FUNDAMENTALS OF

Organic Chemistry

7E

John McMurry
Cornell University

CENGAGE Learning®

Andover • Melbourne • Mexico City • Stamford, CT • Toronto • Hong Kong • New Delhi • Seoul • Singapore • Tokyo

Fundamentals of Organic Chemistry, 7e
John McMurry

ISBN-13: 978-81-315-2097-0
ISBN-10: 81-315-2097-8

Cengage Learning India Private Limited
418, F.I.E., Patparganj
Delhi 110092
India

Tel: 91-11-43641111
Fax: 91-11-43641100
Email: asia.infoindia@cengage.com

Cengage Learning is a leading provider of customized learning solutions with office locations around the globe, including Andover, Melbourne, Mexico City, Stamford (CT), Toronto, Hong Kong, New Delhi, Seoul, Singapore, and Tokyo. Locate your local office at: www.cengage.com/global

Cengage Learning Products are represented in Canada by Nelson Education, Ltd.

For product information, visit our website at **www.cengage.co.in**

Printed in India
Second Indian Reprint 2016

BRIEF CONTENTS

CONTENTS

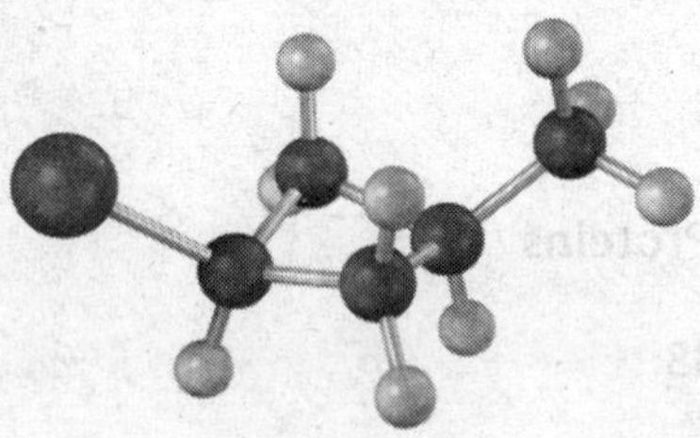

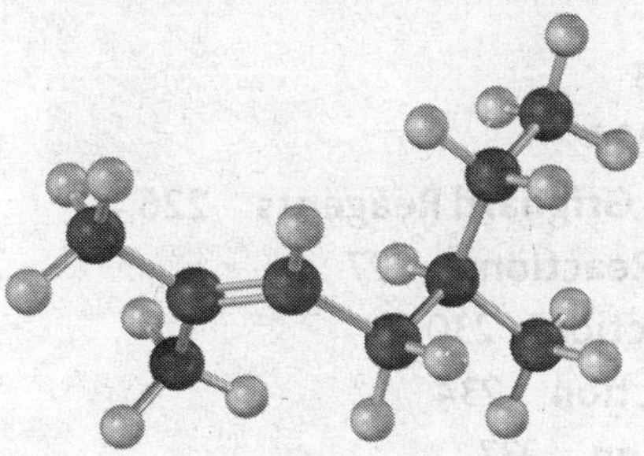

6 Stereochemistry at Tetrahedral Centers

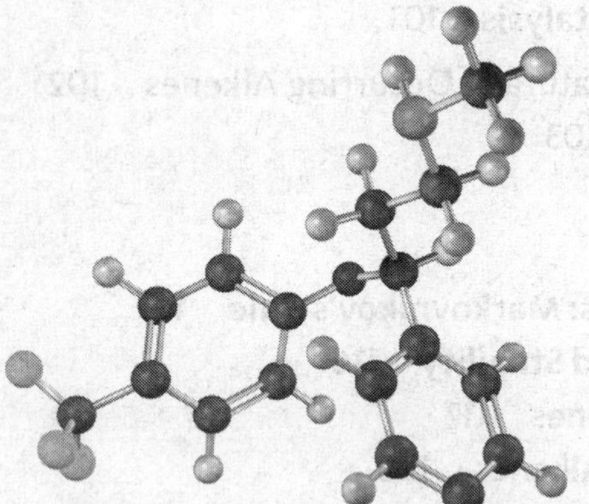

7 Organohalides: Nucleophilic Substitutions and Eliminations

8 Alcohols, Phenols, Ethers, and Their Sulfur Analogs

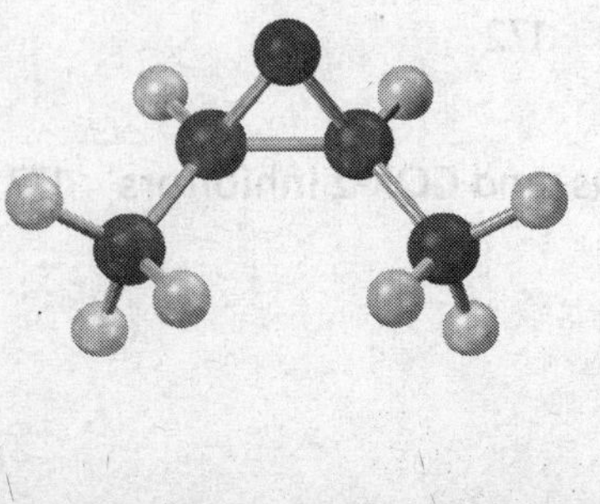

11 Carbonyl Alpha-Substitution Reactions and Condensation Reactions

12 Amines

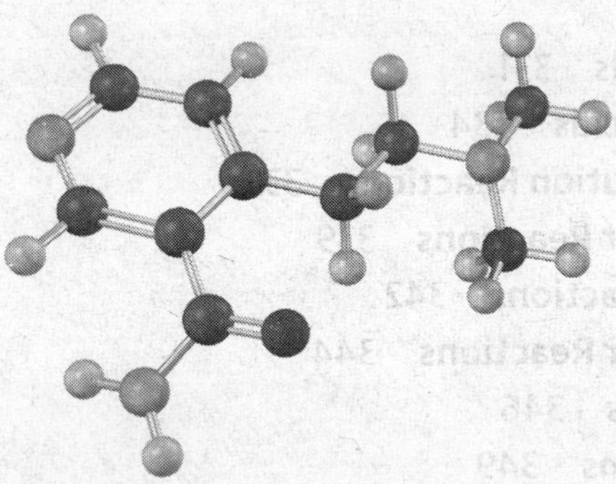

13 Structure Determination

16 Biomolecules: Lipids and Nucleic Acids

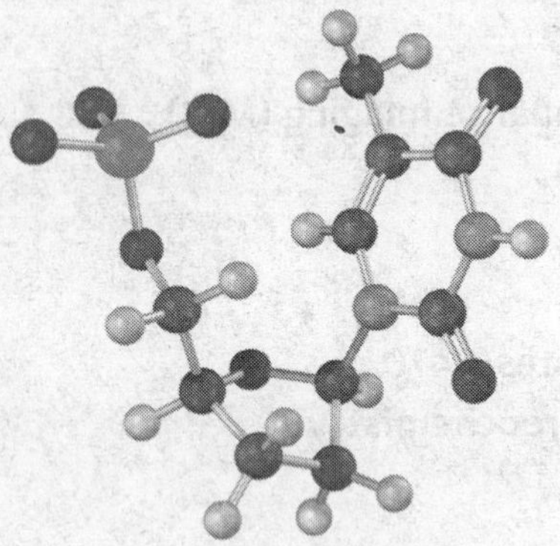

17 The Organic Chemistry of Metabolic Pathways

PREFACE

Organic chemistry is changing rapidly. From its early days dealing primarily with soaps and dyes, organic chemistry has moved to center stage in many fields, from molecular biology to medicine and from agriculture to advanced electronics. Today's organic chemists are learning new languages—particularly those of medicine and molecular biology—to shape the world we live in, and practitioners in many other fields are finding themselves having to learn something of organic chemistry. More than ever before, a fundamental understanding of organic chemistry is critical to addressing complex, interdisciplinary problems.

This seventh edition of *Fundamentals of Organic Chemistry* addresses some of the changes that are occurring by placing a greater emphasis on the applications of organic chemistry, especially applications to medicine and agriculture. Many new examples of biological organic reactions have been added in this edition; *Interlude* boxes at the end of each chapter are rich in the chemistry of drugs and agrochemicals; and problem categories such as "In the Field" and "In the Medicine Cabinet" reinforce the emphasis on applications.

This book is written for a one-semester course in organic chemistry, where content must be comprehensive but to the point. Only those topics needed for a brief course are covered, yet the important pedagogical tools commonly found in larger books are also maintained. In this seventh edition, *Fundamentals of Organic Chemistry* continues its clear explanations, thought-provoking examples and problems, and the trademark vertical format for explaining reaction mechanisms.

The primary organization of this book is by functional group, beginning with the simple (alkanes) and progressing to the more complex. Within the primary organization, there is also an emphasis on explaining the fundamental mechanistic similarities of reactions, and several chapters even have a dual title: Chapter 7 (Organohalides: Nucleophilic Substitutions and Eliminations), Chapter 9 (Aldehydes and Ketones: Nucleophilic Addition Reactions), and Chapter 10 (Carboxylic Acids and Derivatives: Nucleophilic Acyl Substitution Reactions), for instance. Through this approach, memorization is minimized and understanding is maximized.

The first six editions of this text were widely regarded as the clearest and most readable treatments of introductory organic chemistry available. I hope you will find that this seventh edition of *Fundamentals of Organic Chemistry* builds on the strengths of the first six and serves students even better. I have made every effort to make this seventh edition as effective, clear, and readable as possible; to show the beauty, logic, and relevance of organic chemistry; and to make the subject interesting to learn. I welcome all comments on this new edition as well as recommendations for future editions.

FEATURES CONTINUED FROM THE SIXTH EDITION

- Trademarked **vertical reaction mechanisms** give students easy-to-follow descriptions of each step in a reaction pathway. The number of these vertical mechanisms has increased in every edition; see Figure 11.1 on page 375, for example, where the mechanisms of enol formation under both acid-catalyzed and base-catalyzed conditions are compared.

- **Full color** throughout the text highlights the reacting parts of molecules to make it easier to focus on the main parts of a reaction.
- Nearly 100 **electrostatic potential maps** display the polarity patterns in molecules and the importance of these patterns in determining chemical reactivity.
- More than 100 **Visualizing Chemistry problems** challenge students to make the connection between typical line-bond drawings and molecular models.
- Each chapter contains many **Worked Examples** that illustrate how problems can be solved, followed by a similar problem for the student to solve. Each worked-out problem begins with a Strategy discussion that shows how to approach the problem.
- More than 900 **Problems** are included both within the text and at the end of every chapter.
- Current **IUPAC nomenclature** rules, as updated in 1993, are used to name compounds in this text.

CHANGES AND ADDITIONS FOR THE SEVENTH EDITION

The primary reason for preparing a new edition is to keep the book up-to-date, both in its scientific coverage and in its pedagogy. Global changes to the text for this new edition include:

- Writing has been revised at the sentence level.
- Chemical structures have been redrawn.
- Titles have been added to Worked Examples.
- Brief paragraphs titled "Why This Chapter" have been added to chapter introductions to explain the relevance of the chapter material to students.
- Many biologically oriented problems and examples have been added.

Specific changes and additions in individual chapters include:

- **Chapter 1:** A new Section 1.11, Organic Acids and Organic Bases, has been added.
- **Chapter 4:** Coverage of epoxide formation and cleavage has been added to Section 4.6.
- **Chapter 5:** A new *Interlude,* Aspirin, NSAIDs, and COX-2 Inhibitors, has been added.
  Coverage of biologically important aromatic heterocycles has been added to Section 5.9.
- **Chapter 7:** Coverage of alkyl fluoride preparation from alcohols has been added to Section 7.2.
  Coverage of the biologically important E1cB reaction has been added to Section 7.8.
- **Chapter 8:** Coverage of the Grignard reaction has been added to Section 8.3.
  Periodinane oxidation of alcohols has been added to Section 8.4.
  A new *Interlude,* Epoxy Resins and Adhesives, has been added.
- **Chapter 9:** The former Sections 9.6 and 9.11 have been combined in a new Section 9.6, Nucleophilic Addition of Hydride and Grignard Reagents: Alcohol Formation.
  A new *Interlude,* Vitamin C, has been added.

- **Chapter 10:** Coverage of the DCC method of amide synthesis has been added to Section 10.10.
  A new Section 10.12, Biological Carboxylic Acid Derivatives: Thioesters and Acyl Phosphates, has been added.
  Coverage of biodegradable polymers has been added to Section 10.13.
- **Chapter 11:** A new *Interlude,* Barbiturates, has been added.
- **Chapter 12:** Coverage of the azide synthesis of amines has been added to Section 12.4.
  A new *Interlude,* Green Chemistry, has been added.
- **Chapter 13:** The chapter has been reorganized to cover IR before UV.
- **Chapter 14:** A new subsection, Biological Ester Formation: Phosphorylation, has been added to Section 14.7.
  A new Section 14.8, The Eight Essential Monosaccharides, has been added.
- **Chapter 15:** Coverage of major coenzymes has been added to Section 15.9.
  A new *Interlude,* X-Ray Crystallography, has been added.
- **Chapter 16:** All material on nucleic acid chemistry has been updated.
- **Chapter 17:** A new *Interlude,* Statin Drugs, has been added.

BOOK SUPPORT

OWL

OWL Online Web Learning for Organic Chemistry

Authored by Steve Hixson, Peter Lillya, and Peter Samal, all of the University of Massachusetts, Amherst. End-of-chapter questions authored by David W. Brown, Florida Gulf Coast University.

Featuring a modern, intuitive interface, OWL for Organic Chemistry is a customizable, online learning system and assessment tool that reduces faculty workload and facilitates instruction. You can select from various types of assignments—including tutors, simulations, and short answer questions. Questions are numerically, chemically, and contextually parameterized and can accept superscript and subscript as well as structure drawings. With parameterization, OWL for Organic Chemistry offers more than 6000 questions and includes MarvinSketch, an advanced molecular drawing program for drawing gradable structures. In addition, when you become an OWL user, you can expect service that goes far beyond the ordinary.

OWL is continually enhanced with online learning tools to address the various learning styles of today's students such as:

- **Quick Prep** review courses that help students learn essential skills to succeed in General and Organic Chemistry
- **Jmol** molecular visualization program for rotating molecules and measuring bond distances and angles

For more information or to see a demo, please contact your Cengage Learning representative or visit us at www.cengage.com/owl.

ExamView® Computerized Testing

This digital version of the **Test Bank**, revised by Tammy H. Tiner of Texas A&M University, includes a variety of questions per chapter ranging from multiple choice to matching. ISBN-10: 1-4390-5034-1 | ISBN-13: 978-1-4390-5034-7

Instructor's Companion Website
Accessible from www.cengage.com/international, this website provides downloadable files for a library of images from the text as well as WebCT and Blackboard versions of ExamView Computerized Testing.

Study Guide/Solutions Manual, by Susan McMurry
Contains answers to all problems in the text and helps students develop solid problem-solving strategies required for organic chemistry. ISBN-10: 0-538-73676-3 | ISBN-13: 978-0-538-73676-3.

***Pushing Electrons: A Guide for Students of Organic Chemistry*, Third Edition, by Daniel P. Weeks**
Using this brief book, students learn to push electrons to generate resonance structures and write organic mechanisms, an essential skill to learning organic chemistry. ISBN-10: 0-03-020693-6 | ISBN-13: 978-0-03-020693-1

Organic Chemistry Modeling Kits
Cengage Learning offers a variety of organic chemistry model kits as a bundling option for students. Please consult with your Cengage Learning representative for pricing and selection.

Organic Chemistry Laboratory Manuals
Brooks/Cole, Cengage Learning is pleased to offer a choice of organic chemistry laboratory manuals catered to fit your needs. Visit www.cengage.com/chemistry for preset lab manuals. For custom laboratory manuals, visit www.signature-labs.com.

ACKNOWLEDGMENTS

I sincerely thank the many people whose help and suggestions were so valuable in preparing this seventh edition, particularly Sandi Kiselica, Lisa Lockwood, Lisa Weber, and Amee Mosley at Cengage Learning; Dan Fitzgerald at Graphic World Inc., my wife, Susan, who read and improved the entire manuscript; and Professor Tom Lectka at Johns Hopkins University, who made many valuable suggestions. I would also like to thank members of the reviewing panel, who graciously provided many helpful ideas for revising this text: Robert Cameron, Samford University; Alvan C. Hengge, Utah State University; Steven Holmgren, Montana State University; and Richard P. Johnson, University of New Hampshire.

CHAPTER

1

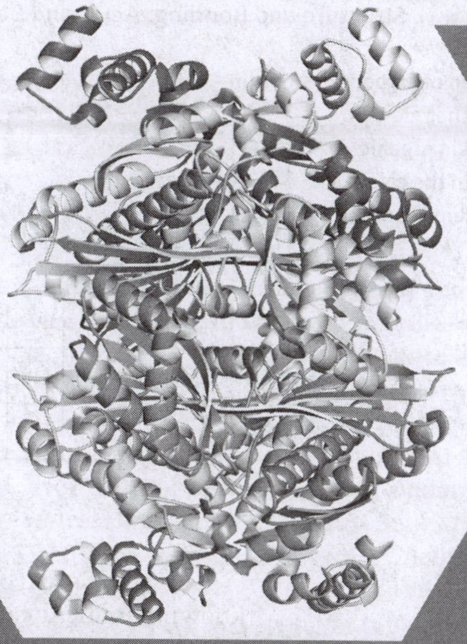

The enzyme HMG-CoA reductase, shown here as a so-called ribbon model, catalyzes a crucial step in the body's synthesis of cholesterol. Understanding how this enzyme functions has led to the development of drugs credited with saving millions of lives.

Structure and Bonding; Acids and Bases

Organic chemistry is all around us. The reactions and interactions of organic molecules allow us to see, smell, fight, and fear. Organic chemistry provides the molecules that feed us, treat our illnesses, protect our crops, and clean our clothes. Anyone with a curiosity about life and living things must have a basic understanding of organic chemistry.

Historically, the term **organic chemistry** dates to the late 1700s, when it was used to mean the chemistry of compounds found in living organisms. Little was known about chemistry at that time, and the behavior of the "organic" substances isolated from plants and animals seemed different from that of the "inorganic" substances found in minerals. Organic compounds were generally low-melting solids and were usually more difficult to isolate, purify, and work with than high-melting inorganic compounds. By the mid-1800s, however, it was clear that there was no fundamental difference between organic and inorganic compounds. The same principles explain the behaviors of all substances, regardless of origin or complexity. The only distinguishing characteristic of organic chemicals is that *all contain the element carbon* (Figure 1.1).

OWL

Online homework for this chapter can be assigned in OWL, an online homework assessment tool.

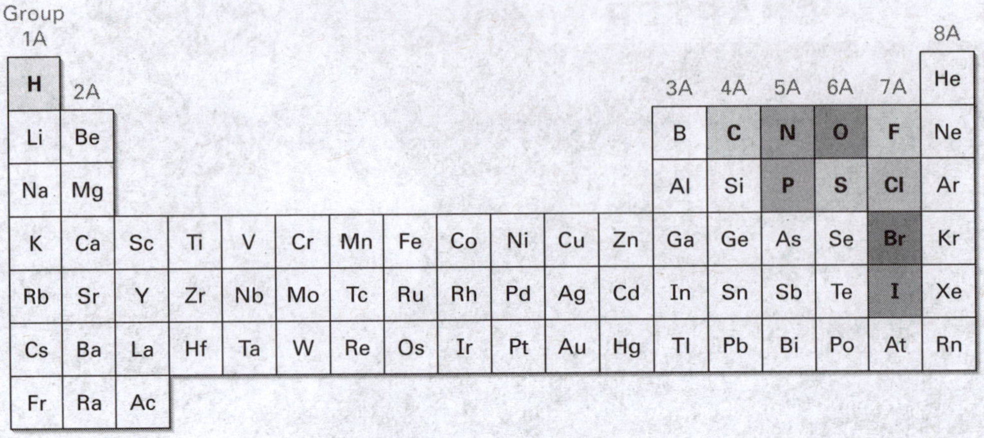

Figure 1.1 The position of carbon in the periodic table. Other elements commonly found in organic compounds are shown in the colors typically used to represent them.

But why is carbon special? Why, of the more than 37 million presently known chemical compounds, do more than 99% of them contain carbon? The answers to these questions come from carbon's electronic structure and its consequent position in the periodic table. As a group 4A element, carbon can share four valence electrons and form four strong covalent bonds. Furthermore, carbon atoms can bond to one another, forming long chains and rings. Carbon, alone of all elements, is able to form an immense diversity of compounds, from the simple methane, with one carbon atom, to the staggeringly complex DNA, which can have more than *100 million* carbons.

Not all carbon compounds are derived from living organisms of course. Modern chemists have developed a remarkably sophisticated ability to design and synthesize new organic compounds in the laboratory—medicines, dyes, polymers, and a host of other substances. Organic chemistry touches the lives of everyone; its study can be a fascinating undertaking.

WHY THIS CHAPTER?

We'll ease into the study of organic chemistry by first reviewing some ideas about atoms, bonds, and molecular geometry that you may recall from your general chemistry course. Much of the material in this chapter is likely to be familiar to you, but some of it may be new and it's a good idea to make sure you understand it before going on.

1.1 Atomic Structure

As you probably know from your general chemistry course, an atom consists of a dense, positively charged *nucleus* surrounded at a relatively large distance by negatively charged *electrons* (Figure 1.2). The nucleus consists of subatomic particles called *neutrons,* which are electrically neutral, and *protons,* which are positively charged. Because an atom is neutral overall, the number of positive protons in the nucleus and the number of negative electrons surrounding the nucleus are the same.

Although extremely small—about 10^{-14} to 10^{-15} meter (m) in diameter—the nucleus nevertheless contains essentially all the mass of the atom. Electrons have negligible mass and circulate around the nucleus at a distance of approximately 10^{-10} m. Thus, the diameter of a typical atom is about

2×10^{-10} m, or 200 picometers (pm), where 1 pm = 10^{-12} m. To give you an idea of how small this is, a thin pencil line is about 3 million carbon atoms wide. Many organic chemists and biochemists still use the unit *angstrom* (Å) to express atomic distances, where 1 Å = 100 pm = 10^{-10} m, but we'll stay with the SI unit picometer in this book.

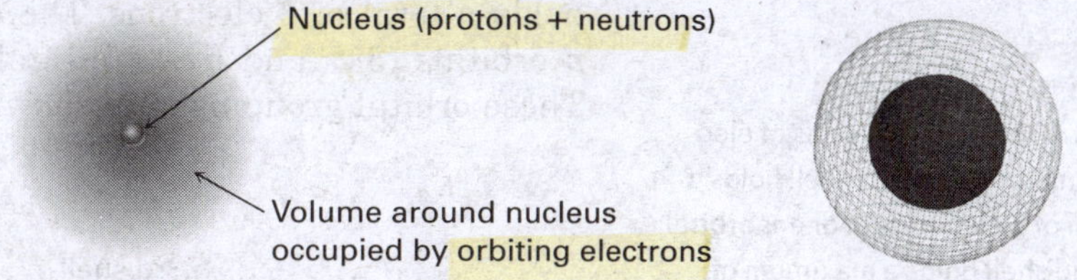

Figure 1.2 A schematic view of an atom. The dense, positively charged nucleus contains most of the atom's mass and is surrounded by negatively charged electrons. The three-dimensional view on the right shows calculated electron-density surfaces. Electron density increases steadily toward the nucleus and is 40 times greater at the blue solid surface than at the gray mesh surface.

A specific atom is described by its *atomic number (Z),* which gives the number of protons (or electrons) it contains, and its *mass number (A),* which gives the total number of protons plus neutrons in its nucleus. All the atoms of a given element have the same atomic number—1 for hydrogen, 6 for carbon, 15 for phosphorus, and so on—but they can have different mass numbers depending on how many neutrons they contain. Atoms with the same atomic number but different mass numbers are called **isotopes.**

The weighted average mass in atomic mass units (amu) of an element's naturally occurring isotopes is called the element's *atomic mass* (or atomic weight)—1.008 amu for hydrogen, 12.011 amu for carbon, 30.974 amu for phosphorus, and so on. Atomic masses of the elements are given in the periodic table in the back of this book.

What about the electrons? How are they distributed in an atom? According to the *quantum mechanical model* of atomic structure, the behavior of a specific electron in an atom can be described by a mathematical expression called a *wave equation*—the same sort of expression used to describe the motion of waves in a fluid. The solution to a wave equation is a *wave function,* or **orbital,** denoted by the Greek letter psi, ψ. An orbital can be thought of as defining a region of space around the nucleus where the electron can most likely be found.

What do orbitals look like? There are four different kinds of orbitals, denoted *s, p, d,* and *f,* each with a different shape. Of the four, we'll be concerned only with *s* and *p* orbitals because these are the most common in organic and biological chemistry. An *s* orbital is spherical, with the nucleus at its center, while a *p* orbital is dumbbell-shaped and can be oriented in space along any of three mutually perpendicular directions, arbitrarily denoted p_x, p_y, and p_z (Figure 1.3). The two parts, or *lobes,* of a *p* orbital have different algebraic signs (+ and −) in the wave function and are separated by a region of zero electron density called a *node.*

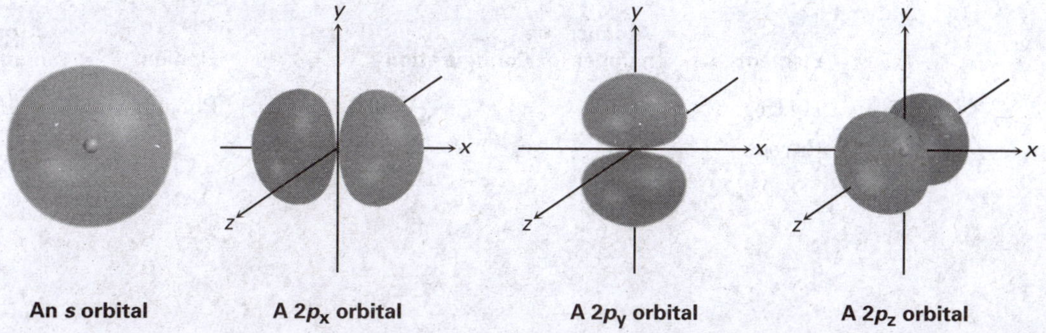

Figure 1.3 Representations of *s* and *p* orbitals. An *s* orbital is spherical, while a *p* orbital is dumbbell-shaped and can be oriented along any of three mutually perpendicular directions. Each *p* orbital has two lobes separated by a node. The two lobes have different algebraic signs in the corresponding wave function, as indicated by the different colors.

Orbitals are organized into different layers around the nucleus of successively larger size and energy. Different layers, or **electron shells,** contain different numbers and kinds of orbitals, and each orbital can be occupied by 2 electrons. The first shell contains only a single *s* orbital, denoted 1*s,* and thus holds only 2 electrons. The second shell contains an *s* orbital (designated 2*s*) and three mutually perpendicular *p* orbitals (each designated 2*p*) and thus holds a total of 8 electrons. The third shell contains an *s* orbital (3*s*), three *p* orbitals (3*p*), and five *d* orbitals (3*d*), for a total capacity of 18 electrons. These orbital groupings are shown in Figure 1.4.

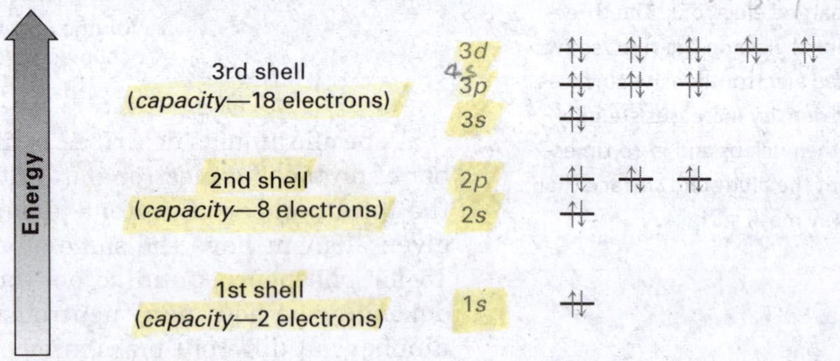

Figure 1.4 The energy levels of electrons in an atom. The first shell holds a maximum of 2 electrons in one 1*s* orbital; the second shell holds a maximum of 8 electrons in one 2*s* and three 2*p* orbitals; the third shell holds a maximum of 18 electrons in one 3*s*, three 3*p*, and five 3*d* orbitals; and so on. The 2 electrons in each orbital are represented by up and down arrows, ↑↓. Although not shown, the energy level of the 4*s* orbital falls between 3*p* and 3*d*.

1.2 Atomic Structure: Electron Configurations

The lowest-energy arrangement, or **ground-state electron configuration,** of an atom is a listing of the orbitals that the atom's electrons occupy. We can predict this arrangement by following three rules.

RULE 1 The orbitals of lowest energy are filled first, according to the order 1*s* → 2*s* → 2*p* → 3*s* → 3*p* → 4*s* → 3*d,* as shown in Figure 1.4.

RULE 2 Only two electrons can occupy an orbital, and they must be of opposite *spin.* (Electrons act in some ways as if they were spinning on an axis, somewhat as the earth spins. This spin can have two orientations, denoted as up ↑ and down ↓.)

RULE 3 If two or more empty orbitals of equal energy are available, one electron occupies each with the spins parallel until all orbitals are half-full.

Some examples of how these rules apply are shown in Table 1.1. Hydrogen, for instance, has only one electron, which must occupy the lowest-energy

Table 1.1 Ground-State Electron Configuration of Some Elements

Element	Atomic number	Configuration		Element	Atomic number	Configuration	
Hydrogen	1	1*s*	↑	Phosphorus	15	3*p*	↑ ↑ ↑
Carbon	6	2*p*	↑ ↑ —			3*s*	↑↓
		2*s*	↑↓			2*p*	↑↓ ↑↓ ↑↓
		1*s*	↑↓			2*s*	↑↓
						1*s*	↑↓

orbital. Thus, hydrogen has a 1s ground-state electron configuration. Carbon has six electrons and the ground-state electron configuration $1s^2\ 2s^2\ 2p^2$. Note that a superscript is used to represent the number of electrons in a particular orbital.

Worked Example 1.1 **Assigning an Electron Configuration to an Element**

Give the ground-state electron configuration of nitrogen.

Strategy Find the atomic number of nitrogen to see how many electrons it has, and then apply the three rules to assign electrons into orbitals according to the energy levels given in Figure 1.4.

Solution Nitrogen has atomic number 7 and thus has seven electrons. The first two electrons go into the lowest-energy orbital ($1s^2$), the next two go into the second-lowest-energy orbital ($2s^2$), and the remaining three go into the next-lowest-energy orbitals ($2p^3$), with one electron in each. Thus, the configuration of nitrogen is $1s^2\ 2s^2\ 2p^3$.

Problem 1.1 How many electrons does each of the following elements have in its outermost electron shell?

(a) Potassium **(b)** Calcium **(c)** Aluminum

Problem 1.2 Give the ground-state electron configuration of the following elements:

(a) Boron **(b)** Phosphorus **(c)** Oxygen **(d)** Argon

1.3 Development of Chemical Bonding Theory

By the mid-1800s, the new science of chemistry was developing rapidly and chemists had begun to probe the forces holding molecules together. In 1858, August Kekulé and Archibald Couper independently proposed that, in all organic compounds, carbon is *tetravalent;* that is, it always forms four bonds when it joins other elements to form chemical compounds. Furthermore, said Kekulé, carbon atoms can bond to one another to form extended chains of linked atoms and chains can double back on themselves to form rings.

Although Kekulé and Couper were correct in describing the tetravalent nature of carbon, chemistry was still viewed in a two-dimensional way until 1874. In that year, Jacobus van't Hoff and Joseph Le Bel added a third dimension to our ideas about organic compounds. They proposed that the four bonds of carbon are not oriented randomly but have specific spatial directions. Van't Hoff went even further and suggested that the four atoms to which carbon is bonded sit at the corners of a regular tetrahedron, with carbon in the center.

A representation of a tetrahedral carbon atom is shown in Figure 1.5. Note the conventions used to show three-dimensionality: solid lines represent bonds in the plane of the page, the heavy wedged line represents a bond coming out of the page toward the viewer, and the dashed line represents a bond receding back behind the page away from the viewer. These representations will be used throughout this text.

Figure 1.5 A representation of van't Hoff's tetrahedral carbon atom. The solid lines represent bonds in the plane of the paper, the heavy wedged line represents a bond coming out of the plane of the page, and the dashed line represents a bond going back behind the plane of the page.

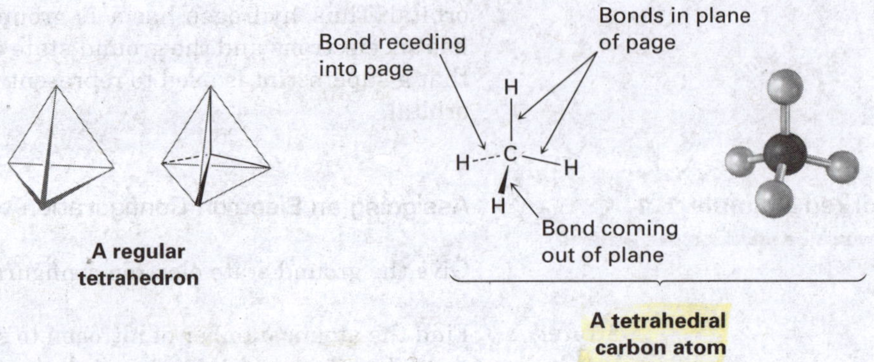

Problem 1.3 Draw a molecule of chloromethane, CH_3Cl, using solid, wedged, and dashed lines to show its tetrahedral geometry.

Problem 1.4 Convert the following molecular model of ethane, C_2H_6, into a structure that uses wedged, normal, and dashed lines to represent three-dimensionality.

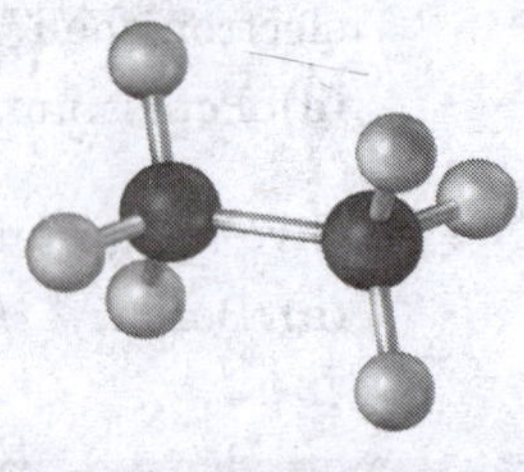

Ethane

1.4 The Nature of Chemical Bonds

Why do atoms bond together, and how can bonds be described electronically? The *why* question is relatively easy to answer: atoms bond together because the compound that results is more stable and lower in energy than the separate atoms. Energy (usually as heat) is always released and flows *out of* the chemical system when a bond forms. Conversely, energy must be put *into* the system to break a bond. Making bonds always releases energy, and breaking bonds always absorbs energy. The *how* question is more difficult. To answer it, we need to know more about the electronic properties of atoms.

We know through observation that eight electrons—an electron *octet*—in an atom's outermost shell, or **valence shell**, impart special stability to the noble-gas elements in group 8A of the periodic table: Ne (2 + *8*); Ar (2 + 8 + *8*); Kr (2 + 8 + 18 + *8*). We also know that the chemistry of main-group elements is governed by their tendency to take on the electron configuration of the nearest noble gas. The alkali metals in group 1A, for example, achieve a noble-gas configuration by losing the single *s* electron from their valence shell to form a cation, while the halogens in group 7A achieve a noble-gas configuration by gaining a *p* electron to fill their valence shell and form an anion. The resultant ions are held together in compounds like $Na^+ Cl^-$ by an electrostatic attraction that we call an *ionic bond.*

How, though, do elements near the middle of the periodic table form bonds? Look at methane, CH_4, the main constituent of natural gas, for example. The bonding in methane is not ionic because it would take too much energy for carbon ($1s^2\ 2s^2\ 2p^2$) to either gain or lose *four* electrons to achieve a noble-gas configuration. As a result, carbon bonds to other atoms, not by gaining or losing electrons, but by *sharing* them. Such a shared-electron bond, first proposed in 1916 by G. N. Lewis, is called a **covalent bond**. The neutral group of atoms held together by covalent bonds is called a **molecule**.

A simple way of indicating the covalent bonds in molecules is to use what are called *Lewis structures,* or **electron-dot structures**, in which the valence-shell electrons of an atom are represented as dots. Thus, hydrogen has one dot representing its 1*s* electron, carbon has four dots ($2s^2\ 2p^2$), oxygen has six dots ($2s^2\ 2p^4$), and so on. A stable molecule results whenever a noble-gas configuration is achieved for all the atoms—eight dots (an octet) for main-group atoms or two dots for hydrogen. Simpler still is the use of *Kekulé structures,* or **line-bond structures**, in which a two-electron covalent bond is indicated as a line drawn between atoms.

Electron-dot structures (Lewis structures)

Line-bond structures (Kekulé structures)

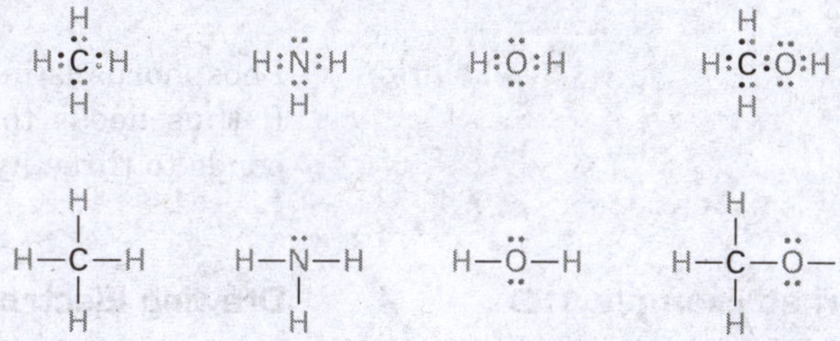

Methane (CH_4) **Ammonia (NH_3)** **Water (H_2O)** **Methanol (CH_3OH)**

The number of covalent bonds an atom forms depends on how many additional valence electrons it needs to reach a noble-gas configuration. Hydrogen has one valence electron (1*s*) and needs one more to reach the helium configuration ($1s^2$), so it forms one bond. Carbon has four valence electrons ($2s^2\ 2p^2$) and needs four more to reach the neon configuration ($2s^2\ 2p^6$), so it forms four bonds. Nitrogen has five valence electrons ($2s^2\ 2p^3$), needs three more, and forms three bonds; oxygen has six valence electrons ($2s^2\ 2p^4$), needs two more, and forms two bonds; and the halogens have seven valence electrons, need one more, and form one bond.

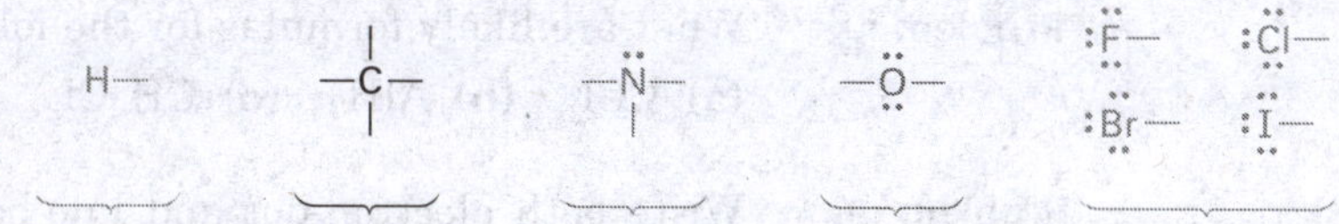

One bond **Four bonds** **Three bonds** **Two bonds** **One bond**

Valence electrons that are not used for bonding are called **lone-pair electrons,** or *nonbonding electrons.* The nitrogen atom in ammonia (NH_3), for instance, shares six valence electrons in three covalent bonds and has its remaining two valence electrons in a nonbonding lone pair. As a time-saving shorthand, nonbonding electrons are often omitted when drawing line-bond structures, but you still have to keep them in mind since they're often crucial in chemical reactions.

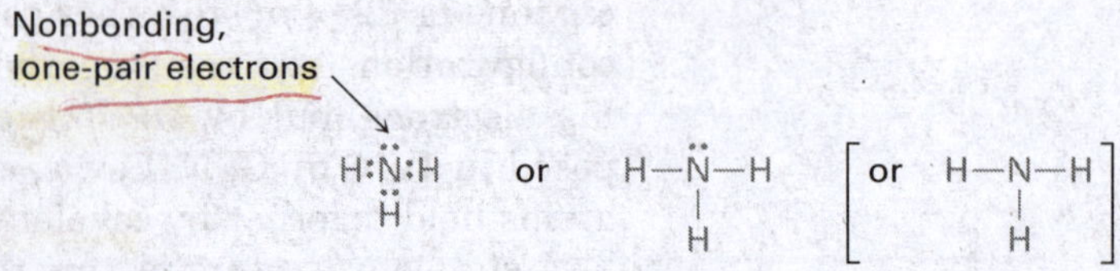

Ammonia

Worked Example 1.2 Predicting the Number of Bonds Formed by an Atom

How many hydrogen atoms does phosphorus bond to in forming phosphine, $PH_?$?

Strategy Identify the periodic group of phosphorus, and tell from that how many electrons (bonds) are needed to make an octet.

Solution Phosphorus is in group 5A of the periodic table and has five valence electrons. It thus needs to share three more electrons to make an octet and therefore bonds to three hydrogen atoms, giving PH_3.

Worked Example 1.3 Drawing Electron-Dot and Line-Bond Structures

Draw both electron-dot and line-bond structures for chloromethane, CH_3Cl.

Strategy Remember that a bond—that is, a pair of shared electrons—is represented as a line between atoms.

Solution Hydrogen has one valence electron, carbon has four valence electrons, and chlorine has seven valence electrons. Thus, chloromethane is represented as

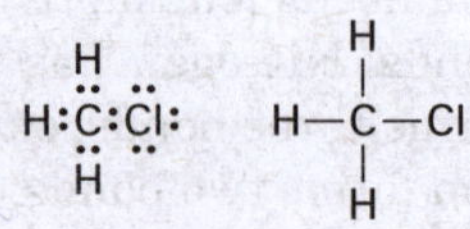

Chloromethane

Problem 1.5 What are likely formulas for the following molecules?

(a) $CCl_?$ **(b)** $AlH_?$ **(c)** $CH_?Cl_2$ **(d)** $SiF_?$

Problem 1.6 Write both electron-dot and line-bond structures for the following molecules, showing all nonbonded electrons:

(a) $CHCl_3$, chloroform **(b)** H_2S, hydrogen sulfide

(c) CH_3NH_2, methylamine

Problem 1.7 Why can't an organic molecule have the formula C_2H_7?

1.5 Forming Covalent Bonds: Valence Bond Theory

How does electron sharing lead to bonding between atoms? According to **valence bond theory**, a covalent bond forms when two atoms approach each other closely and a singly occupied orbital on one atom *overlaps* a singly occupied orbital on the other atom. The electrons are now paired in the overlapping orbitals and are attracted to the nuclei of both atoms, thus bonding the atoms together. In the H_2 molecule, for example, the H–H bond results from the overlap of two singly occupied hydrogen 1*s* orbitals.

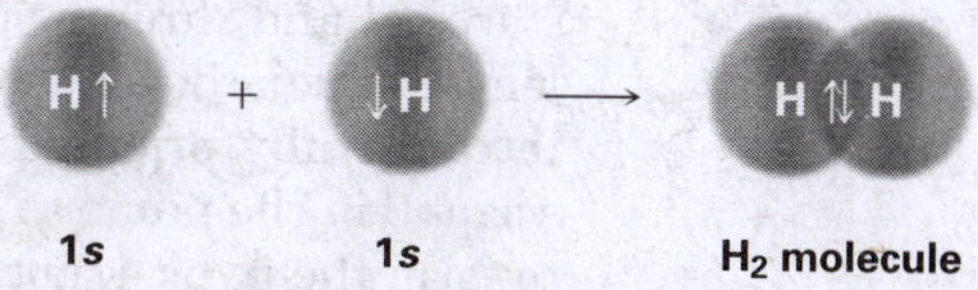

During the bond-forming reaction 2 H· → H_2, 436 kJ/mol (104 kcal/mol) of energy is released. Because the product H_2 molecule has 436 kJ/mol less energy than the starting 2 H· atoms, we say that the product is *more stable* than the reactant and that the new H–H bond has a **bond strength** of 436 kJ/mol. In other words, we would have to put 436 kJ/mol of energy *into* the H–H bond to break the H_2 molecule apart into two H atoms. [For convenience, we'll generally give energies in both the SI unit kilojoules (kJ) and the older unit kilocalories (kcal): 1 kJ = 0.2390 kcal; 1 kcal = 4.184 kJ.]

How close are the two nuclei in the H_2 molecule? If they are too close, they will repel each other because both are positively charged, yet if they are too far apart, they won't be able to share the bonding electrons. Thus, there is an optimum distance between nuclei that leads to maximum stability (Figure 1.6). Called the **bond length**, this distance is 74 pm in the H_2 molecule. Every covalent bond has both a characteristic bond strength and bond length.

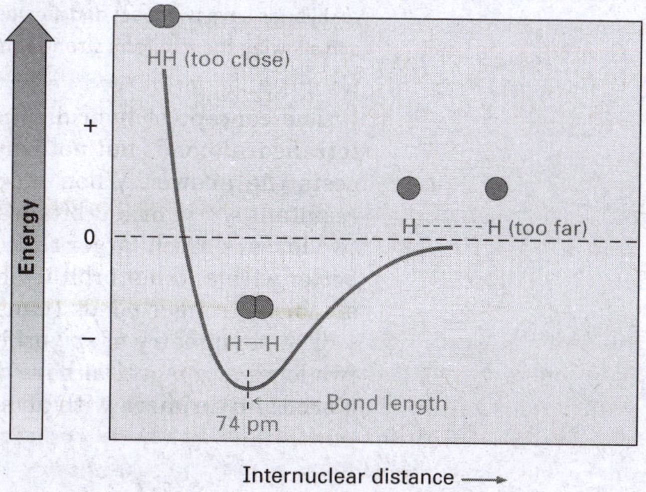

Figure 1.6 A plot of energy versus internuclear distance for two hydrogen atoms. The distance at the minimum energy point is the bond length.

1.6 sp^3 Hybrid Orbitals and the Structure of Methane

The bonding in the H_2 molecule is fairly straightforward, but the situation is more complicated in organic molecules with tetravalent carbon atoms. Take methane, CH_4, for instance. Carbon has four valence electrons ($2s^2\ 2p^2$) and forms four bonds. Because carbon uses two kinds of orbitals for bonding, 2*s* and 2*p*, we might expect methane to have two kinds of C–H bonds. In fact, though, all four C–H bonds in methane are identical and are spatially oriented toward the corners of a regular tetrahedron (Figure 1.5). How can we explain this?

An answer was provided in 1931 by Linus Pauling, who proposed that an *s* orbital and three *p* orbitals can combine, or *hybridize,* to form four equivalent atomic orbitals with tetrahedral orientation. Shown in Figure 1.7, these tetrahedrally oriented orbitals are called **sp^3 hybrids**. Note that the superscript 3 in the name sp^3 tells how many of each type of atomic orbital combine to form the hybrid, not how many electrons occupy it.

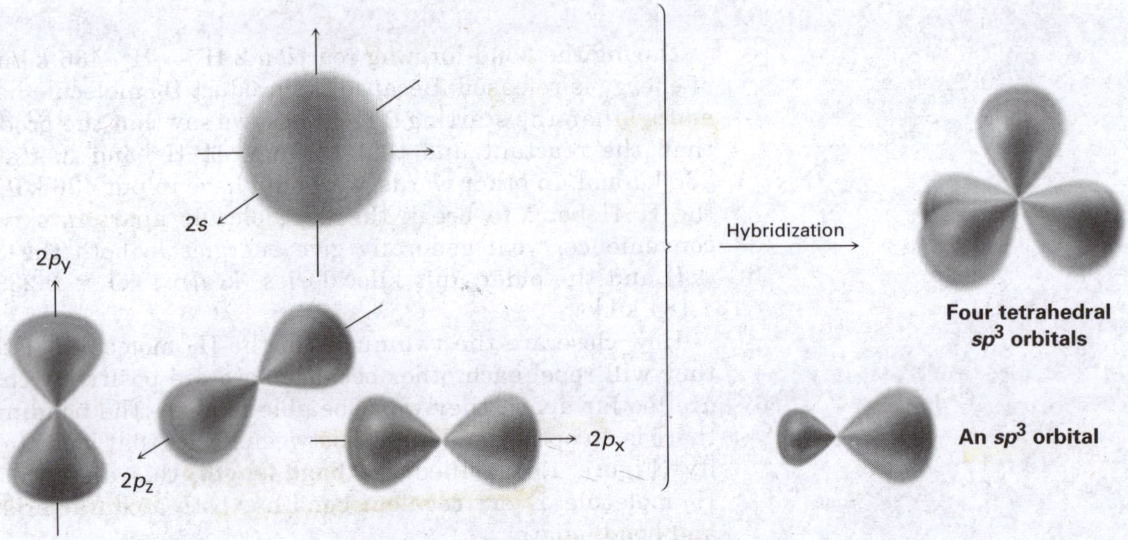

Figure 1.7 Four sp^3 hybrid orbitals (green), oriented to the corners of a regular tetrahedron, are formed by combination of an atomic *s* orbital (red) and three atomic *p* orbitals (red/blue). The sp^3 hybrids have two lobes and are unsymmetrical about the nucleus, giving them a directionality and allowing them to form strong bonds when they overlap an orbital from another atom.

The concept of hybridization explains how carbon forms four equivalent tetrahedral bonds but not why it does so. The shape of the hybrid orbital suggests the answer. When an *s* orbital hybridizes with three *p* orbitals, the resultant sp^3 hybrid orbitals are unsymmetrical about the nucleus. One of the two lobes is much larger than the other (Figure 1.7) and can therefore overlap better with another orbital when it forms a bond. As a result, sp^3 hybrid orbitals form stronger bonds than do unhybridized *s* or *p* orbitals.

The asymmetry of sp^3 orbitals arises because, as noted in Section 1.1, the two lobes of a *p* orbital have different algebraic signs, + and −. Thus, when a *p* orbital hybridizes with an *s* orbital, the positive *p* lobe adds to the *s* orbital but the negative *p* lobe subtracts from the *s* orbital. The resultant hybrid orbital is therefore unsymmetrical about the nucleus and is strongly oriented in one direction.

When each of the four identical sp^3 hybrid orbitals of a carbon atom overlaps with the 1*s* orbital of a hydrogen atom, four identical C–H bonds are formed and methane results. Each C–H bond in methane has a strength of 439 kJ/mol (105 kcal/mol) and a length of 109 pm. Because the four bonds have a specific geometry, we also can define a property called the **bond angle**. The angle formed by each H—C—H is 109.5°, the so-called tetrahedral angle. Methane thus has the structure shown in Figure 1.8.

Figure 1.8 The structure of methane, showing its 109.5° bond angles.

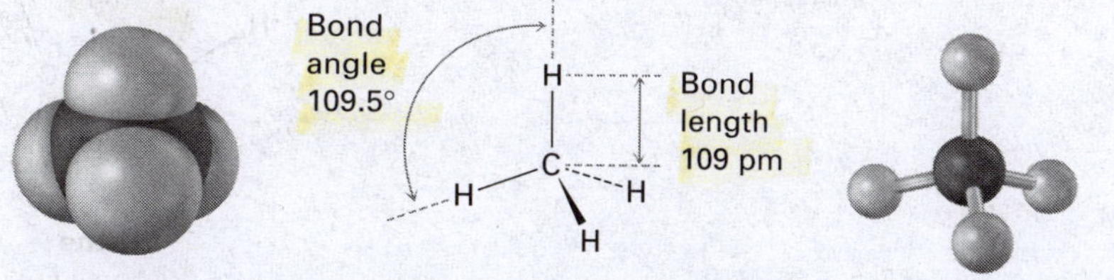

Problem 1.8 Draw a tetrahedral representation of tetrachloromethane, CCl_4, using the standard convention of solid, dashed, and wedged lines.

Problem 1.9 Why do you think a C–H bond (109 pm) is longer than an H–H bond (74 pm)?

1.7 sp^3 Hybrid Orbitals and the Structure of Ethane

The same kind of orbital hybridization that accounts for the methane structure also accounts for the bonding together of carbon atoms into chains and rings to make possible many millions of organic compounds. Ethane, C_2H_6, is the simplest molecule containing a carbon–carbon bond.

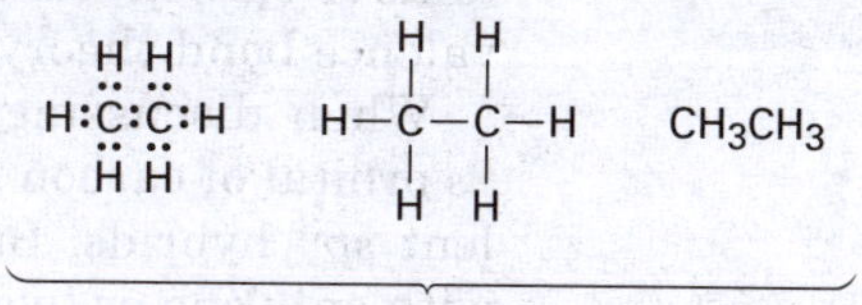

Some representations of ethane

We can picture the ethane molecule by imagining that the two carbon atoms bond to each other by overlap of an sp^3 hybrid orbital from each (Figure 1.9). The remaining three sp^3 hybrid orbitals of each carbon overlap with the 1*s* orbitals of three hydrogens to form the six C–H bonds. The C–H bonds in ethane are similar to those in methane, although a bit weaker—421 kJ/mol (101 kcal/mol) for ethane versus 439 kJ/mol for methane. The C–C bond is 154 pm long and has a strength of 377 kJ/mol (90 kcal/mol). All the bond angles of ethane are near, although not exactly at, the tetrahedral value of 109.5°.

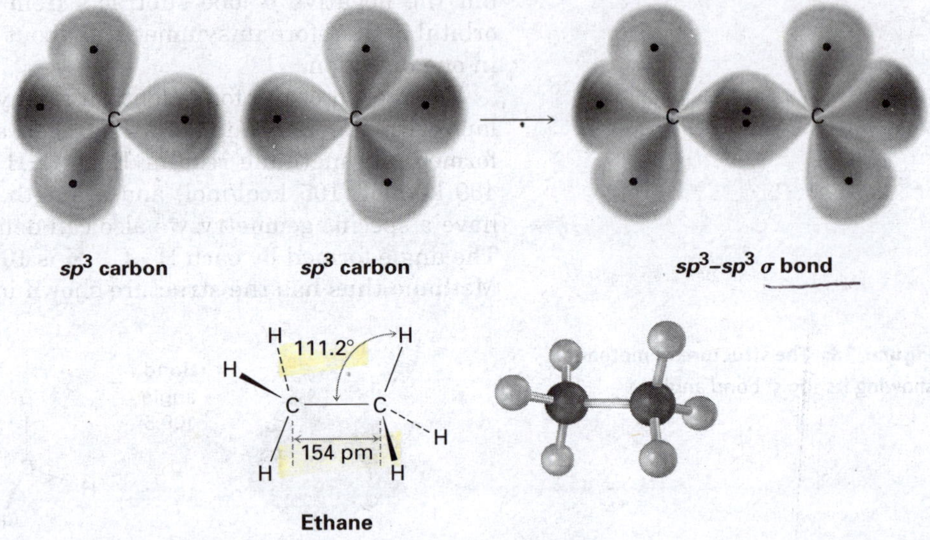

Figure 1.9 The structure of ethane. The carbon–carbon bond is formed by overlap of two carbon sp^3 hybrid orbitals. For clarity, the smaller lobes of the hybrid orbitals are not shown.

Problem 1.10 Draw a line-bond structure for propane, $CH_3CH_2CH_3$. Predict the value of each bond angle, and indicate the overall shape of the molecule.

1.8 Other Kinds of Hybrid Orbitals: sp^2 and sp

The bonds we've seen in methane and ethane are called single bonds because they result from the sharing of one electron pair between bonded atoms. It was recognized more than 100 years ago, however, that in some molecules carbon atoms can also form a double bond by sharing two electron pairs between atoms or a triple bond by sharing three electron pairs. Ethylene, for instance, has the structure $H_2C{=}CH_2$ and contains a carbon–carbon double bond, while acetylene has the structure $HC{\equiv}CH$ and contains a carbon–carbon triple bond. How are multiple bonds described by valence bond theory?

When discussing sp^3 hybrid orbitals in Section 1.6, we said that the 2*s* orbital of carbon combines with all three 2*p* orbitals to form four equivalent sp^3 hybrids. Imagine instead, however, that the 2*s* orbital combines with only one or two of the three available 2*p* orbitals. If the 2*s* orbital combines with only two 2*p* orbitals, three **sp^2 hybrids** result and one unhybridized 2*p* orbital remains unchanged. If the 2*s* orbital combines with only one 2*p* orbital, two ***sp* hybrids** result and two unhybridized 2*p* orbitals remain unchanged.

Like sp^3 hybrids, sp^2 and *sp* hybrid orbitals are unsymmetrical about the nucleus and are strongly oriented in a specific direction so they can form strong bonds. In an sp^2-hybridized carbon atom, for instance, the three sp^2 orbitals lie in a plane at angles of 120° to one another, with the remaining *p* orbital perpendicular to the sp^2 plane (Figure 1.10a). In an *sp*-hybridized carbon atom, the two *sp* orbitals are oriented 180° apart, with the remaining two *p* orbitals perpendicular both to the *sp* hybrids and to each other (Figure 1.10b).

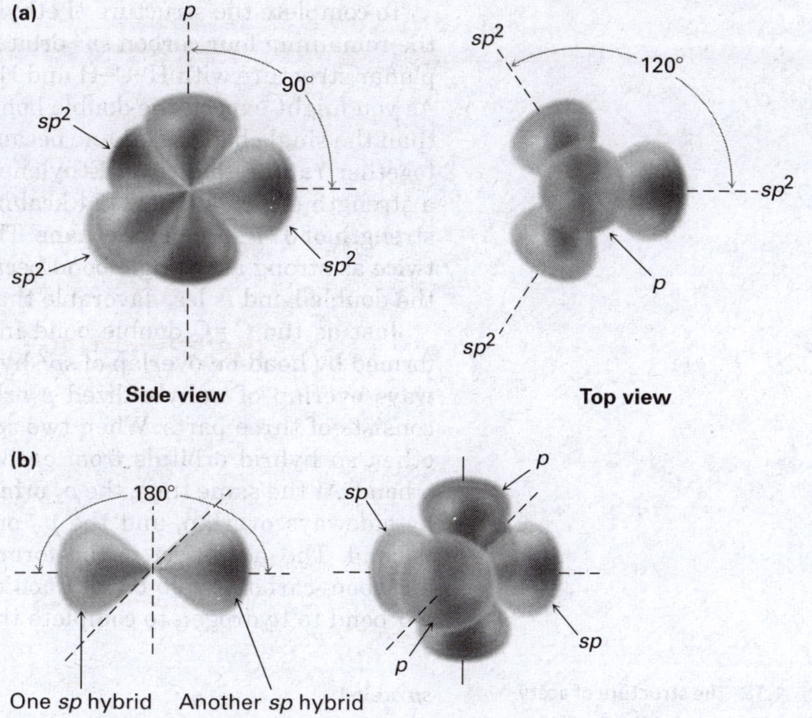

Figure 1.10 (a) An sp^2-hybridized carbon. The three equivalent sp^2 hybrid orbitals (green) lie in a plane at angles of 120° to one another, and a single unhybridized *p* orbital (red/blue) is perpendicular to the sp^2 plane. (b) An *sp*-hybridized carbon atom. The two *sp* hybrid orbitals (green) are oriented 180° away from each other, perpendicular to the two remaining *p* orbitals (red/blue).

When two sp^2-hybridized carbon atoms approach each other, they form a strong bond by sp^2–sp^2 head-on overlap. At the same time, the unhybridized *p* orbitals interact by *sideways* overlap to form a second bond. Head-on overlap gives what is called a **sigma (σ) bond**, while sideways overlap gives a **pi (π) bond**. The combination of sp^2–sp^2 σ overlap and $2p$–$2p$ π overlap results in the net sharing of two electron pairs and the formation of a carbon–carbon double bond (Figure 1.11). Note that the electrons in a σ bond occupy the region centered between nuclei, while the electrons in a π bond occupy regions on either side of a line drawn between nuclei.

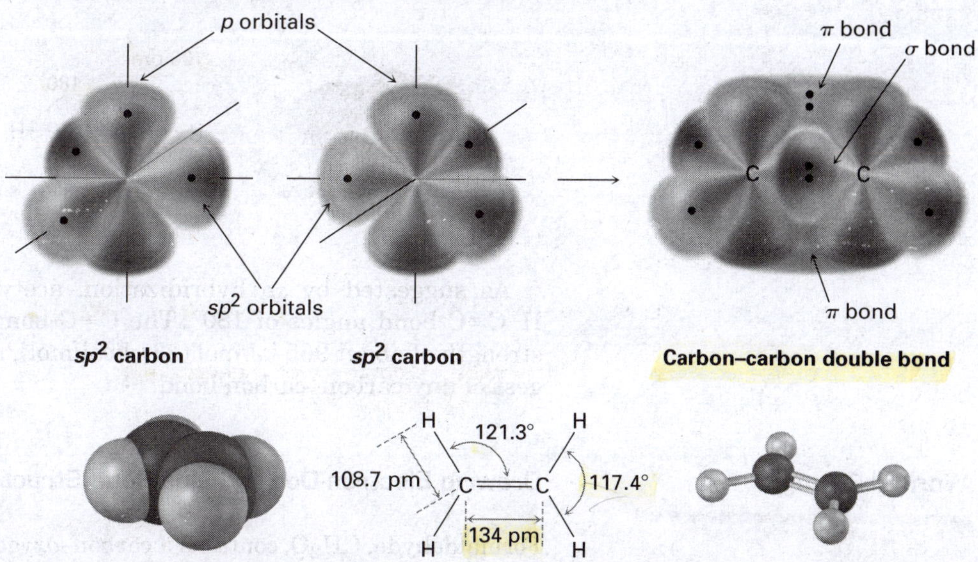

Figure 1.11 The structure of ethylene. Orbital overlap of two sp^2-hybridized carbons forms a carbon–carbon double bond. One part of the double bond results from σ (head-on) overlap of sp^2 orbitals (green), and the other part results from π (sideways) overlap of unhybridized *p* orbitals (red/blue). The π bond has regions of electron density above and below a line drawn between nuclei.

To complete the structure of ethylene, four hydrogen atoms form σ bonds to the remaining four carbon sp^2 orbitals. The resultant ethylene molecule has a planar structure with H–C–H and H–C=C bond angles of approximately 120°. As you might expect, the double bond in ethylene is both shorter and stronger than the single bond in ethane because it has four electrons bonding the nuclei together rather than two. Ethylene has a C=C bond length of 134 pm and a strength of 728 kJ/mol (174 kcal/mol) versus a C–C length of 154 pm and a strength of 377 kJ/mol for ethane. The carbon–carbon double bond is less than twice as strong as a single bond because the sideways overlap in the π part of the double bond is less favorable than the head-on overlap in the σ part.

Just as the C=C double bond in ethylene consists of two parts, a σ part formed by head-on overlap of sp^2 hybrid orbitals and a π part formed by sideways overlap of unhybridized *p* orbitals, the C≡C triple bond in acetylene consists of three parts. When two *sp*-hybridized carbon atoms approach each other, *sp* hybrid orbitals from each overlap head-on to form a strong *sp*–*sp* σ bond. At the same time, the p_z orbitals from each carbon form a p_z–p_z π bond by sideways overlap, and the p_y orbitals overlap similarly to form a p_y–p_y π bond. The net effect is the formation of one σ bond and two π bonds—a carbon–carbon triple bond. Each of the remaining *sp* hybrid orbitals forms a σ bond to hydrogen to complete the acetylene molecule (Figure 1.12).

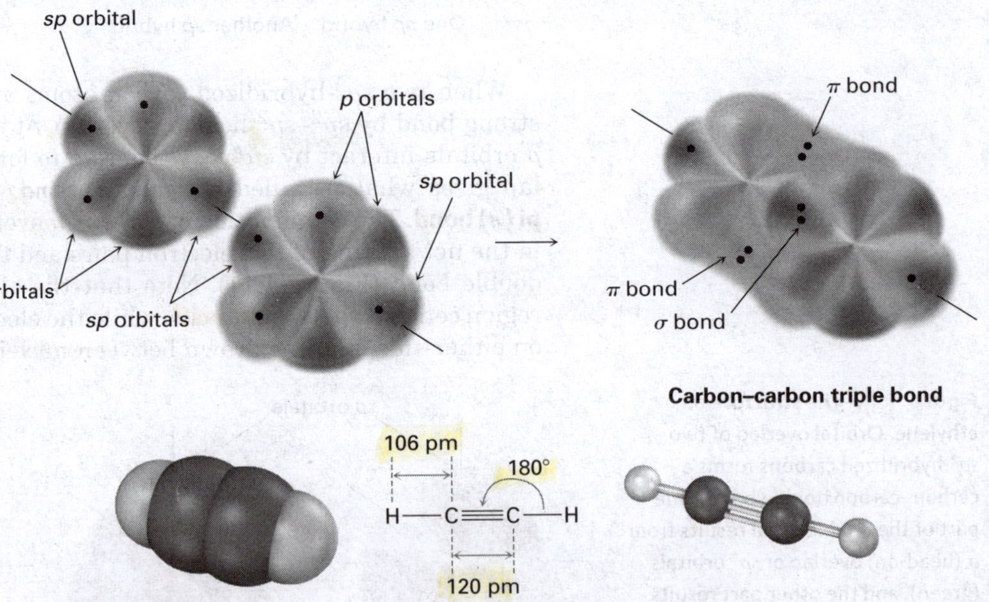

Figure 1.12 The structure of acetylene. The two *sp*-hybridized carbon atoms are joined by one *sp*–*sp* σ bond and two *p*–*p* π bonds.

As suggested by *sp* hybridization, acetylene is a linear molecule with H–C≡C bond angles of 180°. The C≡C bond has a length of 120 pm and a strength of about 965 kJ/mol (231 kcal/mol), making it the shortest and strongest of any carbon–carbon bond.

Worked Example 1.4

Drawing Electron-Dot and Line-Bond Structures

Formaldehyde, CH_2O, contains a carbon–*oxygen* double bond. Draw electron-dot and line-bond structures of formaldehyde, and indicate the hybridization of the carbon atom.

Strategy We know that hydrogen forms one covalent bond, carbon forms four, and oxygen forms two. Trial and error, combined with intuition, must be used to fit the atoms together.

Solution There is only one way that two hydrogens, one carbon, and one oxygen can combine:

Electron-dot structure **Line-bond structure**

Like the carbon atoms in ethylene, the carbon atom in formaldehyde is sp^2-hybridized.

Problem 1.11 Draw both an electron-dot and a line-bond structure for acetaldehyde, CH_3CHO.

Problem 1.12 Draw a line-bond structure for propene, $CH_3CH{=}CH_2$. Indicate the hybridization of each carbon, and predict the value of each bond angle.

Problem 1.13 Draw a line-bond structure for propyne, $CH_3C{\equiv}CH$. Indicate the hybridization of each carbon, and predict a value for each bond angle.

Problem 1.14 Draw a line-bond structure for buta-1,3-diene, $H_2C{=}CH{-}CH{=}CH_2$. Indicate the hybridization of each carbon, and predict a value for each bond angle.

Problem 1.15 Convert the following molecular model of aspirin into a line-bond structure, and identify the hybridization of each carbon atom (gray = C, red = O, ivory = H).

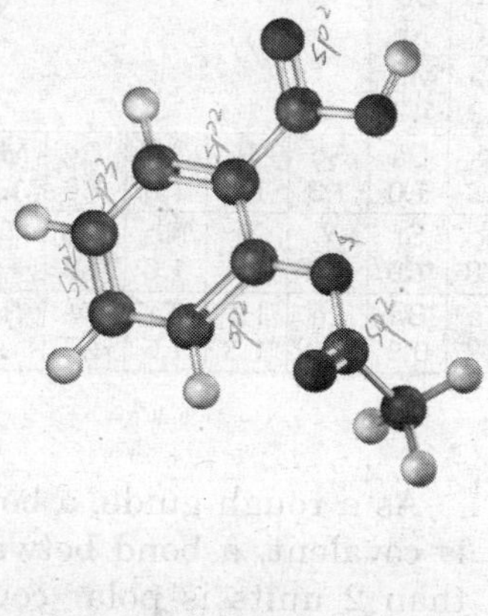

Aspirin (acetylsalicylic acid)

1.9 Polar Covalent Bonds: Electronegativity

Up to this point, we've treated chemical bonds as either ionic or covalent. The bond in sodium chloride, for instance, is ionic. Sodium transfers an electron to chlorine to give Na^+ and Cl^- ions, which are held together in the solid by electrostatic attractions between the unlike charges. The C–C bond in ethane, however, is covalent. The two bonding electrons are shared equally by the two equivalent carbon atoms, resulting in a symmetrical electron distribution in the bond. Most bonds, however, are neither fully ionic nor fully covalent but are somewhere between the two extremes. Such bonds are called **polar covalent bonds,** meaning that the bonding electrons are attracted more strongly by one atom than the other so that the electron distribution between atoms is not symmetrical (Figure 1.13).

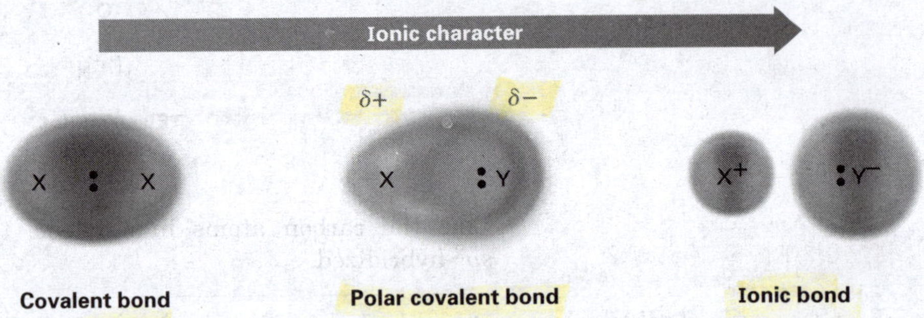

Figure 1.13 The continuum in bonding from covalent to ionic is a result of an unequal distribution of bonding electrons between atoms. The symbol δ (lowercase Greek delta) means *partial* charge, either partial positive (δ+) for the electron-poor atom or partial negative (δ−) for the electron-rich atom.

Bond polarity is due to differences in **electronegativity (EN),** the intrinsic ability of an atom to attract the shared electrons in a covalent bond. As shown in Figure 1.14, electronegativities are based on an arbitrary scale, with fluorine the most electronegative (EN = 4.0) and cesium the least (EN = 0.7). Metals on the left side of the periodic table attract electrons weakly and have lower electronegativities, while oxygen, nitrogen, and halogens on the right side of the periodic table attract electrons strongly and have higher electronegativities. Carbon, the most important element in organic compounds, has an electronegativity value of 2.5.

H 2.1																	He
Li 1.0	Be 1.6											B 2.0	C 2.5	N 3.0	O 3.5	F 4.0	Ne
Na 0.9	Mg 1.2											Al 1.5	Si 1.8	P 2.1	S 2.5	Cl 3.0	Ar
K 0.8	Ca 1.0	Sc 1.3	Ti 1.5	V 1.6	Cr 1.6	Mn 1.5	Fe 1.8	Co 1.9	Ni 1.9	Cu 1.9	Zn 1.6	Ga 1.6	Ge 1.8	As 2.0	Se 2.4	Br 2.8	Kr
Rb 0.8	Sr 1.0	Y 1.2	Zr 1.4	Nb 1.6	Mo 1.8	Tc 1.9	Ru 2.2	Rh 2.2	Pd 2.2	Ag 1.9	Cd 1.7	In 1.7	Sn 1.8	Sb 1.9	Te 2.1	I 2.5	Xe
Cs 0.7	Ba 0.9	La 1.0	Hf 1.3	Ta 1.5	W 1.7	Re 1.9	Os 2.2	Ir 2.2	Pt 2.2	Au 2.4	Hg 1.9	Tl 1.8	Pb 1.9	Bi 1.9	Po 2.0	At 2.1	Rn

Figure 1.14 Electronegativity values and trends. Electronegativity generally increases from left to right across the periodic table and decreases from top to bottom. The values are on an arbitrary scale, with F = 4.0 and Cs = 0.7. Elements in orange are the most electronegative, those in yellow are medium, and those in green are the least electronegative.

As a rough guide, a bond between atoms with similar electronegativities is covalent, a bond between atoms whose electronegativities differ by less than 2 units is polar covalent, and a bond between atoms whose electronegativities differ by 2 units or more is largely ionic. A carbon–hydrogen bond, for instance, is relatively nonpolar because carbon and hydrogen have similar electronegativities. A bond between carbon and a *more* electronegative element such as oxygen or chlorine, however, is polar covalent. The electrons in such a bond are drawn away from carbon toward the more electronegative atom, leaving the carbon with a partial positive charge, denoted δ+, and leaving the more electronegative atom with a partial negative charge, denoted δ− (δ is the lowercase Greek letter delta). An example is the C–O bond in methanol, CH_3OH (Figure 1.15a).

A bond between carbon and a *less* electronegative element is polarized so that carbon bears a partial negative charge and the other atom bears a partial positive charge. An example is the C–Li bond in methyllithium, CH_3Li (Figure 1.15b).

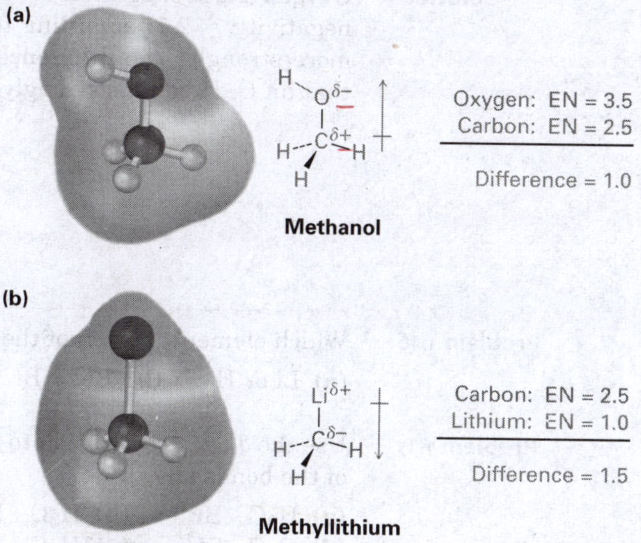

Figure 1.15 **(a)** Methanol, CH_3OH, has a polar covalent C–O bond, and **(b)** methyllithium, CH_3Li, has a polar covalent C–Li bond. The computer-generated representations, called electrostatic potential maps, use color to show calculated charge distributions, ranging from red (electron-rich; δ−) to blue (electron-poor; δ+).

Note in the representations of methanol and methyllithium in Figure 1.15 that a crossed arrow ⟶ is used to indicate the direction of bond polarity. By convention, *electrons are displaced in the direction of the arrow.* The tail of the arrow (which looks like a plus sign) is electron-poor (δ+), and the head of the arrow is electron-rich (δ–).

Note also in Figure 1.15 that charge distributions in a molecule can be displayed visually with what are called *electrostatic potential maps,* which use color to indicate electron-rich (red) and electron-poor (blue) regions. In methanol, oxygen carries a partial negative charge and is colored red, while the carbon and hydrogen atoms carry partial positive charges and are colored blue-green. In methyllithium, lithium carries a partial positive charge (blue), while carbon and the hydrogen atoms carry partial negative charges (red). Electrostatic potential maps are useful because they show at a glance the electron-rich and electron-poor atoms in molecules. We'll make frequent use of these maps throughout the text and will see how electronic structure often correlates with chemical reactivity.

When speaking of an atom's ability to polarize a bond, we often use the term *inductive effect.* An **inductive effect** is simply the shifting of electrons in a σ bond in response to the electronegativity of nearby atoms. Metals, such as lithium and magnesium, inductively donate electrons, whereas reactive nonmetals, such as oxygen and nitrogen, inductively withdraw electrons. Inductive effects play a major role in understanding chemical reactivity, and we'll use them many times throughout this text to explain a variety of chemical phenomena.

Worked Example 1.5 **Predicting the Polarity of Bonds**

Predict the extent and direction of polarization of the O–H bonds in H_2O.

Strategy Look at the electronegativity table in Figure 1.14 to see which atoms attract electrons more strongly.

Solution Oxygen (electronegativity = 3.5) is more electronegative than hydrogen (electronegativity = 2.1) according to Figure 1.14, and it therefore attracts electrons more strongly. The difference in electronegativities (3.5 − 2.1 = 1.4) implies that an O–H bond is strongly polarized.

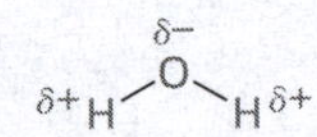

Problem 1.16 Which element in each of the following pairs is more electronegative?

(a) Li or H **(b)** Be or Br **(c)** Cl or I

Problem 1.17 Use the δ+/δ− convention to indicate the direction of expected polarity for each of the bonds shown:

(a) H_3C—Br **(b)** H_3C—NH_2 **(c)** H_2N—H
(d) H_3C—SH **(e)** H_3C—MgBr **(f)** H_3C—F

Problem 1.18 Order the bonds in the following compounds according to their increasing ionic character: CCl_4, $MgCl_2$, $TiCl_3$, Cl_2O.

Problem 1.19 Look at the following electrostatic potential map of chloromethane, and tell the direction of polarization of the C–Cl bond:

Chloromethane

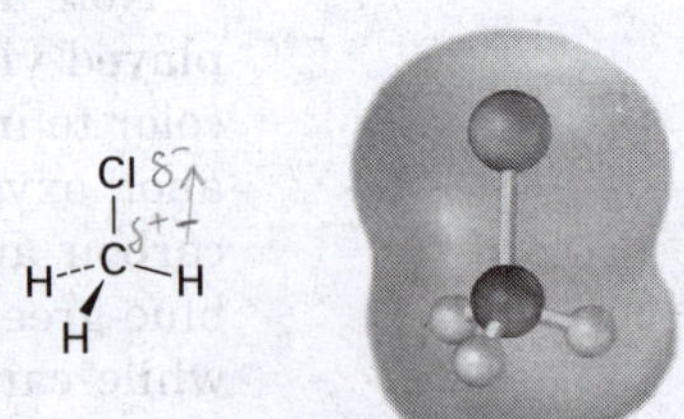

1.10 Acids and Bases: The Brønsted–Lowry Definition

A further important concept related to electronegativity and bond polarity is that of *acidity* and *basicity*. We'll soon see that the acid–base behavior of organic molecules helps explain much of their chemistry. You may recall from a course in general chemistry that two definitions of acidity are frequently used: the *Brønsted–Lowry definition* and the *Lewis definition*. Let's look at the Brønsted–Lowry definition first.

A **Brønsted–Lowry acid** is a substance that donates a hydrogen ion (H^+), and a **Brønsted–Lowry base** is a substance that accepts a hydrogen ion. (The name *proton* is often used as a synonym for H^+ because loss of the valence electron from a neutral hydrogen atom leaves only the hydrogen nucleus—a proton.) When hydrogen chloride gas dissolves in water, for instance, HCl donates a proton and a water molecule accepts the proton, yielding hydronium ion (H_3O^+) and chloride ion (Cl^-). Chloride ion, the product that results when the acid HCl loses a proton, is called the **conjugate base** of the acid, and H_3O^+, the product that results when the base H_2O gains a proton, is called the **conjugate acid** of the base.

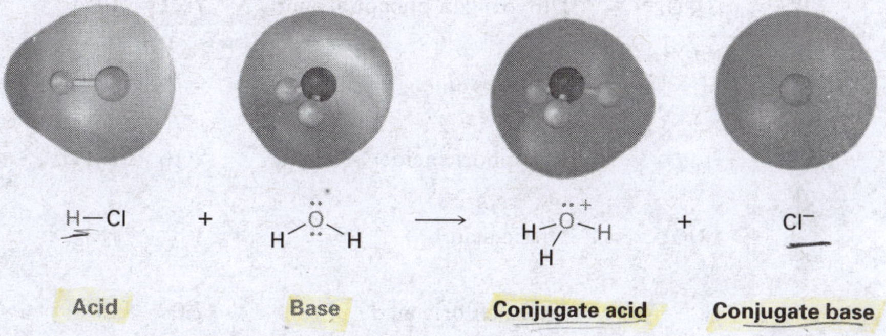

Acids differ in their ability to donate H^+. Stronger acids, such as HCl, react almost completely with water, whereas weaker acids, such as acetic acid (CH_3CO_2H), react only slightly. The exact strength of a given acid HA in water solution can be expressed by its **acidity constant, *Ka***. Remember from general chemistry that the concentration of solvent is ignored in the equilibrium expression and that brackets [ ] around a substance refer to the concentration of the enclosed species in moles per liter.

$$HA + H_2O \rightleftharpoons A^- + H_3O^+$$

$$K_a = \frac{[H_3O^+][A^-]}{[HA]}$$

Stronger acids have their equilibria toward the right and thus have larger acidity constants; weaker acids have their equilibria toward the left and have smaller acidity constants. The range of K_a values for different acids is enormous, running from about 10^{15} for the strongest acids to about 10^{-60} for the weakest. The common inorganic acids such as H_2SO_4, HNO_3, and HCl have K_a's in the range 10^2 to 10^9, while many organic acids have K_a's in the range 10^{-5} to 10^{-15}. As you gain more experience, you'll develop a rough feeling for which acids are "strong" and which are "weak" (remembering that the terms are always relative).

Acid strengths are normally given using pK_a values rather than K_a values, where the **p*K*$_a$** is the negative common logarithm of the K_a.

$$pK_a = -\log K_a$$

A *stronger acid* (larger K_a) has a *smaller* pK_a, and a *weaker acid* (smaller K_a) has a *larger* pK_a. Table 1.2 lists the pK_a's of some common acids in order of their strength.

Table 1.2 Relative Strengths of Some Common Acids and Their Conjugate Bases

	Acid	Name	pK_a	Conjugate base	Name	
Weaker acid	CH_3CH_2OH	Ethanol	16.00	$CH_3CH_2O^-$	Ethoxide ion	Stronger base
	H_2O	Water	15.74	HO^-	Hydroxide ion	
	HCN	Hydrocyanic acid	9.31	CN^-	Cyanide ion	
	$H_2PO_4^-$	Dihydrogen phosphate ion	7.21	HPO_4^{2-}	Hydrogen phosphate ion	
	CH_3CO_2H	Acetic acid	4.76	$CH_3CO_2^-$	Acetate ion	
	H_3PO_4	Phosphoric acid	2.16	$H_2PO_4^-$	Dihydrogen phosphate ion	
	HNO_3	Nitric acid	−1.3	NO_3^-	Nitrate ion	
Stronger acid	HCl	Hydrochloric acid	−7.0	Cl^-	Chloride ion	Weaker base

Notice that the pK_a value shown in Table 1.2 for water is 15.74, which results from the following calculation. Because water is both the acid and the solvent, the equilibrium expression is

$$\underset{\text{(acid)}}{H_2O} + \underset{\text{(solvent)}}{H_2O} \rightleftharpoons OH^- + H_3O^+$$

$$K_a = \frac{[H_3O^+][A^-]}{[HA]} = \frac{[H_3O^+][OH^-]}{[H_2O]} = \frac{[1.0 \times 10^{-7}][1.0 \times 10^{-7}]}{[55.4]} = 1.8 \times 10^{-16}$$

$$pK_a = 15.74$$

The numerator in this expression is the so-called ion-product constant for water, $K_w = [H_3O^+][OH^-] = 1.00 \times 10^{-14}$, and the denominator is the molar concentration of pure water, $[H_2O] = 55.4$ M at 25 °C. The calculation is artificial in that the concentration of "solvent" water is ignored while the concentration of "acid" water is not, but it is nevertheless useful in allowing us to make a comparison of water with other weak acids on a similar footing.

Notice also in Table 1.2 that there is an inverse relationship between the acid strength of an acid and the base strength of its conjugate base. A *strong* acid yields a *weak* conjugate base, and a *weak* acid yields a *strong* conjugate base. To understand this inverse relationship, think about what is happening to the acidic hydrogen in an acid–base reaction: a strong acid is one that loses H^+ easily, meaning that its conjugate base holds the H^+ weakly and is therefore a weak base. A weak acid is one that loses H^+ with difficulty, meaning that its conjugate base *does* hold the proton tightly and is therefore a strong base. The fact that HCl is a strong acid, for example, means that Cl^- does not hold H^+ tightly and is thus a weak base. Water, on the other hand, is a weak acid, meaning that OH^- holds H^+ tightly and is a strong base.

A proton always goes *from* the stronger acid *to* the stronger base in an acid–base reaction. That is, an acid donates a proton to the conjugate base of any acid with a larger pK_a, and the conjugate base of an acid removes a proton from any acid with a smaller pK_a. For example, the data in Table 1.2 indicate that OH^- reacts with acetic acid, CH_3CO_2H, to yield acetate ion, $CH_3CO_2^-$, and H_2O. Because water (pK_a = 15.74) is a weaker acid than acetic acid (pK_a = 4.76), hydroxide ion holds a proton more tightly than acetate ion does.

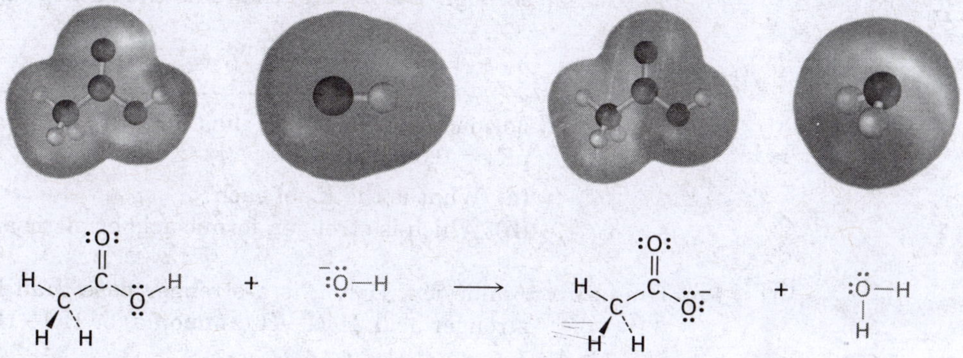

Acetic acid (pK_a = 4.76) | Hydroxide ion | Acetate ion | Water (pK_a = 15.74)

Another way to predict acid–base reactivity is to remember that the product conjugate acid in an acid–base reaction must be weaker and less reactive than the starting acid and that the product conjugate base must be weaker and less reactive than the starting base. In the reaction of acetic acid with hydroxide ion, for example, the product conjugate acid (H_2O) is weaker than the starting acid (CH_3CO_2H) and the product conjugate base ($CH_3CO_2^-$) is weaker than the starting base (OH^-).

$$\underset{\text{Stronger acid}}{CH_3\overset{O}{\overset{\|}{C}}OH} + \underset{\text{Stronger base}}{HO^-} \rightleftharpoons \underset{\text{Weaker acid}}{HOH} + \underset{\text{Weaker base}}{CH_3\overset{O}{\overset{\|}{C}}O^-}$$

Worked Example 1.6 Predicting Acid–Base Reactions

Water has pK_a = 15.74, and acetylene has pK_a = 25. Which of the two is more acidic? Will hydroxide ion react with acetylene?

$$\underset{\text{Acetylene}}{H{-}C{\equiv}C{-}H} + OH^- \xrightarrow{?} H{-}C{\equiv}C{:}^- + H_2O$$

Strategy In comparing two acids, the one with the smaller pK_a is stronger. Thus, water is a stronger acid than acetylene.

Solution Because water loses a proton more easily than acetylene, the HO^- ion has less affinity for a proton than the $HC{\equiv}C{:}^-$ ion. In other words, the anion of acetylene is a stronger base than hydroxide ion, and the reaction will not proceed as written.

Worked Example 1.7 **Calculating K_a from pK_a**

Butanoic acid, the substance responsible for the odor of rancid butter, has pK_a = 4.82. What is its K_a?

Strategy Since pK_a is the negative logarithm of K_a, it's necessary to use a calculator with an ANTILOG or INV LOG function. Enter the value of the pK_a (4.82), change the sign (–4.82), and then find the antilog (1.5×10^{-5}).

Solution $K_a = 1.5 \times 10^{-5}$

Problem 1.20 Formic acid, HCO_2H, has pK_a = 3.75, and picric acid, $C_6H_3N_3O_7$, has pK_a = 0.38.

(a) What is the K_a of each?

(b) Which is stronger, formic acid or picric acid?

Problem 1.21 Amide ion, H_2N^-, is a stronger base than hydroxide ion, HO^-. Which is the stronger acid, H_2N—H (ammonia) or HO—H (water)? Explain.

Problem 1.22 Is either of the following reactions likely to take place according to the pK_a data in Table 1.2?

(a) $HCN + CH_3CO_2^-\ Na^+ \longrightarrow Na^+\ ^-CN + CH_3CO_2H$

(b) $CH_3CH_2OH + Na^+\ ^-CN \longrightarrow CH_3CH_2O^-\ Na^+ + HCN$

1.11 Organic Acids and Organic Bases

Many of the reactions we'll be seeing in future chapters, including essentially all biological reactions, involve organic acids and organic bases. Organic acids are characterized by the presence of a positively polarized hydrogen atom (blue in electrostatic potential maps) and are of two main kinds: those acids such as methanol and acetic acid that contain a hydrogen atom bonded to an electronegative oxygen atom (O—H) and those such as acetone that contain a hydrogen atom bonded to a carbon atom next to a C=O double bond (O=C—C—H). We'll see the reasons for this behavior in Chapters 8 and 11.

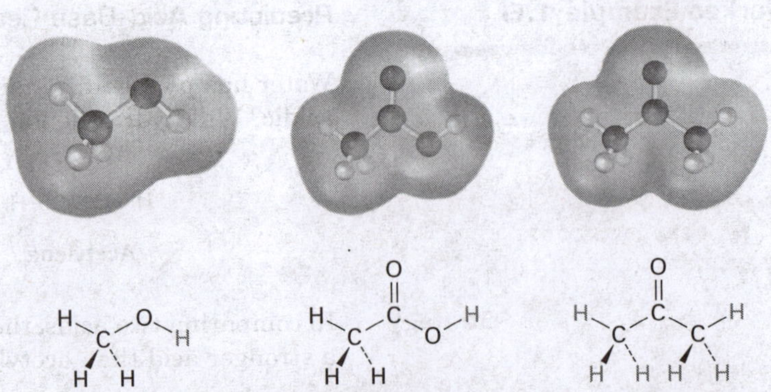

Some organic acids

Methanol (pK_a = 15.54) | Acetic acid (pK_a = 4.76) | Acetone (pK_a = 19.3)

Compounds called *carboxylic acids,* which contain the $-CO_2H$ grouping, are particularly common. They occur abundantly in all living organisms and are involved in almost all metabolic pathways. Acetic acid, pyruvic acid, and citric acid are examples.

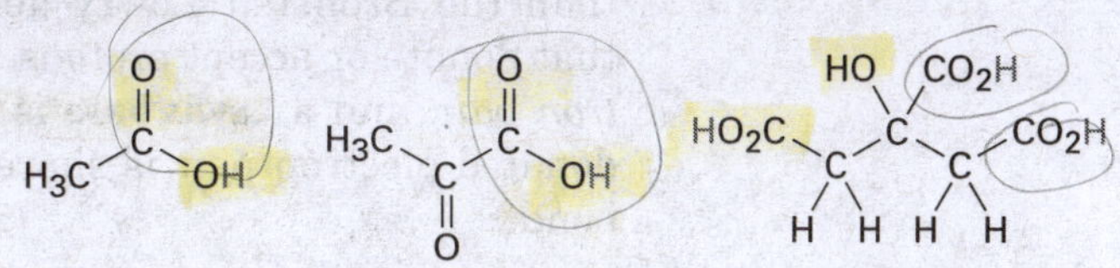

Acetic acid **Pyruvic acid** **Citric acid**

Organic bases are characterized by the presence of an atom (reddish in electrostatic potential maps) with a lone pair of electrons that can bond to H^+. Nitrogen-containing compounds such as methylamine are the most common organic bases, but oxygen-containing compounds can also act as bases when reacting with a sufficiently strong acid. Note that some oxygen-containing compounds can act as both acids and bases depending on the circumstances, just as water can. Methanol and acetone, for instance, act as *acids* when they donate a proton but act as bases when their oxygen atom accepts a proton.

Some organic bases

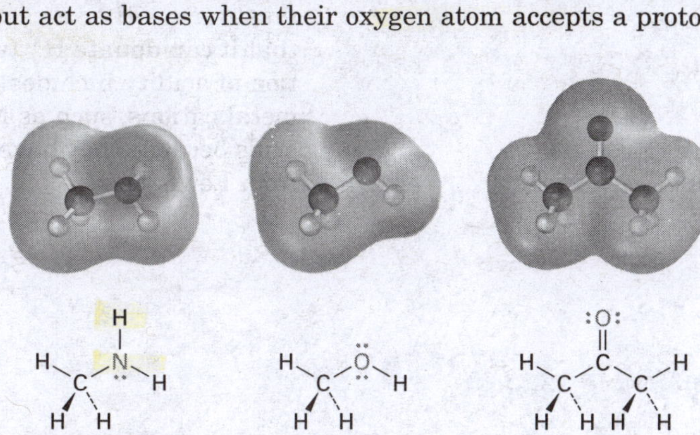

Methylamine **Methanol** **Acetone**

We'll see in Chapter 15 that substances called *amino acids,* so named because they are both amines ($-NH_2$) and carboxylic acids ($-CO_2H$), are the building blocks from which the proteins present in all living organisms are made. Twenty different amino acids go into making up proteins; alanine is an example. Interestingly, alanine and other amino acids exist primarily in a doubly charged form called a *zwitterion* rather than in the uncharged form. The zwitterion form arises because amino acids have both acidic and basic sites within the same molecule and therefore undergo an *internal* acid–base reaction.

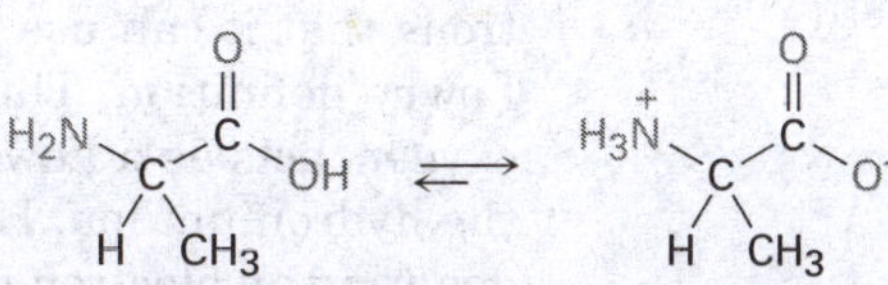

Alanine (uncharged form) **Alanine (zwitterion form)**

1.12 Acids and Bases: The Lewis Definition

The Lewis definition of acids and bases is broader and more encompassing than the Brønsted–Lowry definition because it's not limited to substances that donate or accept protons. A **Lewis acid** is a substance that *accepts an electron pair*, and a **Lewis base** is a substance that *donates an electron pair*. The donated electron pair is shared between the acid and the base in a covalent bond.

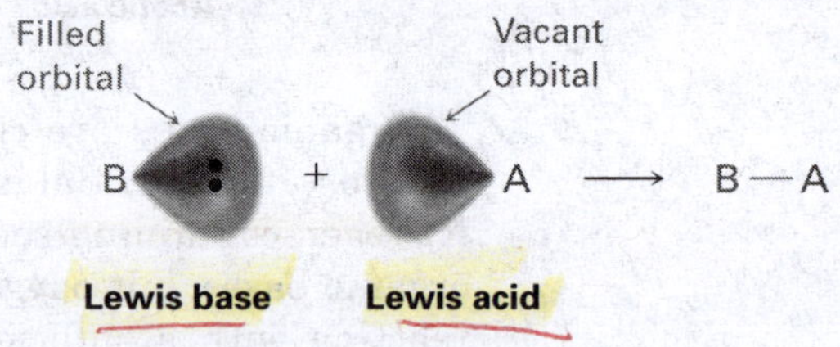

The fact that a Lewis acid is able to accept an electron pair means that it must have either a vacant, low-energy orbital or a polar bond to hydrogen so that it can donate H^+ (which has an empty 1*s* orbital). Thus, the Lewis definition of acidity includes many species in addition to H^+. For example, various metal cations, such as Mg^{2+}, and metal compounds, such as $AlCl_3$, are Lewis acids because they have unfilled valence orbitals and can accept electron pairs from Lewis bases.

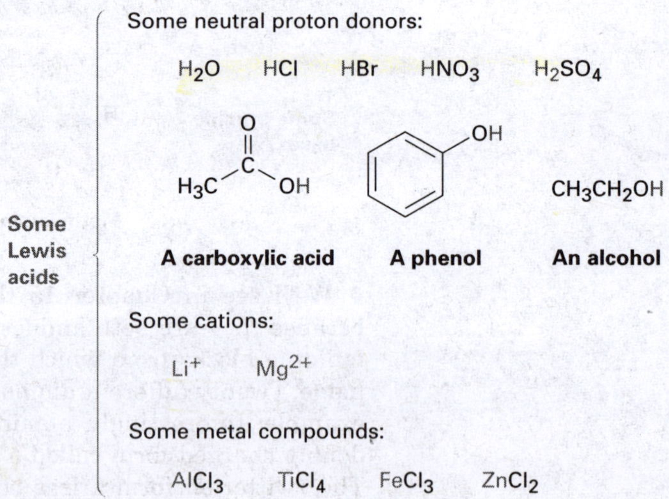

The Lewis definition of a base—a compound with a pair of nonbonding electrons that it can use in bonding to a Lewis acid—is similar to the Brønsted–Lowry definition. Thus, H_2O, with its two pairs of nonbonding electrons on oxygen, acts as a Lewis base by donating an electron pair to an H^+ in forming the hydronium ion, H_3O^+. Similarly, trimethylamine acts as a Lewis base by donating an electron pair on its nitrogen atom to aluminum chloride. In a more general sense, most oxygen- and nitrogen-containing organic compounds can act as Lewis bases because they have lone pairs of electrons.

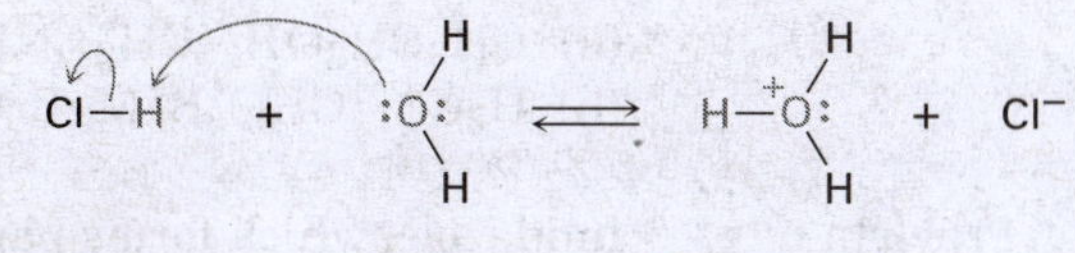

Hydrogen chloride (Lewis acid) **Water** (Lewis base) **Hydronium ion**

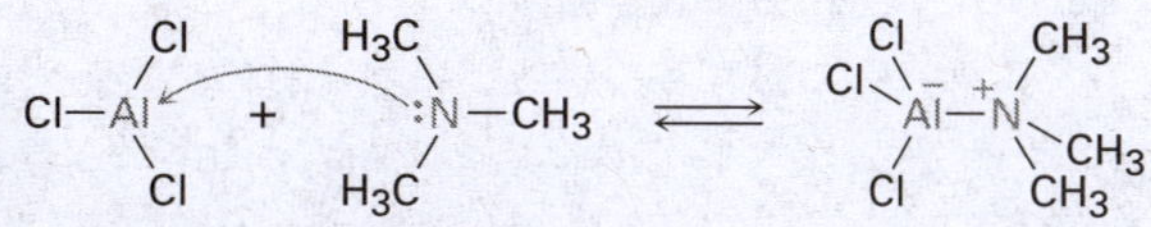

Aluminum trichloride (Lewis acid) **Trimethylamine** (Lewis base)

Look closely at the two acid–base reactions just shown. In the first reaction, the Lewis base water uses an electron pair to abstract H^+ from the polar HCl molecule. In the second reaction, the Lewis base trimethylamine donates an electron pair to a vacant valence orbital of an aluminum atom. In both reactions, the direction of electron-pair flow from the electron-rich Lewis base to the electron-poor Lewis acid is shown using curved arrows. *A curved arrow always means that a pair of electrons moves from the atom at the tail of the arrow to the atom at the head of the arrow.* We'll use this curved-arrow notation frequently in the remainder of this text to indicate electron flow during reactions.

Worked Example 1.8 Using Curved Arrows to Show Electron Flow

Using curved arrows, show how acetaldehyde, CH_3CHO, can act as a Lewis base in a reaction with a strong acid, H^+.

Strategy A Lewis base donates an electron pair to a Lewis acid. We therefore need to locate the electron lone pairs on acetaldehyde and use a curved arrow to show the movement of an electron pair from the oxygen toward a strong acid.

Solution

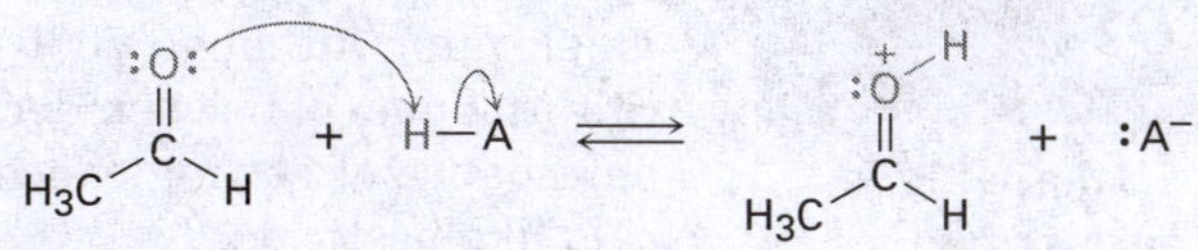

Acetaldehyde

Problem 1.23 Which of the following are likely to act as Lewis acids and which as Lewis bases? Which might act both ways?

(a) CH_3CH_2OH **(b)** $(CH_3)_2NH$ **(c)** $MgBr_2$
(d) $(CH_3)_3B$ **(e)** H_3C^+ **(f)** $(CH_3)_3P$

Problem 1.24 Show how the species in part **(a)** can act as Lewis bases in their reactions with HCl, and show how the species in part **(b)** can act as Lewis acids in their reaction with OH^-:

(a) CH_3CH_2OH, $(CH_3)_2NH$, $(CH_3)_3P$
(b) H_3C^+, $(CH_3)_3B$, $MgBr_2$

Problem 1.25 Imidazole, which forms part of the structure of the amino acid histidine, can act as both an acid and a base. Look at the electrostatic potential map of imidazole, and identify the most acidic hydrogen atom and the most basic nitrogen atom.

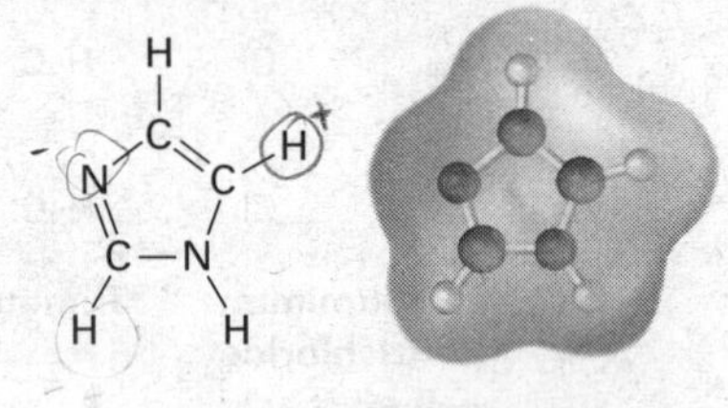

Imidazole

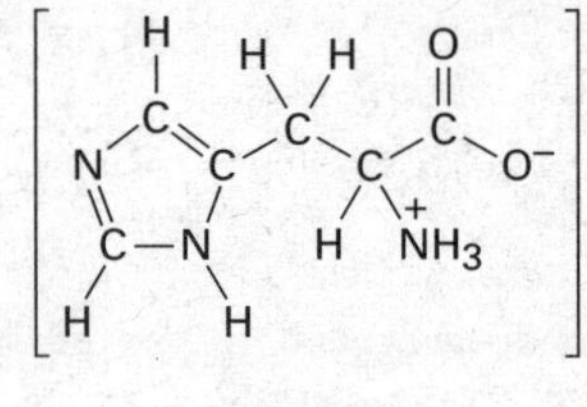

Histidine

INTERLUDE

Organic Foods: Risk versus Benefit

Purestock/Jupiter Images

How dangerous is the pesticide being sprayed on this crop?

Contrary to what you may hear in supermarkets or on television, all foods are organic—complex mixtures of organic molecules. Even so, when applied to food, the word *organic* has come to mean an absence of synthetic chemicals, typically pesticides. How concerned should we be about traces of pesticides in the food we eat? Or toxins in the water we drink? Or pollutants in the air we breathe?

Life is not risk-free—we all take many risks each day without even thinking about it. We decide to ride a bike rather than drive, although there is a ten times greater likelihood per mile of dying in a bicycling accident than in a car. We decide to walk down stairs rather than take an elevator, although 7000 people die from falls each year in the United States. Some of us even decide to smoke cigarettes, although it increases our chance of getting cancer by 50%. But what about risks from chemicals like pesticides?

One thing is certain: without pesticides, whether they target weeds (herbicides), insects (insecticides), or molds and fungi (fungicides), crop production would drop significantly and food prices would increase. Take the herbicide atrazine, for instance. In the United States alone,

continued

INTERLUDE

approximately 100 million pounds of atrazine are used each year to kill weeds in corn, sorghum, and sugar cane fields, greatly improving the yields of these crops. Nevertheless, the use of atrazine continues to be a concern because traces persist in the environment. Indeed, heavy atrazine exposure *can* pose health risks to humans and some animals, but the U.S. Environmental Protection Agency (EPA) is unwilling to ban its use because doing so would result in significantly lower crop yields and increased food costs, and because there is no suitable alternative herbicide available.

How can the potential hazards from a chemical like atrazine be determined? Risk evaluation of chemicals is carried out by exposing test animals, usually mice or rats, to the chemical and then monitoring the animals for signs of harm. To limit the expense and time needed, the amounts administered are typically hundreds or thousands of times greater than those a person might normally encounter. The results obtained in animal tests are then distilled into a single number called an *LD_{50} value*, the amount of substance per kilogram body weight that is a lethal dose for 50% of the test animals. For atrazine, the LD_{50} value is between 1 and 4 g/kg depending on the animal species. Aspirin, for comparison, has an LD_{50} of 1.1 g/kg, and Table 1.3 lists values for some other familiar substances. The lower the value, the more toxic the substance. Note, though, that LD_{50} values tell only about the effects of very heavy exposure for a relatively short time. They say nothing about the risks of long-term exposure, such as whether the substance can cause cancer or interfere with development in the unborn.

Table 1.3 Some LD_{50} Values

Substance	LD_{50} (g/kg)	Substance	LD_{50} (g/kg)
Strychnine	0.005	Chloroform	1.2
Arsenic trioxide	0.015	Iron(II) sulfate	1.5
DDT	0.115	Ethyl alcohol	10.6
Aspirin	1.1	Sodium cyclamate	17

So, should we still use atrazine? All decisions involve tradeoffs, and the answer is rarely obvious. Does the benefit of increased food production outweigh possible health risks of a pesticide? Do the beneficial effects of a new drug outweigh a potentially dangerous side effect in a small number of users? Different people will have different opinions, but an honest evaluation of facts is surely the best way to start. At present, atrazine is approved for continued use in the United States because the EPA believes that the benefits of increased food production outweigh possible health risks. At the same time, though, atrazine use is being phased out in Europe.

Summary and Key Words

The purpose of this chapter has been to get you up to speed—to review some ideas about atoms, bonds, and molecular geometry. As we've seen, **organic chemistry** is the study of carbon compounds. Although a division into inorganic and organic chemistry occurred historically, there is no scientific reason for the division.

An atom is composed of a positively charged nucleus surrounded by negatively charged electrons that occupy specific regions of space called **orbitals**. Different orbitals have different energy levels and shapes. For example, *s* orbitals are spherical and *p* orbitals are dumbbell-shaped.

There are two fundamental kinds of chemical bonds: ionic bonds and **covalent bonds**. The ionic bonds commonly found in inorganic salts result from the electrical attraction of unlike charges. The covalent bonds found in organic molecules result from the sharing of one or more electron pairs between atoms. Electron sharing occurs when two atoms approach and their atomic orbitals overlap. Bonds formed by head-on overlap of atomic orbitals are called **sigma (σ) bonds**, and bonds formed by sideways overlap of *p* orbitals are called **pi (π) bonds**.

In the valence bond description, carbon uses hybrid orbitals to form bonds in organic molecules. When forming only single bonds with tetrahedral geometry, carbon uses four equivalent ***sp*^{3} hybrid orbitals**. When forming double bonds, carbon has three equivalent ***sp*^{2} orbitals** with planar geometry and one unhybridized *p* orbital. When forming triple bonds, carbon has two equivalent ***sp* orbitals** with linear geometry and two unhybridized *p* orbitals.

Organic molecules often have **polar covalent bonds** because of unsymmetrical electron sharing caused by the **electronegativity** of atoms. A carbon–oxygen bond, for instance, is polar because oxygen attracts the bonding electrons more strongly than carbon does. A carbon–metal bond, by contrast, is polarized in the opposite sense because carbon attracts electrons more strongly than metals do.

A **Brønsted–Lowry acid** is a substance that can donate a proton (hydrogen ion, H^+), and a **Brønsted–Lowry base** is a substance that can accept a proton. The strength of an acid is given by its acidity constant, K_a. A **Lewis acid** is a substance that can accept an electron pair. A **Lewis base** is a substance that can donate an unshared electron pair. Most organic molecules that contain oxygen and nitrogen are Lewis bases.

WORKING PROBLEMS

There is no surer way to learn organic chemistry than by working problems. Although careful reading and rereading of this text are important, reading alone isn't enough. You must also be able to use the information you've read and be able to apply your knowledge in new situations. Working problems gives you practice at doing this.

Each chapter in this book provides many problems of different sorts. The in-chapter problems are placed for immediate reinforcement of ideas just learned; the end-of-chapter problems provide additional practice and are of several types. They begin with a short section called "Visualizing Chemistry," which helps you "see" the microscopic world of molecules and provides

continued

practice for working in three dimensions. After the visualizations are many "Additional Problems" that are grouped according to topic.

As you study organic chemistry, take the time to work the problems. Do the ones you can, and ask for help on the ones you can't. If you're stumped by a particular problem, check the accompanying *Study Guide and Solutions Manual* for an explanation that will help clarify the difficulty. Working problems takes effort, but the payoff in knowledge and understanding is immense.

Exercises

Visualizing Chemistry

(Problems 1.1–1.25 appear within the chapter.)

Interactive versions of these problems are assignable in OWL.

1.26 The following model is a representation of citric acid, a substance in the so-called citric acid cycle by which food molecules are metabolized in the body. Only the connections between atoms are shown; multiple bonds are not indicated. Complete the structure by indicating the positions of multiple bonds and lone-pair electrons (gray = C, red = O, ivory = H).

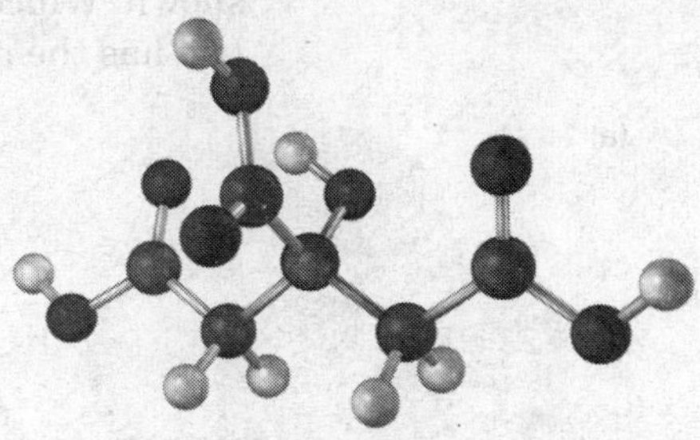

1.27 Convert each of the following molecular models into a line-bond structure, and give the formula of each (gray = C, red = O, blue = N, ivory = H).

(a)

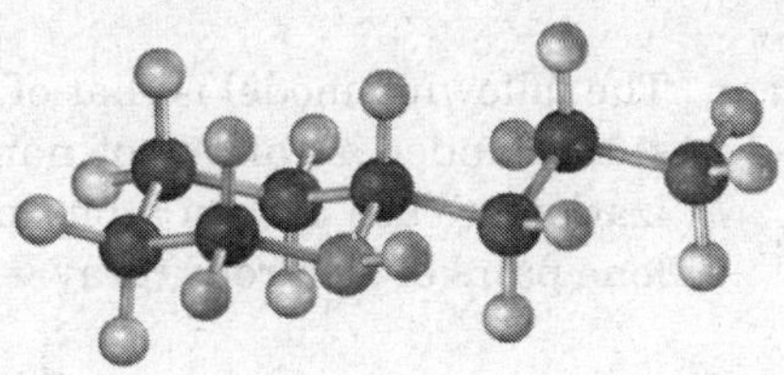

Coniine (the toxic substance in poison hemlock)

(b)

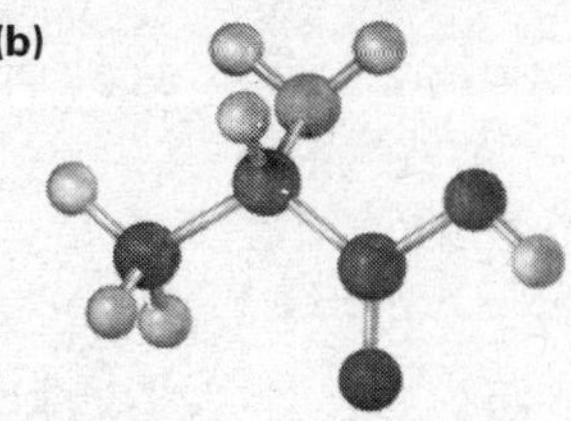

Alanine (an amino acid)

1.28 The following model is that of aspartame, $C_{14}H_{18}N_2O_5$, known commercially under many names, including NutraSweet. Only the connections between atoms are shown; multiple bonds are not indicated. Complete the structure by indicating the positions of multiple bonds (gray = C, red = O, blue = N, ivory = H).

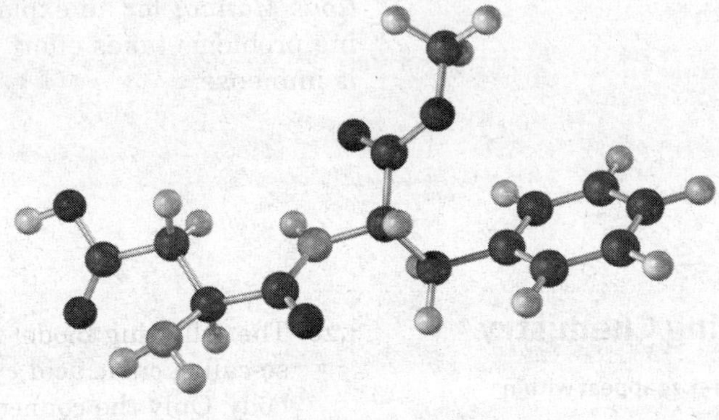

1.29 Electrostatic potential maps of **(a)** acetamide and **(b)** methylamine are shown. Which of the two has the more basic nitrogen atom? Which of the two has the more acidic hydrogen atoms?

(a)

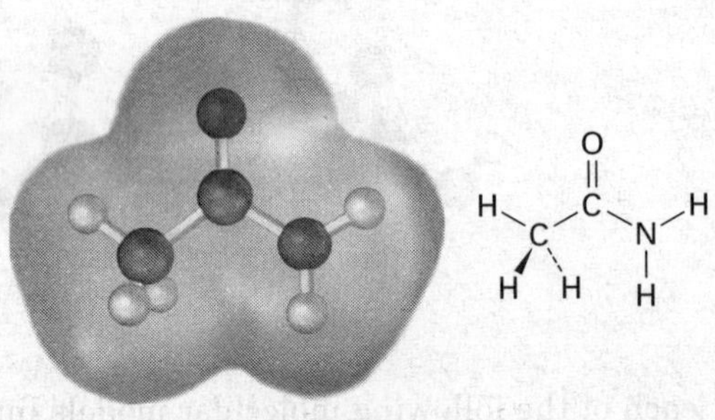

Acetamide

(b)

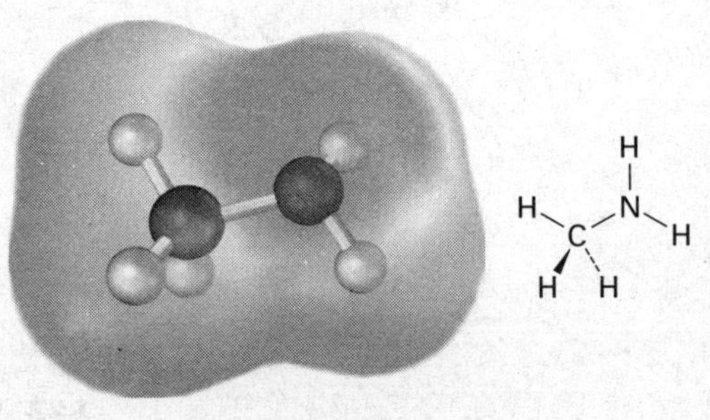

Methylamine

1.30 The following model is that of acetaminophen, a pain reliever sold in drugstores under a variety of names, including Tylenol. Identify the hybridization of each carbon atom in acetaminophen, and tell which atoms have lone pairs of electrons (gray = C, red = O, blue = N, ivory = H).

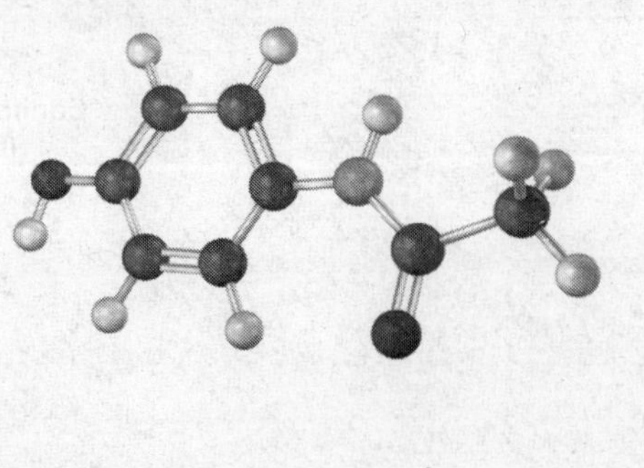

Additional Problems

ELECTRON CONFIGURATIONS

1.31 Give the ground-state electron configuration of the following elements. For example, carbon is $1s^2\ 2s^2\ 2p^2$.
(a) Lithium (b) Sodium (c) Aluminum (d) Sulfur

1.32 How many valence electrons does each of the following atoms have?
(a) Oxygen (b) Magnesium (c) Fluorine

ELECTRON-DOT STRUCTURES

1.33 Fill in any unshared electrons that are missing from the following line-bond structures:

(a) 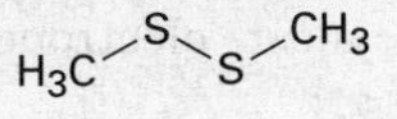
Dimethyl disulfide

(b) 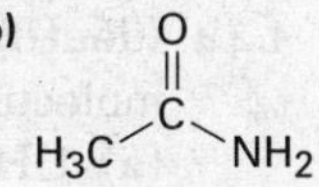
Acetamide

(c) 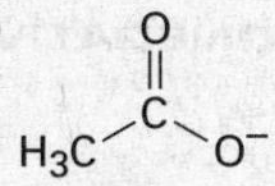
Acetate ion

1.34 Why can't molecules with the following formulas exist?
(a) CH_5 (b) C_2H_6N (c) $C_3H_5Br_2$

1.35 What are the likely formulas of the following molecules?
(a) $AlCl_?$ (b) $CF_2Cl_?$ (c) $NI_?$ (d) $CH_?O$

1.36 Write an electron-dot structure for acetonitrile, $CH_3C{\equiv}N$. How many electrons does the nitrogen atom have in its valence shell? How many are used for bonding, and how many are not used for bonding?

STRUCTURAL FORMULAS

1.37 Convert the following molecular formulas into line-bond structures:
(a) C_3H_8 (b) C_3H_7Br (two possibilities)
(c) C_3H_6 (two possibilities) (d) C_2H_6O (two possibilities)

1.38 Convert the following line-bond structures into molecular formulas:

(a) 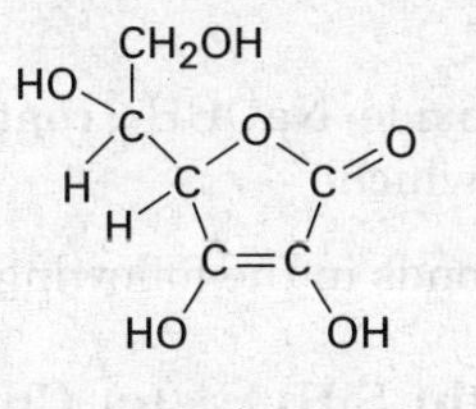
Vitamin C (ascorbic acid)

(b) 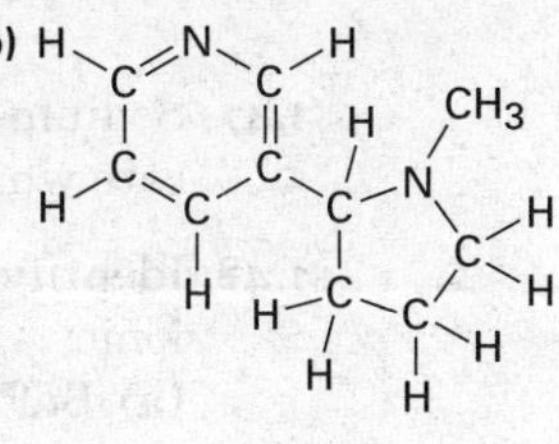
Nicotine

(c) 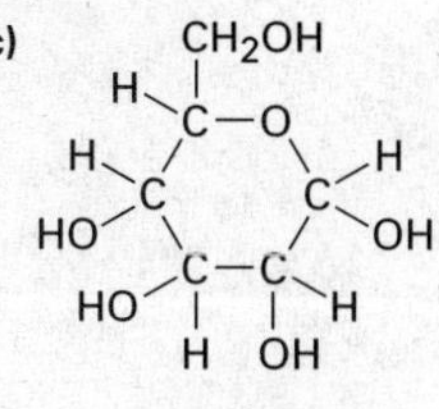
Glucose

1.39 Draw both an electron-dot structure and a line-bond structure for vinyl chloride, C_2H_3Cl, the starting material from which PVC [poly(vinyl chloride)] plastic is made.

1.40 Draw a three-dimensional representation of the OH-bearing carbon atom in ethanol, CH_3CH_2OH.

1.41 There are two structures with the formula C_4H_{10}. Draw them, and tell how they differ.

1.42 Oxaloacetic acid, an important intermediate in food metabolism, has the formula $C_4H_4O_5$ and contains three C=O bonds and two O–H bonds. Propose two possible structures.

1.43 Draw line-bond structures for the following molecules:
(a) Ethyl methyl ether, C_3H_8O, which contains an oxygen atom bonded to two carbons
(b) Butane, C_4H_{10}, which contains a chain of four carbon atoms
(c) Cyclohexene, C_6H_{10}, which contains a ring of six carbon atoms and one carbon–carbon double bond

ELECTRONEGATIVITY

1.44 Identify the most electronegative element in each of the following molecules:
(a) CH_2FCl **(b)** $FCH_2CH_2CH_2Br$
(c) $HOCH_2CH_2NH_2$ **(d)** CH_3OCH_2Li

1.45 Use Figure 1.14 to order the following molecules according to increasing positive character of the carbon atom:

CH_3F, CH_3OH, CH_3Li, CH_3I, CH_3CH_3, CH_3NH_2

1.46 We'll see in the next chapter that organic molecules can be classified according to the *functional groups* they contain, where a functional group is a collection of atoms with a characteristic chemical reactivity. Use the electronegativity values given in Figure 1.14 to predict the polarity of the following functional groups:

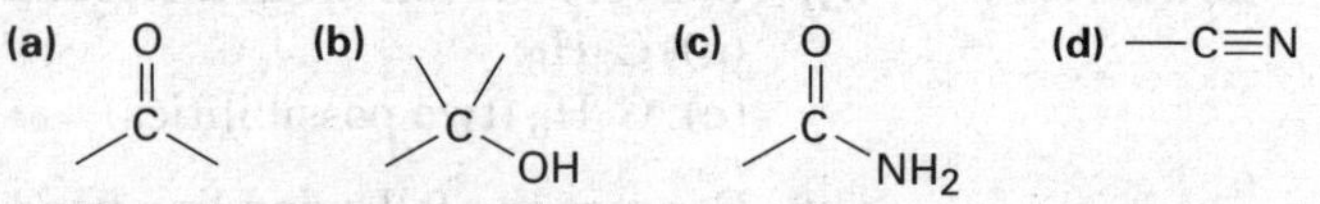

Ketone Alcohol Amide Nitrile

1.47 Sodium methoxide, $NaOCH_3$, contains both ionic and covalent bonds. Indicate which is which.

1.48 Identify the bonds in the following molecules as covalent, polar covalent, or ionic:
(a) BeF_2 **(b)** SiH_4 **(c)** CBr_4

1.49 Indicate which of the bonds in the following molecules are polar covalent, using the symbols $\delta+$ and $\delta-$.
(a) Br_2 **(b)** CH_3Cl **(c)** HF **(d)** CH_3CH_2OH

1.50 Use the electronegativity values in Figure 1.14 to predict which of the indicated bonds in each of the following sets is more polar. Tell the direction of the polarity in each.
(a) Cl—CH_3 or Cl—Cl **(b)** H—CH_3 or H—Cl
(c) HO—CH_3 or $(CH_3)_3$Si—CH_3

HYBRIDIZATION

1.51 Propose structures for molecules that meet the following descriptions:
(a) Contains two sp^2-hybridized carbons and two sp^3-hybridized carbons
(b) Contains only four carbons, all of which are sp^2-hybridized
(c) Contains two sp-hybridized carbons and two sp^2-hybridized carbons

1.52 What kind of hybridization do you expect for each carbon atom in the following molecules?

(a)
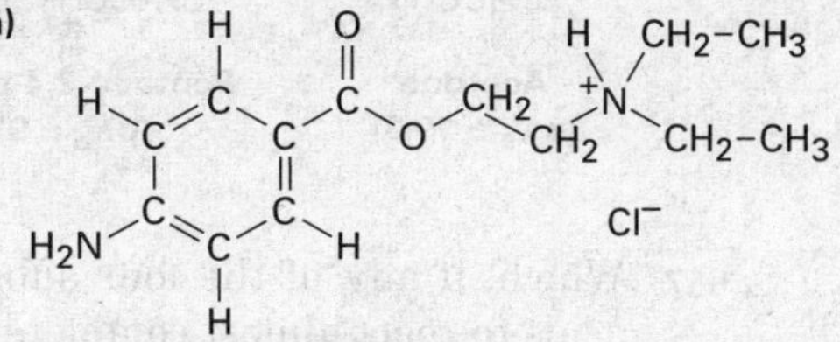

Procaine

(b)
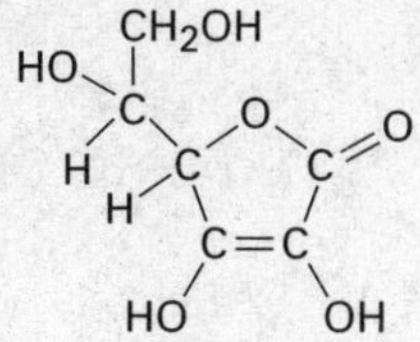

Vitamin C (ascorbic acid)

1.53 What is the hybridization of each carbon atom in benzene? What shape do you expect benzene to have?

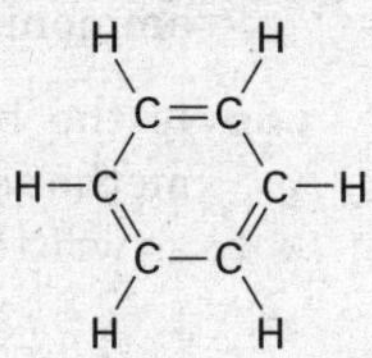

Benzene

1.54 What values do you expect for the indicated bond angles in each of the following molecules, and what kind of hybridization do you expect for the central atom in each?

(a)
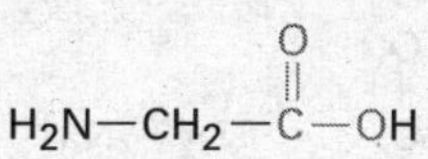

Glycine (an amino acid)

(b)
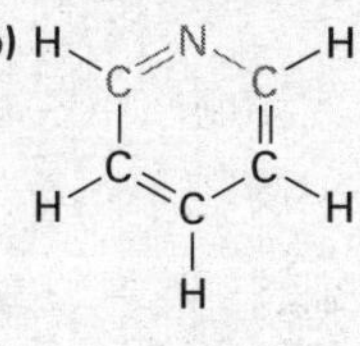

Pyridine

(c)
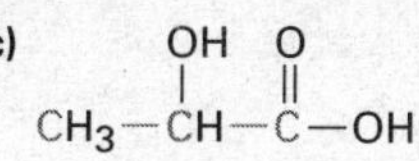

Lactic acid (in sour milk)

1.55 What is the hybridization of each carbon atom in acetonitrile, $CH_3C{\equiv}N$?

ACIDS AND BASES

1.56 Rank the following substances in order of increasing acidity:

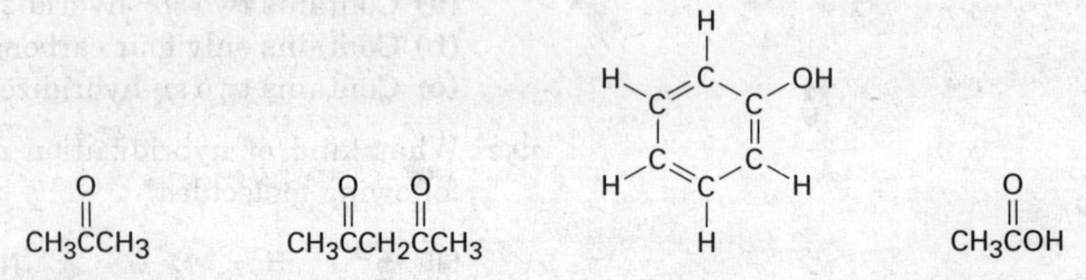

Acetone (pK_a = 19.3) | Pentane-2,4-dione (pK_a = 9) | Phenol (pK_a = 9.9) | Acetic acid (pK_a = 4.76)

1.57 Which, if any, of the four substances in Problem 1.56 are strong enough acids to react almost completely with NaOH? (The pK_a of H_2O is 15.7.)

1.58 Which of the following substances are likely to behave as Lewis acids and which as Lewis bases?

(a) $AlBr_3$ **(b)** $CH_3CH_2NH_2$ **(c)** HF **(d)** CH_3SCH_3

1.59 The ammonium ion (NH_4^+, pK_a = 9.25) has a lower pK_a than the methylammonium ion ($CH_3NH_3^+$, pK_a = 10.66). Which is the stronger base, ammonia (NH_3) or methylamine (CH_3NH_2)? Explain.

1.60 Is the bicarbonate anion (HCO_3^-) a strong enough base to react with methanol (CH_3OH)? In other words, does the following reaction take place as written? (The pK_a of methanol is 15.5; the pK_a of H_2CO_3 is 6.4.)

$$CH_3OH + HCO_3^- \xrightarrow{?} CH_3O^- + H_2CO_3$$

1.61 Ammonia, $H_2N{-}H$, has $pK_a \approx 36$, and acetone has $pK_a \approx 19$. Will the following reaction take place? Explain.

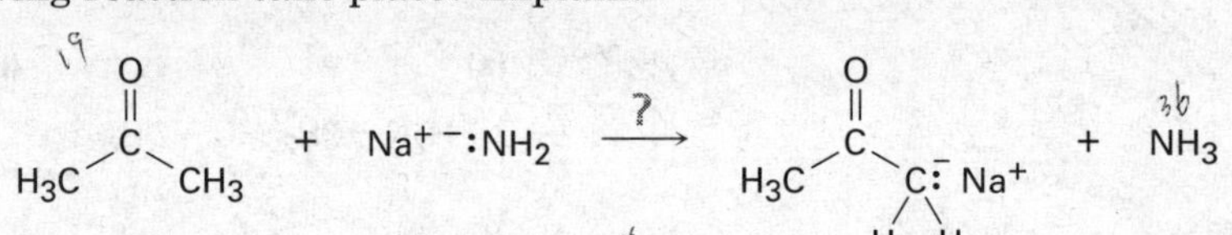

Acetone

1.62 Identify the acids and bases in the following reactions:

(a) $CH_3OH + H^+ \longrightarrow CH_3OH_2^+$

(b) $CH_3OH + NH_2^- \longrightarrow CH_3O^- + NH_3$

(c)

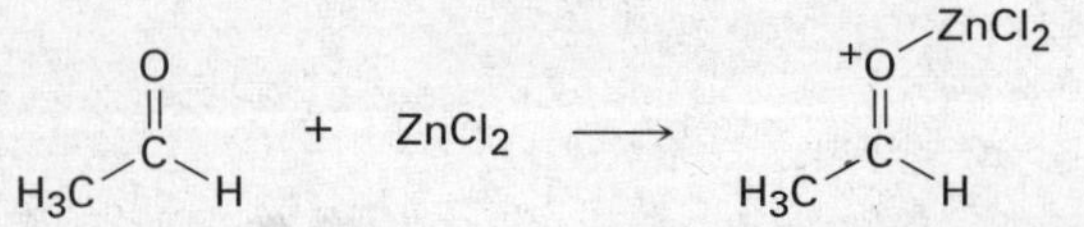

1.63 Predict the structure of the product formed in the reaction of the organic base pyridine with the organic acid acetic acid, and use curved arrows to indicate the direction of electron flow.

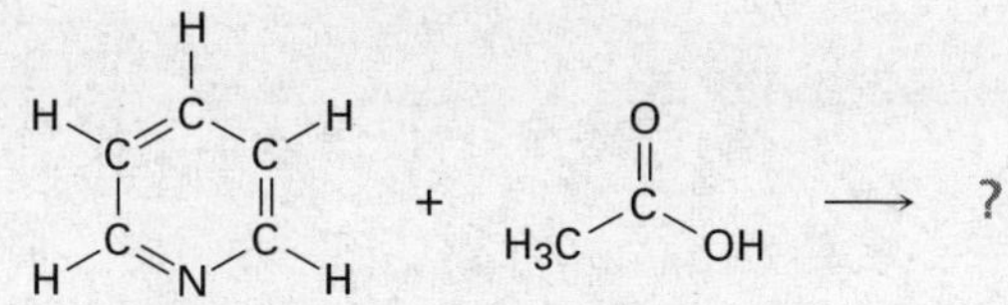

Pyridine **Acetic acid**

General Problems

1.64 Why do you suppose no one has ever been able to make cyclopentyne as a stable molecule?

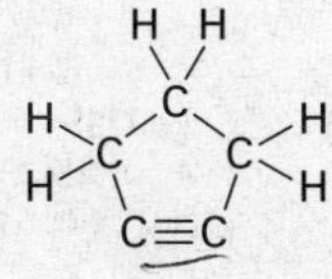

Cyclopentyne

1.65 Draw an orbital picture of allene, $H_2C{=}C{=}CH_2$. What hybridization must the central carbon atom have to form two double bonds? What shape does allene have?

1.66 Draw an electron-dot structure and an orbital picture for carbon dioxide, CO_2. What kind of hybridization does the carbon atom have? What is the relationship between CO_2 and allene (Problem 1.65)?

1.67 Although most stable organic compounds have tetravalent carbon atoms, high-energy species with trivalent carbon atoms also exist. *Carbocations* are one such class of compounds. If the positively charged carbon atom has planar geometry, what hybridization do you think it has? How many valence electrons does the carbon have?

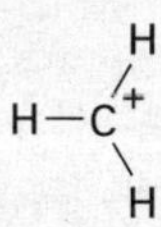

A carbocation

1.68 The ammonium ion, NH_4^+, has a geometry identical to that of methane, CH_4. What kind of hybridization do you think the nitrogen atom has? Explain.

1.69 Complete the electron-dot structure of caffeine, showing all lone-pair electrons, and identify the hybridization of the indicated atoms.

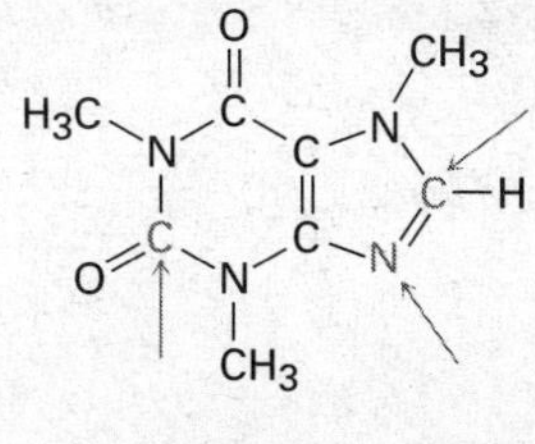

Caffeine

IN THE MEDICINE CABINET **1.70** Nonsteroidal anti-inflammatory drugs, often referred to as NSAIDs, are commonly used to treat minor aches and pains. Four of the most common NSAIDs are:

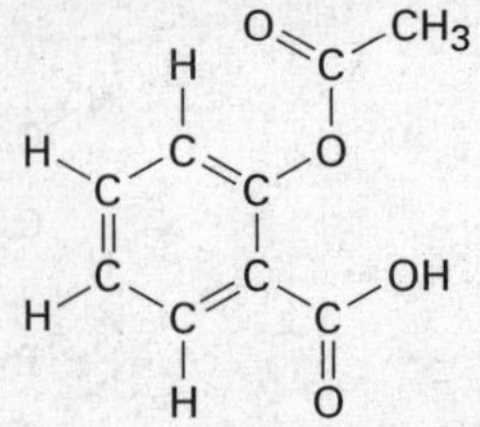

Acetylsalicylic acid (aspirin)

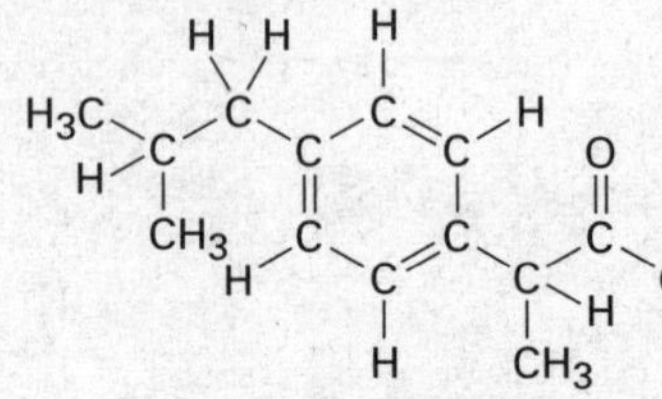

Ibuprofen (Advil, Nuprin, Motrin)

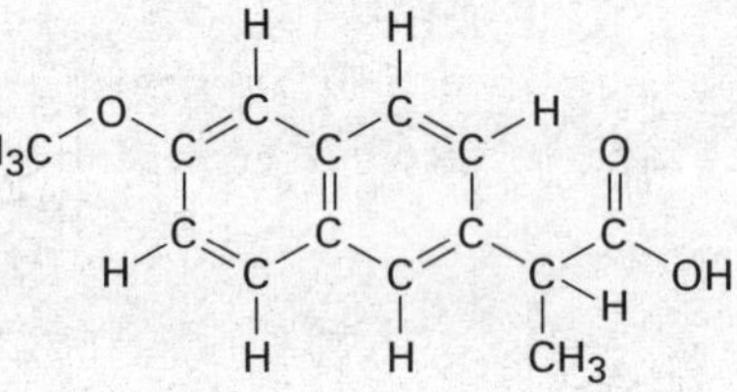

Naproxen (Naprosyn, Aleve)

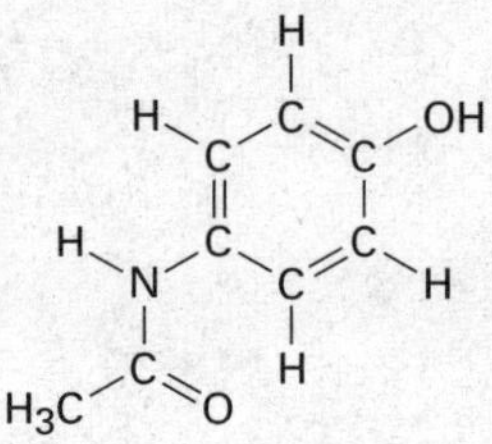

Acetaminophen (Tylenol)

(a) How many sp^3-hybridized carbons are present in aspirin?
(b) How many sp^2-hybridized carbons are present in naproxen?
(c) What is the molecular formula of acetaminophen?
(d) Aspirin, ibuprofen, and naproxen are all believed to target the same enzyme, cyclooxygenase, which produces substances called prostaglandins, that mediate inflammation. Discuss any similarities in the structures of these drugs.

IN THE FIELD **1.71** Herbicides that differ greatly in structure also often differ in how they act.

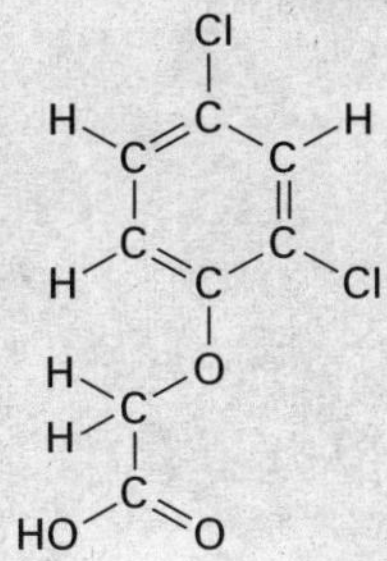

2,4-D (Hi-Dep or Weedar 64)
Disrupts growth regulation signals

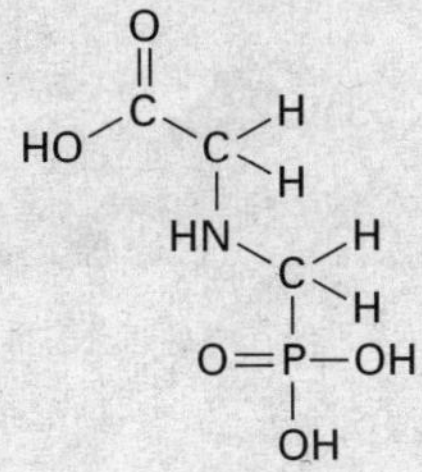

Glyphosate (Roundup)
Inhibits amino acid synthesis

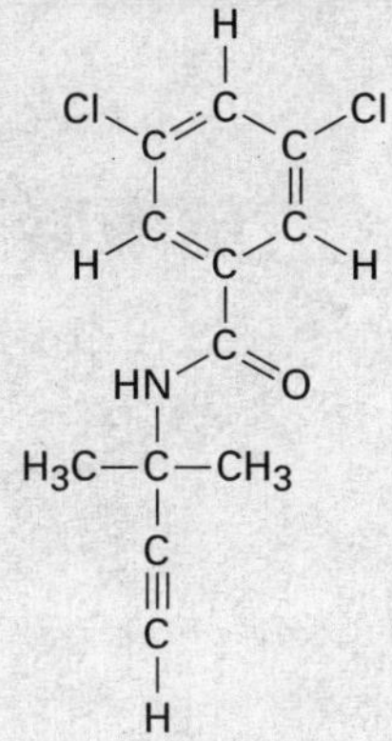

Pronamide (Propyzamide)
Interferes with cell division

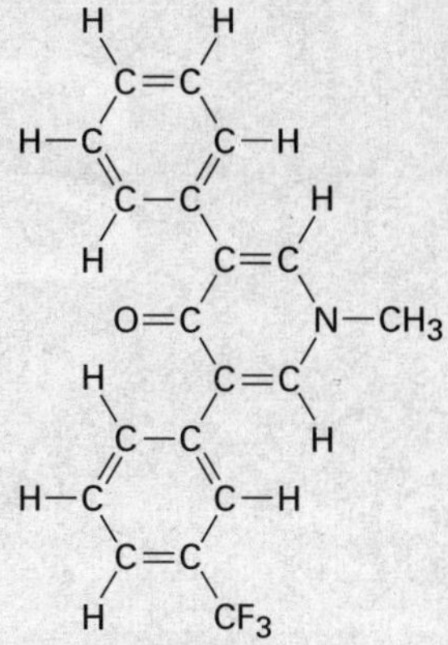

Fluridone (Sonar)
Inhibits pigment biosynthesis

(a) How many sp^3-hybridized carbons are present in 2,4-D?
(b) How many sp^2-hybridized carbons are present in Roundup?
(c) How many sp-hybridized carbons are present in pronamide?
(d) What is the molecular formula of fluridone?

CHAPTER

2

The bristlecone pine is the oldest living organism on Earth. The waxy coating on its needles contains a mixture of organic compounds called alkanes, the subject of this chapter.

Jeff Foott/Getty Images

Alkanes: The Nature of Organic Compounds

There are more than 37 million known organic compounds. Each of these compounds has its own physical properties, such as melting point, and each has its own chemical reactivity. Chemists have learned through years of experience that organic compounds can be classified into families according to their structural features and that the members of a given family often have similar chemical reactivity. Instead of 37 million compounds with random reactivity, there are a few dozen families of compounds whose chemistry is reasonably predictable. We'll study the chemistry of specific families of organic molecules throughout this book, beginning in this chapter with a look at the simplest family, the *alkanes*.

WHY THIS CHAPTER?

Alkanes are relatively unreactive and are rarely involved in chemical reactions, but they nevertheless provide a useful way to introduce some important general ideas. In this chapter, we'll use alkanes to introduce the basic approach to naming organic compounds and to take an initial look at some of the three-dimensional aspects of molecules, a topic of particular importance in understanding biological organic chemistry.

OWL

Online homework for this chapter can be assigned in OWL, an online homework assessment tool.

2.1 Functional Groups

The structural features that make it possible to classify compounds into families are called *functional groups.* A **functional group** is a group of atoms within a molecule that has a characteristic chemical behavior. Chemically, a given functional group behaves almost the same way in every molecule it's a part of. For example, compare ethylene, a plant hormone that causes fruit to ripen, with menthene, a much more complicated molecule found in peppermint oil. Both substances contain a carbon–carbon double-bond functional group, and both therefore react with Br_2 in the same way to give products in which a Br atom has added to each of the double-bond carbons (Figure 2.1). This example is typical: *the chemistry of every organic molecule, regardless of size and complexity, is determined by the functional groups it contains.*

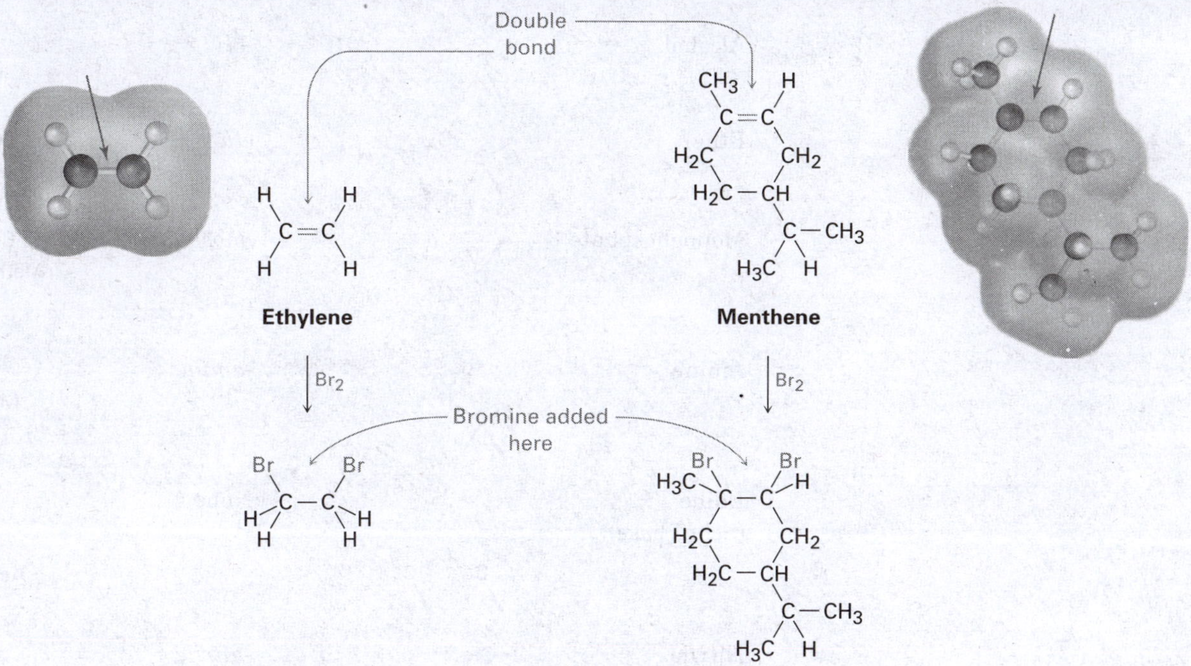

Figure 2.1 The reactions of ethylene and menthene with bromine. In both molecules, the carbon–carbon double-bond functional group reacts with Br_2 in the same way. The size and complexity of the molecules are not important.

Look at Table 2.1, which lists many of the common functional groups and gives simple examples of their occurrence. Some functional groups have only carbon–carbon double or triple bonds; others have halogen atoms; and still others contain oxygen, nitrogen, sulfur, or phosphorus. Much of the chemistry you'll be studying in subsequent chapters is the chemistry of these functional groups.

Table 2.1 Structure of Some Common Functional Groups

Name	Structure*	Name ending	Example
Alkene (double bond)	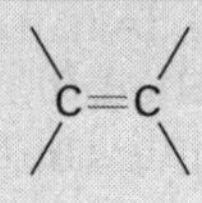	*-ene*	$H_2C{=}CH_2$ Ethene
Alkyne (triple bond)	$—C{\equiv}C—$	*-yne*	$HC{\equiv}CH$ Ethyne
Arene (aromatic ring)	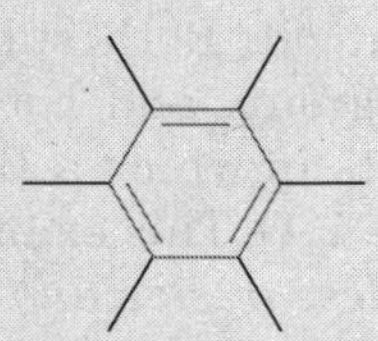	None	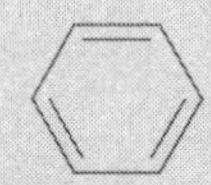 Benzene
Halide	 (X = F, Cl, Br, I)	None	CH_3Cl Chloromethane
Alcohol	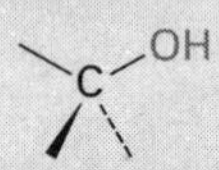	*-ol*	CH_3OH Methanol
Ether	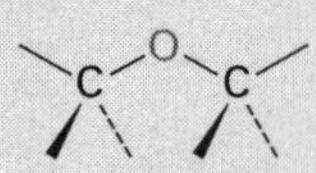	*ether*	CH_3OCH_3 Dimethyl ether
Monophosphate	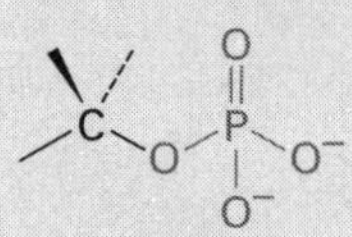	*phosphate*	$CH_3OPO_3^{2-}$ Methyl phosphate
Amine	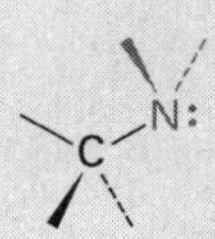	*-amine*	CH_3NH_2 Methylamine
Imine (Schiff base)	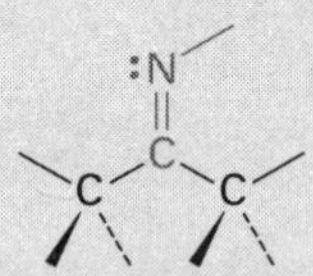	None	NH ‖ CH_3CCH_3 Acetone imine
Nitrile	$—C{\equiv}N$	*-nitrile*	$CH_3C{\equiv}N$ Ethanenitrile
Nitro	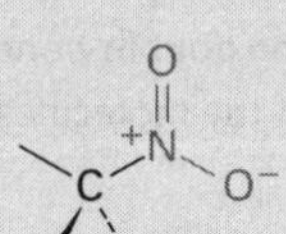	None	CH_3NO_2 Nitromethane

*The bonds whose connections aren't specified are assumed to be attached to carbon or hydrogen atoms in the rest of the molecule.

continued

Table 2.1 Structure of Some Common Functional Groups *(continued)*

Name	Structure*	Name ending	Example
Thiol	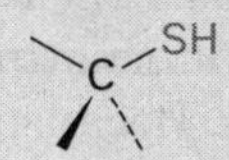	*-thiol*	CH_3SH Methanethiol
Sulfide	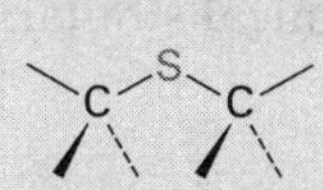	*sulfide*	CH_3SCH_3 Dimethyl sulfide
Disulfide	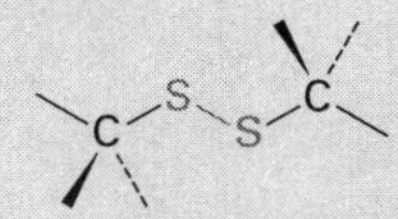	*disulfide*	CH_3SSCH_3 Dimethyl disulfide
Carbonyl	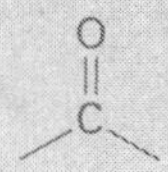		
Aldehyde	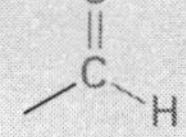	*-al*	$CH_3\overset{\overset{O}{\|}}{C}H$ Ethanal
Ketone	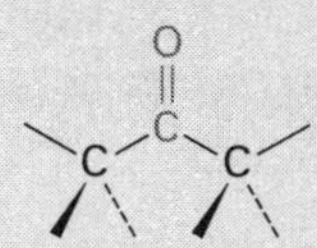	*-one*	$CH_3\overset{\overset{O}{\|}}{C}CH_3$ Propanone
Carboxylic acid	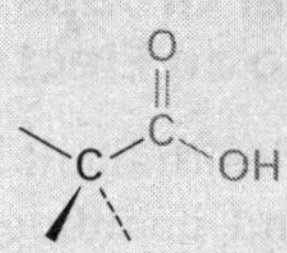	*-oic acid*	$CH_3\overset{\overset{O}{\|}}{C}OH$ Ethanoic acid
Ester	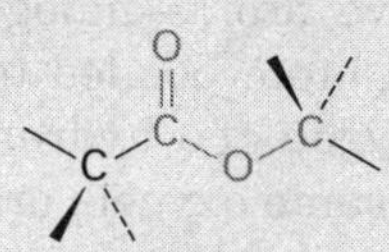	*-oate*	$CH_3\overset{\overset{O}{\|}}{C}OCH_3$ Methyl ethanoate
Amide	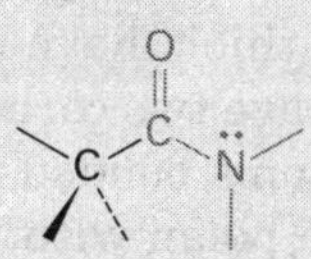	*-amide*	$CH_3\overset{\overset{O}{\|}}{C}NH_2$ Ethanamide
Carboxylic acid anhydride	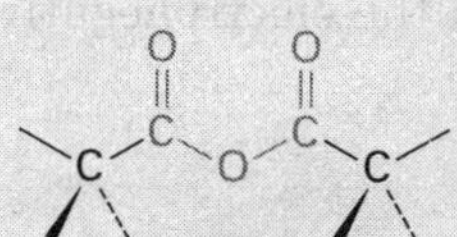	*-oic anhydride*	$CH_3\overset{\overset{O}{\|}}{C}O\overset{\overset{O}{\|}}{C}CH_3$ Ethanoic anhydride
Carboxylic acid chloride	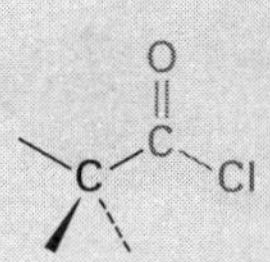	*-oyl chloride*	$CH_3\overset{\overset{O}{\|}}{C}Cl$ Ethanoyl chloride

*The bonds whose connections aren't specified are assumed to be attached to carbon or hydrogen atoms in the rest of the molecule.

Functional Groups with Carbon–Carbon Multiple Bonds

Alkenes, alkynes, and arenes (aromatic compounds) all contain carbon–carbon multiple bonds. *Alkenes* have a double bond, *alkynes* have a triple bond, and *arenes* have alternating double and single bonds in a six-membered ring of carbon atoms. Because of their structural similarities, these compounds also have chemical similarities.

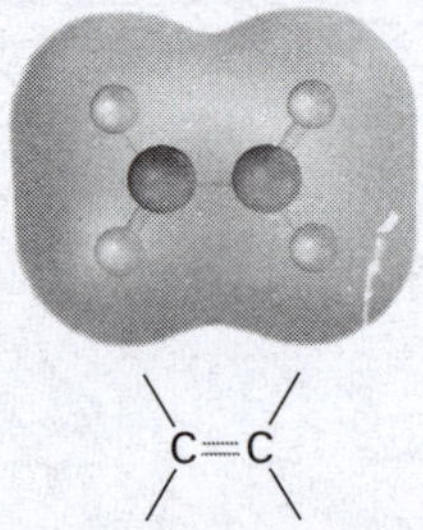

Alkene

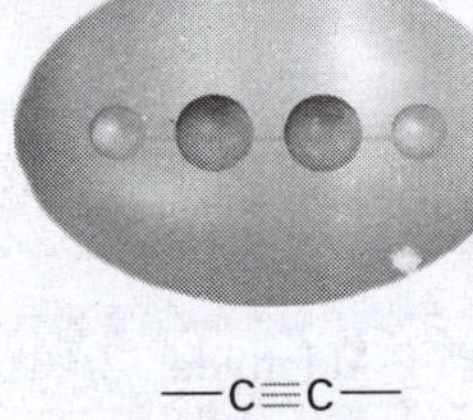

Alkyne

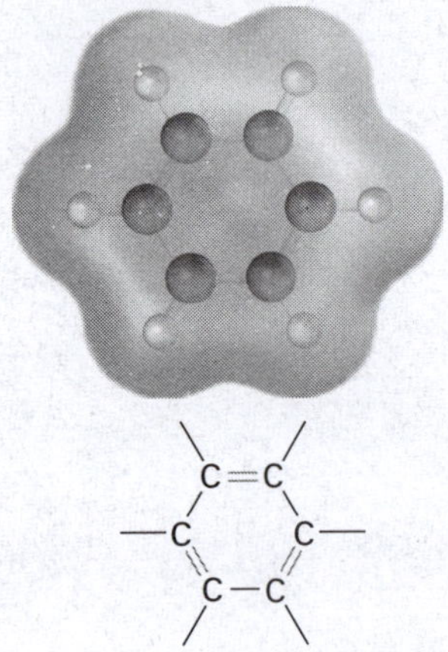

Arene (aromatic ring)

Functional Groups with Carbon Singly Bonded to an Electronegative Atom

Alkyl halides (haloalkanes), alcohols, ethers, alkyl phosphates, amines, thiols, sulfides, and disulfides all have a carbon atom singly bonded to an electronegative atom—halogen, oxygen, nitrogen, or sulfur. Alkyl halides have a carbon atom bonded to halogen (–X), alcohols have a carbon atom bonded to the oxygen of a hydroxyl group (–OH), ethers have two carbon atoms bonded to the same oxygen, organophosphates have a carbon atom bonded to the oxygen of a phosphate group ($-OPO_3^{2-}$), amines have a carbon atom bonded to a nitrogen, thiols have a carbon atom bonded to the sulfur of an –SH group, sulfides have two carbon atoms bonded to the same sulfur, and disulfides have carbon atoms bonded to two sulfurs that are joined together. In all cases, the bonds are polar, with the carbon atom bearing a partial positive charge ($\delta+$) and the electronegative atom bearing a partial negative charge ($\delta-$).

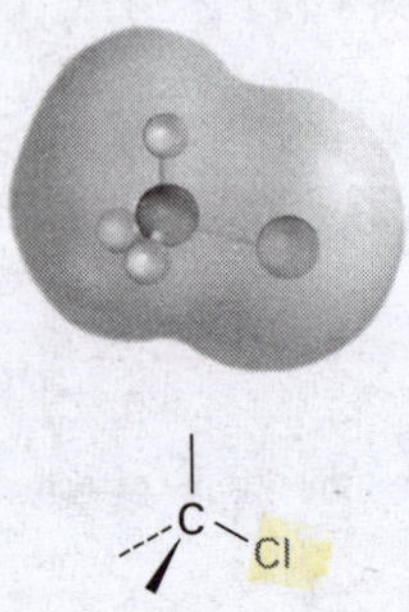

Alkyl halide (haloalkane)

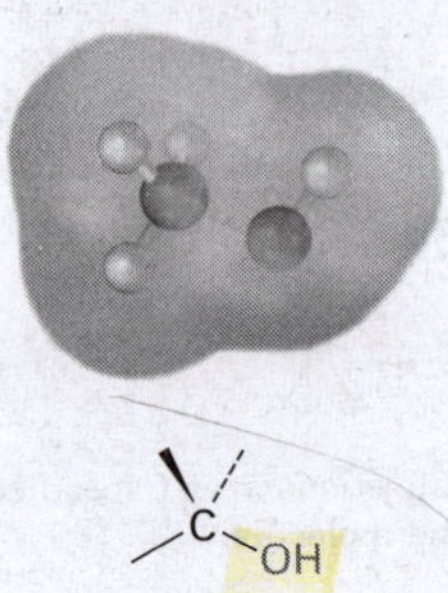

Alcohol

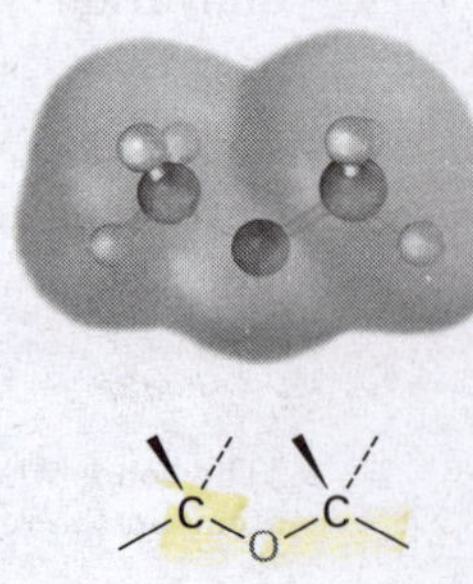

Ether

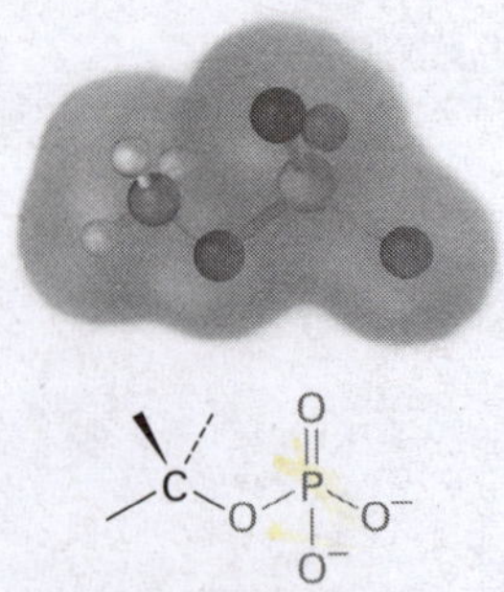

Phosphate

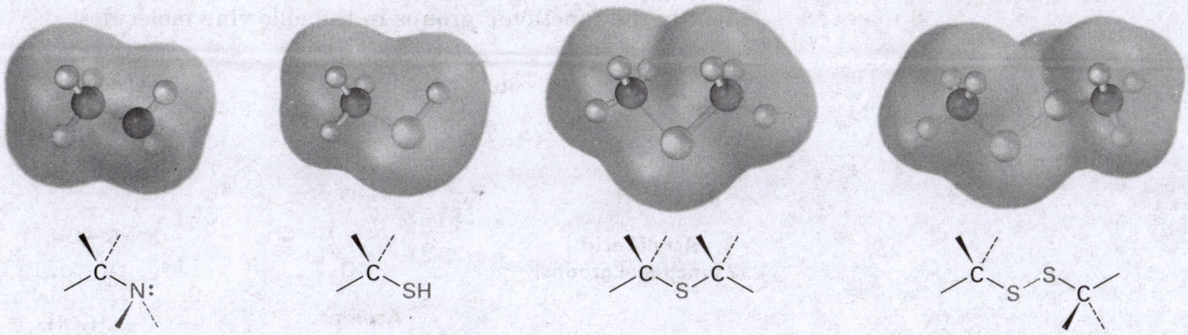

Amine Thiol Sulfide Disulfide

Functional Groups with a Carbon–Oxygen Double Bond (Carbonyl Groups)

The *carbonyl group,* C=O (pronounced car-bo-**neel**) is common to many of the families listed in Table 2.1. Carbonyl groups are present in the great majority of organic compounds and in practically all biological molecules. These compounds behave similarly in many respects but differ depending on the identity of the atoms bonded to the carbonyl-group carbon. Aldehydes have at least one hydrogen bonded to the C=O, ketones have two carbons bonded to the C=O, carboxylic acids have one carbon and one –OH group bonded to the C=O, esters have one carbon and one ether-like oxygen bonded to the C=O, amides have one carbon and one nitrogen bonded to the C=O, acid chlorides have one carbon and one chlorine bonded to the C=O, and so on. The carbonyl carbon atom bears a partial positive charge ($\delta+$), and the oxygen bears a partial negative charge ($\delta-$).

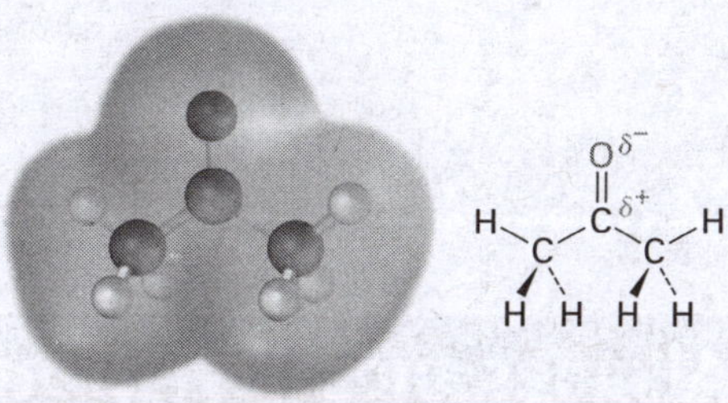

Acetone—a typical carbonyl compound

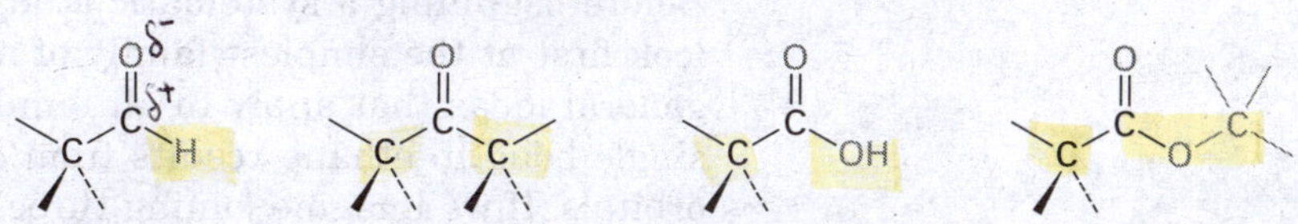

Aldehyde Ketone Carboxylic acid Ester

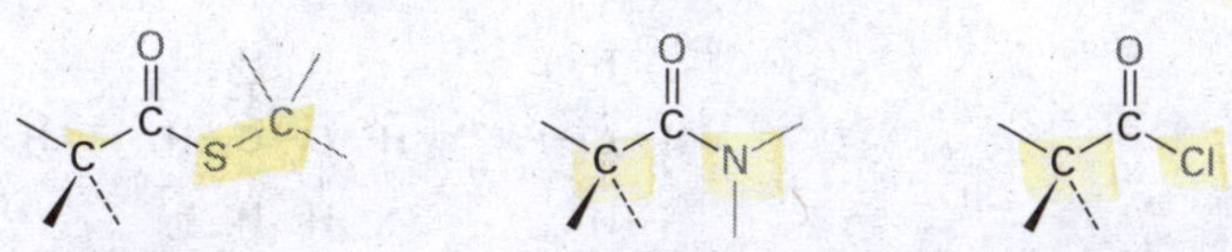

Thioester Amide Acid chloride

Problem 2.1 Identify the functional groups in the following molecules:

(a)

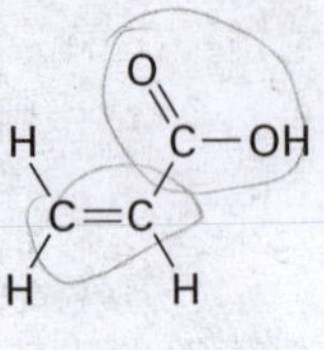

Acrylic acid
(2 functional groups)

(b)

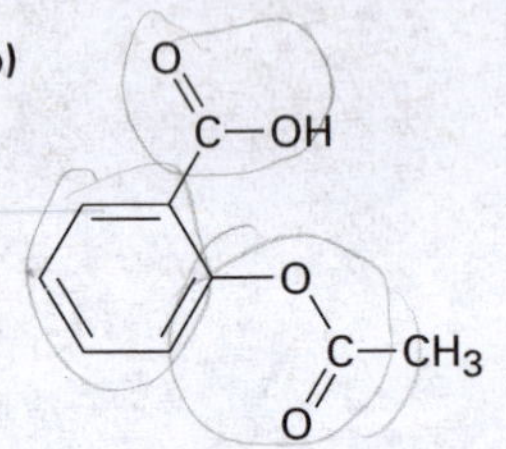

Aspirin
(3 functional groups)

(c)

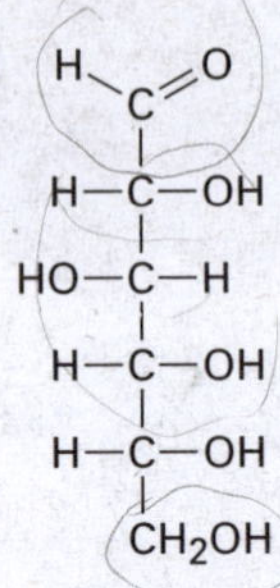

Glucose
(6 functional groups)

Problem 2.2 Propose structures for simple molecules that contain the following functional groups:

(a) Alcohol **(b)** Aromatic ring **(c)** Carboxylic acid
(d) Amine **(e)** Both ketone and amine **(f)** Two double bonds

Problem 2.3 Identify the functional groups in the following model of arecoline, a veterinary drug used to control worms in animals. Convert the drawing into a line-bond structure (gray = C, red = O, blue = N, ivory = H).

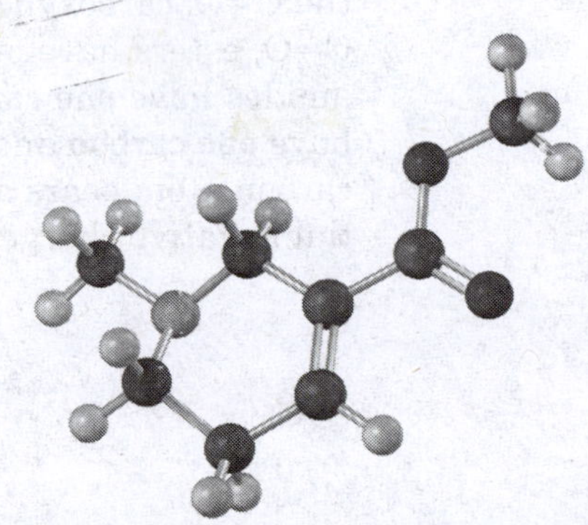

2.2 Alkanes and Alkyl Groups: Isomers

Before beginning a systematic study of the different functional groups, let's look first at the simplest family of molecules—the *alkanes*—to develop some general ideas that apply to all families. We saw in Section 1.7 that the C–C single bond in ethane results from σ (head-on) overlap of carbon sp^3 hybrid orbitals. If we imagine joining three, four, five, or even more carbon atoms by C–C single bonds, we generate the large family of molecules called **alkanes**.

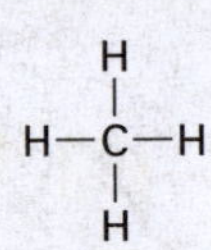

Methane

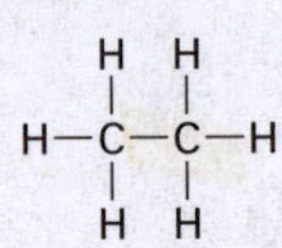

Ethane

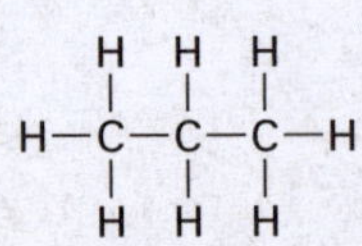

Propane

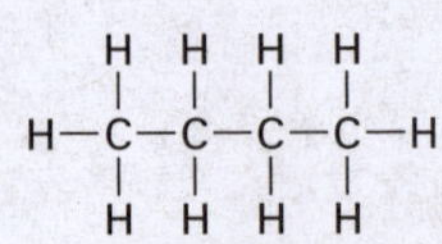

Butane

. . . and so on

Alkanes are often described as *saturated hydrocarbons:* **hydrocarbons** because they contain only carbon and hydrogen atoms; **saturated** because they have only C–C and C–H single bonds and thus contain the maximum possible number of hydrogens per carbon. They have the general formula C_nH_{2n+2}, where n is any integer. Alkanes are also occasionally called **aliphatic** compounds, a word derived from the Greek *aleiphas,* meaning "fat." We'll see in Chapter 16 that animal fats contain long carbon chains similar to alkanes.

$$\mathrm{CH_2O\overset{O}{\overset{\|}{C}}CH_2CH_2CH_2CH_2CH_2CH_2CH_2CH_2CH_2CH_2CH_2CH_2CH_2CH_2CH_2CH_2CH_3}$$
$$\mathrm{CHO\overset{O}{\overset{\|}{C}}CH_2CH_2CH_2CH_2CH_2CH_2CH_2CH_2CH_2CH_2CH_2CH_2CH_2CH_2CH_2CH_2CH_3}$$
$$\mathrm{CH_2O\overset{O}{\overset{\|}{C}}CH_2CH_2CH_2CH_2CH_2CH_2CH_2CH_2CH_2CH_2CH_2CH_2CH_2CH_2CH_2CH_2CH_3}$$

A typical animal fat

Think about the ways that carbon and hydrogen might combine to make alkanes. With one carbon and four hydrogens, only one structure is possible: methane, CH_4. Similarly, there is only one possible combination of two carbons with six hydrogens (ethane, CH_3CH_3) and only one possible combination of three carbons with eight hydrogens (propane, $CH_3CH_2CH_3$). If larger numbers of carbons and hydrogens combine, however, more than one kind of molecule can form. For example, there are two ways that molecules with the formula C_4H_{10} can form: the four carbons can be in a row (butane), or they can branch (isobutane). Similarly, there are three ways in which C_5H_{12} molecules can form, and so on for larger alkanes.

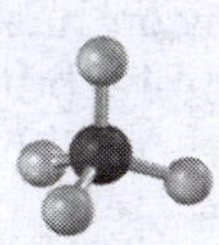

CH_4

Methane, CH_4

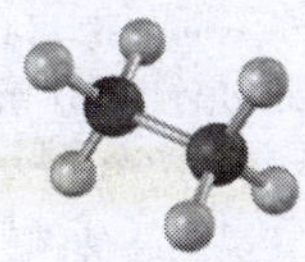

CH_3CH_3

Ethane, C_2H_6

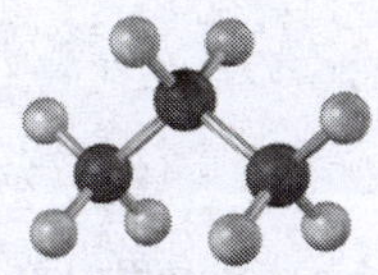

$CH_3CH_2CH_3$

Propane, C_3H_8

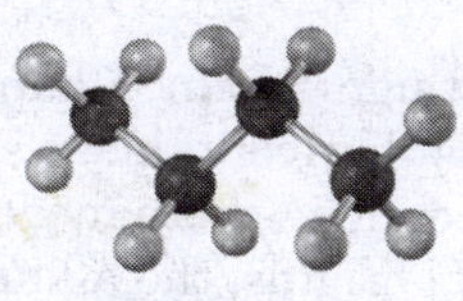

$CH_3CH_2CH_2CH_3$

Butane, C_4H_{10}

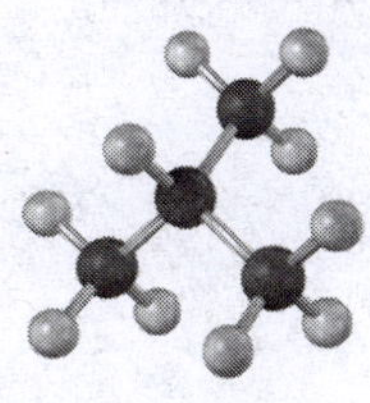

$$\mathrm{CH_3\overset{CH_3}{\overset{|}{C}}HCH_3}$$

Isobutane, C_4H_{10}
(2-methylpropane)

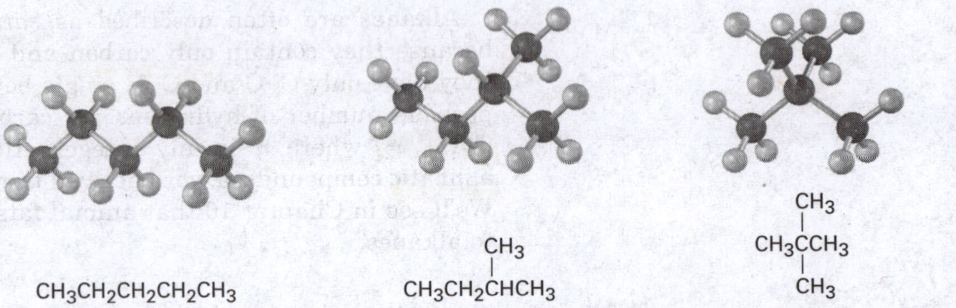

Pentane, C_5H_{12} 2-Methylbutane, C_5H_{12} 2,2-Dimethylpropane, C_5H_{12}

Compounds like butane, whose carbons are connected in a row, are called **straight-chain alkanes, or normal (*n*) alkanes**, whereas compounds with branched carbon chains, such as isobutane (2-methylpropane), are called **branched-chain alkanes**.

Compounds like the two C_4H_{10} molecules and the three C_5H_{12} molecules, which have the same formula but different structures, are called *isomers*, from the Greek *isos* + *meros*, meaning "made of the same parts." **Isomers** have the same numbers and kinds of atoms but differ in the way the atoms are arranged. Compounds like butane and isobutane, whose atoms are connected differently, are called **constitutional isomers**. We'll see shortly that other kinds of isomerism are also possible, even among compounds whose atoms are connected in the same order.

A given alkane can be arbitrarily drawn in many ways. The straight-chain, four-carbon alkane called butane, for instance, can be represented by any of the structures shown in Figure 2.2. These structures don't imply any particular three-dimensional geometry for butane; they only indicate the connections among atoms. In practice, chemists rarely draw all the bonds in a molecule and usually refer to butane by the *condensed structure*, $CH_3CH_2CH_2CH_3$ or $CH_3(CH_2)_2CH_3$. In such representations, the C–C and C–H bonds are "understood" rather than shown. If a carbon has three hydrogens bonded to it, we write CH_3; if a carbon has two hydrogens bonded to it, we write CH_2, and so on. Still more simply, butane can even be represented as n-C_4H_{10}, where *n* signifies *normal*, straight-chain butane.

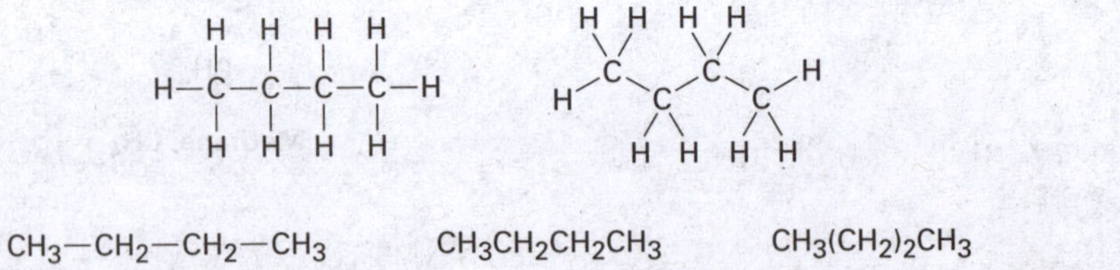

Figure 2.2 Some representations of butane (n-C_4H_{10}). The molecule is the same regardless of how it's drawn. These structures imply only that butane has a continuous chain of four carbon atoms.

Straight-chain alkanes are named according to the number of carbon atoms they contain, as shown in Table 2.2. With the exception of the first four compounds—methane, ethane, propane, and butane—whose names have historical origins, the alkanes are named based on Greek numbers, according to the number of carbons. The suffix *-ane* is added to the end of each name to identify the molecule as an alkane. Thus, pent*ane* is the five-carbon alkane, hex*ane* is the six-carbon alkane, and so on.

If a hydrogen atom is removed from an alkane, the partial structure that remains is called an **alkyl group**. Alkyl groups are named by replacing the *-ane* ending of the parent alkane with an *-yl* ending. For example, removal of

Table 2.2 Names of Straight-Chain Alkanes

Number of carbons (n)	Name	Formula (C_nH_{2n+2})	Number of carbons (n)	Name	Formula (C_nH_{2n+2})
1	Methane	CH_4	9	Nonane	C_9H_{20}
2	Ethane	C_2H_6	10	Decane	$C_{10}H_{22}$
3	Propane	C_3H_8	11	Undecane	$C_{11}H_{24}$
4	Butane	C_4H_{10}	12	Dodecane	$C_{12}H_{26}$
5	Pentane	C_5H_{12}	13	Tridecane	$C_{13}H_{28}$
6	Hexane	C_6H_{14}	20	Icosane	$C_{20}H_{42}$
7	Heptane	C_7H_{16}	30	Triacontane	$C_{30}H_{62}$
8	Octane	C_8H_{18}			

a hydrogen atom from methane, CH_4, generates a *methyl group,* $-CH_3$, and removal of a hydrogen atom from ethane, CH_3CH_3, generates an *ethyl group,* $-CH_2CH_3$. Similarly, removal of a hydrogen atom from the end carbon of any *n*-alkane gives the series of *n*-alkyl groups shown in Table 2.3.

Just as *n*-alkyl groups are generated by removing a hydrogen from an *end* carbon, branched alkyl groups are generated by removing a hydrogen atom from an *internal* carbon. Two 3-carbon alkyl groups and four 4-carbon alkyl groups are possible (Figure 2.3).

Figure 2.3 Alkyl groups generated from straight-chain alkanes.

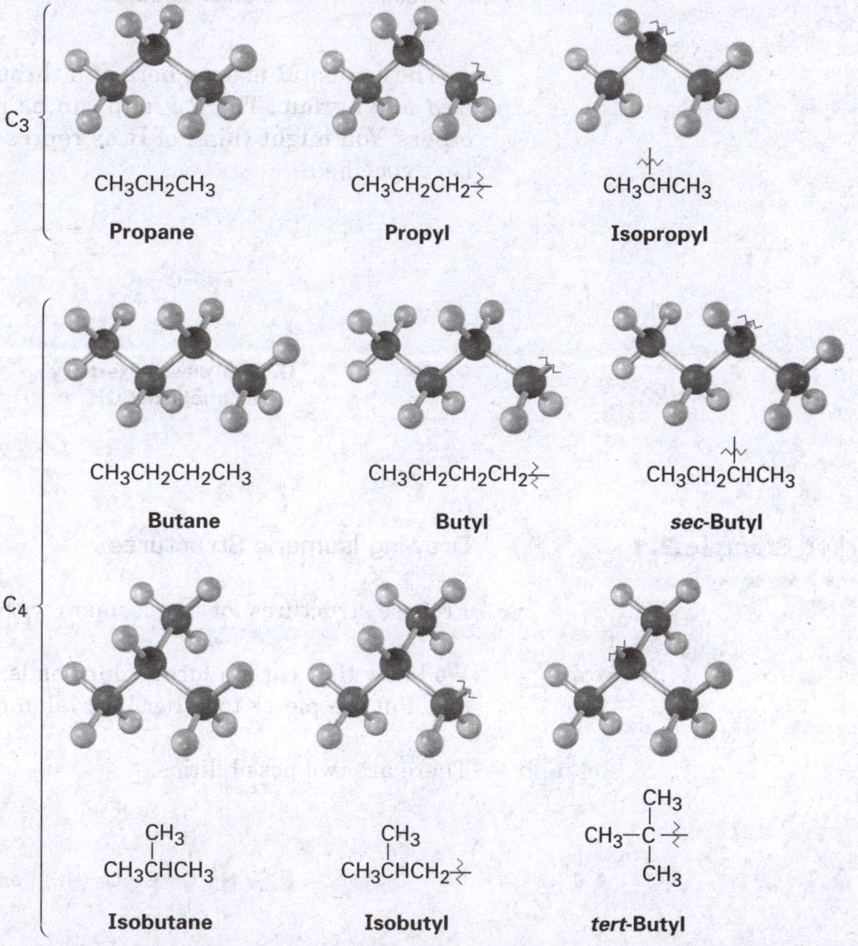

Table 2.3 Some Straight-Chain Alkyl Groups

Alkane	Name	Alkyl group	Name (abbreviation)
CH_4	Methane	$-CH_3$	Methyl (Me)
CH_3CH_3	Ethane	$-CH_2CH_3$	Ethyl (Et)
$CH_3CH_2CH_3$	Propane	$-CH_2CH_2CH_3$	Propyl (Pr)
$CH_3CH_2CH_2CH_3$	Butane	$-CH_2CH_2CH_2CH_3$	Butyl (Bu)
$CH_3CH_2CH_2CH_2CH_3$	Pentane	$-CH_2CH_2CH_2CH_2CH_3$	Pentyl, or amyl

One further word about naming alkyl groups: the prefixes *sec-* (for secondary) and *tert-* (for tertiary) used for the C_4 alkyl groups in Figure 2.3 refer to the number of other carbon atoms attached to the branching carbon atom. There are four possibilities: primary (1°), secondary (2°), tertiary (3°), and quaternary (4°).

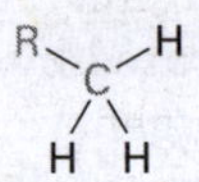

Primary **carbon (1°) is bonded to one other carbon.**

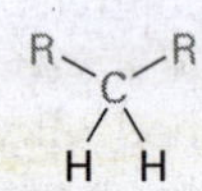

Secondary **carbon (2°) is bonded to two other carbons.**

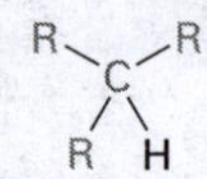

Tertiary **carbon (3°) is bonded to three other carbons.**

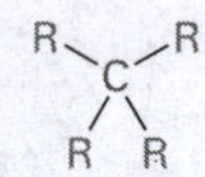

Quaternary **carbon (4°) is bonded to four other carbons.**

The symbol **R** is used here and throughout this text to represent a *generalized* alkyl group. The R group can be methyl, ethyl, or any of a multitude of others. You might think of **R** as representing the **R**est of the molecule, which isn't specified.

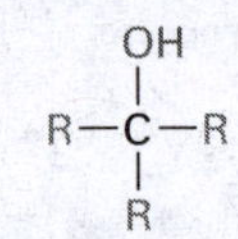

General class of tertiary alcohols, R_3COH

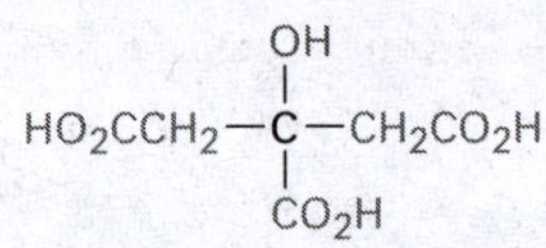

Citric acid—a specific tertiary alcohol

Worked Example 2.1 Drawing Isomeric Structures

Propose structures for two isomers with the formula C_2H_6O.

Strategy We know that carbon forms four bonds, oxygen forms two, and hydrogen forms one. Put the pieces together by trial and error, along with intuition.

Solution There are two possibilities:

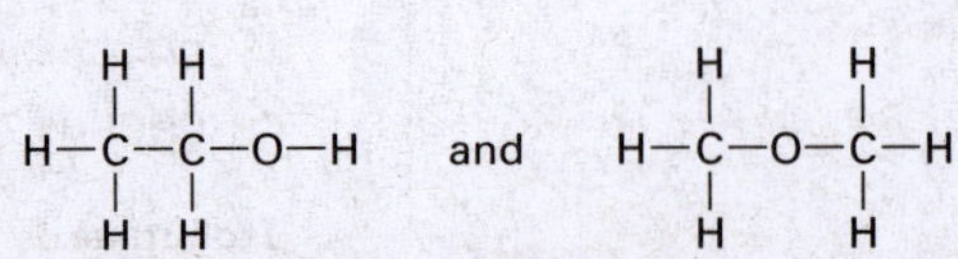

Problem 2.4 Draw structures for the five isomers of C_6H_{14}.

Problem 2.5 Draw structures that meet the following descriptions:

(a) Three isomers with the formula C_8H_{18}
(b) Two isomers with the formula $C_4H_8O_2$

Problem 2.6 Draw the eight possible five-carbon alkyl groups (pentyl isomers).

Problem 2.7 Draw alkanes that meet the following descriptions:

(a) An alkane with two tertiary carbons
(b) An alkane that contains an isopropyl group
(c) An alkane that has one quaternary and one secondary carbon

Problem 2.8 Identify the carbon atoms in the following molecules as primary, secondary, tertiary, or quaternary:

(a) 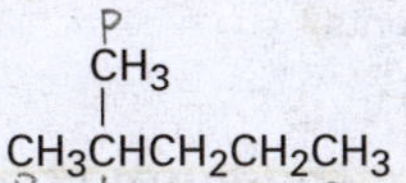

(b) 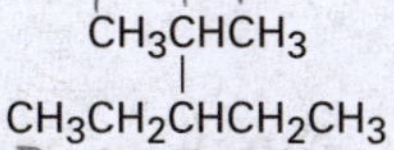

(c) 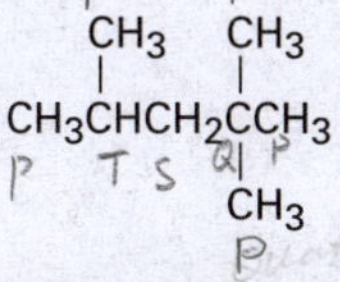

2.3 Naming Branched-Chain Alkanes

In earlier times, when few pure organic chemicals were known, new compounds were named at the whim of their discoverer. Thus, urea (CH_4N_2O) is a crystalline substance isolated from urine, and morphine ($C_{17}H_{19}NO_3$) is an analgesic (painkiller) named after Morpheus, the Greek god of dreams. As the science of organic chemistry slowly grew in the 19th century, so too did the number of known compounds and the need for a systematic method of naming them. The system of naming (*nomenclature*) we'll use in this book is that devised by the International Union of Pure and Applied Chemistry (IUPAC, usually spoken as **eye**-you-pac).

A chemical name typically has four parts in the IUPAC system of nomenclature: prefix, parent, locant, and suffix. The prefix specifies the location and identity of various substituent groups in the molecule, the parent selects a main part of the molecule and tells how many carbon atoms are in that part, the locant gives the location of the primary functional group, and the suffix identifies the primary functional group.

Prefix—Parent—Locant—Suffix

Prefix	Parent	Locant	Suffix
Where and what are the substituents?	**How many carbons?**	**Where is the primary functional group?**	**What is the primary functional group?**

As we cover new functional groups in later chapters, the applicable IUPAC rules of nomenclature will be given. In addition, Appendix A gives an overall view of organic nomenclature and shows how compounds that contain more than one functional group are named. For now, let's see how to name branched-chain alkanes. All but the most complex branched-chain alkanes can be named by following four steps.

STEP 1 **Find the parent hydrocarbon.**

(a) Find the longest continuous carbon chain in the molecule and use the name of that chain as the parent name. The longest chain may not always be obvious; you may have to "turn corners."

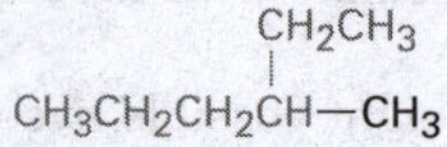

Named as a substituted hexane

(b) If two chains of equal length are present, choose the one with the larger number of branch points as the parent.

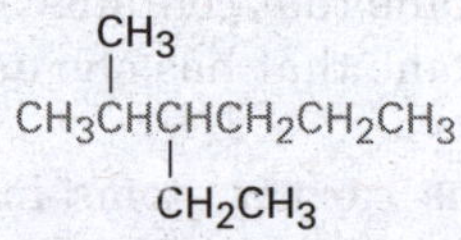

Named as a hexane with *two* substituents

NOT

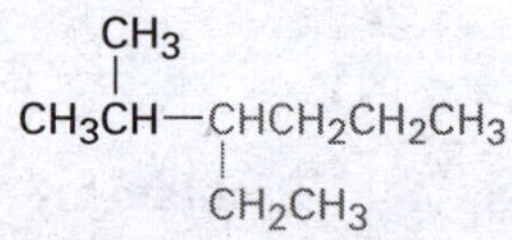

as a hexane with *one* substituent

STEP 2 **Number the atoms in the main chain.**

Beginning at the end nearer the first branch point, number each carbon atom in the parent chain.

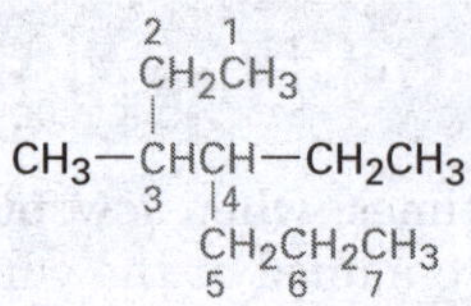

NOT

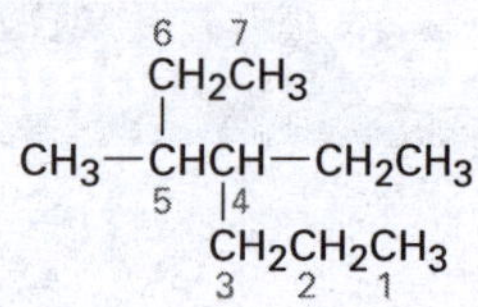

The first branch occurs at C3 in the proper system of numbering but at C4 in the improper system.

STEP 3 **Identify and number the substituents.**

Assign a number, called a *locant,* to each substituent to specify its point of attachment to the parent chain. If there are two substituents on the same carbon, assign them both the same number. There must always be as many numbers in the name as there are substituents.

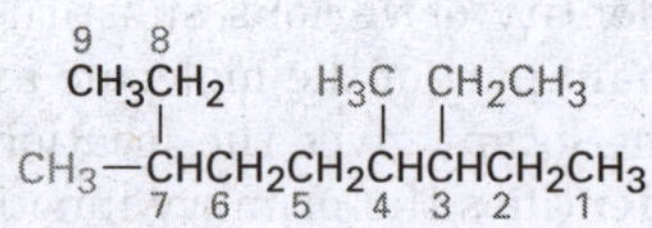

Named as a nonane

Substituents: On C3, CH_2CH_3 (3-ethyl)
On C4, CH_3 (4-methyl)
On C7, CH_3 (7-methyl)

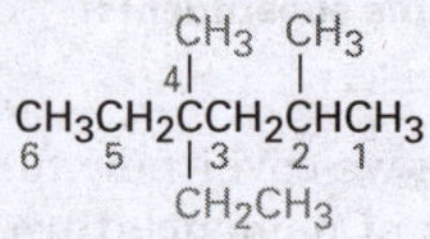

Named as a hexane

Substituents: On C2, CH_3 (2-methyl)
On C4, CH_3 (4-methyl)
On C4, CH_2CH_3 (4-ethyl)

STEP 4 **Write the name as a single word.**

Use hyphens to separate the various prefixes and commas to separate numbers. If two or more different side chains are present, cite them in alphabetical order. If two or more identical side chains are present, use the appropriate multiplier prefixes *di-*, *tri-*, *tetra-*, and so forth. Don't use these prefixes for alphabetizing, though. Full names for some examples follow:

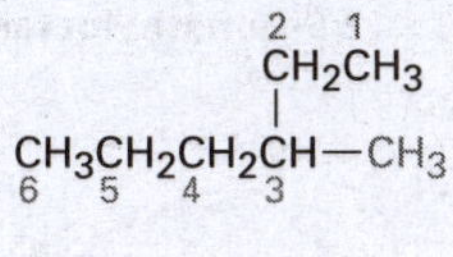

3-Methylhexane

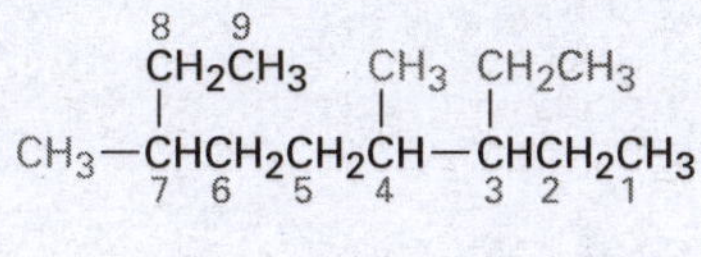

3-Ethyl-4,7-dimethylnonane

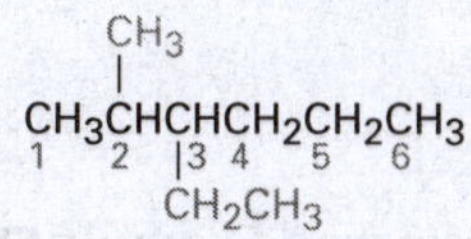

3-Ethyl-2-methylhexane

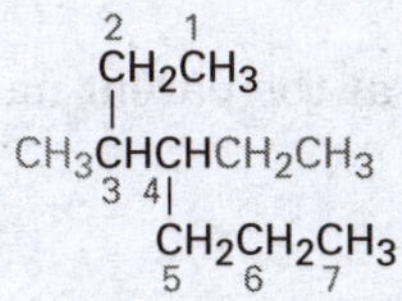

4-Ethyl-3-methylheptane

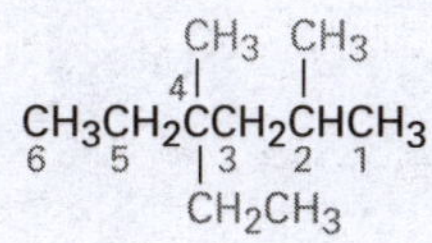

4-Ethyl-2,4-dimethylhexane

For historical reasons, a few simple branched-chain alkyl groups also have nonsystematic, common names, as noted in Figure 2.3.

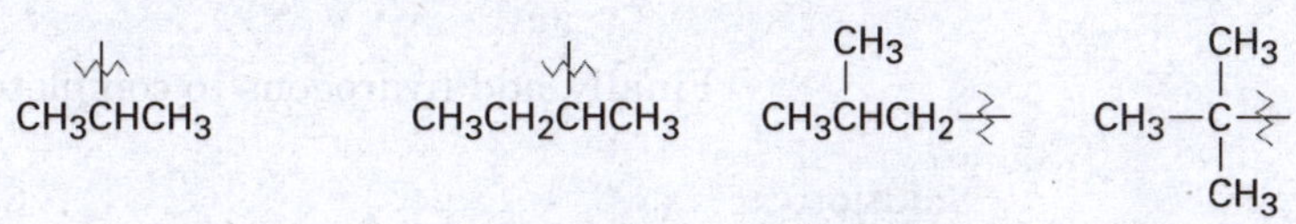

Isopropyl (*i*-Pr) | ***sec*-Butyl (*sec*-Bu)** | **Isobutyl** | ***tert*-Butyl (*t*-Butyl or *t*-Bu)**

3-Carbon alkyl group | **4-Carbon alkyl groups**

When writing the name of an alkane that contains one of these alkyl groups, the nonhyphenated prefix iso- is considered part of the alkyl-group name for alphabetizing purposes, but the hyphenated and italicized prefixes *sec-* and *tert-* are not. Thus, isopropyl and isobutyl are listed alphabetically under *i*, but *sec*-butyl and *tert*-butyl are listed under *b*.

Worked Example 2.2 **Naming an Alkane**

What is the IUPAC name of the following alkane?

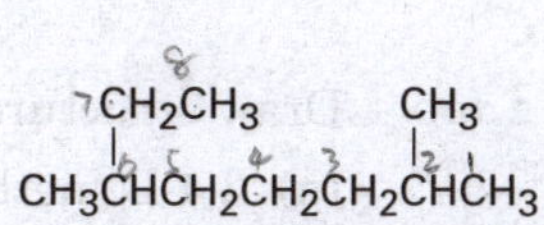

Strategy The molecule has a chain of eight carbons (octane) with two methyl substituents. Numbering from the end nearer the first methyl substituent indicates that the methyls are at C2 and C6.

Solution

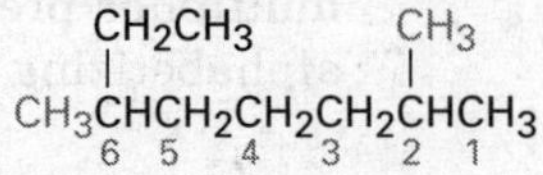

2,6-Dimethyloctane

Worked Example 2.3 Drawing a Structure from a Name

Draw the structure of 3-isopropyl-2-methylhexane.

Strategy First, look at the parent name (hexane) and draw its carbon structure.

C–C–C–C–C–C **Hexane**

Next, find the substituents (3-isopropyl and 2-methyl), and place them on the proper carbons.

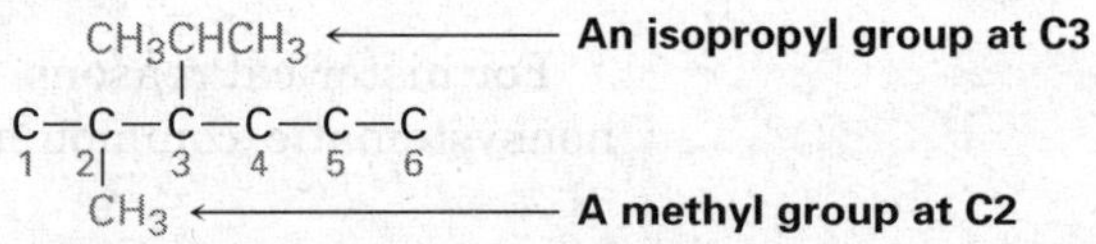

Finally, add hydrogens to complete the structure.

Solution

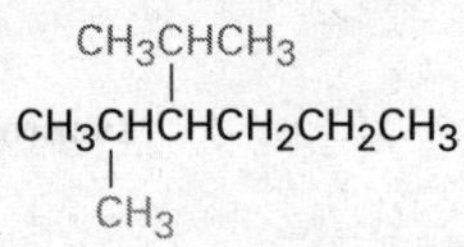

3-Isopropyl-2-methylhexane

Problem 2.9 Give IUPAC names for the following alkanes:

(a) The three isomers of C_5H_{12}

(b)

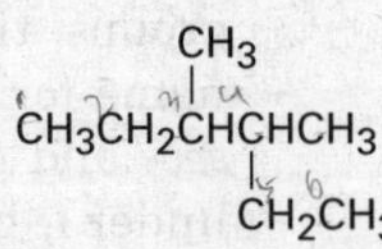

(c)

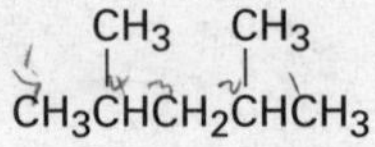

(d)

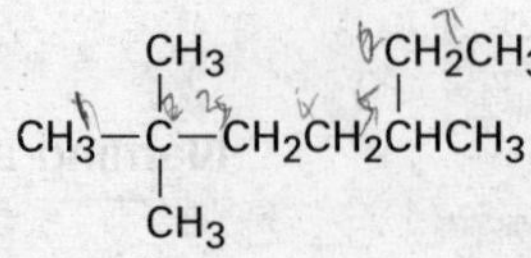

Problem 2.10 Draw structures corresponding to the following IUPAC names:

(a) 3,4-Dimethylnonane **(b)** 3-Ethyl-4,4-dimethylheptane

(c) 2,2-Dimethyl-4-propyloctane **(d)** 2,2,4-Trimethylpentane

Problem 2.11 Name the following alkane:

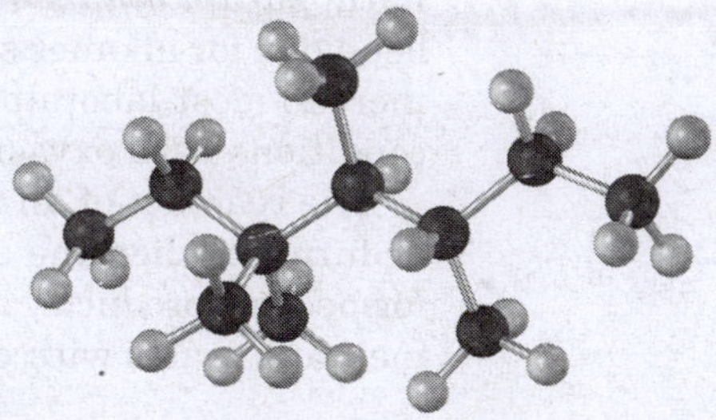

2.4 Properties of Alkanes

Many alkanes occur naturally in the plant and animal world. For example, the waxy coating on cabbage leaves contains nonacosane ($C_{29}H_{60}$), and the wood oil of the Jeffrey pine common to the Sierra Nevada mountains of California contains heptane (C_7H_{16}). By far the major sources of alkanes, however, are the world's natural gas and petroleum deposits. Laid down eons ago, these natural deposits are derived from the decomposition of plant and animal matter, primarily of marine origin. *Natural gas* consists chiefly of methane but also contains ethane, propane, and butane. *Petroleum* is a complex mixture of hydrocarbons that must first be separated into various fractions and then further refined before it can be used.

Petroleum refining begins by fractional distillation of crude oil into three principal cuts, according to their boiling points (bp): straight-run gasoline (bp 20–200 °C), kerosene (bp 175–275 °C), and heating oil, or diesel fuel (bp 250–400 °C). Finally, distillation under reduced pressure yields lubricating oils and waxes, and leaves an undistillable tarry residue of asphalt (Figure 2.4).

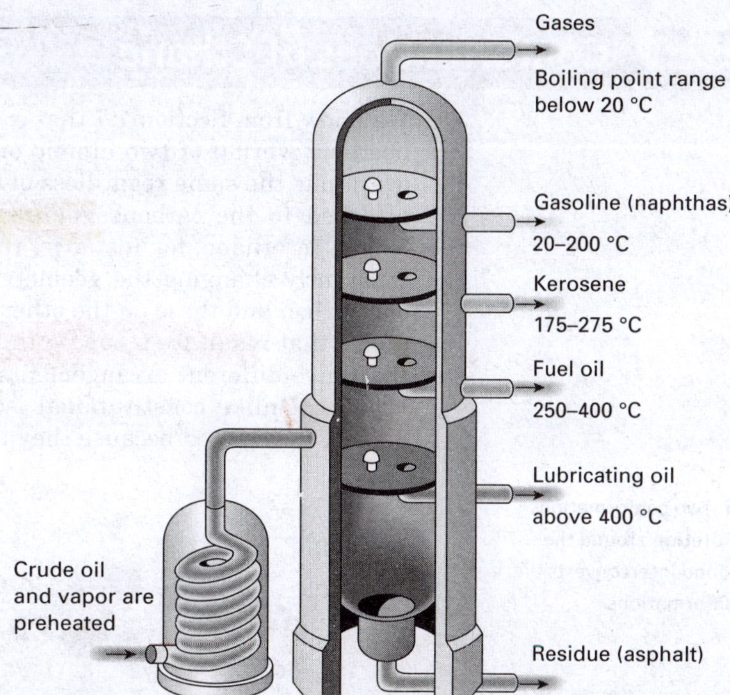

Figure 2.4 Fractional distillation separates petroleum into fractions according to boiling point. The temperature in the tower decreases with increasing height, allowing condensation of the vapors and collection of different components.

Alkanes are sometimes referred to as *paraffins,* a word derived from the Latin *parum affinis,* meaning "slight affinity." This term aptly describes their behavior, for alkanes show little chemical affinity for other substances and are inert to most laboratory reagents. They do, however, react under appropriate conditions with oxygen, chlorine, and a few other substances.

The reaction of an alkane with O_2 occurs during combustion in an engine or furnace when the alkane is used as a fuel. Carbon dioxide and water are formed as products, and a large amount of heat is released. For example, methane reacts with oxygen according to the equation:

$$CH_4 + 2\,O_2 \longrightarrow CO_2 + 2\,H_2O + 890\ \text{kJ}\ (213\ \text{kcal})$$

The reaction of an alkane with Cl_2 occurs when a mixture of the two is irradiated with ultraviolet light (denoted $h\nu$, where ν is the lowercase Greek letter nu). Depending on the relative amounts of the two reactants and on the time allowed for reaction, a sequential replacement of the alkane hydrogen atoms by chlorine occurs, leading to a mixture of chlorinated products. Methane, for instance, reacts with chlorine to yield a mixture of chloromethane (CH_3Cl), dichloromethane (CH_2Cl_2), trichloromethane ($CHCl_3$), and tetrachloromethane (CCl_4).

$$CH_4 + Cl_2 \xrightarrow{h\nu} CH_3Cl + HCl$$
$$CH_3Cl \xrightarrow{Cl_2} CH_2Cl_2 + HCl$$
$$CH_2Cl_2 \xrightarrow{Cl_2} CHCl_3 + HCl$$
$$CHCl_3 \xrightarrow{Cl_2} CCl_4 + HCl$$

2.5 Conformations of Ethane

We know from Section 1.7 that a carbon–carbon single bond results from the head-on overlap of two atomic orbitals. Because the amount of this orbital overlap is the same regardless of the geometric arrangements of other atoms attached to the carbons, rotation is possible around carbon–carbon single bonds. In ethane, for instance, rotation around the C–C bond occurs freely, constantly changing the geometric relationships between the hydrogens on one carbon and those on the other (Figure 2.5). The different arrangements of atoms that result from bond rotation are called **conformations**, and molecules that have different arrangements are called conformational isomers, or *conformers.* Unlike constitutional isomers, however, different conformers can't usually be isolated because they interconvert too rapidly.

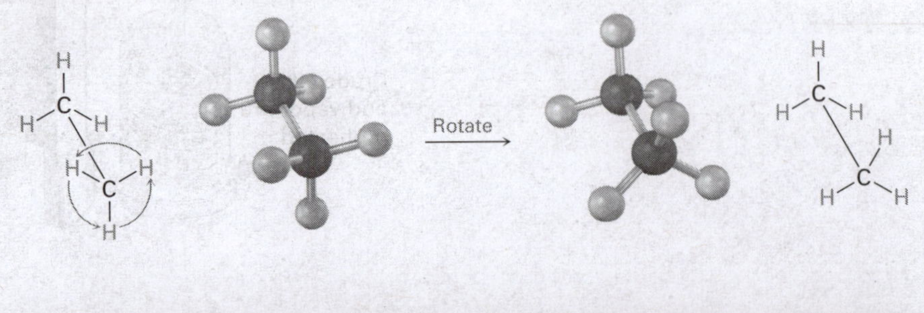

Figure 2.5 Two conformations of ethane. Rotation around the C–C single bond interconverts the different conformations.

Chemists represent different conformations in two ways, as shown in Figure 2.6. A **sawhorse representation** views the C–C bond from an oblique angle and indicates spatial relationships by showing all the C–H bonds. A **Newman projection** views the C–C bond directly end-on and represents the two carbon atoms by a circle. Bonds attached to the front carbon are represented by lines to a dot in the center of the circle, and bonds attached to the rear carbon are represented by lines to the edge of the circle.

Figure 2.6 A sawhorse representation and a Newman projection of ethane. The sawhorse representation views the molecule from an oblique angle, while the Newman projection views the molecule end-on. Note that the molecular model of the Newman projection appears at first to have six atoms attached to a single carbon. Actually, the front carbon, with three attached green atoms, is directly in front of the rear carbon, with three attached red atoms.

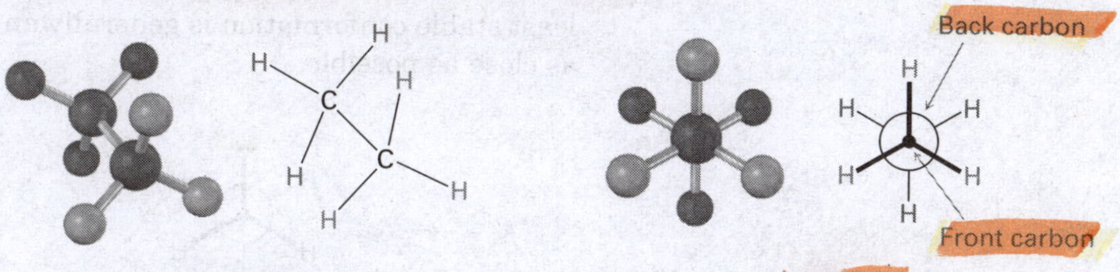

Sawhorse representation

Newman projection

Despite what we've just said, we actually don't observe *perfectly* free rotation in ethane. Experiments show that there is a slight (12 kJ/mol; 2.9 kcal/mol) barrier to rotation and that some conformations are more stable than others. The lowest-energy, most stable conformation is the one in which all six C–H bonds are as far away from one another as possible (**staggered** when viewed end-on in a Newman projection). The highest-energy, least stable conformation is the one in which the six C–H bonds are as close as possible (**eclipsed** in a Newman projection). At any given instant, about 99% of ethane molecules have an approximately staggered conformation, and only about 1% are close to the eclipsed conformation (Figure 2.7).

Figure 2.7 Staggered and eclipsed conformations of ethane. The staggered conformation is lower in energy and more stable by 12.0 kJ/mol.

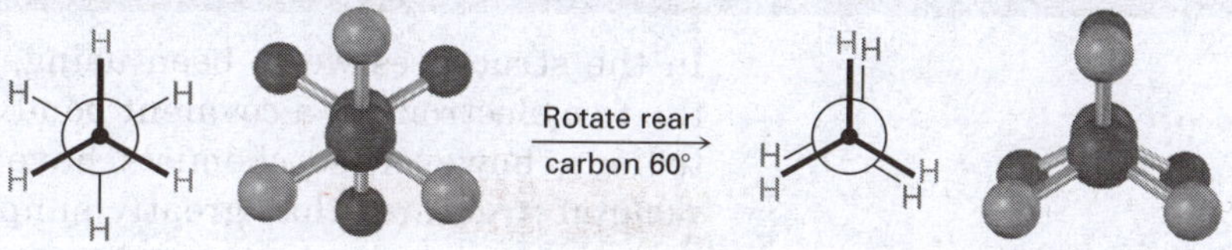

Ethane—staggered conformation

Ethane—eclipsed conformation

What is true for ethane is also true for propane, butane, and all higher alkanes. The most favored conformation for any alkane is the one in which all bonds have staggered arrangements (Figure 2.8).

Figure 2.8 The most stable conformation of any alkane is the one in which the bonds on adjacent carbons are staggered and the carbon chain is fully extended so that large groups are far away from one another, as in this model of decane.

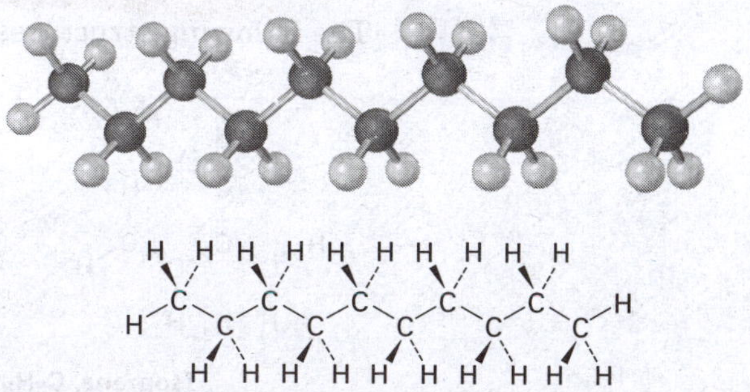

Worked Example 2.4 **Drawing a Newman Projection**

Sight along the C1–C2 bond of 1-chloropropane and draw Newman projections of the most stable and least stable conformations.

Strategy The most stable conformation of a substituted alkane is generally a staggered one in which large groups are as far away from one another as possible. The least stable conformation is generally an eclipsed one in which large groups are as close as possible.

Solution

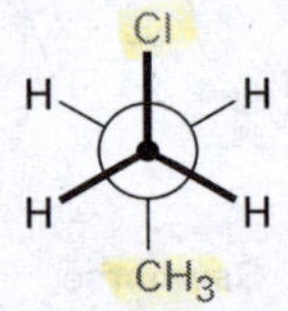

Most stable (staggered)

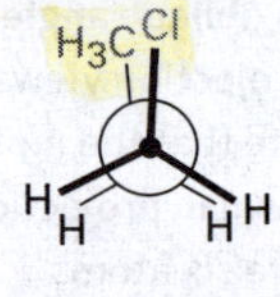

Least stable (eclipsed)

Problem 2.12 Sight along a C–C bond of propane and draw a Newman projection of the most stable conformation. Draw a Newman projection of the least stable conformation.

Problem 2.13 Looking along the C2–C3 bond of butane, there are two different staggered conformations and two different eclipsed conformations. Draw them.

Problem 2.14 Which of the butane conformations you drew in Problem 2.13 do you think is the most stable? Explain.

2.6 Drawing Chemical Structures

In the structures we've been using, a line between atoms has represented the two electrons in a covalent bond. Drawing every bond and every atom is tedious, however, so chemists have devised a shorthand way of drawing **skeletal structures** that greatly simplifies matters. Drawing skeletal structures is straightforward:

- Carbon atoms usually aren't shown. Instead, a carbon atom is assumed to be at the intersection of two lines (bonds) and at the end of each line. Occasionally, a carbon atom might be indicated for emphasis or clarity.
- Hydrogen atoms bonded to carbon aren't shown. Because carbon always has a valence of four, we mentally supply the correct number of hydrogen atoms for each carbon.
- All atoms other than carbon and hydrogen *are* shown.

The following structures give some examples.

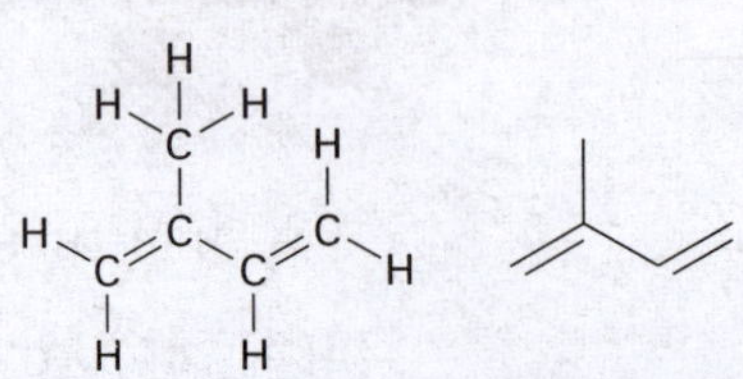

Isoprene, C_5H_8

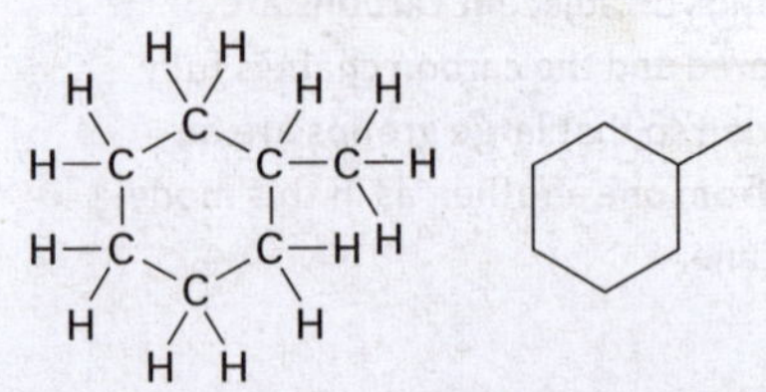

Methylcyclohexane, C_7H_{14}

One further comment: although such groupings as $-CH_3$, $-OH$, and $-NH_2$ are usually written with the C, O, or N atom first and the H atom second, the order of writing is sometimes inverted to H_3C-, $HO-$, and H_2N- if needed to make the bonding connections in a molecule clearer. Larger units such as $-CH_2CH_3$ are not inverted, though; we don't write H_3CH_2C- because it would be confusing. There are, however, no well-defined rules that cover all cases; it's largely a matter of preference.

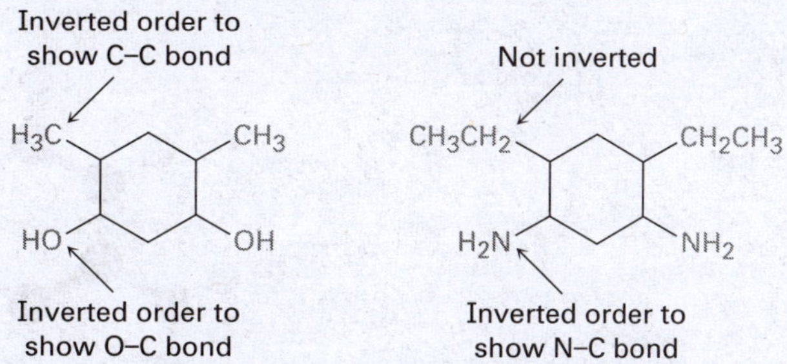

Worked Example 2.5

Interpreting a Skeletal Structure

Carvone, a substance responsible for the odor of spearmint, has the following structure. Tell how many hydrogens are bonded to each carbon, and give the molecular formula of carvone.

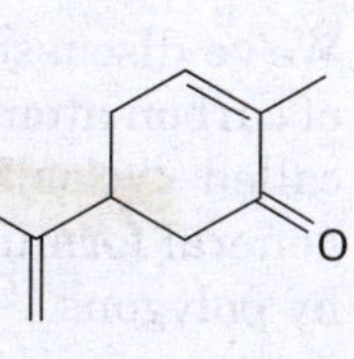

Carvone

Strategy Remember that the end of a line represents a carbon atom with three hydrogens, CH_3; a two-way intersection is a carbon atom with two hydrogens, CH_2; a three-way intersection is a carbon atom with one hydrogen, CH; and a four-way intersection is a carbon atom with no attached hydrogens.

Solution

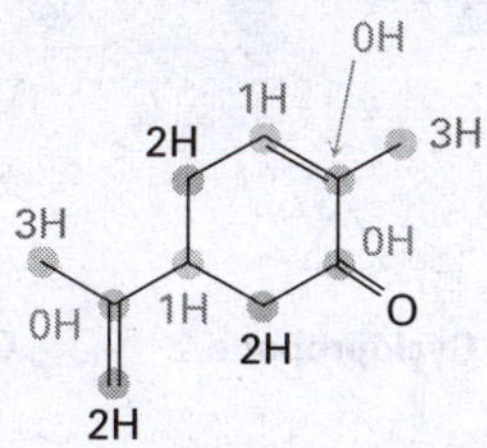

Carvone, $C_{10}H_{14}O$

Problem 2.15 Convert the following skeletal structures into molecular formulas, and tell how many hydrogens are bonded to each carbon:

(a) 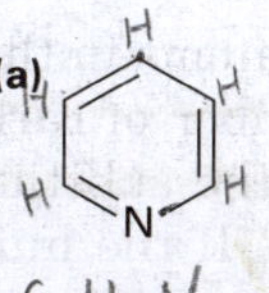

Pyridine

(b) 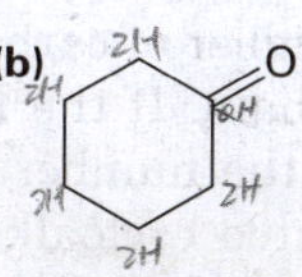

Cyclohexanone

(c) 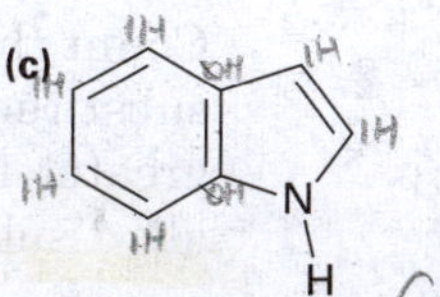

Indole

Problem 2.16 Propose skeletal structures for the following molecular formulas:

(a) C_4H_8 **(b)** C_3H_6O **(c)** C_4H_9Cl

Problem 2.17 The following molecular model is a representation of *para*-aminobenzoic acid (PABA), the active ingredient in many sunscreens. Indicate the positions of the multiple bonds, and draw a skeletal structure (gray = C, red = O, blue = N, ivory = H).

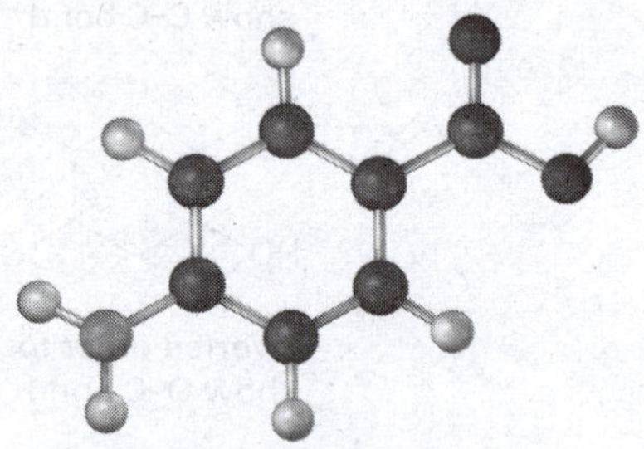

para-Aminobenzoic acid (PABA)

2.7 Cycloalkanes

We've discussed only open-chain alkanes thus far, but compounds with *rings* of carbon atoms are actually more common. Saturated cyclic hydrocarbons are called **cycloalkanes**, or *alicyclic* (*ali*phatic *cyclic*) compounds, and have the general formula $(CH_2)_n$, or C_nH_{2n}. In skeletal drawings, they are represented by polygons.

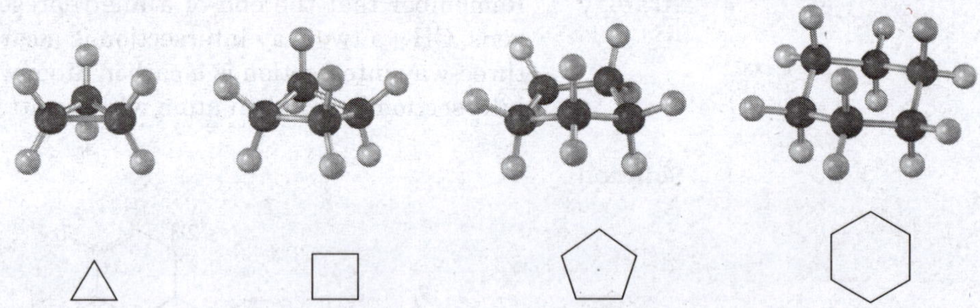

Cyclopropane Cyclobutane Cyclopentane Cyclohexane

Substituted cycloalkanes are named by rules similar to those for open-chain alkanes. For most compounds, there are only two steps.

STEP 1 **Find the parent.**

Count the number of carbon atoms in the ring and the number in the largest substituent chain. If the number of carbon atoms in the ring is equal to or greater than the number in the substituent, the compound is named as an alkyl-substituted cycloalkane. If the number of carbon atoms in the largest substituent is greater than the number in the ring, the compound is named as a cycloalkyl-substituted alkane.

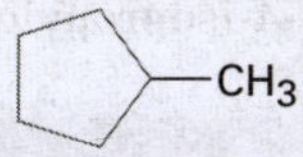

Methylcyclopentane

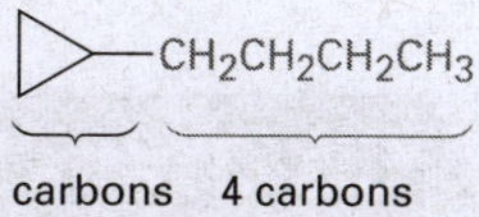

1-Cyclopropylbutane

STEP 2 **Number the substituents, and write the name.**

For substituted cycloalkanes, start at a point of attachment and number around the ring. If two or more substituents are present, begin numbering at the group that has alphabetical priority and proceed around the ring so as to give the second substituent the lowest number.

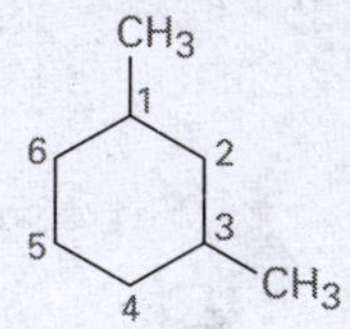

1,3-Dimethylcyclohexane
↑
Lower

NOT

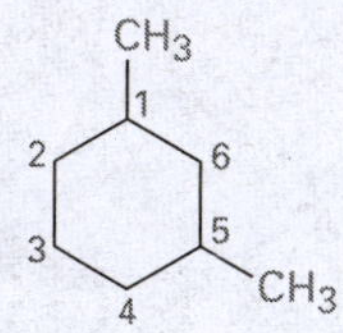

1,5-Dimethylcyclohexane
↑
Higher

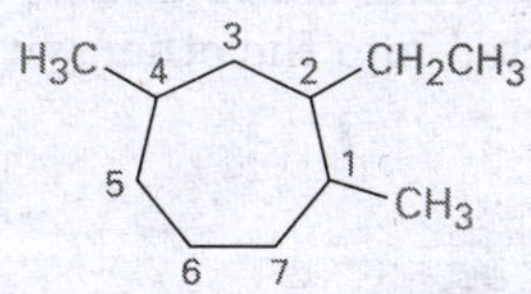

2-Ethyl-1,4-dimethylcycloheptane
↑ ↑
Lower Lower

NOT

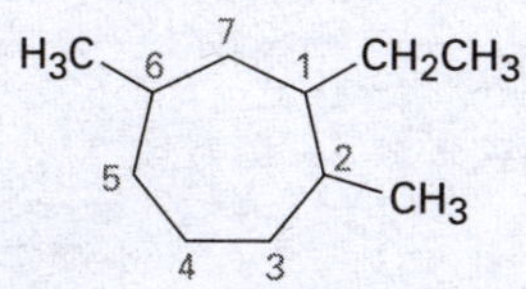

1-Ethyl-2,6-dimethylcycloheptane
↑
Higher

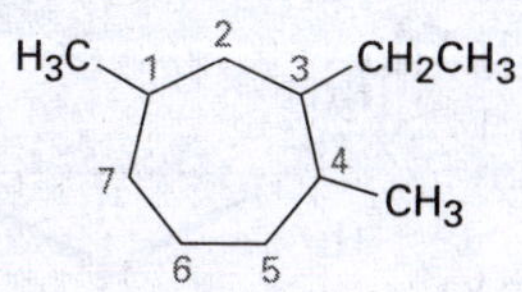

3-Ethyl-1,4-dimethylcycloheptane
↑
Higher

Problem 2.18 Give IUPAC names for the following cycloalkanes:

(a)

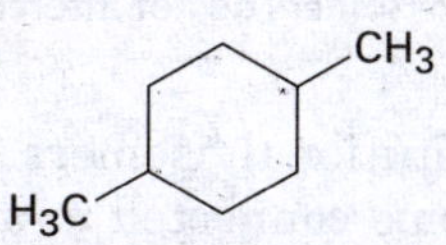

(b)

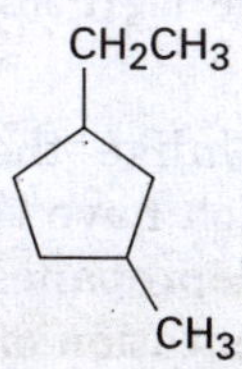

(c)

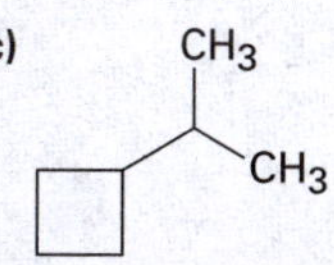

Problem 2.19 Draw structures corresponding to the following IUPAC names:

(a) 1-*tert*-Butyl-2-methylcyclopentane **(b)** 1,1-Dimethylcyclobutane
(c) 1-Ethyl-4-isopropylcyclohexane **(d)** 3-Cyclopropylhexane

2.8 Cis–Trans Isomerism in Cycloalkanes

In many respects, the behavior of cycloalkanes is similar to that of open-chain, acyclic alkanes. Both are nonpolar and chemically inert to most reagents. There are, however, some important differences. One difference is that cycloalkanes are less flexible than their open-chain counterparts. Although open-chain alkanes have nearly free rotation around their C–C single bonds, cycloalkanes have much less freedom of motion. Cyclopropane, for example, must be a rigid, planar molecule. No rotation around a C–C bond can take place in cyclopropane without breaking open the ring (Figure 2.9).

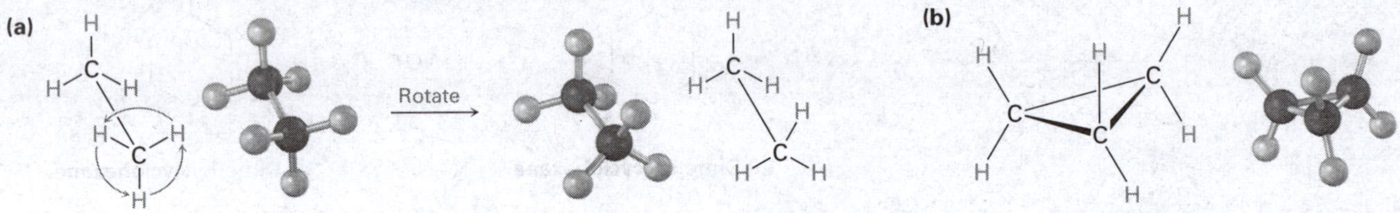

Figure 2.9 **(a)** Rotation occurs around the carbon–carbon bond in ethane, but **(b)** no rotation is possible around the carbon–carbon bonds in cyclopropane without breaking open the ring.

Because of their cyclic structures, cycloalkanes have two sides: a "top" side and a "bottom" side. As a result, isomerism is possible in substituted cycloalkanes. For example, there are two 1,2-dimethylcyclopropane isomers, one with the two methyl groups on the same side of the ring and one with the methyls on opposite sides. Both isomers are stable compounds, and neither can be converted into the other without breaking bonds (Figure 2.10).

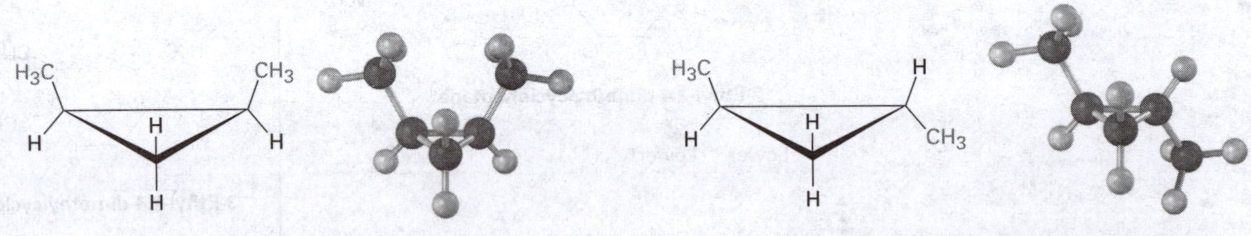

***cis*-1,2-Dimethylcyclopropane** ***trans*-1,2-Dimethylcyclopropane**

Figure 2.10 There are two different 1,2-dimethylcyclopropane isomers: one with the methyl groups on the same side of the ring (cis) and the other with the methyl groups on opposite sides of the ring (trans). The two isomers do not interconvert.

Unlike the constitutional isomers butane and isobutane (Section 2.2), which have their atoms connected in a different order, the two 1,2-dimethylcyclopropanes have the same order of connections but differ in the spatial orientation of the atoms. Such compounds, which have their atoms connected in the same order but differ in three-dimensional orientation, are called *stereochemical* isomers, or **stereoisomers**. More generally, the term **stereochemistry** is used to refer to the three-dimensional aspects of chemical structure and reactivity.

Constitutional isomers (different connections between atoms)

$CH_3-CH(CH_3)-CH_3$ and $CH_3-CH_2-CH_2-CH_3$

Stereoisomers (same connections but different three-dimensional geometry)

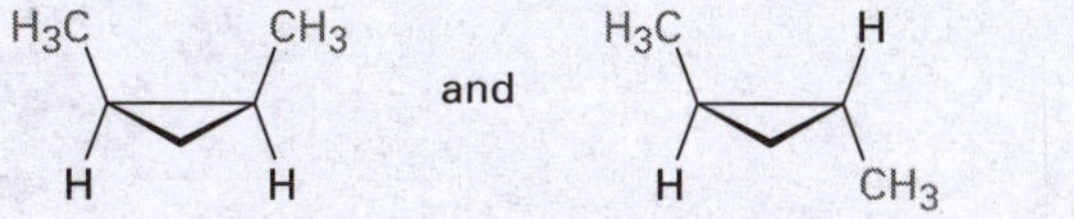

The 1,2-dimethylcyclopropanes are members of a subclass of stereoisomers called **cis–trans isomers**. The prefixes *cis-* (Latin, "on the same side") and *trans-* (Latin, "across") are used to distinguish between them. Cis–trans isomerism is a common occurrence in substituted cycloalkanes and in many cyclic biological molecules.

Worked Example 2.6 Naming Cis–Trans Cycloalkane Isomers

Name the following substances, including the *cis-* or *trans-* prefix:

(a)

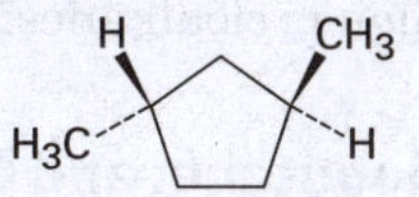

(b)

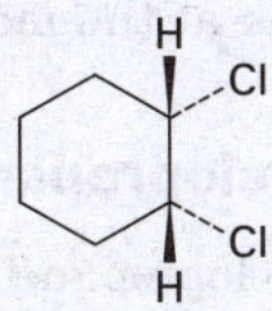

Strategy In these views, the ring is roughly in the plane of the page, a wedged bond protrudes out of the page, and a dashed bond recedes into the page. Two substituents are cis if they are both out of or both into the page, and they are trans if one is out of and one is into the page.

Solution **(a)** *trans*-1,3-Dimethylcyclopentane **(b)** *cis*-1,2-Dichlorocyclohexane

---

Problem 2.20 Draw *cis*-1-chloro-3-methylcyclopentane.

Problem 2.21 Draw both cis and trans isomers of 1,2-dibromocyclobutane.

Problem 2.22 Prostaglandin $F_{2\alpha}$, a hormone that causes uterine contraction during childbirth, has the following structure. Are the two hydroxyl groups (–OH) on the cyclopentane ring cis or trans to each other? What about the two carbon chains attached to the ring?

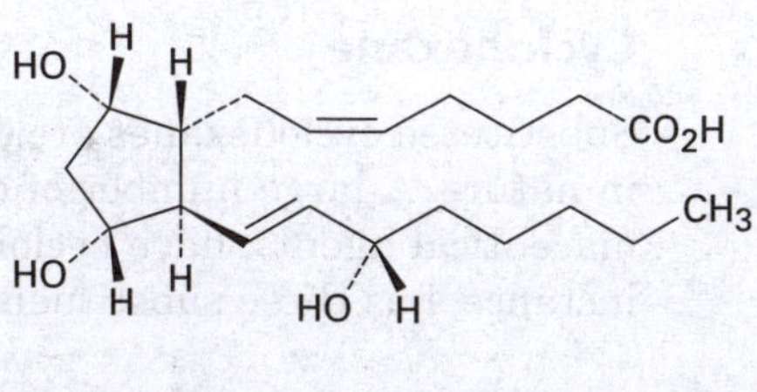

Prostaglandin $F_{2\alpha}$

Problem 2.23 Name the following substances, including the *cis-* or *trans-* prefix (red-brown = Br):

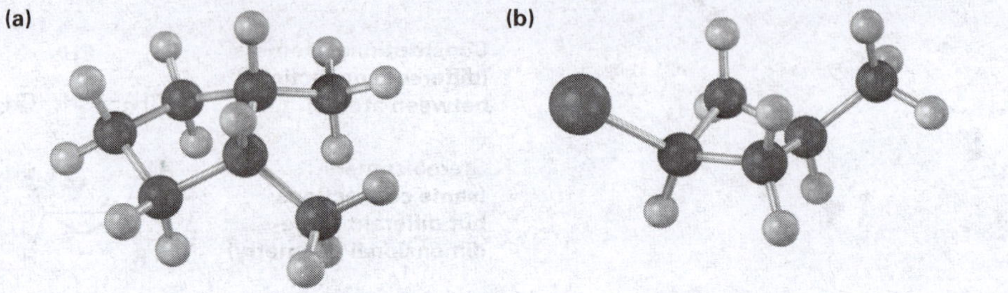

2.9 Conformations of Some Cycloalkanes

In the early days of organic chemistry, cycloalkanes provoked a good deal of consternation among chemists. The problem was that if carbon prefers to have bond angles of 109.5°, how is it possible for cyclopropane and cyclobutane to exist? After all, cyclopropane must have a triangular shape with bond angles near 60°, and cyclobutane must have a roughly square shape with bond angles near 90°. Nonetheless, these compounds *do* exist and are stable. Let's look at the most common cycloalkanes.

Cyclopropane, Cyclobutane, and Cyclopentane

Cyclopropane is a flat, triangular molecule with C–C–C bond angles of 60°, as indicated in Figure 2.11a. The deviation of bond angles from the normal 109.5° tetrahedral value causes an **angle strain** in the molecule that raises its energy and makes it more reactive than unstrained alkanes. All six of the C–H bonds have an eclipsed, rather than staggered, arrangement with their neighbors.

Cyclobutane and cyclopentane are slightly puckered rather than flat, as indicated in Figure 2.11b–c. This puckering makes the C–C–C bond angles a bit smaller than they would otherwise be and increases the angle strain. At the same time, though, the puckering relieves the unfavorable eclipsing interactions of adjacent C–H bonds that would occur if the rings were flat.

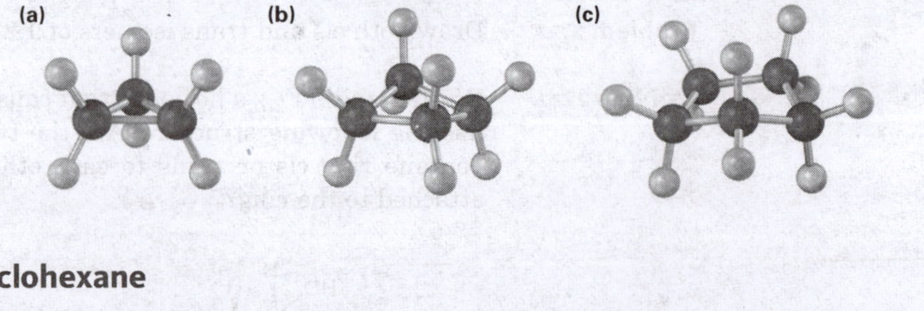

Figure 2.11 The structures of (a) cyclopropane, (b) cyclobutane, and (c) cyclopentane. Cyclopropane is planar, but cyclobutane and cyclopentane are slightly puckered.

Cyclohexane

Substituted cyclohexanes are the most common cycloalkanes and occur widely in nature. A large number of compounds, including steroids and many pharmaceutical agents, have cyclohexane rings. The flavoring agent menthol, for instance, has three substituents on a six-membered ring.

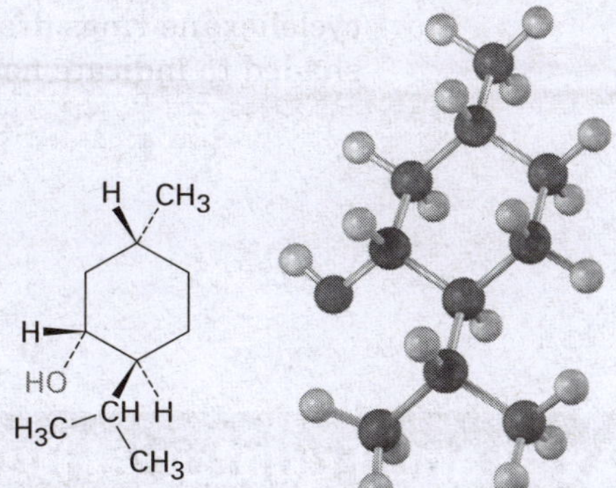

Menthol

Cyclohexane is not flat. Rather, it is puckered into a strain-free, three-dimensional shape called a **chair conformation** because of its similarity to a lounge chair, with a back, a seat, and a footrest (Figure 2.12). All C–C–C bond angles are near 109°, and all adjacent C–H bonds are staggered.

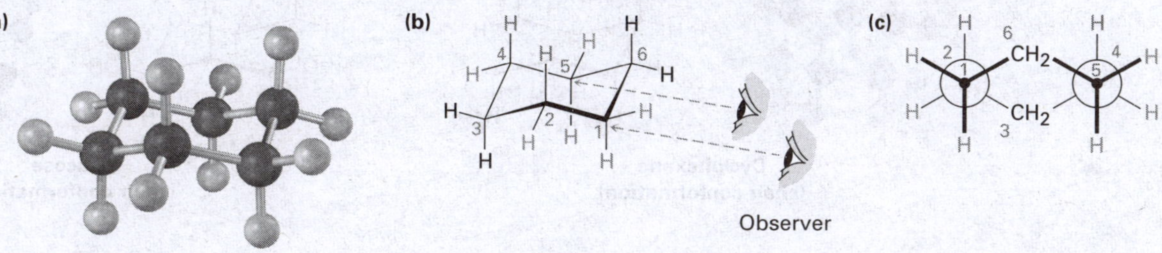

Figure 2.12 The strain-free, chair conformation of cyclohexane. All C–C–C bond angles are close to 109°, and all neighboring C–H bonds are staggered, as evident in the end-on view in (c).

A chair conformation is drawn in three steps.

STEP 1 Draw two parallel lines, slanted downward and slightly offset from each other. This means that four of the cyclohexane carbons lie in a plane.

STEP 2 Place the topmost carbon atom above and to the right of the plane of the other four, and connect the bonds.

STEP 3 Place the bottommost carbon atom below and to the left of the plane of the middle four, and connect the bonds. Note that the bonds to the bottommost carbon atom are parallel to the bonds to the topmost carbon.

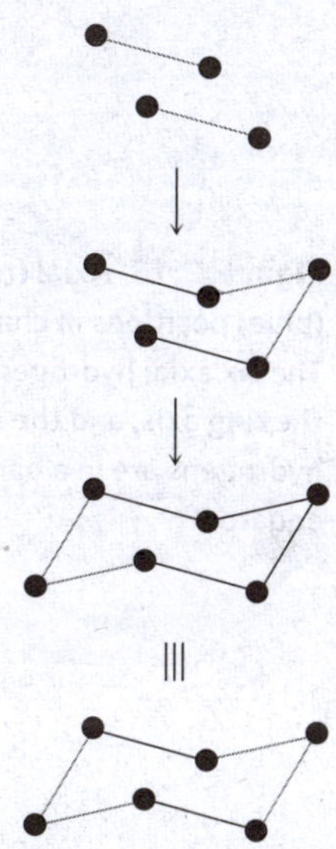

When viewing cyclohexane, it's helpful to remember that the lower bond is in front and the upper bond is in back. If this convention is not defined, an optical illusion can make it appear that the reverse is true. For clarity, all cyclohexane rings drawn in this book will have the front (lower) bond heavily shaded to indicate nearness to the viewer.

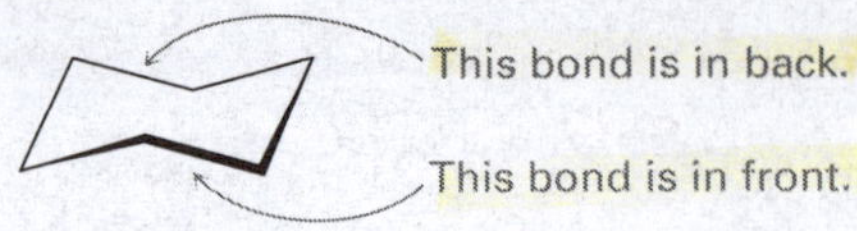

2.10 Axial and Equatorial Bonds in Cyclohexane

The chair conformation of cyclohexane leads to many consequences. We'll see in Section 14.5, for instance, that simple carbohydrates, such as glucose, adopt a conformation based on the cyclohexane chair and that their chemistry is directly affected as a result.

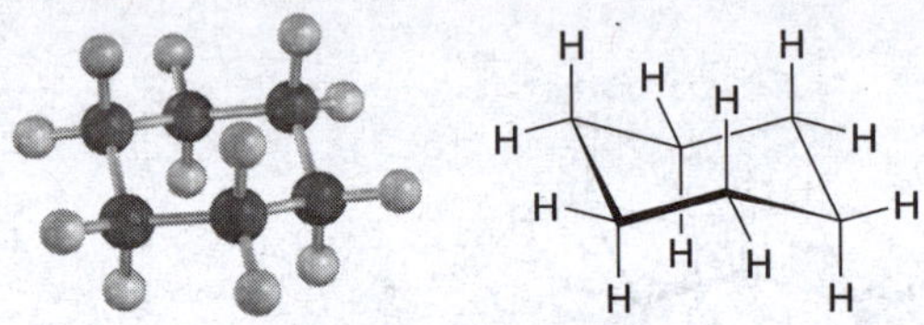

Cyclohexane
(chair conformation)

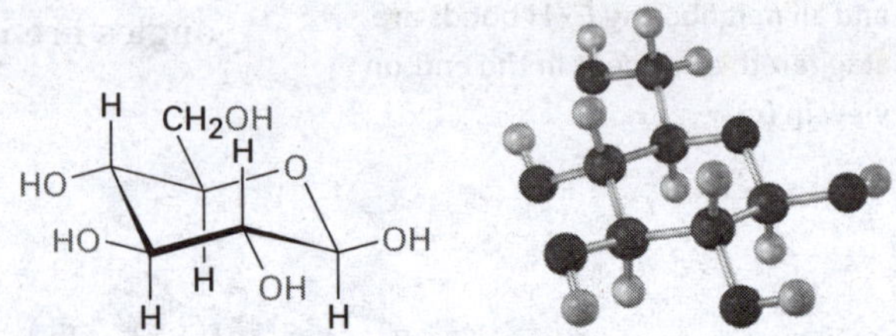

Glucose
(chair conformation)

Another consequence of the chair conformation is that there are two kinds of positions for substituents on the cyclohexane ring: *axial* positions and *equatorial* positions (Figure 2.13). The six **axial positions** are perpendicular to the ring, parallel to the ring axis, and the six **equatorial positions** are in the rough plane of the ring, around the ring equator. Each carbon atom has one axial and one equatorial position, and each side of the ring has three axial and three equatorial positions in an alternating arrangement.

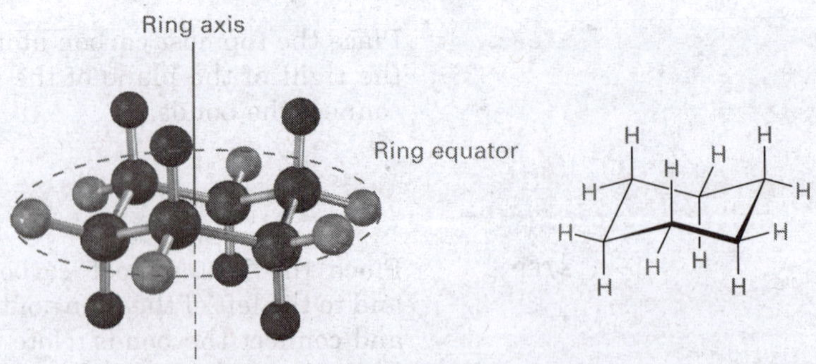

Figure 2.13 Axial (red) and equatorial (blue) positions in chair cyclohexane. The six axial hydrogens are parallel to the ring axis, and the six equatorial hydrogens are in a band around the ring equator.

Note that we haven't used the words *cis* and *trans* in this discussion of cyclohexane conformation. Two hydrogens on the same side of a ring are always cis, regardless of whether they're axial or equatorial and regardless of whether they're adjacent. Similarly, two hydrogens on opposite sides of the ring are always trans.

Axial and equatorial bonds can be drawn by following the procedure in Figure 2.14.

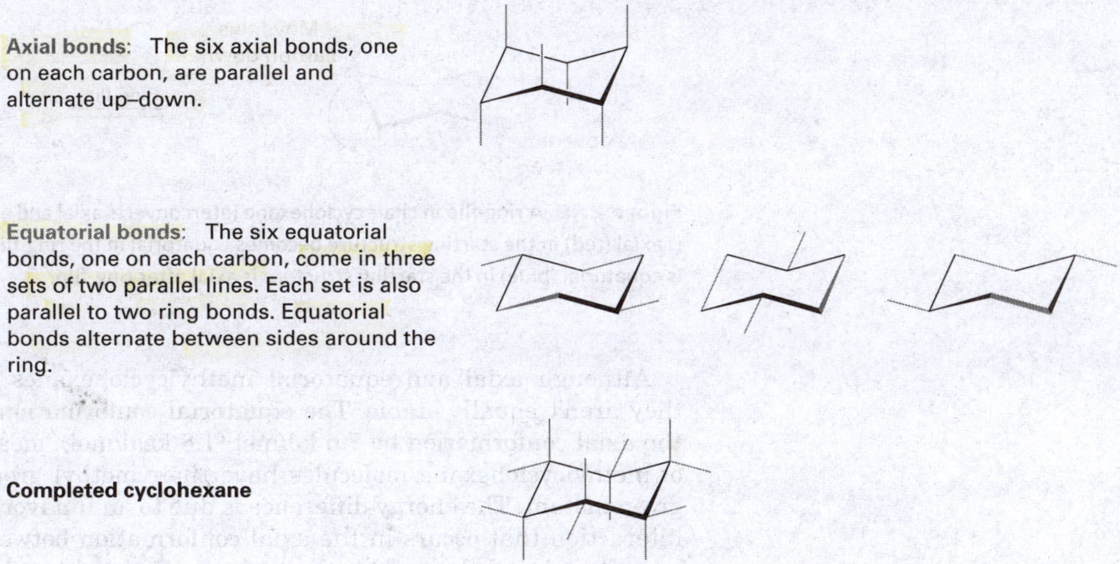

Figure 2.14 A procedure for drawing axial and equatorial bonds in cyclohexane.

Problem 2.24 Draw two chair structures for methylcyclohexane, one with the methyl group axial and one with the methyl group equatorial.

2.11 Conformational Mobility of Cyclohexane

Because chair cyclohexane has two kinds of positions, axial and equatorial, we might expect to find two isomeric forms of a monosubstituted cyclohexane. In fact, we don't. There is only *one* methylcyclohexane, *one* bromocyclohexane, *one* cyclohexanol (hydroxycyclohexane), and so on, because cyclohexane rings are *conformationally mobile* at room temperature. Different chair conformations readily interconvert by a process called a **ring-flip**.

The ring-flip of a chair cyclohexane can be visualized as shown in Figure 2.15 by keeping the middle four carbon atoms in place while folding the two ends in opposite directions. An axial substituent in one chair form becomes an equatorial substituent in the ring-flipped chair form and vice versa. For example, axial methylcyclohexane becomes equatorial methylcyclohexane after ring-flip. Because this interconversion occurs rapidly at room temperature, the individual axial and equatorial isomers can't be isolated.

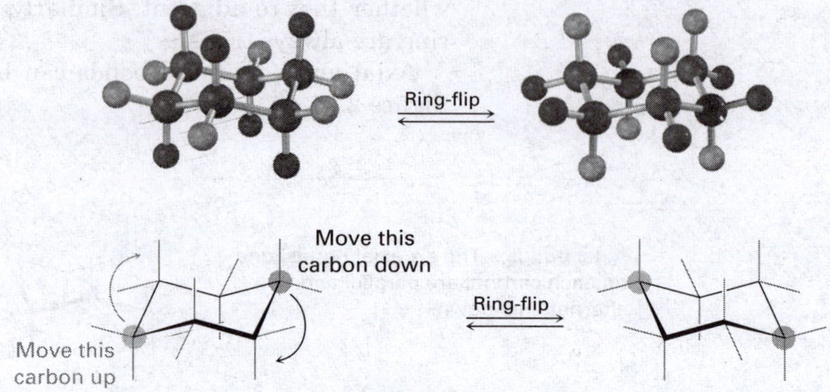

Figure 2.15 A ring-flip in chair cyclohexane interconverts axial and equatorial positions. What is axial (red) in the starting structure becomes equatorial in the ring-flipped structure, and what is equatorial (blue) in the starting structure is axial after ring-flip.

Although axial and equatorial methylcyclohexanes interconvert rapidly, they aren't equally stable. The equatorial conformation is more stable than the axial conformation by 7.6 kJ/mol (1.8 kcal/mol), meaning that about 95% of methylcyclohexane molecules have their methyl group equatorial at any given instant. The energy difference is due to an unfavorable spatial, or *steric,* interaction that occurs in the axial conformation between the methyl group on carbon 1 and the axial hydrogen atoms on carbons 3 and 5. This so-called 1,3-diaxial interaction introduces 7.6 kJ/mol of **steric strain** into the molecule because the axial methyl group and the nearby axial hydrogen are too close together (Figure 2.16).

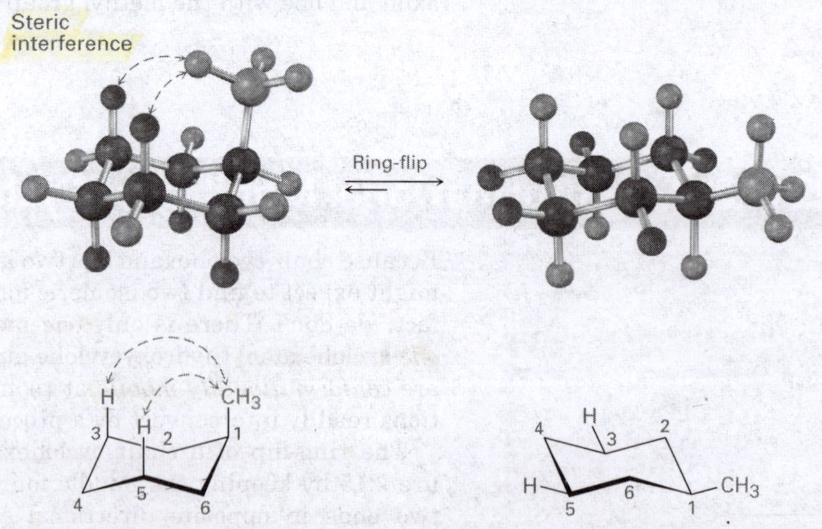

Figure 2.16 Axial versus equatorial methylcyclohexane. The 1,3-diaxial steric interactions in axial methylcyclohexane (easier to see in space-filling models) make the equatorial conformation more stable by 7.6 kJ/mol.

What is true for methylcyclohexane is also true for other monosubstituted cyclohexanes: a substituent is always more stable in an equatorial position than in an axial position. As you might expect, the amount of steric strain increases as the size of the axial substituent group increases.

Worked Example 2.7 Drawing Conformations of Substituted Cyclohexanes

Draw 1,1-dimethylcyclohexane in a chair conformation, indicating which methyl group in your drawing is axial and which is equatorial.

Strategy Draw a chair cyclohexane ring, and then put two methyl groups on the same carbon. The methyl group in the rough plane of the ring is equatorial, and the one directly above or below the ring is axial.

Solution

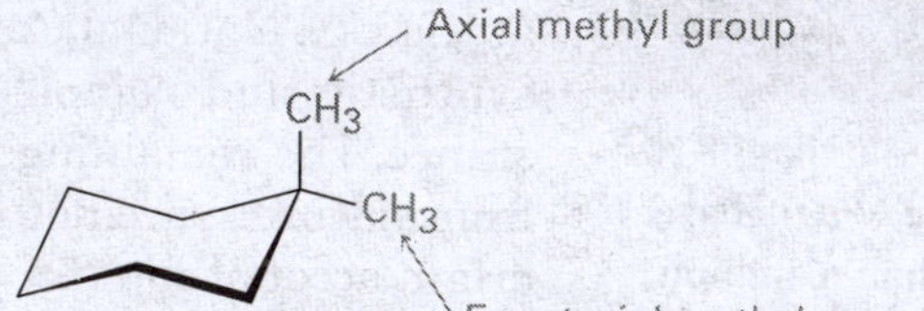

Problem 2.25 Draw two different chair conformations of bromocyclohexane showing all hydrogen atoms. Label all positions as axial or equatorial. Which of the two conformations do you think is more stable?

Problem 2.26 Draw *cis*-1,2-dichlorocyclohexane in a chair conformation, and explain why one chlorine must be axial and one equatorial.

Problem 2.27 Draw *trans*-1,2-dichlorocyclohexane in chair conformation, and explain why both chlorines must be axial or both equatorial.

Problem 2.28 Identify each substituent as axial or equatorial, and tell whether the conformation shown is the more stable or less stable chair form (gray = C, yellow-green = Cl, ivory = H).

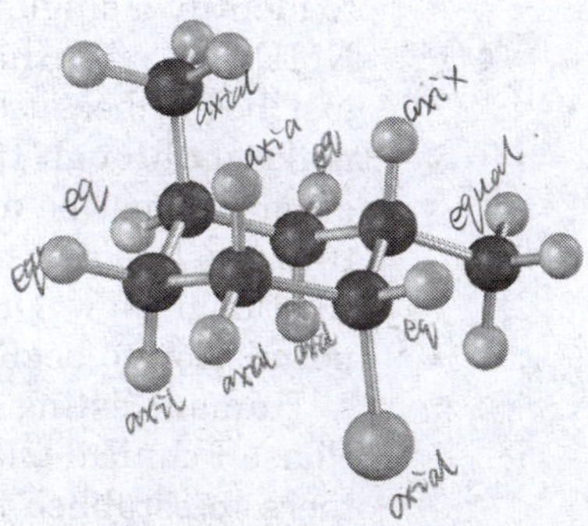

INTERLUDE

Where Do Drugs Come From?

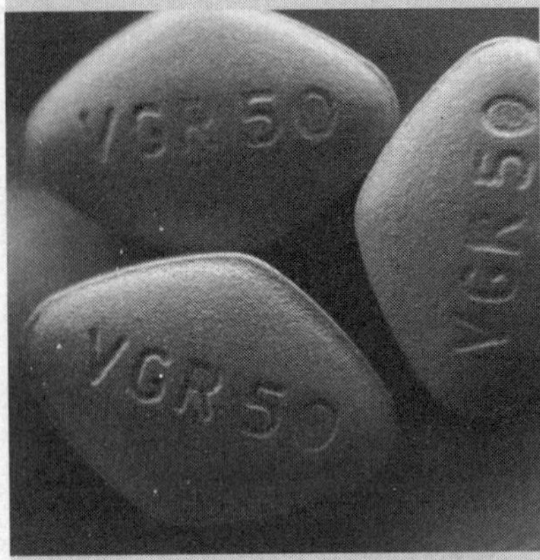

© Charlie Newham/Alamy

Approved for sale in March 1998 to treat male impotency, Viagra has been used by more than 16 million men. It is also used to treat pulmonary hypertension and is currently undergoing study as a treatment for preeclampsia, a complication of pregnancy that is responsible for as many as 70,000 deaths each year. Where do new drugs like this come from?

It has been estimated that major pharmaceutical companies in the United States spend some $33 billion per year on drug research and development, while government agencies and private foundations spend another $28 billion. What does this money buy? For the period 1981 to 2004, the money resulted in a total of 912 new molecular entities (NMEs)—new biologically active chemical substances approved for sale as drugs by the U.S. Food and Drug Administration (FDA). That's an average of only 38 new drugs each year spread over all diseases and conditions, and the number has been steadily falling: in 2004, only 23 NMEs were approved.

Where do the new drugs come from? According to a study carried out at the U.S. National Cancer Institute, only 33% of new drugs are entirely synthetic and completely unrelated to any naturally occurring substance. The remaining 67% take their lead, to a greater or lesser extent, from nature. Vaccines and genetically engineered proteins of biological origin account for 15% of NMEs, but most new drugs come from *natural products,* a catchall term generally taken to mean small molecules found in bacteria, plants, and other living organisms. Unmodified natural products isolated directly from the producing organism account for 28% of NMEs, while natural products that have been chemically modified in the laboratory account for the remaining 24%.

Origin of New Drugs 1981–2004

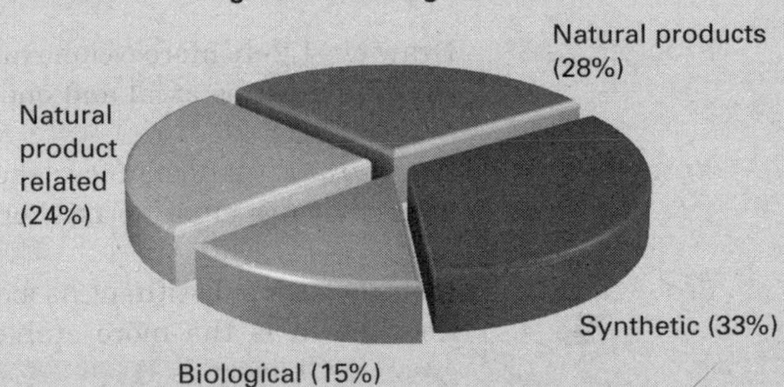

Many years of work go into screening many thousands of substances to identify a single compound that might ultimately gain approval as an NME. But after that single compound has been identified, the work has just begun because it takes an average of 9 to 10 years for a drug to make it through the approval process. First, the safety of the drug in animals must be demonstrated and an economical method of manufacture must be devised. With these preliminaries out of the way, an Investigational New Drug (IND) application is submitted to the FDA for permission to begin testing in humans.

Human testing takes 5 to 7 years and is divided into three phases. Phase I clinical trials are carried out on a small group of healthy volunteers to establish safety and look for side effects. Several months to a year are needed, and only about 70% of drugs pass at this point. Phase

INTERLUDE

II clinical trials next test the drug for 1 to 2 years in several hundred patients with the target disease or condition, looking both for safety and for efficacy, and only about 33% of the original group pass. Finally, phase III trials are undertaken on a large sample of patients to document definitively the drug's safety, dosage, and efficacy. If the drug is one of the 25% of the original group that make it to the end of phase III, all the data are then gathered into a New Drug Application (NDA) and sent to the FDA for review and approval, which can take another 2 years. Ten years have elapsed and at least $500 million has been spent, with only a 20% success rate for the drugs that began testing. Finally, though, the drug will begin to appear in medicine cabinets. The following timeline shows the process.

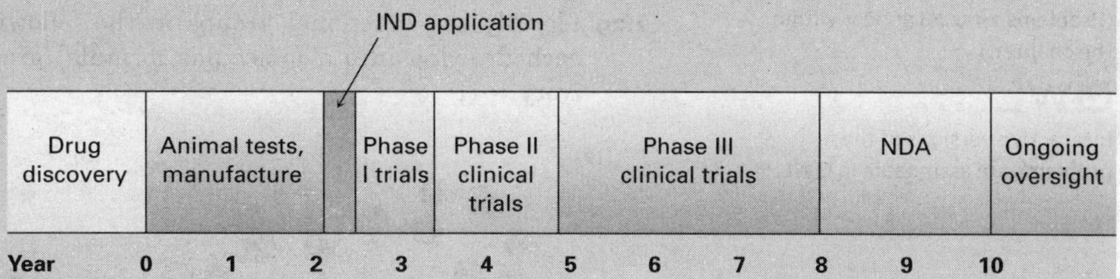

Summary and Key Words

aliphatic 45
alkane 44
alkyl group 46
angle strain 62
axial position 64
branched-chain alkane 46
chair conformation 63
cis–trans isomers 61
conformation 54
constitutional isomers 46
cycloalkane 58
eclipsed conformation 55
equatorial position 64
functional group 39
hydrocarbon 45
isomers 46
Newman projection 55
normal (*n*) alkane 46
ring-flip 65
saturated 45
sawhorse representation 55
skeletal structure 56
staggered conformation 55
stereochemistry 61
stereoisomers 61
steric strain 66
straight-chain alkane 46

Even though alkanes are relatively unreactive and rarely involved in chemical reactions, they nevertheless provide a useful vehicle for introducing some important general ideas. In this chapter, we've used alkanes to introduce the basic approach to naming organic compounds and to take an initial look at some of the three-dimensional aspects of molecules.

A **functional group** is an atom or group of atoms within a larger molecule that has a characteristic chemical reactivity. Because functional groups behave approximately the same way in all molecules in which they occur, the reactions of an organic molecule are largely determined by its functional groups.

Alkanes are a class of **saturated hydrocarbons** having the general formula C_nH_{2n+2}. They contain no functional groups, are chemically rather inert, and can be either **straight-chain** or **branched**. Alkanes are named systematically by a series of IUPAC rules of nomenclature. **Isomers**—compounds that have the same chemical formula but different structures—exist for all but the simplest alkanes. Compounds such as butane and isobutane, which have the same formula but differ in the way their atoms are connected, are called **constitutional isomers**.

Because C–C single bonds are formed by head-on orbital overlap, rotation is possible about them. Alkanes can therefore adopt any of a large number of rapidly interconverting **conformations**. A **staggered conformation** is more stable than an **eclipsed conformation**.

Cycloalkanes contain rings of carbon atoms and have the general formula C_nH_{2n}. Because complete rotation around C–C bonds is not possible in cycloalkanes, conformational mobility is reduced and disubstituted cycloalkanes can exist as **cis–trans stereoisomers**. In a cis isomer, both substituents are on the same side of the ring, whereas in a trans isomer, the substituents are on opposite sides of the ring.

Cyclohexanes are the most common of all rings because of their wide occurrence in nature. Cyclohexane exists in a puckered, strain-free **chair conformation** in which all bond angles are near 109° and all neighboring C–H bonds are staggered. Chair cyclohexane has two kinds of bonding positions: axial and equatorial. **Axial bonds** are directed up and down, parallel to the ring axis; **equatorial bonds** lie in a belt around the ring equator. Chair cyclohexanes can undergo a **ring-flip** that interconverts axial and equatorial positions. Substituents on the ring are more stable in the equatorial than in the axial position.

Exercises

Visualizing Chemistry

(Problems 2.1–2.28 appear within the chapter.)

OWL

Interactive versions of these problems are assignable in OWL.

2.29 Identify the functional groups in the following substances, and convert each drawing into a molecular formula (gray = C, red = O, blue = N, ivory = H).

(a)

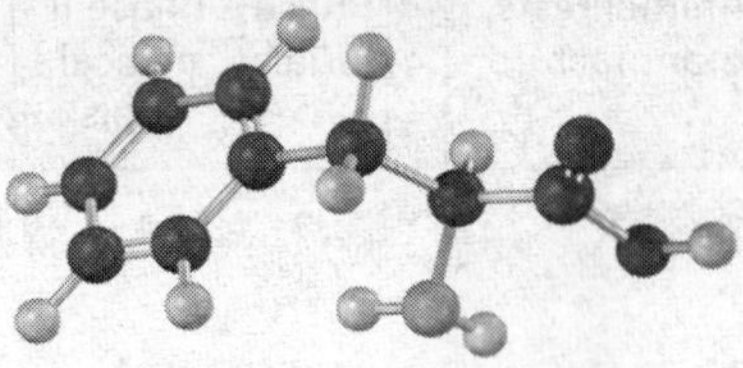

Phenylalanine

(b)

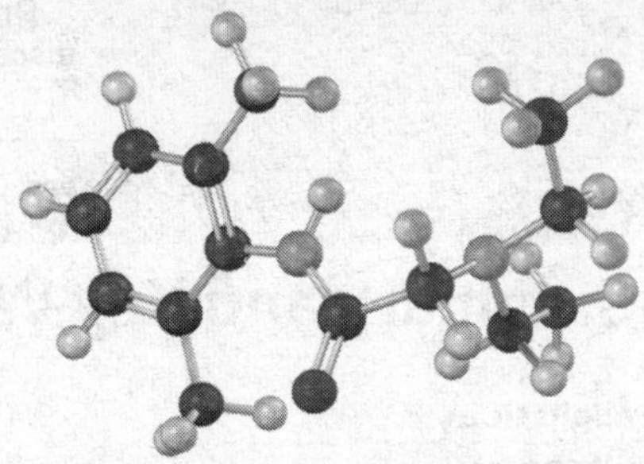

Lidocaine

2.30 Give IUPAC names for the following substances, and convert each drawing into a skeletal structure.

(a)

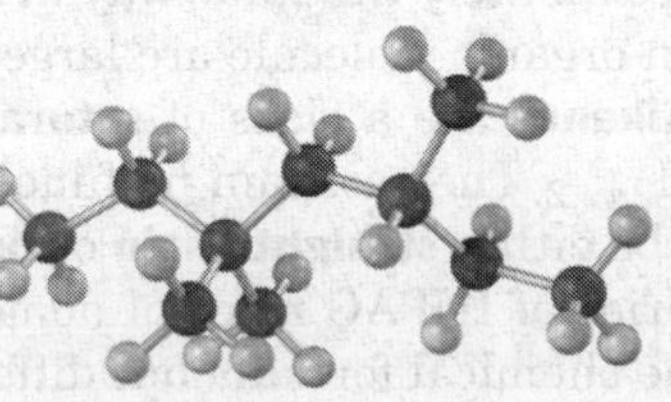

(b)

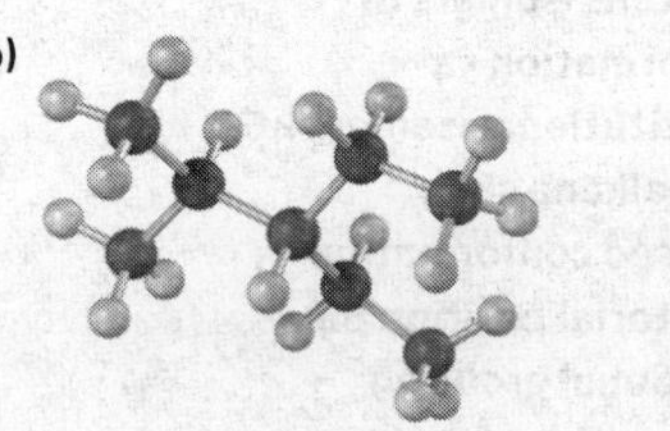

(c)

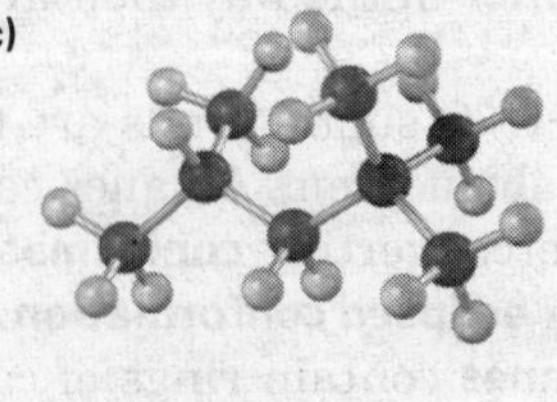

(d)

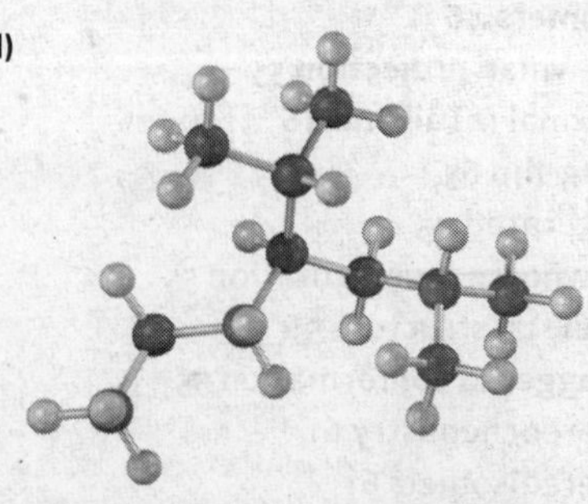

2.31 A trisubstituted cyclohexane with three substituents—red, green, and blue—undergoes a ring-flip to its alternative chair conformation. Identify each substituent as axial or equatorial, and show the positions occupied by the three substituents in the ring-flipped form.

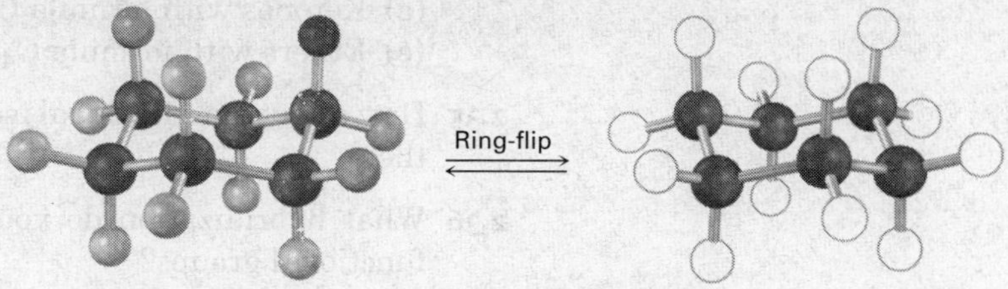

2.32 Glucose exists in two forms having a 36:64 ratio at equilibrium. Draw a skeletal structure of each, describe the difference between them, and tell which of the two you think is more stable (red = O).

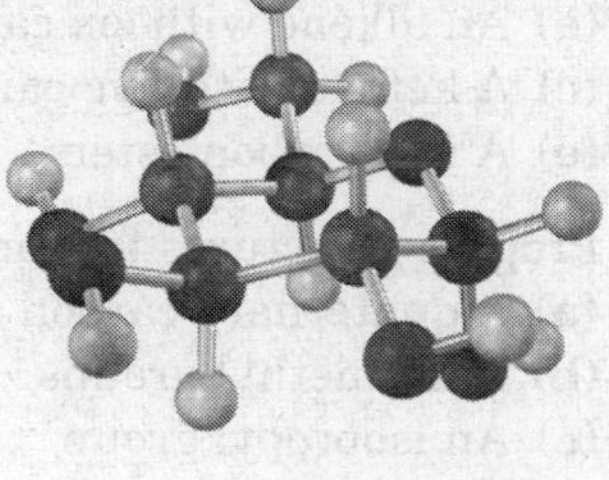

α-Glucose

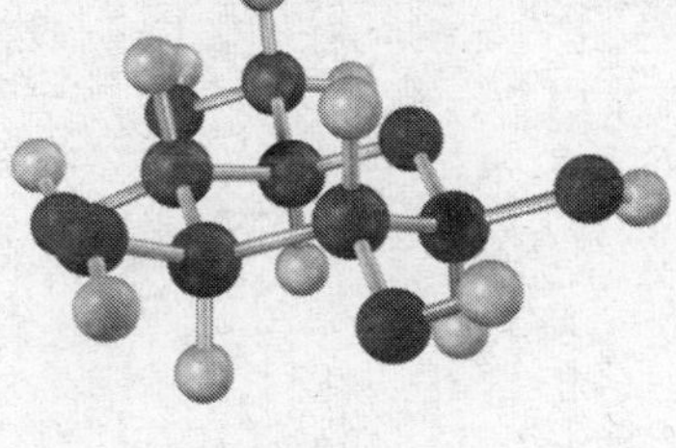

β-Glucose

2.33 The following cyclohexane derivative has three substituents—red, green, and blue. Identify each substituent as axial or equatorial, and identify each pair of relationships (red–blue, red–green, and blue–green) as cis or trans.

Additional Problems

FUNCTIONAL GROUPS AND ISOMERISM

2.34 Write as many structures as you can that fit the following descriptions:
(a) Alcohols with formula $C_4H_{10}O$
(b) Amines with formula $C_5H_{13}N$
(c) Ketones with formula $C_5H_{10}O$
(d) Aldehydes with formula $C_5H_{10}O$
(e) Ethers with formula $C_4H_{10}O$
(f) Esters with formula $C_4H_8O_2$

2.35 How many constitutional isomers are there with the formula C_3H_8O? Draw them.

2.36 What hybridization do you expect for the carbon atom in the following functional groups?
(a) Carboxylic acid chloride **(b)** Thiol **(c)** Imine **(d)** Aldehyde

2.37 Propose suitable structures for the following:
(a) An alkene, C_7H_{14}
(b) A cycloalkene, C_3H_4
(c) A ketone, C_4H_8O
(d) A nitrile, C_5H_9N
(e) A dialkene, C_5H_8
(f) A dialdehyde, $C_4H_6O_2$

2.38 Propose structures for molecules that fit the following descriptions:
(a) An alkene with six carbons
(b) A cycloalkene with five carbons
(c) A ketone with five carbons
(d) An amide with four carbons
(e) A five-carbon ester
(f) An aromatic alcohol

2.39 Propose structures for compounds that contain the following:
(a) A quaternary carbon
(b) Four methyl groups
(c) An isopropyl group
(d) Two tertiary carbons
(e) An amino group ($-NH_2$) bonded to a secondary carbon

2.40 Locate and identify the functional groups in the following molecules. Each intersection of lines and the end of each line represents a carbon atom with the appropriate number of hydrogens attached.

(a)
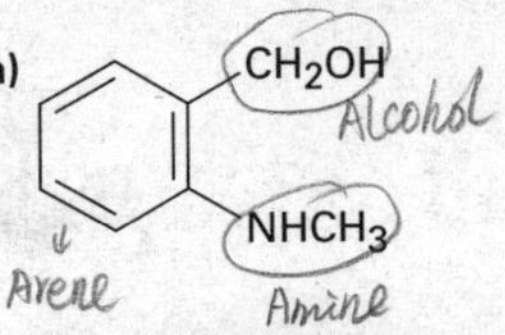

(b)
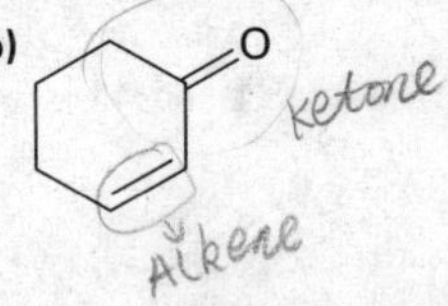

(c)
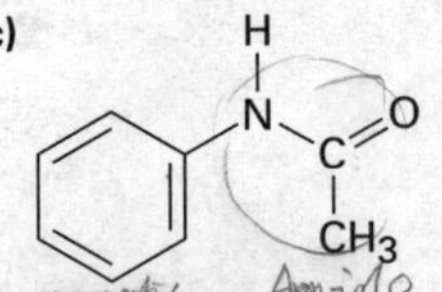

(d)
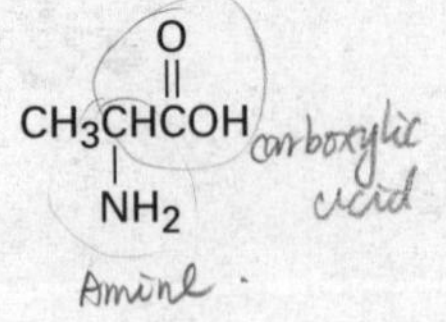

(e)
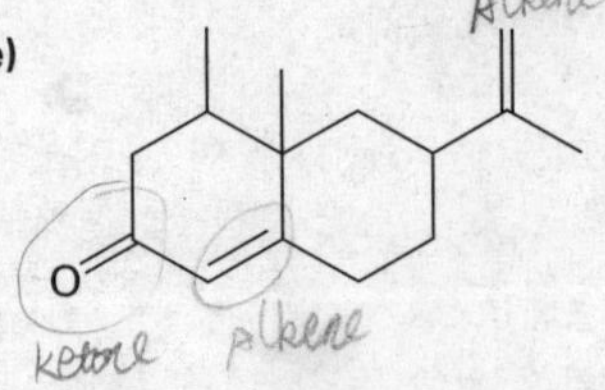

(f)
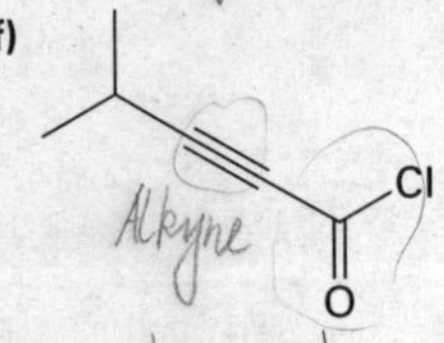

2.41 Draw all monochloro derivatives of 2,5-dimethylhexane.

2.42 Draw all monobromo derivatives of pentane, $C_5H_{11}Br$.

2.43 For each of the following compounds, draw a constitutional isomer with the same functional groups:

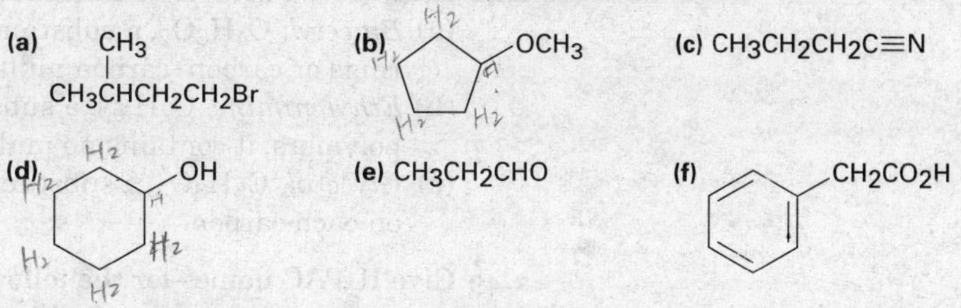

2.44 In each of the following sets, which structures represent the same compound and which represent different compounds?

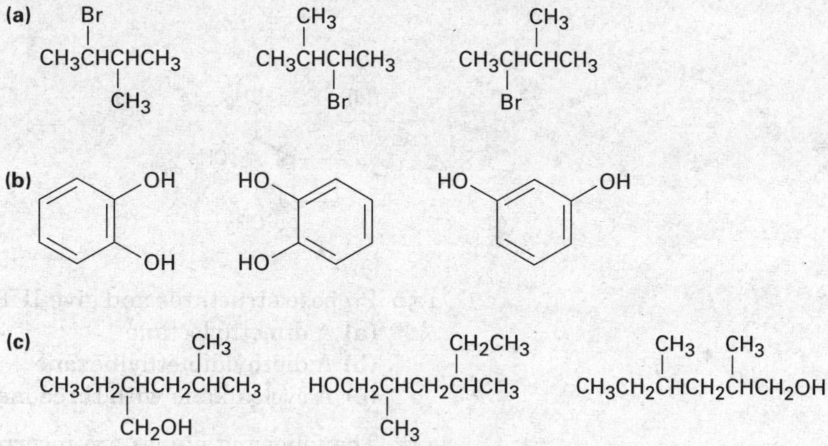

NAMING AND DRAWING CHEMICAL STRUCTURES

2.45 Give IUPAC names for the following alkanes:

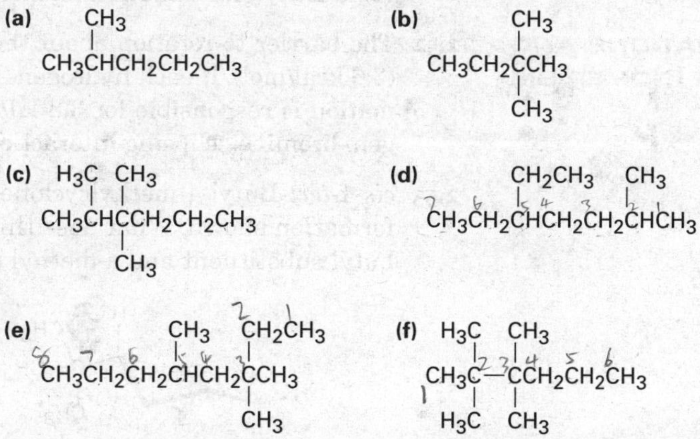

2.46 Draw structures for the following substances:

(a) 2-Methylheptane
(b) 4-Ethyl-2-methylhexane
(c) 4-Ethyl-3,4-dimethyloctane
(d) 2,4,4-Trimethylheptane
(e) 1,1-Dimethylcyclopentane
(f) 4-Isopropyl-3-methylheptane

2.47 Draw and name the five isomers of C_6H_{14}.

2.48 Draw the structures of the following molecules:

(a) *Biacetyl,* $C_4H_6O_2$, a substance with the aroma of butter; it contains no rings or carbon–carbon multiple bonds.

(b) *Ethylenimine,* C_2H_5N, a substance used in the synthesis of melamine polymers; it contains no multiple bonds.

(c) *Glycerol,* $C_3H_8O_3$, a substance used in cosmetics; it has an –OH group on each carbon.

2.49 Give IUPAC names for the following compounds:

(a)

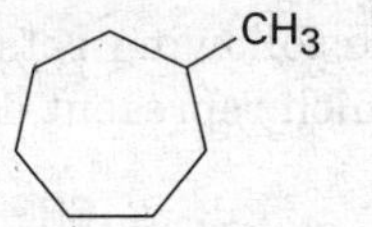

(b)

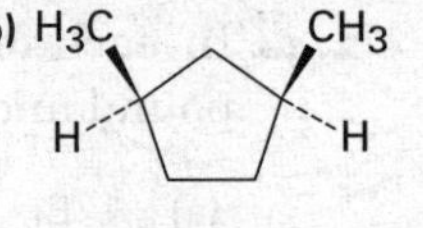

(c)

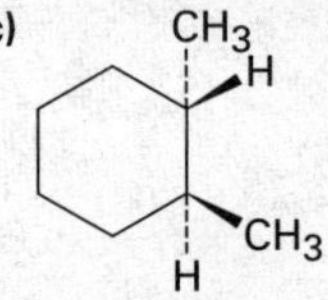

(d)

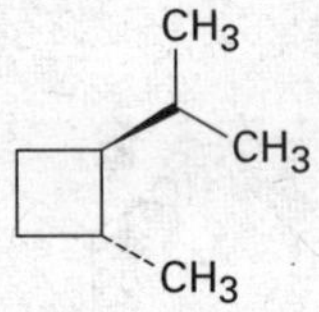

(e)

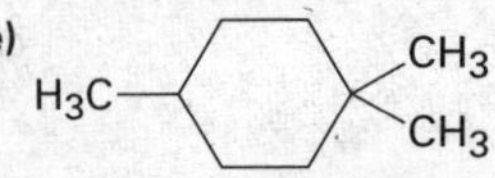

2.50 Propose structures and give IUPAC names for the following:

(a) A dimethyloctane

(b) A diethyldimethylhexane

(c) A cycloalkane with three methyl groups

2.51 The following names are incorrect. Draw the structures represented, and give the proper IUPAC names.

(a) 2,2-Dimethyl-6-ethylheptane **(b)** 4-Ethyl-5,5-dimethylpentane

(c) 3-Ethyl-4,4-dimethylhexane **(d)** 5,5,6-Trimethyloctane

CONFORMATIONS AND CIS-TRANS ISOMERISM

2.52 The barrier to rotation about the C–C bond in bromoethane is 15.0 kJ/mol (3.6 kcal/mol). If each hydrogen–hydrogen interaction in the eclipsed conformation is responsible for 3.8 kJ/mol (0.9 kcal/mol), how much is the hydrogen–bromine eclipsing interaction responsible for?

2.53 *cis*-1-*tert*-Butyl-4-methylcyclohexane exists almost exclusively in the conformation shown. What does this tell you about the relative sizes of a *tert*-butyl substituent and a methyl substituent?

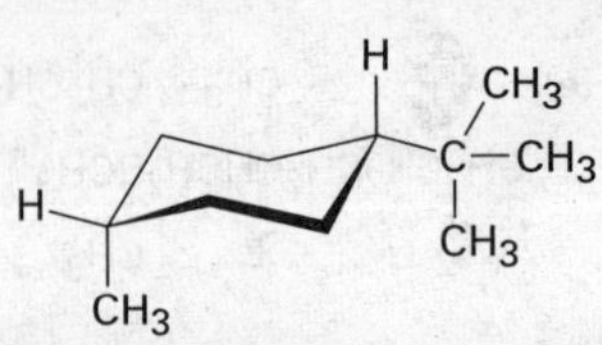

***cis*-1-*tert*-Butyl-4-methylcyclohexane**

2.54 Sighting along the C2–C3 bond of 2-methylbutane, there are two different staggered conformations. Draw them both in Newman projections, tell which is more stable, and explain your choice.

2.55 Sighting along the C2–C3 bond of 2-methylbutane (see Problem 2.54), there are also two possible eclipsed conformations. Draw them both in Newman projections, tell which you think is lower in energy, and explain.

2.56 Which is more stable, *cis*-1,3-dimethylcyclohexane or *trans*-1,3-dimethylcyclohexane? Draw chair conformations of both, and explain your answer.

2.57 Draw *trans*-1,2-dimethylcyclohexane in its more stable chair conformation. Are the methyl groups axial or equatorial?

2.58 Draw *cis*-1,2-dimethylcyclohexane in its more stable chair conformation. Are the methyl groups axial or equatorial? Which is more stable, *cis*-1,2-dimethylcyclohexane or *trans*-1,2-dimethylcyclohexane (Problem 2.57)? Explain.

2.59 Tell whether the following pairs of compounds are identical, constitutional isomers, stereoisomers, or unrelated.

(a) *cis*-1,3-Dibromocyclohexane and *trans*-1,4-dibromocyclohexane

(b) 2,3-Dimethylhexane and 2,3,3-trimethylpentane

(c)

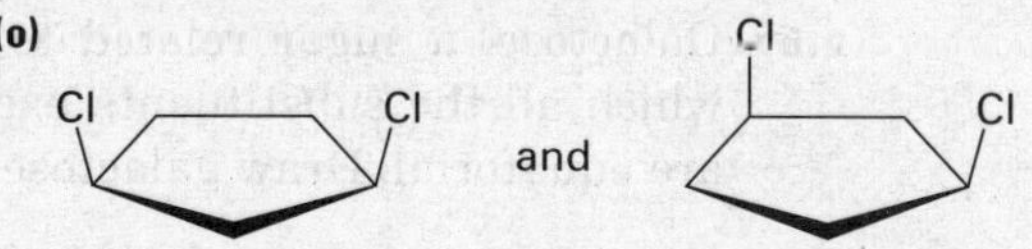

2.60 Draw a stereoisomer of *trans*-1,3-dimethylcyclobutane.

2.61 Draw two constitutional isomers of *cis*-1,2-dibromocyclopentane.

General Problems

2.62 Identify each pair of relationships among the –OH groups in glucose (red–blue, red–green, red–black, blue–green, blue–black, green–black) as cis or trans.

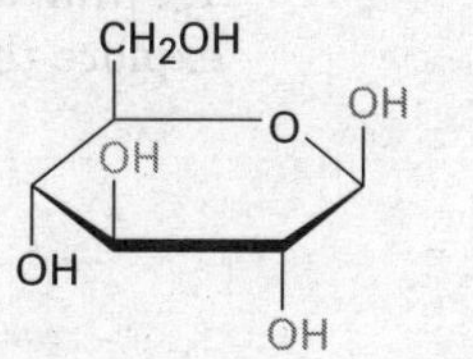

Glucose

2.63 Malic acid, $C_4H_6O_5$, has been isolated from apples. Because malic acid reacts with 2 equivalents of base, it can be formulated as a dicarboxylic acid (that is, it has two $-CO_2H$ groups).

(a) Draw at least five possible structures for malic acid.

(b) If malic acid is also a secondary alcohol (has an –OH group attached to a secondary carbon), what is its structure?

2.64 *N*-Methylpiperidine has the conformation shown. What does this tell you about the relative steric requirements of a methyl group versus an electron lone pair?

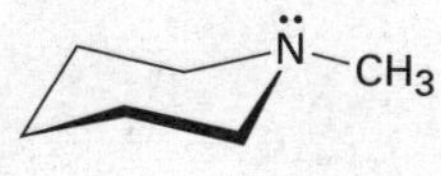

***N*-Methylpiperidine**

2.65 Draw the two chair conformations of menthol, and tell which is more stable.

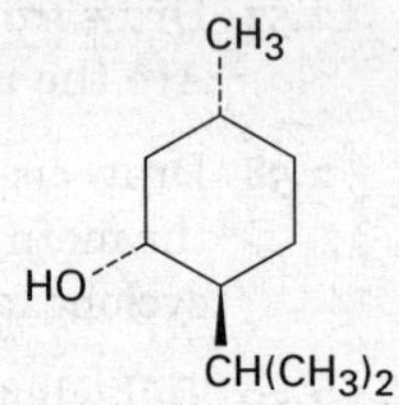

Menthol

2.66 There are four cis–trans isomers of menthol (Problem 2.65), including the one shown. Draw the other three.

2.67 Galactose, a sugar related to glucose, contains a six-membered ring in which all the substituents except the –OH group indicated below in red are equatorial. Draw galactose in its more stable chair conformation.

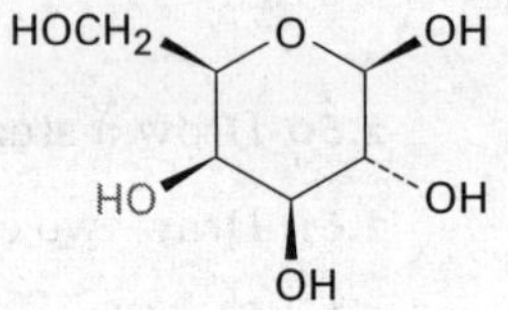

Galactose

2.68 Here's a tough one. There are two different substances named *trans*-1,2-dimethylcyclopentane. What is the relationship between them? (We'll explore this kind of isomerism in Chapter 6.)

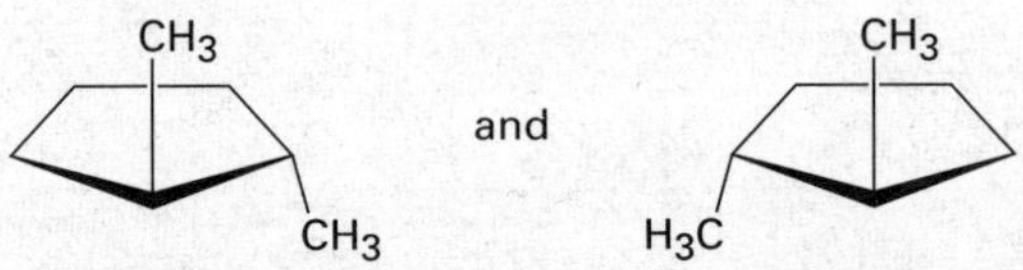

2.69 Draw 1,3,5-trimethylcyclohexane using a hexagon to represent the ring. How many cis–trans stereoisomers are possible?

2.70 One of the two chair structures of *cis*-1-chloro-3-methylcyclohexane is more stable than the other by 15.5 kJ/mol (3.7 kcal/mol). Which is it?

IN THE MEDICINE CABINET

2.71 Hydrocortisone, a naturally occurring hormone produced in the adrenal glands, is often used to treat inflammation, severe allergies, and numerous other conditions. Is the indicated –OH group in the molecule axial or equatorial?

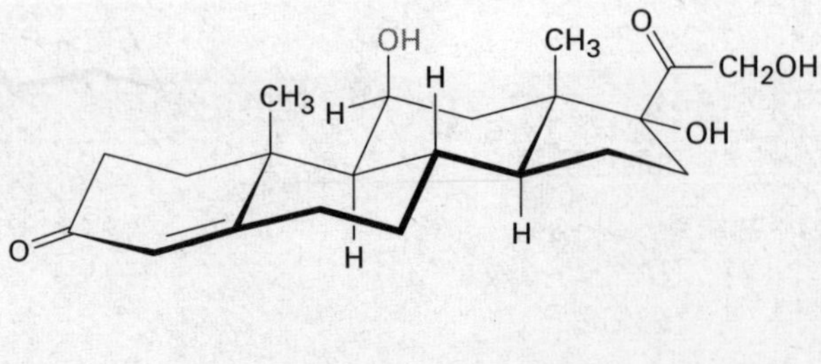

Hydrocortisone

2.72 The so-called statin drugs, such as simvastatin (Zocor), pravastatin (Pravachol), and atorvastatin (Lipitor) are the most widely prescribed drugs in the world, with annual sales estimated at approximately $15 billion.

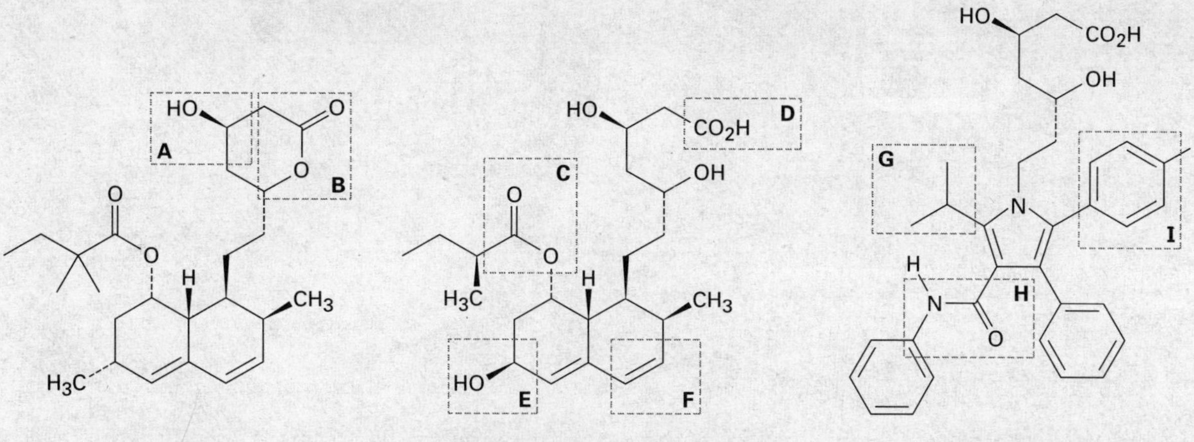

(a) Identify the functional groups (or alkyl groups) labeled **A–I**.
(b) Are the groups **C** and **E** on Pravachol cis or trans?
(c) Why can't groups **G**, **H**, and **I** be identified as cis or trans?

2.73 Amantadine is an antiviral agent that is active against influenza A infection. Draw a three-dimensional representation of amantadine showing the chair cyclohexane rings.

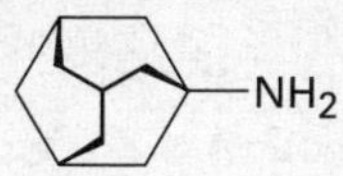

Amantadine

IN THE FIELD

2.74 Metolachlor, a herbicide marketed under the names Bicep, CGA-24705, Dual, Pennant, and Pimagram, is used to control weeds and grasses in fields of plants such as corn, soybeans, cotton, and peanuts. Metolachlor is degraded through oxidation in the environment to produce the water-soluble derivative shown. Identify the three functional groups in metolachlor and the new functional group in the derivative.

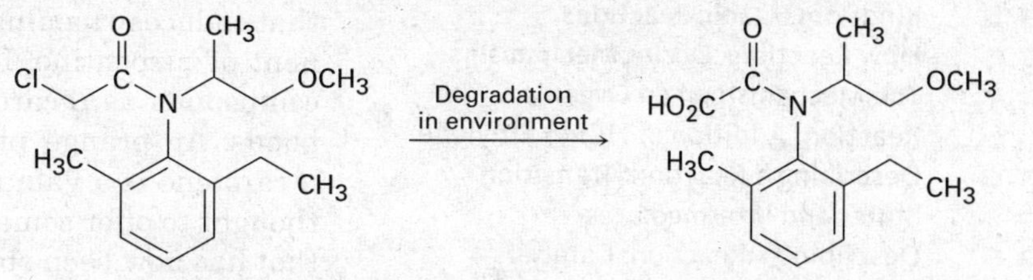

CHAPTER 3

The pink color of flamingo feathers is caused by the presence in the bird's diet of β-carotene, a polyalkene.

Alkenes and Alkynes: The Nature of Organic Reactions

OWL

Online homework for this chapter can be assigned in OWL, an online homework assessment tool.

Alkenes, sometimes called *olefins,* are hydrocarbons that contain a carbon–carbon **double bond**, C=C, and **alkynes** are hydrocarbons that contain a carbon–carbon **triple bond**, C≡C. Alkenes occur abundantly in nature, but alkynes are much less common. Ethylene, for instance, is a plant hormone that induces ripening in fruit, and α-pinene is the major component of turpentine. Life itself would be impossible without such compounds as β-carotene, a *poly*alkene that contains 11 double bonds. An orange pigment responsible for the color of carrots, β-carotene is a valuable dietary source of vitamin A. It was once thought to offer some protection against some types of cancer, but that has now been shown not to be true.

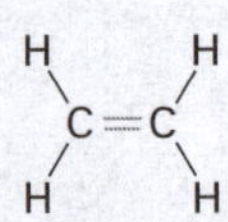

Ethylene

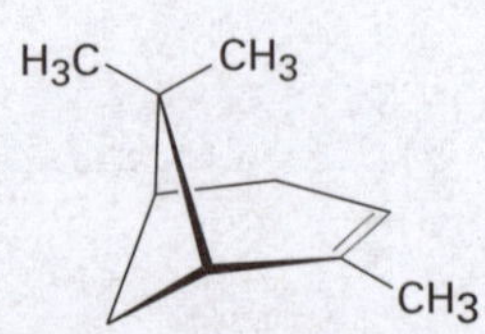

α-Pinene

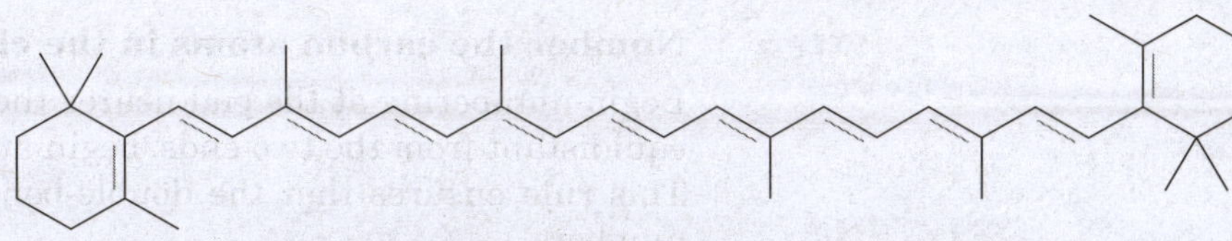

β-Carotene
(orange pigment and vitamin A precursor)

WHY THIS CHAPTER?

Carbon–carbon double bonds are present in most organic and biological molecules, so a good understanding of their behavior is needed. In this chapter, we'll look at some consequences of alkene stereoisomerism and then focus in detail on the broadest and most general class of alkene reactions, the electrophilic addition reaction. Carbon–carbon *triple* bonds, by contrast, occur less commonly than double bonds, so we'll not spend much time on their chemistry.

3.1 Naming Alkenes and Alkynes

Because of their multiple bond, alkenes and alkynes have fewer hydrogens per carbon than related alkanes and are therefore referred to as **unsaturated**. Ethylene, for example, has the formula C_2H_4, and acetylene has the formula C_2H_2, whereas ethane has the formula C_2H_6.

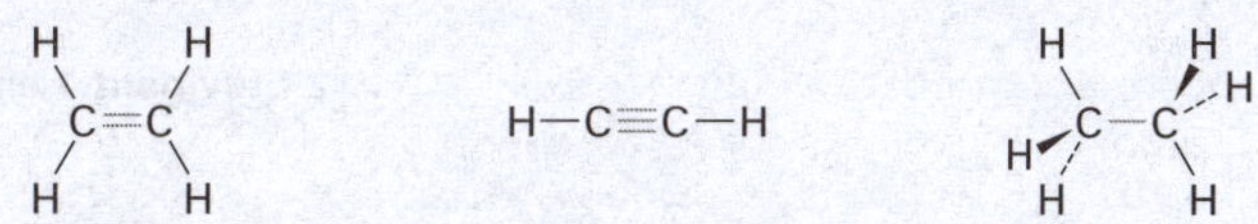

Ethylene: C_2H_4 **Acetylene: C_2H_2** **Ethane: C_2H_6**

(Fewer hydrogens—*unsaturated*) **(More hydrogens—*saturated*)**

Alkenes are named using a series of rules similar to those for alkanes (Section 2.3), with the suffix *-ene* used in place of *-ane* to identify the family. There are three steps.

STEP 1 **Name the parent hydrocarbon.**

Find the longest carbon chain that contains the double bond, and name the compound using the suffix *-ene* in place of *-ane*.

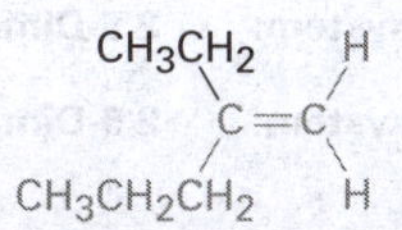

Named as a *pentene* *NOT*

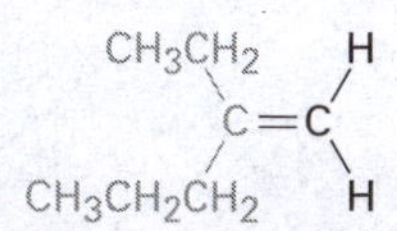

as a hexene, since the double bond is not contained in the six-carbon chain

STEP 2 **Number the carbon atoms in the chain.**

Begin numbering at the end nearer the double bond, or, if the double bond is equidistant from the two ends, begin at the end nearer the first branch point. This rule ensures that the double-bond carbons receive the lowest possible numbers.

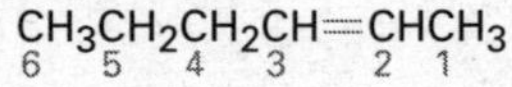

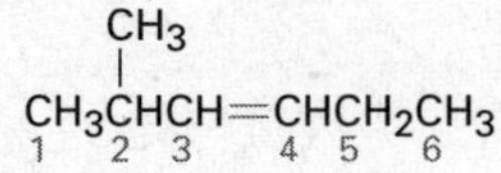

STEP 3 **Write the full name.**

Number the substituents on the main chain according to their position, and list them alphabetically. Indicate the position of the double bond by giving the number of the first alkene carbon and placing that number directly before the *-ene* suffix. If more than one double bond is present, give the position of each and use the appropriate multiplier suffix *-diene, -triene, -tetraene,* and so on.

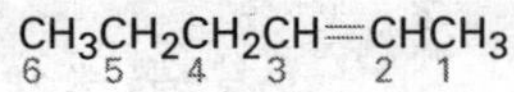

Hex-2-ene

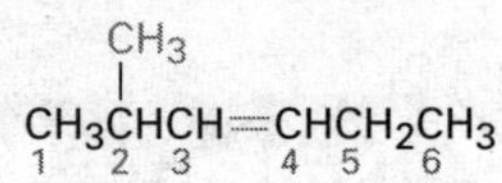

2-Methyl**hex**-3-ene

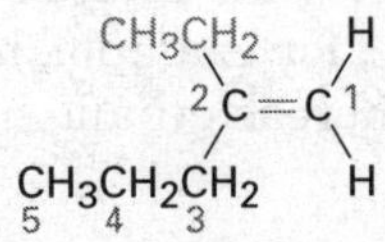

2-Ethyl**pent**-1-ene

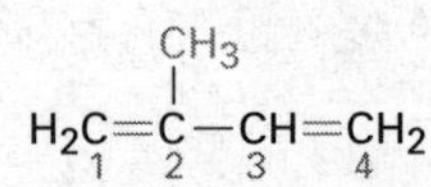

2-Methyl**buta**-1,3-diene

We should also note that IUPAC changed its naming rules in 1993. Prior to that time, the locant, or number locating the position of the double bond, was placed before the parent name rather than before the *-ene* suffix: 2-butene rather than but-2-ene, for instance. Changes always take time to be fully accepted, so the new rules have not yet been adopted universally and some texts have not yet been updated. We'll use the new naming system in this book, although you may encounter the old system elsewhere. Fortunately, the difference between old and new is minor and rarely causes problems.

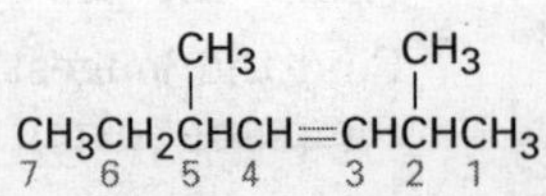

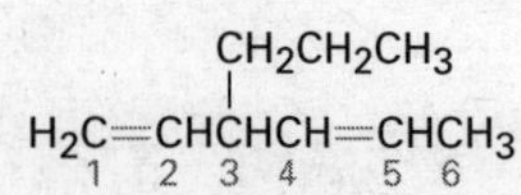

New naming system:	**2,5-Dimethylhept-3-ene**	**3-Propylhexa-1,4-diene**
(Old naming system:	**2,5-Dimethyl-3-heptene**	**3-Propyl-1,4-hexadiene)**

Cycloalkenes are named similarly, but because there is no chain end to begin from, we number the cycloalkene so that the double bond is between C1 and C2 and the first substituent has as low a number as possible. Note that it's not necessary to specify the position of the double bond in the name because it's always between C1 and C2.

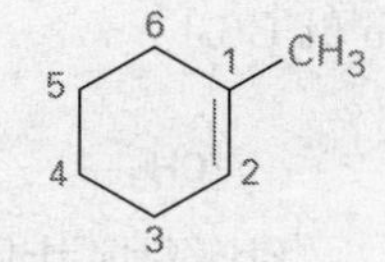

1-Methylcyclohexene

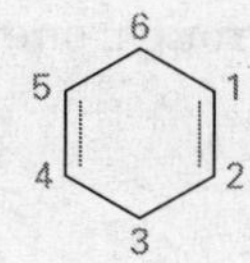

Cyclohexa-1,4-diene
(Old name: 1,4-Cyclohexadiene)

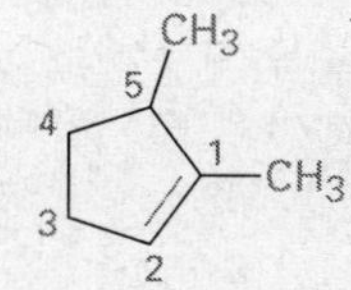

1,5-Dimethylcyclopentene

For historical reasons, there are a few alkenes whose names don't conform to the rules. For example, the alkene corresponding to ethane should be called *ethene,* but the name *ethylene* has been used for so long that it is accepted by IUPAC. Table 3.1 lists some other common names accepted by IUPAC.

Table 3.1 Common Names of Some Alkenes

Compound	Systematic name	Common name
$H_2C{=}CH_2$	Ethene	Ethylene
$CH_3CH{=}CH_2$	Propene	Propylene
$CH_3C(CH_3){=}CH_2$	2-Methylpropene	Isobutylene
$H_2C{=}C(CH_3){-}CH{=}CH_2$	2-Methylbuta-1,3-diene	Isoprene

Alkynes are named in the same way as alkenes, with the suffix *-yne* used in place of *-ene.* Numbering the main chain begins at the end nearer the triple bond so that the triple bond receives as low a number as possible, and the locant is again placed immediately before the *-yne* suffix in the post-1993 naming system.

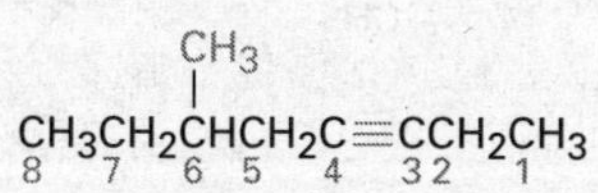

Begin numbering at the end nearer the triple bond.

6-Methyloct-3-yne
(Old name: 6-Methyl-3-octyne)

As with alkyl groups derived from alkanes, *alkenyl* and *alkynyl* groups are also possible.

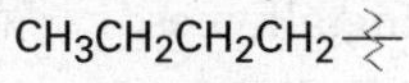

Butyl
(an alkyl group)

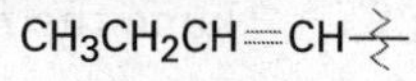

But-1-enyl
(a vinylic group)

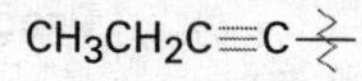

But-1-ynyl
(an alkynyl group)

Worked Example 3.1 Naming an Alkene

What is the IUPAC name of the following alkene?

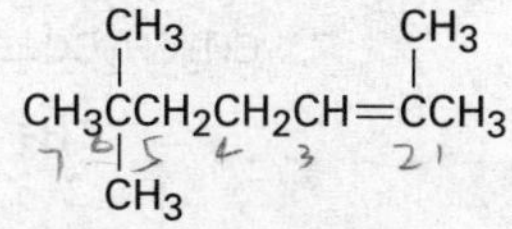

Strategy First, find the longest chain containing the double bond—in this case, a heptene. Next, number the chain beginning at the end nearer the double bond, and identify the substituents at each position. In this case, there are three methyl groups, one at C2 and two at C6.

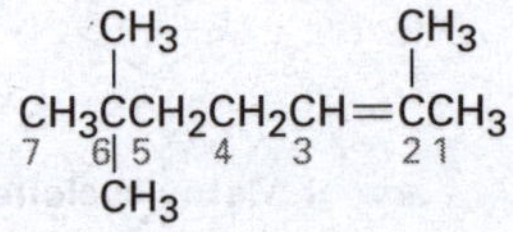

Solution Write the full name, listing the substituents alphabetically and giving the position of each. Identify the position of the double bond by placing the number of the first alkene carbon before the *-ene* suffix: 2,6,6-trimethylhept-2-ene.

Problem 3.1 Give IUPAC names for the following compounds:

(a) 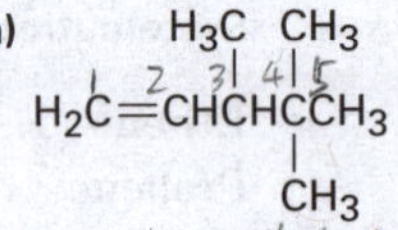

(b) 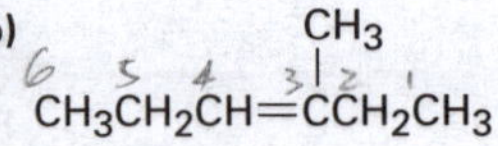

(c) 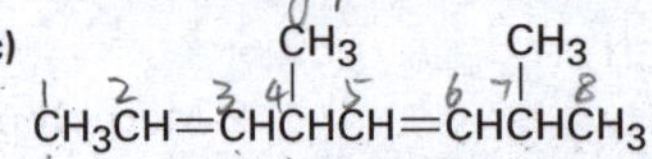

(d) 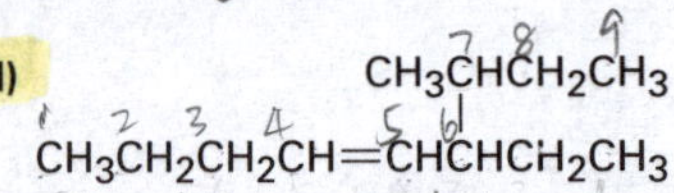

Problem 3.2 Name the following cycloalkenes:

(a) 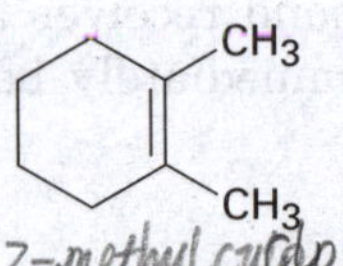

(b) 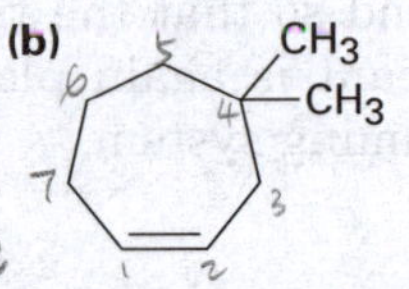

(c) 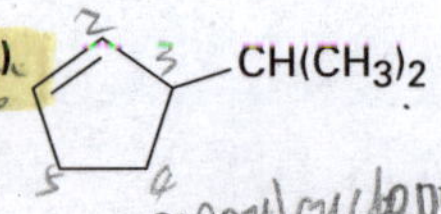

Problem 3.3 Draw structures corresponding to the following IUPAC names:

(a) 2-Methylhex-1-ene

(b) 4,4-Dimethylpent-2-yne

(c) 2-Methylhexa-1,5-diene

(d) 3-Ethyl-2,2-dimethylhept-3-ene

Problem 3.4 Name the following alkynes:

(a) 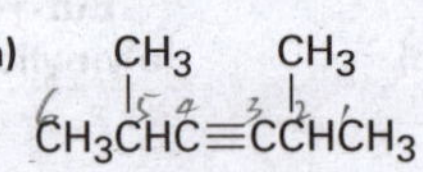

(b) 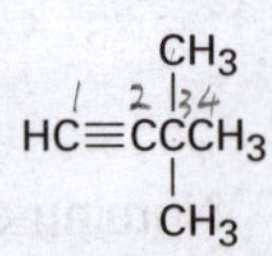

(c) 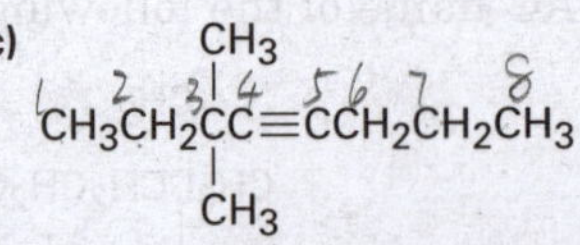

(d) 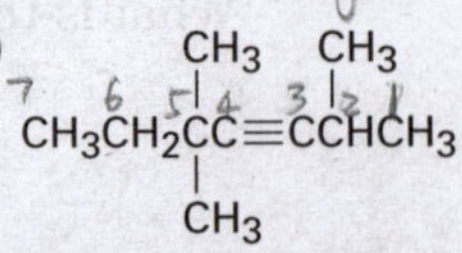

Problem 3.5 Change the following old names to new, post-1993 names, and draw the structure of each compound:

(a) 2,5,5-Trimethyl-2-hexene **(b)** 2,2-Dimethyl-3-hexyne
(c) 2-Methyl-2,5-heptadiene **(d)** 1-Methyl-1,3-cyclopentadiene

3.2 Electronic Structure of Alkenes

We saw in Section 1.8 that the carbon atoms in a double bond have three equivalent sp^2 hybrid orbitals, which lie in a plane at angles of 120° to one another. The fourth carbon orbital is an unhybridized p orbital perpendicular to the sp^2 plane. When two sp^2-hybridized carbon atoms approach each other, they form a σ bond by head-on overlap of sp^2 orbitals and a π bond by sideways overlap of p orbitals. The doubly bonded carbons and the four attached atoms lie in a plane, with bond angles of approximately 120° (Figure 3.1).

We also know from Section 2.5 that rotation can occur around single bonds and that open-chain alkanes like ethane and propane therefore have many rapidly interconverting conformations. The same is not true for double bonds, however. For rotation to take place around a double bond, the π part of the bond must break momentarily (Figure 3.1). Thus, the energy barrier to rotation around a double bond must be at least as great as the strength of the π bond itself, an estimated 350 kJ/mol (84 kcal/mol). Recall that the rotation barrier for a single bond is only about 12 kJ/mol.

Figure 3.1 The π bond must break momentarily for rotation around a carbon–carbon double bond to take place, requiring a large amount of energy.

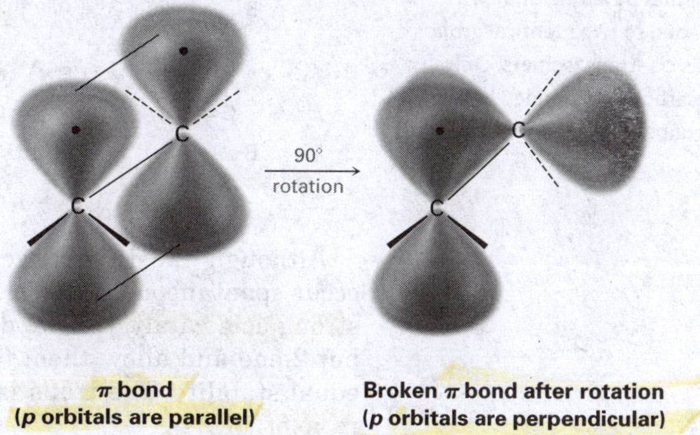

3.3 Cis–Trans Isomers of Alkenes

The lack of rotation around carbon–carbon double bonds is of more than just theoretical interest; it also has chemical consequences. Imagine the situation for a disubstituted alkene such as but-2-ene. (*Disubstituted* means that two substituents other than hydrogen are bonded to the double-bond carbons.) The two methyl groups in but-2-ene can be either on the same side of the double bond or on opposite sides, a situation similar to that in substituted cycloalkanes (Section 2.8).

Because bond rotation can't occur, the two but-2-enes can't spontaneously interconvert and are different chemical compounds. As with disubstituted cycloalkanes, we call such compounds *cis–trans isomers*. The isomer with both substituents on the same side of the double bond is *cis*-but-2-ene, and the isomer with substituents on opposite sides is *trans*-but-2-ene (Figure 3.2).

Figure 3.2 Cis and trans isomers of but-2-ene. The cis isomer has the two methyl groups on the same side of the double bond, and the trans isomer has the methyl groups on opposite sides.

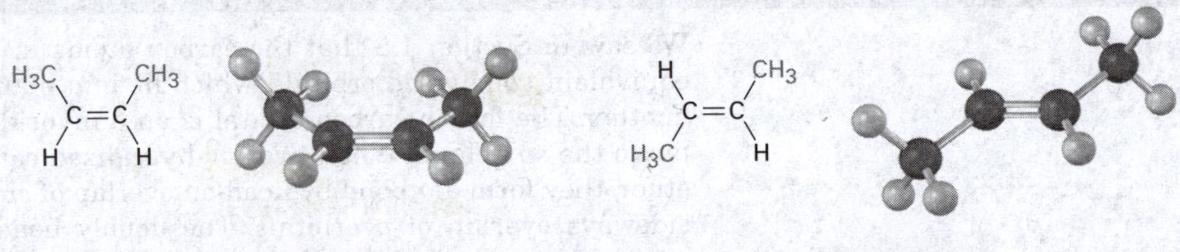

***cis*-But-2-ene** ***trans*-But-2-ene**

Cis–trans isomerism is not limited to disubstituted alkenes. It occurs whenever each double-bond carbon is attached to two different groups. If one of the double-bond carbons is attached to two identical groups, however, then cis–trans isomerism is not possible (Figure 3.3).

Figure 3.3 The requirement for cis–trans isomerism in alkenes. Compounds that have one of their carbons bonded to two identical groups can't exist as cis–trans isomers. Only when both carbons are bonded to two different groups are cis–trans isomers possible.

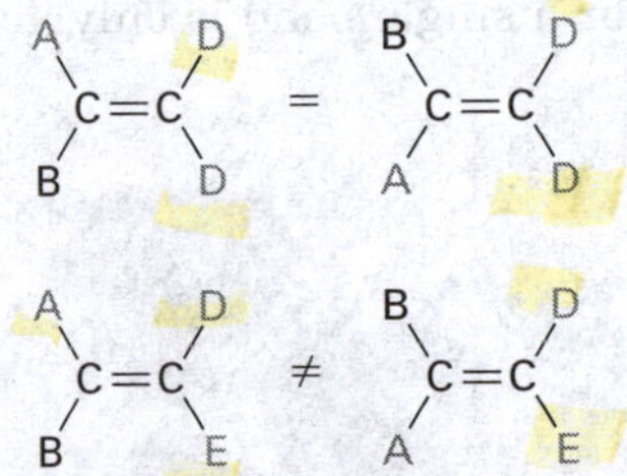

These two compounds are identical; they are not cis–trans isomers.

These two compounds are not identical; they are cis–trans isomers.

Although the interconversion of cis and trans alkene isomers doesn't occur spontaneously, it can be brought about by treating the alkene with a strong acid catalyst. If we do, in fact, interconvert *cis*-but-2-ene with *trans*-but-2-ene and allow them to reach equilibrium, we find that they aren't of equal stability. The trans isomer is more favored than the cis isomer by a ratio of 76:24.

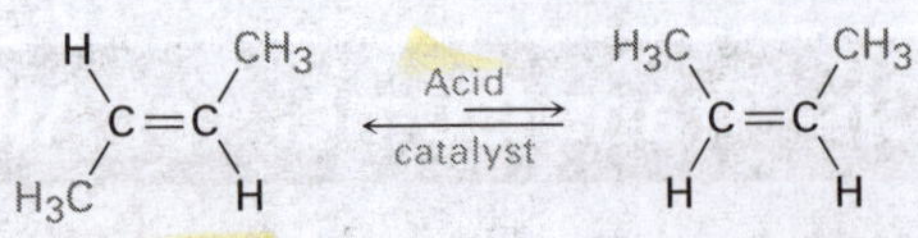

Trans (76%) **Cis (24%)**

Cis alkenes are less stable than their trans isomers because of steric (spatial) interference between the large substituents on the same side of the double bond. This is the same kind of interference, or *steric strain,* that we saw in the axial conformation of methylcyclohexane (Section 2.11).

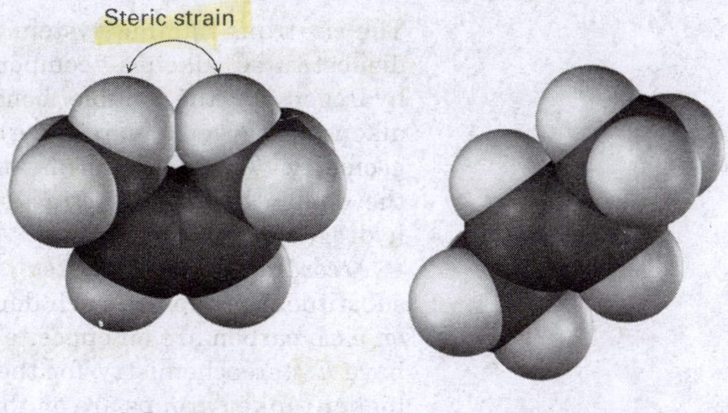

***cis*-But-2-ene** ***trans*-But-2-ene**

Worked Example 3.2 Drawing Cis and Trans Alkene Isomers

Draw the cis and trans isomers of 5-chloropent-2-ene.

Strategy First, draw the molecule without indicating isomers to see the overall structure: $ClCH_2CH_2CH{=}CHCH_3$. Then locate the two substituent groups on the same side of the double bond for the cis isomer and on opposite sides for the trans isomer.

Solution

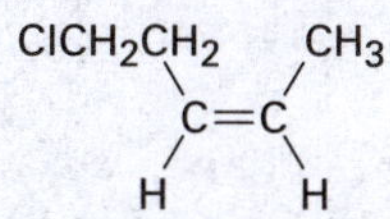

***cis*-5-Chloropent-2-ene**

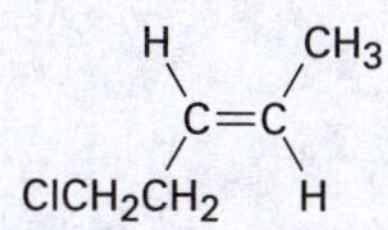

***trans*-5-Chloropent-2-ene**

Problem 3.6 Which of the following compounds can exist as cis–trans isomers? Draw each cis–trans pair.

(a) $CH_3CH{=}CH_2$ **(b)** $(CH_3)_2C{=}CHCH_3$
(c) $ClCH{=}CHCl$ **(d)** $CH_3CH_2CH{=}CHCH_3$
(e) $CH_3CH_2CH{=}C(Br)CH_3$ **(f)** 3-Methylhept-3-ene

Problem 3.7 Name the following alkenes, including the cis or trans designation:

(a)

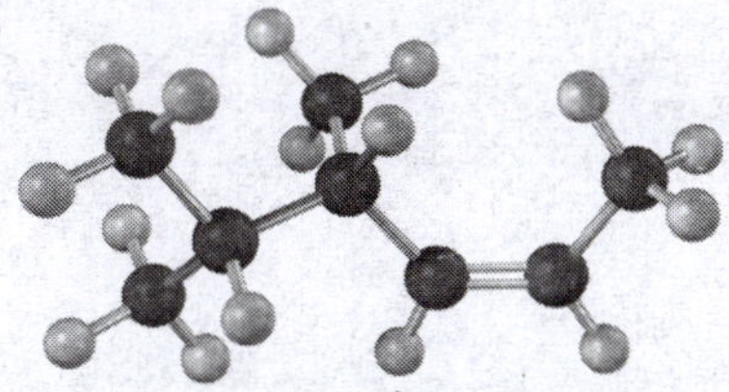

(b)

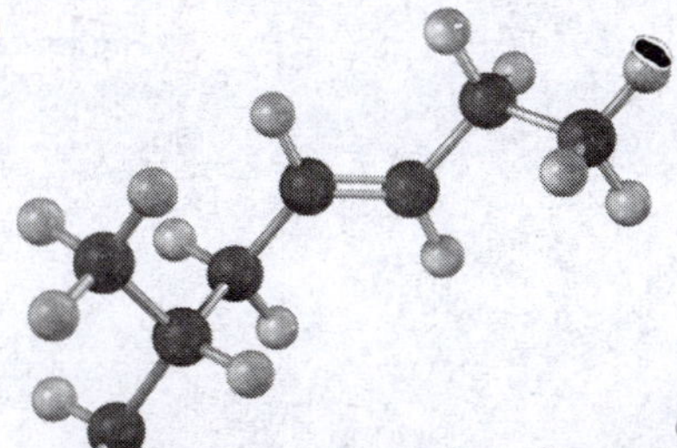

3.4 Sequence Rules: The *E*,*Z* Designation

The cis–trans naming system used in the previous section works only with disubstituted alkenes—compounds that have two substituents other than hydrogen on the double bond. With trisubstituted and tetrasubstituted alkenes, however, a more general method is needed for describing double-bond geometry. (*Trisubstituted* means three substituents other than hydrogen on the double bond, and *tetrasubstituted* means four substituents other than hydrogen.)

According to the ***E*,*Z* system**, a set of *sequence rules* is used to rank the two substituent groups on each double-bond carbon. If the higher-ranked groups on each carbon are on opposite sides of the double bond, the alkene is said to have *E* stereochemistry, for the German *entgegen,* meaning "opposite." If the higher-ranked groups are on the same side, the alkene has *Z* stereochemistry, for the German *zusammen,* meaning "together." (You can remember which is which by noting that the groups are on "ze zame zide" in the *Z* isomer.)

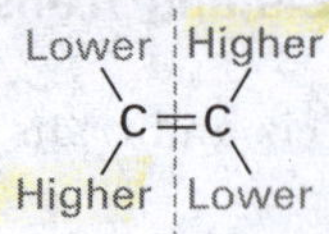

E double bond (Higher-ranked groups are on opposite sides.)

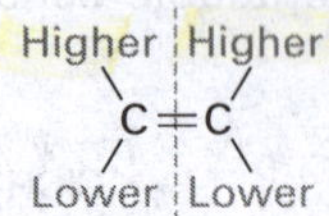

Z double bond (Higher-ranked groups are on the same side.)

Called the *Cahn–Ingold–Prelog rules* after the chemists who proposed them, the sequence rules are as follows:

RULE 1 **Considering the double-bond carbons separately, look at the atoms directly attached to each carbon and rank them according to atomic number**. The atom with the higher atomic number has the higher ranking, and the atom with the lower atomic number (usually hydrogen) has the lower ranking. Thus, the atoms commonly found attached to a double-bond carbon are assigned the following rankings. When different isotopes of the same element are compared, such as deuterium (^{2}H) and protium (^{1}H), the heavier isotope ranks higher than the lighter isotope.

Atomic number	35		17		16		15		8		7		6		(2)		(1)	
Higher ranking	Br	>	Cl	>	S	>	P	>	O	>	N	>	C	>	^{2}H	>	^{1}H	**Lower ranking**

For example:

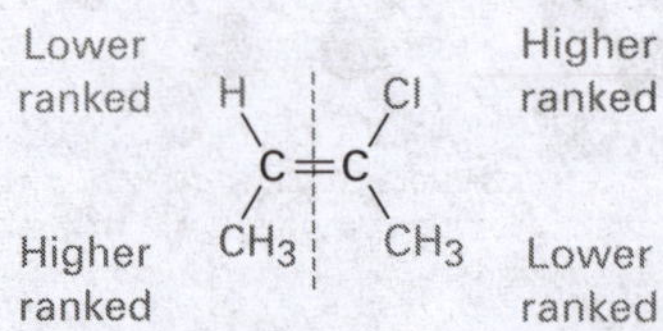

(a) (*E*)-2-Chlorobut-2-ene

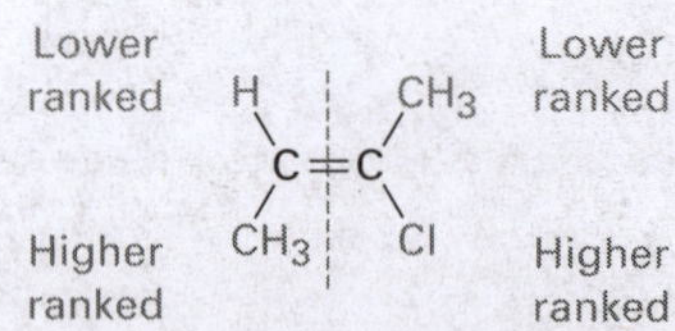

(b) (*Z*)-2-Chlorobut-2-ene

Because chlorine has a higher atomic number than carbon, it ranks higher than a methyl (CH_3) group. Methyl ranks higher than hydrogen, however, and isomer (a) is therefore assigned *E* stereochemistry (higher-ranked groups on opposite sides of the double bond). Isomer (b) has *Z* stereochemistry (higher-ranked groups on "ze zame zide" of the double bond).

RULE 2 **If a decision can't be reached by ranking the first atoms in the substituents, look at the second, third, or fourth atoms away from the double-bond carbons until the first difference is found.** Thus, a $-CH_2CH_3$ substituent and a $-CH_3$ substituent are equivalent by rule 1 because both have carbon as the first atom. By rule 2, however, ethyl ranks higher than methyl because ethyl has a *carbon* as its highest *second* atom, while methyl has only hydrogen as its second atom. Look at the following examples to see how the rule works:

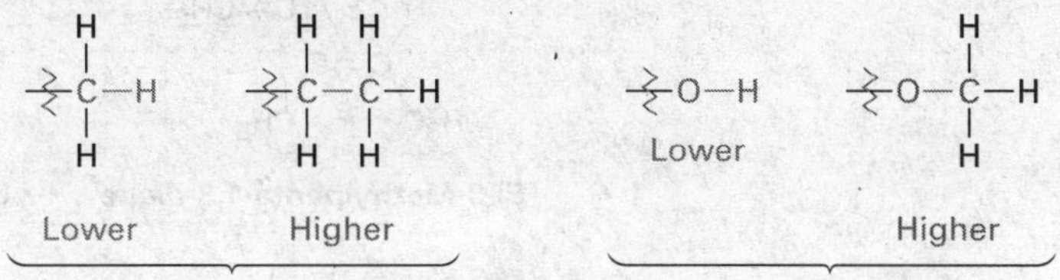

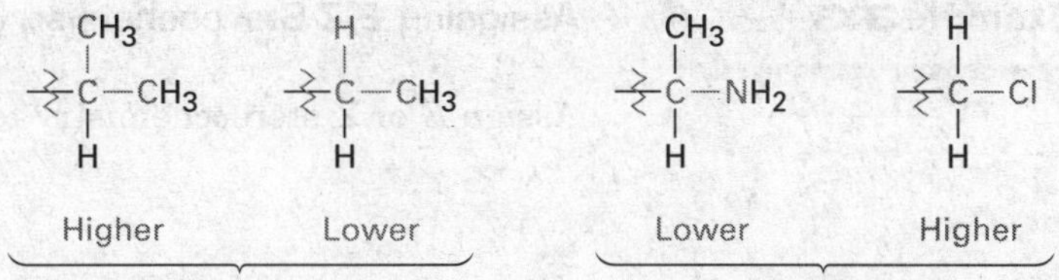

RULE 3 **Multiple-bonded atoms are equivalent to the same number of single-bonded atoms.** For example, an aldehyde substituent (–CH=O), which has a carbon atom *doubly* bonded to *one* oxygen, is equivalent to a substituent having a carbon atom *singly* bonded to *two* oxygens.

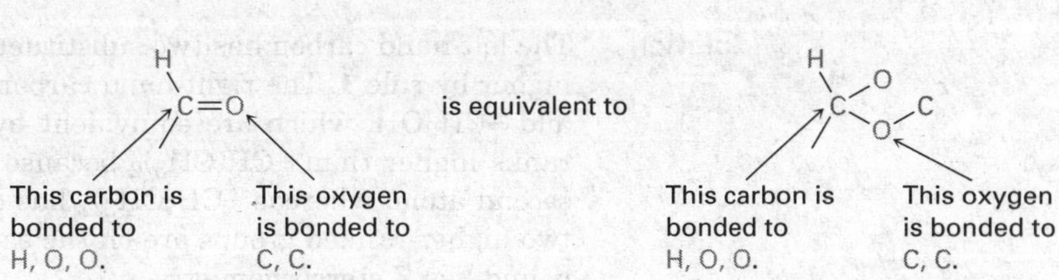

As further examples, the following pairs are equivalent:

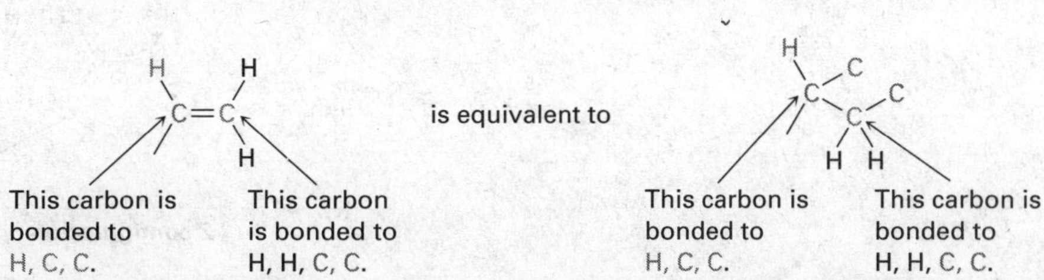

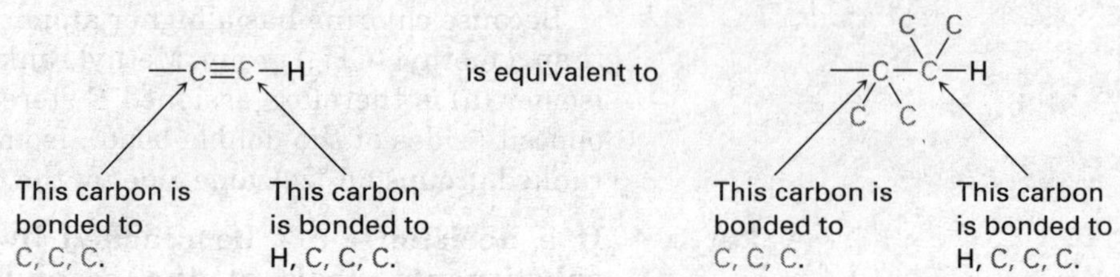

By applying the sequence rules, we can assign the stereochemistry shown in the following examples. Work through each one to convince yourself the assignments are correct.

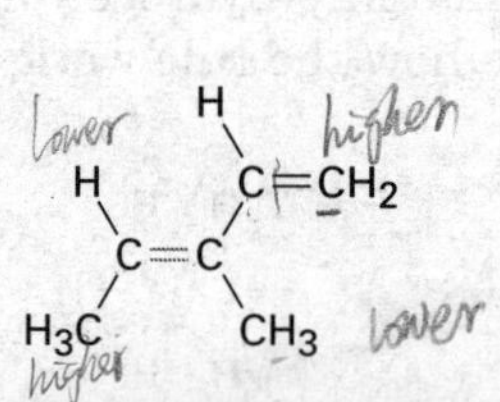

(*E*)-3-Methylpenta-1,3-diene

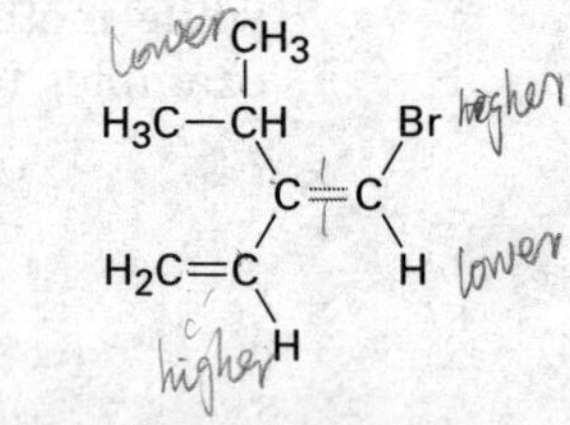

(*E*)-1-Bromo-2-isopropylbuta-1,3-diene

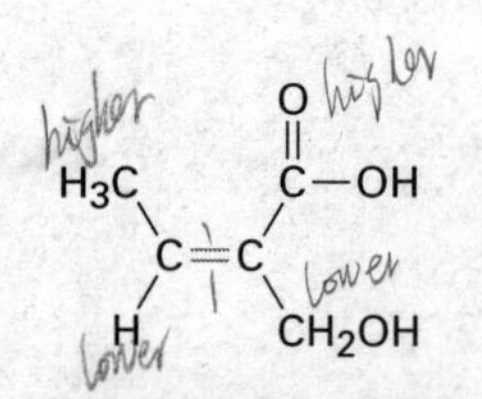

(*Z*)-2-Hydroxymethylbut-2-enoic acid

Worked Example 3.3 Assigning *E*,*Z* Stereochemistry to an Alkene

Assign *E* or *Z* stereochemistry to the double bond in the following compound:

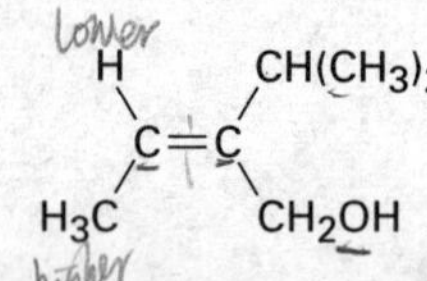

Strategy Look at each double-bond carbon individually, and assign rankings. Then see whether the two higher-ranked groups are on the same or opposite sides of the double bond.

Solution The left-hand carbon has two substituents, $-H$ and $-CH_3$, of which $-CH_3$ ranks higher by rule 1. The right-hand carbon also has two substituents, $-CH(CH_3)_2$ and $-CH_2OH$, which are equivalent by rule 1. By rule 2, however, $-CH_2OH$ ranks higher than $-CH(CH_3)_2$ because $-CH_2OH$ has an *oxygen* as its highest second atom, whereas $-CH(CH_3)_2$ has *carbon* as its highest second atom. The two higher-ranked groups are on the same side of the double bond, so the compound has *Z* stereochemistry.

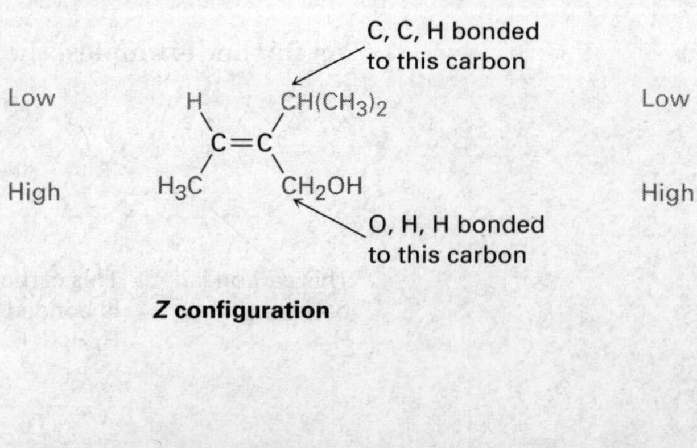

***Z* configuration**

Problem 3.8 Which member in each of the following sets ranks higher?

(a) $-H$ or $-Br$ **(b)** $-Cl$ or $-Br$ **(c)** $-CH_3$ or $-CH_2CH_3$
(d) $-NH_2$ or $-OH$ **(e)** $-CH_2OH$ or $-CH_3$ **(f)** $-CH_2OH$ or $-CH{=}O$

Problem 3.9 Assign *E* or *Z* stereochemistry to the following compounds:

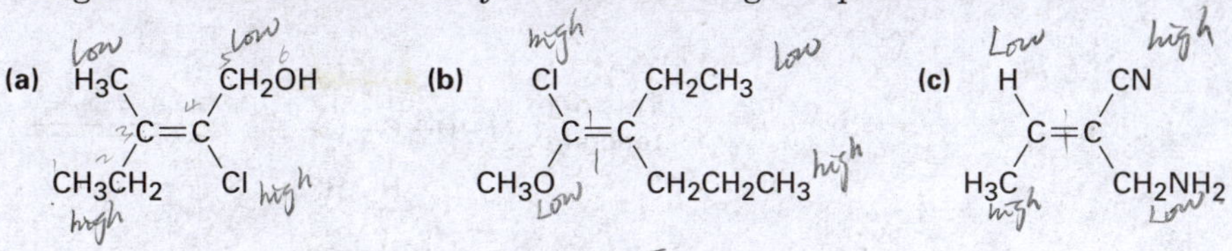

Problem 3.10 Assign *E* or *Z* stereochemistry to the following compound (red = O):

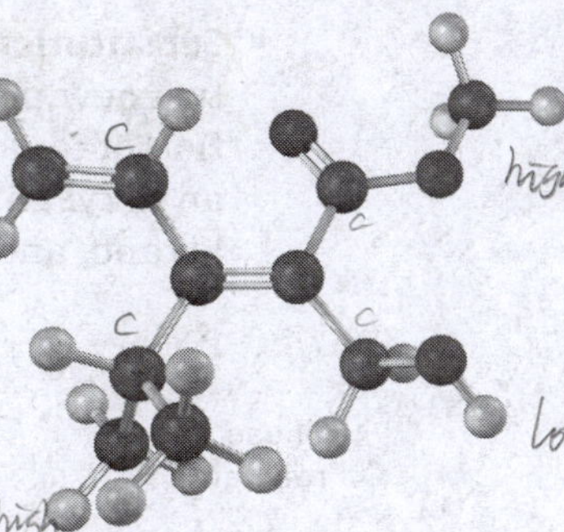

3.5 Kinds of Organic Reactions

Now that we know something about alkenes and alkynes, let's learn about their chemical reactivity. As an introduction, we'll first look at some of the basic principles that underlie all organic reactions. In particular, we'll develop some general notions about why compounds react the way they do, and we'll see some methods that have been developed to help understand how reactions take place.

Organic chemical reactions can be organized either by what kinds of reactions occur or by how reactions occur. Let's look first at the kinds of reactions that take place. There are four particularly broad types of organic reactions: *additions, eliminations, substitutions,* and *rearrangements.*

- **Addition reactions** occur when two reactants add together to form a single new product with no atoms "left over." An example that we'll be studying soon is the reaction of an alkene with HCl to yield an alkyl chloride.

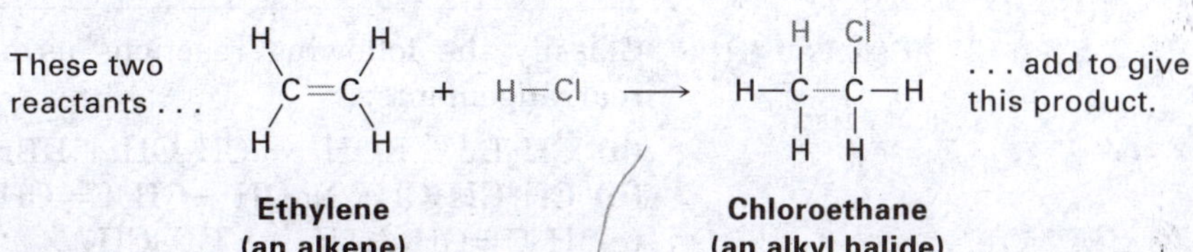

- **Elimination reactions** are, in a sense, the opposite of addition reactions. They occur when a single organic reactant splits into two products, often with formation of a small molecule such as H_2O or HCl. An example is the acid-catalyzed reaction of an alcohol to yield water and an alkene.

This one reactant . . .

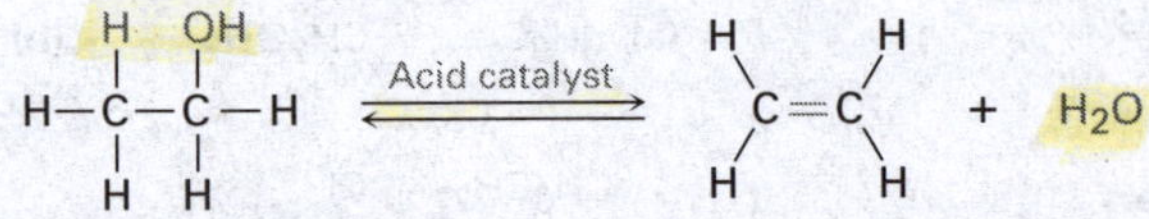

. . . gives these two products.

Ethanol (an alcohol) **Ethylene (an alkene)**

- **Substitution reactions** occur when two reactants exchange parts to give two new products. An example that we saw in Section 2.4 is the reaction of an alkane with Cl_2 in the presence of ultraviolet light to yield an alkyl chloride. A –Cl group substitutes for the –H group of the alkane, and two new products result.

These two reactants . . .

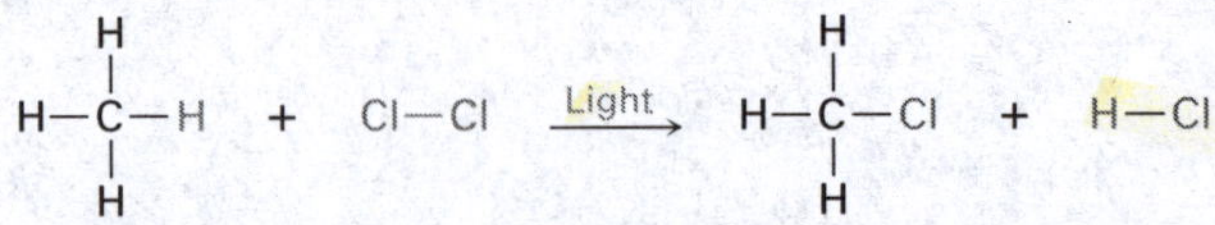

. . . give these two products.

Methane (an alkane) **Chloromethane (an alkyl halide)**

- **Rearrangement reactions** occur when a single organic reactant undergoes a reorganization of bonds and atoms to yield a single isomeric product. An example that we saw in Section 3.3 is the conversion of *cis*-but-2-ene into its isomer *trans*-but-2-ene by treatment with an acid catalyst.

This reactant . . .

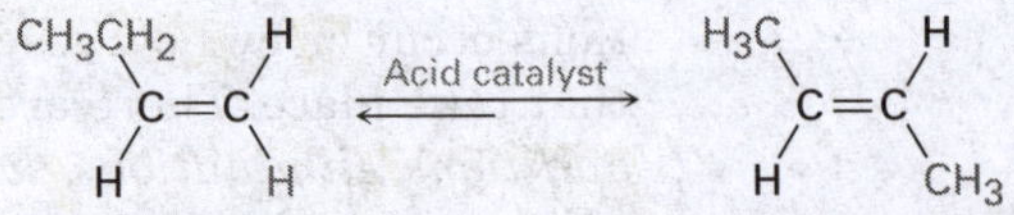

. . . gives this isomeric product.

But-1-ene **But-2-ene**

Problem 3.11 Classify the following reactions as additions, eliminations, substitutions, or rearrangements:

(a) $CH_3Br + KOH \rightarrow CH_3OH + KBr$

(b) $CH_3CH_2Cl + NaOH \rightarrow H_2C{=}CH_2 + NaCl$

(c) $H_2C{=}CH_2 + H_2 \rightarrow CH_3CH_3$

3.6 How Reactions Occur: Mechanisms

Having looked at the kinds of reactions that take place, let's now see *how* reactions occur. An overall description of how a reaction occurs is called a **reaction mechanism**. A mechanism describes what takes place at each stage of a chemical transformation—which bonds are broken and in what order, which bonds are formed and in what order, and what the relative rates of the steps are.

All chemical reactions involve bond-breaking in the reactant molecules and bond-making in the product molecules, which means that the electrons in those bonds must move about and reorganize. Fundamentally, a covalent two-electron bond can break in two ways: a bond can break in an electronically *symmetrical* way so that one electron remains with each product fragment, or a bond can break in an electronically *unsymmetrical* way so that both electrons remain with one product fragment, leaving the other fragment with a vacant orbital. The symmetrical cleavage is said to be *homolytic*, and the unsymmetrical cleavage is said to be *heterolytic*.

We'll develop the point in more detail later, but you might note for now that the movement of *one* electron in the symmetrical process is indicated using a half-headed, or "fishhook," arrow (⌒), while the movement of *two* electrons in an unsymmetrical process is indicated using a full-headed curved arrow (↷), as noted previously in Section 1.12.

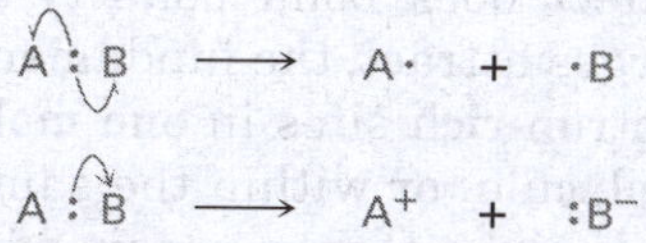

Symmetrical bond-breaking (radical): one bonding electron stays with each product.

Unsymmetrical bond-breaking (polar): two bonding electrons stay with one product.

Just as there are two ways in which a bond can break, there are two ways in which a covalent two-electron bond can form. A bond can form in an electronically symmetrical way if one electron is donated to the new bond by each reactant or in an electronically unsymmetrical way if both bonding electrons are donated by one reactant.

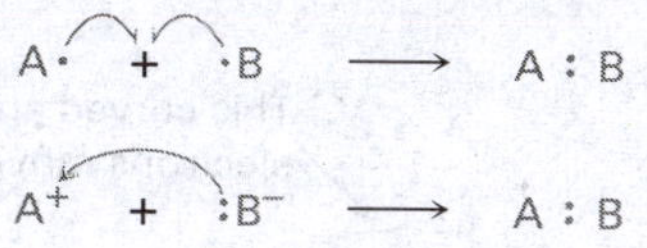

Symmetrical bond-making (radical): one bonding electron is donated by each reactant.

Unsymmetrical bond-making (polar): two bonding electrons are donated by one reactant.

Processes that involve symmetrical bond-breaking and bond-making are called **radical reactions**. A **radical**, often called a *free radical,* is a neutral chemical species that contains an odd number of electrons and thus has a single, unpaired electron in one of its orbitals. Processes that involve unsymmetrical bond breaking and making are called **polar reactions**. Polar reactions involve species that have an even number of electrons and thus have only electron pairs in their orbitals. Polar processes are by far the more common reaction type in both organic and biological chemistry, and much of this book is devoted to their description.

To see how polar reactions occur, we first need to look more deeply into the effects of bond polarity on organic molecules. We saw in Section 1.9 that certain bonds in a molecule, particularly the bonds in functional groups, are often polar. When carbon bonds to a more electronegative atom, such as chlorine or oxygen, the bond is polarized so that the carbon bears a partial positive charge ($\delta+$) and the electronegative atom bears a partial negative charge ($\delta-$). When carbon bonds to a less electronegative atom, such as a metal, the opposite polarity results. As always, electrostatic potential maps show electron-rich regions of a molecule in red and electron-poor regions in blue.

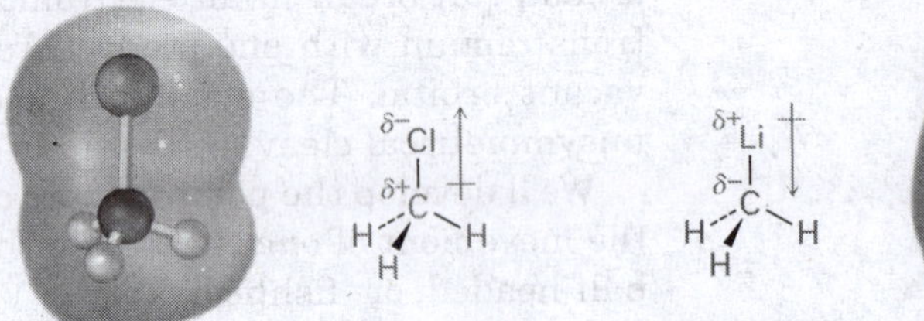

Chlorometane Methyllithium

What effect does bond polarity have on chemical reactions? Because unlike charges attract, the fundamental characteristic of all polar reactions is that electron-rich sites in one molecule react with electron-poor sites in another molecule (or within the same molecule). Bonds are made when an electron-rich atom shares a pair of electrons with an electron-poor atom, and bonds are broken when one atom leaves with both electrons from the former bond.

As noted previously, chemists normally indicate the movement of an electron pair during a polar reaction by using a curved, full-headed arrow. A curved arrow shows where electrons move when reactant bonds are broken and product bonds are formed. It means that an electron pair moves *from* the atom (or bond) at the tail of the arrow *to* the atom at the head of the arrow during the reaction.

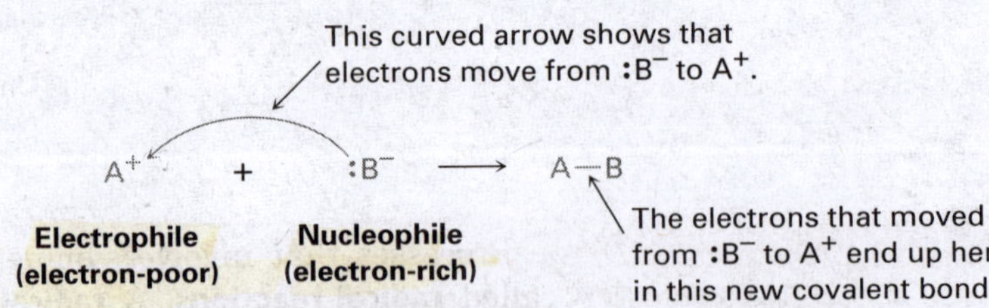

In referring to the electron-rich and electron-poor species involved in polar reactions, chemists use the words *nucleophile* and *electrophile*. A **nucleophile** is a substance that is "nucleus loving" and thus attracted to a positive charge. A nucleophile has a negatively polarized, electron-rich atom and can form a bond by donating an electron pair to a positively polarized, electron-poor atom. Nucleophiles can be either neutral or negatively charged and usually have lone-pairs of electrons: ammonia, water, hydroxide ion, and chloride ion are examples. An **electrophile**, by contrast, is "electron-loving." An electrophile has a positively polarized, electron-poor atom and can form a bond by accepting a pair of electrons from a nucleophile. Electrophiles can be either neutral or positively charged. Acids (H^+ donors), alkyl halides, and carbonyl compounds are examples (Figure 3.4).

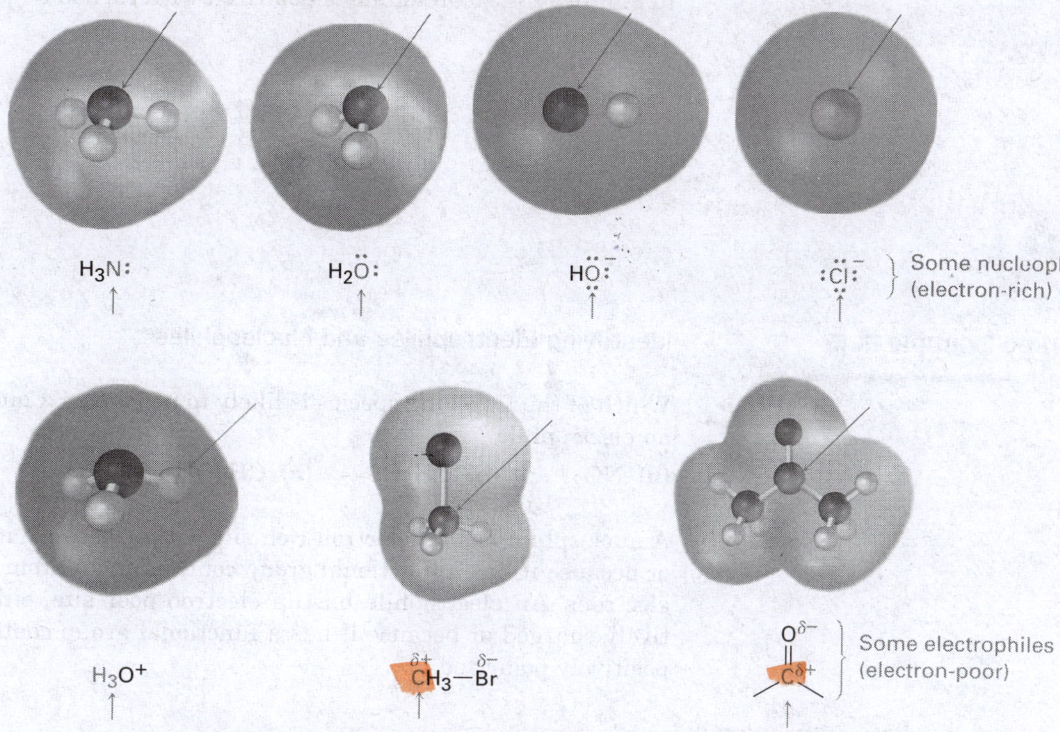

Figure 3.4 Some nucleophiles and electrophiles. Electrostatic potential maps identify the nucleophilic (red; negative) and electrophilic (blue; positive) atoms.

Note that neutral compounds can often react either as nucleophiles or as electrophiles, depending on the circumstances. After all, if a compound is neutral yet has an electron-*rich* nucleophilic site within it, it must also have a corresponding electron-*poor* electrophilic site. Water, for instance, acts as an electrophile when it donates H^+ but acts as a nucleophile when it donates a nonbonding pair of electrons. Similarly, a carbonyl compound acts as an electrophile when it reacts at its positively polarized carbon atom, yet acts as a nucleophile when it donates a pair of electrons from its negatively polarized oxygen atom.

If the definitions of nucleophiles and electrophiles sound similar to those given in Section 1.12 for Lewis acids and Lewis bases, that's because there is indeed a correlation. Lewis bases are electron donors and behave as nucleophiles, whereas Lewis acids are electron acceptors and behave as electrophiles. Thus, much of organic chemistry is explainable in terms of acid–base reactions. The main difference is that the words *acid* and *base* are used broadly in all fields of chemistry, while the words *nucleophile* and *electrophile* are used primarily in organic chemistry when bonds to carbon are involved.

Worked Example 3.4 Predicting the Polarity of a Bond

What is the direction of bond polarity in the amine functional group, $C{-}NH_2$?

Strategy Look at the electronegativity values in Figure 1.14 on page 16 to see which atoms withdraw electrons more strongly.

Solution Nitrogen (EN $= 3.0$) is more electronegative than carbon (EN $= 2.5$) according to Figure 1.14, so an amine is polarized with carbon $\delta+$ and nitrogen $\delta-$.

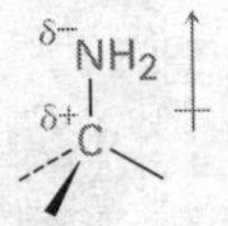

An amine

Worked Example 3.5 Identifying Electrophiles and Nucleophiles

Which of the following species is likely to behave as a nucleophile and which as an electrophile?

(a) NO_2^+ **(b)** CH_3O^- **(c)** CH_3OH

Strategy A nucleophile has an electron-rich site, either because it is negatively charged or because it has a functional group containing an atom that has a lone pair of electrons. An electrophile has an electron-poor site, either because it is positively charged or because it has a functional group containing an atom that is positively polarized.

Solution

(a) NO_2^+ (nitronium ion) is likely to be an electrophile because it is positively charged.

(b) CH_3O^- (methoxide ion) is likely to be a nucleophile because it is negatively charged.

(c) CH_3OH (methyl alcohol) can be either a nucleophile, because it has two lone pairs of electrons on oxygen, or an electrophile, because it has polar C–O and O–H bonds.

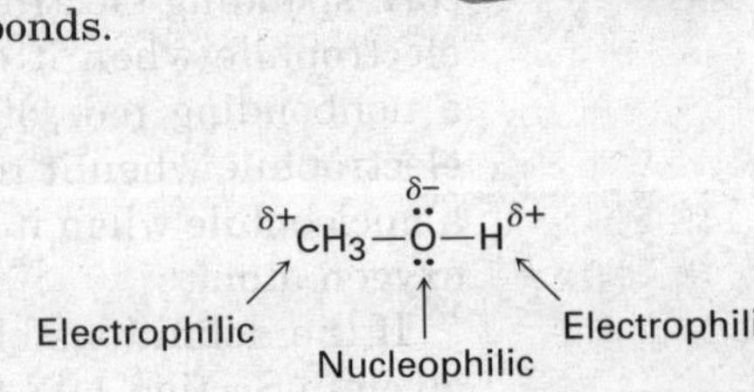

Problem 3.12 What is the direction of bond polarity in the following functional groups? (See Figure 1.14 on page 16 for electronegativity values.)

(a) Aldehyde **(b)** Ether
(c) Ester **(d)** Alkylmagnesium bromide, R—MgBr

Problem 3.13 Which of the following are most likely to behave as electrophiles, and which as nucleophiles? Explain.

(a) NH_4^+ **(b)** $C{\equiv}N^-$ **(c)** Br^+ **(d)** CH_3NH_2 **(e)** $H{-}C{\equiv}C{-}H$

Problem 3.14 An electrostatic potential map of boron trifluoride is shown. Is BF_3 likely to be an electrophile or a nucleophile? Draw an electron-dot structure for BF_3, and explain the result.

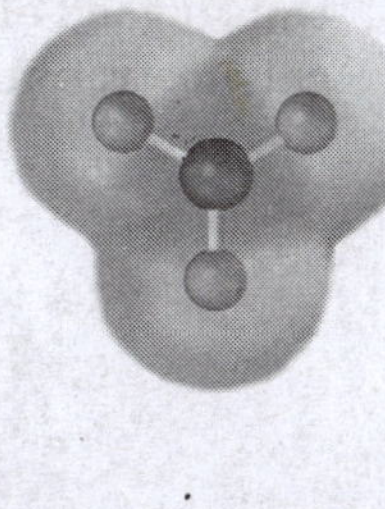

BF_3

3.7 The Mechanism of an Organic Reaction: Addition of HCl to Ethylene

Let's look in detail at a typical polar reaction, the addition reaction of ethylene with HCl. When ethylene is treated with hydrogen chloride at room temperature, chloroethane is produced. Overall, the reaction can be formulated as:

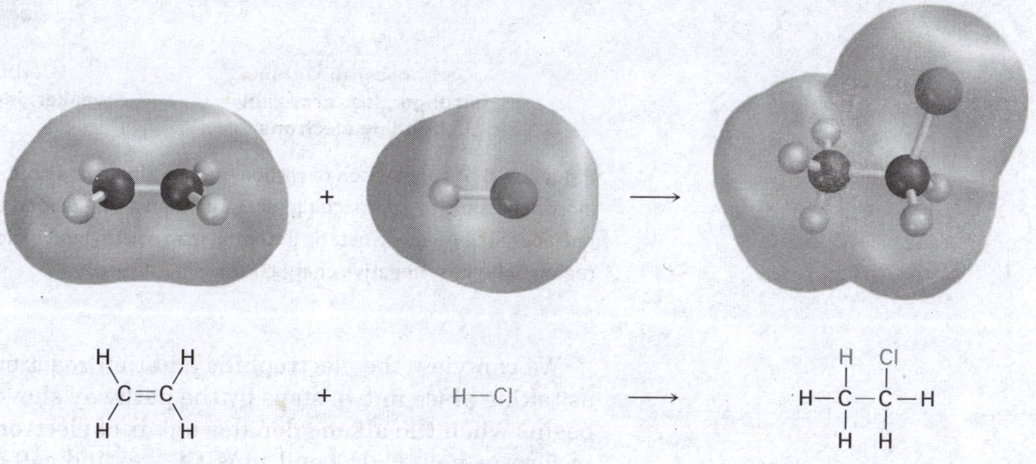

Ethylene (nucleophile) **Hydrogen chloride (electrophile)** **Chloroethane**

The reaction is an example of a general polar reaction type known as an *electrophilic addition reaction* and can be understood using the general ideas discussed in the previous section. Let's begin by looking at the nature of the two reactants.

What do we know about ethylene? We know from Sections 1.8 and 3.2 that a carbon–carbon double bond results from orbital overlap of two sp^2-hybridized carbon atoms. The σ part of the double bond results from sp^2–sp^2 overlap, and the π part results from p–p overlap.

What kind of chemical reactivity might we expect of a C=C bond? Unlike the valence electrons in alkanes, which are relatively inaccessible because they are tied up in strong, nonpolar C–C and C–H σ bonds between nuclei, the

π electrons in alkenes are accessible to external reagents because they are located above and below the plane of the double bond rather than between the nuclei (Figure 3.5). Furthermore, an alkene π bond is much weaker than an alkane σ bond, so an alkene is more reactive. As a result, C=C bonds behave as nucleophiles in much of their chemistry. That is, alkenes typically react by donating an electron pair from the double bond to form a new bond with an electron-poor, electrophilic partner.

What about HCl? As a strong acid, HCl is a powerful proton (H^+) donor and thus a good electrophile. The reaction of HCl with ethylene is therefore a typical electrophile–nucleophile combination as in all polar reactions.

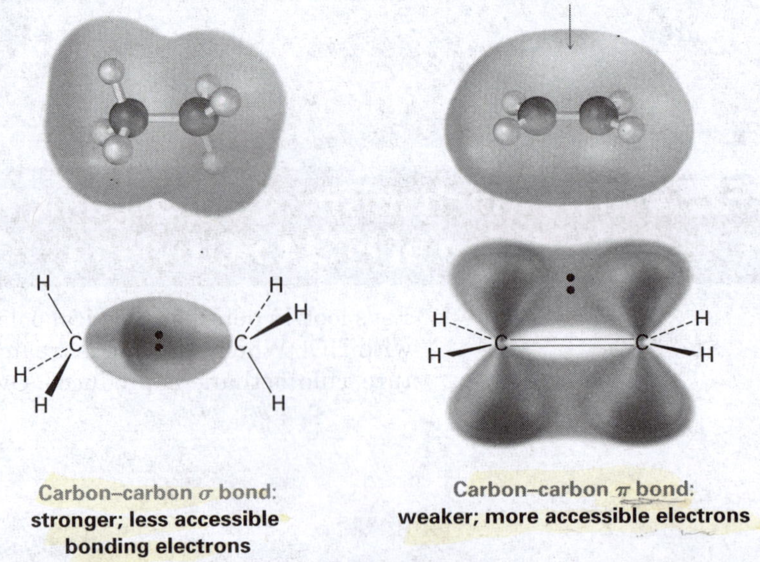

Figure 3.5 A comparison of carbon–carbon single and double bonds. A double bond is both more accessible to approaching reactants than a single bond and more electron-rich (more nucleophilic). An electrostatic potential map of ethylene indicates that the double bond is the region of highest negative charge (red).

We can view the electrophilic addition reaction between ethylene and HCl as taking place in two steps by the pathway shown in Figure 3.6. The reaction begins when the alkene donates a pair of electrons from its C=C bond to HCl to form a new C–H bond plus Cl^-, as indicated by the path of the curved arrows in the first step of Figure 3.6. One curved arrow begins at the middle of the double bond (the source of the electron pair) and points to the hydrogen atom in HCl (the atom to which a bond will form). This arrow indicates that a new C–H bond forms using electrons from the former C=C bond. A second curved arrow begins in the middle of the H–Cl bond and points to the Cl, indicating that the H–Cl bond breaks and the electrons remain with the Cl atom, giving Cl^-.

When one of the alkene carbon atoms bonds to the incoming hydrogen, the other carbon atom, having lost its share of the double-bond electrons, now has only six valence electrons and is left with a positive charge. This positively charged species—a carbon-cation, or **carbocation**—is itself an electrophile that can accept an electron pair from nucleophilic Cl^- anion in a second step, forming a C–Cl bond and yielding the neutral addition product. Once again, a curved arrow in Figure 3.6 shows the electron-pair movement from Cl^- to the positively charged carbon.

MECHANISM

Figure 3.6 The mechanism of the electrophilic addition of HCl to ethylene. The reaction takes place in two steps and involves an intermediate carbocation.

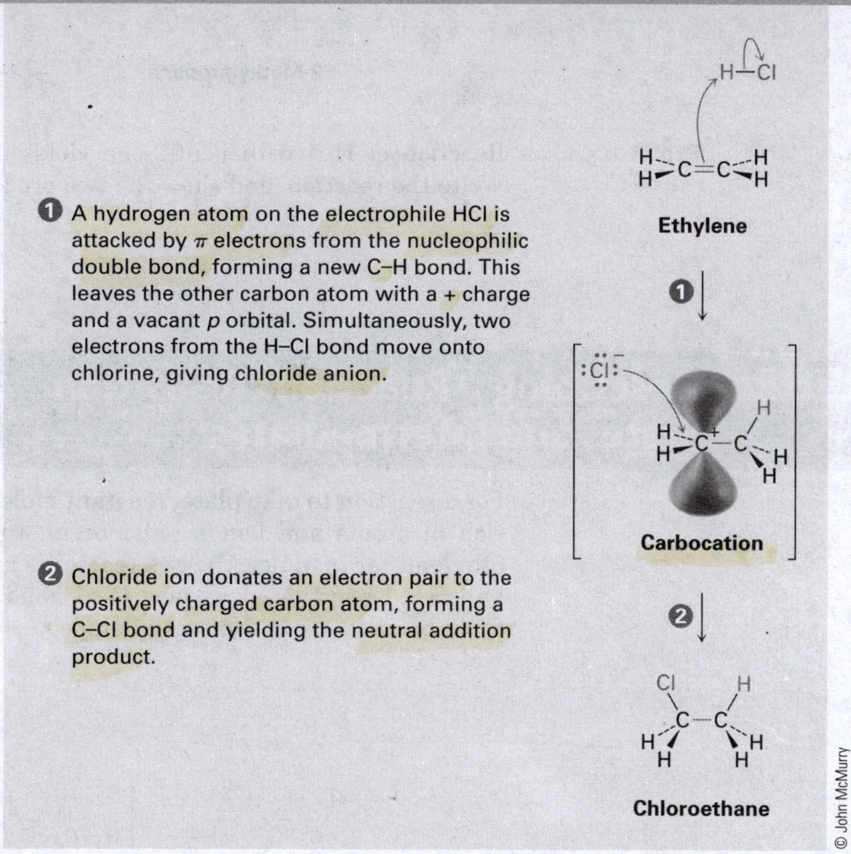

Worked Example 3.6 Predicting the Product of an Electrophilic Addition Reaction

What product would you expect from reaction of HCl with cyclohexene?

Strategy HCl adds to the double-bond functional group in cyclohexene in exactly the same way it adds to ethylene, yielding an addition product.

Solution

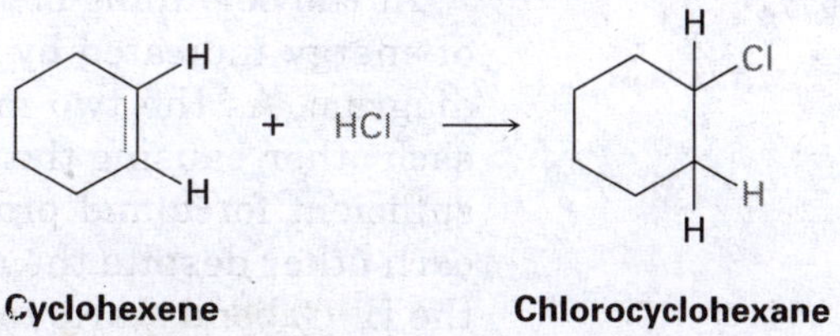

Problem 3.15 Reaction of HCl with 2-methylpropene yields 2-chloro-2-methylpropane. What is the structure of the carbocation formed during the reaction? Show the mechanism of the reaction.

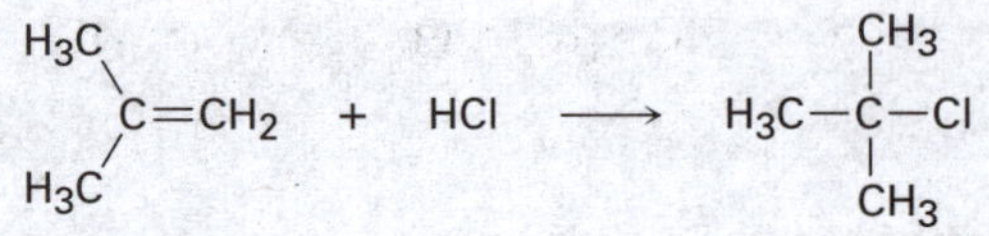

2-Methylpropene **2-Chloro-2-methylpropane**

Problem 3.16 Reaction of HCl with pent-2-ene yields a mixture of two addition products. Write the reaction, and show the two products.

3.8 Describing a Reaction: Transition States and Intermediates

For a reaction to take place, reactant molecules must collide and a reorganization of atoms and bonds must occur. In the addition reaction of HCl and ethylene, for instance, the two reactants approach each other, the C=C π bond and H–Cl bond break, a new C–H bond forms in the first step, and a new C–Cl bond forms in the second step.

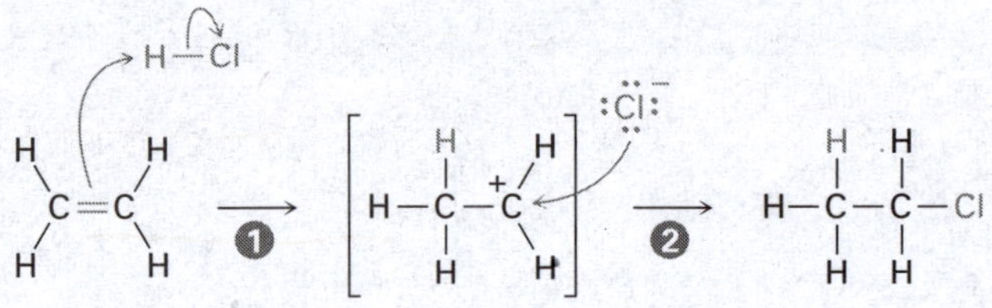

Carbocation

To depict graphically the energy changes that occur during a reaction, chemists use *energy diagrams* of the sort shown in Figure 3.7. The vertical axis of the diagram represents the total energy of all reactants, and the horizontal axis, called the *reaction coordinate,* represents the progress of the reaction from beginning to end.

At the beginning of the reaction, ethylene and HCl have the total amount of energy indicated by the reactant level at point **A** on the left side of the diagram. As the two molecules crowd together, their electron clouds repel each other, causing the energy level to rise. If the collision has occurred with sufficient force and proper orientation, the reactants continue to approach each other despite the repulsion until the new C–H bond starts to form and the H–Cl bond starts to break. At some point (**B** on the diagram), a structure of maximum energy is reached, a structure called the **transition state**. The transition state represents the highest-energy structure involved in this step of the reaction and can't be isolated or directly observed. Nevertheless, we can imagine it to be a kind of activated complex of the two reactants in which the C=C π bond is partially broken and the new C–H bond is partially formed.

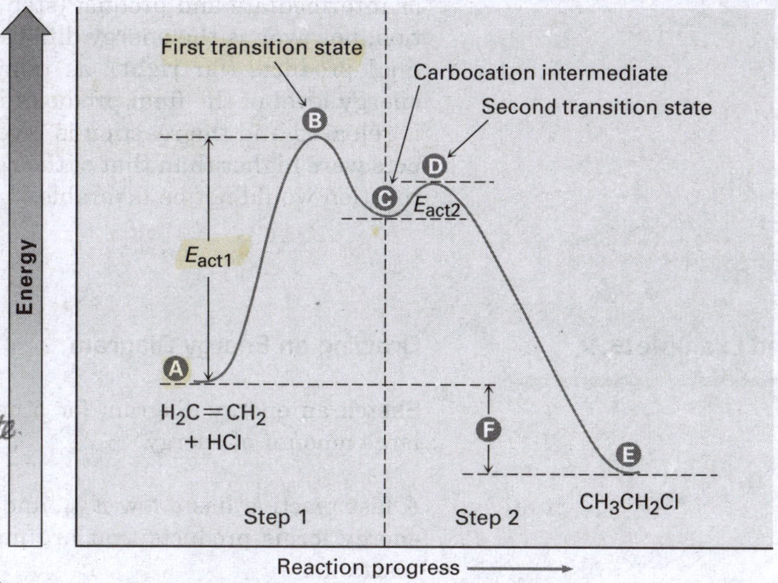

Figure 3.7 An energy diagram for the reaction of ethylene with HCl. Two separate steps are involved, each with its own activation energy, transition state, and energy change. The energy minimum between the two steps represents the carbocation reaction intermediate.

The energy difference between reactants **A** and transition state **B** is called the **activation energy, E_{act}**, and is a measure of how rapidly the reaction occurs. A large activation energy results in a slow reaction because few of the reacting molecules collide with enough energy to reach the transition state. A small activation energy results in a rapid reaction because almost all reacting molecules are energetic enough to climb to the transition state. As an analogy, think about hikers climbing over a mountain pass. If the pass is a high one, the hikers need a lot of energy and surmount the barrier slowly. If the pass is low, however, the hikers need less energy and reach the top quickly.

Most organic reactions have activation energies in the range 40 to 125 kJ/mol (10–30 kcal/mol). Reactions with activation energies less than 80 kJ/mol take place at or below room temperature, while reactions with higher activation energies often require heating to give the molecules enough energy to climb the activation barrier.

Once the high-energy transition state **B** has been reached, energy is released as the new C–H bond forms fully, so the curve in Figure 3.7 turns downward until it reaches a minimum at point **C**, representing the energy level of the carbocation. We call the carbocation, which is formed transiently during the course of the multistep reaction, a **reaction intermediate**. As soon as the carbocation intermediate is formed in the first step, it immediately reacts with Cl^- in a second step to give the final product, chloroethane. This second step has its own activation energy, E_{act2}, and its own transition state (**D**), which we can think of as an activated complex between the electrophilic carbocation intermediate and nucleophilic Cl^- anion in which the new C–Cl bond is partially formed. Finally, the curve turns downward as the C–Cl bond forms fully to give the final addition product at point **E**.

Each individual step in the reaction has its own energy change, represented by the difference in levels between reactant and intermediate (step 1) or intermediate and product (step 2).The *overall* energy change for the reaction, however, is the energy difference between initial reactants (far left) and final products (far right), as represented by **F** in Figure 3.7. Because the energy level of the final products is lower than that of the reactants, energy is released and the reaction is favorable. If the energy level of the final products were higher than that of the reactants, energy would be absorbed and the reaction would not be favorable.

Worked Example 3.7 Drawing an Energy Diagram

Sketch an energy diagram for a one-step reaction that is fast and releases a large amount of energy.

Strategy A fast reaction has a low E_{act}, and a reaction that releases a large amount of energy forms products that are much lower in energy and more stable than reactants.

Solution

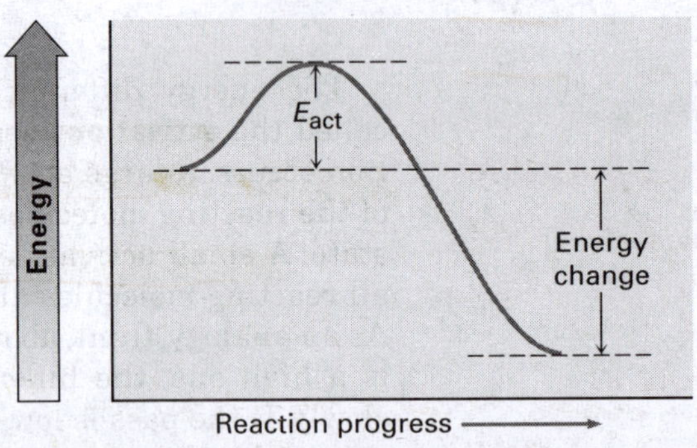

Problem 3.17 Which reaction is faster, one with E_{act} = 60 kJ/mol or one with E_{act} = 80 kJ/mol?

Problem 3.18 Sketch an energy diagram to represent each of the following situations.

(a) A reaction that releases energy and takes place in one step

(b) A reaction that absorbs energy and takes place in one step

Problem 3.19 Draw an energy diagram for a two-step reaction whose first step absorbs energy and whose second step releases energy. Label the intermediate.

3.9 Describing a Reaction: Catalysis

How fast a reaction occurs depends on the value of its activation energy, E_{act}, as noted in the previous section. Unfortunately, there is no way to predict the size of the activation energy for a reaction, and it may well happen that the E_{act} of a process is too large for the reaction to occur easily, even at high temperature. The only solution in such a situation is to find a way to change the reaction mechanism to an alternative pathway that occurs through different steps with lower activation energies.

A **catalyst** is a substance that increases the rate of a chemical transformation by providing an alternative mechanism. The catalyst *does* take part in the reaction, but it is regenerated at some point and thus undergoes no net change. An example that we'll see in the next chapter, for instance, is the use of a metal catalyst such as palladium to effect the reaction of an alkene with H_2 gas and produce an alkane. In the absence of palladium, an alkene undergoes no reaction with H_2 gas even at high temperature, but in the presence of palladium, reaction occurs rapidly at room temperature. Called a *hydrogenation reaction*, the process is used industrially to convert liquid vegetable oil to solid cooking fats.

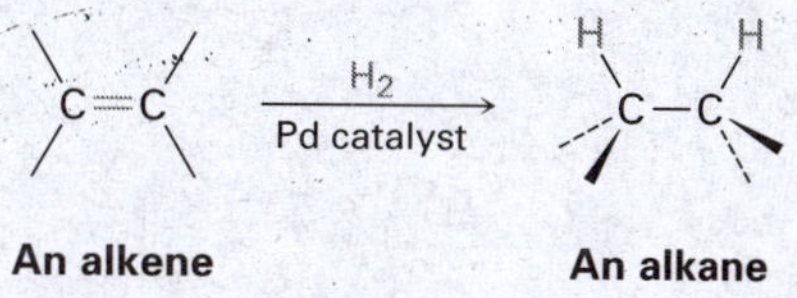

The hundreds of thousands of biological reactions that take place in living organisms almost all involve catalysts. Biological reactions use the same mechanisms as reactions that take place in the laboratory and can be described in similar ways, but they are constrained by the fact that they must have low enough activation energies to occur at moderate temperatures. This constraint is met through the use of large, structurally complex catalysts called *enzymes*, which provide reaction mechanisms that proceed through a series of small steps rather than through one or two large steps. Thus, a typical energy diagram for an enzyme-catalyzed biological reaction might look like that in Figure 3.8.

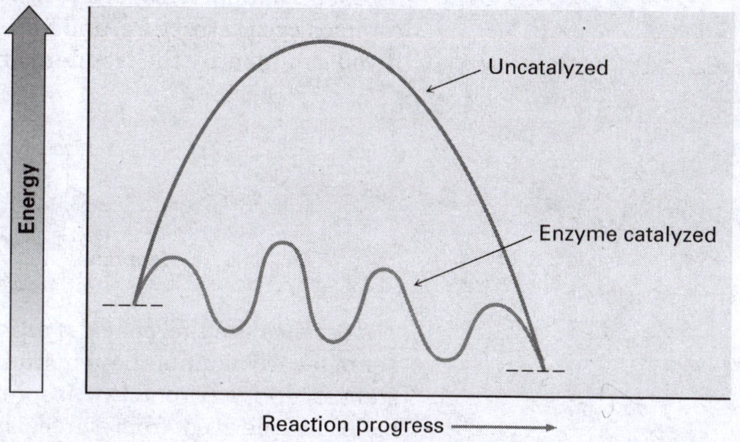

Figure 3.8 An energy diagram for a typical, enzyme-catalyzed biological reaction (blue curve) versus an uncatalyzed laboratory reaction (red curve). The biological reaction takes place in several steps, each of which has a relatively small activation energy.

INTERLUDE

Terpenes: Naturally Occurring Alkenes

© Biosphoto/Noun/Peter Arnold Inc.

The wonderful fragrance of leaves from the California bay laurel is due primarily to myrcene, a simple terpene.

It has been known for centuries that codistillation of many plant materials with steam produces a fragrant mixture of liquids called *essential oils.* For hundreds of years, such plant extracts have been used as medicines, spices, and perfumes. The investigation of essential oils also played a major role in the emergence of organic chemistry as a science during the 19th century.

Chemically, plant essential oils consist largely of mixtures of compounds known as *terpenoids*—small organic molecules with an immense diversity of structure. More than 35,000 different terpenoids are known. Some are open-chain molecules, and others contain rings; some are hydrocarbons, and others contain oxygen. Hydrocarbon terpenoids, in particular, are known as *terpenes,* and all contain double bonds. For example:

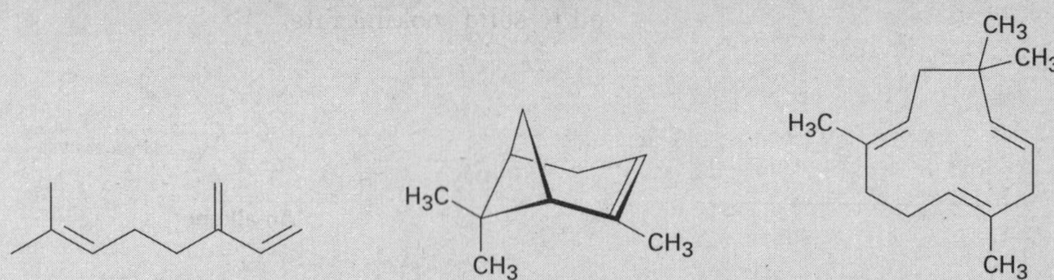

Myrcene (oil of bay) **α-Pinene (turpentine)** **Humulene (oil of hops)**

Regardless of their apparent structural differences, all terpenoids are related. According to a formalism called the *isoprene rule,* they can be thought of as arising from head-to-tail joining of 5-carbon isoprene units (2-methylbuta-1,3-diene). Carbon 1 is the head of the isoprene unit, and carbon 4 is the tail. For example, myrcene contains two isoprene units joined head to tail, forming an 8-carbon chain with two 1-carbon branches. α-Pinene similarly contains two isoprene units assembled into a more complex cyclic structure, and humulene contains three isoprene units. See if you can identify the isoprene units in α-pinene and humulene.

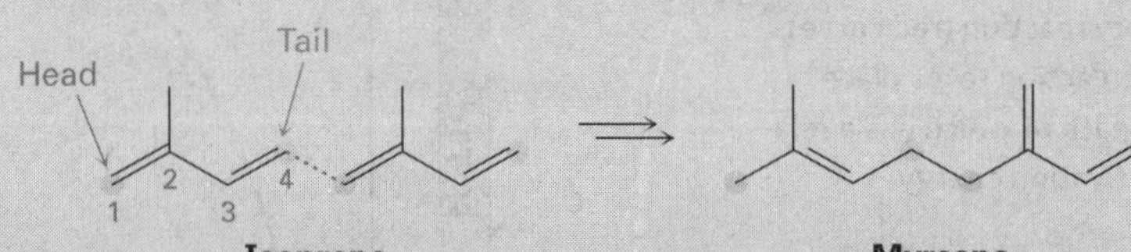

Isoprene **Myrcene**

Terpenes (and terpenoids) are further classified according to the number of 5-carbon units they contain. Thus, *monoterpenes* are 10-carbon substances derived from two isoprene units, *sesquiterpenes* are 15-carbon molecules derived from three isoprene units, *diterpenes* are 20-carbon

continued

INTERLUDE

substances derived from four isoprene units, and so on. Monoterpenes and sesquiterpenes are found primarily in plants, but the higher terpenoids occur in both plants and animals, and many have important biological roles. The triterpenoid lanosterol, for example, is the precursor from which all steroid hormones are made.

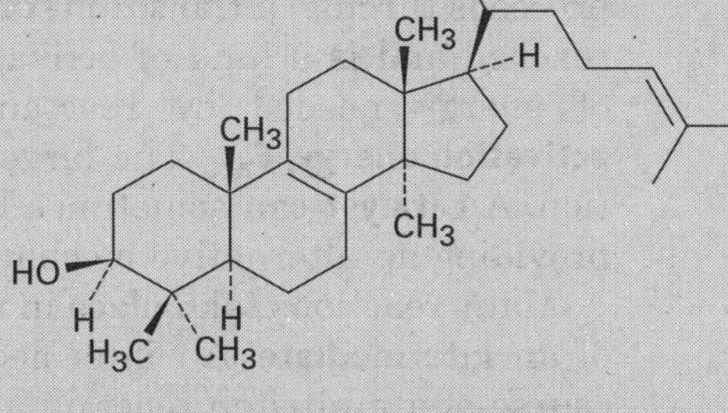

Lanosterol (a triterpene, C_{30})

Isoprene itself is not the true biological precursor of terpenoids. Nature instead uses two "isoprene equivalents"—isopentenyl diphosphate and dimethylallyl diphosphate—which are themselves made by two different routes depending on the organism. Lanosterol, in particular, is biosynthesized from acetic acid by a complex pathway that has been worked out in great detail.

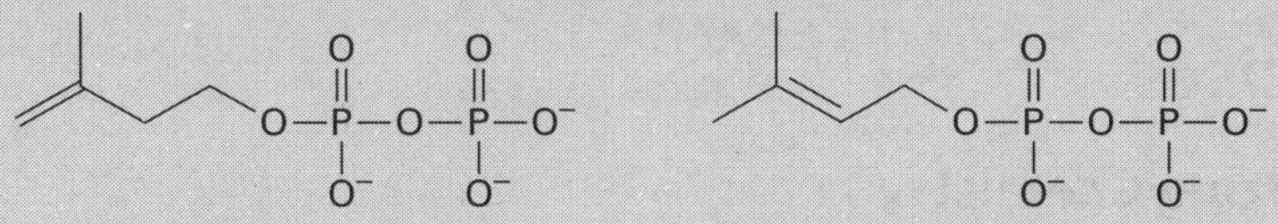

Isopentenyl diphosphate **Dimethylallyl diphosphate**

Summary and Key Words

activation energy, E_{act} 99
addition reaction 89
alkene 78
alkyne 78
carbocation 96
catalyst 101
electrophile 93
elimination reaction 90
E,Z system 86
nucleophile 92
polar reaction 91
radical 91
radical reaction 91
reaction intermediate 99
reaction mechanism 91
rearrangement reaction 90
substitution reaction 90
transition state 99
unsaturated 79

All chemical reactions, whether in the laboratory or in living organisms, follow the same "rules." To understand both organic and biological chemistry, it's necessary to know not just *what* occurs but also *why* and *how* chemical reactions take place. In this chapter, we've taken a brief look at the fundamental kinds of organic reactions. We've looked first at alkene stereoisomerism and then we've used alkene chemistry as a vehicle to see why reactions occur and how they can be described.

Alkenes are hydrocarbons that contain a carbon–carbon double bond, C=C, and **alkynes** are hydrocarbons that contain a carbon–carbon triple bond, C≡C. Because they contain fewer hydrogens than related alkanes, alkenes are often referred to as **unsaturated**.

A double bond consists of two parts: a σ bond formed by head-on overlap of two sp^2 orbitals and a π bond formed by sideways overlap of two p orbitals. Because rotation around the double bond is not possible, substituted alkenes can exist as cis–trans isomers. The geometry of a double bond can be described by the ***E,Z* system** as either *E (entgegen)* or *Z (zusammen)* by application of a series of sequence rules that rank the substituent groups on the double-bond carbons.

A full description of how a reaction occurs is called its **mechanism**. There are two kinds of organic mechanisms: polar and radical. **Polar reactions**, the most common kind, involve even-electron species and occur when an electron-rich reagent, or **nucleophile**, donates an electron pair to an electron-poor reagent, or **electrophile**, in forming a new bond. **Radical reactions** involve odd-electron species and occur when each reactant donates one electron in forming a new bond.

A reaction can be described pictorially by using an energy diagram, which follows the course of the reaction from reactant to product. Every reaction proceeds through a **transition state**, which represents the highest-energy point reached and is a kind of activated complex between reactants. The amount of energy needed by reactants to reach the transition state is the **activation energy, E_{act}**. The larger the activation energy, the slower the reaction. A **catalyst** can sometimes be used to increase the rate of a reaction by providing an alternative mechanistic pathway.

Many reactions take place in more than one step and involve the formation of an **intermediate**. An intermediate is a species that is formed during the course of a multistep reaction and that lies in an energy minimum between two transition states. Intermediates are more stable than transition states but are often too reactive to be isolated.

Exercises

Visualizing Chemistry

(Problems 3.1–3.19 appear within the chapter.)

OWL

Interactive versions of these problems are assignable in OWL.

3.20 Assign E or Z stereochemistry to each of the following alkenes, and convert each drawing into a skeletal structure (red = O, yellow-green = Cl).

(a)

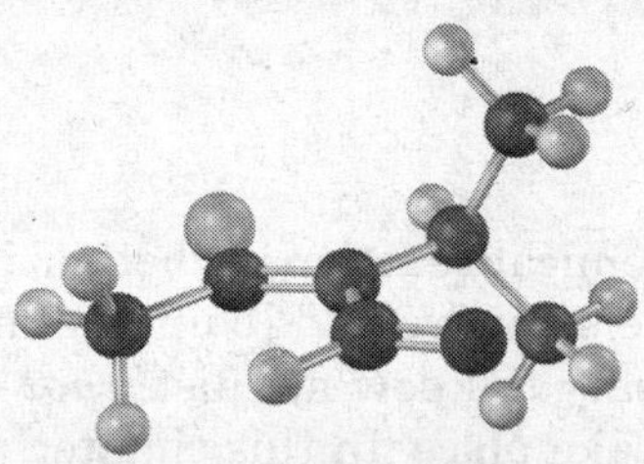

(b)

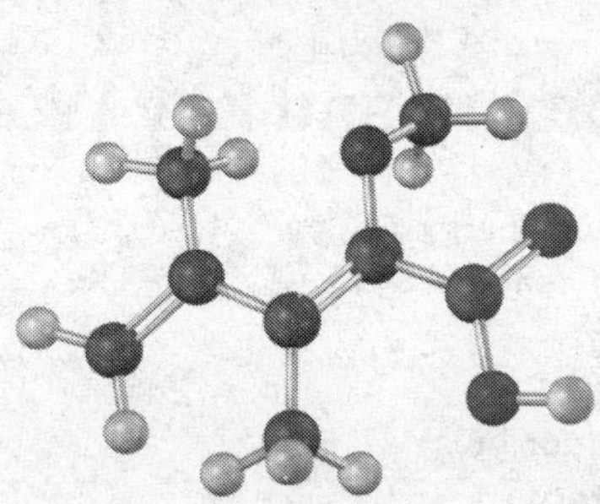

3.21 Name the following alkenes, and convert each drawing into a skeletal structure.

(a)

(b)

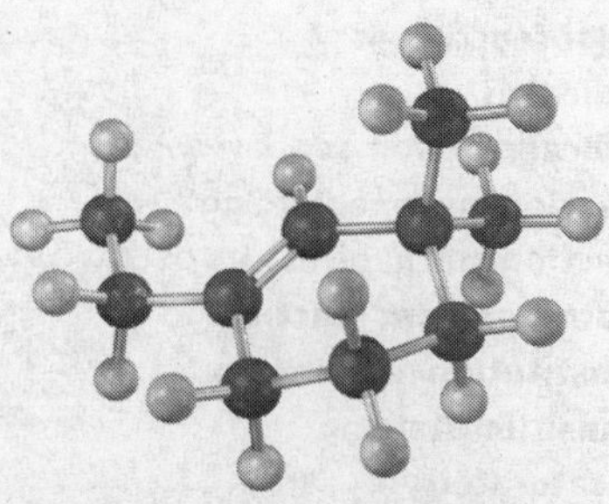

3.22 Electrostatic potential maps of **(a)** formaldehyde (CH_2O) and **(b)** methanethiol (CH_3SH) are shown. Is the formaldehyde carbon atom likely to be electrophilic or nucleophilic? What about the methanethiol sulfur atom? Explain.

(a)

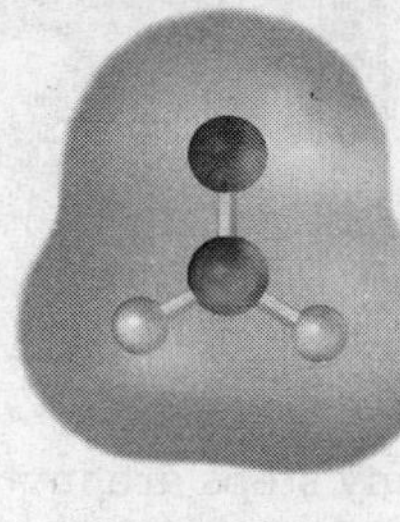

Formaldehyde

(b)

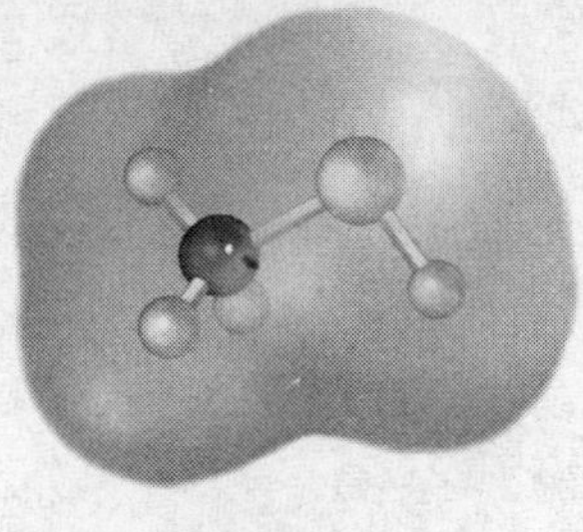

Methanethiol

3.23 The following alkyl chloride can be prepared by addition of HCl to two different alkenes. Name and draw the structures of both (yellow-green = Cl).

3.24 The following carbocation is a possible intermediate in the electrophilic addition of HCl with two different alkenes. Write structures for both.

Additional Problems

3.25 Predict the direction of polarization of the functional groups in each of the following molecules.

(a) $CH_3CH_2C{\equiv}N$

(b)

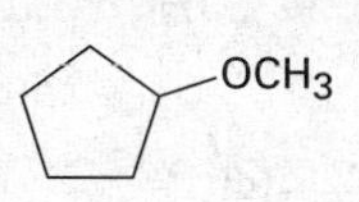

(c) $CH_3\overset{O}{\overset{\|}{C}}CH_2\overset{O}{\overset{\|}{C}}OCH_3$

(d)

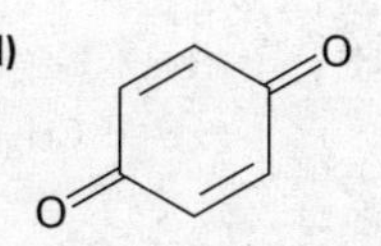

(e)

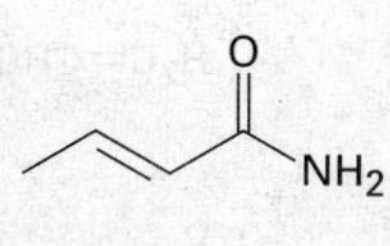

(f)

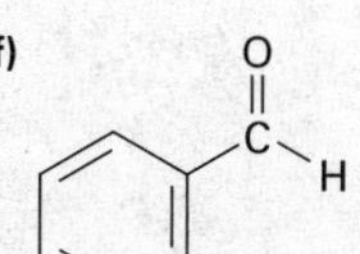

3.26 Look at the following energy diagram for an enzyme-catalyzed reaction:

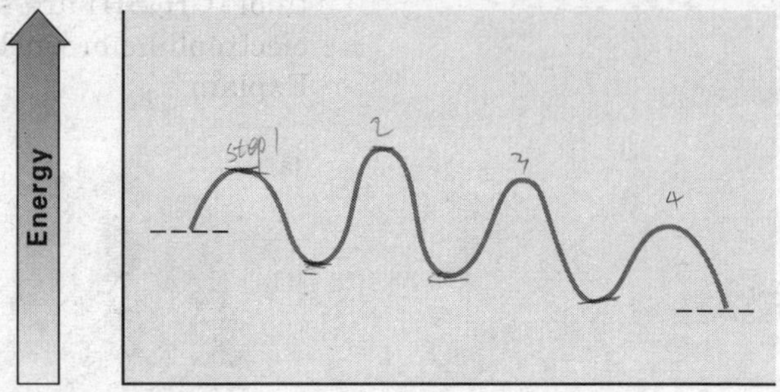

(a) How many steps are involved?
(b) Which is the fastest step, and which is the slowest?

3.27 Which of the following are likely to behave as electrophiles and which as nucleophiles?
(a) Cl^- **(b)** $N(CH_3)_3$ **(c)** Hg^{2+} **(d)** CH_3S^- **(e)** CH_3^+

3.28 Identify the likely electrophilic and nucleophilic sites in each of the following molecules:

(a)
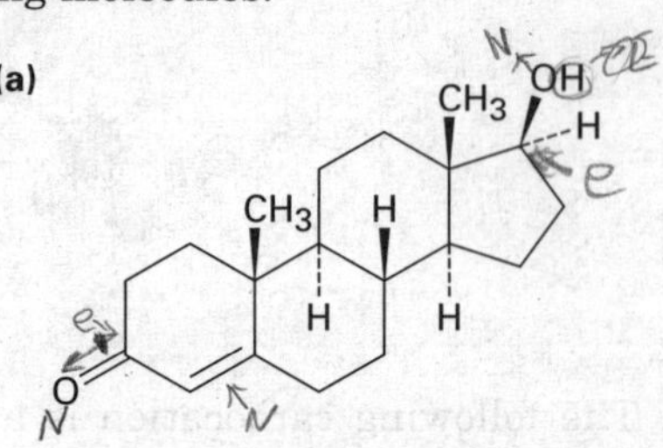

Testosterone

(b)
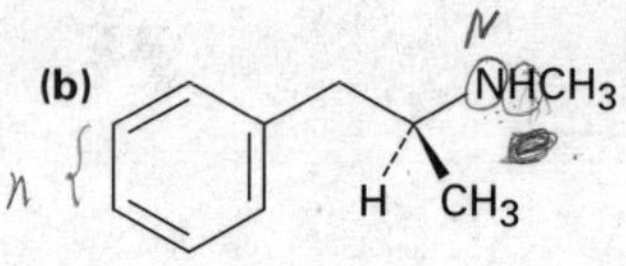

Amphetamine

NAMING ALKENES, ALKYNES, AND CYCLOALKENES

3.29 Name the following cycloalkenes:

(a)
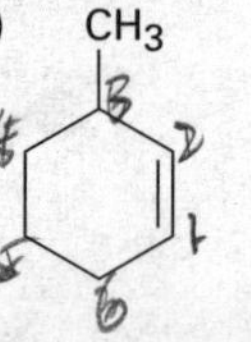

(b)
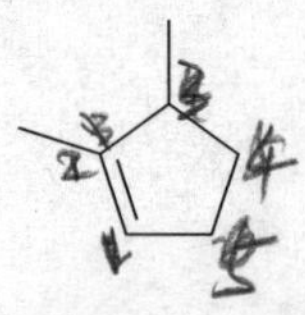

(c)
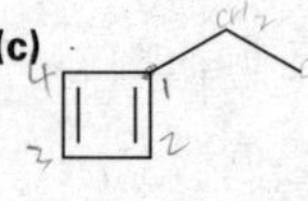

(d)
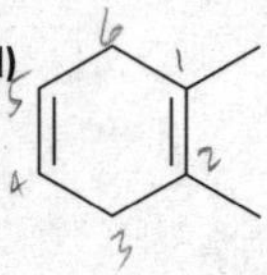

3.30 Name the following alkenes:

(a)
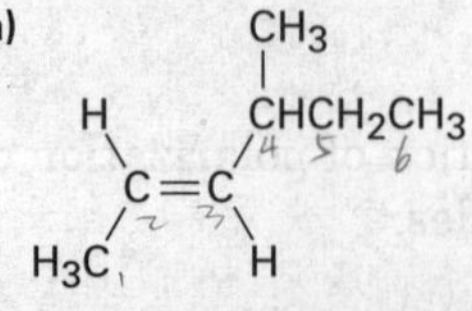

(b)
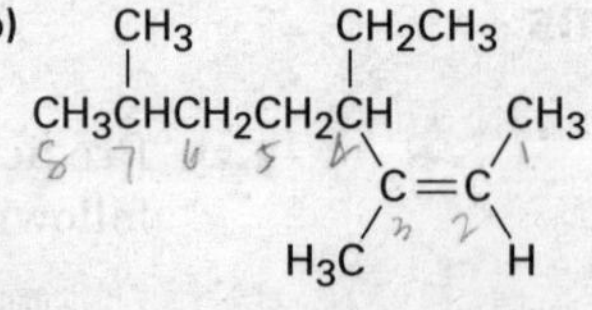

(c)
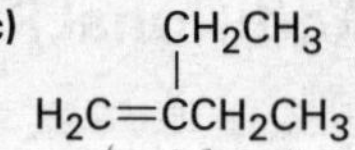

(d)
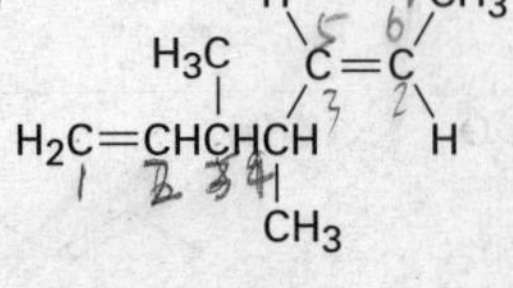

(e)
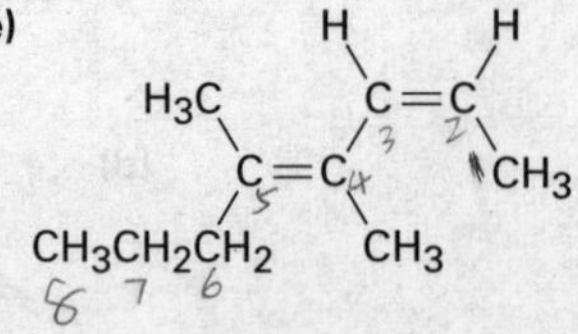

(f) $H_2C{=}C{=}CHCH_3$

3.31 The following names are incorrect. Draw each molecule, tell why its name is wrong, and give its correct name.
(a) 1-Methylcyclopent-2-ene (b) 1-Methylpent-1-ene
(c) 6-Ethylcycloheptene (d) 3-Methyl-2-ethylcyclohexene

3.32 Draw structures corresponding to the following IUPAC names:
(a) 3-Propylhept-2-ene (b) 2,4-Dimethylhex-2-ene
(c) Octa-1,5-diene (d) 4-Methylpenta-1,3-diene
(e) *cis*-4,4-Dimethylhex-2-ene (f) *(E)*-3-Methylhept-3-ene

3.33 Correct the following pre-1993 names to current names, and draw each structure:
(a) 2,5-Dimethyl-3-hexyne (b) *(Z)*-3-Methyl-2-pentene

3.34 Draw the structures of the following cycloalkenes:
(a) *cis*-4,5-Dimethylcyclohexene (b) 3,3,4,4-Tetramethylcyclobutene

DOUBLE-BOND ISOMERS

3.35 Neglecting cis–trans isomers, there are five substances with the formula C_4H_8. Draw and name them.

3.36 Which of the molecules you drew in Problem 3.35 show cis–trans isomerism? Draw and name their cis–trans isomers.

3.37 Rank the following sets of substituents according to the sequence rules:
(a) $-CH_3$, $-Br$, $-H$, $-I$
(b) $-OH$, $-OCH_3$, $-H$, $-CO_2H$
(c) $-CH_3$, $-CO_2H$, $-CH_2OH$, $-CHO$
(d) $-CH_3$, $-CH{=}CH_2$, $-CH_2CH_3$, $-CH(CH_3)_2$

3.38 Which of the following molecules show cis–trans isomerism?

(a) 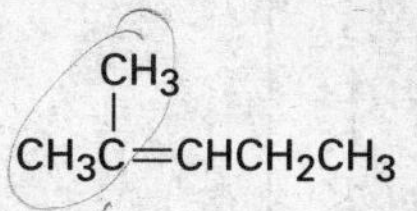

(b) 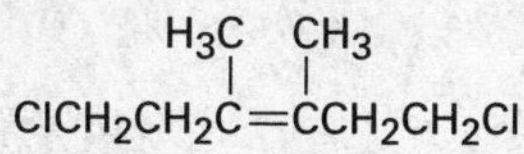

(c) 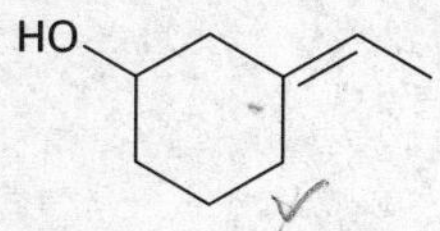

3.39 Rank the following pairs of substituents according to the sequence rules:

(a) $-CH_2CH_3$ $-CH_2CH_2CH_3$

(b) 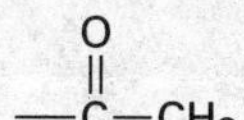 $-CH_2OH$

(c) 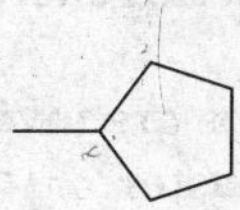 $-CH(CH_3)_2$

(d) 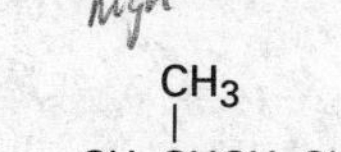 $-CH_2CH_2CH_2Br$

3.40 Draw and name molecules that meet the following descriptions:
(a) An alkene, C_6H_{12}, that does not show cis–trans isomerism
(b) The *E* isomer of a trisubstituted alkene, C_6H_{12}
(c) A cycloalkene, C_7H_{12}, with a tetrasubstituted double bond

3.41 Cyclodecene can exist in both cis and trans forms, but cyclohexene cannot. Explain.

3.42 Draw and name all possible stereoisomers of hepta-2,4-diene.

3.43 Assign *E* or *Z* stereochemistry to the following alkenes:

(a) 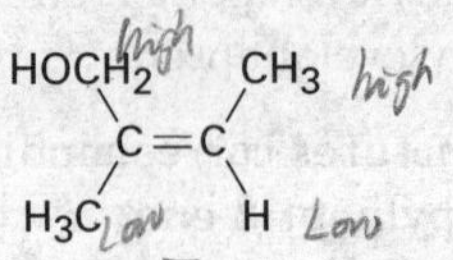

(b) 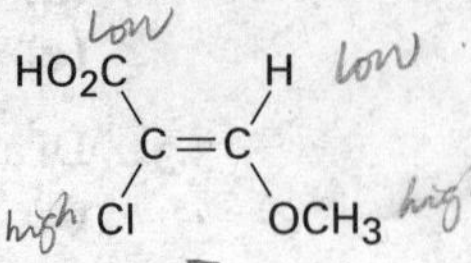

(c) 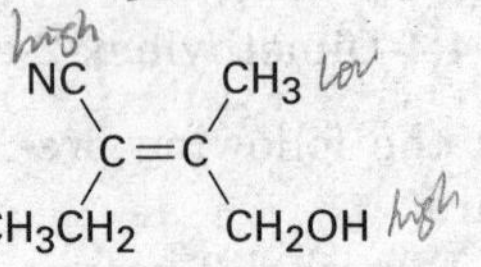

(d) 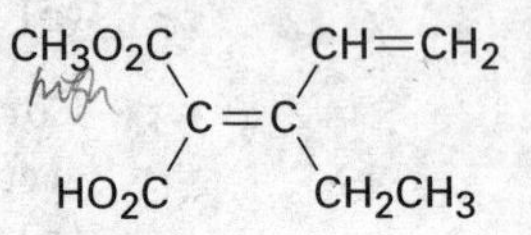

3.44 Menthene, a hydrocarbon found in mint plants, has the IUPAC name 1-isopropyl-4-methylcyclohexene. What is the structure of menthene?

3.45 Draw and name the six C_5H_{10} alkene isomers, including *E,Z* isomers.

ENERGY DIAGRAMS

3.46 Draw an energy diagram for a two-step reaction that releases energy and whose first step is faster than its second step. Label the parts of the diagram corresponding to reactants, products, transition states, intermediate, activation energies, and overall energy change.

3.47 Describe the difference between a transition state and a reaction intermediate.

3.48 Consider the energy diagram shown:

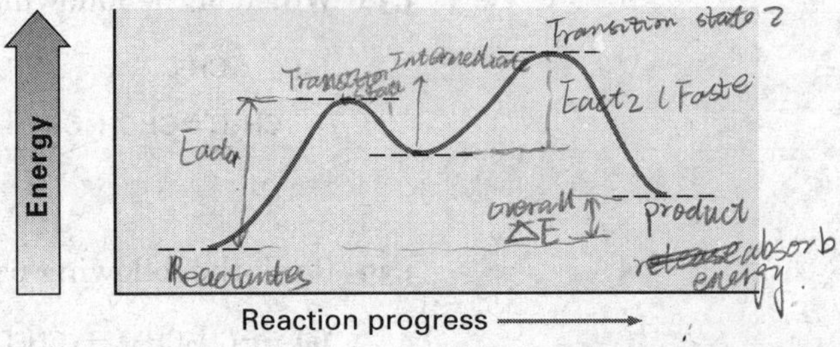

(a) Indicate the overall energy change for the reaction. Is it positive or negative?
(b) How many steps are involved in the reaction?
(c) Which step is faster?
(d) How many transition states are there? Label them.

3.49 Draw an energy diagram for a two-step reaction whose second step is faster than its first step.

3.50 Draw an energy diagram for a reaction whose products and reactants are of equal stability.

3.51 If a reaction has $E_{act} = 15$ kJ/mol, is it likely to be fast or slow at room temperature? Explain.

General Problems

3.52 α-Farnesene is a constituent of the natural waxy coating found on apples. What is its IUPAC name?

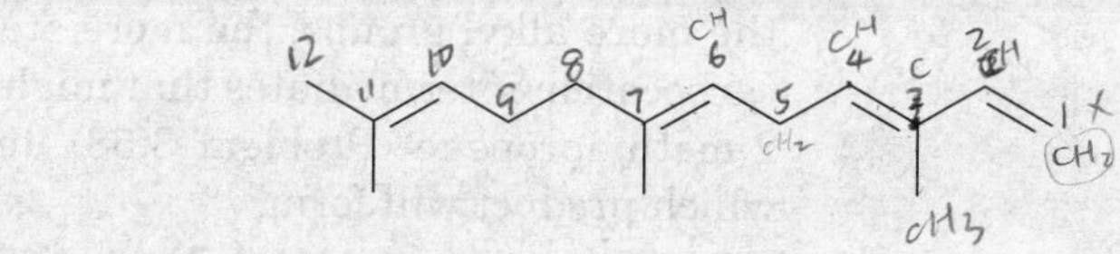

α-Farnesene

3.53 Indicate E or Z stereochemistry for each of the double bonds in α-farnesene (see Problem 3.52).

3.54 Methoxide ion (CH_3O^-) reacts with bromoethane in a single step according to the following equation:

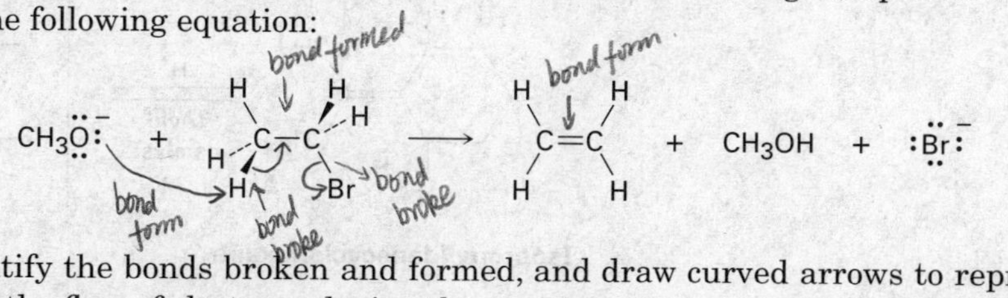

Identify the bonds broken and formed, and draw curved arrows to represent the flow of electrons during the reaction.

3.55 Name the following cycloalkenes:

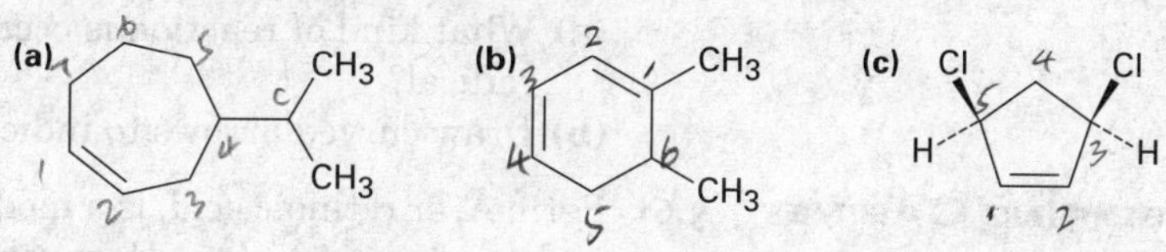

3.56 Follow the flow of electrons indicated by the curved arrows in each of the following reactions, and predict the products that result:

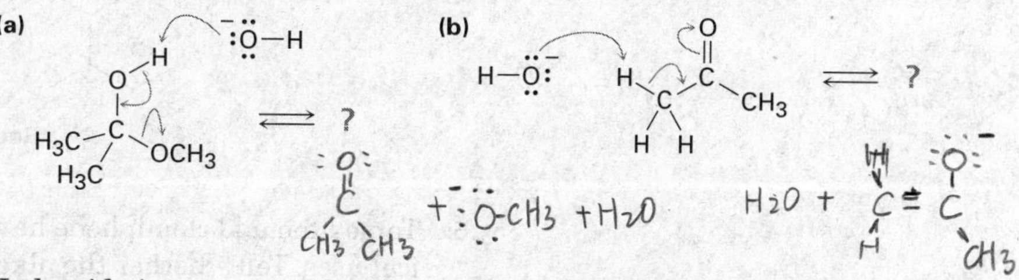

3.57 Hydroxide ion reacts with chloromethane in a single step according to the following equation:

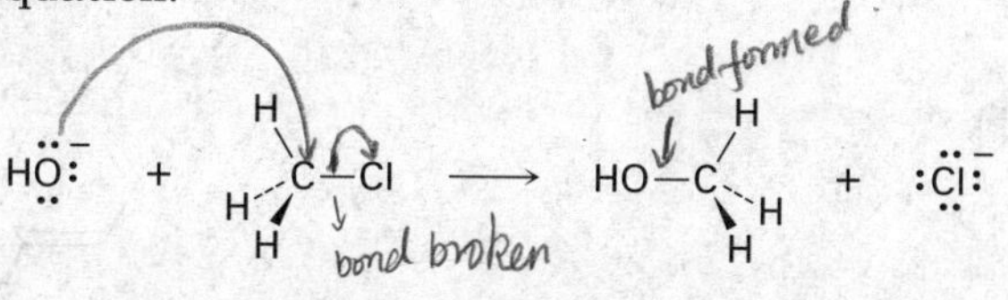

Identify the bonds broken and formed, and draw curved arrows to represent the flow of electrons during the reaction.

3.58 Reaction of 2-methylpropene with HCl might, in principle, lead to a mixture of two products. Draw them.

3.59 We'll see in the next chapter that the stability of carbocations depends on the number of alkyl groups attached to the positively charged carbon—the more alkyl groups, the more stable the cation. Draw the two possible carbocation intermediates that might be formed in the reaction of HCl with 2-methylpropene (Problem 3.58), tell which is more stable, and predict which product will form.

3.60 When isopropylidenecyclohexane is treated with strong acid at room temperature, isomerization occurs by the mechanism shown below to yield 1-isopropylcyclohexene:

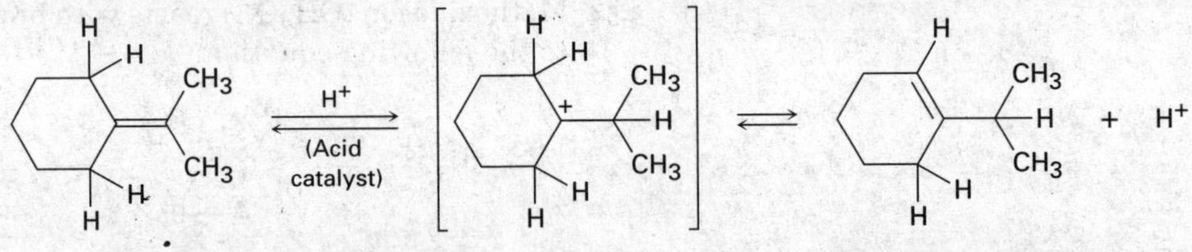

Isopropylidenecyclohexane

1-Isopropylcyclohexene

At equilibrium, the product mixture contains about 30% isopropylidenecyclohexane and about 70% 1-isopropylcyclohexene.

(a) What kind of reaction is occurring? Is the mechanism polar or radical?

(b) Draw curved arrows to indicate electron flow in each step.

IN THE MEDICINE CABINET

3.61 Retin A, or retinoic acid, is a medication commonly used to reduce wrinkles and treat severe acne. How many different isomers arising from double bond isomerizations are possible?

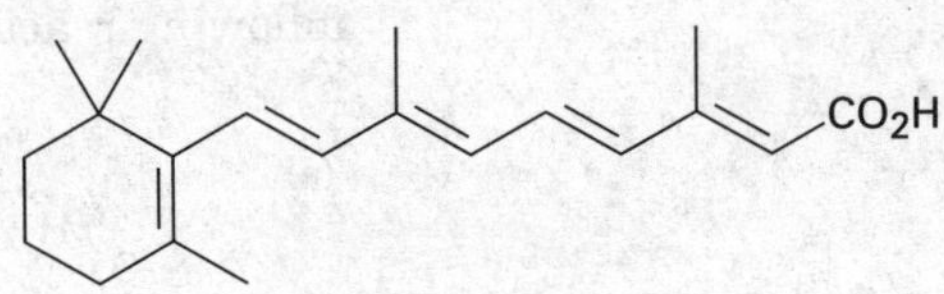

Retin A (retinoic acid)

3.62 Tamoxifen and clomiphene have similar structures but very different medical uses. Tell whether the alkene double bond in each is *E* or *Z*.

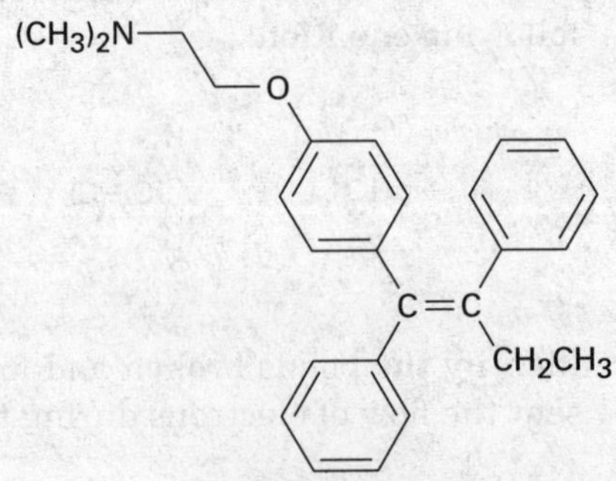

Tamoxifen (anticancer)

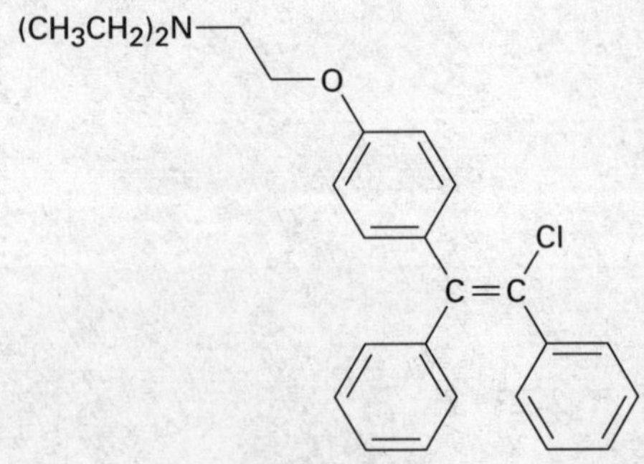

Clomiphene (fertility treatment)

IN THE FIELD

3.63 Lycopene, the pigment that gives tomatoes their red color, is a terpene derived formally by the joining together of numerous isoprene units (see the *Interlude* in this chapter). Start at one end of the molecule, and identify all the contiguous isoprene groupings.

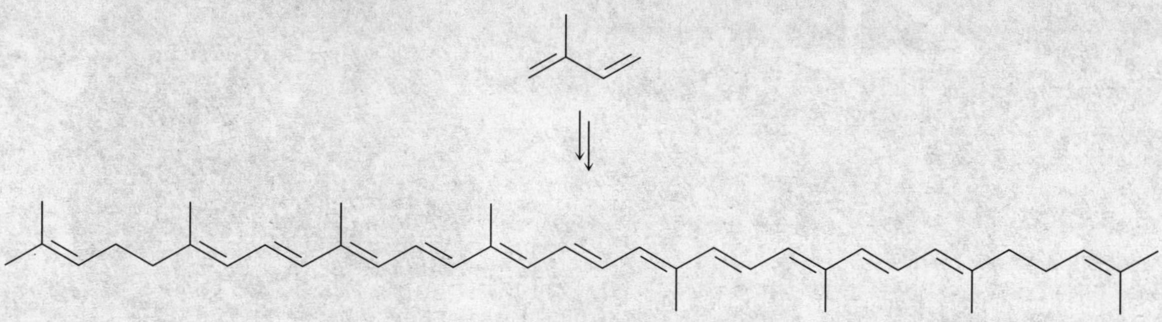

Lycopene

CHAPTER

4

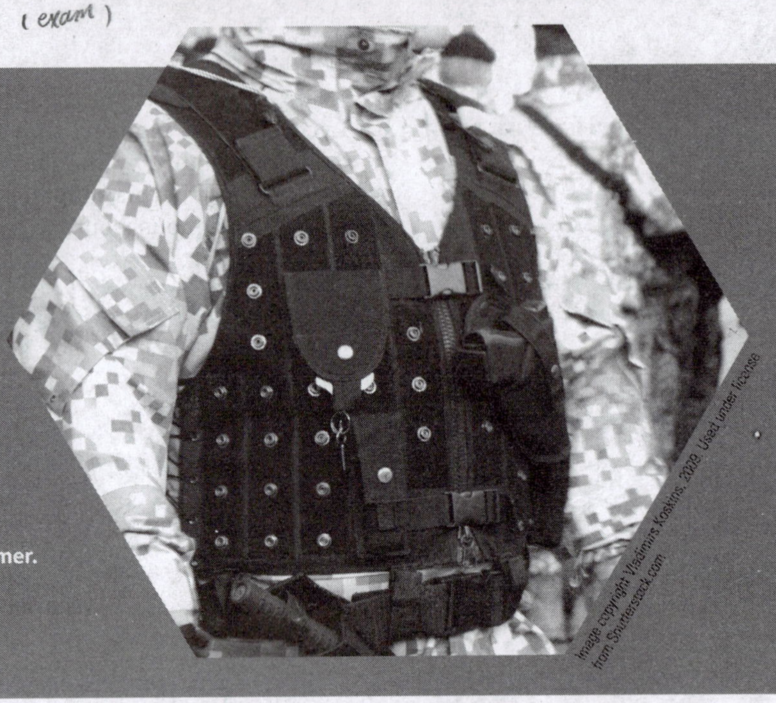

The Spectra fiber in the bulletproof vests used by police and military is made of ultra high molecular weight polyethylene, a simple alkene polymer.

Reactions of Alkenes and Alkynes

Much of the background needed to understand organic reactions has been covered, and it's now time to begin a systematic description of the major functional groups. We'll start in this chapter with a study of the alkene and alkyne families of compounds, and we'll see that the most important reaction of these two functional groups is the addition to the C=C and C≡C multiple bonds of various reagents X–Y to yield saturated products. In fact, all the reactions we'll discuss in this chapter follow the same pattern.

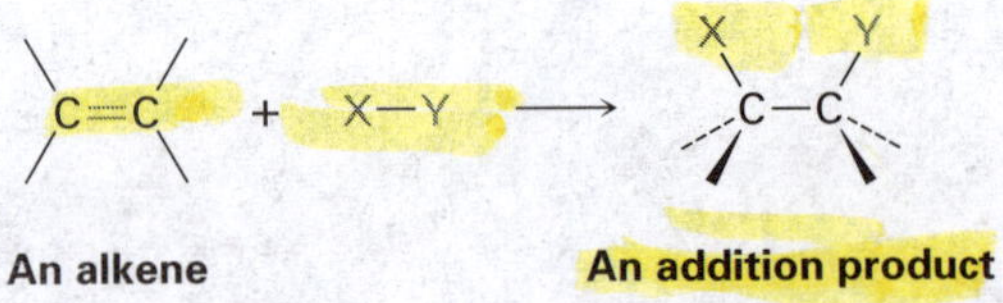

WHY THIS CHAPTER?

Both in this chapter on alkenes and in future chapters on other functional groups, we'll discuss a variety of reactions but will focus on the general principles and patterns of reactivity that tie organic chemistry together. There are no shortcuts; you have to know the reactions to understand organic chemistry.

OWL

Online homework for this chapter can be assigned in OWL, an online homework assessment tool.

4.1 Addition of HX to Alkenes: Markovnikov's Rule

We saw in Section 3.7 that alkenes react with HCl to yield alkyl chloride addition products. For example, ethylene reacts with HCl to give chloroethane. The reaction takes place in two steps and involves a carbocation intermediate.

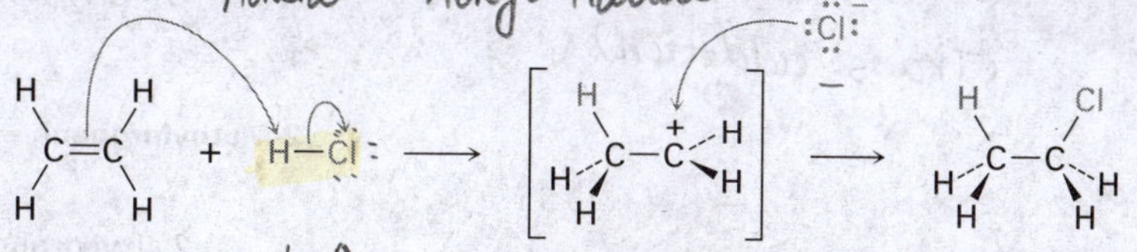

Ethylene **Carbocation intermediate** **Chloroethane**

The addition of halogen acids, HX, to alkenes is a general reaction that allows chemists to prepare a variety of halo-substituted alkane products. Thus, HCl, HBr, and HI all add to alkenes.

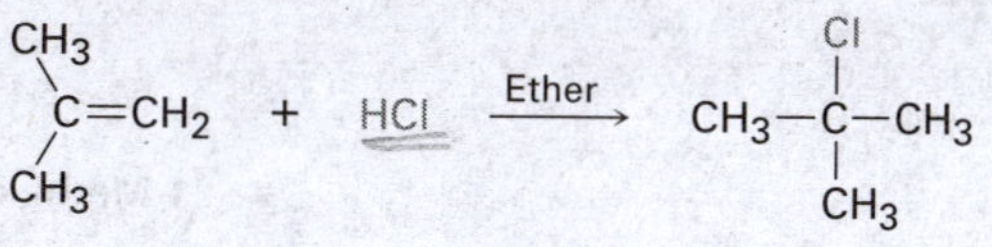

2-Methylpropene **2-Chloro-2-methylpropane**

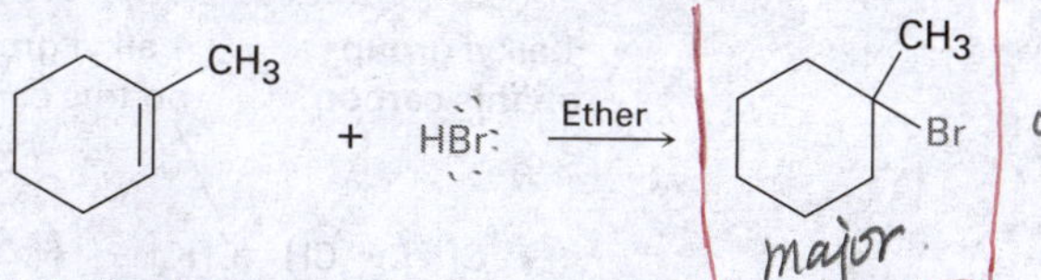

1-Methylcyclohexene **1-Bromo-1-methylcyclohexane**

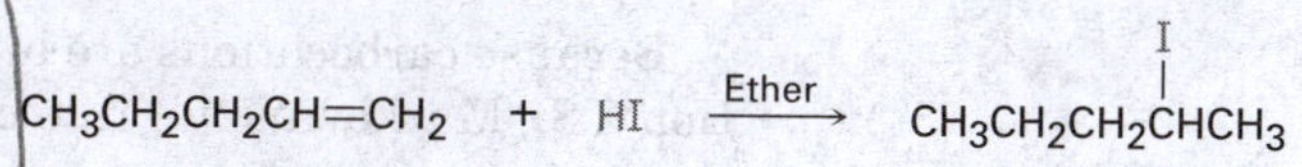

Pent-1-ene **2-Iodopentane**

Look carefully at the three reactions just shown. In each case, an unsymmetrically substituted alkene has given a single addition product rather than the mixture that might have been expected. For example, 2-methylpropene *might* have reacted with HCl to give 1-chloro-2-methylpropane in addition to 2-chloro-2-methylpropane, but it didn't. We say that such reactions are **regiospecific** (**ree**-jee-oh-specific) when only one of the two possible orientations of addition occurs.

A regiospecific reaction:

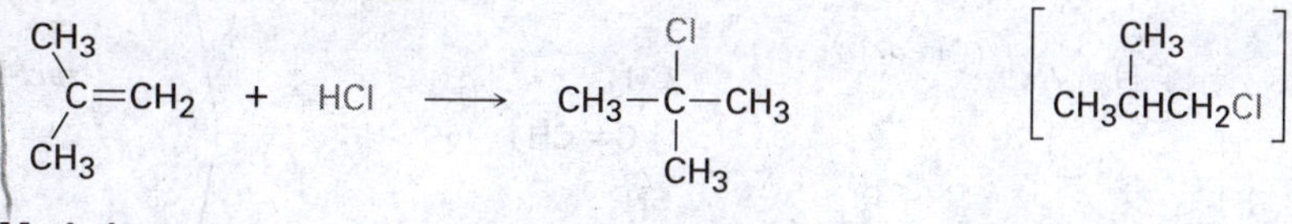

2-Methylpropene **2-Chloro-2-methylpropane (sole product)** **1-Chloro-2-methylpropane (*NOT* formed)**

After looking at the results of many such reactions, the Russian chemist Vladimir Markovnikov proposed in 1869 what has become known as **Markovnikov's rule**:

MARKOVNIKOV'S RULE In the addition of HX to an alkene, the H attaches to the carbon with fewer alkyl substituents and the X attaches to the carbon with more alkyl substituents.

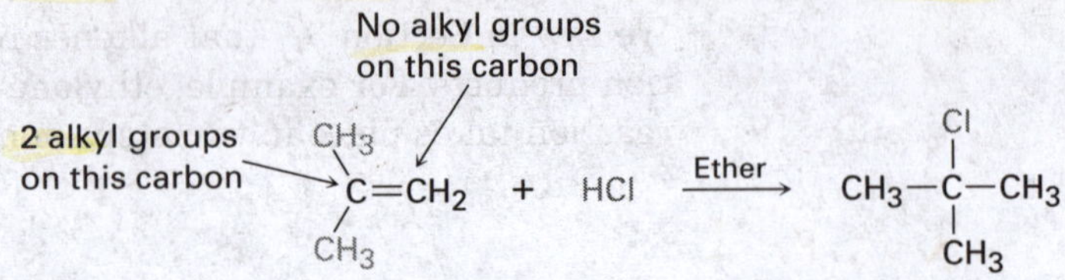

2-Methylpropene **2-Chloro-2-methylpropane**

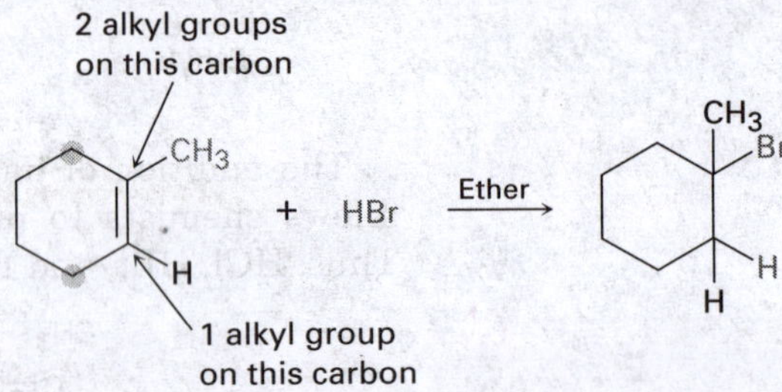

1-Methylcyclohexene **1-Bromo-1-methylcyclohexane**

When both double-bond carbon atoms have the same degree of substitution, a mixture of addition products results.

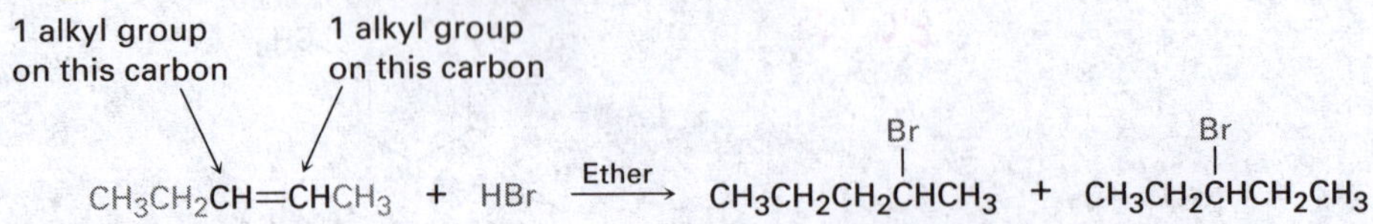

Pent-2-ene **2-Bromopentane** **3-Bromopentane**

Because carbocations are involved as intermediates in these reactions (Section 3.8), Markovnikov's rule can be restated.

MARKOVNIKOV'S RULE (RESTATED) In the addition of HX to an alkene, the more highly substituted carbocation is formed as the intermediate rather than the less highly substituted one.

For example, addition of H^+ to 2-methylpropene yields the intermediate *tertiary* carbocation rather than the alternative primary carbocation. Why should this be?

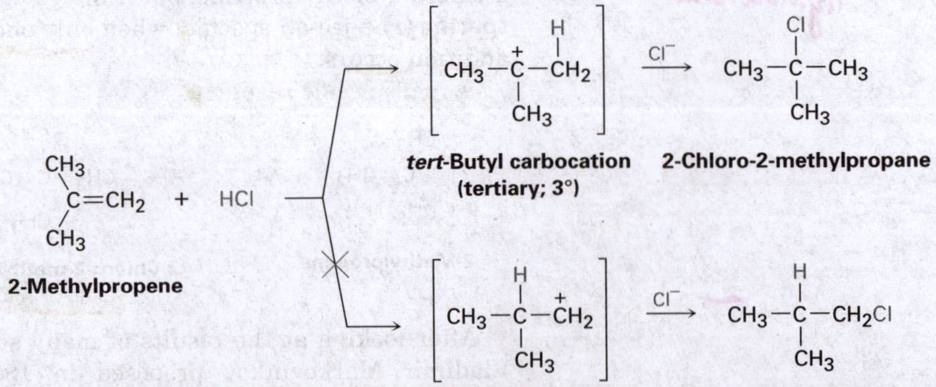

2-Methylpropene

***tert*-Butyl carbocation (tertiary; 3°)** **2-Chloro-2-methylpropane**

Isobutyl carbocation (primary; 1°) **1-Chloro-2-methylpropane *(NOT formed)***

Worked Example 4.1 Predicting the Product of an Alkene Addition Reaction

What product would you expect from the reaction of HCl with 1-ethylcyclopentene?

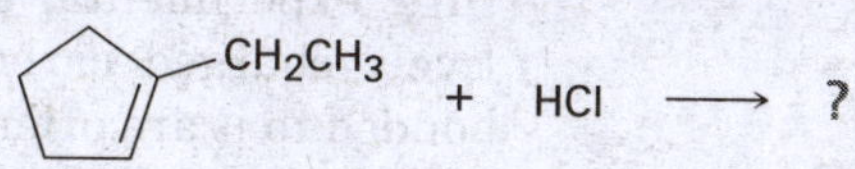

Strategy When solving a problem that asks you to predict a reaction product, begin by looking at the functional group(s) in the reactants and deciding what kind of reaction is likely to occur. In the present instance, the reactant is an alkene that will probably undergo an electrophilic addition reaction with HCl. Next, recall what you know about electrophilic addition reactions, and use your knowledge to predict the product. You know that electrophilic addition reactions follow Markovnikov's rule, so H^+ will add to the double-bond carbon that has one alkyl group (C2 on the ring) and Cl will add to the double-bond carbon that has two alkyl groups (C1 on the ring).

Solution The expected product is 1-chloro-1-ethylcyclopentane.

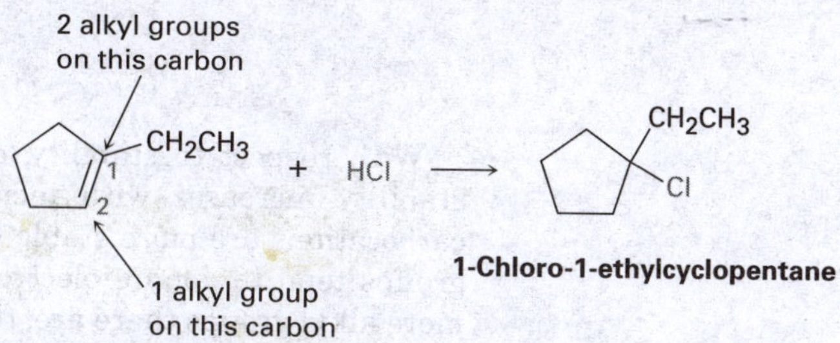

Problem 4.1 Predict the products of the following reactions:

(a)
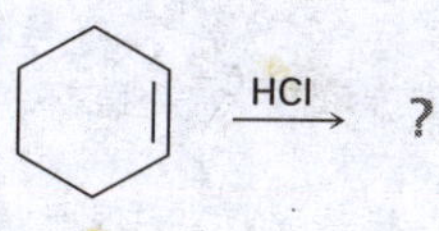

(b)
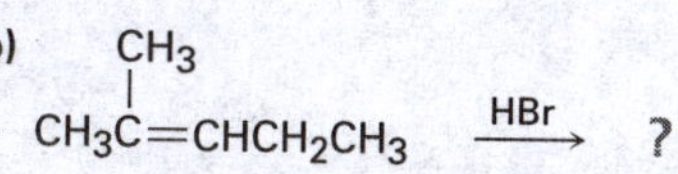

(c)
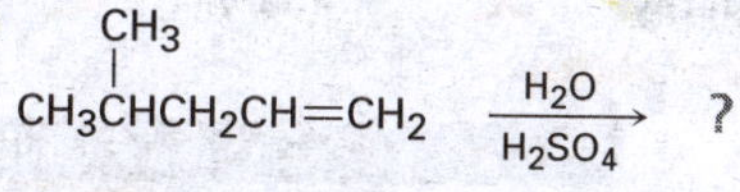

(Addition of H_2O occurs.)

(d)
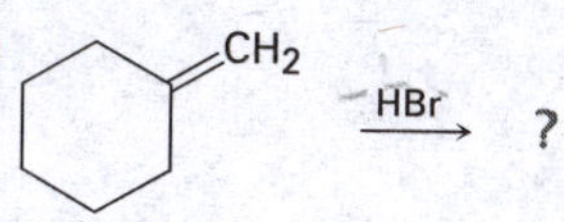

Problem 4.2 What alkenes would you start with to prepare the following alkyl halides?

(a)
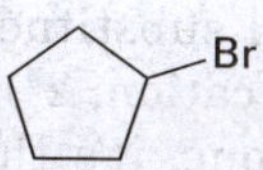

(b)
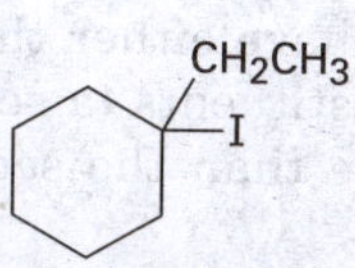

(c)
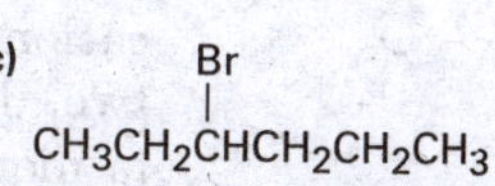

(d)
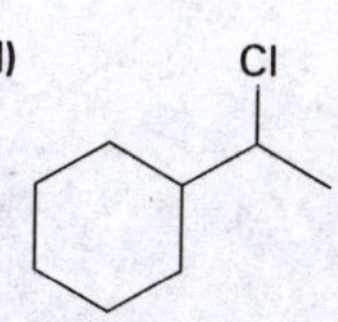

4.2 Carbocation Structure and Stability

To understand why Markovnikov's rule works, we need to learn more about the structure and stability of substituted carbocations. With respect to structure, experimental evidence shows that carbocations are *planar.* The positively charged carbon atom is sp^2-hybridized, and the three substituents bonded to it are oriented to the corners of an equilateral triangle, as indicated in Figure 4.1. Because there are only six valence electrons on carbon and all six are used in the three σ bonds, the *p* orbital extending above and below the plane is unoccupied.

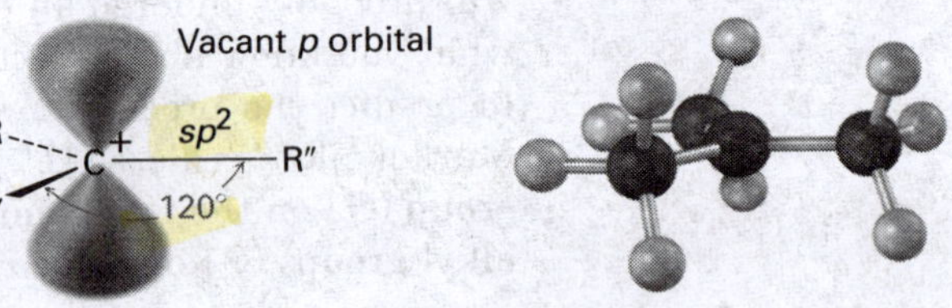

Figure 4.1 The structure of a carbocation. The trivalent carbon is sp^2-hybridized and has an unoccupied *p* orbital perpendicular to the plane of the carbon and three attached groups.

With respect to stability, experimental evidence shows that carbocation stability increases with increasing substitution. More highly substituted carbocations are more stable than less highly substituted ones because alkyl groups tend to donate electrons to the positively charged carbon atom. The more alkyl groups there are, the more electron donation there is and the more stable the carbocation.

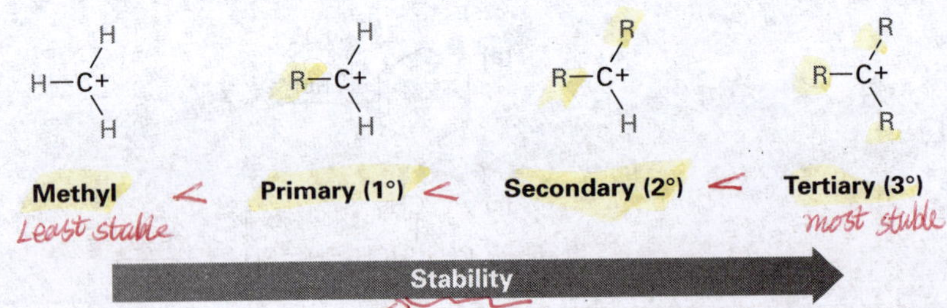

With this knowledge, we can now explain Markovnikov's rule. In the reaction of 1-methylcyclohexene with HBr, for instance, the intermediate carbocation might have either three alkyl substituents (a tertiary cation, 3°) or two alkyl substituents (a secondary cation, 2°). Because the tertiary cation is more stable than the secondary one, it's the tertiary cation that forms as the reaction intermediate, thus leading to the observed tertiary alkyl bromide product.

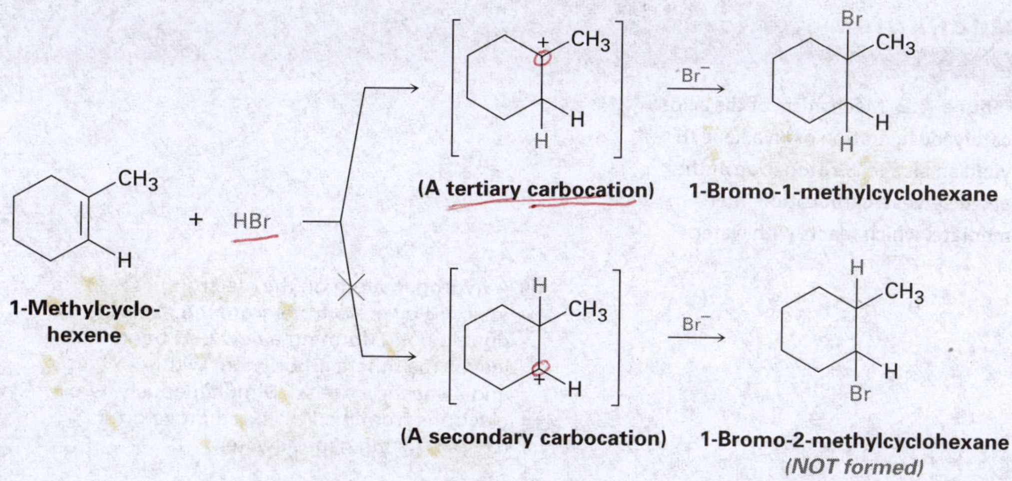

Problem 4.3 Show the structures of the carbocation intermediates you would expect in the following reactions:

(a)

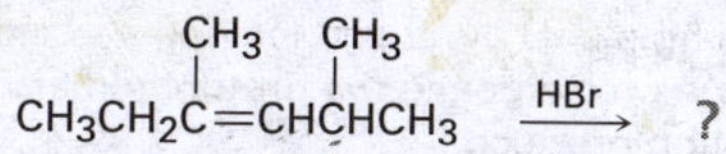

(b)

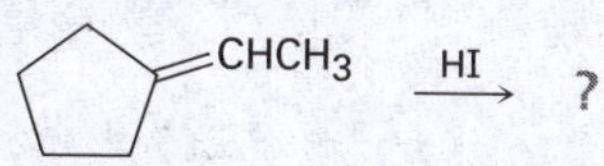

4.3 Addition of Water to Alkenes

Just as HCl and HBr add to alkenes to yield alkyl halides, H_2O adds to alkenes to yield alcohols, ROH, a process called *hydration*. Industrially, more than 300,000 tons of ethanol are produced each year in the United States by this method.

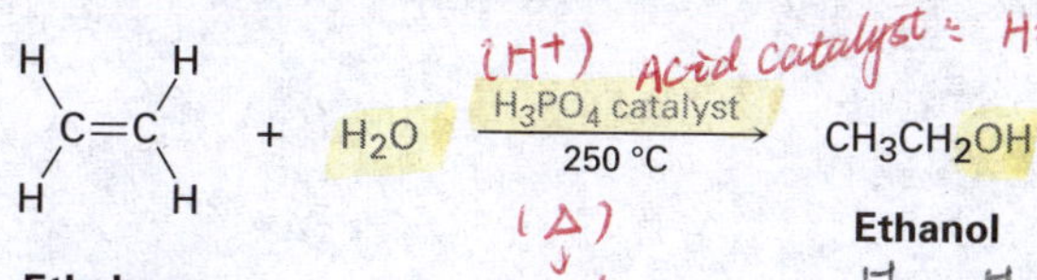

Hydration takes place on treatment of the alkene with water and a strong acid catalyst by a mechanism similar to that of HX addition. Thus, protonation of the alkene double bond yields a carbocation intermediate, which reacts with water as nucleophile to yield a protonated alcohol (ROH_2^+). Loss of H^+ from this protonated alcohol then gives the neutral alcohol and regenerates the acid catalyst (Figure 4.2). The addition of water to an unsymmetrical alkene follows Markovnikov's rule just as addition of HX does, giving the more highly substituted alcohol as product.

MECHANISM

Figure 4.2 Mechanism of the acid-catalyzed hydration of an alkene to yield an alcohol. Protonation of the alkene gives a carbocation intermediate, which reacts with water.

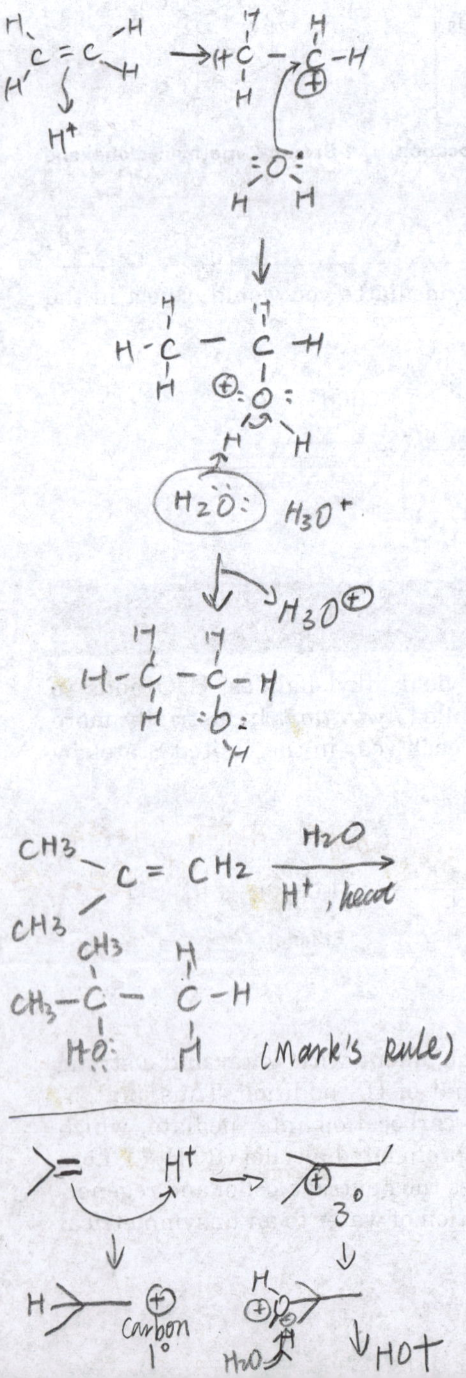

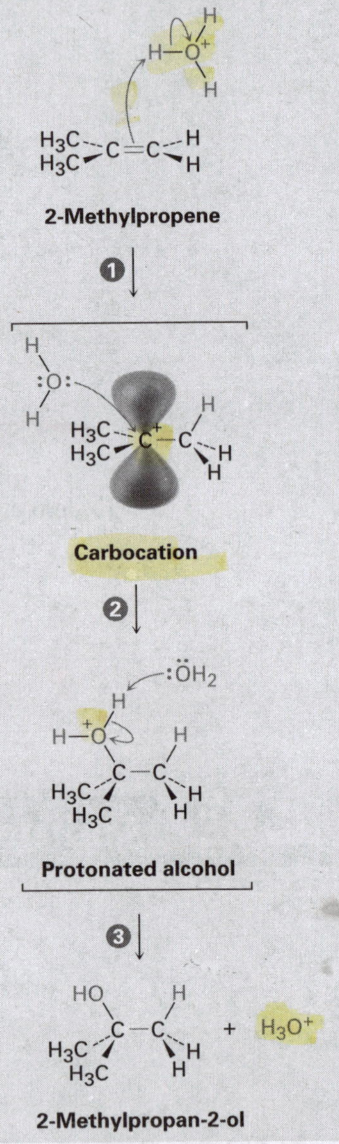

❶ A hydrogen atom on the electrophile H_3O^+ is attacked by π electrons from the nucleophilic double bond, forming a new C–H bond. This leaves the other carbon atom with a + charge and a vacant *p* orbital. Simultaneously, two electrons from the H–O bond move onto oxygen, giving neutral water.

❷ The nucleophile H_2O donates an electron pair to the positively charged carbon atom, forming a C–O bond and leaving a positive charge on oxygen in the protonated alcohol addition product.

❸ Water acts as a base to remove H^+, regenerating H_3O^+ and yielding the neutral alcohol addition product.

Unfortunately, the reaction conditions required for hydration are severe: the hydration of ethylene to produce ethanol, for instance, requires a phosphoric acid catalyst and reaction temperatures of up to 250 °C. As a result, sensitive molecules are sometimes destroyed. To get around this difficulty, chemists have devised several alternative methods of alkene hydration that take place under nonacidic conditions at room temperature, but we'll not discuss them here.

Hydration of carbon–carbon double bonds also occurs in various biological pathways, although not by the carbocation mechanism. Instead, biological hydrations usually require that the double bond be adjacent to a carbonyl group (C=O) for reaction to proceed. Fumarate, for instance, is hydrated to give malate as one step in the citric acid cycle of food metabolism. We'll see the function of the nearby carbonyl group in Section 9.10 but might note for now that the reaction occurs through a mechanism that involves formation of an anion intermediate followed by protonation by an acid HA.

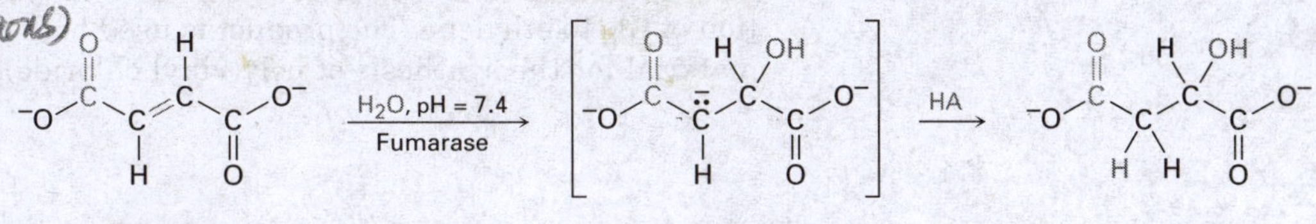

Fumarate **Anion intermediate** **Malate**

Worked Example 4.2 Predicting the Product of an Alkene Hydration Reaction

What product would you expect from acid-catalyzed addition of water to methylenecyclopentane?

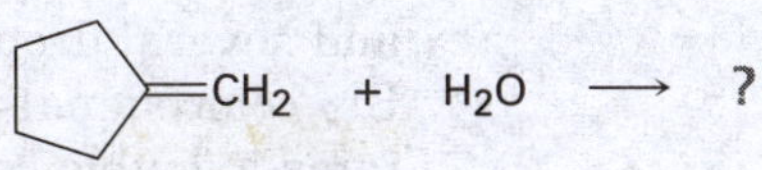

Methylenecyclopentane

Strategy According to Markovnikov's rule, H^+ adds to the carbon that already has more hydrogens (the $=CH_2$ carbon) and OH adds to the carbon that has fewer hydrogens (the ring carbon). Thus, the product will be a tertiary alcohol.

Solution

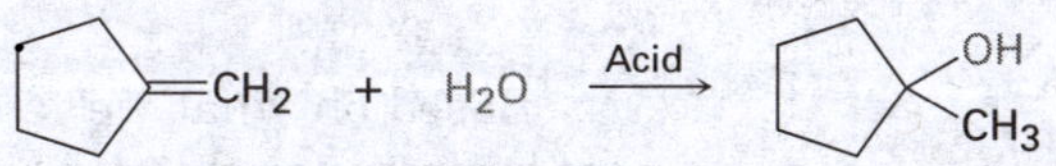

Problem 4.4 What product would you expect to obtain from the acid-catalyzed addition of water to the following alkenes?

(a)

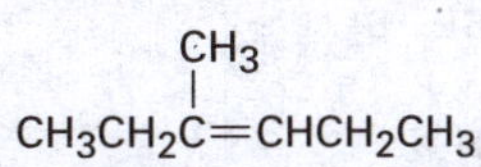

(b) 1-Methylcyclopentene **(c)** 2,5-Dimethylhept-2-ene

Problem 4.5 What alkenes might the following alcohols be made from?

(a)

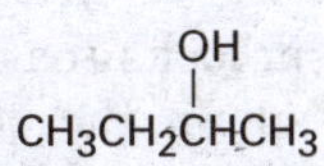

(b)

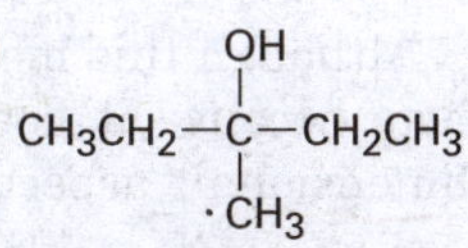

(c)

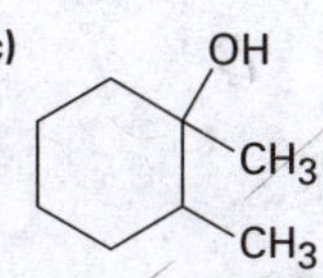

4.4 Addition of Halogens to Alkenes

Many other substances besides HX and H_2O add to alkenes. Bromine and chlorine, for instance, add readily to yield 1,2-dihaloalkanes, a process called *halogenation.* More than 10 million tons of 1,2-dichloroethane (also called ethylene dichloride) are synthesized each year in the United States by addition of Cl_2 to ethylene. The product is used both as a solvent and as a starting material for the synthesis of poly(vinyl chloride), PVC.

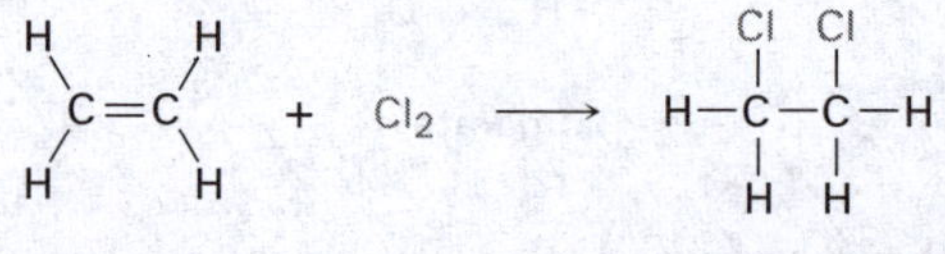

Ethylene **1,2-Dichloroethane (ethylene dichloride)**

Addition of Br_2 also acts as a simple and rapid laboratory test for unsaturation. A sample of unknown structure is dissolved in dichloromethane, CH_2Cl_2, and several drops of Br_2 are added. Immediate disappearance of the reddish Br_2 color signals a positive test and indicates that the sample molecule contains a double bond.

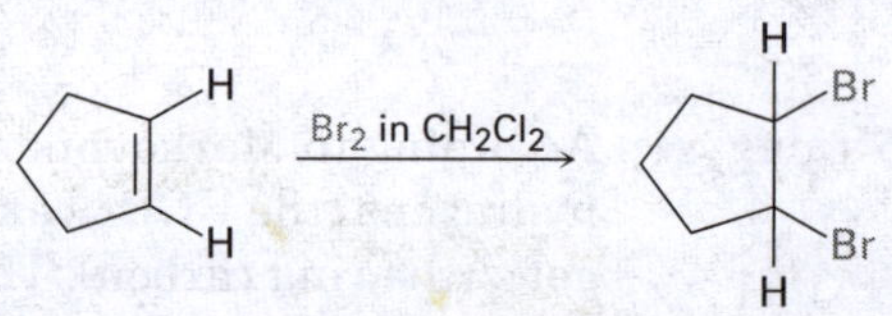

Cyclopentene **1,2-Dibromocyclopentane (95%)**

Based on what we've seen thus far, a possible mechanism for the reaction of bromine (or chlorine) with an alkene might involve electrophilic addition of Br^+ to the alkene, giving a carbocation that could undergo further reaction with Br^- to yield the dibromo addition product.

Possible mechanism?

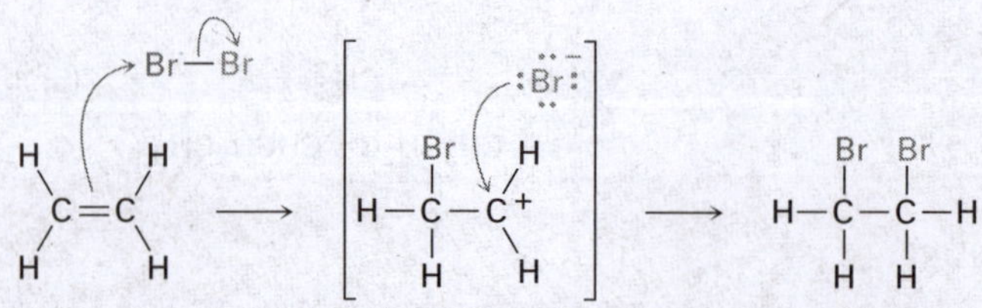

Possible mechanism?

Although this mechanism looks reasonable, it's not consistent with known facts because it doesn't explain the stereochemical (Section 2.8), or three-dimensional, aspects of halogen addition. That is, the mechanism doesn't explain what product stereoisomers are formed. When the halogenation reaction is carried out on a cycloalkene, such as cyclopentene, only *trans*-1,2-dibromocyclopentane is formed rather than the mixture of cis and trans products that might have been expected if a planar carbocation intermediate were involved. We say that the reaction occurs with **anti stereochemistry**, meaning that the two bromine atoms come from opposite faces of the double bond approximately 180° apart.

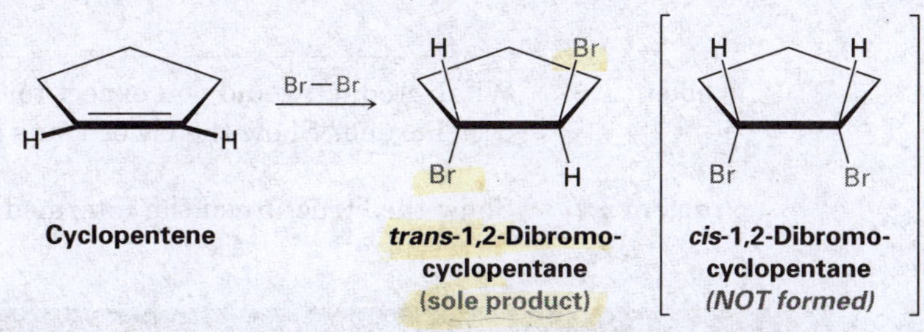

The stereochemical result is best explained by imagining that the reaction intermediate is not a true carbocation but is instead a *bromonium ion,* R_2Br^+, formed in a single step by addition of Br^+ to the double bond. Since the bromine atom effectively "shields" one side of the molecule, reaction with Br^- ion in the second step occurs from the opposite, more accessible side to give the anti product (Figure 4.3).

Figure 4.3 **Mechanism of the addition of Br_2 to an alkene. A bromonium ion intermediate is formed, shielding one face of the double bond and resulting in trans stereochemistry for the addition product.**

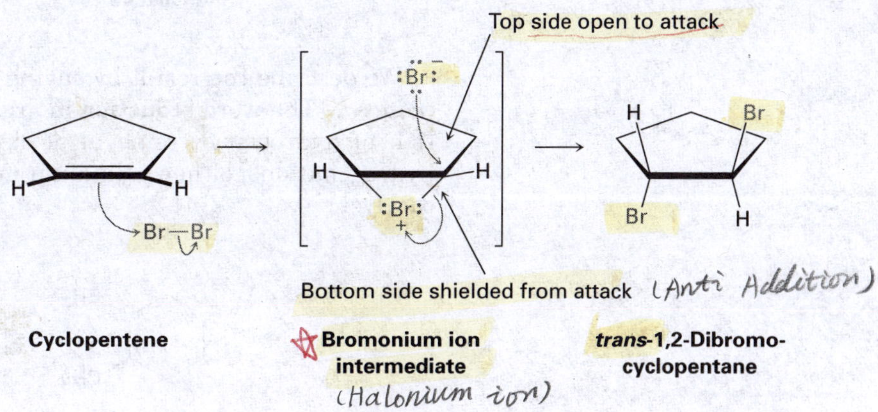

The addition of halogens to carbon–carbon double bonds also occurs in nature just as it does in the laboratory but is limited primarily to marine organisms, which live in a halide-rich environment. The biological halogenation reactions are carried out by enzymes called *haloperoxidases,* which use H_2O_2 to oxidize Br^- or Cl^- ions to a biological equivalent of Br^+ or Cl^+. Electrophilic addition to the double bond of a substrate molecule then yields a bromonium or chloronium ion intermediate just as in the laboratory, and reaction with another halide ion completes the process. Halomon, for example, an anticancer pentahalide isolated from red alga, is thought to arise by a route that involves twofold addition of BrCl through the corresponding bromonium ions.

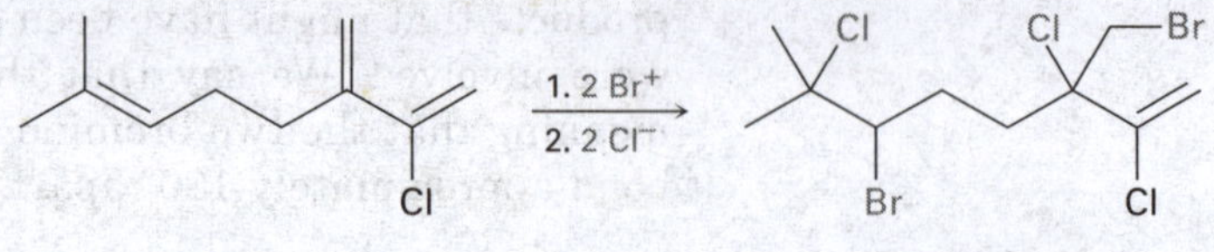

Halomon

Problem 4.6 What product would you expect to obtain from addition of Br_2 to 1,2-dimethylcyclohexene? Show the cis or trans stereochemistry of the product.

Problem 4.7 Show the structure of the intermediate bromonium ion formed in Problem 4.6.

4.5 Reduction of Alkenes: Hydrogenation

Addition of H_2 to the C=C bond occurs when an alkene is exposed to an atmosphere of hydrogen gas in the presence of a metal catalyst to yield an alkane.

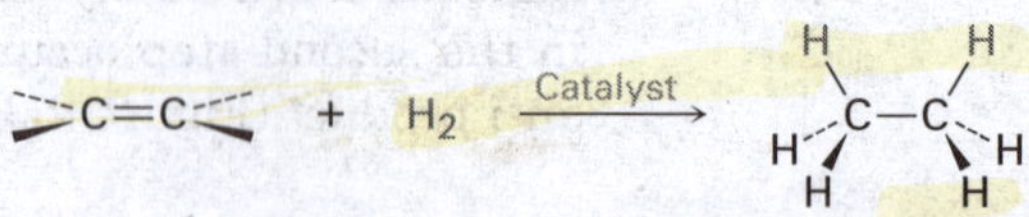

An alkene An alkane

We describe the result by saying that the double bond is **hydrogenated**, or *reduced.* (The word **reduction** in organic chemistry usually refers to the addition of hydrogen or removal of oxygen from a molecule.) For most alkene hydrogenations, either palladium metal or platinum (as PtO_2) is used as the catalyst.

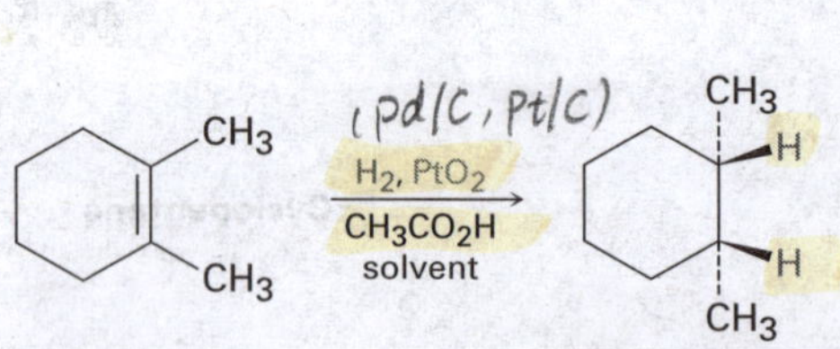

1,2-Dimethylcyclohexene *cis*-1,2-Dimethylcyclohexane (82%)

cis addition

Catalytic hydrogenation of alkenes, unlike most other organic reactions, is a *heterogeneous* process, rather than a homogeneous one. That is, the hydrogenation reaction occurs on the surface of solid catalyst particles rather than in solution. Following initial adsorption of H_2 onto the catalyst surface, complexation between catalyst and alkene then occurs as a vacant orbital on the metal interacts with the filled alkene π orbital. Next, hydrogen is inserted into the double bond, and the saturated product diffuses away from the catalyst (Figure

4.4). The reaction occurs with **syn stereochemistry** (the opposite of *anti*), meaning that both hydrogens add to the double bond from the same side.

MECHANISM

Figure 4.4 Mechanism of alkene hydrogenation. The reaction takes place with syn stereochemistry on the surface of insoluble catalyst particles.

1. Molecular hydrogen adsorbs to the catalyst surface and dissociates into hydrogen atoms.
2. The alkene adsorbs to the catalyst surface, using its π bond to complex to the metal atoms.
3. A hydrogen atom is transferred from the metal to one of the alkene carbon atoms, forming a partially reduced intermediate with a C–H bond and carbon–metal σ bond.
4. A second hydrogen is transferred from the metal to the second carbon, giving the alkane product and regenerating the catalyst. Because both hydrogens are transferred to the same face of the alkene, the reduction has syn stereochemistry.

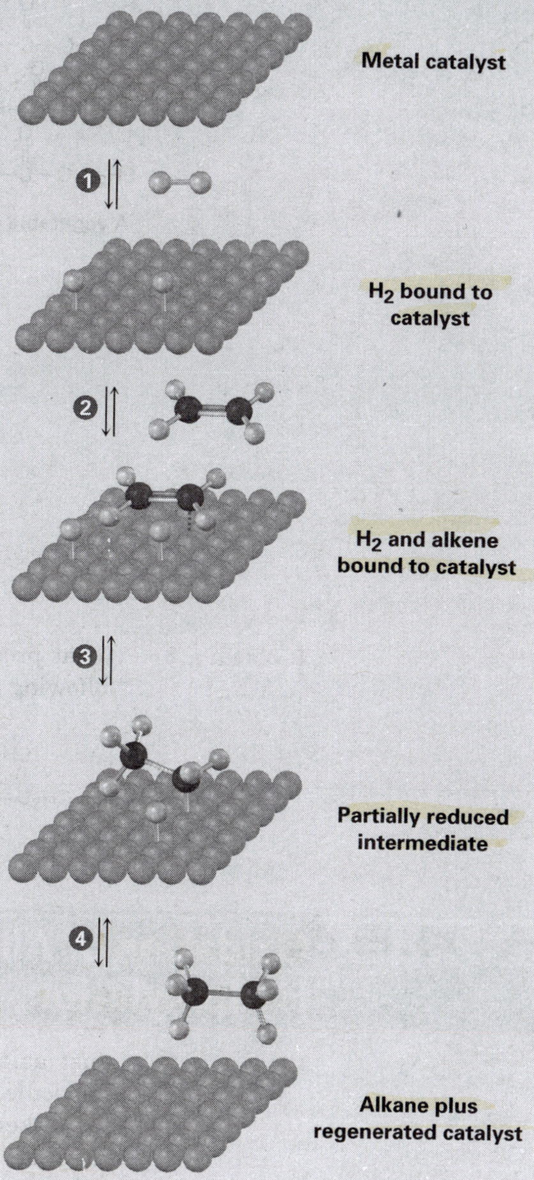

In addition to its usefulness in the laboratory, catalytic hydrogenation is also important in the food industry, where unsaturated vegetable oils are reduced to produce the saturated fats used in margarine and cooking products. As we'll see in Section 16.1, vegetable oils are triesters of glycerol, $HOCH_2CH(OH)CH_2OH$, with three long-chain carboxylic acids called *fatty acids.* The fatty acids are generally polyunsaturated, and their double bonds have cis stereochemistry. Complete hydrogenation yields the corresponding saturated fatty acids, but incomplete hydrogenation often results in partial cis–trans isomerization of a remaining double bond. When eaten and digested, the free trans fatty acids are released, raising blood cholesterol levels and contributing to potential coronary problems.

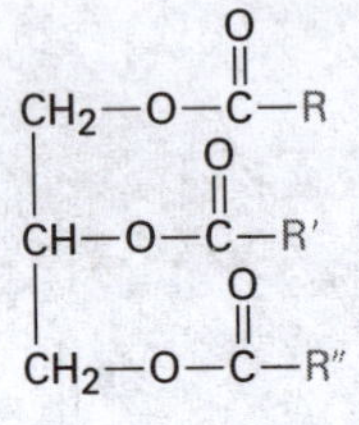

A vegetable oil

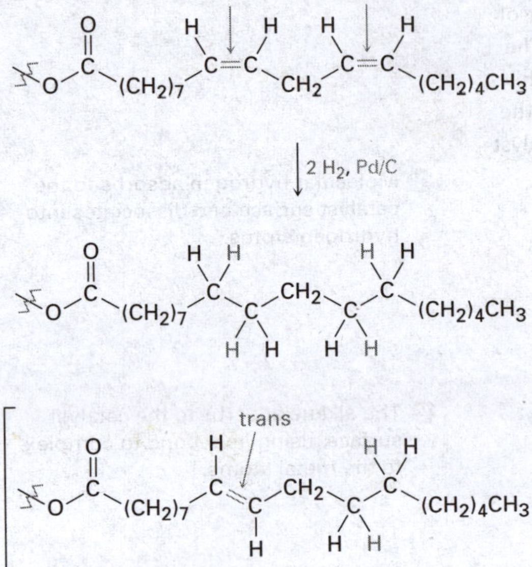

A polyunsaturated fatty acid in vegetable oil

A saturated fatty acid in margarine

A trans fatty acid

Problem 4.8 What product would you expect to obtain from catalytic hydrogenation of the following alkenes?

(a) 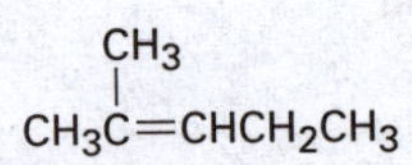

(b) 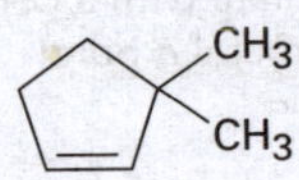

4.6 Oxidation of Alkenes: Epoxidation, Hydroxylation, and Cleavage

Just as the word *reduction* usually refers to the addition of *hydrogen* to a molecule, the word **oxidation** usually means the addition of *oxygen*. For example, alkenes are oxidized to give *epoxides* on treatment with a peroxyacid, RCO_3H, such as *meta*-chloroperoxybenzoic acid. An **epoxide**, also called an *oxirane*, is a cyclic ether with an oxygen atom in a three-membered ring.

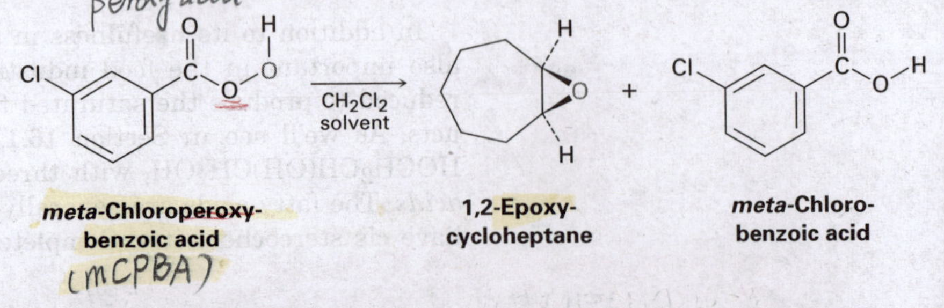

Cycloheptene | *meta*-Chloroperoxybenzoic acid | 1,2-Epoxycycloheptane | *meta*-Chlorobenzoic acid

Epoxides undergo an acid-catalyzed ring-opening reaction with water (a *hydrolysis*) to give the corresponding dialcohol, or *diol*, also called a *glycol*. The net result of the two-step alkene epoxidation/hydrolysis is thus a **hydroxylation**—the addition of an –OH group to each of the two double-bond carbons. In fact, more than 3 million tons of ethylene glycol, $HOCH_2CH_2OH$, most of it used for automobile antifreeze, are produced each year in the United States by epoxidation of ethylene followed by hydrolysis.

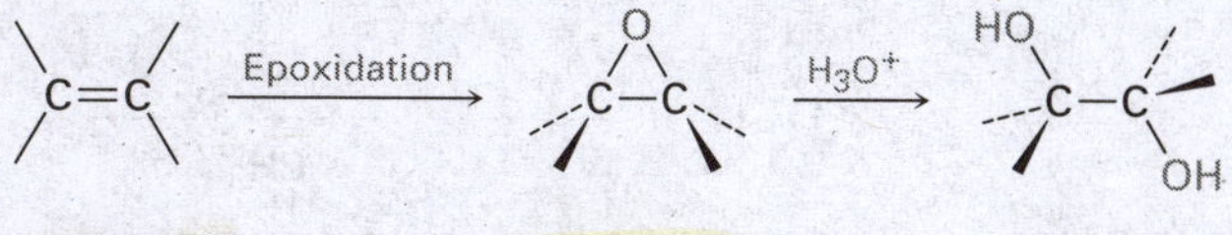

An alkene **An epoxide** **A 1,2-diol**

Acid-catalyzed epoxide opening takes place by protonation of the epoxide to increase its reactivity, followed by nucleophilic addition of water. This nucleophilic addition is analogous to the final step of alkene bromination that we saw in Section 4.4, in which a cyclic bromonium ion was opened by a reaction with bromide ion. As a result, a *trans*-1,2-diol results when an epoxycycloalkane is opened by aqueous acid, just as a *trans*-1,2-dibromide results when a cycloalkene is brominated.

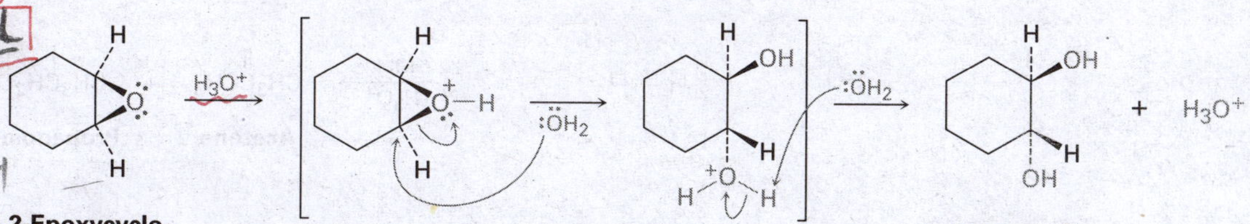

1,2-Epoxycyclohexane

***trans*-Cyclohexane-1,2-diol (86%)**

The hydroxylation of an alkene can also be carried out in a single step by reaction of the alkene with potassium permanganate, $KMnO_4$, in basic solution. For example, cyclohexene gives *cis*-cyclohexane-1,2-diol.

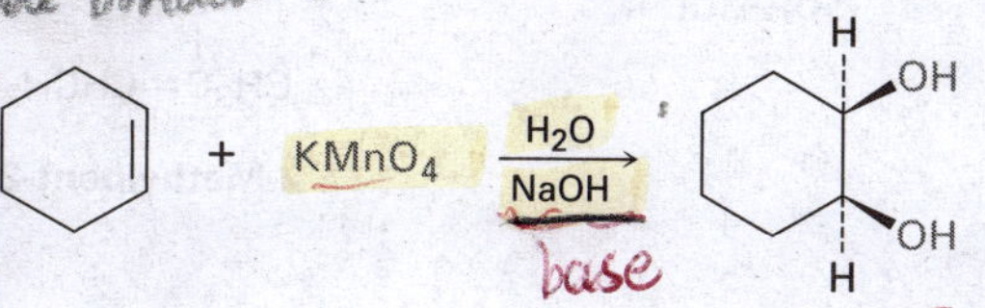

Cyclohexene ***cis*-Cyclohexane-1,2-diol (37%)**

When oxidation of the alkene is carried out with $KMnO_4$ in acidic rather than basic solution, *cleavage* of the double bond occurs and carbonyl-containing products are obtained. If the double bond is tetrasubstituted, the two carbonyl-containing products are ketones; if a hydrogen is present on the double bond, one of the carbonyl-containing products is a carboxylic acid; and if two hydrogens are present on one carbon, CO_2 is formed.

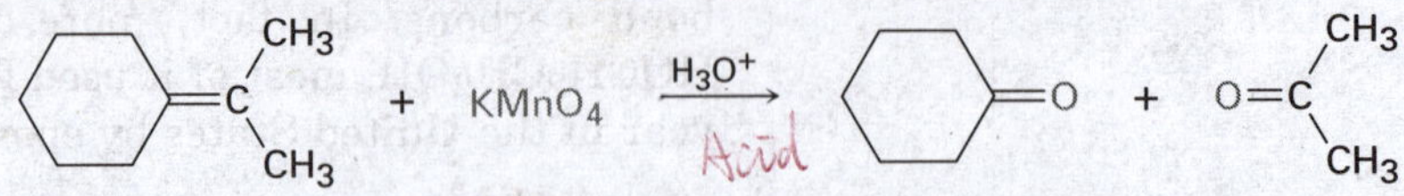

Isopropylidenecyclohexane → Cyclohexanone + Acetone (two ketones)

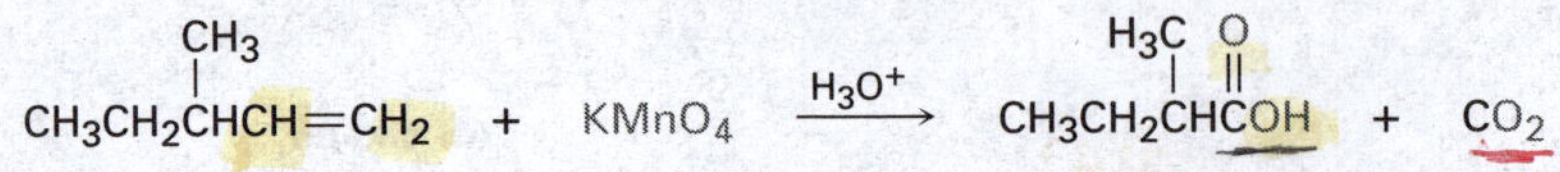

3-Methylpent-1-ene → 2-Methylbutanoic acid (45%)

Worked Example 4.3 Predicting the Structure of a Reactant Given the Products

What alkene gives a mixture of acetone and propanoic acid on reaction with acidic $KMnO_4$?

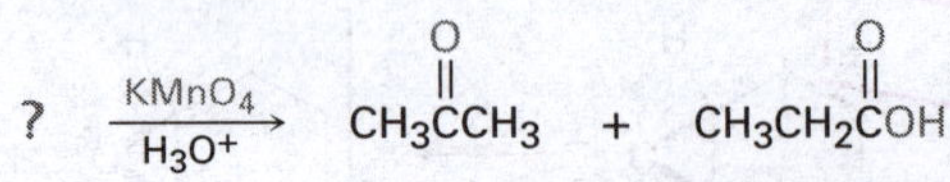

Acetone + Propanoic acid

Strategy When solving a problem that asks how to prepare a given product, *always work backward.* Look at the product, identify the functional group(s) it contains, and ask yourself, "How can I prepare that functional group?" In the present instance, the products are a ketone and a carboxylic acid, which can be prepared by reaction of an alkene with acidic $KMnO_4$. To find the starting alkene that gives the cleavage products shown, remove the oxygen atoms from the two products, join the fragments with a double bond, and replace the –OH by –H.

Solution

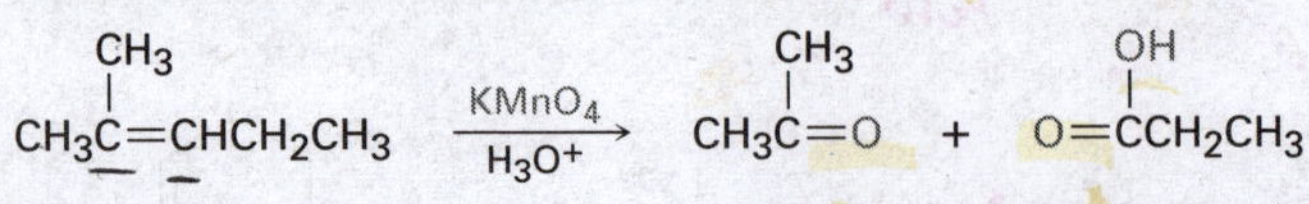

2-Methylpent-2-ene → Acetone + Propanoic acid

Problem 4.9 Predict the product of the reaction of 1,2-dimethylcyclohexene with the following:

(a) $KMnO_4$, H_3O^+ **(b)** $KMnO_4$, OH^-, H_2O

Problem 4.10 Propose structures for alkenes that yield the following products on treatment with acidic $KMnO_4$:

(a) $(CH_3)_2C{=}O + CO_2$ **(b)** 2 equiv $CH_3CH_2CO_2H$

4.7 Addition of Radicals to Alkenes: Polymers

No other group of synthetic chemicals has had as great an impact on our day-to-day lives as *polymers.* From carpeting to clothing to foam coffee cups, we are literally surrounded by polymers.

A **polymer** is a large—sometimes *very* large—molecule built up by repetitive bonding together of many smaller molecules, called **monomers.** Nature makes wide use of biological polymers. Cellulose, for example, is a polymer built of repeating sugar monomers; proteins are polymers built of repeating amino acid monomers; and nucleic acids are polymers built of repeating nucleotide monomers.

Cellulose—a glucose polymer

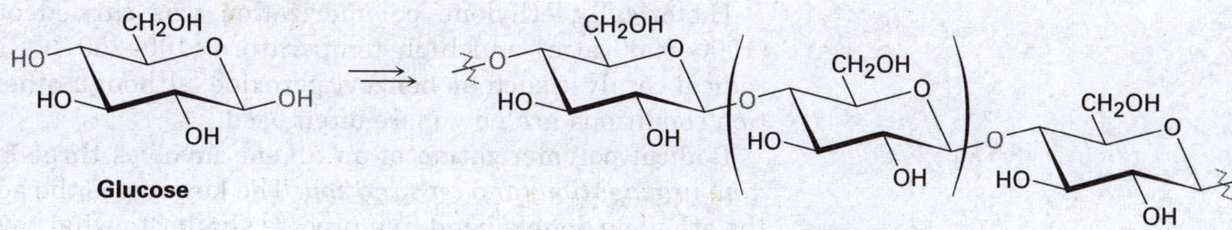

Glucose

Cellulose

Protein—an amino acid polymer

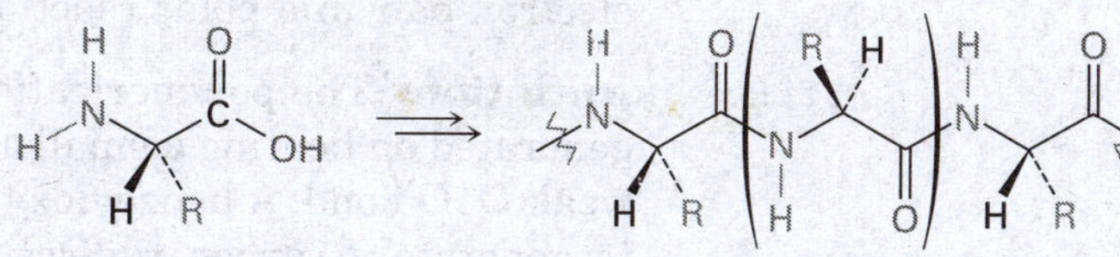

An amino acid

A protein

Nucleic acid—a nucleotide polymer

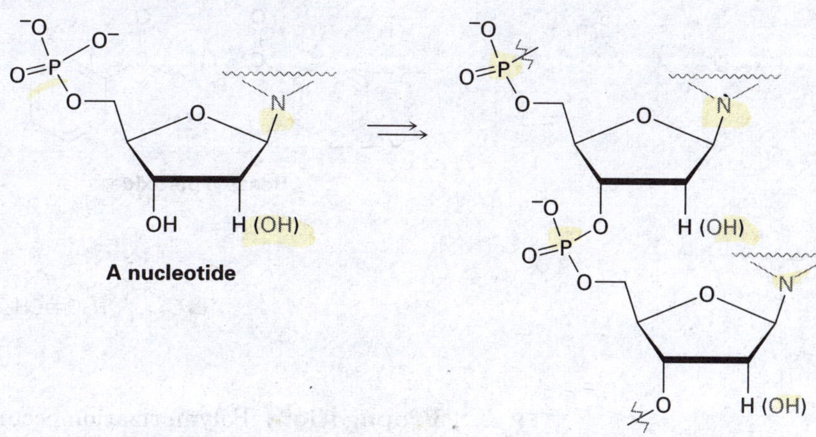

A nucleotide

A nucleic acid

The simplest synthetic polymers are those that result when an alkene is treated with a small amount of a suitable polymerization catalyst. Ethylene, for example, yields polyethylene, an enormous alkane that may have up to

200,000 monomer units incorporated into a gigantic hydrocarbon chain. Approximately 19 million tons per year of polyethylene are manufactured in the United States alone.

Polyethylene—a synthetic alkene polymer

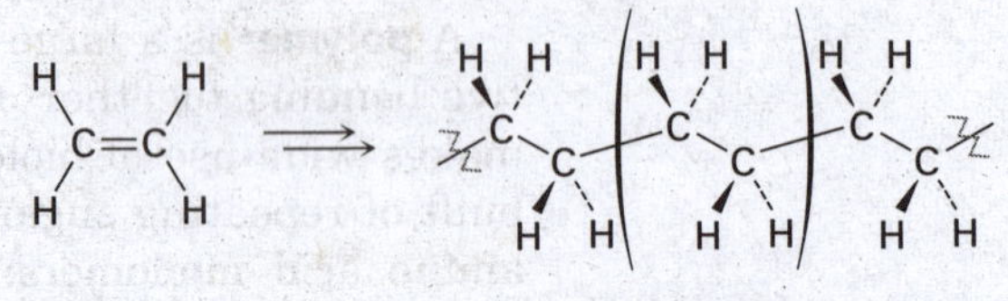

Ethylene **Polyethylene**

Historically, ethylene polymerization was carried out at high pressure (1000–3000 atm) and high temperature (100–250 °C) in the presence of a radical catalyst such as benzoyl peroxide, although other catalysts and reaction conditions are now more often used.

Radical polymerization of an alkene involves three kinds of steps: *initiation, propagation,* and *termination.* The key step is the addition of a radical to the ethylene double bond in a process similar to what takes place in the addition of an electrophile to an alkene (Section 3.7). In writing the mechanism, a curved half-arrow, or "fishhook," is used to show the movement of a single electron, as opposed to the full curved arrow used to show the movement of an electron pair in a polar reaction.

STEP 1 **Initiation:** The polymerization reaction is initiated when a few radicals are generated on heating a small amount of benzoyl peroxide catalyst to break the weak O–O bond. A benzoyloxy radical then adds to the C=C bond of ethylene to generate a carbon radical. One electron from the carbon–carbon double bond pairs up with the odd electron on the benzoyloxy radical to form a C–O bond, and the other electron remains on carbon.

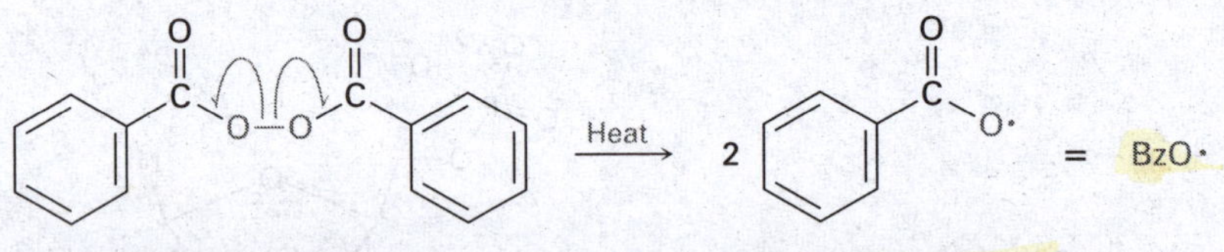

Benzoyl peroxide **Benzoyloxy radical**

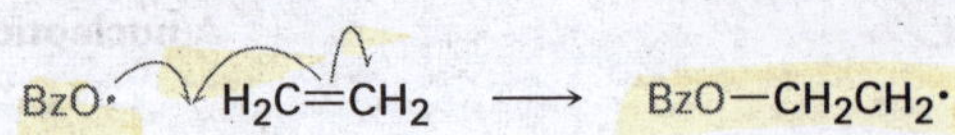

STEP 2 **Propagation:** Polymerization occurs when the carbon radical formed in the initiation step adds to another ethylene molecule to yield another radical. Repetition of the process for hundreds or thousands of times builds the polymer chain.

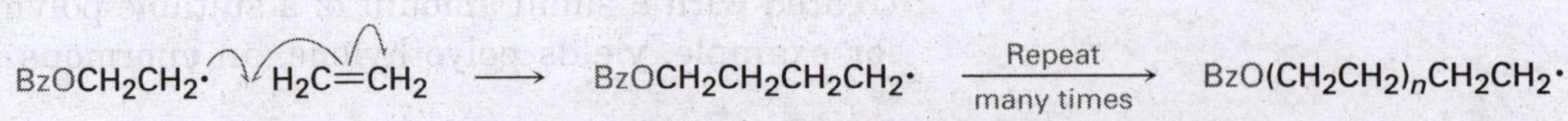

STEP 3 **Termination:** The polymerization process is eventually ended by a reaction that consumes the radical. Combination of two growing chains is one possible chain-terminating reaction.

$$2\ R{-}CH_2CH_2\cdot \longrightarrow R{-}CH_2CH_2CH_2CH_2{-}R$$

Ethylene is not unique in its ability to form a polymer. Many substituted ethylenes, called *vinyl monomers,* undergo polymerization, yielding polymers with substituent groups regularly spaced along the polymer chain. Propylene, for example, yields polypropylene.

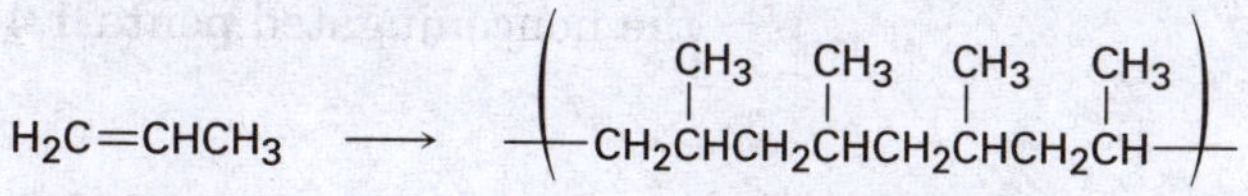

Propylene **Polypropylene**

Table 4.1 shows some commercially important vinyl monomers and lists some industrial uses of the different polymers that result.

Table 4.1 Some Alkene Polymers and Their Uses

Monomer	Formula	Trade or common name of polymer	Uses
Ethylene	$H_2C{=}CH_2$	Polyethylene	Packaging, bottles
Propene (propylene)	$H_2C{=}CHCH_3$	Polypropylene	Moldings, rope, carpets
Chloroethylene (vinyl chloride)	$H_2C{=}CHCl$	Poly(vinyl chloride) Tedlar	Insulation, films, pipes
Styrene	$H_2C{=}CHC_6H_5$	Polystyrene	Foam, moldings
Tetrafluoroethylene	$F_2C{=}CF_2$	Teflon	Gaskets, nonstick coatings
Acrylonitrile	$H_2C{=}CHCN$	Orlon, Acrilan	Fibers
Methyl methacrylate	CH_3 $H_2C{=}CCO_2CH_3$	Plexiglas, Lucite	Paint, sheets, moldings
Vinyl acetate	$H_2C{=}CHOCOCH_3$	Poly(vinyl acetate)	Paint, adhesives, foams

Worked Example 4.4 **Predicting the Structure of a Polymer**

Show the structure of poly(vinyl chloride), a polymer made from $H_2C{=}CHCl$, by drawing several repeating units.

Strategy Imagine breaking the carbon–carbon double bond in the monomer unit, and then form single bonds by connecting numerous units together.

Solution The general structure of poly(vinyl chloride) is

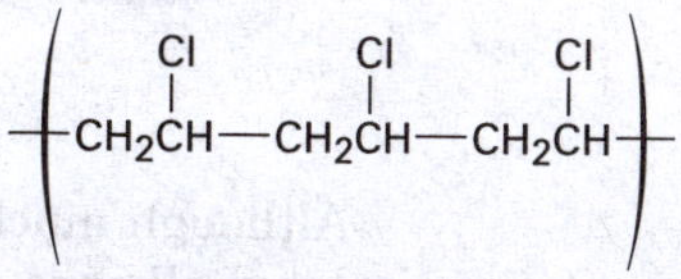

Problem 4.11 Show the structure of Teflon by drawing several repeating units. The monomer unit is tetrafluoroethylene, $F_2C{=}CF_2$.

4.8 Conjugated Dienes

The unsaturated compounds we've looked at thus far have had only one double bond, but many compounds have numerous sites of unsaturation. If the different unsaturations are well separated in a molecule, they react independently, but if they're close together, they may interact with one another. In particular, compounds that have alternating single and double bonds—so-called **conjugated** compounds—have some distinctive characteristics. The conjugated diene buta-1,3-diene, for instance, behaves quite differently from the nonconjugated penta-1,4-diene in some respects.

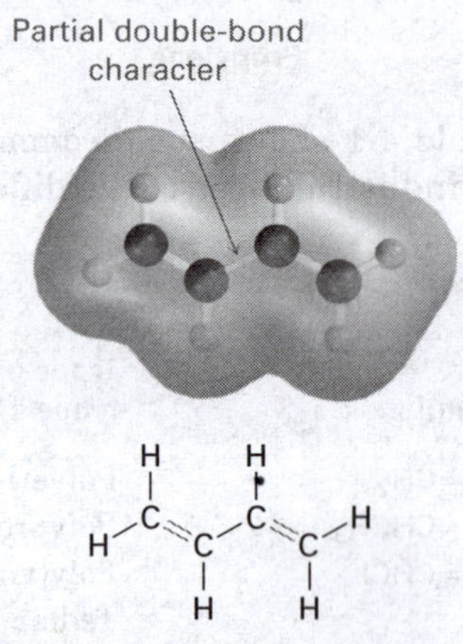

Buta-1,3-diene (conjugated)

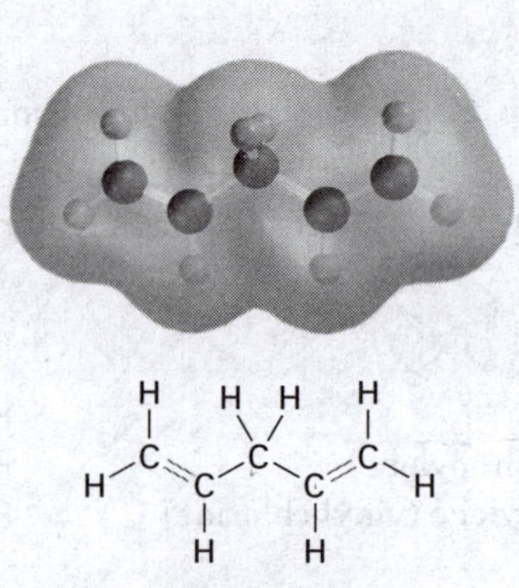

Penta-1,4-diene (nonconjugated)

What's so special about conjugated dienes that we need to look at them separately? The orbital view of buta-1,3-diene shown in Figure 4.5 provides a clue to the answer. *There is an electronic interaction between the two double bonds of a conjugated diene* because of *p* orbital overlap across the central single bond. This interaction of *p* orbitals across a single bond gives conjugated dienes some unusual properties.

Figure 4.5 An orbital view of buta-1,3-diene. Each of the four carbon atoms has a *p* orbital, allowing for an electronic interaction across the C2–C3 single bond.

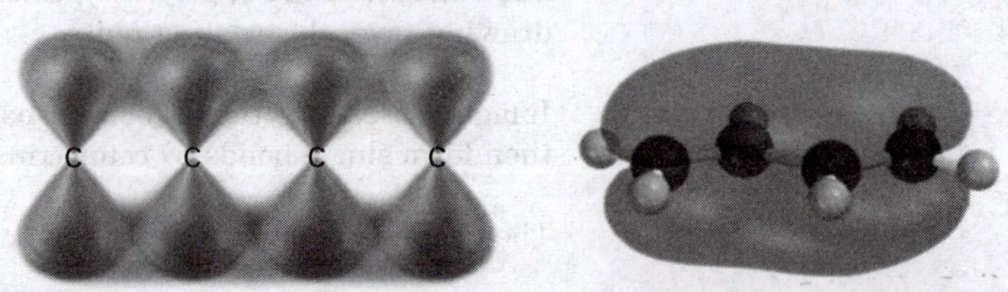

Although much of the chemistry of conjugated dienes is similar to that of isolated alkenes, there is a striking difference in their electrophilic addition reactions with HX and X_2. When HX adds to an isolated alkene, Markovnikov's rule usually predicts the formation of a single product, but when HX adds to a conjugated diene, mixtures of products are often obtained. Reaction of HBr with buta-1,3-diene, for instance, yields two products.

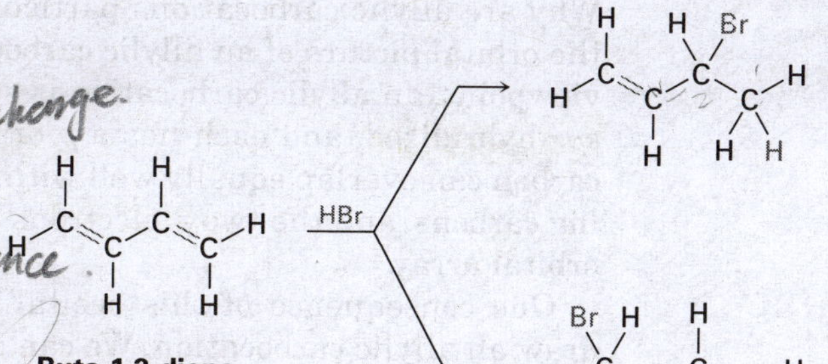

Buta-1,3-diene

3-Bromobut-1-ene (71%; 1,2-addition)

1-Bromobut-2-ene (29%; 1,4-addition)

3-Bromobut-1-ene is the typical Markovnikov product of **1,2-addition**, but 1-bromobut-2-ene appears unusual. The double bond in this product has moved to a position between C2 and C3, and HBr has added to C1 and C4, a result described as **1,4-addition**. In the same way, Br_2 adds to buta-1,3-diene to give a mixture of 3,4-dibromobut-1-ene and 1,4-dibromobut-2-ene.

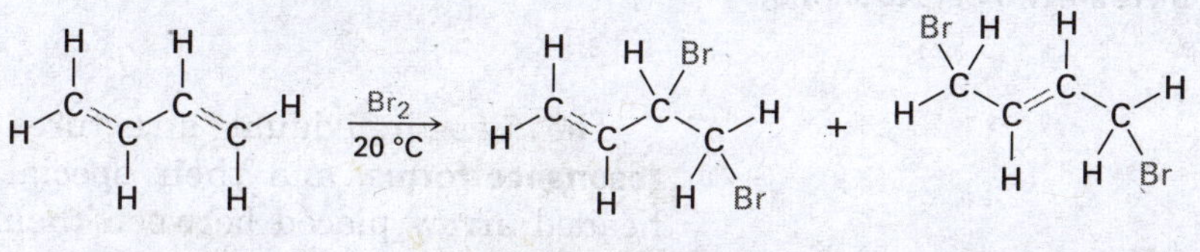

Buta-1,3-diene

3,4-Dibromobut-1-ene (55%; 1,2-addition)

1,4-Dibromobut-2-ene (45%; 1,4-addition)

How can we account for the formation of the 1,4-addition product? The answer is that an *allylic carbocation* is involved as an intermediate in the reaction, where the word **allylic** means "next to a double bond." When H^+ adds to an electron-rich π bond of buta-1,3-diene, two carbocation intermediates are possible—a primary nonallylic carbocation and a secondary allylic carbocation. Allylic carbocations are more stable and therefore form faster than less stable, nonallylic carbocations.

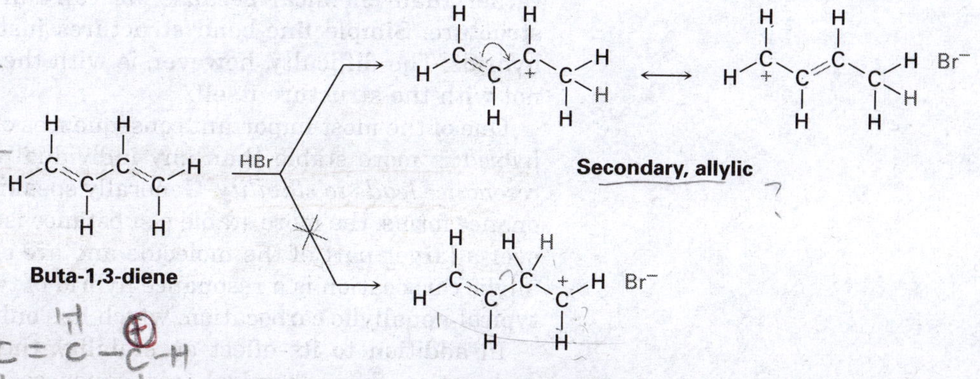

Buta-1,3-diene

Secondary, allylic

Primary, nonallylic (NOT formed)

4.9 Stability of Allylic Carbocations: Resonance

Why are allylic carbocations particularly stable? To see the answer, look at the orbital picture of an allylic carbocation in Figure 4.6. From an electronic viewpoint, an allylic carbocation is symmetrical. All three carbon atoms are sp^2-hybridized, and each has a p orbital. Thus, the p orbital on the central carbon can overlap equally well with p orbitals on *either* of the two neighboring carbons, and the two p electrons are free to move about the entire three-orbital array.

One consequence of this orbital picture is that there are two ways to draw an allylic carbocation. We can draw it with the vacant p orbital on the right and the double bond on the left, or we can draw it with the vacant p orbital on the left and the double bond on the right. *Neither structure is correct by itself; the true structure of the allylic carbocation is somewhere between the two.*

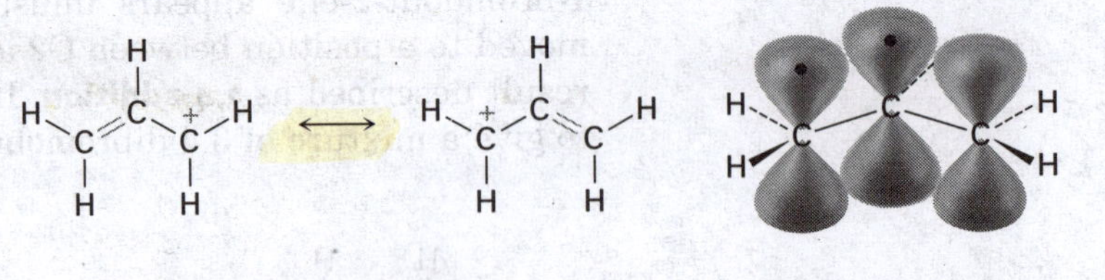

Figure 4.6 An orbital picture of an allylic carbocation. The vacant p orbital on the positively charged carbon can overlap the double-bond p orbitals. As a result, there are two ways to draw the structure.

The two individual structures of an allylic carbocation are called **resonance forms**, and their special relationship is indicated by a double-headed arrow placed between them. The only difference between the resonance forms is the position of the bonding electrons. The atoms themselves remain in exactly the same place in both resonance forms, the connections between atoms are the same, and the three-dimensional shapes of the resonance forms are the same.

A good way to think about resonance is to realize that a species like an allylic carbocation is no different from any other. An allylic carbocation doesn't jump back and forth between two resonance forms, spending part of its time looking like one and the rest of its time looking like the other. Rather, an allylic carbocation has a single, unchanging structure called a **resonance hybrid** that is a blend of the two individual forms and has characteristics of both. The only "problem" with the allylic carbocation is visual rather than chemical because we can't draw it using a single line-bond structure. Simple line-bond structures just don't work well for resonance hybrids. The difficulty, however, is with the *representation* of the structure, not with the structure itself.

One of the most important consequences of resonance is that the resonance hybrid is more stable than any individual resonance form. In other words, *resonance leads to stability.* Generally speaking, the larger the number of resonance forms, the more stable a substance is because electrons are spread out over a larger part of the molecule and are closer to more nuclei. Because an allylic carbocation is a resonance hybrid of two forms, it is more stable than a typical nonallylic carbocation, which has only one form.

In addition to its effect on stability, the resonance picture of an allylic carbocation has chemical consequences. When the allylic carbocation produced by protonation of buta-1,3-diene reacts with Br^- ion to complete the addition, reaction can occur at either C1 or C3, because both share the positive charge. The result is a mixture of 1,2- and 1,4-addition products.

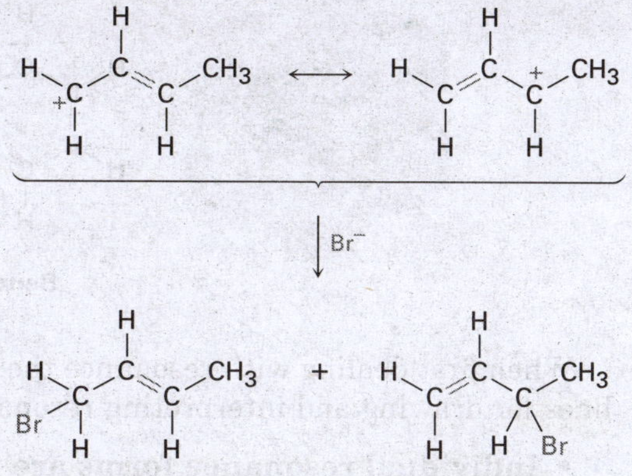

1,4-Addition (29%) 1,2-Addition (71%)

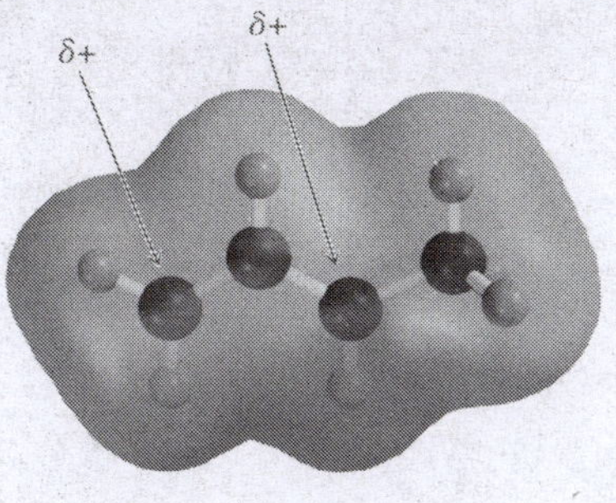

Problem 4.12 Buta-1,3-diene reacts with Br_2 to yield a mixture of 1,2- and 1,4-addition products. Show the structures of both.

4.10 Drawing and Interpreting Resonance Forms

Resonance is an extremely useful concept for explaining a variety of chemical phenomena. In the acetate ion, for instance, the lengths of the two C–O bonds are identical. Although there is no single line-bond structure that can account for this equivalence of C–O bonds, resonance theory accounts for it nicely. The acetate ion is simply a resonance hybrid of two resonance forms, with both oxygens sharing the π electrons and the negative charge equally.

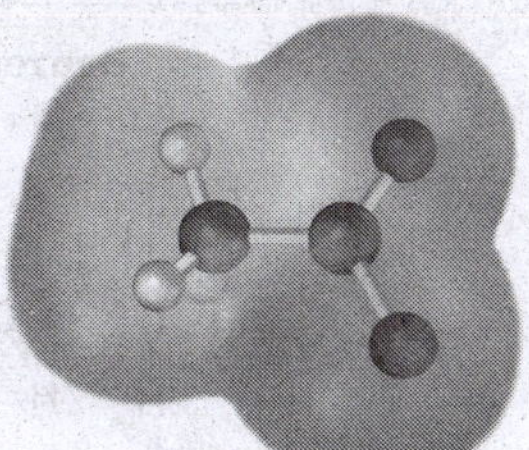

Acetate ion—two resonance forms

As another example, we'll see in the next chapter that the six carbon–carbon bonds in aromatic compounds like benzene are equivalent because benzene is a resonance hybrid of two forms. Each form has alternating single and double bonds, and neither form is correct by itself; the true benzene structure is a hybrid of the two forms.

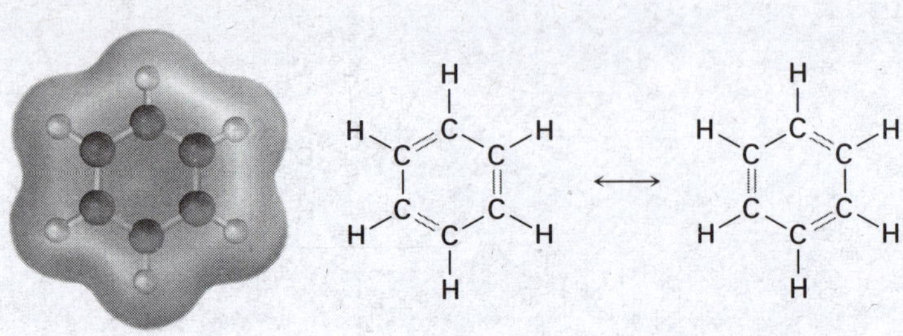

Benzene (two resonance forms)

When first dealing with resonance theory, it's useful to have a set of guidelines for drawing and interpreting resonance forms.

- **Individual resonance forms are imaginary, not real.** The real structure is a composite, or hybrid, of the different forms. Substances like the allylic carbocation, the acetate ion, and benzene are no different from any other: they have single, unchanging structures. The only difference between these and other substances is in the way they must be represented on paper.
- **Resonance forms differ only in the placement of their π or nonbonding electrons.** Neither the position nor the hybridization of any atom changes from one resonance form to another. In benzene, for example, the π electrons in the double bonds move, but the six carbon and six hydrogen atoms remain in the same place. By contrast, two structures such as penta-1,3-diene and penta-1,4-diene are *not* resonance structures because their hydrogen atoms don't occupy the same positions. Instead, the two dienes are constitutional isomers.

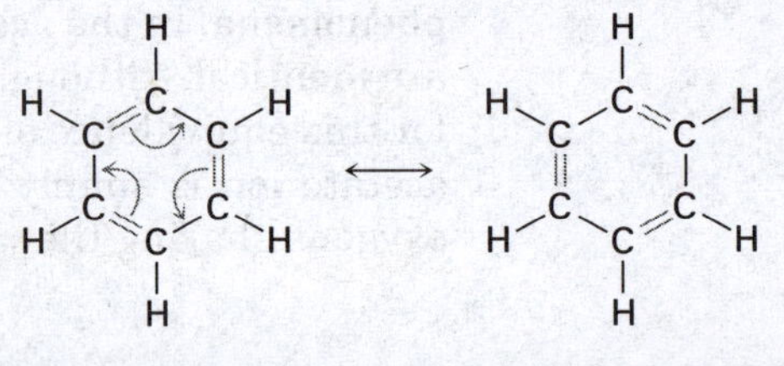

Benzene—two resonance forms

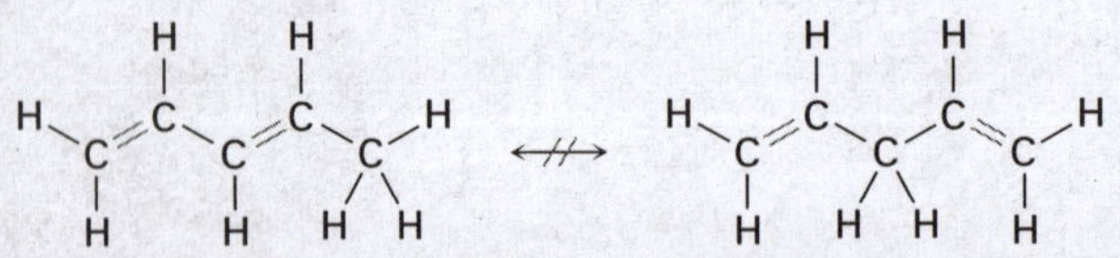

Penta-1,3-diene　　Penta-1,4-diene

Constitutional isomers

- **Different resonance forms of a substance don't have to be equivalent.** For example, the allylic carbocation obtained by reaction of buta-1,3-diene with H^+ is unsymmetrical. One end of the π electron system has a methyl substituent, and the other end is unsubstituted. Even though the two resonance forms aren't equivalent, both contribute to the overall resonance hybrid.

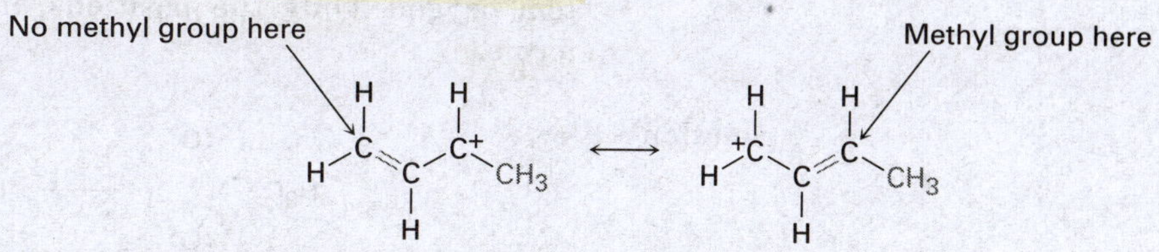

When two resonance forms are not equivalent, the actual structure of the resonance hybrid is closer to the more stable form than to the less stable form. Thus, we might expect the butenyl carbocation to look a bit more like a secondary carbocation than like a primary one.

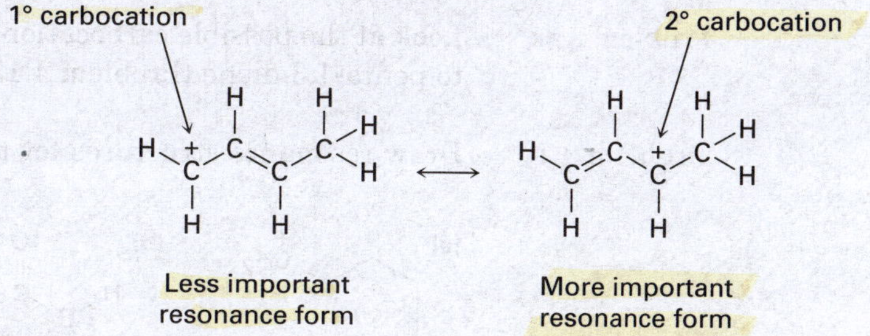

- **Resonance forms must be valid electron-dot structures and obey normal rules of valency.** A resonance form is like any other structure: the octet rule still applies. For example, one of the following structures for the acetate ion is not a valid resonance form because the carbon atom has five bonds and ten valence electrons.

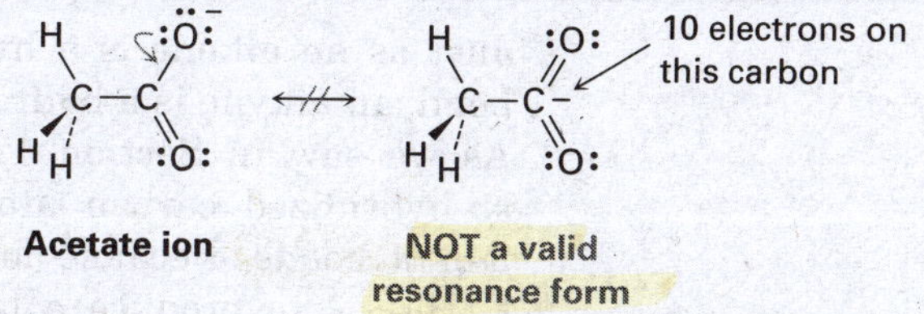

- **Resonance leads to stability.** The greater the number of resonance forms, the more stable the substance. We've already seen, for example, that an allylic carbocation is more stable than a nonallylic one. In the same way, we'll see in the next chapter that a benzene ring is more stable than a cyclic alkene.

Worked Example 4.5 Using Resonance Structures

Use resonance structures to explain why the two N–O bonds of nitromethane are equivalent.

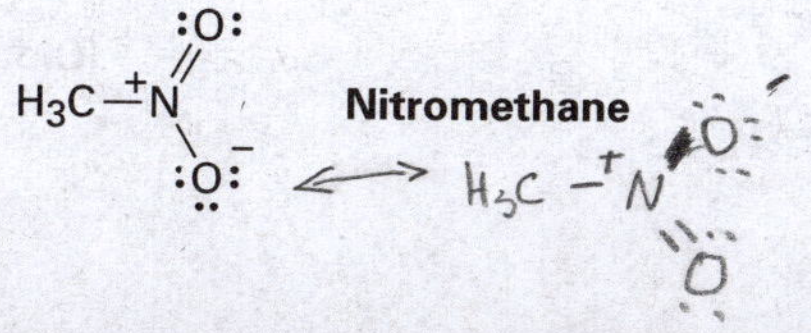

Strategy Resonance forms differ only in the placement of π (multiple-bond) and nonbonding electrons. Nitromethane has two equivalent resonance forms, which can be drawn by showing the double bond either to the top oxygen or to the bottom oxygen. Only the positions of the electrons are different in the two forms.

Solution

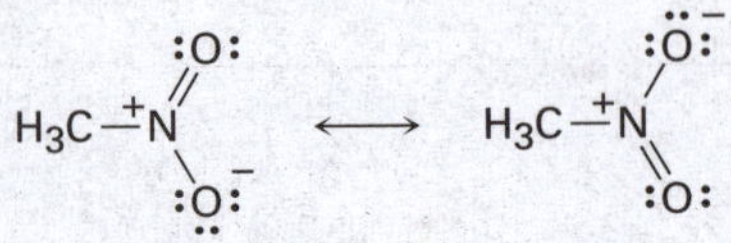

Nitromethane

Problem 4.13 Give the structure of all possible monoadducts of HCl and penta-1,3-diene, $CH_3CH{=}CH{-}CH{=}CH_2$.

Problem 4.14 Look at the possible carbocation intermediates produced during addition of HCl to penta-1,3-diene (Problem 4.13), and predict which is the most stable.

Problem 4.15 Draw resonance structures for the following species:

(a)

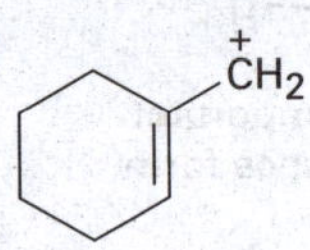

(b)

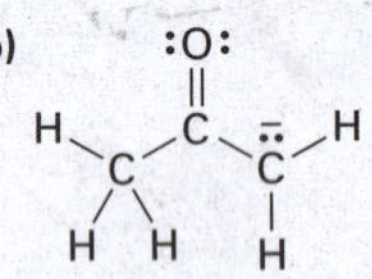

(c)

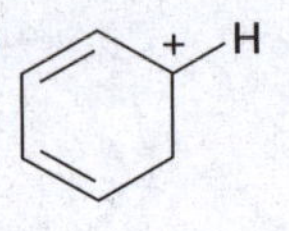

4.11 Alkynes and Their Reactions

Just as an alkene is a hydrocarbon that contains a carbon–carbon *double* bond, an alkyne is a hydrocarbon that contains a carbon–carbon *triple* bond. As we saw in Section 1.8, a C≡C bond results from the overlap of two *sp*-hybridized carbon atoms and consists of one *sp–sp* σ bond and two *p–p* π bonds. Because four hydrogens must be removed from an alkane, C_nH_{2n+2}, to produce a triple bond, the general formula for an alkyne is C_nH_{2n-2}. Alkynes occur much less commonly than alkenes, so we'll look at them only briefly.

As we saw in Section 3.1, alkynes are named using the suffix *-yne,* and the position of the triple bond is indicated by its number in the chain. Numbering begins at the chain end nearer the triple bond so that the triple bond receives as low a number as possible, and the number is placed immediately before the *-yne* suffix in the post-1993 naming system.

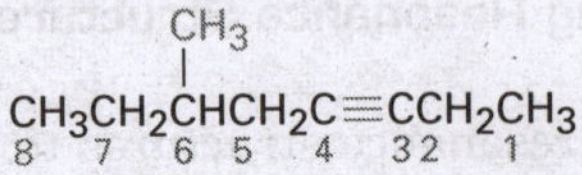

Begin numbering at the end nearer the triple bond.

6-Methyloct-3-yne

(Old name: 6-Methyl-3-octyne)

Compounds containing both double and triple bonds are called *enynes* (not ynenes). Numbering of the hydrocarbon chain starts from the end nearer the first multiple bond, whether double or triple. If there is a choice in numbering, double bonds receive lower numbers than triple bonds. For example,

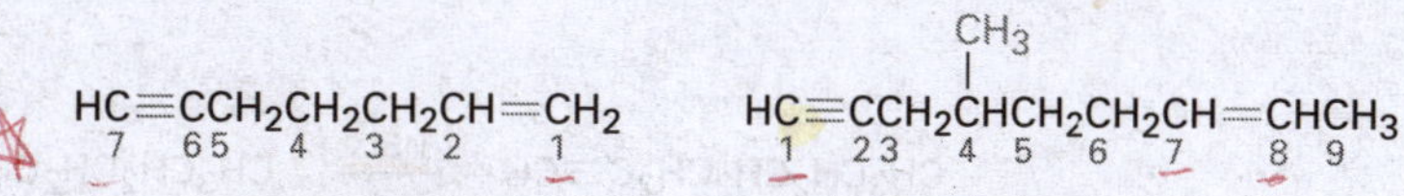

Hept-1-en-6-yne **4-Methylnon-7-en-1-yne**

(Old name: 1-Hepten-6-yne) **(Old name: 4-Methyl-7-nonen-1-yne)**

Problem 4.16 Give IUPAC names for the following compounds:

(a) 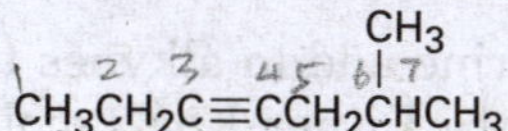 **(b)** 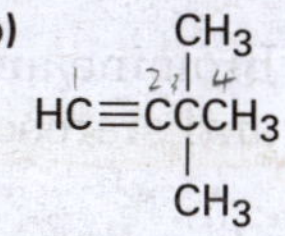

(c) 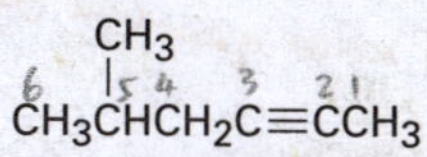 **(d)** 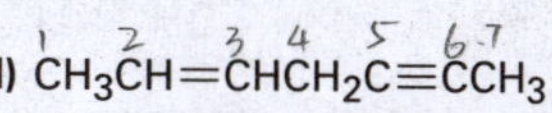

Alkyne Reactions: Addition of H_2

Alkynes are converted into alkanes by reduction with 2 molar equivalents of H_2 over a palladium catalyst. The reaction proceeds through an alkene intermediate, and the reaction can be stopped at the alkene stage if the right catalyst is used. The catalyst most often used for this purpose is the Lindlar catalyst, a specially prepared form of palladium metal. Because hydrogenation occurs with syn stereochemistry, alkynes give cis alkenes when reduced. For example,

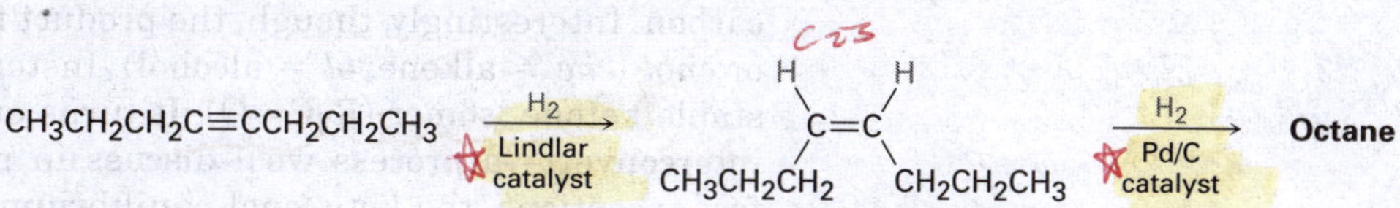

Oct-4-yne ***cis*-Oct-4-ene**

Alkyne Reactions: Addition of HX

Alkynes give electrophilic addition products on reaction with HCl, HBr, and HI just as alkenes do. Although the reaction can usually be stopped after addition of 1 molar equivalent of HX to yield a *vinylic* halide (**vinylic** means "on the C=C double bond"), an excess of HX leads to formation of a dihalide product. As the following example indicates, the regioselectivity of addition to a monosubstituted alkyne usually follows Markovnikov's rule. The H atom adds to the terminal carbon of the triple bond, and the X atom adds to the internal, more highly substituted carbon.

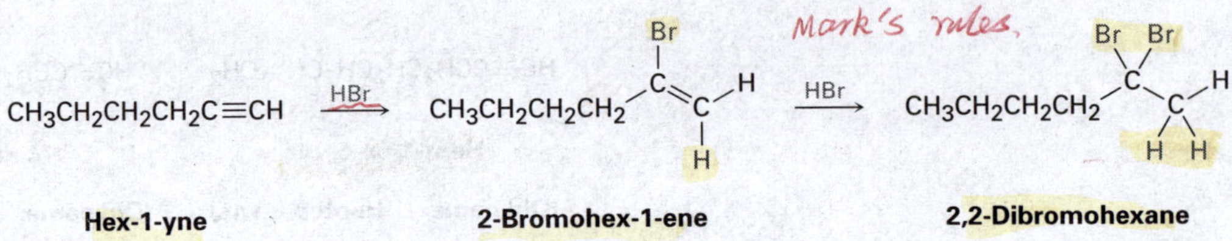

Alkyne Reactions: Addition of X_2

Bromine and chlorine add to alkynes to give dihalide addition products with anti stereochemistry. Either 1 or 2 molar equivalents can be added.

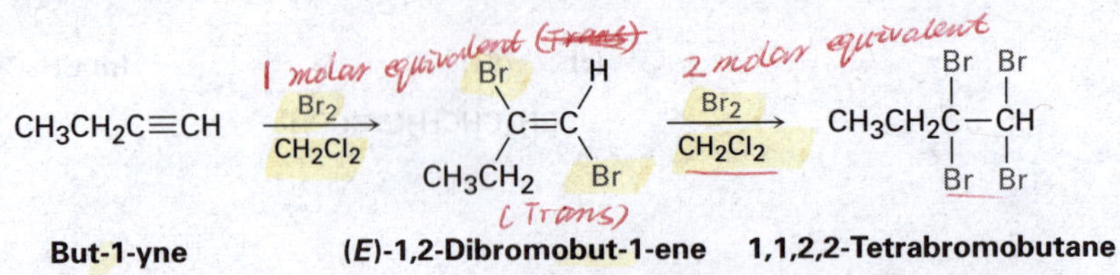

Alkyne Reactions: Addition of H_2O

Addition of water takes place when an alkyne is treated with aqueous sulfuric acid in the presence of mercuric sulfate catalyst. Markovnikov regioselectivity is found for the hydration reaction, with the H attaching to the less substituted carbon and the OH attaching to the more substituted carbon. Interestingly, though, the product is not the expected vinylic alcohol, or *enol* (*ene* = alkene; *ol* = alcohol). Instead, the enol rearranges to a more stable ketone isomer ($R_2C{=}O$). It turns out that enols and ketones rapidly interconvert—a process we'll discuss in more detail in Section 11.1. With few exceptions, the keto–enol equilibrium heavily favors the ketone. Enols are rarely isolated.

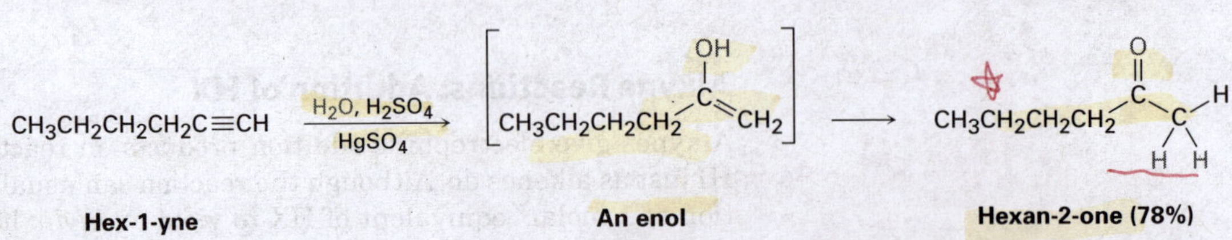

A mixture of both possible ketones results when an internal alkyne (R—C≡C—R′) is hydrated, but only a single product is formed from reaction of a terminal alkyne (R—C≡C—H).

An internal alkyne

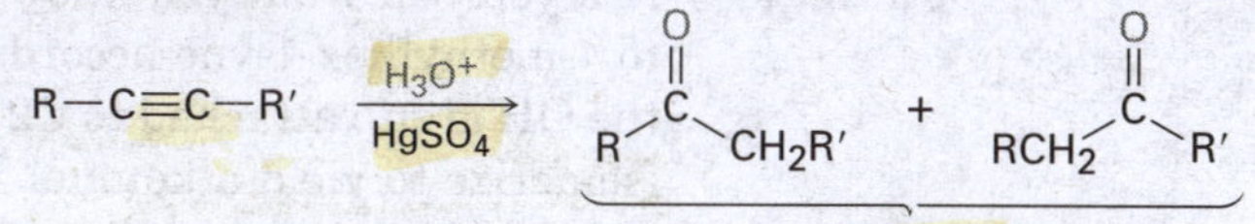

Mixture

A terminal alkyne

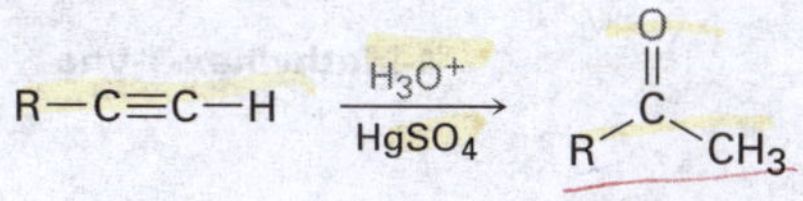

A methyl ketone

Alkyne Reactions: Formation of Acetylide Anions

The most striking difference between the chemistry of alkenes and alkynes is that terminal alkynes (R—C≡C—H) are weakly acidic, with $pK_a \approx 25$ (Section 1.10). Alkenes, by contrast, are far less acidic ($pK_a \approx 44$). When a terminal alkyne is treated with a strong base such as sodium amide, $NaNH_2$, the terminal hydrogen is removed and an **acetylide anion** is formed.

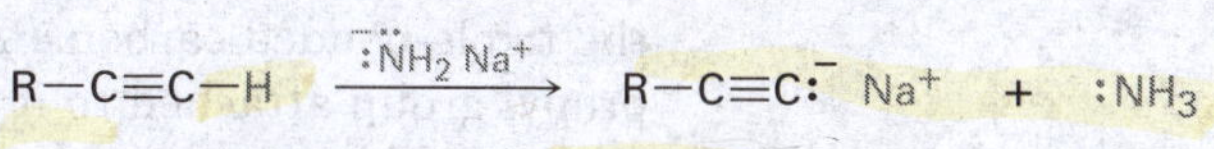

A terminal alkyne **An acetylide anion**

The presence of an unshared electron pair on the negatively charged alkyne carbon makes acetylide anions both basic and nucleophilic. As a result, acetylide anions react with alkyl halides such as bromomethane to substitute for the halogen and yield a new alkyne product. We won't study the mechanism of this substitution reaction until Chapter 7 but will note for now that it is a very useful method for preparing larger alkynes from smaller precursors. Terminal alkynes can be prepared by reaction of acetylene itself, and internal alkynes can be prepared by further reaction of a terminal alkyne.

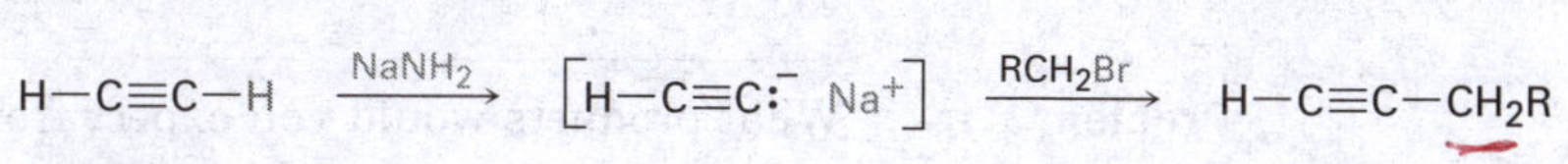

Acetylene **A terminal alkyne**

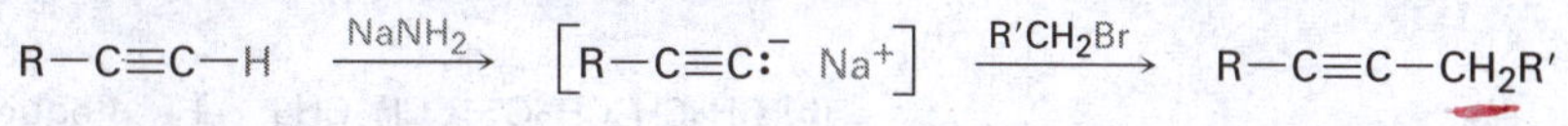

A terminal alkyne **An internal alkyne**

The one limitation to the reaction of an acetylide anion with an alkyl halide is that only primary alkyl halides, RCH_2X, can be used, for reasons that will be discussed in Chapter 7.

Worked Example 4.6 Predicting the Product of an Alkyne Hydration Reaction

What product would you obtain by hydration of 4-methylhex-1-yne?

Strategy Ask yourself what you know about alkyne addition reactions. Addition of water to 4-methylhex-1-yne according to Markovnikov's rule will yield a product with the OH group attached to C2 rather than C1. This initially formed enol will then isomerize to yield a ketone.

Solution

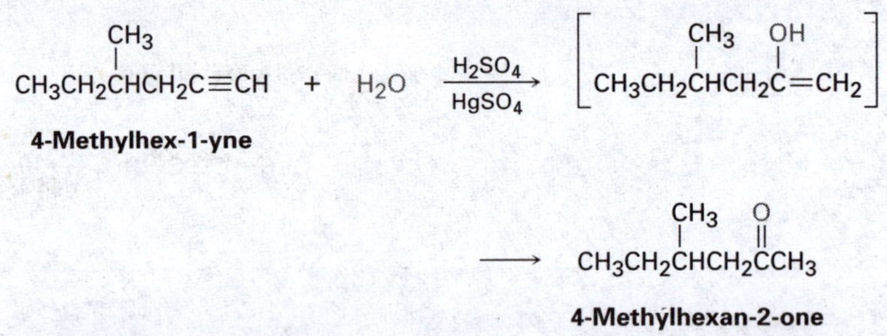

Worked Example 4.7 Synthesizing an Alkyne

What alkyne and what alkyl halide would you use to prepare pent-1-yne?

Strategy As always when synthesizing a compound, work the problem backward. Draw the structure of the target molecule, and identify the alkyl group(s) attached to the triple-bonded carbons. In the present case, one of the alkyne carbons has a propyl group attached to it and the other has a hydrogen attached. Thus, pent-1-yne could be prepared by treatment of acetylene with $NaNH_2$ to yield sodium acetylide, followed by reaction with 1-bromopropane.

Solution

$$\underset{\textbf{Acetylene}}{H{-}C{\equiv}C{-}H} + :\ddot{N}H_2^-\ Na^+ \longrightarrow \underset{\textbf{Sodium acetylide}}{H{-}C{\equiv}C{:}^-\ Na^+} + :NH_3$$

$$H{-}C{\equiv}C{:}^-\ Na^+ + \underset{\textbf{1-Bromopropane}}{CH_3CH_2CH_2Br} \longrightarrow \underset{\textbf{Pent-1-yne}}{H{-}C{\equiv}C{-}CH_2CH_2CH_3}$$

This propyl group comes from 1-bromopropane.

Problem 4.17 What products would you expect from the following reactions?

(a) $CH_3CH_2CH_2C{\equiv}CH + 1\ \text{equiv}\ Cl_2 \longrightarrow ?$

(b) $CH_3CH_2CH_2C{\equiv}CCH_2CH_3 + 1\ \text{equiv}\ HBr \longrightarrow ?$

(c)

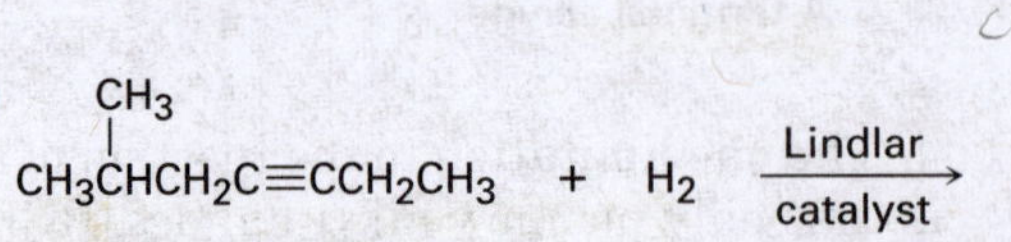

Problem 4.18 What product would you obtain by hydration of oct-4-yne?

Problem 4.19 What alkynes would you start with to prepare the following ketones by a hydration reaction?

(a) $CH_3CH_2CH_2\overset{\overset{O}{\|}}{C}CH_3$ (b) $CH_3CH_2CH_2\overset{\overset{O}{\|}}{C}CH_2CH_3$

Problem 4.20 Show the alkyne and alkyl halide from which the following products can be obtained. Where two routes look feasible, list both.

(a) $CH_3\underset{}{\overset{\overset{CH_3}{|}}{C}}HCH_2CH_2C{\equiv}CH$ (b) $CH_3CH_2CH_2C{\equiv}CCH_3$ (c) $CH_3\overset{\overset{CH_3}{|}}{C}HC{\equiv}CCH_3$

INTERLUDE

Natural Rubber

Crude rubber is harvested from the rubber tree, *Hevea brasiliensis*.

Rubber—an unusual name for an unusual substance—is a naturally occurring alkene polymer produced by more than 400 different plants. The major source is the so-called rubber tree, *Hevea brasiliensis,* from which the crude material is harvested as it drips from a slice made through the bark. The name *rubber* was coined by Joseph Priestley, the discoverer of oxygen and early researcher of rubber chemistry, for the simple reason that one of rubber's early uses was to rub out pencil marks on paper.

Unlike polyethylene and other simple alkene polymers, natural rubber is a polymer of a conjugated diene, *isoprene,* or 2-methylbuta-1,3-diene. The polymerization takes place by 1,4-addition (Section 4.8) of isoprene monomers to the growing chain, leading to formation of a polymer that still contains double bonds spaced regularly at four-carbon intervals. As the following structure shows, these double bonds have *Z* configuration.

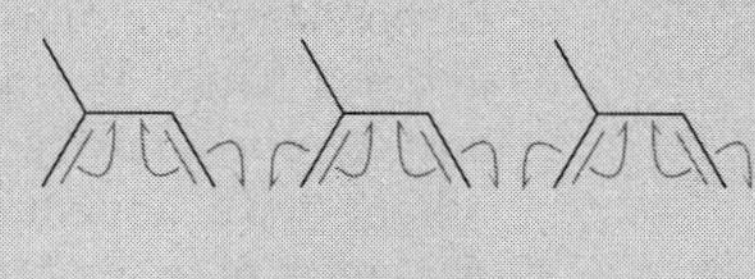

Many isoprene units

↓

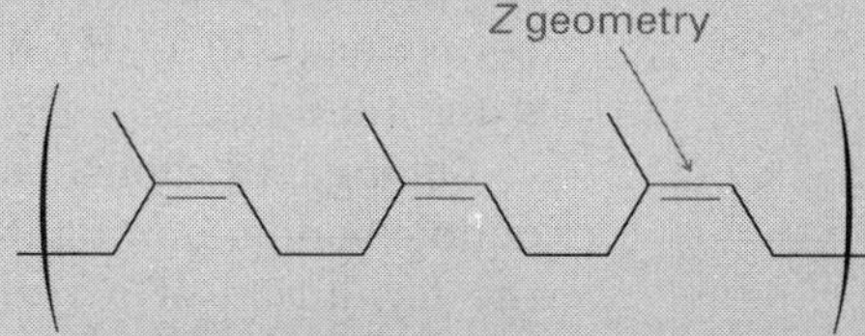

A segment of natural rubber

continued

INTERLUDE

Crude rubber, called *latex,* is collected from the tree as an aqueous dispersion that is washed, dried, and coagulated by warming in air. The resultant polymer has chains that average about 5000 monomer units in length and have molecular weights of 200,000 to 500,000 amu. This crude coagulate is too soft and tacky to be useful until it is hardened by heating with elemental sulfur, a process called *vulcanization.* By mechanisms that are still not fully understood, vulcanization cross-links the rubber chains together by forming carbon–sulfur bonds between them, thereby hardening and stiffening the polymer. The exact degree of hardening can be varied, yielding material soft enough for automobile tires or hard enough for bowling balls (*ebonite*).

The remarkable ability of rubber to stretch and then contract to its original shape is due to the irregular shapes of the polymer chains caused by the double bonds. These double bonds introduce bends and kinks into the polymer chains, thereby preventing neighboring chains from nestling together. When stretched, the randomly coiled chains straighten out and orient along the direction of the pull but are kept from sliding over one another by the cross-links. When the stretch is released, the polymer reverts to its original random state.

Summary and Key Words

acetylide anion 139
1,2-addition 131
1,4-addition 131
allylic 131
anti stereochemistry 121
conjugation 130
epoxide 124
hydrogenation 122
hydroxylation 125
Markovnikov's rule 113
monomer 127
oxidation 124
polymer 127
reduction 122
regiospecific 113
resonance form 132
resonance hybrid 132
syn stereochemistry 123
vinylic 137

With the background needed to understand organic reactions now covered, this chapter has begun the systematic description of major functional groups. The chemistry of alkenes is dominated by addition reactions of electrophiles. When HX reacts with an alkene, **Markovnikov's rule** predicts that the H will add to the carbon that has fewer alkyl substituents and the X group will add to the carbon that has more alkyl substituents. Many electrophiles besides HX add to alkenes. Thus, Br_2 and Cl_2 add to give 1,2-dihalide addition products having **anti stereochemistry**. Addition of H_2O (*hydration*) takes place on reaction of the alkene with aqueous acid, and addition of H_2 (**hydrogenation**) occurs in the presence of a metal catalyst such as platinum or palladium. **Oxidation** of alkenes is often carried out using potassium permanganate, $KMnO_4$. Under basic conditions, $KMnO_4$ reacts with alkenes to yield cis **1,2-diols**. Under neutral or acidic conditions, $KMnO_4$ cleaves double bonds to yield carbonyl-containing products. Alkenes can also be converted into **epoxides** by reaction with a peroxy acid (RCO_3H), and epoxides can be hydrolyzed by aqueous acid to yield trans 1,2-diols.

Conjugated dienes, such as buta-1,3-diene, contain alternating single and double bonds. Conjugated dienes undergo **1,4-addition** of electrophiles through the formation of a resonance-stabilized **allylic** carbocation intermediate. No single line-bond representation can depict the true structure of an allylic carbocation. Rather, the true structure is a **resonance hybrid** intermediate between two contributing resonance forms. The only difference between two **resonance forms** is in the location of double-bond and lone-pair electrons. The atoms remain in the same places in both structures.

Many simple alkenes undergo **polymerization** when treated with a radical catalyst. **Polymers** are large molecules built up by the repetitive bonding together of many small **monomer** units.

Alkynes are hydrocarbons that contain a carbon–carbon triple bond. Much of the chemistry of alkynes is similar to that of alkenes. For example, alkynes react with 1 equivalent of HBr and HCl to yield **vinylic** halides, and with 1 equivalent of Br_2 and Cl_2 to yield 1,2-dihalides. Alkynes can also be hydrated by reaction with aqueous sulfuric acid in the presence of mercuric sulfate catalyst. The reaction leads to an intermediate enol that immediately isomerizes to a ketone. Alkynes can be hydrogenated with the Lindlar catalyst to yield a cis alkene. Terminal alkynes are weakly acidic and can be converted into **acetylide anions** by treatment with a strong base. Reaction of the acetylide anion with a primary alkyl halide then gives an internal alkyne.

Summary of Reactions

Note: No stereochemistry is implied unless specifically stated or indicated with wedged, solid, and dashed lines.

1. Reactions of alkenes
   (a) Addition of HX, where X = Cl, Br, or I (Sections 4.1–4.2)

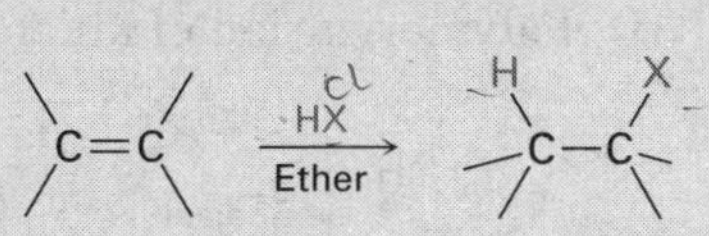

   (b) Addition of H_2O (Section 4.3)

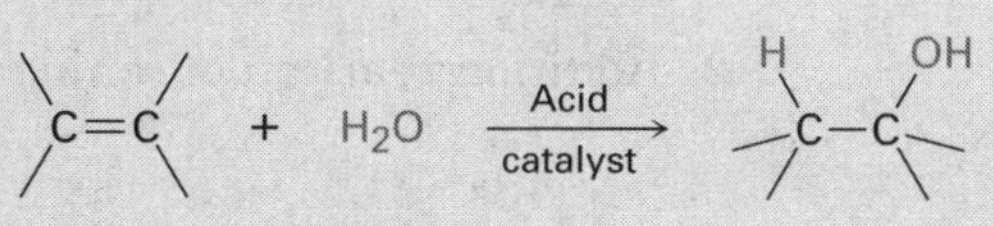

   (c) Addition of X_2, where X = Cl, Br (Section 4.4)

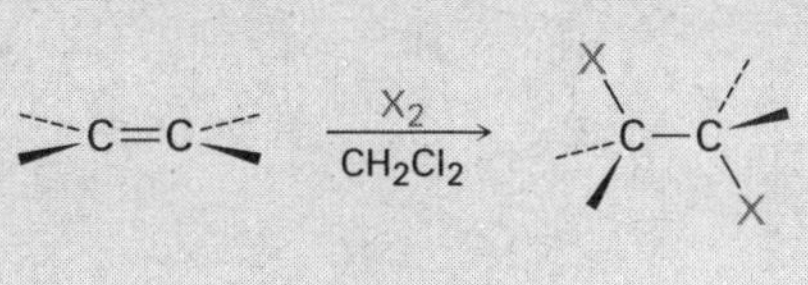

   (d) Addition of H_2 (Section 4.5)

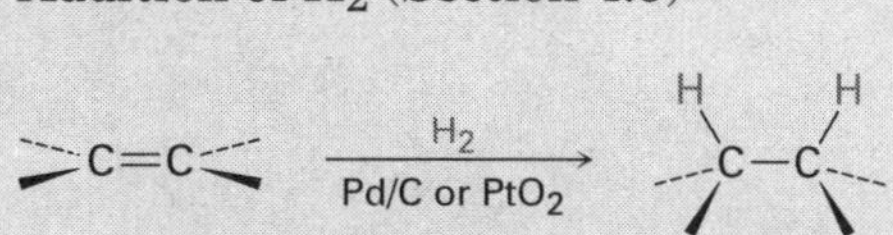

continued

(e) Epoxidation (Section 4.6)

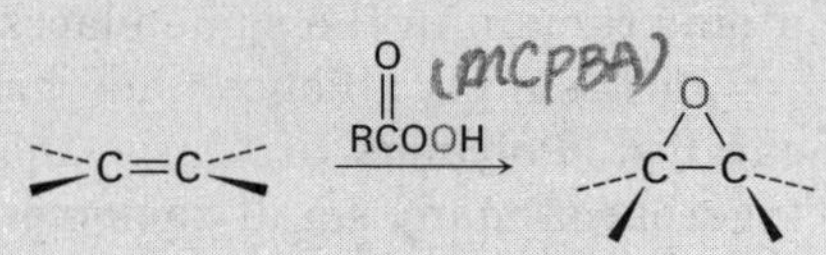

(f) Hydroxylation by acid-catalyzed epoxide hydrolysis (Section 4.6)

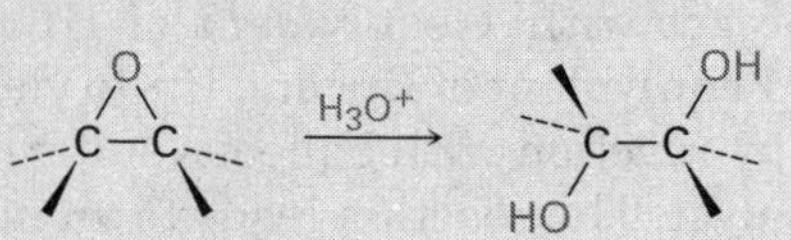

(g) Hydroxylation with $KMnO_4$ (Section 4.6)

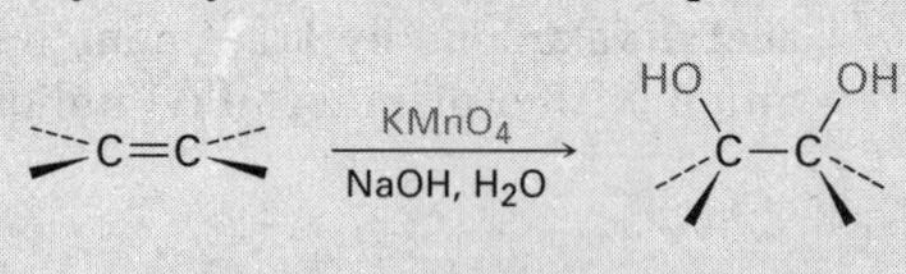

Syn addition

(h) Oxidative cleavage of alkenes with acidic $KMnO_4$ (Section 4.6)

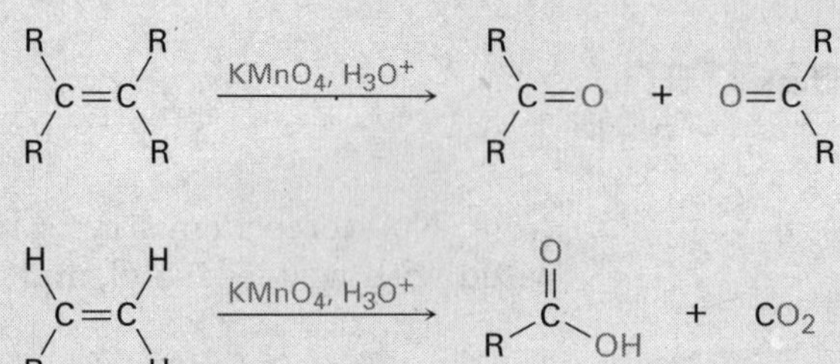

(i) Polymerization of alkenes (Section 4.7)

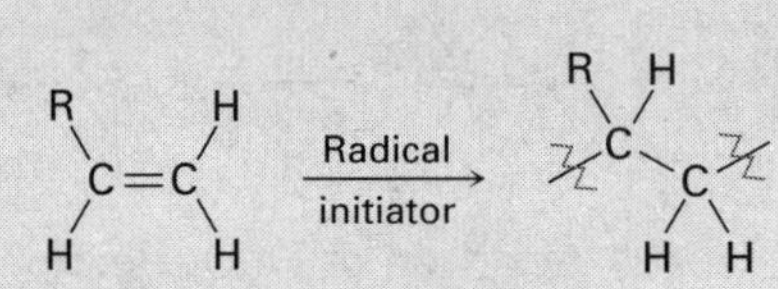

2. Addition reaction of conjugated dienes (Section 4.8)

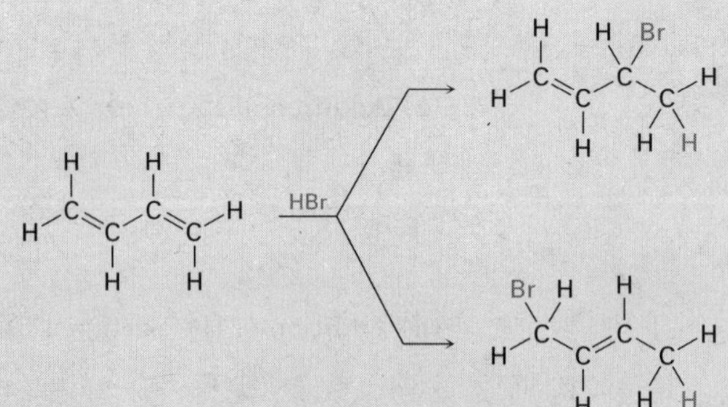

3. Reactions of alkynes (Section 4.11)

(a) Addition of H_2

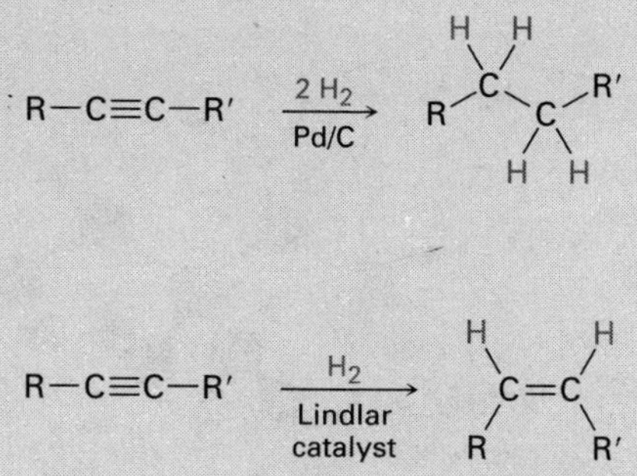

A cis alkene

(b) Addition of HX, where X = Cl, Br, or I

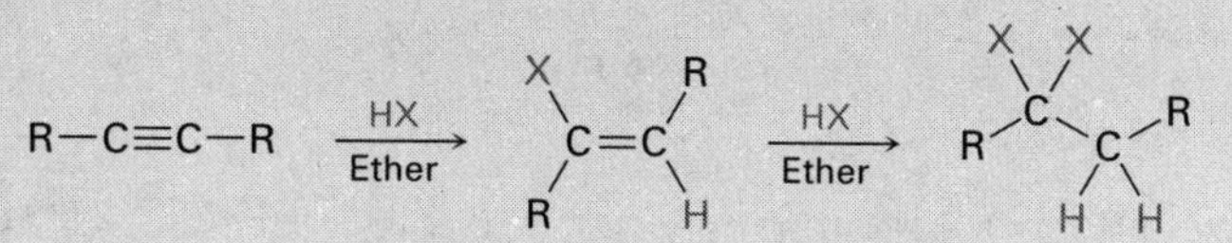

(c) Addition of X_2, where X = Cl, Br

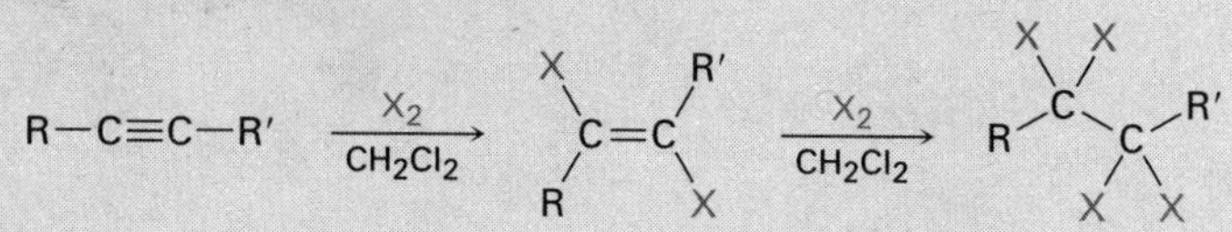

(d) Addition of H_2O

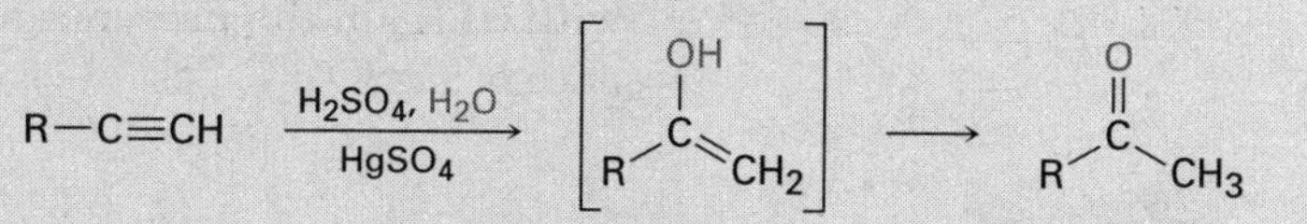

An enol A methyl ketone

(e) Acetylide anion formation

$$R-C\equiv C-H \xrightarrow[NH_3]{NaNH_2} R-C\equiv C:^- \; Na^+ \; + \; NH_3$$

(f) Reaction of acetylide anions with alkyl halides

$$HC\equiv CH \xrightarrow{NaNH_2} HC\equiv C^- \; Na^+ \xrightarrow{RCH_2Br} HC\equiv CCH_2R$$

Acetylene — **A terminal alkyne**

$$RC\equiv CH \xrightarrow{NaNH_2} RC\equiv C^- \; Na^+ \xrightarrow{R'CH_2Br} RC\equiv CCH_2R'$$

A terminal alkyne — **An internal alkyne**

Exercises

Visualizing Chemistry

(Problems 4.1–4.20 appear within the chapter.)

OWL

Interactive versions of these problems are assignable in OWL.

4.21 What alkenes would give the following alcohols on hydration? (Red = O).

(a)

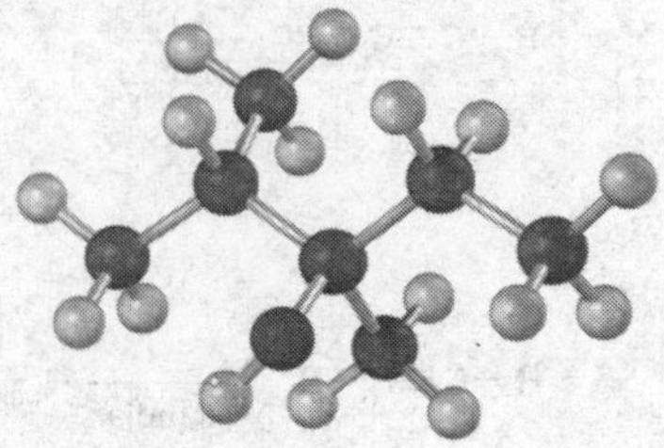

(b)

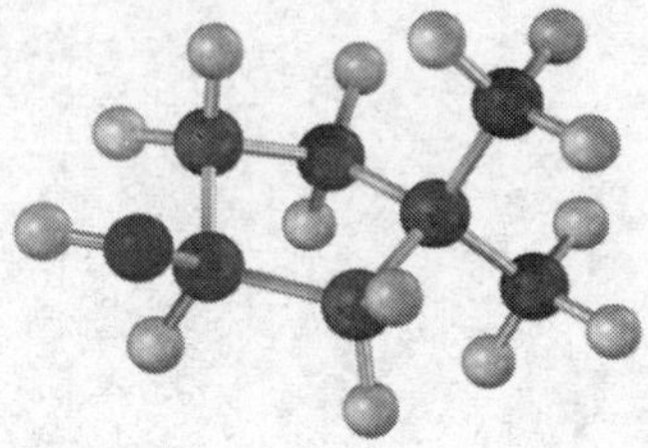

4.22 Name the following alkenes, and predict the products of their reaction with (i) $KMnO_4$ in aqueous acid and (ii) $KMnO_4$ in aqueous NaOH:

(a)

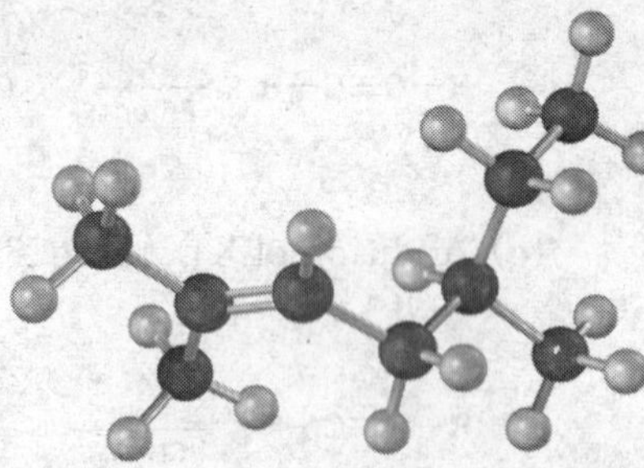

(b)

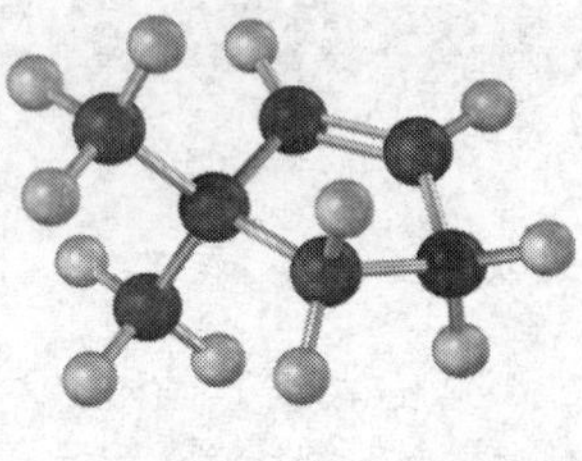

4.23 Name the following alkynes, and predict the products of their reaction with (i) H_2 in the presence of a Lindlar catalyst and (ii) H_3O^+ in the presence of $HgSO_4$:

(a)

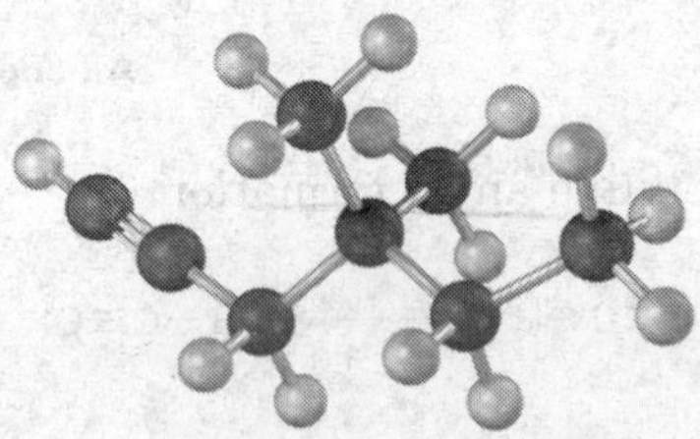

(b)

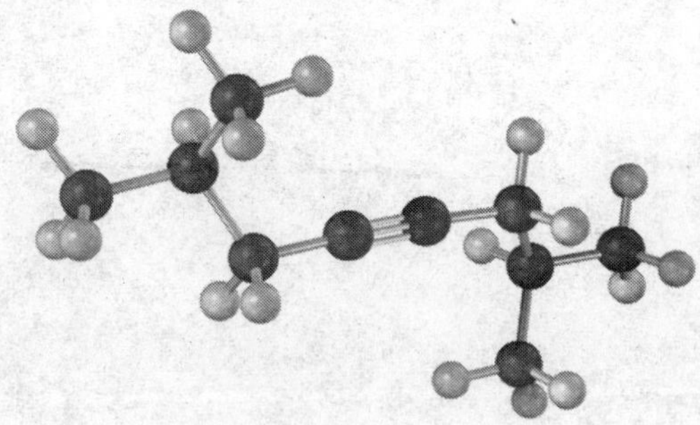

4.24 From what alkene was the following 1,2-diol made, and what method was used, epoxide hydrolysis or $KMnO_4$ in basic solution? (Red = O).

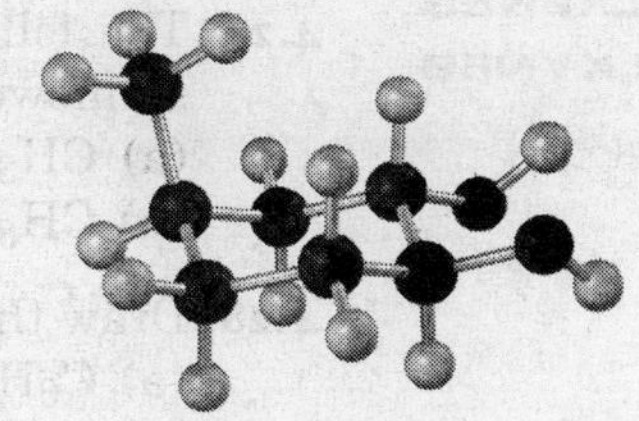

4.25 The following model is that of an allylic carbocation intermediate formed by protonation of a conjugated diene with HBr. Show the structure of the diene and the structures of the final reaction products.

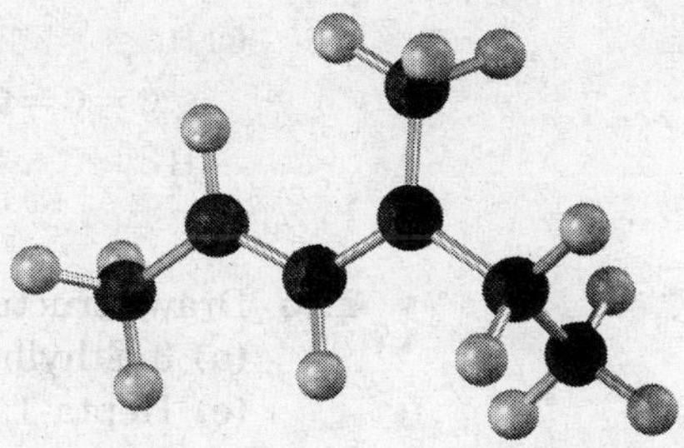

4.26 From what alkyne might each of the following substances have been made? (Red = O, yellow-green = Cl).

(a)

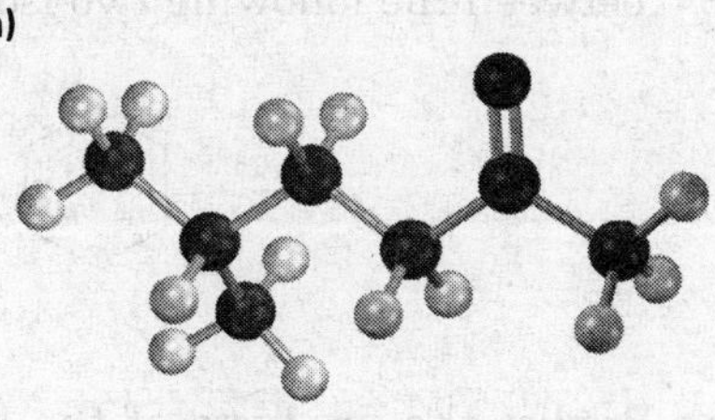

(b)

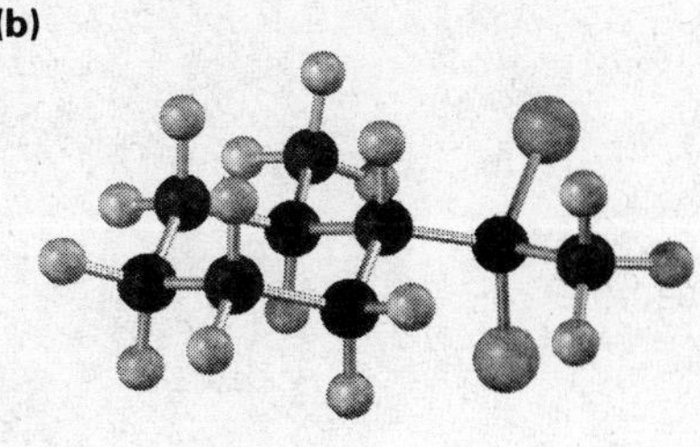

Additional Problems

NAMING ALKENES AND ALKYNES

4.27 The following two hydrocarbons have been isolated from plants in the sunflower family. Name them according to IUPAC rules.
(a) $CH_3CH{=}CHC{\equiv}CC{\equiv}CCH{=}CHCH{=}CHCH{=}CH_2$ (all trans)
(b) $CH_3C{\equiv}CC{\equiv}CC{\equiv}CC{\equiv}CC{\equiv}CCH{=}CH_2$

4.28 Draw three possible structures for each of the following formulas:
(a) C_6H_8 **(b)** C_6H_8O

4.29 Give IUPAC names for the following compounds:

(a)

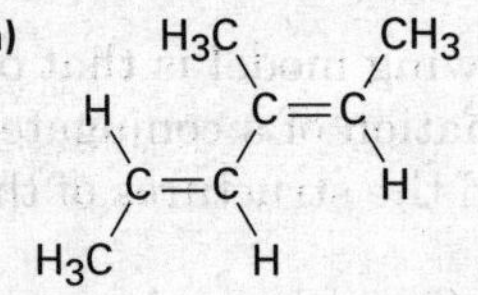

(b)

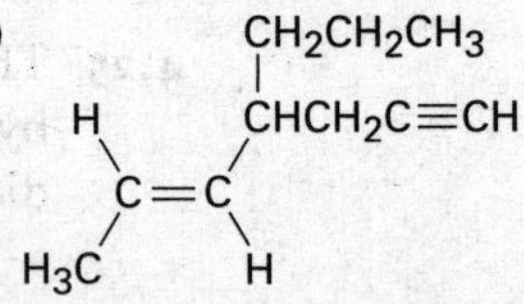

(c)

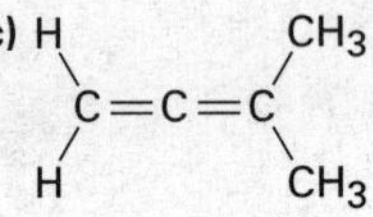

(d)

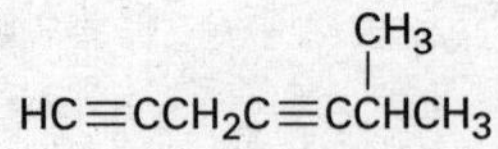

4.30 Draw structures corresponding to the following IUPAC names:
(a) 3-Ethylhept-1-yne **(b)** 3,5-Dimethylhex-4-en-1-yne
(c) Hepta-1,5-diyne **(d)** 1-Methylcyclopenta-1,3-diene

4.31 Draw and name all the possible pentyne isomers, C_5H_8.

4.32 Draw and name the six possible diene isomers of formula C_5H_8. Which of the six are conjugated dienes?

PREDICT THE PRODUCTS

4.33 Using an oxidative cleavage reaction, explain how you would distinguish between the following two isomeric cyclohexadienes:

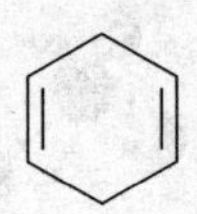

and

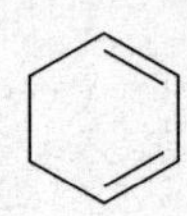

4.34 Predict the products of the following reactions. Indicate regioselectivity where relevant. (The aromatic ring is inert to all the indicated reagents.)

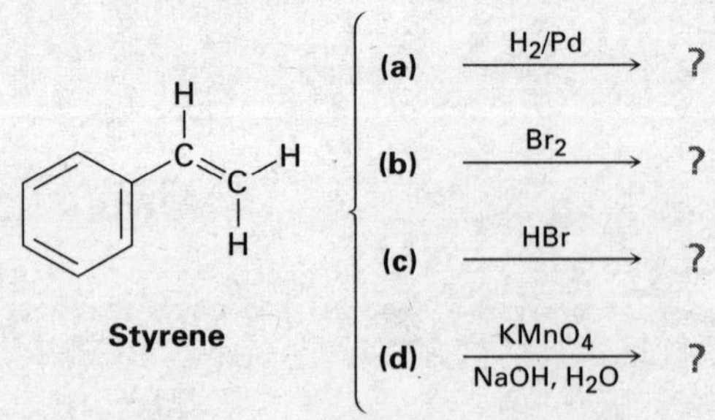

4.35 Predict the products of the following reactions on dec-5-yne:

(a) $\xrightarrow{H_2\text{, Lindlar catalyst}}$? (b) $\xrightarrow{2\text{ equiv }Br_2}$? (c) $\xrightarrow{H_2O,\ H_2SO_4,\ HgSO_4}$?

4.36 Predict the products of the following reactions on hex-1-yne:

(a) $\xrightarrow{1\text{ equiv HBr}}$? (b) $\xrightarrow{1\text{ equiv }Cl_2}$? (c) $\xrightarrow{H_2\text{, Lindlar catalyst}}$?

4.37 Formulate the reaction of cyclohexene with (i) Br_2 and (ii) *meta*-chloroperoxybenzoic acid followed by H_3O^+. Show the reaction intermediates and the final products with correct cis or trans stereochemistry.

4.38 Suggest structures for alkenes that give the following reaction products. There may be more than one answer for some cases.

(a) ? $\xrightarrow{H_2\text{/Pd catalyst}}$ 2-Methylhexane

(b) ? $\xrightarrow{Br_2\text{ in }CH_2Cl_2}$ 2,3-Dibromo-5-methylhexane

(c) ? $\xrightarrow{HBr}$ 2-Bromo-3-methylheptane

(d) ? $\xrightarrow[H_2O]{KMnO_4,\ OH^-}$ $CH_3CH(CH_3)CH_2CH(OH)CH(OH)CH_2CH_3$

4.39 What products would you expect to obtain from reaction of cyclohexa-1,3-diene with each of the following?

(a) 1 mol Br_2 in CH_2Cl_2 (b) 1 mol HCl
(c) 1 mol DCl (D = deuterium, ^2H) (d) 2 mol H_2 over a Pd catalyst

PREDICT THE REACTANTS

4.40 Give the structure of an alkene that yields the following keto acid on reaction with $KMnO_4$ in aqueous acid:

$$? \xrightarrow[H_3O^+]{KMnO_4} HOC(=O)CH_2CH_2CH_2CH_2C(=O)CH_3$$

4.41 Draw the structure of a hydrocarbon that reacts with only 1 equivalent of H_2 on catalytic hydrogenation and gives only pentanoic acid, $CH_3CH_2CH_2CH_2CO_2H$, on treatment with acidic $KMnO_4$. Write the reactions involved.

4.42 What alkynes would you hydrate to obtain the following ketones?

(a) 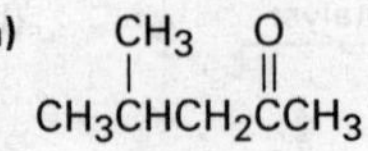

(b) 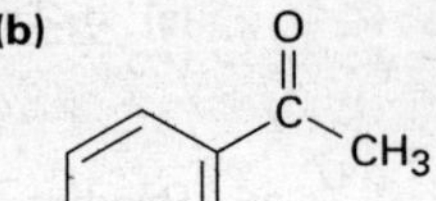

4.43 Draw the structure of a hydrocarbon that reacts with 2 equivalents of H_2 on catalytic hydrogenation and gives only succinic acid on reaction with acidic $KMnO_4$.

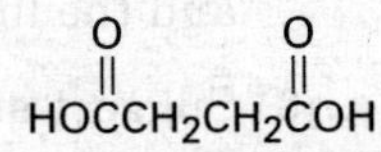

Succinic acid

4.44 What alkenes would you hydrate to obtain the following alcohols?

(a) 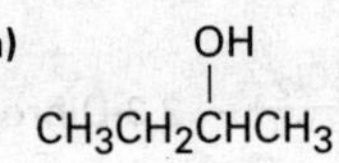

(b) 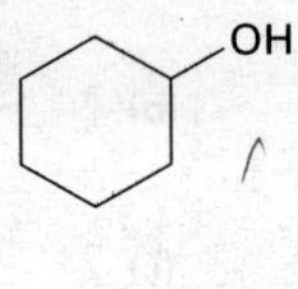

(c) 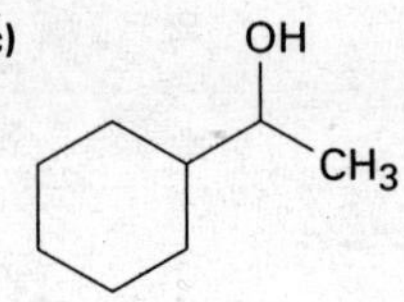

SYNTHESIS

4.45 Using but-1-yne as the only organic starting material, along with any inorganic reagents needed, how would you synthesize the following compounds? (More than one step may be needed.)

(a) Butane (b) 1,1,2,2-Tetrachlorobutane

(c) 2-Bromobutane (d) Butan-2-one ($CH_3CH_2COCH_3$)

4.46 In planning the synthesis of a compound, it's as important to know what *not* to do as to know what to do. What is wrong with each of the following reactions?

(a) 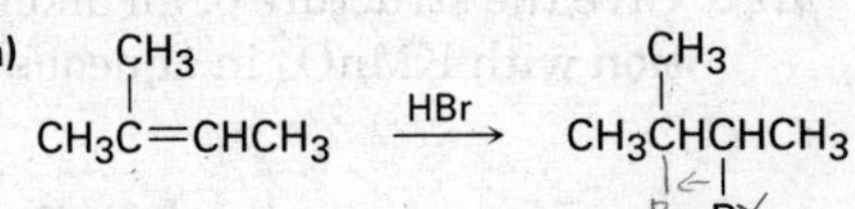

(b) 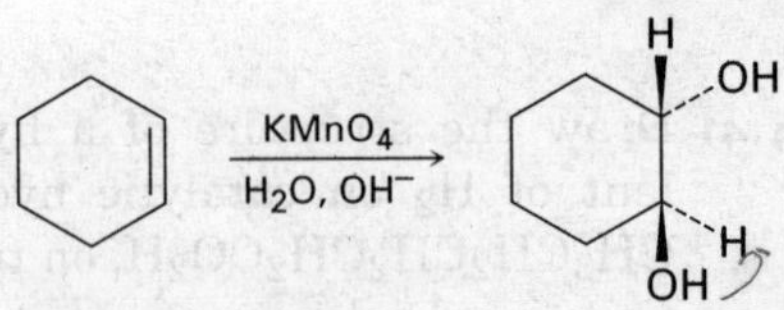

(c) 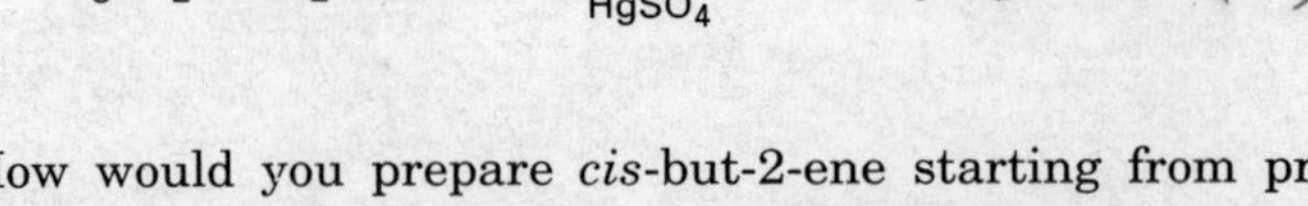

4.47 How would you prepare *cis*-but-2-ene starting from propyne, an alkyl halide, and any other reagents needed? (This problem can't be worked in a single step. You'll have to carry out more than one reaction.)

RESONANCE

4.48 Draw three additional resonance structures for the benzyl cation.

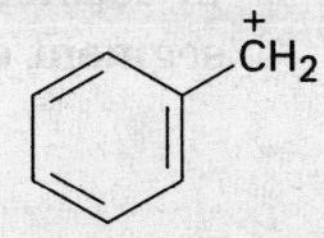

Benzyl cation

4.49 In light of your answer to Problem 4.48, what product would you expect from the following reaction? Explain.

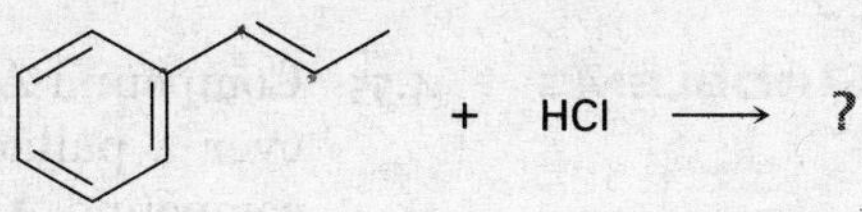

4.50 One of the following pairs of structures represents resonance forms, and one does not. Explain which is which.

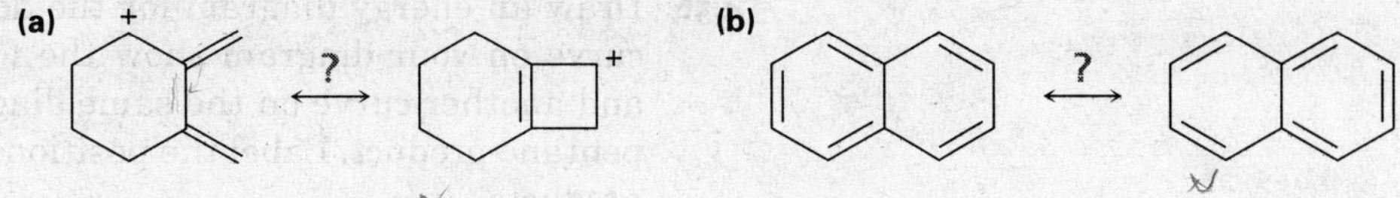

4.51 Draw the indicated number of additional resonance structures for each of the following substances:

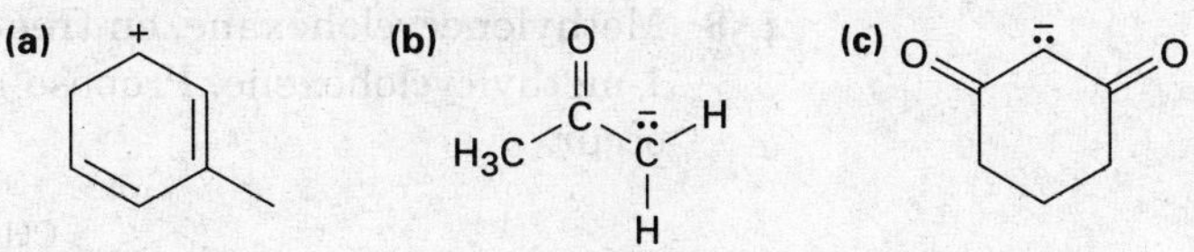

(two) (one) (two)

POLYMERS

4.52 What monomer unit might be used to prepare the following polymer?

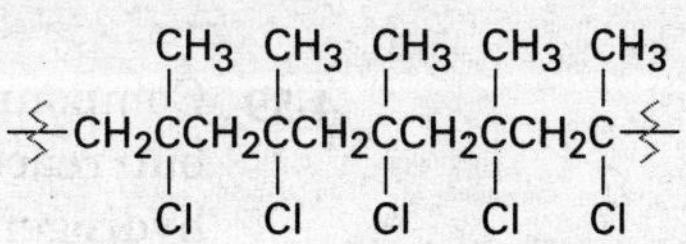

4.53 Plexiglas, a clear plastic used to make many molded articles, is made by polymerization of methyl methacrylate. Draw a representative segment of Plexiglas.

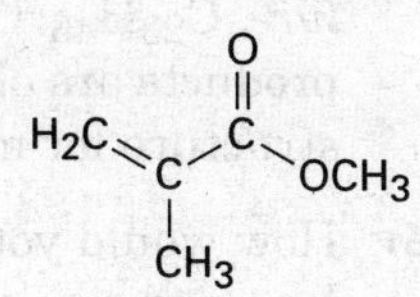

Methyl methacrylate

4.54 Poly(vinyl pyrrolidone), prepared by from *N*-vinylpyrrolidone, is used both in cosmetics and as a synthetic blood substitute. Draw a representative segment of the polymer.

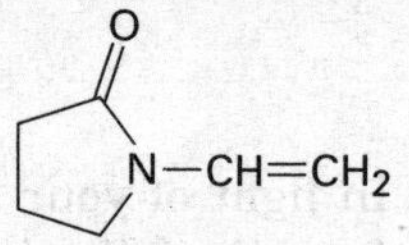

***N*-Vinylpyrrolidone**

GENERAL PROBLEMS

4.55 Compound A, C_9H_{12}, absorbs 3 equivalents of H_2 on catalytic reduction over a palladium catalyst to give **B**, C_9H_{18}. On reaction with $KMnO_4$, compound **A** gives, among other things, a ketone that was identified as cyclohexanone. On treatment with $NaNH_2$ in NH_3, followed by addition of iodomethane, compound **A** gives a new hydrocarbon **C**, $C_{10}H_{14}$. What are the structures of **A**, **B**, and **C**?

4.56 Draw an energy diagram for the addition of HBr to pent-1-ene. Let one curve on your diagram show the formation of 1-bromopentane product and another curve on the same diagram show the formation of 2-bromopentane product. Label the positions for all reactants, intermediates, and products.

4.57 Make sketches of what you imagine the transition-state structures to look like in the reaction of HBr with pent-1-ene (Problem 4.56).

4.58 Methylenecyclohexane, on treatment with strong acid, isomerizes to yield 1-methylcyclohexene. Propose a mechanism by which the reaction might occur.

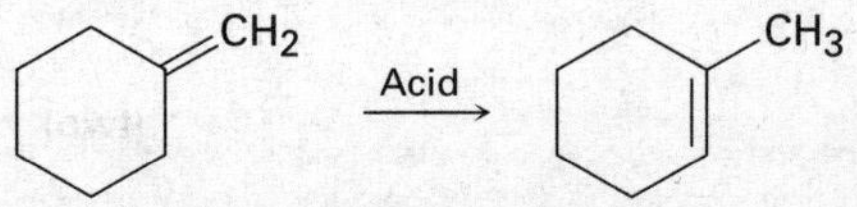

Methylenecyclohexane **1-Methylcyclohexene**

4.59 Compound **A** has the formula C_8H_8. It reacts rapidly with acidic $KMnO_4$ but reacts with only 1 equivalent of H_2 over a palladium catalyst. On hydrogenation under conditions that reduce aromatic rings, A reacts with 4 equivalents of H_2, and hydrocarbon **B**, C_8H_{16}, is produced. The reaction of **A** with $KMnO_4$ gives CO_2 and a carboxylic acid **C**, $C_7H_6O_2$. What are the structures of **A**, **B**, and **C**? Write all the reactions.

4.60 The sex attractant of the common housefly is a hydrocarbon named *muscalure*, $C_{23}H_{46}$. On treatment of muscalure with aqueous acidic $KMnO_4$, two products are obtained, $CH_3(CH_2)_{12}CO_2H$ and $CH_3(CH_2)_7CO_2H$. Propose a structure for muscalure.

4.61 How would you synthesize muscalure (Problem 4.60) starting from acetylene and any alkyl halides needed? (The double bond in muscalure is cis.)

4.62 Reaction of 2-methylpropene with CH_3OH in the presence of H_2SO_4 catalyst yields methyl *tert*-butyl ether, $CH_3OC(CH_3)_3$, by a mechanism analogous to that of acid-catalyzed alkene hydration. Write the mechanism.

4.63 Prelaureatin, a substance isolated from marine algae, is thought to arise from laurediol by the following route. Propose a mechanism.

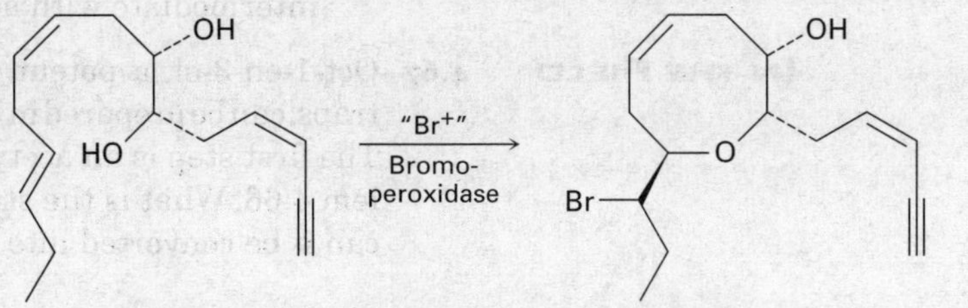

Laurediol **Prelaureatin**

4.64 Hydroxylation of *cis*-but-2-ene with basic $KMnO_4$ yields a different product than hydroxylation of *trans*-but-2-ene. Draw the structure, show the stereochemistry of each product, and explain the result. We'll explore the stereochemistry of the products in more detail in Chapter 6.

4.65 α-Terpinene, $C_{10}H_{16}$, is a pleasant-smelling hydrocarbon that has been isolated from oil of marjoram. On hydrogenation over a palladium catalyst, α-terpinene reacts with 2 mol equiv of hydrogen to yield a new hydrocarbon, $C_{10}H_{20}$. On reaction with acidic $KMnO_4$, α-terpinene yields oxalic acid and 6-methylheptane-2,5-dione. Propose a structure for α-terpinene.

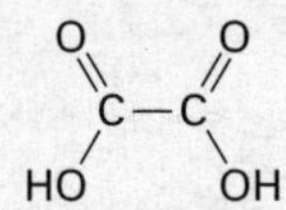

Oxalic acid

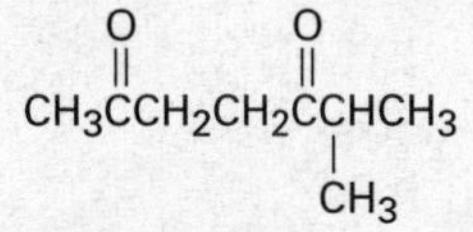

6-Methylheptane-2,5-dione

IN THE MEDICINE CABINET

4.66 The oral contraceptive agent Mestranol is synthesized by addition of acetylide ion to a carbonyl group.

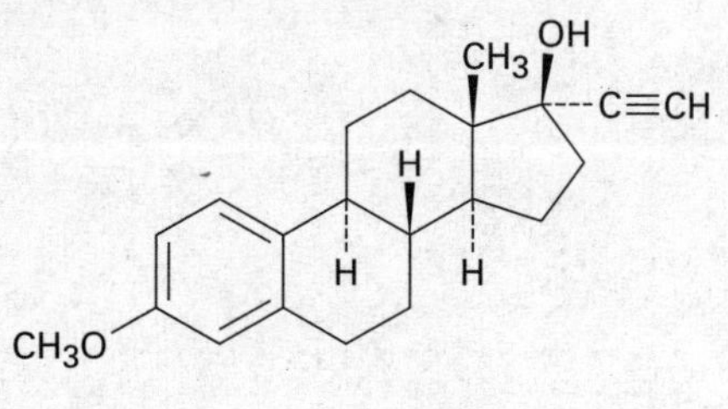

Mestranol

(a) To understand the acetylide-addition reaction, first draw a resonance structure of the C=O double bond based on electronegativity values, indicating where the + and − charges belong in the circles on the following abbreviated structure:

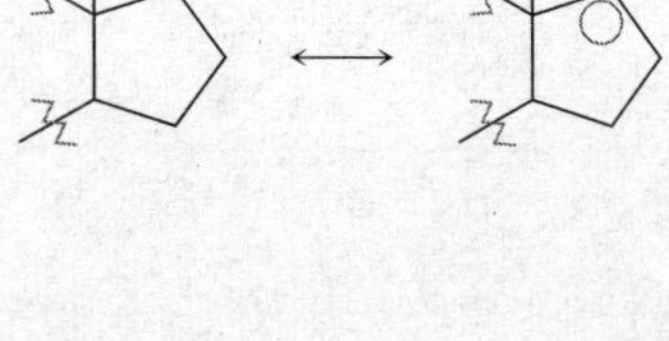

(b) Then draw a two-step mechanism for the addition of acetylide ion to this resonance structure and subsequent protonation of the intermediate with acid.

IN THE FIELD

4.67 Oct-1-en-3-ol, a potent mosquito attractant commonly used in mosquito traps, can be prepared in two steps from hexanal, $CH_3CH_2CH_2CH_2CH_2CHO$. The first step is an acetylide-addition reaction like that described in Problem 4.66. What is the structure of the product from the first step, and how can it be converted into oct-1-en-3-ol?

$$CH_3CH_2CH_2CH_2CH_2\overset{\displaystyle OH}{\overset{|}{C}}HCH{=}CH_2$$

Oct-1-en-3-ol

4.68 As we saw in the *Interlude* at the end of this chapter, natural rubber is a polymer of 2-methylbuta-1,3-diene that contains *Z* double bonds. Synthetic rubber, by contrast, is similar to natural rubber but contains *E* double bonds. Draw the structure of a representative section of synthetic rubber.

CHAPTER

5

A fennel plant is an aromatic herb used in cooking. A phenyl group (pronounced exactly the same way) is the characteristic structural unit of "aromatic" organic compounds.

Aromatic Compounds

In the early days of organic chemistry, the word *aromatic* was used to describe fragrant substances such as benzene (from coal distillate), benzaldehyde (from cherries, peaches, and almonds), and toluene (from tolu balsam). It was soon realized, however, that substances classed as aromatic differed from most other organic compounds in their chemical behavior.

Today, the association of aromaticity with fragrance has long been lost, and we now use the word **aromatic** to refer to the class of compounds that contain six-membered benzene-like rings with three double bonds. Many valuable compounds are aromatic in part, such as the steroidal hormone estrone and the cholesterol-lowering drug atorvastatin, marketed as Lipitor. Benzene itself causes a depressed white blood cell count (leukopenia) on prolonged exposure and should not be used as a laboratory solvent.

OWL

Online homework for this chapter can be assigned in OWL, an online homework assessment tool.

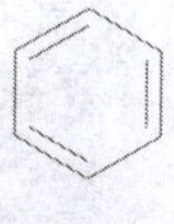

Benzene

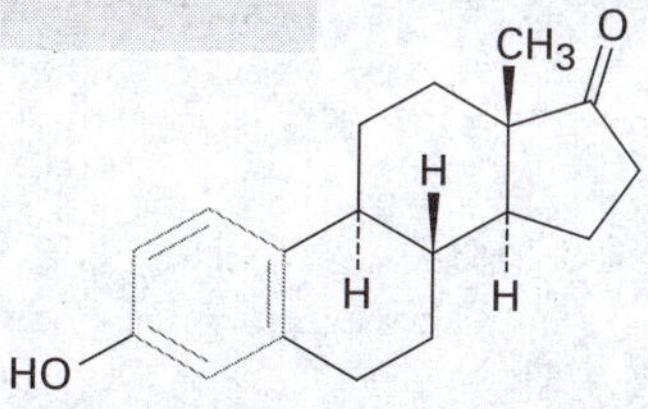

Estrone

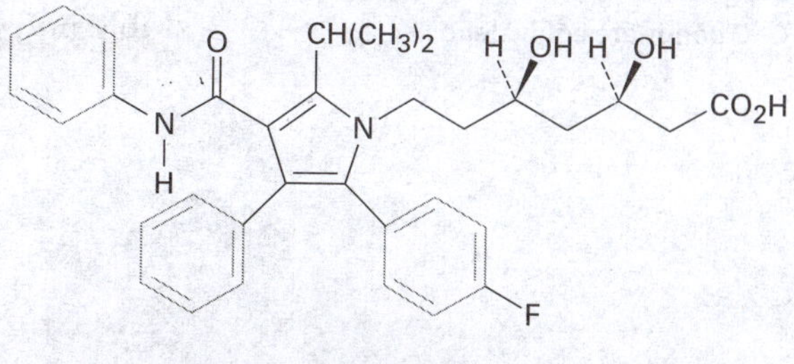

Atorvastatin (Lipitor)

WHY THIS CHAPTER?

Aromatic rings are a common part of many organic structures and are particularly important in nucleic acid chemistry and in the chemistry of several amino acids. In this chapter, we'll find out how and why aromatic compounds are different from such apparently related compounds as alkenes.

5.1 Structure of Benzene

Benzene (C_6H_6) has eight fewer hydrogens than the corresponding six-carbon alkane (C_6H_{14}) and is clearly unsaturated, usually being represented as a six-membered ring with alternating double and single bond. Yet it has been known since the mid-1800s that benzene is much less reactive than typical alkenes and fails to undergo typical alkene addition reactions. Cyclohexene, for instance, reacts rapidly with Br_2 and gives the addition product 1,2-dibromocyclohexane, but benzene reacts only slowly with Br_2 and gives the *substitution* product C_6H_5Br.

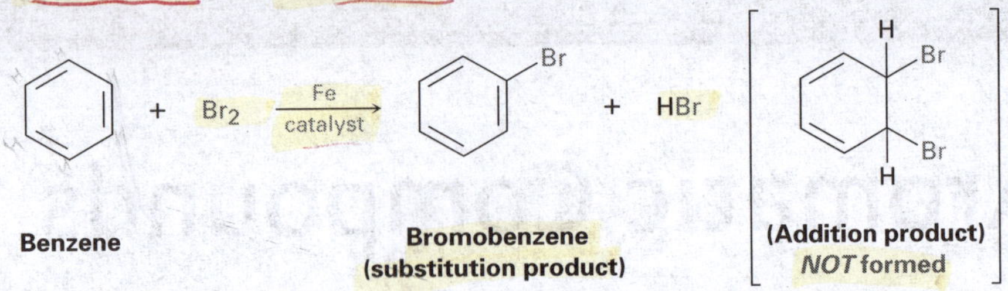

Further evidence for the unusual nature of benzene is that all its carbon–carbon bonds have the same length—139 pm—intermediate between typical single (154 pm) and double (134 pm) bonds. In addition, the electron density in all six carbon–carbon bonds is identical, as shown by an electrostatic potential map (Figure 5.1a). Thus, benzene is a planar molecule with the shape of a regular hexagon. All C–C–C bond angles are 120°, all six carbon atoms are sp^2-hybridized, and each carbon has a p orbital perpendicular to the plane of the six-membered ring.

Because all six carbon atoms and all six p orbitals in benzene are equivalent, it's impossible to define three localized π bonds in which a given p orbital overlaps only one neighboring p orbital. Rather, each p orbital overlaps equally well with both neighboring p orbitals, leading to a picture of benzene in which all six π electrons are free to move about the entire ring (Figure 5.1b). In resonance terms (Sections 4.9 and 4.10), benzene is a hybrid of two equivalent forms. Neither form is correct by itself; the true structure of benzene is somewhere in between the two resonance forms but is impossible to draw with our usual conventions. Because of this resonance, benzene is more stable and less reactive than a typical alkene.

Figure 5.1 **(a)** An electrostatic potential map of benzene and **(b)** an orbital picture. Each of the six carbon atoms has a *p* orbital that can overlap equally well with neighboring *p* orbitals on both sides. The π electrons are thus shared around the ring in two doughnut-shaped clouds, and all C–C bonds are equivalent.

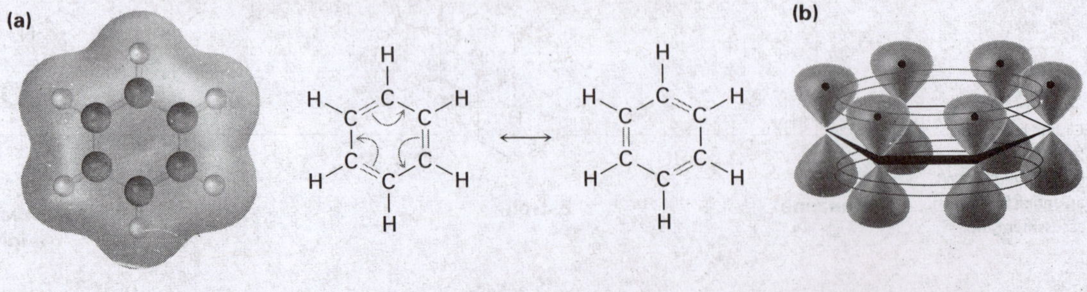

Problem 5.1 Line-bond structures appear to imply that there are two different isomers of 1,2-dibromobenzene, one with the bromine-bearing carbon atoms joined by a double bond and one with the bromine-bearing carbons joined by a single bond. In fact, though, there is only one 1,2-dibromobenzene. Explain.

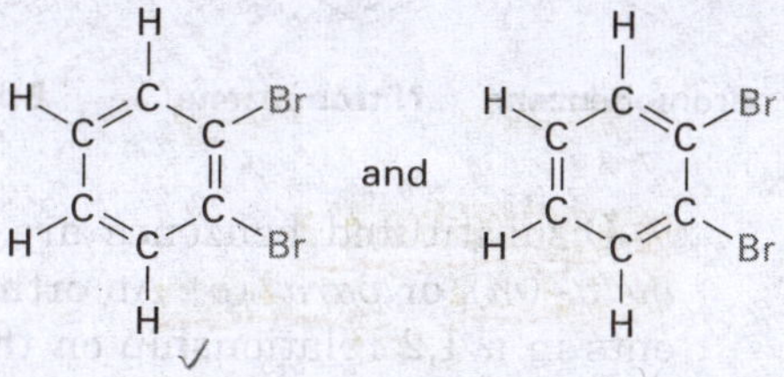

1,2-Dibromobenzene

5.2 Naming Aromatic Compounds

Aromatic substances, more than any other class of organic compounds, have acquired a large number of common names. IUPAC rules discourage the use of most such names but allow some of the more widely used ones to be retained (Table 5.1). Thus, methylbenzene is commonly known as *toluene,* hydroxybenzene as *phenol,* aminobenzene as *aniline,* and so on.

Table 5.1 Common Names of Some Aromatic Compounds

Structure	Name	Structure	Name
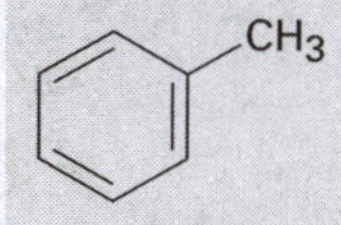	Toluene (bp 111 °C)	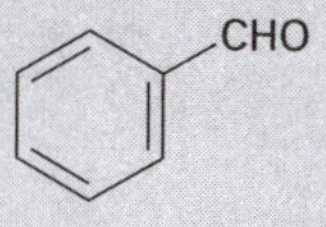	Benzaldehyde (bp 178 °C)
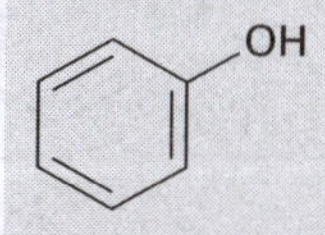	Phenol (mp 43 °C)	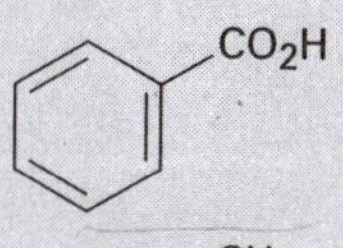	Benzoic acid (mp 122 °C)
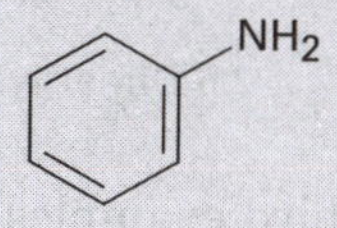	Aniline (bp 184 °C)	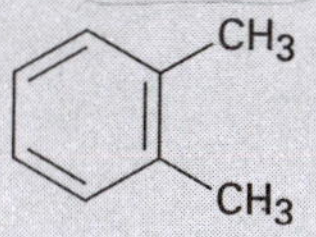	*ortho*-Xylene (bp 144 °C)
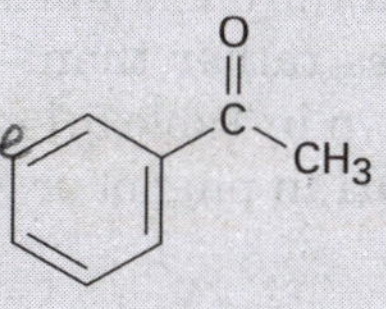	Acetophenone (mp 21 °C)	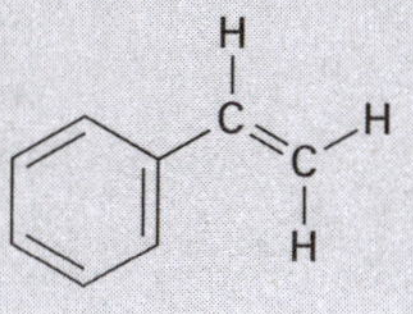	Styrene (bp 145 °C)

Monosubstituted benzenes are systematically named in the same manner as other hydrocarbons, with *-benzene* as the parent name. Thus, C_6H_5Br is bromobenzene, $C_6H_5NO_2$ is nitrobenzene, and $C_6H_5CH_2CH_3$ is ethylbenzene. The name **phenyl**, pronounced **fen-nil** and sometimes abbreviated as Ph or Φ (Greek phi), is used for the $-C_6H_5$ unit when the benzene ring is considered as a substituent. In addition, a generalized aromatic substituent is called an

aryl group, abbreviated as Ar, and the name **benzyl** is used for the $C_6H_5CH_2$– group.

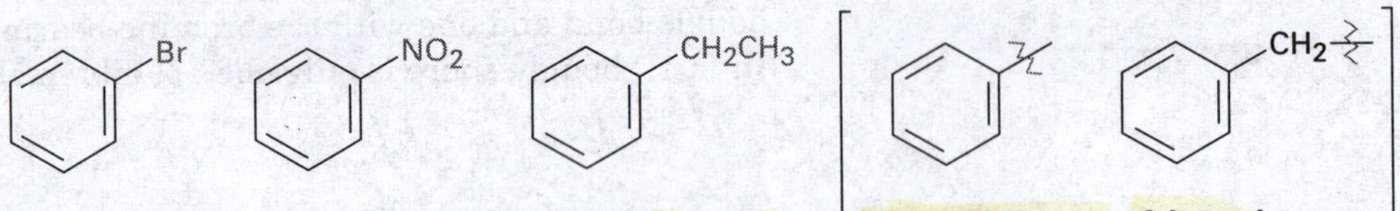

Bromobenzene **Nitrobenzene** **Ethylbenzene** **A phenyl group** **A benzyl group**

Disubstituted benzenes are named using one of the prefixes *ortho-* (*o*), *meta-* (*m*), or *para-* (*p*). An ortho-disubstituted benzene has its two substituents in a 1,2 relationship on the ring; a meta-disubstituted benzene has its two substituents in a 1,3 relationship; and a para-disubstituted benzene has its substituents in a 1,4 relationship.

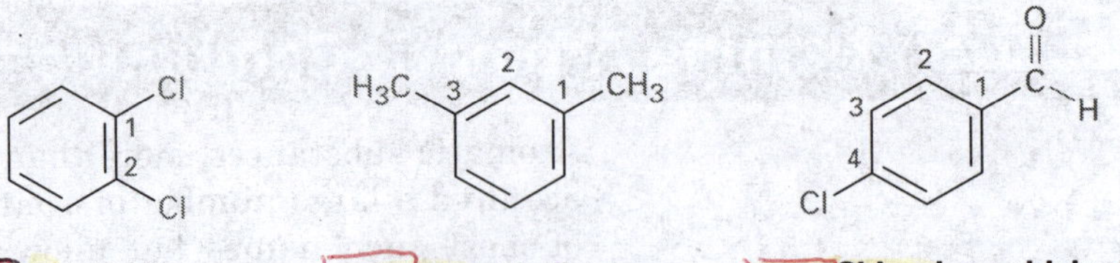

***ortho*-Dichlorobenzene** **1,2 disubstituted** **_meta_-Dimethylbenzene (_meta_-xylene)** **1,3 disubstituted** **_para_-Chlorobenzaldehyde** **1,4 disubstituted**

As with cycloalkanes (Section 2.7), benzenes with more than two substituents are named by choosing a point of attachment as carbon 1 and numbering the substituents on the ring so that the second substituent has as low a number as possible. The substituents are listed alphabetically when writing the name.

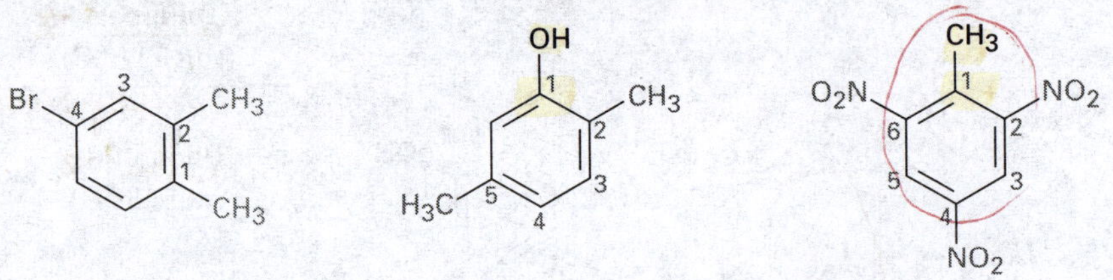

4-Bromo-1,2-dimethylbenzene **2,5-Dimethylphenol** **2,4,6-Trinitrotoluene (TNT)**

Note in the second and third examples shown that *-phenol* and *-toluene* are used as the parent names rather than *-benzene.* Any of the monosubstituted aromatic compounds shown in Table 5.1 can be used as a parent name, with the principal substituent (–OH in phenol or $-CH_3$ in toluene) considered as C1.

Worked Example 5.1 Naming an Aromatic Compound

What is the IUPAC name of the following compound?

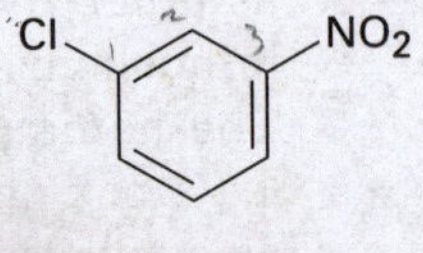

Solution Because the nitro group ($-NO_2$) and chloro group are on carbons 1 and 3, they have a meta relationship. Citing the two substituents in alphabetical order gives the IUPAC name *m*-chloronitrobenzene.

Problem 5.2 Tell whether the following compounds are ortho, meta, or para disubstituted:

(a) 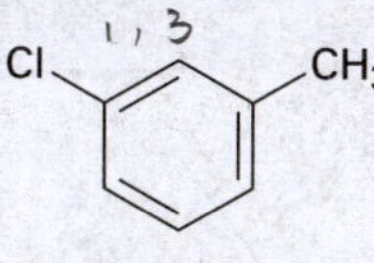
(b) 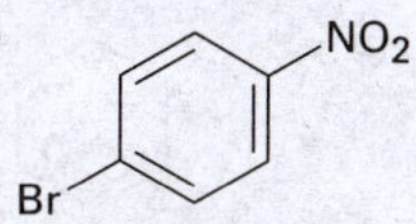
(c) 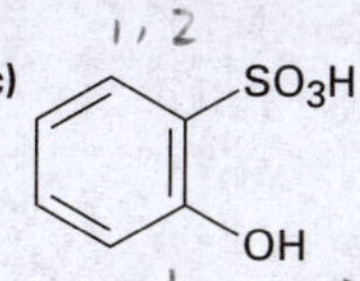

Problem 5.3 Give IUPAC names for the following compounds:

(a) 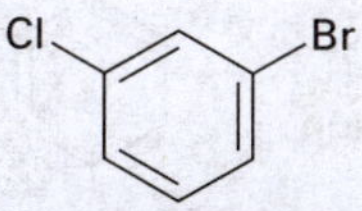
(b) 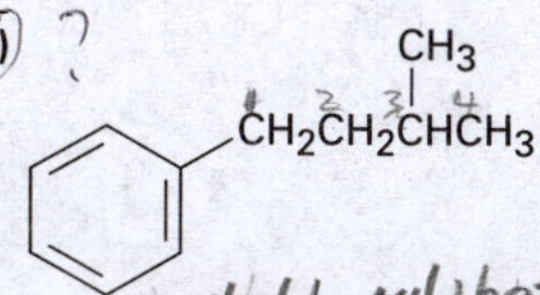
(c) 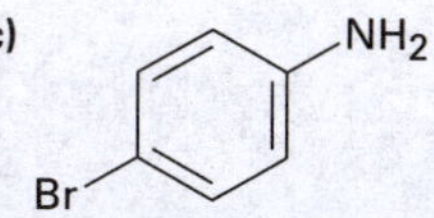
(d) 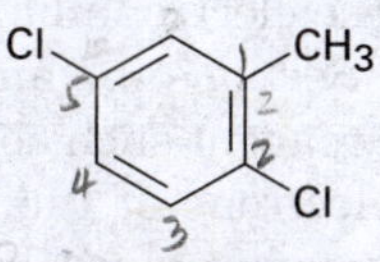
(e) 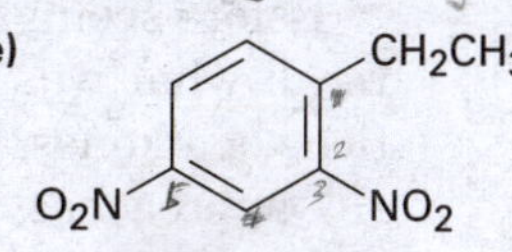
(f) 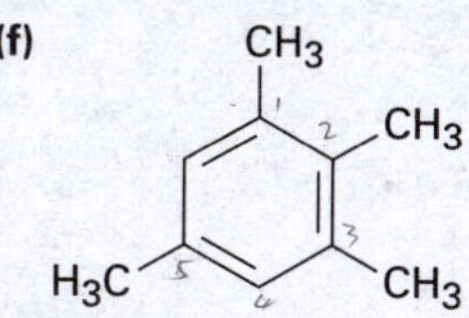

Problem 5.4 Draw structures corresponding to the following IUPAC names:

(a) *p*-Bromochlorobenzene (b) *p*-Bromotoluene
(c) *m*-Chloroaniline (d) 1-Chloro-3,5-dimethylbenzene

5.3 Electrophilic Aromatic Substitution Reactions: Bromination

The most common reaction of aromatic compounds is **electrophilic aromatic substitution**, a process in which an electrophile (E^+) reacts with an aromatic ring and substitutes for one of the hydrogens.

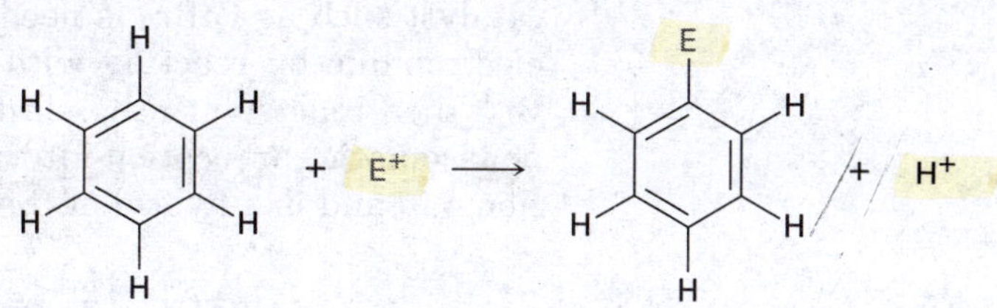

Many different substituents can be introduced onto the aromatic ring by electrophilic substitution. To list some possibilities, an aromatic ring can be substituted by a halogen (–Cl, –Br, –I), a nitro group ($-NO_2$), a sulfonic acid group ($-SO_3H$), an alkyl group (–R), or an acyl group (–COR). Starting from only a few simple materials, it's possible to prepare many thousands of substituted aromatic compounds (Figure 5.2).

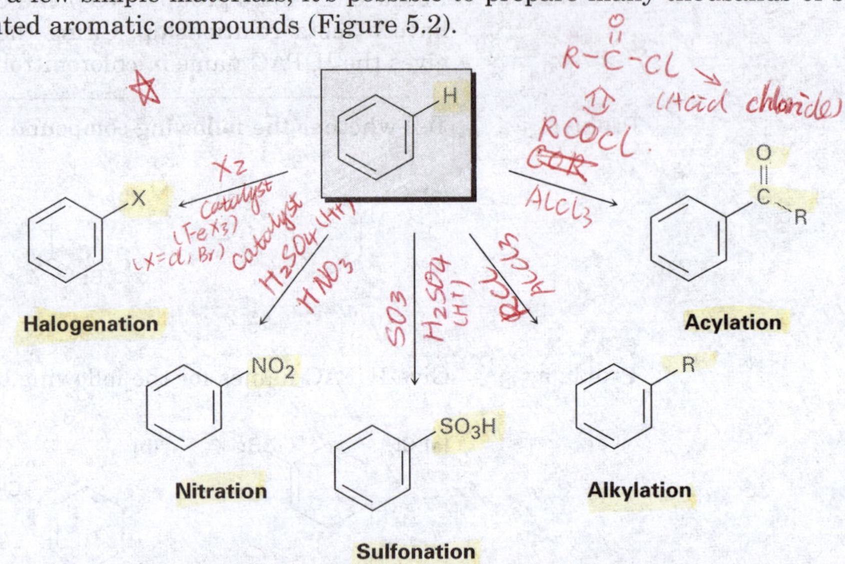

Figure 5.2 Some electrophilic aromatic substitution reactions.

Before seeing how these electrophilic substitution reactions occur, let's briefly recall what was said in Sections 3.7 and 3.8 about electrophilic addition reactions of alkenes. When a reagent such as HCl adds to an alkene, the electrophilic H^+ approaches the π electrons of the double bond and forms a bond to one carbon, leaving a positive charge at the other carbon. This carbocation intermediate then reacts with the nucleophilic Cl^- ion to yield the addition product.

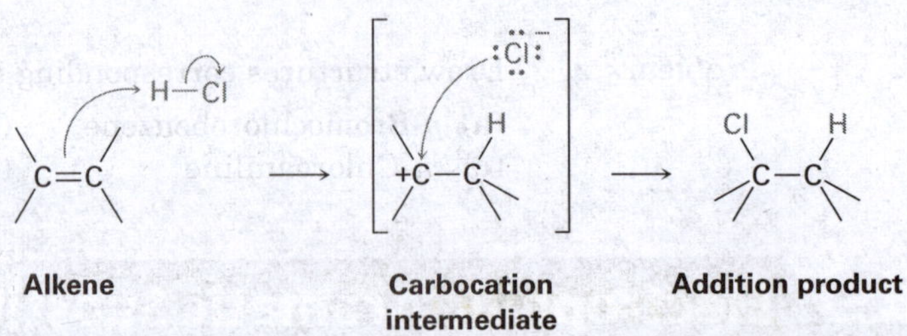

An electrophilic aromatic substitution reaction begins in a similar way, but there are a number of differences. One difference is that aromatic rings are less reactive toward electrophiles than alkenes are. For example, Br_2 in CH_2Cl_2 solution reacts instantly with most alkenes but does not react with benzene at room temperature. For bromination of benzene to take place, a catalyst such as $FeBr_3$ is needed. The catalyst makes the Br_2 molecule more electrophilic by reacting with it to give $FeBr_4^-$ and Br^+. The electrophilic Br^+ then reacts with the electron-rich (nucleophilic) benzene ring to yield a nonaromatic carbocation intermediate. This carbocation is doubly allylic (Section 4.9) and is a hybrid of three resonance forms.

$$Br{-}Br \quad + \quad FeBr_3 \quad \longrightarrow \quad Br^+ \; ^-FeBr_4$$

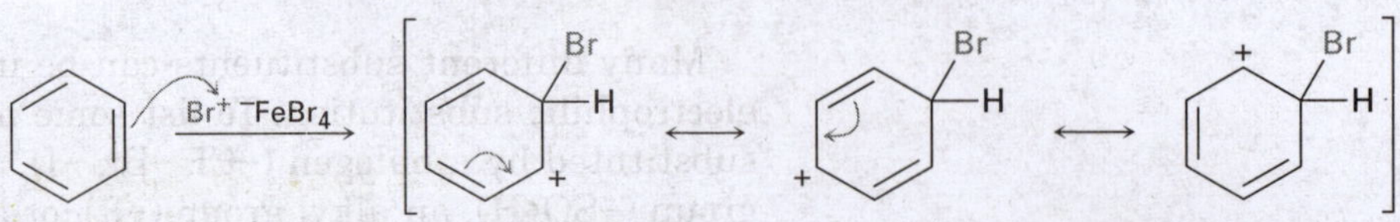

Although more stable than a typical nonallylic carbocation because of resonance, the intermediate in electrophilic aromatic substitution is much less stable than the starting benzene ring itself. Thus, reaction of an electrophile with a benzene ring has a relatively high activation energy and is rather slow.

Another difference between alkene addition reactions and aromatic substitution reactions occurs after the electrophile has added to the benzene ring and the carbocation intermediate has formed. Instead of adding Br^- to give an addition product, the carbocation intermediate loses H^+ from the bromine-bearing carbon to give a substitution product. The net effect is the substitution of H^+ by Br^+ by the overall mechanism shown in Figure 5.3.

MECHANISM

Figure 5.3 The mechanism of the electrophilic bromination of benzene. The reaction occurs in two steps and involves a resonance-stabilized carbocation intermediate.

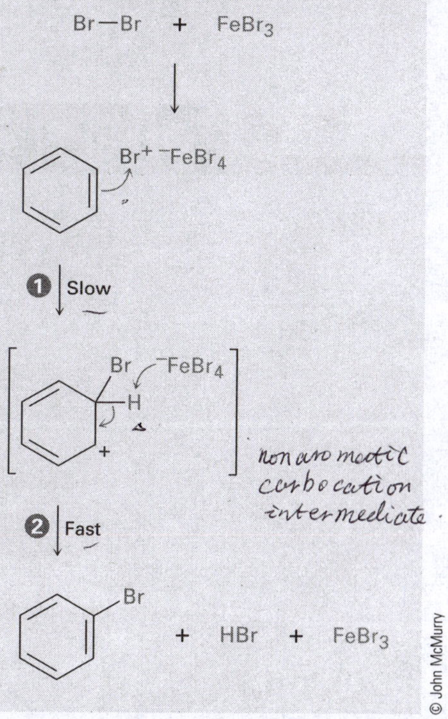

❶ An electron pair from the benzene ring attacks the positively polarized bromine, forming a new C–Br bond and leaving a nonaromatic carbocation intermediate.

❷ A base removes H^+ from the carbocation intermediate, and the neutral substitution product forms as two electrons from the C–H bond move to re-form the aromatic ring.

Why does the reaction of Br_2 with benzene take a different course than its reaction with an alkene? The answer is straightforward: if *addition* occurred, the resonance stabilization of the aromatic ring would be lost and the overall reaction would be energetically unfavorable. When *substitution* occurs, though, the resonance stability of the aromatic ring is retained and the reaction is favorable. An energy diagram for the overall process is shown in Figure 5.4.

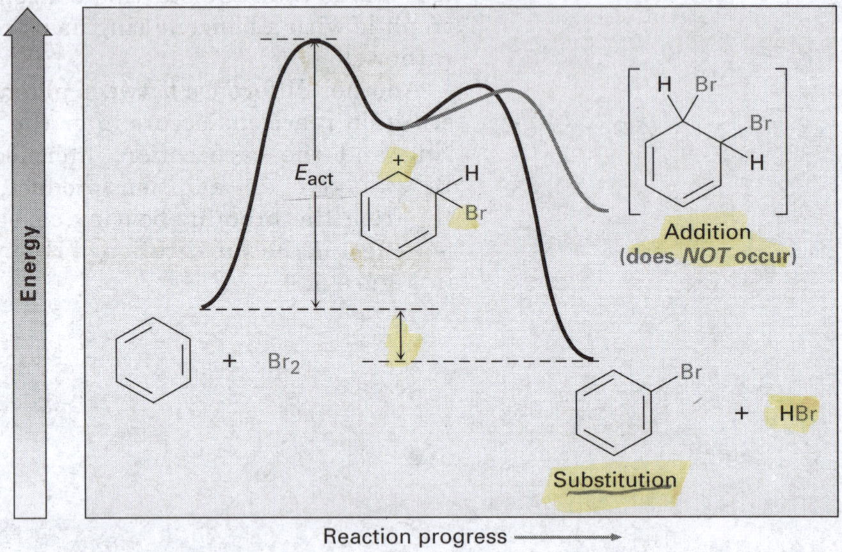

Figure 5.4 An energy diagram for the electrophilic bromination of benzene. The reaction occurs in two steps and releases energy.

Problem 5.5 There are three products that might form on reaction of toluene (methylbenzene) with Br_2. Draw and name them.

5.4 Other Electrophilic Aromatic Substitution Reactions

Many other electrophilic aromatic substitutions occur by the same general mechanism as bromination. Let's look at some briefly.

Chlorination and Iodination

Aromatic rings react with Cl_2 in the presence of $FeCl_3$ catalyst to yield chlorobenzenes, just as they react with Br_2 and $FeBr_3$ to give bromobenzenes. This kind of reaction is used in the synthesis of numerous pharmaceutical agents, including the antiallergy medication loratadine, marketed as Claritin.

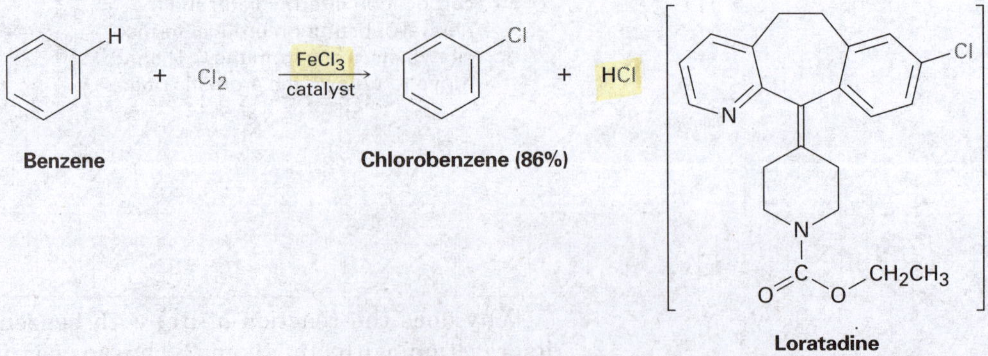

Electrophilic aromatic halogenations also occur in the biological synthesis of many naturally occurring molecules, particularly those produced by marine organisms. In humans, the best-known example occurs in the thyroid gland during the biosynthesis of thyroxine, a thyroid hormone involved in regulating growth and metabolism. The amino acid tyrosine is first iodinated by thyroid peroxidase, and two of the iodinated tyrosine molecules then couple. The electrophilic iodinating agent is an I^+ species, perhaps hypoiodous acid (HIO), that is formed from iodide ion by oxidation with H_2O_2.

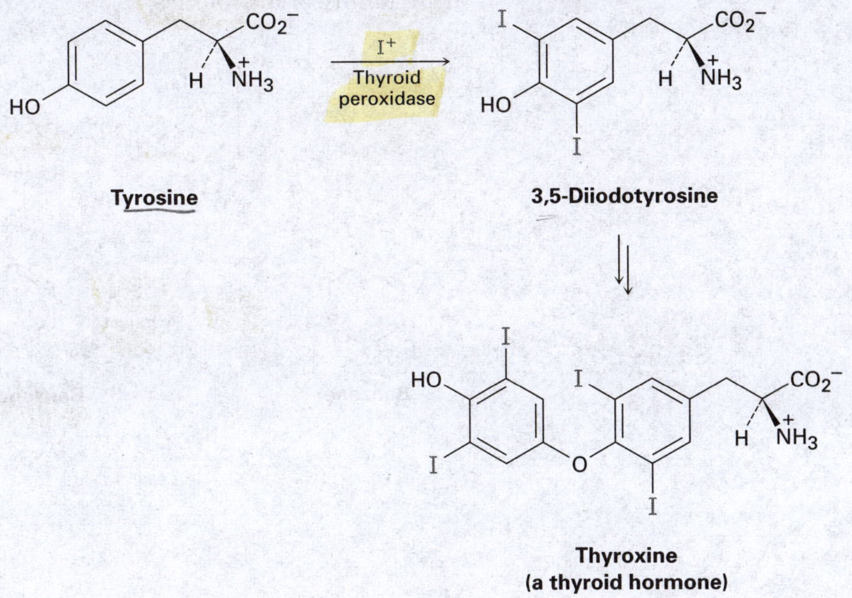

Nitration

Aromatic rings are nitrated by reaction with a mixture of concentrated nitric and sulfuric acids. The electrophile is the nitronium ion, NO_2^+, which is formed from HNO_3 by protonation and loss of water and which reacts with benzene in much the same way Br^+ does.

Aromatic nitration does not occur in nature but is particularly important in the laboratory because the nitro-substituted product can be reduced by reagents such as iron, tin, or $SnCl_2$ to yield an amino-substituted product, or *arylamine,* $ArNH_2$. Attachment of an amino group to an aromatic ring by the two-step nitration/reduction sequence is a key part of the industrial synthesis of many dyes and pharmaceutical agents.

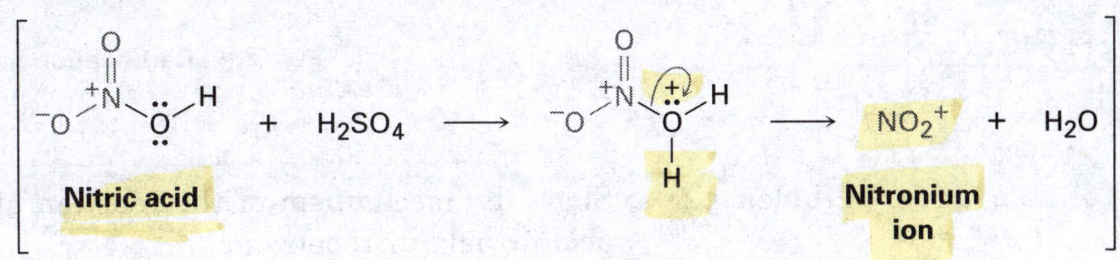

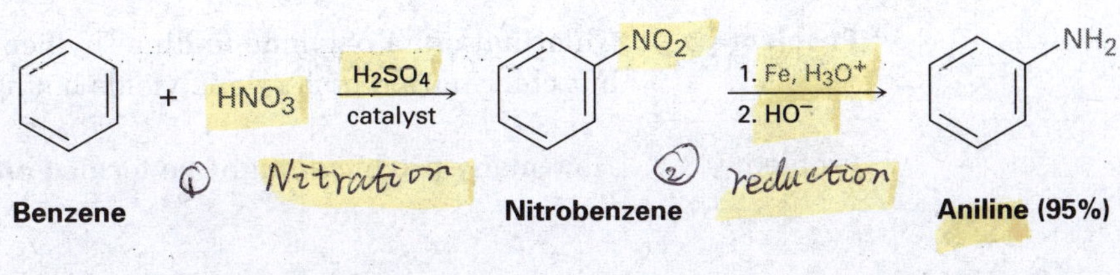

Sulfonation

Aromatic rings are sulfonated by reaction with so-called fuming sulfuric acid, a mixture of SO_3 and H_2SO_4. The reactive electrophile is HSO_3^+, and substitution occurs by the usual two-step mechanism seen for bromination. Aromatic sulfonation is a key step in the synthesis of such compounds as the sulfa drug family of antibiotics.

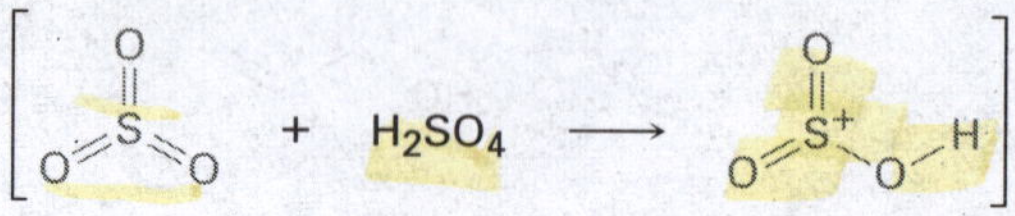

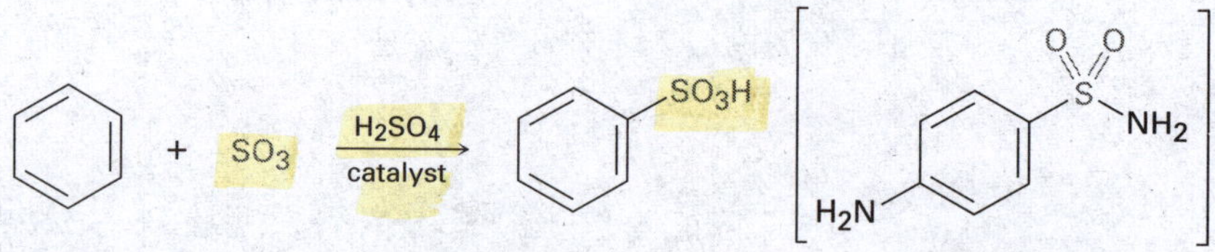

Worked Example 5.2

Writing the Mechanism of an Electrophilic Aromatic Substitution Reaction

Show the mechanism of the reaction of benzene with fuming sulfuric acid to yield benzenesulfonic acid.

Strategy The reaction of benzene with fuming sulfuric acid to yield benzenesulfonic acid is a typical electrophilic aromatic substitution reaction, which occurs by the usual two-step mechanism. An electrophile first adds to the aromatic ring, and H^+ is then lost. In sulfonation reactions, the electrophile is HSO_3^+.

Solution

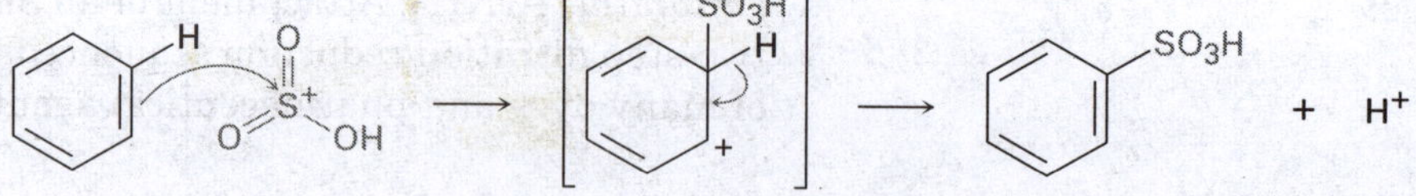

Carbocation intermediate

Problem 5.6 Show the mechanism of the reaction of benzene with nitric acid and sulfuric acid to yield nitrobenzene.

✓ **Problem 5.7** Chlorination of *o*-xylene (*o*-dimethylbenzene) yields a mixture of two products, but chlorination of *p*-xylene yields a single product. Explain.

✓ **Problem 5.8** How many products might be formed on chlorination of *m*-xylene?

5.5 The Friedel–Crafts Alkylation and Acylation Reactions

One of the most useful electrophilic aromatic substitution reactions is *alkylation*—the introduction of an alkyl group onto the benzene ring. Called the **Friedel–Crafts alkylation reaction** after its discoverers, the reaction is carried out by treating the aromatic compound with an alkyl chloride, RCl, in the presence of $AlCl_3$ to generate a carbocation electrophile, R^+. Aluminum chloride catalyzes the reaction by helping the alkyl halide to dissociate in much the same way that $FeBr_3$ catalyzes aromatic brominations by helping Br_2 dissociate (Section 5.3). Loss of H^+ then completes the reaction (Figure 5.5).

MECHANISM

Figure 5.5 Mechanism of the Friedel–Crafts alkylation reaction. The electrophile is a carbocation, generated by $AlCl_3$-assisted ionization of an alkyl chloride.

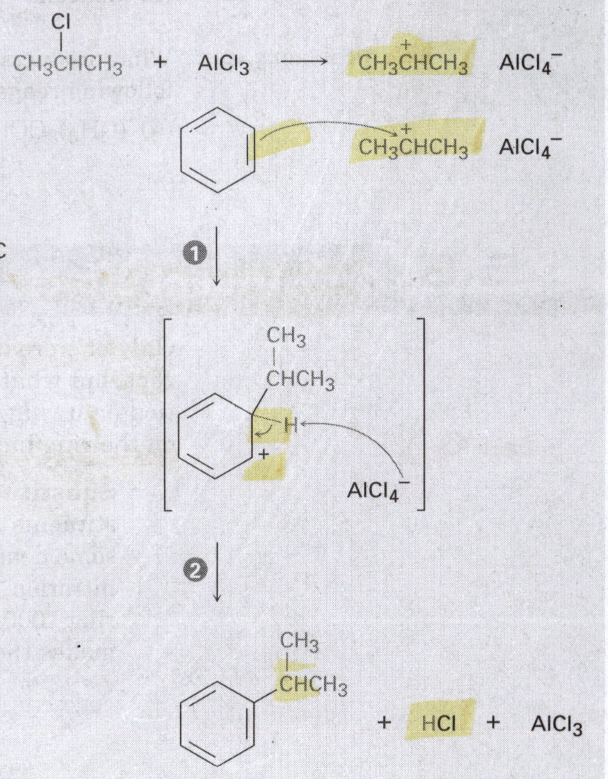

1. An electron pair from the aromatic ring attacks the carbocation, forming a C–C bond and yielding a new carbocation intermediate.
2. Loss of a proton then gives the neutral alkylated substitution product.

Despite its utility, the Friedel–Crafts alkylation reaction has several limitations. For one, only *alkyl* halides can be used. Aromatic (aryl) halides such as chlorobenzene don't react. In addition, Friedel–Crafts reactions don't succeed on aromatic rings that are already substituted by the groups $-NO_2$, $-C{\equiv}N$, $-SO_3H$, or $-COR$. Such aromatic rings are much less reactive than benzene for reasons we'll discuss in the next two sections.

Closely related to the Friedel–Crafts alkylation reaction is the **Friedel–Crafts acylation reaction**. When an aromatic compound is treated with a carboxylic acid chloride (RCOCl) in the presence of $AlCl_3$, an **acyl** (**a**-sil) **group** (R—C=O) is introduced onto the ring. For example, reaction of benzene with acetyl chloride yields the ketone acetophenone.

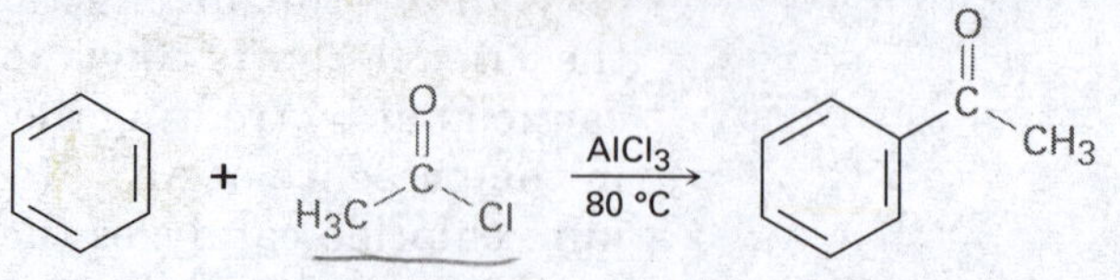

Benzene **Acetyl chloride** **Acetophenone (95%)**

Problem 5.9 What products would you expect to obtain from the reaction of the following compounds with chloroethane and $AlCl_3$?

(a) Benzene **(b)** *p*-Xylene

Problem 5.10 What products would you expect to obtain from the reaction of benzene with the following reagents?

(a) $(CH_3)_3CCl$, $AlCl_3$ **(b)** CH_3CH_2COCl, $AlCl_3$

5.6 Substituent Effects in Electrophilic Aromatic Substitution

Only one product can form when an electrophilic substitution occurs on benzene, but what would happen if we were to carry out an electrophilic substitution on a ring that already has a substituent? A substituent already present on the ring has two effects:

- **Substituents affect the *reactivity* of an aromatic ring**. Some substituents activate a ring, making it more reactive than benzene, and some deactivate a ring, making it less reactive than benzene. In aromatic nitration, for instance, the presence of an –OH substituent makes the ring 1000 times more reactive than benzene, while an $-NO_2$ substituent makes the ring more than 10 million times less reactive.

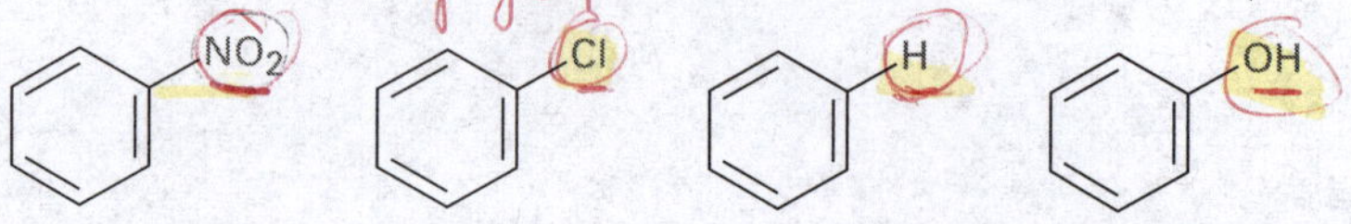

Relative rate of nitration	6×10^{-8}	0.033	1	1000

Reactivity →

- **Substituents affect the *orientation* of a reaction**. The three possible disubstituted products—ortho, meta, and para—are usually not formed in equal amounts. Instead, the nature of the substituent already present on the ring determines the position of the second substitution. An –OH group directs further substitution toward the ortho and para positions, for instance, while a –CN directs further substitution primarily toward the meta position.

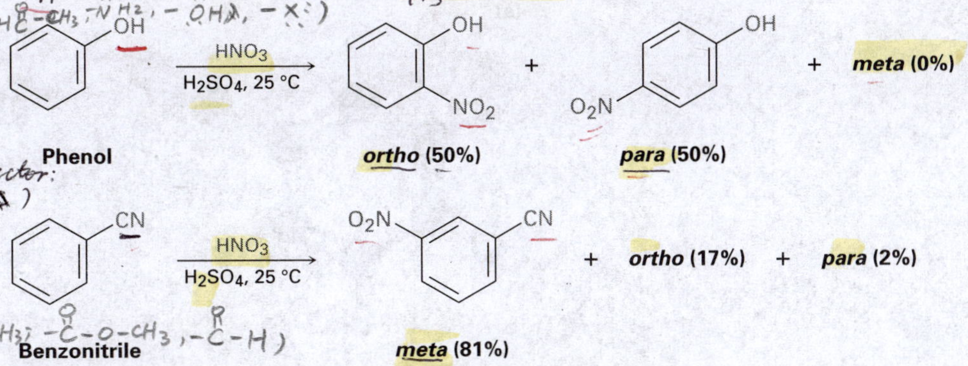

Substituents can be classified into three groups, as shown in Figure 5.6: *meta-directing deactivators*, *ortho- and para-directing deactivators*, and *ortho- and para-directing activators*. There are no meta-directing activators. Note how the directing effect of a group correlates with its reactivity. All meta-directing groups are deactivating, and all ortho- and para-directing groups other than halogen are activating. The halogens are unique in being ortho- and para-directing but deactivating.

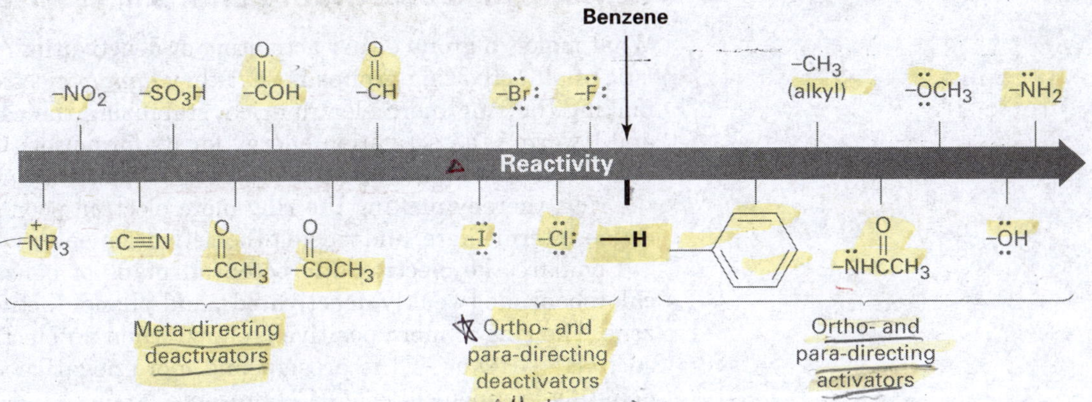

Figure 5.6 Substituent effects in electrophilic aromatic substitutions. All activating groups are ortho- and para-directing, and all deactivating groups other than halogen are meta-directing. The halogens are unique in being deactivating but ortho- and para-directing.

Worked Example 5.3 Predicting Relative Reactivity in Electrophilic Aromatic Substitution Reactions

Which would you expect to react faster in an electrophilic aromatic substitution reaction, chlorobenzene or ethylbenzene? Explain.

Strategy Look at Figure 5.6, and compare the relative reactivities of chloro and alkyl groups.

Solution A chloro substituent is deactivating, whereas an alkyl group is activating. Thus, ethylbenzene is more reactive than chlorobenzene.

Problem 5.11 Use Figure 5.6 to rank the compounds in each of the following groups in order of their reactivity toward electrophilic aromatic substitution:

(a) Nitrobenzene, phenol (hydroxybenzene), toluene
(b) Phenol, benzene, chlorobenzene, benzoic acid
(c) Benzene, bromobenzene, benzaldehyde, aniline (aminobenzene)

Problem 5.12 Draw and name the products you would expect to obtain by reaction of the following substances with Cl_2 and $FeCl_3$ (blue = N, reddish brown = Br):

(a)

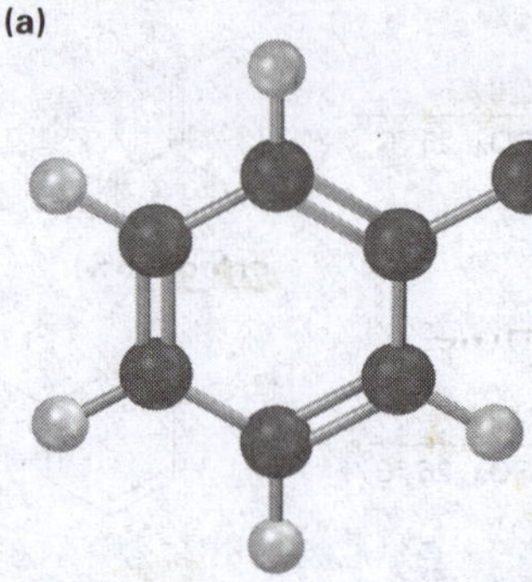

(b)

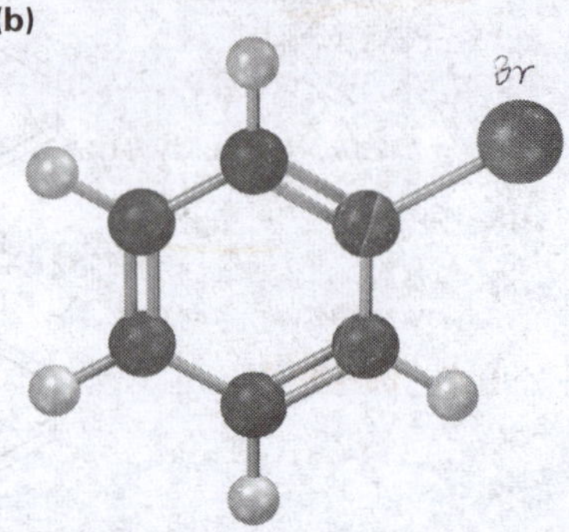

5.7 An Explanation of Substituent Effects

We saw in the previous section that substituents affect both the reactivity of an aromatic ring and the orientation of further aromatic substitutions. Let's look at the effects separately.

Activating and Deactivating Effects in Aromatic Rings

What makes a group either activating or deactivating? The common characteristic of all activating groups is that they *donate* electrons to the ring, thereby making the ring more electron-rich, stabilizing the carbocation intermediate, and lowering the activation energy for its formation. Conversely, the common characteristic of all deactivating groups is that they *withdraw* electrons from the ring, thereby making the ring more electron-poor, destabilizing the carbocation intermediate, and raising the activation energy for its formation.

Compare the electrostatic potential maps of benzaldehyde (deactivated), chlorobenzene (weakly deactivated), and phenol (activated) with that of benzene. The ring is more positive (yellow) when an electron-withdrawing group such as –CHO or –Cl is present and more negative (red) when an electron-donating group such as –OH is present.

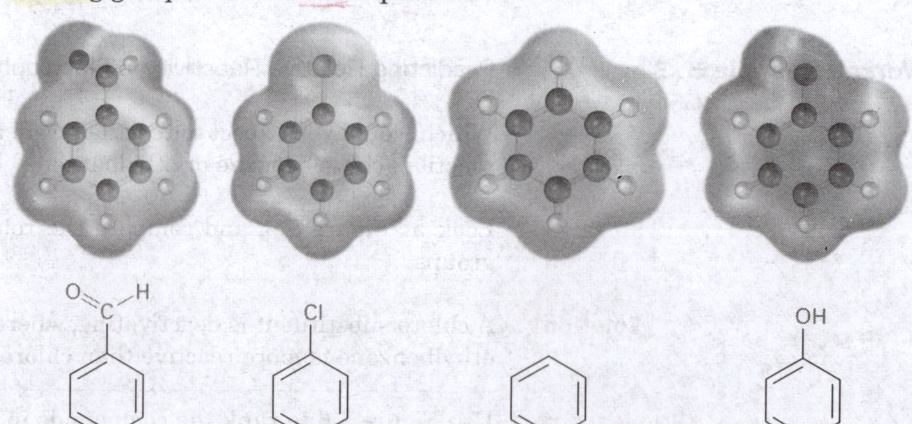

Benzaldehyde **Chlorobenzene** **Benzene** **Phenol**

Electron donation or withdrawal may occur by either an inductive effect (Section 1.9) or a resonance effect (Section 4.10). An inductive effect is the withdrawal or donation of electrons through a σ bond due to an electronegativity difference between the ring and the attached substituent atom, while a resonance effect is the withdrawal or donation of electrons through a π bond due to the overlap of a p orbital on the substituent with a p orbital on the aromatic ring.

Orienting Effects in Aromatic Rings: Ortho and Para Directors

Let's look at the nitration of phenol as an example of how ortho- and para-directing substituents work. In the first step, reaction with the electrophilic nitronium ion (NO_2^+) can occur either ortho, meta, or para to the –OH group, giving the carbocation intermediates shown in Figure 5.7. The ortho and para intermediates are more stable than the meta intermediate because they have more resonance forms—four rather than three—including a particularly favorable one that allows the positive charge to be stabilized by electron donation from the substituent oxygen atom. Because the ortho and para intermediates are more stable than the meta intermediate, they are formed faster.

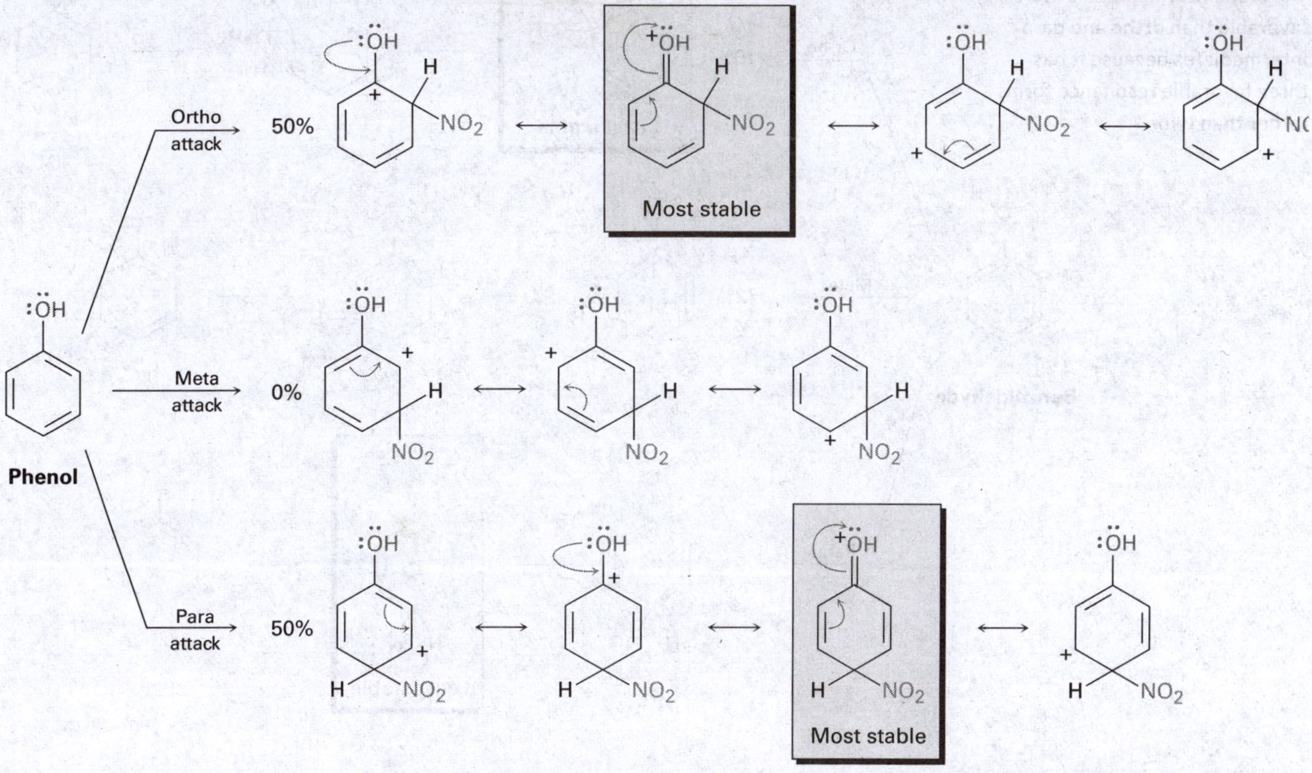

Figure 5.7 Carbocation intermediates in the nitration of phenol. The ortho and para intermediates are more stable than the meta intermediate because they have more resonance forms, including a particularly favorable one that involves electron donation from the oxygen atom.

In general, any substituent that has a lone pair of electrons on the atom directly bonded to the aromatic ring allows an electron-donating resonance interaction to occur and thus acts as an ortho and para director.

Ortho/para directors

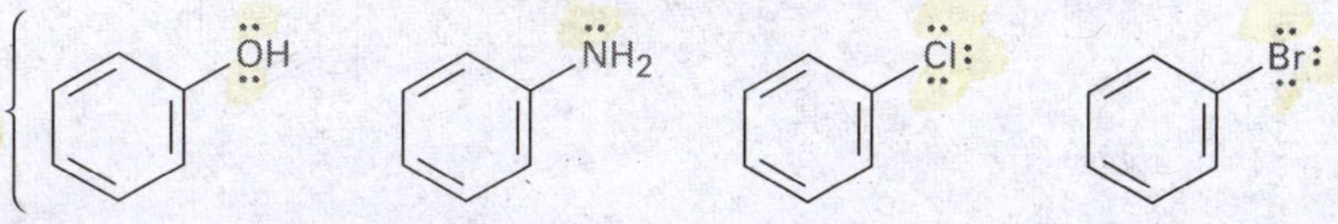

Orienting Effects in Aromatic Rings: Meta Directors

The influence of meta-directing substituents can be explained using the same kinds of arguments used for ortho and para directors. Look at the chlorination of benzaldehyde, for instance (Figure 5.8). Of the three possible carbocation intermediates, the meta intermediate has three favorable resonance forms, while the ortho and para intermediates have only two. In both ortho and para intermediates, the third resonance form is particularly unfavorable because it places the positive charge directly on the carbon that bears the aldehyde group, where it is disfavored by a repulsive interaction with the positively polarized carbon atom of the C=O group. Hence, the meta intermediate is more favored and is formed faster than the ortho and para intermediates.

Figure 5.8 Intermediates in the chlorination of benzaldehyde. The meta intermediate is more favorable than ortho and para intermediates because it has three favorable resonance forms rather than two.

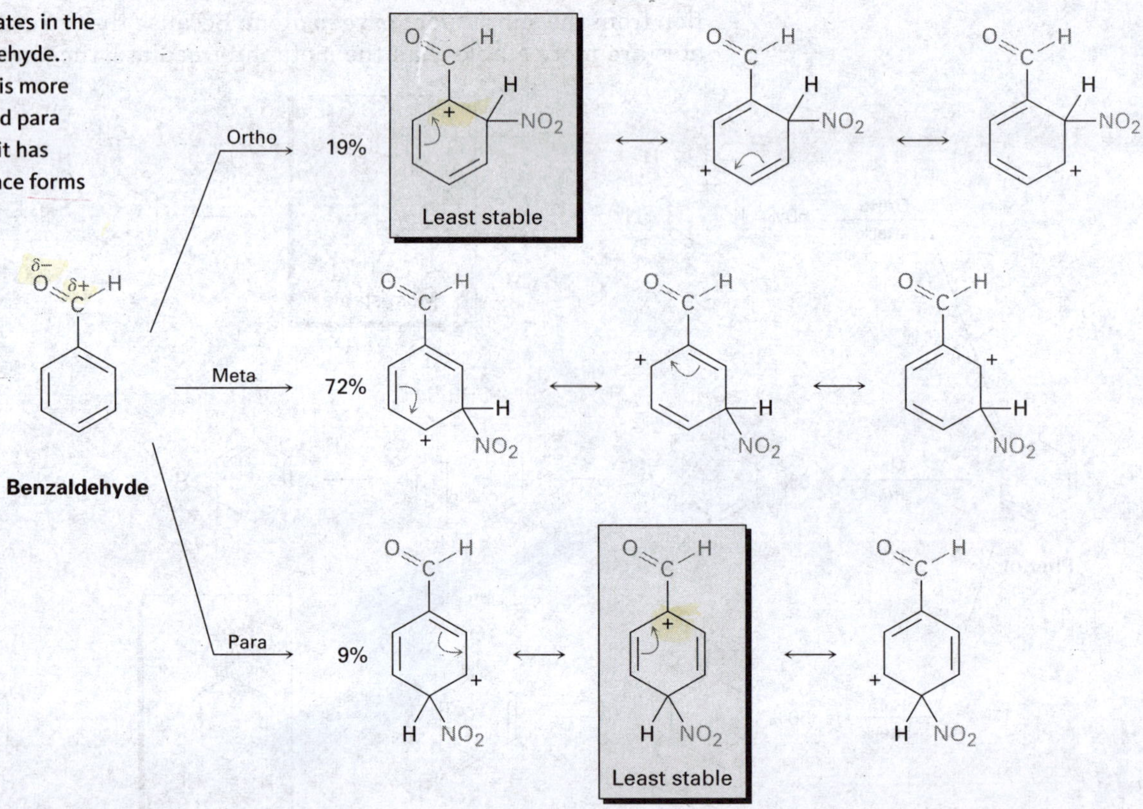

In general, any substituent that has a positively polarized atom ($\delta+$) directly attached to the ring makes one of the resonance forms of the ortho and para intermediates unfavorable, and thus acts as a meta director.

Meta directors

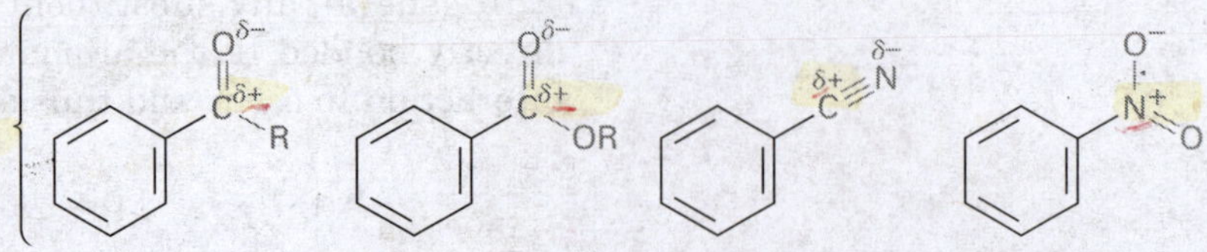

Worked Example 5.4 **Predicting the Product of an Electrophilic Aromatic Substitution Reaction**

What product(s) would you expect from bromination of aniline, $C_6H_5NH_2$?

Strategy Look at Figure 5.6 to see whether the $-NH_2$ substituent is ortho- and para-directing or meta-directing. Because an amino group has a lone pair of electrons on the nitrogen atom, it is ortho- and para-directing and we expect to obtain a mixture of *o*-bromoaniline and *p*-bromoaniline.

Solution

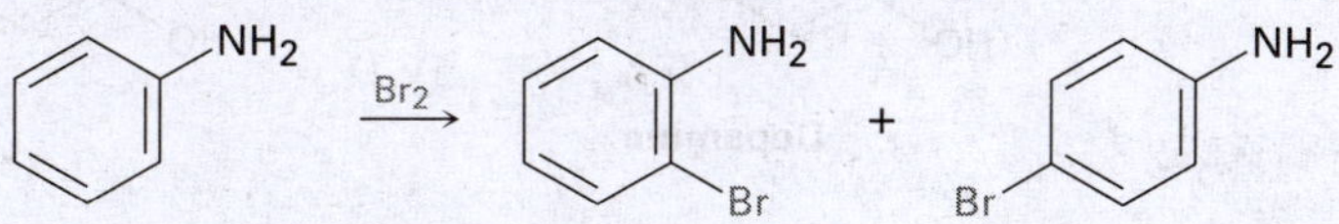

Aniline ***o*-Bromoaniline** ***p*-Bromoaniline**

Problem 5.13 What product(s) would you expect from sulfonation of the following compounds?

(a) Nitrobenzene **(b)** Bromobenzene **(c)** Toluene
(d) Benzoic acid **(e)** Benzonitrile

Problem 5.14 Draw resonance structures of the three possible carbocation intermediates to show how a methoxyl group ($-OCH_3$) directs bromination toward ortho and para positions.

Problem 5.15 Draw resonance structures of the three possible carbocation intermediates to show how an acetyl group, $CH_3C{=}O$, directs bromination toward the meta position.

5.8 Oxidation and Reduction of Aromatic Compounds

Despite its unsaturation, a benzene ring does not usually react with strong oxidizing agents such as $KMnO_4$. (Recall from Section 4.6 that $KMnO_4$ cleaves alkene C=C bonds.) Alkyl groups attached to the aromatic ring are readily attacked by oxidizing agents, however, and are converted into carboxyl groups ($-CO_2H$). For example, butylbenzene is oxidized by $KMnO_4$ to give benzoic acid. The mechanism of this reaction is complex and involves attack on the side-chain C–H bonds at the position next to the aromatic ring (the **benzylic position**) to give radical intermediates.

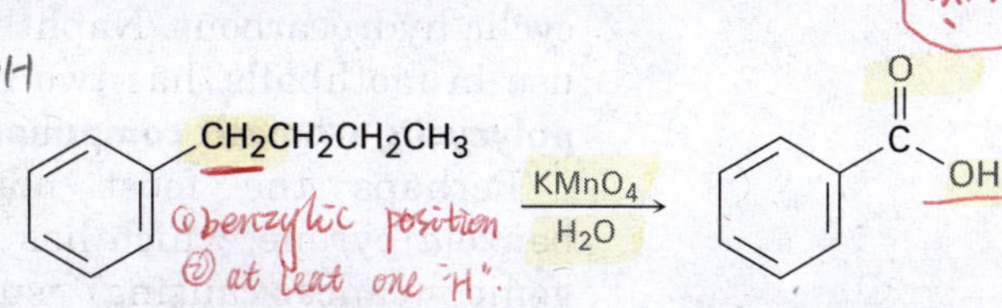

Butylbenzene **Benzoic acid (85%)**

Analogous oxidations occur in various biological pathways. The neurotransmitter norepinephrine, for instance, is biosynthesized from dopamine by a benzylic hydroxylation reaction. The process is catalyzed by the copper-containing enzyme dopamine β-monooxygenase and occurs by a radical mechanism.

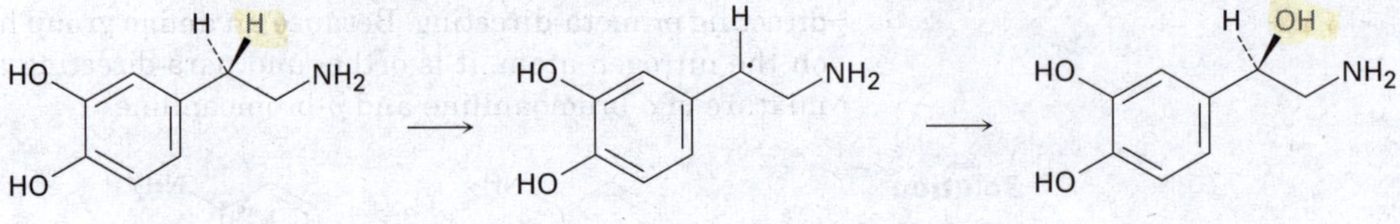

Dopamine **Norepinephrine**

Just as aromatic rings are usually inert to oxidation, they are also inert to reduction under typical alkene hydrogenation conditions. Only if high temperatures and pressures are used does reduction of an aromatic ring occur. For example, *o*-dimethylbenzene (*o*-xylene) gives *cis*-1,2-dimethylcyclohexane if reduced at high pressure.

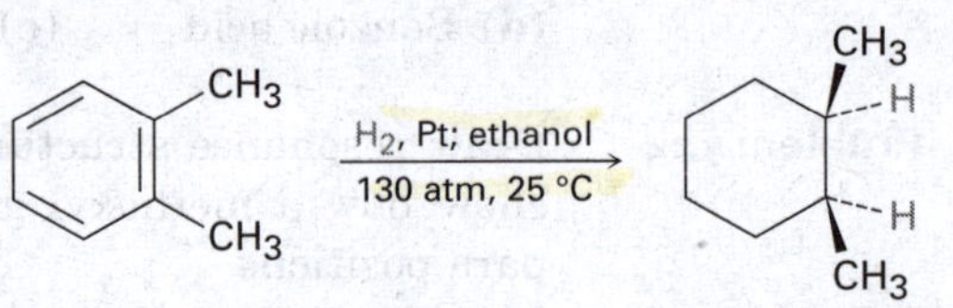

***o*-Xylene** ***cis*-1,2-Dimethylcyclohexane**

Problem 5.16 What aromatic products would you expect to obtain from oxidation of the following substances with $KMnO_4$?

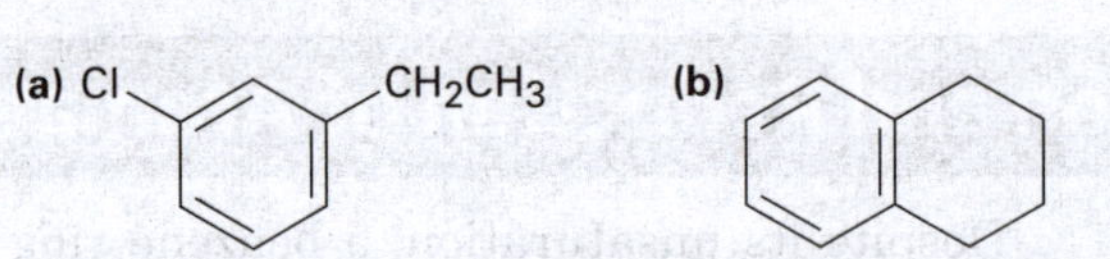

Tetralin

5.9 Other Aromatic Compounds

The concept of aromaticity—the unusual chemical stability present in cyclic conjugated molecules like benzene—can be extended beyond simple monocyclic hydrocarbons. Naphthalene, for instance, a substance familiar for its use in mothballs, has two benzene-like rings fused together and is thus a **polycyclic aromatic compound**.

Perhaps the most notorious polycyclic aromatic hydrocarbon is benzo[*a*]pyrene, which has five benzene-like rings and is a major carcinogenic (cancer-causing) substance found in chimney soot, cigarette smoke, and well-done barbecued meat. Once in the body, benzo[*a*]pyrene is metabolically converted into a diol epoxide that binds to DNA, where it induces mutations.

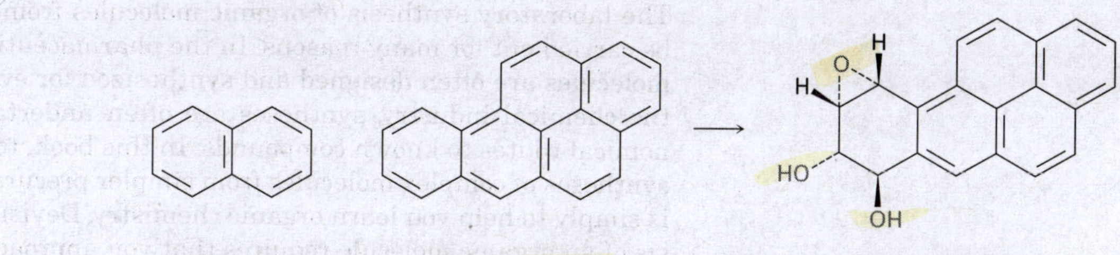

Naphthalene Benzo[a]pyrene A diol epoxide

In addition to monocyclic and polycyclic aromatic hydrocarbons, some aromatic compounds are **heterocycles**—cyclic compounds that contain atoms of two or more elements in their rings. Nitrogen, sulfur, and oxygen atoms are all found along with carbon in various aromatic compounds. We'll see in Chapter 12, for instance, that the nitrogen-containing heterocycles pyridine, pyrimidine, pyrrole, and imidazole are aromatic, even though they aren't hydrocarbons and even though two of them have five-membered rather than six-membered rings (Figure 5.9). They are aromatic because they all, like benzene, contain a cyclic conjugated array of six π electrons. Pyridine and pyrimidine have one π electron on each of their six ring atoms. Pyrrole and imidazole have one π electron on each of four ring atoms and an additional two π electrons (the lone pair) on their N–H nitrogen.

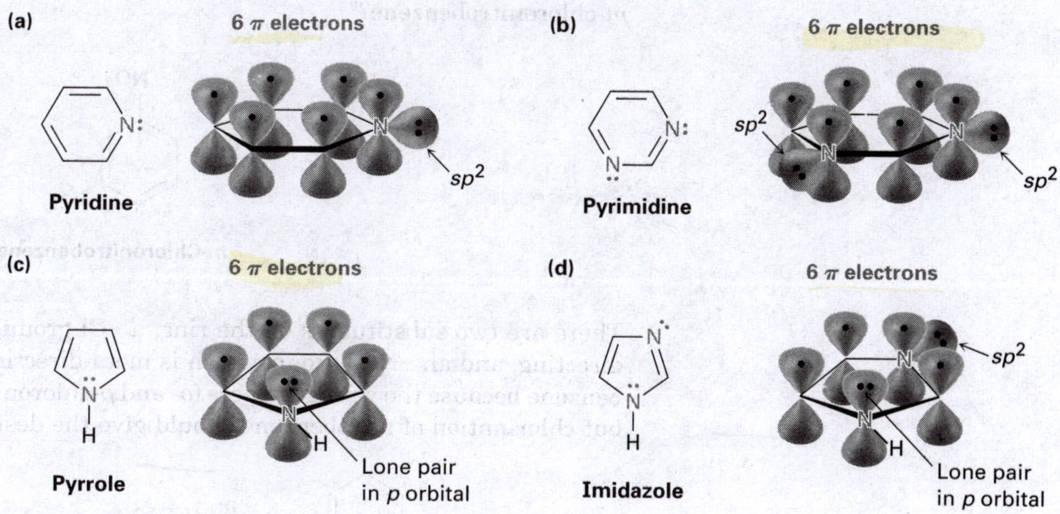

Figure 5.9 Orbital views of the nitrogen-containing compounds **(a)** pyridine, **(b)** pyrimidine, **(c)** pyrrole, and **(d)** imidazole. All are aromatic because, like benzene, they contain a cyclic conjugated system of six π electrons. Pyridine and pyrimidine have one π electron on each of their six ring atoms. Pyrrole and imidazole have one π electron on each of four ring atoms and an additional two π electrons (the lone pair) on their N–H nitrogen.

Problem 5.17 There are three resonance structures of naphthalene, of which only one is shown. Draw the other two.

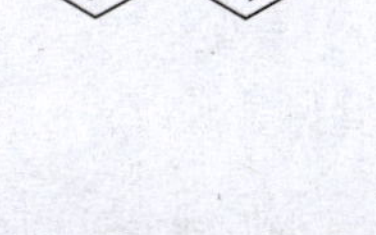

Naphthalene

5.10 Organic Synthesis

The laboratory synthesis of organic molecules from simple precursors might be carried out for many reasons. In the pharmaceutical industry, new organic molecules are often designed and synthesized for evaluation as medicines. In the chemical industry, syntheses are often undertaken to devise more economical routes to known compounds. In this book, too, we'll sometimes devise syntheses of complex molecules from simpler precursors, but the purpose here is simply to help you learn organic chemistry. Devising a route for the synthesis of an organic molecule requires that you approach chemical problems in a logical way, draw on your knowledge of organic reactivity, and organize that knowledge into a workable plan.

The only trick to devising an organic synthesis is to *work backward*. Look at the product and ask yourself, "What is the immediate precursor of that product?" Having found an immediate precursor, work backward again, one step at a time, until a suitable starting material is found. Let's try some examples.

Worked Example 5.5 Synthesizing a Substituted Aromatic Compound

Synthesize *m*-chloronitrobenzene starting from benzene.

Strategy Work backward by first asking, "What is an immediate precursor of *m*-chloronitrobenzene?"

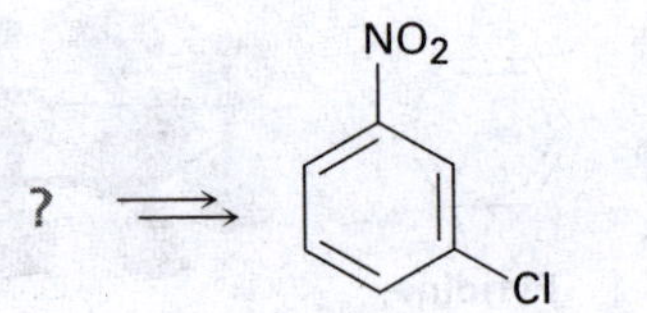

There are two substituents on the ring, a –Cl group, which is ortho- and para-directing, and an $-NO_2$ group, which is meta-directing. We can't nitrate chlorobenzene because the wrong isomers (*o*- and *p*-chloronitrobenzenes) would result, but chlorination of nitrobenzene should give the desired product.

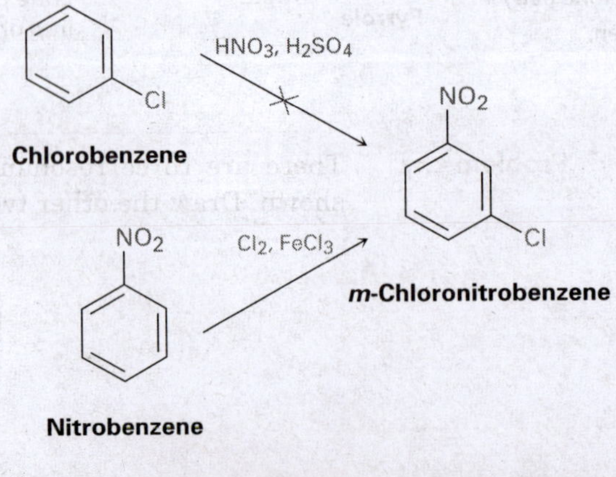

"What is an immediate precursor of nitrobenzene?" Benzene, which can be nitrated.

Solution We've solved the problem in two steps:

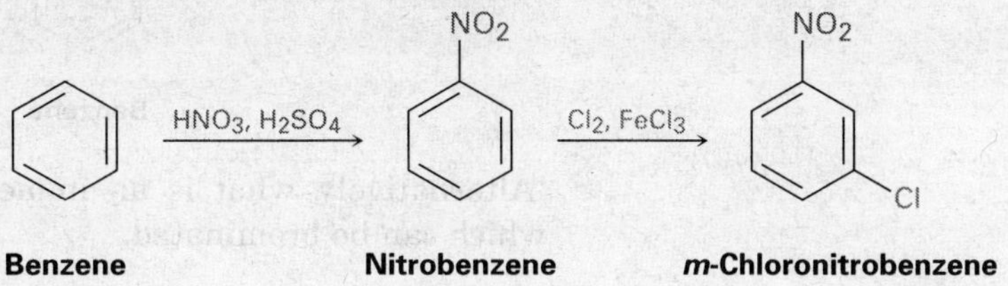

Worked Example 5.6

Synthesizing a Substituted Aromatic Compound

Synthesize *p*-bromobenzoic acid starting from benzene.

Strategy Work backward by first asking, "What is an immediate precursor of *p*-bromobenzoic acid?"

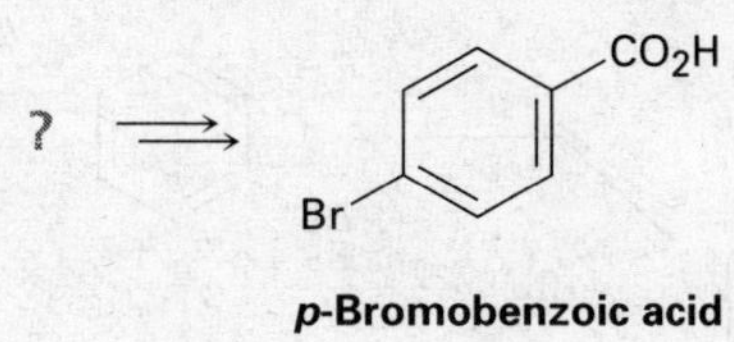

There are two substituents on the ring, a $-CO_2H$ group, which is meta-directing, and a $-Br$ atom, which is ortho- and para-directing. We can't brominate benzoic acid because the wrong isomer (*m*-bromobenzoic acid) would be formed. We've seen, however, that oxidation of alkylbenzene side chains yields benzoic acids. An immediate precursor of our target molecule might therefore be *p*-bromotoluene.

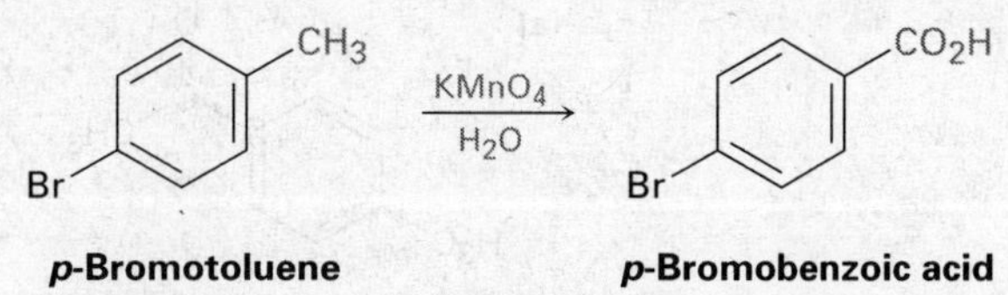

"What is an immediate precursor of *p*-bromotoluene?" Perhaps toluene, because the methyl group would direct bromination to the ortho and para positions, and we could then separate isomers. Alternatively, bromobenzene might be an immediate precursor because we could carry out a Friedel–Crafts alkylation and obtain the para product. Both methods are satisfactory.

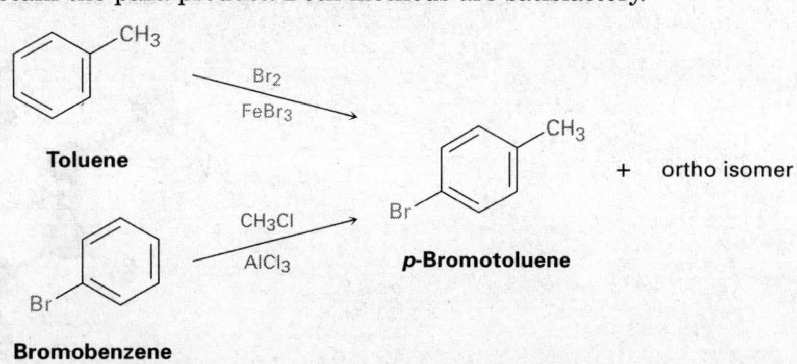

"What is an immediate precursor of toluene?" Benzene, which can be methylated in a Friedel–Crafts reaction.

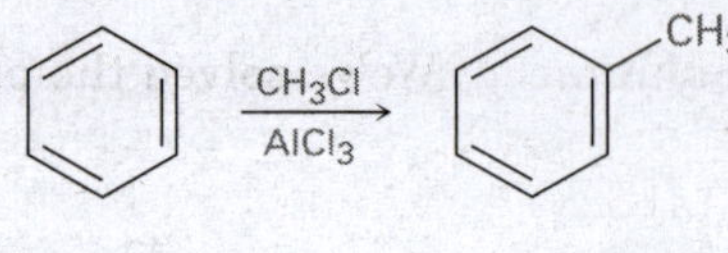

Benzene **Toluene**

"Alternatively, what is an immediate precursor of bromobenzene?" Benzene, which can be brominated.

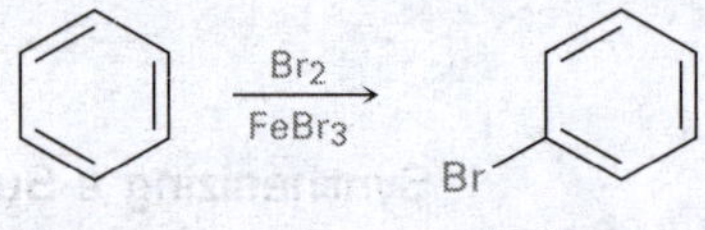

Benzene **Bromobenzene**

Solution Our backward synthetic (*retrosynthetic*) analysis has provided two workable routes from benzene to *p*-bromobenzoic acid.

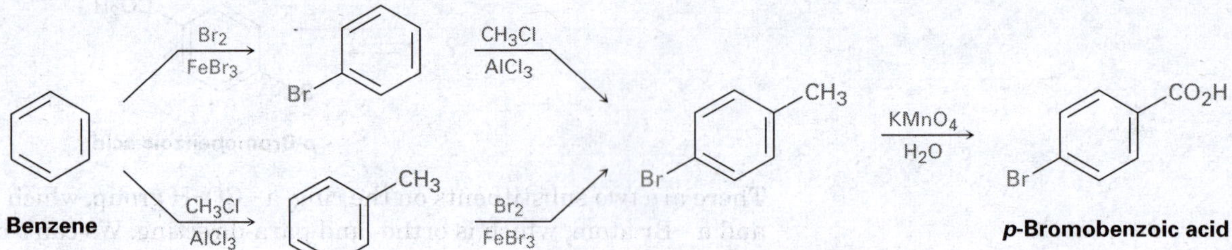

Benzene ***p*-Bromobenzoic acid**

Problem 5.18 Propose syntheses of the following substances starting from benzene:

(a)

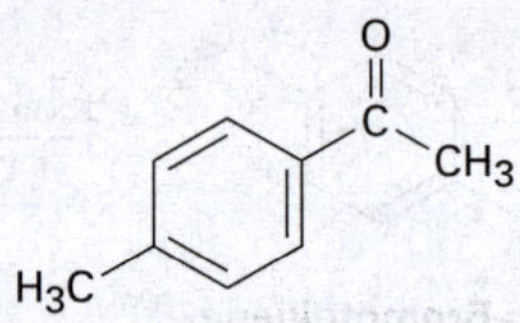

(b)

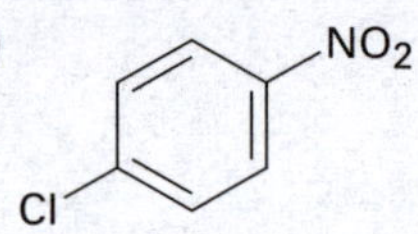

Problem 5.19 Synthesize the following substances from benzene:

(a) *o*-Bromotoluene **(b)** 2-Bromo-1,4-dimethylbenzene

Problem 5.20 How would you prepare the following substance from benzene? (Yellow-green = Cl.)

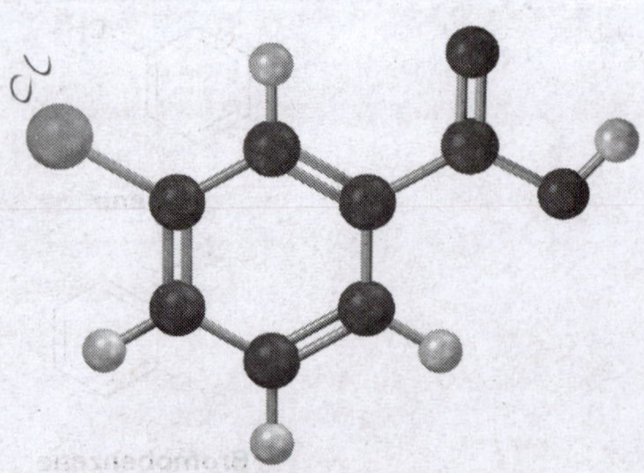

INTERLUDE

Aspirin, NSAIDs, and COX-2 Inhibitors

Long-distance runners sometimes call ibuprofen "the fifth basic food group" because of its ability to control aches and pains.

Whatever the cause—tennis elbow, a sprained ankle, or a wrenched knee—pain and inflammation seem to go together. They are, however, different in their origin, and powerful drugs are available for treating each separately. Codeine, for example, is a powerful *analgesic,* or pain reliever, used in the management of debilitating pain, while cortisone and related steroids are potent *anti-inflammatory* agents, used for treating arthritis and other crippling inflammations. For minor pains and inflammation, both problems are often treated at the same time by using a common, over-the-counter medication called an *NSAID,* or nonsteroidal anti-inflammatory drug.

The most common NSAID is aspirin, or acetylsalicylic acid, whose use goes back to the late 1800s. It had been known from before the time of Hippocrates in 400 BC that fevers could be lowered by chewing the bark of willow trees. The active agent in willow bark was found in 1827 to be an aromatic compound called *salicin,* which could be converted by reaction with water into salicyl alcohol and then oxidized to give salicylic acid. Salicylic acid turned out to be even more effective than salicin for reducing fever and to have analgesic and anti-inflammatory action as well. Unfortunately, it also turned out to be too corrosive to the walls of the stomach for everyday use. Conversion of the phenol –OH group into an acetate ester, however, yielded acetylsalicylic acid, which proved just as potent as salicylic acid but less corrosive to the stomach.

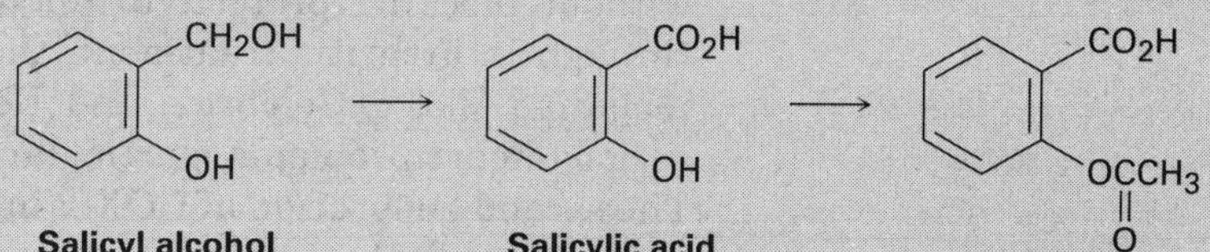

Salicyl alcohol **Salicylic acid** **Acetylsalicylic acid (aspirin)**

Although extraordinary in its powers, aspirin is also more dangerous than commonly believed. A dose of only about 15 g can be fatal to a small child, and aspirin can cause stomach bleeding and allergic reactions in long-term users. Even more serious is a condition called *Reye's syndrome,* a potentially fatal reaction to aspirin sometimes seen in children recovering from the flu. As a result of these problems, numerous other NSAIDs have been developed in the last several decades, most notably ibuprofen and naproxen.

Like aspirin, both ibuprofen and naproxen are relatively simple aromatic compounds containing a side-chain carboxylic acid group. Ibuprofen, sold under the names Advil, Nuprin, Motrin, and others, has

continued

INTERLUDE

roughly the same potency as aspirin but is less prone to cause stomach upset. Naproxen, sold under the names Aleve and Naprosyn, also has about the same potency as aspirin but remains active in the body six times longer.

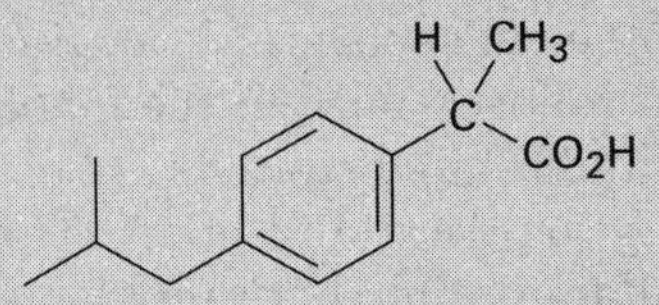

Ibuprofen
(Advil, Nuprin, Motrin)

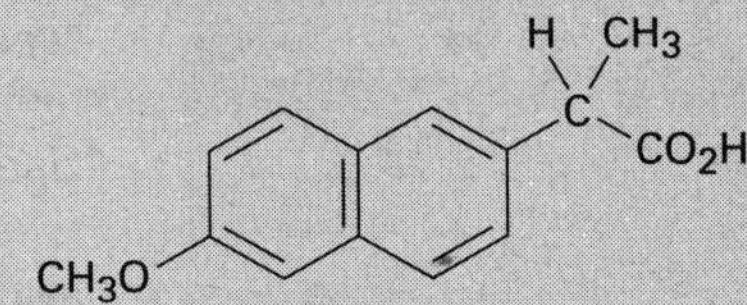

Naproxen
(Aleve, Naprosyn)

Aspirin and other NSAIDs function by blocking the cyclooxygenase (COX) enzymes that carry out the body's synthesis of compounds called prostaglandins (Section 6.3). There are two forms of the enzyme: COX-1, which carries out the normal physiological production of prostaglandins, and COX-2, which mediates the body's response to arthritis and other inflammatory conditions. Unfortunately, both COX-1 and COX-2 enzymes are blocked by aspirin, ibuprofen, and other NSAIDs, thereby shutting down not only the response to inflammation but also various protective functions, including the control mechanism for production of acid in the stomach.

Medicinal chemists have devised a number of drugs that act as selective inhibitors of the COX-2 enzyme. Inflammation is thereby controlled without blocking protective functions. Originally heralded as a breakthrough in arthritis treatment, the first generation of COX-2 inhibitors, including Vioxx, Celebrex, and Bextra, turned out to cause potentially serious heart problems, particularly in elderly or compromised patients. The second generation of COX-2 inhibitors now under development promises to be safer but will be closely scrutinized for side effects before gaining approval.

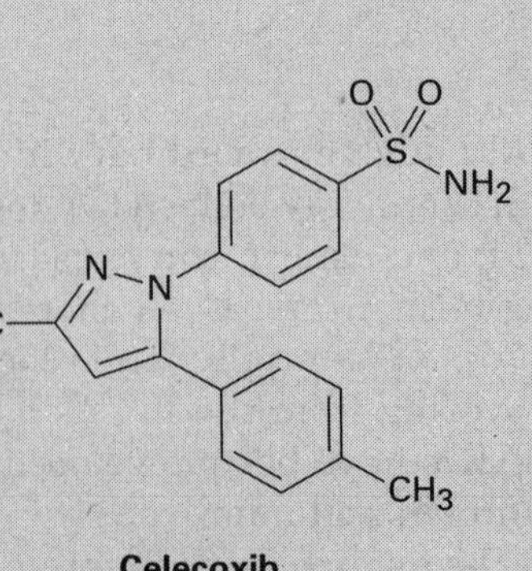

Celecoxib
(Celebrex)

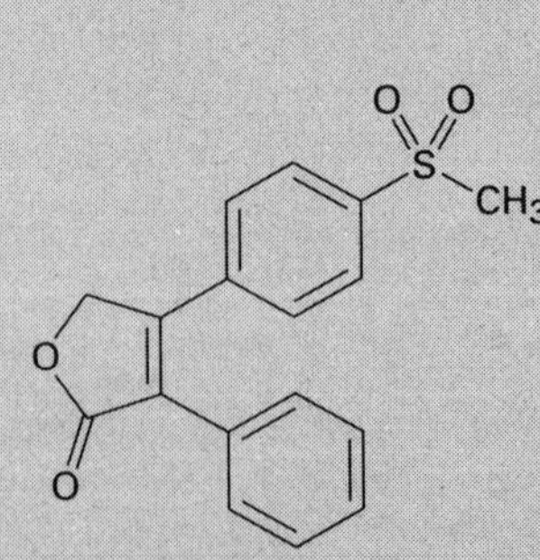

Rofecoxib
(Vioxx)

Summary and Key Words

acyl group 166
aromatic 155
aryl group 158
benzyl group 158
benzylic position 171
electrophilic aromatic substitution reaction 159
Friedel–Crafts acylation reaction 165
Friedel–Crafts alkylation reaction 165
heterocycle 173
phenyl group 157
polycyclic aromatic compound 172

Aromatic rings are a common part of many biological molecules and pharmaceutical agents and are particularly important in nucleic acid chemistry. In this chapter, we've seen how and why aromatic compounds are different from such apparently related compounds as alkenes, and we've seen some of their most common reactions.

The word **aromatic** refers to the class of compounds structurally related to benzene. Aromatic compounds are named according to IUPAC rules, with disubstituted benzenes referred to as either ortho (1,2 disubstituted), meta (1,3 disubstituted), or para (1,4 disubstituted). Benzene is a resonance hybrid of two equivalent forms, neither of which is correct by itself. The true structure of benzene is intermediate between the two.

The most common reaction of aromatic compounds is **electrophilic aromatic substitution**. In this two-step polar reaction, the π electrons of the aromatic ring first attack the electrophile to yield a resonance-stabilized carbocation intermediate, which then loses H^+ to give a substituted aromatic product. Bromination, chlorination, iodination, nitration, sulfonation, **Friedel–Crafts alkylation**, and **Friedel–Crafts acylation** can all be carried out. Friedel–Crafts alkylation is particularly useful for preparing a variety of alkylbenzenes but is limited because only alkyl halides can be used and strongly deactivated rings do not react.

Substituents on the aromatic ring affect both the reactivity of the ring toward further substitution and the orientation of that further substitution. Substituents can be classified as either activators or deactivators, and as either ortho and para directors or meta directors.

The side chains of alkylbenzenes have unique reactivity because of the neighboring aromatic ring. Thus, an alkyl group attached to the aromatic ring can be degraded to a carboxyl group ($-CO_2H$) by oxidation with aqueous $KMnO_4$. In addition, the aromatic ring can be reduced to yield a cyclohexane on catalytic hydrogenation at high pressure.

In addition to substituted benzenes, polycyclic hydrocarbons such as naphthalene are also aromatic, and nitrogen-containing **heterocycles** such as pyridine, pyrimidine, pyrrole, and imidazole are aromatic because their rings have the same six-π-electron electronic structure as benzene.

Summary of Reactions

1. Electrophilic aromatic substitution
   (a) Bromination (Section 5.3)

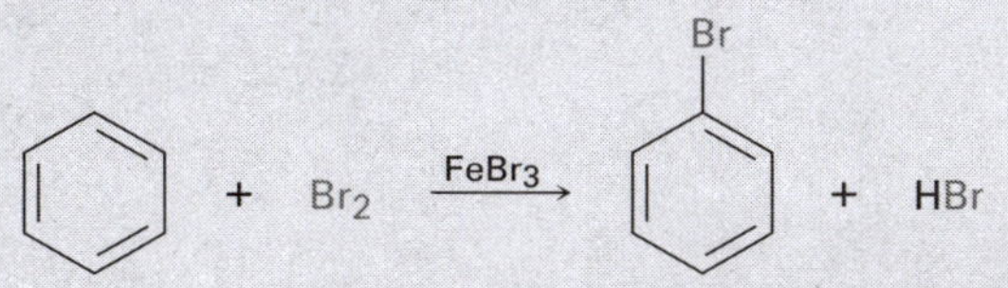

continued

(b) Chlorination (Section 5.4)

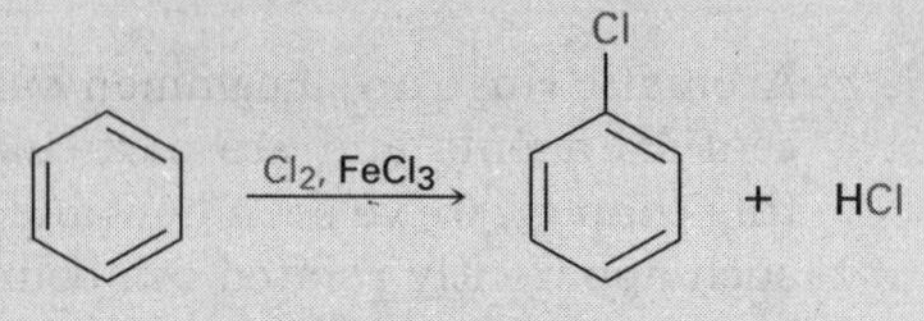

(c) Nitration (Section 5.4)

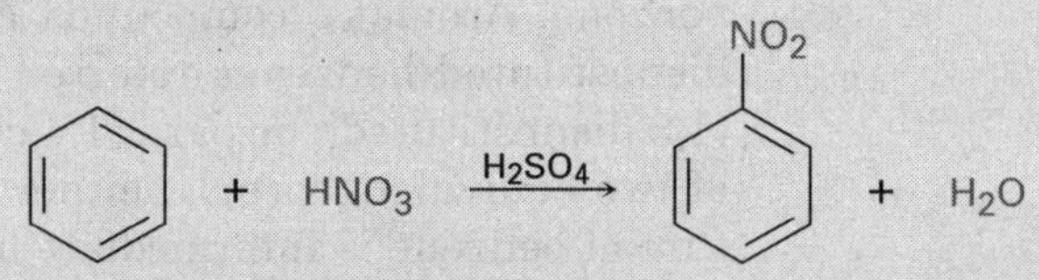

(d) Sulfonation (Section 5.4)

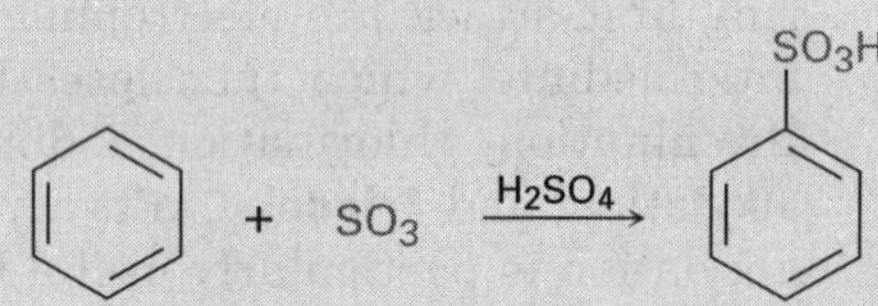

(e) Friedel–Crafts alkylation (Section 5.5)

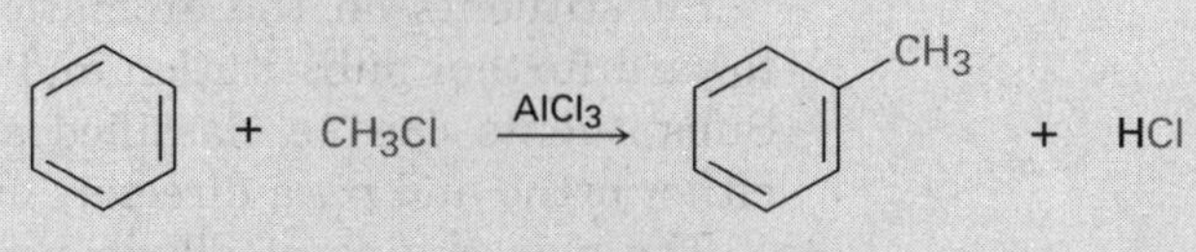

(f) Friedel–Crafts acylation (Section 5.5)

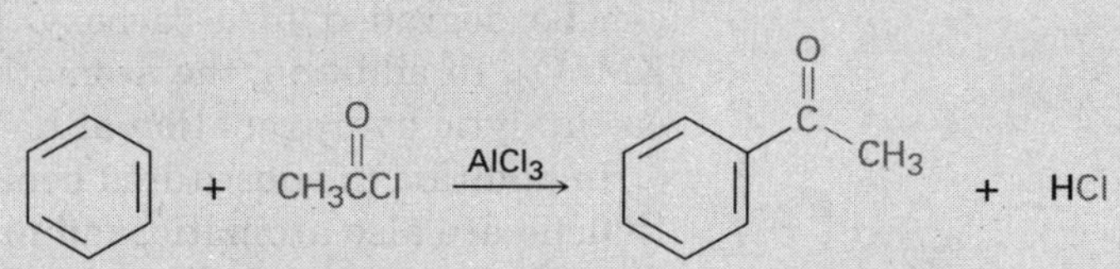

2. Oxidation of aromatic side chains (Section 5.8)

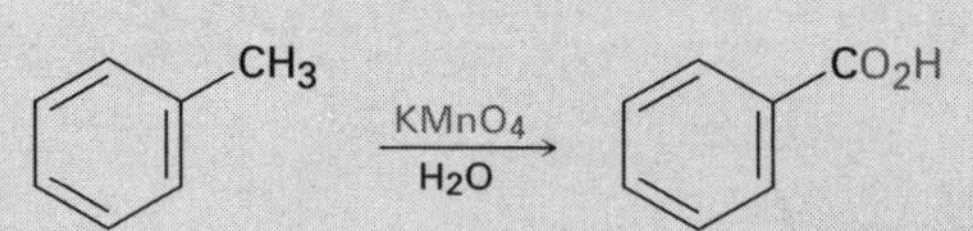

3. Hydrogenation of aromatic rings (Section 5.8)

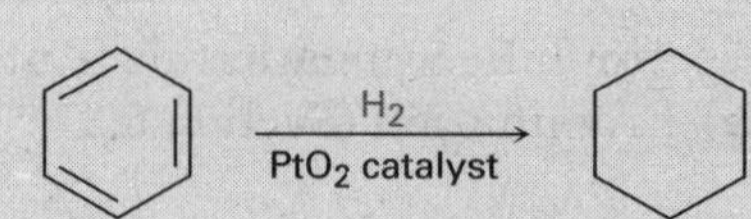

Exercises

Visualizing Chemistry

(Problems 5.1–5.20 appear within the chapter.)

OWL

Interactive versions of these problems are assignable in OWL.

5.21 The following structure represents a carbocation. Draw two resonance structures, indicating the positions of the double bonds.

5.22 Give IUPAC names for the following substances (red = O, blue = N).

(a)

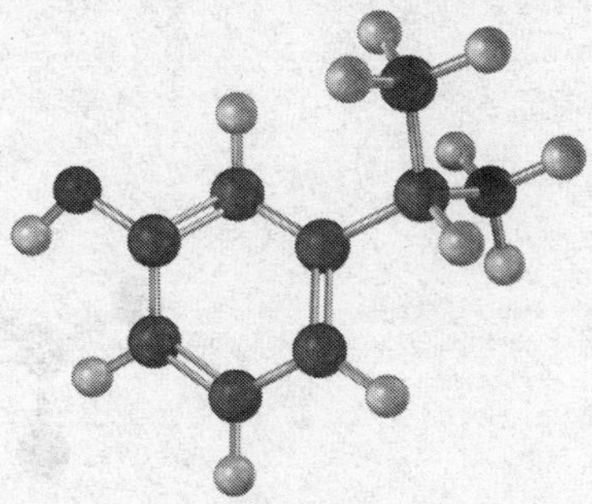

(b)

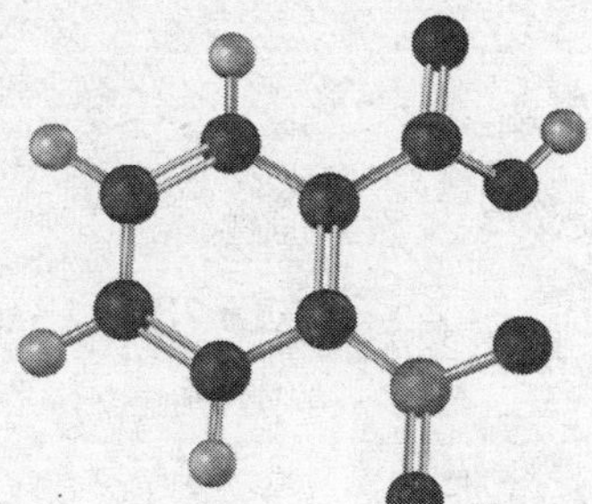

5.23 Draw and name the product from reaction of each of the following substances with (i) Br_2, $FeBr_3$ and (ii) CH_3COCl, $AlCl_3$ (red = O):

(a)

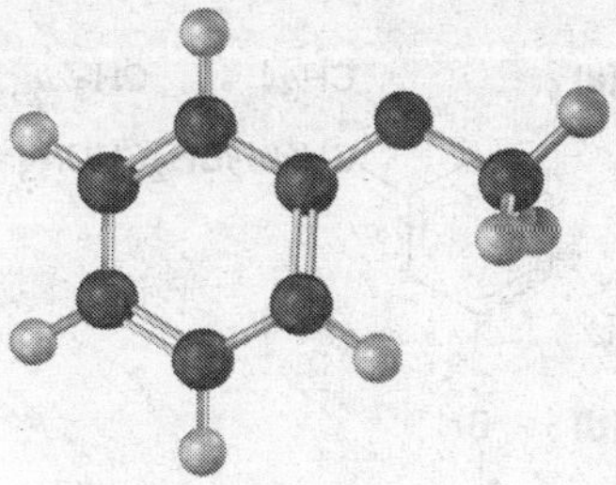

(b)

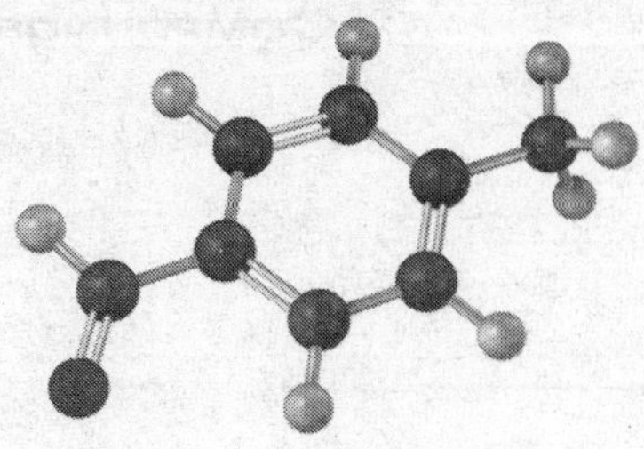

5.24 The following compound can't be synthesized using the methods discussed in this chapter. Why not?

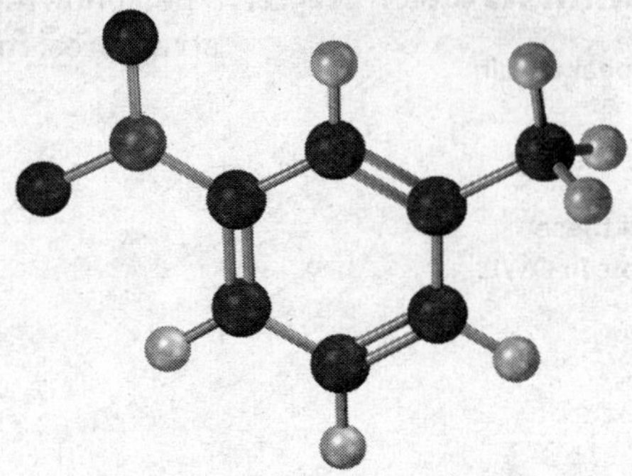

5.25 How would you synthesize the following compound starting from benzene? More than one step is needed.

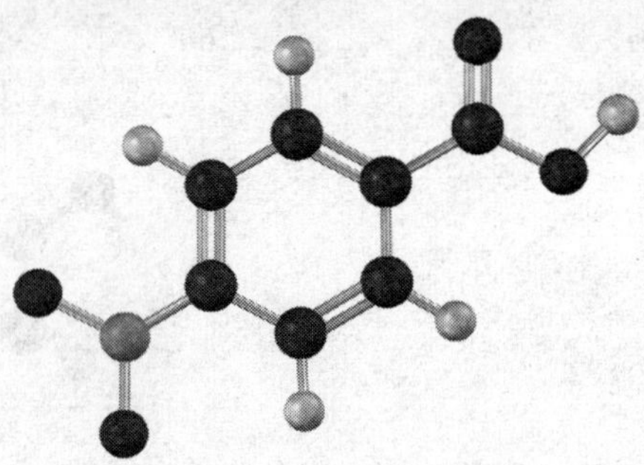

Additional Problems

Naming Aromatic Compounds

5.26 Give IUPAC names for the following compounds:

(a)

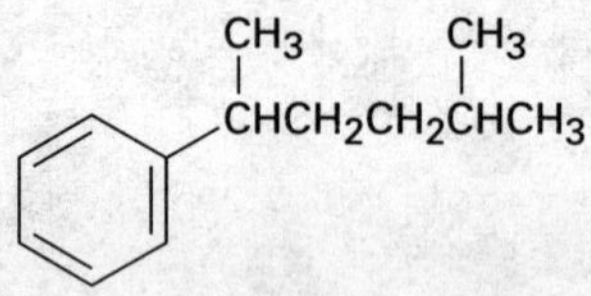

(b)

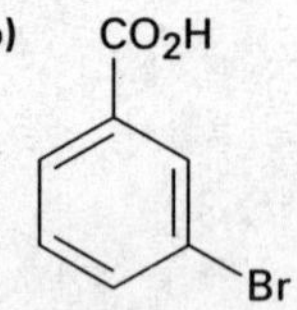

(c)

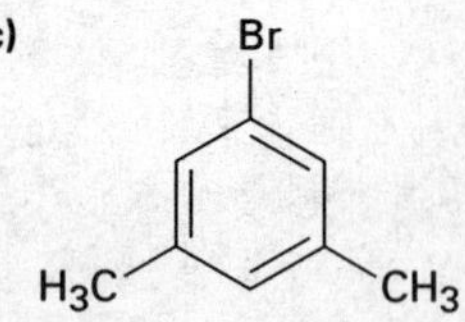

(d)

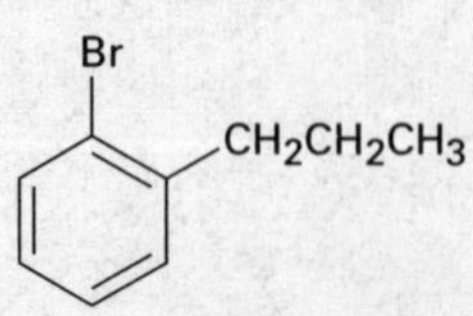

(e)

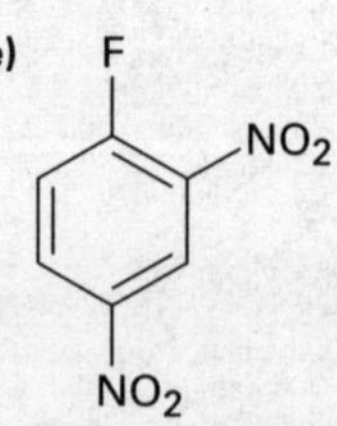

(f)

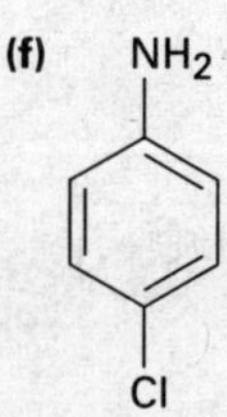

5.27 Draw and name all aromatic compounds with the formula C_7H_7Cl.

5.28 Draw and name all isomeric bromodimethylbenzenes.

5.29 Draw structures corresponding to the following names:
(a) *m*-Bromophenol (b) Benzene-1,3,5-triol
(c) *p*-Iodonitrobenzene (d) 2,4,6-Trinitrotoluene (TNT)
(e) *o*-Aminobenzoic acid (f) 3-Methyl-2-phenylhexane

REACTIONS AND SUBSTITUENT EFFECTS

5.30 Predict the major product(s) of mononitration of the following substances:
(a) Bromobenzene (b) Benzonitrile (cyanobenzene)
(c) Benzoic acid (d) Nitrobenzene
(e) Phenol (f) Benzaldehyde

5.31 Which of the substances listed in Problem 5.30 react faster than benzene and which react slower?

5.32 Formulate the reaction of benzene with 2-chloro-2-methylpropane in the presence of $AlCl_3$ catalyst to give *tert*-butylbenzene.

5.33 Identify each of the following groups as an activator or deactivator and as an *o,p*-director or *m*-director:

(a) 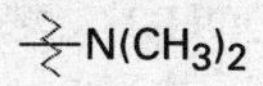

(b) 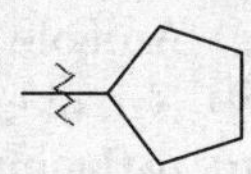
(c) 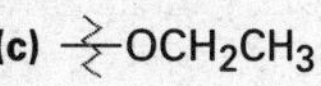

(d) 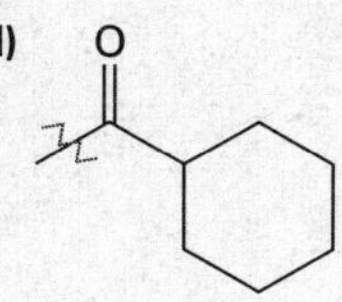

5.34 Propose structures for aromatic hydrocarbons meeting the following descriptions:
(a) C_9H_{12}; can give only one product on aromatic bromination
(b) C_8H_{10}; can give three products on aromatic chlorination
(c) $C_{10}H_{14}$; can give two products on aromatic nitration

5.35 Predict the major product(s) of the following reactions:

(a) 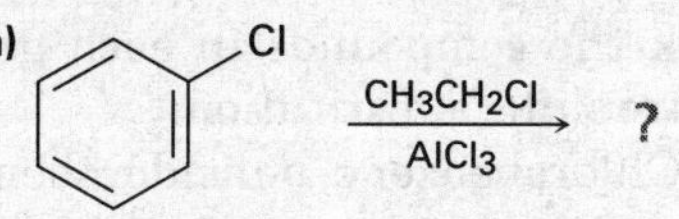

(b) 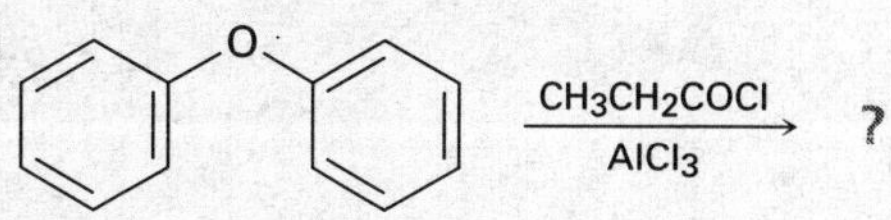

(c) 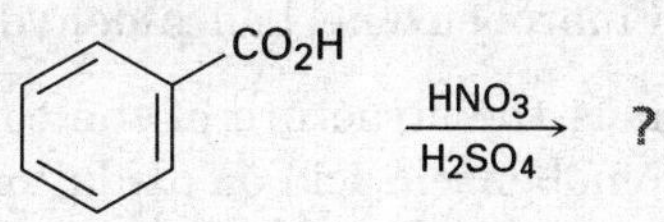

(d) 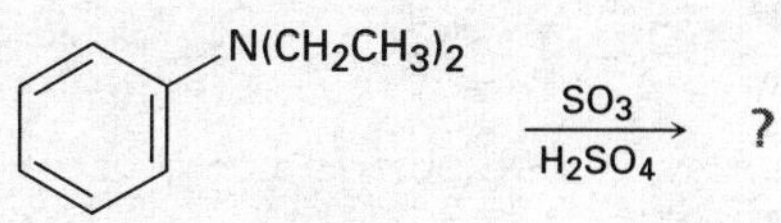

5.36 In some cases, the Friedel–Crafts acylation reaction can occur *intramolecularly,* that is, within the same molecule. Predict the product of the following reaction:

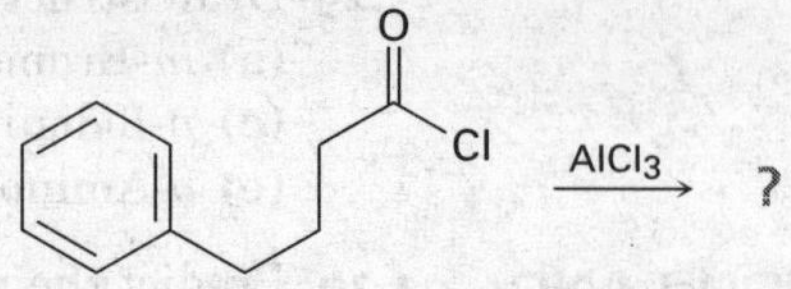

5.37 Draw the three additional resonance structures of anthracene.

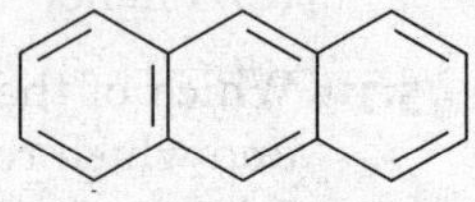

Anthracene

5.38 Rank the following aromatic compounds in the expected order of their reactivity toward Friedel–Crafts acylation. Which compounds are unreactive?
(a) Bromobenzene
(b) Toluene
(c) Anisole ($C_6H_5OCH_3$)
(d) Nitrobenzene
(e) *p*-Bromotoluene

5.39 The orientation of electrophilic aromatic substitution on a disubstituted benzene ring is usually controlled by whichever of the two groups already on the ring is the more powerful activator. Name and draw the structure(s) of the major product(s) of electrophilic chlorination of the following substances:
(a) *m*-Nitrophenol **(b)** *o*-Methylphenol **(c)** *p*-Chloronitrobenzene

5.40 Predict the major product(s) you would expect to obtain from sulfonation of the following substances (see Problem 5.39):
(a) *o*-Chlorotoluene **(b)** *m*-Bromophenol **(c)** *p*-Nitrotoluene

5.41 Rank the compounds in each group according to their reactivity toward electrophilic substitution:
(a) Chlorobenzene, *o*-dichlorobenzene, benzene
(b) *p*-Bromonitrobenzene, nitrobenzene, phenol
(c) Fluorobenzene, benzaldehyde, *o*-dimethylbenzene

5.42 What is the structure of the compound with formula C_8H_9Br that gives *p*-bromobenzoic acid on oxidation with $KMnO_4$?

MECHANISMS

5.43 Propose a mechanism to explain the fact that deuterium (D, ^2H) slowly replaces hydrogen (^1H) in the aromatic ring when benzene is treated with D_2SO_4.

5.44 Show the steps involved in the Friedel–Crafts reaction of benzene with CH_3Cl.

5.45 Use resonance structures of the possible carbocation intermediates to explain why bromination of biphenyl occurs at the ortho and para positions rather than at the meta positions.

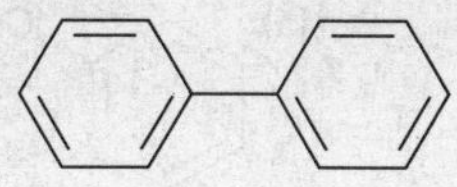

Biphenyl

5.46 In light of your answer to Problem 5.45, at what position and on which ring would you expect nitration of 4-bromobiphenyl to occur?

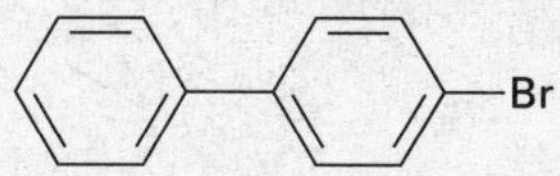

4-Bromobiphenyl

SYNTHESIS

5.47 Explain by drawing resonance structures of the intermediate carbocations why naphthalene undergoes electrophilic aromatic substitution at C1 rather than at C2.

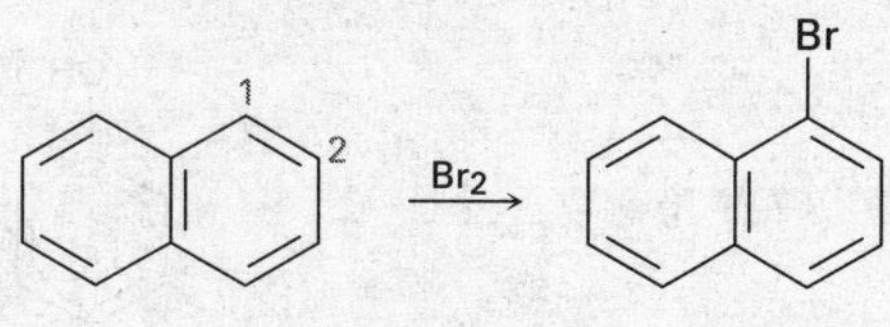

5.48 Starting with benzene, how would you synthesize the following substances? Assume that you can separate ortho and para isomers if necessary.
(a) *m*-Bromobenzenesulfonic acid **(b)** *o*-Chlorobenzenesulfonic acid
(c) *p*-Chlorotoluene

5.49 Starting from any aromatic hydrocarbon of your choice, how would you synthesize the following substances? Ortho and para isomers can be separated if necessary.
(a) *o*-Nitrobenzoic acid **(b)** *p-tert*-Butylbenzoic acid

GENERAL PROBLEMS

5.50 We said in Section 4.9 that an allylic carbocation is stabilized by resonance. Draw resonance structures to account for the similar stabilization of a benzylic carbocation.

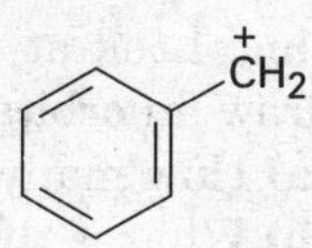

A benzylic carbocation

5.51 Addition of HBr to 1-phenylpropene yields (1-bromopropyl)benzene as the only product. Propose a mechanism for the reaction, and explain why none of the other regioisomer is produced (see Problem 5.50).

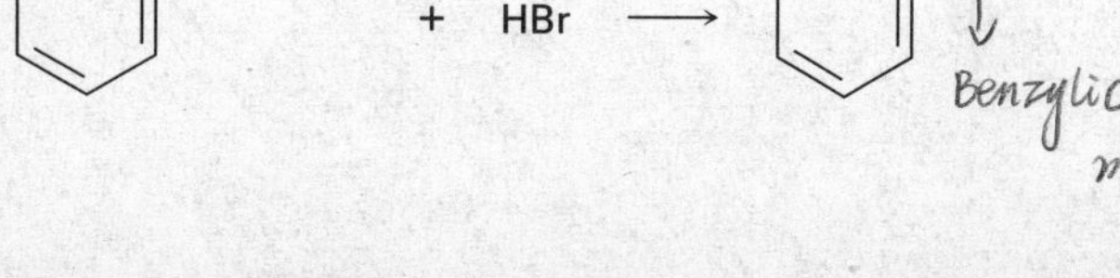

5.52 Starting with toluene, how would you synthesize the three nitrobenzoic acids?

5.53 The following syntheses have flaws in them. What is wrong with each?

(a)

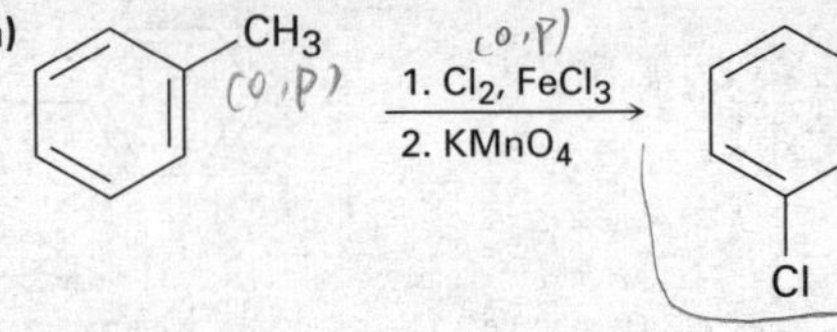

(b)

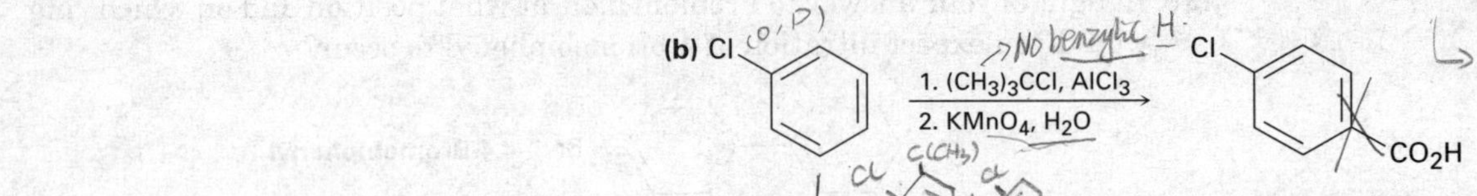

5.54 Carbocations generated by reaction of an alkene with a strong acid catalyst can react with aromatic rings in a Friedel–Crafts reaction. Propose a mechanism to account for the industrial synthesis of the food preservative BHT from *p*-cresol and 2-methylpropene:

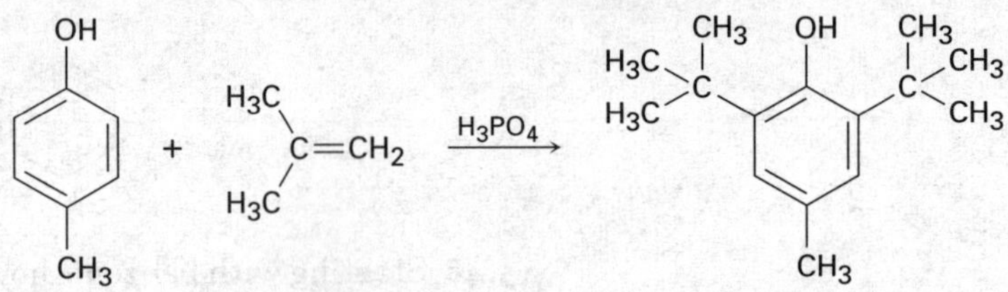

***p*-Cresol** **BHT**

5.55 Would you expect the trimethylammonium group to be an activating or deactivating substituent? Explain.

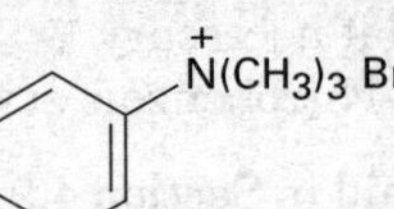

Phenyltrimethylammonium bromide

5.56 Indole is an aromatic compound that has a benzene ring fused to a pyrrole ring. Look at the electronic structure of pyrrole in Figure 5.9c, and then draw an orbital picture of indole.

(a) How many π electrons does indole have?

(b) What is the electronic relationship of indole to naphthalene?

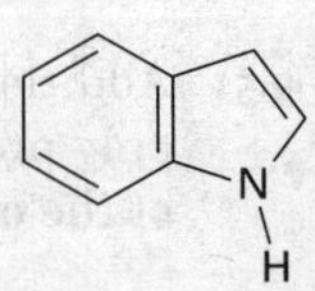

Indole

5.57 Benzene can be hydroxylated by reaction with H_2O_2 in the presence of an acid catalyst. What is the likely structure of the reactive electrophile? Propose a mechanism for the reaction.

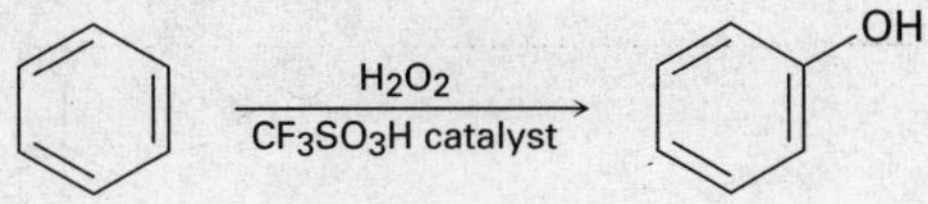

5.58 You know the mechanism of HBr addition to alkenes, and you know the effects of various substituent groups on aromatic substitution. Use this knowledge to predict which of the following two alkenes reacts faster with HBr. Explain your answer by drawing resonance structures of the carbocation intermediates.

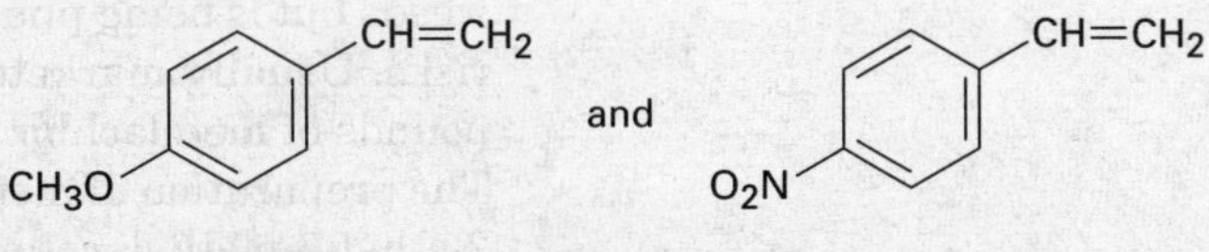

5.59 Identify the reagents represented by the letters **a** through **d** in the following scheme:

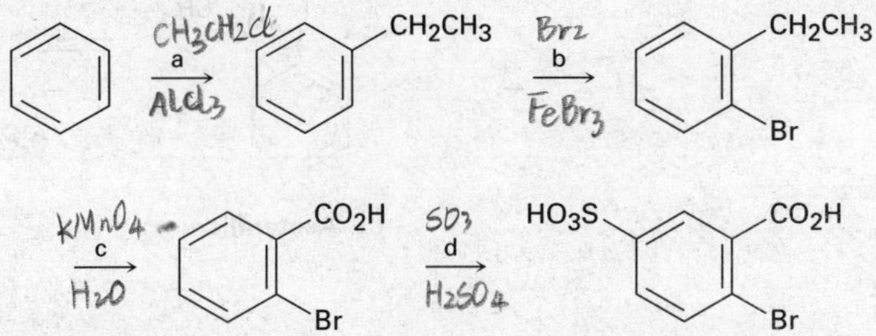

IN THE MEDICINE CABINET

5.60 Ribavirin, an antiviral agent used against hepatitis C and viral pneumonia, contains a 1,2,4-triazole ring. Look at the electronic structure of imidazole in Figure 5.9d, and then explain why the ring is aromatic.

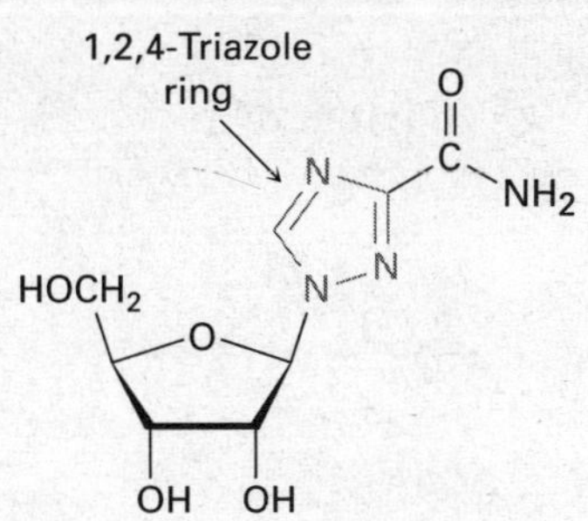

Ribavirin

IN THE FIELD **5.61** Synthesis of the herbicide 2,4-D begins with chlorination of phenol followed by reaction of the product with NaOH and chloroacetic acid. Name the chlorinated intermediate, and use resonance structures to explain the pattern of chlorination in the first step.

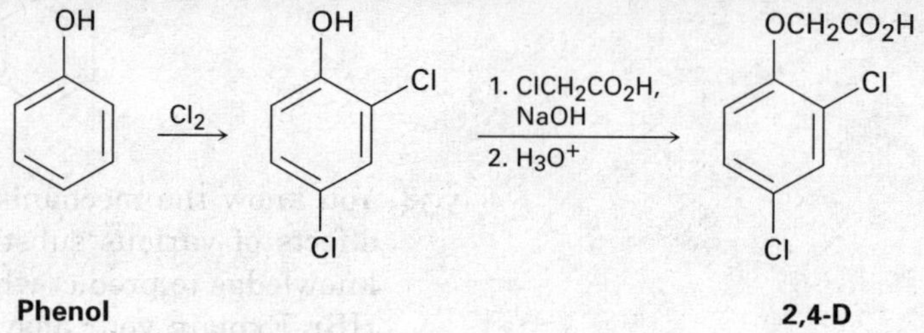

5.62 The herbicide metolachlor is broadly used in the United States to control weeds but is being phased out in Europe because of possible environmental risks. Usually marketed under the name Dual, approximately 50 million pounds of metolachlor are applied on crops each year in the United States. The preparation of metolachlor begins with the conversion of acetanilide to 2-ethyl-6-methylacetanilide. How would you accomplish this conversion?

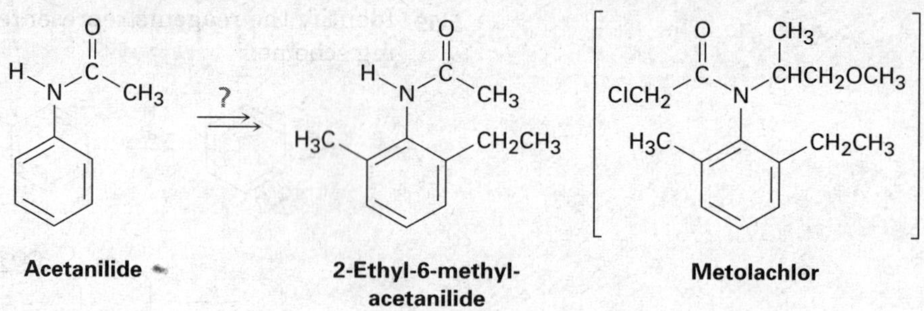

CHAPTER

6

Like the mountain whose image is reflected in a lake, many organic molecules also have mirror-image counterparts.

Stereochemistry at Tetrahedral Centers

OWL

Online homework for this chapter can be assigned in OWL, an online homework assessment tool.

Are you right-handed or left-handed? You may not spend much time thinking about it, but handedness plays a surprisingly large role in your daily activities. Many musical instruments, such as oboes and clarinets, have a handedness to them; the last available softball glove always fits the wrong hand; left-handed people write in a "funny" way. The fundamental reason for these difficulties is that our hands aren't identical; rather, they're *mirror images.* When you hold a *left* hand up to a mirror, the image you see looks like a *right* hand. Try it.

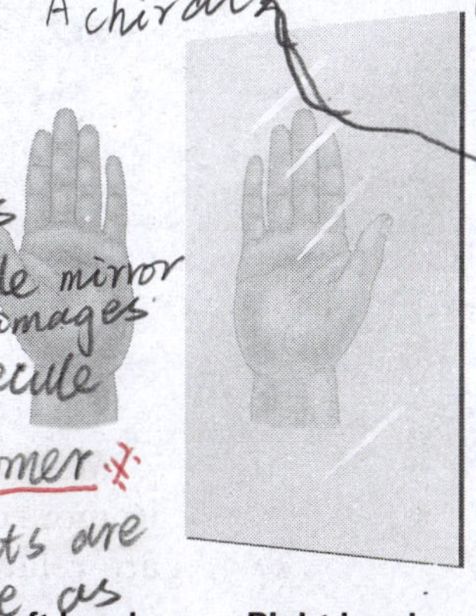

Handedness is also important in organic and biological chemistry, where it primarily arises as a consequence of the tetrahedral stereochemistry of sp^3-hybridized carbon atoms. Many drugs and almost all the molecules in our bodies—amino acids, carbohydrates, nucleic acids, and many more—are handed. Furthermore, it is molecular handedness that makes possible the precise interactions between enzymes and their substrates that are involved in the hundreds of thousands of chemical reactions on which life is based.

WHY THIS CHAPTER?

Understanding the causes and consequences of molecular handedness is crucial to understanding organic and biological chemistry. The subject can seem a bit complex at first, but the material covered in this chapter nevertheless forms the basis for much of the remainder of the book.

6.1 Enantiomers and the Tetrahedral Carbon

What causes molecular handedness? To see how molecular handedness arises, look at generalized molecules of the type CH_3X, CH_2XY, and CHXYZ shown in Figure 6.1. On the left are three molecules, and on the right are their images reflected in a mirror. The CH_3X and CH_2XY molecules are identical to their mirror images and thus are not handed. If you make a molecular model of each molecule and its mirror image, you'll find that you can superimpose one on the other. The CHXYZ molecule, by contrast, is *not* identical to its mirror image. You can't superimpose a model of the molecule on a model of its mirror image for the same reason that you can't superimpose a left hand on a right hand: they simply aren't the same.

Figure 6.1 Tetrahedral carbon atoms and their mirror images. Molecules of the type CH_3X and CH_2XY are identical to their mirror images, but a molecule of the type CHXYZ is not. A CHXYZ molecule is related to its mirror image in the same way that a right hand is related to a left hand.

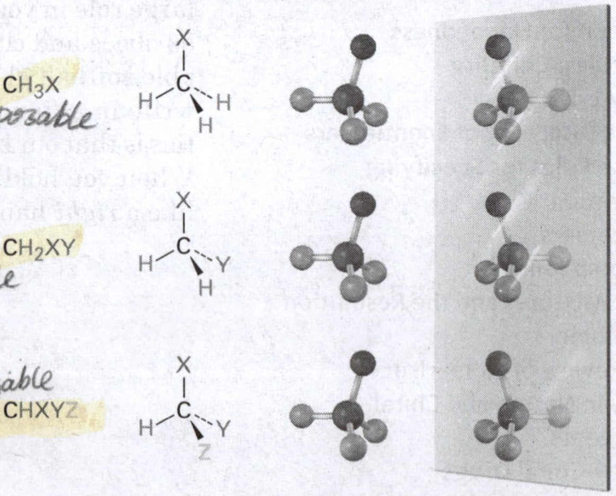

Molecules that are not identical to their mirror images are kinds of stereoisomers called **enantiomers** (Greek *enantio,* meaning "opposite"). Enantiomers are related to each other as a right hand is related to a left hand and result whenever a tetrahedral carbon is bonded to four different substituents (one need not be H). For example, lactic acid (2-hydroxypropanoic acid) exists as a pair of enantiomers because there are four different groups (–H, –OH, $–CH_3$, and $–CO_2H$) bonded to the central carbon atom. Both are found in sour milk, but only the (+) enantiomer occurs in muscle tissue.

Lactic acid: a molecule of general formula CHXYZ

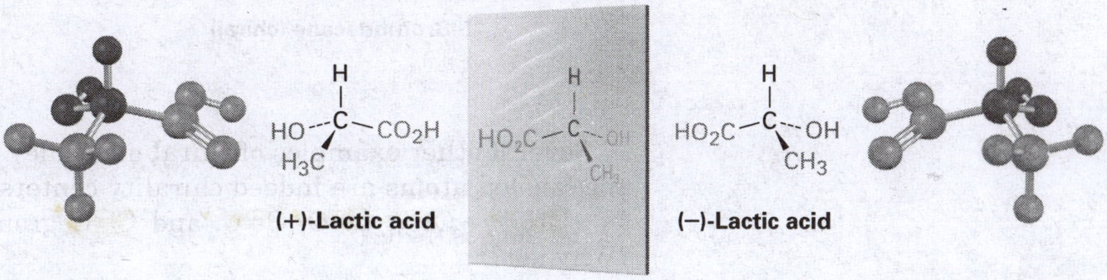

No matter how hard you try, you can't superimpose a molecule of (+)-lactic acid on a molecule of (−)-lactic acid. If any two groups match up, say –H and $–CO_2H$, the remaining two groups don't match (Figure 6.2).

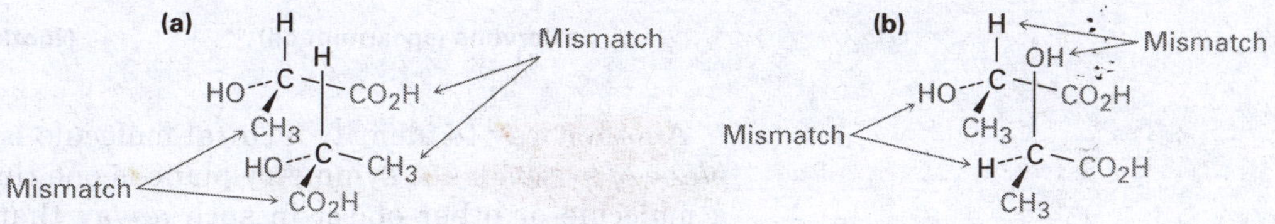

Figure 6.2 Attempts at superimposing the mirror-image forms of lactic acid: **(a)** When the –H and –OH substituents match up, the $–CO_2H$ and $–CH_3$ substituents don't. **(b)** When $–CO_2H$ and $–CH_3$ match up, –H and –OH don't. Regardless of how the molecules are oriented, they aren't identical.

6.2 The Reason for Handedness in Molecules: Chirality

A molecule that is not identical to its mirror image is said to be **chiral** (**ky**-ral, from the Greek *cheir,* meaning "hand"). You can't take a chiral molecule and its enantiomer and place one on the other so that all atoms coincide.

How can you predict whether a given molecule is or is not chiral? By far the most common (although not the only) cause of chirality in an organic molecule is the presence of a carbon atom bonded to four different groups—for example, the central carbon atom in lactic acid. Such carbons are referred to as *stereocenters,* or **chirality centers**. Note that *chirality* is a property of the entire molecule, whereas a chirality *center* is the *cause* of chirality.

Detecting chirality centers in a complex molecule takes practice, because it's not always apparent that four different groups are bonded to a given carbon. The differences don't necessarily appear right next to the chirality center. For example, 5-bromodecane is a chiral molecule because four different groups are bonded to C5 (marked by an asterisk). A butyl substituent is very *similar* to a pentyl substituent, but it isn't identical. The difference isn't apparent until four carbons away from the chirality center, but there's still a difference.

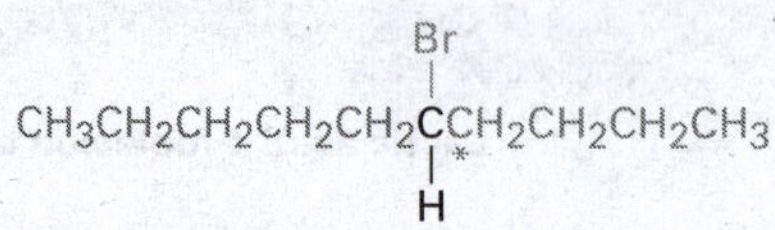

5-Bromodecane (chiral)

Substituents on carbon 5

–H

–Br

$-CH_2CH_2CH_2CH_3$ (butyl)

$-CH_2CH_2CH_2CH_2CH_3$ (pentyl)

Several other examples of chiral molecules follow. Check for yourself that the labeled atoms are indeed chirality centers. You might note that carbons in $-CH_2-$, $-CH_3$, C=O, C=C, and C≡C groups *can't* be chirality centers. (Why not?)

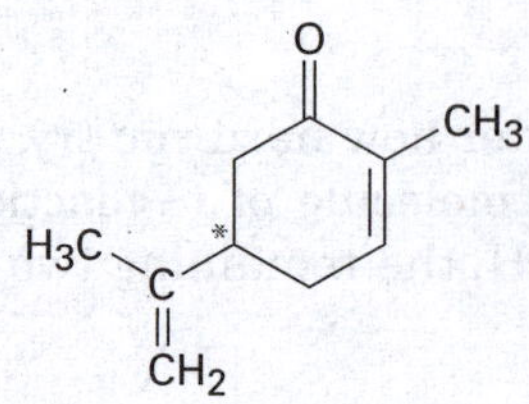

Carvone (spearmint oil)

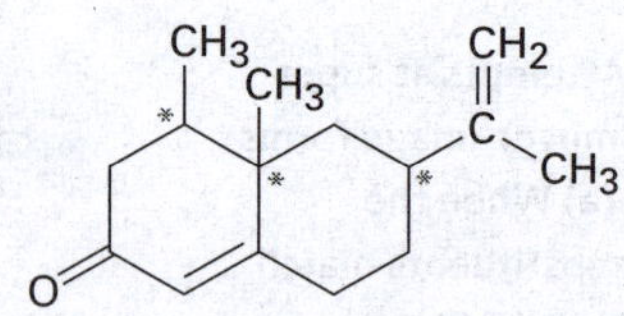

Nootkatone (grapefruit oil)

Another way to identify a chiral molecule is to look for the presence of a *plane of symmetry.* A symmetry plane is one that cuts through the middle of a molecule or other object in such a way that one half of the molecule or object is a mirror image of the other half. A laboratory flask, for example, has a plane of symmetry. If you were to cut the flask in half, one half would be an exact mirror image of the other half. A hand, however, does not have a plane of symmetry. One "half" of a hand is not a mirror image of the other "half" (Figure 6.3).

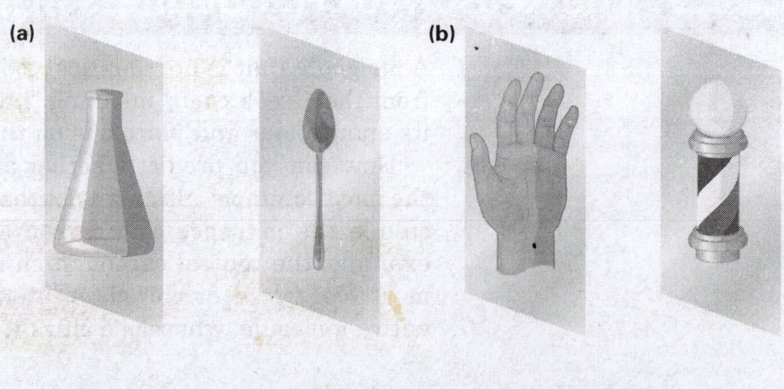

Figure 6.3 The meaning of *symmetry plane.* **(a)** Objects like a flask or spoon have planes of symmetry passing through them that make the right and left halves mirror images. **(b)** Objects like a hand or barber pole have no symmetry plane; the right "half" is not a mirror image of the left half.

A molecule that has a plane of symmetry in any of its possible conformations must be identical to its mirror image and hence must be nonchiral, or **achiral**. Thus, propanoic acid, $CH_3CH_2CO_2H$, contains a plane of symmetry when lined up as shown in Figure 6.4 and is achiral. Lactic acid, $CH_3CH(OH)CO_2H$, however, has no plane of symmetry in any conformation and is chiral.

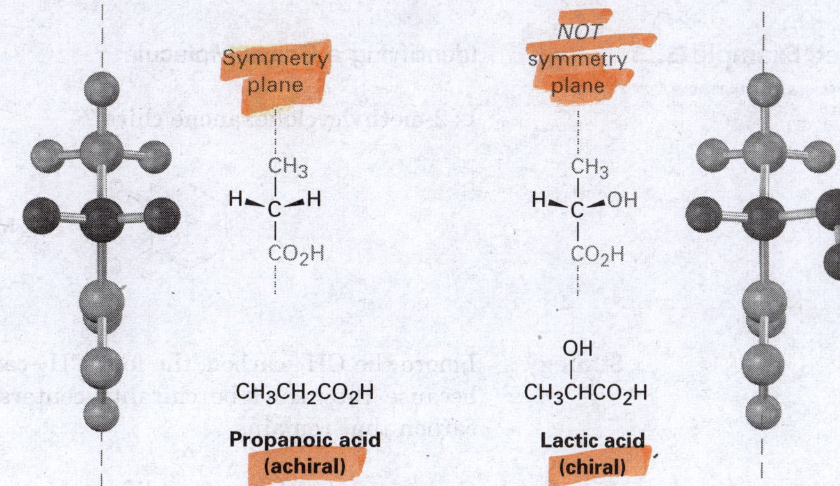

Figure 6.4 The achiral propanoic acid molecule versus the chiral lactic acid molecule. Propanoic acid has a plane of symmetry that makes one side of the molecule a mirror image of the other side. Lactic acid has no such symmetry plane.

Worked Example 6.1

Drawing a Chiral Molecule

Draw the structure of a chiral alcohol.

Strategy An alcohol is a compound that contains the –OH functional group. To make an alcohol chiral, we need to have four different groups bonded to a single carbon atom, say –H, –OH, $-CH_3$, and $-CH_2CH_3$.

Solution

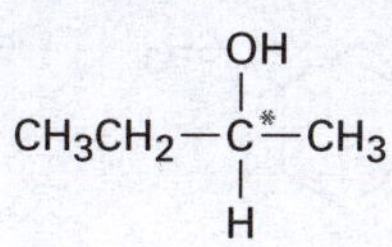

Butan-2-ol (chiral)

Worked Example 6.2

Identifying a Chiral Molecule

Is 3-methylhexane chiral?

Strategy Draw the structure of 3-methylhexane and cross out all the CH_2 and CH_3 carbons because they can't be chirality centers. Then look closely at any carbon that remains to see if it's bonded to four different groups.

Solution Carbon 3 is bonded to –H, $-CH_3$, $-CH_2CH_3$, and $-CH_2CH_2CH_3$, so the molecule is chiral.

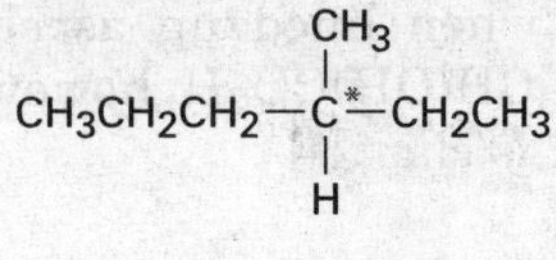

3-Methylhexane (chiral)

Worked Example 6.3

Identifying a Chiral Molecule

Is 2-methylcyclohexanone chiral?

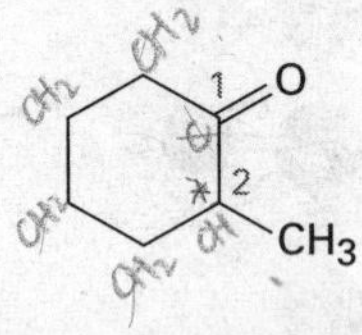

2-Methylcyclohexanone

Strategy Ignore the CH_3 carbon, the four CH_2 carbons in the ring, and the C=O carbon because they can't be chirality centers. Then look carefully at C2, the only carbon that remains.

Solution Carbon 2 is bonded to four different groups: a $-CH_3$ group, an –H atom, a –C=O carbon in the ring, and a $-CH_2-$ ring carbon, so 2-methylcyclohexanone is chiral.

Problem 6.1 Which of the following objects are chiral?

(a) Soda can **(b)** Screwdriver **(c)** Screw **(d)** Shoe

Problem 6.2 Which of the following molecules are chiral?

(a) 3-Bromopentane **(b)** 1,3-Dibromopentane

(c) 3-Methylhex-1-ene **(d)** *cis*-1,4-Dimethylcyclohexane

Problem 6.3 Which of the following molecules are chiral? Identify the chirality center(s) in each.

(a)

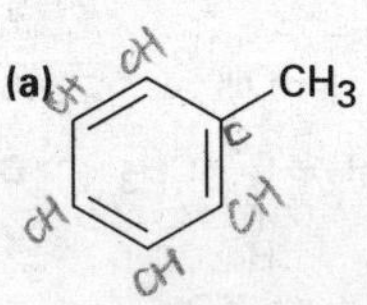

Toluene

(b)

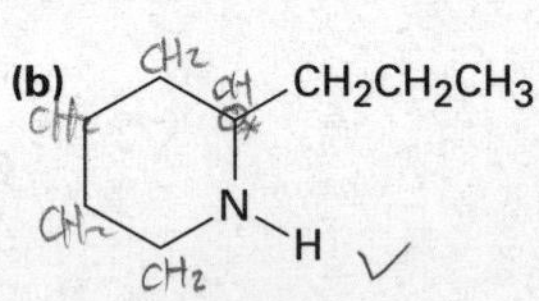

Coniine (from poison hemlock)

(c)

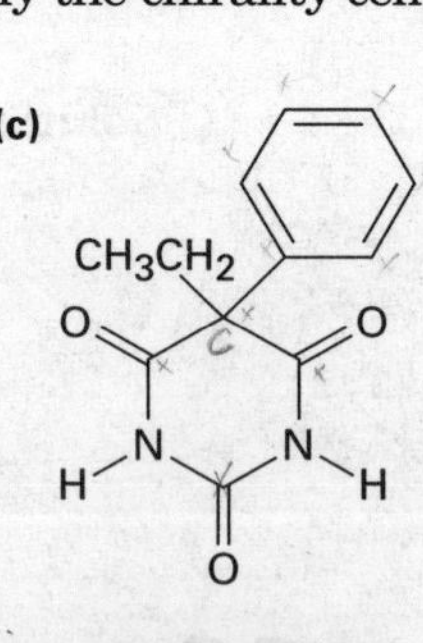

Phenobarbital (tranquilizer)

Problem 6.4 Alanine, an amino acid found in proteins, is chiral. Draw the two enantiomers of alanine using the standard convention of solid, wedged, and dashed lines.

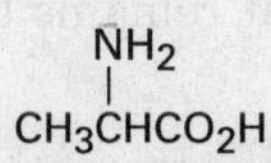

Alanine

Problem 6.5 Identify the chirality centers in the following molecules (yellow-green = Cl, pale yellow = F):

(a)

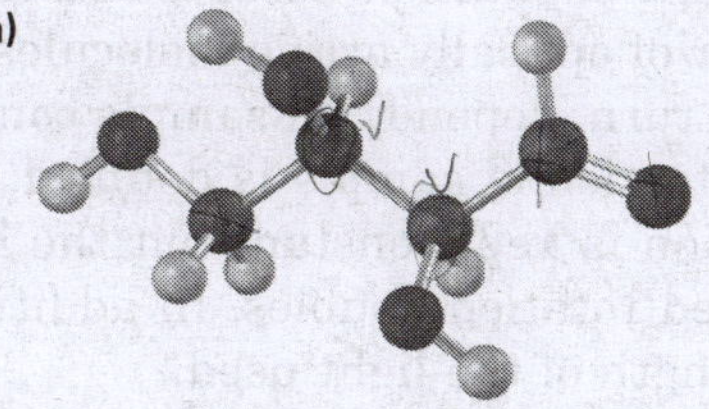

Threose
(a sugar)

(b)

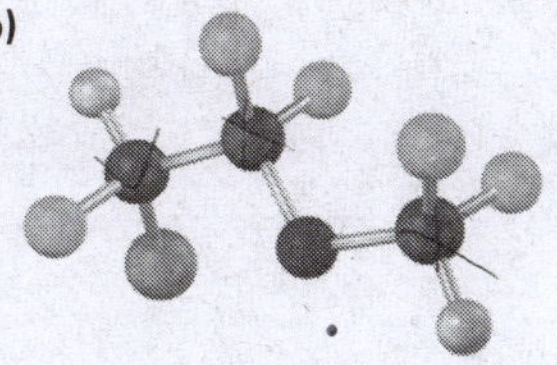

Enflurane
(an anesthetic)

6.3 Optical Activity

The study of chirality originated in the early 19th century during investigations by the French physicist Jean-Baptiste Biot into the nature of *plane-polarized light.* A beam of ordinary light consists of electromagnetic waves that oscillate in an infinite number of planes at right angles to the direction of light travel. When a beam of ordinary light is passed through a device called a *polarizer,* however, only the light waves oscillating in a single plane pass through and the light is said to be plane-polarized. Light waves in all other planes are blocked out.

Biot made the remarkable observation that when a beam of plane-polarized light passes through a solution of certain organic molecules, such as sugar or camphor, the plane of polarization is *rotated* through an angle α. Not all organic substances exhibit this property, but those that do are said to be **optically active**.

The angle of rotation can be measured with an instrument called a *polarimeter,* represented in Figure 6.5. A solution of optically active organic molecules is placed in a sample tube, plane-polarized light is passed through the tube, and rotation of the polarization plane occurs. The light then goes through a second polarizer called the *analyzer.* By rotating the analyzer until the light passes through *it,* we can find the new plane of polarization and can tell to what extent rotation has occurred.

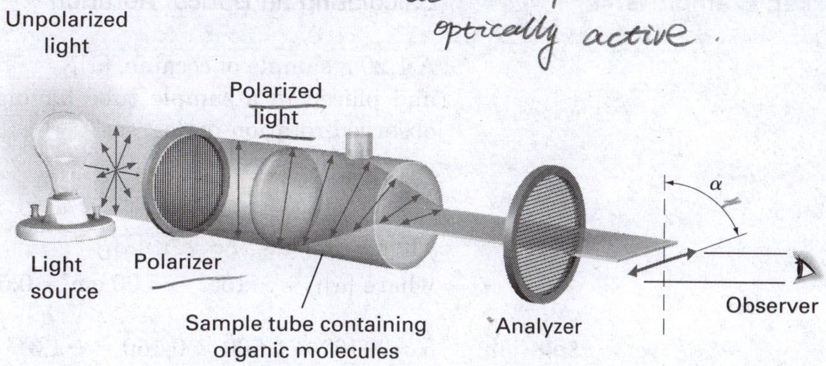

Figure 6.5 Schematic representation of a polarimeter. Plane-polarized light passes through a solution of optically active molecules, which rotate the plane of polarization.

In addition to determining the extent of rotation, we can also find the direction. From the vantage point of the observer looking directly at the analyzer, some optically active molecules rotate polarized light to the left (counterclockwise) and are said to be **levorotatory**, whereas others rotate polarized light to the right (clockwise) and are said to be **dextrorotatory**. By convention, rotation to the left is given a minus sign (−), and rotation to the right is given a plus sign (+). (−)-Morphine, for example, is levorotatory, and (+)-sucrose is dextrorotatory.

The extent of rotation observed in a polarimetry experiment depends on the number of optically active molecules encountered by the light beam. This number, in turn, depends on sample concentration and sample pathlength. If the concentration of sample is doubled, the observed rotation doubles. If the concentration is kept constant but the length of the sample tube is doubled, the observed rotation doubles. In addition, the angle of rotation depends on the wavelength of the light used.

To express optical rotations in a meaningful way so that comparisons can be made, we have to choose standard conditions. The **specific rotation, $[\alpha]_D$,** of a compound is defined as the observed rotation when light of 589.6 nanometer (nm; 1 nm = 10^{-9} m) wavelength is used with a sample pathlength l of 1 decimeter (dm; 1 dm = 10 cm) and a sample concentration c of 1 g/cm^3.

$$[\alpha]_D = \frac{\text{Observed rotation (degrees)}}{\text{Pathlength, } l \text{ (dm)} \times \text{ Concentration, } c \text{ (g/cm}^3\text{)}} = \frac{\alpha}{l \times c}$$

When optical rotation data are expressed in this standard way, the specific rotation, $[\alpha]_D$, is a physical constant characteristic of a given optically active compound. For example, (+)-lactic acid has $[\alpha]_D = +3.82$, and (−)-lactic acid has $[\alpha]_D = -3.82$. That is, the two enantiomers rotate plane-polarized light to the same extent but in opposite directions. Note that the units of specific rotation are [(deg · cm^2)/g] but that values are usually expressed without the units. Some additional examples are listed in Table 6.1.

Table 6.1 Specific Rotation of Some Organic Molecules

Compound	$[\alpha]_D$	Compound	$[\alpha]_D$
Penicillin V	233	Cholesterol	−31.5
Sucrose	+66.47	Morphine	−132
Camphor	+44.26	Cocaine	−16
Chloroform	0	Acetic acid	0

Worked Example 6.4 **Calculating an Optical Rotation**

A 1.20 g sample of cocaine, $[\alpha]_D = -16$, was dissolved in 7.50 mL of chloroform and placed in a sample tube having a pathlength of 5.00 cm. What was the observed rotation in degrees?

Strategy Since $[\alpha]_D = \dfrac{\alpha}{l \times c}$

Then $\alpha = l \times c \times [\alpha]_D$

where $[\alpha]_D = -16$; $l = 5.00$ cm $= 0.500$ dm; $c = 1.20$ g/7.50 cm^3 $= 0.160$ g/cm^3

Solution $\alpha = -16° \times 0.500 \times 0.160 = -1.3°$

Problem 6.6 Is cocaine (Worked Example 6.4) dextrorotatory or levorotatory?

Problem 6.7 A 1.50 g sample of coniine, the toxic extract of poison hemlock, was dissolved in 10.0 mL of ethanol and placed in a sample cell with a 5.00 cm pathlength. The observed rotation at the sodium D line was +1.21°. Calculate $[\alpha]_D$ for coniine.

6.4 Pasteur's Discovery of Enantiomers

Little was done after Biot's discovery of optical activity until 1848, when Louis Pasteur began work on a study of crystalline tartaric acid salts derived from wine. On recrystallizing a concentrated solution of sodium ammonium tartrate below 28 °C, Pasteur made the surprising observation that two distinct kinds of crystals precipitated. Furthermore, the two kinds of crystals were mirror images and were related in the same way that a right hand is related to a left hand.

Working carefully with tweezers, Pasteur was able to separate the crystals into two piles, one of "right-handed" crystals and one of "left-handed" crystals, as shown in Figure 6.6. Although the original sample, a 50:50 mixture of right and left, was optically inactive, *solutions of crystals from each of the sorted piles were optically active* and their specific rotations were equal in amount but opposite in sign.

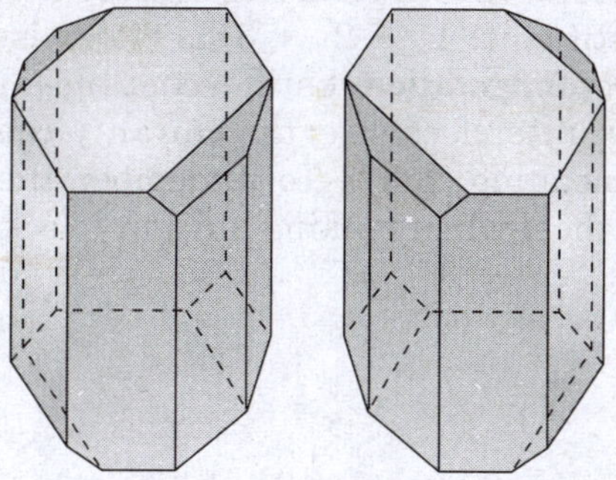

Figure 6.6 Drawings of sodium ammonium tartrate crystals taken from Pasteur's original sketches. One of the crystals is dextrorotatory in solution, and the other is levorotatory.

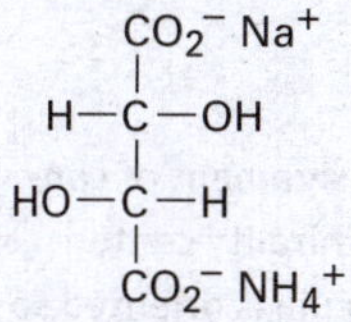

Sodium ammonium tartrate

Pasteur was far ahead of his time. Although the structural theory of Kekulé had not yet been proposed, Pasteur explained his results by speaking of the molecules themselves, saying, "There is no doubt that [in the *dextro* tartaric acid] there exists an asymmetric arrangement having a nonsuperimposable image. It is no less certain that the atoms of the *levo* acid possess precisely the inverse asymmetric arrangement." Pasteur's vision was extraordinary, for it was not until 25 years later that his theories regarding the asymmetry of chiral molecules were confirmed.

Today, we would describe Pasteur's work by saying that he had discovered enantiomers. Enantiomers, also called *optical isomers,* have identical physical properties, such as melting points and boiling points, but differ in the direction in which their solutions rotate plane-polarized light.

6.5 Sequence Rules for Specifying Configuration

Drawings provide visual representations of stereochemistry, but a verbal method for specifying the three-dimensional arrangement, or **configuration**, of substituents around a chirality center is also necessary. The method used employs the same sequence rules given in Section 3.4 for specifying *E* and *Z* alkene stereochemistry. Let's briefly review these sequence rules and then see how they're used to specify the configuration of a chirality center. For a more thorough review, you should reread Section 3.4.

RULE 1 **Look at the four atoms directly attached to the chirality center, and rank them according to atomic number.** The atom with the highest atomic number has the highest ranking (first), and the atom with the lowest atomic number (usually hydrogen) has the lowest ranking (fourth).

RULE 2 **If a decision can't be reached by ranking the first atoms in the substituent, look at the second, third, or fourth atoms away from the chirality center until the first difference is found.**

RULE 3 **Multiple-bonded atoms are equivalent to the same number of single-bonded atoms.**

Having ranked the four groups attached to a chirality center, we describe the stereochemical configuration around the carbon by orienting the molecule so that the group with the lowest ranking (4) points directly back, away from us. We then look at the three remaining substituents, which now appear to radiate toward us like the spokes on a steering wheel (Figure 6.7). If a curved arrow drawn from the highest to second-highest to third-highest ranked substituent ($1 \rightarrow 2 \rightarrow 3$) is clockwise, we say that the chirality center has the ***R* configuration** (Latin *rectus*, meaning "right"). If an arrow from $1 \rightarrow 2 \rightarrow 3$ is counterclockwise, the chirality center has the ***S* configuration** (Latin *sinister*, meaning "left"). To remember these assignments, think of a car's steering wheel when making a *R*ight (clockwise) turn.

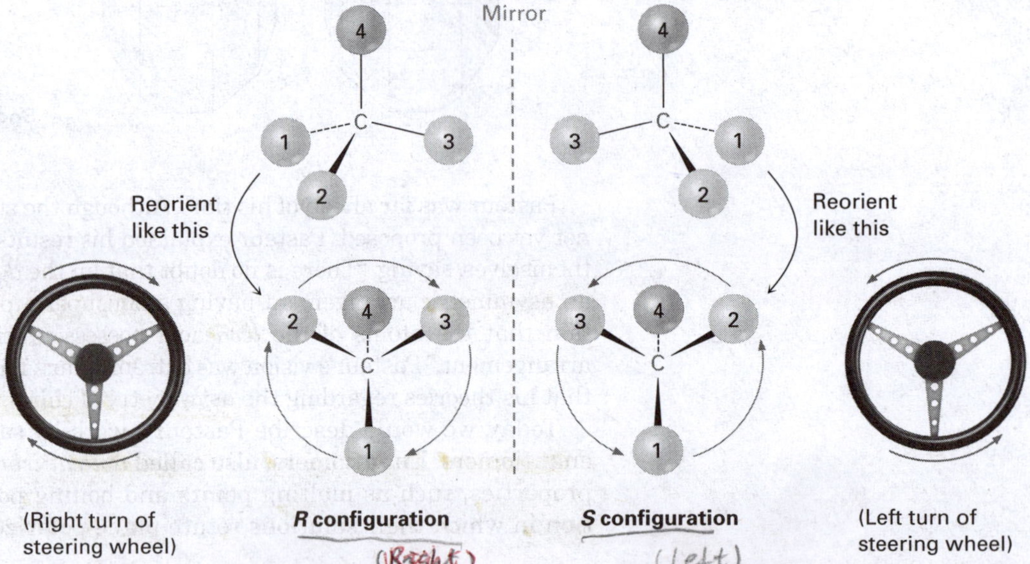

Figure 6.7 Assignment of configuration to a chirality center. When the molecule is oriented so that the lowest-ranked group (4) is toward the rear, the remaining three groups radiate toward the viewer like the spokes of a steering wheel. If the direction of travel $1 \rightarrow 2 \rightarrow 3$ is clockwise (right turn), the center has the *R* configuration. If the direction of travel $1 \rightarrow 2 \rightarrow 3$ is counterclockwise (left turn), the center is *S*.

Look at (−)-lactic acid in Figure 6.8 for an example of how to assign configuration. Sequence rule 1 says that –OH is ranked 1 and –H is ranked 4, but it doesn't allow us to distinguish between $-CH_3$ and $-CO_2H$ because both groups have carbon as their first atom. Sequence rule 2, however, says that $-CO_2H$ ranks higher than $-CH_3$ because O (the highest second atom in $-CO_2H$) outranks H (the highest second atom in $-CH_3$). Now, turn the molecule so that the fourth-ranked group (–H) is oriented toward the rear, away from the observer. Since a curved arrow from 1 (–OH) to 2 ($-CO_2H$) to 3 ($-CH_3$) is clockwise (right turn of the steering wheel), (−)-lactic acid has the *R* configuration. Applying the same procedure to (+)-lactic acid leads to the opposite assignment.

Figure 6.8 Assigning configuration to (a) (*R*)-(−)-lactic acid and (b) (*S*)-(+)-lactic acid.

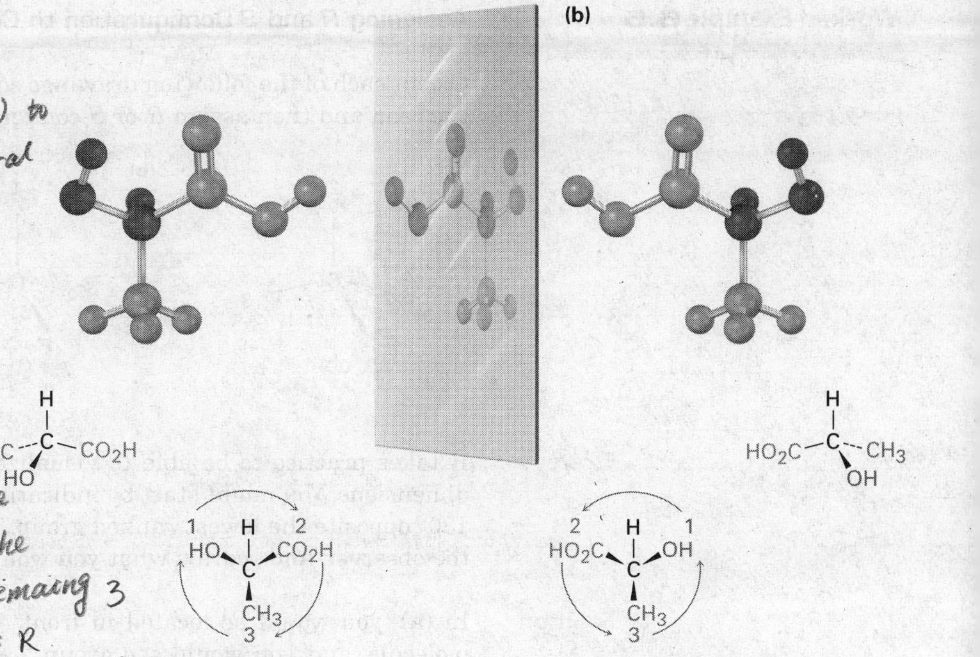

R configuration (−)-Lactic acid

S configuration (+)-Lactic acid

Further examples are provided by naturally occurring (−)-glyceraldehyde and (+)-alanine, both of which have an *S* configuration as shown in Figure 6.9. Note that the sign of optical rotation, (+) or (−), is not related to the *R*,*S* designation. (*S*)-Glyceraldehyde happens to be levorotatory (−), and (*S*)-alanine happens to be dextrorotatory (+), but there is no simple correlation between *R*,*S* configuration and direction or magnitude of optical rotation.

Figure 6.9 Assigning configuration to (a) (−)-glyceraldehyde and (b) (+)-alanine. Both happen to have the *S* configuration, although one is levorotatory and the other is dextrorotatory.

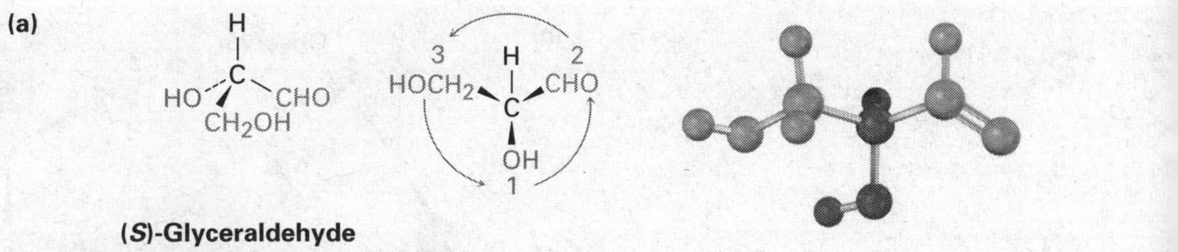

(*S*)-Glyceraldehyde
[(*S*)-(−)-2,3-Dihydroxypropanal]
$[\alpha]_D = -8.7$

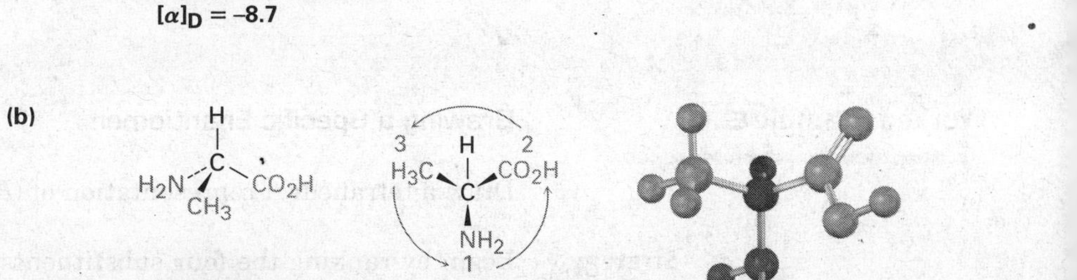

(*S*)-Alanine
[(*S*)-(+)-2-Aminopropanoic acid]
$[\alpha]_D = +8.5$

Worked Example 6.5 Assigning *R* and *S* Configuration to Chirality Centers

Orient each of the following drawings so that the lowest-ranked group is toward the rear, and then assign *R* or *S* configuration:

(a)

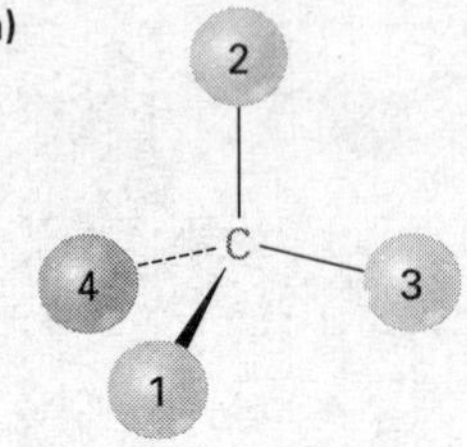

(b)

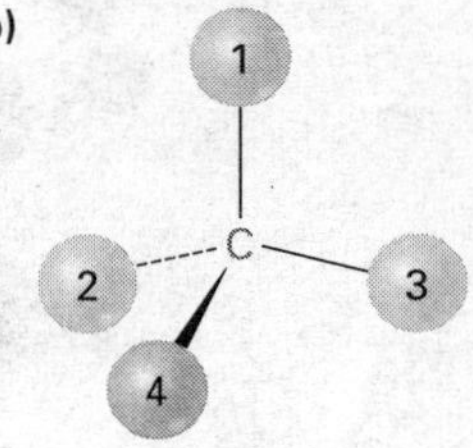

Strategy It takes practice to be able to visualize and orient a chirality center in three dimensions. You might start by indicating where the observer must be located—180° opposite the lowest-ranked group. Then imagine yourself in the position of the observer, and redraw what you would see.

Solution In **(a)**, you would be located in front of the page toward the top right of the molecule, and you would see group 2 to your left, group 3 to your right, and group 1 below you. This corresponds to an *R* configuration.

(a)

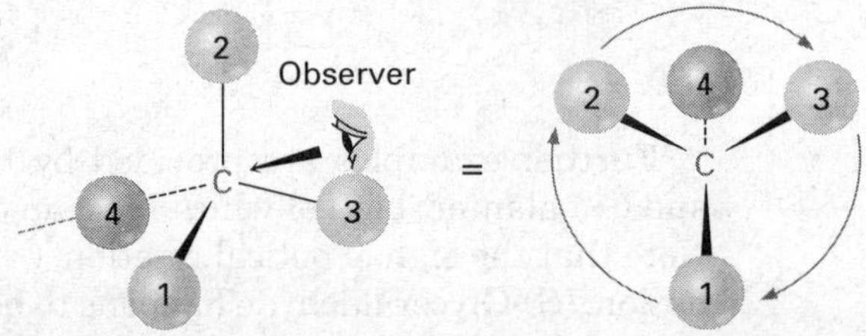

***R* configuration**

In **(b)**, you would be located behind the page toward the top left of the molecule from your point of view, and you would see group 3 to your left, group 1 to your right, and group 2 below you. This also corresponds to an *R* configuration.

(b)

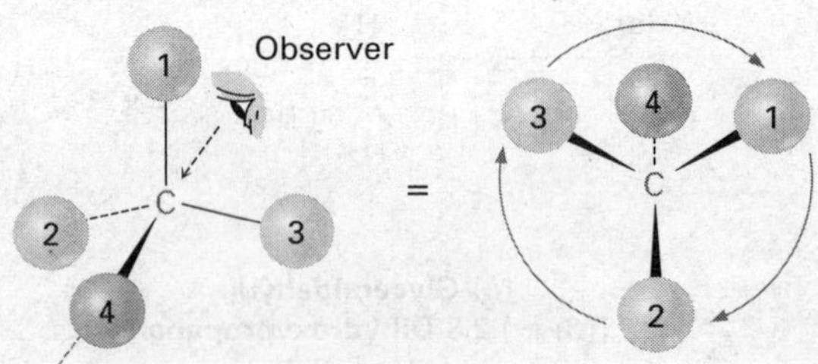

***R* configuration**

Worked Example 6.6 Drawing a Specific Enantiomer

Draw a tetrahedral representation of (*R*)-2-chlorobutane.

Strategy Begin by ranking the four substituents bonded to the chirality center: (1) –Cl, (2) $-CH_2CH_3$, (3) $-CH_3$, (4) –H. To draw a tetrahedral representation of the molecule, orient the lowest-ranked group (–H) away from you and imagine that the other three groups are coming out of the page toward you. Then place the remaining three substituents such that the direction of travel 1 → 2 → 3 is clockwise (right turn), and tilt the molecule toward you to bring the rear hydrogen into view. Using molecular models is a great help in working problems of this sort.

Solution

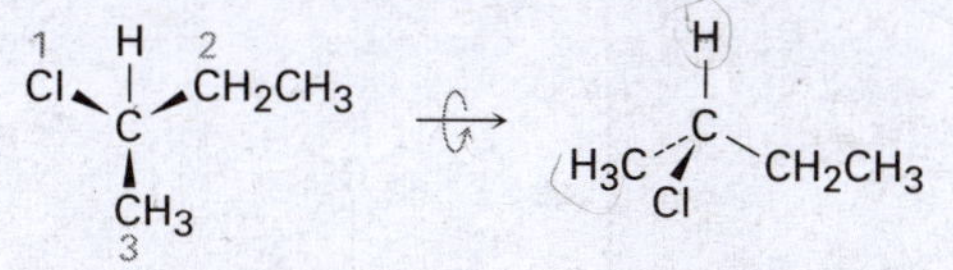

(R)-2-Chlorobutane

Problem 6.8 Rank the substituents in each of the following sets:

(a) $-H$, $-OH$, $-CH_2CH_3$, $-CH_2CH_2OH$
(b) $-CO_2H$, $-CO_2CH_3$, $-CH_2OH$, $-OH$
(c) $-CN$, $-CH_2NH_2$, $-CH_2NHCH_3$, $-NH_2$
(d) $-SH$, $-CH_2SCH_3$, $-CH_3$, $-SSCH_3$

Problem 6.9 Assign R,S configurations to the following molecules:

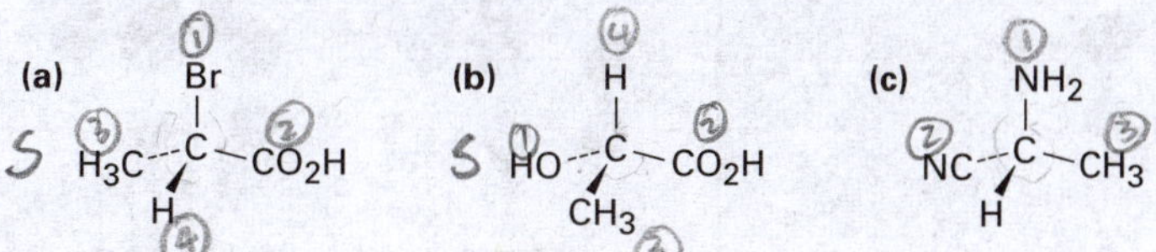

Problem 6.10 Draw a tetrahedral representation of (S)-pentan-2-ol (2-hydroxypentane).

Problem 6.11 Assign R or S configuration to the chirality center in the following molecular model of the amino acid methionine (red = O, blue = N, yellow = S):

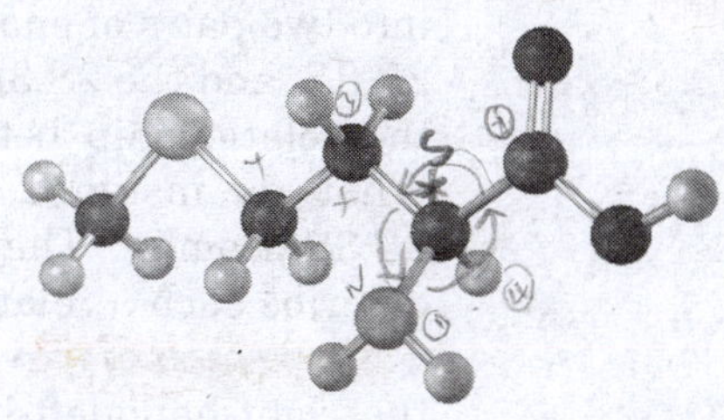

6.6 Diastereomers

Molecules like lactic acid and glyceraldehyde are relatively simple to deal with because each has only one chirality center and only two stereoisomers. The situation becomes more complex, however, for molecules that have more than one chirality center. As a general rule, a molecule with n chirality centers can have up to 2^n stereoisomers (although it may have fewer). Take the amino acid threonine (2-amino-3-hydroxybutanoic acid), for instance. Because threonine has two chirality centers (C2 and C3), there are $2^2 = 4$ possible stereoisomers, as shown in Figure 6.10. Check for yourself that the *R,S* configurations are correct.

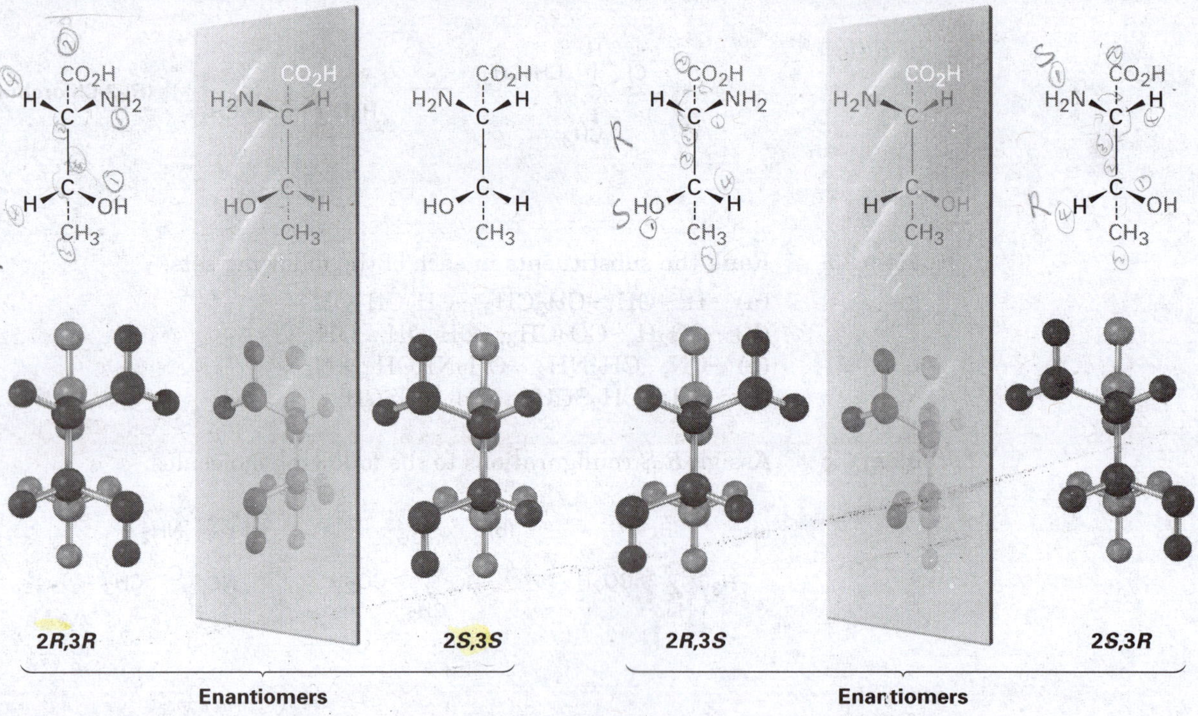

Figure 6.10 The four stereoisomers of 2-amino-3-hydroxybutanoic acid.

The four stereoisomers of 2-amino-3-hydroxybutanoic acid can be grouped into two pairs of enantiomers. The 2*S*,3*S* stereoisomer is the mirror image of 2*R*,3*R*, and the 2*S*,3*R* stereoisomer is the mirror image of 2*R*,3*S*. But what is the relationship between any two molecules that are not mirror images? What, for instance, is the relationship between the 2*R*,3*R* isomer and the 2*R*,3*S* isomer? They are stereoisomers, yet they aren't enantiomers. To describe such a relationship, we need a new term—*diastereomer.*

Diastereomers are stereoisomers that are not mirror images. Since we used the right-hand/left-hand analogy to describe the relationship between two enantiomers, we might extend the analogy by saying that the relationship between diastereomers is like that of hands from different people. Your hand and your friend's hand look *similar,* but they aren't identical and they aren't mirror images. The same is true of diastereomers: they're similar, but they aren't identical and they aren't mirror images.

Note carefully the difference between enantiomers and diastereomers: enantiomers have opposite configurations at *all* chirality centers, whereas diastereomers have opposite configurations at *some* (one or more) chirality centers but the same configuration at others. A full description of the four threonine stereoisomers is given in Table 6.2. Of the four, only the 2*S*,3*R* isomer occurs naturally in plants and animals and is an essential human nutrient. This result is typical: most biological molecules are chiral, and often only one stereoisomer is found in nature.

Table 6.2 Relationships among the Four Stereoisomers of Threonine

Stereoisomer	Enantiomer	Diastereomer
2*R*,3*R*	2*S*,3*S*	2*R*,3*S* and 2*S*,3*R*
2*S*,3*S*	2*R*,3*R*	2*R*,3*S* and 2*S*,3*R*
2*R*,3*S*	2*S*,3*R*	2*R*,3*R* and 2*S*,3*S*
2*S*,3*R*	2*R*,3*S*	2*R*,3*R* and 2*S*,3*S*

Problem 6.12 Assign *R* or *S* configuration to each chirality center in the following molecules:

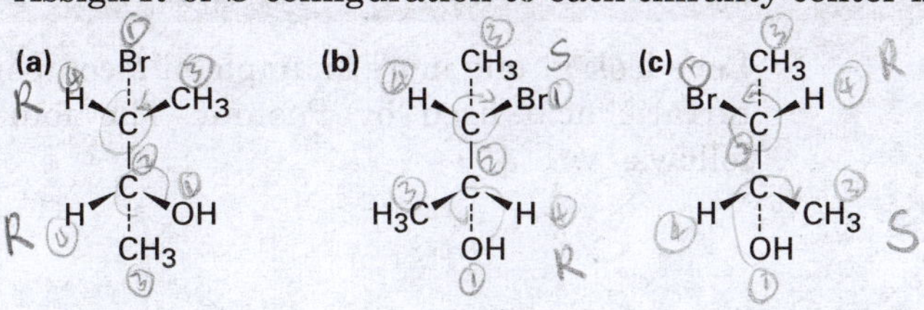

Problem 6.13 Which of the compounds in Problem 6.12 are enantiomers, and which are diastereomers?

Problem 6.14 Of the following molecules **(a)** through **(d)**, one is D-erythrose 4-phosphate, an intermediate in the Calvin photosynthetic cycle by which plants incorporate CO_2 into carbohydrates. If D-erythrose 4-phosphate has *R* stereochemistry at both chirality centers, which of the structures is it? Which of the remaining three structures is the enantiomer of D-erythrose 4-phosphate, and which are diastereomers?

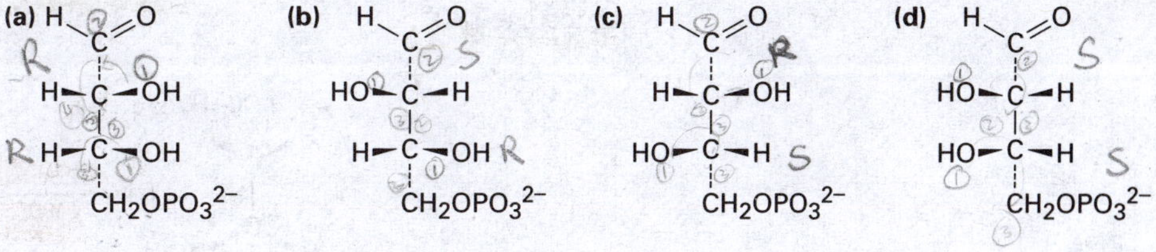

Problem 6.15 Nandrolone is an anabolic steroid used to build muscle mass. How many chirality centers does nandrolone have? How many stereoisomers of nandrolone are possible in principle?

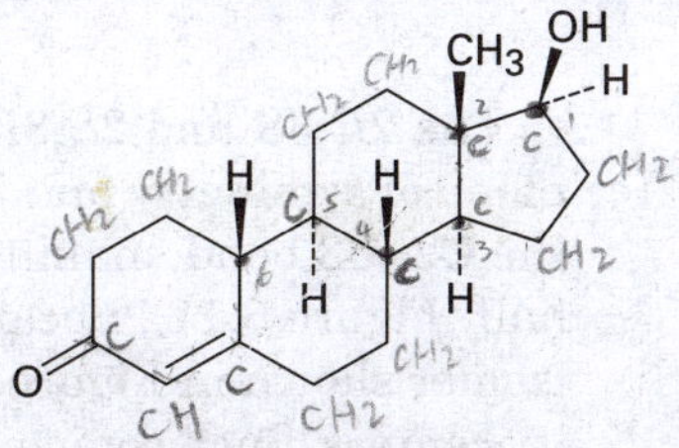

Nandrolone

Problem 6.16 Assign *R*,*S* configuration to each chirality center in the following molecular model of the amino acid isoleucine (red = O, blue = N):

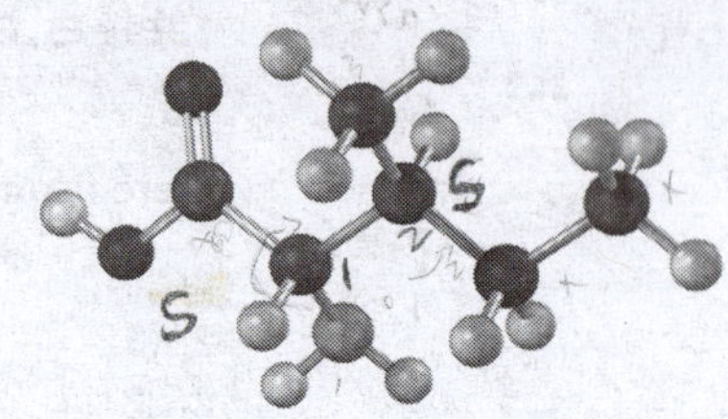

6.7 Meso Compounds

Let's look at one more example of a compound with two chirality centers: the tartaric acid used by Pasteur. The four stereoisomers can be drawn as follows:

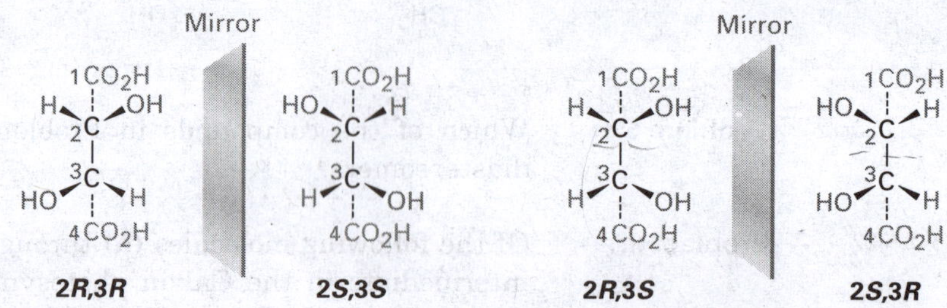

The mirror-image 2*R*,3*R* and 2*S*,3*S* structures are not identical and therefore represent an enantiomeric pair. A careful look, however, shows that the 2*R*,3*S* and 2*S*,3*R* structures *are* identical, as can be seen by rotating one structure 180°.

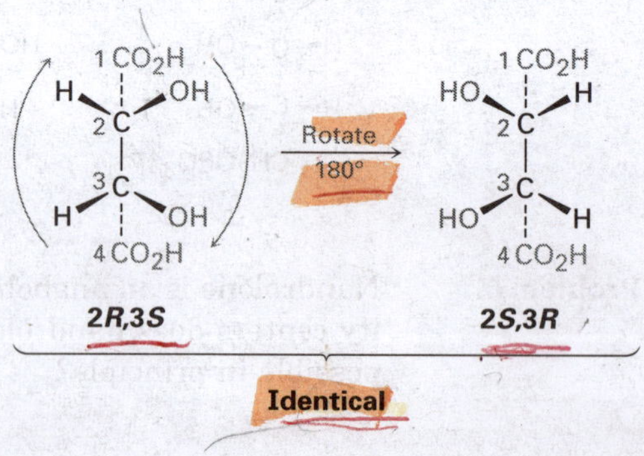

The 2*R*,3*S* and 2*S*,3*R* structures are identical because the molecule has a plane of symmetry and is therefore achiral. The symmetry plane cuts through the C2–C3 bond, making one half of the molecule a mirror image of the other half (Figure 6.11). Because of the plane of symmetry, the tartaric acid stereoisomer shown in Figure 6.11 is achiral, despite the fact that it has two chirality centers. Such compounds that are achiral, yet contain chirality centers, are

called **meso (me-zo) compounds**. Thus, tartaric acid exists in three stereoisomeric forms: two enantiomers and one meso form.

Figure 6.11 A symmetry plane cutting through the C2–C3 bond of *meso*-tartaric acid makes the molecule achiral even though it contains two chirality centers.

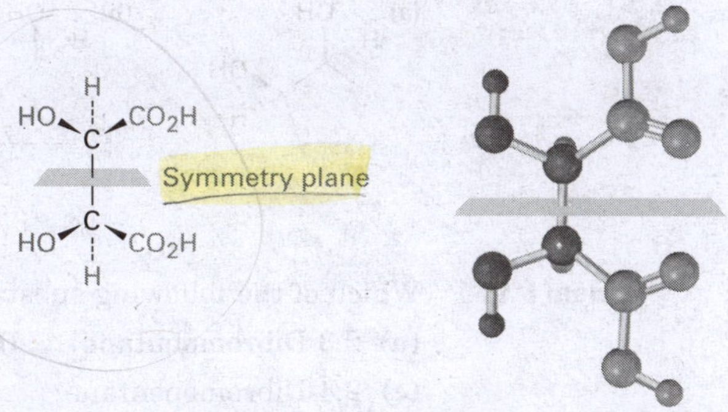

Some physical properties of the three stereoisomers of tartaric acid are shown in Table 6.3. The (+) and (−) enantiomers have identical melting points, solubilities, and densities but differ in the sign of their rotation of plane-polarized light. The meso isomer, by contrast, is diastereomeric with the (+) and (−) forms. It is therefore a different compound altogether and has different physical properties.

Table 6.3 Some Properties of the Stereoisomers of Tartaric Acid

Stereoisomer	Melting point (°C)	$[\alpha]_D$	Density (g/cm^3)	Solubility at 20 °C (g/100 mL H_2O)
(+)	168–170	+12	1.7598	139.0
(−)	168–170	−12	1.7598	139.0
Meso	146–148	0	1.6660	125.0

Worked Example 6.7 Identifying Meso Compounds

Does *cis*-1,2-dimethylcyclobutane have any chirality centers? Is it chiral?

Strategy To see whether a chirality center is present, look for a carbon atom bonded to four different groups. To see whether the molecule is chiral, look for a symmetry plane. Not all molecules with chirality centers are chiral; meso compounds are an exception.

Solution A look at the structure of *cis*-1,2-dimethylcyclobutane shows that both methyl-bearing ring carbons (C1 and C2) are chirality centers. Overall, though, the compound is achiral because there is a symmetry plane bisecting the ring between C1 and C2. Thus, *cis*-1,2-dimethylcyclobutane is a meso compound.

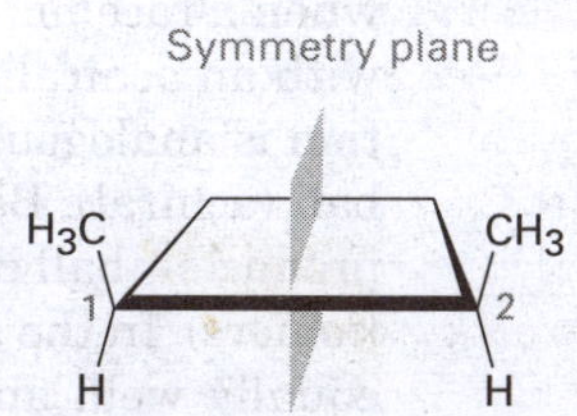

Problem 6.17 Which of the following structures represent meso compounds?

(a)
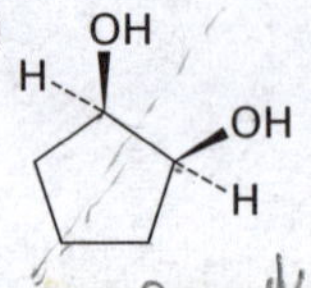

(b)
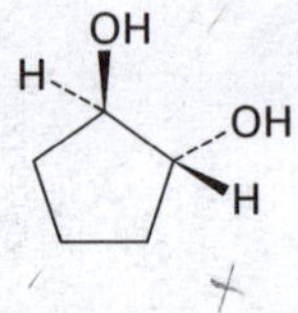

(c)
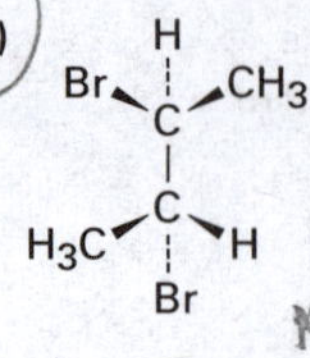

Problem 6.18 Which of the following substances have meso forms? Draw them.

(a) 2,3-Dibromobutane (b) 2,3-Dibromopentane

(c) 2,4-Dibromopentane

6.8 Racemic Mixtures and the Resolution of Enantiomers

To end this discussion of stereoisomerism, let's return for a final look at Pasteur's discovery of enantiomers, described in Section 6.4. Pasteur took an optically inactive tartaric acid salt and found that he could crystallize from it two optically active forms having the 2*R*,3*R* and 2*S*,3*S* configurations. But what was the optically inactive form he started with? It couldn't have been *meso*-tartaric acid, because *meso*-tartaric acid is a different compound and can't interconvert with the two chiral enantiomers without breaking and re-forming bonds.

The answer is that Pasteur started with a 50:50 *mixture* of the two chiral tartaric acid enantiomers. Such a mixture is called a *racemic mixture,* or **racemate** (ra-suh-mate). Racemic mixtures are often denoted by the symbol (±) or by the prefix *d,l* to indicate that they contain equal amounts of dextrorotatory and levorotatory enantiomers. Such mixtures show no optical activity because the (+) rotation from one enantiomer exactly cancels the (−) rotation from the other. Through good luck, Pasteur was able to separate, or **resolve,** the racemate into its (+) and (−) enantiomers. Unfortunately, the crystallization technique he used doesn't work for most racemic mixtures, so other methods are required.

The most common method of resolution uses an acid–base reaction between a racemic mixture of chiral carboxylic acid (RCO_2H) and an amine base (RNH_2) to yield an ammonium salt.

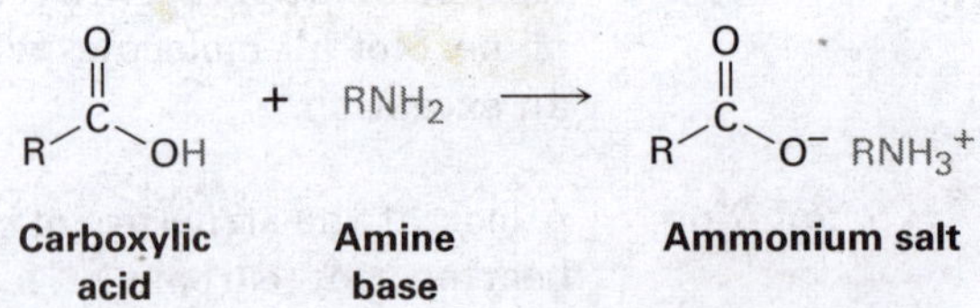

To understand how this method of resolution works, let's see what happens when a racemic mixture of chiral acids, such as (+)- and (−)-lactic acids, reacts with an achiral amine base, such as methylamine. Stereochemically, the situation is analogous to what happens when left and right hands (chiral) pick up a ball (achiral). Both left and right hands pick up the ball equally well, and the products—ball in right hand versus ball in left hand—are mirror images (enantiomers). In the same way, both (+)- and (−)-lactic acid react with methylamine equally well, and the product is a racemic mixture of enantiomeric methylammonium (+)-lactate and methylammonium (−)-lactate (Figure 6.12).

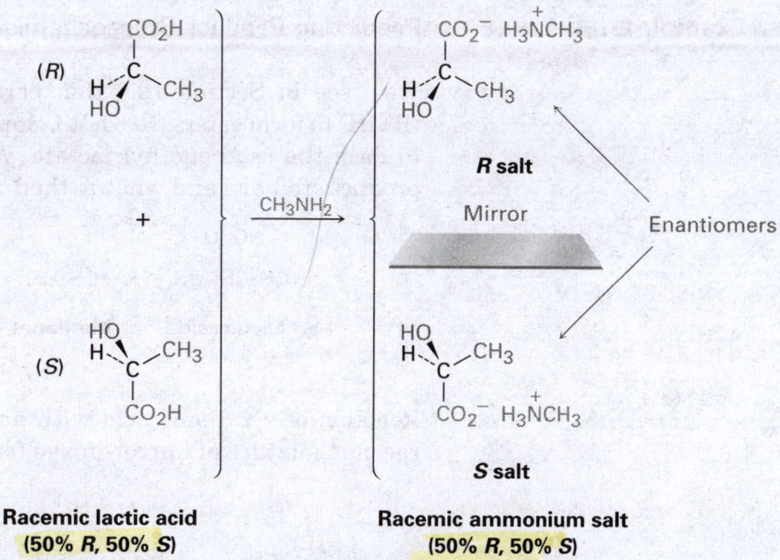

Figure 6.12 Reaction of racemic lactic acid with achiral methylamine leads to a racemic mixture of enantiomeric ammonium salts.

Now let's see what happens when the racemic mixture of (+)- and (−)-lactic acids reacts with a *single* enantiomer of a *chiral* amine base, such as (*R*)-1-phenylethylamine. Stereochemically, this situation is analogous to what happens when left and right hands (chiral) put on a right-handed glove (*also chiral*). The left and right hands don't put on the right-handed glove in the same way. The products—right hand in right glove versus left hand in right glove—are not mirror images; they're similar but different.

In the same way, (+)- and (−)-lactic acids react with (*R*)-1-phenylethylamine to give two different products (Figure 6.13). (*R*)-Lactic acid reacts with (*R*)-1-phenylethylamine to give the *R*,*R* salt, whereas (*S*)-lactic acid reacts with the same *R* amine to give the *S*,*R* salt. *The two salts are diastereomers.* They have different chemical and physical properties, and it may therefore be possible to separate them by crystallization or some other means. Once separated, acidification of the two diastereomeric salts with HCl then allows us to isolate the two pure enantiomers of lactic acid and recover the chiral amine for further use. That is, the original racemic mixture has been resolved.

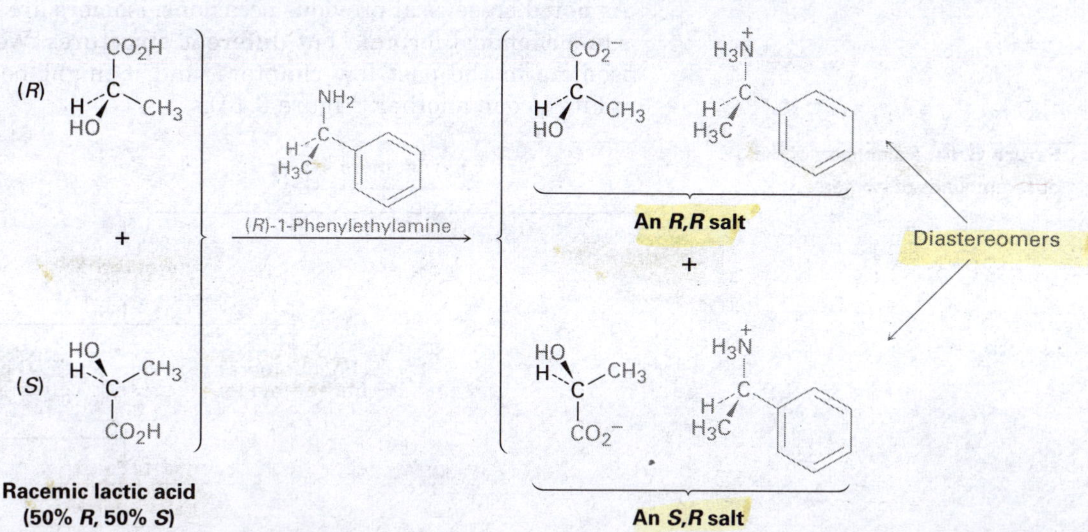

Figure 6.13 Reaction of racemic lactic acid with (*R*)-1-phenylethylamine yields a mixture of diastereomeric ammonium salts, which have different properties and can be separated.

Worked Example 6.8 Predicting Product Stereochemistry

We'll see in Section 10.6 that carboxylic acids (RCO_2H) react with alcohols (R′OH) to form esters (RCO_2R'). Suppose that (±)-lactic acid reacts with CH_3OH to form the ester methyl lactate. What stereochemistry would you expect the products to have and what is their relationship?

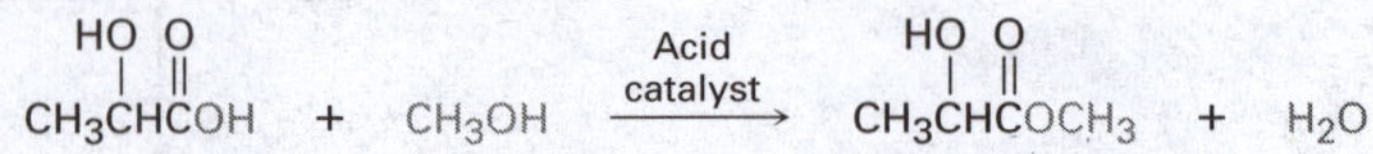

Lactic acid **Methanol** **Methyl lactate**

Solution Reaction of a racemic acid with an achiral alcohol such as methanol yields a racemic mixture of mirror-image (enantiomeric) products:

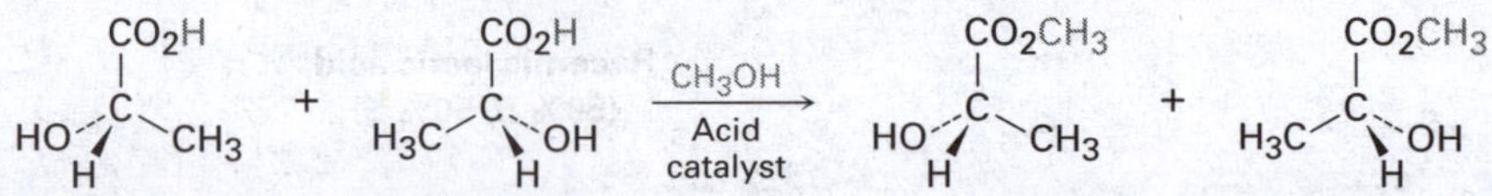

(*S*)-Lactic acid **(*R*)-Lactic acid** **Methyl (*S*)-lactate** **Methyl (*R*)-lactate**

Problem 6.19 Suppose that acetic acid (CH_3CO_2H) reacts with (*S*)-butan-2-ol to form an ester (see Worked Example 6.8). What stereochemistry would you expect the product(s) to have, assuming that the singly bonded oxygen atom comes from the alcohol rather than the acid?

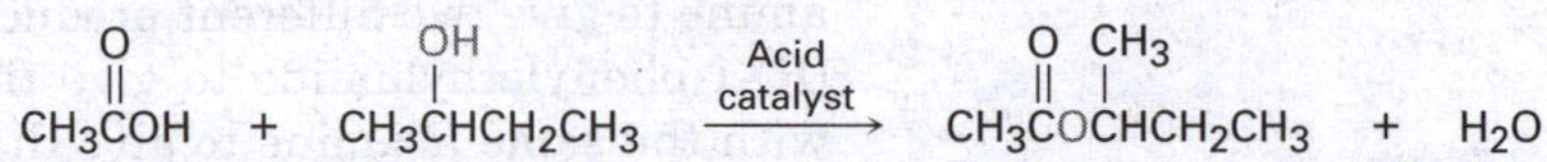

Acetic acid **Butan-2-ol** ***sec*-Butyl acetate**

6.9 A Brief Review of Isomerism

As noted on several previous occasions, isomers are compounds that have the same chemical formula but different structures. We've seen several kinds of isomers in the past few chapters, and it might be helpful to see how they relate to one another (Figure 6.14).

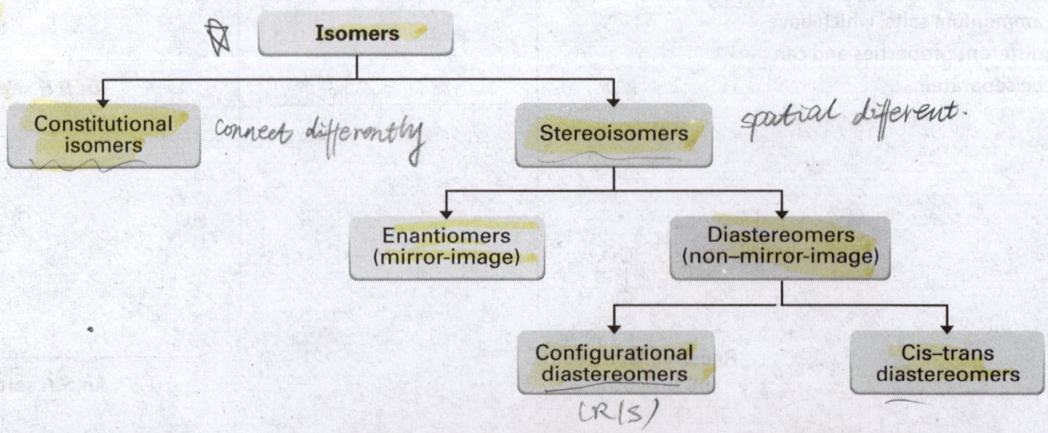

Figure 6.14 A summary of the different kinds of isomers.

There are two fundamental types of isomerism, both of which we've now encountered: constitutional isomerism and stereoisomerism.

- **Constitutional isomers** (Section 2.2) are compounds whose atoms are connected differently. Among the kinds of constitutional isomers we've seen are skeletal, functional, and positional isomers.

Different carbon skeletons

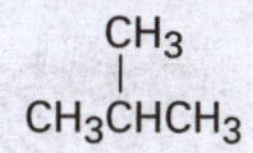

2-Methylpropane

and

$CH_3CH_2CH_2CH_3$

Butane

Different functional groups

CH_3CH_2OH

Ethyl alcohol

and

CH_3OCH_3

Dimethyl ether

Different position of functional groups

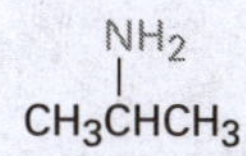

Isopropylamine

and

$CH_3CH_2CH_2NH_2$

Propylamine

- **Stereoisomers** (Section 2.8) are compounds whose atoms are connected in the same way but with a different spatial arrangement. Among the kinds of stereoisomers we've seen are enantiomers, diastereomers, and cis–trans isomers (both in alkenes and in cycloalkanes). Actually, cis–trans isomers are just special kinds of diastereomers because they are non–mirror-image stereoisomers.

Enantiomers (nonsuperimposable mirror-image stereoisomers)

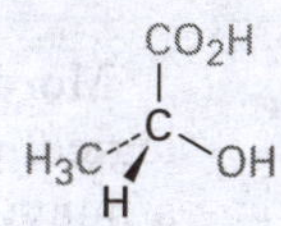

(*R*)-Lactic acid

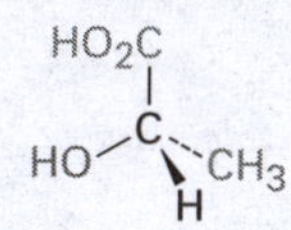

(*S*)-Lactic acid

Diastereomers (nonsuperimposable non–mirror-image stereoisomers)

Configurational diastereomers

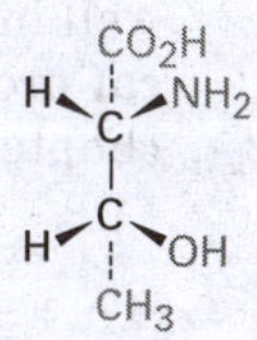

2*R*,3*R*-2-Amino-3-hydroxybutanoic acid

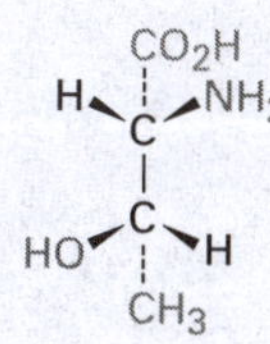

2*R*,3*S*-2-Amino-3-hydroxybutanoic acid

Cis–trans diastereomers (substituents on same side or opposite side of double bond or ring)

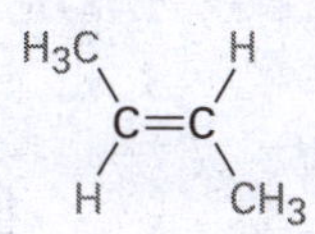

***trans*-But-2-ene**

and

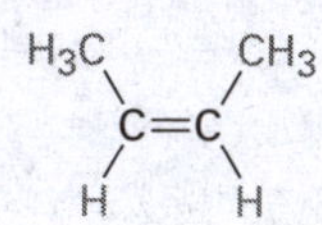

***cis*-But-2-ene**

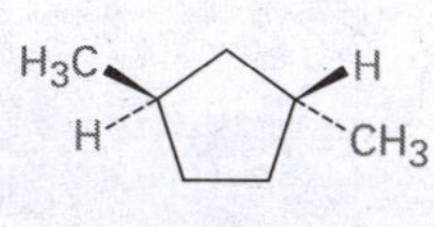

***trans*-1,3-Dimethylcyclopentane**

and

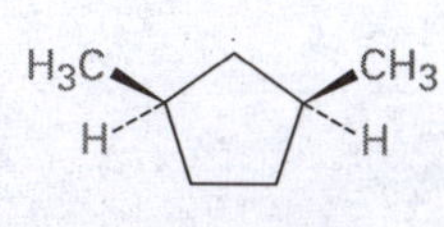

***cis*-1,3-Dimethylcyclopentane**

Problem 6.20 What kinds of isomers are the following pairs?

(a) (*S*)-5-Chlorohex-2-ene and chlorocyclohexane

(b) (2*R*,3*R*)-Dibromopentane and (2*S*,3*R*)-dibromopentane

6.10 Chirality in Nature and Chiral Environments

Although the different enantiomers of a chiral molecule have the same physical properties, they almost always have different biological properties. For example, the (+) enantiomer of limonene has the odor of oranges and lemons, but the (−) enantiomer has the odor of pine trees.

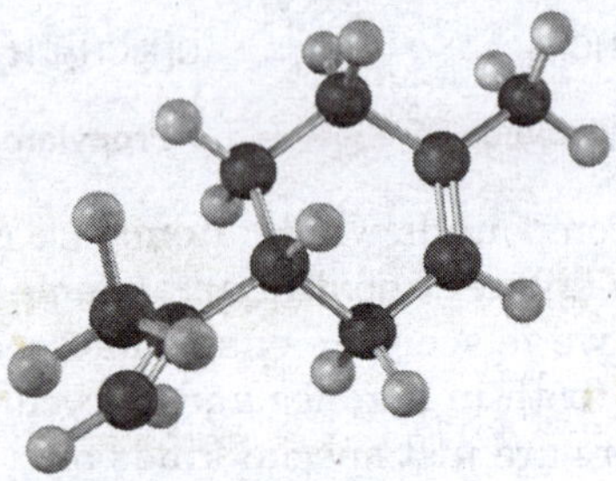

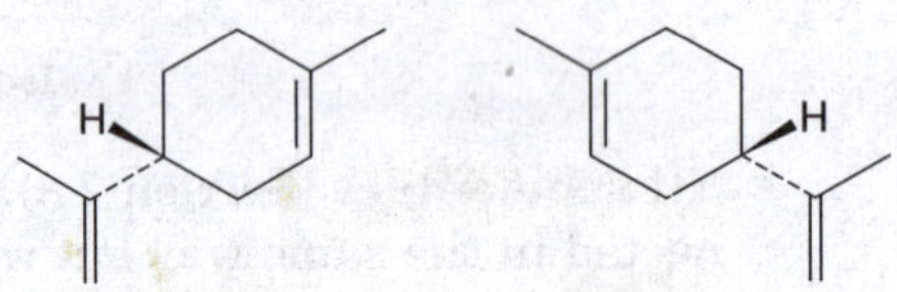

(+)-Limonene (in citrus fruits)

(−)-Limonene (in pine trees)

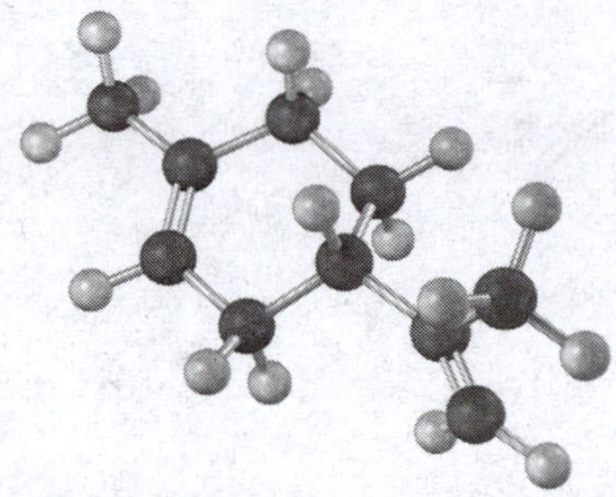

More dramatic examples of how a change in chirality can affect the biological properties of a molecule are found in many drugs, such as fluoxetine, a heavily prescribed medication sold under the trade name Prozac. Racemic fluoxetine is an extraordinarily effective antidepressant, but it has no activity against migraine. The pure *S* enantiomer, however, works remarkably well in preventing migraine. Other examples of how chirality affects biological properties are given in the *Interlude* "Chiral Drugs" at the end of this chapter.

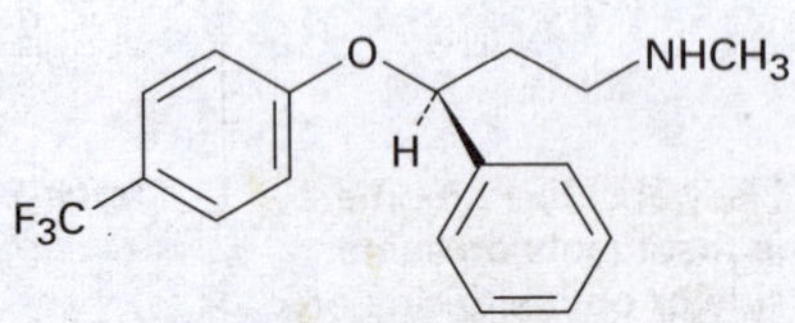

(*S*)-Fluoxetine (prevents migraine)

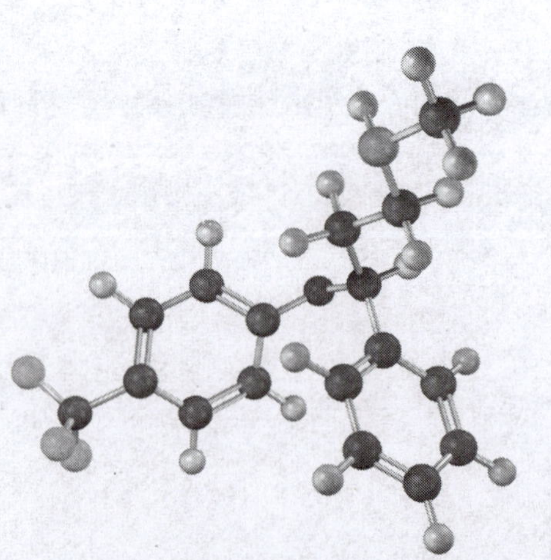

Why do different enantiomers have different biological properties? To have a biological effect, a substance typically must fit into an appropriate receptor in the body that has an exactly complementary shape. But because biological receptors are chiral, only one enantiomer of a chiral substrate can fit in, just as only a right hand will fit into a right-handed glove. The mirror-image enantiomer will be a misfit, like a left hand in a right-handed glove. A representation of the interaction between a chiral molecule and a chiral biological receptor is shown in Figure 6.15. One enantiomer fits the receptor perfectly, but the other does not.

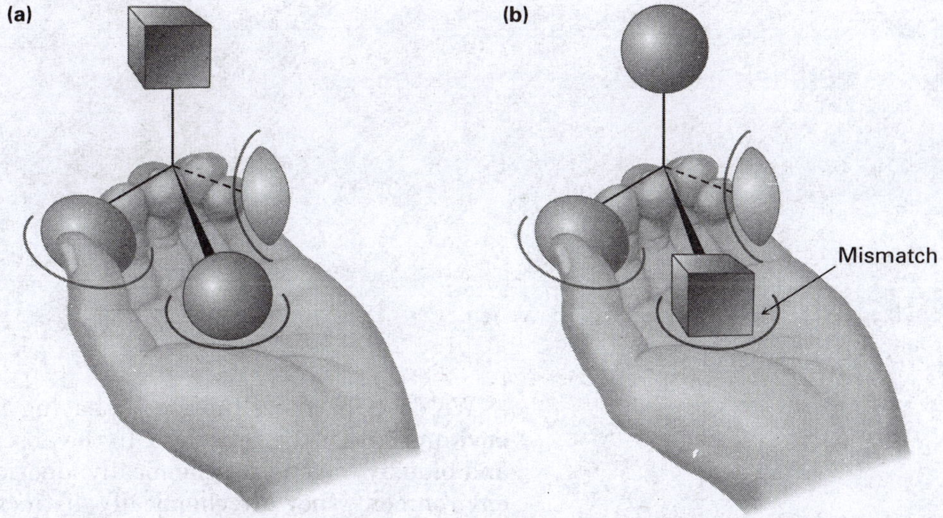

Figure 6.15 Imagine that a left hand interacts with a chiral object, much as a biological receptor interacts with a chiral molecule. **(a)** One enantiomer fits into the hand perfectly: green thumb, red palm, and gray pinkie finger, with the blue substituent exposed. **(b)** The other enantiomer, however, can't fit into the hand. When the green thumb and gray pinkie finger interact appropriately, the palm holds a blue substituent rather than a red one, with the red substituent exposed.

The hand-in-glove fit of a chiral substrate into a chiral receptor is relatively straightforward, but it's less obvious how selective reactions can also occur on achiral molecules. Take the reaction of ethanol (CH_3CH_2OH) with the biochemical oxidizing agent NAD^+ and the enzyme yeast alcohol dehydrogenase to yield acetaldehyde (CH_3CHO). Even though ethanol is achiral, the oxidation reaction occurs with specific removal of only one of the two apparently equivalent $-CH_2-$ hydrogen atoms (Figure 6.16a).

We can understand this result by imagining that the chiral receptor on yeast alcohol dehydrogenase again has three binding sites (Figure 6.16b). When an achiral substrate interacts with the receptor, green (OH) and gray (CH_3) substituents of the substrate are held appropriately but only the blue (H_b) hydrogen substituent is also held while the red hydrogen (H_a) is specifically exposed for removal in the oxidation reaction.

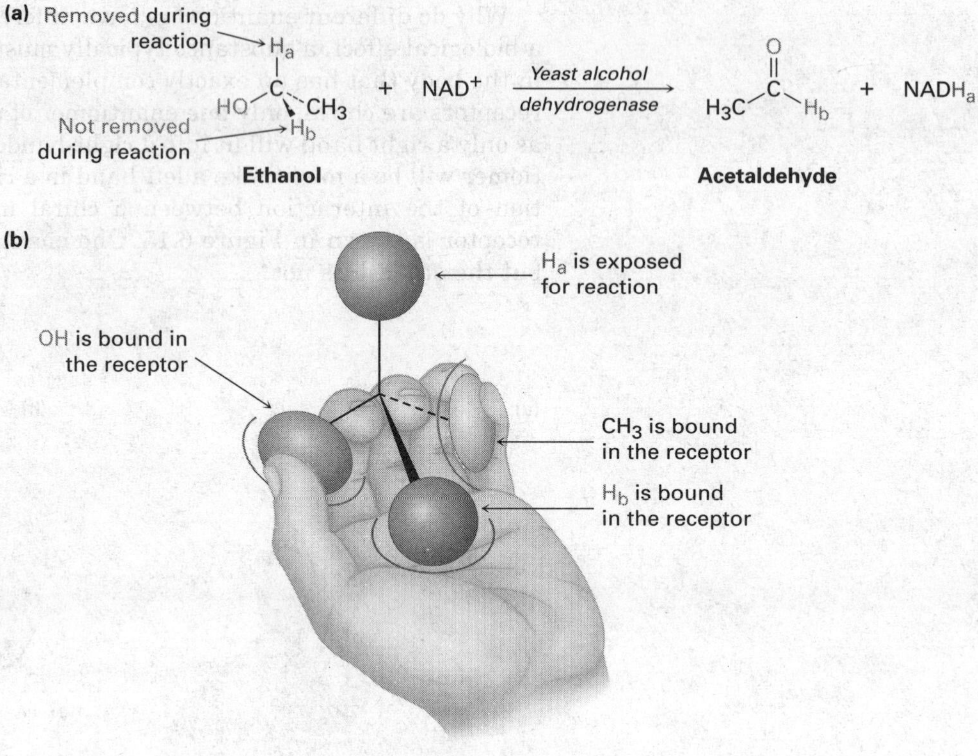

Figure 6.16 When the achiral substrate molecule ethanol is held in a chiral environment on binding to a biological receptor, the two seemingly identical hydrogens are distinguishable. Thus, only a specific hydrogen (red) is removed in an oxidation reaction.

We describe the situation by saying that the receptor provides a **chiral environment** for the substrate. In the absence of a chiral environment, the red and blue hydrogens are chemically identical, but in the presence of the chiral environment, they are chemically distinctive (Figure 6.16b) so that only one of them (red) is exposed for reaction. In effect, the chiral environment transfers its own chirality to the achiral substrate.

INTERLUDE

Chiral Drugs

The hundreds of different pharmaceutical agents approved for use by the U.S. Food and Drug Administration come from many sources. Many drugs are isolated directly from plants or bacteria, and others are made by chemical modification of naturally occurring compounds, but an estimated 33% are made entirely in the laboratory and have no relatives in nature.

Those drugs that come from natural sources, either directly or after chemical modification, are usually chiral and are generally found only as a single enantiomer rather than as a racemate. Penicillin V, for example, an antibiotic isolated from the *Penicillium* mold, has the 2*S*,5*R*,6*R* configuration. Its enantiomer, which does not occur naturally but can be made in the laboratory, has no antibiotic activity.

continued

INTERLUDE

The *S* enantiomer of ibuprofen soothes the aches and pains of athletic injuries. The *R* enantiomer has no effect.

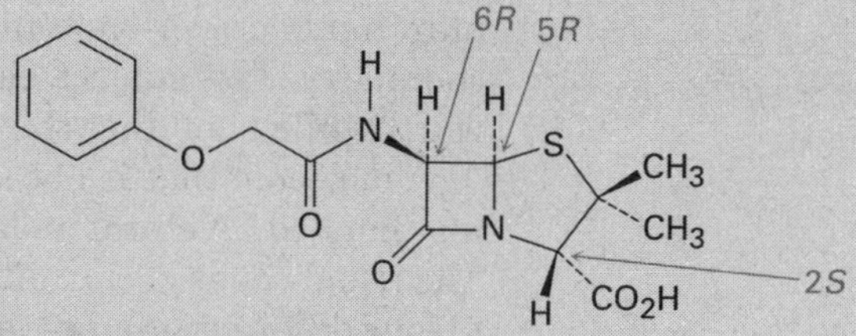

Penicillin V (2*S*,5*R*,6*R* configuration)

In contrast to drugs from natural sources, those drugs that are made entirely in the laboratory either are achiral or, if chiral, are often produced and sold as racemic mixtures. Ibuprofen, for example, has one chirality center and is sold commercially under such trade names as Advil, Nuprin, and Motrin as a 50:50 mixture of *R* and *S* enantiomers. It turns out, however, that only the *S* enantiomer is active as an analgesic and anti-inflammatory agent. The *R* enantiomer of ibuprofen is inactive, although it is slowly converted in the body to the active *S* form.

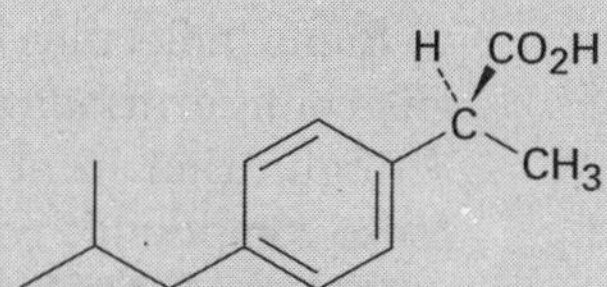

(*S*)-Ibuprofen
(an active analgesic agent)

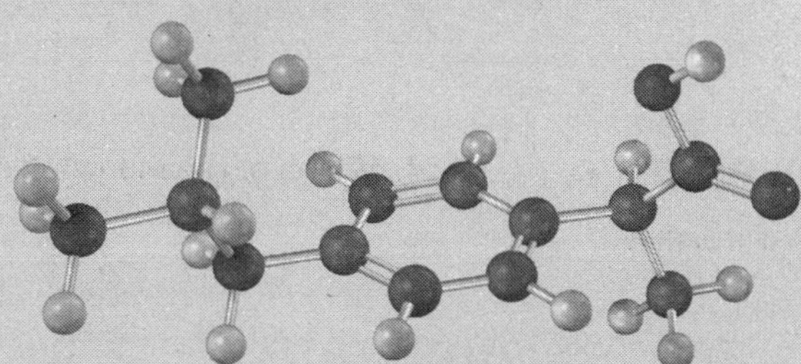

Not only is it chemically wasteful to synthesize and administer an enantiomer that does not serve the intended purpose, many instances are now known where the presence of the "wrong" enantiomer in a racemic mixture either affects the body's ability to utilize the "right" enantiomer or has unintended pharmacological effects of its own. The presence of (*R*)-ibuprofen in the racemic mixture, for instance, slows the rate at which the *S* enantiomer takes effect in the body, from 12 minutes to 38 minutes.

To get around this problem, pharmaceutical companies attempt to devise methods of *enantioselective synthesis,* which allow them to prepare only a single enantiomer rather than a racemic mixture. Viable methods have been developed for the preparation of (*S*)-ibuprofen, which is now being marketed in Europe.

Summary and Key Words

achiral 193
chiral 191
chiral environment 212
chirality center 191
configuration 197
dextrorotatory 195
diastereomers 202
enantiomers 190
levorotatory 195
meso compound 205
optical activity 195
R configuration 198
racemate (racemic mixture) 206
resolution 206
S configuration 198
specific rotation, $[\alpha]_D$ 196

In this chapter, we've looked at some of the causes and consequences of molecular handedness—a topic crucial to understanding organic and biological chemistry. The subject can be a bit complex, but it is so important that it's worthwhile spending the time needed to become familiar with it.

A molecule that is not identical to its mirror image is said to be **chiral**, meaning "handed." A chiral molecule is one that does not contain a plane of symmetry. The usual cause of chirality is the presence of a tetrahedral carbon atom bonded to four different groups—a so-called **chirality center**. Chiral compounds can exist as a pair of mirror-image stereoisomers called **enantiomers**, which are related to each other as a right hand is related to a left hand. When a beam of plane-polarized light is passed through a solution of a pure enantiomer, the plane of polarization is rotated, and the compound is said to be **optically active**.

The three-dimensional **configuration** of a chirality center is specified as either ***R*** or ***S***. Sequence rules are used to rank the four substituents on the chiral carbon, and the molecule is then oriented so that the lowest-ranked group points directly away from the viewer. If a curved arrow drawn in the direction of decreasing rank for the remaining three groups is clockwise, the chirality center has the *R* configuration. If the direction is counterclockwise, the chirality center has the *S* configuration.

Some molecules have more than one chirality center. Enantiomers have opposite configurations at all chirality centers, whereas **diastereomers** have the same configuration in at least one center but opposite configurations at the others. **Meso compounds** contain chirality centers but are achiral overall because they contain a plane of symmetry. **Racemates** are 50:50 mixtures of (+) and (−) enantiomers. Racemic mixtures and individual diastereomers differ in both their physical properties and their biological properties and can often be **resolved**.

Exercises

Visualizing Chemistry

(Problems 6.1–6.20 appear within the chapter.)

OWL
Interactive versions of these problems are assignable in OWL.

6.21 Which of the following structures are identical? (Red = O, yellow-green = Cl.)

(a)

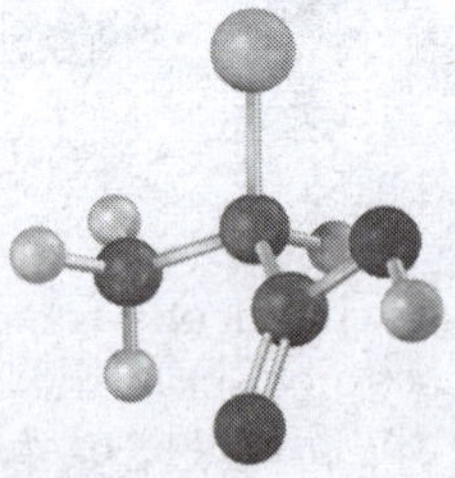

(b)

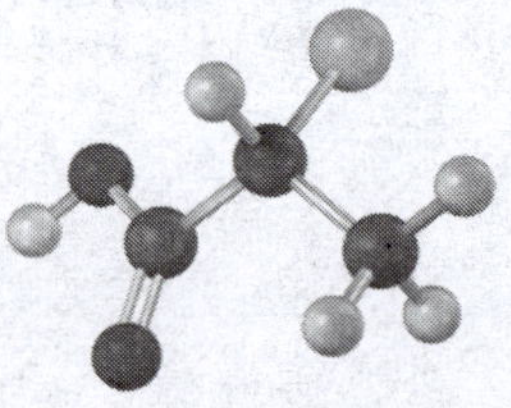

(c)

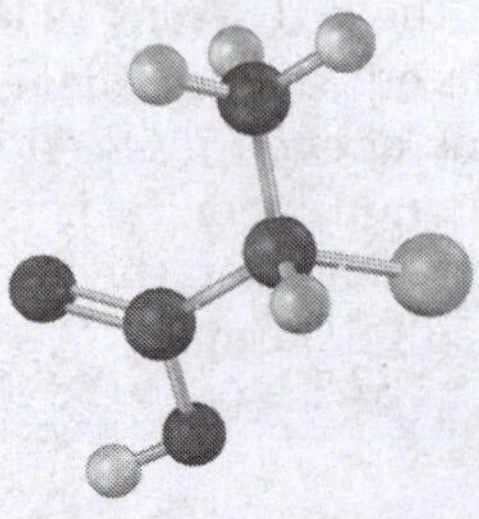

(d)

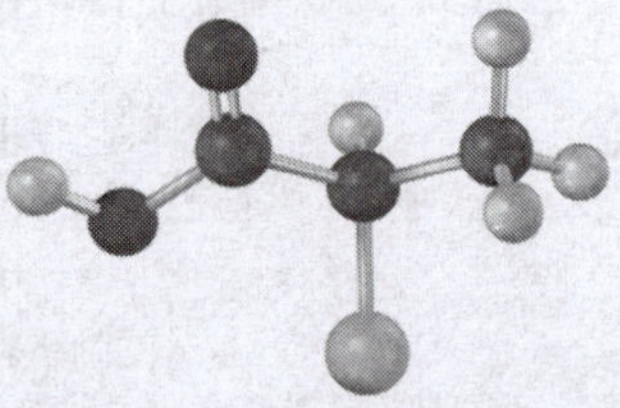

6.22 Assign R or S configuration to each chirality center in pseudoephedrine, an over-the-counter decongestant found in cold remedies (red = O, blue = N).

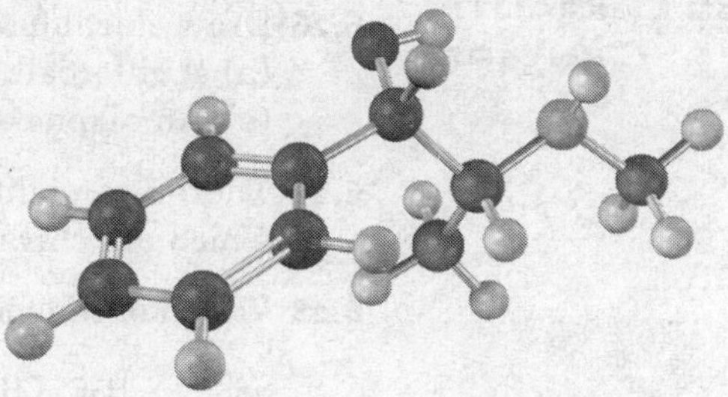

6.23 Assign R or S configuration to the chirality centers in the following molecules (red = O, blue = N):

(a)

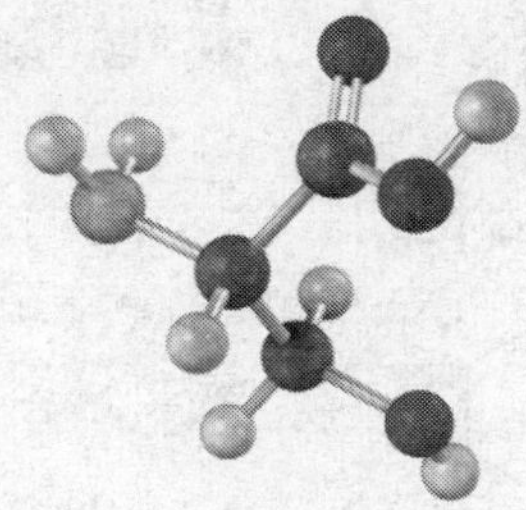

Serine

(b)

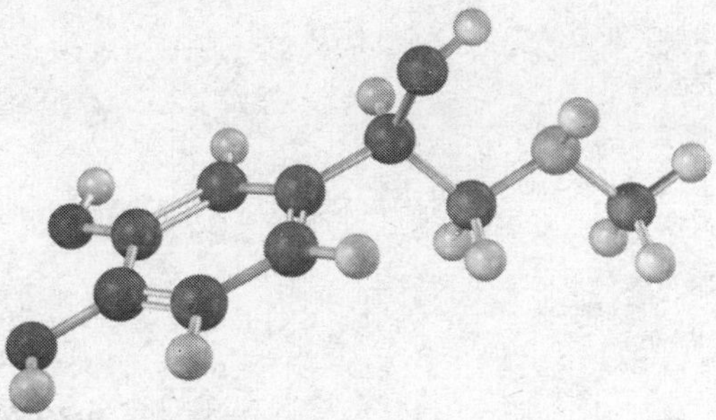

Adrenaline

6.24 Orient each of the following drawings so that the lowest-ranked group is toward the rear, and then assign R or S configuration:

(a)

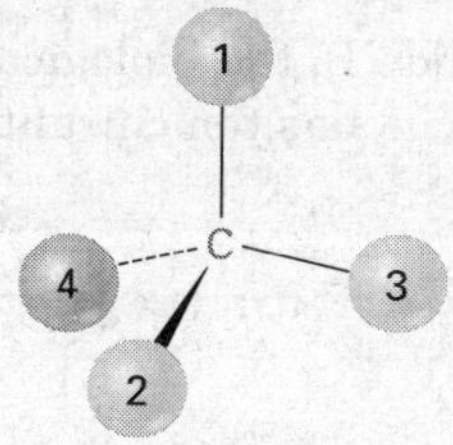

(b)

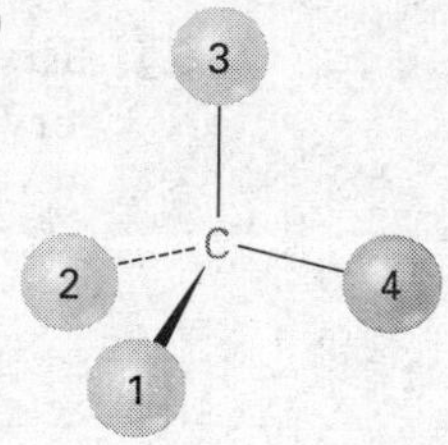

(c)

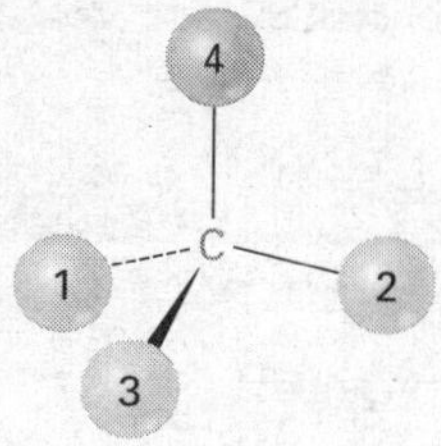

6.25 Which, if any, of the following structures represent meso compounds? (Red = O, blue = N, yellow-green = Cl.)

(a)

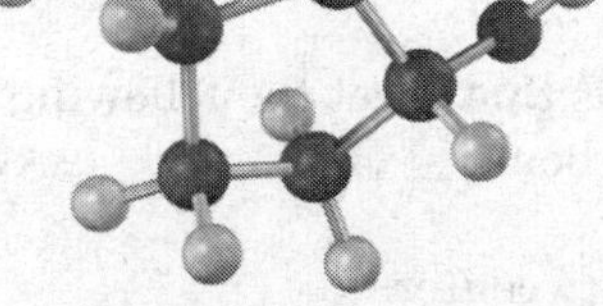

(b)

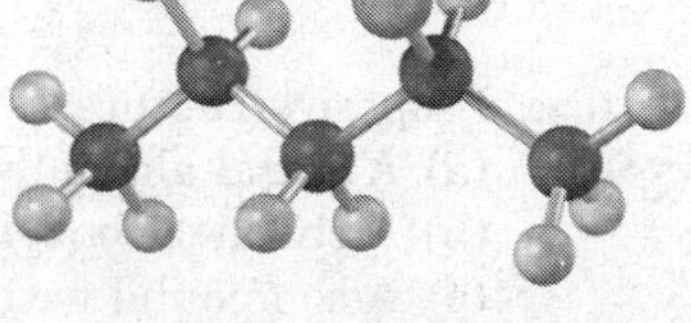

(c)

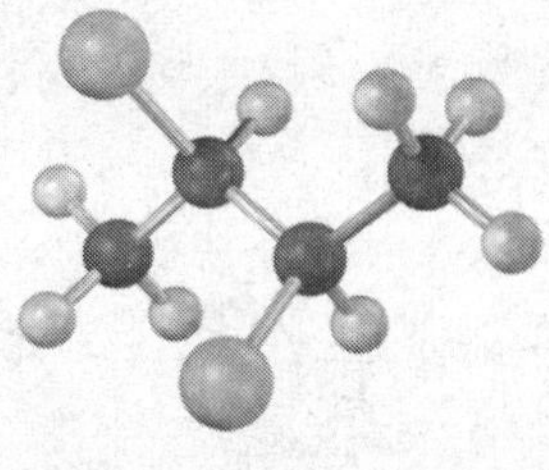

Additional Problems

Identifying Chirality Centers

6.26 Draw chiral molecules that meet the following descriptions:
(a) A chloroalkane, $C_5H_{11}Cl$ **(b)** An alcohol, $C_6H_{14}O$
(c) An alkene, C_6H_{12} **(d)** An alkane, C_8H_{18}

6.27 There are eight alcohols with the formula $C_5H_{12}O$. Draw them, and tell which are chiral.

6.28 Which of the following compounds are chiral? Label all chirality centers.

(a)

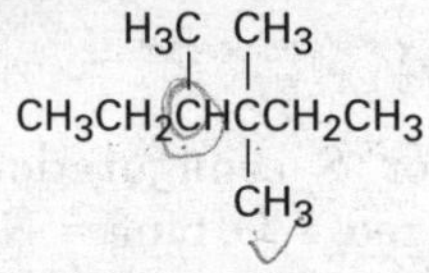

(b)

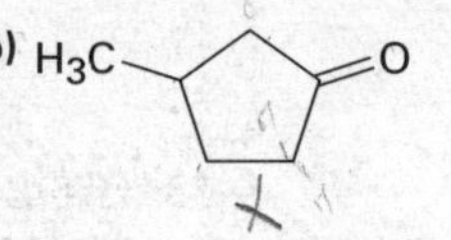

(c)

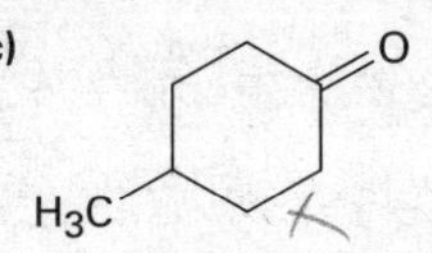

(d)

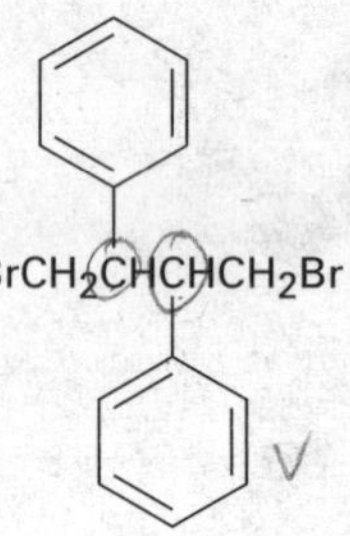

(e)

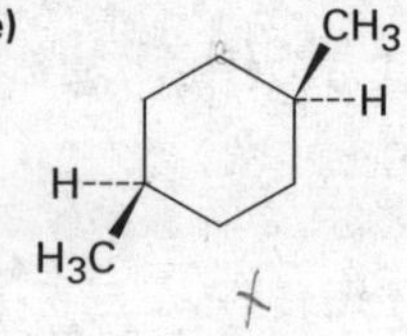

(f)

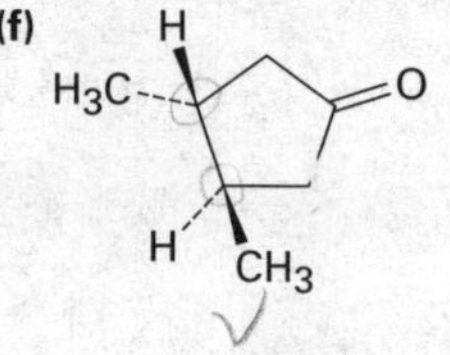

6.29 Which of the following objects are chiral?
(a) A basketball **(b)** A fork **(c)** A wine glass
(d) A golf club **(e)** A monkey wrench **(f)** A snowflake

6.30 Which of the following compounds are chiral?
(a) 2,4-Dimethylheptane **(b)** 5-Ethyl-3,3-dimethylheptane
(c) *cis*-1,3-Dimethylcyclohexane **(d)** 4,5-Dimethylocta-2,6-diene

6.31 Erythronolide B, the biological precursor of the broad-spectrum antibiotic erythromycin, has ten chirality centers. Identify them with asterisks.

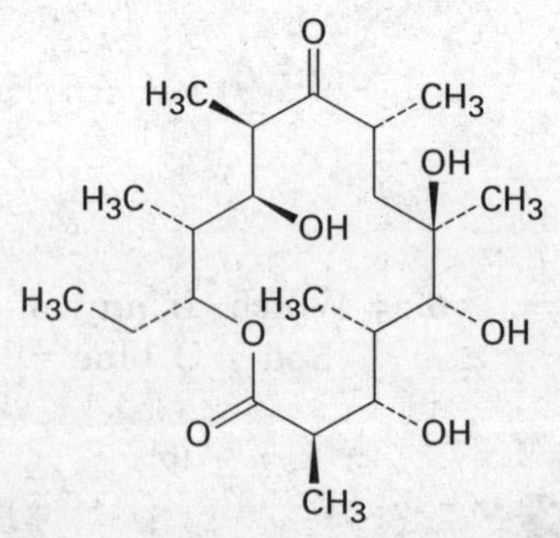

Erythronolide B

6.32 Propose structures for compounds that meet the following descriptions:
(a) A chiral alcohol with four carbons
(b) A chiral carboxylic acid
(c) A compound with two chirality centers

Optical Activity

6.33 Polarimeters are so sensitive that they can measure rotations to the thousandth of a degree, an important advantage when only small amounts of a sample are available. For example, when 7.00 mg of ecdysone, an insect hormone that controls molting in the silkworm moth, was dissolved in 1.00 mL of chloroform in a cell with a 2.00 cm pathlength, an observed rotation of +0.087° was found. Calculate $[\alpha]_D$ for ecdysone.

6.34 Naturally occurring (*S*)-serine has $[\alpha]_D = -6.83$. What specific rotation do you expect for (*R*)-serine?

6.35 Cholic acid, the major steroid found in bile, was found to have a rotation of +2.22° when a 3.00 g sample was dissolved in 5.00 mL of alcohol in a sample tube with a 1.00 cm pathlength. Calculate $[\alpha]_D$ for cholic acid.

Assigning R, S Configuration to Chirality Centers

6.36 Draw tetrahedral representations of both enantiomers of the amino acid serine. Tell which of your structures is *S* and which is *R*.

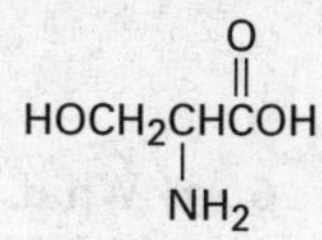

Serine

6.37 One enantiomer of lactic acid is shown below. Is it *R* or *S*? Draw its mirror image in the standard tetrahedral representation.

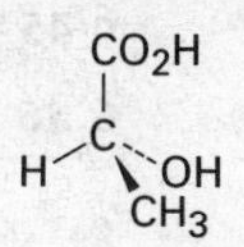

6.38 Rank the substituents in each of the following sets:

(a) –H, –OH, $–OCH_3$, $–CH_3$

(b) –Br, $–CH_3$, $–CH_2Br$, –Cl

(c) $–CH{=}CH_2$, $–CH(CH_3)_2$, $–C(CH_3)_3$, $–CH_2CH_3$

(d) $–CO_2CH_3$, $–COCH_3$, $–CH_2OCH_3$, $–OCH_3$

6.39 Assign *R* or *S* configuration to the chirality centers in the following molecules:

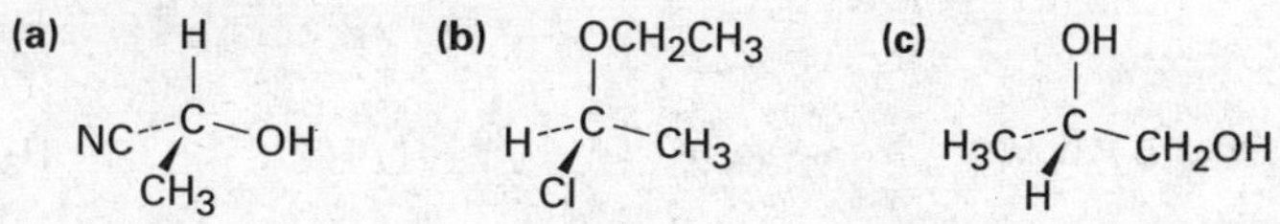

6.40 Rank the substituents in each of the following sets:

(a)

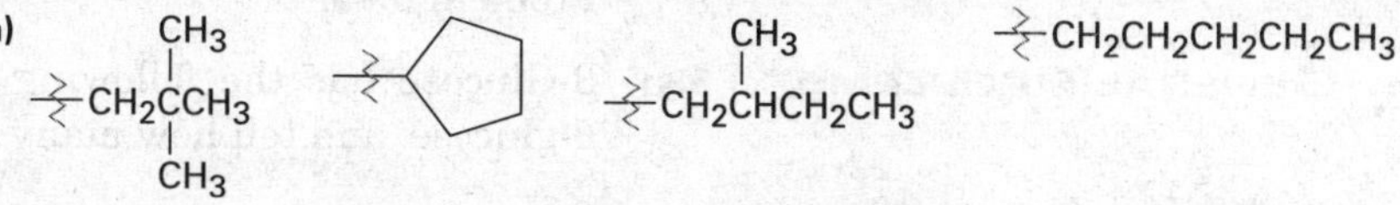

(b) –SH $–NH_2$ $–SO_3H$ $–OCH_2CH_2OH$

6.41 Assign R or S configuration to each chirality center in the following biological molecules:

(a) 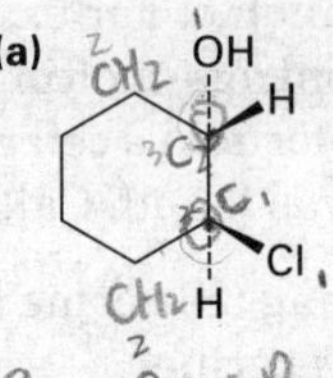 **(b)** 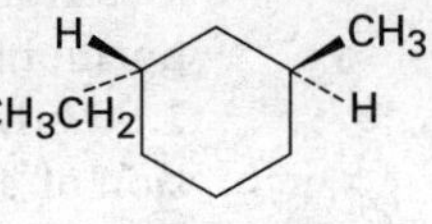 **(c)** 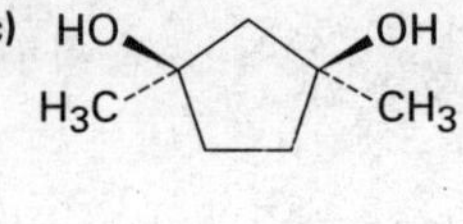

6.42 Assign R or S configuration to the chirality centers in the following molecules:

(a) 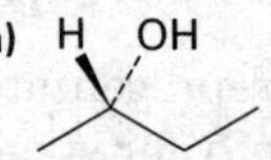 **(b)** 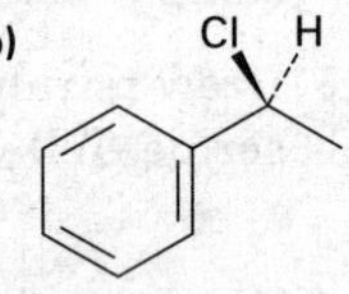 **(c)** 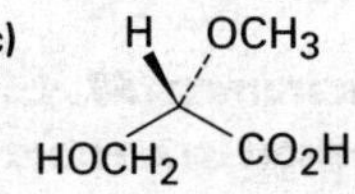

STEREOCHEMICAL RELATIONSHIPS

6.43 What is the stereochemical configuration of the enantiomer of (2S,4R)-dibromooctane?

6.44 What are the stereochemical configurations of the two diastereomers of (2S,4R)-dibromooctane?

6.45 Draw a Newman projection of *meso*-tartaric acid.

6.46 Draw Newman projections of (2R,3R)- and (2S,3S)-tartaric acid, and compare them to the projection you drew in Problem 6.45 for the meso form.

6.47 Draw examples of the following:
(a) A meso compound with the formula C_8H_{18}
(b) A compound with two chirality centers, one R and the other S

6.48 What is the relationship between the specific rotations of (2R,3R)-pentane-2,3-diol and (2S,3S)-pentane-2,3-diol? Between (2R,3S)-pentane-2,3-diol and (2R,3R)-pentane-2,3-diol?

6.49 Tell whether the following Newman projection of 2-chlorobutane is R or S. (You might want to review Section 2.5.)

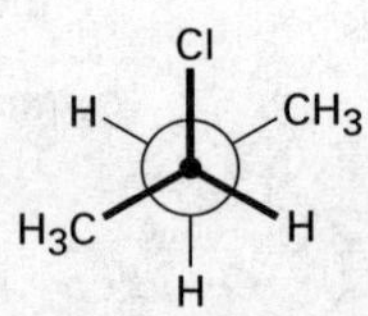

6.50 Draw a Newman projection that is enantiomeric with the one shown in Problem 6.49.

GENERAL PROBLEMS

6.51 β-Glucose has the following structure. Identify the chirality centers in β-glucose, and tell how many stereoisomers of glucose are possible.

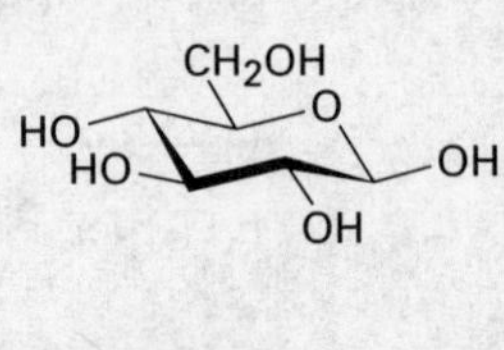

β-Glucose

6.52 Draw the meso form of each of the following molecules, and indicate the plane of symmetry in each:

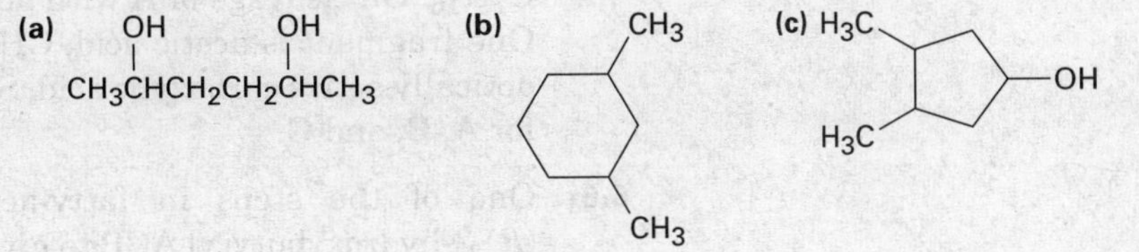

6.53 Ribose, an essential part of ribonucleic acid (RNA), has the following structure:

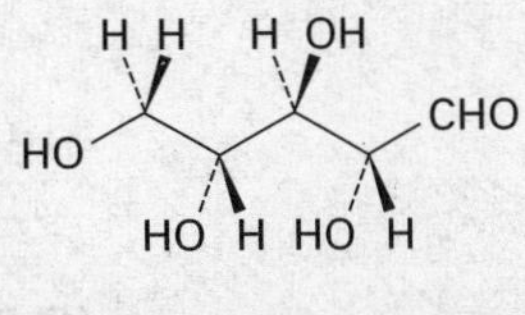

Ribose

How many chirality centers does ribose have? Identify them with asterisks. How many stereoisomers of ribose are there?

6.54 Draw the structure of the enantiomer of ribose (Problem 6.53).

6.55 Draw the structure of a diastereomer of ribose (Problem 6.53).

6.56 Draw the two cis–trans stereoisomers of 1,2-dimethylcyclopentane, assign *R,S* configurations to the chirality centers, and indicate whether the stereoisomers are chiral or meso.

6.57 Assign *R* or *S* configuration to the chirality centers in ascorbic acid (vitamin C).

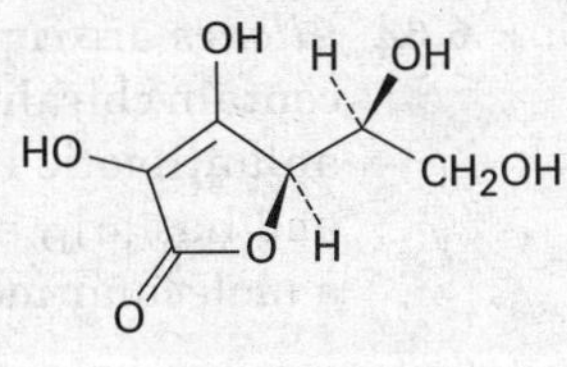

Ascorbic acid

6.58 Draw a tetrahedral representation of (*R*)-3-chloropent-1-ene.

6.59 We saw in Section 4.6 that alkenes undergo reaction with peroxycarboxylic acids to give epoxides. For example, *cis*-but-2-ene gives 2,3-epoxybutane:

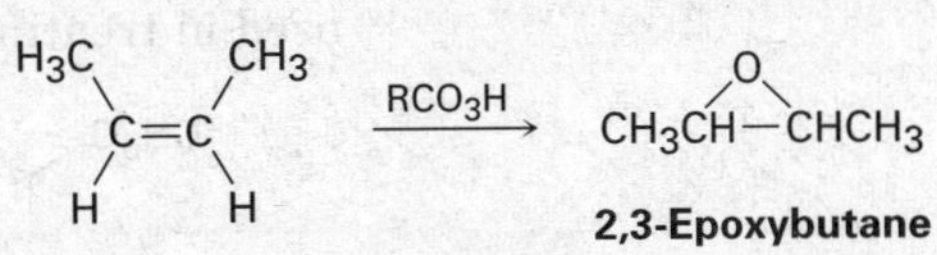

2,3-Epoxybutane

Assuming that both C–O bonds form from the same side of the double bond (syn stereochemistry; Section 4.5), show the stereochemistry of the product. Is the epoxide chiral? Is it optically active?

6.60 Compound **A**, C_7H_{14}, is optically active. On catalytic reduction over a palladium catalyst, 1 equivalent of H_2 is absorbed, yielding compound **B**, C_7H_{16}. On cleavage of **A** with acidic $KMnO_4$, two fragments are obtained. One fragment is acetic acid, CH_3CO_2H, and the other fragment, **C**, is an optically active carboxylic acid. Show the reactions, and propose structures for **A**, **B**, and **C**.

6.61 One of the steps in fatty-acid biosynthesis is the dehydration of (*R*)-3-hydroxybutyryl ACP to give *trans*-crotonyl ACP. The reaction occurs with anti stereochemistry, meaning that the OH and H groups lost during the reaction depart from opposite sides of the molecule. Which hydrogen is lost, H_a or H_b?

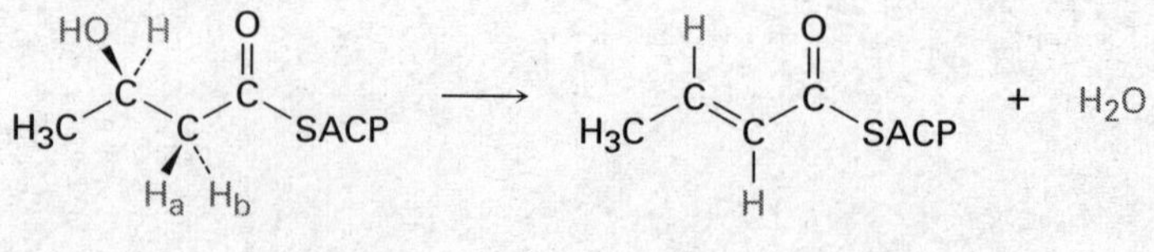

(*R*)-3-Hydroxybutyryl ACP **_trans_-Crotonyl ACP**

6.62 On catalytic hydrogenation over a platinum catalyst, ribose (Problem 6.53) is converted into ribitol. Is ribitol optically active or inactive? Explain.

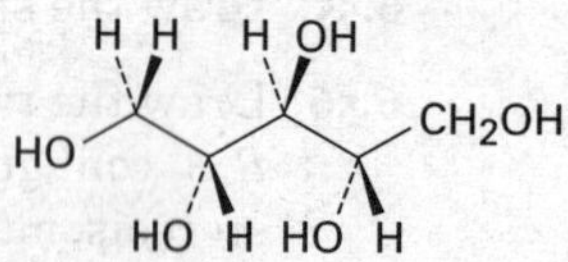

Ribitol

6.63 Draw the structure of (*R*)-2-methylcyclohexanone.

6.64 *Allenes* are compounds with adjacent C=C bonds. Even though they don't contain chirality centers, many allenes are chiral. For example, mycomycin, an antibiotic isolated from the bacterium *Nocardia acidophilus*, is chiral and has $[\alpha]_D = -130$. Can you explain why mycomycin is chiral? Making a molecular model should be helpful.

$$HC{\equiv}C{-}C{\equiv}C{-}CH{=}C{=}CH{-}CH{=}CH{-}CH{=}CH{-}CH_2CO_2H$$

Mycomycin

IN THE MEDICINE CABINET

6.65 Compound **A** is a precursor used for synthesizing dopa, whose *S* isomer is used in treating Parkinson's disease.

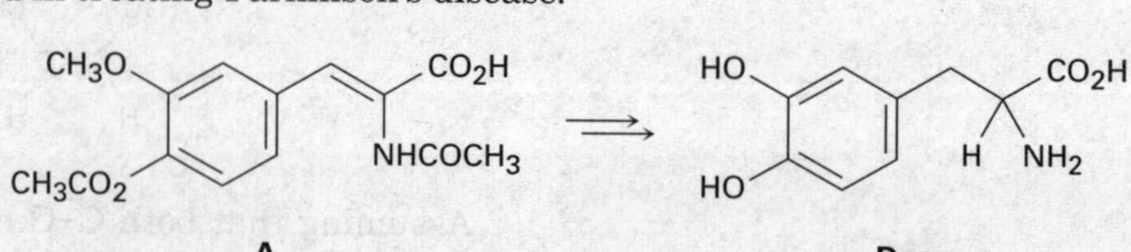

A **Dopa**

(a) The first step in the synthesis is catalytic hydrogenation of the carbon–carbon double bond in **A** to yield two enantiomeric hydrogenation products. Draw them, and assign *R*,*S* configuration to each.

(b) Following hydrogenation, several additional transformations are carried out to yield dopa. Do you expect the enantiomers of dopa to have similar physical properties?

(c) Do you expect the enantiomers of dopa to perform equally well as drugs?

(d) The Monsanto process, commercialized in 1974, carries out the double-bond hydrogenation using a chiral catalyst that produces only a single enantiomer, which is subsequently converted into (*S*)-dopa. Show the stereochemistry of (*S*)-dopa, and explain how a chiral hydrogenation catalyst can produce a single enantiomer as product.

IN THE FIELD **6.66** Metolachlor, a herbicide previously encountered in Problems 2.74 and 5.62, kills weeds by inhibiting the enzyme fatty-acid elongase, which is needed to make a waxy coating on plant leaves. With the enzyme activity inhibited, the wax is not produced and the weed dies.

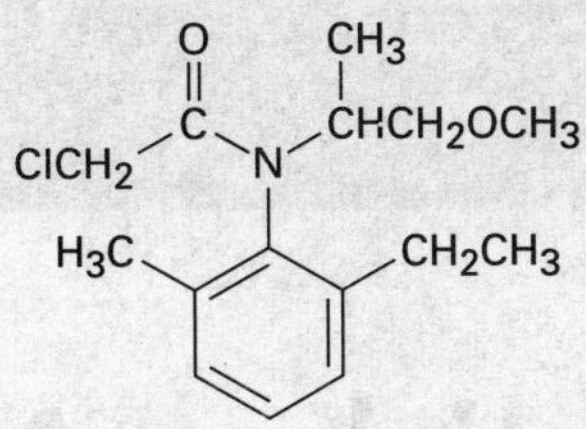

Metolachlor

(a) Only the *S* enantiomer of metolachlor inhibits fatty-acid elongase. Draw it.

(b) Why do the *R* and *S* enantiomers have different activities?

CHAPTER 7

The gases released during volcanic eruptions contain large amounts of organohalides, including chloromethane, chloroform, dichlorodifluoromethane, and many others.

Organohalides: Nucleophilic Substitutions and Eliminations

Now that we've covered the chemistry of hydrocarbons, it's time to start looking at more complex substances that contain elements in addition to just C and H. We'll begin by discussing the chemistry of **organohalides**, compounds that contain one or more halogen atoms.

Halogen-substituted organic compounds are widespread in nature, and more than 5000 organohalides have been found in algae and various other marine organisms. Chloromethane, for instance, is released in large amounts by ocean kelp, as well as by forest fires and volcanoes. Halogen-containing compounds also have a vast array of industrial applications, including their use as solvents, inhaled anesthetics, refrigerants, and pesticides. The modern electronics industry, for example, relies on halogenated solvents such as trichloroethylene for cleaning semiconductor chips and other components.

OWL

Online homework for this chapter can be assigned in OWL, an online homework assessment tool.

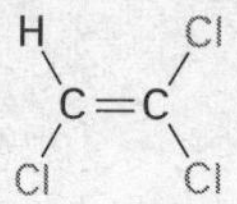

Trichloroethylene (a solvent)

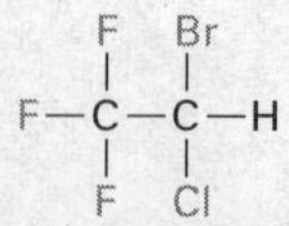

Halothane (an inhaled anesthetic)

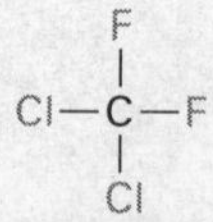

Dichlorodifluoromethane (a refrigerant)

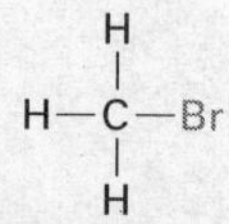

Bromomethane (a fumigant)

A large variety of organohalides are known. The halogen might be bonded to an alkynyl group (C≡C—X), a vinylic group (C=C—X), an aromatic ring (Ar—X), or an alkyl group. We'll be concerned in this chapter primarily with **alkyl halides**, compounds with a halogen atom bonded to a saturated, sp^3-hybridized carbon atom.

WHY THIS CHAPTER?

Alkyl halides (R—X) are encountered much less frequently than their oxygen-containing relative alcohols (R—OH), but some of the *kinds* of reactions they undergo—nucleophilic substitutions and eliminations—*are* encountered frequently. Thus, alkyl halide chemistry acts as a relatively simple model for many mechanistically similar but structurally more complex reactions. We'll begin with a look at how to name and prepare alkyl halides, and we'll then make a detailed study of their substitution and elimination reactions—two of the most important and well-studied reaction types in organic chemistry.

7.1 Naming Alkyl Halides

Although commonly called *alkyl halides,* halogen-substituted alkanes are named systematically as *haloalkanes* (Section 2.3), treating the halogen as a substituent on a parent alkane chain. There are three steps.

STEP 1 **Find the longest chain, and name it as the parent**. If a multiple bond is present, the parent chain must contain it.

STEP 2 **Number the carbons of the parent chain beginning at the end nearer the first substituent, whether alkyl or halo**. Assign each substituent a number according to its position on the chain. If there are substituents the same distance from both ends, begin numbering at the end nearer the substituent with alphabetical priority.

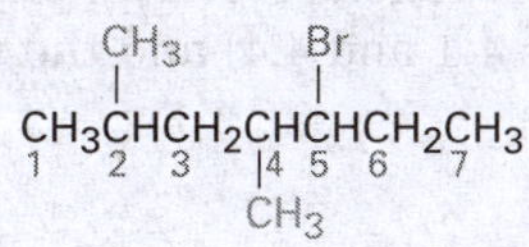

5-Bromo-2,4-dimethyl**heptane**

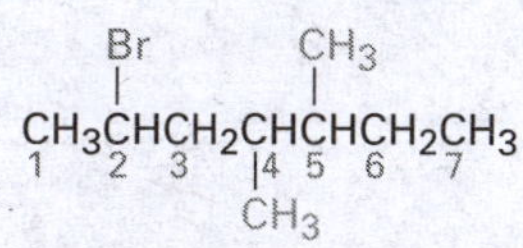

2-Bromo-4,5-dimethyl**heptane**

STEP 3 **Write the name**. List all substituents in alphabetical order, and use one of the prefixes *di-*, *tri-*, and so forth if more than one of the same substituent is present.

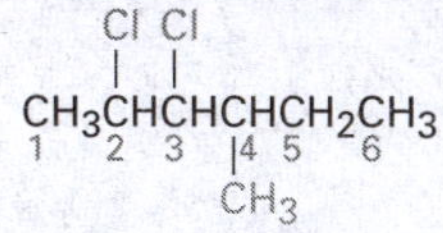

2,3-Dichloro-4-methyl**hexane**

In addition to their systematic names, many simple alkyl halides are also named by identifying first the alkyl group and then the halogen. For example,

CH_3I can be called either iodomethane or methyl iodide. Such names are well entrenched in the chemical literature and in daily usage, but they won't be used in this book.

CH_3I

Iodomethane (or methyl iodide)

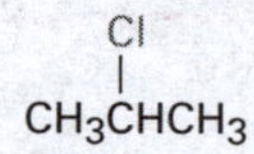

2-Chloropropane (or isopropyl chloride)

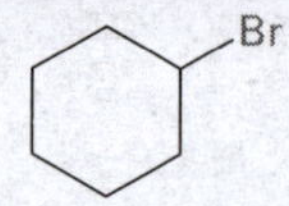

Bromocyclohexane (or cyclohexyl bromide)

Problem 7.1 Give the IUPAC names of the following alkyl halides:

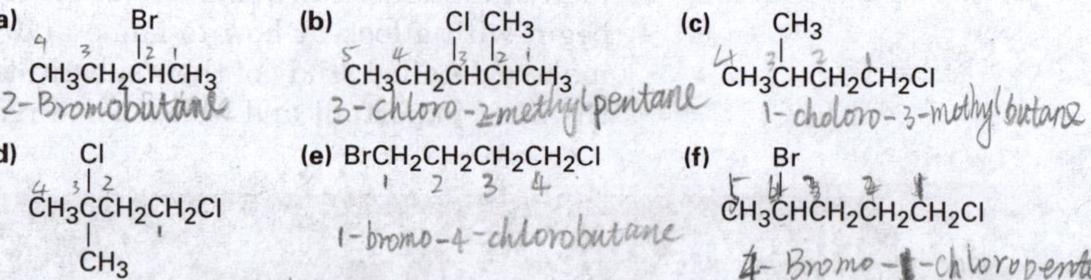

Problem 7.2 Draw structures corresponding to the following names:

(a) 2-Chloro-3,3-dimethylhexane **(b)** 3,3-Dichloro-2-methylhexane
(c) 3-Bromo-3-ethylpentane **(d)** 2-Bromo-5-chloro-3-methylhexane

7.2 Preparing Alkyl Halides

We've already seen several methods for preparing alkyl halides, including the addition reactions of HX and X_2 with alkenes in electrophilic addition reactions (Sections 4.1 and 4.4) and the reaction of an alkane with Cl_2 (Section 2.4).

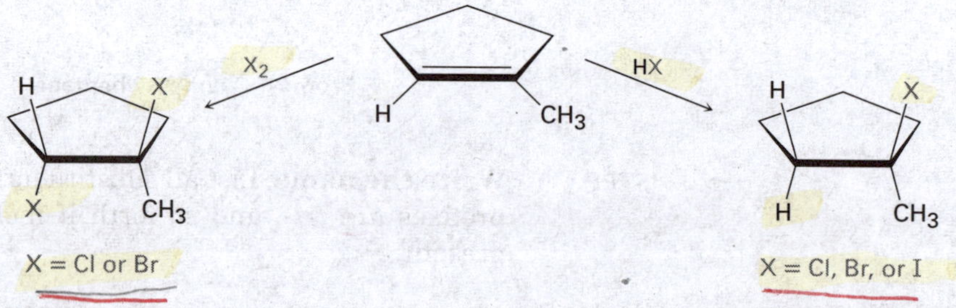

$$CH_4 + Cl_2 \xrightarrow{h\nu} CH_3Cl + HCl$$

Methane **Chloromethane**

The most generally useful method for preparing alkyl halides is to make them from alcohols, which themselves are easily obtained from carbonyl compounds. The reaction can often be carried out simply by treating the alcohol with HCl or HBr. 1-Methylcyclohexanol, for example, is converted into 1-chloro-1-methylcyclohexane by treating with HCl.

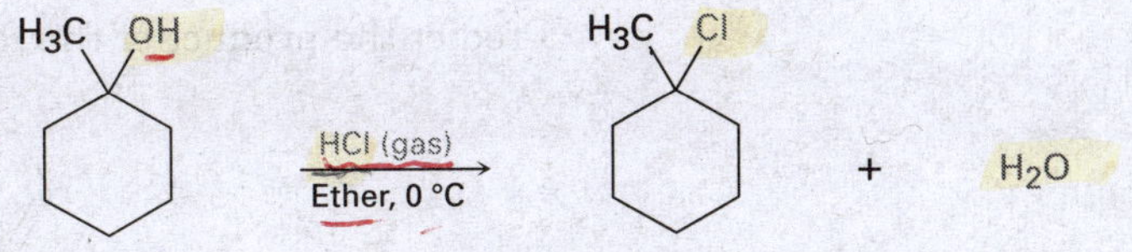

1-Methylcyclohexanol **1-Chloro-1-methylcyclohexane (90%)**

For reasons that will be discussed in Section 7.6, the HX reaction works best with tertiary alcohols. Primary and secondary alcohols react much more slowly.

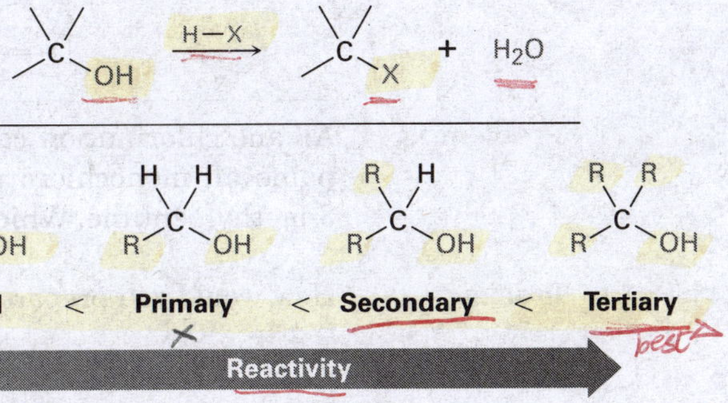

Primary and **secondary** alcohols are best converted into alkyl halides by treatment with either thionyl chloride ($SOCl_2$) or phosphorus tribromide (PBr_3). These reactions normally take place in high yield.

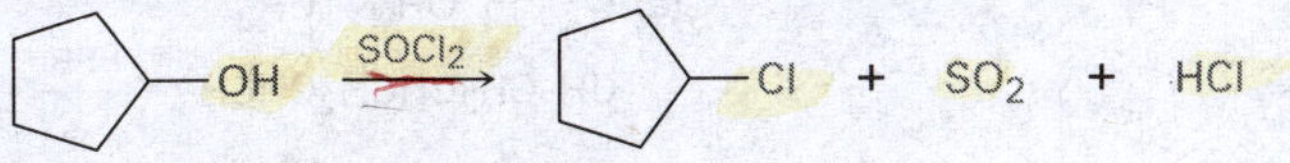

Cyclopentanol **Chlorocyclopentane**

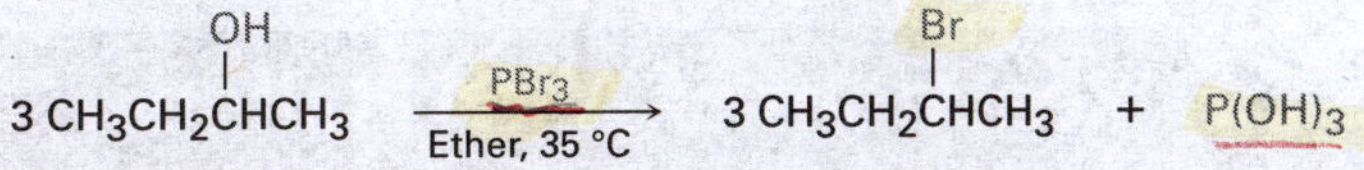

Butan-2-ol **2-Bromobutane (86%)**

Alkyl fluorides can also be prepared from alcohols. Numerous alternative reagents are used for the reaction, including diethylaminosulfur trifluoride [$(CH_3CH_2)_2NSF_3$] and HF–pyridine, where pyridine is the nitrogen-containing analog of benzene.

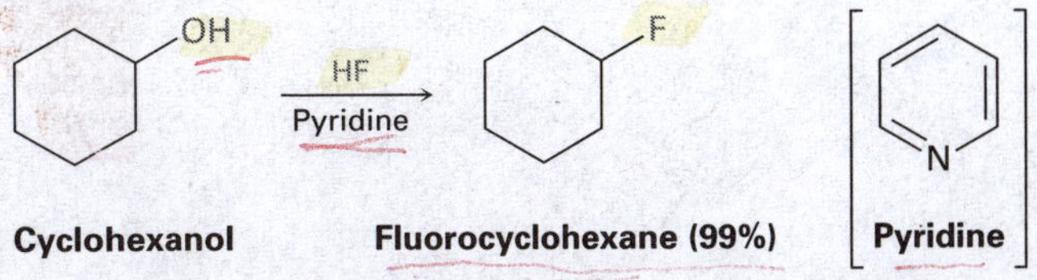

Cyclohexanol **Fluorocyclohexane (99%)** **Pyridine**

Worked Example 7.1 **Synthesizing an Alkyl Halide**

Predict the product of the following reaction:

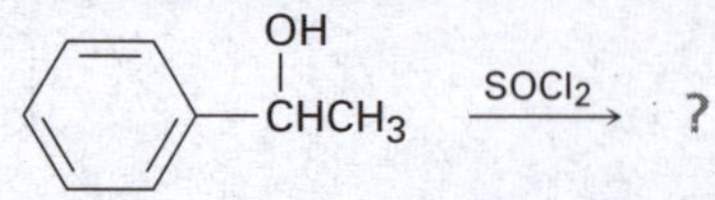

Strategy A big part of learning organic chemistry is remembering reactions. Ask yourself what you know about alcohols, and then recall that alcohols yield alkyl chlorides on treatment with $SOCl_2$.

Solution

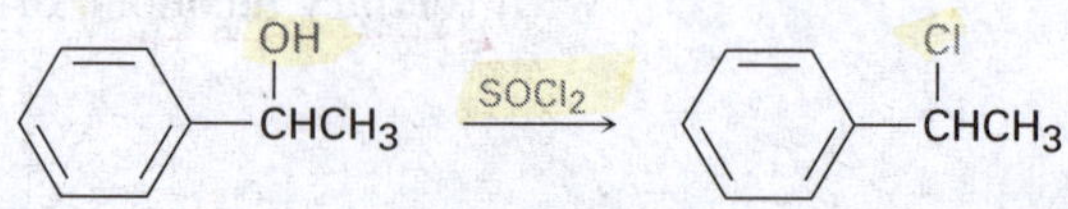

Problem 7.3 Alkane chlorination can occur at any position in the alkane chain. Draw and name all monochloro products you might obtain from radical chlorination of 3-methylpentane. Which, if any, are chiral?

Problem 7.4 How would you prepare the following alkyl halides from the appropriate alcohols?

(a) 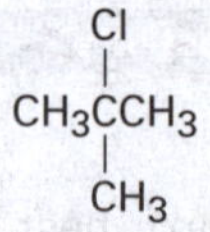 **(b)** 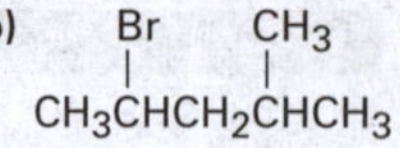 **(c)** 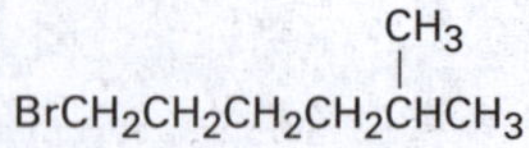 **(d)** 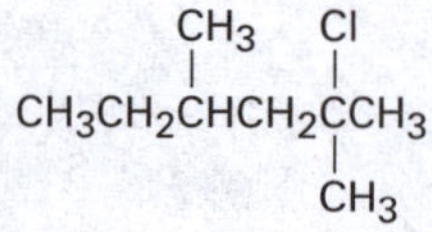

Problem 7.5 Predict the products of the following reactions:

(a) 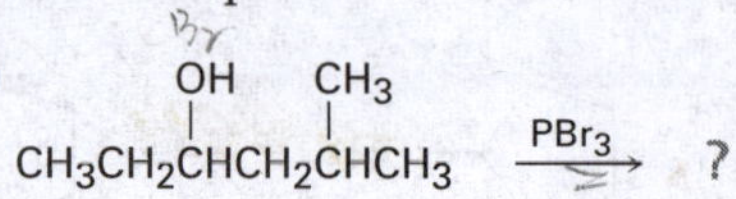 **(b)** 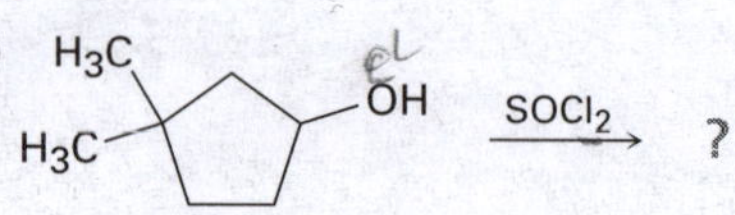

7.3 Reactions of Alkyl Halides: Grignard Reagents

Alkyl halides, RX, react with magnesium metal in ether solvent to yield alkylmagnesium halides, RMgX. The products, called **Grignard reagents** after their discoverer, Victor Grignard, are examples of *organometallic* compounds because they contain a carbon–metal bond. In addition to alkyl halides, alkenyl (vinylic) and aryl (aromatic) halides also react with magnesium to give Grignard reagents. The halogen can be Cl, Br, or I, but not F.

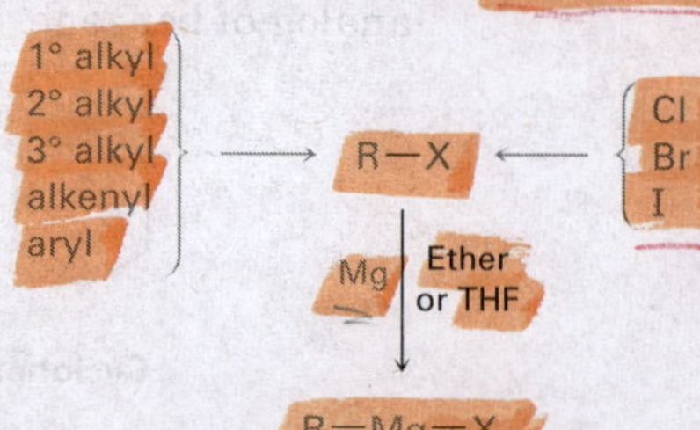

As you might expect from the discussion of electronegativity in Section 1.9, a carbon–magnesium bond is polarized, making the carbon both nucleophilic and basic. An electrostatic potential map of methylmagnesium iodide, for instance, indicates the electron-rich (red) character of the carbon bonded to magnesium.

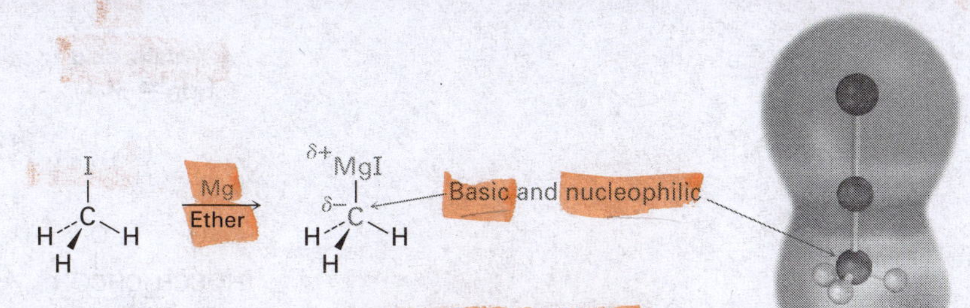

Iodomethane   Methylmagnesium iodide

A Grignard reagent is formally the magnesium salt, $R_3C^{-+}MgX$, of a carbon acid, $R_3C—H$, and is thus a carbon anion, or **carbanion**. But because hydrocarbons are such weak acids, carbon anions are very strong bases. Grignard reagents must therefore be protected from atmospheric moisture to prevent their being protonated and destroyed in an acid–base reaction: $R—Mg—X + H_2O \rightarrow R—H + HO—Mg—X$.

Grignard reagents themselves don't occur in living organisms, but they are useful carbon-based nucleophiles in important laboratory reactions, as we'll see in the next chapter. In addition, they act as a simple model for other, more complex carbon-based nucleophiles that *are* important in biological chemistry. We'll see examples in Chapter 17.

7.4 Nucleophilic Substitution Reactions

Because they are electrophiles, alkyl halides do one of two things when they react with nucleophiles/bases, such as hydroxide ion: either they undergo *substitution* of the X group by the nucleophile or *elimination* of HX to yield an alkene.

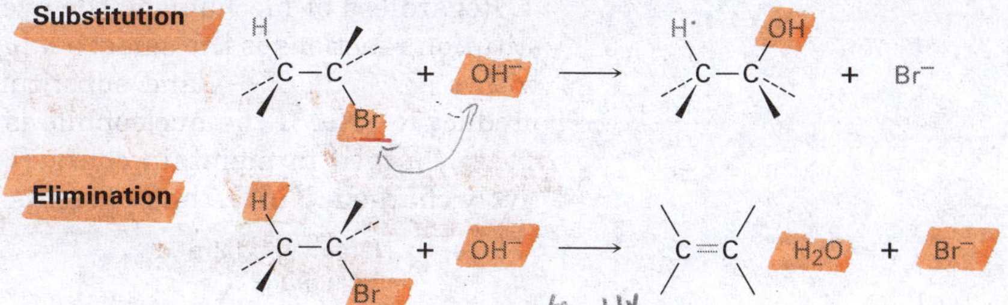

Let's look first at substitution reactions. The discovery of the nucleophilic substitution reaction of alkyl halides dates back to 1896 when the German chemist Paul Walden discovered that (+)- and (−)-malic acids could be interconverted. When Walden treated (−)-malic acid with PCl_5, he isolated (+)-chlorosuccinic acid. This, on reaction with wet Ag_2O, gave (+)-malic acid. Similarly, reaction of (+)-malic acid with PCl_5 gave (−)-chlorosuccinic acid, which was converted into (−)-malic acid when treated with wet Ag_2O. The full cycle of reactions reported by Walden is shown in Figure 7.1.

Figure 7.1 Walden's cycle of reactions interconverting (+)- and (−)-malic acids.

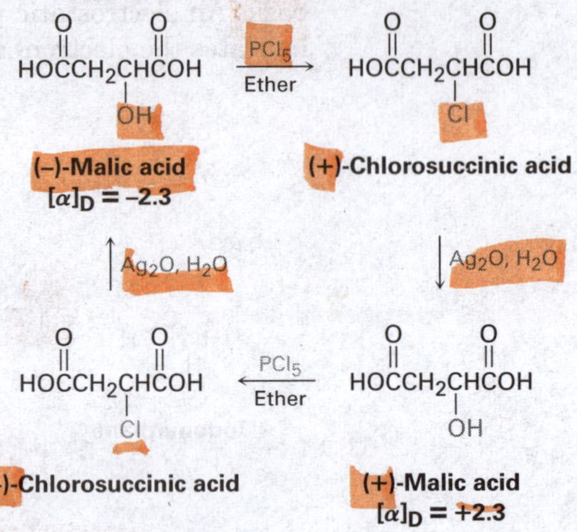

At the time, the results were astonishing. Since (−)-malic acid was converted into (+)-malic acid, *some reactions in the cycle must have occurred with an inversion, or change, in the configuration of the chirality center.* But which ones, and how? Remember from Section 6.5 that you can't tell the configuration of a chirality center from the sign of optical rotation.

Today we refer to the transformations taking place in Walden's cycle as **nucleophilic substitution reactions** because each step involves the substitution of one nucleophile (chloride ion, Cl^-, or hydroxide ion, OH^-) by another. Nucleophilic substitution reactions are one of the most common and versatile reaction types in organic chemistry.

$$\text{R—X} + \text{Nu:}^- \longrightarrow \text{R—Nu} + \text{X:}^-$$

Following the work of Walden, further investigations were undertaken during the 1920s and 1930s to clarify the mechanism of nucleophilic substitution reactions and to find out how inversions of configuration occur. These investigations showed that nucleophilic substitutions occur by two major pathways, named the *S_N1 reaction* and the *S_N2 reaction*. In both cases, the "S_N" part of the name stands for *substitution, nucleophilic*. The meanings of *1* and *2* are discussed in the next two sections.

Regardless of mechanism, the overall change during all nucleophilic substitution reactions is the same: a *nucleophile* (symbolized Nu: or Nu:$^-$) reacts with a *substrate* R—X and substitutes for a *leaving group* X:$^-$ to yield the product R—Nu. If the nucleophile is neutral (Nu:), then the product is positively charged to maintain charge conservation. If the nucleophile is negatively charged (Nu:$^-$), the product is neutral.

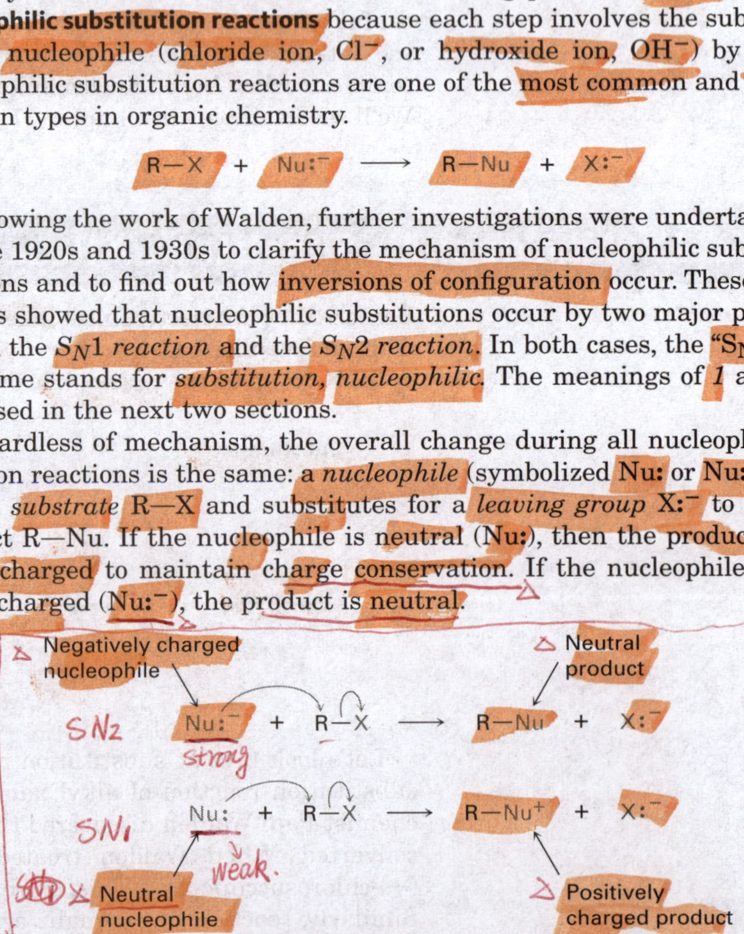

A wide array of substances can be prepared using nucleophilic substitution reactions. In fact, we've already seen examples in previous chapters. The reaction of an acetylide anion with an alkyl halide (Section 4.11), for instance, is an S_N2 reaction in which the acetylide nucleophile replaces halide. Table 7.1 lists other examples.

$$\underset{\textbf{An acetylide anion}}{R{-}C{\equiv}C{:}^-} + CH_3Br \xrightarrow[\text{reaction}]{S_N2} R{-}C{\equiv}C{-}CH_3 + Br^-$$

Table 7.1 Some Nucleophilic Substitution Reactions with Bromomethane

$$Nu{:}^- + CH_3Br \rightarrow CH_3Nu + Br^-$$

Nucleophile		Product	
Formula	**Name**	**Formula**	**Name**
H_2O	Water	$CH_3OH_2^+$	Methylhydronium ion
$CH_3CO_2^-$	Acetate	$CH_3CO_2CH_3$	Methyl acetate
NH_3	Ammonia	$CH_3NH_3^+$	Methylammonium ion
Cl^-	Chloride	CH_3Cl	Chloromethane
HO^-	Hydroxide	CH_3OH	Methanol
CH_3O^-	Methoxide	CH_3OCH_3	Dimethyl ether
I^-	Iodide	CH_3I	Iodomethane
^-CN	Cyanide	CH_3CN	Acetonitrile
HS^-	Hydrosulfide	CH_3SH	Methanethiol

Worked Example 7.2

Predicting the Product of a Substitution Reaction

What is the substitution product from reaction of 1-chloropropane with NaOH?

Strategy Write the two reactants, and identify the nucleophile (in this instance, OH^-) and the leaving group (in this instance, Cl^-). Then, replace the –Cl group by –OH and write the complete equation.

Solution

$$\underset{\textbf{1-Chloropropane}}{CH_3CH_2CH_2Cl} + Na^+ \, ^-OH \longrightarrow \underset{\textbf{Propan-1-ol}}{CH_3CH_2CH_2OH} + Na^+ \, ^-Cl$$

Worked Example 7.3

Using a Substitution Reaction in a Synthesis

How would you prepare propane-1-thiol, $CH_3CH_2CH_2SH$, using a nucleophilic substitution reaction?

Strategy Identify the group in the product that is introduced by nucleophilic substitution. In this case, the product contains an –SH group, so it might be prepared by reaction of SH^- (hydrosulfide ion) with an alkyl halide such as 1-bromopropane.

Solution

$$\underset{\textbf{1-Bromopropane}}{CH_3CH_2CH_2Br} + Na^+ \, ^-SH \longrightarrow \underset{\textbf{Propane-1-thiol}}{CH_3CH_2CH_2SH} + Na^+ \, ^-Br$$

Problem 7.6 What substitution products would you expect to obtain from the following reactions?

(a)

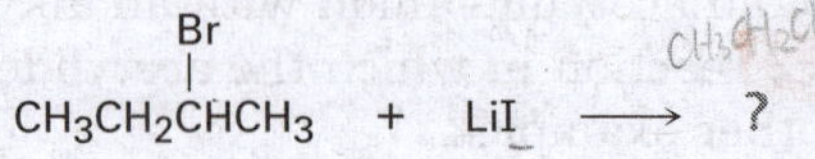

(b)

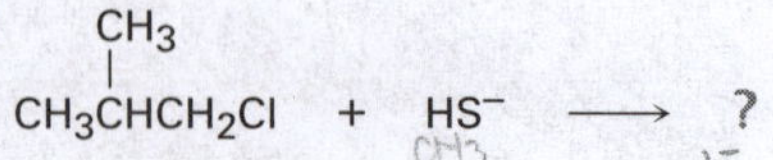

(c)

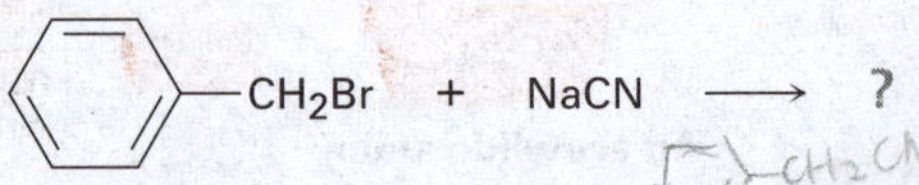

Problem 7.7 How might you prepare the following substances by using nucleophilic substitution reactions?

(a) $CH_3CH_2CH_2CH_2OH$ **(b)** $(CH_3)_2CHCH_2CH_2N_3$

7.5 Substitutions: The S_N2 Reaction

An **S_N2 reaction** takes place in a single step without intermediates when the entering nucleophile approaches the substrate from a direction 180° away from the leaving group. As the nucleophile comes in on one side of the molecule, an electron pair on the nucleophile Nu:⁻ forces out the leaving group X:⁻, which departs from the other side of the molecule and takes with it the electron pair from the C–X bond. In the transition state for the reaction, the new Nu–C bond is partially forming at the same time the old C–X bond is partially breaking, and the negative charge is shared by both the incoming nucleophile and the outgoing leaving group. The mechanism is shown in Figure 7.2 for the reaction of OH^- with (*S*)-2-bromobutane.

MECHANISM

Figure 7.2 The mechanism of the S_N2 reaction. The reaction takes place in a single step when the incoming nucleophile approaches from a direction 180° away from the leaving halide ion, thereby inverting the stereochemistry at carbon.

① The nucleophile ⁻OH uses its lone-pair electrons to attack the alkyl halide carbon 180° away from the departing halogen. This leads to a transition state with a partially formed C–OH bond and a partially broken C–Br bond.

② The stereochemistry at carbon is inverted as the C–OH bond forms fully and the bromide ion departs with the electron pair from the former C–Br bond.

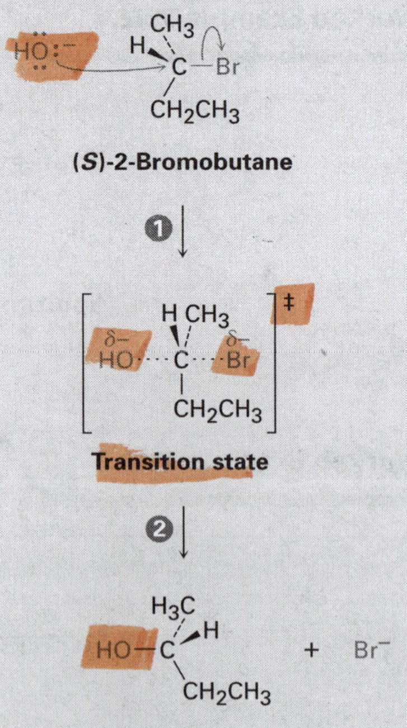

Let's see what evidence there is for this mechanism and what the chemical consequences are.

Rates of S_N2 Reactions

In every chemical reaction, there is a direct relationship between the rate at which the reaction occurs and the concentrations of the reactants. The S_N2 reaction of CH_3Br with OH^- to yield CH_3OH, for instance, takes place in a single step when substrate and nucleophile collide and react. At a given concentration of reactants, the reaction takes place at a certain rate. If we double the concentration of OH^-, the frequency of collision between the two reactants doubles and we find that the reaction rate also doubles. Similarly, if we double the concentration of CH_3Br, the reaction rate doubles. Thus, the origin of the "2" in S_N2: S_N2 reactions are said to be **bimolecular** because the rate of the reaction depends on the concentrations of *two* substances—alkyl halide and nucleophile.

$$HO{:}^- + CH_3{-}Br{:} \longrightarrow HO{-}CH_3 + {:}Br{:}^-$$

Problem 7.8 What effects would the following changes have on the rate of the S_N2 reaction between CH_3I and sodium acetate?

(a) The CH_3I concentration is tripled.

(b) Both CH_3I and CH_3CO_2Na concentrations are doubled.

Stereochemistry of S_N2 Reactions

Look carefully at the mechanism of the S_N2 reaction shown in Figure 7.2. As the incoming nucleophile attacks the substrate and begins pushing out the leaving group on the opposite side, the configuration of the molecule *inverts* (Figure 7.3). (*S*)-2-Bromobutane gives (*R*)-butan-2-ol, for example, by an inversion of configuration that occurs through a planar transition state.

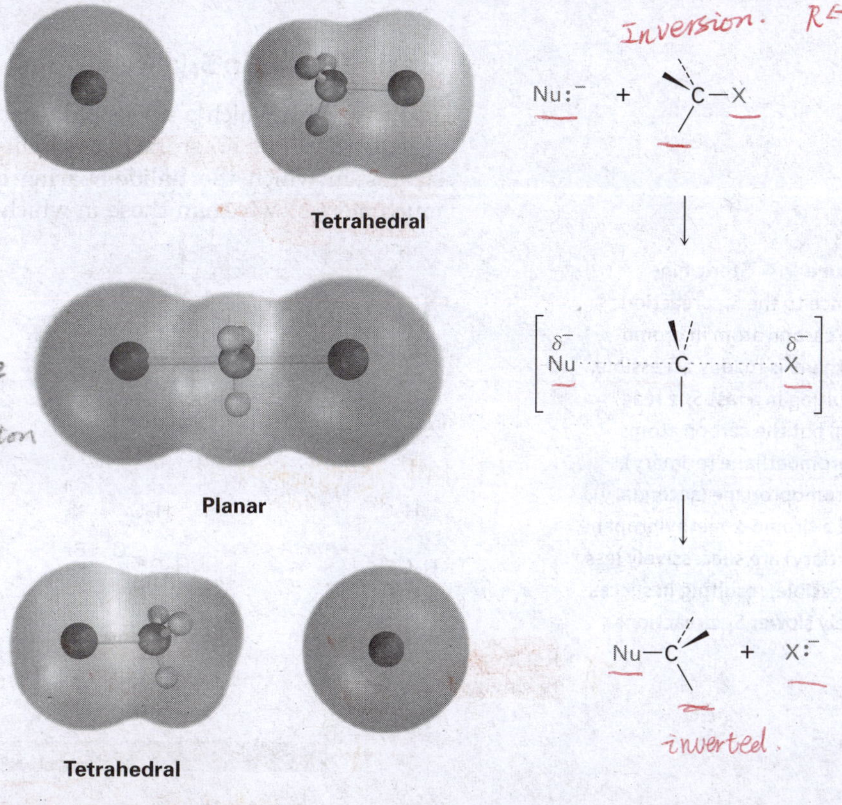

Figure 7.3 The transition state of the S_N2 reaction has a planar arrangement of the carbon atom and the remaining three groups. Electrostatic potential maps show that the negative charge (red) is shared by the incoming nucleophile and the leaving group in the transition state. (The dotted red lines indicate partial bonding.)

Worked Example 7.4 Predicting the Product of a Substitution Reaction

What product would you expect to obtain from the S_N2 reaction of (*S*)-2-iodooctane with sodium cyanide, NaCN?

Strategy Identify the nucleophile (cyanide ion) and the leaving group (iodide ion). Then carry out the substitution, inverting the configuration at the chirality center. (*S*)-2-Iodooctane reacts with CN^- to yield (*R*)-2-methyloctanenitrile.

Solution

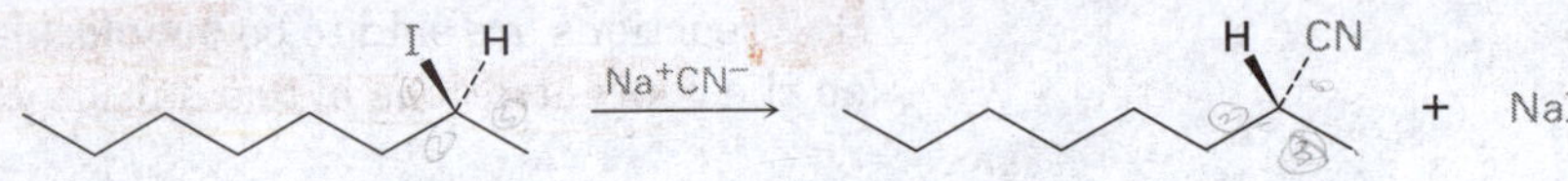

(*S*)-2-Iodooctane → (*R*)-2-Methyloctanenitrile

Problem 7.9 What product would you expect to obtain from the S_N2 reaction of (*S*)-2-bromohexane with sodium acetate, CH_3CO_2Na? Show the stereochemistry of both product and reactant.

Problem 7.10 Assign configuration to the following substance, and draw the structure of the product that would result on nucleophilic substitution reaction with HS^- (reddish brown = Br):

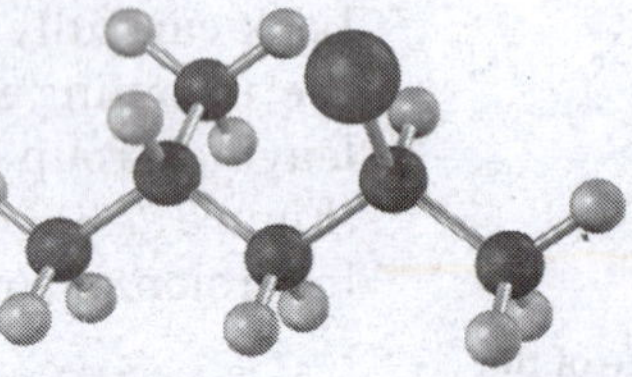

Steric Effects in S_N2 Reactions

The ease with which a nucleophile can approach a substrate to carry out an S_N2 reaction depends on steric accessibility to the halide-bearing carbon. Bulky substrates, in which the halide-bearing carbon atom is difficult to approach, react much more slowly than those in which the carbon is more accessible (Figure 7.4).

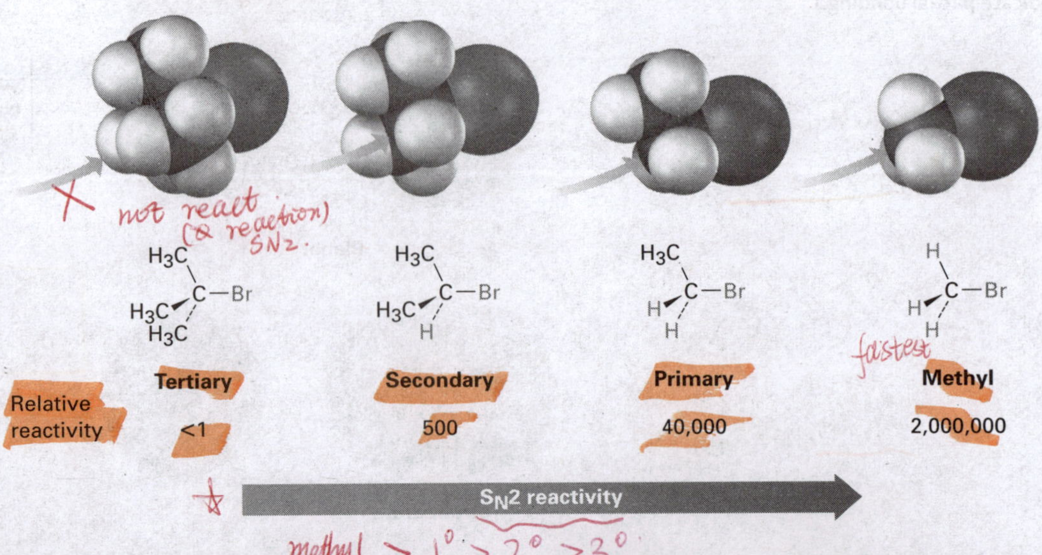

Figure 7.4 Steric hindrance to the S_N2 reaction. The carbon atom in bromomethane is readily accessible, resulting in a fast S_N2 reaction, but the carbon atoms in bromoethane (primary), 2-bromopropane (secondary), and 2-bromo-2-methylpropane (tertiary) are successively less accessible, resulting in successively slower S_N2 reactions.

Methyl halides (CH_3X) are the most reactive substrates, followed by primary alkyl halides (RCH_2—X) such as ethyl and propyl. Alkyl branching next to the leaving group slows the reaction greatly for secondary halides (R_2CH—X), and further branching effectively halts the reaction for tertiary halides (R_3C—X).

Vinylic ($R_2C{=}CRX$) and aryl (Ar—X) halides are not shown on this reactivity list because they are completely unreactive toward S_N2 displacements. This lack of reactivity is due to steric hindrance: the incoming nucleophile would have to burrow through part of the molecule to carry out a displacement.

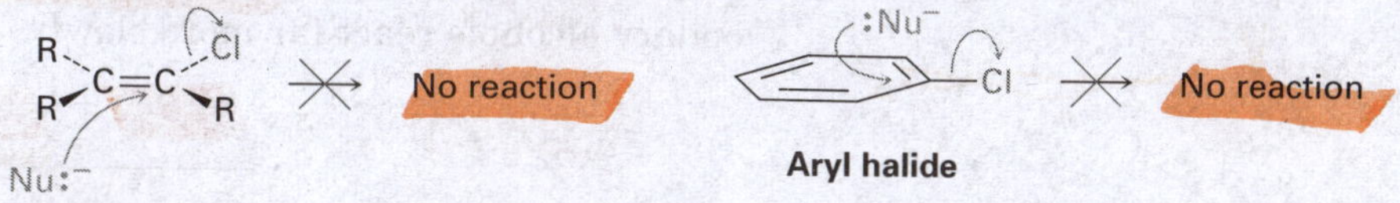

Vinylic halide   Aryl halide

Worked Example 7.5 Predicting the Rates of Substitution Reactions

Which would you expect to be faster, the S_N2 reaction of OH^- ion with 1-bromopentane or with 2-bromopentane?

Strategy Decide which substrate is less hindered. Since 1-bromopentane is a 1° halide and 2-bromopentane is a 2° halide, reaction with the less hindered 1-bromopentane is faster.

Solution

Primary: $CH_3CH_2CH_2CH_2CH_2Br$ — **1-Bromopentane**

Secondary: $CH_3CH_2CH_2CH(Br)CH_3$ — **2-Bromopentane**

Problem 7.11 Which of the following S_N2 reactions would you expect to be faster?

(a) Reaction of CN^- (cyanide ion) with $CH_3CH(Br)CH_3$ or with $CH_3CH_2CH_2Br$?

(b) Reaction of I^- with $(CH_3)_2CHCH_2Cl$ or with $H_2C{=}CHCl$?

The Leaving Group in S_N2 Reactions

Another variable that can affect the S_N2 reaction is the nature of the leaving group displaced by the attacking nucleophile. Because the leaving group is expelled with a negative charge in most S_N2 reactions, the best leaving groups are those that give the most stable anions (anions of strong acids). A halide ion (I^-, Br^-, or Cl^-) is the most common leaving group, although others are also possible. Anions such as F^-, OH^-, OR^-, and NH_2^- are rarely found as leaving groups.

	OH^-, NH_2^-, OR^-	F^-	Cl^-	Br^-	I^-
Relative reactivity	<<1	1	200	10,000	30,000

Leaving group reactivity →

Problem 7.12 Rank the following compounds in order of their expected reactivity toward S_N2 reaction: CH_3I, CH_3F, CH_3Br.

7.6 Substitutions: The S_N1 Reaction

Most nucleophilic substitutions take place by the S_N2 pathway just discussed, but an alternative called the **S_N1 reaction** can also occur. In general, S_N1 reactions take place only on *tertiary* substrates and only under neutral or acidic conditions in a hydroxylic solvent such as water or alcohol. We saw in Section 7.2, for example, that alkyl halides can be prepared from alcohols by treatment with HCl or HBr. Tertiary alcohols react rapidly, but primary and secondary alcohols react far more slowly.

$$R{-}Br + H_2O \longrightarrow R{-}OH + HBr$$

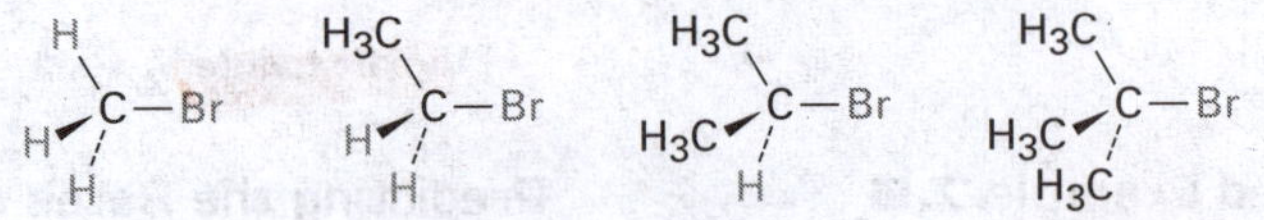

	Methyl	Primary	Secondary	Tertiary
Relative reactivity	< 1	1	12	1,200,000

Reactivity →

What's going on here? Clearly, a nucleophilic substitution reaction is taking place—a halogen is replacing a hydroxyl group—yet the reactivity order 3° > 2° > 1° is backward from the normal S_N2 order. Furthermore, an –OH group is being replaced, although we said in the previous section that OH^- is a poor leaving group. These reactions can't be taking place by the S_N2 mechanism we've been discussing but are instead taking place by the S_N1 mechanism shown in Figure 7.5.

Unlike what occurs in an S_N2 reaction, where the leaving group is displaced at the same time that the incoming nucleophile approaches, an S_N1 reaction occurs by spontaneous loss of the leaving group *before* the incoming nucleophile approaches. Loss of the leaving group gives a carbocation intermediate, which then reacts with the nucleophile in a second step to yield the substitution product.

This two-step mechanism explains why tertiary alcohols react with HBr so much more rapidly than primary or secondary ones do: S_N1 reactions can occur only when stable carbocation intermediates are formed. The more stable the carbocation intermediate, the faster the S_N1 reaction. Thus, the reactivity order of alcohols with HBr is the same as the stability order of carbocations (Section 4.2).

Rates of S_N1 Reactions

Unlike an S_N2 reaction, whose rate depends on the concentrations of both substrate and nucleophile, the rate of an S_N1 reaction depends only on the concentration of the substrate and is independent of the nucleophile concentration. Thus, the origin of the "1" in S_N1: S_N1 reactions are **unimolecular** because the rate of the reaction depends on the concentration of only *one* substance—the substrate. The observation that S_N1 reactions are unimolecular means that the substrate must undergo a spontaneous reaction without involvement of the nucleophile, exactly what the mechanism shown in Figure 7.5 accounts for.

MECHANISM

Figure 7.5 The mechanism of the S_N1 reaction of *tert*-butyl alcohol with HBr to yield an alkyl halide. Neutral water is the leaving group.

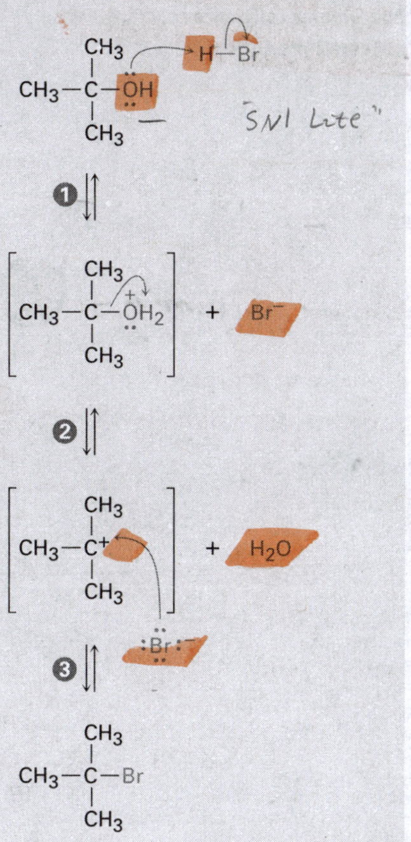

1. The –OH group is first protonated by HBr.
2. Spontaneous dissociation of the protonated alcohol occurs in a slow, rate-limiting step to yield a carbocation intermediate plus water.
3. The carbocation intermediate reacts with bromide ion in a fast step to yield the neutral substitution product.

Carbocation

Problem 7.13 What effect would the following changes have on the rate of the S_N1 reaction of *tert*-butyl alcohol with HBr?

(a) The HBr concentration is tripled.

(b) The HBr concentration is halved, and the *tert*-butyl alcohol concentration is doubled.

Stereochemistry of S_N1 Reactions

Because an S_N1 reaction occurs through a carbocation intermediate, as shown in Figure 7.5, its stereochemistry is different from that of an S_N2 reaction. Carbocations, as we've seen, are planar, sp^2-hybridized, and achiral. The positively charged carbon can therefore react with a nucleophile equally well from either side, leading to a 50:50 (racemic) mixture of enantiomers (Figure 7.6). In other words, if we carry out an S_N1 reaction on a single enantiomer of a chiral substrate and go through an achiral carbocation intermediate, the molecule momentarily loses its chirality so the product is optically inactive.

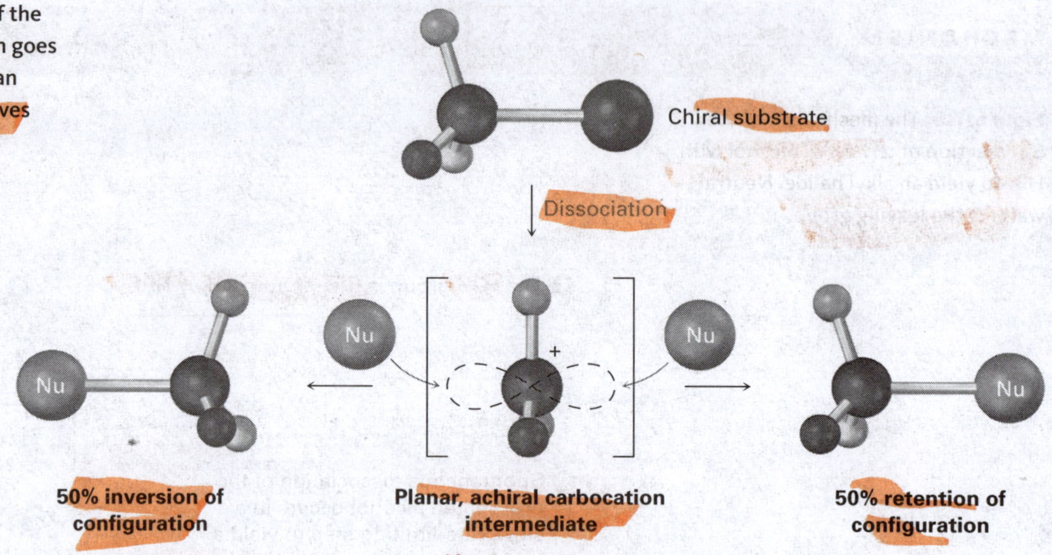

Figure 7.6 A Stereochemistry of the S_N1 reaction. Because the reaction goes through an achiral intermediate, an enantiomerically pure reactant gives a racemic product.

An example of an S_N1 reaction on a chiral substrate occurs on treatment of (*R*)-1-phenylbutan-1-ol with HCl: a racemic alkyl chloride product is formed.

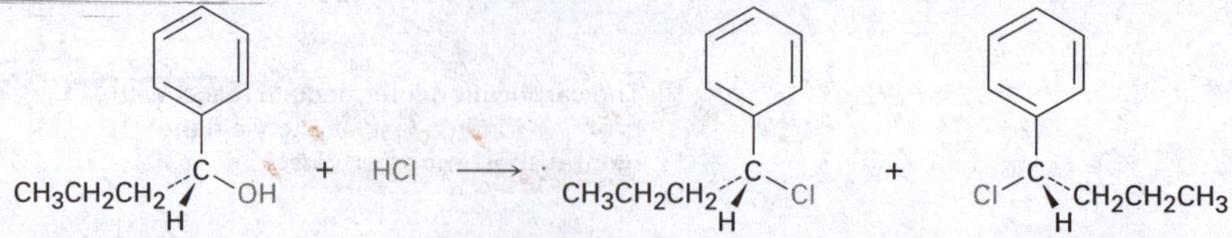

(*R*)-Phenylbutan-1-ol — **(*R*)-1-Phenyl-1-chlorobutane (50%, retention)** — **(*S*)-1-Phenyl-1-chlorobutane (50%, inversion)**

Worked Example 7.6 Predicting the Stereochemistry of a Substitution Reaction

What stereochemistry would you expect for the S_N1 reaction of (*R*)-3-bromo-3-methylhexane with methanol to yield 3-methoxy-3-methylhexane?

Strategy First draw the starting alkyl halide, showing its correct stereochemistry. Then replace the –Br with a methoxy group ($-OCH_3$) to give the racemic product.

Solution

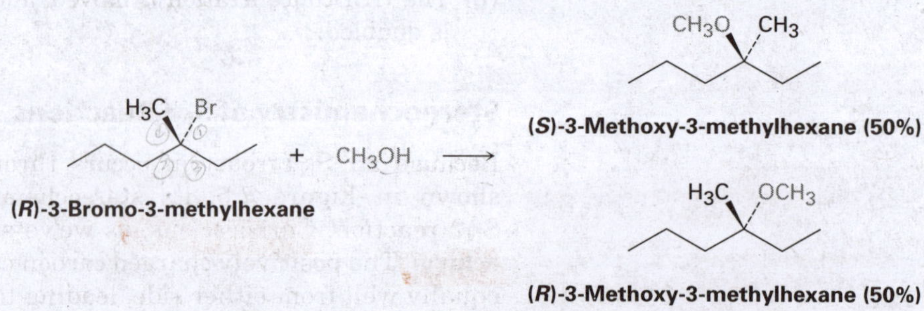

Problem 7.14 What product would you expect from the S_N1 reaction of (*S*)-3-methyloctan-3-ol [(*S*)-3-hydroxy-3-methyloctane] with HBr? Show the stereochemistry of both starting material and product.

Problem 7.15 Assign configuration to the following substrate, and show the stereochemistry and identity of the product you would obtain by S_N1 reaction with H_2O (reddish brown = Br):

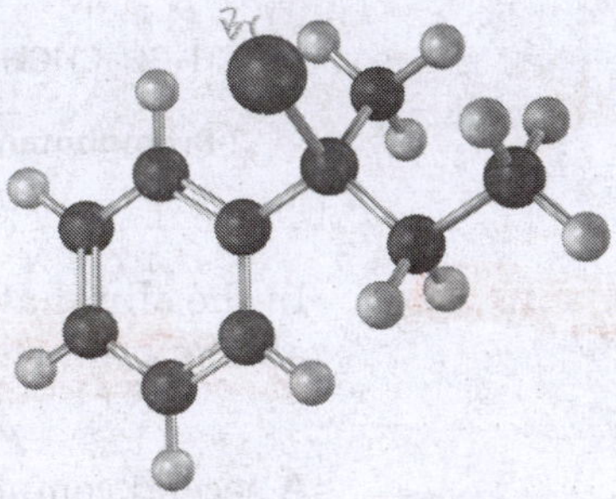

The Leaving Group in S_N1 Reactions

The best leaving groups in S_N1 reactions are those that give the most stable anions, just as in S_N2 reactions. Note that if an S_N1 reaction is carried out under acidic conditions, as occurs when a tertiary alcohol reacts with HX to yield an alkyl halide (Figure 7.5), neutral water is the leaving group. The S_N1 reactivity order of leaving groups is

$$HO^- < Cl^- < Br^- < I^- \approx H_2O$$

Leaving group reactivity →

7.7 Eliminations: The E2 Reaction

Thus far, we've looked only at substitution reactions, but in fact two kinds of reactions can happen when a nucleophile/base reacts with an alkyl halide. The nucleophile/base can either substitute for the halide in an S_N1 or S_N2 reaction, or it can cause elimination of HX, leading to formation of an alkene.

Substitution

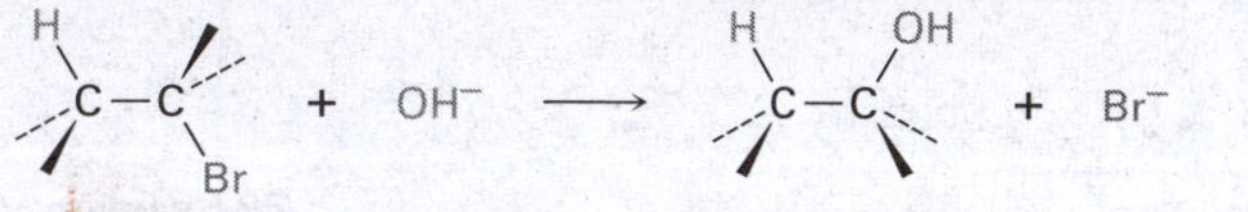

Elimination

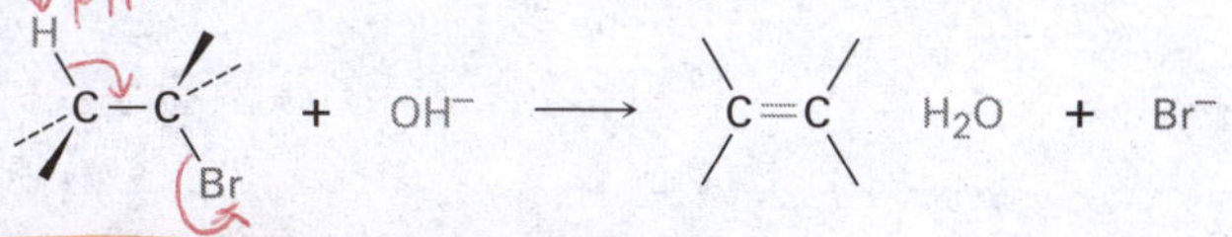

The elimination of HX from an alkyl halide is a very useful method for preparing alkenes, but the topic is complex for several reasons. One complication is the problem of regiochemistry. What products result by loss of HX from an unsymmetrical halide? In fact, elimination reactions almost always give mixtures of alkene products, and the best we can usually do is to predict which will be the major product.

According to **Zaitsev's rule**, a predictive guideline formulated in 1875 by the Russian chemist Alexander Zaitsev, base-induced elimination reactions generally give the more highly substituted alkene product—that is, the alkene with the larger number of alkyl substituents on the double bond. Treatment of 2-bromobutane with KOH in ethanol, for instance, gives primarily but-2-ene (disubstituted; two alkyl group substituents on the double-bond carbons) rather than but-1-ene (monosubstituted; one alkyl group substituent on the double-bond carbons).

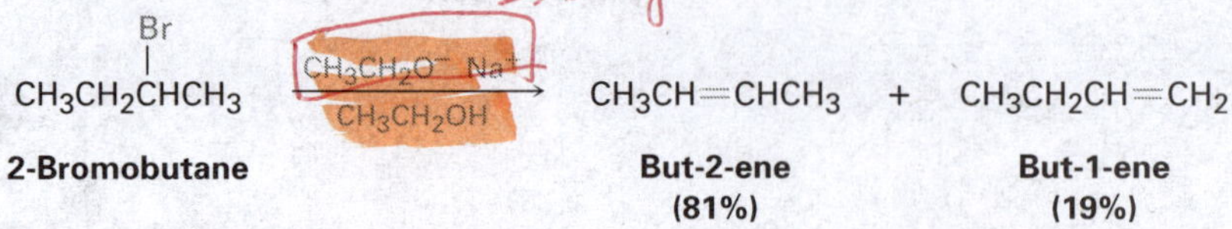

ZAITSEV'S RULE

In the elimination of HX from an alkyl halide, the more highly substituted alkene product predominates.

A second complication is that eliminations can take place by several different mechanisms, just as substitutions can. We'll consider the three most common mechanisms—the E1, E2, and E1cB reactions—which differ in the timing of C–H and C–X bond-breaking. In the E1 reaction, the C–X bond breaks first to give a carbocation intermediate that undergoes subsequent base abstraction of H^+. In the E2 reaction, base-induced C–H bond cleavage is simultaneous with C–X bond cleavage, giving the alkene in a single step. In the E1cB reaction (cB for "conjugate base"), base abstraction of the proton occurs first, giving a carbanion ($R:^-$) intermediate that loses X^- in a subsequent step. All three mechanisms occur frequently in the laboratory, but the E1cB mechanism predominates in biological pathways.

E1 Reaction: C–X bond breaks first to give a carbocation intermediate, followed by base removal of a proton to yield the alkene.

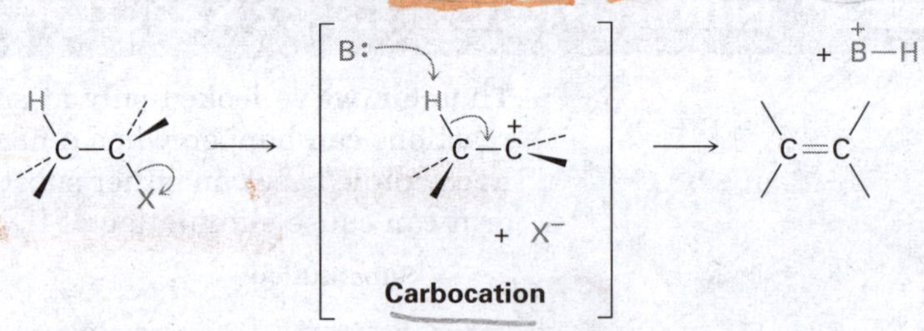

E2 Reaction: C–H and C–X bonds break simultaneously, giving the alkene in a single step without intermediates.

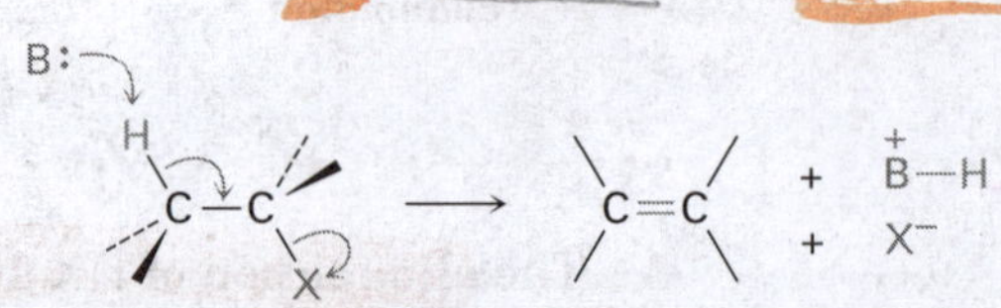

E1cB Reaction: C–H bond breaks first, giving a carbanion intermediate that loses X^- to form the alkene.

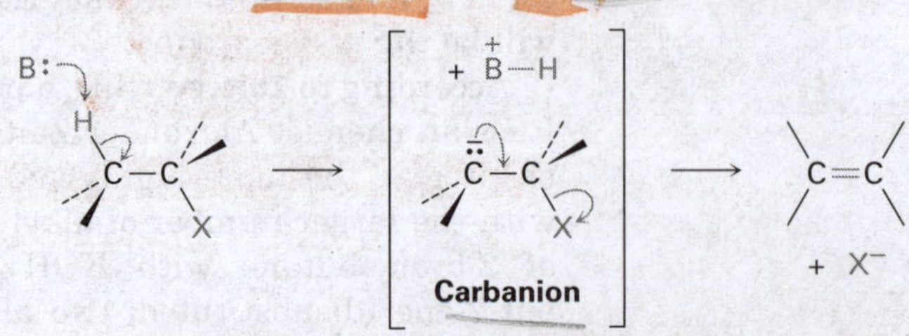

Let's look first at the most common elimination pathway—the **E2 reaction** (for *elimination, bimolecular*). The process takes place when an alkyl halide is treated with a strong base, such as hydroxide ion or alkoxide ion (RO^-), and occurs by the mechanism shown in Figure 7.7.

MECHANISM

Figure 7.7 Mechanism of the E2 reaction. The reaction takes place in a single step through a transition state in which the double bond begins to form at the same time the H and X groups are leaving. Red dotted lines indicate partial bonding in the transition state.

1 Base (B:) attacks a neighboring hydrogen and begins to remove the H at the same time as the alkene double bond starts to form and the X group starts to leave.

2 Neutral alkene is produced when the C–H bond is fully broken and the X group has departed with the C–X bond electron pair.

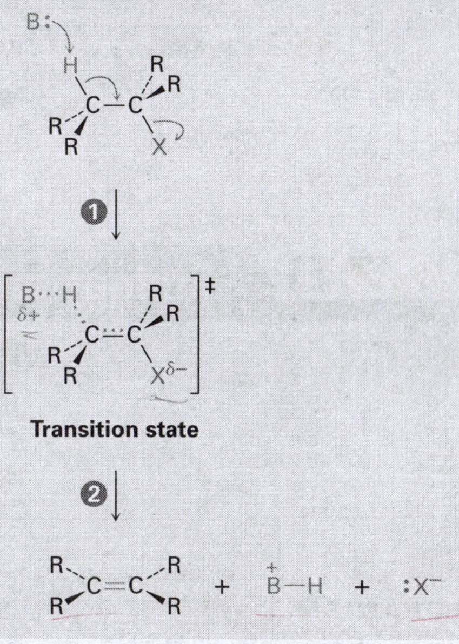

Like the S_N2 reaction discussed in Section 7.5, the E2 reaction takes place in one step without intermediates. As the attacking base begins to abstract H^+ from a carbon atom next to the leaving group, the C–H and C–X bonds begin to break and the C=C double bond begins to form. When the leaving group departs, it takes with it the two electrons from the former C–X bond.

Worked Example 7.7

Predicting the Product of an Elimination Reaction

What product would you expect from reaction of 1-chloro-1-methylcyclohexane with KOH in ethanol?

Strategy

Treatment of an alkyl halide with a strong base such as KOH yields an alkene. To find the products in a specific case, draw the structure of the reactant and locate the hydrogen atoms on neighboring carbons. Then generate the potential alkene products by removing HX in as many ways as possible. The major product will be the one that has the most highly substituted double bond—in this case, 1-methylcyclohexene.

Solution

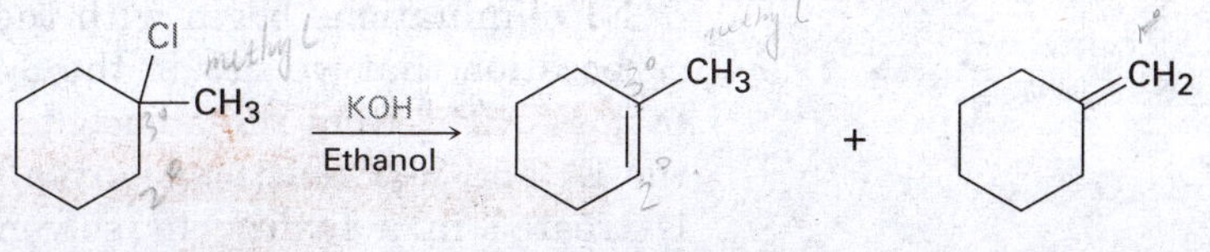

1-Chloro-1-methylcyclohexane | 1-Methylcyclohexene (major) | Methylenecyclohexane (minor)

Problem 7.16 Ignoring double-bond stereochemistry, what products would you expect from elimination reactions of the following alkyl halides?

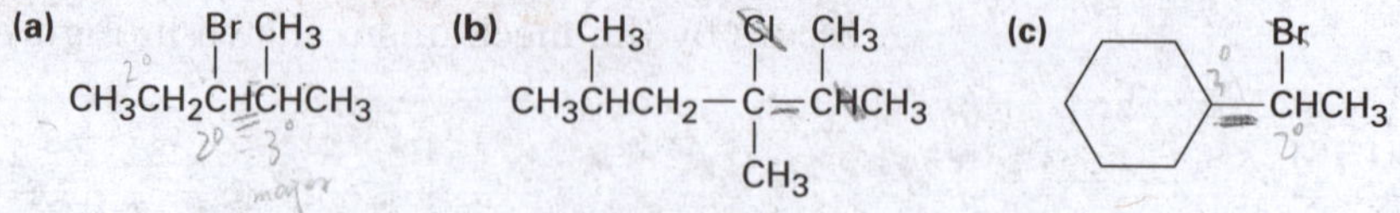

Problem 7.17 What alkyl halides might the following alkenes have been made from?

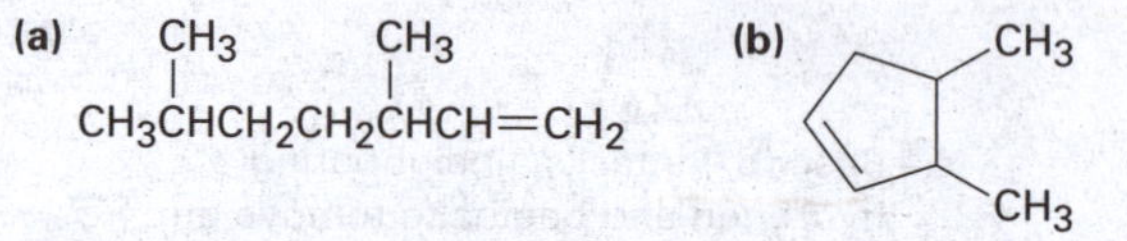

7.8 Eliminations: The E1 and E1cB Reactions

The E1 Reaction

Just as the E2 reaction is analogous to the S_N2 reaction, the S_N1 reaction has a close analog called the **E1 reaction** (for *elimination, unimolecular*). The E1 mechanism is shown in Figure 7.8 for the elimination of HCl from 2-chloro-2-methylpropane.

MECHANISM

Figure 7.8 Mechanism of the E1 reaction. Two steps are involved, and a carbocation intermediate is present.

❶ Spontaneous dissociation of the tertiary alkyl chloride yields an intermediate carbocation in a slow, rate-limiting step.

❷ Loss of a neighboring H^+ in a fast step yields the neutral alkene product. The electron pair from the C–H bond goes to form the alkene π bond.

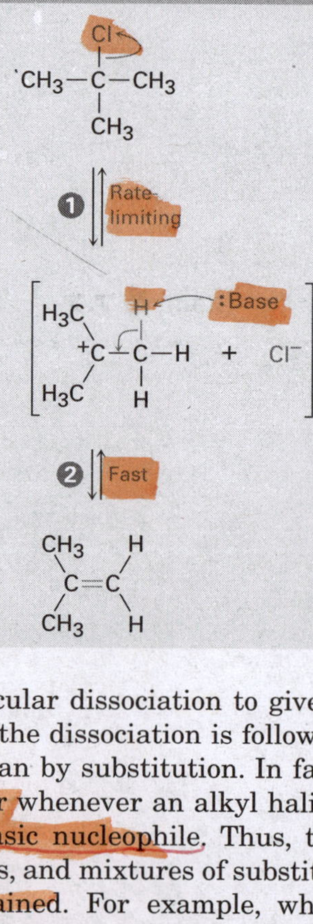

E1 eliminations begin with the same unimolecular dissociation to give a carbocation that we saw in the S_N1 reaction, but the dissociation is followed by loss of H^+ from the adjacent carbon rather than by substitution. In fact, the E1 and S_N1 reactions normally occur together whenever an alkyl halide is treated in a hydroxylic solvent with a nonbasic nucleophile. Thus, the best E1 substrates are also the best S_N1 substrates, and mixtures of substitution and elimination products are usually obtained. For example, when

2-chloro-2-methylpropane is warmed to 65 °C in 80% aqueous ethanol, a 64:36 mixture of 2-methylpropan-2-ol (S_N1) and 2-methylpropene (E1) results.

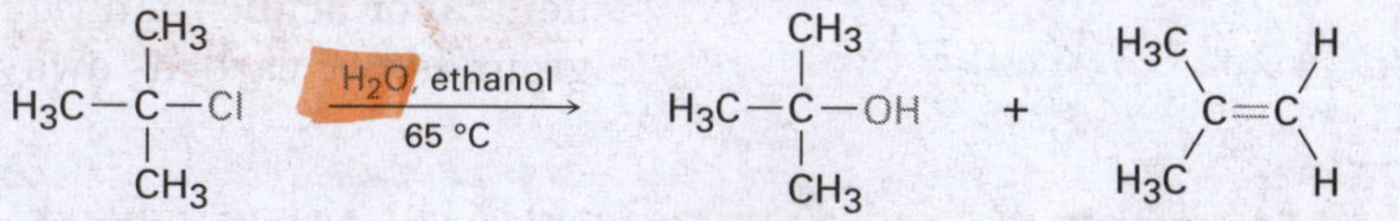

2-Chloro-2-methylpropane | **2-Methylpropan-2-ol (64%)** | **2-Methylpropene (36%)**

The E1cB Reaction

In contrast to the E1 reaction, which involves a carbocation intermediate, the **E1cB reaction** takes place through a *carbanion* intermediate. Base-induced abstraction of a proton gives an anion, which immediately expels a leaving group on the adjacent carbon. The reaction is particularly common in substrates that have a poor leaving group, such as –OH, two carbons removed from a carbonyl group, HO—C—CH—C=O. The poor leaving group disfavors the alternative E1 and E2 possibilities, and the carbonyl group makes the adjacent hydrogen unusually acidic by resonance stabilization of the anion intermediate. We'll look at this acidifying effect of a carbonyl group in Section 11.4. Note that the carbon–carbon double bond in the product is conjugated to the carbonyl, C=C—C=O, a situation similar to that in conjugated dienes (Section 4.8).

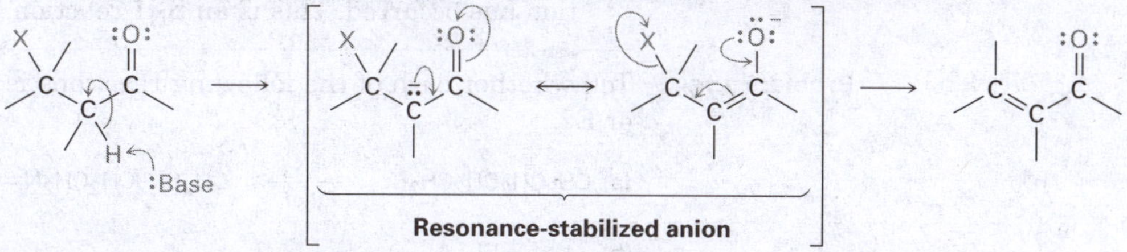

Problem 7.18 What effect on the rate of an E1 reaction of 2-chloro-2-methylpropane would you expect if the concentration of the alkyl halide were tripled?

7.9 A Summary of Reactivity: S_N1, S_N2, E1, E1cB, and E2

Now that we've seen five different kinds of nucleophilic substitution/elimination reactions, you may be wondering how to predict what will take place in any given case. Will substitution or elimination occur? Will the reaction be unimolecular or bimolecular? There are no rigid answers to these questions, but it's possible to recognize some trends and make some generalizations.

- **Primary alkyl halide (RCH_2X)** S_N2 substitution occurs if a nucleophile such as I^-, Br^-, RS^-, NH_3, or CN^-, is used; E2 elimination occurs if a strong base such as OH^- or an alkoxide ion (RO^-) is used; and E1cB elimination occurs if the leaving group is two carbons away from a carbonyl group (HO—C—CH—C=O).
- **Secondary alkyl halide (R_2CHX)** S_N2 substitution predominates if a weakly basic nucleophile is used; E2 elimination predominates if a strong base is used; and E1cB elimination takes place if the leaving group is two carbons away from a carbonyl group (HO—C—CH—C=O).

- **Tertiary alkyl halide (R_3CX)** E2 elimination occurs when a base is used, but S_N1 substitution and E1 elimination occur together under neutral or acidic conditions. E1cB elimination takes place if the leaving group is two carbons away from a carbonyl group (HO—C—CH—C=O).

Worked Example 7.8 **Predicting the Mechanism of a Reaction**

Tell whether each of the following reactions is likely to be S_N1, S_N2, E1, E1cB, or E2:

(a) 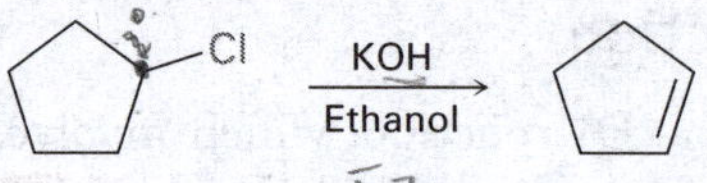

(b) 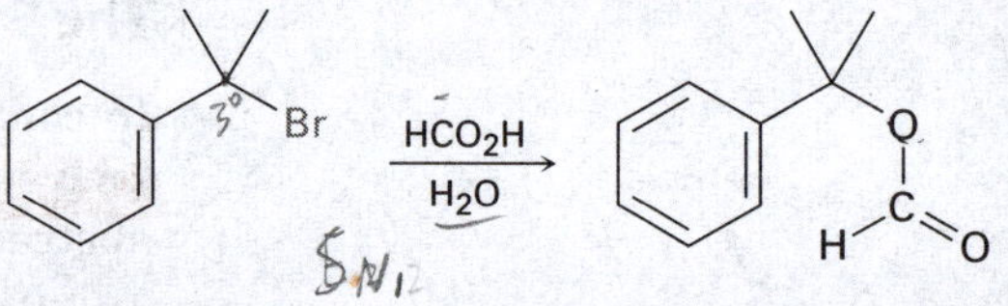

Strategy Look to see whether the substrate is primary, secondary, or tertiary, and determine whether substitution or elimination has occurred. Then apply the generalizations summarized above.

Solution (a) The substrate is a secondary alkyl halide, a strong base is used, and an elimination has occurred. This is an E2 reaction.

(b) The substrate is a tertiary halide, an acidic solvent is used, and a substitution has occurred. This is an S_N1 reaction.

Problem 7.19 Tell whether each of the following reactions is likely to be S_N1, S_N2, E1, E1cB, or E2:

(a) $CH_3CH_2CH_2CH_2Br \xrightarrow[\text{Ether}]{NaN_3} CH_3CH_2CH_2CH_2N{=}N{=}N$

(b) $CH_3CH_2CH(Cl)CH_2CH_3 \xrightarrow[\text{Ethanol}]{KOH} CH_3CH_2CH{=}CHCH_3$

(c) 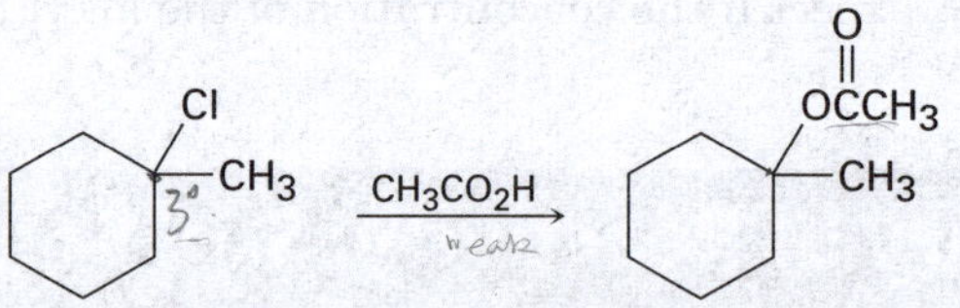

(d) 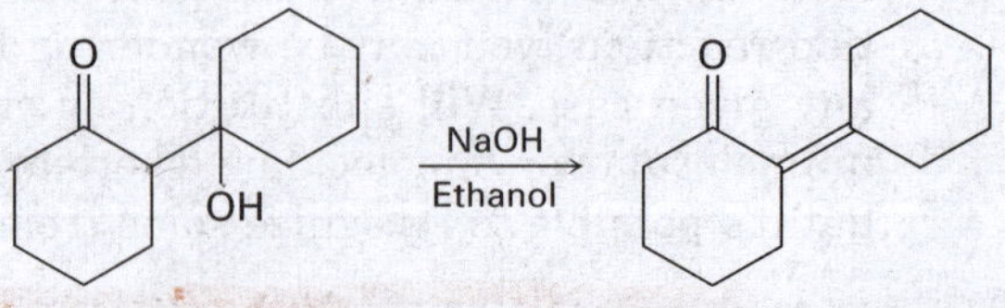

7.10 Substitution and Elimination Reactions in Living Organisms

All chemistry, whether carried out in flasks by chemists or in cells by living organisms, follows the same rules. Biological reactions therefore occur by the same addition, substitution, elimination, and rearrangement mechanisms encountered in laboratory reactions. Both S_N1 and S_N2 reactions, for instance, are well-known in biological chemistry.

Among the most common biological substitution reactions is *methylation*, the transfer of a $-CH_3$ group from an electrophilic donor to a nucleophile. A laboratory chemist might choose CH_3I for such a reaction, but living organisms use the complex molecule *S*-adenosylmethionine (abbreviated SAM) as the biological methyl donor. Because the sulfur atom in *S*-adenosylmethionine has a positive charge (a *sulfonium* ion, R_3S^+), it is a highly reactive leaving group for S_N2 displacements on the methyl carbon. In the biosynthesis of epinephrine (adrenaline) from norepinephrine, for instance, the nucleophilic nitrogen atom of norepinephrine attacks the electrophilic methyl carbon atom of *S*-adenosylmethionine in an S_N2 reaction, displacing *S*-adenosylhomocysteine.

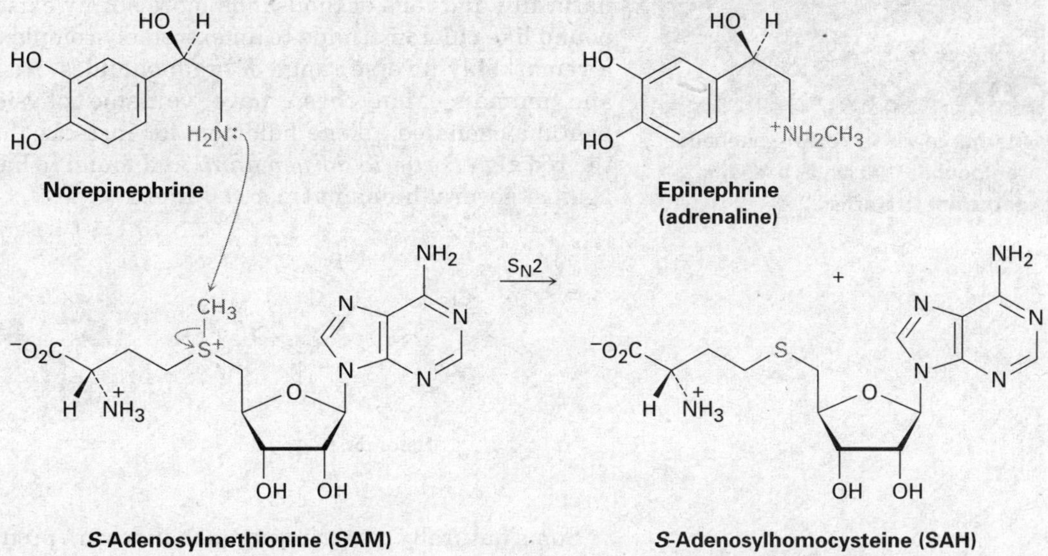

After dealing only with simple halides such as CH_3I up to this point, it's a shock to encounter a molecule as complex as *S*-adenosylmethionine. From a chemical standpoint, however, CH_3I and *S*-adenosylmethionine do exactly the same thing: both transfer a methyl group by an S_N2 reaction.

Eliminations, like substitutions, also occur frequently in biological pathways, with the E1cB mechanism particularly common. The substrate is usually an alcohol, and the H atom removed is usually adjacent to a carbonyl group, just as in laboratory reactions. A typical example occurs during the biosynthesis of fats when a 3-hydroxybutyryl thioester is dehydrated to the corresponding unsaturated (crotonyl) thioester. The base in this reaction is an amine functional group in the enzyme.

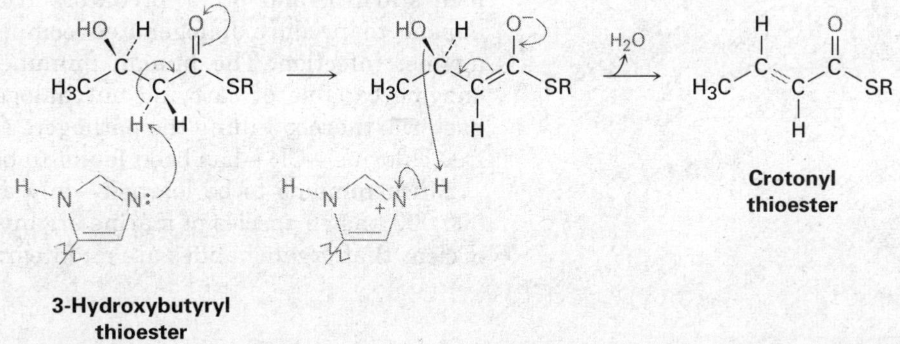

INTERLUDE

Naturally Occurring Organohalides

Marine corals secrete organohalide compounds that act as a feeding deterrent to starfish.

As recently as 1970, only about 30 naturally occurring organohalides were known. It was simply assumed that chloroform, halogenated phenols, chlorinated aromatic compounds called PCBs, and other such substances found in the environment were industrial pollutants. Now, a bit more than a third of a century later, the situation is quite different. More than 5000 organohalides have been found to occur naturally, and tens of thousands more surely exist. From a simple compound like chloromethane to an extremely complex one like vancomycin, a remarkably diverse range of organohalides exists in plants, bacteria, and animals. Many even have valuable physiological activity. The pentahalogenated alkene halomon, for instance, has been isolated from the red alga *Portieria hornemannii* and found to have anticancer activity against several human tumor cell lines.

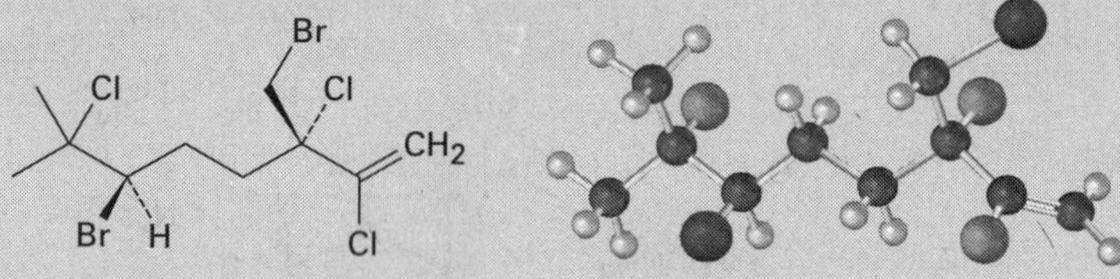

Halomon

Some naturally occurring organohalides are produced in massive quantities. Forest fires, volcanoes, and marine kelp release up to *5 million tons* of CH_3Cl per year, for example, while annual industrial emissions total about 26,000 tons. Termites are thought to release as much as 10^8 kg of chloroform per year. A detailed examination of the Okinawan acorn worm *Ptychodera flava* found that the 64 million worms living in a 1 km^2 study area excreted nearly 8000 pounds per year of bromophenols and bromoindoles, compounds previously thought to be nonnatural pollutants.

Why do organisms produce organohalides, many of which are undoubtedly toxic? The answer seems to be that many organisms use organohalogen compounds for self-defense, either as feeding deterrents, as irritants to predators, or as natural pesticides. Marine sponges, coral, and sea hares, for example, release foul-tasting organohalides that deter fish, starfish, and other predators from eating them. Even humans appear to produce halogenated compounds as part of their defense against infection. The human immune system contains a peroxidase enzyme capable of carrying out halogenation reactions on fungi and bacteria, thereby killing the pathogen. And most remarkable of all, even free chlorine—Cl_2—has been found to be present in humans.

Much remains to be learned—only a few hundred of the more than 500,000 known species of marine organisms have been examined—but it is clear that organohalides are an integral part of the world around us.

Summary and Key Words

alkyl halide 223
bimolecular reaction 231
carbanion 227
E1 reaction 240
E1cB reaction 241
E2 reaction 239
Grignard reagent 226
nucleophilic substitution reaction 228
organohalide 222
S_N1 reaction 234
S_N2 reaction 230
unimolecular reaction 234
Zaitsev's rule 237

Alkyl halides are not often found in terrestrial organisms, but the kinds of reactions they undergo are among the most important and well-studied reaction types in organic chemistry. In this chapter, we saw how to name and prepare alkyl halides, and we made a detailed study of their substitution and elimination reactions.

Alkyl halides are usually prepared from alcohols by treatment either with HX (for tertiary alcohols) or with $SOCl_2$ or PBr_3 (for primary and secondary alcohols). Alkyl halides react with magnesium metal to form organomagnesium halides, called **Grignard reagents (RMgX)**. Because Grignard reagents are both nucleophilic and basic, they react with acids to yield hydrocarbons.

Treatment of an alkyl halide with a nucleophile/base results either in substitution or elimination. **Nucleophilic substitution reactions** occur by two mechanisms: S_N2 and S_N1. In the **S_N2 reaction,** the entering nucleophile attacks the substrate from a direction 180° away from the leaving group, resulting in an umbrella-like inversion of configuration at the carbon atom. S_N2 reactions are strongly inhibited by increasing steric bulk of the reagents and are favored only for primary substrates and simple secondary substrates. In the **S_N1 reaction,** the substrate spontaneously dissociates to a carbocation, which reacts with a nucleophile in a second step. As a consequence, S_N1 reactions take place with racemization of configuration at the carbon atom and are favored only for tertiary substrates.

Elimination reactions occur commonly by three mechanisms—E2, E1, and E1cB—which differ in the timing of C–X and C–H bond-breaking. In the **E2 reaction**, a base abstracts a hydrogen at the same time that the adjacent halide group departs. The E2 reaction usually gives a mixture of alkene products in which the more highly substituted alkene predominates (**Zaitsev's rule**). In the **E1 reaction**, C–X bond-breaking occurs first. The substrate spontaneously dissociates to form a carbocation, which subsequently loses H^+ from a neighboring carbon. In the **E1cB reaction**, C–H bond-breaking occurs first. A base removes H^+ to give a **carbanion** intermediate, followed by loss of the leaving halide group from the adjacent carbon. Biological elimination reactions typically occur by this E1cB mechanism.

Summary of Reactions

1. Synthesis of alkyl halides from alcohols (Section 7.2)
   (a) Reaction of alcohols with HX, where X = Cl, Br

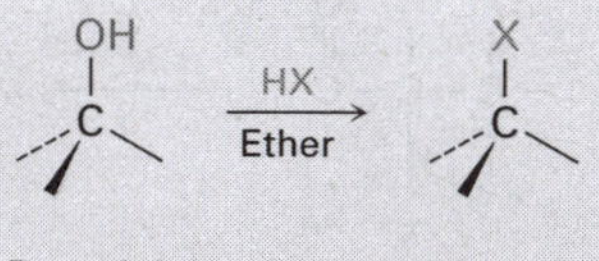

Reactivity order: 3° > 2° > 1°

continued

(b) Reaction of primary and secondary alcohols with $SOCl_2$ and PBr_3

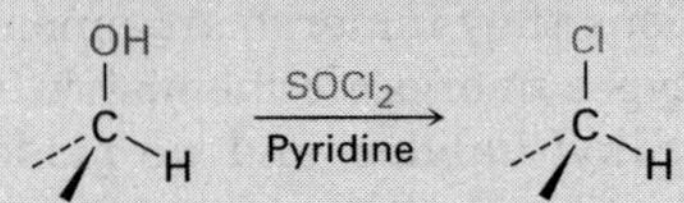

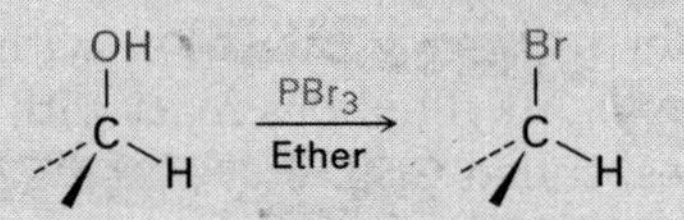

2. Reactions of alkyl halides

(a) Formation of Grignard reagents (Section 7.3)

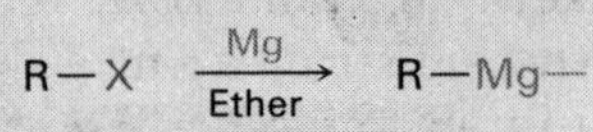

(b) S_N2 reaction: backside attack of nucleophile on alkyl halide (Section 7.5)

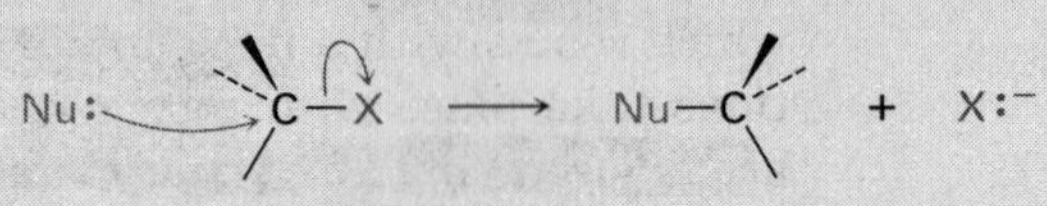

(c) S_N1 reaction: carbocation intermediate is involved (Section 7.6)

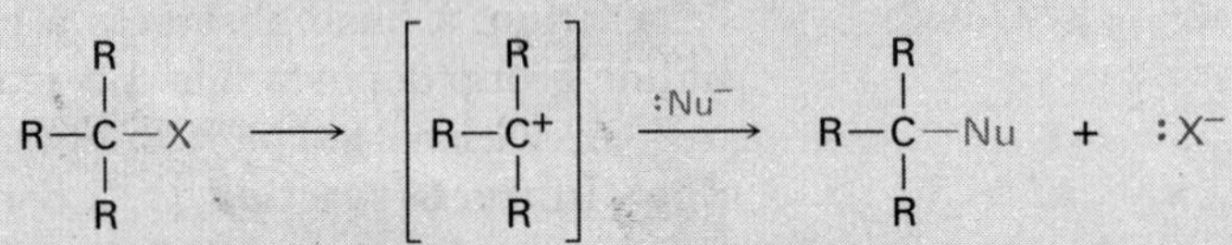

(d) E2 reaction (Section 7.7)

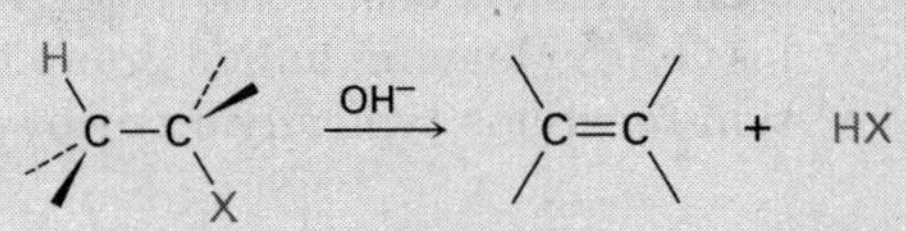

(e) E1 reaction (Section 7.8)

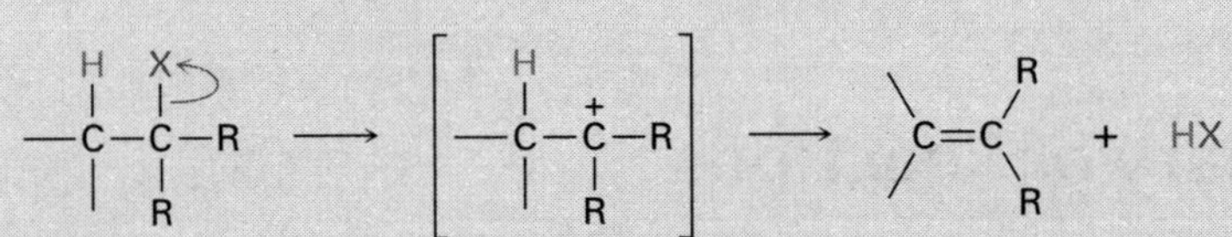

(f) E1cB reaction (Section 7.8)

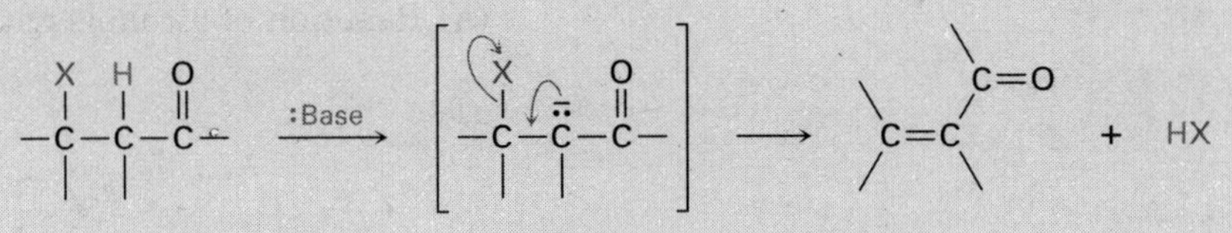

Exercises

Visualizing Chemistry

(Problems 7.1–7.19 appear within the chapter.)

OWL

Interactive versions of these problems are assignable in OWL.

7.20 Draw the structure of the product you expect from E2 reaction of the following molecule with NaOH (yellow-green = Cl):

7.21 From what alkyl bromide was the following alkyl acetate made by S_N2 reaction? Write the reaction, showing all stereochemistry.

7.22 Assign *R* or *S* configuration to the following molecule, write the product you would expect from S_N2 reaction with NaCN, and assign *R* or *S* configuration to the product (red = O, yellow-green = Cl):

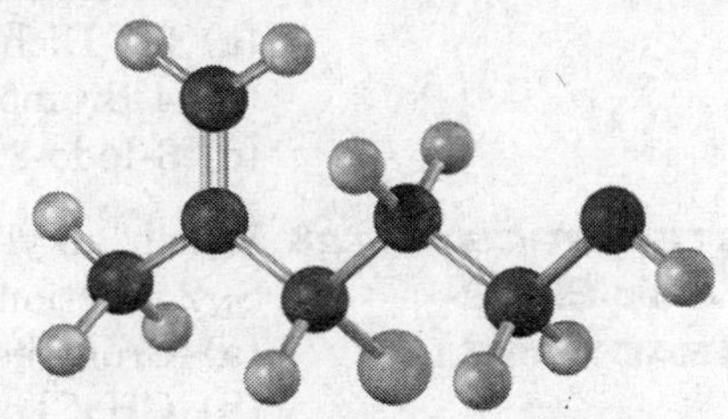

7.23 Write the product you would expect from reaction of each of the following alkyl halides with (i) $Na^+ \ ^-SCH_3$ and (ii) NaOH (yellow-green = Cl):

(a)

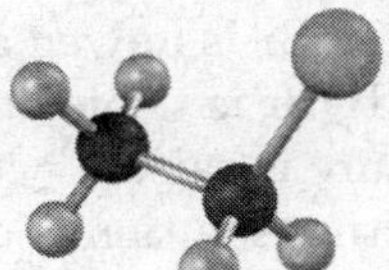

(b)

(c)

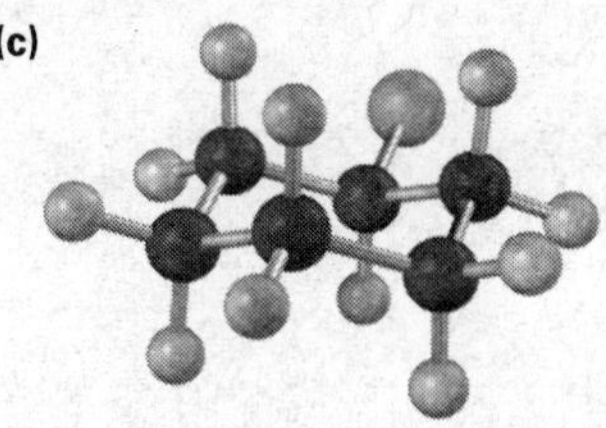

7.24 The following alkyl bromide can be prepared by reaction of the alcohol (*S*)-pentan-2-ol with PBr_3. Name the compound, assign (*R*) or (*S*) stereochemistry, and tell whether the reaction of the alcohol with PBr_3 occurs with retention of the same stereochemistry or with a change in stereochemistry (reddish brown = Br).

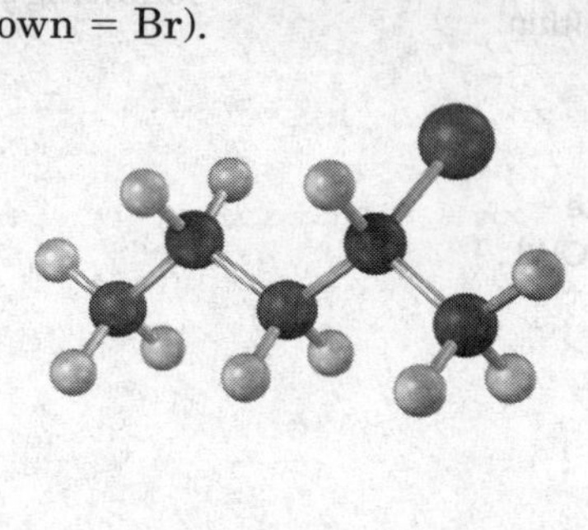

Additional Problems

NAMING ALKYL HALIDES

7.25 Draw and name the monochlorination products you might obtain by reaction of 2-methylpentane with Cl_2. Which of the products are chiral?

7.26 Name the following alkyl halides:

(a) 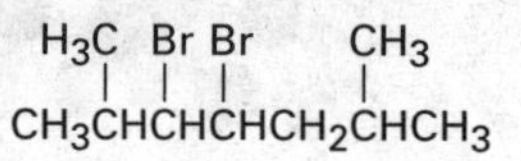

(b) 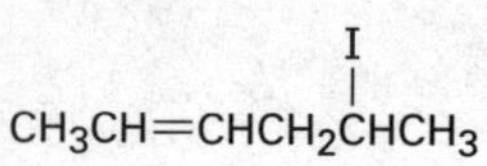

(c) 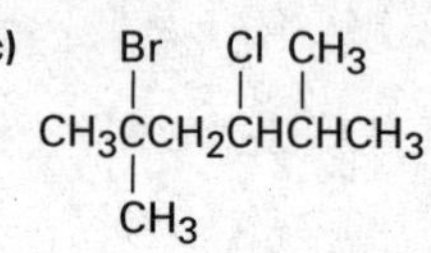

(d) 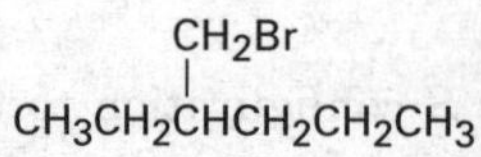

(e) $ClCH_2CH_2CH_2C{\equiv}CCH_2Br$

7.27 Draw structures corresponding to the following IUPAC names:

(a) 2,3-Dichloro-4-methylhexane

(b) 4-Bromo-4-ethyl-2-methylhexane

(c) 3-Iodo-2,2,4,4-tetramethylpentane

CHARACTERISTICS OF S_N1 AND S_N2 REACTIONS

7.28 Which alkyl halide in each of the following pairs will react faster in an S_N2 reaction with OH^-?

(a) Bromobenzene or benzyl bromide, $C_6H_5CH_2Br$

(b) CH_3Cl or $(CH_3)_3CCl$

(c) $CH_3CH{=}CHBr$ or $H_2C{=}CHCH_2Br$

7.29 What effect would you expect the following changes to have on the S_N2 reaction of CH_3Br and CN^- to give CH_3CN?

(a) The concentration of CH_3Br is tripled and that of CN^- is halved.

(b) The concentration of CH_3Br is halved and that of CN^- is tripled.

(c) The concentration of CH_3Br is tripled and that of CN^- is doubled.

(d) The reaction temperature is raised.

(e) The volume of the reacting solution is doubled by addition of more solvent.

7.30 Describe the effects of the following variables on both S_N2 and S_N1 reactions:

(a) Substrate structure **(b)** Leaving group

7.31 Which ion in each of the following pairs is a better leaving group?

(a) F^- or Br^- **(b)** Cl^- or NH_2^- **(c)** OH^- or I^-

7.32 Order the following compounds with respect to both S_N1 and S_N2 reactivity:

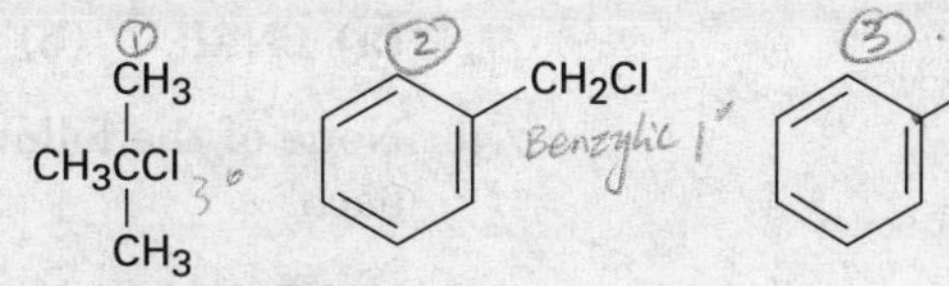

7.33 Order each set of compounds with respect to S_N2 reactivity:

(a)

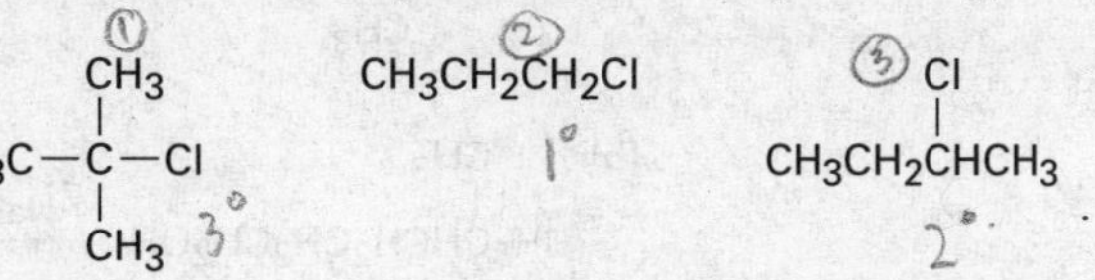

(b)

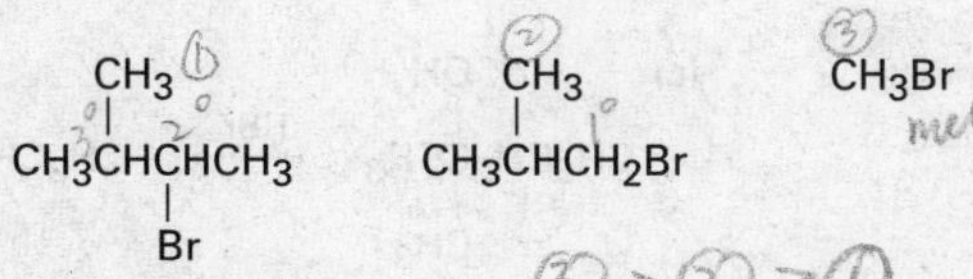

7.34 What effect would you expect the following changes to have on the S_N1 reaction of $(CH_3)_3CBr$ with CH_3OH to give $(CH_3)_3COCH_3$?

(a) The concentration of $(CH_3)_3CBr$ is doubled and that of CH_3OH is halved.

(b) The concentration of $(CH_3)_3CBr$ is halved and that of CH_3OH is doubled.

(c) The concentrations of both $(CH_3)_3CBr$ and CH_3OH are tripled.

(d) The reaction temperature is lowered.

SYNTHESIS

7.35 Predict the product(s) of the following reactions:

(a)

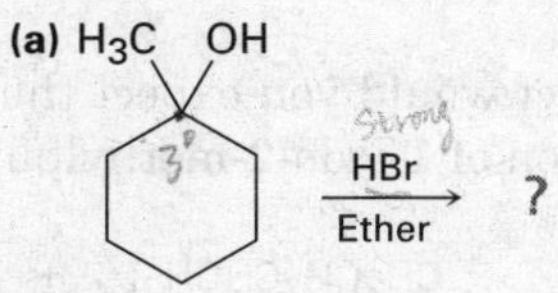

(b) $CH_3CH_2CH_2CH_2OH \xrightarrow{SOCl_2} ?$

(c)

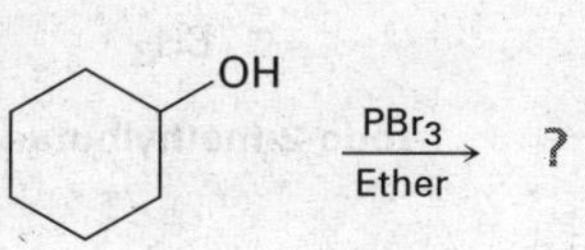

(d) $CH_3CH_2CH(Br)CH_3 \xrightarrow[\text{Ether}]{\text{Mg}} \text{A} \xrightarrow{H_2O} \text{B}$

7.36 How would you prepare the following compounds, starting with cyclopentene and any other reagents needed?

(a) Chlorocyclopentane

(b) Cyclopentanol

(c) Cyclopentylmagnesium chloride

(d) Cyclopentane

7.37 What products do you expect from reaction of 1-bromopropane with the following reagents?

(a) NaI **(b)** NaCN **(c)** NaOH **(d)** Mg **(e)** $NaOCH_3$

7.38 None of the following reactions take place as written. What is wrong with each?

(a)

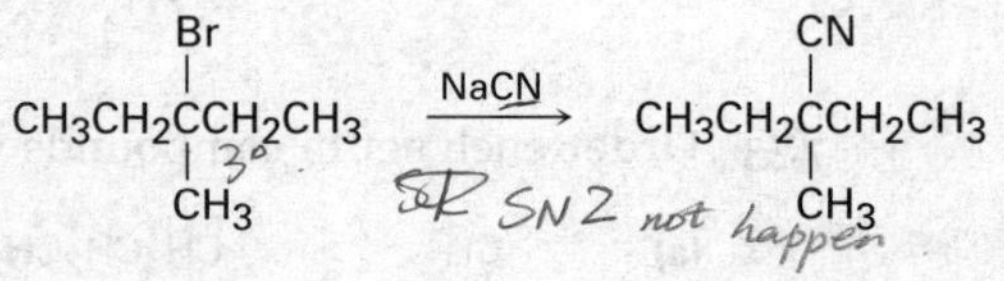

(b)

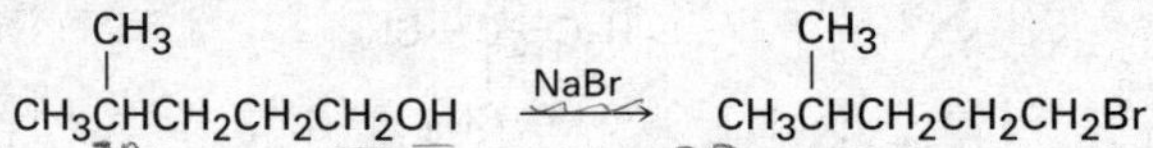

(c)

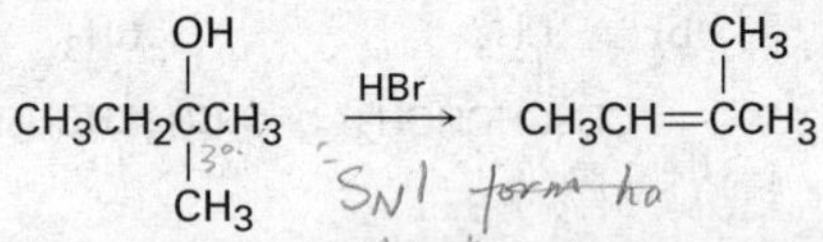

7.39 How might you prepare the following molecules using a nucleophilic substitution reaction at some step?

(a) CH_3CH_2Br **(b)** $CH_3CH_2CH_2CH_2CN$ **(c)**

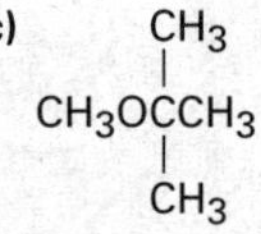

(d) $CH_3CH_2CH_2N{=}\overset{+}{N}{=}N^-$ **(e)** CH_3CH_2SH **(f)**

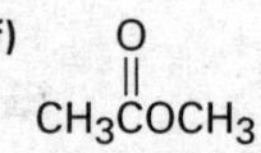

SUBSTITUTIONS AND ELIMINATIONS

7.40 What effect would you expect the following changes to have on the rate of the reaction of 1-iodo-2-methylbutane with CN^-?

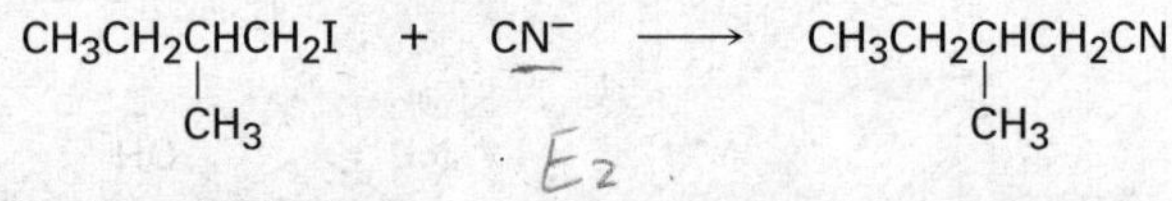

1-Iodo-2-methylbutane

(a) CN^- concentration is halved and 1-iodo-2-methylbutane concentration is doubled.

(b) Both CN^- and 1-iodo-2-methylbutane concentrations are tripled.

7.41 Propose a structure for an alkyl halide that can give a mixture of three alkenes on E2 reaction.

7.42 Heating either *tert*-butyl chloride or *tert*-butyl bromide with ethanol yields the same reaction mixture: approximately 80% *tert*-butyl ethyl ether [$(CH_3)_3COCH_2CH_3$] and 20% 2-methylpropene. Explain why the identity of the leaving group has no effect on the product mixture.

7.43 Predict the major alkene product from the following eliminations:

(a)

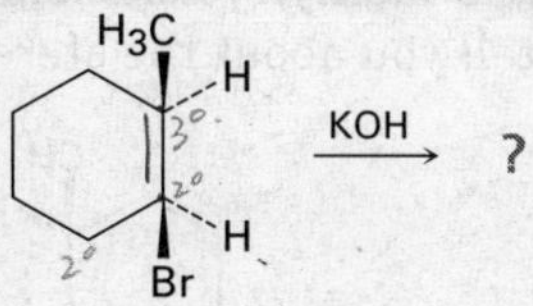

(b)

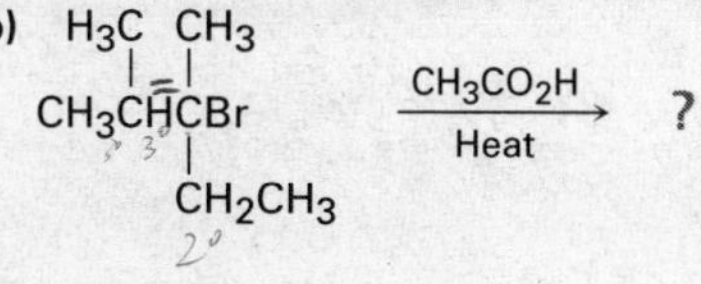

7.44 What effect would you expect on the rate of reaction of ethyl alcohol with 2-iodo-2-methylbutane if the concentration of the alkyl halide is tripled?

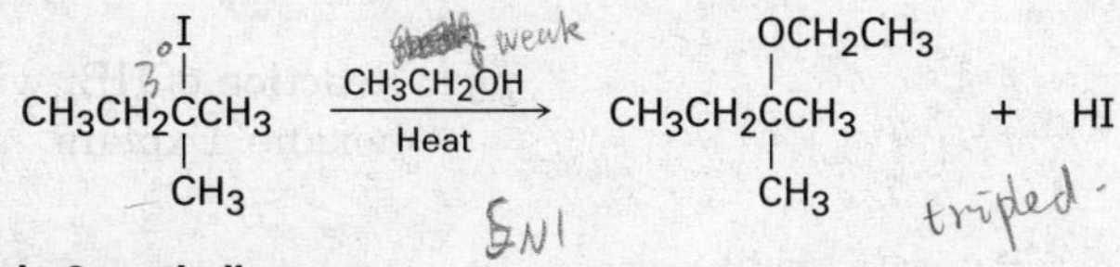

2-Iodo-2-methylbutane

7.45 Treatment of an alkyl chloride C_4H_9Cl with strong base gives a mixture of three isomeric alkene products. What is the structure of the alkyl chloride, and what are the structures of the three products?

7.46 Identify the following reactions as either S_N1, S_N2, E1, or E2:

(a)

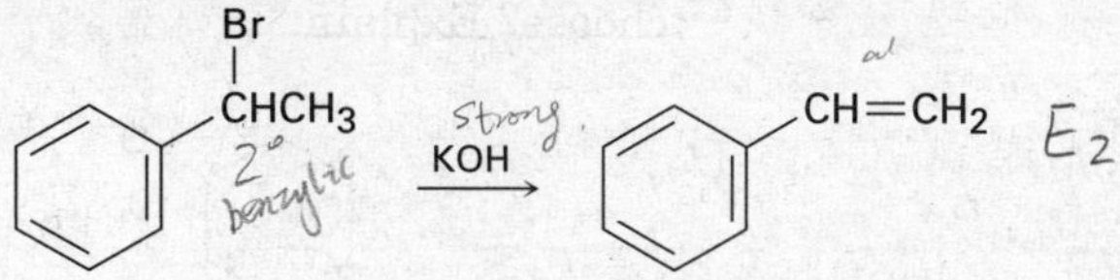

(b)

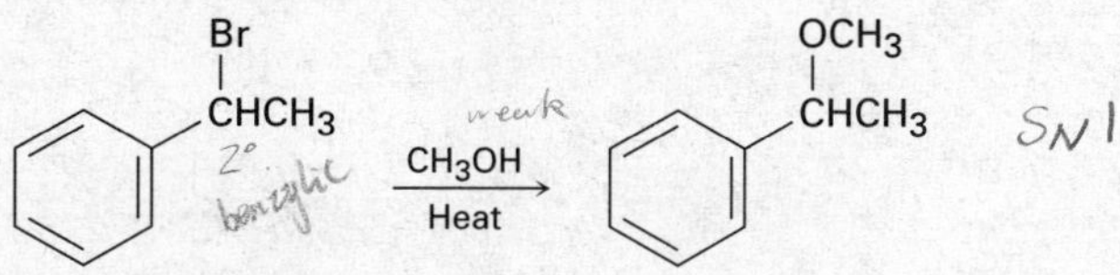

GENERAL PROBLEMS

7.47 Draw resonance structures of the benzyl radical $C_6H_5CH_2\cdot$ to account for the fact that radical chlorination of toluene with Cl_2 occurs exclusively on the methyl group rather than on the aromatic ring.

7.48 Predict the product and give the stereochemistry of reactions of the following nucleophiles with (*R*)-2-bromooctane:
(a) CN^- **(b)** $CH_3CO_2^-$ **(c)** Br^-

7.49 Draw all isomers of C_4H_9Br, name them, and arrange them in order of decreasing reactivity in the S_N2 reaction.

7.50 Although the radical chlorination of alkanes with Cl_2 is usually unselective and gives a mixture of products, chlorination of propene, $CH_3CH{=}CH_2$, occurs almost exclusively on the methyl group rather than on a double-bond carbon. Draw resonance structures of the allyl radical $CH_2{=}CHCH_2\cdot$ to account for this result.

7.51 *trans*-1-Bromo-2-methylcyclohexane yields the non-Zaitsev elimination product 3-methylcyclohexene on treatment with KOH. What does this result tell you about the stereochemistry of E2 reactions?

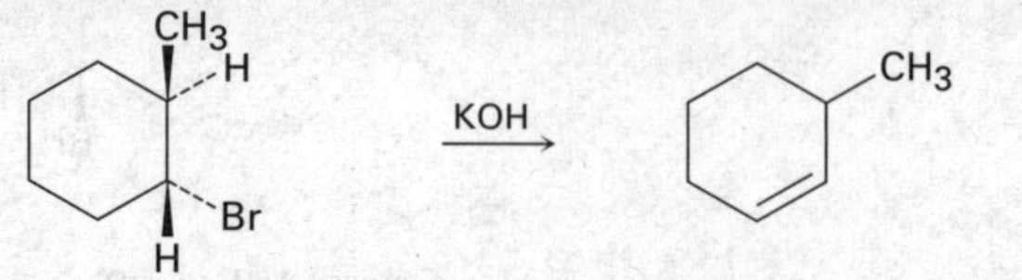

***trans*-1-Bromo-2-methylcyclohexane** **3-Methylcyclohexene**

7.52 Reaction of HBr with (*R*)-3-methylhexan-3-ol yields (±)-3-bromo-3-methylhexane. Explain.

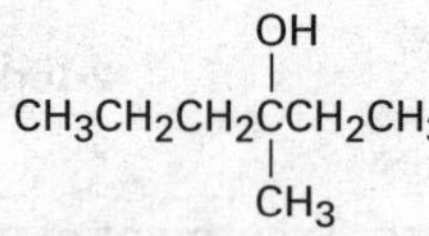

3-Methylhexan-3-ol

7.53 Ethers can be prepared by S_N2 reaction of an alkoxide ion with an alkyl halide: R—O$^-$ + R′—Br → R—O—R′ + Br$^-$. Suppose you wanted to prepare cyclohexyl methyl ether. Which of the following two routes would you choose? Explain.

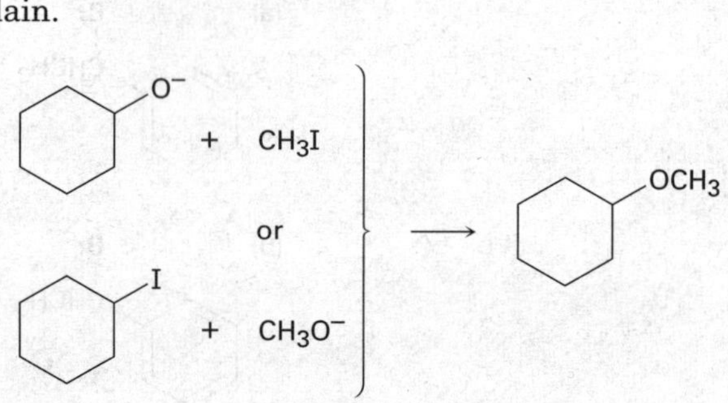

7.54 How could you prepare diethyl ether, $CH_3CH_2OCH_2CH_3$, starting from ethyl alcohol and any inorganic reagents needed? More than one step is needed. (See Problem 7.53.)

7.55 The S_N2 reaction can occur *intramolecularly,* meaning within the same molecule. What product would you expect from treatment of 4-bromobutan-1-ol with base?

$$BrCH_2CH_2CH_2CH_2OH \xrightarrow{\text{Base}} BrCH_2CH_2CH_2CH_2O^-\ Na^+ \longrightarrow\ ?$$

7.56 How could you prepare cyclohexane starting from 3-bromocyclohexene? More than one step is needed.

7.57 How can you explain the fact that treatment of (*R*)-2-bromohexane with NaBr yields racemic 2-bromohexane?

7.58 Why do you suppose it's not possible to prepare a Grignard reagent from a bromoalcohol such as 4-bromopentan-1-ol?

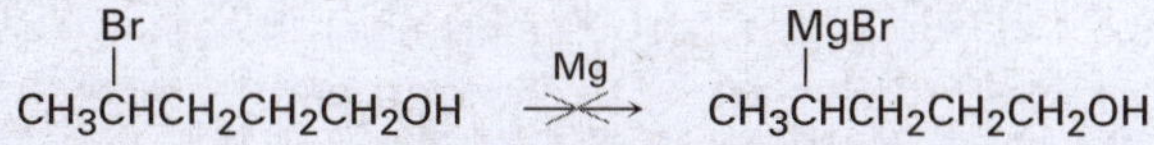

7.59 (*S*)-Butan-2-ol slowly racemizes to give (±)-butan-2-ol on standing in dilute sulfuric acid. Propose a mechanism to account for this observation.

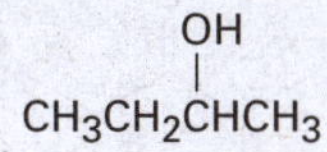

Butan-2-ol

7.60 The following reaction is an important step in the laboratory synthesis of proteins. Propose a mechanism.

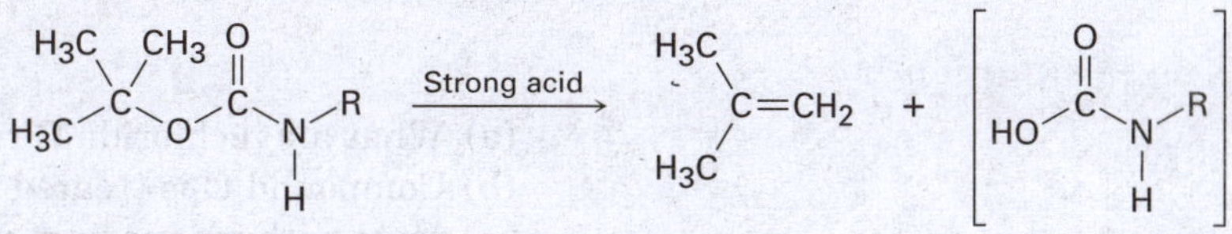

7.61 Metabolism of *S*-adenosylhomocysteine (Section 7.10) involves the following step. Propose a mechanism.

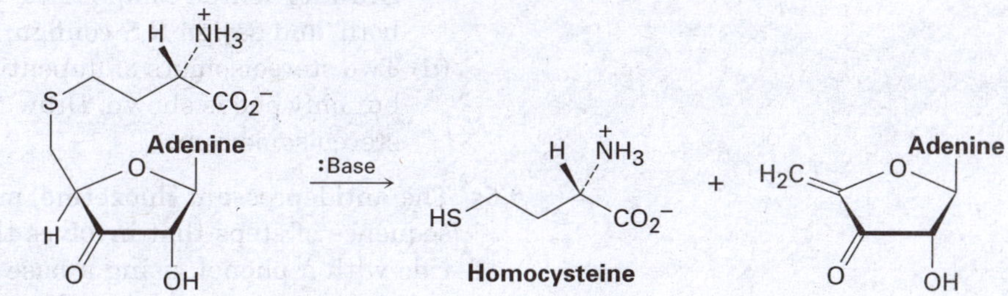

7.62 Compound **A** is optically inactive and has the formula $C_{16}H_{16}Br_2$. On treatment with strong base, **A** gives hydrocarbon **B**, $C_{16}H_{14}$, which absorbs 2 equivalents of H_2 when reduced over a palladium catalyst. Hydrocarbon **B** also reacts with acidic $KMnO_4$ to give two carbonyl-containing products. One product, **C**, is a carboxylic acid with the formula $C_7H_6O_2$. The other product is oxalic acid, HO_2CCO_2H. Formulate the reactions involved, and suggest structures for **A**, **B**, and **C**.

7.63 One step in the urea cycle for ridding the body of ammonia is the conversion of argininosuccinate to the amino acid arginine plus fumarate. Propose a mechanism for the reaction, and show the structure of arginine.

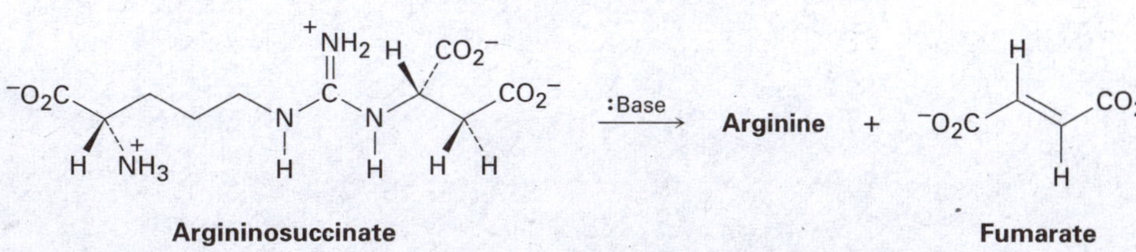

IN THE MEDICINE CABINET **7.64** The antipsychotic drug flupentixol is prepared by the following scheme:

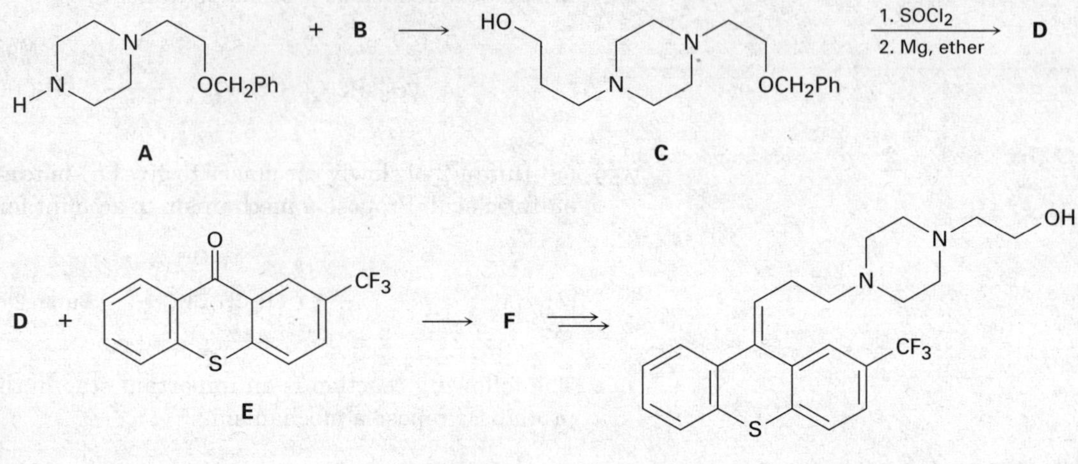

Flupentixol

(a) What alkyl chloride **B** reacts with amine **A** to form **C**?

(b) Compound **C** is treated with $SOCl_2$, and the product is allowed to react with magnesium metal to give a Grignard reagent **D**. What is the structure of **D**?

(c) We'll see in Section 9.6 that Grignard reagents add to ketones, such as **E**, to give tertiary alcohols, such as **F**. Because of the newly formed chirality center, compound **F** exists as a pair of enantiomers. Draw both, and assign *R,S* configuration.

(d) Two stereoisomers of flupentixol are subsequently formed from **F**, but only one is shown. Draw the other isomer, and identify the type of stereoisomerism.

7.65 The antidepressant fluoxetine, marketed as Prozac, can be prepared by a sequence of steps that involves the substitution reaction of an alkyl chloride with a phenol, using a base to convert the phenol into its phenoxide anion.

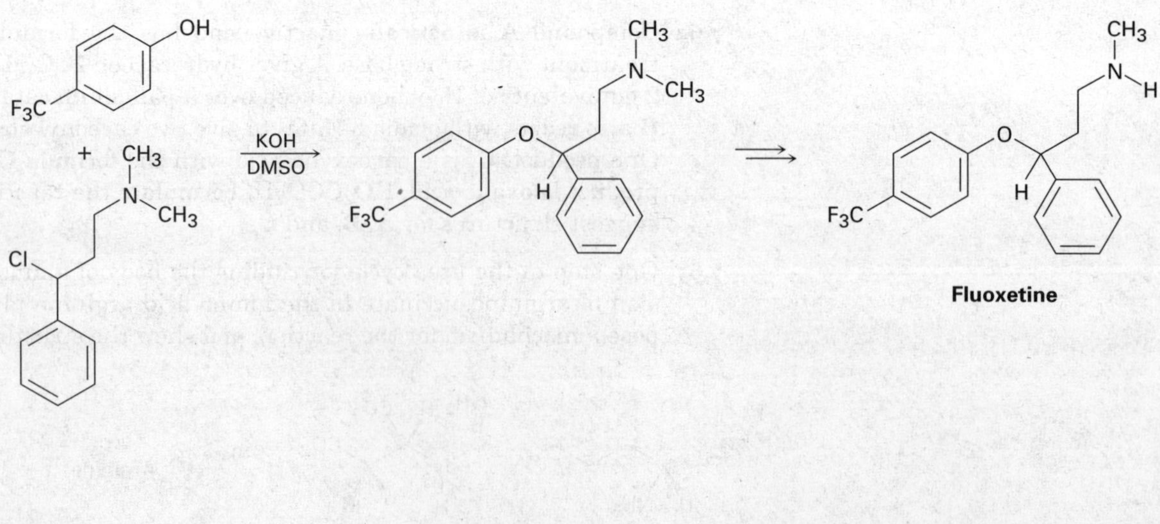

Fluoxetine

(a) Identify the nucleophile and electrophile in the reaction.

(b) The rate of the substitution reaction depends on concentrations of both the alkyl chloride and phenol. Is this an S_N1 or an S_N2 reaction?

(c) The physiologically active enantiomer of fluoxetine has (*S*) stereochemistry. Based on your answer in part **(b)**, draw the structure of the alkyl chloride you would need, showing the correct stereochemistry.

IN THE FIELD

7.66 All five of the following herbicides contain a trifluoromethyl group attached to a benzene ring. This group is commonly included in many bioactive molecules because it slows down oxidative decomposition of the aromatic ring, making the $-CF_3$ containing compounds more stable than their $-CH_3$ analogs.

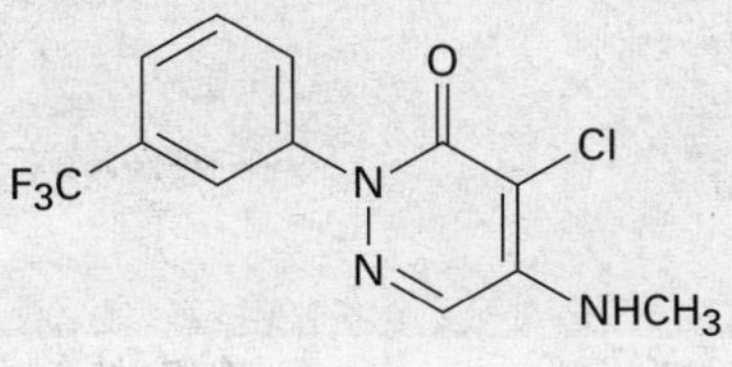

Norflurazon

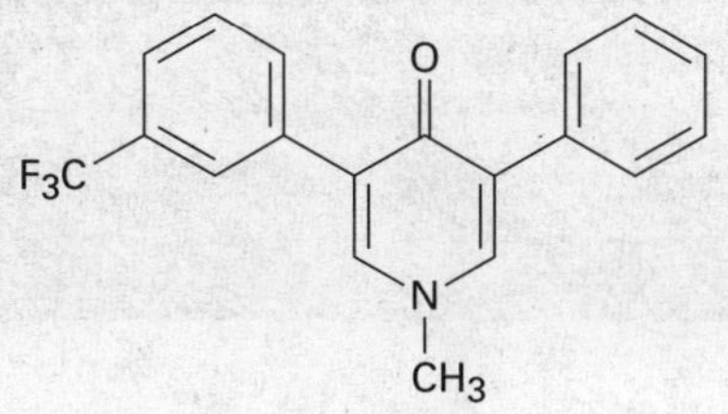

Fluridone

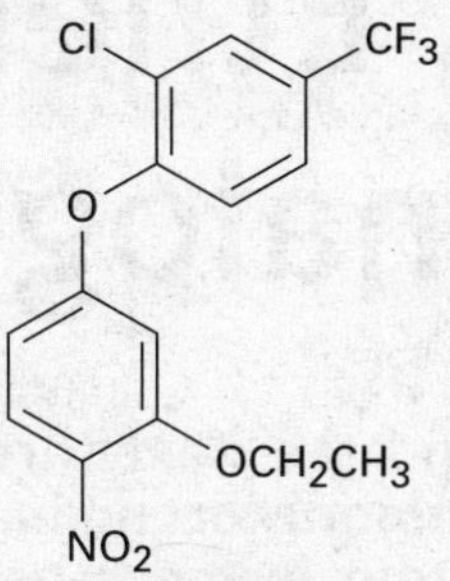

Oxyfluorfen

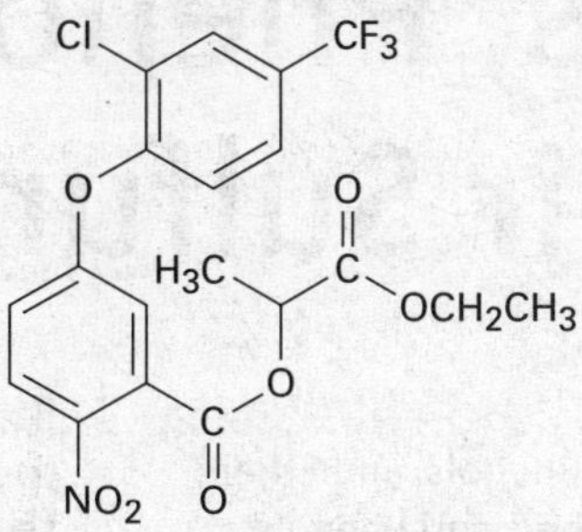

Lactofen

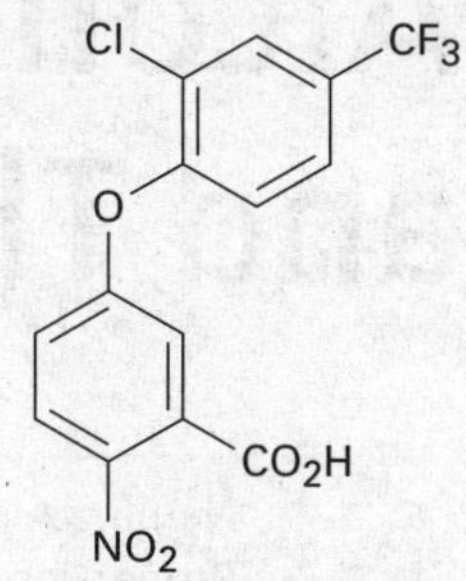

Acifluorfen

(a) Oxidation is defined as the loss of electrons. Is it likely to be easier to remove electrons from a benzene ring with a $-CF_3$ or $-CH_3$ group attached? Why?

(b) Two of the compounds shown inhibit pigment synthesis in plants, while three inhibit lipid synthesis. Group the compounds by their likely mechanism of action. What criteria did you use to make these groupings?

CHAPTER

8

The Harger Drunkometer was introduced in 1938 to help keep drunk drivers off the road. Drunk driving is still a problem, but methods for detecting the amount of alcohol consumed have improved greatly in the last 70 years.

Alcohols, Phenols, Ethers, and Their Sulfur Analogs

In this and the next three chapters, we'll focus on the most commonly occurring of all functional groups—those that contain *oxygen*—beginning in this chapter with a look at compounds that contain C–O single bonds. An **alcohol** is a compound that has a hydroxyl group bonded to a saturated, *sp*^{3}-hybridized carbon atom, R—OH; a **phenol** has a hydroxyl group bonded to an aromatic ring, Ar—OH; and an **ether** has an oxygen atom bonded to two organic groups, R—O—R′. The corresponding sulfur analogs are called **thiols** (R—SH), **thiophenols** (Ar—SH), and **sulfides** (R—S—R′).

Alcohols, phenols, and ethers occur widely in nature and have many industrial, pharmaceutical, and biological applications. Ethanol, for instance, is a fuel additive, an industrial solvent, and a beverage; phenol is a general disinfectant, commonly called carbolic acid; and diethyl ether—the familiar "ether" of medical use—is frequently used as a reaction solvent and was once popular as an inhaled anesthetic.

WHY THIS CHAPTER?

Up to this point, we've focused on developing some general ideas of organic reactivity and on looking at the chemistry of hydrocarbons and alkyl halides. With that background, it's now time to begin a study of oxygen-containing

OWL

Online homework for this chapter can be assigned in OWL, an online homework assessment tool.

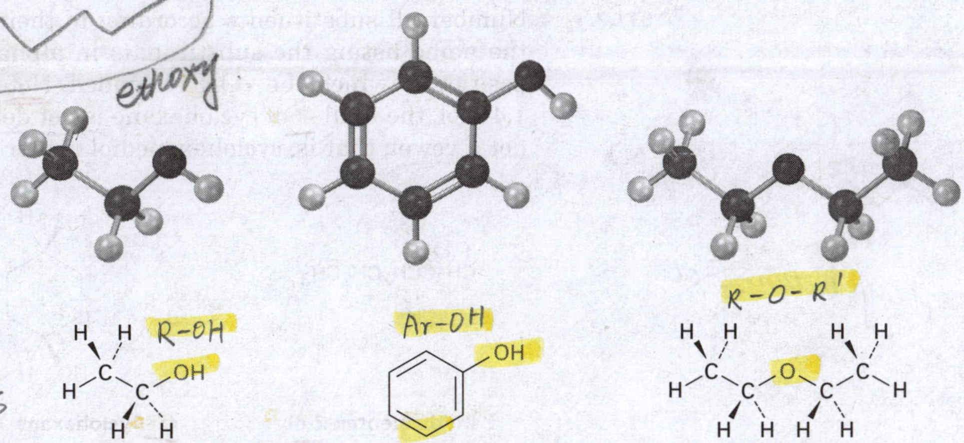

Ethanol **Phenol** **Diethyl ether**

functional groups, which lie at the heart of organic and biological chemistry. In fact, practically every one of the hundreds of thousands of different compounds in your body contains oxygen. We'll look at compounds with C–O single bonds in this chapter and then move on to carbonyl compounds in Chapters 9 to 11.

8.1 Naming Alcohols, Phenols, and Ethers

Alcohols

Alcohols are classified as primary (1°), secondary (2°), or tertiary (3°), depending on the number of carbon substituents bonded to the hydroxyl-bearing carbon.

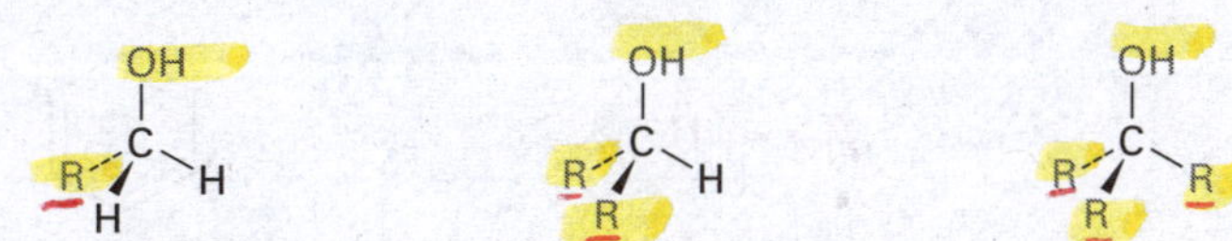

A primary (1°) alcohol **A secondary (2°) alcohol** **A tertiary (3°) alcohol**

Simple alcohols are named in the IUPAC system as derivatives of the parent alkane, using the suffix *-ol*. As always, the locant indicating the position of the hydroxyl group in the parent chain is placed immediately before the *-ol* suffix in the newer, post-1993 naming system.

STEP 1 Select the longest carbon chain containing the hydroxyl group, and replace the *-e* ending of the corresponding alkane with *-ol*. The *-e* is deleted to prevent the occurrence of two adjacent vowels: propanol rather than propaneol, for example.

STEP 2 Number the carbons of the parent chain beginning at the end nearer the hydroxyl group.

STEP 3 Number all substituents according to their position on the chain, and write the name listing the substituents in alphabetical order and identifying the position to which the –OH is bonded. Note that in naming *cis*-cyclohexane-1,4-diol, the final *-e* of cyclohexane is not deleted because the next letter (d) is not a vowel; that is, cyclohexanediol rather than cyclohexandiol.

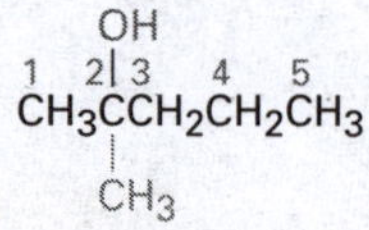

2-Methylpentan-2-ol

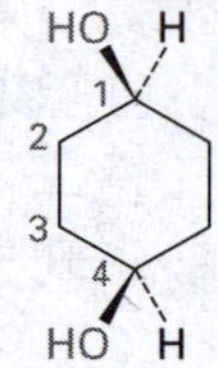

***cis*-Cyclohexane-1,4-diol**

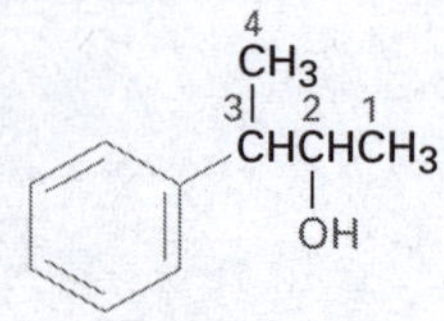

3-Phenylbutan-2-ol

Some well-known alcohols also have common names that are accepted by IUPAC. For example:

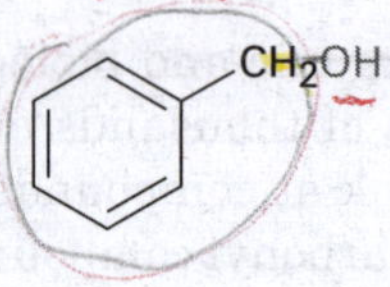

Benzyl alcohol (phenylmethanol)

$H_2C{=}CHCH_2OH$

Allyl alcohol (prop-2-en-1-ol)

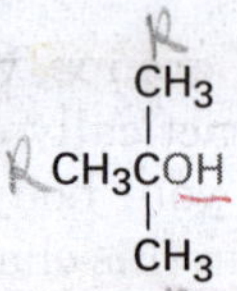

***tert*-Butyl alcohol (2-methylpropan-2-ol)**

$HOCH_2CH_2OH$

Ethylene glycol (ethane-1,2-diol)

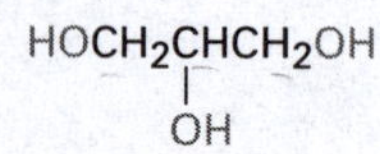

Glycerol (propane-1,2,3-triol)

Phenols

The word *phenol* is used both as the name of a specific substance (hydroxybenzene) and as the family name for all hydroxy-substituted aromatic compounds. Substituted phenols are named as described previously in Section 5.2 for aromatic compounds, with *-phenol* used as the parent name rather than *-benzene*.

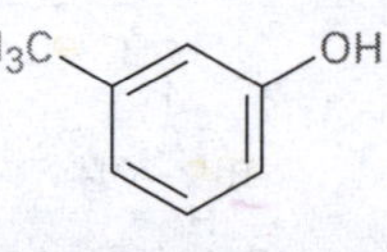

***m*-Methylphenol (*m*-cresol)**

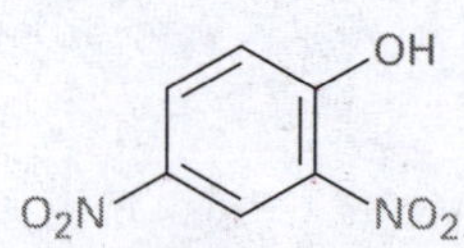

2,4-Dinitrophenol

Ethers

Simple ethers that contain no other functional groups are named by identifying the two organic groups and adding the word *ether*.

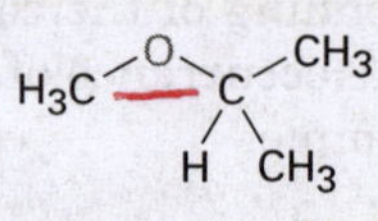

Isopropyl methyl ether

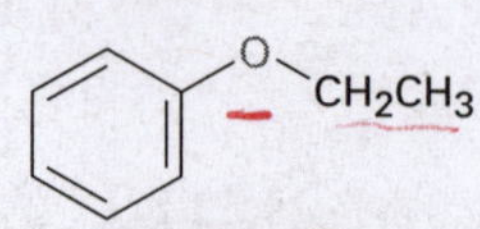

Ethyl phenyl ether

If other functional groups are present, the ether part is named as an *alkoxy* substituent. For example:

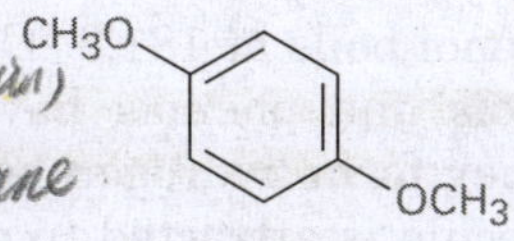

p-Dimethoxybenzene

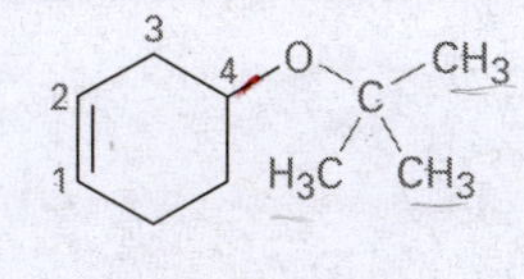

4-*tert*-Butoxycyclohex-1-ene

Problem 8.1 Give IUPAC names for the following compounds:

(a)

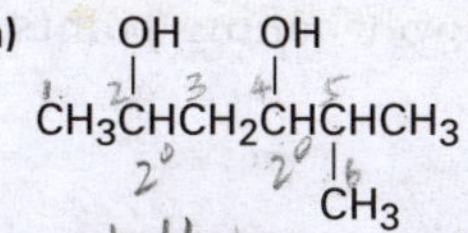

(b)

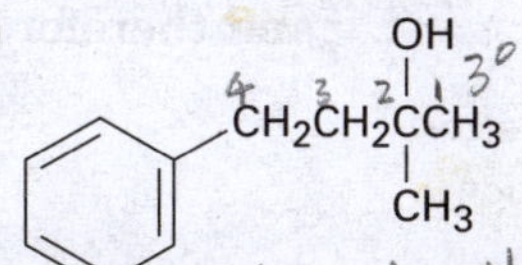

(c)

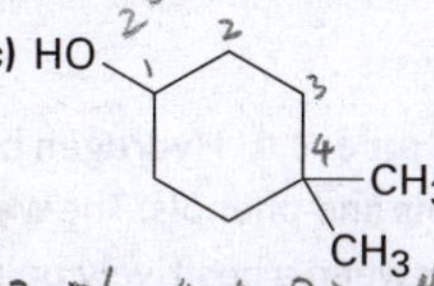

(d)

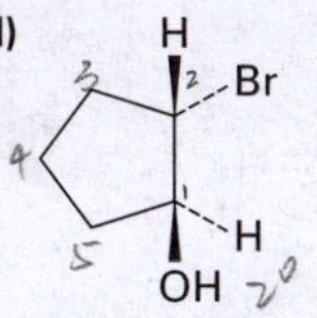

(e)

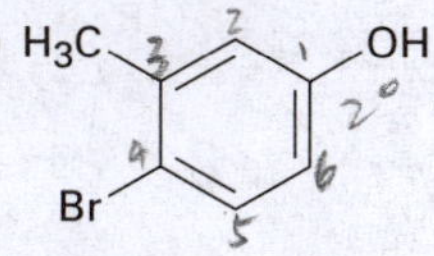

(f)

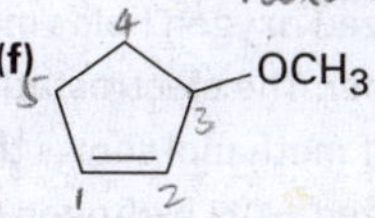

Problem 8.2 Identify the alcohols in Problem 8.1 as primary, secondary, or tertiary.

Problem 8.3 Draw structures corresponding to the following IUPAC names:

(a) 2-Methylhexan-2-ol **(b)** Hexane-1,5-diol
(c) 2-Ethylbut-2-en-1-ol **(d)** Cyclohex-3-en-1-ol
(e) *o*-Bromophenol **(f)** 2,4,6-Trinitrophenol

Problem 8.4 Name the following ethers by IUPAC rules:

(a)

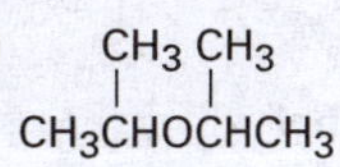

(b)

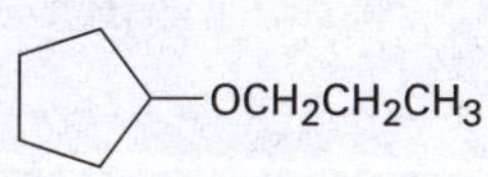

(c)

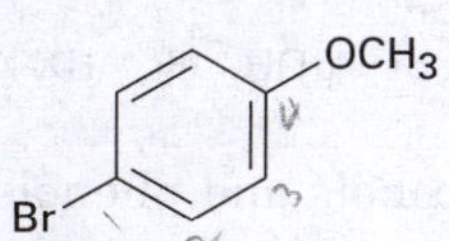

(d)

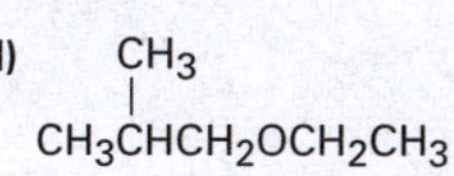

8.2 Properties of Alcohols and Phenols: Hydrogen Bonding and Acidity

Alcohols, phenols, and ethers can be thought of as organic derivatives of water in which one or both of the hydrogens have been replaced by organic parts: H—O—H becomes R—O—H, Ar —O—H, or R—O—R′. Thus, all three classes of compounds have nearly the same geometry as water. The C–O–H or C–O–C bond angles are approximately tetrahedral—109° in methanol and 112° in dimethyl ether, for instance—and the oxygen atoms are sp^3-hybridized.

Also like water, alcohols and phenols have higher boiling points than might be expected. Propan-1-ol and butane have similar molecular weights, for instance, yet propan-1-ol boils at 97.2 °C and butane boils at −0.5 °C. Similarly, phenol boils at 181.9 °C but toluene boils at 110.6 °C.

Alcohols and phenols have unusually high boiling points because, like water, they form hydrogen bonds. The positively polarized –OH hydrogen of one molecule is attracted to a lone pair of electrons on the negatively polarized oxygen of another molecule, resulting in a weak force that holds the molecules together (Figure 8.1). These forces must be overcome for a molecule to break free from the liquid and enter the vapor, so the boiling temperature is raised. Ethers, because they lack hydroxyl groups, can't form hydrogen bonds and therefore have lower boiling points.

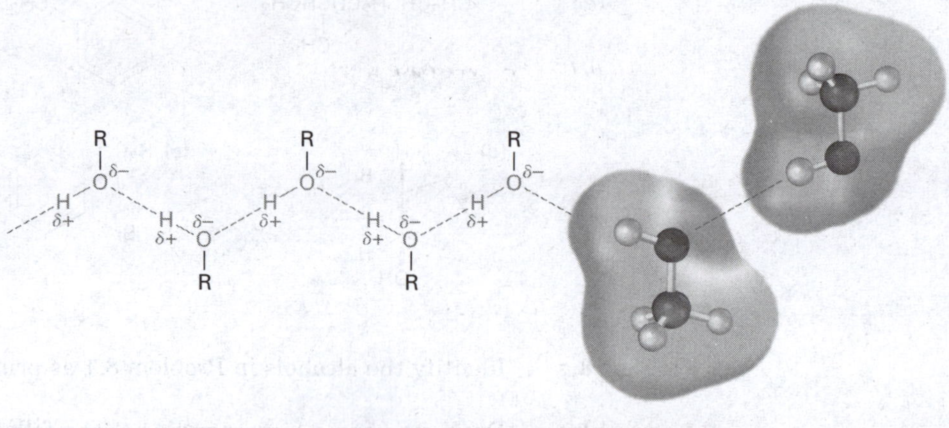

Figure 8.1 Hydrogen bonding in alcohols and phenols. The weak attraction between a positively polarized –OH hydrogen and a negatively polarized oxygen holds molecules together. The electrostatic potential map of methanol shows the positively polarized –OH hydrogen (blue) and the negatively polarized oxygen (red).

Another similarity with water is that alcohols and phenols are both weakly basic and weakly acidic. As weak Lewis bases, alcohols and phenols are reversibly protonated by strong acids to yield oxonium ions, ROH_2^+.

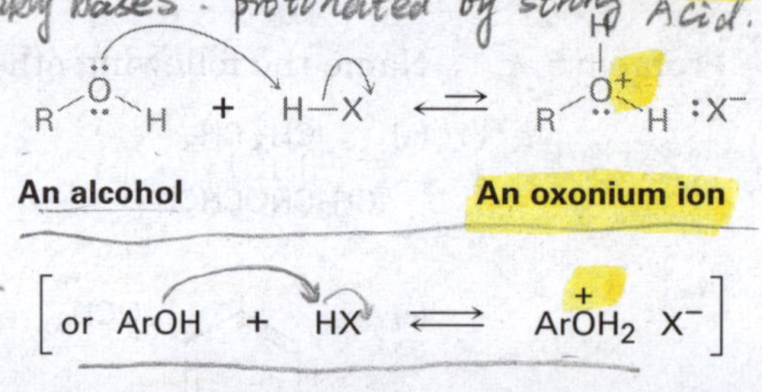

As weak acids, alcohols and phenols dissociate to a slight extent in dilute aqueous solution by donating a proton to water, generating H_3O^+ and an **alkoxide ion (RO^-)** or a **phenoxide ion (ArO^-)**.

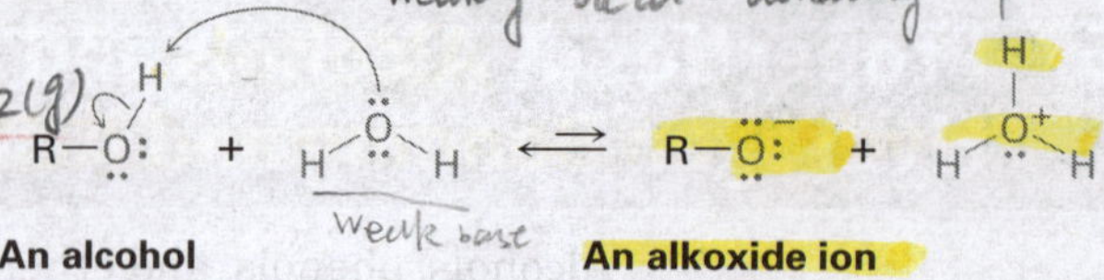

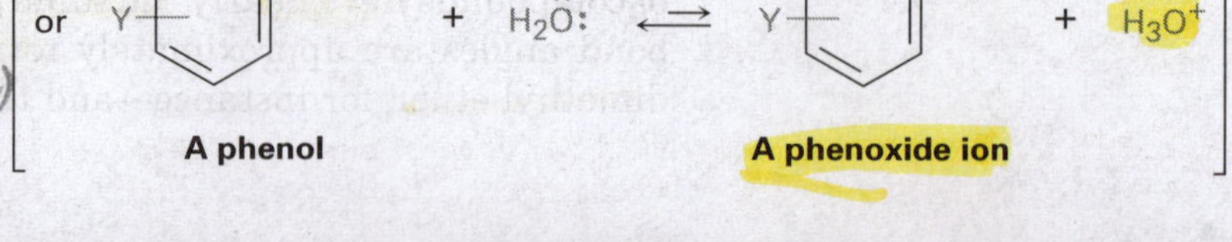

Recall from the earlier discussion of acidity in Sections 1.10 and 1.11 that the strength of any acid HA in water can be expressed by an acidity constant, K_a.

$$K_a = \frac{[A^-][H_3O^+]}{[HA]} \qquad pK_a = -\log K_a$$

Compounds with a smaller K_a and larger pK_a are less acidic, whereas compounds with a larger K_a and smaller pK_a are more acidic. Table 8.1 gives the pK_a values of some common alcohols and phenols.

Table 8.1 Acidity of Some Alcohols and Phenols

Compound	pK_a	
$(CH_3)_3COH$	18.00	Weaker acid
CH_3CH_2OH	16.00	↓
H_2O	15.74	↓
CH_3OH	15.54	↓
p-Methylphenol	10.17	↓
Phenol	9.89	↓
p-Nitrophenol	7.15	Stronger acid

The data in Table 8.1 show that alcohols are about as acidic as water. Thus, they don't react with weak bases such as amines or bicarbonate ion and they react to only a limited extent with metal hydroxides, such as NaOH. They do, however, react with alkali metals to yield alkoxides that are themselves strong bases. Alkoxides are named commonly by replacing the *-ane* suffix of the corresponding alkane with *-oxide:* methane gives methoxide, for instance. They are named systematically by adding the *-ate* suffix to the name of the alcohol. Methanol becomes methanolate, for instance.

$$2\ CH_3OH + 2\ Na \longrightarrow 2\ CH_3O^-\ Na^+ + H_2$$

Methanol → **Sodium methanolate (sodium methoxide)**

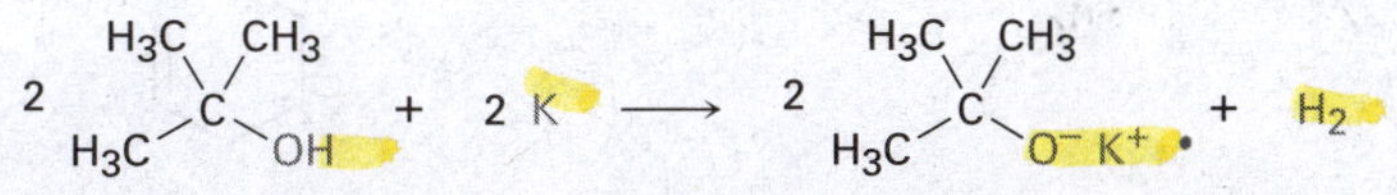

2-Methylpropan-2-ol (*tert*-butyl alcohol) → **Potassium 2-methylpropan-2-olate (potassium *tert*-butoxide)**

Phenols are about a million times more acidic than alcohols and are therefore soluble in dilute aqueous NaOH.

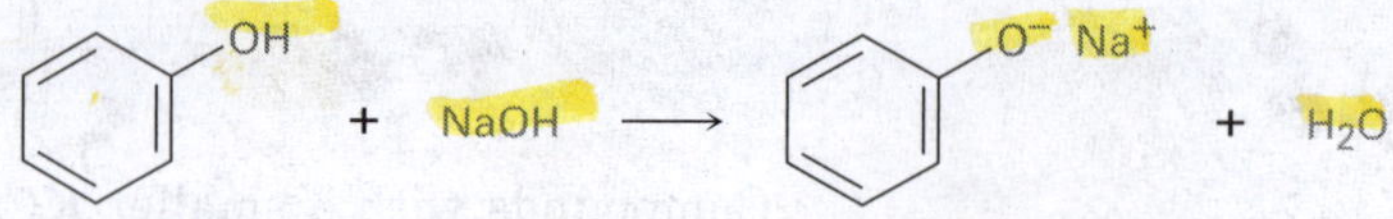

Phenol

Sodium phenoxide (sodium phenolate)

Phenols are more acidic than alcohols because the phenoxide anion is resonance-stabilized by the aromatic ring. Sharing the negative charge over the ring increases the stability of the phenoxide anion and thus increases the tendency of the corresponding phenol to dissociate (Figure 8.2).

Figure 8.2 A resonance-stabilized phenoxide ion is more stable than an alkoxide ion. Electrostatic potential maps show how the negative charge is more concentrated on oxygen in the methoxide ion but is less concentrated on the phenoxide oxygen.

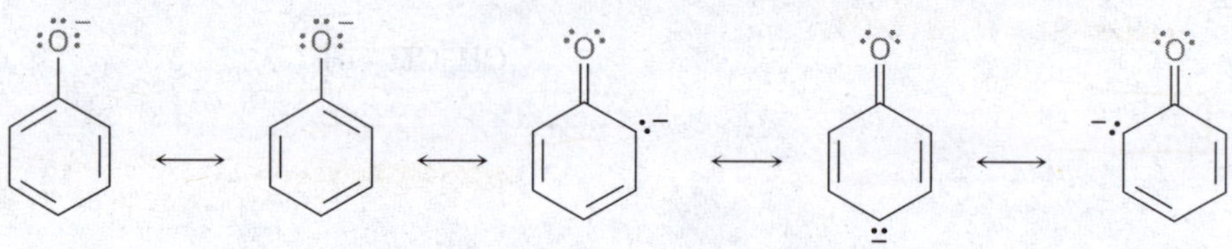

Phenoxide ion

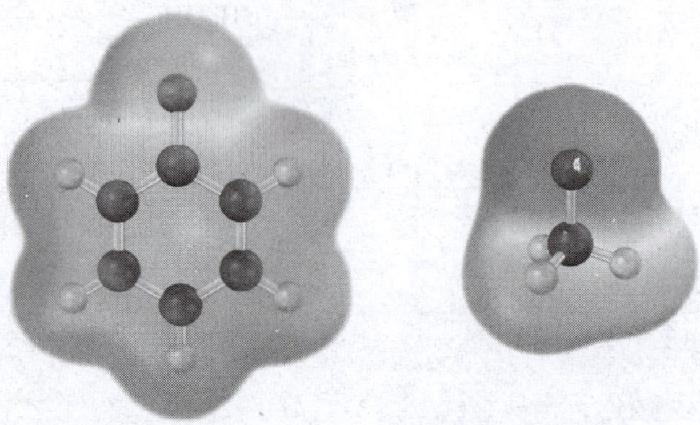

Methoxide ion

8.3 Synthesis of Alcohols from Carbonyl Compounds

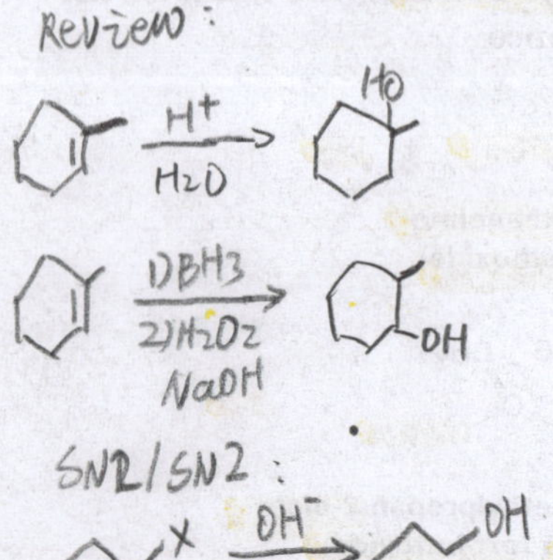

Alcohols occupy a central position in organic chemistry. They can be prepared from many other kinds of compounds (alkenes, alkyl halides, ketones, aldehydes, and esters, among others), and they can be transformed into an equally wide assortment of compounds. We saw in Section 4.3, for instance, that alcohols can be prepared by hydration of alkenes.

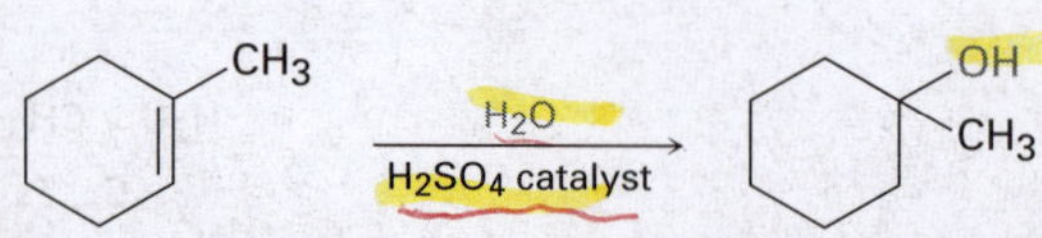

1-Methylcyclohexene

1-Methylcyclohexanol

Reduction of Carbonyl Compounds

The most general method for preparing alcohols, both in the laboratory and in living organisms, is by the reduction of a carbonyl compound. Just as reduction of an alkene adds hydrogen to a C=C bond to give an alkane (Section 4.5), reduction of a carbonyl compound adds hydrogen to a C=O bond to give an alcohol. All kinds of carbonyl compounds can be reduced, including aldehydes, ketones, carboxylic acids, and esters.

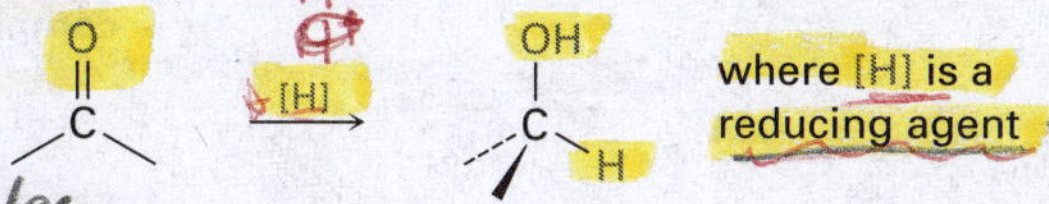

A carbonyl compound **An alcohol**

Reduction of Aldehydes and Ketones

Aldehydes are reduced to give primary alcohols, and ketones are reduced to give secondary alcohols.

An aldehyde **A primary alcohol** **A ketone** **A secondary alcohol**

Many reducing reagents are available, but sodium borohydride, $NaBH_4$, is usually chosen because of its safety. It is a white, crystalline solid that can be weighed in the open atmosphere and used in either water or alcohol solution.

Aldehyde reduction

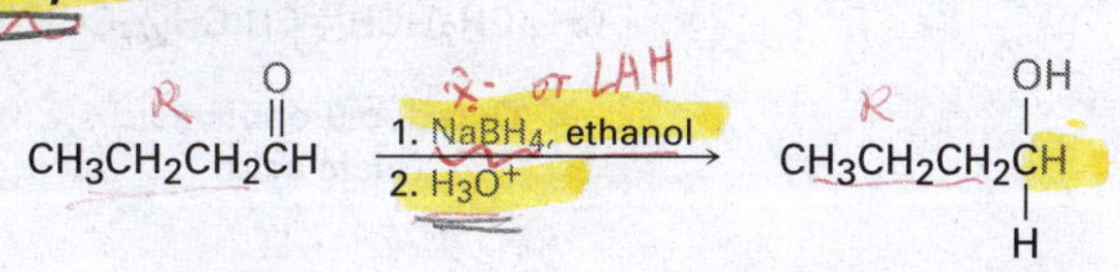

Butanal **Butan-1-ol (85%) (a 1° alcohol)**

Ketone reduction

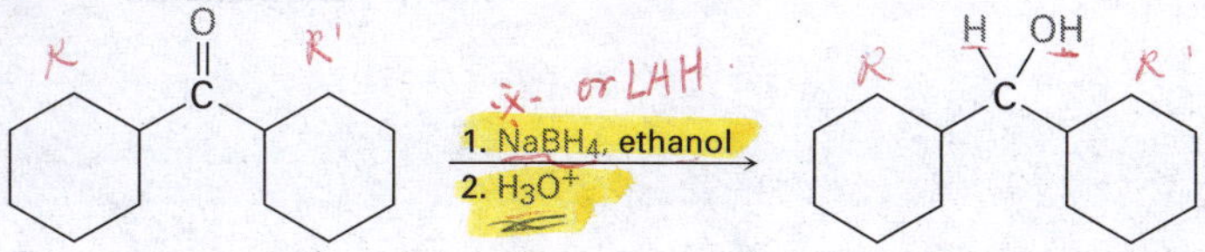

Dicyclohexyl ketone **Dicyclohexylmethanol (88%) (a 2° alcohol)**

We'll defer a discussion of the mechanisms of these reductions until Section 9.6 but might note for now that they involve the addition of a nucleophilic hydride ion (:H^-) to the positively polarized, electrophilic carbon atom of the carbonyl group. The initial product is an alkoxide ion, which is protonated by addition of H_3O^+ in a second step to yield the alcohol product.

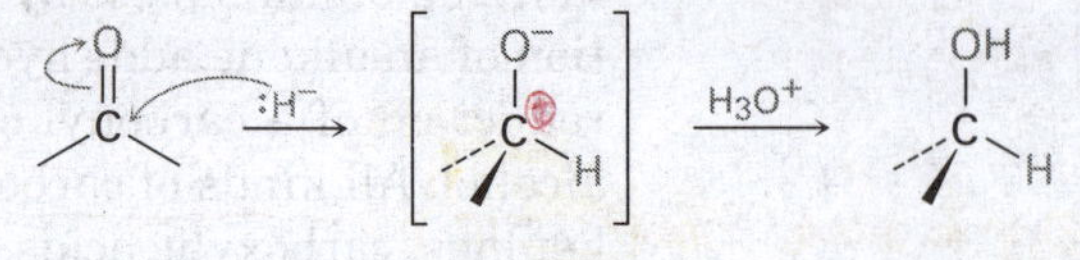

A carbonyl compound — An alkoxide ion intermediate — An alcohol

Reduction of Carboxylic Acids and Esters

Esters and carboxylic acids are reduced to give primary alcohols.

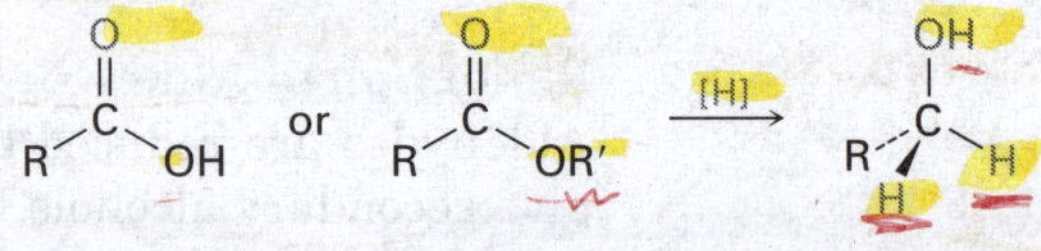

A carboxylic acid — An ester — A primary alcohol

These reactions aren't as rapid as the reductions of aldehydes and ketones, so the more powerful reducing agent lithium aluminum hydride ($LiAlH_4$) is used rather than $NaBH_4$. ($LiAlH_4$ will also reduce aldehydes and ketones.) Note that only one hydrogen is added to the carbonyl carbon atom during the reduction of an aldehyde or ketone, but two hydrogens are added to the carbonyl carbon during reduction of an ester or carboxylic acid.

Carboxylic acid reduction

$$CH_3(CH_2)_7CH{=}CH(CH_2)_7\overset{O}{\overset{\|}{C}}OH \xrightarrow[2.\ H_3O^+]{1.\ LiAlH_4,\ ether} CH_3(CH_2)_7CH{=}CH(CH_2)_7CH_2OH$$

Octadec-9-enoic acid (oleic acid) → **Octadec-9-en-1-ol (87%)**

Ester reduction

$$CH_3CH_2CH{=}CH\overset{O}{\overset{\|}{C}}OCH_3 \xrightarrow[2.\ H_3O^+]{1.\ LiAlH_4,\ ether} CH_3CH_2CH{=}CHCH_2OH + CH_3OH$$

Methyl pent-2-enoate → **Pent-2-en-1-ol (91%)**

Worked Example 8.1 — Predicting the Product of a Reduction Reaction

Predict the product of the following reaction:

$$CH_3CH_2CH_2\overset{O}{\overset{\|}{C}}CH_2CH_3 \xrightarrow[2.\ H_3O^+]{1.\ NaBH_4} ?$$

Strategy Ketones are reduced by treatment with $NaBH_4$ to yield secondary alcohols. Thus, reduction of hexan-3-one yields hexan-3-ol.

Solution

$$\underset{\textbf{Hexan-3-one}}{CH_3CH_2CH_2\overset{O}{\overset{\|}{C}}CH_2CH_3} \xrightarrow[\text{2. } H_3O^+]{\text{1. } NaBH_4} \underset{\textbf{Hexan-3-ol}}{CH_3CH_2CH_2\overset{OH}{\overset{|}{C}H}CH_2CH_3}$$

Worked Example 8.2 Synthesizing an Alcohol by Reduction of a Carbonyl Compound

What carbonyl compound(s) might you reduce to obtain the following alcohols?

(a)

$$CH_3CH_2\overset{CH_3}{\overset{|}{C}H}CH_2\overset{OH}{\overset{|}{C}H}CH_3$$

(b)

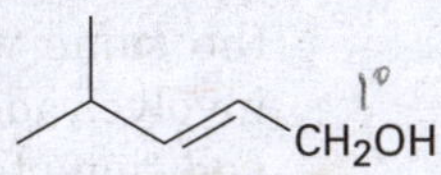

Strategy Identify the target alcohol as primary, secondary, or tertiary. A primary alcohol can be prepared by reduction of an aldehyde, an ester, or a carboxylic acid; a secondary alcohol can be prepared by reduction of a ketone; and a tertiary alcohol can't be prepared by reduction.

Solution **(a)** The target molecule is a secondary alcohol, which can be prepared by reduction of a ketone with $NaBH_4$ (or $LiAlH_4$).

$$CH_3CH_2\overset{CH_3}{\overset{|}{C}H}CH_2\overset{O}{\overset{\|}{C}}CH_3 \xrightarrow[\text{2. } H_3O^+]{\text{1. } NaBH_4 \text{ or } LiAlH_4} CH_3CH_2\overset{CH_3}{\overset{|}{C}H}CH_2\overset{OH}{\overset{|}{C}H}CH_3$$

(b) The target molecule is a primary alcohol, which can be prepared by reduction of an aldehyde, an ester, or a carboxylic acid with $LiAlH_4$.

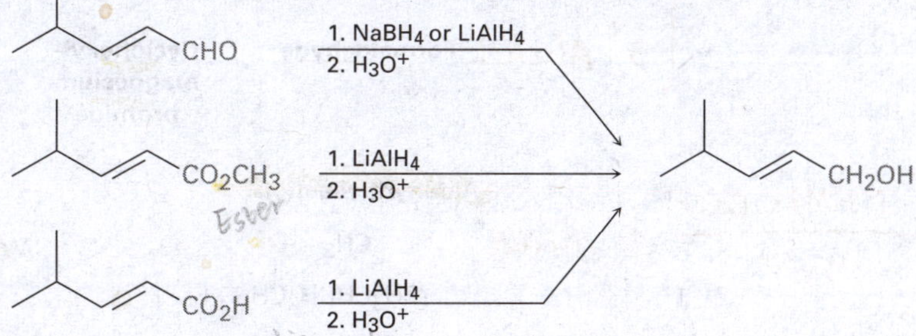

Problem 8.5 How would you carry out the following reactions?

(a)

$$CH_3\overset{O}{\overset{\|}{C}}CH_2CH_2\overset{O}{\overset{\|}{C}}OCH_3 \xrightarrow{?} CH_3\overset{OH}{\overset{|}{C}H}CH_2CH_2\overset{O}{\overset{\|}{C}}OCH_3$$

(b)

$$CH_3\overset{O}{\overset{\|}{C}}CH_2CH_2\overset{O}{\overset{\|}{C}}OCH_3 \xrightarrow{?} CH_3\overset{OH}{\overset{|}{C}H}CH_2CH_2CH_2OH$$

Problem 8.6 What carbonyl compounds give the following alcohols on reduction with $LiAlH_4$? Show all possibilities.

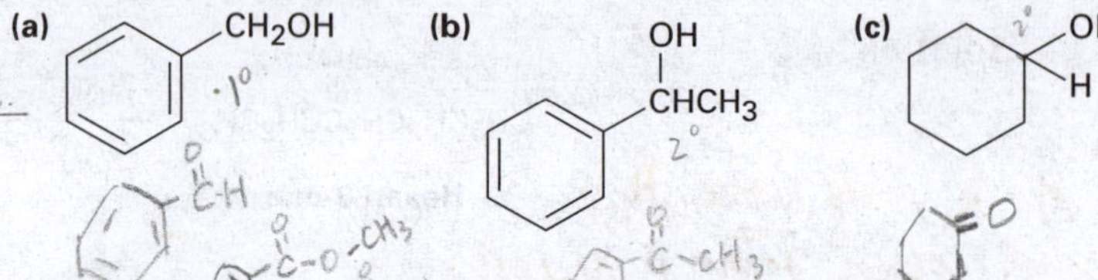

Grignard Reactions of Carbonyl Compounds

Grignard reagents (RMgX), prepared by reaction of organohalides with magnesium (Section 7.3), react with carbonyl compounds to yield alcohols in much the same way that hydride reducing agents do. Just as carbonyl reduction involves addition of a hydride ion to the C=O bond, Grignard reaction involves addition of a carbanion ($R:^{-}\ ^{+}MgX$).

$$R{-}X + Mg \longrightarrow \overset{\delta-}{R}{-}\overset{\delta+}{MgX}$$

A Grignard reagent

R = 1°, 2°, or 3° alkyl, aryl, or vinylic
X = Cl, Br, I

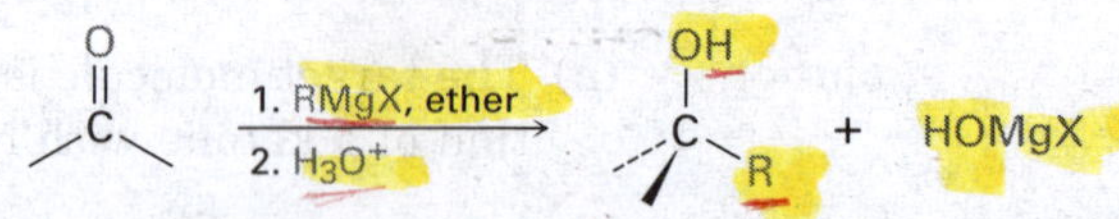

Formaldehyde, $H_2C{=}O$, reacts with Grignard reagents giving primary alcohols, aldehydes give secondary alcohols, and ketones give tertiary alcohols.

Formaldehyde reaction

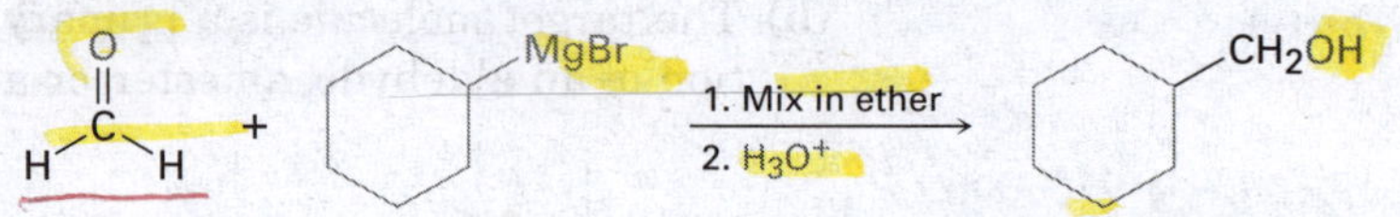

Formaldehyde **Cyclohexylmagnesium bromide** **Cyclohexylmethanol (65%) (a 1° alcohol)**

Aldehyde reaction

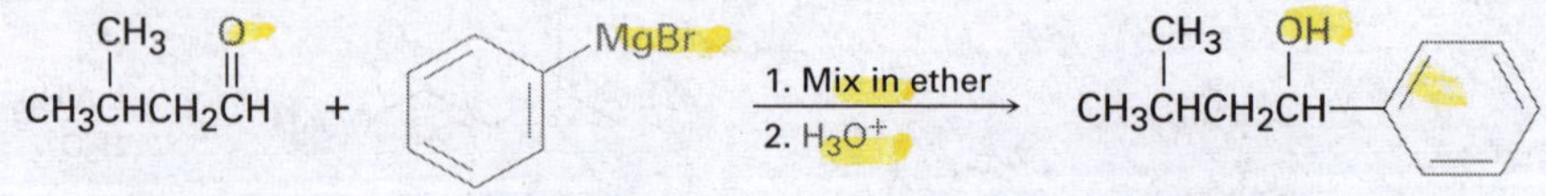

3-Methylbutanal **Phenylmagnesium bromide** **3-Methyl-1-phenylbutan-1-ol (73%) (a 2° alcohol)**

Ketone reaction

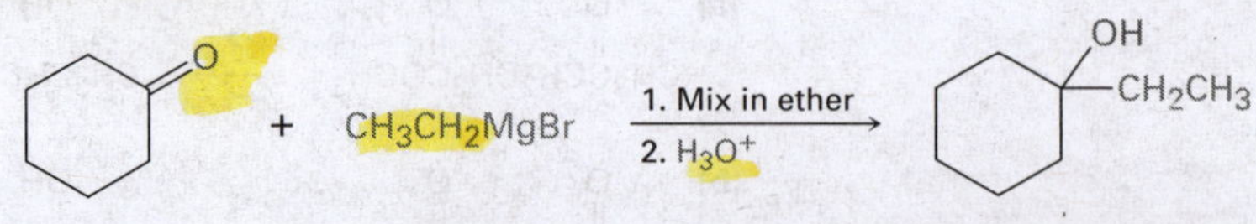

Cyclohexanone **Ethylmagnesium bromide** **1-Ethylcyclohexanol (89%) (a 3° alcohol)**

Esters react with Grignard reagents to yield tertiary alcohols in which two of the substituents bonded to the hydroxyl-bearing carbon have come from the Grignard reagent, just as esters are reduced by $LiAlH_4$ with addition of two hydrogens. As with the reduction of carbonyl compounds, we'll defer a discussion of the mechanism of Grignard additions until Section 9.6.

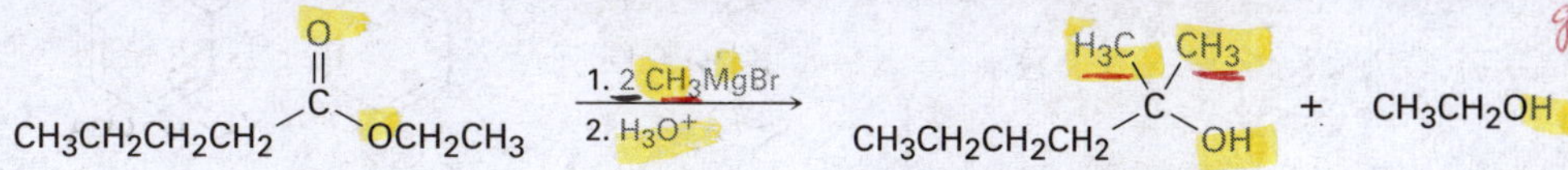

Ethyl pentanoate → 2-Methylhexan-2-ol (85%) (a 3° alcohol)

Worked Example 8.3 Using a Grignard Reaction to Synthesize an Alcohol

How could you use the reaction of a Grignard reagent with a carbonyl compound to synthesize 2-methylpentan-2-ol?

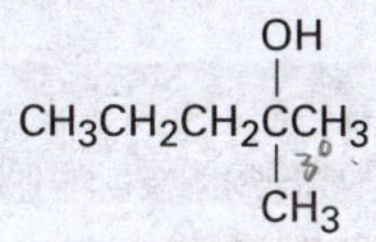

2-Methylpentan-2-ol

Strategy Identify the alcohol as 1°, 2°, or 3°. If the alcohol is secondary, the starting carbonyl compound must be an aldehyde. If the alcohol is tertiary and the three groups are all different, the starting carbonyl compound must be a ketone. If the alcohol is tertiary and two of the three groups are identical, the starting carbonyl compound might be either a ketone or an ester.

Solution In the present instance, the product is a tertiary alcohol with two methyl groups and one propyl group. Starting from a ketone, the possibilities are addition of methylmagnesium bromide to pentan-2-one and addition of propylmagnesium bromide to acetone.

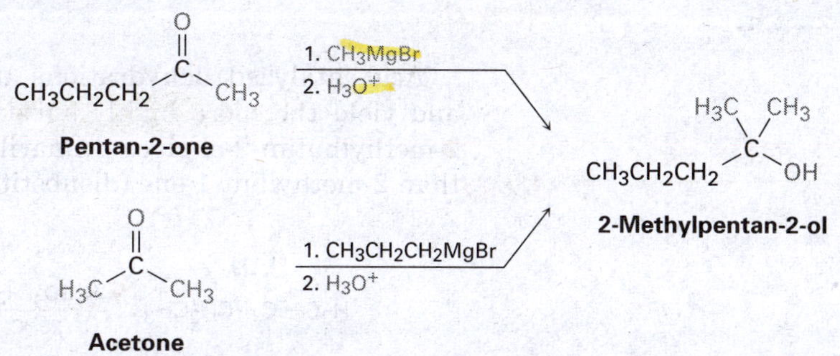

Starting from an ester, the only possibility is addition of methylmagnesium bromide to an ester of butanoic acid, such as methyl butanoate.

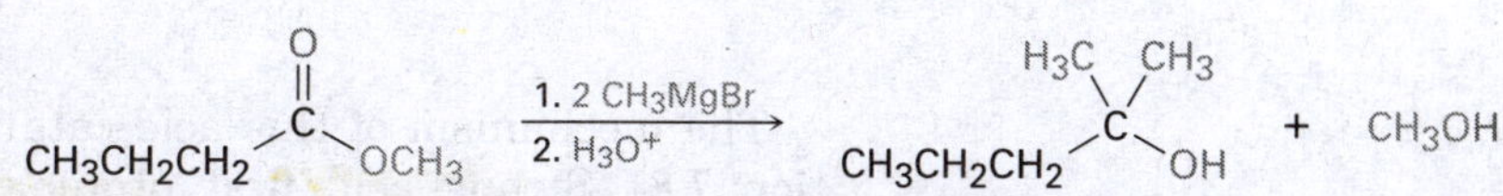

Methyl butanoate → 2-Methylpentan-2-ol

Problem 8.7 Show the products obtained from addition of CH_3MgBr to the following carbonyl compounds:

(a)

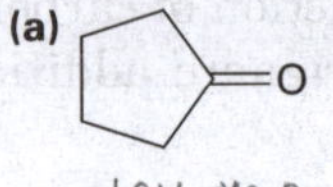

(b)

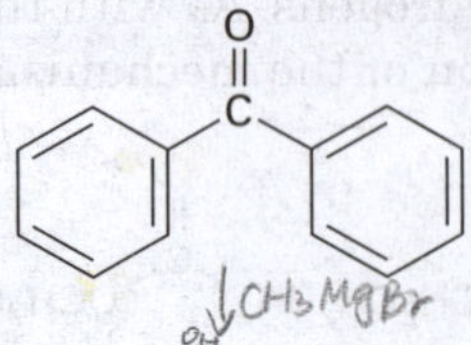

(c) $CH_3CH_2CH_2\overset{\overset{O}{\|}}{C}CH_2CH_3$

Problem 8.8 How might you use a Grignard reaction to prepare the following alcohols:

(a) $CH_3\overset{\overset{OH}{|}}{\underset{\underset{CH_3}{|}}{C}}CH_3$

(b)

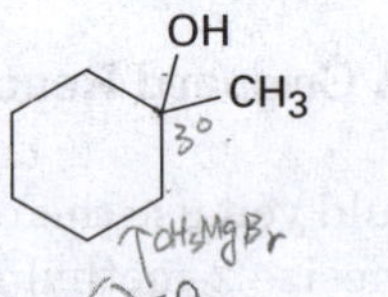

(c) $CH_3CH_2\overset{\overset{OH}{|}}{\underset{\underset{CH_3}{|}}{C}}CH_2CH_3$

8.4 Reactions of Alcohols

Dehydration of Alcohols

Alcohols undergo *dehydration*—the elimination of H_2O—to give alkenes. A method that works particularly well for secondary and tertiary alcohols is treatment with a strong acid. For example, when 1-methylcyclohexanol is treated with aqueous sulfuric acid, dehydration occurs to yield 1-methylcyclohexene.

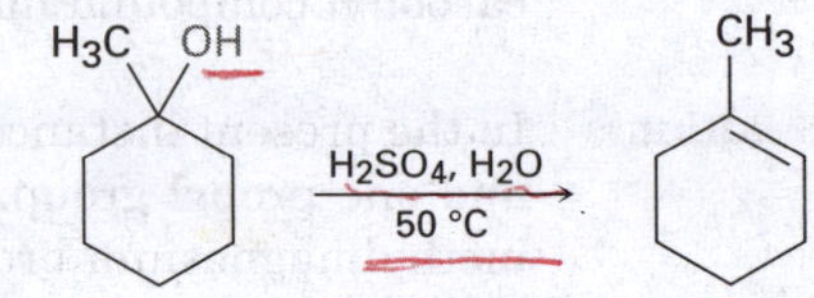

1-Methylcyclohexanol 1-Methylcyclohexene (91%)

Acid-catalyzed dehydrations usually follow Zaitsev's rule (Section 7.7) and yield the more highly substituted alkene as the major product. Thus, 2-methylbutan-2-ol gives primarily 2-methylbut-2-ene (trisubstituted) rather than 2-methylbut-1-ene (disubstituted).

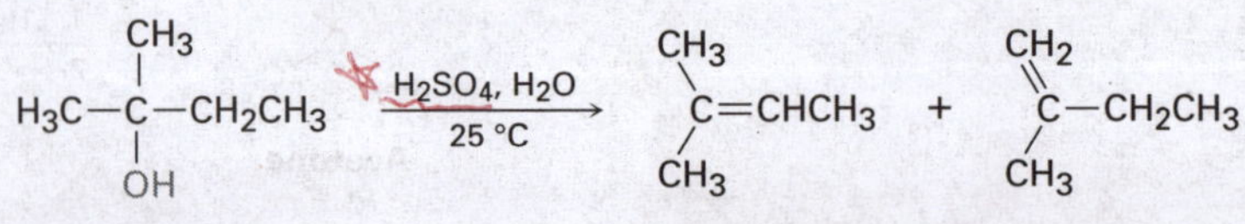

2-Methylbutan-2-ol **2-Methylbut-2-ene (trisubstituted)** Major product **2-Methylbut-1-ene (disubstituted)** Minor product

The mechanism of this acid-catalyzed dehydration is an E1 process (Section 7.8). Strong acid first protonates the alcohol oxygen, the protonated intermediate spontaneously loses water to generate a carbocation, and loss of H^+ from a neighboring carbon atom then yields the alkene product (Figure 8.3).

MECHANISM

Figure 8.3 Mechanism of the acid-catalyzed dehydration of a tertiary alcohol to yield an alkene. The process is an E1 reaction and involves a carbocation intermediate.

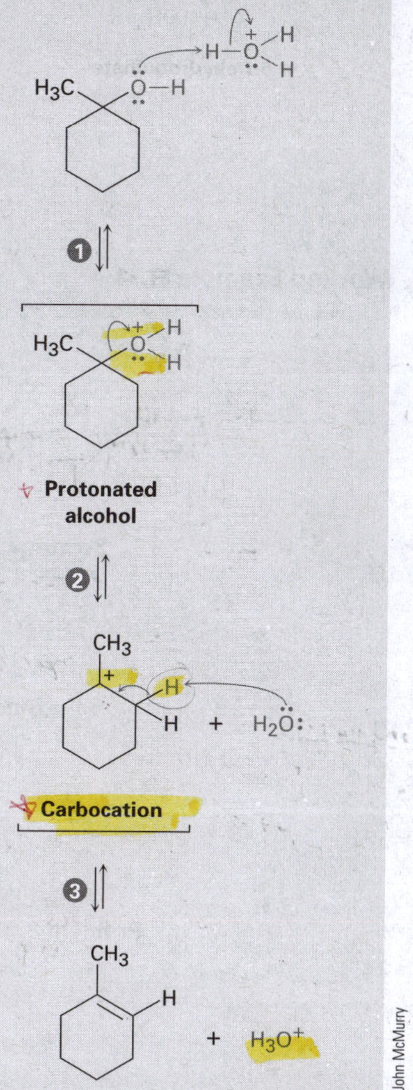

❶ Two electrons from the oxygen atom bond to H^+, yielding a protonated alcohol intermediate.

❷ The carbon–oxygen bond breaks, and the two electrons from the bond stay with oxygen, leaving a carbocation intermediate.

❸ Two electrons from a neighboring carbon–hydrogen bond form the alkene π bond, and H^+ (a proton) is eliminated.

As noted previously in Section 7.10, biological dehydrations are also common and usually occur by an E1cB mechanism on a substrate in which the –OH group is two carbons away from a carbonyl group. An example occurs in the biosynthesis of the aromatic amino acid tyrosine. A base (:B) first abstracts a proton from the carbon adjacent to the carbonyl group, and the anion intermediate then expels the –OH group with simultaneous protonation by an acid (HA) to form water.

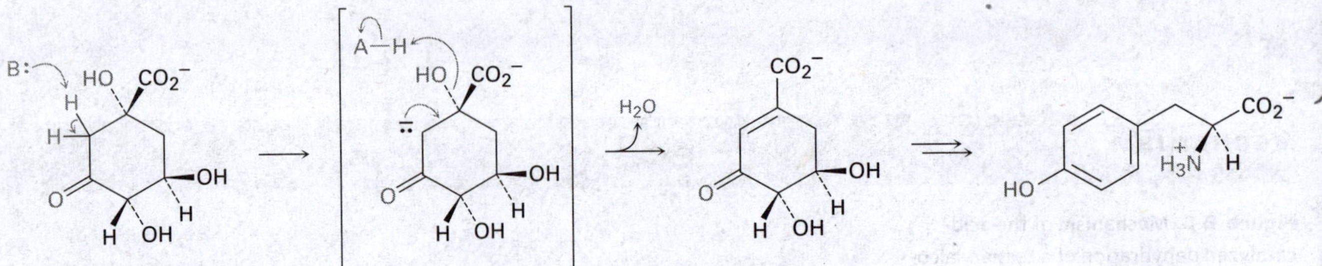

5-Dehydroquinate — Anion intermediate — 5-Dehydroshikimate — Tyrosine

Worked Example 8.4 Predicting the Product of a Dehydration Reaction

Predict the major product of the following reaction:

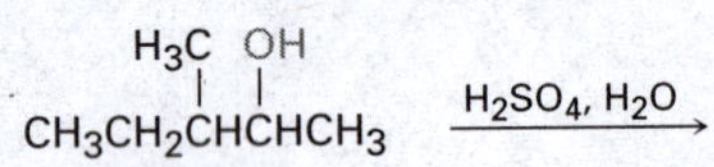

Strategy Treatment of an alcohol with H_2SO_4 leads to dehydration and formation of the more highly substituted alkene product (Zaitsev's rule). Thus, dehydration of 3-methylpentan-2-ol yields 3-methylpent-2-ene as the major product rather than 3-methylpent-1-ene.

Solution

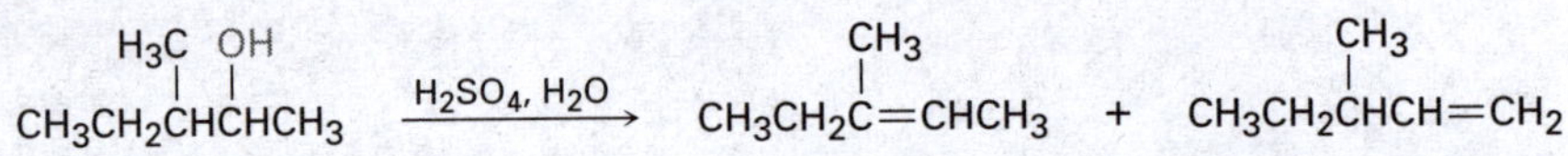

3-Methylpentan-2-ol — 3-Methylpent-2-ene (major) — 3-Methylpent-1-ene (minor)

Problem 8.9 Predict the products you would expect from the following reactions. Indicate the major product in each case.

(a) 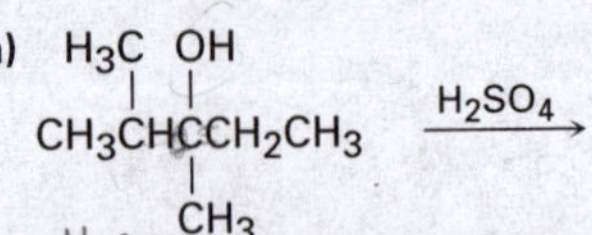

(b) 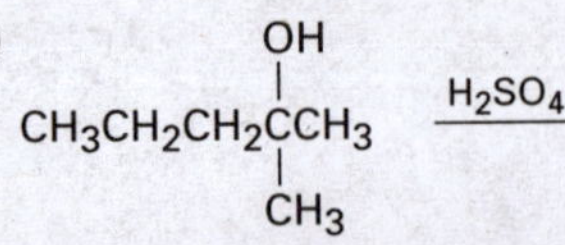

Problem 8.10 What alcohols might the following alkenes be made from?

(a) 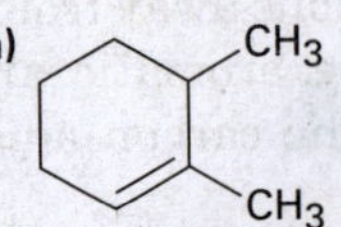

(b) $CH_3CH_2CH{=}CHCH_2CH_2CH_3$

Oxidation of Alcohols

Perhaps the most valuable reaction of alcohols is their oxidation to yield carbonyl compounds—the opposite of the reduction of carbonyl compounds to give alcohols. Primary alcohols yield aldehydes or carboxylic acids, and secondary alcohols yield ketones, but tertiary alcohols don't normally react with oxidizing agents.

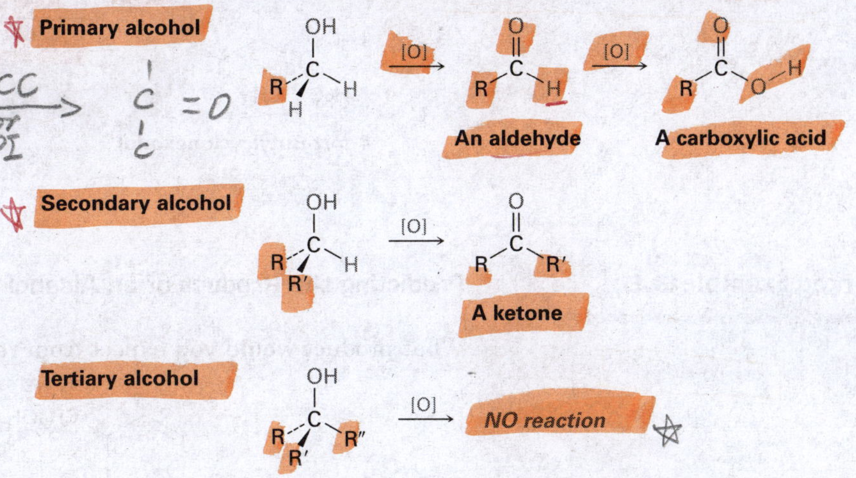

Primary alcohols are oxidized either to aldehydes or to carboxylic acids, depending on the reagent used. Older methods were often based on Cr(VI) reagents, such as CrO_3 or $Na_2Cr_2O_7$, but the most common current choice for preparing an aldehyde from a primary alcohol in the laboratory is to use a *periodinane,* which contains an iodine atom in the +5 oxidation state. This reagent is too expensive for large-scale use in industry, however.

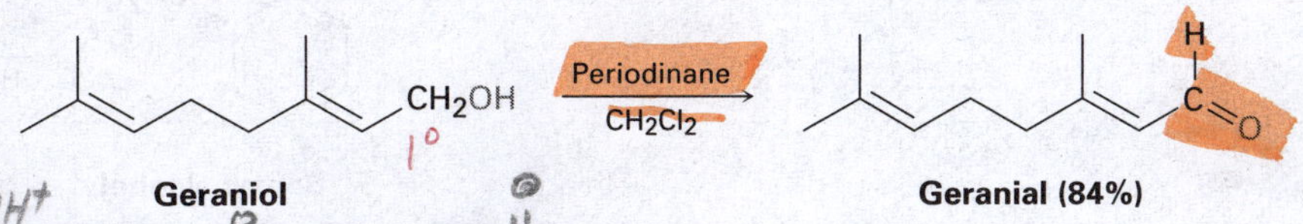

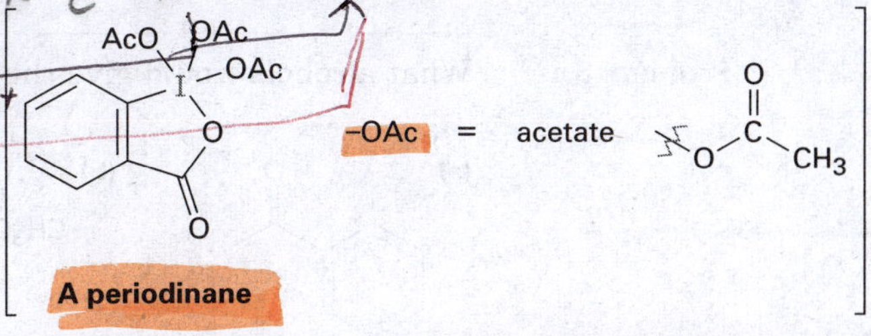

Many oxidizing agents, such as chromium trioxide (CrO_3) and sodium dichromate ($Na_2Cr_2O_7$) in aqueous acid solution, oxidize primary alcohols to carboxylic acids. Although aldehydes are intermediates in these oxidations, they usually can't be isolated because further oxidation takes place too rapidly.

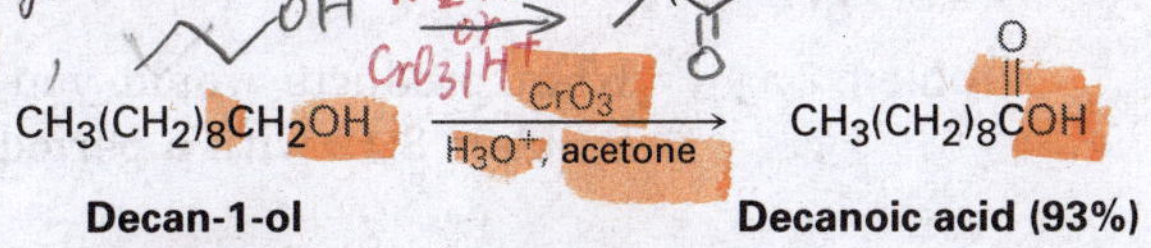

Secondary alcohols are oxidized to produce ketones. For a sensitive or costly alcohol, a periodinane is often used. For a large-scale oxidation, however, an inexpensive reagent such as CrO_3 or $Na_2Cr_2O_7$ in aqueous acetic acid is more economical.

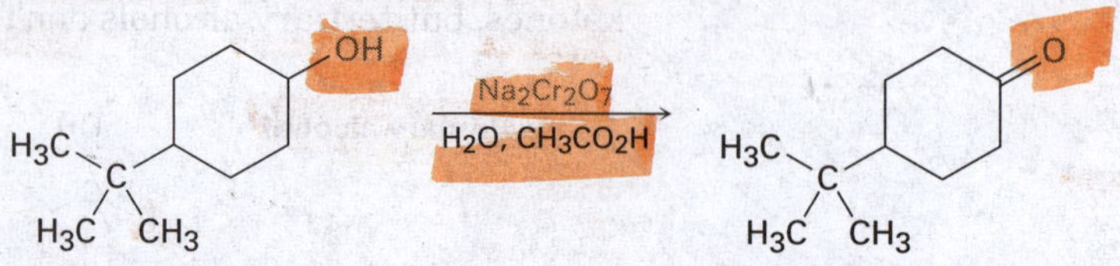

4-*tert*-Butylcyclohexanol — 4-*tert*-Butylcyclohexanone (91%)

Worked Example 8.5 — Predicting the Product of an Alcohol Oxidation

What product would you expect from reaction of benzyl alcohol with CrO_3?

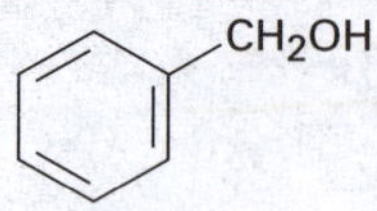

Benzyl alcohol

Strategy Treatment of a primary alcohol with CrO_3 yields a carboxylic acid. Thus, oxidation of benzyl alcohol yields benzoic acid.

Solution

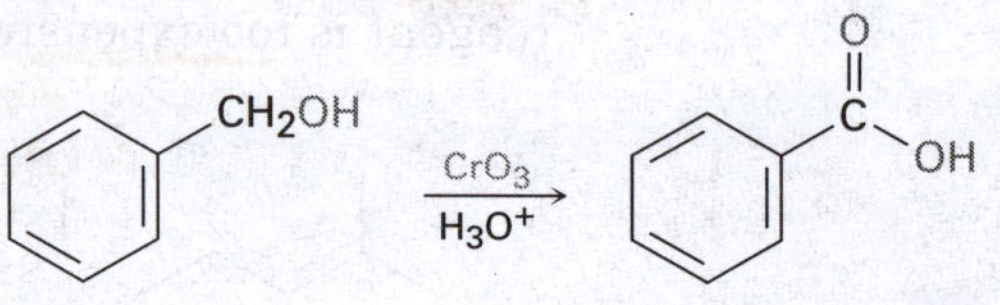

Benzyl alcohol — Benzoic acid

Problem 8.11 What alcohols would give the following products on oxidation?

(a) 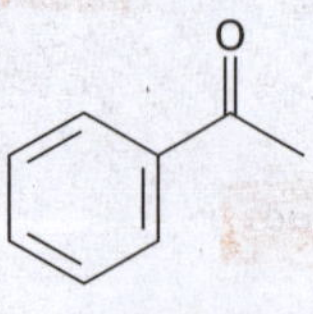 (b) 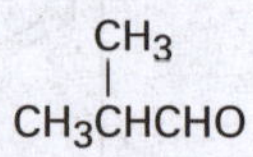 (c) 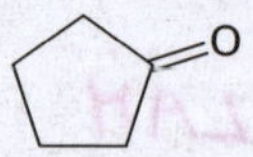

Problem 8.12 What products would you expect to obtain from oxidation of the following alcohols with CrO_3?

(a) Cyclohexanol **(b)** Hexan-1-ol **(c)** Hexan-2-ol

Problem 8.13 What products would you expect to obtain from oxidation of the alcohols in Problem 8.12 with a periodinane?

Conversion into Ethers

Alcohols are converted into ethers by formation of the corresponding alkoxide ion followed by reaction with an alkyl halide, a reaction known as the *Williamson ether synthesis.* As noted in Section 8.2, the alkoxide ion needed in the reaction can be prepared either by reaction of an alcohol with an alkali metal or by reaction with a strong base such as sodium hydride, NaH.

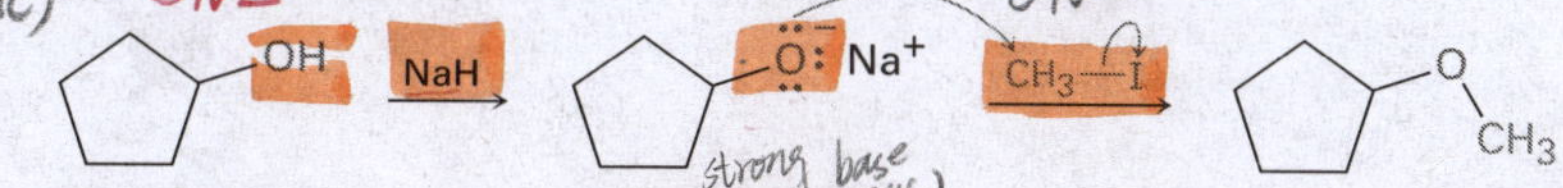

Cyclopentanol **Alkoxide ion** **Cyclopentyl methyl ether (74%)**

Mechanistically, the Williamson synthesis is an S_N2 reaction (Section 7.5) and occurs by nucleophilic substitution of halide ion by the alkoxide ion. As with all S_N2 reactions, primary alkyl halides work best because competitive E2 elimination of HX can occur with more hindered substrates. Unsymmetrical ethers are therefore best prepared by reaction of the more hindered alkoxide partner with the less hindered alkyl halide partner, rather than vice versa. *tert*-Butyl methyl ether, for example, is best prepared by reaction of *tert*-butoxide ion with iodomethane, rather than by reaction of methoxide ion with 2-chloro-2-methylpropane.

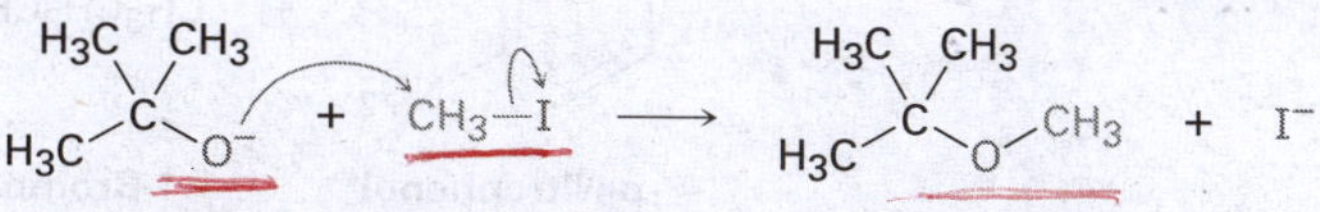

***tert*-Butoxide** **Iodomethane** ***tert*-Butyl methyl ether**

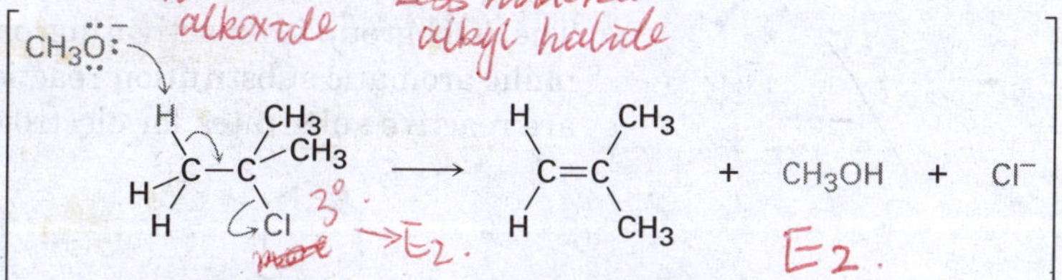

2-Chloro-2-methylpropane **2-Methylpropene**

Problem 8.14 How would you prepare the following ethers?

(a) $CH_3OCH_2CH_2CH_3$ **(b)**

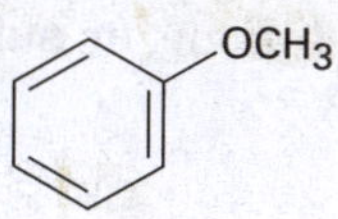

(c)

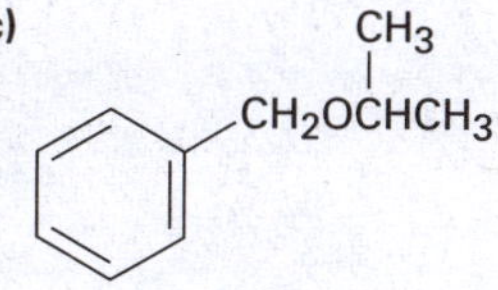

Problem 8.15 Rank the following alkyl halides in order of their expected reactivity toward an alkoxide ion in the Williamson ether synthesis: bromoethane, 2-bromopropane, chloroethane, 2-chloro-2-methylpropane.

Problem 8.16 Draw the structure of the carbonyl compound(s) from which the following alcohol might have been prepared, and show the products you would obtain by treatment of the alcohol with (i) Na metal, followed by CH_3I, (ii) $SOCl_2$, and (iii) CrO_3.

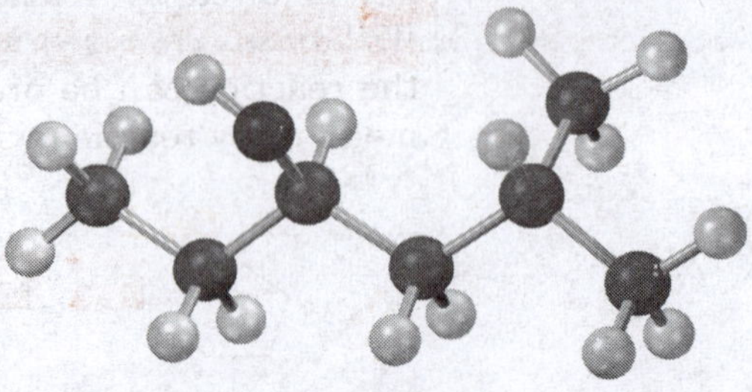

8.5 Reactions of Phenols

Alcohol-Like Reactions of Phenols

Phenols and alcohols behave very differently despite the fact that both have –OH groups. Phenols can't be dehydrated by treatment with acid and can't be converted into halides by treatment with HX. Phenols can, however, be converted into ethers by S_N2 reaction with alkyl halides in the presence of base. Williamson ether synthesis with phenols occurs easily because phenols are more acidic than alcohols and are more readily converted into their anions.

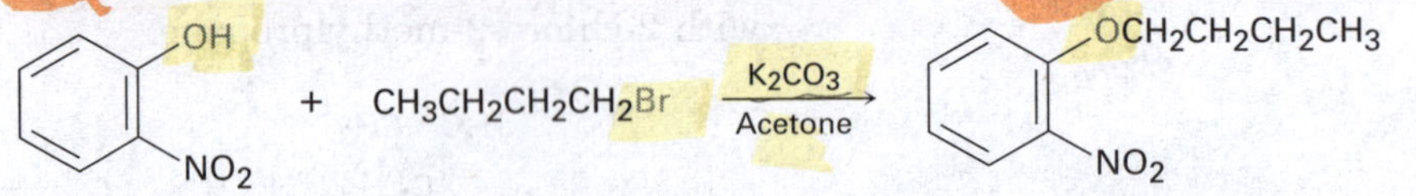

o-Nitrophenol **1-Bromobutane** **Butyl o-nitrophenyl ether (80%)**

Electrophilic Aromatic Substitution Reactions of Phenols

The –OH group is an activating, ortho- and para-directing substituent in electrophilic aromatic substitution reactions (Sections 5.6 and 5.7). As a result, phenols are reactive substrates for electrophilic halogenation, nitration, and sulfonation.

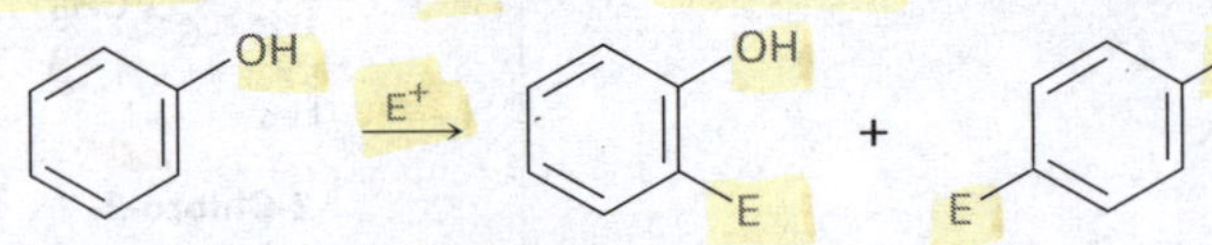

Oxidation of Phenols: Quinones

Phenols don't undergo oxidation in the same way that alcohols do because they don't have a hydrogen atom on the hydroxyl-bearing carbon. Instead, oxidation of a phenol with sodium dichromate yields a cyclohexa-2,5-diene-1,4-dione, or **quinone**.

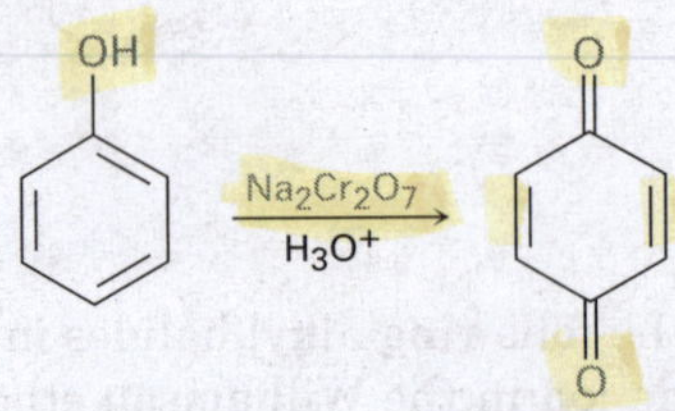

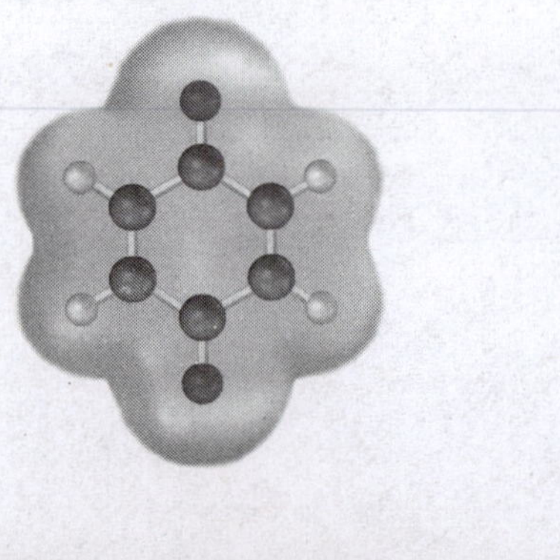

Phenol **Benzoquinone (79%)**

Quinones are an interesting and valuable class of compounds because of their oxidation–reduction properties. They can be easily reduced to **hydroquinones** (*p*-dihydroxybenzenes) by $NaBH_4$ or $SnCl_2$, and hydroquinones can be easily oxidized back to quinones by $Na_2Cr_2O_7$.

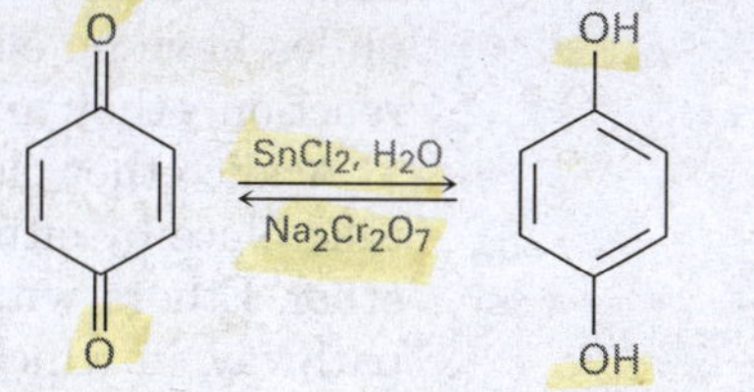

Benzoquinone **Hydroquinone**

The oxidation–reduction properties of quinones are crucial to the functioning of living cells, where compounds called *ubiquinones* act as biochemical oxidizing agents to mediate the electron-transfer processes involved in energy production. Ubiquinones, also called *coenzymes Q,* are components of the cells of all aerobic organisms, from the simplest bacterium to humans. They are so named because of their ubiquitous occurrence in nature.

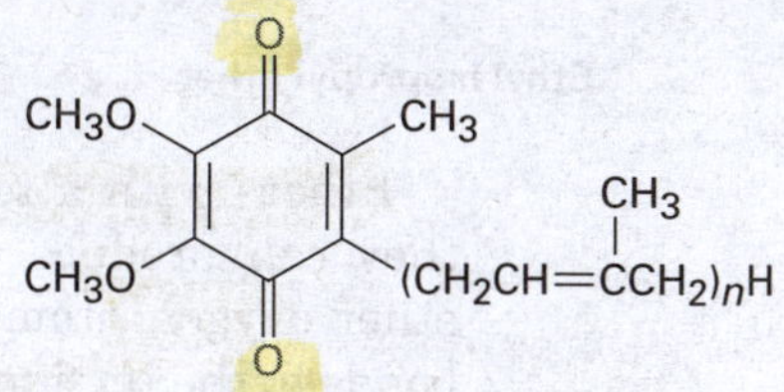

Ubiquinones (n = 1–10)

Ubiquinones function within the mitochondria of cells to mediate the respiration process in which electrons are transported from the biological reducing agent NADH to molecular oxygen. Through a complex series of steps, the ultimate result is a cycle whereby NADH is oxidized to NAD^+, O_2 is reduced to water, and energy is produced. Ubiquinone acts only as an intermediary and is itself unchanged.

Step 1

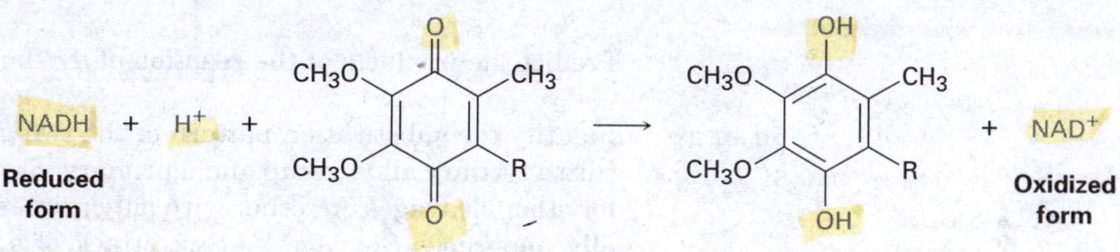

Step 2

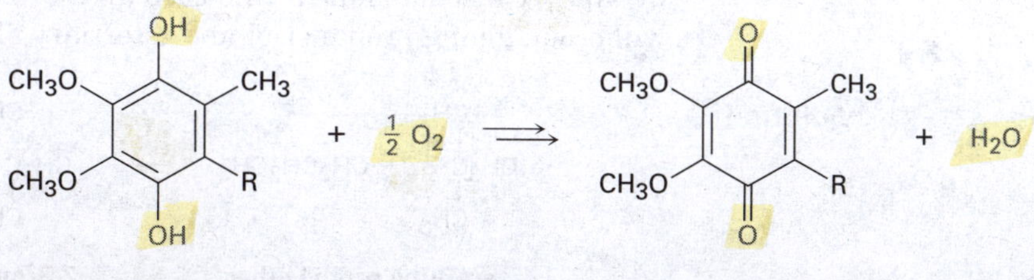

Net change: $NADH + \frac{1}{2} O_2 + H^+ \longrightarrow NAD^+ + H_2O$

8.6 Reactions of Ethers

Ethers are unreactive to most common reagents, a property that accounts for their frequent use as reaction solvents. Halogens, mild acids, bases, and nucleophiles have no effect on most ethers. In fact, ethers undergo only one general reaction—they are cleaved by strong acids such as aqueous HI or HBr.

Acidic ether cleavages are typical nucleophilic substitution reactions, which take place by either an S_N1 or S_N2 pathway, depending on the structure of the ether. Ethers with only primary and secondary alkyl groups react by an S_N2 pathway, in which nucleophilic halide ion attacks the protonated ether at the less highly substituted site. The ether oxygen atom stays with the more hindered alkyl group, and the halide bonds to the less hindered group. For example, ethyl isopropyl ether yields isopropyl alcohol and iodoethane on cleavage by HI.

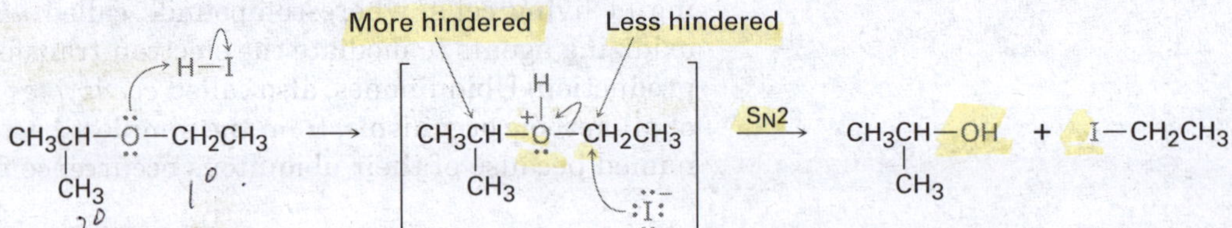

Ethyl isopropyl ether — Isopropyl alcohol — Iodoethane

Ethers with a tertiary alkyl group cleave by an S_N1 mechanism because they can produce stable intermediate carbocations. In such reactions, the ether oxygen atom stays with the less hindered alkyl group and the halide bonds to the tertiary group. Like most S_N1 reactions, the cleavage is rapid and often takes place at room temperature or below.

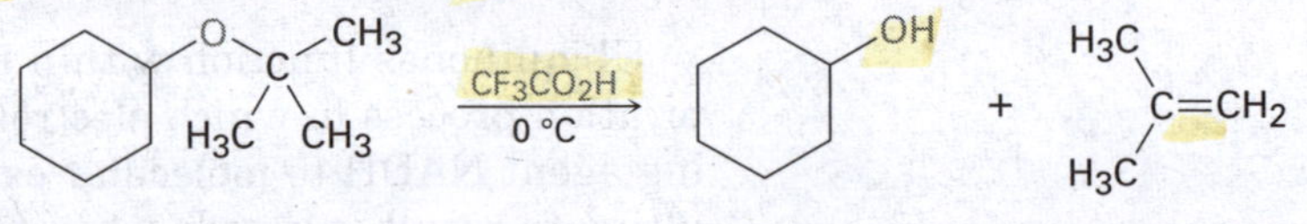

tert-Butyl cyclohexyl ether — Cyclohexanol (90%) — 2-Methylpropene

Worked Example 8.6 Predicting the Product of an Ether Cleavage

Predict the products of the reaction of *tert*-butyl propyl ether with HBr.

Strategy Identify the substitution pattern of the two groups attached to oxygen—in this case a tertiary alkyl group and a primary alkyl group. Then recall the guidelines for ether cleavages. An ether with only primary and secondary alkyl groups usually undergoes cleavage by S_N2 attack of a nucleophile on the less hindered alkyl group, but an ether with a tertiary alkyl group usually undergoes cleavage by an S_N1 mechanism. In this case an S_N1 cleavage of the tertiary C–O bond will occur, giving propan-1-ol and a tertiary alkyl bromide.

Solution

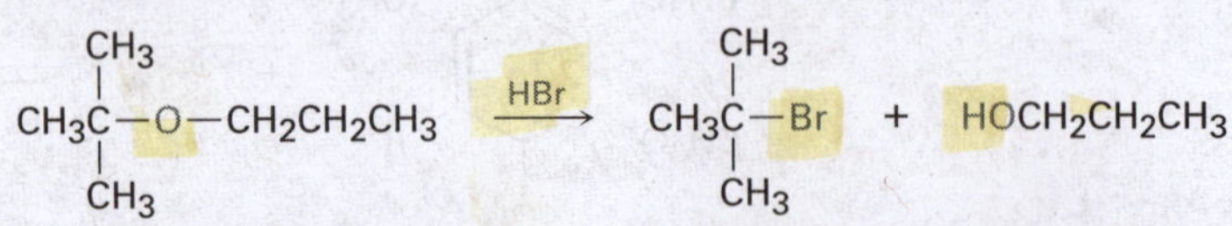

tert-Butyl propyl ether — 2-Bromo-2-methylpropane — Propan-1-ol

Problem 8.17 What products do you expect from the reaction of the following ethers with HI?

(a)

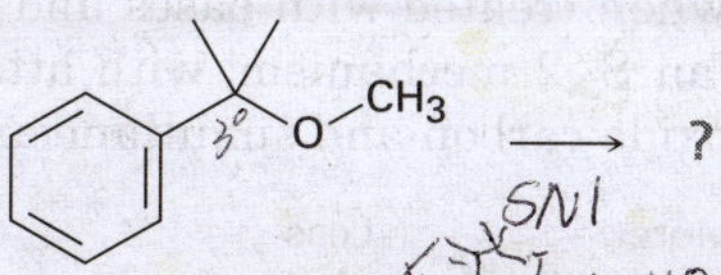

(b)

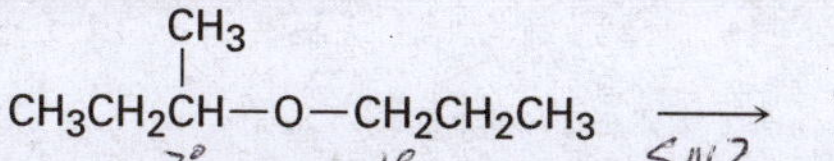

8.7 Cyclic Ethers: Epoxides

For the most part, cyclic ethers behave like acyclic ethers. The chemistry of the ether functional group is the same whether it's in an open chain or in a ring. Thus, the cyclic ether tetrahydrofuran (THF) is often used as a solvent because of its inertness.

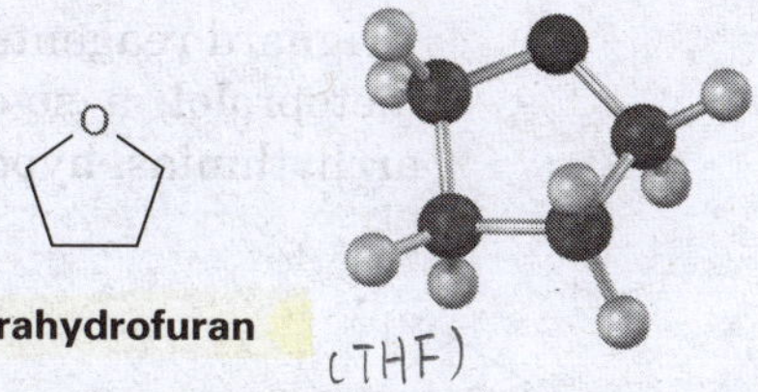

Tetrahydrofuran

The three-membered-ring ethers, called *epoxides,* make up the one group of cyclic ethers that behave differently from open-chain ethers. The strain of the three-membered ring makes epoxides much more reactive than other ethers. As we saw in Section 4.6, epoxides are prepared by reaction of an alkene with a peroxyacid, RCO_3H, usually *m*-chloroperoxybenzoic acid.

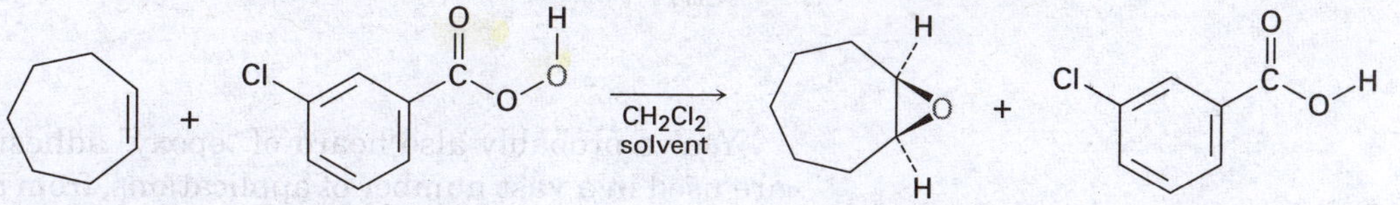

Cycloheptene | ***meta*-Chloroperoxybenzoic acid** | **1,2-Epoxycycloheptane** | ***meta*-Chlorobenzoic acid**

Epoxide rings are cleaved by treatment with acid just like other ethers, but they react under much milder conditions because of the strain of the three-membered ring. Dilute aqueous acid at room temperature converts an epoxide to a 1,2-diol (Section 4.6), and halogen acids HX convert an epoxide into a halo alcohol called a *halohydrin.* Both reactions take place by S_N2 attack of the nucleophile (H_2O or X^-) on the protonated epoxide so that the product has trans stereochemistry.

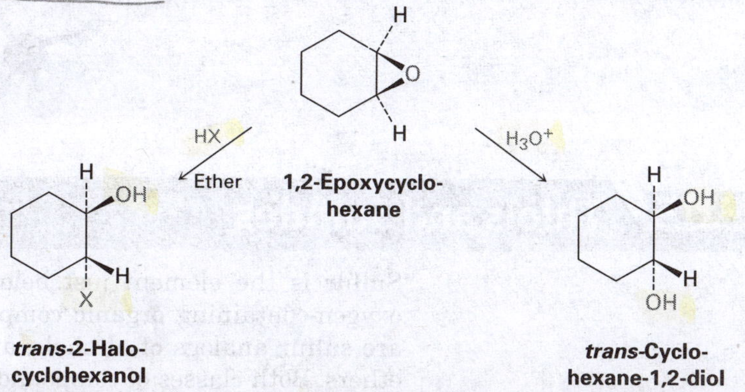

1,2-Epoxycyclohexane

***trans*-2-Halocyclohexanol** | ***trans*-Cyclohexane-1,2-diol**

In addition to their reaction with acids, epoxides also undergo rapid ring-opening when treated with bases and other nucleophiles. The reaction takes place by an S_N2 mechanism, with attack of the nucleophile at the less hindered epoxide carbon and simultaneous cleavage of the C–O bond.

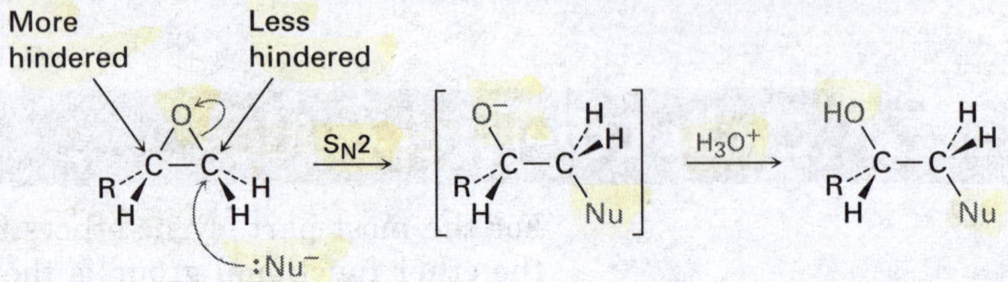

Many different nucleophiles can be used for epoxide opening, including hydroxide ion (HO^-), alkoxide ions (RO^-), amines (RNH_2 or R_2NH), and Grignard reagents (RMgX). An example occurs in the commercial synthesis of metoprolol, a so-called beta-blocker that is used for treatment of cardiac arrhythmias, hypertension, and heart attacks.

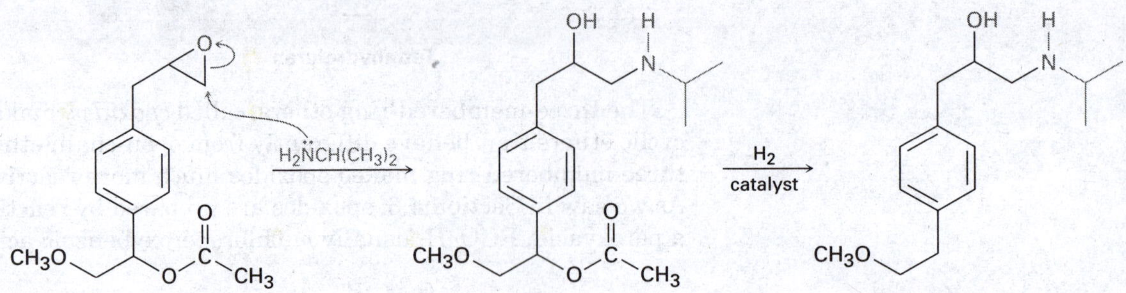

Metoprolol

You've probably also heard of "epoxy" adhesives, resins, and coatings that are used in a vast number of applications, from aircraft construction to simple home repairs. The *Interlude* at the end of this chapter tells more about them and describes how they work.

Problem 8.18 Show the structure of the product you would obtain by treatment of the following epoxide with aqueous acid. What is the stereochemistry of the product if the ring-opening takes place by normal backside S_N2 attack?

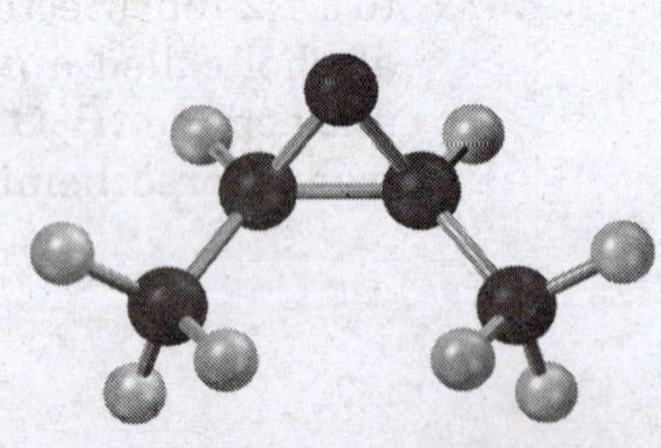

8.8 Thiols and Sulfides

Sulfur is the element just below oxygen in the periodic table, and many oxygen-containing organic compounds have sulfur analogs. *Thiols* (R—SH) are sulfur analogs of alcohols, and sulfides (R—S—R′) are sulfur analogs of ethers. Both classes of compounds are widespread in living organisms.

Thiols are named in the same way as alcohols, with the suffix *-thiol* used in place of *-ol*. The –SH group itself is referred to as a **mercapto group**.

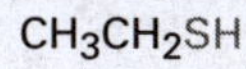

CH_3CH_2SH

Ethanethiol

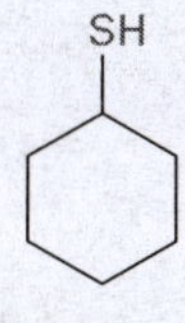

Cyclohexanethiol

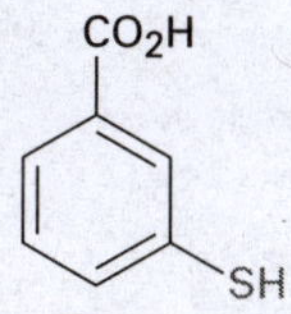

m-Mercaptobenzoic acid

Sulfides are named in the same way as ethers, with *sulfide* used in place of *ether* for simple compounds and with *alkylthio* used in place of *alkoxy* for more complex substances.

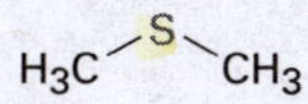

Dimethyl sulfide

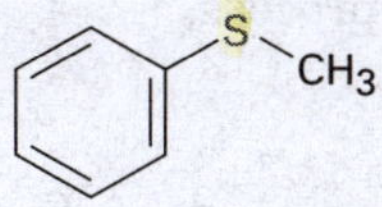

Methyl phenyl sulfide

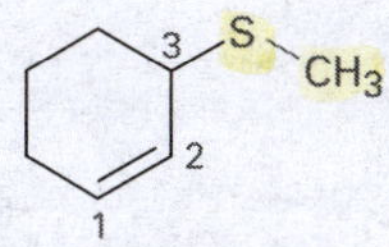

3-(Methylthio)cyclohexene

Thiols can be prepared from the corresponding alkyl halide by S_N2 displacement with a sulfur nucleophile such as hydrosulfide anion, SH^-.

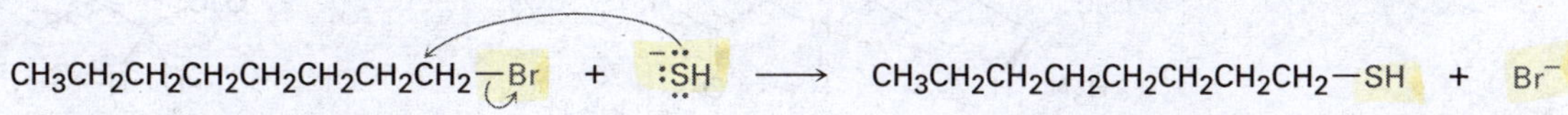

$CH_3CH_2CH_2CH_2CH_2CH_2CH_2CH_2{-}Br + {:}SH^- \longrightarrow CH_3CH_2CH_2CH_2CH_2CH_2CH_2CH_2{-}SH + Br^-$

1-Bromooctane — Octane-1-thiol (83%)

Sulfides are prepared by treating a primary or secondary alkyl halide with a *thiolate ion,* RS^-, the sulfur analog of an alkoxide ion. Reaction occurs by an S_N2 mechanism analogous to the Williamson ether synthesis (Section 8.4). Thiolate anions are among the best nucleophiles known, so these reactions usually work well.

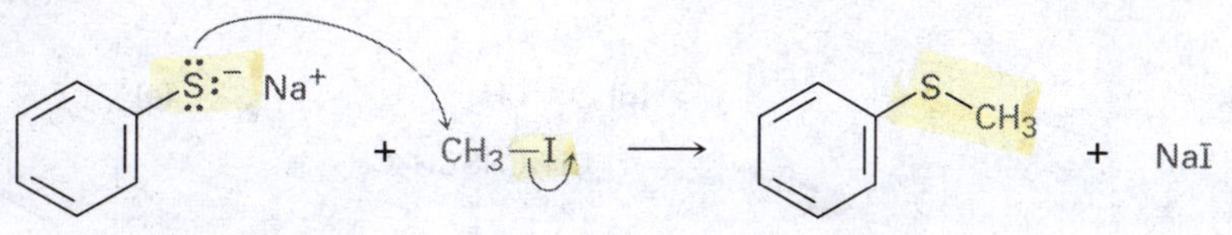

Sodium benzenethiolate — Methyl phenyl sulfide (96%)

Physically, the most unforgettable characteristic of thiols is their appalling odor. Skunk scent, in fact, is due primarily to the simple thiols 3-methylbutane-1-thiol and but-2-ene-1-thiol. Chemically, thiols can be oxidized by mild reagents such as bromine or iodine to yield *disulfides,* R—S—S—R, and disulfides can be reduced back to thiols by treatment with zinc metal and acetic acid.

$$\underset{\text{A thiol}}{2\,R{-}SH} \underset{Zn,\,H^+}{\overset{I_2}{\rightleftharpoons}} \underset{\text{A disulfide}}{R{-}S{-}S{-}R} + 2\,HI$$

This thiol–disulfide interconversion is a key part of numerous biological processes. We'll see in Section 15.4, for instance, that disulfide formation is involved in defining the structure and three-dimensional conformations of proteins, where disulfide "bridges" often form cross-links between thiol-containing amino acids in the protein chains. Disulfide formation is also involved in the process by which cells protect themselves from oxidative degradation. A cellular component called *glutathione* removes potentially harmful oxidants and is itself oxidized to glutathione disulfide in the process.

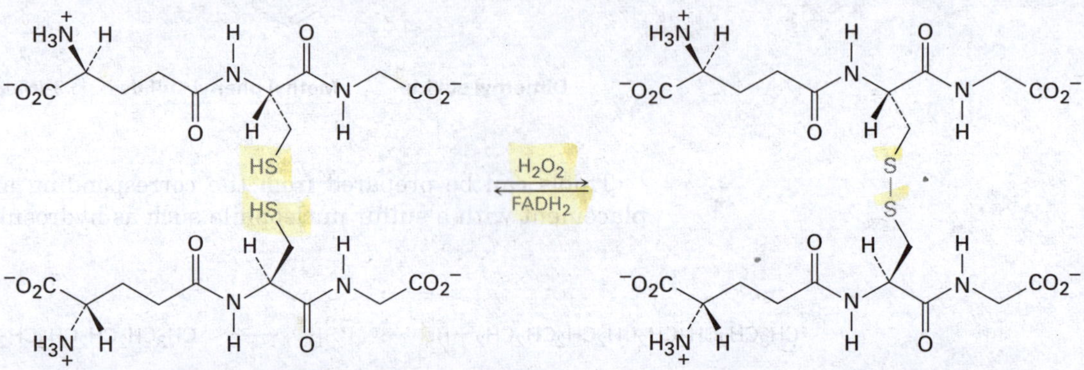

Glutathione (GSH) | Glutathione disulfide (GSSG)

Problem 8.19 Name the following compounds:

(a) 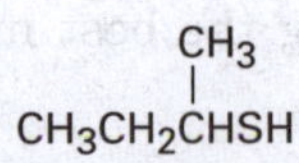

(b) 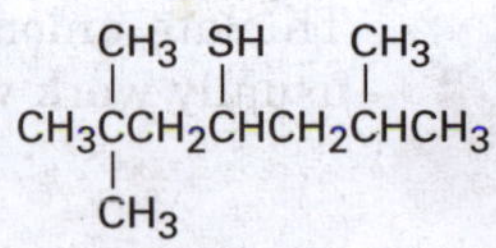

(c) 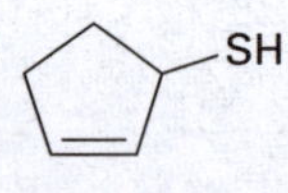

(d) 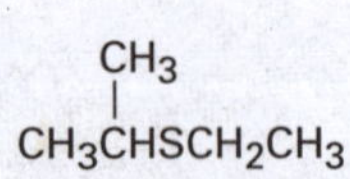

(e) 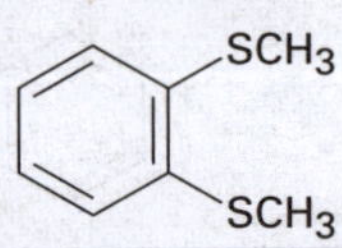

Problem 8.20 But-2-ene-1-thiol is a component of skunk spray. How would you synthesize this substance from but-2-en-1-ol? From methyl but-2-enoate, $CH_3CH{=}CHCO_2CH_3$? More than one step is required in both instances.

INTERLUDE

Epoxy Resins and Adhesives

Kayaks are often made of a high-strength polymer coated with epoxy resin.

Few nonchemists know exactly what an epoxide is, but practically everyone has used an "epoxy glue" for household repairs or an epoxy resin for a protective coating. Worldwide, approximately $15 billion of epoxies are used annually for a vast number of adhesive and coating applications including many in the aerospace industry. Much of the new Boeing 787 Dreamliner, for instance, is held together with epoxy-based adhesives.

Epoxy resins and adhesives generally consist of two components that are mixed just prior to use. One component is a liquid "prepolymer," and the second is a "curing agent" that reacts with the prepolymer and causes it to solidify.

The most widely used epoxy resins and adhesives are based on a prepolymer made from bisphenol A and epichlorohydrin. On treatment with base, bisphenol A is converted into its anion, which acts as a nucleophile in an S_N2 reaction with epichlorohydrin. Each epichlorohydrin molecule can react with two molecules of bisphenol A, once by S_N2 displacement of chloride ion and once by nucleophilic opening of the epoxide ring. At the same time, each bisphenol A molecule can react with two epichlorohydrins, leading to a long polymer chain. Each end of a prepolymer chain has an unreacted epoxy group, and each chain has numerous secondary alcohol groups spaced regularly along its midsection.

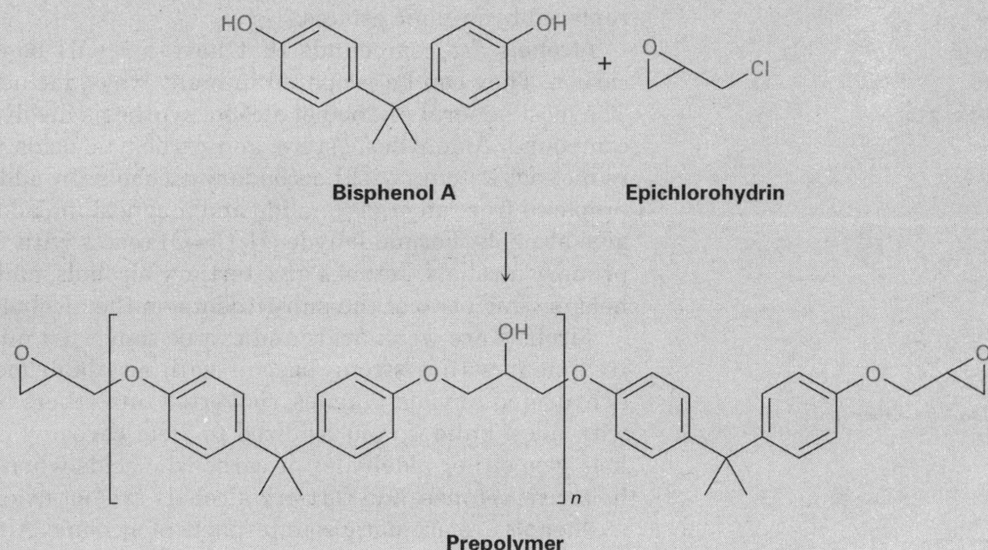

When the epoxide is to be used, a basic curing agent such as a tertiary amine, R_3N, is added to cause the individual prepolymer chains to link together. This "cross-linking" of chains is simply a base-catalyzed, S_N2 epoxide ring-opening of an –OH group in the middle of one chain with

continued

INTERLUDE

an epoxide group on the end of another chain. The result of such cross linking is formation of a vast, three-dimensional tangle that has enormous strength and chemical resistance.

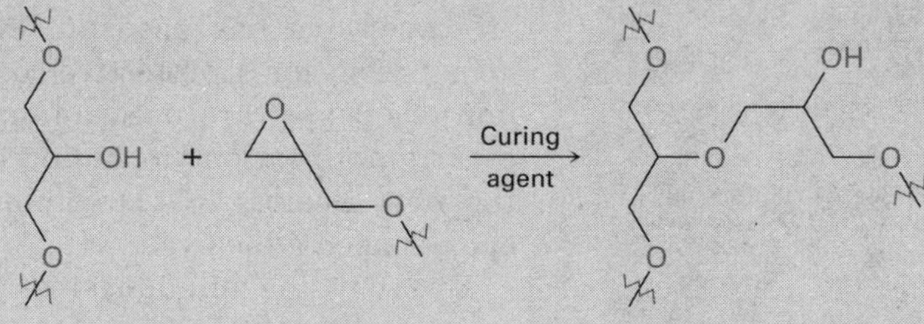

Summary and Key Words

alcohol 256
alkoxide ion 260
ether 256
hydroquinone 275
mercapto group 279
phenol 256
phenoxide ion 260
quinone 274
sulfide 256
thiol 256
thiophenol 256

In past chapters, we focused on developing general ideas of organic reactivity, looking at the chemistry of hydrocarbons and alkyl halides. With that accomplished, we have now begun in this chapter to study the oxygen-containing functional groups that lie at the heart of organic chemistry. To understand the chemistry of living organisms, in particular, it's necessary to understand oxygen-containing functional groups. Alcohols, phenols, and ethers are organic derivatives of water in which one or both of the water hydrogens have been replaced by organic groups.

Alcohols are compounds that have an –OH bonded to an sp^3-hybridized carbon. They can be prepared in many ways, including hydration of alkenes. The most general method of alcohol synthesis involves reduction of a carbonyl compound. Aldehydes, esters, and carboxylic acids yield primary alcohols on reduction; ketones yield secondary alcohols. In addition, Grignard reagents, prepared from an organohalide and magnesium, add to carbonyl compound to give alcohols. Formaldehyde ($H_2C{=}O$) reacts with Grignard reagents to give primary alcohols, ketones give tertiary alcohols, and esters give tertiary alcohols in which two of the substituents on the alcohol carbon are identical.

Alcohols are weak acids and can be converted into their **alkoxide anions** by treatment with a strong base or with an alkali metal. Alcohols can also be dehydrated to yield alkenes, converted into ethers by reaction of their anions with alkyl halides, and oxidized to yield carbonyl compounds. Primary alcohols give either aldehydes or carboxylic acids when oxidized, secondary alcohols give ketones, and tertiary alcohols are not oxidized.

Phenols are aromatic counterparts of alcohols. Although similar to alcohols in some respects, phenols are more acidic than alcohols because **phenoxide anions** are stabilized by resonance. Phenols undergo electrophilic aromatic substitution reactions readily and can be oxidized to yield **quinones**. Quinones, in turn, can be reduced to give *p*-dihydroxybenzenes, called **hydroquinones**.

Ethers have two organic groups bonded to the same oxygen atom. They are prepared by S_N2 reaction of an alkoxide ion with a primary alkyl halide—the Williamson ether synthesis. Ethers are inert to most reagents but are cleaved by the strong acids HBr and HI. Epoxides—cyclic ethers with an oxygen atom in a three-membered ring—differ from other ethers in their ease of cleavage. The high reactivity of the strained three-membered ether ring allows epoxides to react with aqueous acid, yielding diols and with HX yielding halohydrins. In addition, epoxides react with nucleophiles such as amines and alkoxides.

Sulfides (R—S—R′) and **thiols (R—SH)** are sulfur analogs of ethers and alcohols. Thiols are prepared by S_N2 reaction of an alkyl halide with HS^-, and sulfides are prepared by further alkylation of the thiol with an alkyl halide.

Summary of Reactions

1. Synthesis of alcohols (Section 8.3)
   (a) Reduction of carbonyl compounds
   (1) Aldehydes

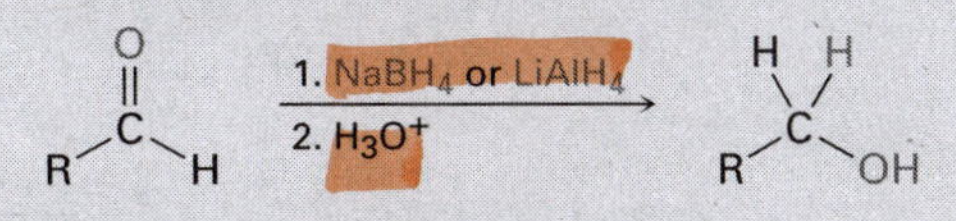

Primary alcohol

(2) Ketones

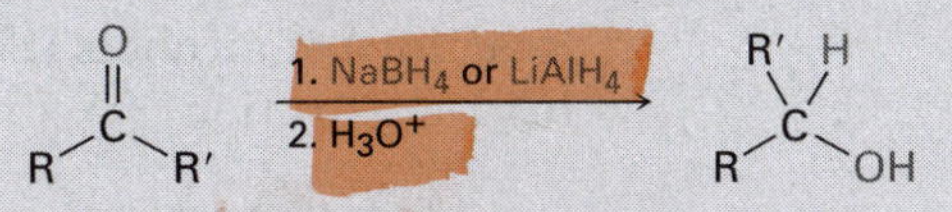

Secondary alcohol

(3) Esters

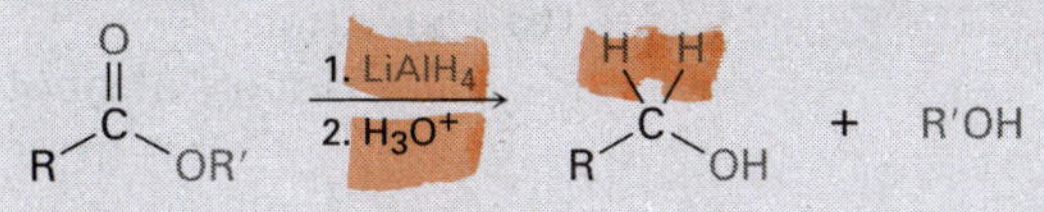

Primary alcohol

(4) Carboxylic acids

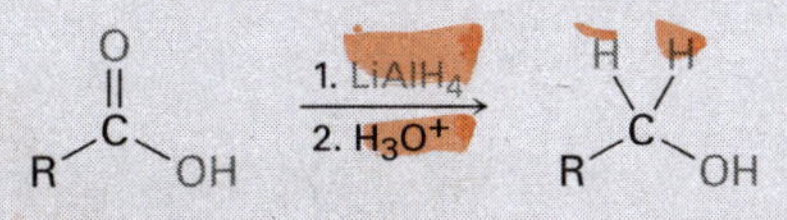

Primary alcohol

continued

(b) Grignard reaction with carbonyl compounds

(1) Formaldehyde

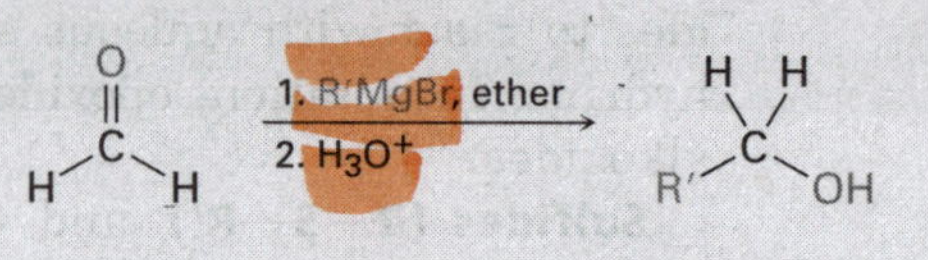

Primary alcohol

(2) Aldehydes

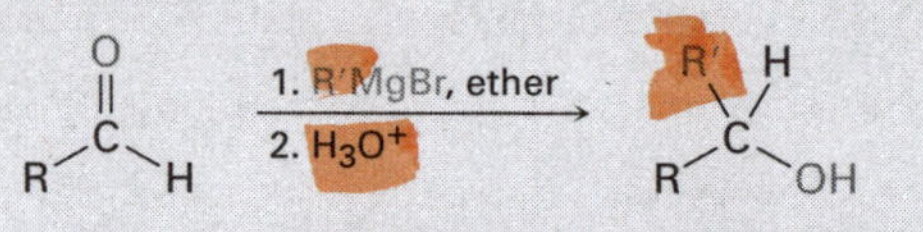

Secondary alcohol

(3) Ketones

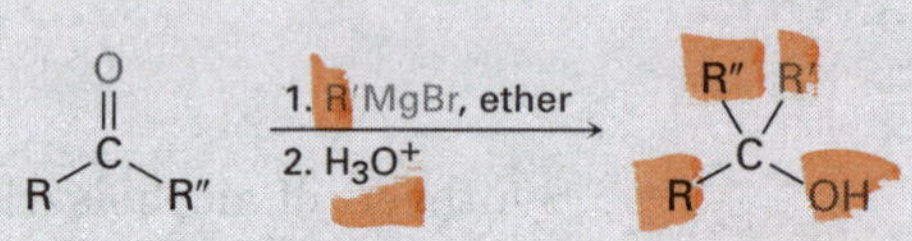

Tertiary alcohol

(4) Esters

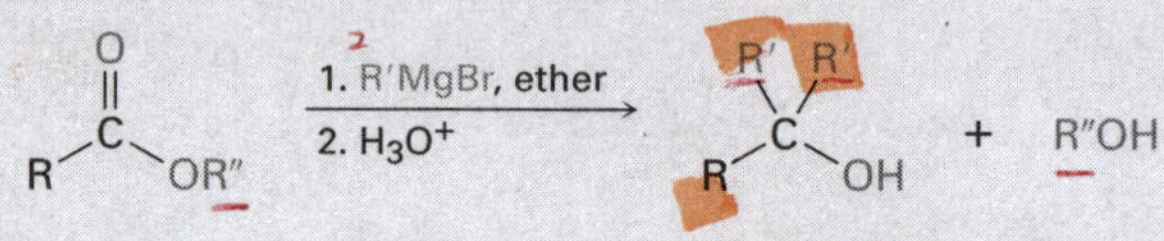

Tertiary alcohol

2. Reactions of alcohols (Section 8.4)

(a) Dehydration

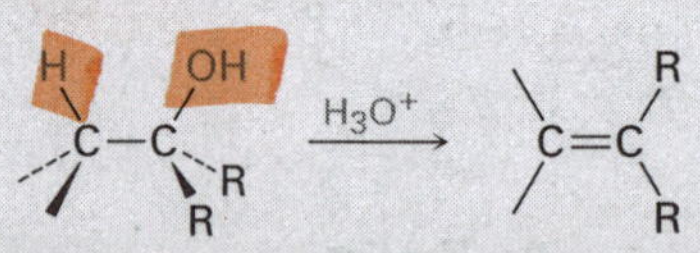

(b) Oxidation

(1) Primary alcohols

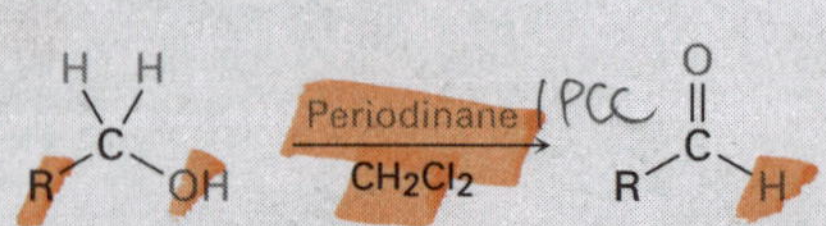

Aldehyde

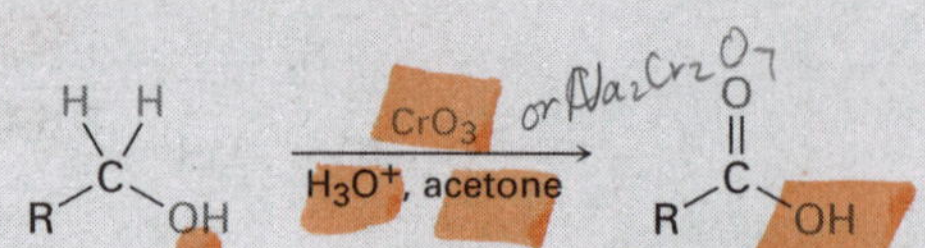

Carboxylic acid

(2) Secondary alcohols

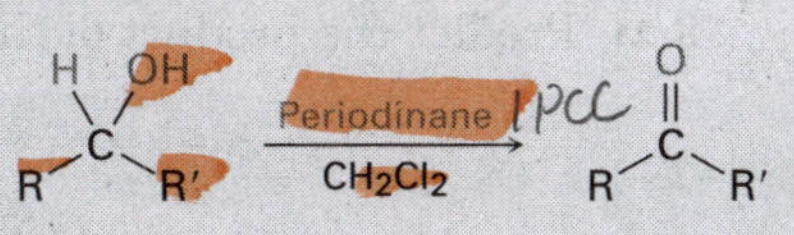

Ketone

3. Synthesis of ethers (Section 8.4)
   Williamson ether synthesis

$$RO^- + R'CH_2X \longrightarrow ROCH_2R' + X^-$$

4. Reactions of ethers
   (a) Acidic cleavage with HBr or HI (Section 8.6)

$$R{-}O{-}R' \xrightarrow[H_2O]{HX} RX + R'OH$$

   (b) Epoxide opening with aqueous acid (Section 8.7)

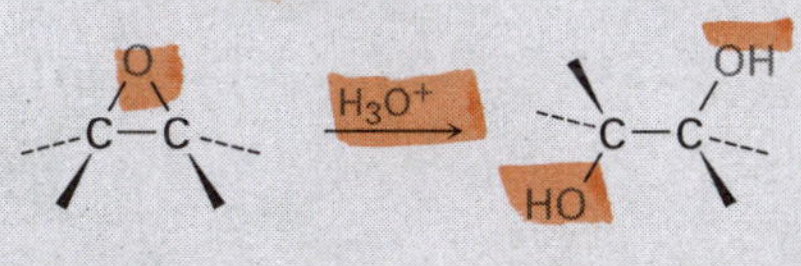

   (c) Epoxide opening with nucleophiles (Section 8.7)

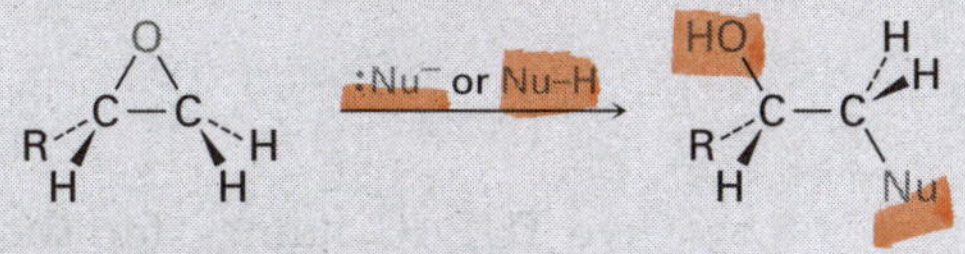

5. Reactions of phenols (Section 8.5)
   Oxidation to give quinones

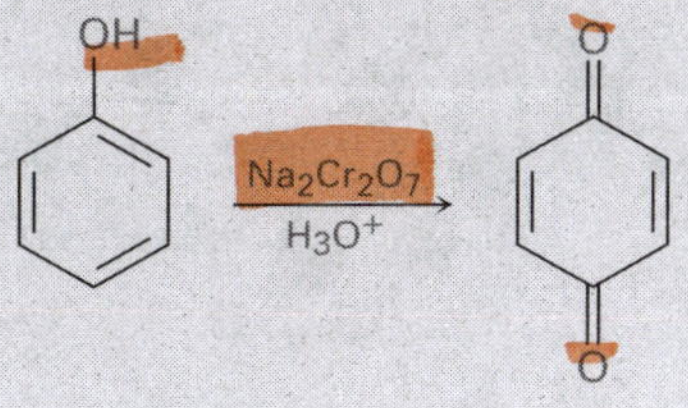

6. Synthesis of thiols (Section 8.8)

$$RCH_2Br \xrightarrow[2.\ H_2O,\ NaOH]{1.\ HS^-} RCH_2SH$$

7. Synthesis of sulfides (Section 8.8)

$$RS^- + R'CH_2Br \longrightarrow RSCH_2R' + Br^-$$

Exercises

Visualizing Chemistry

(Problems 8.1–8.20 appear within the chapter.)

OWL

Interactive versions of these problems are assignable in OWL.

8.21 Predict the product of the following reaction (red = O):

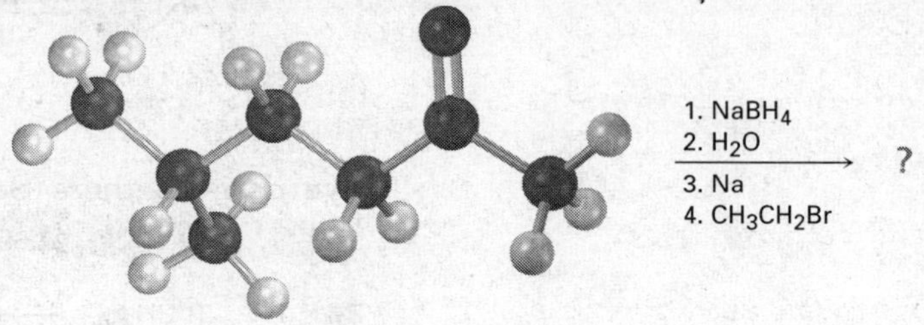

8.22 Show the product you would obtain by reaction of the following compound with:
(a) $NaBH_4$ **(b)** CH_3CH_2MgBr

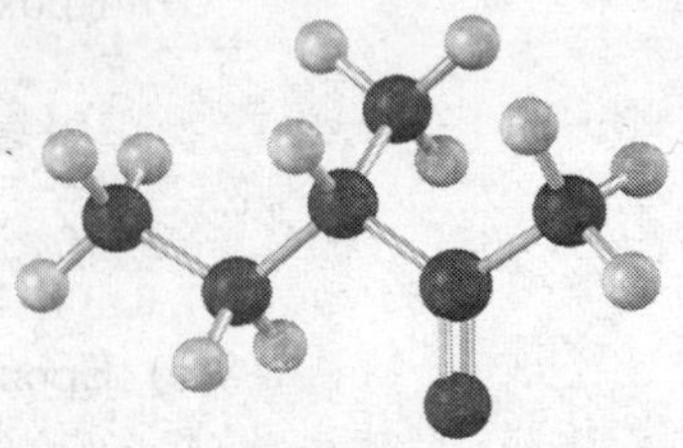

8.23 Give IUPAC names for the following compounds (red = O, blue = N, yellow = S):

(a)

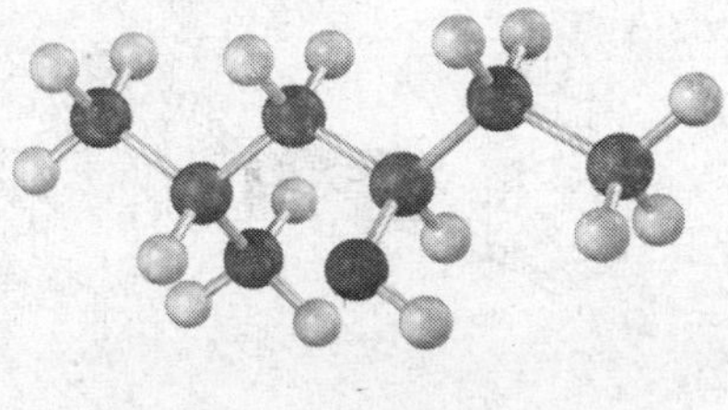

(b)

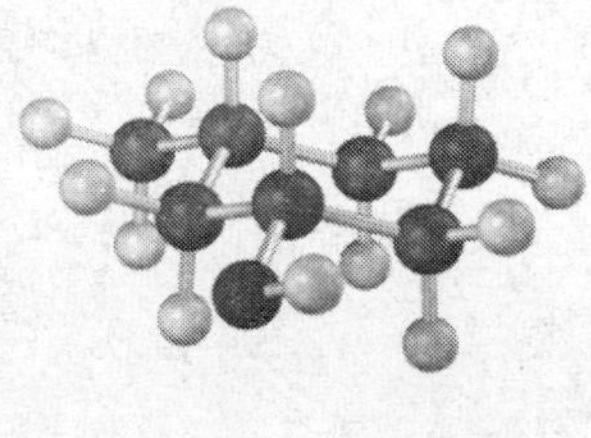

(c)

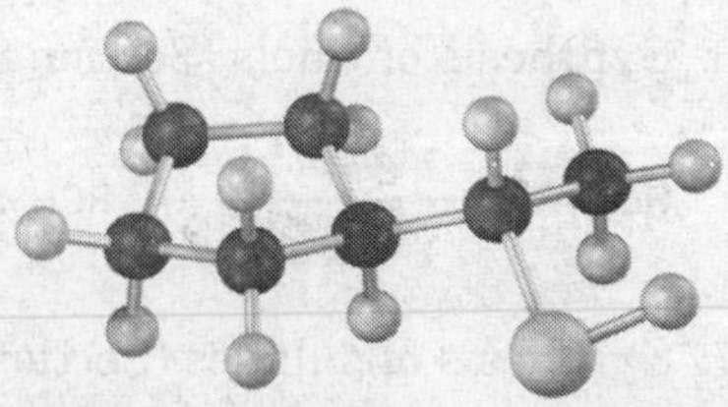

(d)

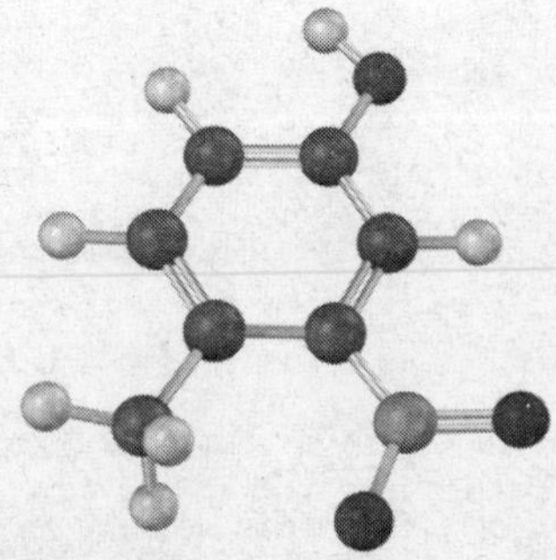

8.24 Show the product, including stereochemistry, that would result from reaction of the following epoxide with HBr:

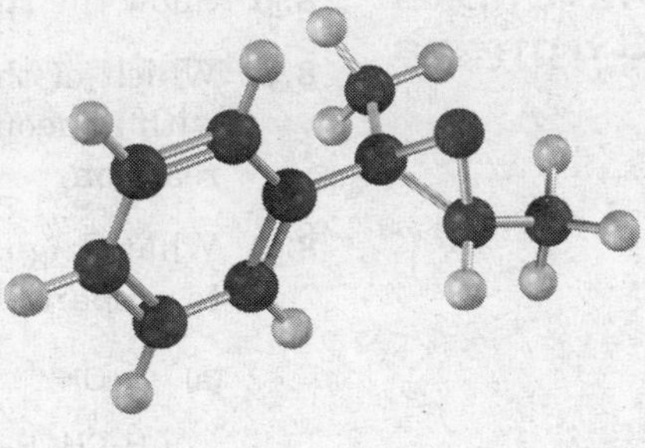

8.25 Show the structures of the carbonyl compounds that would give the following alcohols on reduction. Show also the structure of the products that would result by treating the alcohols with a periodinane and with aqueous acidic CrO_3.

(a)

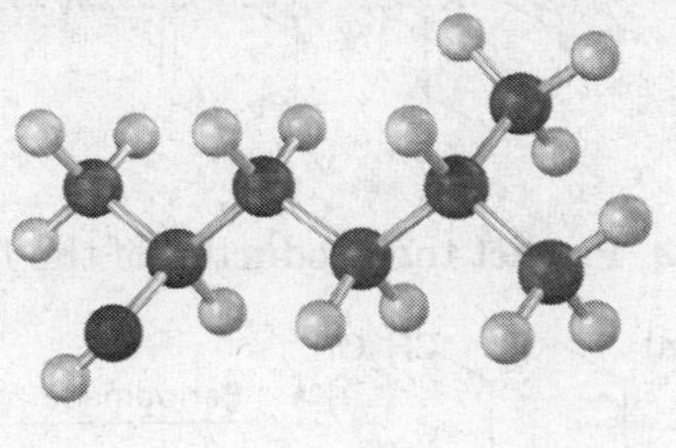

(b)

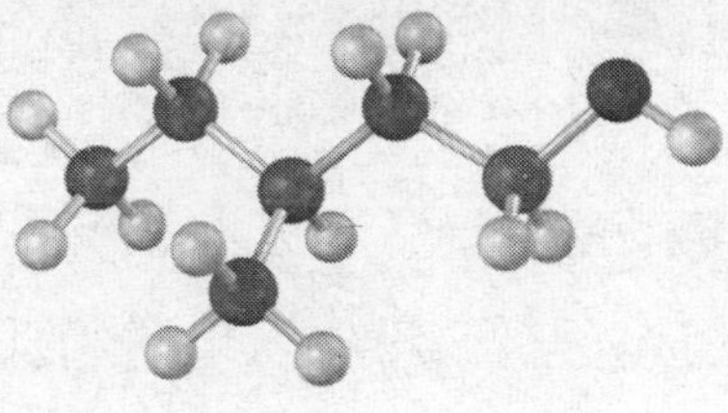

Additional Problems

Naming Alcohols and Ethers

8.26 Draw and name the eight isomeric alcohols with the formula $C_5H_{12}O$.

8.27 Which of the eight alcohols you identified in Problem 8.26 are chiral?

8.28 Name the following compounds:

(a)

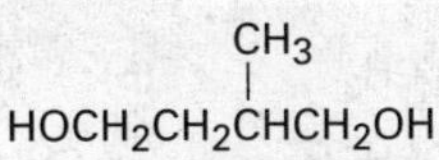

(b)

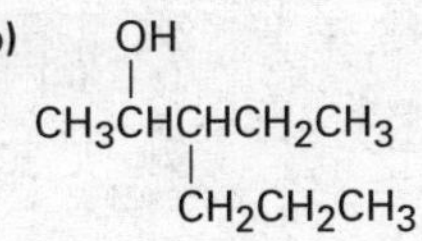

(c)

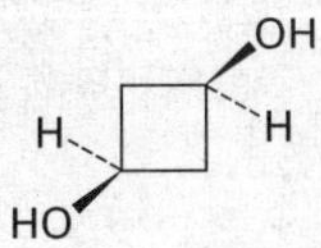

(d)

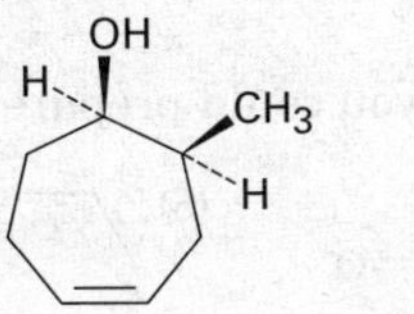

(e)

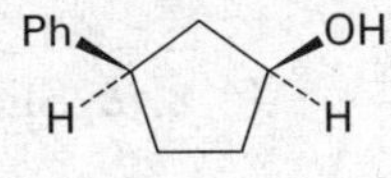

(f)

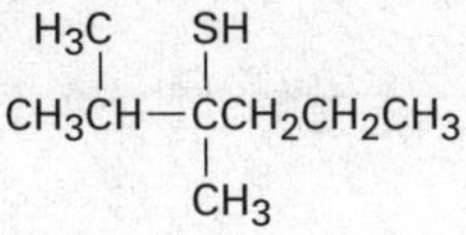

8.29 Draw structures corresponding to the following IUPAC names:

(a) Ethyl isopropyl ether
(b) 3,4-Dimethoxybenzoic acid
(c) 2-Methylheptane-2,5-diol
(d) *trans*-3-Ethylcyclohexanol
(e) 4-Allyl-2-methoxyphenol (eugenol, from oil of cloves)

8.30 Draw and name the six ethers that are isomeric with the alcohols you drew in Problem 8.26. Which are chiral?

REACTIONS AND SYNTHESIS

8.31 Show the HI cleavage products of the ethers you drew in Problem 8.30.

8.32 Which of the eight alcohols you identified in Problem 8.26 would react with aqueous acidic CrO_3? Show the products you would expect from each reaction.

8.33 What Grignard reagent and what carbonyl compound might you start with to prepare the following alcohols?

(a) 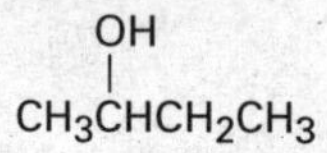

(b) 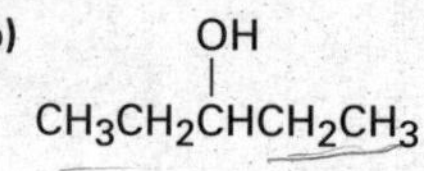

(c) 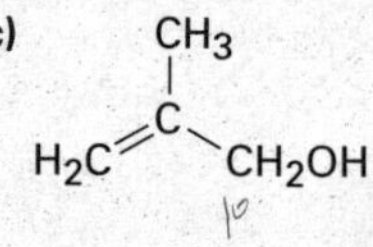

(d) 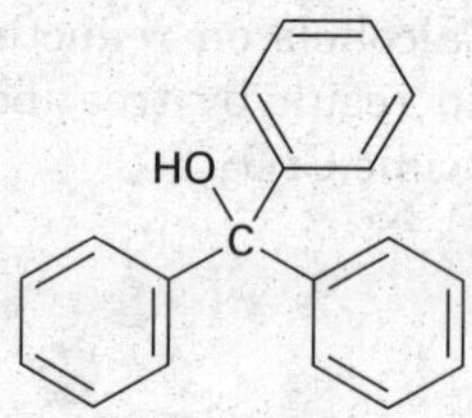

(e) 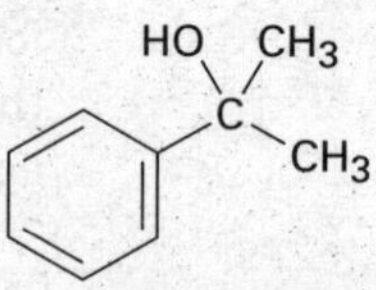

(f) 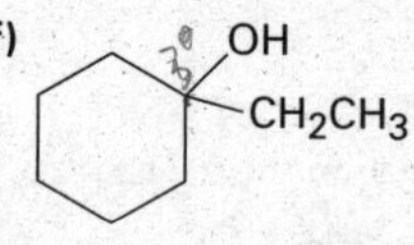

8.34 Predict the product(s) of the following transformations:

(a) 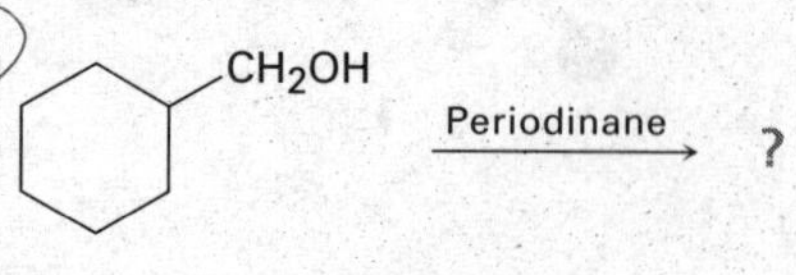

(b) 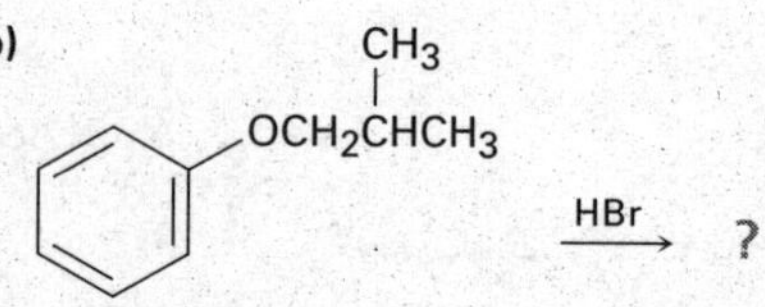

(c) 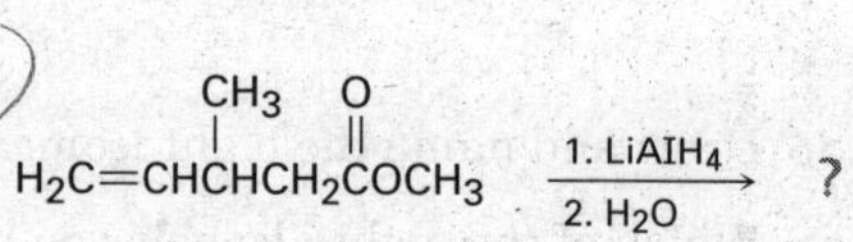

(d) 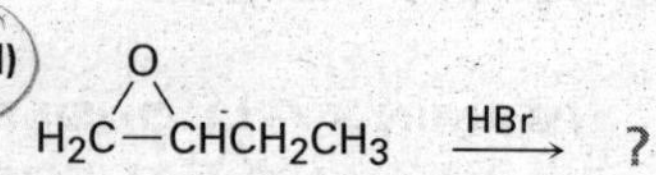

8.35 Predict the likely products of reaction of the following ethers with HI:

(a) 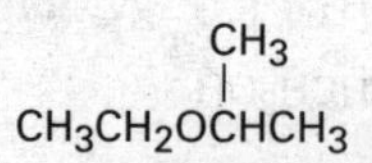

(b) 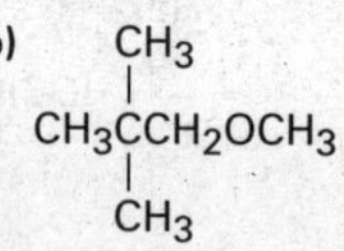

8.36 Show how you could prepare the following substances from cyclohexanol:

(a) 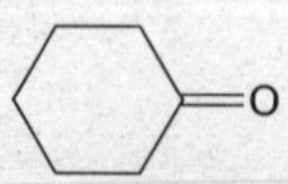

(b) 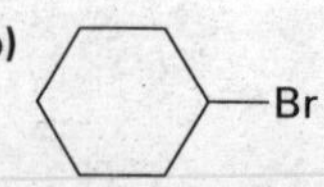

(c) 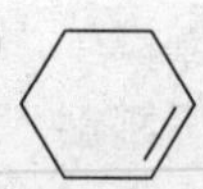

(d) 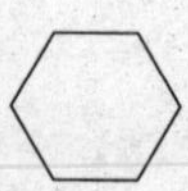

8.37 Show how you could prepare the following substances from propan-1-ol:

(a) 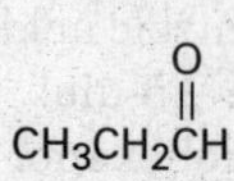

(b) 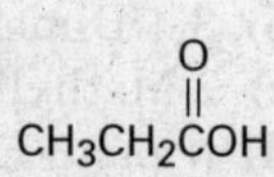

(c) $CH_3CH_2CH_2O^-Na^+$

(d) $CH_3CH_2CH_2Cl$

8.38 What alcohols would you oxidize to obtain the following products?

(a) 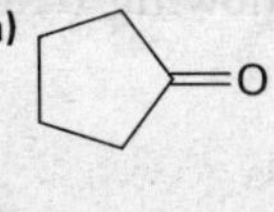 (b) 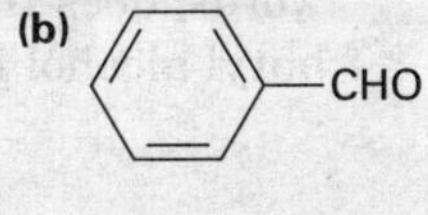 (c) 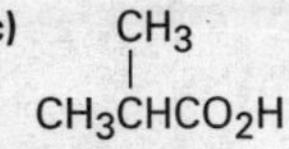

8.39 Predict the product(s) of the following reactions:

(a) 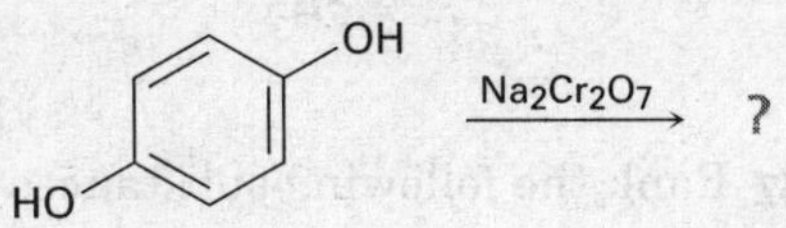

(b) 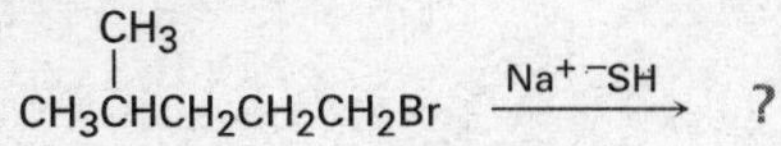

(c) 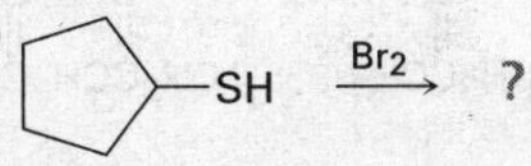

8.40 What carbonyl compounds would you reduce to prepare the following alcohols? List all possibilities.

(a) 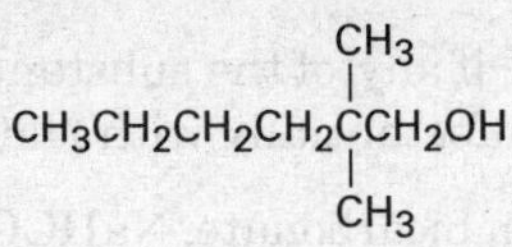 (b) 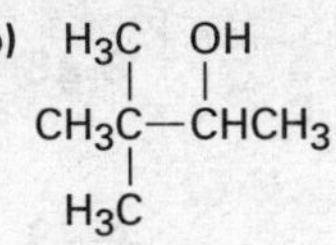 (c) 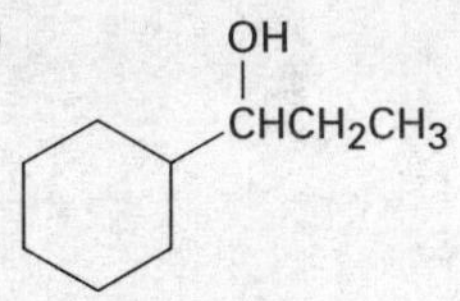

8.41 Show the alcohols you would obtain by reduction of the following carbonyl compounds:

(a) 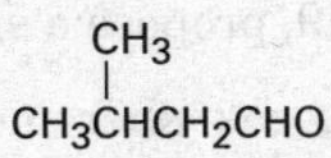 (b) 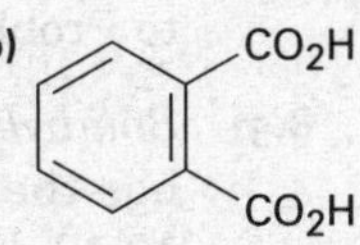 (c) 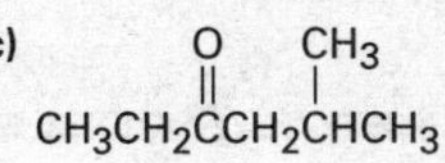

8.42 Give the structures of the major products you would obtain from reaction of phenol with the following reagents:

(a) Br_2 (1 mol) **(b)** Br_2 (3 mol)
(c) NaOH, then CH_3I **(d)** $Na_2Cr_2O_7$, H_3O^+

8.43 How would you prepare the following compounds from 2-phenylethanol?

(a) Benzoic acid
(b) Ethylbenzene
(c) 1-Bromo-2-phenylethane
(d) Phenylacetic acid ($C_6H_5CH_2CO_2H$)
(e) Phenylacetaldehyde ($C_6H_5CH_2CHO$)

8.44 What products would you obtain from reaction of 1-methylcyclohexanol with the following reagents?

(a) HBr
(b) H_2SO_4
(c) CrO_3
(d) Na
(e) Product of part (d), then CH_3I

8.45 What products would you obtain from reaction of butan-1-ol with the following reagents?

(a) PBr_3 **(b)** CrO_3, H_3O^+ **(c)** Na **(d)** A periodinane

ACIDITY

8.46 Is *tert*-butoxide anion a strong enough base to react with water? In other words, does the following reaction take place as written? (The pK_a of *tert*-butyl alcohol is 18.)

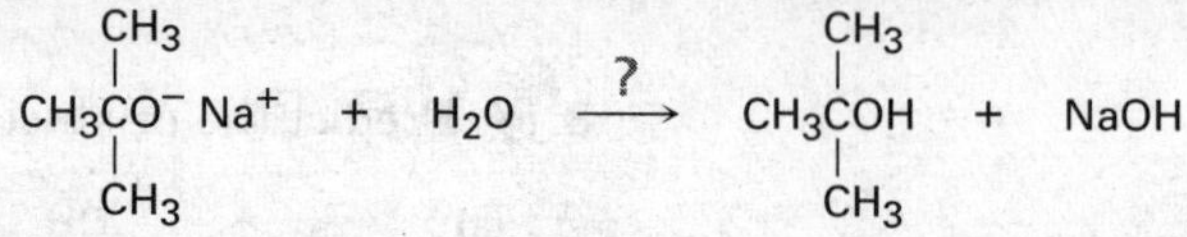

8.47 Rank the following substances in order of increasing acidity:

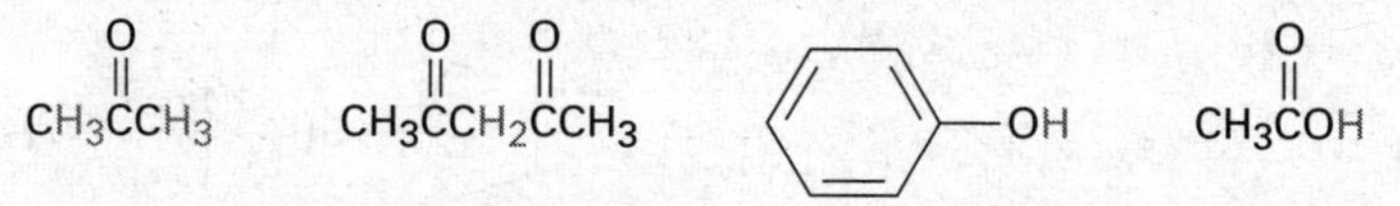

Acetone	Pentane-2,4-dione	Phenol	Acetic acid
(pK_a = 19)	(pK_a = 9)	(pK_a = 9.9)	(pK_a = 4.7)

8.48 Which, if any, of the substances in Problem 8.47 are strong enough acids to react substantially with NaOH? (The pK_a of H_2O is 15.7.)

8.49 Sodium bicarbonate, $NaHCO_3$, is the sodium salt of carbonic acid (H_2CO_3), pK_a = 6.4. Which of the substances shown in Problem 8.47 will react with sodium bicarbonate?

8.50 Assume you have two unlabeled bottles, one that contains phenol (pK_a = 9.9) and one that contains acetic acid (pK_a = 4.7). In light of your answer to Problem 8.49, propose a simple way to tell what is in each bottle.

GENERAL PROBLEMS

8.51 *Bombykol,* the sex pheromone secreted by the female silkworm moth, has the formula $C_{16}H_{30}O$ and the systematic name (10*E*,12*Z*)-hexadeca-10,12-dien-1-ol. Draw bombykol showing correct stereochemistry for the two double bonds.

8.52 When 4-chlorobutane-1-thiol is treated with a strong base such as sodium hydride, NaH, tetrahydrothiophene is produced. Suggest a mechanism for this reaction.

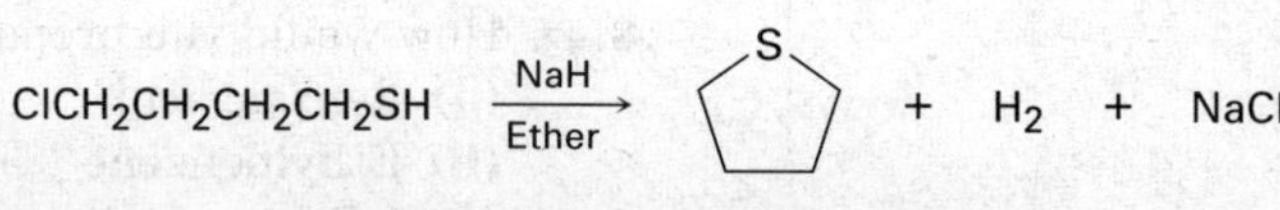

Tetrahydrothiophene

8.53 *Carvacrol* is a naturally occurring substance isolated from oregano, thyme, and marjoram. What is its IUPAC name?

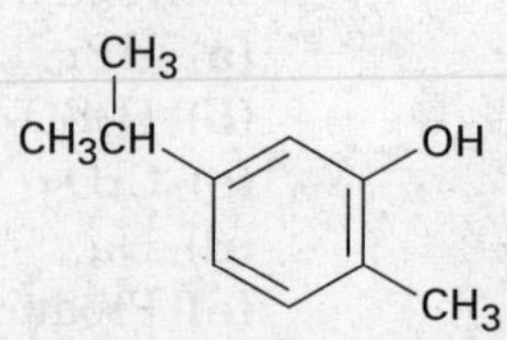

Carvacrol

8.54 How would you prepare the following ethers?

(a) 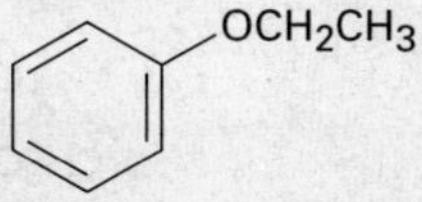 **(b)** 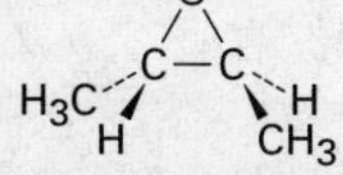

8.55 It's found experimentally that a substituted cyclohexanol with an axial –OH group reacts with aqueous CrO_3 more rapidly than its isomer with an equatorial –OH group. Draw both *cis-* and *trans*-4-*tert*-butylcyclohexanol, and predict which oxidizes faster. (The large *tert*-butyl group is equatorial in both.)

8.56 *tert*-Butyl ethers can be prepared by the reaction of an alcohol with 2-methylpropene in the presence of an acid catalyst. Propose a mechanism.

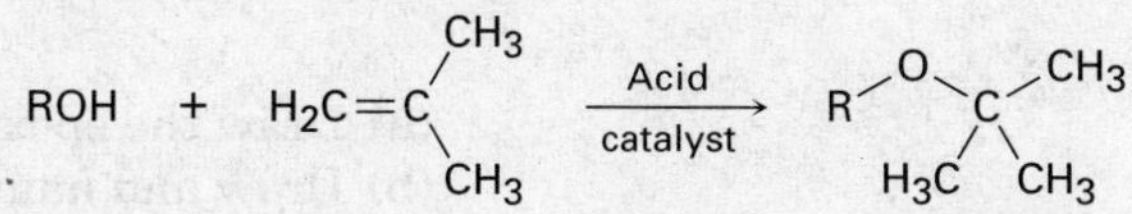

8.57 *tert*-Butyl ethers react with trifluoroacetic acid, CF_3CO_2H, to yield an alcohol and 2-methylpropene. Tell what kind of reaction is occurring, and propose a mechanism.

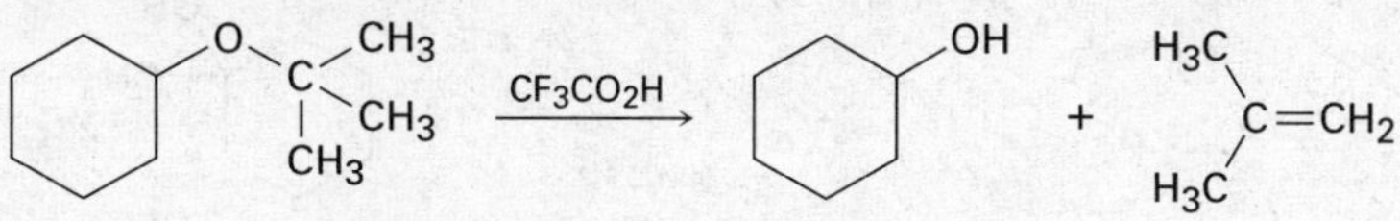

8.58 Why can't the Williamson ether synthesis be used to prepare diphenyl ether?

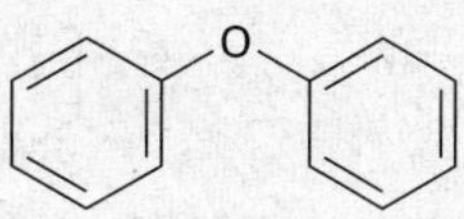

Diphenyl ether

8.59 Because all hamsters look pretty much alike, pairing and mating is governed by chemical means of communication rather than by physical attraction. Investigations have shown that dimethyl disulfide, CH_3SSCH_3, is secreted by female hamsters as a sex attractant for males. How would you synthesize dimethyl disulfide in the laboratory if you wanted to trick your hamster?

8.60 Epoxides react with Grignard reagents to yield alcohols. Propose a mechanism.

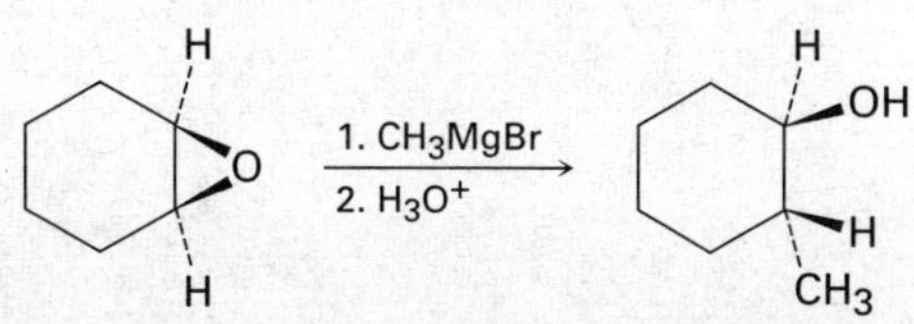

8.61 Reduction of butan-2-one with $NaBH_4$ yields butan-2-ol. Explain why the product is chiral but not optically active.

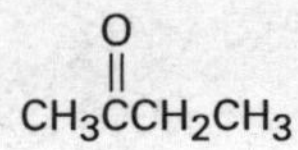

Butan-2-one

8.62 Methyl phenyl ether can be cleaved to yield iodomethane and lithium phenoxide when heated with LiI. Propose a mechanism for this reaction.

8.63 Imagine that you have treated (2*R*,3*R*)-2,3-epoxy-3-methylpentane with aqueous acid to carry out a ring-opening reaction.

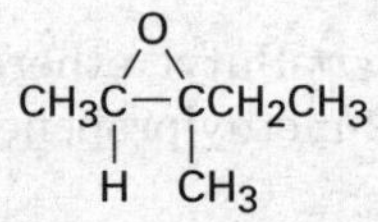

2,3-Epoxy-3-methylpentane (no stereochemistry implied)

(a) Draw the epoxide, showing stereochemistry.
(b) Draw and name the product, showing stereochemistry.
(c) Is the product chiral? Explain.
(d) Is the product optically active? Explain.

8.64 Identify the reagents **a** through **d** in the following scheme:

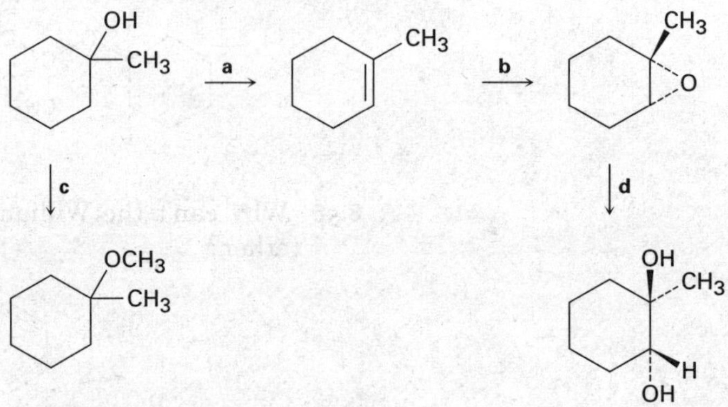

8.65 What cleavage product would you expect from reaction of tetrahydrofuran with hot aqueous HI?

Tetrahydrofuran

IN THE MEDICINE CABINET

8.66 Captopril is a drug commonly used to treat high blood pressure and heart failure. It functions by decreasing the concentration of chemicals that constrict the blood vessels, so blood flows more smoothly and the heart can pump blood more efficiently.

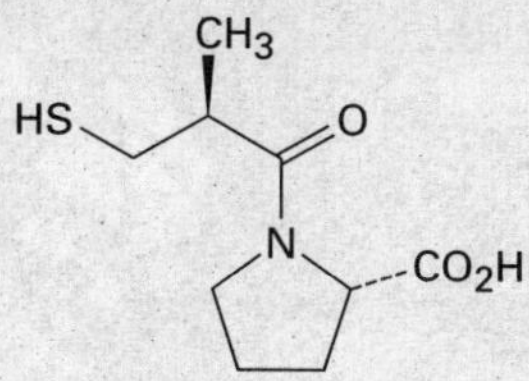

Captopril

(a) Assign *R*,*S* configuration to the two chirality centers in captopril.
(b) What three functional groups are present in captopril?
(c) Draw the disulfide that results from oxidation of captopril.

8.67 Nonoxynol 9 is a potent spermicide made by reacting ethylene oxide with *p*-nonylphenoxide. Propose a mechanism for this multistep reaction.

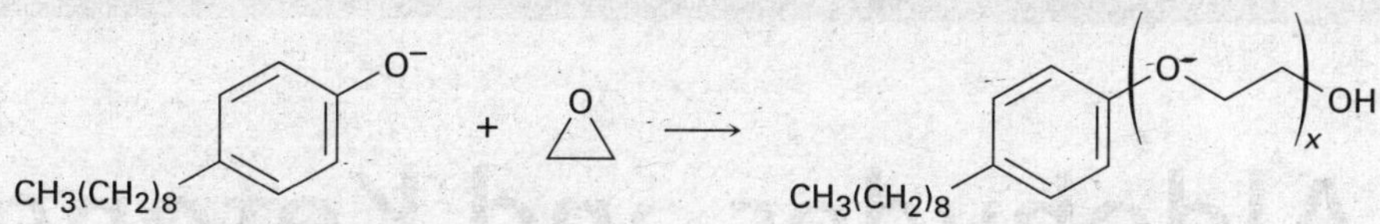

***p*-Nonylphenoxide** **Ethylene oxide** **Nonoxynol 9**

IN THE FIELD

8.68 The herbicide 2,4,5-T (2,4,5-trichlorophenoxyacetic acid) can be prepared by heating a mixture of 2,4,5-trichlorophenol and $ClCH_2CO_2H$ with NaOH. Show the mechanism of the reaction.

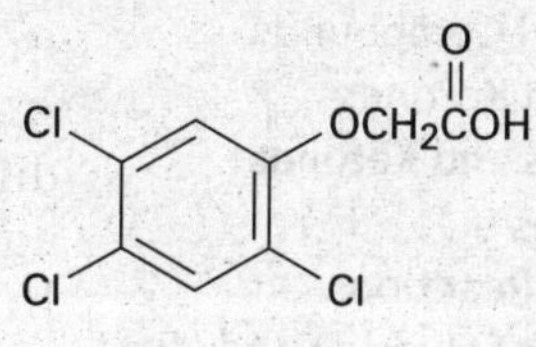

2,4,5-T

CHAPTER

9

Few flowers are more beautiful or more fragrant than roses. Their perfumed odor is due to several simple organic compounds, including the ketone β-damascenone.

Aldehydes and Ketones: Nucleophilic Addition Reactions

OWL

Online homework for this chapter can be assigned in OWL, an online homework assessment tool.

In this and the next two chapters, we'll discuss the most important functional group in both organic and biological chemistry—the carbonyl group, C=O. There are many different kinds of carbonyl compounds, but we'll begin our coverage by looking at just two: **aldehydes (RCHO)** and **ketones (R_2CO)**. In nature, many substances found in living organisms are aldehydes or ketones. The aldehyde pyridoxal phosphate, for instance, is a coenzyme involved in a large number of metabolic reactions, and the ketone hydrocortisone is a steroid hormone secreted by the adrenal glands to regulate fat, protein, and carbohydrate metabolism.

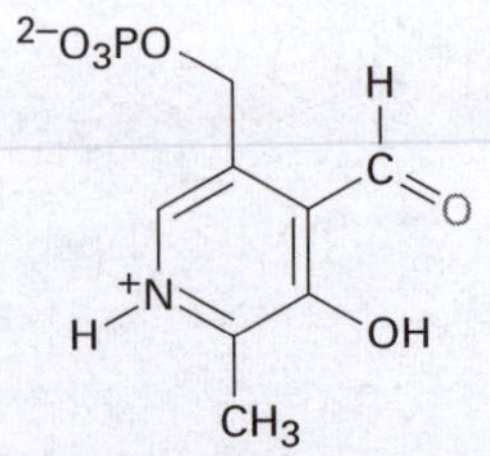

Pyridoxal phosphate (PLP)

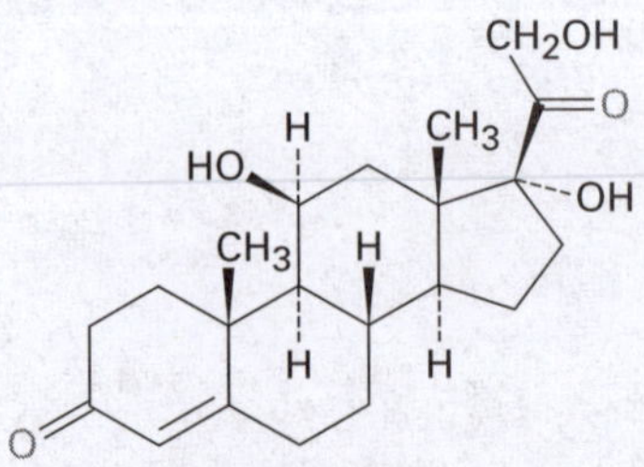

Hydrocortisone

In the chemical industry, simple aldehydes and ketones are produced in large amounts for use as solvents and as starting materials to prepare a host of other compounds. More than 23 million tons per year of formaldehyde, $H_2C{=}O$, are produced worldwide for use in building insulation materials and in the adhesive resins that bind particle board and plywood. Acetone, $(CH_3)_2C{=}O$, is widely used as an industrial solvent, with approximately 3.3 million tons per year produced worldwide.

WHY THIS CHAPTER?

Much of organic chemistry is the chemistry of carbonyl compounds. Aldehydes and ketones, in particular, are intermediates in the synthesis of many pharmaceutical agents, in almost all biological pathways, and in numerous industrial processes. We'll look in this chapter at some of their most important reactions.

9.1 The Nature of Carbonyl Compounds

It's useful to classify carbonyl compounds into two general categories based on the kinds of chemistry they undergo. In one category are aldehydes and ketones; in the other are carboxylic acids and their derivatives. The C=O groups in aldehydes and ketones are bonded to atoms (H and C) that aren't electronegative enough to stabilize a negative charge and therefore can't act as leaving groups in nucleophilic substitution reactions. The C=O groups in carboxylic acids and their derivatives are bonded to atoms (oxygen, halogen, nitrogen, and so forth) that *can* stabilize a negative charge and therefore can act as leaving groups in substitution reactions.

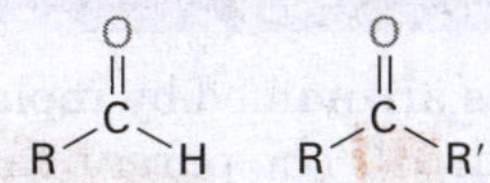

Aldehyde **Ketone**

The –R′ and –H in these compounds can't act as leaving groups in nucleophilic substitution reactions.

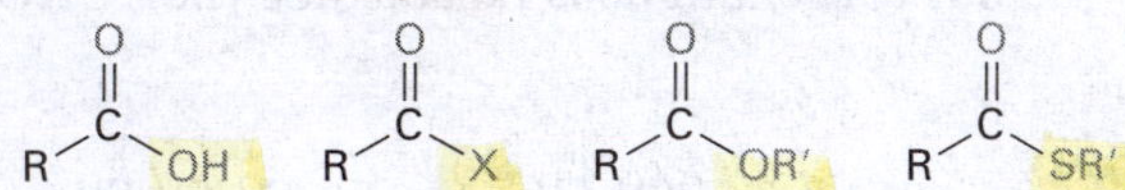

Carboxylic acid **Acid halide** **Ester** **Thioester**

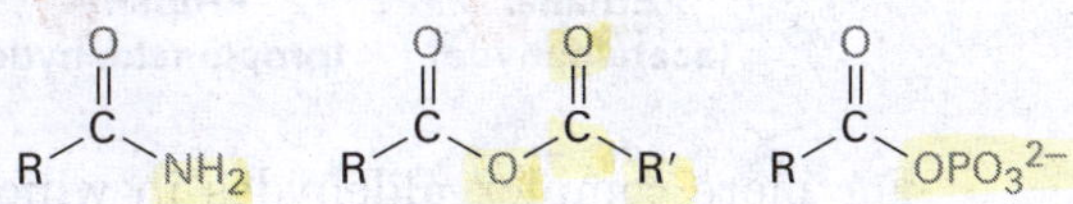

Amide **Acid anhydride** **Acyl phosphate**

The –OH, –X, –OR′, –SR, $-NH_2$, –OCOR′, and $-OPO_3^{2-}$ in these compounds can act as leaving groups in nucleophilic substitution reactions.

The carbon–oxygen double bond of carbonyl groups is similar in some respects to the carbon–carbon double bond of alkenes (Figure 9.1). The carbonyl carbon atom is sp^2-hybridized and forms three σ bonds. The fourth valence electron remains in a carbon *p* orbital and forms a π bond to oxygen by overlap with an oxygen *p* orbital. The oxygen also has two nonbonding pairs of electrons, which occupy its remaining two orbitals. Like alkenes, carbonyl compounds are planar about the double bond and have bond angles of approximately 120°.

Figure 9.1 Electronic structure of the carbonyl group.

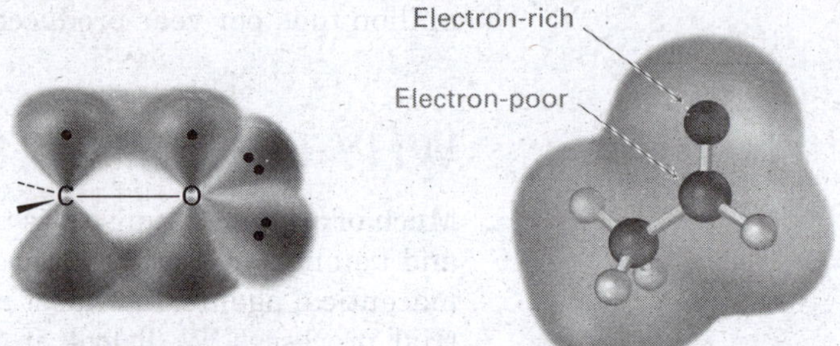

Carbonyl group

As indicated by the electrostatic potential map in Figure 9.1, the carbon–oxygen double bond is polarized because of the high electronegativity of oxygen relative to carbon. Thus, the carbonyl carbon is positively polarized and electrophilic (Lewis acidic), while the carbonyl oxygen is negatively polarized and nucleophilic (Lewis basic). We'll see in this and the next two chapters that most carbonyl-group reactions are the result of this bond polarity.

Problem 9.1 Propose structures for molecules that meet the following descriptions:

(a) A ketone, $C_5H_{10}O$ **(b)** An aldehyde, $C_6H_{10}O$
(c) A keto aldehyde, $C_6H_{10}O_2$ **(d)** A cyclic ketone, C_5H_8O

9.2 Naming Aldehydes and Ketones

Aldehydes are named by replacing the terminal *-e* of the corresponding alkane name with *-al*. The parent chain must contain the –CHO group, and numbering begins at the –CHO carbon, which is always C1. In the following examples, note that the longest chain in 2-ethyl-4-methylpentanal is actually a hexane, but this chain does not include the –CHO group and thus is not the parent.

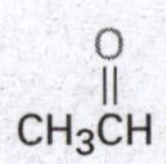

Ethanal (acetaldehyde)

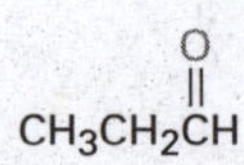

Propanal (propionaldehyde)

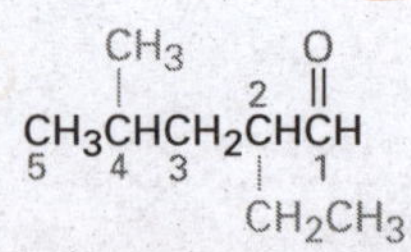

2-Ethyl-4-methylpentanal

For more complex aldehydes in which the –CHO group is attached to a ring, the suffix *-carbaldehyde* is used.

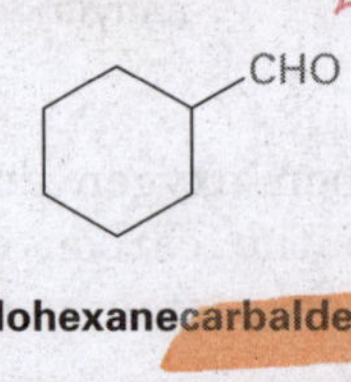

Cyclohexanecarbaldehyde

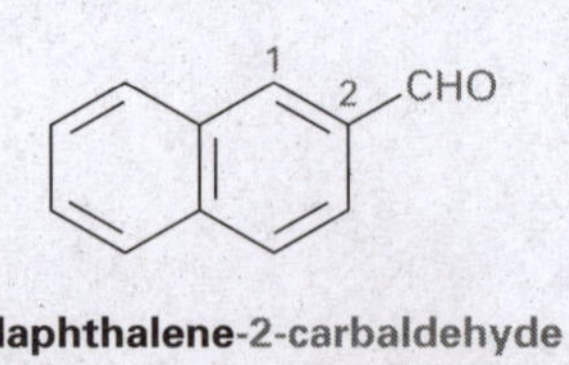

Naphthalene-2-carbaldehyde

A few simple and well-known aldehydes have common names that are recognized by IUPAC. Several that you might encounter are listed in Table 9.1.

Table 9.1 Common Names of Some Simple Aldehydes

Formula	Common name	Systematic name
HCHO	Formaldehyde	Methanal
CH_3CHO	Acetaldehyde	Ethanal
$H_2C{=}CHCHO$	Acrolein	Propenal
$CH_3CH{=}CHCHO$	Crotonaldehyde	But-2-enal
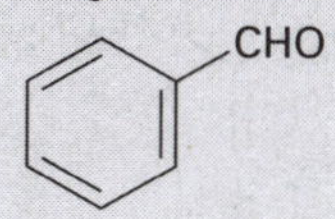	Benzaldehyde	Benzenecarbaldehyde

Ketones are named by replacing the terminal *-e* of the corresponding alkane name with *-one.* The parent chain is the longest one that contains the ketone group, and numbering begins at the end nearer the carbonyl carbon. As with alkenes (Section 3.1) and alcohols (Section 8.1), the numerical locant is placed immediately before the *-one* suffix in newer IUPAC recommendations. For example:

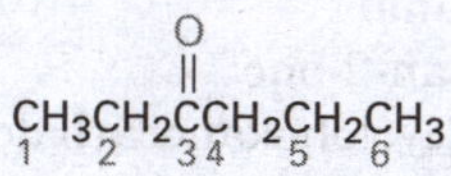

Hexan-3-one

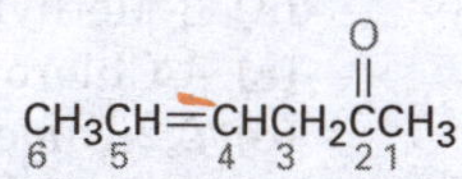

Hex-4-en-2-one

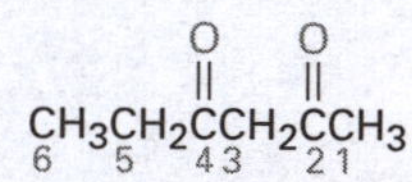

Hexane-2,4-dione

A few ketones also have common names.

Acetone

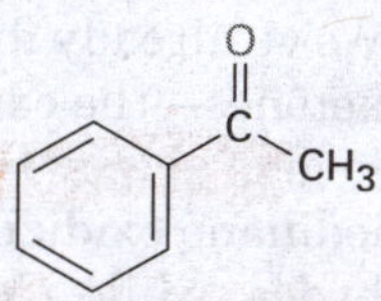

Acetophenone

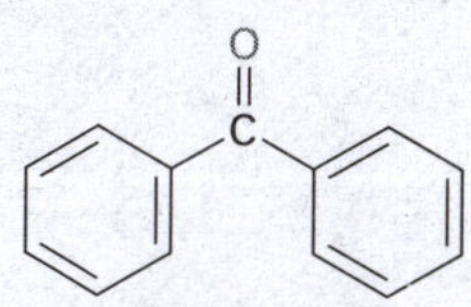

Benzophenone

When it's necessary to refer to the –COR group as a substituent, the general term **acyl group** is used. More specifically, $-COCH_3$ is an *acetyl* group, –CHO is a *formyl* group, –COAr is an *aroyl* group, and $-COC_6H_5$ is a *benzoyl* group.

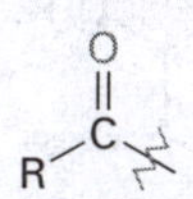

An acyl group (R = alkyl, alkenyl)

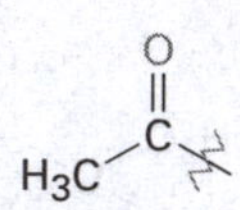

Acetyl

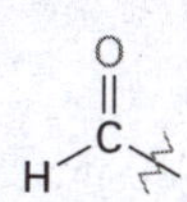

Formyl

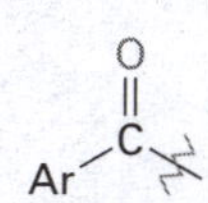

Aroyl (Ar = aromatic)

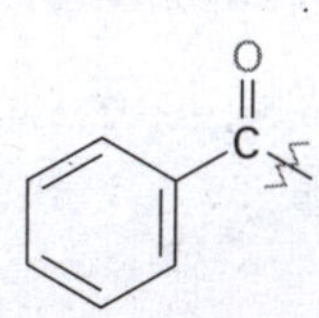

Benzoyl

Occasionally, the doubly bonded oxygen is considered a substituent, and the prefix *oxo-* is used. For example:

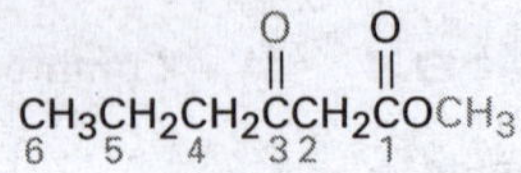

Methyl 3-oxohexanoate

Problem 9.2 Name the following aldehydes and ketones:

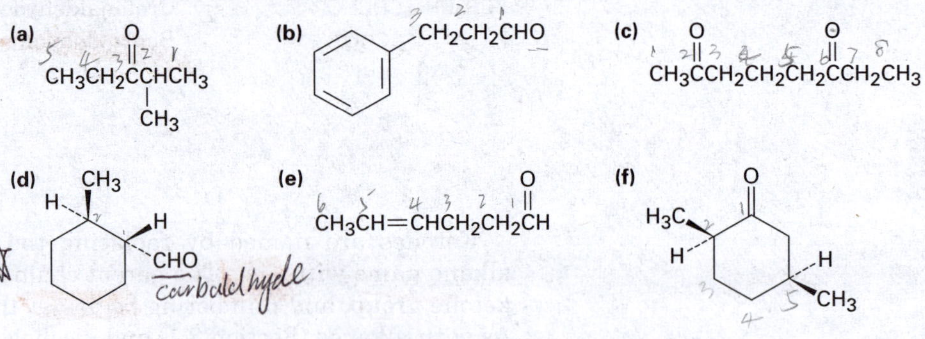

Problem 9.3 Draw structures corresponding to the following IUPAC names:

(a) 3-Methylbutanal
(b) 3-Methylbut-3-enal
(c) 4-Chloropentan-2-one
(d) Phenylacetaldehyde
(e) 2,2-Dimethylcyclohexanecarbaldehyde
(f) Cyclohexane-1,3-dione

9.3 Synthesis of Aldehydes and Ketones

We've already discussed one of the best methods for preparing aldehydes and ketones—the oxidation of alcohols (Section 8.4). Primary alcohols are oxidized to give aldehydes, and secondary alcohols are oxidized to give ketones. A periodinane oxidizing agent in dichloromethane is often chosen for making aldehydes, while CrO_3 and $Na_2Cr_2O_7$ in aqueous acid are frequently used for making ketones.

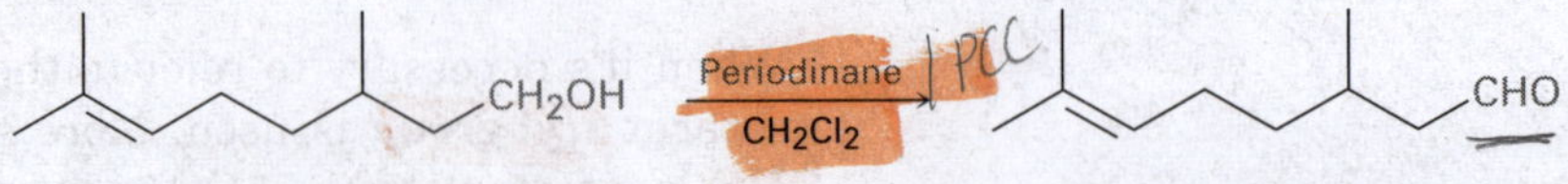

Citronellol **Citronellal (82%)**

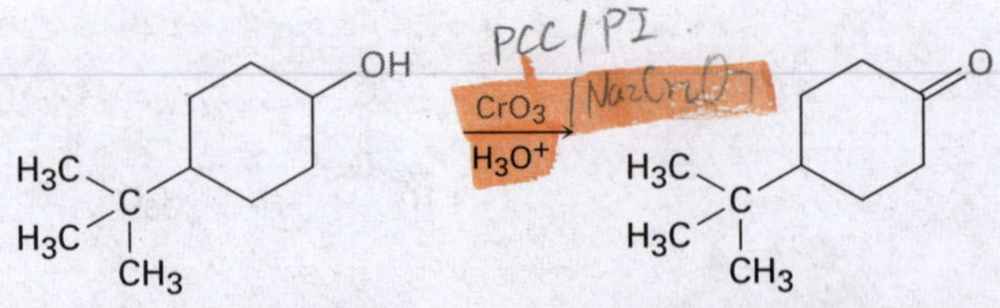

4-*tert*-Butylcyclohexanol **4-*tert*-Butylcyclohexanone (90%)**

Other methods we've seen for preparing ketones include the hydration of a terminal alkyne to yield a methyl ketone (Section 4.11) and the Friedel–Crafts acylation of an aromatic ring (Section 5.5).

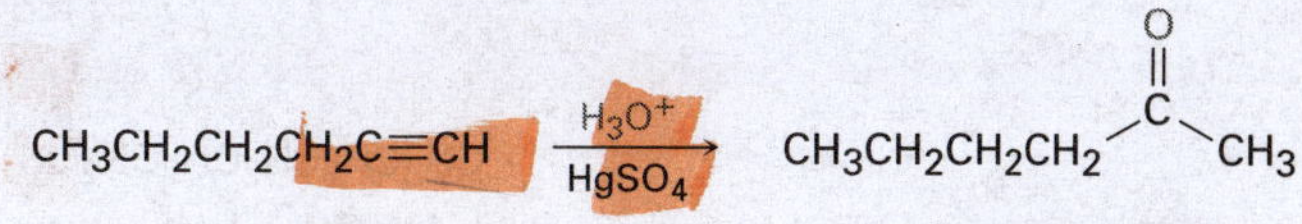

Hex-1-yne **Hexan-2-one (78%)**

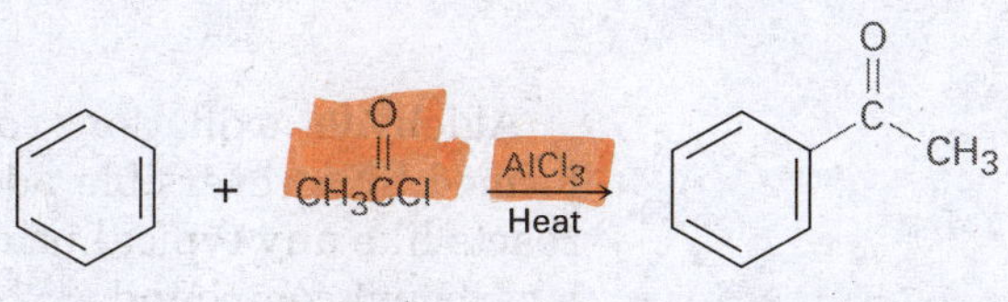

Benzene **Acetyl chloride** **Acetophenone (95%)**

Problem 9.4 How could you prepare pentanal from the following starting materials?

(a) Pentan-1-ol **(b)** $CH_3CH_2CH_2CH_2CO_2H$ **(c)** Dec-5-ene

Problem 9.5 How could you prepare hexan-2-one from the following starting materials?

(a)

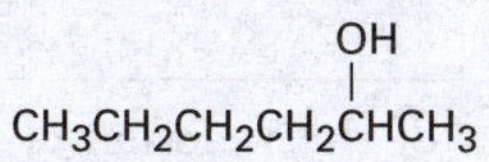

(b) $CH_3CH_2CH_2CH_2C{\equiv}CH$

(c)

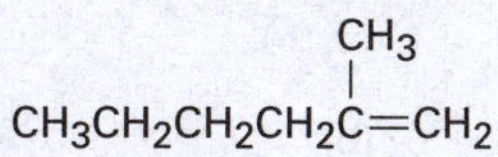

Problem 9.6 How would you carry out the following transformations? More than one step may be required.

(a) Hex-3-ene → Hexan-3-one **(b)** Benzene → 1-Phenylethanol

9.4 Oxidation of Aldehydes

Aldehydes are easily oxidized to yield carboxylic acids, RCHO → RCO_2H, but ketones are unreactive toward oxidation. This reactivity difference is a consequence of structure: aldehydes have a –CHO proton that can be removed during oxidation, but ketones do not.

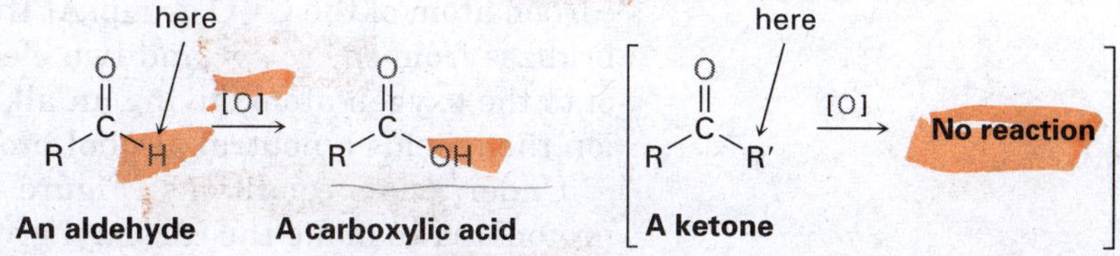

An aldehyde **A carboxylic acid** **A ketone**

Many oxidizing agents convert aldehydes into carboxylic acids, but CrO_3 in aqueous acid is a common choice. The oxidation takes place rapidly at room temperature.

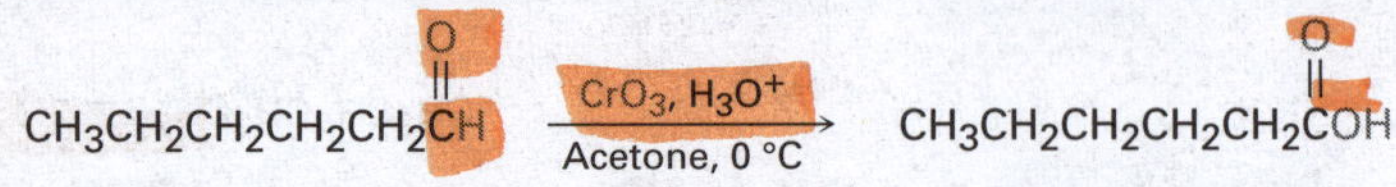

Hexanal — Hexanoic acid (85%)

Aldehyde oxidations occur through intermediate *hydrates,* which are formed by a reversible addition of water to the carbonyl group. The hydrate reacts like any typical primary or secondary alcohol and is rapidly oxidized to a carbonyl compound.

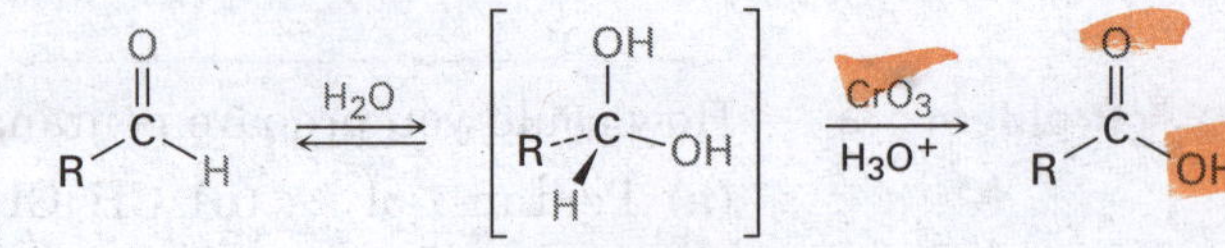

An aldehyde — A hydrate — A carboxylic acid

Problem 9.7 Predict the products of the reaction of the following substances with CrO_3 in aqueous acid:

(a) 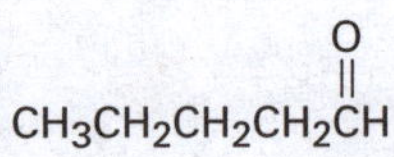

(b) 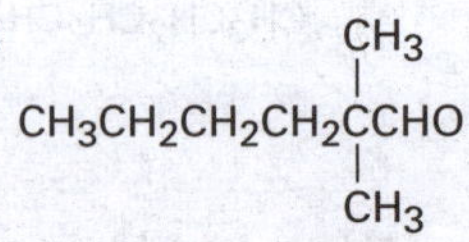

(c) 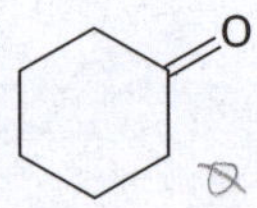

9.5 Nucleophilic Addition Reactions

The most common reaction of aldehydes and ketones is the **nucleophilic addition reaction**, in which a nucleophile adds to the electrophilic carbon of the carbonyl group. As shown in Figure 9.2, the reaction can take place under either basic or acidic conditions.

Under basic conditions (Figure 9.2a), the nucleophile is negatively charged ($:Nu^-$) and uses a pair of its electrons to form a bond to the electrophilic carbon atom of the C=O group. At the same time, the C=O carbon atom rehybridizes from sp^2 to sp^3 and two electrons from the C=O π bond are pushed onto the oxygen atom, giving an alkoxide ion. Addition of H^+ to the alkoxide ion then yields a neutral alcohol product.

Under acidic conditions (Figure 9.2b), the carbonyl oxygen atom is first protonated to make the carbonyl group more strongly electrophilic. A neutral nucleophile (:Nu—H) then uses a pair of electrons to bond to the carbon atom of the C=O group, and two electrons from the C=O π bond move onto the oxygen atom. The positive charge on oxygen is thereby neutralized, while the nucleophile gains a positive charge. Finally, a deprotonation gives the neutral alcohol addition product and regenerates the acid catalyst.

MECHANISM

(a) Basic conditions

1. A negatively charged nucleophile :Nu⁻ adds to the electrophilic carbon and pushes π electrons from the C=O bond onto oxygen, giving an alkoxide ion.
2. The alkoxide ion is protonated, either by added acid H–A or by solvent, to give a neutral alcohol addition product.

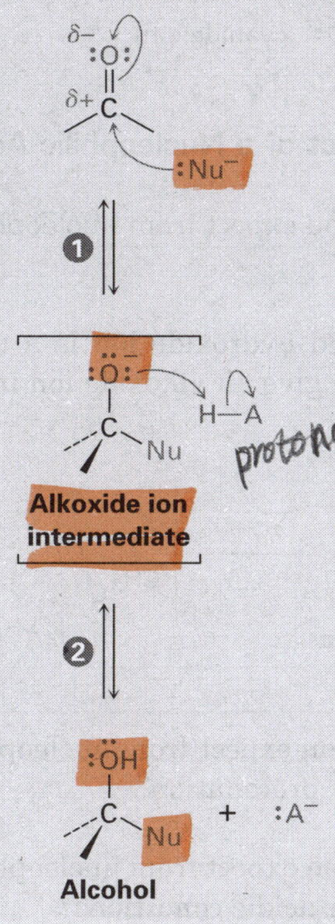

(b) Acidic conditions

1. The carbonyl oxygen is protonated by an acid H–A, making the carbon more strongly electrophilic.
2. A neutral nucleophile :Nu–H adds to the electrophilic carbon, pushing the π electrons from the C=O onto oxygen. The oxygen becomes neutral, and the nucleophile gains the + charge.
3. A base deprotonates the intermediate, giving the neutral alcohol addition product and regenerating the acid catalyst H–A.

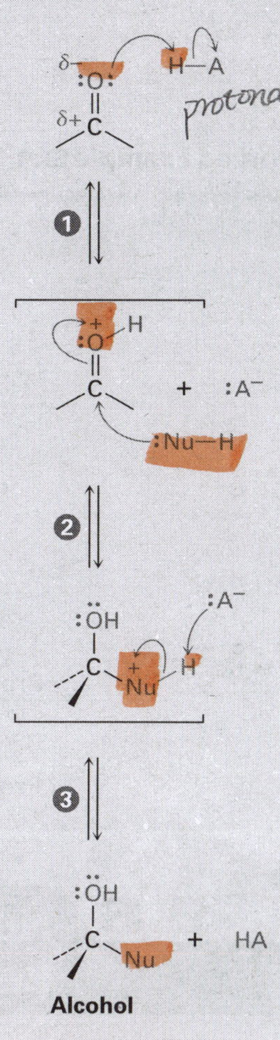

Figure 9.2 General mechanism of a nucleophilic addition reaction of aldehydes and ketones under both basic and acidic conditions. **(a)** Under basic conditions, a negatively charged nucleophile adds to the carbonyl group to give an alkoxide ion intermediate, which is subsequently protonated. **(b)** Under acidic conditions, protonation of the carbonyl group occurs first, followed by addition of a neutral nucleophile and subsequent deprotonation.

Note that the main difference between the basic and acidic reactions is in the timing of the protonation. Under basic conditions, the nucleophile is negatively charged and protonation occurs last. Under acidic conditions, the nucleophile is neutral and protonation occurs first. Examples of some common nucleophiles are

Some negatively charged nucleophiles (basic conditions)		Some neutral nucleophiles (acidic conditions)	
$H\ddot{O}:^-$	(hydroxide ion)	$H\ddot{O}H$	(water)
$H:^-$	(hydride ion)	$R\ddot{O}H$	(alcohol)
$R_3C:^-$	(carbanion)	$:NH_3$	(ammonia)
$R\ddot{O}:^-$	(alkoxide ion)	$R\ddot{N}H_2$	(amine)
$N{\equiv}C:^-$	(cyanide ion)		

Worked Example 9.1

Predicting the Product of a Nucleophilic Addition Reaction

What product would you expect from nucleophilic addition of aqueous hydroxide ion to acetaldehyde?

Strategy The negatively charged hydroxide ion is a nucleophile, which can add to the C=O carbon atom and give an alkoxide ion intermediate. Protonation will then yield a hydrate.

Solution

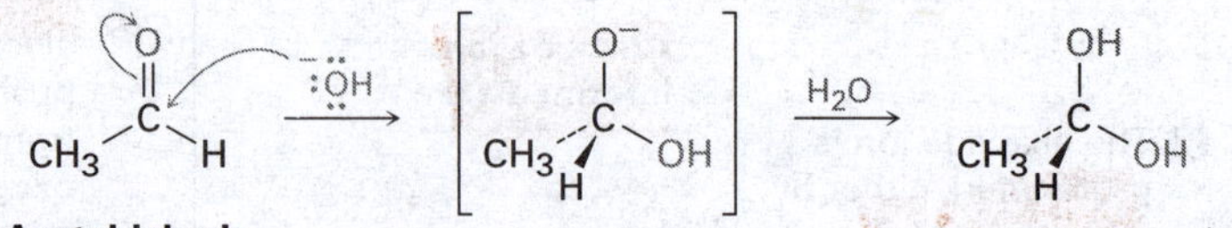

Acetaldehyde

Problem 9.8 What product would you expect from nucleophilic addition of cyanide ion, CN^-, to acetone, followed by protonation?

Problem 9.9 What product would you expect from nucleophilic addition of methanol, CH_3OH, to benzaldehyde under acidic conditions?

9.6 Nucleophilic Addition of Hydride and Grignard Reagents: Alcohol Formation

Addition of Hydride Reagents: Reduction

As we saw in Section 8.3, the most common method for preparing alcohols, both in the laboratory and in living organisms, is by the reduction of carbonyl compounds. Aldehydes are reduced with sodium borohydride ($NaBH_4$) to give primary alcohols, and ketones are similarly reduced to give secondary alcohols.

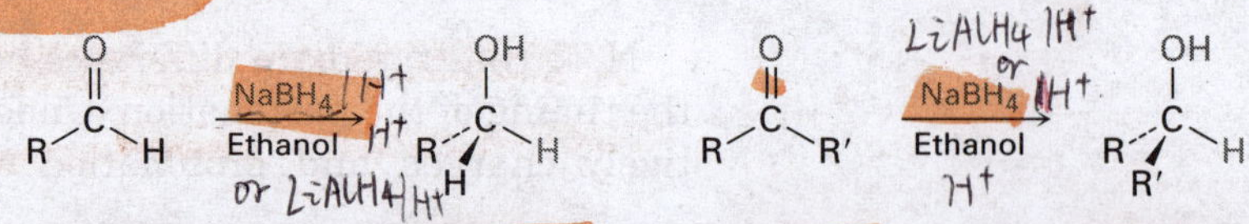

Aldehyde — 1° Alcohol — Ketone — 2° Alcohol

Carbonyl reduction occurs by a typical nucleophilic addition mechanism under basic conditions, as shown previously in Figure 9.2a. The nucleophile is a negatively charged hydride ion (:H^-) supplied by $NaBH_4$, and the initially formed alkoxide ion intermediate is protonated by ethanol solvent. The reaction is irreversible because the reverse process would require expulsion of a very poor leaving group.

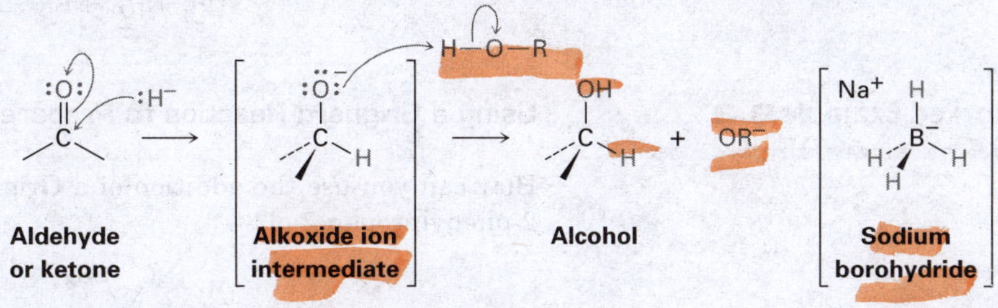

Addition of Grignard Reagents

Just as aldehydes and ketones undergo nucleophilic addition with hydride ion to give alcohols, they undergo a similar addition with Grignard reagents, R:$^-$ $^+$MgX (Section 8.3). Aldehydes give secondary alcohols on reaction with Grignard reagents in ether solution, and ketones give tertiary alcohols.

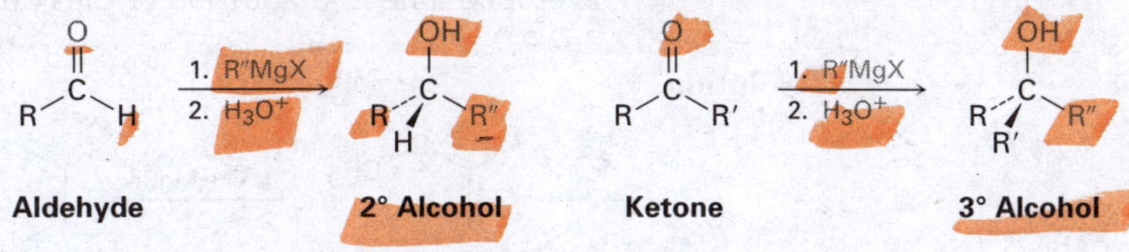

Like the reaction with hydride ion, a Grignard reaction takes place by a typical nucleophilic addition mechanism under basic conditions. The nucleophile is a carbanion (R:$^-$) from the Grignard reagent, which adds to the C=O bond and produces a tetrahedrally hybridized magnesium alkoxide intermediate. Protonation by addition of aqueous acid in a separate step then gives the neutral alcohol. Like reduction, the Grignard reaction is irreversible.

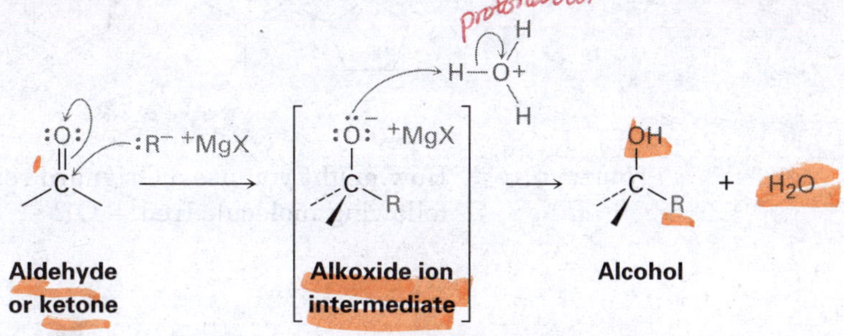

Although widely applicable, the Grignard reaction also has limitations. For example, a Grignard reagent can't be prepared from an organohalide that has other reactive functional groups in the same molecule. Some functional groups—carbonyls, for instance—cause the Grignard reagent to add to itself. Other groups—alcohols, for instance—destroy the Grignard reagent by protonation (Section 7.3). In general, Grignard reagents can't be prepared from compounds that contain the following functional groups:

–CHO, –COR, –CONR$_2$, –C≡N, –NO$_2$, –SO$_2$R } A Grignard reagent reacts with these groups.

–OH, –NH$_2$, –NHR, –SH, –CO$_2$H } A Grignard reagent is protonated by these groups.

Worked Example 9.2 Using a Grignard Reaction to Prepare an Alcohol

How can you use the addition of a Grignard reagent to a ketone to synthesize 2-phenylpropan-2-ol?

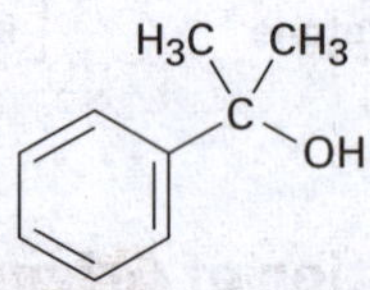

2-Phenylpropan-2-ol

Strategy Look at the product, and identify the groups bonded to the alcohol carbon atom. In this instance, there are two methyl groups ($-CH_3$) and one phenyl ($-C_6H_5$). One of the three must come from a Grignard reagent, and the remaining two must come from a ketone. Thus, the possibilities are addition of CH_3MgBr to acetophenone and addition of C_6H_5MgBr to acetone.

Solution

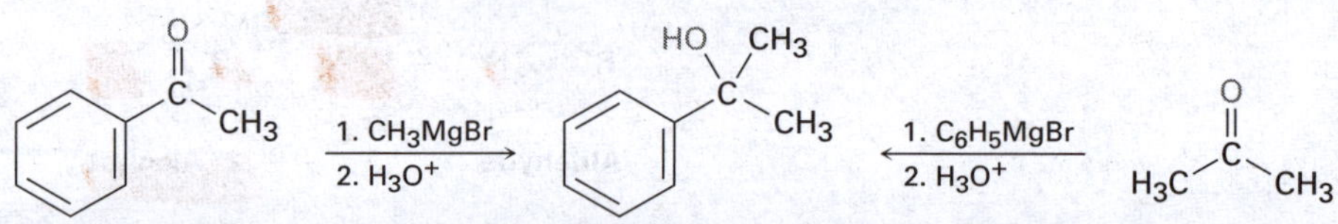

Acetophenone **2-Phenylpropan-2-ol** **Acetone**

Problem 9.10 How might you prepare the following alcohols from an aldehyde or ketone? Show all possibilities.

(a)

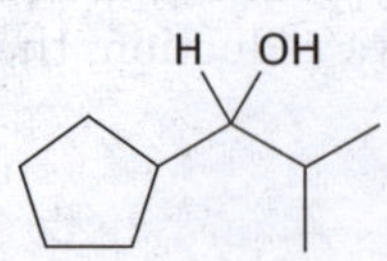

(b)

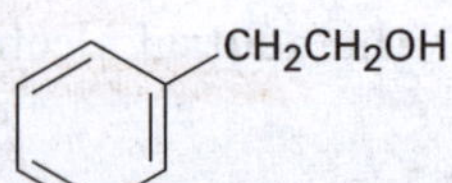

(c)

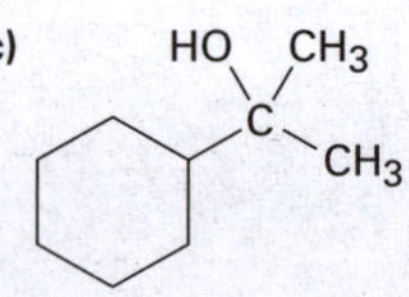

Problem 9.11 How might you use a Grignard reaction of an aldehyde or ketone to prepare the following molecule (red = O)?

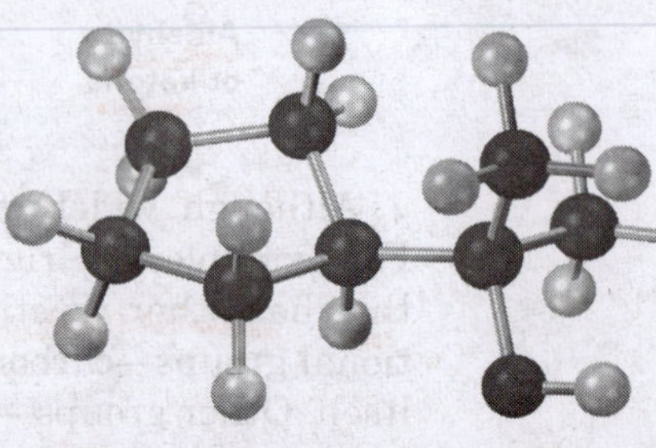

9.7 Nucleophilic Addition of Water: Hydrate Formation

Aldehydes and ketones undergo a nucleophilic addition reaction with water to yield the corresponding carbonyl hydrates, sometimes called **geminal (gem) diols** from the Latin *geminus,* meaning "twin." The reaction is reversible, and the gem diol product can eliminate water to regenerate a ketone or aldehyde.

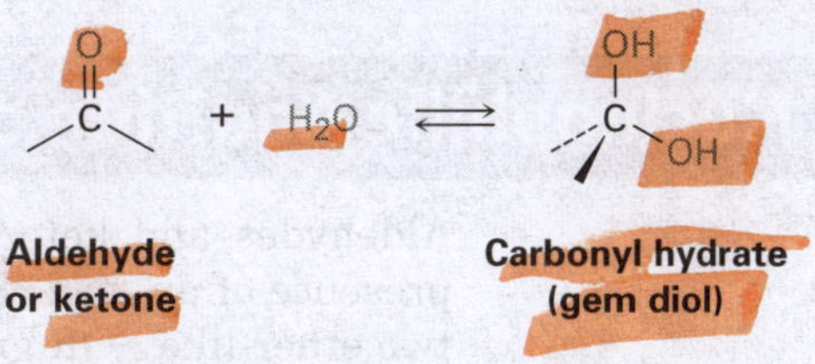

The position of the equilibrium between hydrate and aldehyde/ketone depends on the structure of the carbonyl compound. The equilibrium strongly favors the carbonyl compound in most cases, with the hydrate favored only for a few simple aldehydes. An aqueous solution of acetone, for instance, consists of about 0.1% hydrate and 99.9% ketone, whereas an aqueous solution of formaldehyde (CH_2O) consists of 99.9% hydrate and 0.1% aldehyde.

The nucleophilic addition of water to aldehydes and ketones is slow in pure water but is catalyzed by both base and acid. The base-catalyzed addition reaction takes place by the mechanism shown previously in Figure 9.2a, with the negatively charged hydroxide ion as the nucleophile and water as the proton source in the final step.

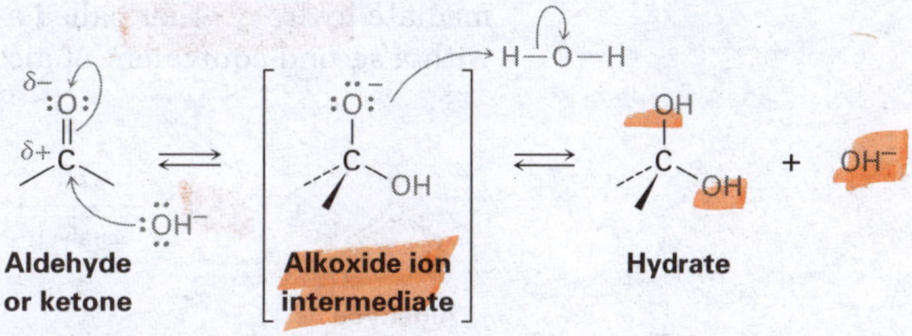

The acid-catalyzed reaction takes place by the mechanism shown previously in Figure 9.2b. The acid catalyst first protonates the basic oxygen atom of the carbonyl group to increase its reactivity, water adds as the nucleophile, and deprotonation gives the gem diol product.

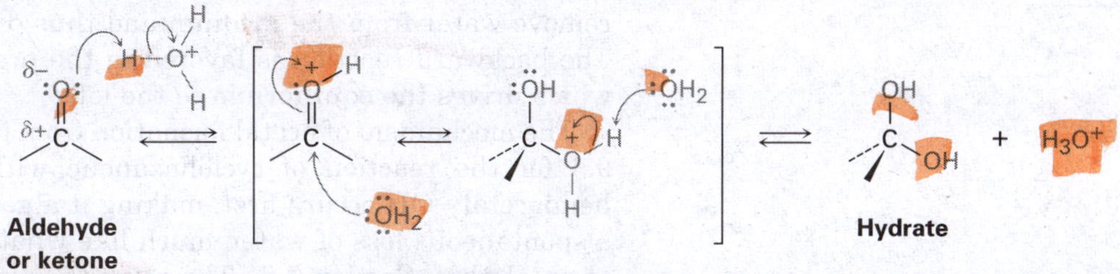

Note the crucial difference between the base-catalyzed and acid-catalyzed processes: the *base*-catalyzed reaction takes place rapidly because hydroxide ion is a much better *nucleophile* than neutral water. The *acid*-catalyzed reaction takes place rapidly because the protonated carbonyl compound is a much better *electrophile* than the neutral compound.

Problem 9.12 The oxygen in water is primarily (99.8%) ^{16}O, but water enriched with the heavy isotope ^{18}O is also available. When a ketone or aldehyde is dissolved in $H_2^{18}O$, the isotopic label becomes incorporated into the carbonyl group: $R_2C{=}O + H_2^{18}O \rightarrow R_2C{=}^{18}O + H_2O$. Explain.

9.8 Nucleophilic Addition of Alcohols: Acetal Formation

Aldehydes and ketones undergo a reversible reaction with alcohols in the presence of an acid catalyst to yield **acetals, $R_2C(OR')_2$**, compounds that have two ether-like –OR groups bonded to the same carbon.

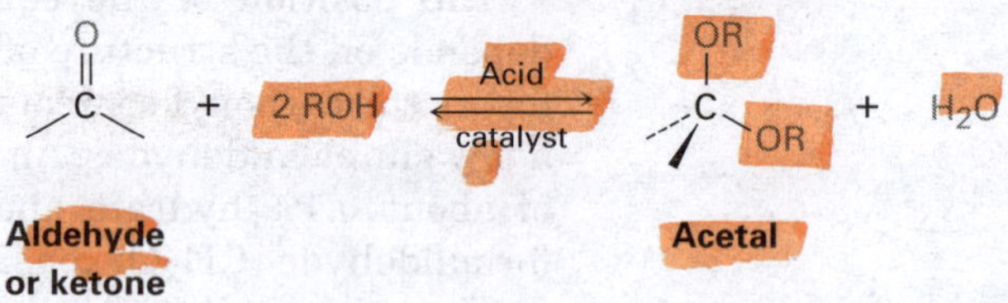

Acetal formation involves the acid-catalyzed nucleophilic addition of an alcohol to an aldehyde or ketone in a process analogous to that of the acid-catalyzed addition of water discussed in the previous section. The initial nucleophilic addition step occurs by the usual mechanism (Figure 9.2b) and yields an intermediate hydroxy ether called a **hemiacetal**. The hemiacetal then reacts further with a second equivalent of alcohol and gives the acetal plus water.

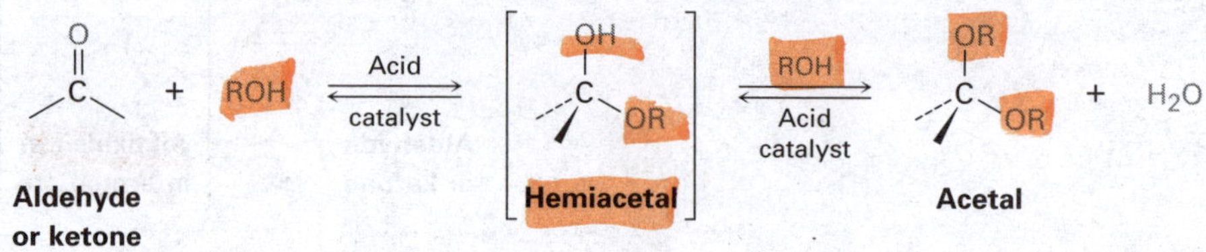

As with hydrate formation, all the steps during acetal formation are reversible, and the reaction can be made to go either forward (from carbonyl compound to acetal) or backward (from acetal to carbonyl compound), depending on the reaction conditions. The forward reaction is favored by conditions that remove water from the medium and thus drive the equilibrium to the right. The backward reaction is favored in the presence of a large excess of water, which drives the equilibrium to the left.

The mechanism of acetal formation from the hemiacetal is shown in Figure 9.3 for the reaction of cyclohexanone with methanol. Protonation of the hemiacetal –OH occurs first, making it a good leaving group and facilitating a spontaneous loss of water, much like what happens during the E1 reaction of an alcohol (Section 7.8). The resonance-stabilized cation that results then undergoes nucleophilic addition of alcohol to the C=O bond, followed by deprotonation to give the acetal.

MECHANISM

Figure 9.3 Mechanism of formation of an acetal from a hemiacetal. Protonation and loss of water give an intermediate cation, which undergoes a nucleophilic addition reaction with methanol.

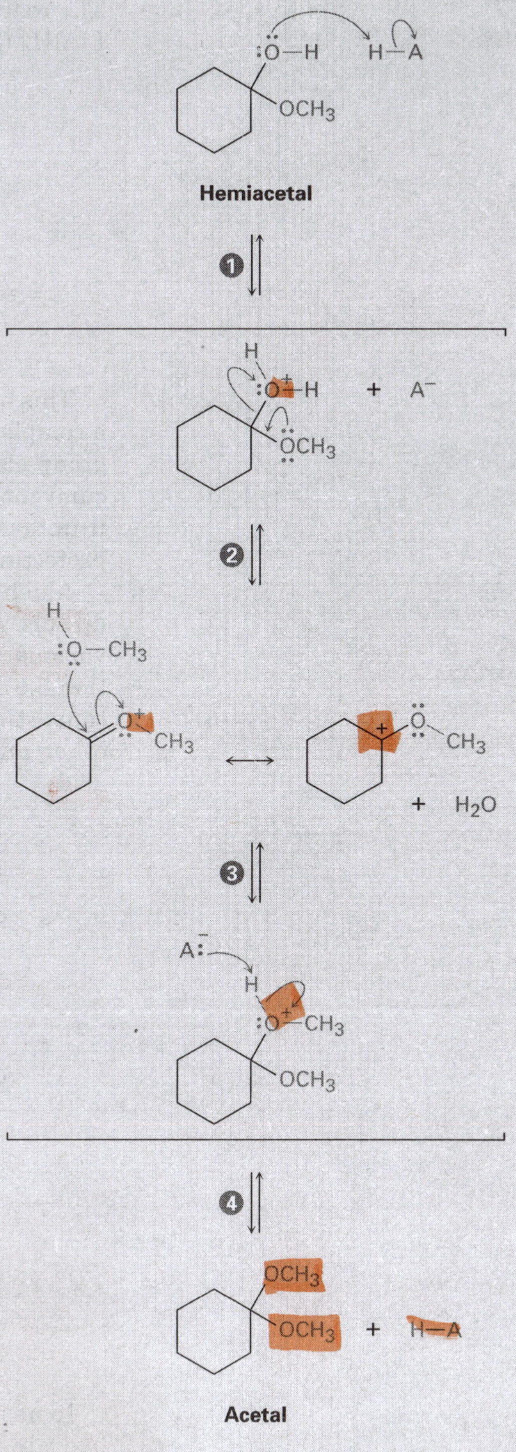

1. The –OH of the hemiacetal is protonated by an acid H–A, making it a good leaving group.

2. An electron pair on the –OCH_3 group moves toward carbon, expelling water and giving a C=$\overset{+}{O}CH_3$ bond with a positively charged, trivalent oxygen.

3. Nucleophilic addition of methanol to the C=O bond pushes the π electrons toward oxygen and neutralizes the positive charge.

4. Deprotonation by the base :A^- gives the neutral acetal and regenerates the acid catalyst.

Acetals are valuable to organic chemists because they can serve as **protecting groups** for aldehydes and ketones. To see what this means, imagine that you need to reduce an ester group selectively in the presence of a keto group. The reaction can't be done in a single step because treating a keto ester with $LiAlH_4$ (Section 8.3) would reduce both groups.

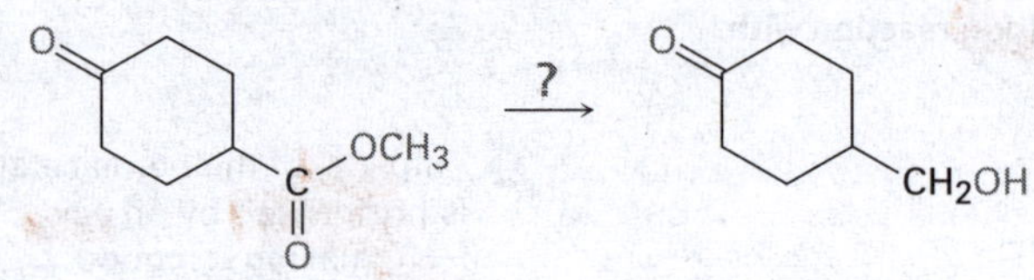

This situation isn't unusual; it often happens that one functional group in a complex molecule interferes with intended chemistry on another functional group elsewhere in the molecule. In such situations, it's often possible to circumvent the problem by *protecting* the interfering functional group to render it unreactive, then carrying out the desired reaction, and then removing the protecting group.

Aldehydes and ketones are usually protected by converting them into acetals. Acetals, like other ethers, are stable to bases, reducing agents, and various nucleophiles, but they can be cleaved by treatment with acid (Section 8.6). Thus, you can selectively reduce the ester group in a keto ester by converting the keto group into an acetal, reducing the ester with $LiAlH_4$ in ether, and removing the acetal protecting group by treatment with aqueous acid.

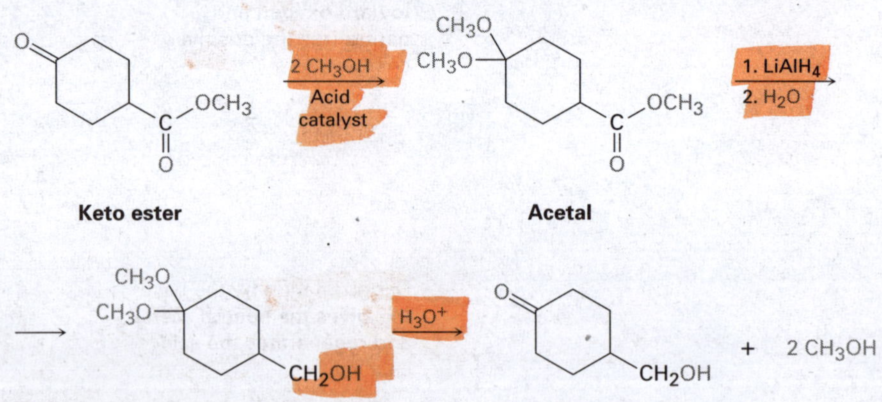

In nature, acetal and hemiacetal groups are particularly common in carbohydrate chemistry. Glucose, for instance, is a polyhydroxy aldehyde that undergoes a spontaneous *internal* nucleophilic addition reaction and exists primarily as a cyclic hemiacetal. Numerous glucose molecules then join together with acetal links to form either cellulose or starch. We'll study this and other reactions of carbohydrates in Chapter 14.

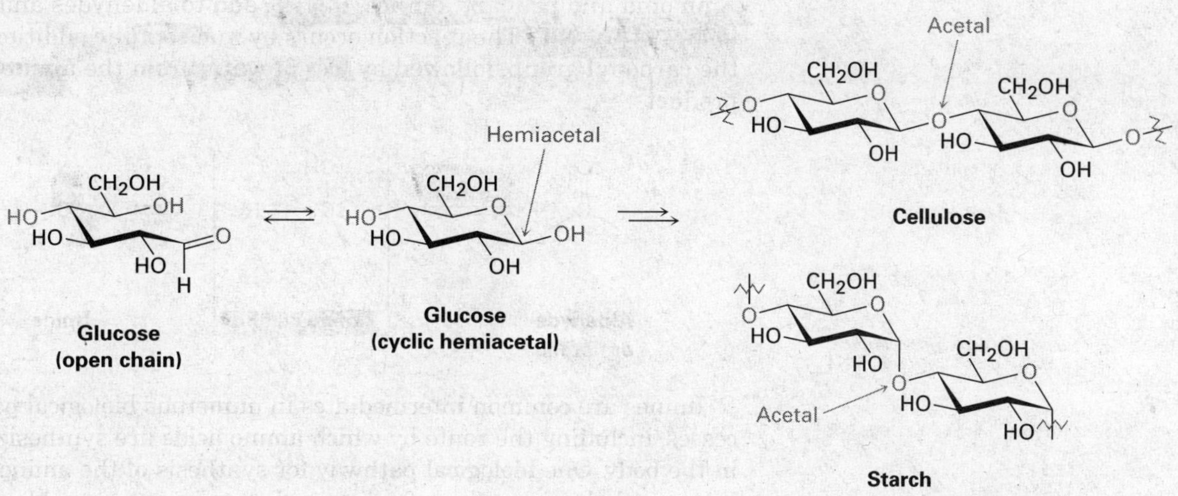

Worked Example 9.3 Predicting the Structure of an Acetal

What product would you obtain from the acid-catalyzed reaction of 2-methylcyclopentanone with methanol?

Strategy A ketone reacts with an alcohol in the presence of acid to yield an acetal. To find the product, replace the oxygen of the ketone with two $-OCH_3$ groups from the alcohol.

Solution

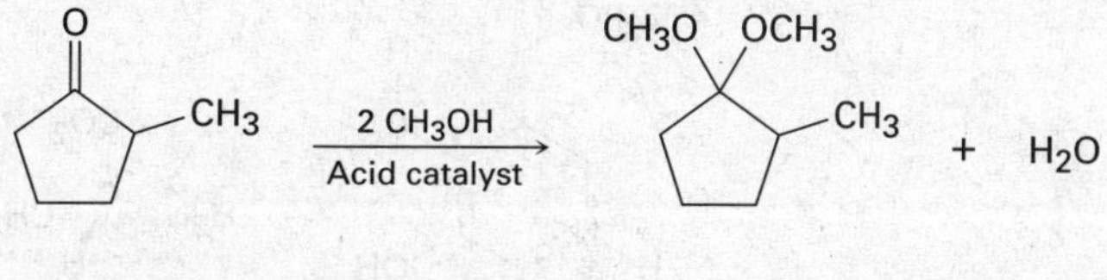

Problem 9.13 Write the mechanism of the acid-catalyzed reaction of methanol with cyclohexanone to give a hemiacetal.

Problem 9.14 When an aldehyde or ketone is treated with a diol such as ethylene glycol (ethane-1,2-diol) and an acid catalyst, a *cyclic* acetal is formed. Draw the structure of the product you would obtain from benzaldehyde and ethylene glycol.

Problem 9.15 Show how you might carry out the following transformation. (A protection step is needed.)

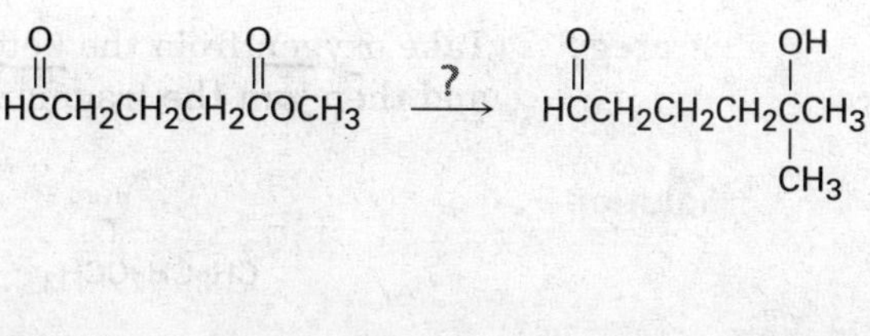

9.9 Nucleophilic Addition of Amines: Imine Formation

Ammonia and primary amines, $R'NH_2$, add to aldehydes and ketones to yield **imines, $R_2C{=}NR'$**. The reaction occurs by nucleophilic addition of the amine to the carbonyl group, followed by loss of water from the amino alcohol addition product.

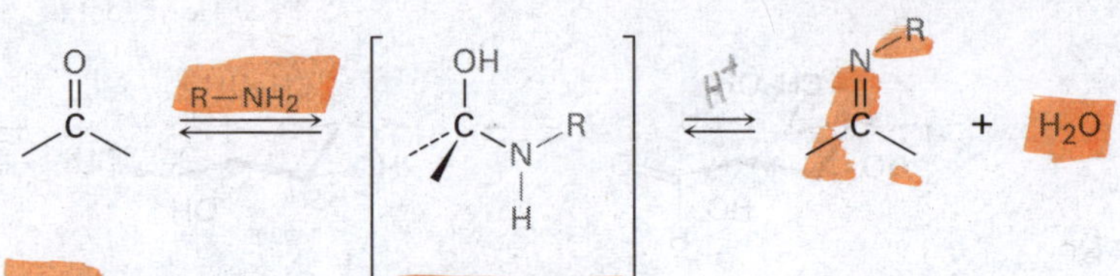

Aldehyde or ketone **Amino alcohol** **Imine**

Imines are common intermediates in numerous biological pathways and processes, including the route by which amino acids are synthesized and degraded in the body. One biological pathway for synthesis of the amino acid alanine, for instance, is by formation of an imine between pyruvic acid and ammonia, followed by reduction.

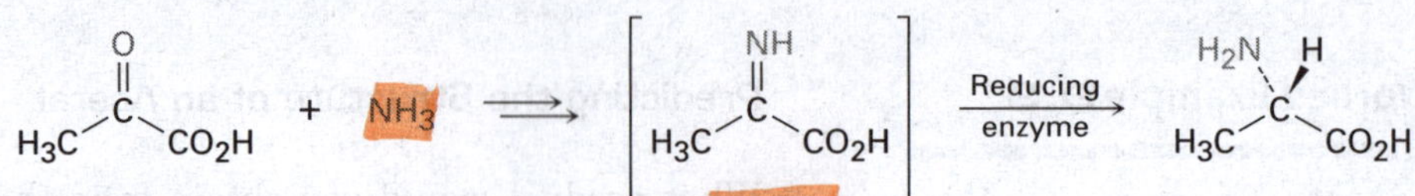

Pyruvic acid **Imine intermediate** **Alanine**

The pathway for biological degradation of alanine involves reaction with the aldehyde pyridoxal phosphate, a derivative of vitamin B_6, to yield an imine that is then further degraded.

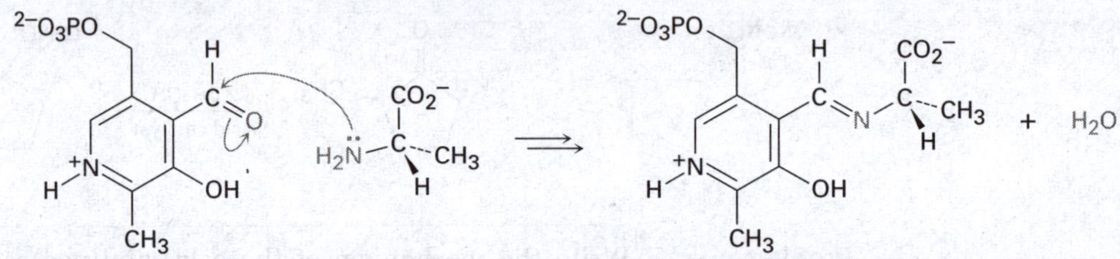

Pyridoxal phosphate **Alanine** **An imine**

Worked Example 9.4 Predicting the Product of Imine Formation

What product do you expect from the reaction of butan-2-one with hydroxylamine, NH_2OH?

Strategy Take oxygen from the ketone and two hydrogens from the amine to form water, and then join the fragments that remain.

Solution

$$CH_3CH_2C(=O)CH_3 + H_2NOH \longrightarrow CH_3CH_2C(=NOH)CH_3 + H_2O$$

Problem 9.16 Write the products you would obtain from treatment of cyclohexanone with the following:

(a) CH_3NH_2 **(b)** CH_3CH_2OH, H^+ **(c)** $NaBH_4$

Problem 9.17 Show how the following molecule can be prepared from a carbonyl compound and an amine (blue = N):

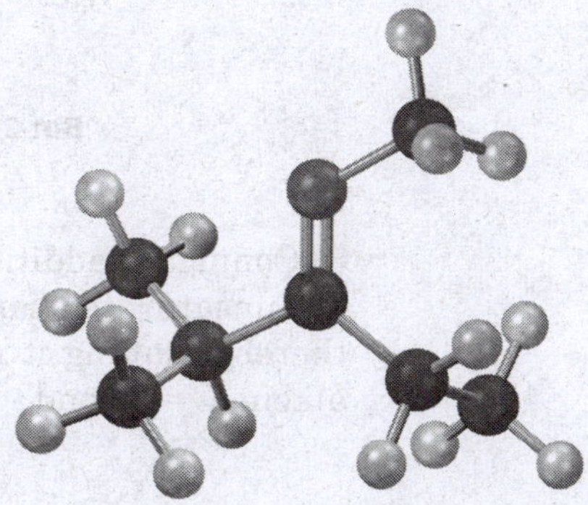

9.10 Conjugate Nucleophilic Addition Reactions

The reactions we've seen to this point have involved addition of a nucleophile directly to the carbonyl group in what is called a **1,2-addition**. Closely related to this direct addition is the *conjugate addition*, or **1,4-addition**, of a nucleophile to the C=C bond of an α,β-unsaturated aldehyde or ketone. (The carbon atom next to a carbonyl group is often called the α carbon, the next one is the β carbon, and so on.) Thus, an α,β-unsaturated aldehyde or ketone has its C=C and C=O bonds conjugated, much as a conjugated diene does (Section 4.8).

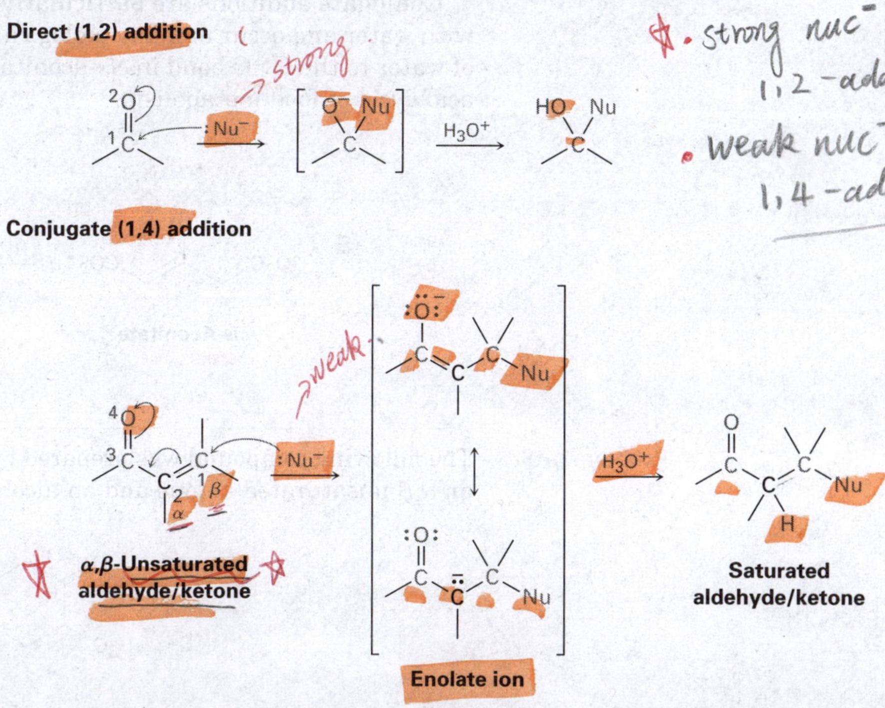

The initial product of conjugate addition is a resonance-stabilized **enolate ion**, which typically undergoes protonation on the α carbon to give a saturated aldehyde or ketone product. For example, methylamine reacts with but-3-en-2-one to give an amino ketone addition product.

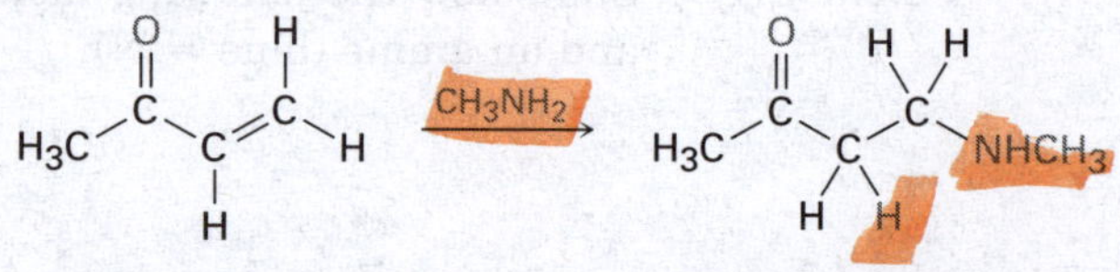

But-3-en-2-one   Conjugate addition product

Conjugate addition occurs because the electronegative oxygen atom of the α,β-unsaturated carbonyl compound withdraws electrons from the β carbon, thereby making it more electron-poor and more electrophilic than a typical alkene C=C bond.

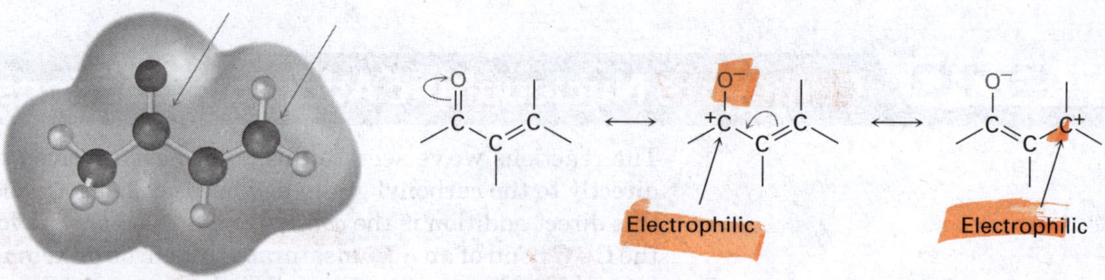

Conjugate additions are particularly common with amine nucleophiles and with water and occur in many biological pathways. An example is the addition of water to the C=C bond in *cis*-aconitate to give isocitrate, a step in the citric acid cycle of food metabolism.

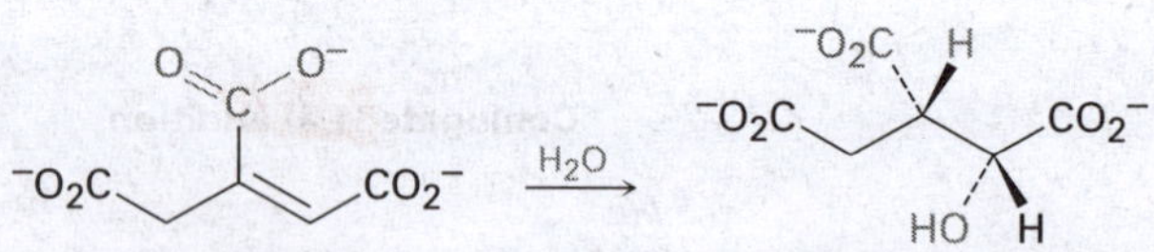

cis-Aconitate   Isocitrate

Problem 9.18 The following compound was prepared by a conjugate addition reaction between an α,β-unsaturated ketone and an alcohol. Identify the two reactants.

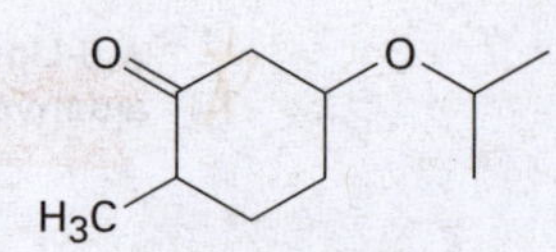

INTERLUDE

Vitamin C

© Bettmann/CORBIS

John Franklin's expedition in 1845 to chart the Northwest Passage between Atlantic and Pacific oceans resulted in the death of all 129 men aboard his two ships *Terror* and *Erebus*. Many of the men died of scurvy, a bleeding disease caused by a vitamin C deficiency.

Vitamin C, or ascorbic acid, is surely the best known of all vitamins. It was the first vitamin to be discovered (1928), the first to be structurally characterized (1933), and the first to be synthesized in the laboratory (1933). Over 200 million pounds of vitamin C are synthesized worldwide each year—more than the total amount of all other vitamins combined. In addition to its use as a vitamin supplement, vitamin C is used as a food preservative, a "flour improver" in bakeries, and an animal food additive.

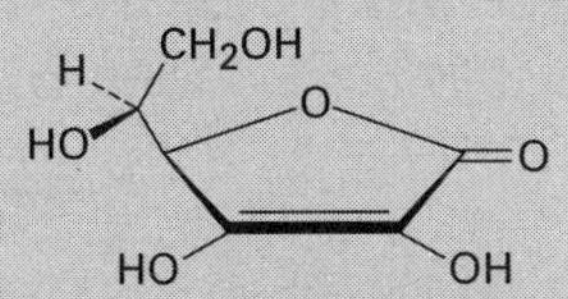

Vitamin C (ascorbic acid)

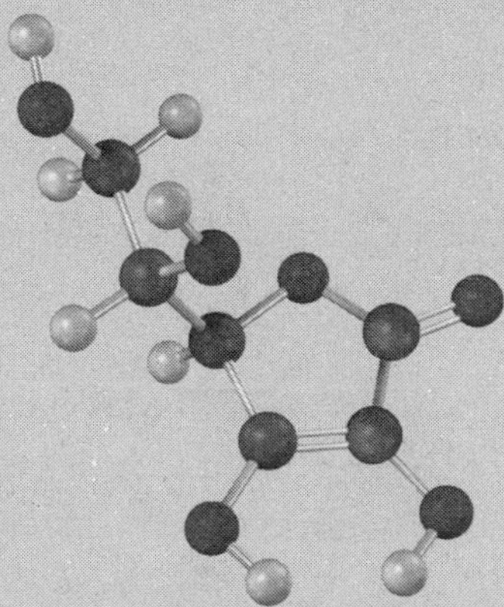

Vitamin C is perhaps most famous for its antiscorbutic properties, meaning that it prevents the onset of scurvy, a bleeding disease affecting those with a deficiency of fresh vegetables and citrus fruits in their diet. Sailors in the Age of Exploration were particularly susceptible to scurvy, and the death toll was high. The Portuguese explorer Vasco da Gama lost more than half his crew to scurvy during his 2-year voyage around the Cape of Good Hope in 1497–1499. Even as late as 1845, all 129 men aboard John Franklin's expedition to find a Northwest Passage died, many of scurvy.

In more recent times, large doses of vitamin C have been claimed to prevent the common cold, cure infertility, delay the onset of symptoms in AIDS, and inhibit the development of gastric and cervical cancers. None of these claims have been backed by medical evidence, however. In the largest study yet done of the effect of vitamin C on the common cold, a meta-analysis of more than 100 separate trials covering 40,000 people found no difference in the incidence of colds between those who took supplemental vitamin C regularly and those who did not. When taken *during* a cold, however, vitamin C does appear to decrease the cold's duration by 8%.

The industrial preparation of vitamin C involves an unusual blend of biological and laboratory organic chemistry. The Hoffmann-La Roche company synthesizes ascorbic acid from glucose through the five-step route shown in Figure 9.4. Glucose, a pentahydroxy aldehyde, is first reduced to sorbitol, which is then oxidized by the microorganism *Acetobacter suboxydans*. No chemical reagent is known that is selective enough to oxidize

continued

INTERLUDE

only one of the six alcohol groups in sorbitol, so an enzymatic reaction is used. Treatment with acetone and an acid catalyst then converts four of the other hydroxyl groups into acetal linkages, and the remaining hydroxyl group is chemically oxidized to a carboxylic acid by reaction with aqueous NaOCl (household bleach). Hydrolysis with acid then removes the two acetal groups and causes an internal ester-forming reaction to take place to give ascorbic acid. Each of the five steps takes place in better than 90% yield.

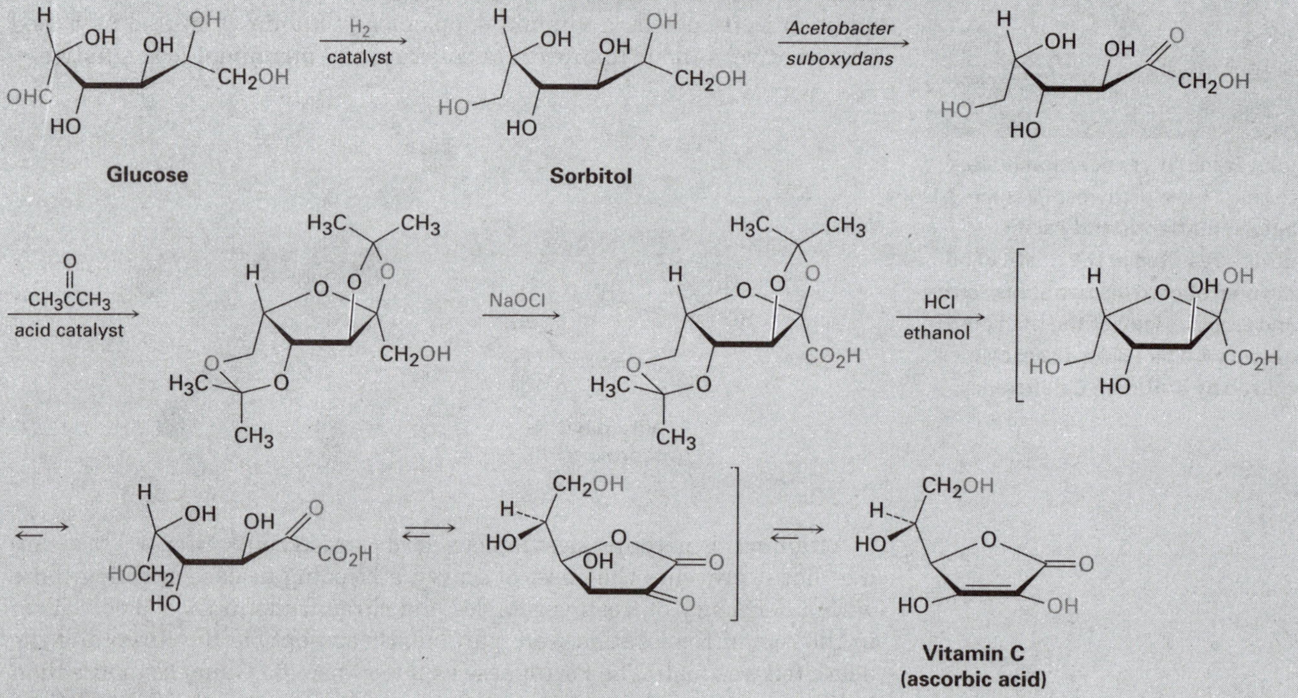

Summary and Key Words

acetal 306
acyl group 297
1,2-addition reaction 311
1,4-addition reaction 311
aldehyde (RCHO) 294
carbonyl group 294
enolate ion 312
geminal (gem) diol 305
hemiacetal 306
imine 310
ketone ($R_2C{=}O$) 294
nucleophilic addition reaction 300
protecting group 308

Aldehydes and ketones are among the most important of all compounds, both in the chemical industry and in biological chemistry. In this chapter, we've looked at some of their typical reactions.

A carbon–oxygen double bond is structurally similar to a carbon–carbon double bond. The carbonyl carbon atom is sp^2-hybridized and forms both an sp^2 σ bond and a p π bond to oxygen. Carbonyl groups are strongly polarized because of the electronegativity of oxygen.

Aldehydes are usually prepared by oxidation of primary alcohols, and ketones are often prepared by oxidation of secondary alcohols. Aldehydes and ketones behave similarly in much of their chemistry. Both undergo **nucleophilic addition reactions**, which are useful for preparing a variety of products. For example, aldehydes and ketones undergo addition of hydride ion ($H:^-$) to give alcohol reduction products. Similarly, addition of the carbanion from Grignard reagents ($R:^{-}\,^{+}MgX$) gives alcohol products.

Reversible addition of an aldehyde or ketone with water yields a hydrate, also called a **gem diol**. Similarly, aldehydes and ketones react reversibly with alcohols to yield first **hemiacetals** and then **acetals**. Acetals are particularly useful as carbonyl **protecting groups**. Ammonia and primary amines add to aldehydes and ketones to give **imines**, **$R_2C{=}NR'$**.

Closely related to the direct 1,2-addition of nucleophiles to aldehydes and ketones is the conjugate **1,4-addition** of nucleophiles to the C=C double bond of α,β-unsaturated aldehydes and ketones. Both direct and conjugate addition reactions are common in biological pathways.

Summary of Reactions

Nucleophilic addition reactions of aldehydes and ketones

1. Reaction with hydride reagents to yield alcohols (Section 9.6)

   (a) Reaction of aldehydes to yield primary alcohols

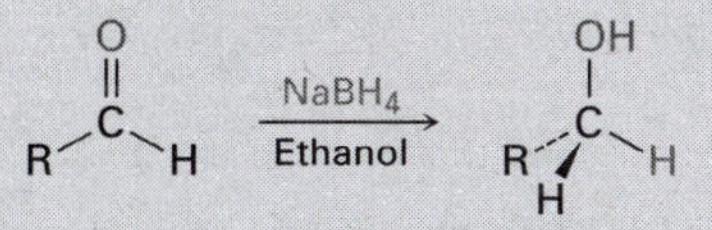

   (b) Reaction of ketones to yield secondary alcohols

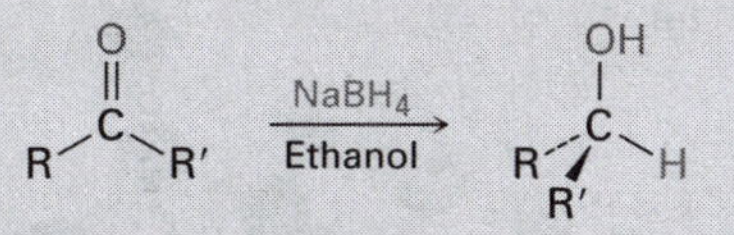

2. Reaction with Grignard reagents to yield alcohols (Section 9.6)

   (a) Reaction of aldehydes to yield secondary alcohols

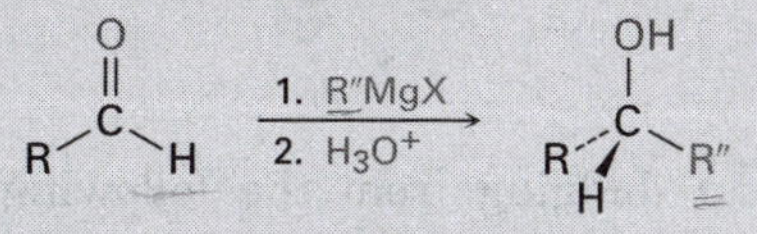

   (b) Reaction of ketones to yield tertiary alcohols

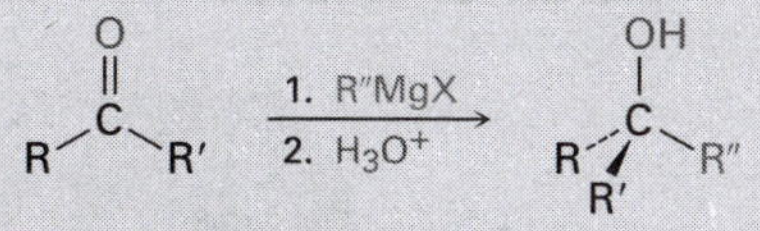

3. Reaction with alcohols to yield acetals (Section 9.8)

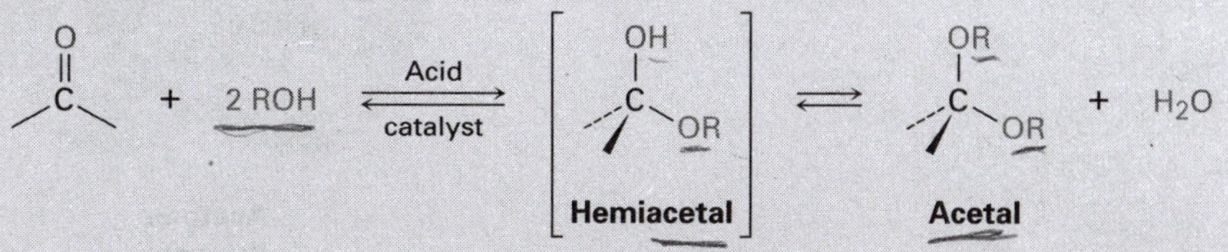

continued

4. Reaction with amines to yield imines (Section 9.9)

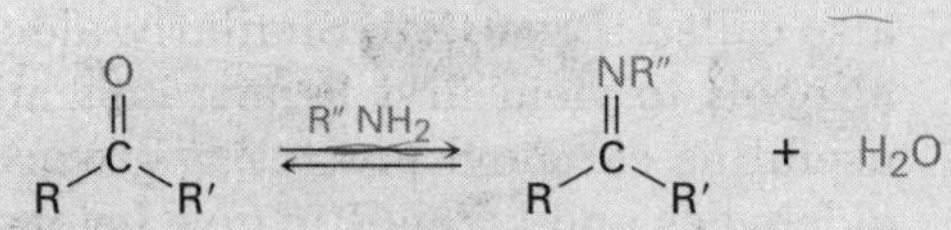

5. Conjugate (1,4) nucleophilic addition reaction (Section 9.10)

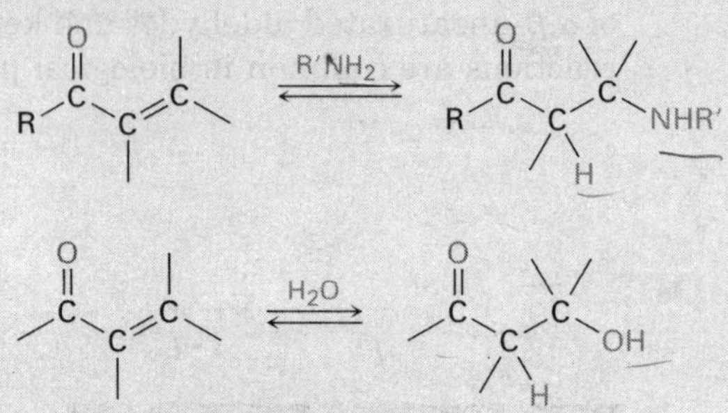

Exercises

Visualizing Chemistry

(Problems 9.1–9.18 appear within the chapter.)

OWL Interactive versions of these problems are assignable in OWL.

9.19 Identify the kinds of carbonyl groups in the following molecules (red = O, blue = N):

(a)

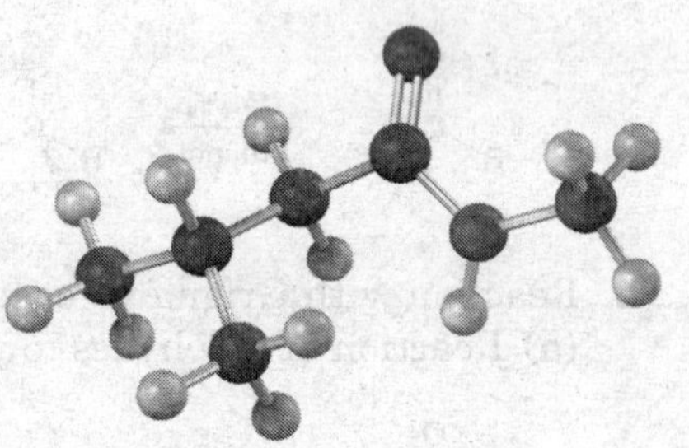

(b)

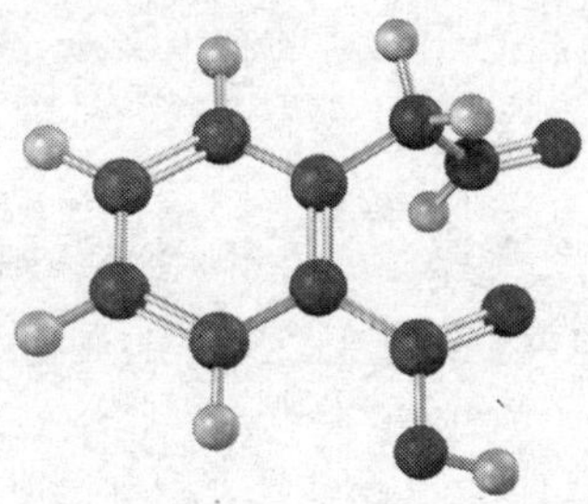

9.20 Judging from the following electrostatic potential maps, which kind of carbonyl compound has the more electrophilic carbonyl carbon atom, a ketone or an acid chloride? Which has the more nucleophilic carbonyl oxygen atom? Explain.

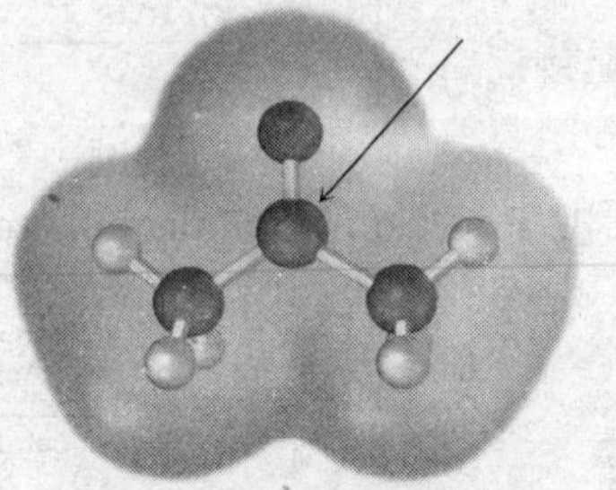

Acetone (ketone)

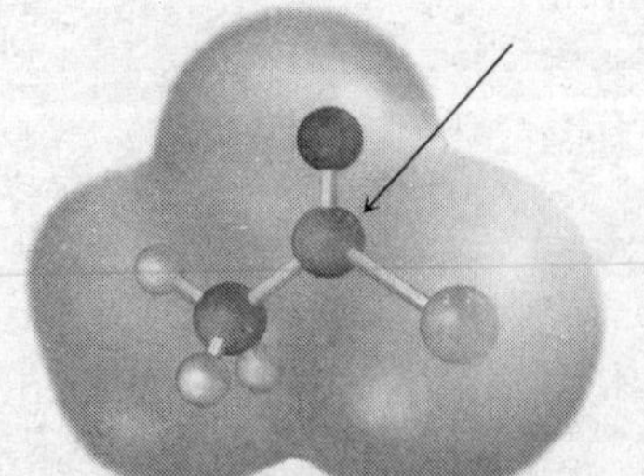

Acetyl chloride (acid chloride)

9.21 Compounds called *cyanohydrins* result from the nucleophilic addition of HCN to an aldehyde or ketone. Draw and name the carbonyl compound that the following cyanohydrin was prepared from (red = O, blue = N):

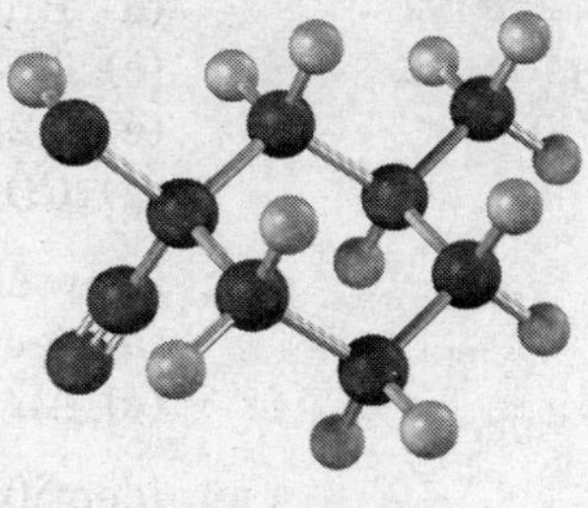

9.22 The following model represents the product resulting from addition of a nucleophile to an aldehyde or ketone. Identify the reactants, and write the reaction (red = O, blue = N).

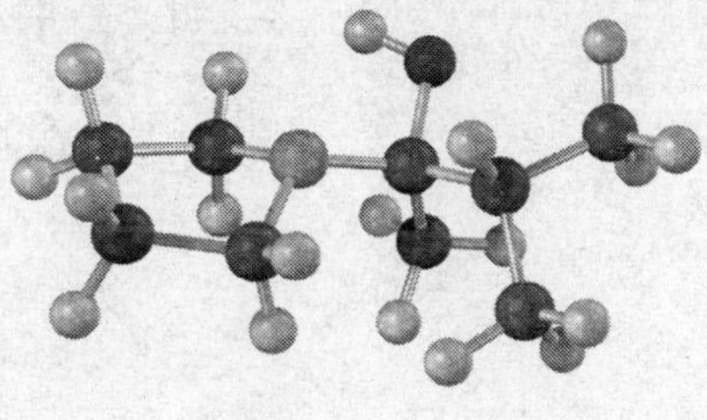

9.23 Identify the reactants from which the following molecules were prepared. If an acetal, identify the carbonyl compound and the alcohol; if an imine, identify the carbonyl compound and the amine; if an alcohol, identify the carbonyl compound and the Grignard reagent (red = O, blue = N):

(a)

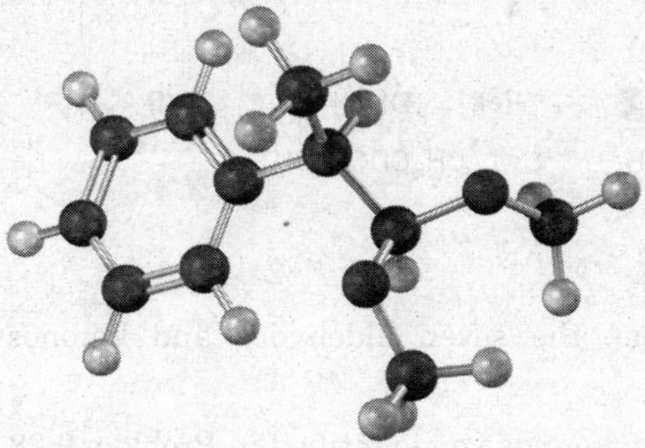

(b)

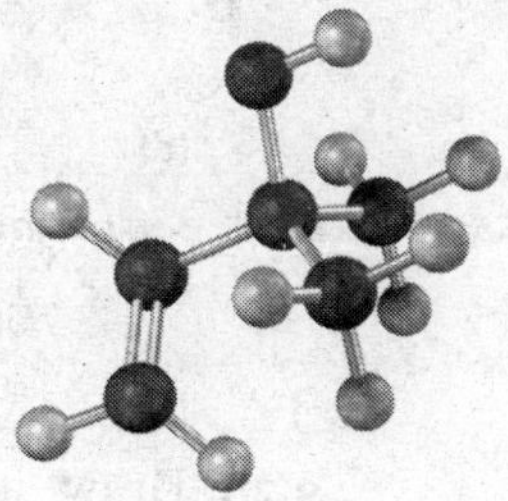

(c)

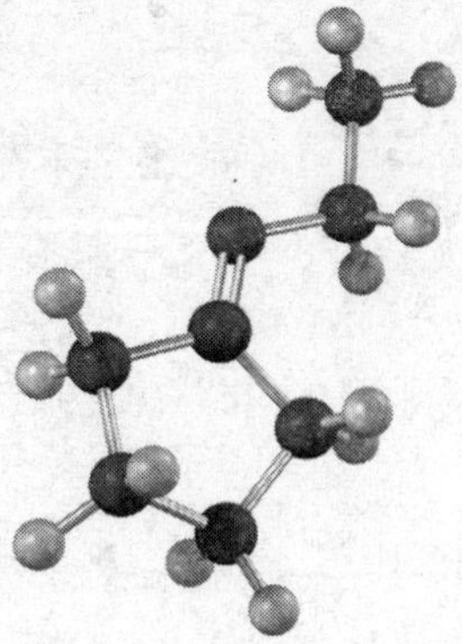

Additional Problems

Identifying and Naming Carbonyl Compounds

9.24 Draw structures corresponding to the following names:

(a) Bromoacetone
(b) 3-Methylbutan-2-one
(c) 3,5-Dinitrobenzaldehyde
(d) 3,5-Dimethylcyclohexanone
(e) 2,2,4,4-Tetramethylpentan-3-one
(f) Butanedial
(g) (*S*)-2-Hydroxypropanal
(h) 3-Phenylprop-2-enal

9.25 Draw structures of molecules that meet the following descriptions:

(a) A cyclic ketone, C_6H_8O
(b) A diketone, $C_6H_{10}O_2$
(c) An aryl ketone, $C_9H_{10}O$
(d) A 2-bromo aldehyde, C_5H_9BrO

9.26 Identify the different kinds of carbonyl functional groups in the following molecules:

(a) 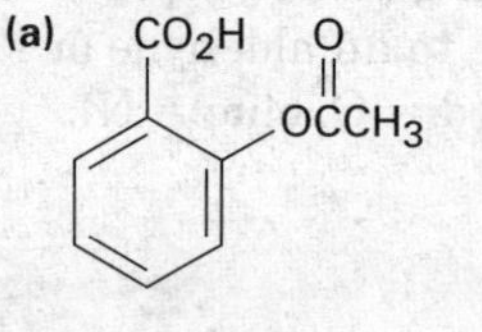

Aspirin

(b) 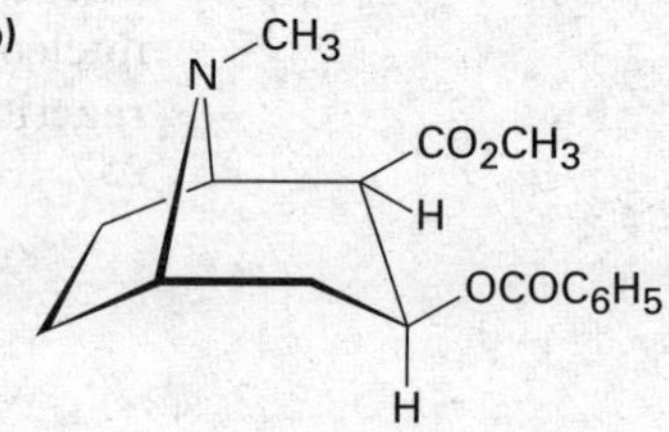

Cocaine

(c) 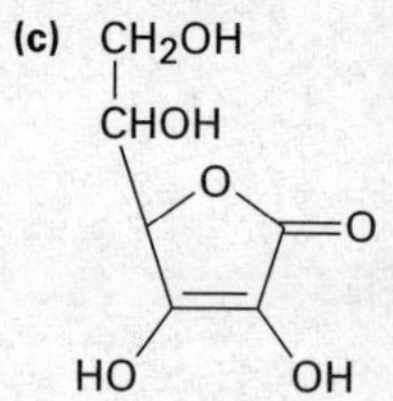

Ascorbic acid (vitamin C)

9.27 Give IUPAC names for the following structures:

(a) 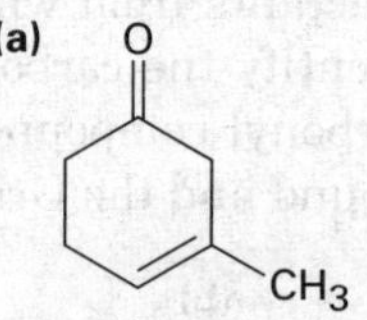

(b) 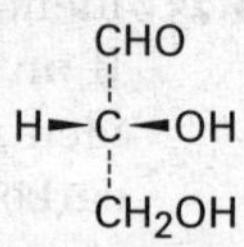

(c) 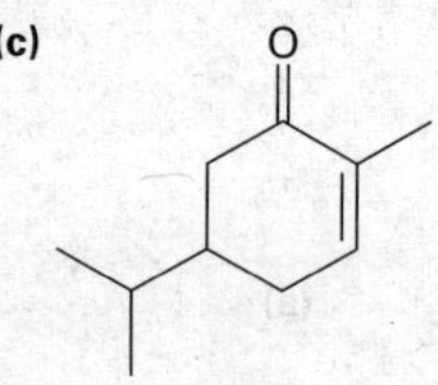

(d) 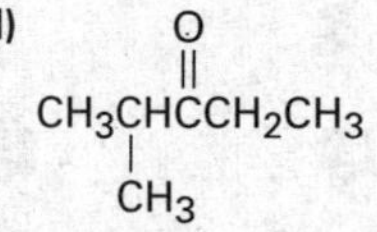

(e) 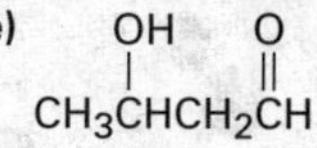

(f) 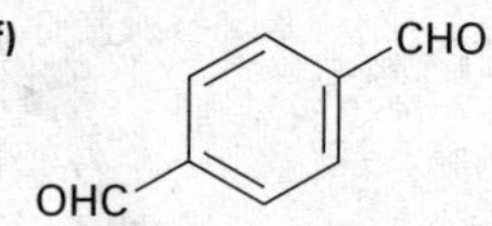

9.28 Draw and name the seven aldehydes and ketones with the formula $C_5H_{10}O$.

9.29 Which of the compounds you identified in Problem 9.28 are chiral?

Reactions

9.30 Predict the products of the reaction of phenylacetaldehyde, $C_6H_5CH_2CHO$, with the following reagents:

(a) $NaBH_4$, then H_3O^+
(b) aqueous acidic CrO_3
(c) NH_2OH
(d) CH_3MgBr, then H_3O^+
(e) CH_3OH, H^+ catalyst

9.31 Answer Problem 9.30 for the reaction of acetophenone, $C_6H_5COCH_3$, with the same reagents.

9.32 Reaction of butan-2-one with HCN yields a *cyanohydrin* product [$R_2C(OH)CN$] having a new chirality center. Explain why the product is not optically active.

9.33 In light of your answer to Problem 9.32, what stereochemistry would you expect the product from the reaction of phenylmagnesium bromide with butan-2-one to have?

9.34 Show the products that result from the reaction of phenylmagnesium bromide with the following reagents:
(a) CH_2O **(b)** Benzophenone ($C_6H_5COC_6H_5$) **(c)** Pentan-3-one

9.35 Identify the nucleophile that has added to acetone to give the following products:

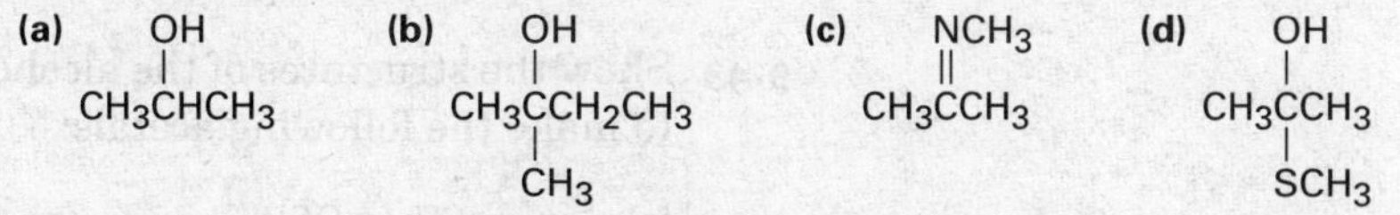

9.36 Show the structures of the intermediate hemiacetals and the final acetals that result from the following reactions:

(a)

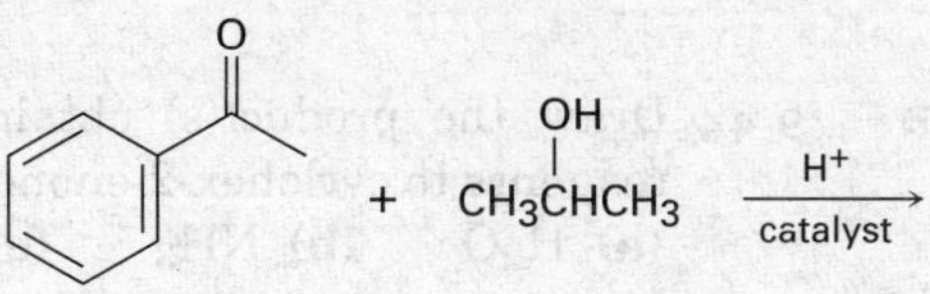

(b)

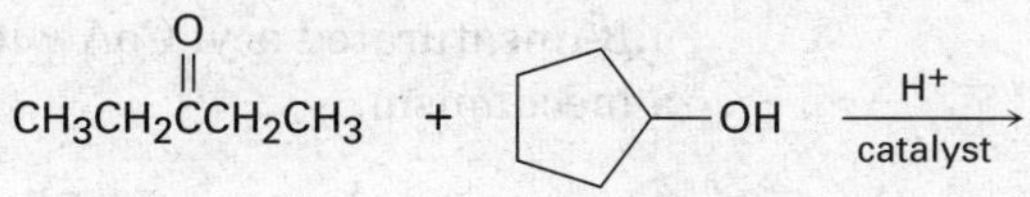

SYNTHESIS

9.37 Use a Grignard reaction on an aldehyde or ketone to synthesize the following compounds:
(a) Pentan-2-ol **(b)** 1-Phenylbutan-2-ol
(c) 1-Ethylcyclohexanol **(d)** Diphenylmethanol

9.38 Starting from cyclohex-2-enone and any other reagents needed, how would you prepare the following substances? More than one step may be required.

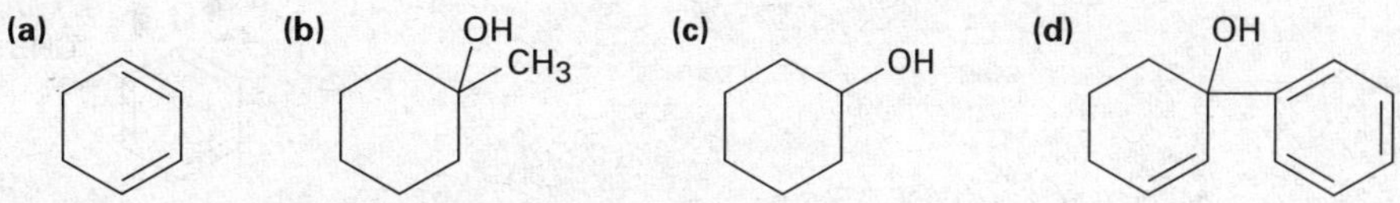

9.39 How could you make the following alcohols using a Grignard reaction of an aldehyde or ketone? Show all possibilities.

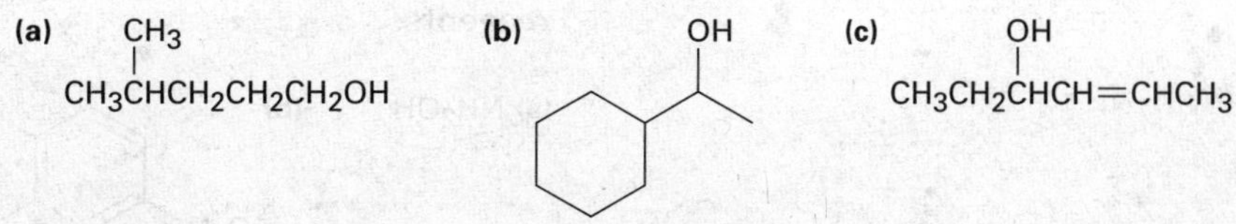

9.40 Which of the alcohols shown in Problem 9.39 could you make by reduction of a carbonyl compound? What carbonyl compound would you use in each case?

9.41 How would you synthesize the following compounds from cyclohexanone?

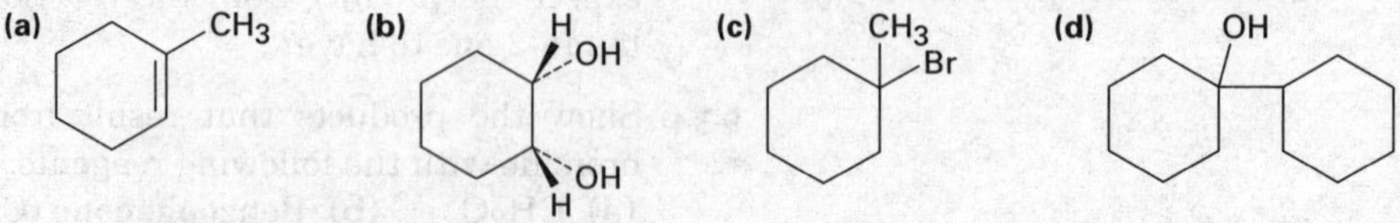

9.42 How could you convert bromobenzene into benzoic acid, $C_6H_5CO_2H$? (More than one step is required.)

9.43 Show the structures of the alcohols and aldehydes or ketones you would use to make the following acetals:

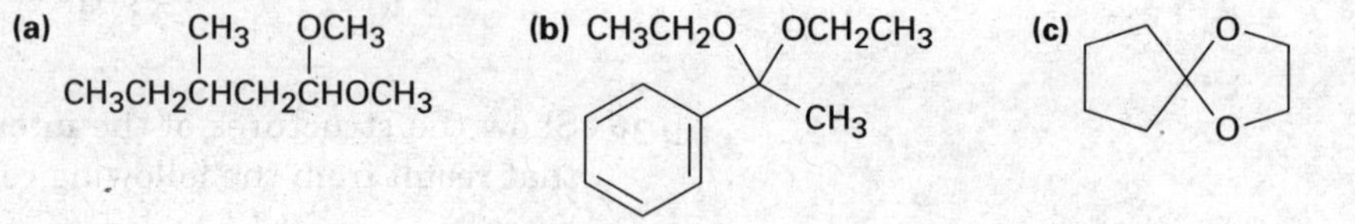

GENERAL PROBLEMS

9.44 Draw the product(s) obtained by conjugate addition of the following reagents to cyclohex-2-enone:
(a) H_2O **(b)** NH_3 **(c)** CH_3OH **(d)** CH_3CH_2SH

9.45 One of the steps in the metabolism of fats is the reaction of an α,β-unsaturated acyl CoA with water to give a β-hydroxyacyl CoA. Propose a mechanism.

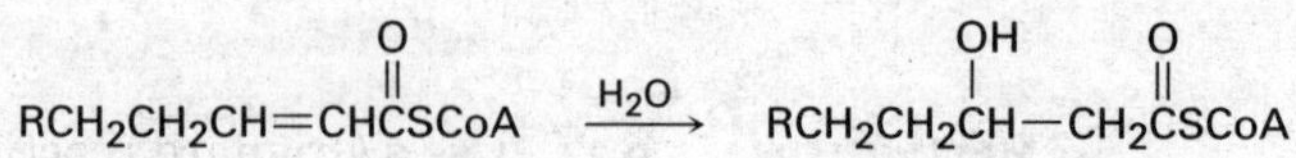

Unsaturated acyl CoA **β-Hydroxyacyl CoA**

9.46 How can you explain the observation that the S_N2 reaction of (dibromomethyl)benzene with NaOH yields benzaldehyde rather than (dihydroxymethyl)benzene?

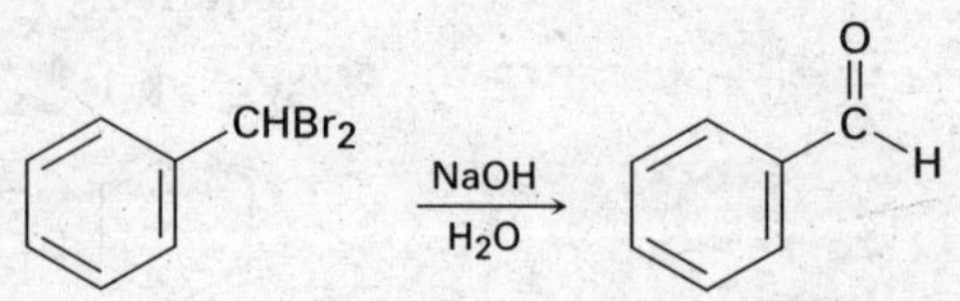

(Dibromomethyl)benzene **Benzaldehyde**

9.47 Show the products from the reaction of pentan-2-one with the following reagents:

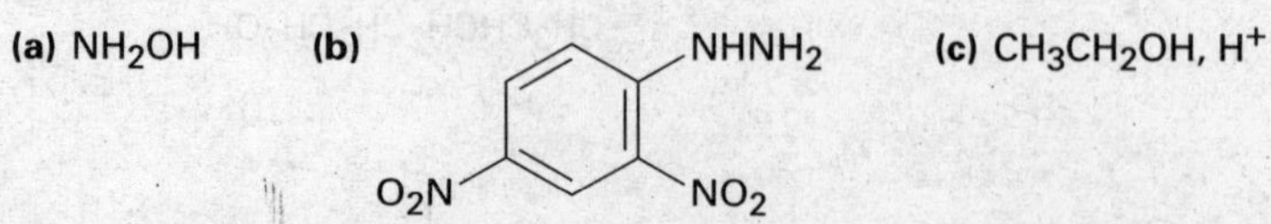

9.48 When glucose is treated with $NaBH_4$, reaction occurs to yield *sorbitol,* a commonly used food additive. Show how this reduction occurs.

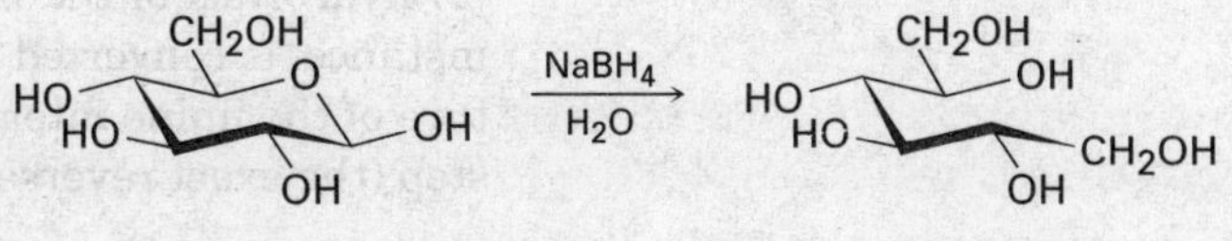

Glucose **Sorbitol**

9.49 Identify the reagents **a** through **d** in the following scheme:

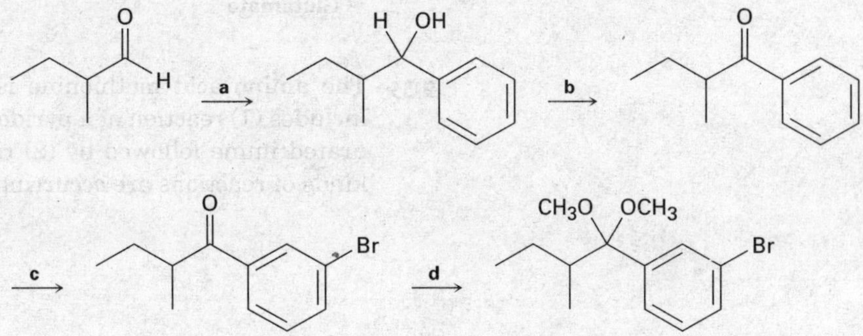

9.50 Ketones react with dimethylsulfonium methylide to yield epoxides by a mechanism that involves (1) an initial nucleophilic addition followed by (2) an intramolecular S_N2 substitution. Show the mechanism.

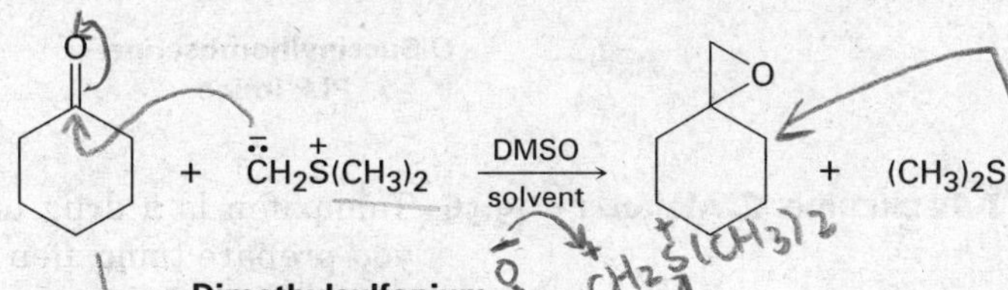

Dimethylsulfonium methylide

9.51 Carvone is the major constituent of spearmint oil. What products would you expect from the reaction of carvone with the following reagents?

(a) $LiAlH_4$, then H_3O^+ **(b)** C_6H_5MgBr, then H_3O^+
(c) H_2, Pd catalyst **(d)** CH_3OH, H^+

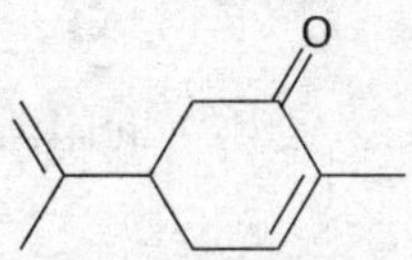

Carvone

9.52 Treatment of an aldehyde or ketone with hydrazine, H_2NNH_2, yields an *azine,* $R_2C{=}N{-}N{=}CR_2$. Propose a mechanism.

9.53 Treatment of an aldehyde or ketone with a thiol (RSH) in the presence of an acid catalyst yields a *thioacetal,* $R'_2C(SR)_2$. To what other reaction is this thioacetal formation analogous? Propose a mechanism for the reaction.

9.54 One of the biological pathways by which an amine is converted to a ketone involves two steps: (1) enzymatic oxidation of the amine to give an imine and (2) hydrolysis of the imine to give a ketone plus ammonia. Glutamate, for instance, is converted by this process into α-ketoglutarate. Show the structure of the imine intermediate, and propose a mechanism for the hydrolysis step (the exact reverse of imine formation).

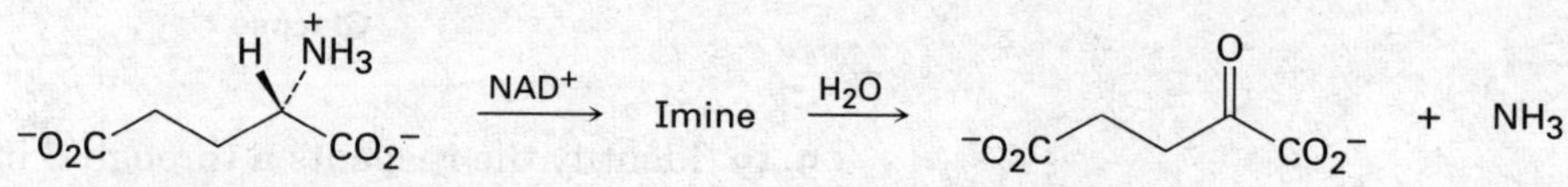

Glutamate **α-Ketoglutarate**

9.55 The amino acid methionine is biosynthesized by a multistep route that includes (1) reaction of a pyridoxal phosphate (PLP) imine to give an unsaturated imine followed by (2) reaction with the amino acid cysteine. What kinds of reactions are occurring in the two steps?

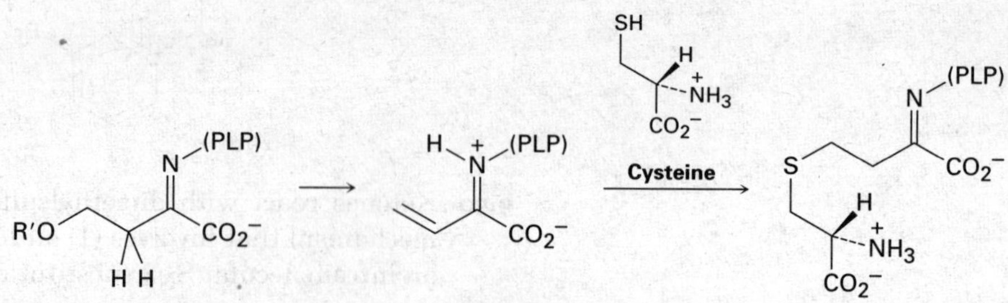

O-Succinylhomoserine–PLP imine **Unsaturated imine**

IN THE MEDICINE CABINET

9.56 Tamoxifen is a drug used in the treatment of breast cancer. How could you prepare tamoxifen from benzene, the following ketone, and any other reagents needed?

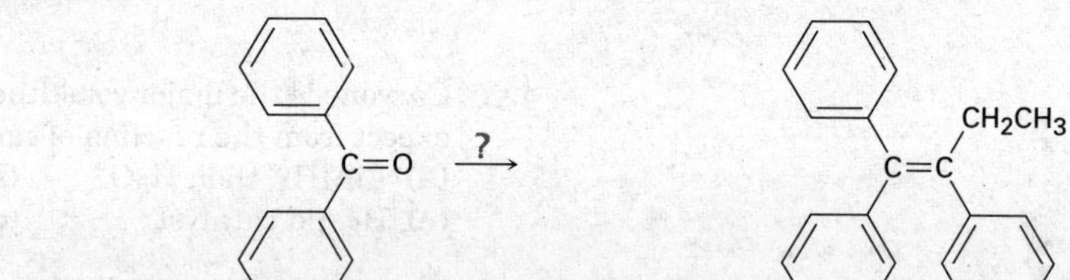

Tamoxifen

9.57 Pralidoxime iodide is a general antidote for poisoning by many insecticides. The drug is made in two steps starting with pyridine-2-carbaldehyde.

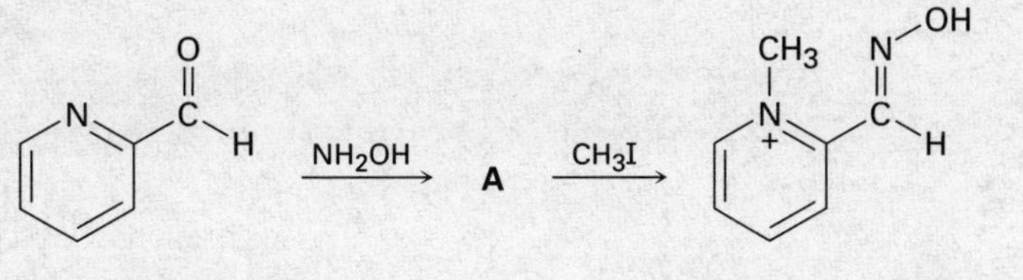

(a) Show the mechanism of the reaction of hydroxylamine (NH_2OH) with pyridine-2-carbaldehyde, and give the structure of **A**.

(b) Reaction of **A** with iodomethane to give pralidoxime iodide is an S_N2 reaction. Show the mechanism.

IN THE FIELD

9.58 Synthesis of the herbicide metolachlor, seen previously in Problems 2.74, 5.62, and 6.66, begins with an oxidation followed by an imine formation:

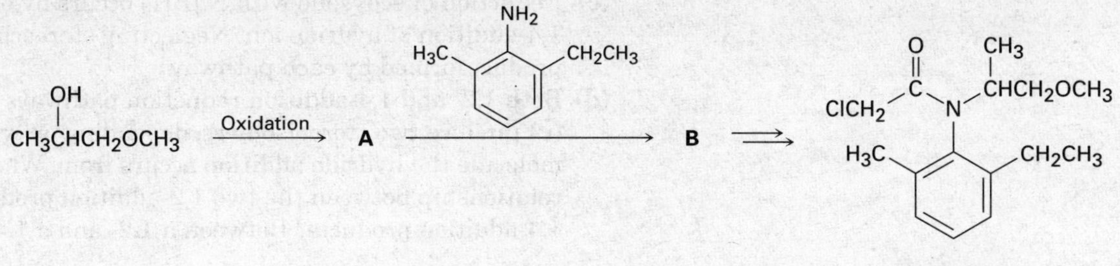

(a) The starting ether can be obtained by reacting an epoxide with sodium methoxide, $Na^{+-}OCH_3$. Propose a structure for the epoxide (commonly called propylene oxide).

(b) What is the structure of intermediate **A**?

(c) What oxidizing agent would you use to form **A**?

(d) What is the structure of imine **B**?

9.59 Many insecticides function by blocking cellular receptors for the insect molting hormone ecdysone.

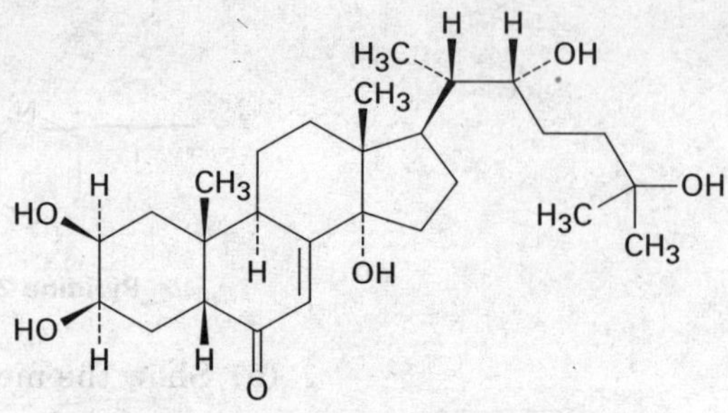

Ecdysone

(a) Categorize each of the hydroxyl groups in ecdysone as primary, secondary, or tertiary.
(b) How many chirality centers does ecdysone have?
(c) Reduction of ecdysone with $NaBH_4$ occurs by both 1,2- and 1,4-addition of hydride ion. Neglecting stereochemistry, show the product formed by each pathway.
(d) Both 1,2- and 1,4-addition reduction pathways described in part **(c)** produce two stereoisomers, depending on which side of the molecule the hydride addition occurs from. What term describes the relationship between the two 1,2-addition products? Between the two 1,4-addition products? Between a 1,2- and a 1,4-addition product?

CHAPTER

10

The burning sensation produced by touching or eating chili peppers is due to capsaicin, a carboxylic acid derivative called an amide.

Carboxylic Acids and Derivatives: Nucleophilic Acyl Substitution Reactions

Carboxylic acids and their derivatives are the most abundant of all organic compounds, both in the laboratory and in living organisms. Although there are many different kinds of carboxylic acid derivatives, we'll be concerned only with some of the most common ones: **acid halides**, **acid anhydrides**, **esters**, **amides**, and related compounds called **nitriles**. In addition, *acyl phosphates* and *thioesters* are acid derivatives of particular importance in numerous biological processes. The common structural feature of all these compounds is that they contain an acyl group bonded to an electronegative atom or substituent that can act as a leaving group in substitution reactions.

OWL

Online homework for this chapter can be assigned in OWL, an online homework assessment tool.

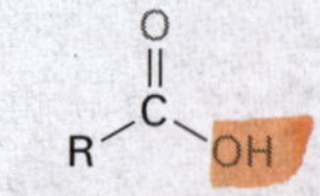

Carboxylic acid

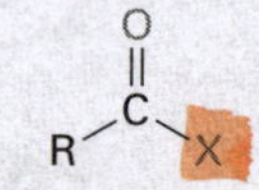

Acid halide (X = Cl, Br)

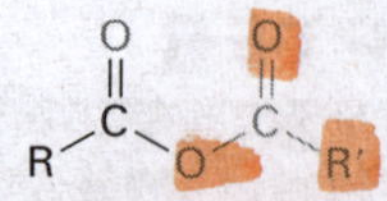

Acid anhydride

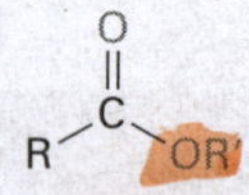

Ester

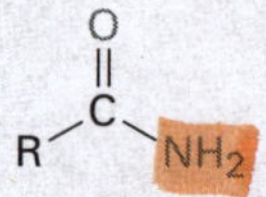

Amide

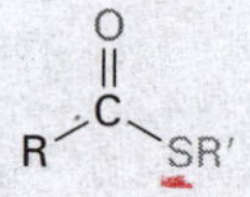

Thioester

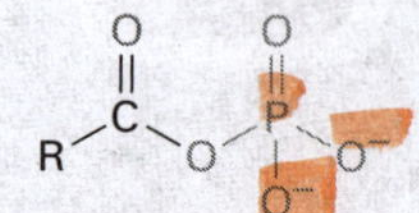

Acyl phosphate

R—C≡N

Nitrile

WHY THIS CHAPTER?

Because carboxylic acids and their derivatives are involved in so many industrial processes and most biological pathways, an understanding of their properties and behavior is fundamental to understanding organic and biological chemistry. In this chapter, we'll first discuss carboxylic acids themselves and will then explore in detail the most common reaction of carboxylic acid derivatives—the nucleophilic acyl substitution reaction.

10.1 Naming Carboxylic Acids and Derivatives

Carboxylic Acids: RCO_2H

Simple open-chain carboxylic acids are named by replacing the terminal *-e* of the corresponding alkane name with *-oic acid.* The $-CO_2H$ carbon is numbered C1.

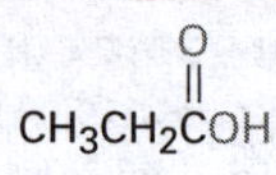

Propanoic acid

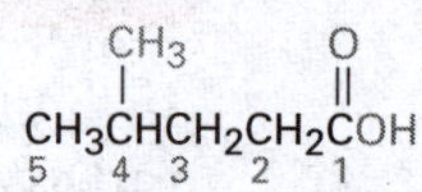

4-Methylpentanoic acid

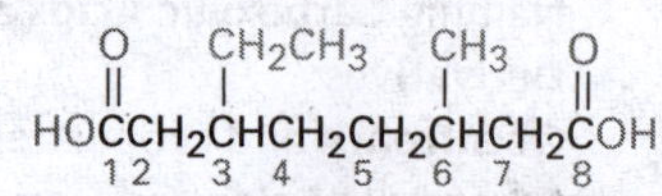

3-Ethyl-6-methyloctanedioic acid

Compounds that have a $-CO_2H$ group (a **carboxyl group**) bonded to a ring are named using the suffix *-carboxylic acid.* The carboxyl carbon is attached to C1 on the ring and is not itself numbered.

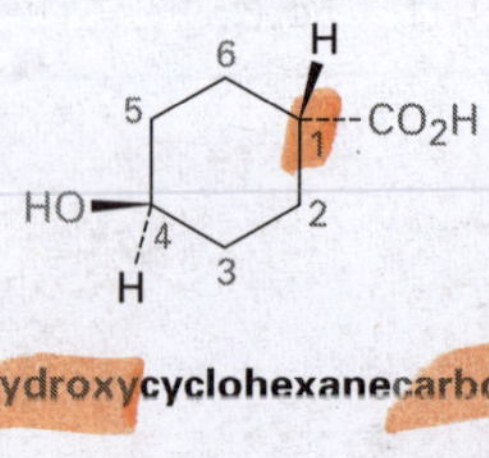

trans-4-Hydroxycyclohexanecarboxylic acid

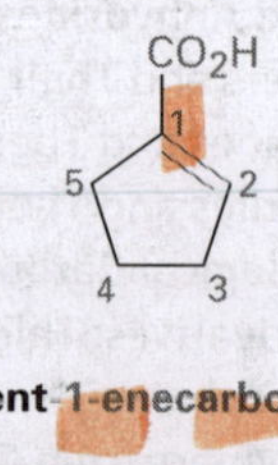

Cyclopent-1-enecarboxylic acid

Because many carboxylic acids were among the first organic compounds to be isolated and purified, a large number of acids have common names (Table 10.1). We'll use systematic names in this book, with the exception of formic (methanoic) acid, HCO_2H, and acetic (ethanoic) acid, CH_3CO_2H, whose names are so well known that it makes little sense to refer to them in any other way. Also listed in Table 10.1 are the names for acyl groups (R–C=O) derived from the parent acids by removing –OH. Except for the eight acyl groups at the top of Table 10.1, whose common names have a *-yl* ending, all others are named systematically with an *-oyl* ending.

Table 10.1 Some Common Names of Carboxylic Acids and Acyl Groups

Structure	Name	Acyl group
HCO_2H	Formic	Formyl
CH_3CO_2H	Acetic	Acetyl
$CH_3CH_2CO_2H$	Propionic	Propionyl
$CH_3CH_2CH_2CO_2H$	Butyric	Butyryl
HO_2CCO_2H	Oxalic	Oxalyl
$HO_2CCH_2CO_2H$	Malonic	Malonyl
$HO_2CCH_2CH_2CO_2H$	Succinic	Succinyl
$HO_2CCH_2CH_2CH_2CO_2H$	Glutaric	Glutaryl
$HO_2CCH_2CH_2CH_2CH_2CO_2H$	Adipic	Adipoyl
$H_2C{=}CHCO_2H$	Acrylic	Acryloyl
$HO_2CCH{=}CHCO_2H$	Maleic (cis)	Maleoyl
	Fumaric (trans)	Fumaroyl
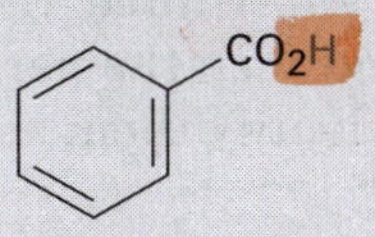	Benzoic	Benzoyl

Acid Halides: RCOX

Acid halides are named by identifying first the acyl group, as in Table 10.1, and then the halide. Those cyclic carboxylic acids that take a *-carboxylic acid* ending use *-carbonyl* for the name ending of the corresponding acyl group. For example:

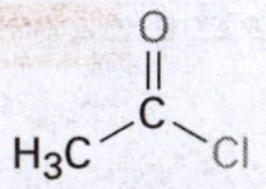

Acetyl chloride (from acetic acid)

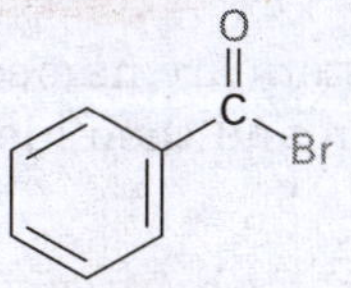

Benzoyl bromide (from benzoic acid)

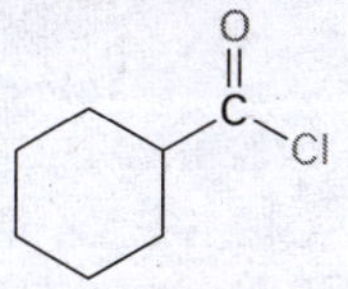

Cyclohexanecarbonyl chloride (from cyclohexanecarboxylic acid)

Acid Anhydrides: RCO_2COR'

Symmetrical anhydrides from simple carboxylic acids and cyclic anhydrides from dicarboxylic acids are named by replacing the word *acid* with *anhydride*.

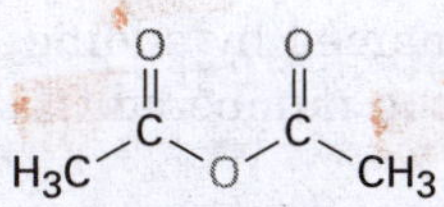

Acetic anhydride

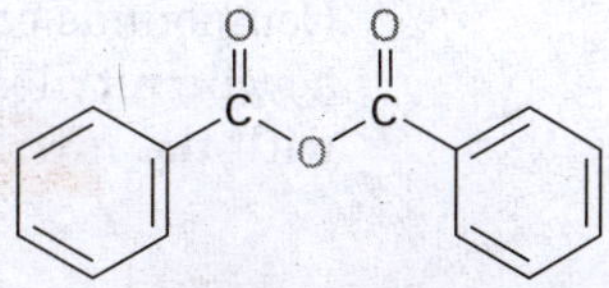

Benzoic anhydride

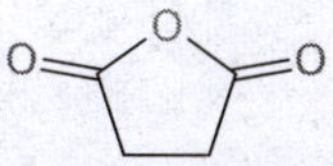

Succinic anhydride

Unsymmetrical anhydrides—those prepared from two different carboxylic acids—are named by citing the two acids alphabetically and then adding *anhydride.*

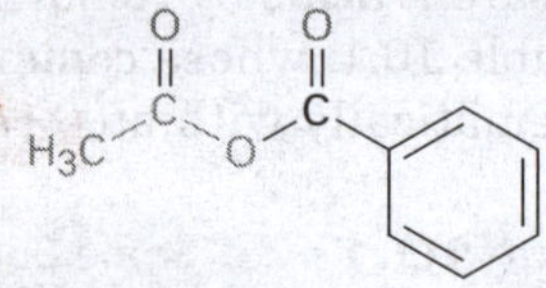

Acetic benzoic anhydride

Amides: $RCONH_2$

Amides with an unsubstituted $-NH_2$ group are named by replacing the *-oic acid* or *-ic acid* ending with *-amide,* or by replacing the *-carboxylic acid* ending with *-carboxamide.*

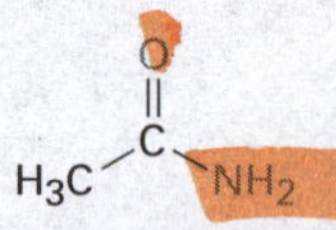

Acetamide

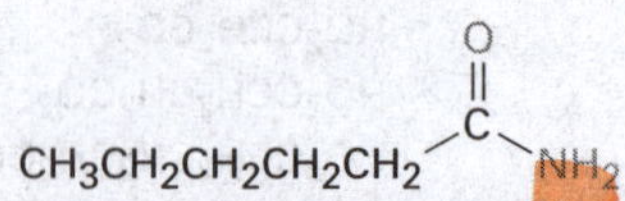

Hexanamide

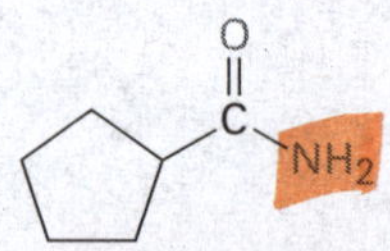

Cyclopentane-carboxamide

If the nitrogen atom is substituted, the amide is named by first identifying the substituent groups and then the parent amide. The substituents are preceded by the letter *N* to identify them as being directly attached to nitrogen.

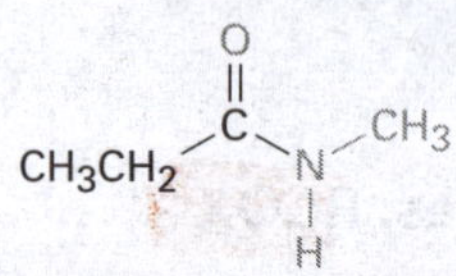

***N*-Methylpropanamide**

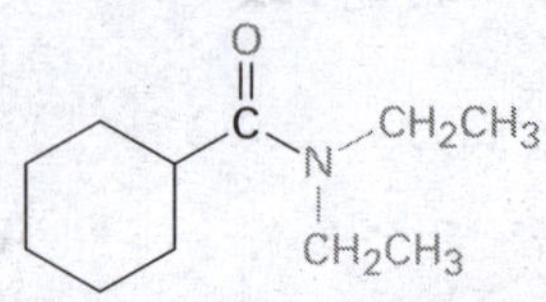

***N,N*-Diethylcyclohexanecarboxamide**

Esters: RCO_2R'

Esters are named by first giving the name of the alkyl group attached to oxygen and then identifying the carboxylic acid, with *-ic acid* replaced by *-ate.*

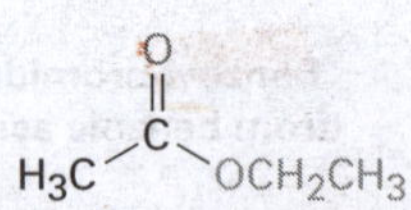

Ethyl acetate

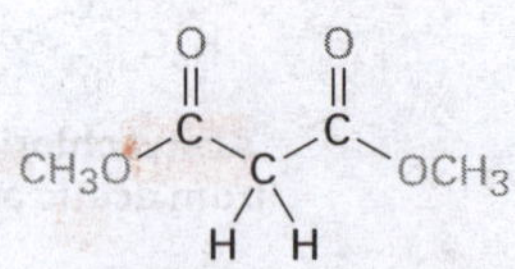

Dimethyl malonate

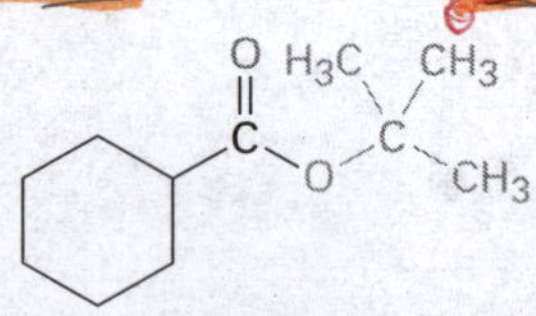

***tert*-Butyl cyclohexane-carboxylate**

Nitriles: R—C≡N

Compounds containing the $-C{\equiv}N$ functional group are called *nitriles*. Simple acyclic nitriles are named by adding *-nitrile* as a suffix to the alkane name, with the nitrile carbon numbered C1.

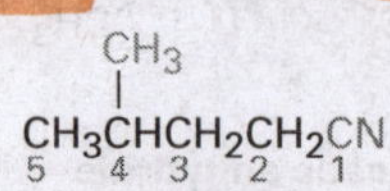

4-Methylpentanenitrile

More complex nitriles are named as derivatives of carboxylic acids by replacing the *-ic acid* or *-oic acid* ending with *-onitrile,* or by replacing the *-carboxylic acid* ending with *-carbonitrile.* In this system, the nitrile carbon atom is attached to C1 but is not itself numbered.

$CH_3C\equiv N$

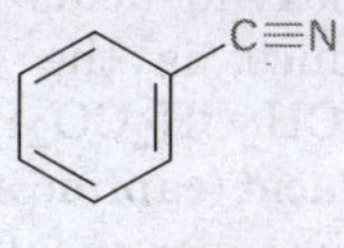

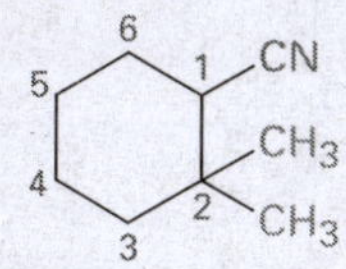

Acetonitrile (from acetic acid) **Benzonitrile (from benzoic acid)** **2,2-Dimethylcyclohexanecarbonitrile (from 2,2-dimethylcyclohexane-carboxylic acid)**

Problem 10.1 Give IUPAC names for the following carboxylic acids:

(a) 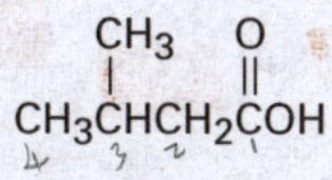 **(b)** 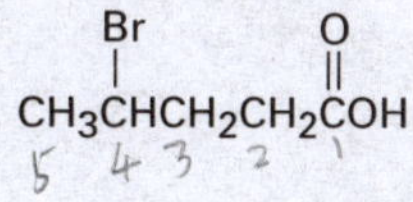 **(c)** 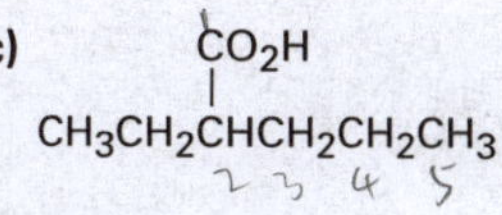

(d) 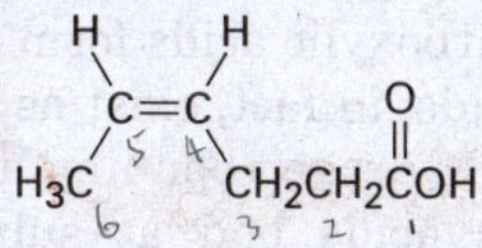 **(e)** 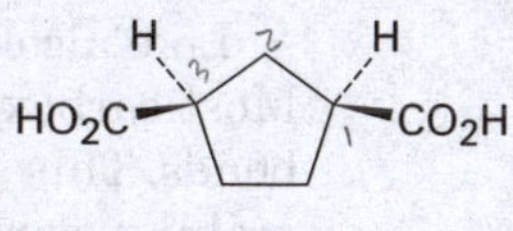

Problem 10.2 Draw structures corresponding to the following names:

(a) 2,3-Dimethylhexanoic acid **(b)** 4-Methylpentanoic acid
(c) *o*-Hydroxybenzoic acid **(d)** *trans*-Cyclobutane-1,2-dicarboxylic acid

Problem 10.3 Give IUPAC names for the following acyl derivatives:

(a) 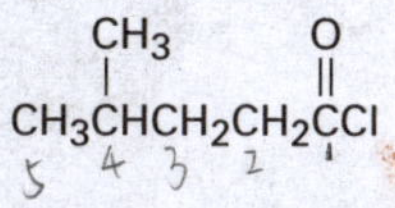 **(b)** 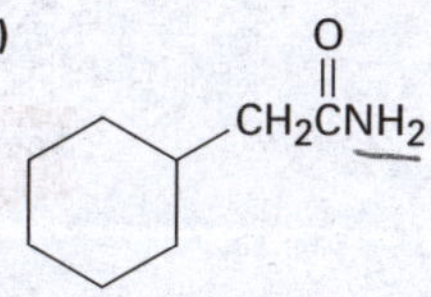 **(c)** 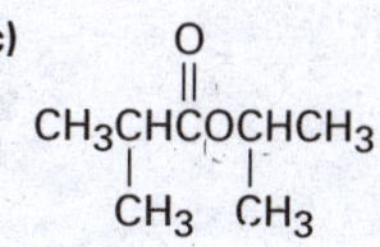

(d) 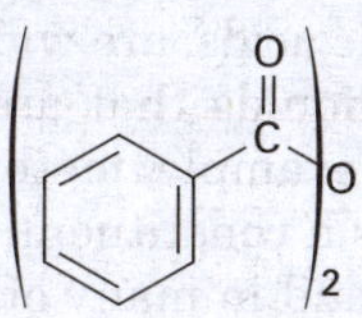 **(e)** 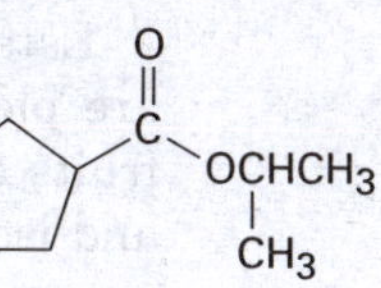 **(f)** 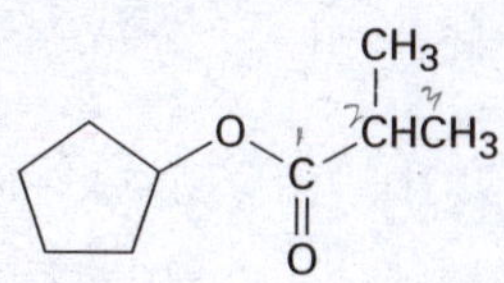

(g) 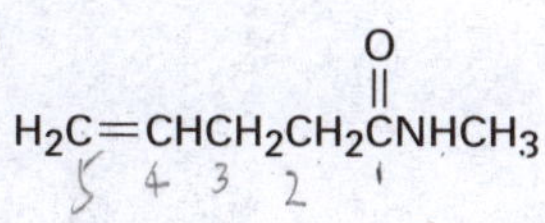 **(h)** 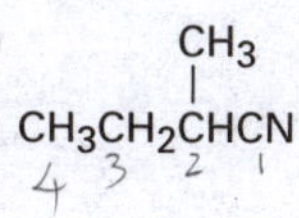

Problem 10.4 Draw structures corresponding to the following names:

(a) 2,2-Dimethylpropanoyl chloride **(b)** *N*-Methylbenzamide
(c) 5,5-Dimethylhexanenitrile **(d)** *tert*-Butyl butanoate
(e) *trans*-2-Methylcyclohexanecarboxamide **(f)** *p*-Methylbenzoic anhydride
(g) *cis*-3-Methylcyclohexanecarbonyl bromide **(h)** *p*-Bromobenzonitrile

10.2 Occurrence and Properties of Carboxylic Acids and Derivatives

Carboxylic acids are everywhere in nature. Acetic acid, CH_3CO_2H, for instance, is the principal organic component of vinegar; butanoic acid, $CH_3CH_2CH_2CO_2H$, is responsible for the rancid odor of sour butter; and hexanoic acid (caproic acid), $CH_3(CH_2)_4CO_2H$, is responsible for the aroma of goats (Latin *caper*, meaning "goat") and dirty socks.

Approximately 5 million tons of acetic acid are produced each year worldwide for a variety of purposes, including preparation of the vinyl acetate polymer used in paints and adhesives. About 20% of the acetic acid synthesized industrially is obtained by oxidation of acetaldehyde. Much of the remaining 80% is prepared by the rhodium-catalyzed reaction of methanol with carbon monoxide.

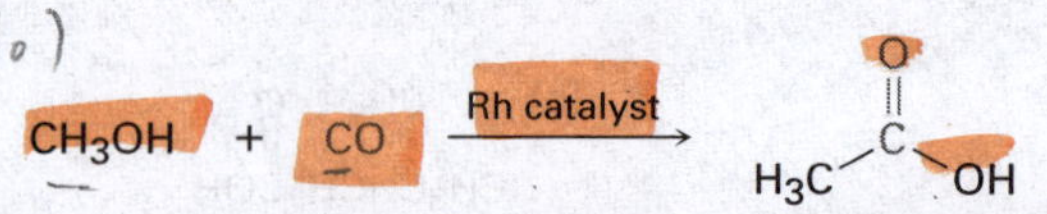

Like alcohols, carboxylic acids form strong intermolecular hydrogen bonds. Most carboxylic acids, in fact, exist as *dimers* held together by two hydrogen bonds. This strong hydrogen bonding has a noticeable effect on boiling points, making carboxylic acids boil at substantially higher temperatures than alkanes or alcohols of similar molecular weight. Acetic acid, for instance has a boiling point of 117.9 °C, versus 78.3 °C for ethanol.

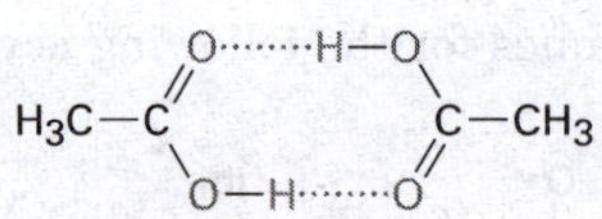

Acetic acid dimer

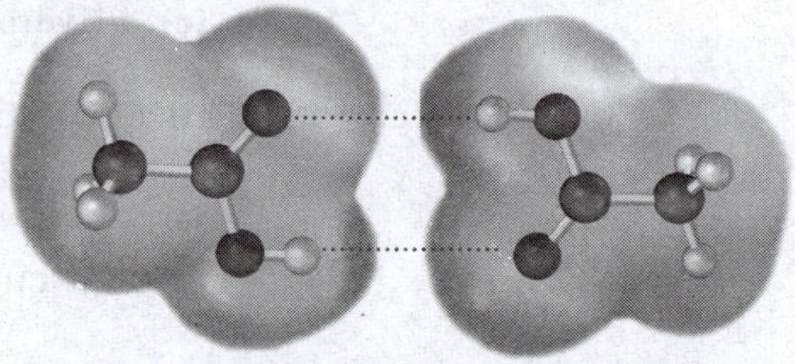

Esters, like carboxylic acids, are widespread in nature. Many simple esters are pleasant-smelling liquids that are responsible for the fragrant odors of fruits and flowers. For example, methyl butanoate is found in pineapple oil, and isopentyl acetate is a constituent of banana oil. The ester linkage is also present in animal fats and in many other biologically important molecules.

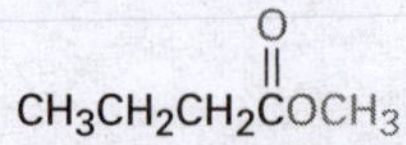

Methyl butanoate (from pineapples)

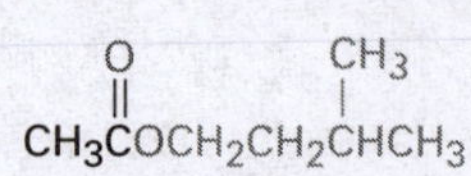

Isopentyl acetate (from bananas)

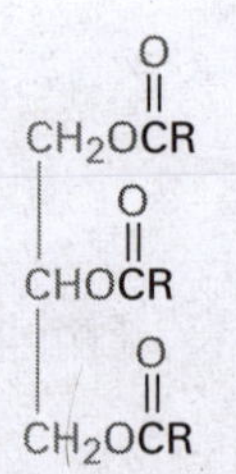

A fat
(R = C_{11-17} chains)

The chemical industry uses esters for a variety of purposes: ethyl acetate is a commonly used solvent, and dialkyl phthalates are used as plasticizers to keep polymers from becoming brittle. You might be aware that there is current concern about possible toxicity of phthalates at high concentrations, although a recent assessment by the U.S. Food and Drug Administration found the risk to be minimal for most people, with the possible exception of male infants.

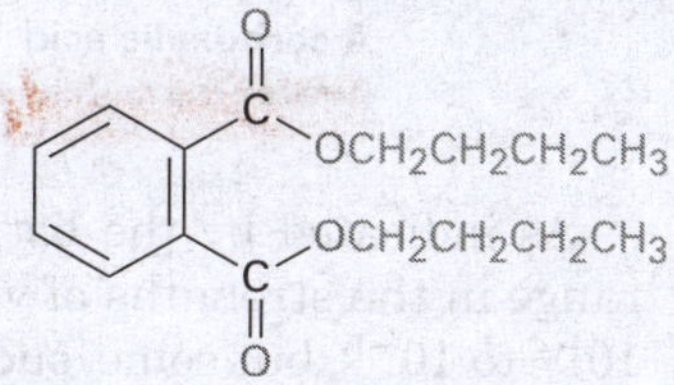

Dibutyl phthalate (a plasticizer)

Amides, like acids and esters, are abundant in living organisms—proteins, nucleic acids, and many pharmaceuticals have amide functional groups. The reason for this abundance of amides is that they are the least reactive of the common acid derivatives and are thus stable to the temperatures and aqueous conditions found in living organisms.

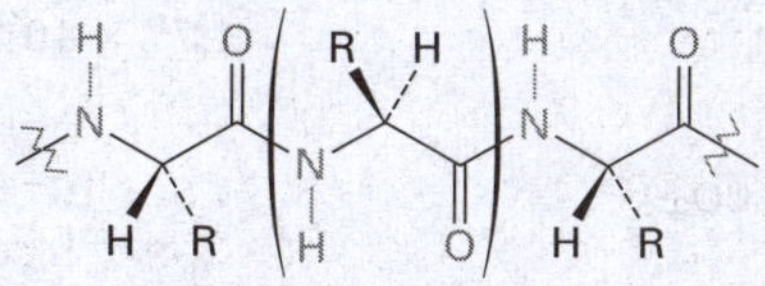

A protein segment

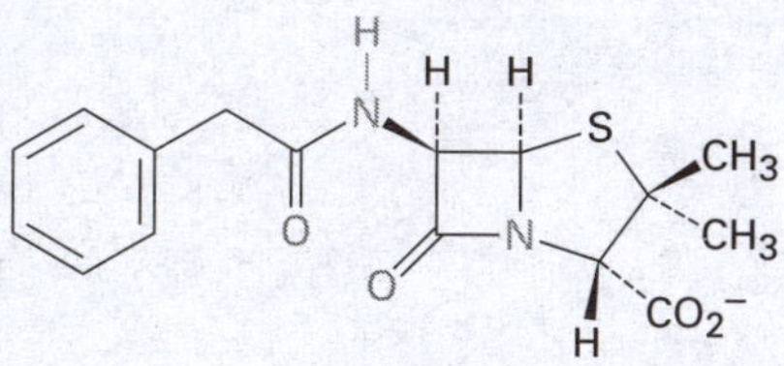

Benzylpenicillin (penicillin G)

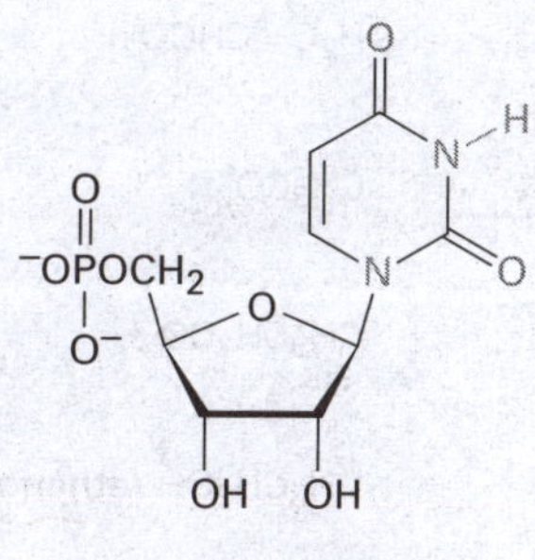

Uridine 5′-phosphate (a ribonucleotide)

Acid chlorides and anhydrides are frequently used in chemical laboratories but are not found in nature because of their high reactivity.

10.3 Acidity of Carboxylic Acids

The most obvious property of carboxylic acids is implied by their name: carboxylic acids are *acidic.* Acetic acid, for example, has $K_a = 1.75 \times 10^{-5}$ ($pK_a = 4.76$). In practical terms, a K_a value near 10^{-5} means that only about

1% of the molecules in a 0.1 M aqueous solution are dissociated. Because of their acidity, carboxylic acids react with bases such as NaOH to give water-soluble metal **carboxylates**, $RCO_2^-\ Na^+$.

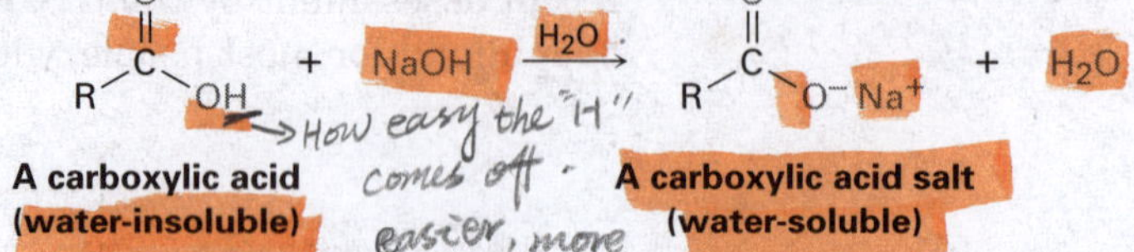

As indicated by the list of K_a values in Table 10.2, there is a considerable range in the strengths of various carboxylic acids. For most, K_a is in the range 10^{-4} to 10^{-5}, but some, such as trifluoroacetic acid ($K_a = 0.59$) are much stronger. The electron-withdrawing fluorine substituents stabilize the carboxylate ion by sharing the negative charge and thus favor dissociation of the acid.

Table 10.2 Acid Strengths of Some Carboxylic Acids

Structure	K_a	pK_a	
CF_3CO_2H	0.59	0.23	Stronger acid
HCO_2H	1.77×10^{-4}	3.75	↑
$HOCH_2CO_2H$	1.5×10^{-4}	3.84	
$C_6H_5CO_2H$	6.46×10^{-5}	4.19	
$H_2C{=}CHCO_2H$	5.6×10^{-5}	4.25	
CH_3CO_2H	1.75×10^{-5}	4.76	
$CH_3CH_2CO_2H$	1.34×10^{-5}	4.87	
CH_3CH_2OH (ethanol)	(1.00×10^{-16})	(16.00)	Weaker acid

Although much weaker than mineral acids, carboxylic acids are nevertheless much stronger acids than alcohols and phenols. The K_a of ethanol, for example, is approximately 10^{-16}, making ethanol a weaker acid than acetic acid by a factor of 10^{11}.

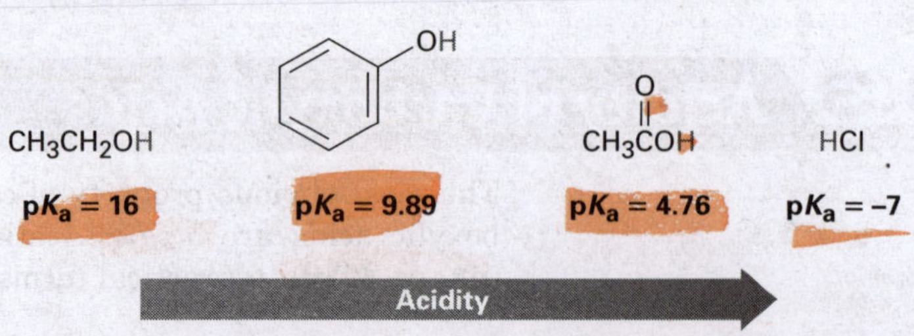

Why are carboxylic acids so much more acidic than alcohols even though both contain –OH groups? To answer this question, compare the relative stabilities of carboxylate anions versus alkoxide anions (Figure 10.1). In an alkoxide ion, the negative charge is localized on one oxygen atom, but in a carboxylate ion, the negative charge is spread out over both oxygen atoms because a carboxylate anion is a resonance hybrid of two equivalent structures (Section 4.10). Because a carboxylate ion is more stable than an alkoxide ion, it is lower in energy and is present to a greater extent at equilibrium.

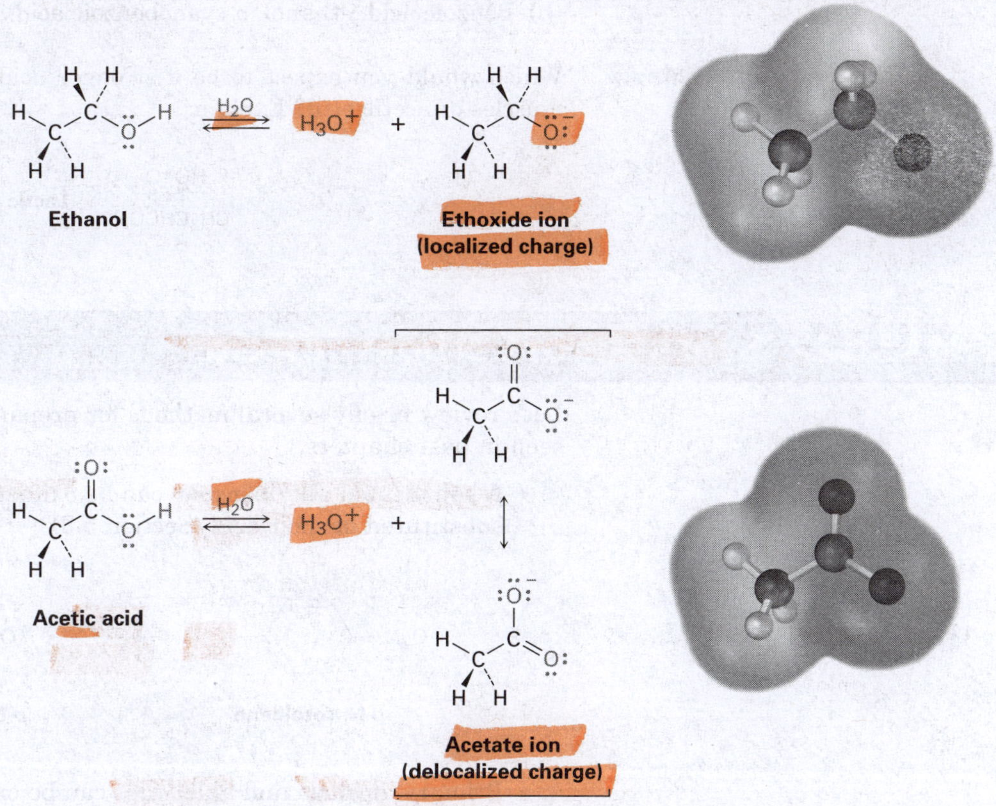

Figure 10.1 An alkoxide ion has its charge localized on one oxygen atom and is less stable, while a carboxylate ion has the charge spread equally over both oxygens and is therefore more stable.

Worked Example 10.1

Predicting Acid Strength

Which would you expect to be the stronger acid, benzoic acid or *p*-nitrobenzoic acid?

Solution The more stabilized the carboxylate anion, the stronger the acid. We know from its effect on aromatic substitution (Section 5.7) that a nitro group is electron-withdrawing and can stabilize a negative charge. Thus, a *p*-nitrobenzoate ion is more stable than a benzoate ion, and *p*-nitrobenzoic acid is stronger than benzoic acid.

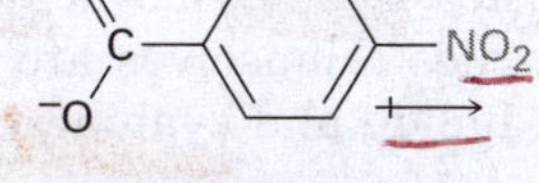

Nitro group withdraws electrons from ring and stabilizes negative charge.

Problem 10.5 Draw structures for the products of the following reactions:

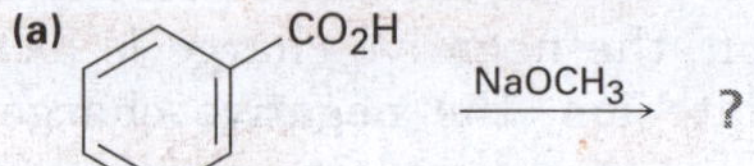

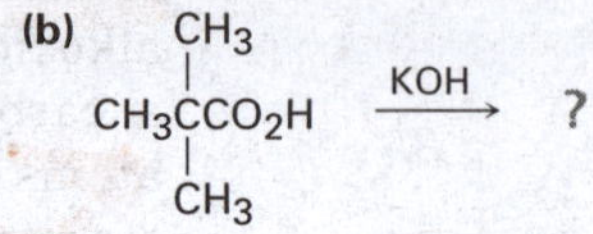

Problem 10.6 Rank the following compounds in order of increasing acidity:

(a) Sulfuric acid, methanol, phenol, *p*-nitrophenol, acetic acid
(b) Benzoic acid, ethanol, *p*-cyanobenzoic acid

Problem 10.7 Which would you expect to be a stronger acid, the lactic acid found in tired muscles or acetic acid? Explain.

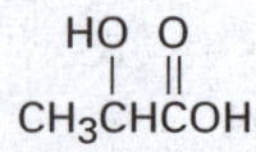

Lactic acid

10.4 Synthesis of Carboxylic Acids

Let's review briefly several methods for preparing carboxylic acids that we've seen in past chapters.

- A substituted alkylbenzene can be oxidized with $KMnO_4$ to give a substituted benzoic acid (Section 5.8).

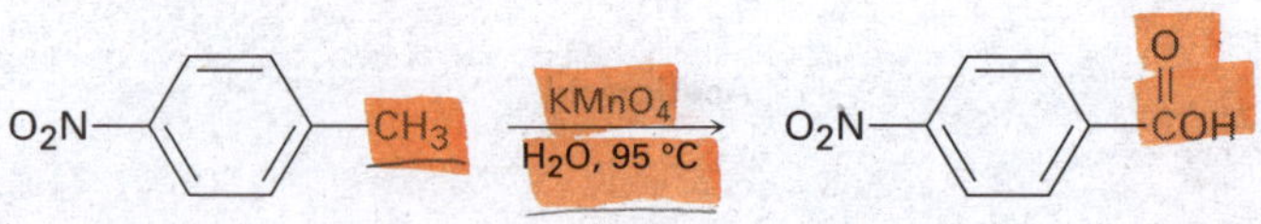

p-Nitrotoluene **p-Nitrobenzoic acid (88%)**

- Primary alcohols and aldehydes can be oxidized with aqueous CrO_3 or $Na_2Cr_2O_7$ to give carboxylic acids (Sections 8.4 and 9.4).

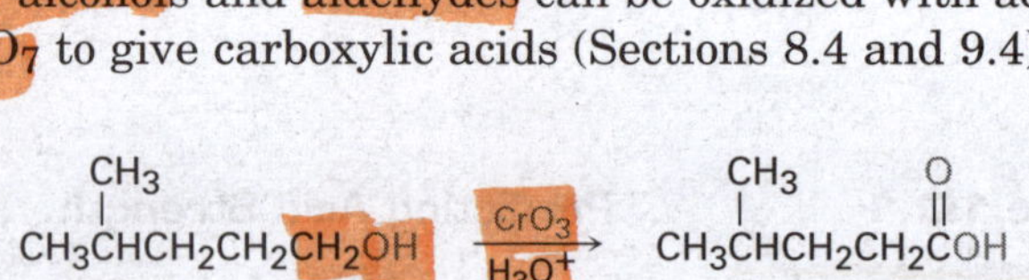

4-Methylpentan-1-ol **4-Methylpentanoic acid**

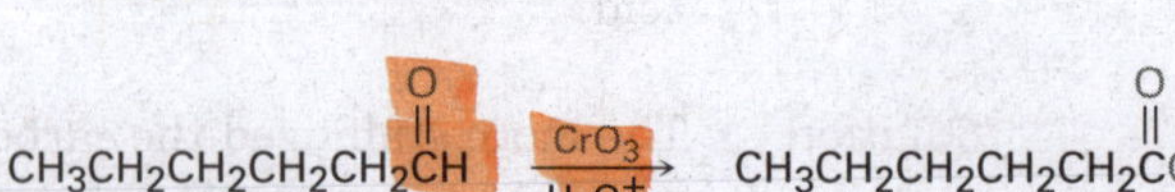

Hexanal **Hexanoic acid**

In addition to the preceding two methods, there are numerous other ways to prepare carboxylic acids. For instance, carboxylic acids can be prepared from nitriles, R—C≡N, by a *hydrolysis* reaction with hot aqueous acid or base. Since nitriles themselves are usually prepared by an S_N2 reaction between an alkyl halide and cyanide ion, CN^-, the two-step sequence of cyanide ion displacement followed by nitrile hydrolysis is a good method for converting an alkyl halide into a carboxylic acid: RBr → RC≡N → RCO_2H. As with all S_N2 reactions, the method works best with primary alkyl halides, although secondary alkyl halides can sometimes be used (Section 7.5).

An example of nitrile hydrolysis occurs in the commercial synthesis of the antiarthritis drug fenoprofen, a nonsteroidal anti-inflammatory agent (see Chapter 5 *Interlude*) marketed under the name Nalfon.

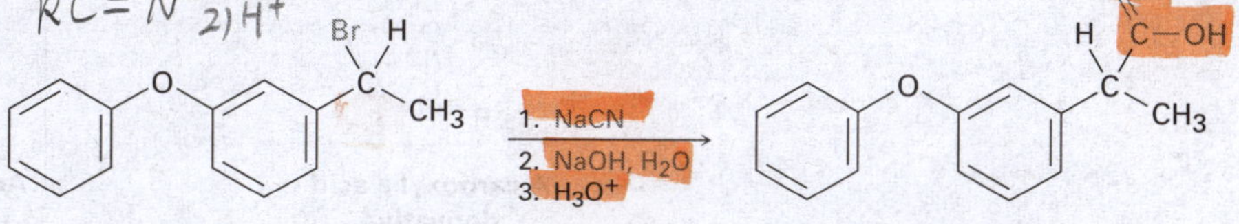

Fenoprofen
(an antiarthritis agent)

Problem 10.8 Show the steps in the conversion of iodomethane to acetic acid by the nitrile hydrolysis route. Would a similar route work for the conversion of iodobenzene to benzoic acid? Explain.

10.5 Nucleophilic Acyl Substitution Reactions

We saw in Chapter 9 that the addition of a nucleophile to the polar C=O bond is a general feature of aldehyde and ketone chemistry. Carboxylic acids and their derivatives also react with nucleophiles, but the ultimate product is different from that of the aldehyde/ketone reaction. Instead of undergoing protonation to yield an alcohol, the initially formed alkoxide intermediate expels one of the substituents originally bonded to the carbonyl carbon, leading to the formation of a new carbonyl compound by a **nucleophilic acyl substitution reaction** (Figure 10.2).

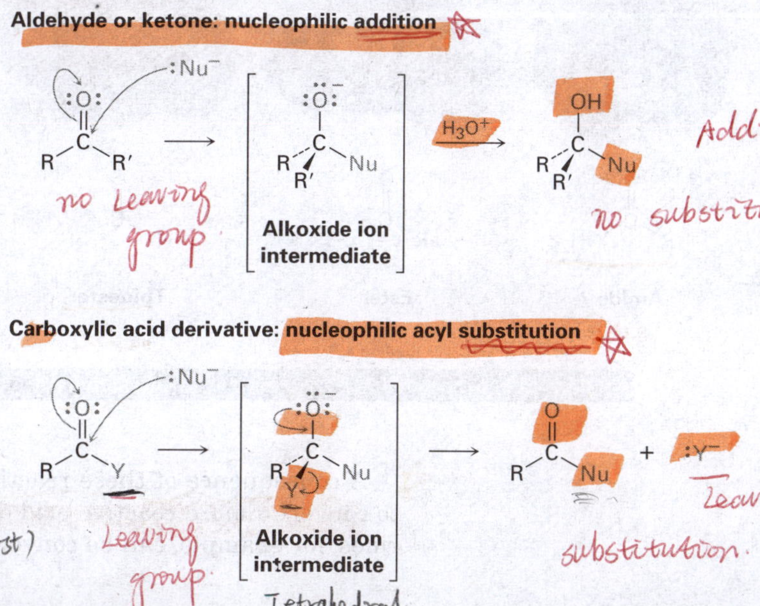

Figure 10.2 The general mechanisms of nucleophilic addition and nucleophilic acyl substitution reactions. Both reactions begin with the addition of a nucleophile to a polar C=O bond to give a tetrahedral, alkoxide ion intermediate. The intermediate formed from an aldehyde or ketone is protonated to give an alcohol, but the intermediate formed from a carboxylic acid derivative expels a leaving group to give a new carbonyl compound.

The different behavior toward nucleophiles of aldehydes/ketones and carboxylic acid derivatives is a consequence of structure. Carboxylic acid derivatives have an acyl carbon bonded to a group –Y that can leave as a stable anion. As soon as addition of a nucleophile occurs, the group leaves and a new carbonyl compound forms. Aldehydes and ketones have no such leaving group, however, and therefore don't undergo substitution.

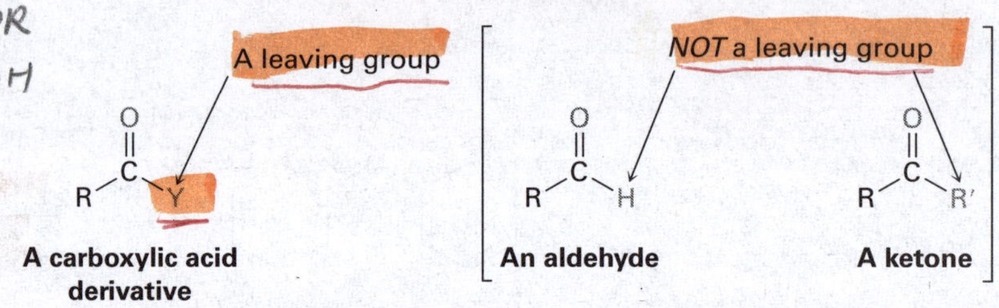

Both the initial addition step and the subsequent elimination step can affect the overall rate of a nucleophilic acyl substitution reaction, but the addition step is generally the rate-limiting one. Thus, any factor that makes the carbonyl group more reactive toward nucleophiles favors the substitution process.

As a general rule, the more electron-poor the C=O carbon, the more readily the compound reacts with nucleophiles. Thus, acid chlorides are the most reactive compounds because the electronegative chlorine atom strongly withdraws electrons from the carbonyl carbon, whereas amides are the least reactive compounds. Although the differences are subtle, electrostatic potential maps indicate the relative reactivities of various carboxylic acid derivatives by the relative blueness on the C=O carbons. Note that thioesters, RCOSR′, which are commonly found in biological molecules, have a reactivity intermediate between that of esters and acid anhydrides. Thioesters are thus stable enough to exist in living organisms but are reactive enough to be useful.

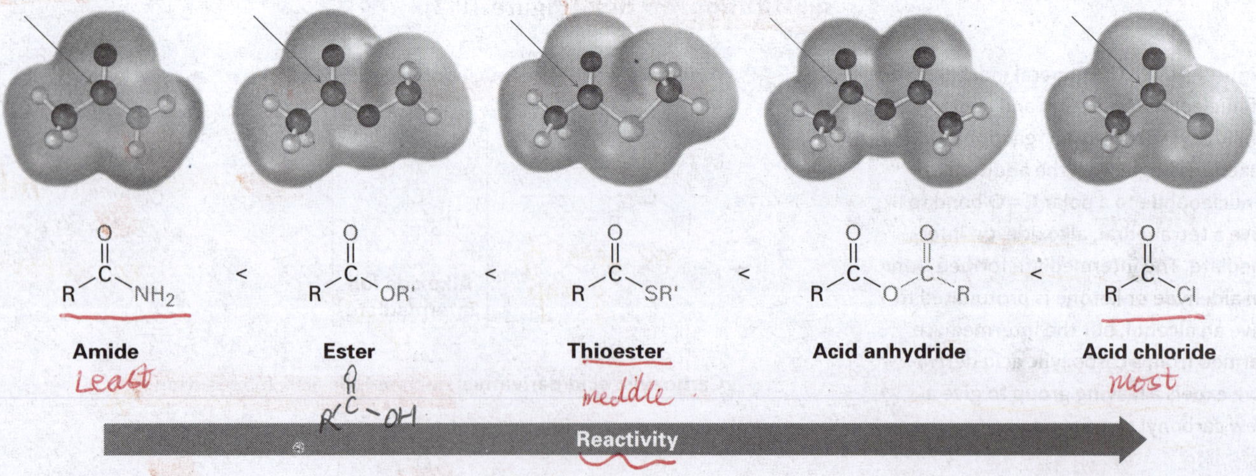

A consequence of these reactivity differences is that it's usually possible to convert a more reactive acid derivative into a less reactive one. Acid chlorides, for example, can be converted into esters and amides, but amides and esters can't be converted into acid chlorides. Remembering the reactivity order is therefore a useful way to keep track of a large number of reactions (Figure 10.3).

Figure 10.3 **Interconversions of carboxylic acid derivatives. More reactive compounds can be converted into less reactive ones, but not vice versa.**

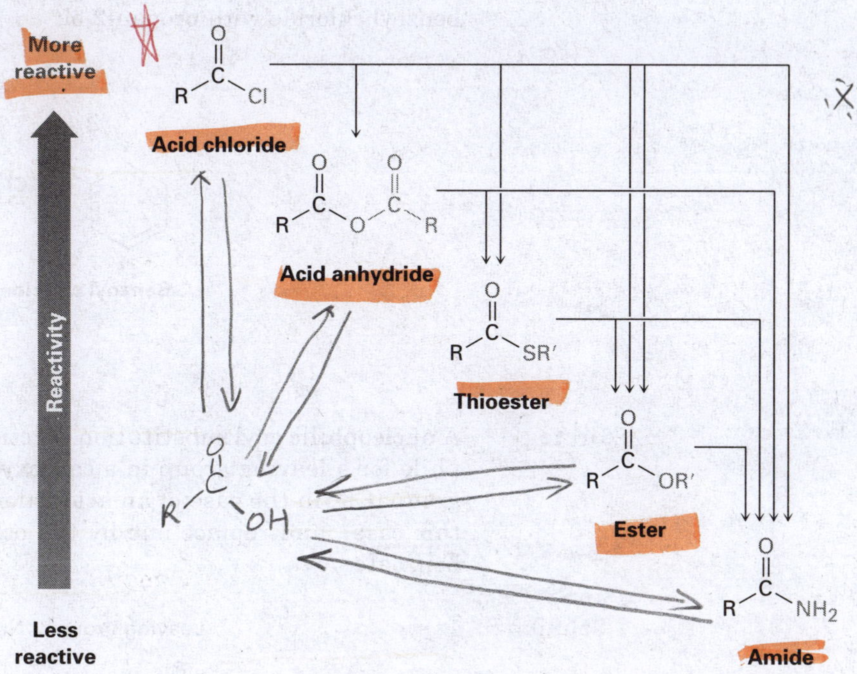

In studying the chemistry of carboxylic acids and derivatives in the next few sections, we'll be concerned largely with the reactions of just a few nucleophiles and will see that the same kinds of reactions keep occurring (Figure 10.4).

- **Hydrolysis:** Reaction with water to yield a carboxylic acid
- **Alcoholysis:** Reaction with an alcohol to yield an ester
- **Aminolysis:** Reaction with ammonia or an amine to yield an amide
- **Reduction:** Reaction with a hydride reducing agent to yield an alcohol
- **Grignard reaction:** Reaction with an organomagnesium reagent to yield an alcohol

Figure 10.4 **Some general reactions of carboxylic acid derivatives.**

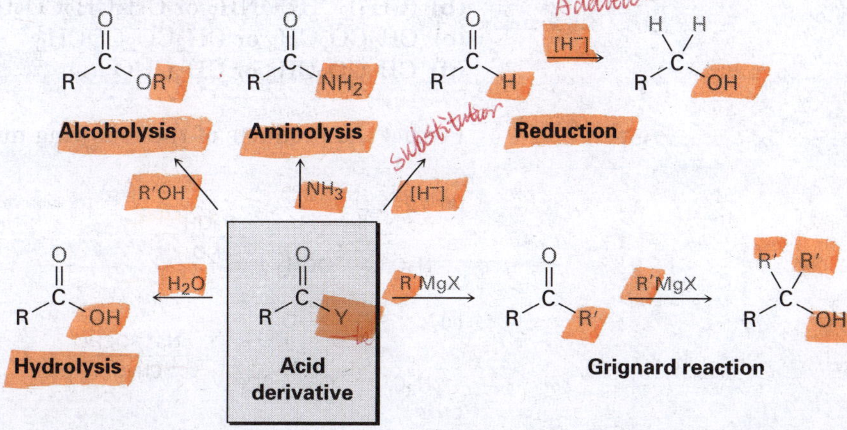

Worked Example 10.2 Predicting the Product of a Nucleophilic Acyl Substitution Reaction

Predict the product of the following nucleophilic acyl substitution reaction of benzoyl chloride with propan-2-ol:

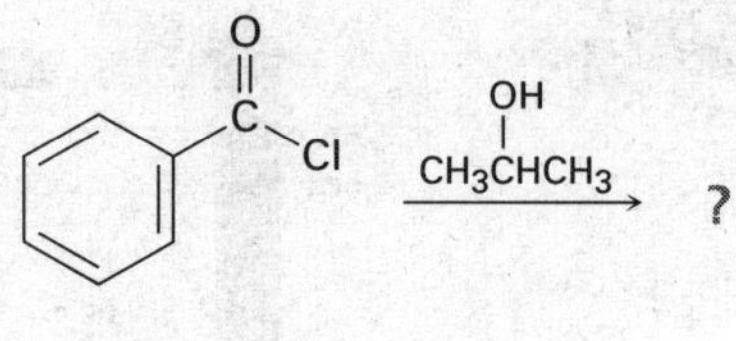

Benzoyl chloride

Strategy A nucleophilic acyl substitution reaction involves the substitution of a nucleophile for a leaving group in a carboxylic acid derivative. Identify the leaving group (Cl^- in the case of an acid chloride) and the nucleophile (an alcohol in this case), and replace one by the other. The product is the ester isopropyl benzoate.

Solution

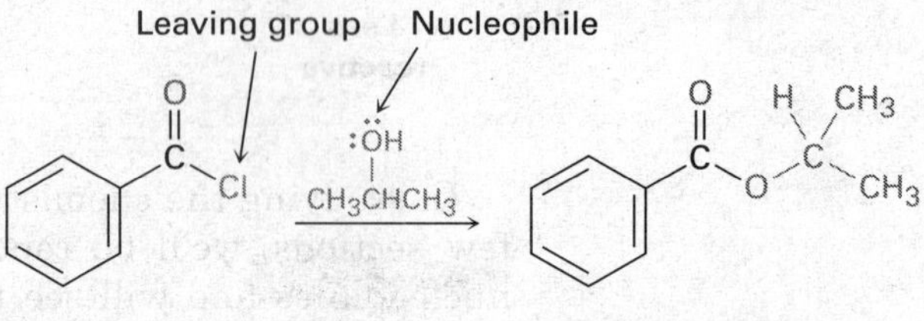

Benzoyl chloride **Isopropyl benzoate**

Problem 10.9 Which compound in each of the following sets is more reactive in nucleophilic acyl substitution reactions?

(a) CH_3COCl or $CH_3CO_2CH_3$
(b) $(CH_3)_2CHCONH_2$ or $CH_3CH_2CO_2CH_3$
(c) $CH_3CO_2CH_3$ or $CH_3CO_2COCH_3$
(d) $CH_3CO_2CH_3$ or CH_3CHO

Problem 10.10 Predict the products of the following nucleophilic acyl substitution reactions:

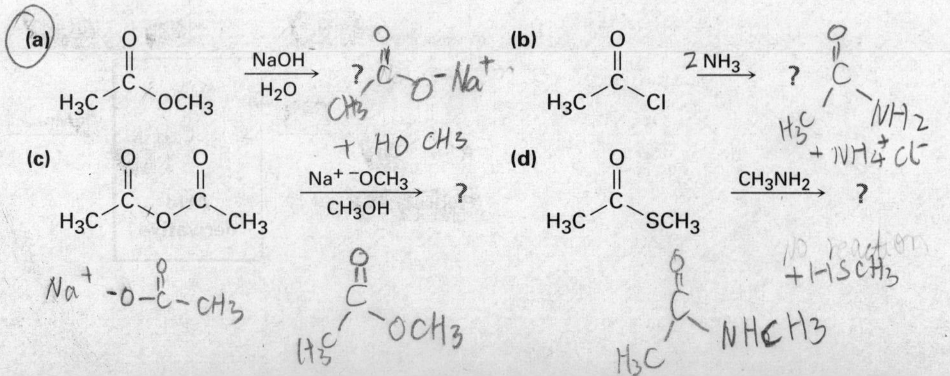

10.6 Carboxylic Acids and Their Reactions

The direct nucleophilic acyl substitution of a carboxylic acid is difficult because –OH is a poor leaving group (Section 7.5). Thus, it's usually necessary to enhance the reactivity of the acid, either by using a strong acid catalyst to protonate the carboxyl and make it a better acceptor or by converting the –OH into a better leaving group. Under the right conditions, however, acid chlorides, anhydrides, esters, and amides can all be prepared from carboxylic acids.

Conversion of Acids into Acid Chlorides ($RCO_2H \rightarrow RCOCl$)

Carboxylic acids are converted into acid chlorides by treatment with thionyl chloride, $SOCl_2$. The reaction occurs by a nucleophilic acyl substitution pathway in which the carboxylic acid is first converted into an acyl chlorosulfite intermediate, thereby replacing the –OH of the acid with a much better leaving group. The chlorosulfite then reacts with a nucleophilic chloride ion.

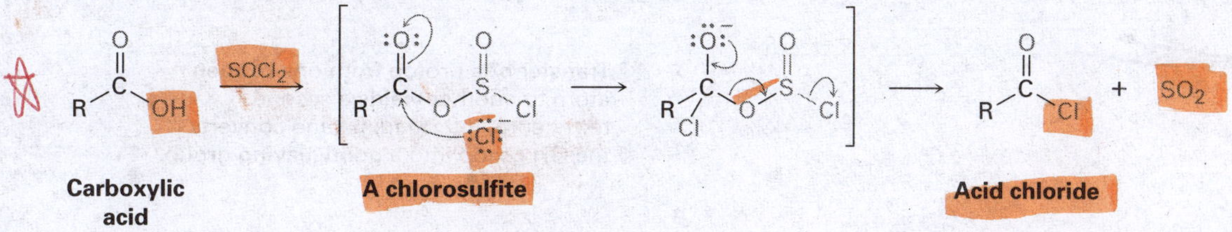

Conversion of Acids into Esters ($RCO_2H \rightarrow RCO_2R'$)

Perhaps the most useful reaction of carboxylic acids is their conversion into esters by reaction with an alcohol—the substitution of –OH by –OR. Called the **Fischer esterification reaction**, the simplest method involves heating the carboxylic acid with an acid catalyst in an alcohol solvent.

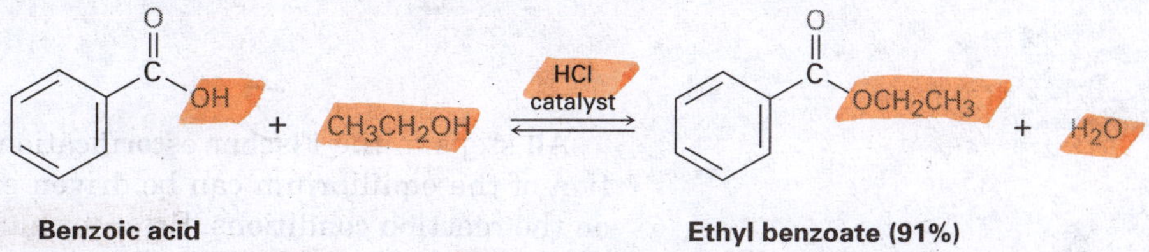

As shown in Figure 10.5, the acid catalyst first protonates an oxygen atom of the $-CO_2H$ group, which gives the carboxylic acid a positive charge and makes it more reactive toward nucleophiles. An alcohol molecule then adds to the protonated carboxylic acid, and subsequent loss of water yields the ester product.

MECHANISM

Figure 10.5 Mechanism of the Fischer esterification reaction of a carboxylic acid to yield an ester. The reaction is an acid-catalyzed nucleophilic acyl substitution.

1. Protonation of the carbonyl oxygen activates the carboxylic acid . . .
2. . . . toward nucleophilic attack by alcohol, yielding a tetrahedral intermediate.
3. Transfer of a proton from one oxygen atom to another yields a second tetrahedral intermediate and converts the OH group into a good leaving group.
4. Loss of a proton and expulsion of H_2O regenerates the acid catalyst and gives the ester product.

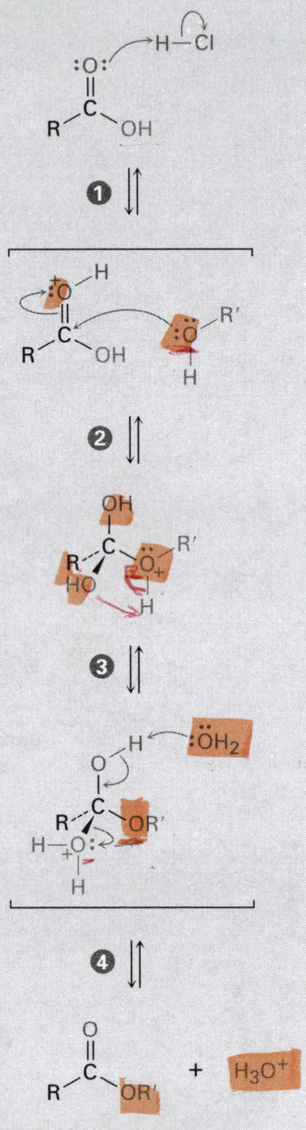

All steps in the Fischer esterification reaction are reversible, and the position of the equilibrium can be driven either forward or backward depending on the reaction conditions. Ester formation is favored when alcohol is used as the solvent, but carboxylic acid is favored when the solvent is water.

Conversion of Acids into Amides ($RCO_2H \rightarrow RCONH_2$)

Amides are carboxylic acid derivatives in which the acid –OH group has been replaced by a nitrogen substituent, $-NH_2$, –NHR, or $-NR_2$. Amides are difficult to prepare directly from acids by substitution with an amine because amines are bases, which convert acidic carboxyl groups into their unreactive carboxylate anions. Thus, the –OH must be activated by making it a better, nonacidic leaving group. In practice, amides are usually prepared by treating the carboxylic acid with dicyclohexylcarbodiimide (DCC) to activate it, followed by addition of the amine. We'll see in Section 15.7 that this DCC method for preparing amides is particularly useful for the laboratory synthesis of proteins from amino acids.

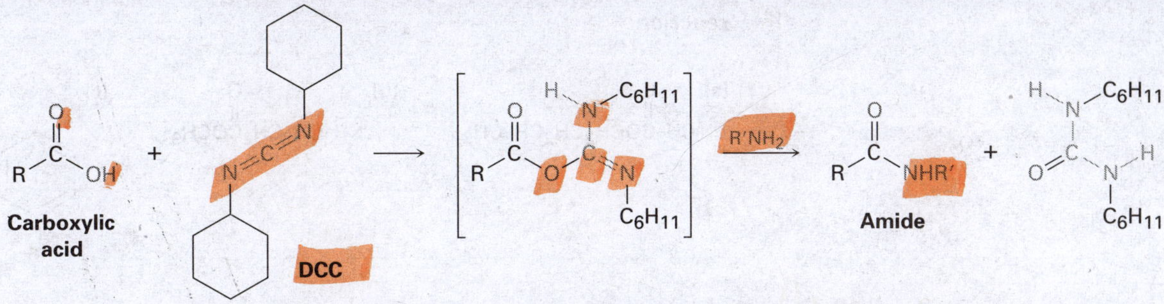

Conversion of Acids into Alcohols ($RCO_2H \rightarrow RCH_2OH$)

As we saw in Section 8.3, carboxylic acids are reduced by lithium aluminum hydride ($LiAlH_4$) to yield primary alcohols. The reaction occurs by initial substitution of the acid –OH group by –H to give an aldehyde intermediate that is further reduced to the alcohol.

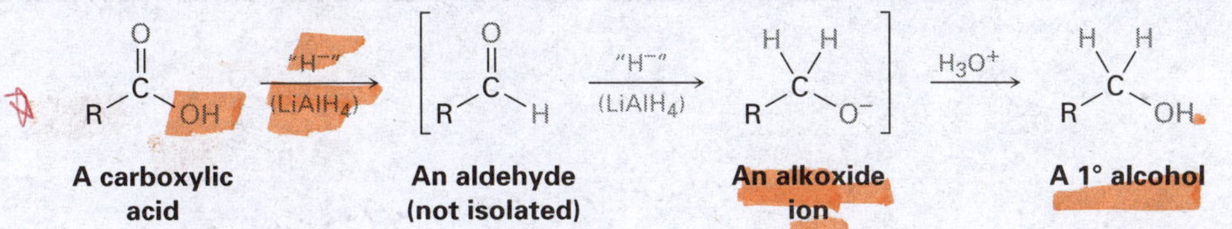

Worked Example 10.3 Synthesizing an Ester from an Acid

How might you prepare the following ester using a Fischer esterification reaction?

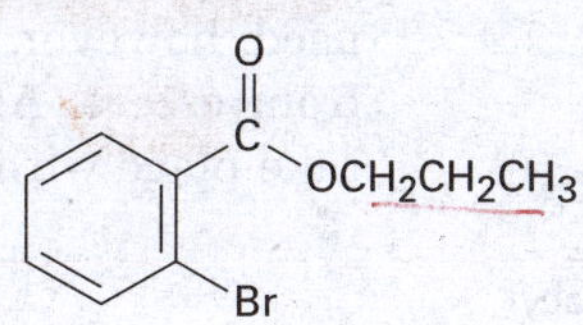

Strategy Begin by identifying the two parts of the ester. The acyl part comes from the carboxylic acid, and the –OR part comes from the alcohol. In this case, the target molecule is propyl *o*-bromobenzoate, so it can be prepared by treating *o*-bromobenzoic acid with propan-1-ol.

Solution

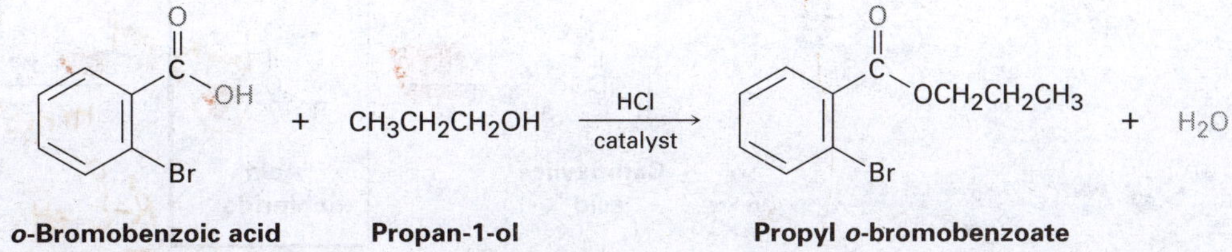

Problem 10.11 What products would you obtain by treating benzoic acid with the following reagents?

(a) $SOCl_2$ (b) CH_3OH, HCl (c) $LiAlH_4$ (d) NaOH

Problem 10.12 Show how you might prepare the following esters using a Fischer esterification reaction:

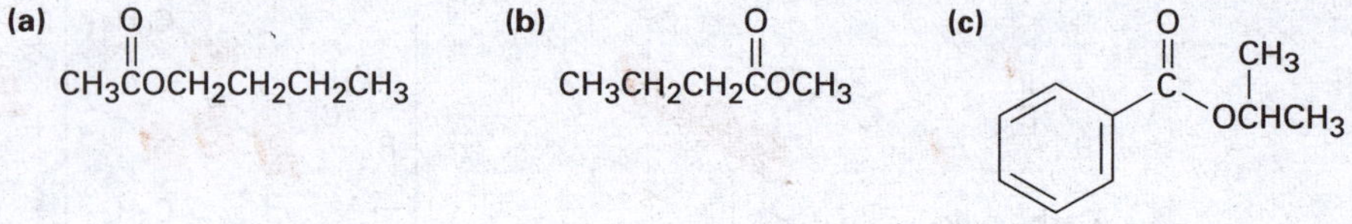

10.7 Acid Halides and Their Reactions

Acid chlorides are prepared from carboxylic acids by reaction with thionyl chloride, $SOCl_2$, as we saw in the previous section.

Acid halides are among the most reactive of the various carboxylic acid derivatives and can be converted into many other kinds of substances. The halogen can be replaced by –OH to yield an acid, by –OR to yield an ester, or by $-NH_2$ to yield an amide. In addition, acid halides can be reduced by $LiAlH_4$ to give primary alcohols or allowed to react with Grignard reagents to give tertiary alcohols (Figure 10.6). Neither of these latter two processes is often used, however, because the product alcohols can be made more conveniently from esters. Although illustrated only for acid chlorides, similar reactions take place with other acid halides.

Figure 10.6 Some nucleophilic acyl substitution reactions of acid chlorides.

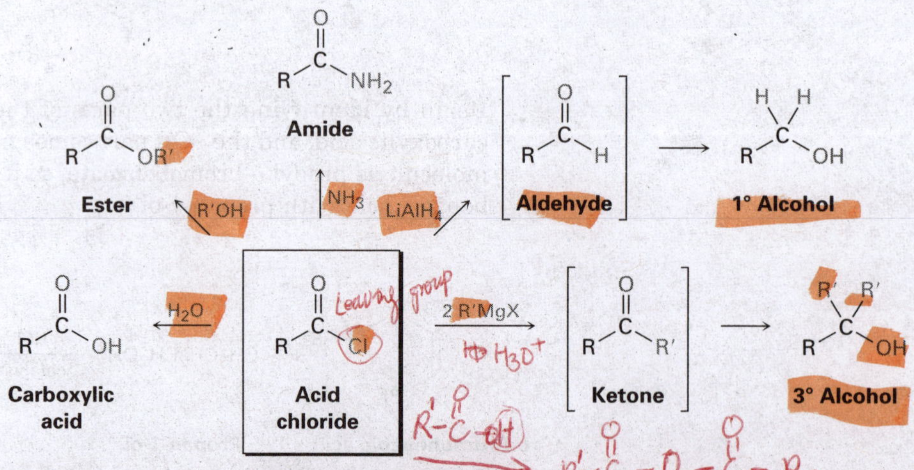

Conversion of Acid Chlorides into Acids ($RCOCl \rightarrow RCO_2H$)

Acid chlorides react with water to yield carboxylic acids—the substitution of –Cl by –OH. This hydrolysis reaction is a typical nucleophilic acyl substitution process and is initiated by attack of the nucleophile water on the acid chloride carbonyl group. The initially formed tetrahedral intermediate undergoes loss of HCl to yield the product.

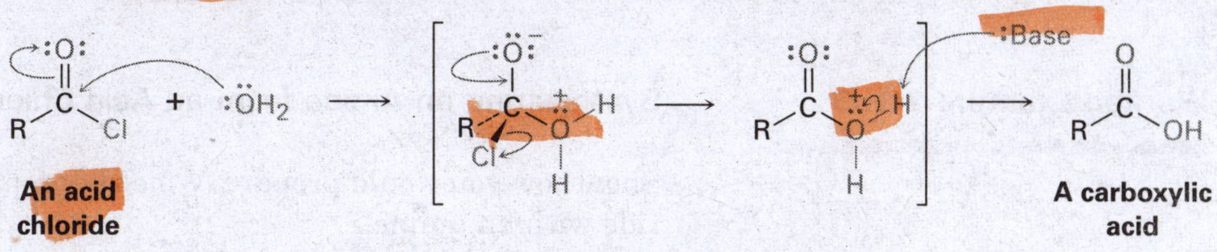

Conversion of Acid Chlorides into Esters ($RCOCl \rightarrow RCO_2R'$)

Acid chlorides react with alcohols to yield esters in a reaction analogous to their reaction with water to yield acids.

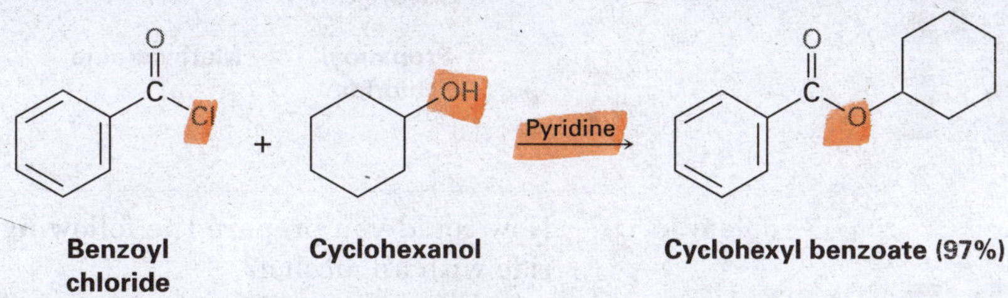

Because HCl is generated as a by-product, the reaction is usually carried out in the presence of an amine base such as pyridine (see Section 12.6), which reacts with the HCl as it's formed and prevents it from causing side reactions.

Conversion of Acid Chlorides into Amides ($RCOCl \rightarrow RCONH_2$)

Acid chlorides react rapidly with ammonia and with amines to give amides—the substitution of –Cl by $–NR_2$. Both monosubstituted and disubstituted amines can be used. For example, 2-methylpropanamide is prepared by reaction of 2-methylpropanoyl chloride with ammonia. Note that one extra equivalent of ammonia is added to react with the HCl generated.

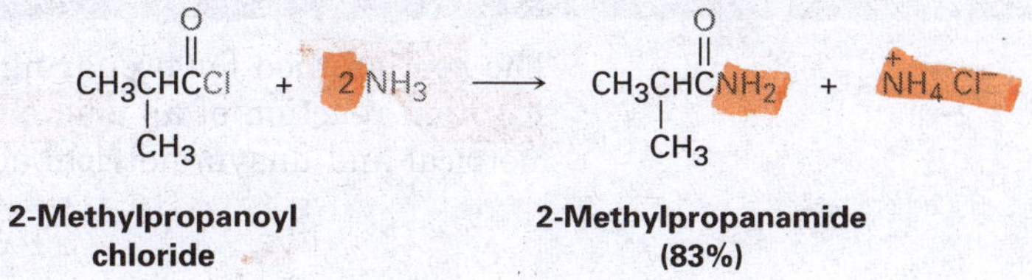

Worked Example 10.4

Synthesizing an Ester from an Acid Chloride

Show how you could prepare ethyl benzoate by reaction of an acid chloride with an alcohol.

Strategy As its name implies, ethyl benzoate can be made by reaction of *ethyl* alcohol with the acid chloride of *benzoic* acid.

Solution

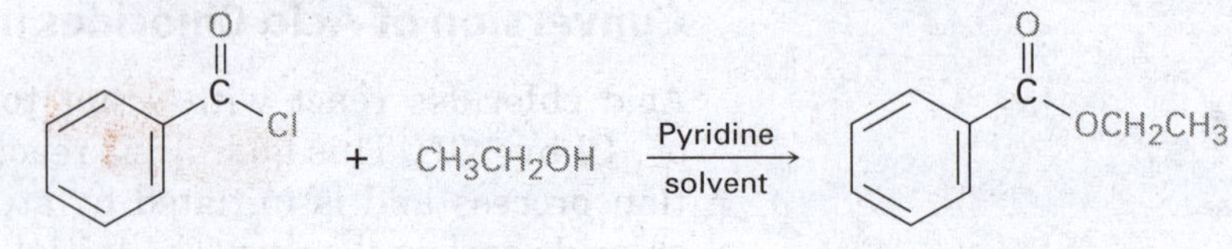

Benzoyl chloride **Ethanol** **Ethyl benzoate**

Worked Example 10.5 Synthesizing an Amide from an Acid Chloride

Show how you would prepare *N*-methylpropanamide by reaction of an acid chloride with an amine.

Strategy The name of the product gives a hint as to how it can be prepared. Reaction of *methyl*amine with *propanoyl* chloride gives *N*-methylpropanamide.

Solution

$$CH_3CH_2\overset{O}{\overset{\|}{C}}Cl \;+\; 2\,CH_3NH_2 \longrightarrow CH_3CH_2\overset{O}{\overset{\|}{C}}NHCH_3 \;+\; CH_3NH_3^{+}\,Cl^{-}$$

Propanoyl chloride **Methylamine** ***N*-Methylpropanamide**

Problem 10.13 How could you prepare the following esters using the reaction of an acid chloride with an alcohol?

(a) $CH_3CH_2CO_2CH_3$ **(b)** $CH_3CO_2CH_2CH_3$ **(c)** Cyclohexyl acetate

Problem 10.14 Write the steps in the mechanism of the reaction between ammonia and 2-methylpropanoyl chloride to yield 2-methylpropanamide.

Problem 10.15 What amines would react with what acid chlorides to give the following amide products?

(a) $CH_3CH_2CONH_2$ **(b)** $(CH_3)_2CHCH_2CONHCH_3$
(c) *N,N*-Dimethylpropanamide **(d)** *N,N*-Diethylbenzamide

10.8 Acid Anhydrides and Their Reactions

The best method for preparing acid anhydrides is by a nucleophilic acyl substitution reaction of an acid chloride with a carboxylic acid anion. Both symmetrical and unsymmetrical acid anhydrides can be prepared in this way.

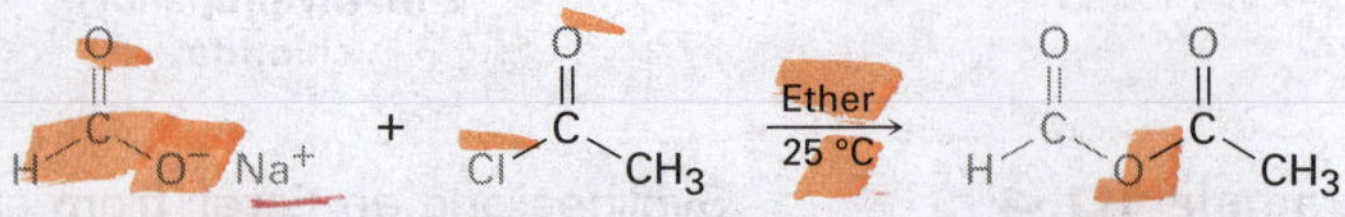

Sodium formate **Acetyl chloride** **Acetic formic anhydride (64%)**

The chemistry of acid anhydrides is similar to that of acid chlorides. Thus, acid anhydrides react with water to form acids, with alcohols to form esters, and with amines to form amides (Figure 10.7). They also undergo reduction with $LiAlH_4$ to give primary alcohols and Grignard reaction to give tertiary alcohols, but neither of these reactions is often used since the alcohol products can be made more conveniently from esters.

Figure 10.7 Some reactions of acid anhydrides.

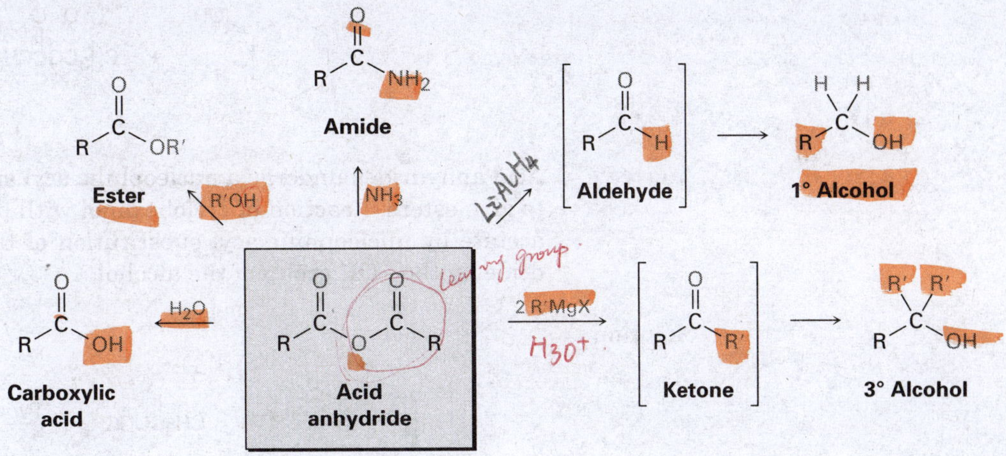

Acetic anhydride is often used to prepare acetate esters of complex alcohols and to prepare substituted acetamides from amines. For example, aspirin (an ester) is prepared by reaction of acetic anhydride with *o*-hydroxybenzoic acid. Similarly, acetaminophen (an amide; the active ingredient in Tylenol) is prepared by reaction of acetic anhydride with *p*-hydroxyaniline.

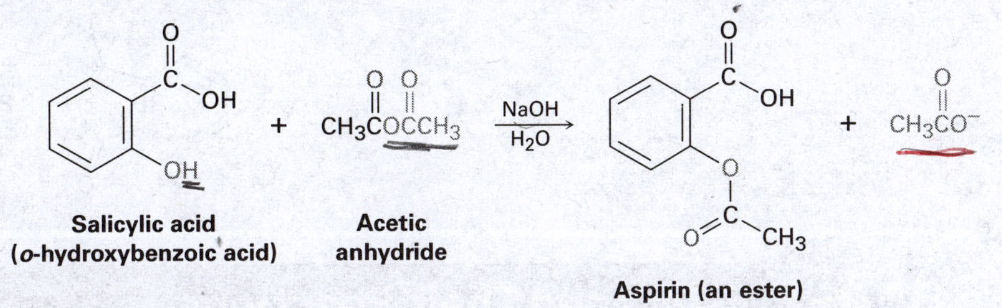

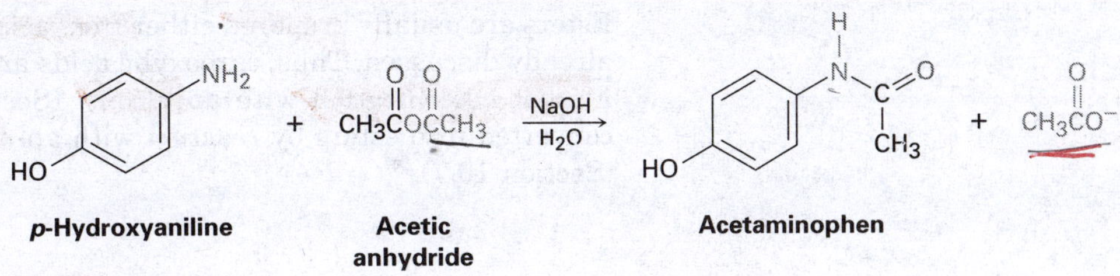

Notice in both of these examples that only "half" of the anhydride molecule is used; the other half acts as the leaving group during the nucleophilic acyl substitution step and produces carboxylate anion as a by-product. Thus, anhydrides are inefficient to use, and acid chlorides are normally used instead.

Worked Example 10.6 Predicting the Product of a Nucleophilic Acyl Substitution Reaction

What is the product of the following reaction?

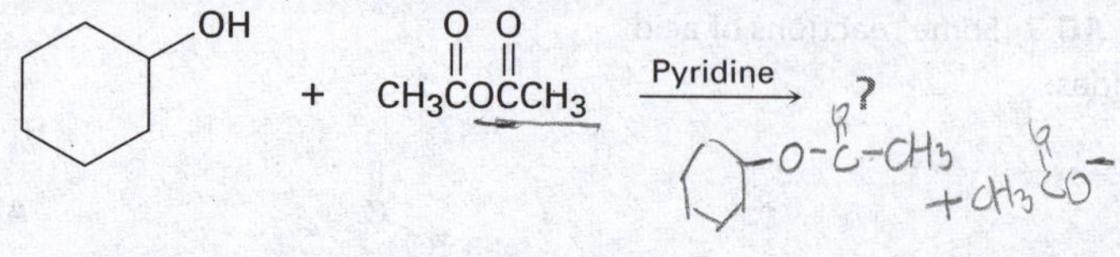

Strategy Acid anhydrides undergo a nucleophilic acyl substitution reaction with alcohols to give esters. Reaction of cyclohexanol with acetic anhydride yields cyclohexyl acetate by nucleophilic acyl substitution of the $-OCOCH_3$ group of the anhydride by the −OR group of the alcohol.

Solution

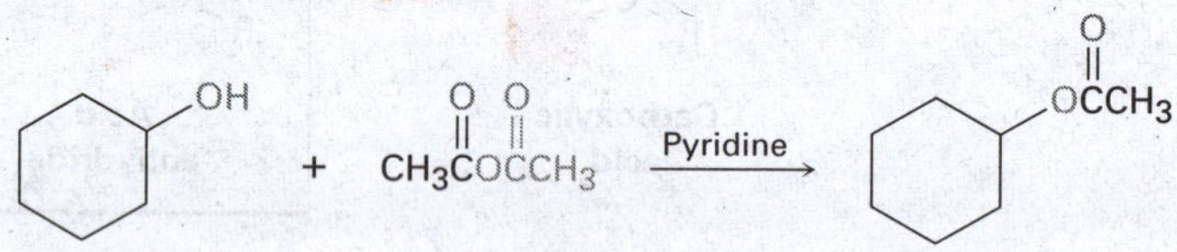

Cyclohexanol **Cyclohexyl acetate**

Problem 10.16 Write the steps in the mechanism of the reaction between *p*-hydroxyaniline and acetic anhydride to prepare acetaminophen.

Problem 10.17 What product would you expect to obtain from the reaction of 1 equivalent of methanol with a cyclic anhydride such as phthalic anhydride?

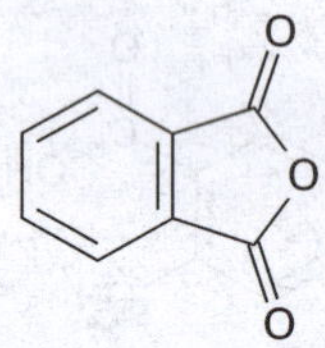

Phthalic anhydride

10.9 Esters and Their Reactions

Esters are usually prepared either from acids or acid chlorides by the methods already discussed. Thus, carboxylic acids are converted directly into esters by Fischer esterification with an alcohol (Section 10.6), and acid chlorides are converted into esters by reaction with an alcohol in the presence of pyridine (Section 10.7).

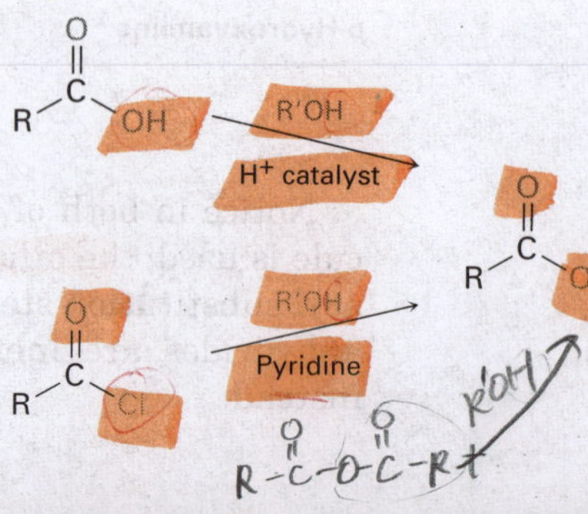

Esters show the same kinds of chemistry we've seen for other acyl derivatives, but they're less reactive toward nucleophiles than acid chlorides or anhydrides. Figure 10.8 shows some general reactions of esters.

Figure 10.8 Some reactions of esters.

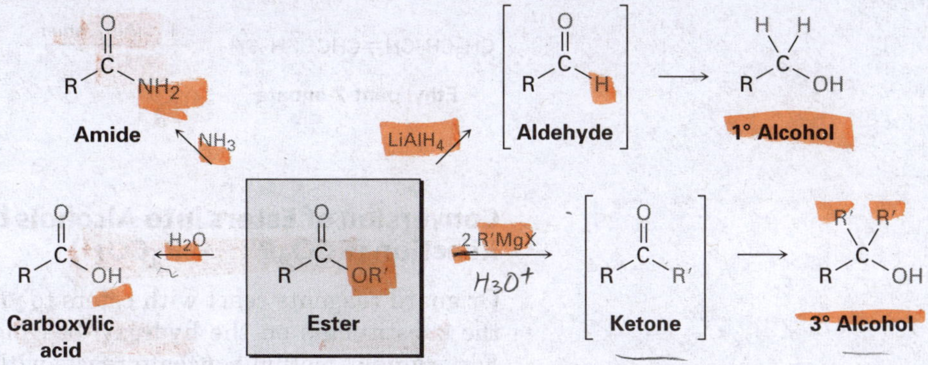

Conversion of Esters into Acids ($RCO_2R' \rightarrow RCO_2H$)

Esters are hydrolyzed either by aqueous base or by aqueous acid to yield a carboxylic acid plus an alcohol.

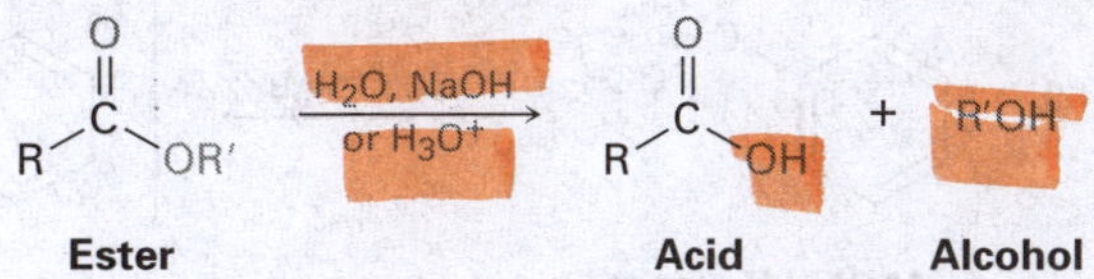

Hydrolysis in basic solution is called *saponification,* after the Latin word *sapo,* "soap." (As we'll see in Section 16.2, soap is made by the base-induced ester hydrolysis of animal fat.) Ester hydrolysis occurs by a typical nucleophilic acyl substitution pathway in which OH^- nucleophile adds to the ester carbonyl group, yielding a tetrahedral alkoxide intermediate. Loss of RO^- then gives a carboxylic acid, which is deprotonated to give the acid carboxylate plus alcohol.

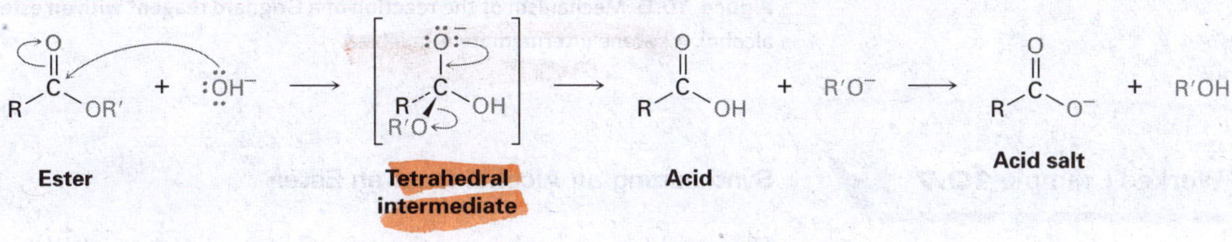

Conversion of Esters into Alcohols by Reduction ($RCO_2R' \rightarrow RCH_2OH$)

Esters are reduced to primary alcohols by treatment with $LiAlH_4$ (Section 8.3). The reaction occurs by an initial nucleophilic acyl substitution reaction in which hydride ion adds to the carbonyl group followed by elimination of an alkoxide ion to give an aldehyde intermediate. Further reduction of the aldehyde by a typical nucleophilic addition process gives the primary alcohol.

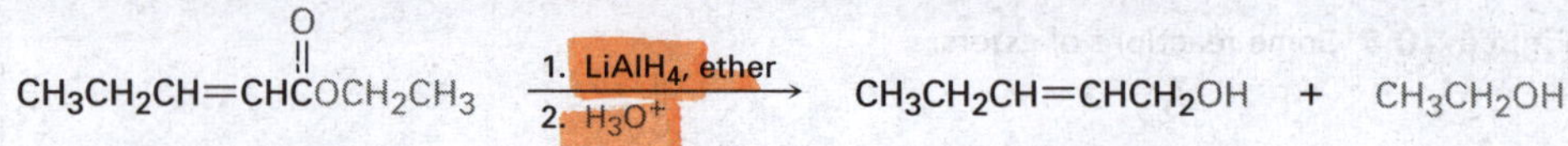

Ethyl pent-2-enoate — Pent-2-en-1-ol (91%)

Conversion of Esters into Alcohols by Grignard Reaction ($RCO_2R' \rightarrow R_3COH$)

Grignard reagents react with esters to yield tertiary alcohols in which two of the substituents on the hydroxyl-bearing carbon are identical (Section 8.3). For example, methyl benzoate reacts with 2 equivalents of CH_3MgBr to yield 2-phenylpropan-2-ol. The reaction occurs by nucleophilic addition of a Grignard reagent to the ester, elimination of alkoxide ion to give an intermediate ketone, and further nucleophilic addition to the ketone to yield the tertiary alcohol (Figure 10.9).

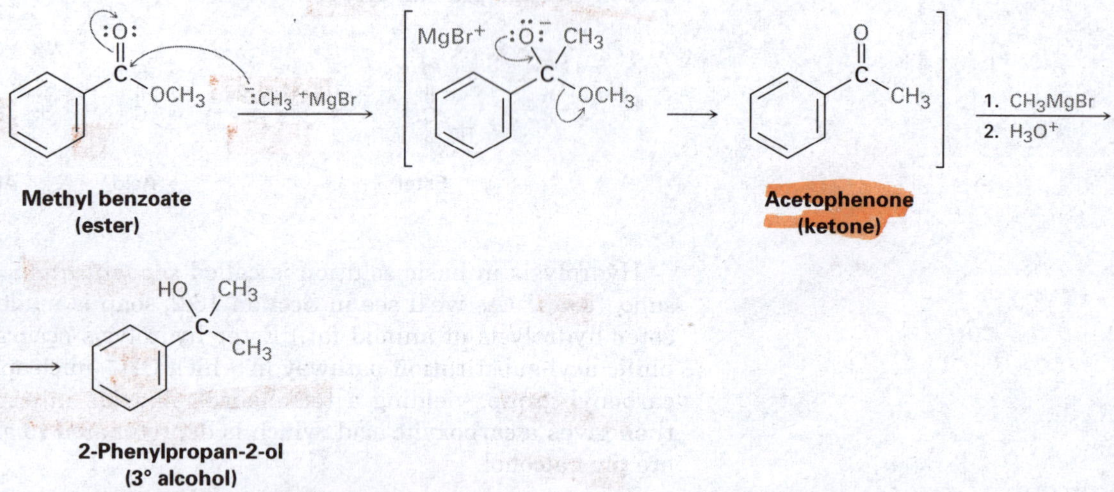

Figure 10.9 Mechanism of the reaction of a Grignard reagent with an ester to yield a tertiary alcohol. A ketone intermediate is involved.

Worked Example 10.7 — **Synthesizing an Alcohol from an Ester**

How could you use the reaction of a Grignard reagent with an ester to prepare 1,1-diphenylpropan-1-ol?

Strategy The product of the reaction between a Grignard reagent and an ester is a tertiary alcohol in which the alcohol carbon and one of the attached groups have come from the ester and the remaining two groups bonded to the alcohol carbon have come from the Grignard reagent. Since 1,1-diphenylpropan-1-ol has two phenyl groups and one ethyl group bonded to the alcohol carbon, it must be prepared from reaction of a phenylmagnesium halide with an ester of propanoic acid.

Solution

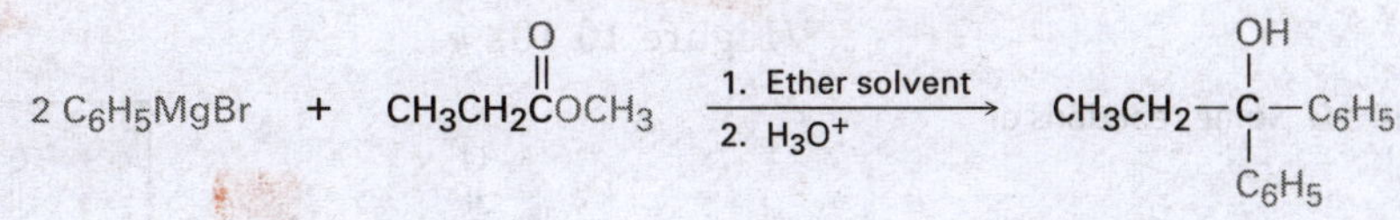

1,1-Diphenylpropan-1-ol

Problem 10.18 Show the products of hydrolysis of the following esters:

(a) 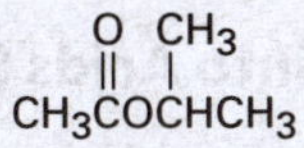 **(b)** 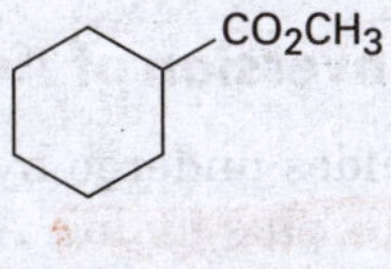

Problem 10.19 Why do you suppose the saponification of esters is not reversible? In other words, why doesn't treatment of a carboxylic acid with an alkoxide ion give an ester?

Problem 10.20 Show the products you would obtain by reduction of the following esters with $LiAlH_4$:

(a) 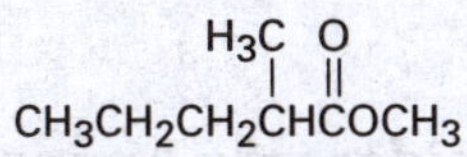 **(b)** 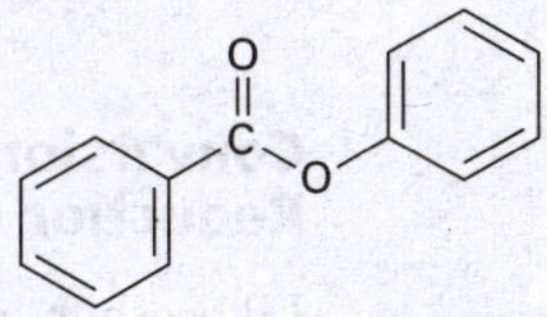

Problem 10.21 What ester and what Grignard reagent might you use to prepare the following alcohols?

(a) 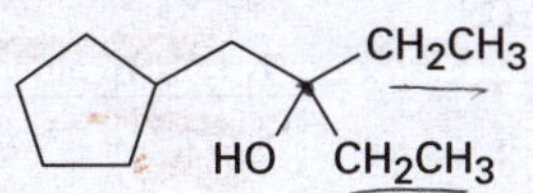 **(b)** 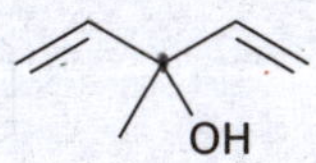

10.10 Amides and Their Reactions

Amides are usually prepared by reaction of an acid chloride with an amine (Section 10.7). They are much less reactive than acid chlorides, acid anhydrides, and esters, and the amide bond is stable enough to link different amino acids together to form proteins.

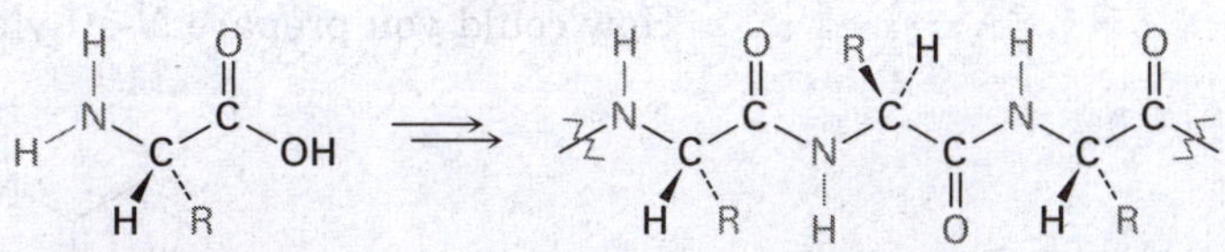

An amino acid **A protein**

The most common reactions of amides are their hydrolysis to give carboxylic acids and their reduction with $LiAlH_4$. Interestingly, though, the reduction product of an amide is an *amine* rather than the expected alcohol (Figure 10.10).

Figure 10.10 Some reactions of amides.

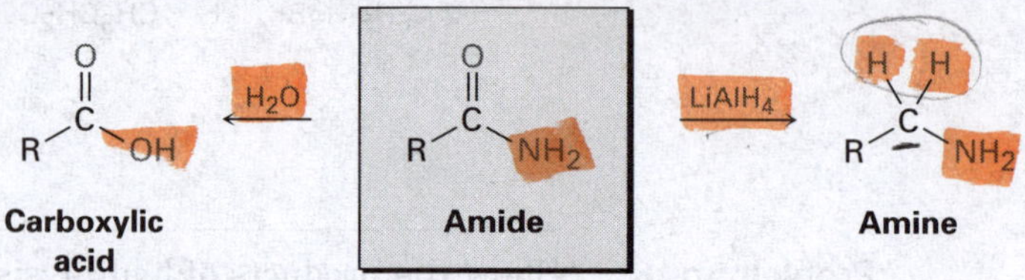

Conversion of Amides into Acids ($RCONH_2 \rightarrow RCO_2H$)

Amides undergo hydrolysis in either aqueous acid or base to yield carboxylic acids plus amine. Although the reaction is slow and requires prolonged heating, the overall transformation is a typical nucleophilic acyl substitution of $-OH$ for $-NH_2$. In biochemistry, the reaction is particularly useful for hydrolyzing proteins to their constituent amino acids.

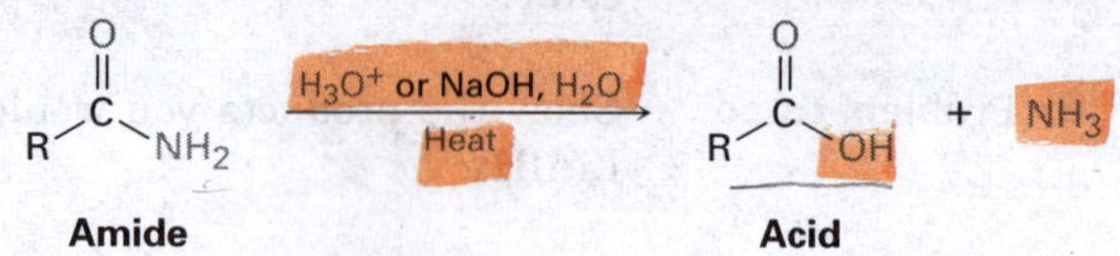

Conversion of Amides into Amines by Reduction ($RCONH_2 \rightarrow RCH_2NH_2$)

Like other carboxylic acid derivatives, amides are reduced by $LiAlH_4$. The product of this reduction, however, is an amine rather than an alcohol. For example:

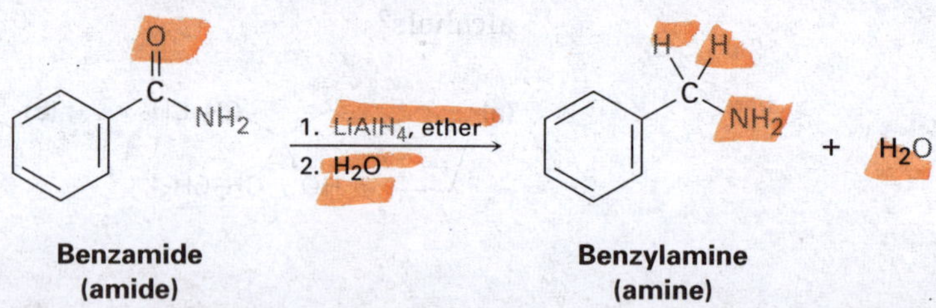

The effect of amide reduction is to convert the amide carbonyl group into a methylene group ($C{=}O \rightarrow CH_2$). This kind of reaction is specific for amides and does not occur with other carboxylic acid derivatives.

Worked Example 10.8

Synthesizing an Amine from an Amide

How could you prepare *N*-ethylaniline by reduction of an amide with $LiAlH_4$?

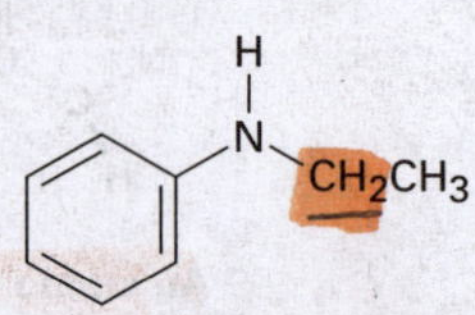

***N*-Ethylaniline**

Strategy Reduction of an amide with $LiAlH_4$ yields an amine. To find the starting material for synthesis of *N*-ethylaniline, look for a CH_2 position next to the nitrogen atom and replace that CH_2 by C=O. In this case, the amide is *N*-phenylacetamide.

Solution

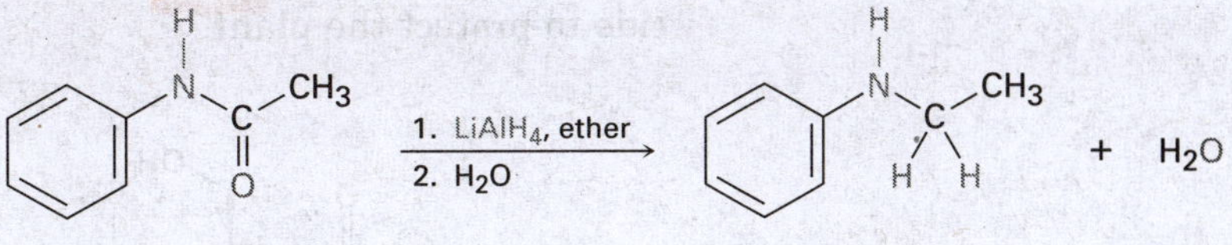

N-Phenylacetamide **N-Ethylaniline**

Problem 10.22 How would you convert *N*-ethylbenzamide into the following substances?

(a) Benzoic acid
(b) Benzyl alcohol
(c) *N*-Ethylbenzylamine, $C_6H_5CH_2NHCH_2CH_3$

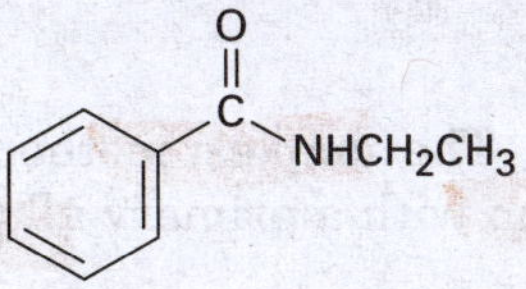

N-Ethylbenzamide

Problem 10.23 The reduction of an amide with $LiAlH_4$ to yield an amine occurs with both acyclic and cyclic amides (*lactams*). What product would you obtain from reduction of 5,5-dimethylpyrrolidin-2-one with $LiAlH_4$?

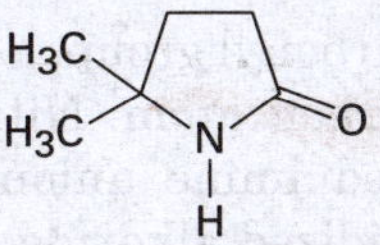

5,5-Dimethylpyrrolidin-2-one (a lactam)

10.11 Nitriles and Their Reactions

Nitriles, R—C≡N, are analogous to carboxylic acids in that both have a carbon atom with three bonds to electronegative atoms and both contain a multiple bond.

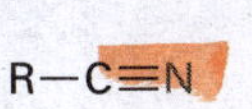

A nitrile—three bonds to nitrogen

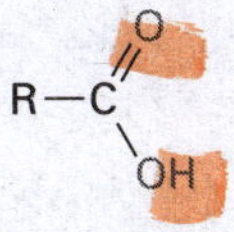

An acid—three bonds to two oxygens

Nitriles occur less frequently in living organisms than do acid derivatives, although more than 1000 examples are known. Cyanocycline A, for instance, has been isolated from the bacterium *Streptomyces lavendulae* and found to have both antimicrobial and antitumor activity. Lotaustralin, isolated from the cassava plant, contains a sugar with an acetal carbon, one oxygen of which is bonded to a nitrile-bearing carbon (Sugar—O—C—CN). On hydrolysis of the acetal, hydrogen cyanide is released, thereby acting as a natural insecticide to protect the plant.

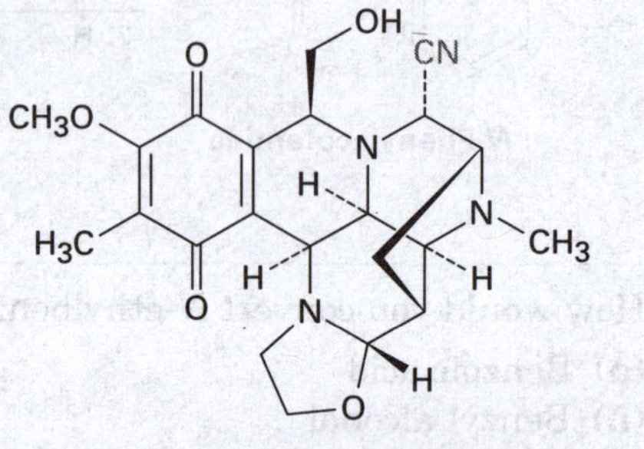

Cyanocycline A

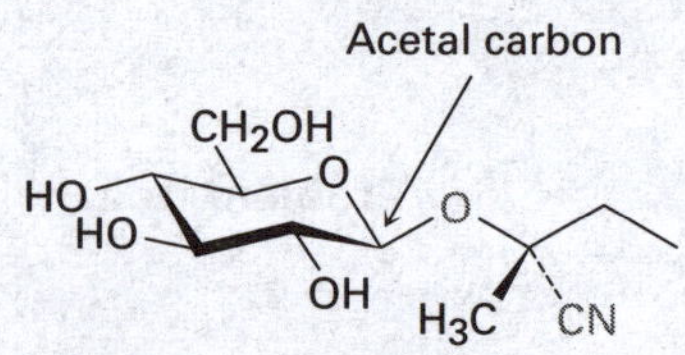

Lotaustralin
(a cyanogenic glycoside)

The simplest method of preparing nitriles is by the S_N2 reaction of cyanide ion with a primary alkyl halide, as discussed in Section 7.5.

$$RCH_2Br + Na^+ \ ^-CN \xrightarrow[\text{reaction}]{S_N2} RCH_2C{\equiv}N + NaBr$$

Reactions of Nitriles

Like a carbonyl group, a nitrile group is strongly polarized and has an electrophilic carbon atom. Nitriles therefore react with nucleophiles to yield sp^2-hybridized imine anions in a reaction analogous to the formation of an sp^3-hybridized alkoxide ion by nucleophilic addition to a carbonyl group. The imine anion then goes on to yield further products.

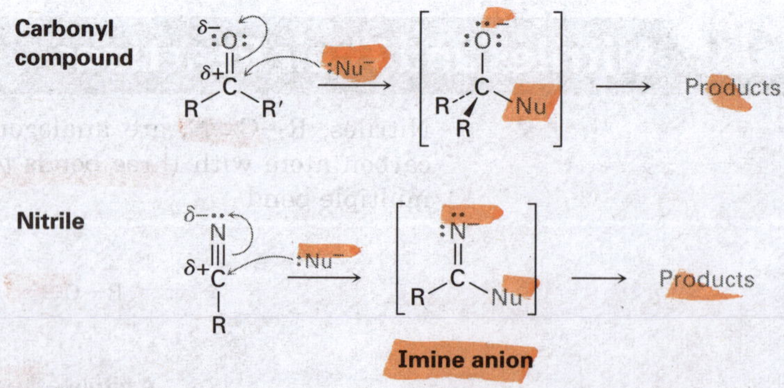

Among the most useful reactions of nitriles are hydrolysis to give a carboxylic acid plus ammonia, reduction to yield an amine, and Grignard reaction to give a ketone (Figure 10.11).

Figure 10.11 Some reactions of nitriles.

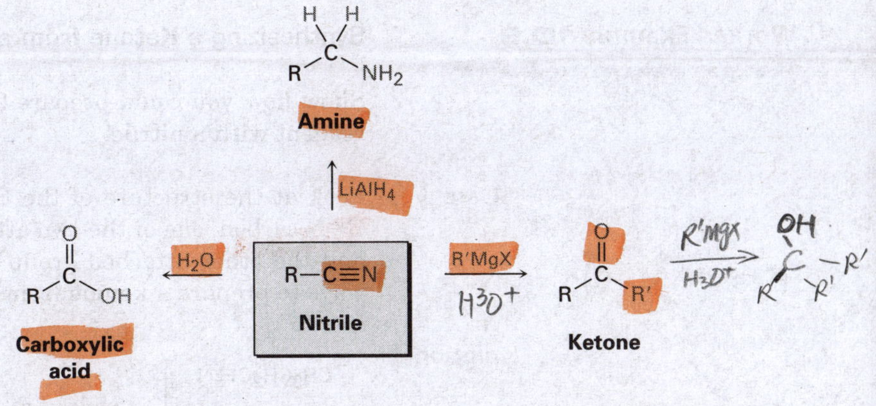

Conversion of Nitriles into Carboxylic Acids ($RCN \rightarrow RCO_2H$)

Nitriles are hydrolyzed in either acidic or basic solution to yield carboxylic acids and ammonia (or an amine). For example, benzonitrile gives benzoic acid.

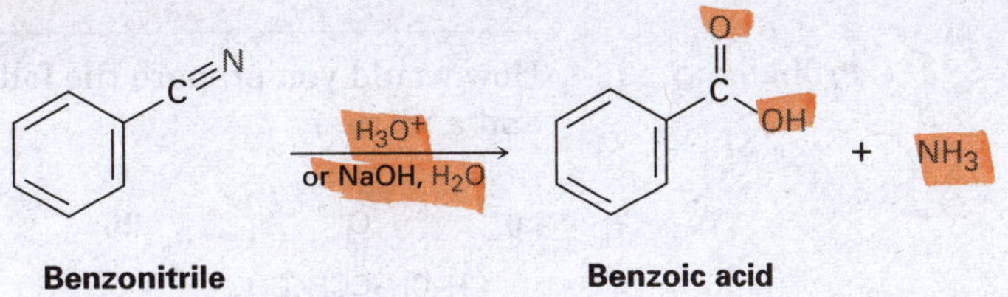

Conversion of Nitriles into Amines by Reduction ($RCN \rightarrow RCH_2NH_2$)

Reduction of a nitrile with $LiAlH_4$ gives a primary amine, RNH_2, just as reduction of an ester gives a primary alcohol, ROH. For example:

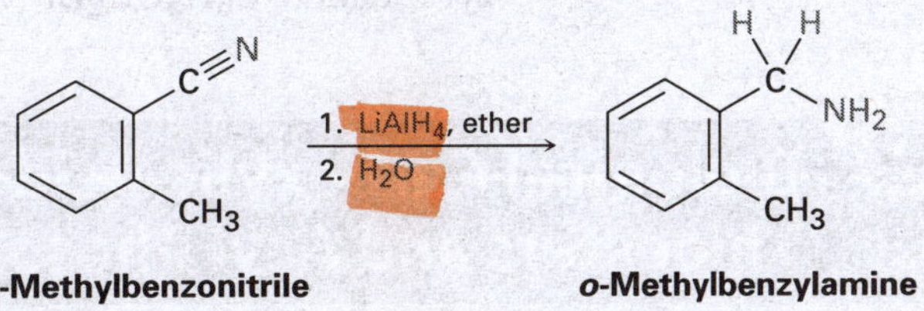

Conversion of Nitriles into Ketones by Reaction with Grignard Reagents

Grignard reagents, RMgX, add to nitriles to give intermediate imine anions that can be hydrolyzed to yield ketones. For example, benzonitrile reacts with ethylmagnesium bromide to give propiophenone.

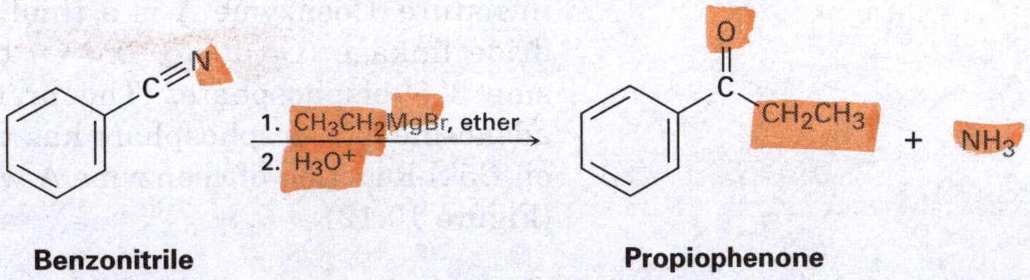

Worked Example 10.9 Synthesizing a Ketone from a Nitrile

Show how you could prepare 2-methylpentan-3-one by reaction of a Grignard reagent with a nitrile.

Strategy Look at the structure of the target ketone. The C=O carbon comes from the C≡N carbon, one of the two attached groups comes from the Grignard reagent, and the other attached group was present in the nitrile. Thus, there are two ways to prepare a ketone from a nitrile by Grignard addition.

Solution

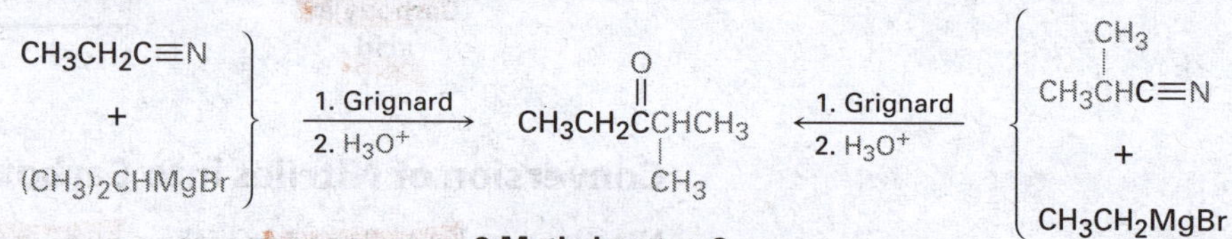

2-Methylpentan-3-one

Problem 10.24 How would you prepare the following ketones by reaction of a Grignard reagent and a nitrile?

(a)

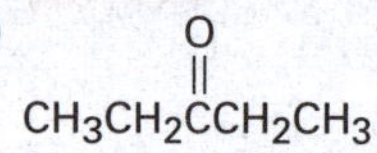

(b)

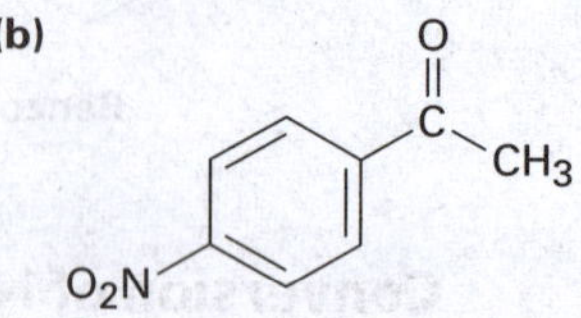

Problem 10.25 How would you prepare 1-phenylbutan-2-one, $C_6H_5CH_2COCH_2CH_3$, from benzyl bromide, $C_6H_5CH_2Br$? More than one step is needed.

10.12 Biological Carboxylic Acid Derivatives: Thioesters and Acyl Phosphates

As mentioned in the chapter introduction, the substrate for nucleophilic acyl substitution reactions in living organisms is generally either a **thioester (RCOSR′)** or an **acyl phosphate** ($RCO_2PO_3^{2-}$ or $RCO_2PO_3R'^-$). Both are intermediate in reactivity between acid chlorides and esters. Thus, they are stable enough to exist in living organisms but reactive enough to undergo acyl substitution.

Acetyl coenzyme A, abbreviated acetyl CoA, is the most common thioester in nature. Coenzyme A is a thiol (RSH) that contains a phosphoric anhydride linkage (O=P—O—P=O) between phosphopantetheine and adenosine 3′,5′-bisphosphate. (The prefix *bis*- means "two" and indicates that adenosine 3′,5′-bisphosphate has two phosphate groups, one on C3′ and one on C5′.) Reaction of coenzyme A with an acyl phosphate gives the acyl CoA (Figure 10.12).

Figure 10.12 Formation of the thioester acetyl CoA by nucleophilic acyl substitution reaction of coenzyme A (CoA) with acetyl adenylate.

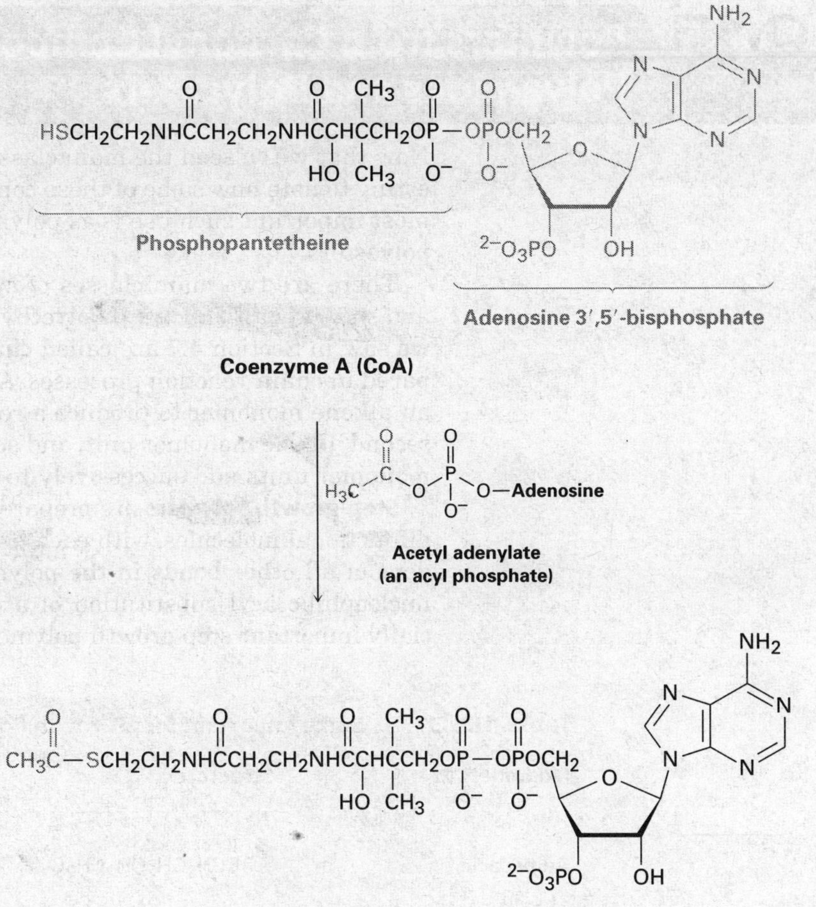

Acetyl CoA

Once formed, an acyl CoA is a substrate for numerous nucleophilic acyl substitution reactions. For example, *N*-acetylglucosamine, a component of cartilage and other connective tissues, is synthesized by an aminolysis reaction between glucosamine and acetyl CoA.

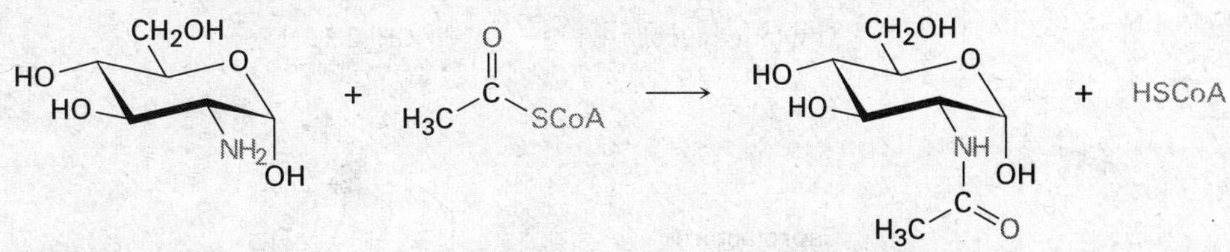

Glucosamine (an amine) **N-Acetylglucosamine (an amide)**

Problem 10.26 Write the mechanism of the reaction shown in Figure 10.12 between coenzyme A and acetyl adenylate to give acetyl CoA.

10.13 Polymers from Carbonyl Compounds: Polyamides and Polyesters

Now that we've seen the main classes of carboxylic acid derivatives, it's interesting to note how some of these compounds are used in daily life. Surely their most important such use is as polymers, particularly polyamides (*nylons*) and polyesters.

There are two main classes of synthetic polymers: *chain-growth polymers* and *step-growth polymers.* Polyethylene and other alkene polymers like those we saw in Section 4.7 are called **chain-growth polymers** because they are prepared in chain-reaction processes. An initiator first adds to the double bond of an alkene monomer to produce a reactive intermediate, which then adds to a second alkene monomer unit, and so on. The polymer chain lengthens as more monomer units add successively to the end of the growing chain.

Step-growth polymers are prepared by polymerization reactions between two difunctional molecules, with each new bond formed in a discrete step, independent of all other bonds in the polymer. The key bond-forming step is often a nucleophilic acyl substitution of a carboxylic acid derivative. Some commercially important step-growth polymers are shown in Table 10.3.

Table 10.3 Some Important Step-Growth Polymers and Their Uses

Monomers	Structure	Polymer	Uses
Adipic acid	$HO\overset{O}{\overset{\|}{C}}CH_2CH_2CH_2CH_2\overset{O}{\overset{\|}{C}}OH$	Nylon 66	Fibers, clothing, tire cord
+			
Hexamethylenediamine	$H_2NCH_2CH_2CH_2CH_2CH_2CH_2NH_2$		
Dimethyl terephthalate	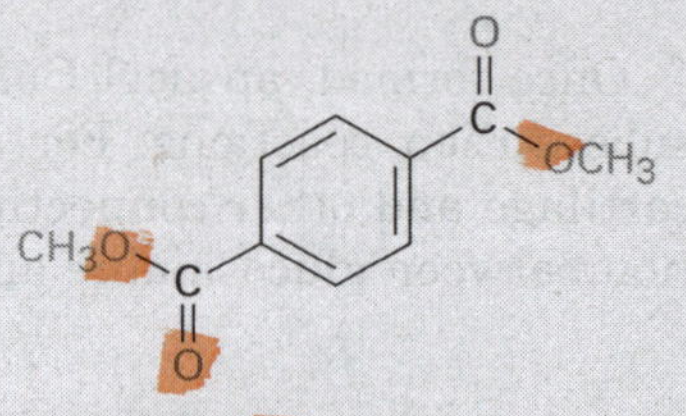	Dacron, Mylar, Terylene	Fibers, clothing, films, tire cord
+			
Ethylene glycol	$HOCH_2CH_2OH$		
Caprolactam	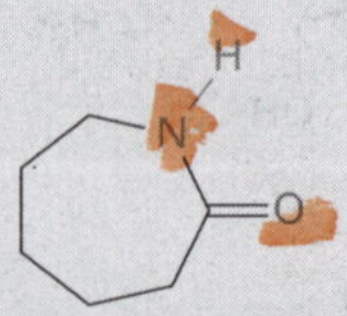	Nylon 6, Perlon	Fibers, castings

Polyamides

The best-known step-growth polymers are the *polyamides,* or **nylons**, prepared by reaction of a diamine with a diacid. For example, nylon 66 is prepared by reaction of adipic acid (hexanedioic acid) with hexamethylenediamine (hexane-1,6-diamine) at 280 °C. The designation "66" tells the number of carbon atoms in the diamine (the first 6) and the diacid (the second 6).

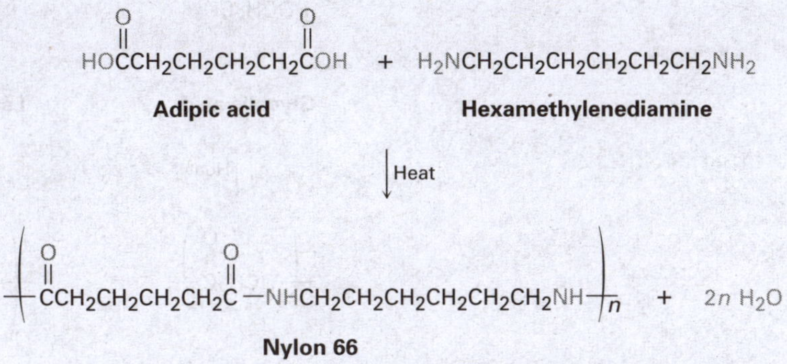

Nylons are used both in engineering applications and in making fibers. A combination of high impact strength and abrasion resistance makes nylon an excellent metal substitute for bearings and gears. As fiber, nylon is used in a variety of applications, from clothing to tire cord to ropes.

Polyesters

Just as a polyamide is made by reaction between a diacid and a diamine, a **polyester** is made by reaction between a diacid and a dialcohol. The most generally useful polyester is that made by reaction between dimethyl terephthalate (dimethyl benzene-1,4-dicarboxylate) and ethylene glycol (ethane-1,2-diol). The product is used under the trade name Dacron to make clothing fiber and tire cord, and under the name Mylar to make recording tape. The tensile strength of poly(ethylene terephthalate) film is nearly equal to that of steel.

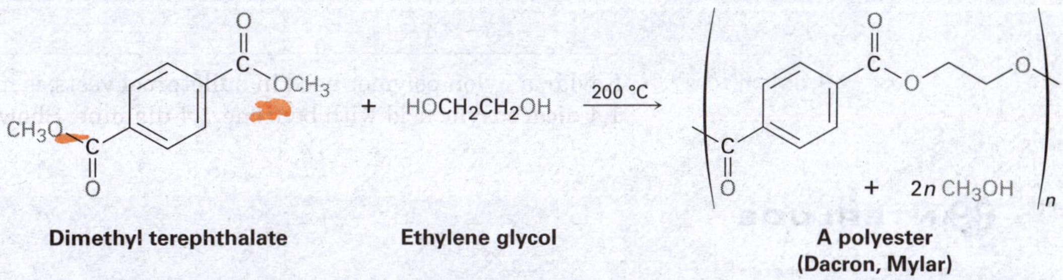

Biodegradable Polymers

Because plastics are too often thrown away rather than recycled, much work has been carried out on developing *biodegradable* polymers, which can be broken down rapidly in landfills by soil microorganisms. Among the most common biodegradable polymers are poly(glycolic acid) (PGA), poly(lactic acid) (PLA), and polyhydroxybutyrate (PHB). All are polyesters and are therefore susceptible to hydrolysis of their ester links. As an example, biodegradable sutures made of poly(glycolic acid) are hydrolyzed and absorbed by the body within 90 days after surgery.

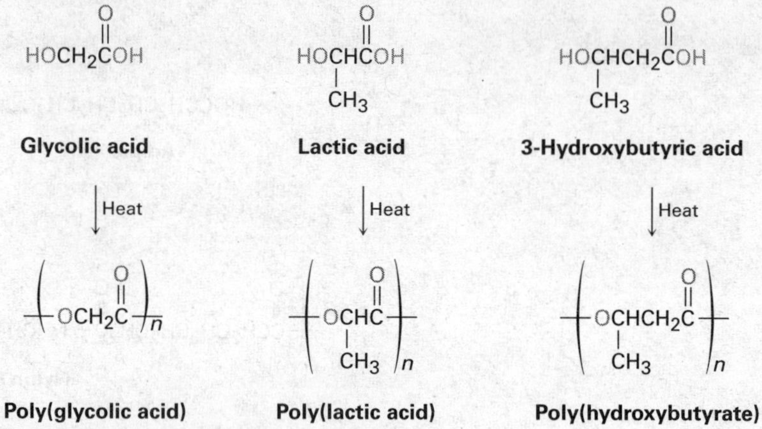

Worked Example 10.10 Predicting the Structure of a Polymer

Draw the structure of Qiana, a polyamide made by high-temperature reaction of hexanedioic acid with cyclohexane-1,4-diamine.

Strategy Reaction of a diacid with a diamine gives a polyamide.

Solution

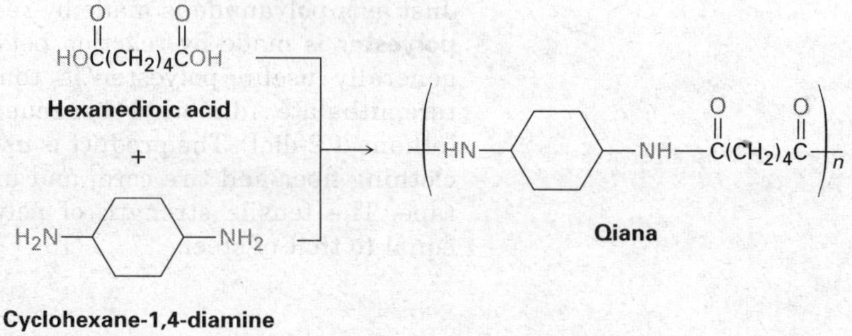

Problem 10.27 Kevlar, a nylon polymer used in bulletproof vests, is made by reaction of benzene-1,4-dicarboxylic acid with benzene-1,4-diamine. Show the structure of Kevlar.

INTERLUDE

β-Lactam Antibiotics

You should never underestimate the value of hard work and logical thinking, but it's also true that blind luck plays a role in most real scientific breakthroughs. What has been called "the supreme example of luck in all scientific history" occurred in the late summer of 1928, when the

continued

INTERLUDE

Andrew McClenaghan/Photo Researchers, Inc.

***Penicillium* mold growing in a petri dish.**

Scottish bacteriologist Alexander Fleming went on vacation, leaving in his lab a culture plate recently inoculated with the bacterium *Staphylococcus aureus.*

While Fleming was away, an extraordinary chain of events occurred. First, a 9-day cold spell lowered the laboratory temperature to a point where the *Staphylococcus* on the plate could not grow. During this time, spores from a colony of the mold *Penicillium notatum* being grown on the floor below wafted up into Fleming's lab and landed in the culture plate. The temperature then rose, and both *Staphylococcus* and *Penicillium* began to grow. On returning from vacation, Fleming discarded the plate into a tray of antiseptic, intending to sterilize it. Evidently, though, the plate did not sink deeply enough into the antiseptic, because when Fleming happened to glance at it a few days later, what he saw changed the course of history. He noticed that the growing *Penicillium* mold appeared to dissolve the colonies of staphylococci.

Fleming realized that the *Penicillium* mold must be producing a chemical that killed the *Staphylococcus* bacteria, and he spent several years trying to isolate the substance. Finally, in 1939, the Australian pathologist Howard Florey and the German refugee Ernst Chain managed to isolate the active substance, called *penicillin.* The dramatic ability of penicillin to cure infections in mice was soon demonstrated, and successful tests in humans followed shortly thereafter. By 1943, penicillin was being produced on a large scale for military use in World War II, and by 1944 it was being used on civilians. Fleming, Florey, and Chain shared the 1945 Nobel Prize in Medicine.

Now called benzylpenicillin, or penicillin G, the substance first discovered by Fleming is but one member of a large class of so-called β-lactam antibiotics, compounds with a four-membered lactam (cyclic amide) ring. The four-membered lactam ring is fused to a five-membered, sulfur-containing ring, and the carbon atom next to the lactam carbonyl group is bonded to an acylamino substituent, RCONH–. This acylamino side chain can be varied in the laboratory to provide many hundreds of penicillin analogs with different biological activity profiles. Ampicillin, for instance, has an α-aminophenylacetamido substituent [$PhCH(NH_2)CONH$–].

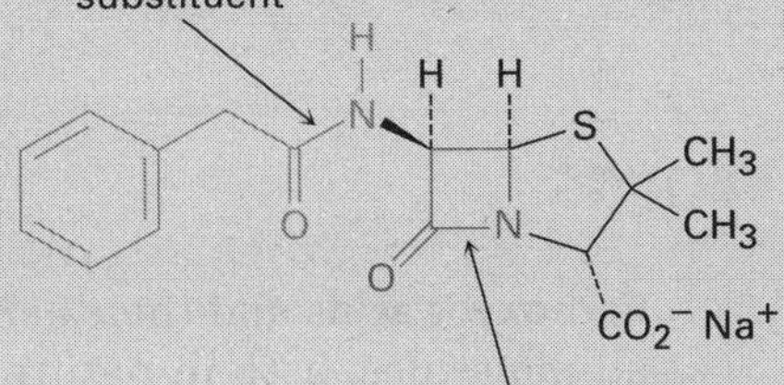

Benzylpenicillin (penicillin G)

Closely related to the penicillins are the *cephalosporins,* a group of β-lactam antibiotics that contain an unsaturated six-membered, sulfur-containing ring. Cephalexin, marketed under the trade name Keflex, is an

continued

INTERLUDE

example. Cephalosporins generally have much greater antibacterial activity than penicillins, particularly against resistant strains of bacteria.

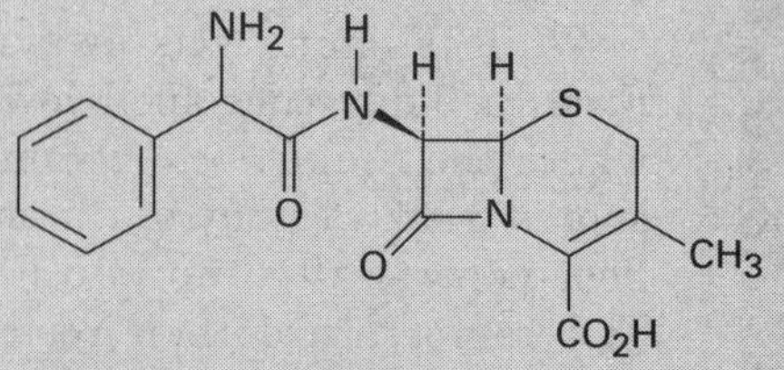

Cephalexin (a cephalosporin)

The biological activity of penicillins and cephalosporins is due to the presence of the strained, unusually reactive β-lactam ring, which undergoes an irreversible nucleophilic acyl substitution reaction with the transpeptidase enzyme needed to synthesize and repair bacterial cell walls. With the enzyme thus inactivated, the cell walls remain either incomplete or weakened, so the bacteria die.

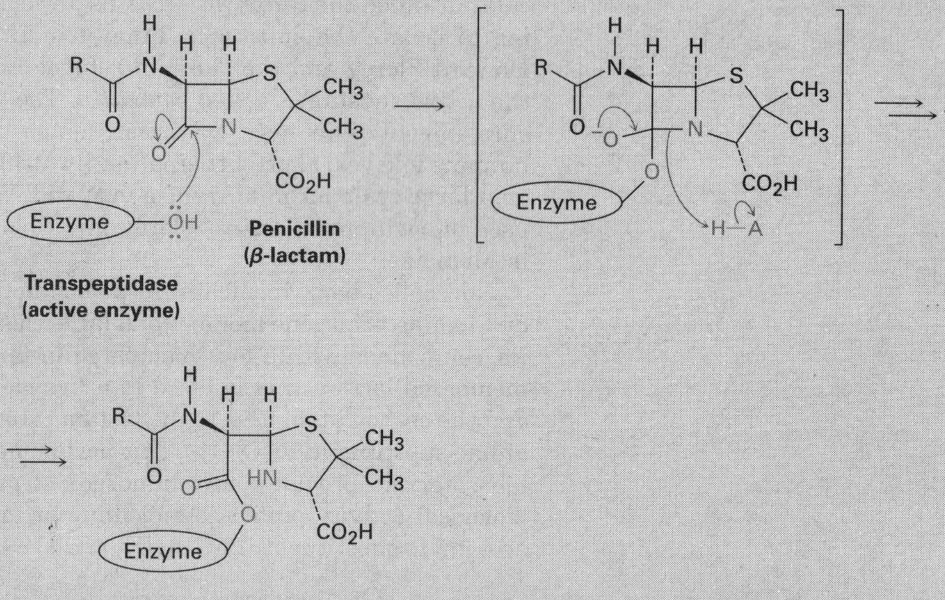

Transpeptidase (active enzyme)

Penicillin (β-lactam)

(inactive enzyme)

Summary and Key Words

acid anhydride (RCO_2COR') 325
acid halide (RCOCl) 325
acyl phosphate ($RCO_2PO_3^{2-}$) 354
amide ($RCONH_2$) 325
carboxyl group 326
carboxylate ion 332
carboxylic acid (RCO_2H) 325
chain-growth polymer 356
ester (RCO_2R') 325

Carboxylic acids and their derivatives are among the most widely occurring of all molecules, both in nature and in the chemical laboratory. In this chapter, we covered the chemistry necessary for understanding them and thus also necessary for understanding the chemistry of living organisms.

The distinguishing characteristic of carboxylic acids is their acidity. Although weaker than mineral acids like HCl, carboxylic acids are much more acidic than alcohols because carboxylate ions are stabilized by resonance.

Carboxylic acids can be transformed into a variety of carboxylic acid derivatives in which the acid –OH group has been replaced by another substituent.

- Fischer esterification reaction 339
- nitrile (RCN) 325
- nucleophilic acyl substitution reaction 335
- nylon 356
- polyester 357
- step-growth polymer 356
- thioester (RCOSR′) 354

Acid chlorides, **acid anhydrides**, **esters**, and **amides** are the most common. The chemistry of all these derivatives is similar and is dominated by a single general reaction type: the **nucleophilic acyl substitution reaction**. These substitutions take place by addition of a nucleophile to the polar carbonyl group of the acid derivative, followed by expulsion of a leaving group. The reactivity order of acid derivatives is acid halide > acid anhydride > ester > amide.

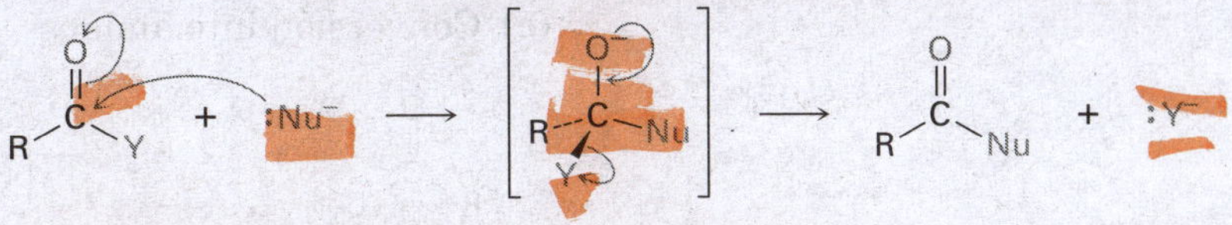

The most common reactions of carboxylic acid derivatives are substitution by water (*hydrolysis*) to yield an acid, by an alcohol (*alcoholysis*) to yield an ester, by an amine (*aminolysis*) to yield an amide, by hydride ion to yield an alcohol (*reduction*), and by an organometallic reagent to yield an alcohol (*Grignard reaction*).

Nitriles, R—C≡N, are related to carboxylic acid derivatives because they undergo nucleophilic additions to the polar C≡N bond in the same way carbonyl compounds do. The most important reactions of nitriles are their hydrolysis to yield carboxylic acids, their reduction to yield primary amines, and their reaction with Grignard reagents to yield ketones.

Step-growth polymers, such as polyamides and polyesters, are prepared by reactions between difunctional molecules. Polyamides (nylons) are formed by reaction between a diacid and a diamine; polyesters are formed from a diacid and a diol.

Summary of Reactions

1. Reactions of carboxylic acids (Section 10.6)

(a) Conversion into acid chlorides

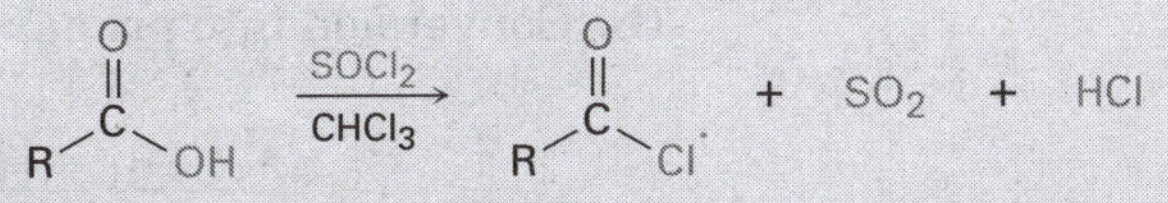

(b) Conversion into esters (Fischer esterification)

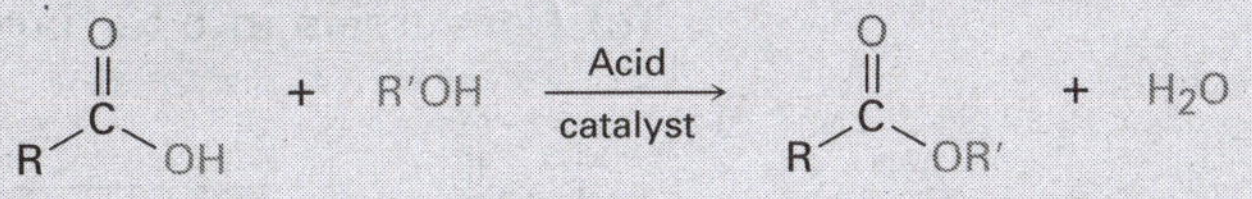

(c) Conversion into amides

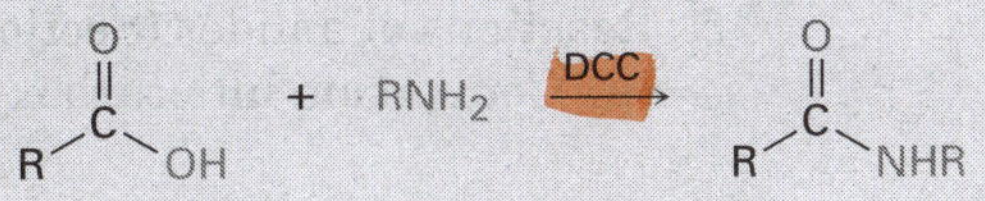

2. Reactions of acid halides (Section 10.7)

(a) Conversion into carboxylic acids

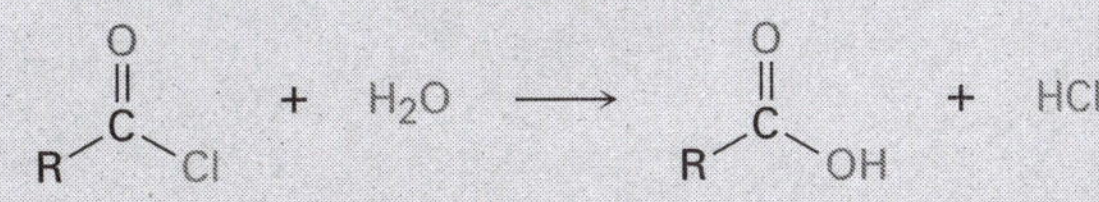

continued

(b) Conversion into esters

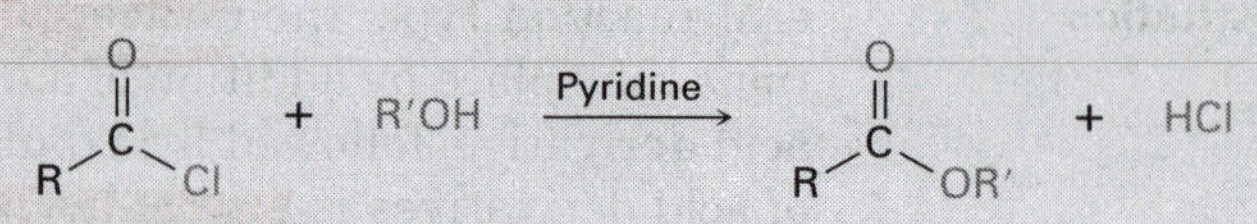

(c) Conversion into amides

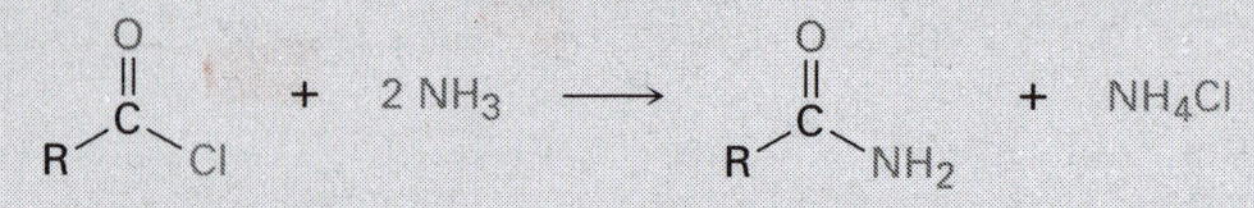

3. Reactions of acid anhydrides (Section 10.8)
   (a) Conversion into esters

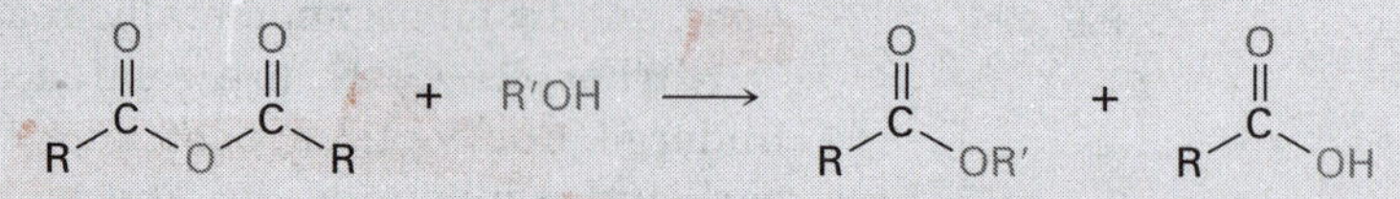

   (b) Conversion into amides

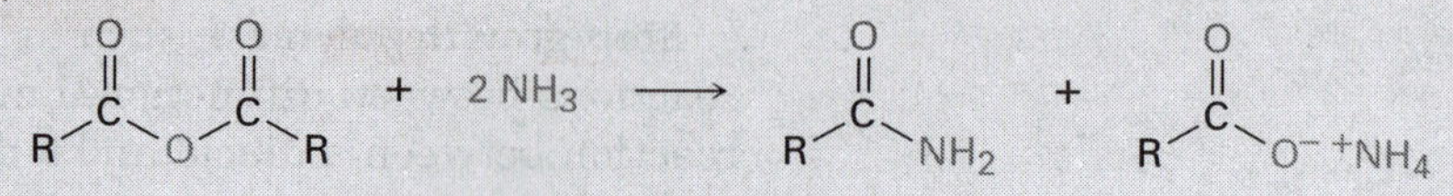

4. Reactions of esters (Section 10.9)
   (a) Conversion into acids

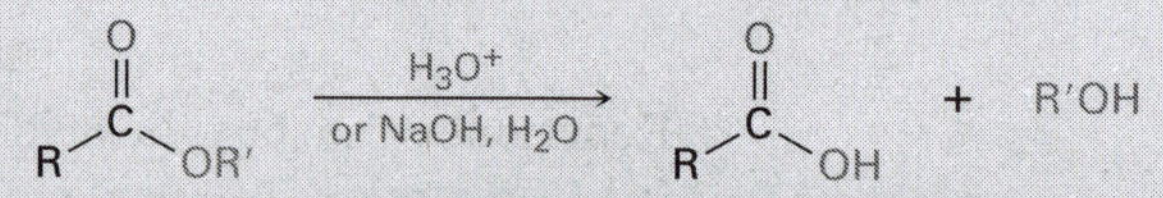

   (b) Conversion into primary alcohols by reduction

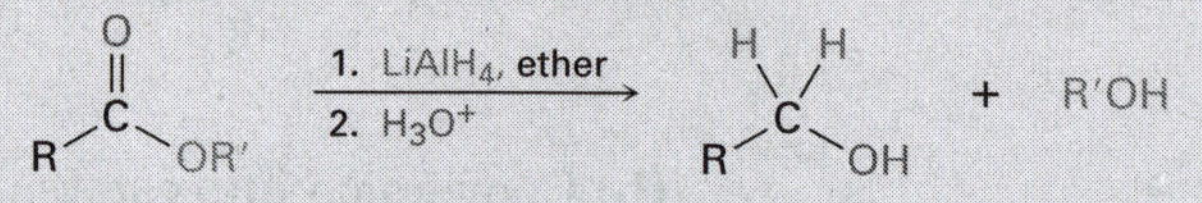

   (c) Conversion into tertiary alcohols by Grignard reaction

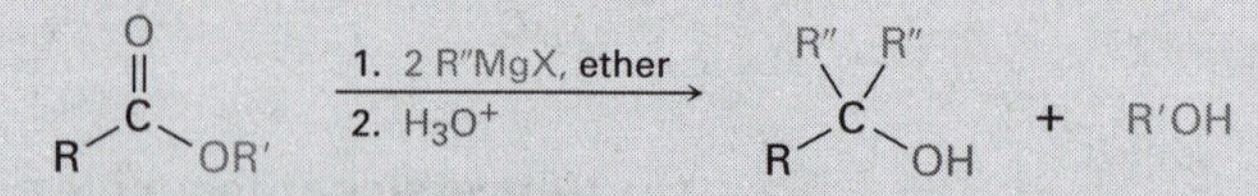

5. Reactions of amides (Section 10.10)
   (a) Conversion into carboxylic acids

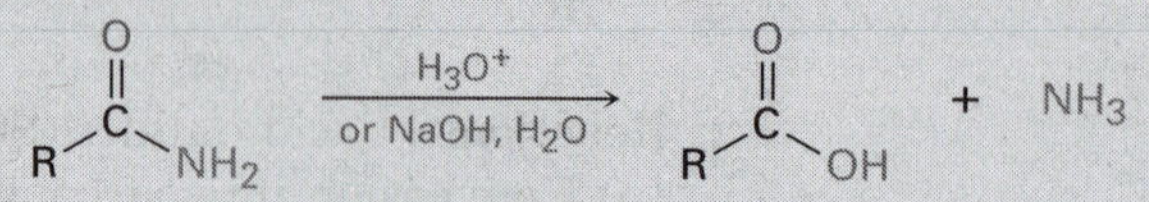

(b) Conversion into amines by reduction

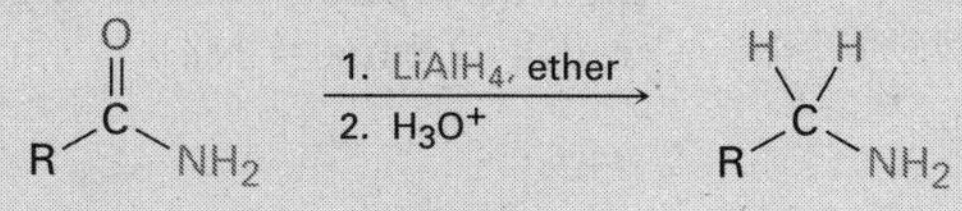

6. Reactions of nitriles (Section 10.11)

(a) Conversion into carboxylic acids

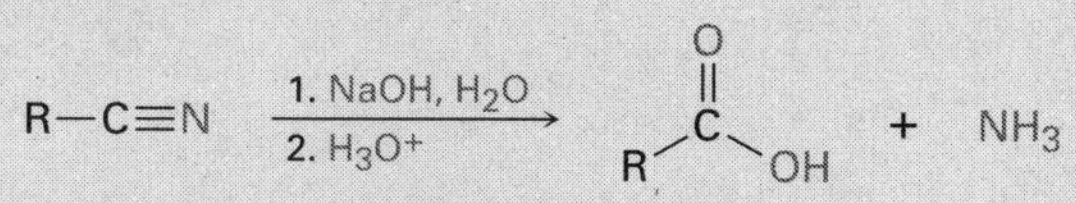

(b) Conversion into amines by reduction

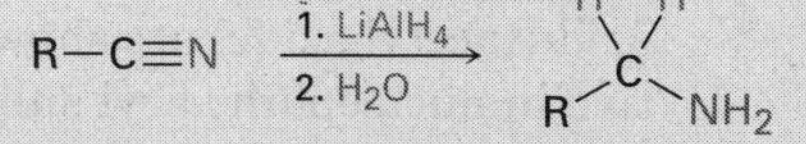

(c) Conversion into ketones by Grignard reaction

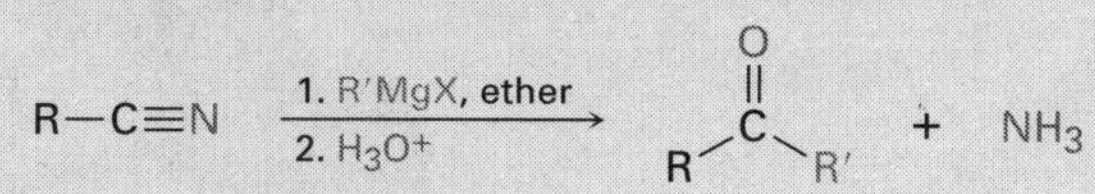

Exercises

Visualizing Chemistry

(Problems 10.1–10.27 appear within the chapter.)

OWL

Interactive versions of these problems are assignable in OWL.

10.28 Name the following compounds (red = O, blue = N):

(a)

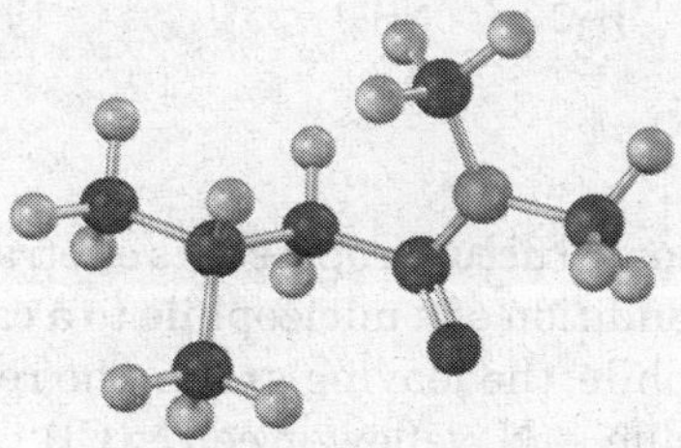

(b)

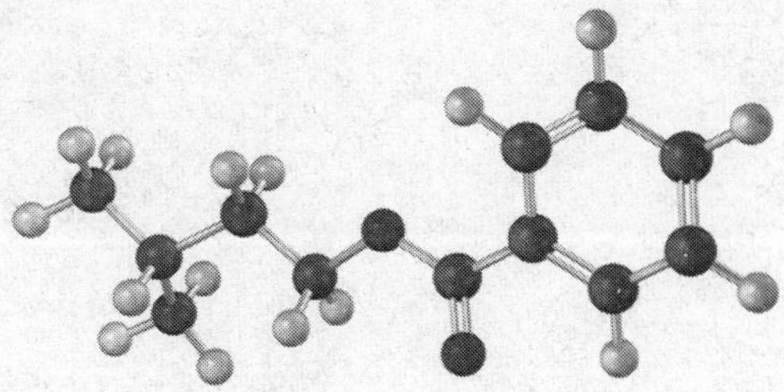

10.29 Show how you could prepare each of the following compounds starting with an appropriate carboxylic acid and any other reagents needed (red = O, blue = N, reddish brown = Br):

(a)

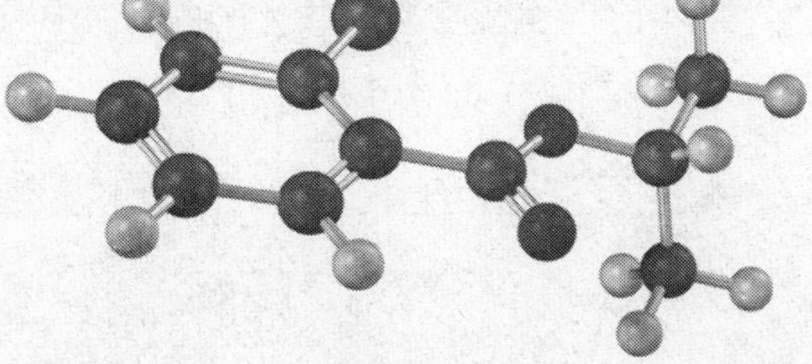

(b)

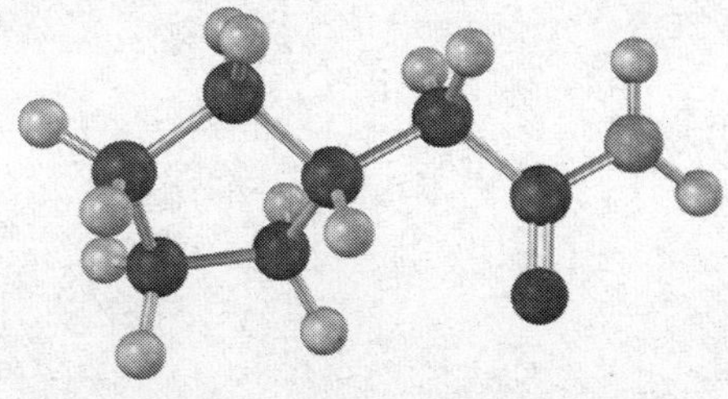

10.30 The following structure represents a tetrahedral alkoxide ion intermediate formed by addition of a nucleophile to a carboxylic acid derivative. Identify the nucleophile, the leaving group, the reactant, and the ultimate product (red = O, blue = N).

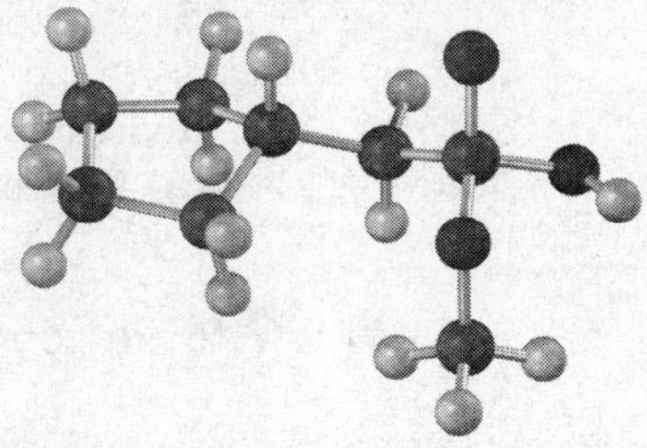

10.31 Electrostatic potential maps of a typical amide (acetamide) and an acyl azide (acetyl azide) are shown. Which of the two do you think is more reactive in nucleophilic acyl substitution reactions? Explain.

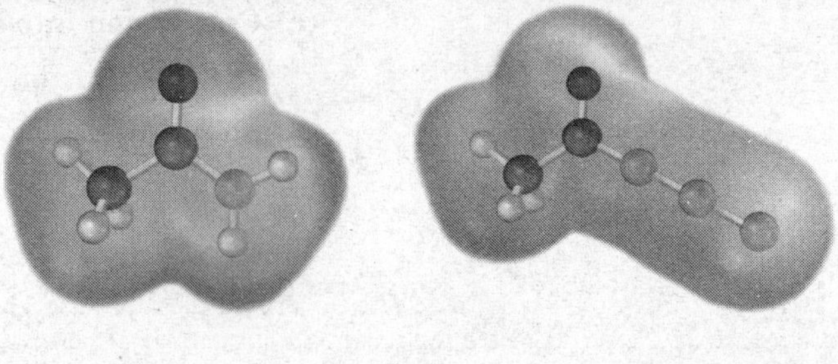

Acetamide

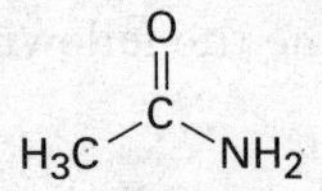

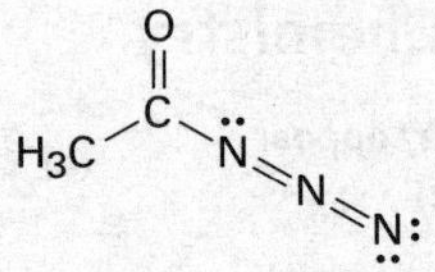

Acetyl azide

10.32 The following structure represents a tetrahedral alkoxide ion intermediate formed by addition of a nucleophile to a carboxylic acid derivative. Identify the nucleophile, the leaving group, the reactant, and the ultimate product (red = O, blue = N, yellow-green = Cl).

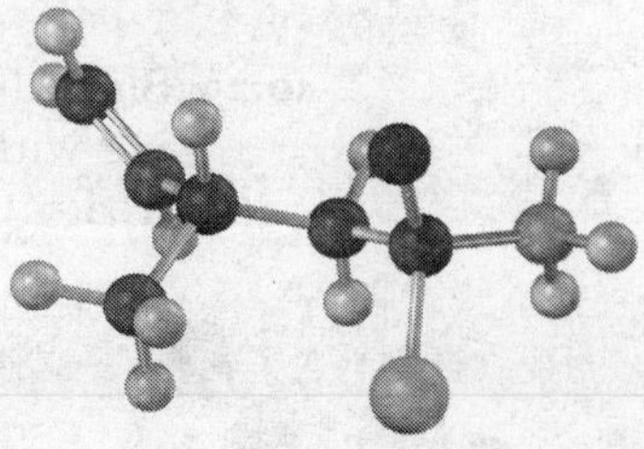

Additional Problems

Naming Carboxylic Acids and Derivatives

10.33 Draw structures corresponding to the following IUPAC names:
(a) 4,5-Dimethylheptanoic acid
(b) *cis*-Cyclohexane-1,2-dicarboxylic acid
(c) Heptanedioic acid
(d) Triphenylacetic acid
(e) 2,2-Dimethylhexanamide
(f) Phenylacetamide
(g) Cyclobut-2-enecarbonitrile
(h) Ethyl cyclohexanecarboxylate

10.34 Give IUPAC names for the following carboxylic acids:

(a) 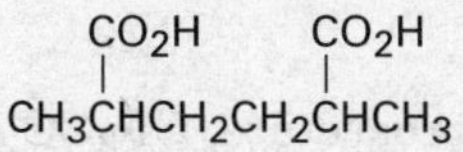

(b) 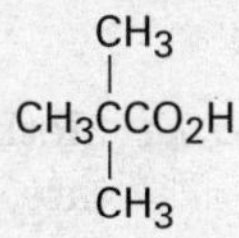

(c) 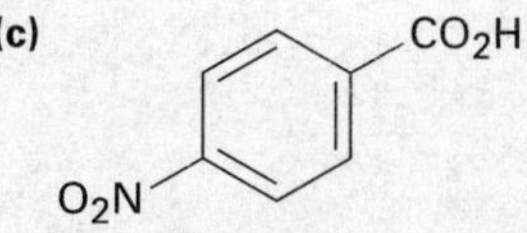

(d) 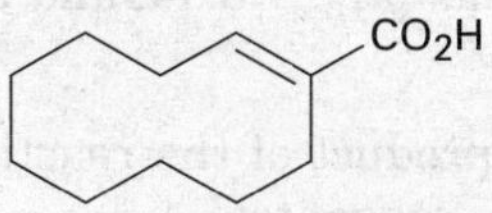

(e) 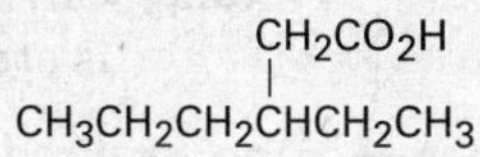

(f) 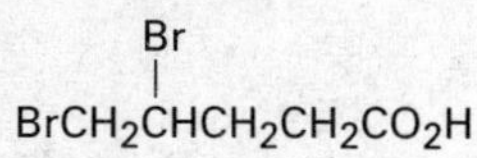

10.35 Draw and name compounds that meet the following descriptions:
(a) Three acid chlorides, C_6H_9ClO
(b) Three amides, $C_7H_{11}NO$
(c) Three nitriles, C_5H_7N
(d) Three esters, $C_5H_8O_2$

10.36 Give IUPAC names for the following carboxylic acid derivatives:

(a) 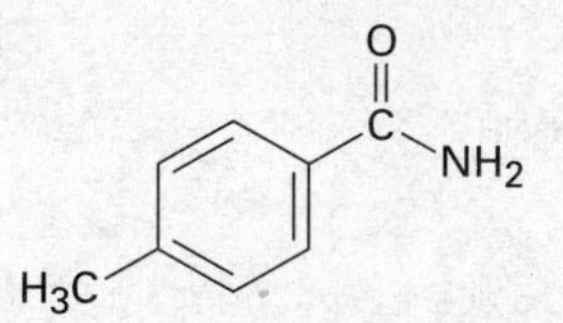

(b) 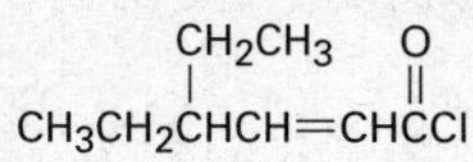

(c) 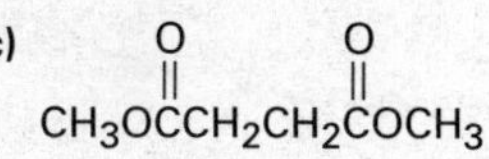

(d) 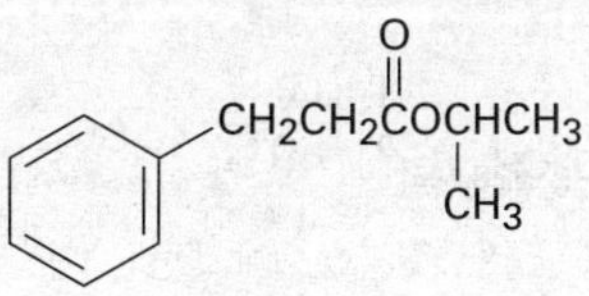

(e) 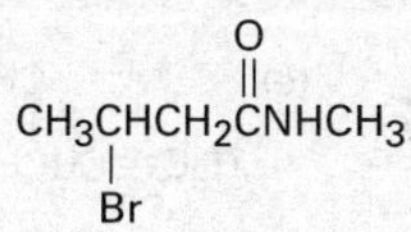

(f) 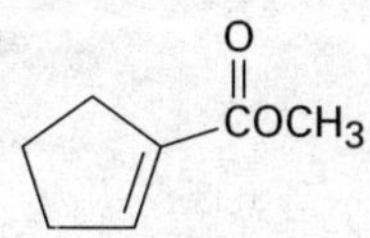

10.37 Draw and name the eight carboxylic acids with formula $C_6H_{12}O_2$. Which are chiral?

Reactivity and Acidity

10.38 The following reactivity order has been found for the saponification of alkyl acetates by aqueous NaOH:

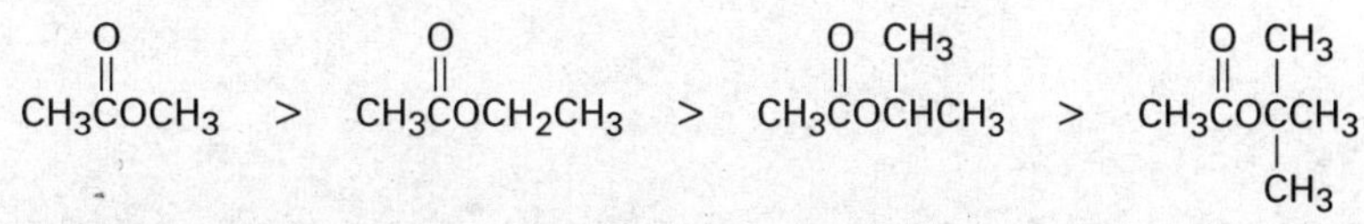

How can you explain this reactivity order?

10.39 How can you explain the fact that 2-chlorobutanoic acid has $pK_a = 2.86$, 3-chlorobutanoic acid has $pK_a = 4.05$, 4-chlorobutanoic acid has $pK_a = 4.52$, and butanoic acid itself has $pK_a = 4.82$?

10.40 Order the compounds in each of the following sets with respect to increasing acidity:
(a) Acetic acid, chloroacetic acid, trifluoroacetic acid
(b) Benzoic acid, *p*-bromobenzoic acid, *p*-nitrobenzoic acid
(c) Acetic acid, phenol, cyclohexanol

10.41 Rank the following compounds in order of their reactivity toward nucleophilic acyl substitution:

(a) CH_3COCH_3 (C=O) **(b)** CH_3CCl (C=O) **(c)** CH_3CNH_2 (C=O) **(d)** CH_3COCCH_3 (both C=O)

10.42 Methyl trifluoroacetate, $CF_3CO_2CH_3$, is more reactive than methyl acetate, $CH_3CO_2CH_3$, in nucleophilic acyl substitution reactions. Explain.

10.43 Citric acid has $pK_a = 3.14$, and tartaric acid has $pK_a = 2.98$. Which acid is stronger?

REACTIONS

10.44 Predict the product of the reaction of *p*-methylbenzoic acid with each of the following reagents:
(a) $LiAlH_4$ **(b)** CH_3OH, HCl **(c)** $SOCl_2$ **(d)** NaOH, then CH_3I

10.45 Predict the product(s) of the following reactions:

(a)
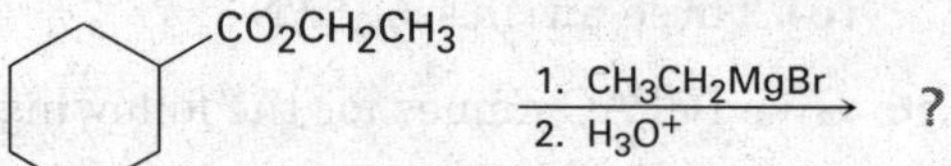

(b)
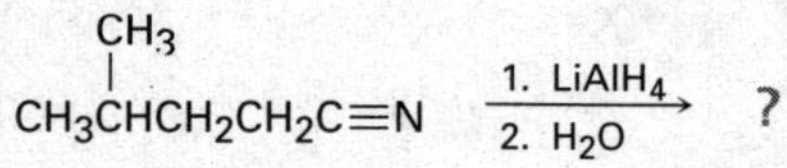

(c)
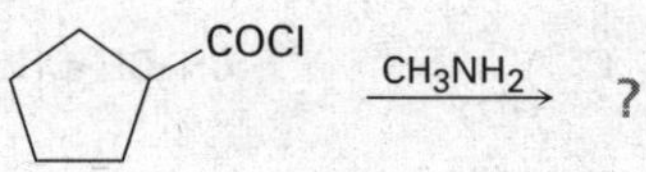

(d)
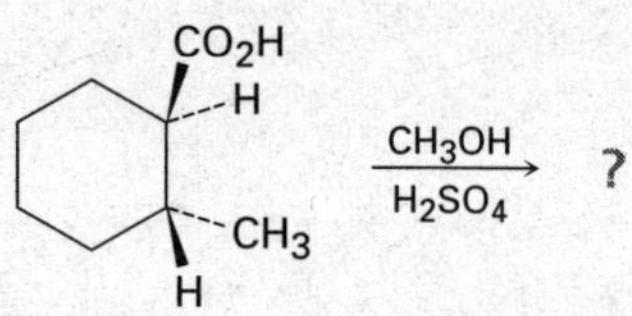

(e)
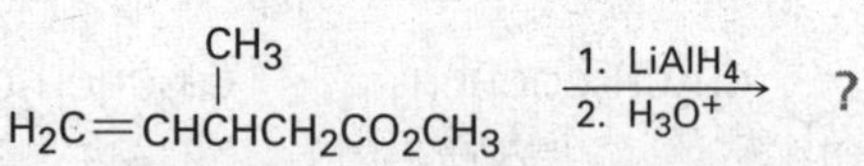

(f)
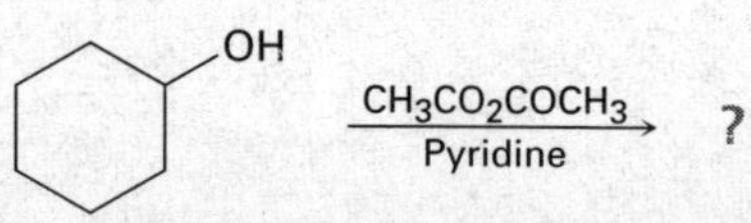

(g)
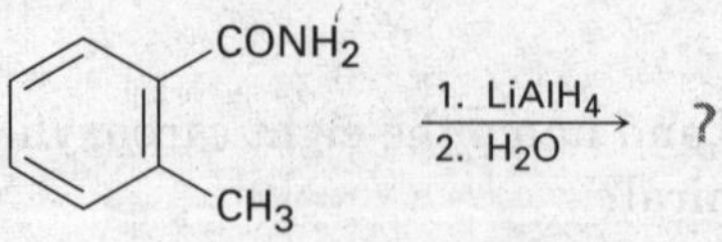

(h)
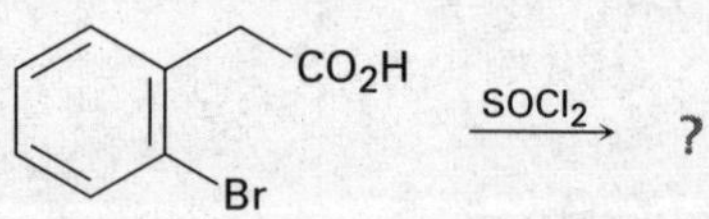

10.46 Acid chlorides undergo reduction with $LiAlH_4$ in the same way that esters do to yield primary alcohols. What are the products of the following reactions?

(a) 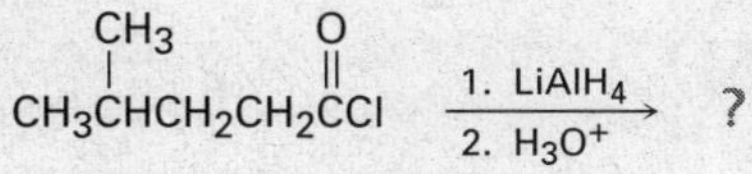

(b) 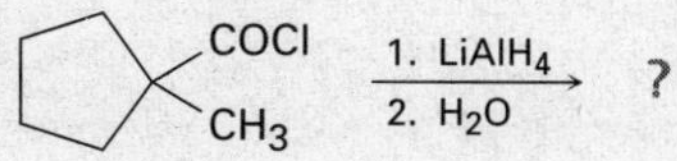

10.47 A chemist in need of 2,2-dimethylpentanoic acid decided to synthesize some by reaction of 2-chloro-2-methylpentane with NaCN, followed by hydrolysis of the product. After carrying out the reaction sequence, however, none of the desired product could be found. What do you suppose went wrong?

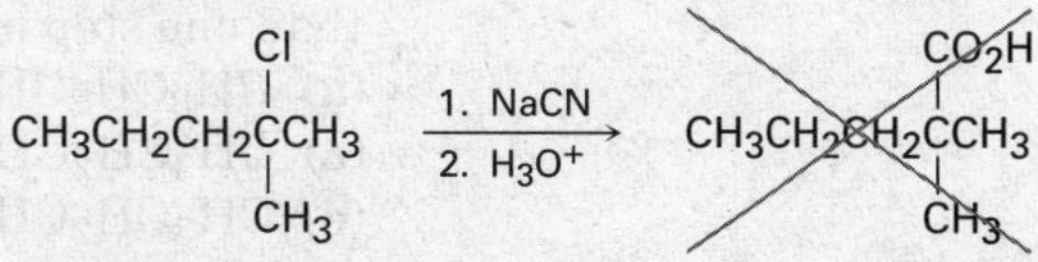

10.48 What product would you expect from the reaction of a cyclic ester such as butyrolactone with $LiAlH_4$?

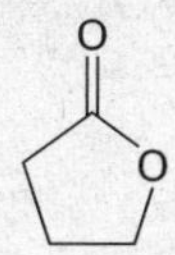

Butyrolactone

10.49 How can you explain the observation that an attempted Fischer esterification of 2,4,6-trimethylbenzoic acid with methanol/HCl is unsuccessful? No ester is obtained, and the starting acid is recovered unchanged.

10.50 If 5-hydroxypentanoic acid is treated with an acid catalyst, an intramolecular esterification reaction occurs. What is the structure of the product? (*Intramolecular* means within the same molecule.)

$HOCH_2CH_2CH_2CH_2CO_2H$ **5-Hydroxypentanoic acid**

MECHANISMS

10.51 The reaction of an acid chloride with $LiAlH_4$ to yield a primary alcohol (Problem 10.46) takes place in two steps. The first step is a nucleophilic acyl substitution of H^- for Cl^- to yield an aldehyde, and the second step is nucleophilic addition of H^- to the aldehyde to yield an alcohol. Write the mechanism of the reduction of CH_3COCl.

10.52 Reaction of a carboxylic acid with trifluoroacetic anhydride leads to an unsymmetrical anhydride that rapidly reacts with alcohol to give an ester.

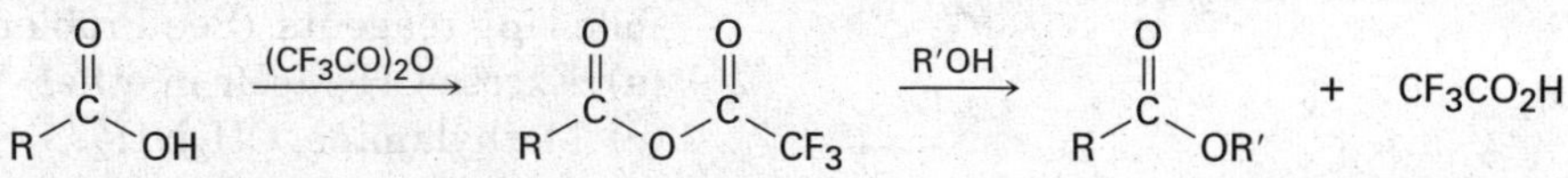

(a) Propose a mechanism for formation of the unsymmetrical anhydride.
(b) Why is the unsymmetrical anhydride unusually reactive?
(c) Why does the unsymmetrical anhydride react as indicated rather than giving a trifluoroacetate ester plus carboxylic acid?

10.53 Acid chlorides undergo reaction with Grignard reagents at −78 °C to yield ketones. Propose a mechanism for the reaction.

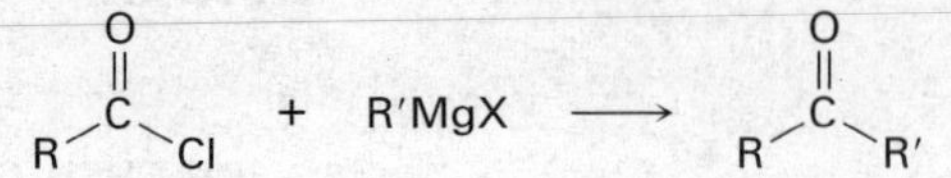

10.54 If the reaction of an acid chloride with a Grignard reagent (Problem 10.53) is carried out at room temperature, a tertiary alcohol is formed.
(a) Propose a mechanism for this reaction.
(b) What are the products of the reaction of CH_3MgBr with the acid chlorides given in Problem 10.46?

SYNTHESIS

10.55 How might you prepare the following products from butanoic acid? More than one step may be needed.
(a) $CH_3CH_2CH_2CH_2OH$ **(b)** $CH_3CH_2CH_2CHO$
(c) $CH_3CH_2CH_2CH_2Br$ **(d)** $CH_3CH_2CH_2CH_2CN$
(e) $CH_3CH_2CH{=}CH_2$ **(f)** $CH_3CH_2CH_2CH_2NH_2$

10.56 Show how you might prepare the anti-inflammatory agent ibuprofen starting from isobutylbenzene. More than one step is needed.

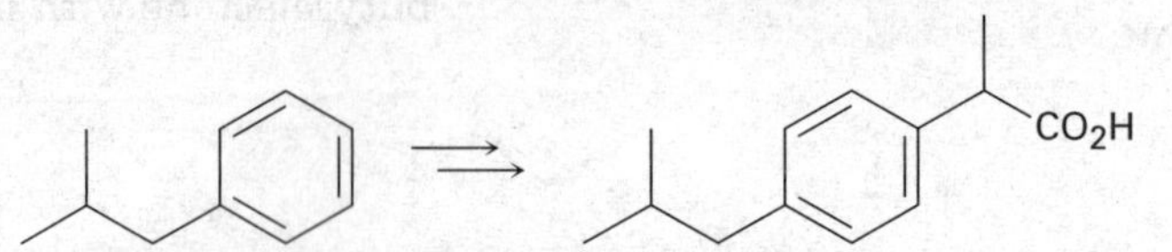

Isobutylbenzene **Ibuprofen**

10.57 How can you prepare acetophenone (methyl phenyl ketone) from the following starting materials? More than one step may be needed.
(a) Benzonitrile **(b)** Bromobenzene
(c) Methyl benzoate **(d)** Benzene

GENERAL PROBLEMS

10.58 What esters and what Grignard reagents would you use to make the following alcohols? Show all possibilities.

(a) 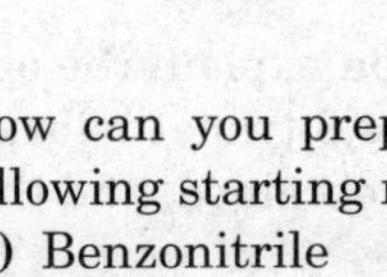

(b) 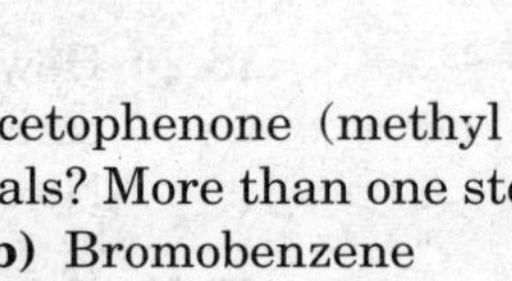

10.59 When dimethyl carbonate, $CH_3OCO_2CH_3$, is treated with phenylmagnesium bromide, triphenylmethanol is formed. Explain.

10.60 Predict the product, if any, of reaction between propanoyl chloride and the following reagents. (See Problems 10.46 and 10.53.)
(a) Excess CH_3MgBr in ether **(b)** NaOH in H_2O
(c) Methylamine, CH_3NH_2 **(d)** $LiAlH_4$
(e) Cyclohexanol **(f)** Sodium acetate

10.61 Answer Problem 10.60 for reaction between methyl propanoate and the listed reagents.

10.62 When *methyl* acetate is heated in pure ethanol containing a small amount of HCl catalyst, *ethyl* acetate results. Explain.

10.63 What product would you expect to obtain on treatment of the cyclic ester butyrolactone with excess phenylmagnesium bromide? (See Problem 10.48.)

10.64 In the *iodoform reaction,* a triiodomethyl ketone reacts with aqueous NaOH to yield a carboxylate ion and iodoform (triiodomethane). Propose a mechanism for this reaction.

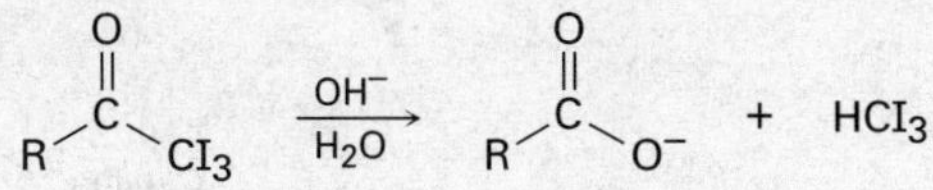

10.65 *N,N*-Diethyl-*m*-toluamide (DEET) is the active ingredient in many insect repellents. How might you synthesize DEET from *m*-bromotoluene?

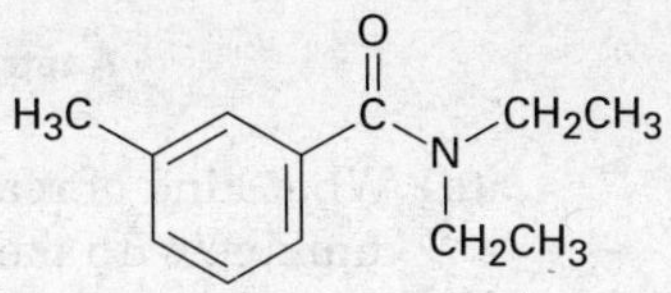

***N,N*-Diethyl-*m*-toluamide**

10.66 How could you make the following esters:

(a) 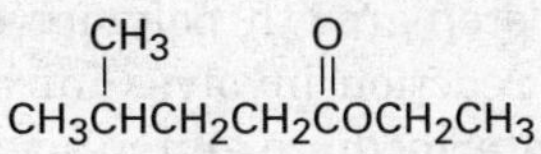

(b) 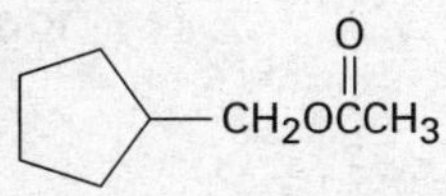

10.67 *tert*-Butoxycarbonyl azide, an important reagent used in protein synthesis, is prepared by treating *tert*-butoxycarbonyl chloride with sodium azide. Propose a mechanism for this reaction.

10.68 What products would you obtain on saponification of the following esters?

(a) 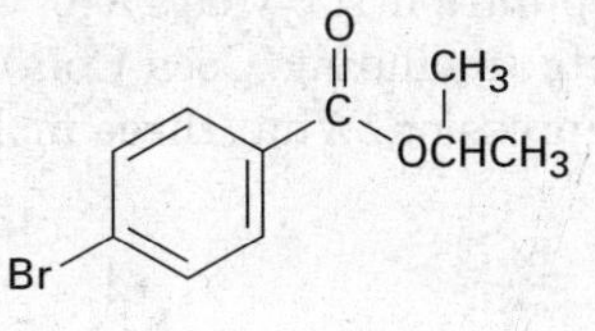

(b) Cyclohexyl propanoate

10.69 A *dendrimer*, unlike a linear polymer, is a highly branched, tree-shaped molecule. An example is formed by reaction of ethylenediamine ($H_2NCH_2CH_2NH_2$) with methyl acrylate ($H_2C{=}CHCO_2CH_3$). The synthesis begins with initial formation of a tetraester, which is allowed to react with more diamine to give a tetraamino tetraamide. The tetraamino tetraamide then reacts again with excess methyl acrylate to form an octaester, which reacts again with excess diamine, and so on.

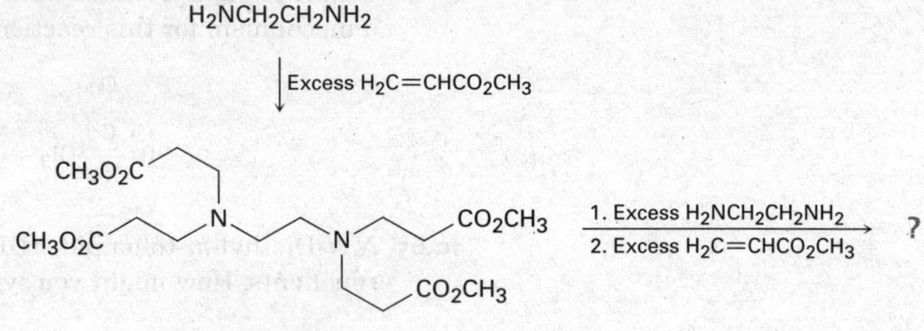

A tetraester

(a) What kind of reaction is involved in the initial reaction of the diamine with methyl acrylate?

(b) What kind of reaction is involved in the reaction of the tetraester with diamine? Show the product.

10.70 The step-growth polymer called nylon 6 is prepared from caprolactam. The reaction involves initial reaction of caprolactam with water to give an intermediate amino acid, followed by heating to form the polymer. Propose mechanisms for both steps, and show the structure of nylon 6.

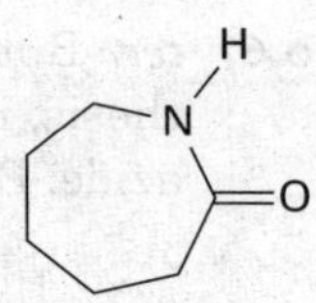

Caprolactam

10.71 Draw a representative segment of the polyester that would result from reaction of pentanedioic acid ($HO_2CCH_2CH_2CH_2CO_2H$) and pentane-1,5-diol.

IN THE MEDICINE CABINET

10.72 Yesterday's drug can be today's poison. Cocaine enjoyed a much better reputation 100 years ago, when it was used as a stimulant in many products (including Coca-Cola) as well as in drops to treat toothaches and depression. What three molecules are produced by hydrolysis of cocaine?

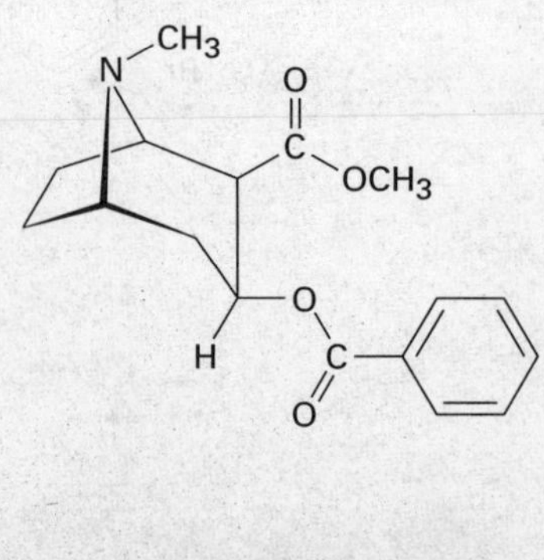

Cocaine

10.73 Cocaine's addictive properties led researchers to look for less addictive alternatives to relieve pain. Lidocaine, for instance, has many of the structural features of cocaine (Problem 10.72) but doesn't have the same risk. Lidocaine is prepared by the reaction sequence shown. Indicate the type of reaction in each step, and draw a mechanism.

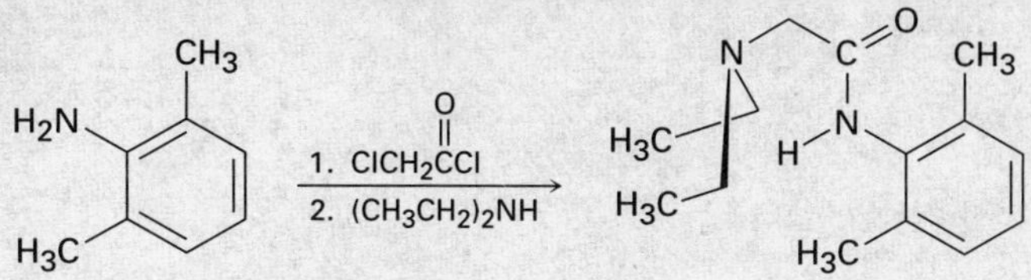

Lidocaine

IN THE FIELD

10.74 Carbaryl, an insecticide marketed under the trade name Sevin, is prepared by reaction between a phenol and methyl isocyanate, $H_3C{-}N{=}C{=}O$. Propose a mechanism for the reaction.

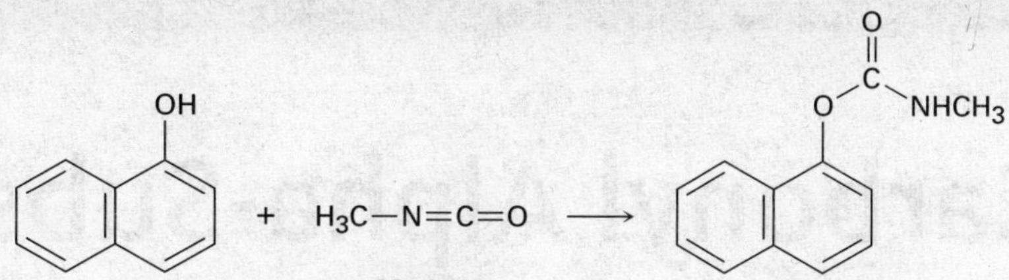

Methyl isocyanate

Carbaryl

10.75 Parathion was one of the first organophosphate insecticides. Assuming that the P=S group shows the same chemistry as a C=O group and that *p*-nitrophenoxide ion is a good leaving group, show how parathion can inactivate its target enzyme (acetylcholine esterase) by forming a stable enzyme intermediate.

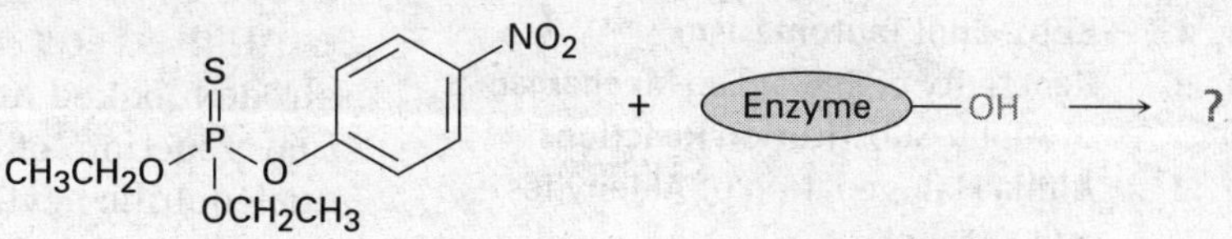

Parathion

CHAPTER

11

The lives of rock climbers depend on their ropes, typically made of a nylon polymer prepared by a carbonyl condensation reaction.

Carbonyl Alpha-Substitution Reactions and Condensation Reactions

Most of the chemistry of carbonyl compounds can be explained by just four fundamental reactions. We've already looked in detail at two of the four: the nucleophilic addition reaction of aldehydes and ketones (Chapter 9) and the nucleophilic acyl substitution reaction of carboxylic acid derivatives (Chapter 10). In this chapter, we'll look at the remaining two: the *alpha-substitution reaction* and the *carbonyl condensation reaction.*

Alpha-substitution reactions occur at the position *next to* the carbonyl group—the α position—and result in the substitution of an α hydrogen atom by an electrophile (E) through either an *enol* or *enolate ion* intermediate.

OWL

Online homework for this chapter can be assigned in OWL, an online homework assessment tool.

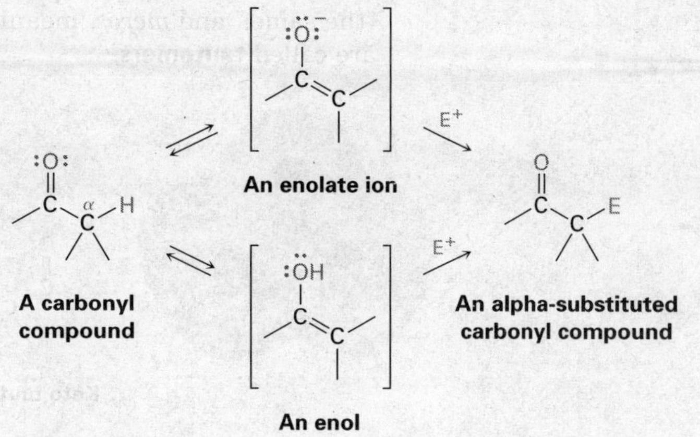

Carbonyl condensation reactions take place between two carbonyl partners and involve a *combination* of α substitution and nucleophilic addition steps. One partner is converted into its enolate ion and undergoes an α-substitution reaction when it carries out a nucleophilic addition to the second partner. The product is a β-hydroxy carbonyl compound.

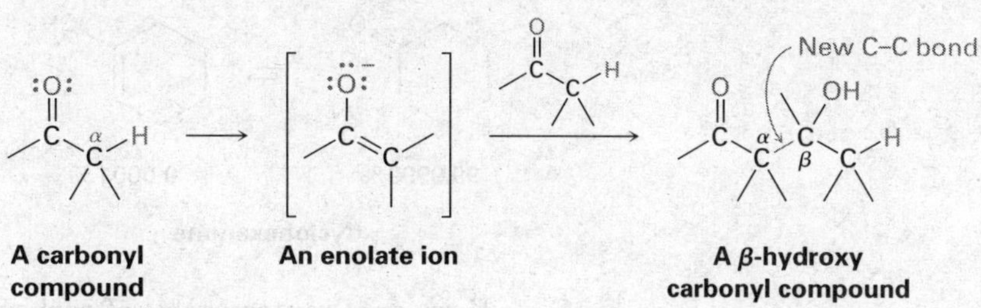

WHY THIS CHAPTER?

Many laboratory schemes, pharmaceutical syntheses, and biochemical pathways make frequent use of carbonyl α-substitution and carbonyl condensation reactions. Their great value is that they are two of the few general methods available for forming carbon–carbon bonds, thereby making it possible to build larger molecules from smaller ones. In this chapter, we'll see how these reactions occur.

11.1 Keto–Enol Tautomerism

A carbonyl compound with a hydrogen atom on its α carbon rapidly equilibrates with its corresponding **enol** (*ene* + *ol;* unsaturated alcohol) isomer. This spontaneous interconversion between two isomers, usually with the change in position of a hydrogen, is called *tautomerism,* from the Greek *tauto,* meaning "the same," and *meros,* meaning "part." The individual keto and enol isomers are called **tautomers**.

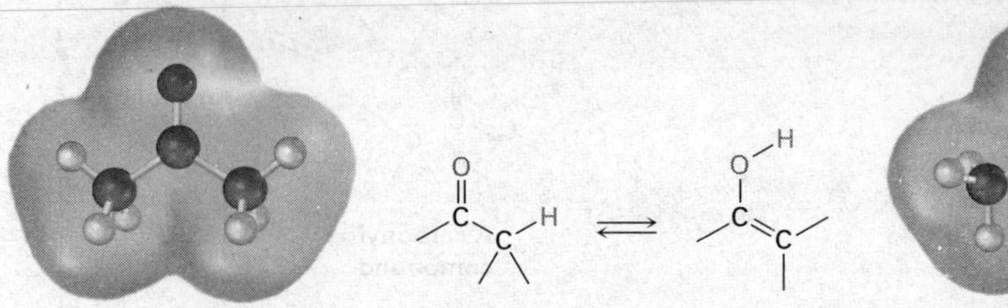

Keto tautomer **Enol tautomer**

Most monocarbonyl compounds exist almost entirely in the keto form at equilibrium, and it's usually difficult to isolate the pure enol. Cyclohexanone, for example, contains only about 0.0001% of its enol tautomer at room temperature, and the amount of enol tautomer is even less for carboxylic acids, esters, and amides. Only when the enol can be stabilized by conjugation or by intramolecular hydrogen bond formation does the enol sometimes predominate. Thus, pentane-2,4-dione is about 76% enol tautomer.

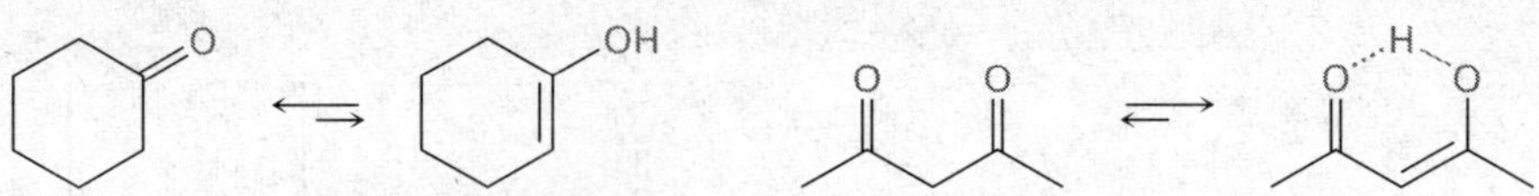

99.9999 % 0.0001 % 24 % 76 %

Cyclohexanone **Pentane-2,4-dione**

Keto–enol tautomerism of carbonyl compounds is catalyzed by both acids and bases. Acid catalysis involves protonation of the carbonyl oxygen atom (a Lewis base) to give an intermediate cation that loses H^+ from the α carbon to yield the enol (Figure 11.1a).

Base-catalyzed enol formation occurs because the presence of a carbonyl group makes the hydrogens on the α carbon weakly acidic. Thus, a carbonyl compound can act as an acid and donate one of its α hydrogens to a sufficiently strong base. The resultant resonance-stabilized anion, an **enolate ion**, is then protonated to yield a neutral compound. If protonation of the enolate ion takes place on the α carbon, the keto tautomer is regenerated and no net change occurs. If, however, protonation takes place on the oxygen atom, then an enol tautomer is formed (Figure 11.1b).

Note that only the protons on the α position of carbonyl compounds are acidic. The protons at beta (β), gamma (γ), delta (δ), and other positions aren't acidic because the resulting anions can't be resonance-stabilized by the carbonyl group.

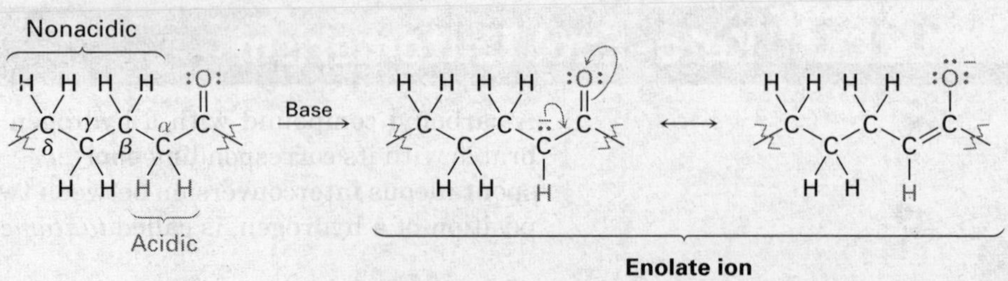

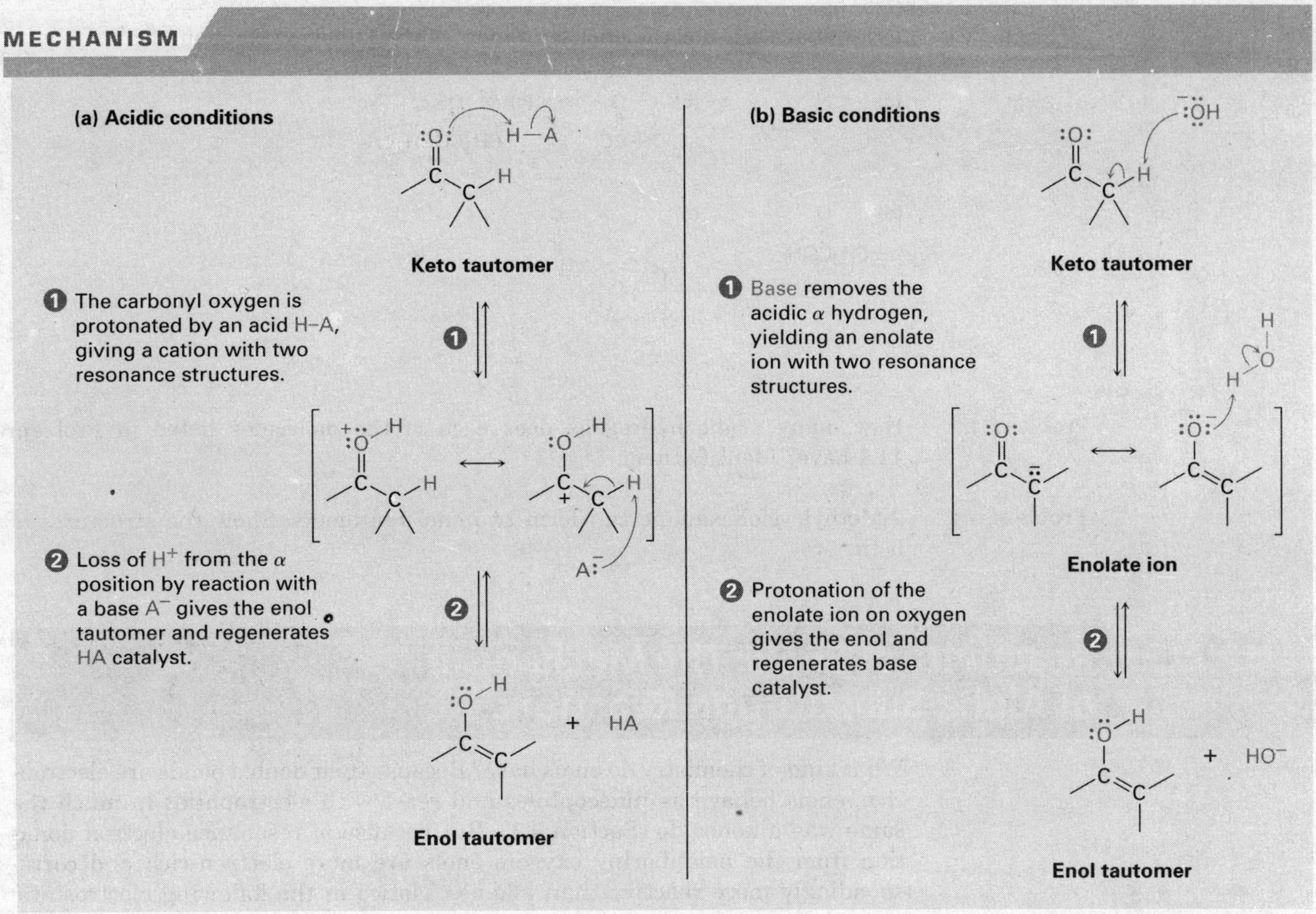

Figure 11.1 Mechanism of enol formation under both acid-catalyzed and base-catalyzed conditions. (a) Acid catalysis involves initial protonation of the carbonyl oxygen followed by removal of H^+ from the α position. (b) Base catalysis involves initial deprotonation of the α position to give an enolate ion, followed by reprotonation on oxygen.

Worked Example 11.1 Drawing the Structure of an Enol Tautomer

Show the structure of the enol tautomer of butanal.

Strategy Form the enol by removing a hydrogen from the carbon next to the carbonyl carbon. Then draw a resonance structure that has a double bond between the two carbons, and replace the hydrogen on the carbonyl oxygen.

Solution

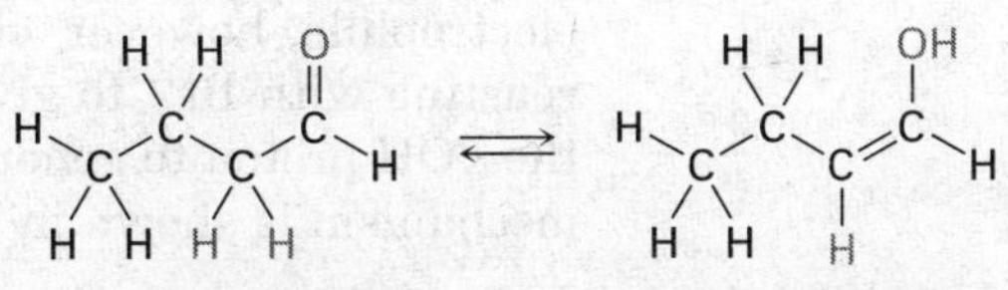

Butanal Enol tautomer

Problem 11.1 Draw structures for the enol tautomers of the following compounds:

(a) 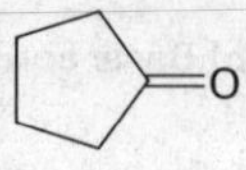

(b) 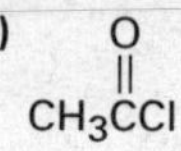

(c) 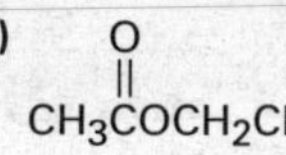

(d) 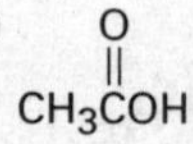

(e) 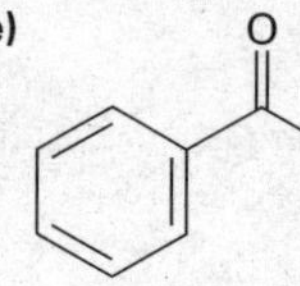

Problem 11.2 How many acidic hydrogens does each of the molecules listed in Problem 11.1 have? Identify them.

Problem 11.3 2-Methylcyclohexanone can form two enol tautomers. Show the structures of both.

11.2 Reactivity of Enols: The Mechanism of Alpha-Substitution Reactions

What kind of chemistry do enols have? Because their double bonds are electron-rich, enols behave as nucleophiles and react with electrophiles in much the same way alkenes do (Section 4.1). But because of resonance electron donation from the neighboring oxygen, enols are more electron-rich and correspondingly more reactive than alkenes. Notice in the following electrostatic potential map that there is a substantial amount of electron density (yellow-red) on the α carbon.

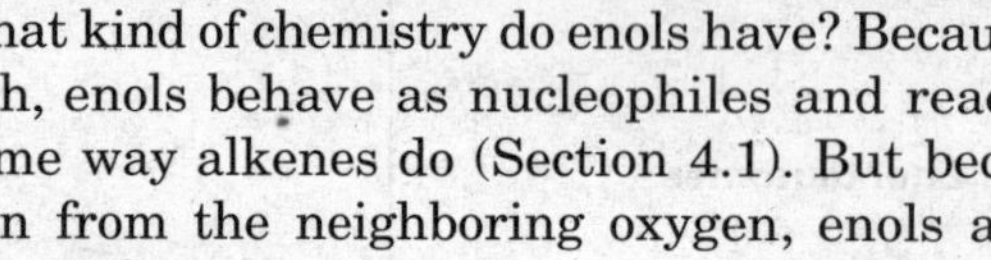

Enol tautomer

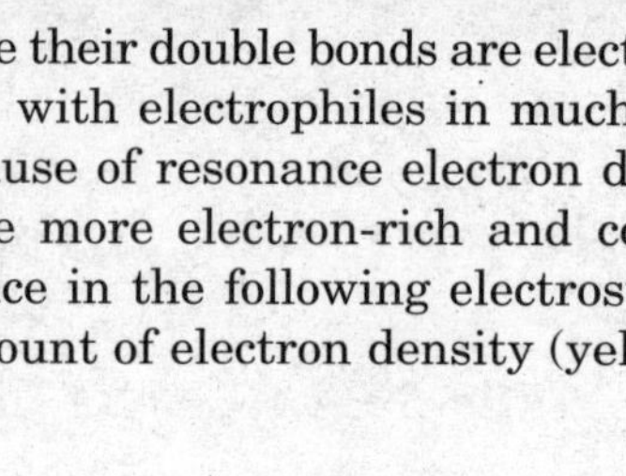

When an *alkene* reacts with an electrophile, such as Br_2, addition of Br^+ occurs to give an intermediate cation, and subsequent reaction with Br^- gives the addition product (Section 4.4). When an *enol* reacts with an electrophile, however, only the initial addition step is the same. Instead of reacting with Br^- to give an addition product, the intermediate cation loses the –OH proton to generate an α-substituted carbonyl compound. The general mechanism is shown in Figure 11.2.

MECHANISM

Figure 11.2 The general mechanism of a carbonyl α-substitution reaction with an electrophile, E^+.

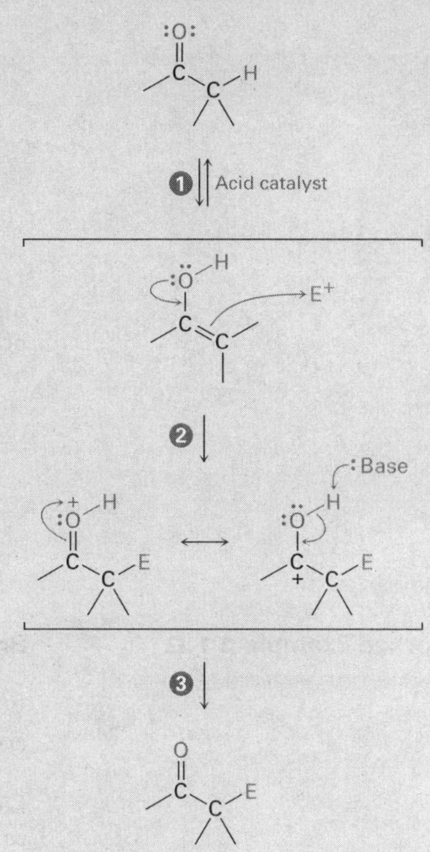

1. Acid-catalyzed enol formation occurs by the usual mechanism.

2. An electron pair from the enol oxygen attacks an electrophile (E^+), forming a new bond and leaving a cation intermediate that is stabilized by resonance between two forms.

3. Loss of a proton from oxygen yields the neutral alpha-substitution product as a new C=O bond is formed.

11.3 Alpha Halogenation of Aldehydes and Ketones

Aldehydes and ketones are halogenated at their α positions by reaction with Cl_2, Br_2, or I_2 in acidic solution. Bromine in acetic acid solvent is most often used. The reaction is a typical α-substitution process that proceeds through an enol intermediate.

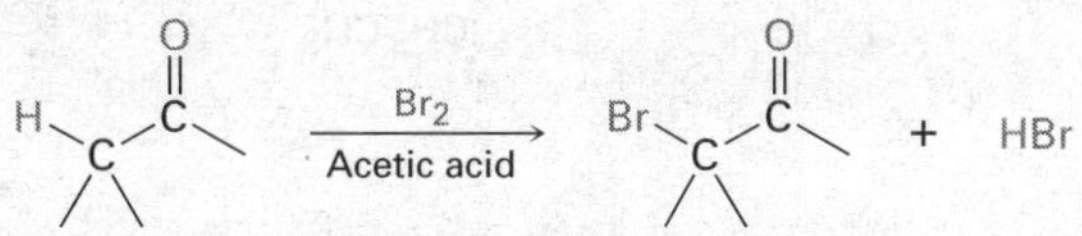

α-Bromo ketones are useful because they undergo elimination of HBr on treatment with base to yield α,β-unsaturated ketones. For example, 2-bromo-2-methylcyclohexanone gives 2-methylcyclohex-2-enone when heated in the organic base pyridine. The reaction takes place by an E2 elimination pathway (Section 7.7) and is a good way to introduce a C=C bond into a molecule.

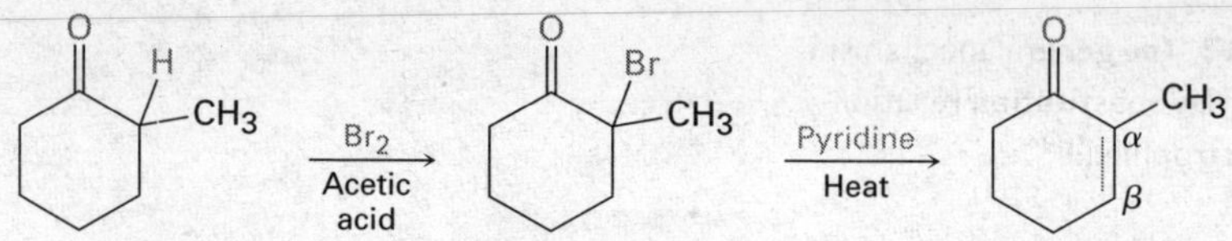

2-Methylcyclohexanone — 2-Bromo-2-methylcyclohexanone — 2-Methylcyclohex-2-enone (63%)

Ketone halogenation also occurs in biological systems, particularly in marine alga, where dibromoacetaldehyde, bromoacetone, 1,1,1-tribromoacetone, and other related compounds have been found.

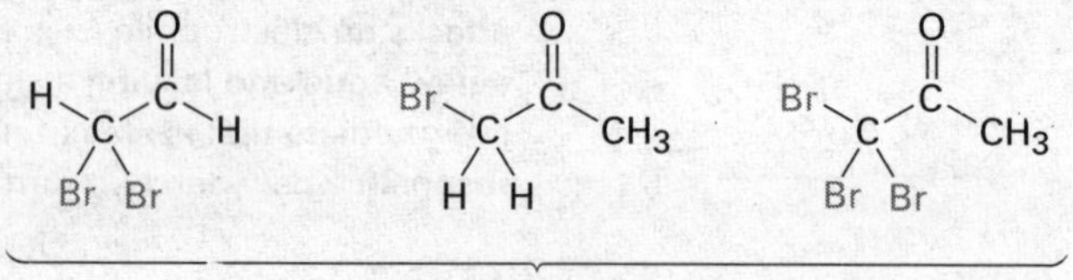

From the Hawaiian alga *Asparagopsis taxiformis*

Worked Example 11.2

Brominating a Ketone

What product would you obtain from the reaction of cyclopentanone with Br_2 in acetic acid?

Strategy Locate the α hydrogens in the starting ketone, and replace one of them by Br to carry out an α-substitution reaction.

Solution

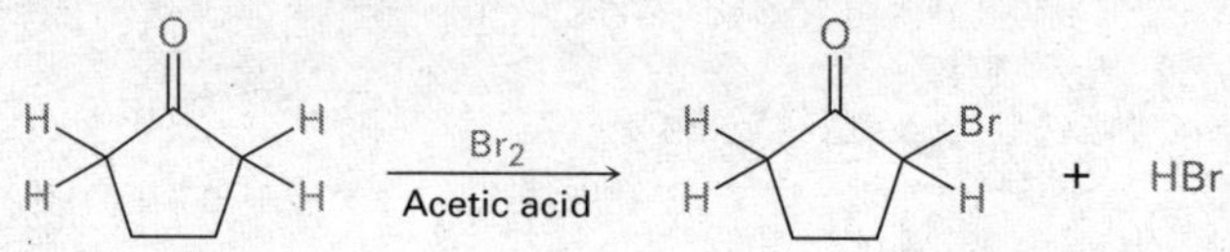

Cyclopentanone — 2-Bromocyclopentanone

Problem 11.4 Show the products you would obtain by bromination of the following ketones:

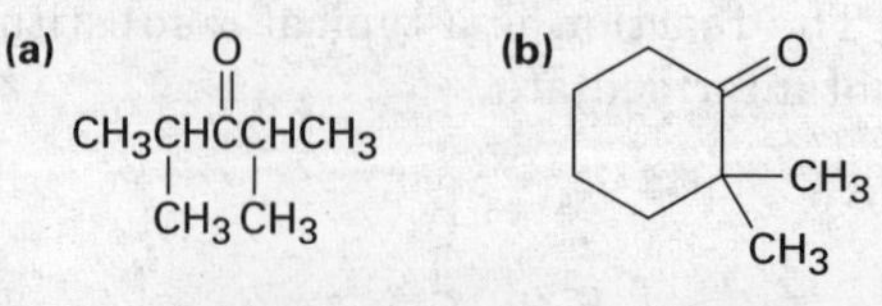

Problem 11.5 Show how you might prepare pent-1-en-3-one from pentan-3-one:

Pentan-3-one — Pent-1-en-3-one

11.4 Acidity of Alpha Hydrogen Atoms: Enolate Ion Formation

As noted in Section 11.1, a hydrogen on the α position of a carbonyl compound is weakly acidic and can be removed by a strong base to give an enolate ion. In comparing acetone ($pK_a = 19.3$) with ethane ($pK_a \approx 60$), for instance, the presence of the carbonyl group increases the acidity of the ketone over the alkane by a factor of 10^{40}.

Acetone ($pK_a = 19.3$)

Ethane ($pK_a \approx 60$)

Why are carbonyl compounds acidic? The reason can be seen by looking at the orbital picture of an enolate ion in Figure 11.3. Abstraction of a proton from a carbonyl compound occurs when an α C–H bond is oriented roughly parallel to the p orbitals of the carbonyl group. The α carbon atom of the enolate ion is sp^2-hybridized and has a p orbital that overlaps the neighboring carbonyl p orbitals. Thus, the negative charge is shared by the electronegative oxygen atom, and the enolate ion is stabilized by resonance.

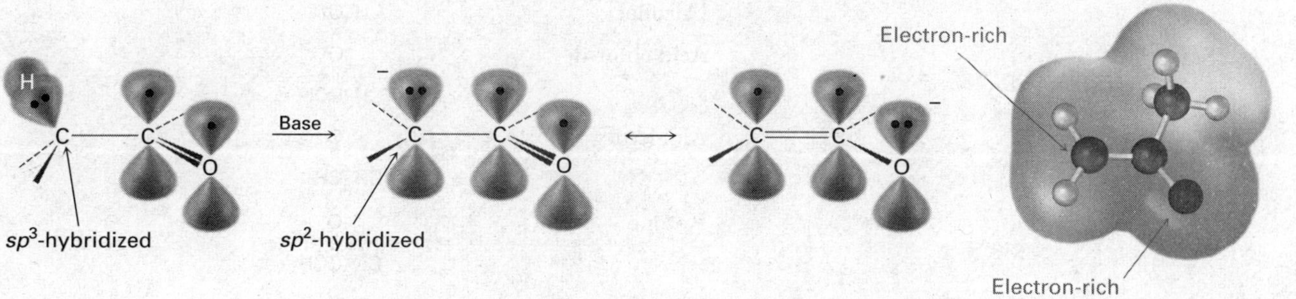

Figure 11.3 Mechanism of enolate ion formation by abstraction of an α hydrogen from a carbonyl compound. The enolate ion is stabilized by resonance and the negative charge (red) is shared by the oxygen and the α carbon atom, as indicated by the electrostatic potential map.

Because carbonyl compounds are only weakly acidic, a strong base is needed to form an enolate ion. If an alkoxide ion, such as sodium ethoxide, is used, ionization of acetone takes place only to the extent of about 0.1% because acetone ($pK_a = 19.3$) is a weaker acid than ethanol ($pK_a = 16$). If, however, a more powerful base such as sodium amide ($Na^+ \ ^-{:}NH_2$, the sodium salt of ammonia) is used, then a carbonyl compound is completely converted into its enolate ion. In practice, the strong base lithium diisopropylamide, abbreviated LDA, is commonly used. LDA is similar in its basicity to sodium amide but is more soluble in organic solvents because of its two alkyl groups.

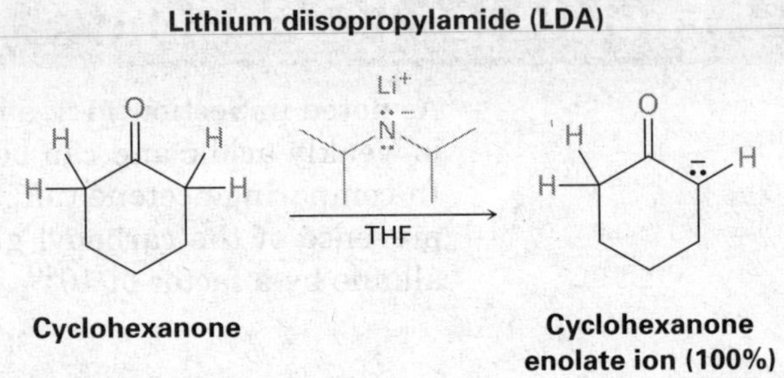

Table 11.1 lists the approximate pK_a values of some different types of carbonyl compounds and shows how these values compare with other common acids. Note that nitriles, too, are acidic and can be converted into enolate-like anions.

Table 11.1 Acidity Constants for Some Carbonyl Compounds

Functional group	Example	pK_a
Carboxylic acid	CH_3COH (C=O)	5
1,3-Diketone	$CH_3CCH_2CCH_3$ (two C=O)	9
3-Keto ester	$CH_3CCH_2COCH_3$ (two C=O)	11
1,3-Diester	$CH_3OCCH_2COCH_3$ (two C=O)	13
[Alcohol	CH_3OH	16]
Acid chloride	CH_3CCl (C=O)	16
Aldehyde	CH_3CH (C=O)	17
Ketone	CH_3CCH_3 (C=O)	19
Thioester	CH_3CSCH_3 (C=O)	21
Ester	CH_3COCH_3 (C=O)	25
Nitrile	$CH_3C{\equiv}N$	25
N,N-Dialkylamide	$CH_3CN(CH_3)_2$ (C=O)	30

When a C–H bond is flanked by *two* carbonyl groups, its acidity is enhanced even more. Thus, Table 11.1 shows that 1,3-diketones (called β-diketones), 3-keto esters (β-keto esters), and 1,3-diesters (malonic esters) are more acidic than water. The enolate ions derived from these β-dicarbonyl compounds are stabilized by sharing of the negative charge by both neighboring carbonyl oxygens. The enolate ion from pentane-2,4-dione, for instance, has three resonance forms.

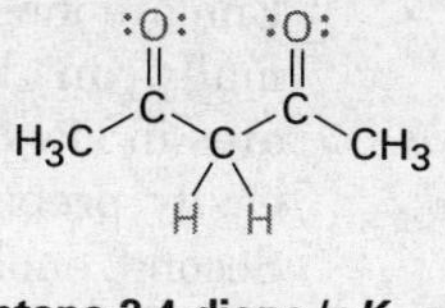

Pentane-2,4-dione (pK_a = 9)

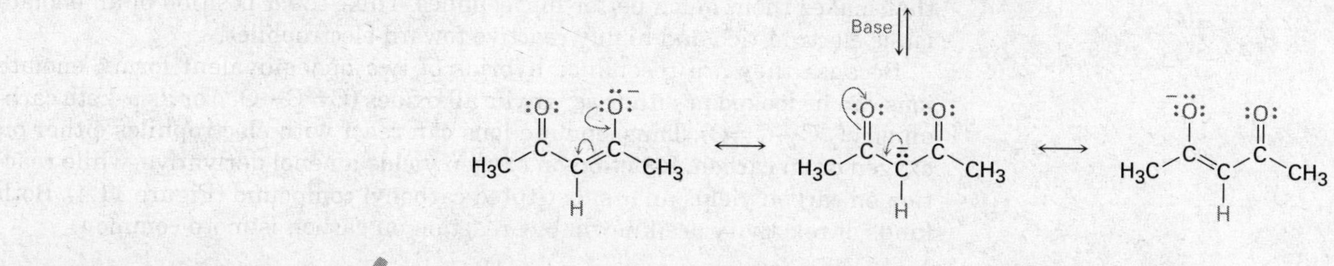

Worked Example 11.3

Forming an Enolate Ion from a Ketone

Draw structures of the two enolate ions you could obtain by reaction of 3-methylcyclohexanone with a strong base.

Strategy Locate the acidic hydrogens, and remove them one at a time to generate the possible enolate ions. In this case, 3-methylcyclohexanone can be deprotonated either at C2 or at C6.

Solution

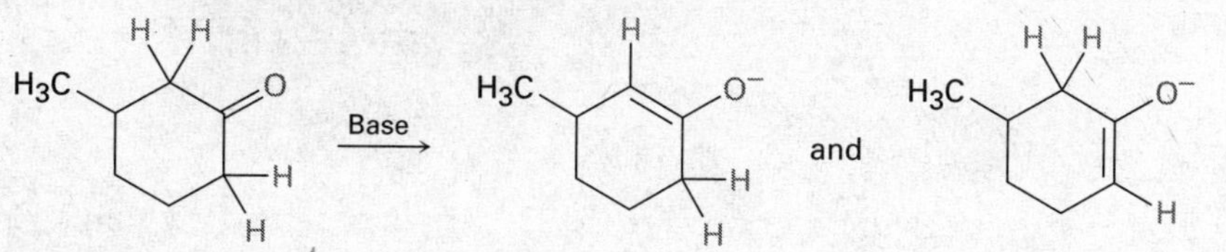

Problem 11.6 Identify all acidic hydrogens in the following molecules:

(a) CH_3CH_2CHO
(b) $(CH_3)_3CCOCH_3$
(c) CH_3CO_2H
(d) $CH_3CH_2CH_2C{\equiv}N$
(e) Cyclohexane-1,3-dione

Problem 11.7 Show the enolate ions you would obtain by deprotonation of the following carbonyl compounds:

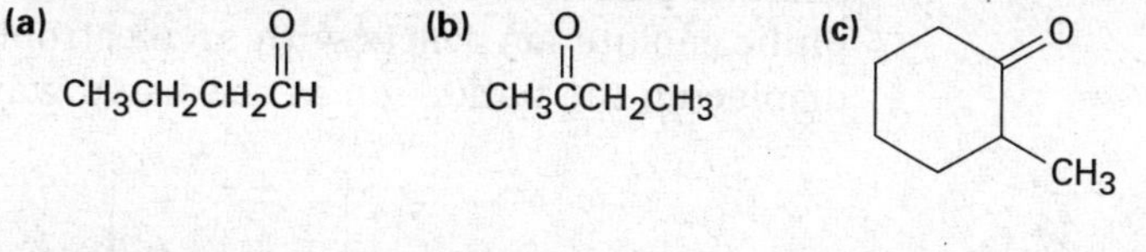

Problem 11.8 Draw three resonance forms for the most stable enolate ion you would obtain by deprotonation of methyl 3-oxobutanoate.

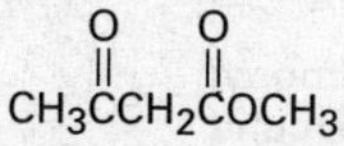

Methyl 3-oxobutanoate

11.5 Reactivity of Enolate Ions

Enolate ions are more useful than enols for two reasons. First, pure enols normally can't be isolated but are instead generated only as short-lived intermediates in low concentration. By contrast, stable solutions of pure enolate ions are easily prepared from carbonyl compounds by treatment with a strong base. Second, enolate ions are more reactive than enols and undergo many reactions that enols don't. Whereas enols are neutral, enolate ions have a negative charge that makes them much better nucleophiles. Thus, the α position of an enolate ion is electron-rich and highly reactive toward electrophiles.

Because they are resonance hybrids of two nonequivalent forms, enolate ions can be looked at either as vinylic alkoxides (C═C—O⁻) or as α-keto carbanions (⁻C—C═O). Thus, enolate ions can react with electrophiles either on oxygen or on carbon. Reaction on oxygen yields an enol derivative, while reaction on carbon yields an α-substituted carbonyl compound (Figure 11.4). Both kinds of reactivity are known, but reaction on carbon is more common.

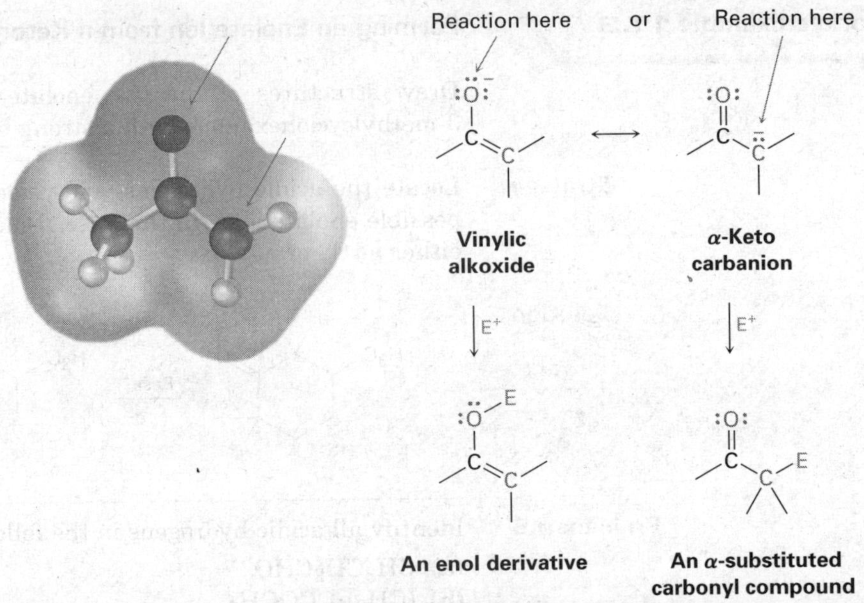

Figure 11.4 The electrostatic potential map of acetone enolate ion shows how the negative charge is delocalized over both the oxygen and the α carbon. As a result, two modes of reaction of an enolate ion with an electrophile E^+ are possible. Reaction on carbon to yield an α-substituted carbonyl product is more common.

11.6 Alkylation of Enolate Ions

Perhaps the most useful reaction of enolate ions is their **alkylation** by treatment with an alkyl halide, thereby forming a new C–C bond and joining two smaller pieces into one larger molecule. Alkylation occurs when the nucleophilic enolate ion reacts with an electrophilic alkyl halide in an S_N2 reaction, displacing the halide ion in the usual way (Section 7.5).

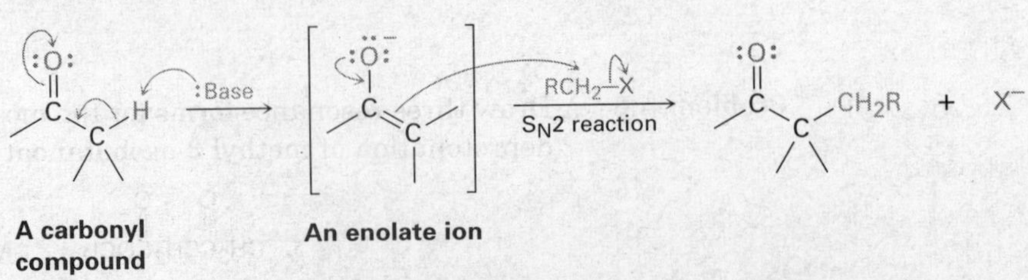

Like all S_N2 reactions, alkylations are successful only when a primary alkyl halide (RCH_2X) or methyl halide (CH_3X) is used, because a competing E2 elimination occurs if a secondary or tertiary halide is used. The leaving group halide can be Cl^-, Br^-, or I^-.

One of the best-known carbonyl alkylation reactions is the **malonic ester synthesis**, a method for preparing a carboxylic acid from an alkyl halide while lengthening the carbon chain by two atoms.

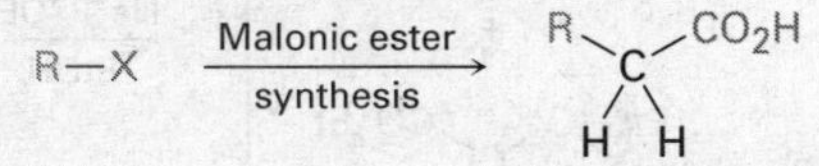

Diethyl propanedioate, commonly called diethyl malonate or *malonic ester*, is relatively acidic ($pK_a = 13$) because its α hydrogen atoms are flanked by two carbonyl groups. Thus, malonic ester is easily converted into its enolate ion by reaction with sodium ethoxide in ethanol. The enolate ion, in turn, is readily alkylated by treatment with an alkyl halide, yielding an α-substituted malonic ester. Note in the following examples that the abbreviation "Et" is used for an ethyl group, $-CH_2CH_3$.

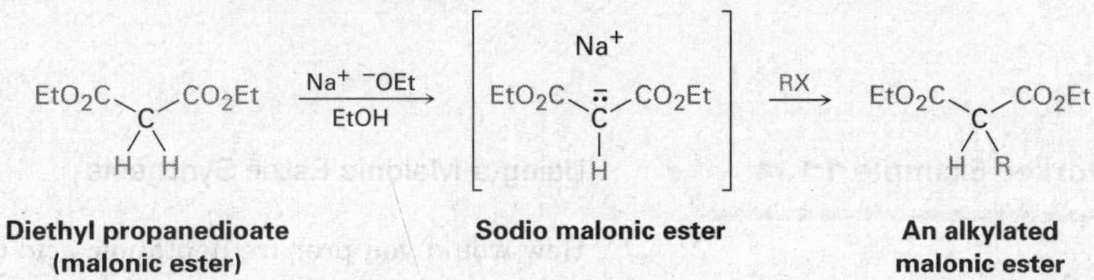

The product of a malonic ester alkylation has one acidic α hydrogen remaining, so the alkylation process can be repeated to yield a dialkylated malonic ester.

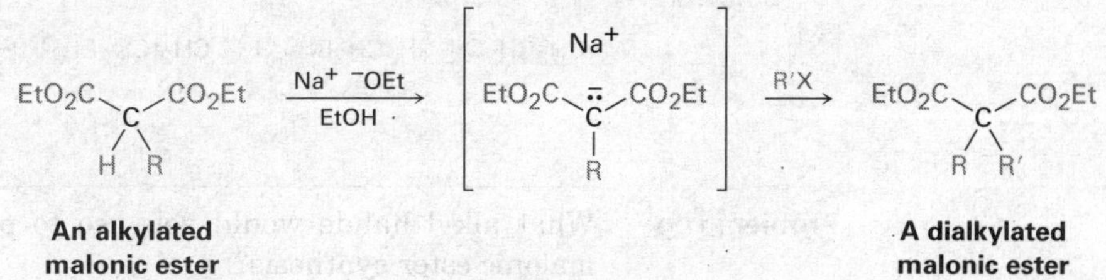

On heating with aqueous hydrochloric acid, the alkylated (or dialkylated) malonic ester undergoes hydrolysis of its two ester groups followed by *decarboxylation* (loss of CO_2) to yield a substituted monocarboxylic acid. Decarboxylation is a unique feature of compounds like malonic acids that have a second carbonyl group two atoms away from the $-CO_2H$ and is not a general reaction of carboxylic acids.

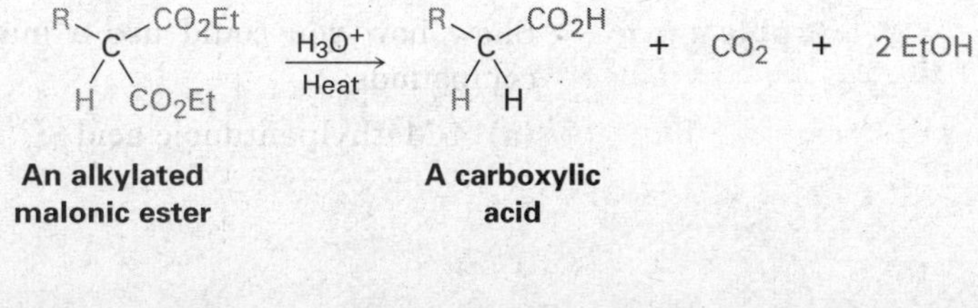

The overall result of the malonic ester synthesis is to convert an alkyl halide into a carboxylic acid with a carbon chain that has been lengthened by two atoms ($RX \rightarrow RCH_2CO_2H$).

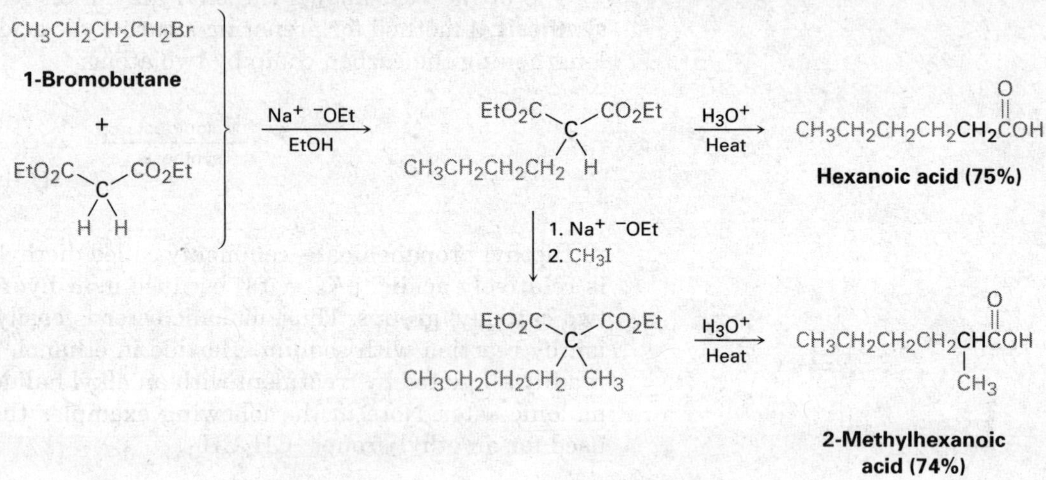

Worked Example 11.4 Using a Malonic Ester Synthesis

How would you prepare heptanoic acid by a malonic ester synthesis?

Strategy The malonic ester synthesis converts an alkyl halide into a carboxylic acid having two more carbon atoms. Thus, a seven-carbon acid chain must be derived from a five-carbon alkyl halide such as 1-bromopentane.

Solution

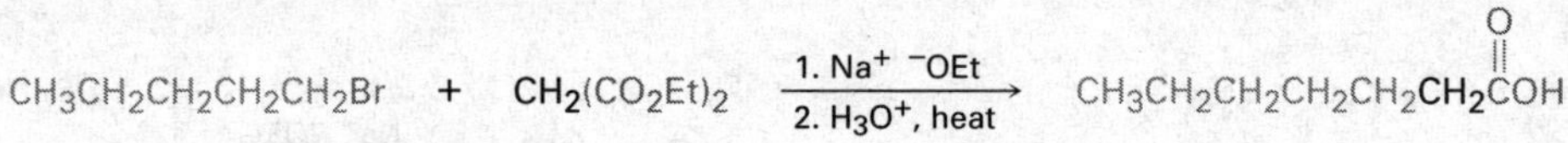

Problem 11.9 What alkyl halide would you use to prepare the following compounds by a malonic ester synthesis?

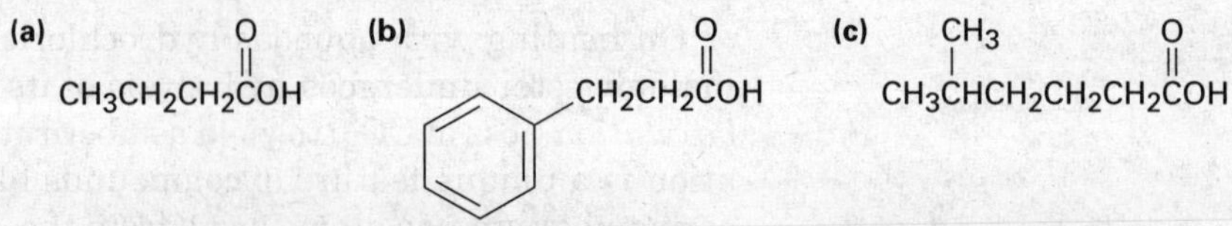

Problem 11.10 Show how you could use a malonic ester synthesis to prepare the following compounds:

(a) 4-Methylpentanoic acid **(b)** 2-Methylpentanoic acid

Problem 11.11 Show how you could use a malonic ester synthesis to prepare the following compound:

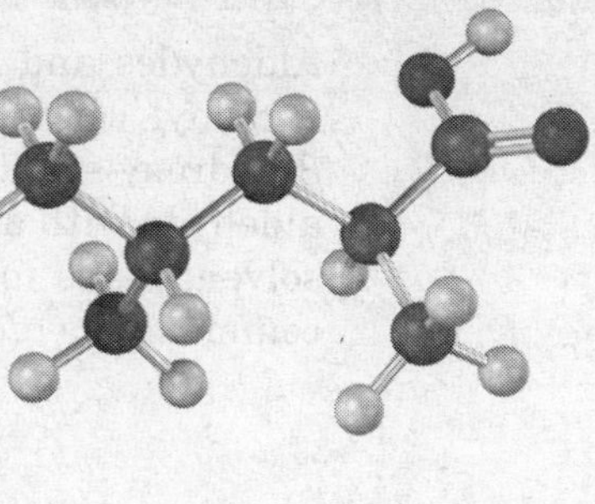

11.7 Carbonyl Condensation Reactions

By this point, we've seen that carbonyl compounds can behave as either electrophiles or nucleophiles. In an α-substitution reaction, the carbonyl compound behaves as a nucleophile after being converted into an enol or enolate ion. In a nucleophilic addition reaction or a nucleophilic acyl substitution reaction, however, the carbonyl group behaves as an electrophile by accepting electrons from an attacking nucleophile.

Carbonyl condensation reactions, the fourth and last general category of carbonyl-group reactions we'll study, involve *both* kinds of reactivity. These reactions take place between two carbonyl partners and involve a combination of nucleophilic addition and α-substitution steps. One partner (the nucleophilic donor) is converted into an enolate ion and undergoes an α-substitution reaction, while the other partner (the electrophilic acceptor) undergoes a nucleophilic addition reaction. There are numerous variations of carbonyl condensation reactions, depending on the two carbonyl partners, but the general mechanism remains the same.

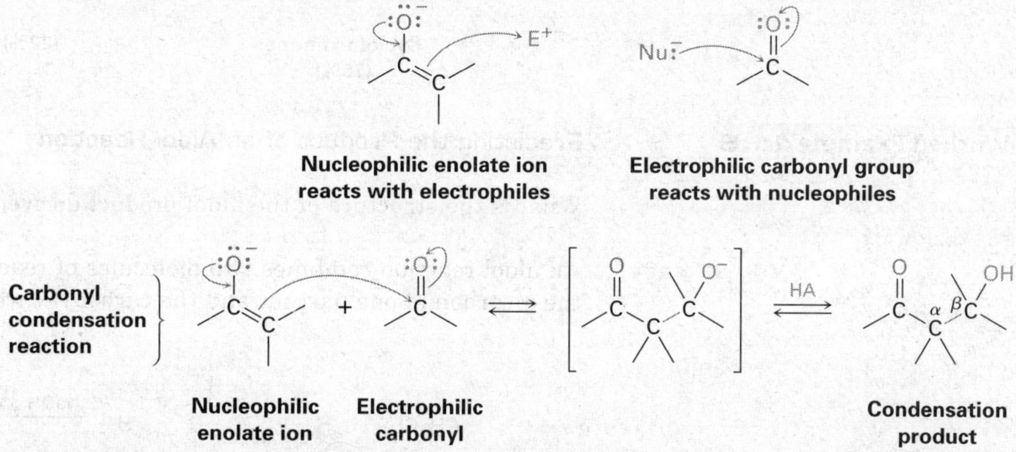

11.8 Condensations of Aldehydes and Ketones: The Aldol Reaction

Aldehydes and ketones with an α hydrogen atom undergo a base-catalyzed carbonyl condensation reaction called the **aldol reaction**. The product is a β-hydroxy-substituted carbonyl compound. For example, treatment of acetaldehyde with a base such as sodium ethoxide or sodium hydroxide in a protic solvent leads to rapid and reversible formation of 3-hydroxybutanal, known commonly as *aldol* (*ald*ehyde + alcoh*ol*).

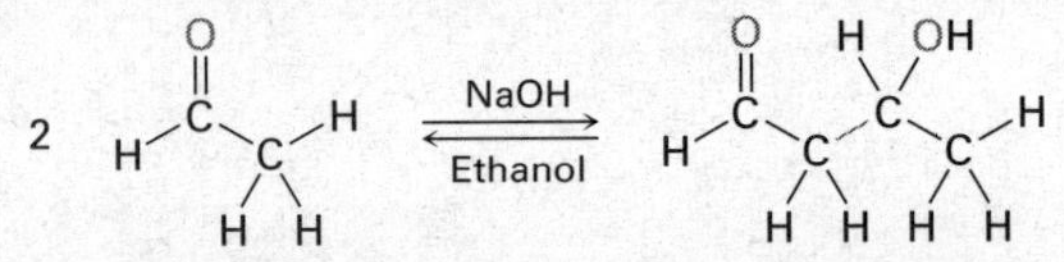

Acetaldehyde **3-Hydroxybutanal (aldol)**

The exact position of the aldol equilibrium depends both on reaction conditions and on substrate structure. The equilibrium generally favors condensation product for aldol reaction of aldehydes with no α substituent (RCH_2CHO) but favors reactant for α-substituted aldehydes (R_2CHCHO) and for ketones.

Aldehydes

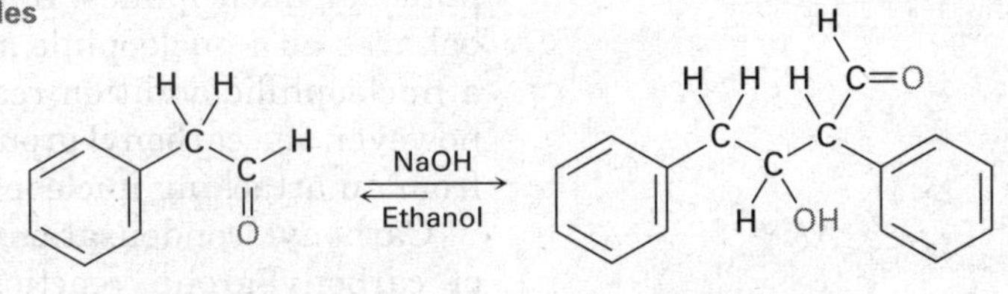

Phenylacetaldehyde (10%) **(90%)**

Ketones

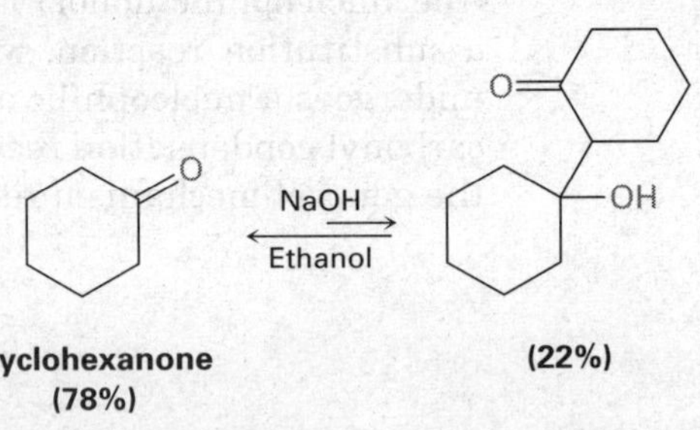

Cyclohexanone (78%) **(22%)**

Worked Example 11.5 Predicting the Product of an Aldol Reaction

What is the structure of the aldol product derived from propanal?

Strategy An aldol reaction combines two molecules of reactant, forming a bond between the α carbon of one partner and the carbonyl carbon of the second partner.

Solution

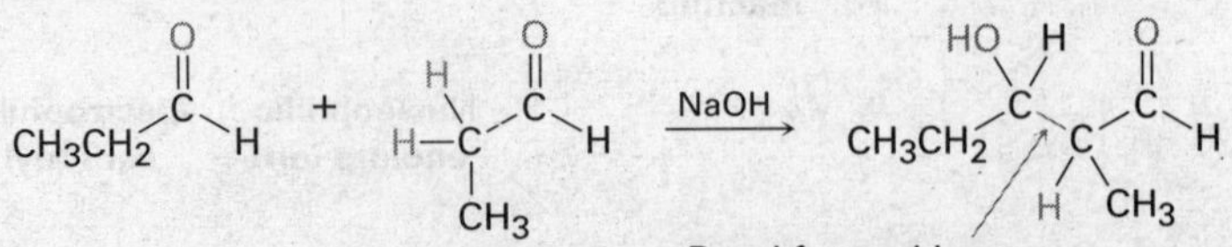

Problem 11.12 Which of the following compounds can undergo the aldol reaction, and which cannot? Explain.

(a) 2,2-Dimethylcyclohexanone
(b) Benzaldehyde
(c) 2,2,6,6-Tetramethylcyclohexanone
(d) Formaldehyde

Problem 11.13 Show the product of the aldol reaction of the following compounds:

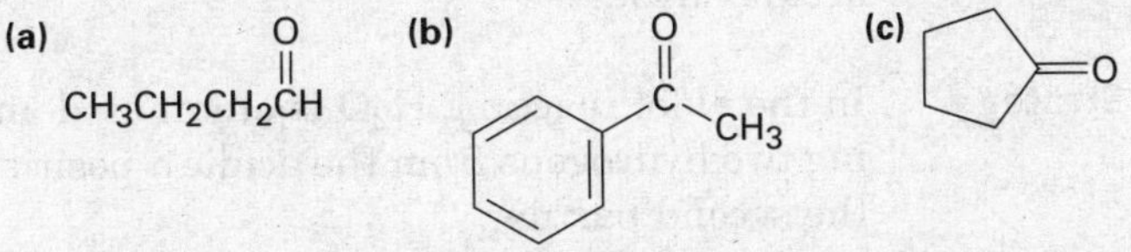

11.9 Dehydration of Aldol Products: Synthesis of Enones

The β-hydroxy aldehydes and β-hydroxy ketones formed in aldol reactions are easily dehydrated to yield α,β-unsaturated products, or conjugated **enones** (*ene + one*). In fact, it's this loss of water that gives the aldol *condensation* its name, because water condenses out of the reaction.

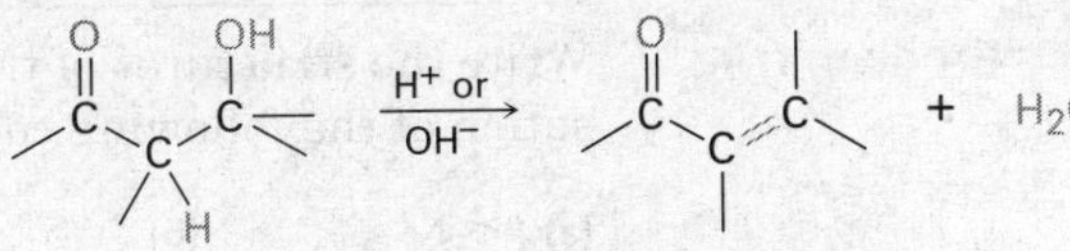

A β-hydroxy ketone or aldehyde — A conjugated enone

Most alcohols are resistant to dehydration by dilute acid or base because hydroxide ion is a poor leaving group (Section 8.4), but β-hydroxy carbonyl compounds dehydrate easily because of the carbonyl group. Under *basic* conditions, an α hydrogen is abstracted and the resultant enolate ion expels the OH^- leaving group in an E1cB reaction (Section 7.8). Under *acidic* conditions, an enol is formed, the –OH group is protonated, and H_2O is then expelled in an E1 or E2 reaction.

Base-catalyzed

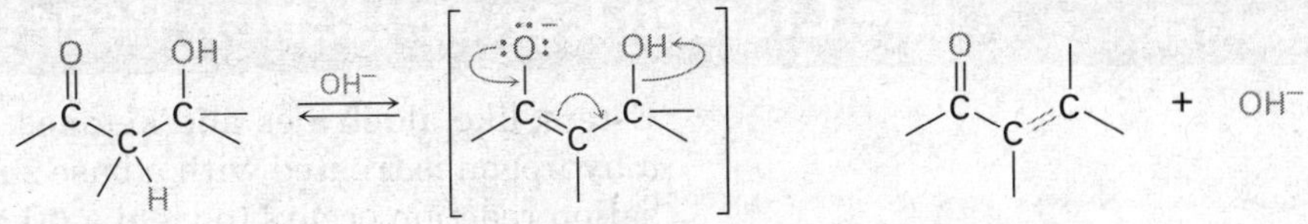

Enolate ion

Acid-catalyzed

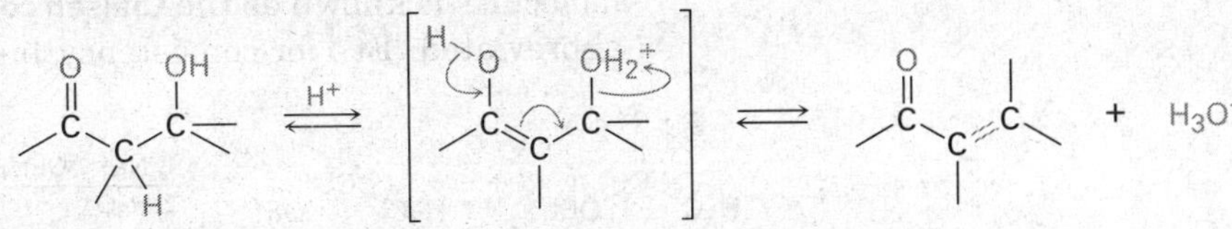

Enol

The reaction conditions needed for aldol dehydration are often only a bit more vigorous (slightly higher temperature, for instance) than the conditions needed for the aldol condensation itself. As a result, conjugated enones are often obtained directly from aldol reactions without isolating the intermediate β-hydroxy carbonyl compounds.

Worked Example 11.6 Dehydrating an Aldol Condensation Product

What is the structure of the enone obtained from aldol condensation of acetaldehyde?

Strategy In the aldol reaction, H_2O is eliminated and a double bond is formed by removing two hydrogens from the acidic α position of one partner and the oxygen from the second partner.

Solution

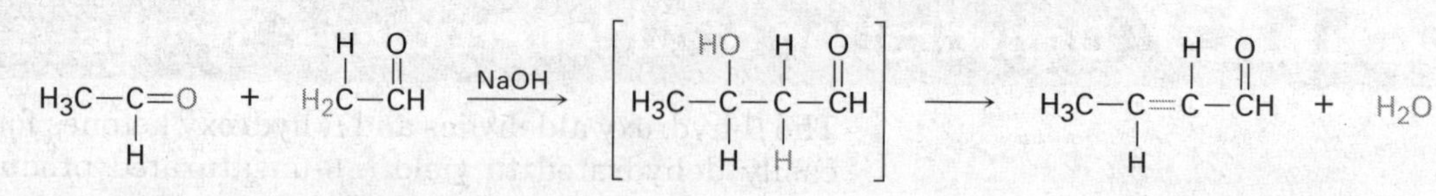

But-2-enal

Problem 11.14 Write the structures of the enone products you would obtain from aldol condensation of the following compounds:

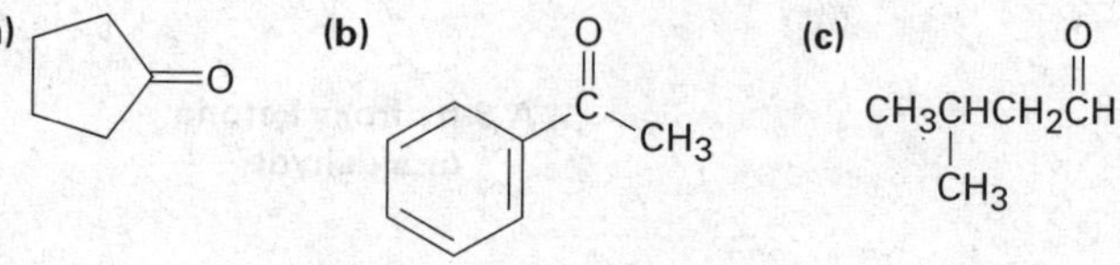

Problem 11.15 Aldol condensation of butan-2-one leads to a mixture of two enones (ignoring double-bond stereochemistry). Draw them.

11.10 Condensations of Esters: The Claisen Condensation Reaction

Esters, like aldehydes and ketones, are weakly acidic. When an ester with an α hydrogen is treated with a base such as sodium ethoxide, a carbonyl condensation reaction occurs to yield a β-keto ester. For example, ethyl acetate yields ethyl acetoacetate on treatment with base. This reaction between two ester molecules is known as the **Claisen condensation reaction**. We'll use ethyl esters, abbreviated "Et," for consistency, but other esters also work.

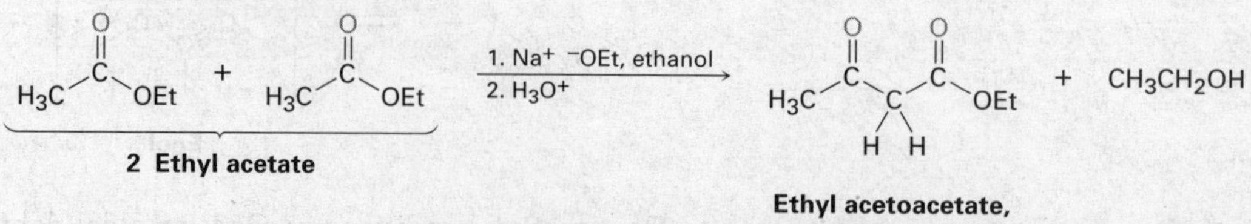

Ethyl acetoacetate, a β-keto ester (75%)

The mechanism of the Claisen condensation is similar to that of the aldol condensation and involves the nucleophilic addition of an ester enolate ion to the carbonyl group of a second ester molecule (Figure 11.5). The only

MECHANISM

Figure 11.5 The mechanism of the Claisen condensation reaction.

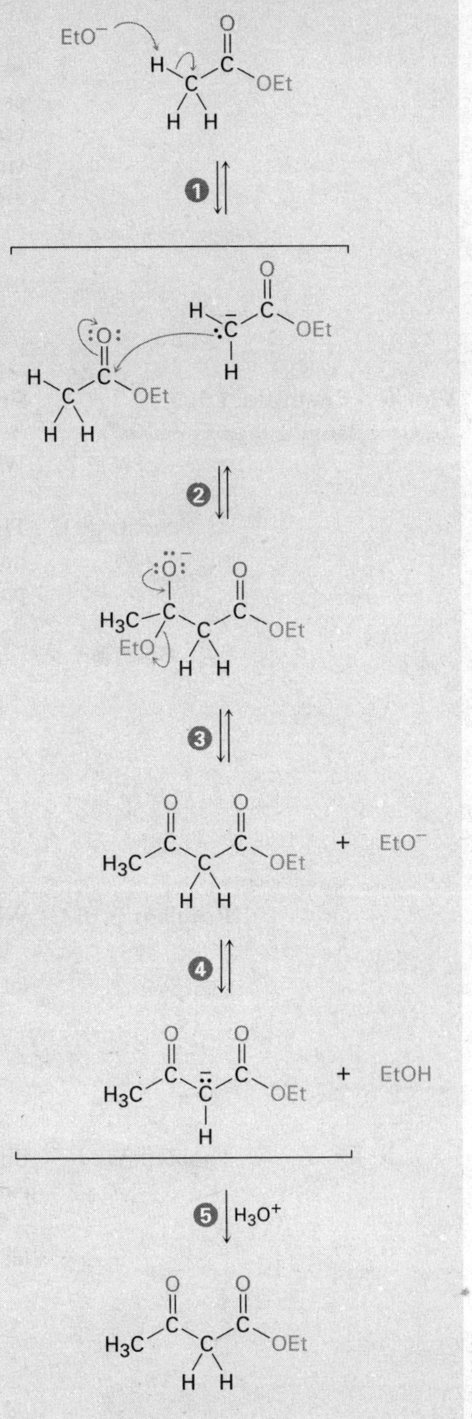

1. Base abstracts an acidic alpha hydrogen atom from an ester molecule, yielding an ester enolate ion.
2. The enolate ion adds in a nucleophilic addition reaction to a second ester molecule, giving a tetrahedral alkoxide intermediate.
3. The tetrahedral intermediate expels ethoxide ion to yield a new carbonyl compound, ethyl acetoacetate.
4. But ethoxide ion is a strong enough base to deprotonate ethyl acetoacetate, shifting the equilibrium and driving the overall reaction to completion.
5. Protonation of the enolate ion by addition of aqueous acid in a separate step yields the final β-keto ester product.

difference between the aldol condensation of an aldehyde or ketone and the Claisen condensation of an ester involves the fate of the initially formed intermediate. The alkoxide intermediate in the aldol reaction is protonated to give an alcohol product—exactly the behavior previously seen for aldehydes and ketones (Section 9.5). The alkoxide intermediate in the Claisen reaction, however, expels an alkoxide leaving group to yield a nucleophilic acyl substitution product—exactly the behavior previously seen for esters (Section 10.9).

If the starting ester has more than one acidic α hydrogen, the β-keto ester product contains a highly acidic, doubly activated hydrogen atom that can be removed by base. This deprotonation of the product means that a full equivalent of base rather than a catalytic amount must be used in the reaction. The deprotonation drives the equilibrium completely to the product side so that high yields are usually obtained in Claisen condensations.

Worked Example 11.7

Predicting the Product of a Claisen Condensation

What product would you obtain from Claisen condensation of ethyl propanoate?

Strategy The Claisen condensation of an ester results in the loss of one molecule of alcohol and the formation of a β-keto ester product in which an acyl group of one partner bonds to the α carbon of the second partner.

Solution

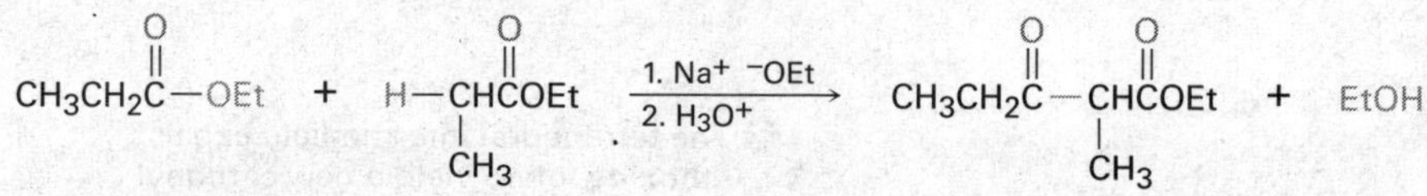

2 Ethyl propanoate — Ethyl 2-methyl-3-oxopentanoate

Problem 11.16 Which of the following esters can't undergo Claisen condensation? Explain.

(a) 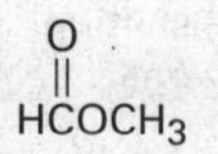 (b) 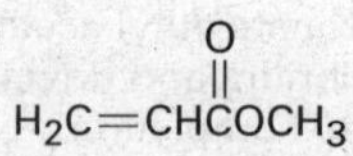 (c) 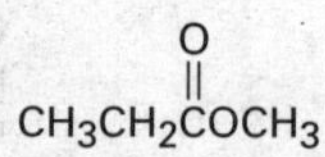

Problem 11.17 Show the products you would obtain by Claisen condensation of the following esters:

(a) 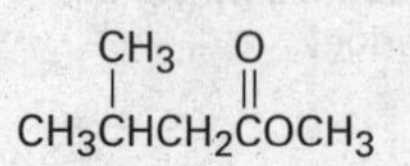 (b) 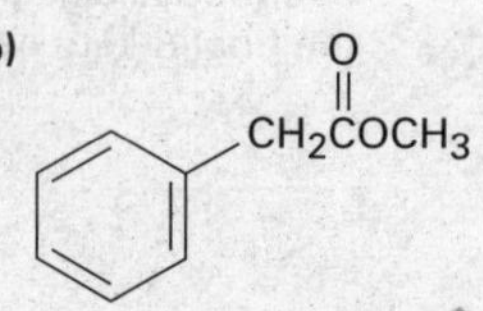 (c) 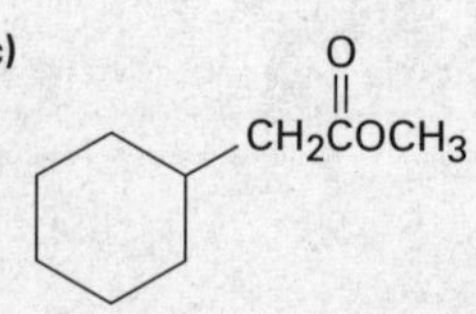

11.11 Some Biological Carbonyl Reactions

Biochemistry *is* carbonyl chemistry. Almost every metabolic process used by living organisms involves one or more of the four fundamental carbonyl-group reactions we've seen in the past three chapters. The digestion and metabolic breakdown of all the major classes of food molecules—fats, carbohydrates, and proteins—take place by nucleophilic addition reactions, nucleophilic acyl substitutions, α substitutions, and carbonyl condensations. Conversely, hormones and other crucial biological molecules are built up from smaller precursors by these same carbonyl-group reactions.

Take *glycolysis,* for example, the metabolic pathway by which organisms convert glucose to pyruvate as the first step in extracting energy from carbohydrates.

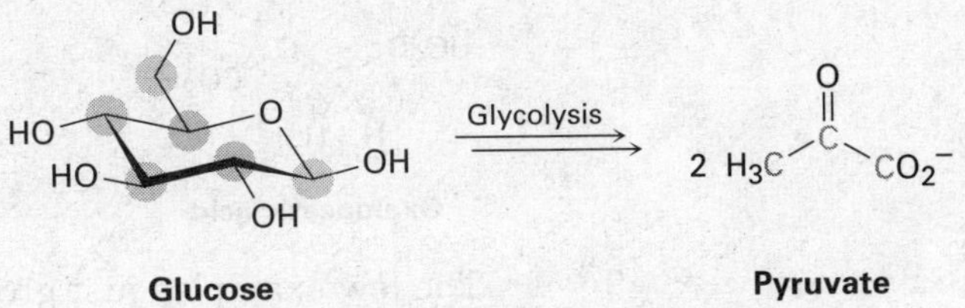

Glycolysis is a ten-step process that begins with conversion of glucose from its cyclic hemiacetal form to its open-chain aldehyde form—a retro nucleophilic addition reaction. The aldehyde then undergoes tautomerization to yield an enol, which undergoes yet another tautomerization to give the ketone fructose.

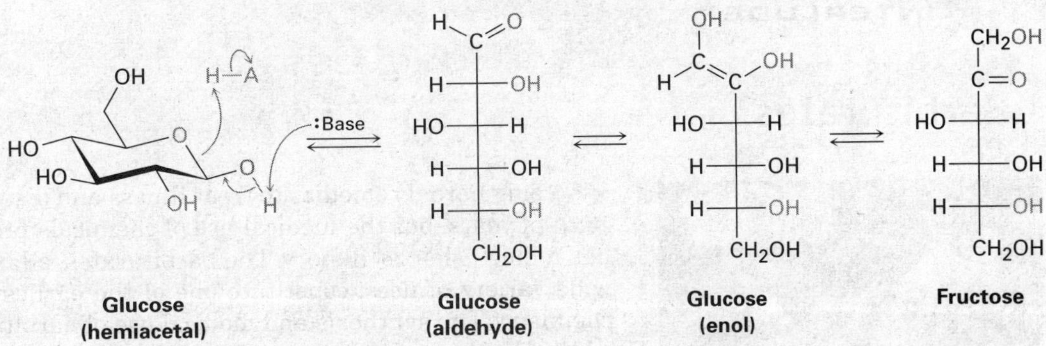

Fructose, a β-hydroxy ketone, is then cleaved into two three-carbon molecules—one ketone and one aldehyde—by a retro aldol reaction. Still further carbonyl-group reactions then occur until pyruvate results.

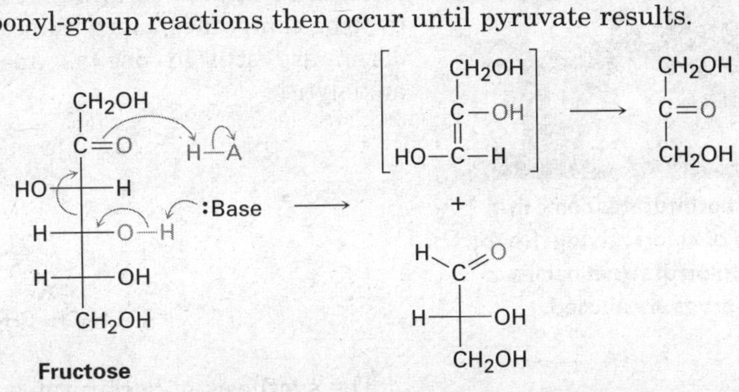

As another example of a biological carbonyl reaction, nature uses the two-carbon acetate fragment of acetyl CoA as a major building block for synthesis.

Acetyl CoA can act not only as an electrophilic acceptor, being attacked by nucleophiles at the carbonyl group, but also as a nucleophilic donor by loss of its acidic α hydrogen. Once formed, the enol or enolate ion of acetyl CoA can add to another carbonyl group in a condensation reaction. For example, citric acid is biosynthesized by nucleophilic addition of acetyl CoA to the ketone carbonyl group of oxaloacetic acid (2-oxobutanedioic acid) in an aldol-like reaction.

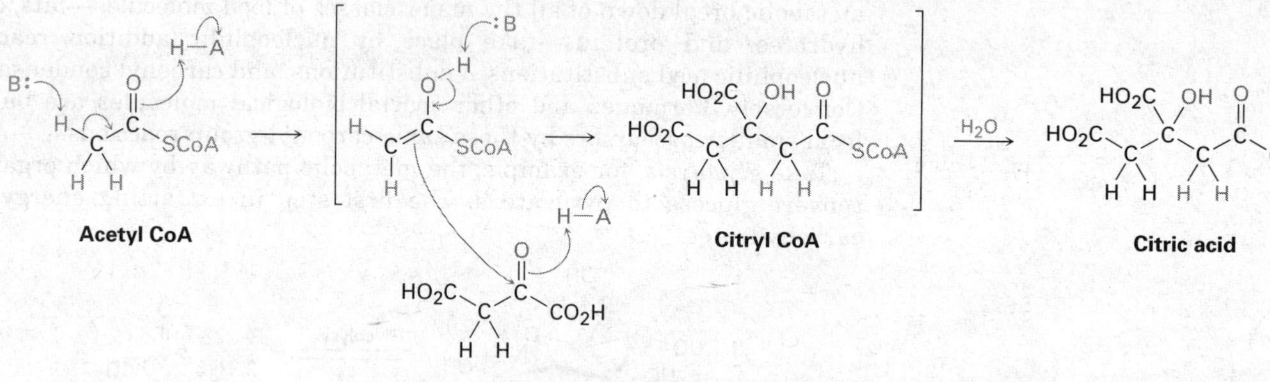

Acetyl CoA

Citryl CoA

Citric acid

Oxaloacetic acid

The few examples just given are only an introduction; we'll look at several of the major metabolic pathways and see more carbonyl-group reactions in Chapter 17. A good grasp of carbonyl chemistry is crucial to an understanding of biochemistry.

INTERLUDE

Barbiturates

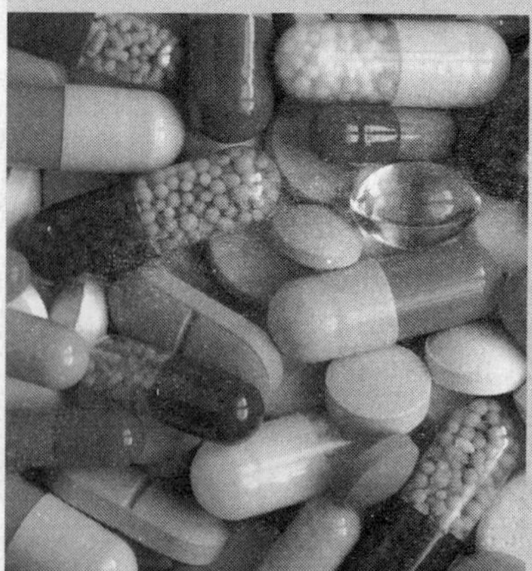

Different barbiturates come in a multitude of colors, giving rise to similarly colorful street names when the drugs are abused.

Using herbal remedies to treat illness and disease goes back thousands of years, but the medical use of chemicals prepared in the laboratory has a much shorter history. The barbiturates, a large class of drugs with a wide variety of uses, constitute one of the earliest successes of medicinal chemistry. The synthesis and medical use of barbiturates goes back to 1904 when the Bayer company in Germany first marketed a compound called barbital, trade named Veronal, as a treatment for insomnia. Since that time, more than 2500 different barbiturate analogs have been synthesized by drug companies, more than 50 have been used medicinally, and about a dozen are still in use as anesthetics, anticonvulsants, sedatives, and anxiolytics.

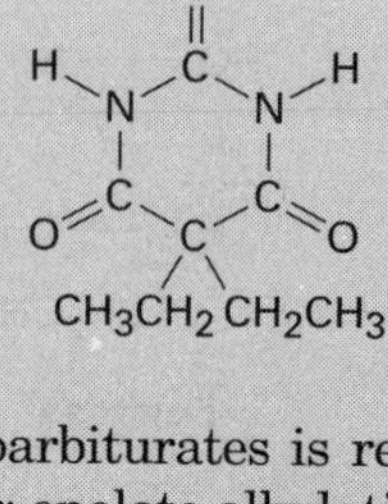

Barbital (Veronal), the first barbiturate

The synthesis of barbiturates is relatively simple and relies on reactions that are now familiar: enolate alkylations and nucleophilic acyl substitutions.

continued

INTERLUDE

Starting with diethyl malonate, or malonic ester, alkylation of the corresponding enolate ion with simple alkyl halides provides a wealth of different disubstituted malonic esters. Reaction with urea (H_2NCONH_2) then gives the product barbiturates by a twofold nucleophilic acyl substitution reaction of the ester groups with the $-NH_2$ groups of urea (Figure 11.6). Amobarbital (Amytal), pentobarbital (Nembutal), and secobarbital (Seconal) are typical examples.

Figure 11.6 The synthesis of barbiturates relies on malonic ester alkylations and nucleophilic acyl substitution reactions. More than 2500 different barbiturates have been synthesized over the past 100 years. In addition to their legal medical uses, some barbiturates are also used illegally as street drugs under many colorful names.

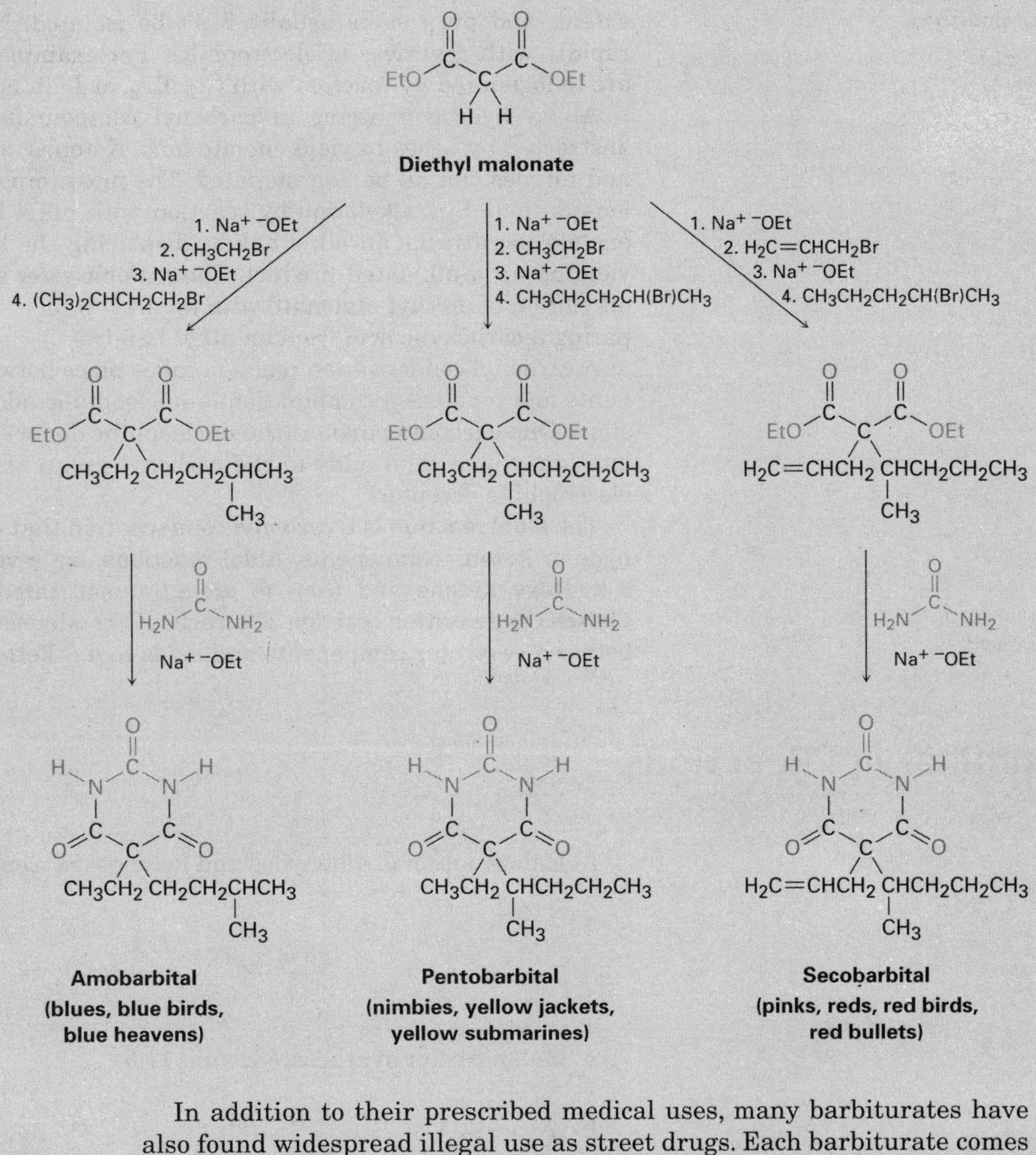

In addition to their prescribed medical uses, many barbiturates have also found widespread illegal use as street drugs. Each barbiturate comes as a tablet of regulated size, shape, and color, and their street names often mimic those colors. Although still used today, most barbiturates have been replaced by safer, more potent alternatives with markedly different structures.

Summary and Key Words

aldol reaction 386
alkylation reaction 382
alpha-substitution reaction 372
carbonyl condensation reaction 373
Claisen condensation reaction 388
enol 373
enolate ion 374
enone 387
malonic ester synthesis 383
tautomers 374

Alpha substitution reactions and **carbonyl condensation** reactions are two of the four fundamental reaction types in carbonyl-group chemistry. Both are used in biosynthetic pathways and in the chemical laboratory for building up larger molecules from smaller precursors. In this chapter, we saw how and why these reactions occur.

Alpha-substitution reactions take place via **enol** or **enolate ion** intermediates and result in the replacement of an α hydrogen atom by another substituent. Carbonyl compounds are in rapid equilibrium with their enols, a process known as *tautomerism*. Enol **tautomers** are normally present to only a small extent, and pure enols usually can't be isolated. Nevertheless, enols react rapidly with a variety of electrophiles. For example, aldehydes and ketones are halogenated by reaction with Cl_2, Br_2, or I_2 in acetic acid solution.

Alpha hydrogen atoms in carbonyl compounds are acidic and can be abstracted by bases to yield enolate ions. Ketones, aldehydes, esters, amides, and nitriles can all be deprotonated. The most important reaction of enolate ions is their S_N2 **alkylation** by reaction with alkyl halides. The nucleophilic enolate ion attacks an alkyl halide, displacing the leaving halide group and yielding an α-alkylated product. The **malonic ester synthesis**, which involves alkylation of diethyl malonate with an alkyl halide, is a good method for preparing a carboxylic acid from an alkyl halide.

A carbonyl condensation reaction takes place between two carbonyl components and involves a combination of nucleophilic addition and α-substitution steps. One carbonyl partner (the nucleophilic donor) is converted into its enolate ion, which then adds to the carbonyl group of the second partner (the electrophilic acceptor).

The **aldol reaction** is a carbonyl condensation that occurs between two aldehyde or ketone components. Aldol reactions are reversible, leading first to a β-hydroxy ketone and then to an α,β-unsaturated ketone, or **enone**. The **Claisen condensation reaction** is a carbonyl condensation reaction that occurs between two ester components and leads to a β-keto ester product.

Summary of Reactions

1. Halogenation of aldehydes and ketones (Section 11.3)

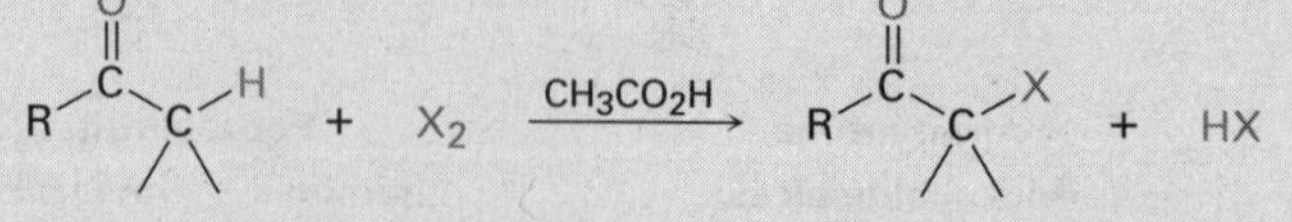

2. Malonic ester synthesis (Section 11.6)

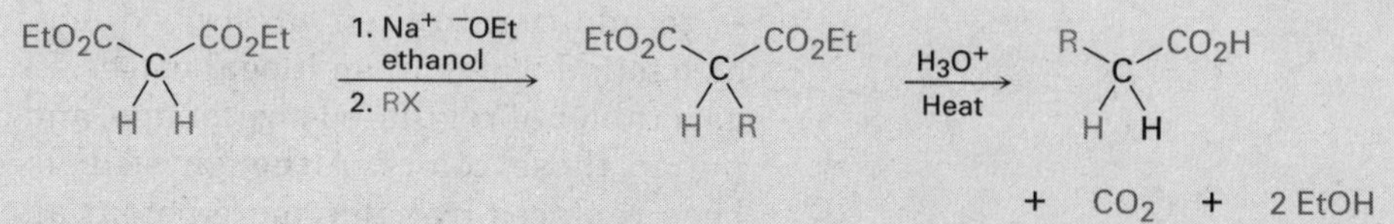

3. Aldol reaction of aldehydes and ketones (Section 11.8)

$$2\ \mathrm{RCH_2\overset{O}{\overset{\|}{C}}H} \xrightleftharpoons{\text{NaOH, ethanol}} \mathrm{RCH_2\overset{OH}{\overset{|}{C}}H\underset{R}{\underset{|}{C}}H\overset{O}{\overset{\|}{C}}H}$$

4. Claisen condensation reaction of esters (Section 11.10)

$$2\ \mathrm{RCH_2\overset{O}{\overset{\|}{C}}OR'} \xrightleftharpoons{\text{Na}^+\ ^-\text{OEt, ethanol}} \mathrm{RCH_2\overset{O}{\overset{\|}{C}}{-}\underset{R}{\underset{|}{C}}H\overset{O}{\overset{\|}{C}}OR'} + \mathrm{HOR'}$$

Exercises

Visualizing Chemistry

(Problems 11.1–11.17 appear within the chapter.)

OWL

Interactive versions of these problems are assignable in OWL.

11.18 The following structure represents an intermediate formed by addition of an ester enolate ion to a second ester molecule. Identify the reactant, the leaving group, and the product.

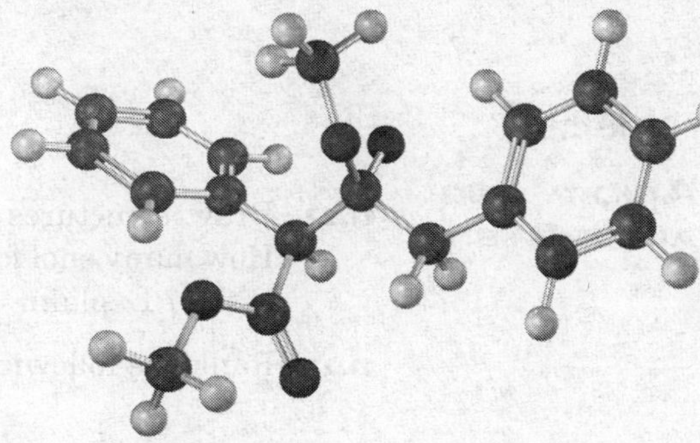

11.19 What aldehydes or ketones might the following enones have been prepared from by aldol reaction?

(a)

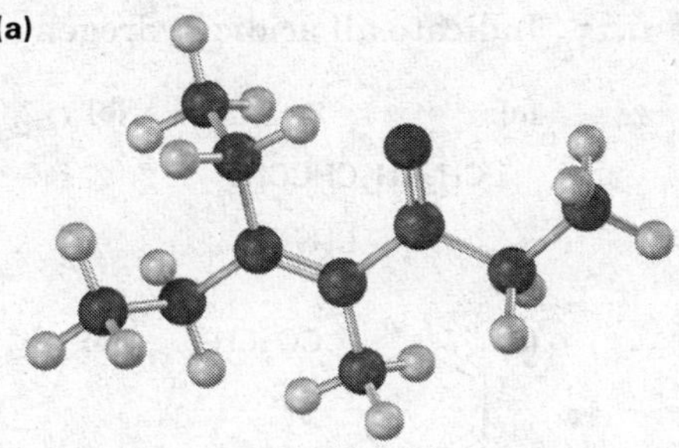

(b)

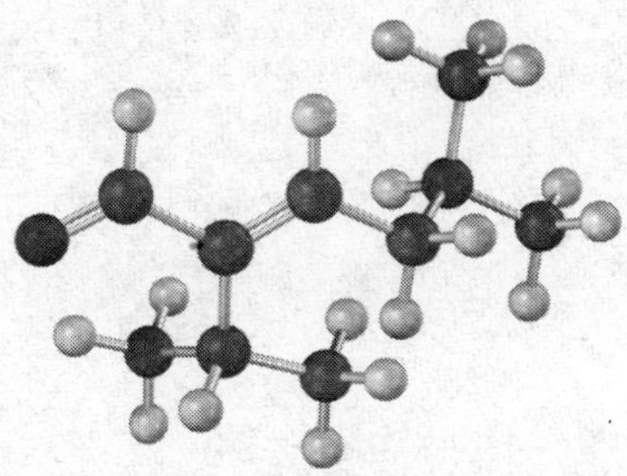

11.20 Show the steps in preparing the following molecule using a malonic ester synthesis:

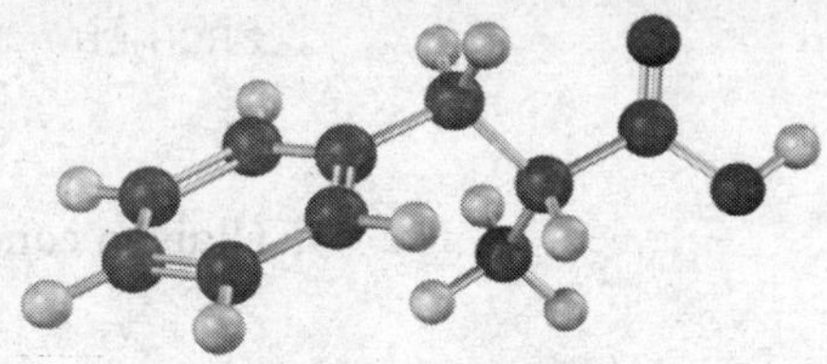

11.21 The following molecule was formed by an *intramolecular* aldol reaction of a *dicarbonyl* compound. Show the structure of the dicarbonyl reactant.

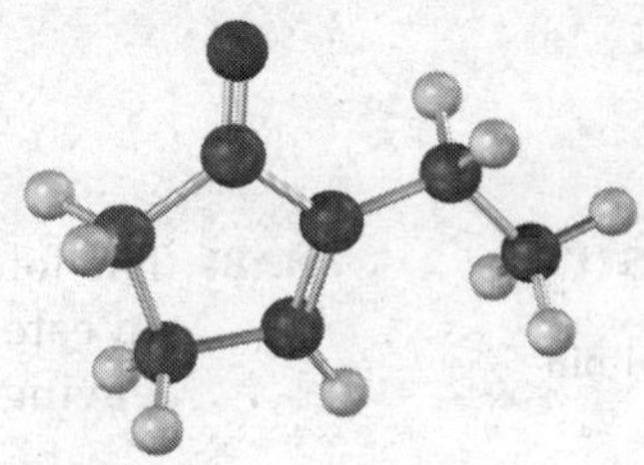

Additional Problems

Acidity and Tautomerism

11.22 Draw structures for the monoenol tautomers of cyclohexane-1,3-dione. How many enol forms are possible, and which would you expect to be most stable? Explain.

11.23 Rank the following compounds in order of increasing acidity:

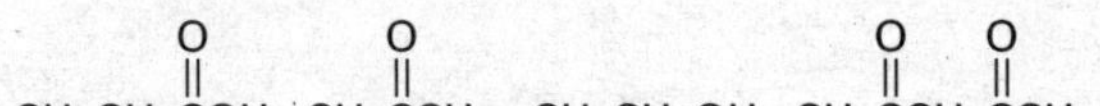

11.24 Why do you suppose pentane-2,4-dione is 76% enolized at equilibrium although acetone is enolized only to the extent of about 0.0001%?

11.25 Indicate all acidic hydrogen atoms in the following molecules:

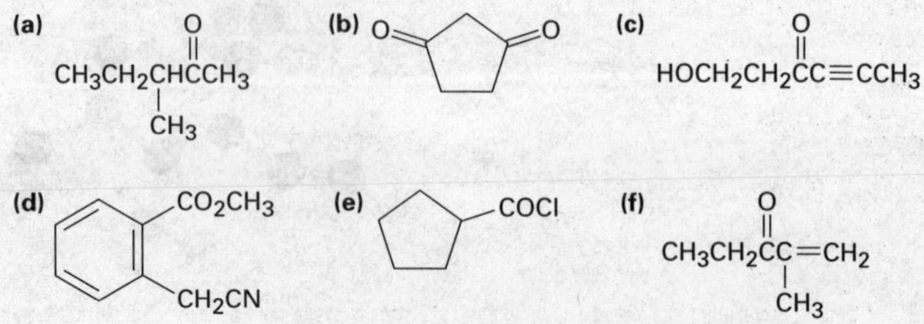

11.26 Why is an enolate ion generally more reactive than a neutral enol?

11.27 Write resonance structures for the following anions:

(a) $CH_3\overset{\overset{O}{\|}}{C}\ddot{\bar{C}}H\overset{\overset{O}{\|}}{C}CH_3$ **(b)** $:\bar{C}H_2C{\equiv}N$ **(c)** $CH_3CH{=}CH\ddot{\bar{C}}H\overset{\overset{O}{\|}}{C}CH_3$

(d) $N{\equiv}C\ddot{\bar{C}}HCO_2C_2H_5$

MECHANISMS **11.28** How can you account for the fact that *cis*- and *trans*-4-*tert*-butyl-2-methylcyclohexanone are interconverted by base treatment? Which of the two isomers is more stable, and why? (See Section 2.10.)

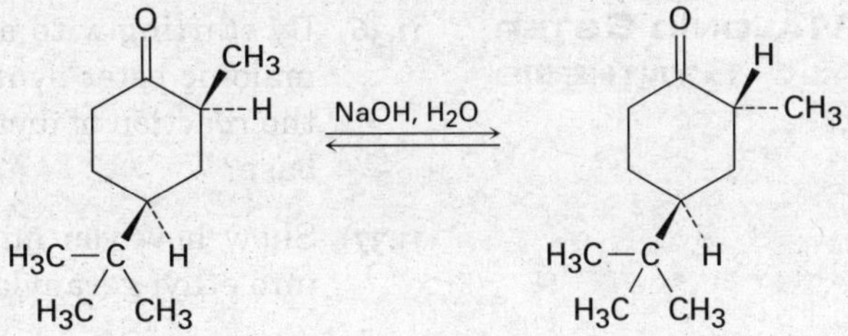

11.29 How do the mechanisms of base-catalyzed enolization and acid-catalyzed enolization differ?

11.30 Nonconjugated β,γ-unsaturated ketones such as cyclohex-3-enone are in an acid-catalyzed equilibrium with their conjugated α,β-unsaturated isomers. Propose a mechanism for the acid-catalyzed interconversion.

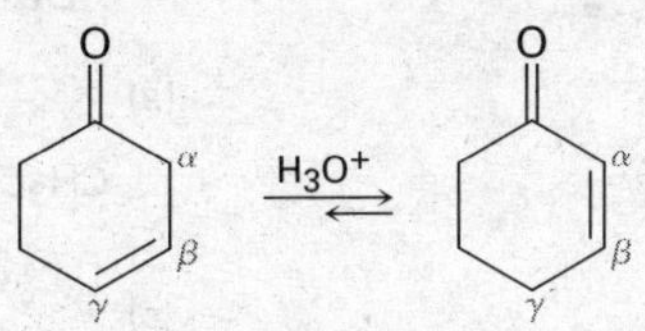

11.31 The α,β to β,γ interconversion of unsaturated ketones (see Problem 11.30) is catalyzed by base as well as by acid. Propose a mechanism.

11.32 One consequence of the base-catalyzed α,β to β,γ isomerization of unsaturated ketones (see Problem 11.31) is that C5-substituted cyclopent-2-enones can be interconverted with C2-substituted cyclopent-2-enones. Propose a mechanism for this isomerization.

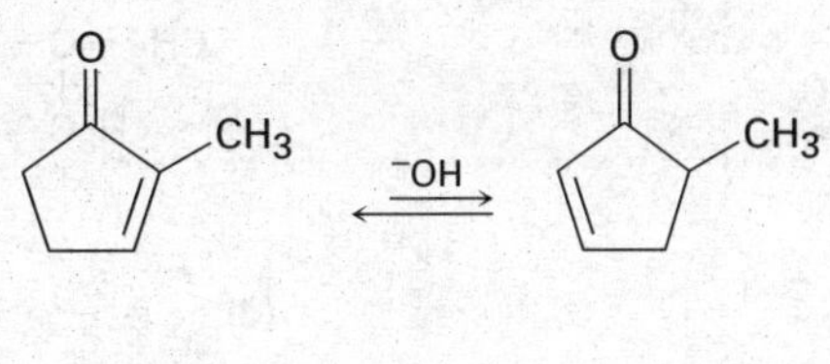

11.33 When optically active (R)-2-methylcyclohexanone is treated with aqueous HCl or NaOH, racemic 2-methylcyclohexanone is produced. Explain.

11.34 When optically active (R)-3-methylcyclohexanone is treated with aqueous HCl or NaOH, no racemization occurs. Instead, the optically active ketone is recovered unchanged. How can you reconcile this observation with your answer to Problem 11.33?

11.35 When acetone is treated with acid in deuterated water, D_2O, deuterium becomes incorporated into the molecule. Propose a mechanism.

$$\mathrm{CH_3\overset{\overset{O}{\|}}{C}CH_3 + D_2O \overset{DCl}{\rightleftharpoons} CH_3\overset{\overset{O}{\|}}{C}CH_2D}$$

Malonic Ester Synthesis

11.36 By starting with a *dihalide*, cyclic compounds can be prepared using the malonic ester synthesis. What product would you expect to obtain from the reaction of diethyl malonate, 1,4-dibromobutane, and 2 equivalents of base?

11.37 Show how you might convert geraniol, the chief constituent of rose oil, into ethyl geranylacetate.

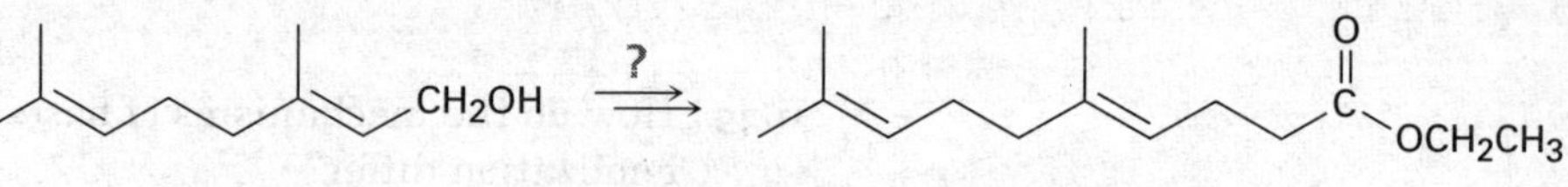

Geraniol **Ethyl geranylacetate**

11.38 Which of the following esters can be prepared by a malonic ester synthesis? Show what reagents you would use.

(a) $\mathrm{CH_3CH_2CH_2CH_2\overset{\overset{O}{\|}}{C}OCH_2CH_3}$

(b) $\mathrm{CH_3\underset{}{\overset{\overset{CH_3}{|}}{C}}HCH_2\overset{\overset{O}{\|}}{C}OCH_2CH_3}$

(c) $\mathrm{CH_3CH_2\underset{\underset{CH_3}{|}}{C}H\overset{\overset{O}{\|}}{C}OCH_2CH_3}$

(d) $\mathrm{CH_3-\underset{\underset{H_3C}{|}}{\overset{\overset{H_3C}{|}}{C}}-\overset{\overset{O}{\|}}{C}OCH_2CH_3}$

Aldol Condensation

11.39 Which of the following compounds would you expect to undergo aldol condensation? Draw the product in each case.

(a) $\mathrm{CH_3-\underset{\underset{H_3C}{|}}{\overset{\overset{H_3C}{|}}{C}}-\overset{\overset{O}{\|}}{C}H}$

(b) 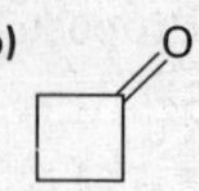

(c) 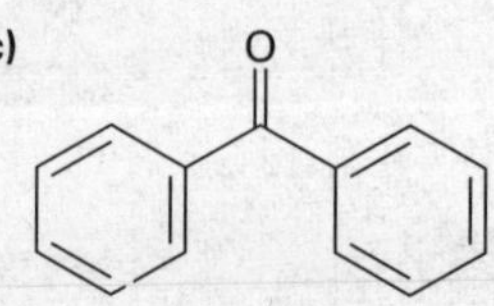

(d) $\mathrm{CH_3(CH_2)_8\overset{\overset{O}{\|}}{C}H}$

(e) 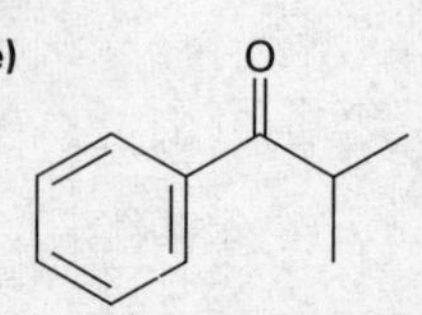

(f) $\mathrm{H\overset{\overset{O}{\|}}{C}H}$

11.40 The aldol condensation reaction can be carried out intramolecularly by treatment of a diketone with base. What diketone would you start with to prepare 3-methylcyclohex-2-enone? Show the reaction.

11.41 What product would you expect to obtain from intramolecular aldol condensation of hexanedial, $OHCCH_2CH_2CH_2CH_2CHO$? (See Problem 11.40.)

11.42 How might you prepare the following compounds using an aldol condensation reaction?

(a)

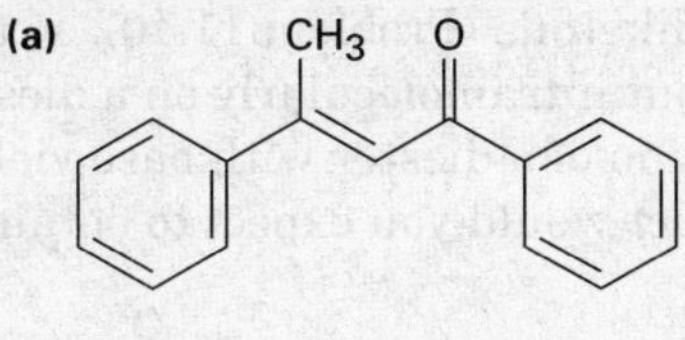

(b) $CH_3C(CH_3){=}CHC(O)CH_3$

(c) $CH_3CH_2CH_2CH{=}C(CH_2CH_3)CHO$

(d)

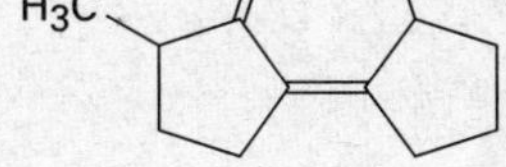

CLAISEN CONDENSATION

11.43 If a mixture of ethyl acetate and ethyl benzoate is treated with base, a mixture of two Claisen condensation products is obtained. Show their structures, and explain.

11.44 The Claisen condensation is reversible. That is, a β-keto ester can be cleaved by base into two fragments. Show the mechanism by which the following cleavage occurs:

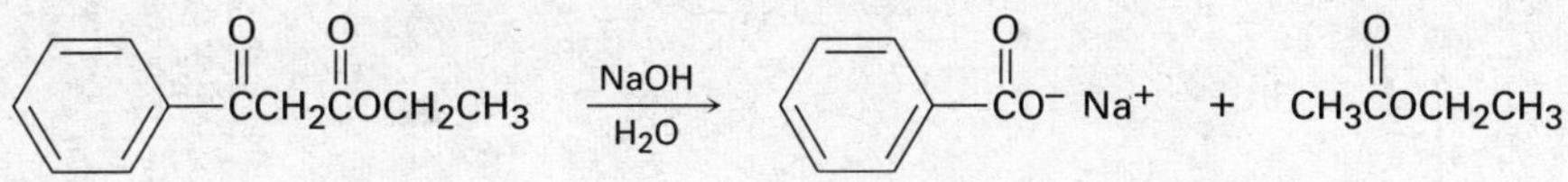

11.45 If a 1 : 1 mixture of ethyl acetate and ethyl propanoate is treated with base under Claisen condensation conditions, a mixture of four β-keto ester products is obtained. Show their structures, and explain.

GENERAL PROBLEMS

11.46 The *acetoacetic ester synthesis* is closely related to the malonic ester synthesis, but involves alkylation with the anion of ethyl açetoacetate rather than diethyl malonate. Treatment of the ethyl acetoacetate anion with an alkyl halide, followed by decarboxylation, yields a ketone product:

$$CH_3\overset{O}{\overset{\|}{C}}CH_2\overset{O}{\overset{\|}{C}}OCH_2CH_3 \xrightarrow[\text{2. RX} \quad \text{3. } H_3O^+]{\text{1. } Na^+ \ ^-OCH_2CH_3} CH_3\overset{O}{\overset{\|}{C}}CH_2{-}R + CO_2 + HOCH_2CH_3$$

How would you prepare the following compounds using an acetoacetic ester synthesis?

(a)

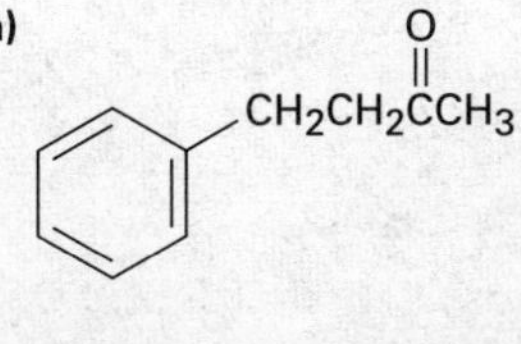

(b) $CH_3CH(CH_3)CH_2CH_2C(O)CH_3$

(c) $CH_3CH_2CH_2CH(CH_3)C(O)CH_3$

11.47 Which of the following compounds can't be prepared by an acetoacetic ester synthesis (see Problem 11.46)? Explain.

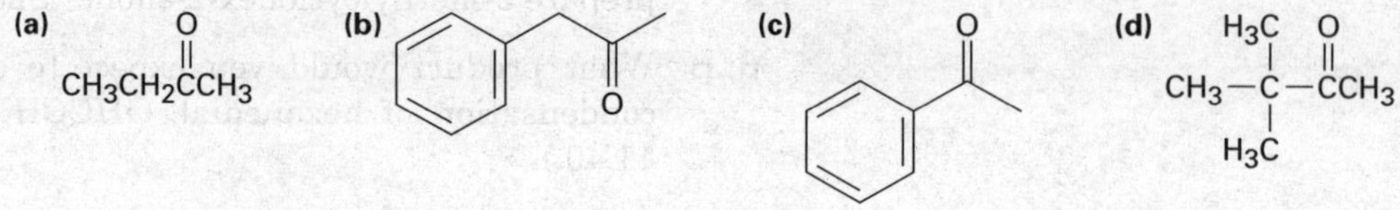

11.48 Just as the aldol condensation can be carried out intramolecularly on a diketone (Problem 11.40), so too can the Claisen condensation be carried out intramolecularly on a diester. Called the *Dieckmann cyclization,* reaction of a diester with base yields a cyclic β-keto ester product. What product would you expect to obtain in the following reaction?

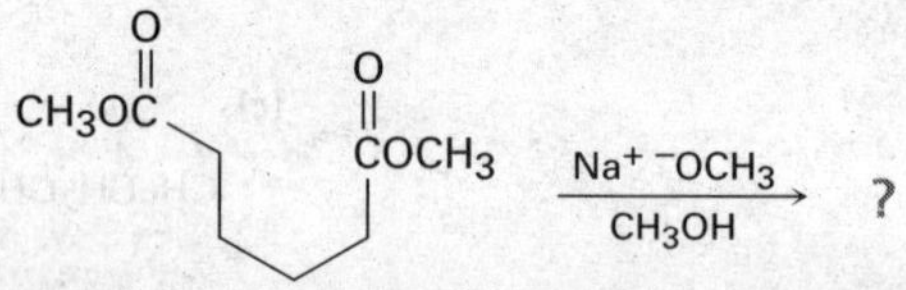

11.49 The cyclic β-keto ester formed in a Dieckmann cyclization reaction (Problem 11.48) can be converted by treatment with base into an anion and alkylated in a process much like that of the acetoacetic ester synthesis (Problem 11.46). Show the product of the following reaction:

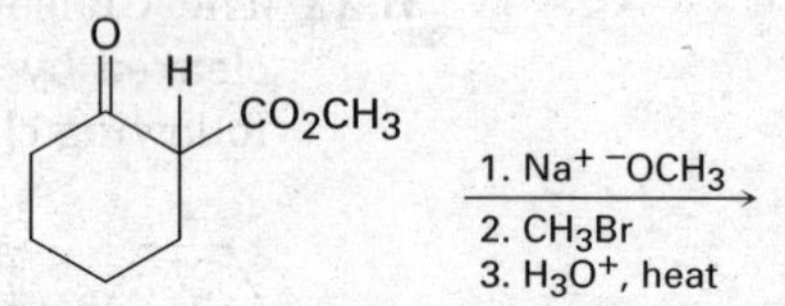

11.50 Butan-1-ol is synthesized commercially from acetaldehyde by a three-step route that involves an aldol reaction followed by two reductions. Show how you might carry out this transformation.

11.51 Cinnamaldehyde, the aromatic constituent of cinnamon oil, can be synthesized by a mixed aldol-like reaction between benzaldehyde and acetaldehyde. Formulate the reaction. What other product would you expect to obtain?

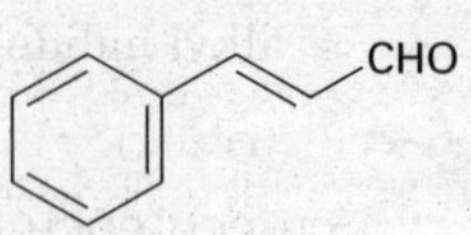

Cinnamaldehyde

11.52 Monoalkylated acetic acids (RCH_2CO_2H) and dialkylated acetic acids (R_2CHCO_2H) can be prepared by malonic ester synthesis, but trialkylated acetic acids (R_3CCO_2H) can't be prepared. Explain.

11.53 Amino acids can be prepared by reaction of alkyl halides with diethyl acetamidomalonate, followed by heating the initial alkylation product with aqueous HCl. Show how you would prepare alanine, $CH_3CH(NH_2)CO_2H$, one of the 20 amino acids found in proteins.

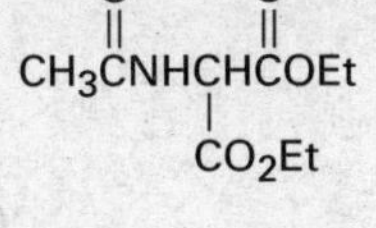

Diethyl acetamidomalonate

11.54 Leucine, one of the 20 amino acids found in proteins, is metabolized by a pathway that includes the following base-catalyzed step. Propose a mechanism.

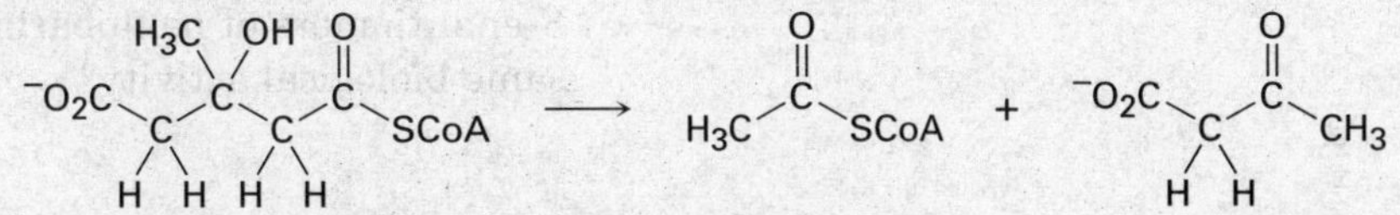

3-Hydroxy-3-methylglutaryl CoA **Acetyl CoA** **Acetoacetate**

11.55 Treatment of an α,β-unsaturated carbonyl compound with base yields an anion by removal of H^+ from the γ carbon. Why are hydrogens on the γ carbon atom acidic?

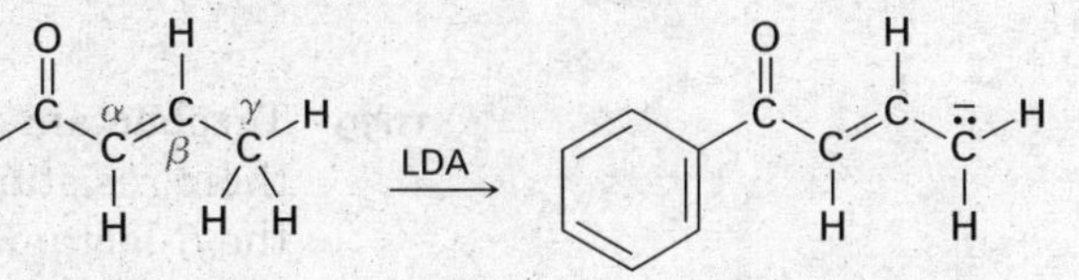

11.56 Using curved arrows, propose a mechanism for the following decarboxylation reaction, one of the steps in the biosynthesis of the amino acid tyrosine.

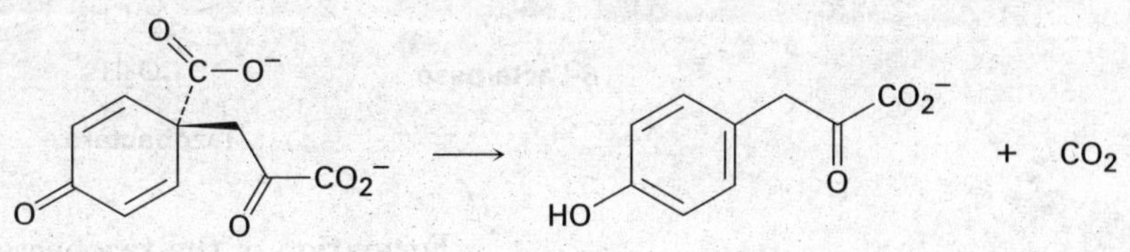

11.57 The following reaction occurs in two steps: (1) An intramolecular aldol reaction to yield a cyclic β-hydroxy ketone and (2) a *retro* aldol-like reaction. Write both steps and show their mechanisms.

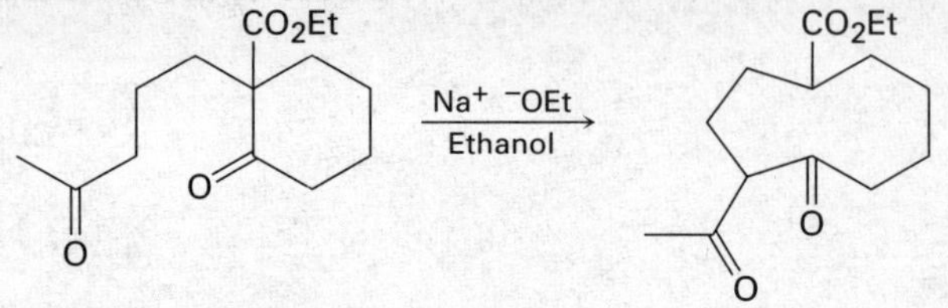

IN THE MEDICINE CABINET **11.58** Pentobarbital, first marketed as Nembutal and still used as an anticonvulsant and sedative, is administered as a racemate. Draw the *R*- and *S*-enantiomers of pentobarbital. Are both enantiomers likely to have the same biological activity?

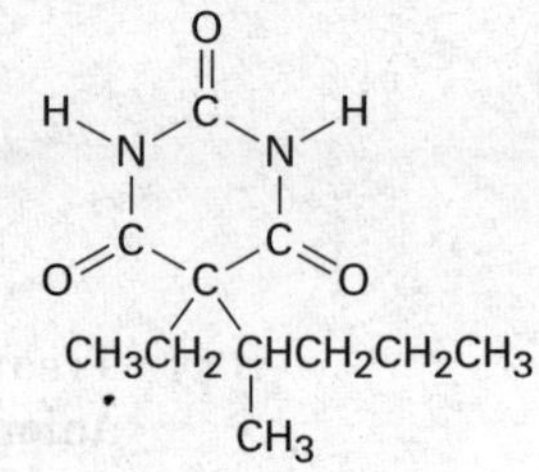

Pentobarbital

11.59 Tazobactam inhibits the β-lactamase enzyme that gives many bacteria their resistance to penicillins. It functions by reacting irreversibly with the β-lactamase, thereby trapping and deactivating it.

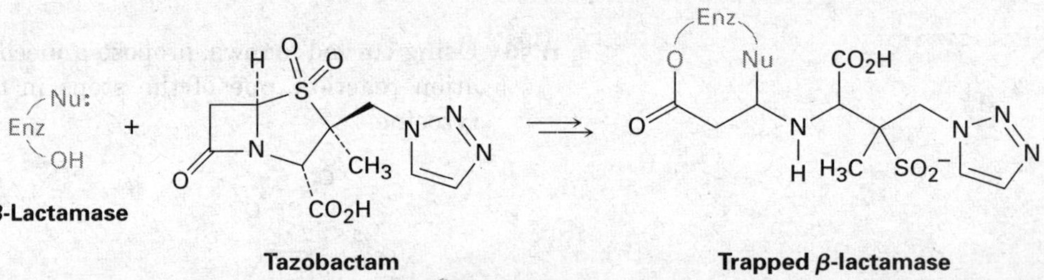

Formation of the tazobactam-trapped β-lactamase proceeds through the following steps. Show a mechanism for each.

(a) Nucleophilic acyl substitution by a β-lactamase –OH group with tazobactam opens the four-membered cyclic amide (β-lactam) ring.

(b) Formation of an imine (C=N) occurs with expulsion of $-SO_2^-$ as leaving group.

(c) Rearrangement of the imine gives an α,β-unsaturated ester.

(d) Intramolecular conjugate addition by the Nu: group of β-lactamase to the α,β-unsaturated ester gives the final product.

IN THE FIELD **11.60** Carboxin, marketed as Vitavax, is a fungicide used on corn and wheat. A synthesis of carboxin is shown:

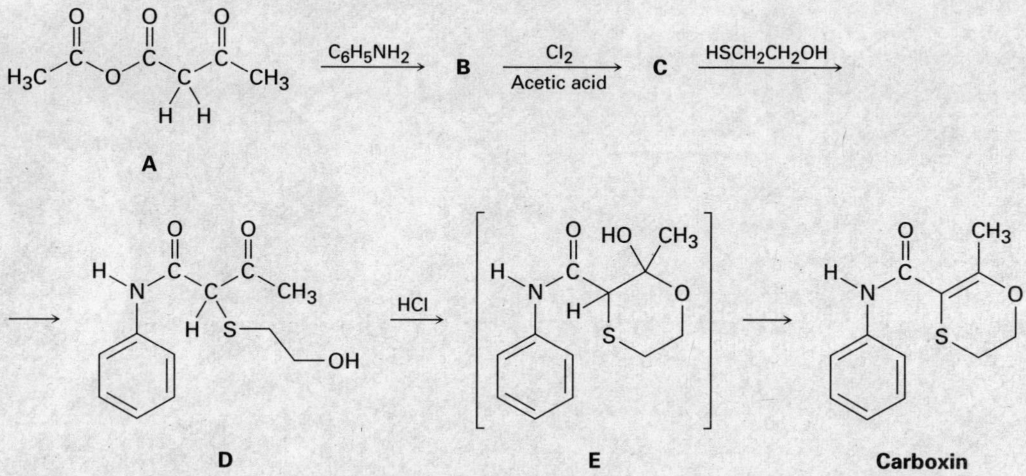

(a) Reaction of anhydride **A** with aniline ($C_6H_5NH_2$) gives an amide, **B**. Show the structure of **B** and the mechanism for its formation.

(b) Acid-catalyzed chlorination of **B** gives **C**. Show the structure of **C** and the mechanism for its formation.

(c) What kind of mechanism is involved in the reaction of **C** with $HSCH_2CH_2OH$ to give **D**?

(d) Formation of carboxin from **D** occurs through intermediate **E**. What new functional group appears in this intermediate? Show a mechanism for formation of **E**.

(e) The conversion of **E** to carboxin occurs through an enol intermediate. Show a mechanism for this reaction.

CHAPTER 12

The characteristic and unmistakable odor of rotting fish is due to a mixture of simple alkylamines.

Amines

Amines are organic derivatives of ammonia in the same way that alcohols and ethers are organic derivatives of water. Like ammonia, amines contain a nitrogen atom with a lone pair of electrons, making amines both basic and nucleophilic.

Amines occur widely in both plants and animals. Trimethylamine, for instance, occurs in animal tissues and is partially responsible for the distinctive odor of fish; nicotine is found in tobacco; and cocaine is a stimulant found in the South American coca bush. In addition, amino acids are the building blocks from which all proteins are made, and cyclic amine bases are constituents of nucleic acids.

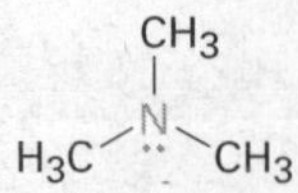

Trimethylamine

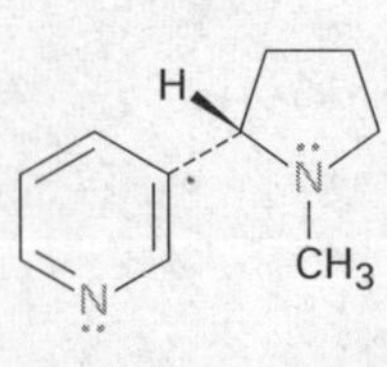

Nicotine

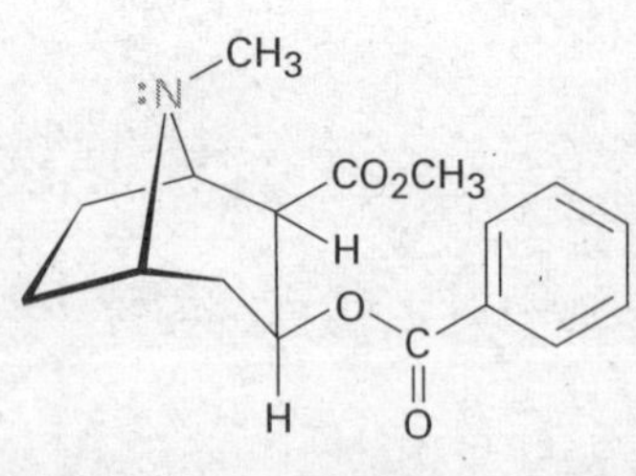

Cocaine

WHY THIS CHAPTER?

OWL

Online homework for this chapter can be assigned in OWL, an online homework assessment tool.

By the end of this chapter, we will have seen all the common functional groups. Of those groups, amines and carbonyl compounds are the most abundant and have the richest chemistry. In addition to the proteins and nucleic acids already mentioned, the majority of pharmaceutical agents contain amine functional groups, and many of the common coenzymes necessary for biological catalysis are amines.

12.1 Naming Amines

Amines can be either alkyl-substituted (**alkylamines**) or aryl-substituted (**arylamines**) and can be classed as *primary* (RNH_2), *secondary* (R_2NH), or *tertiary* (R_3N), depending on the number of organic substituents attached to nitrogen. For example, methylamine (CH_3NH_2) is a primary alkylamine and trimethylamine [$(CH_3)_3N$] is a tertiary alkylamine. Note that this usage of the terms primary, secondary, and tertiary is different from our previous usage. When we speak of a tertiary alcohol or alkyl halide, we refer to the degree of substitution at the alkyl *carbon* atom, but when we speak of a tertiary amine, we refer to the degree of substitution at the *nitrogen* atom.

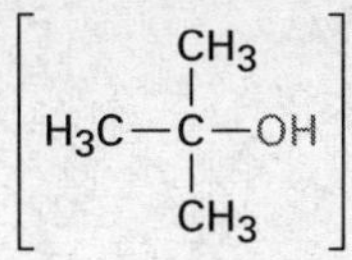

***tert*-Butyl alcohol** **(a tertiary alcohol)**

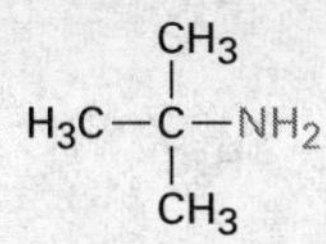

***tert*-Butylamine** **(a primary alkylamine)**

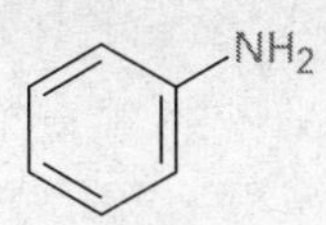

Aniline **(a primary arylamine)**

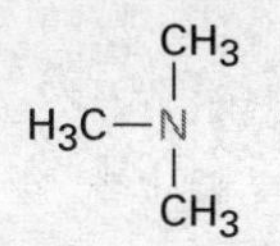

Trimethylamine **(a tertiary alkylamine)**

Compounds containing a nitrogen atom with four attached groups also exist, but the nitrogen atom must carry a formal positive charge. Such compounds are called **quaternary ammonium salts**.

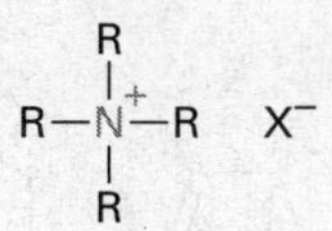

A quaternary ammonium salt

Primary amines, RNH_2, are named in the IUPAC system by adding the suffix *-amine* to the name of the organic substituent.

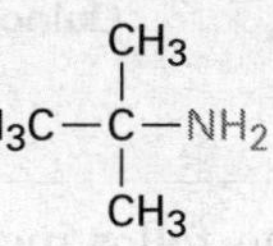

***tert*-Butylamine**

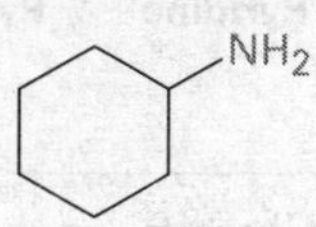

Cyclohexylamine

$H_2NCH_2CH_2CH_2CH_2NH_2$

Butane-1,4-diamine

Amines with more than one functional group are named by considering the $-NH_2$ as an *amino* substituent on the parent molecule.

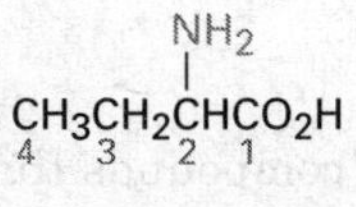

2-Aminobutanoic acid

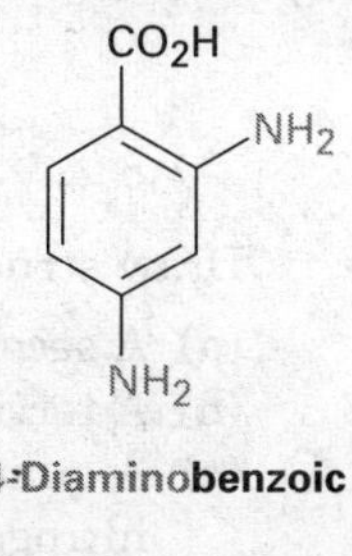

2,4-Diaminobenzoic acid

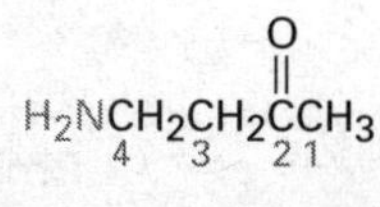

4-Aminobutan-2-one

Symmetrical secondary and tertiary amines are named by adding the prefix *di-* or *tri-* to the alkyl group.

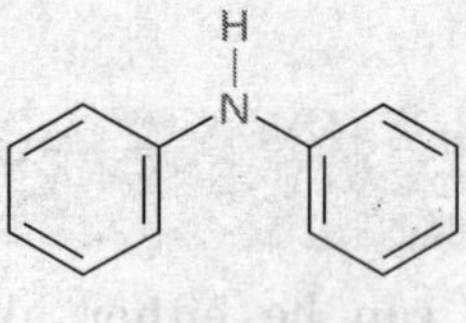

Diphenylamine

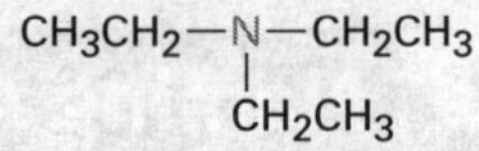

Triethylamine

Unsymmetrically substituted secondary and tertiary amines are named as *N*-substituted primary amines. The largest organic group is chosen as the parent, and the other groups are considered as *N*-substituents on the parent (*N* because they're attached to nitrogen).

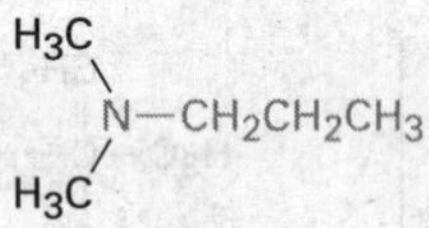

***N,N*-Dimethylpropylamine**

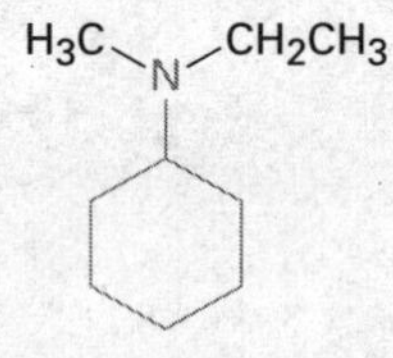

***N*-Ethyl-*N*-methylcyclohexylamine**

Heterocyclic amines—compounds in which the nitrogen atom occurs as part of a ring—are also common, and each different heterocyclic ring system has its own parent name. The heterocyclic nitrogen atom is always numbered as position 1.

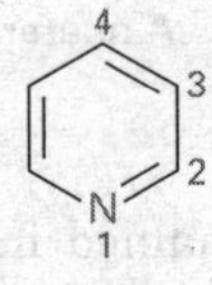

Pyridine

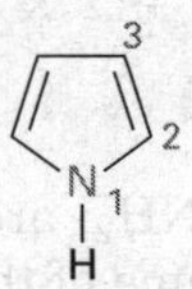

Pyrrole

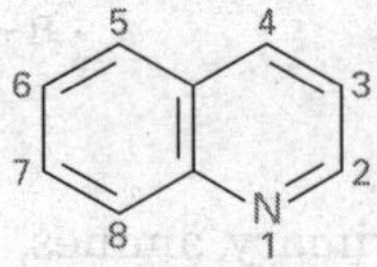

Quinoline

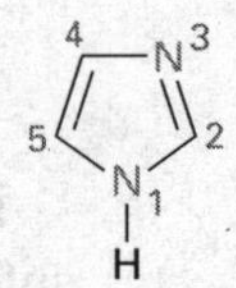

Imidazole

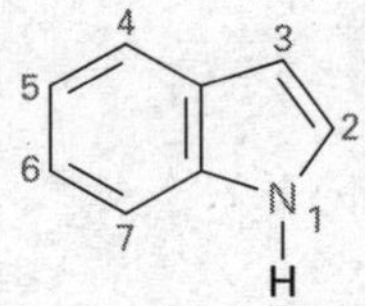

Indole

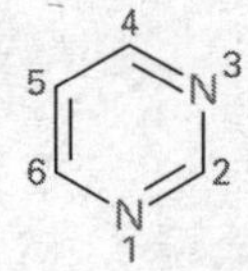

Pyrimidine

Problem 12.1 Classify each of the following compounds as either a primary, secondary, or tertiary amine:

(a)

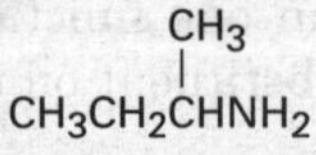

(b)

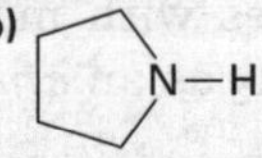

(c)

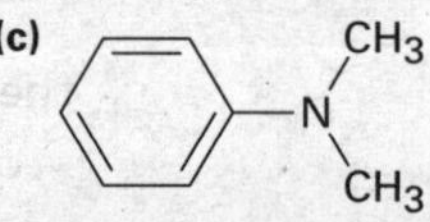

Problem 12.2 Draw structures of compounds that meet the following descriptions:

(a) A secondary amine with one isopropyl group

(b) A tertiary amine with one phenyl group and one ethyl group

(c) A quaternary ammonium salt with four different groups bonded to nitrogen

Problem 12.3 Name the following compounds:

(a) 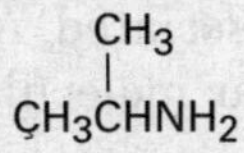

(b) $CH_3CH_2NHCH_2CH_3$

(c) 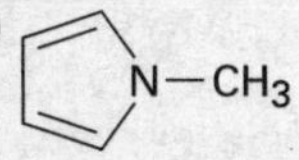

(d) 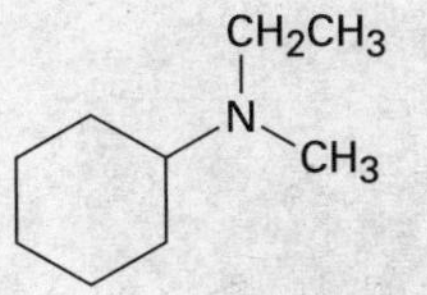

(e) 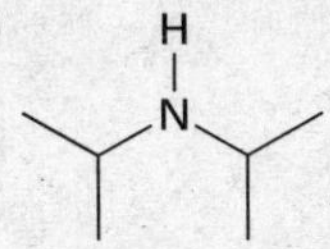

(f) 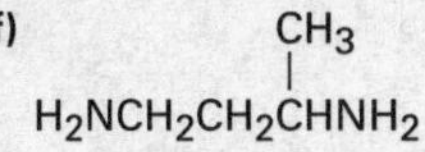

Problem 12.4 Draw structures corresponding to the following IUPAC names:

(a) Triethylamine **(b)** *N*-Methylaniline
(c) Tetraethylammonium bromide **(d)** *p*-Bromoaniline
(e) *N*-Ethyl-*N*-methylcyclopentylamine

12.2 Structure and Properties of Amines

The bonding in alkylamines is similar to the bonding in ammonia. The nitrogen atom is sp^3-hybridized, with the three substituents occupying three corners of a regular tetrahedron and the lone pair of electrons occupying the fourth corner. As you might expect, the C–N–C bond angles are very close to the 109° tetrahedral value—108° in trimethylamine, for example.

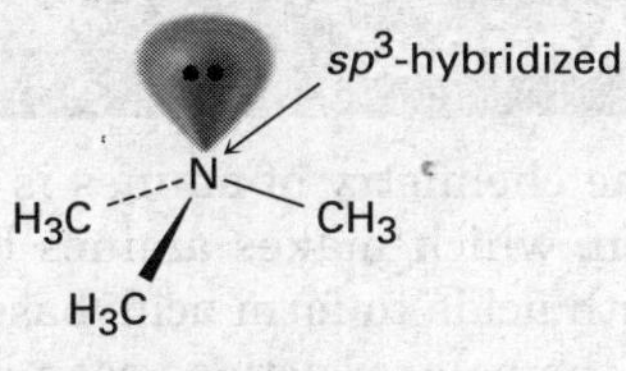

Trimethylamine

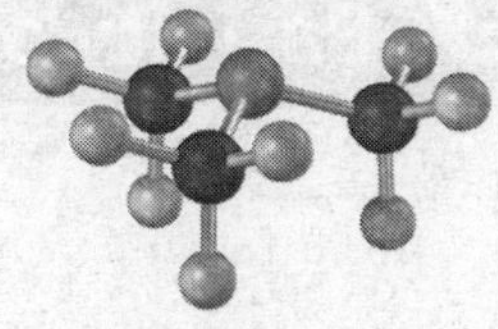

Alkylamines have a variety of applications in the chemical industry as starting materials for the preparation of insecticides and pharmaceuticals. Labetalol, for instance, a so-called beta-blocker used for the treatment of high blood pressure, is prepared by S_N2 reaction of an epoxide with a primary amine. The substance marketed for drug use is a mixture of all four possible stereoisomers, but the biological activity derives primarily from the (*R,R*) isomer.

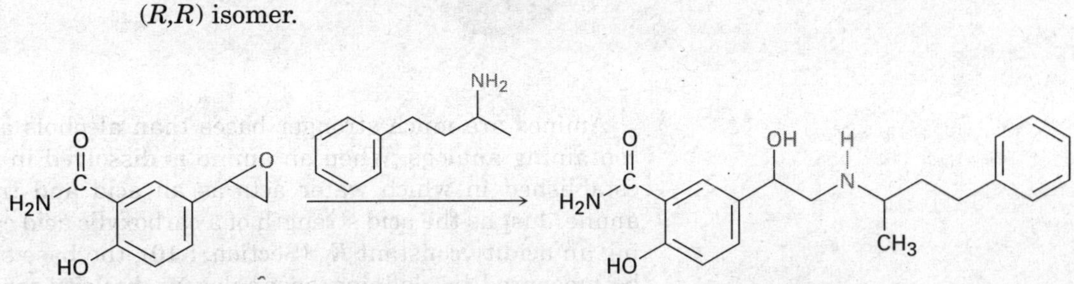

Labetalol

Like alcohols, amines with fewer than five carbon atoms are generally water-soluble. Also like alcohols, primary and secondary amines form hydrogen bonds and are highly associated. As a result, amines have higher boiling points than alkanes of similar molecular weight.

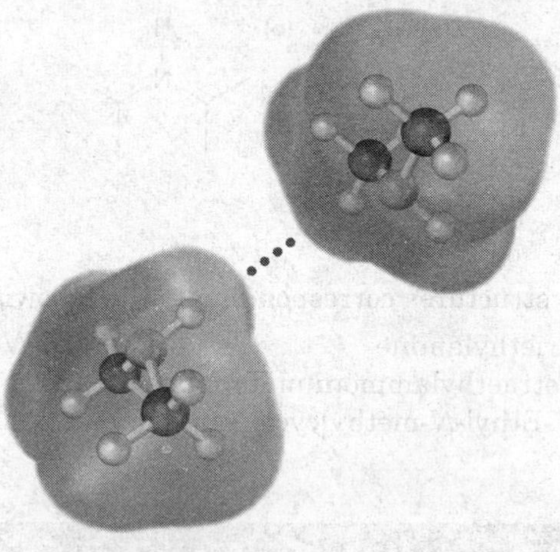

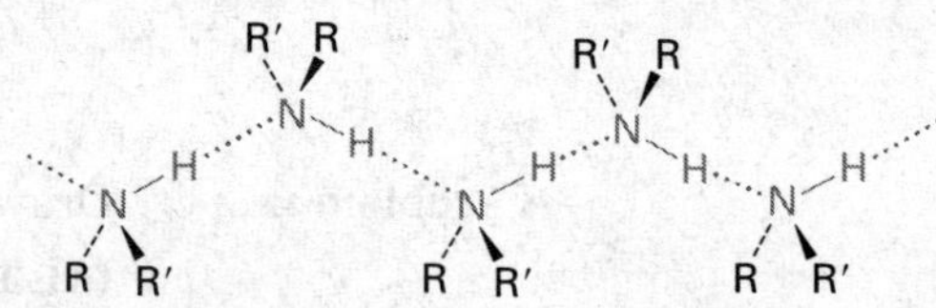

One other characteristic property of amines is their odor. Low-molecular-weight amines such as trimethylamine have a distinctive fish-like aroma, while diamines such as putrescine (butane-1,4-diamine) have odors as putrid as their common names suggest.

12.3 Basicity of Amines

The chemistry of amines is dominated by the lone pair of electrons on nitrogen, which makes amines both basic and nucleophilic. They therefore react with acids to form acid–base salts, and they react with electrophiles in many of the polar reactions seen in past chapters.

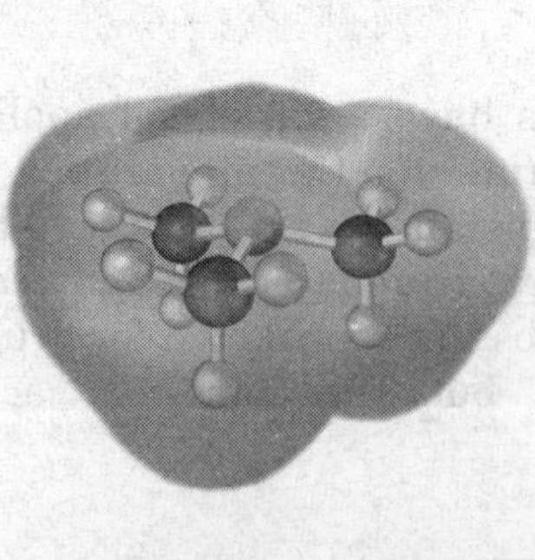

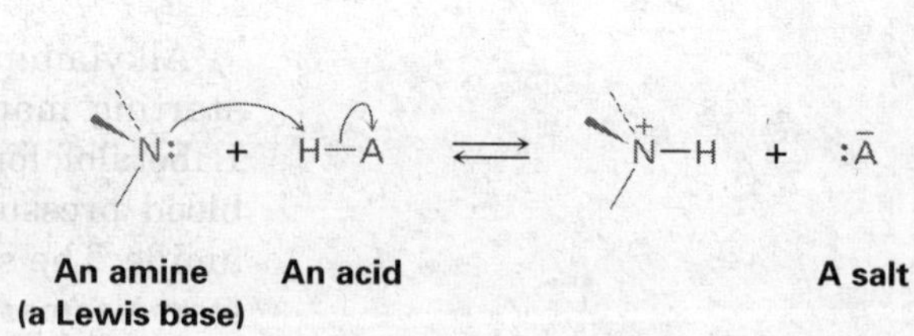

Amines are much stronger bases than alcohols and ethers, their oxygen-containing analogs. When an amine is dissolved in water, an equilibrium is established in which water acts as an acid and transfers a proton to the amine. Just as the acid strength of a carboxylic acid can be measured by defining an acidity constant K_a (Section 1.10), the base strength of an amine can be measured by defining an analogous **basicity constant K_b**. The larger the

K_b and the smaller the pK_b, the more favorable the proton-transfer equilibrium and the stronger the base.

For the reaction

$$RNH_2 + H_2O \rightleftharpoons RNH_3^+ + OH^-$$

$$K_b = \frac{[RNH_3^+]\,[OH^-]}{[RNH_2]}$$

$$pK_b = -\log K_b$$

Table 12.1 gives the pK_b values of some common amines. As indicated, substitution has relatively little effect on alkylamine basicity; most simple alkylamines have pK_b's in the narrow range 3 to 4. Arylamines, however, are weaker bases than alkylamines by a factor of about 10^6, as is the heterocyclic amine pyridine.

Table 12.1 Basicity of Some Common Amines

Name	Structure	pK_b
Ammonia	NH_3	4.74
Primary alkylamine		
Methylamine	CH_3NH_2	3.36
Ethylamine	$CH_3CH_2NH_2$	3.25
Secondary alkylamine		
Diethylamine	$(CH_3CH_2)_2NH$	3.02
Tertiary alkylamine		
Triethylamine	$(CH_3CH_2)_3N$	3.24
Arylamine		
Aniline	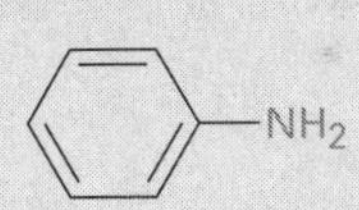	9.37
Heterocyclic amine		
Pyridine	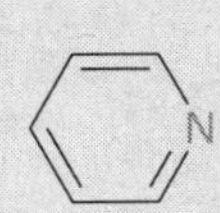	8.75

The decreased basicity of arylamines relative to alkylamines is due to the fact that the nitrogen lone-pair electrons in an arylamine are shared by orbital overlap with the *p* orbitals of the aromatic ring through five resonance forms and are less available for bonding to an acid.

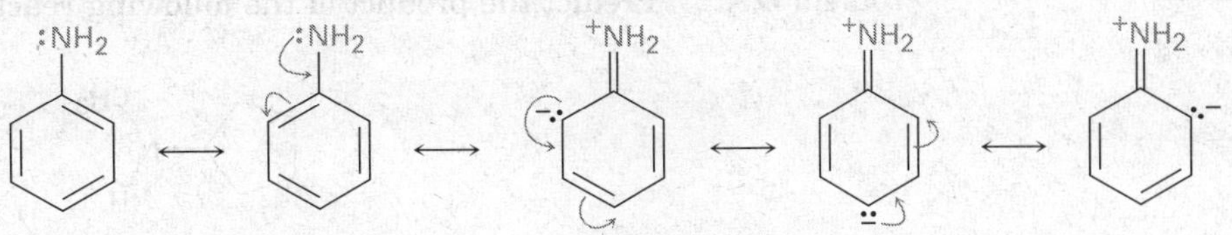

In contrast to amines, *amides* ($RCONH_2$) are nonbasic. Amides aren't protonated by aqueous acids, and they are poor nucleophiles. The main reason for this decreased basicity of amides relative to amines is that the nitrogen lone-pair electrons are shared by orbital overlap with the neighboring carbonyl-group π orbital. In resonance terms, amides are more stable and less reactive than amines because they are hybrids of two resonance forms. This amide resonance stabilization is lost when the nitrogen atom is protonated, however, so protonation is disfavored. Electrostatic potential maps show clearly this decreased electron density on the amide nitrogen.

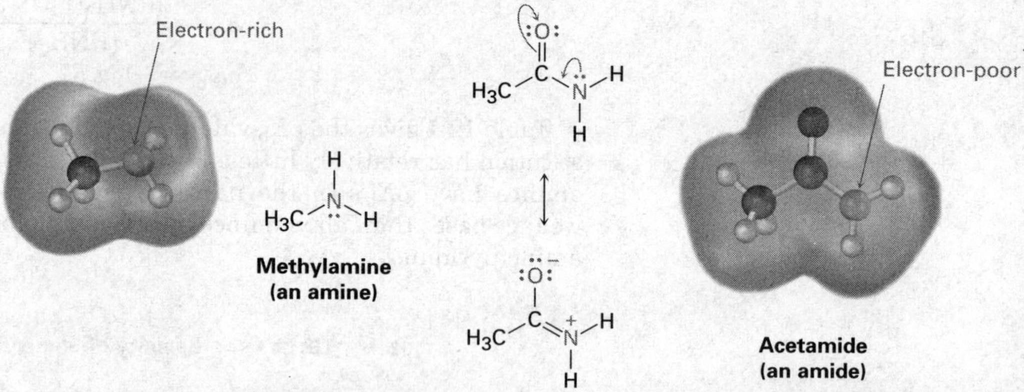

It's often possible to take advantage of its basicity to purify an amine. If a mixture of an amine (basic) and a ketone (neutral) is dissolved in an organic solvent and aqueous HCl is added, the basic amine dissolves in the acidic water as its ammonium ion, while the neutral ketone remains in the organic solvent. Separation of the water layer and neutralization of the ammonium ion by addition of NaOH then provides the pure amine (Figure 12.1).

Figure 12.1 Separation and purification of an amine from a mixture.

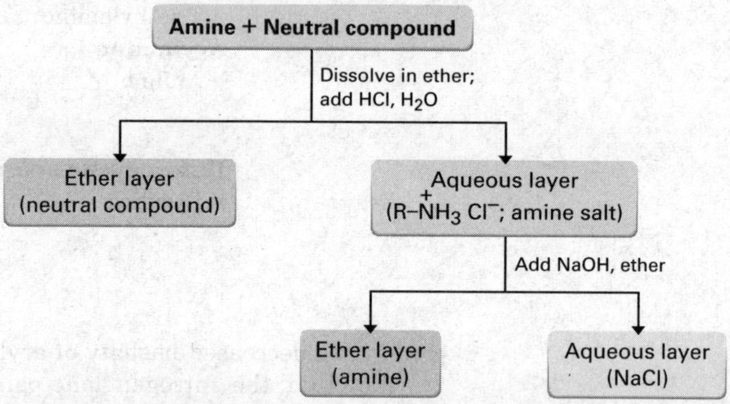

Problem 12.5 Predict the product of the following reaction:

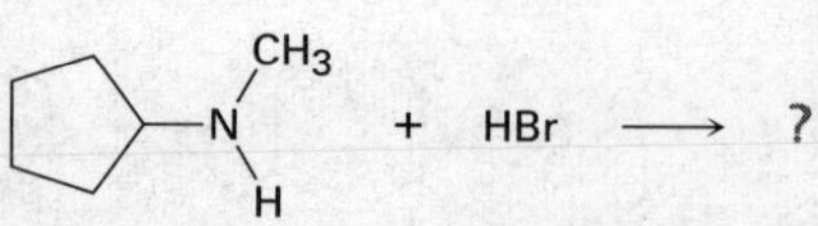

Problem 12.6 Which compound in each of the following pairs is more basic?

(a) $CH_3CH_2NH_2$ or $CH_3CH_2CONH_2$ **(b)** NaOH or $C_6H_5NH_2$
(c) CH_3NHCH_3 or $CH_3NHC_6H_5$ **(d)** CH_3OCH_3 or $(CH_3)_3N$

12.4 Synthesis of Amines

Reduction of Nitriles and Amides

We've already seen how amines can be prepared by reduction of amides (Section 10.10) and nitriles (Section 10.11) with $LiAlH_4$. The two-step sequence of S_N2 reaction of an alkyl halide with cyanide ion, followed by reduction, is a good method for converting an alkyl halide into a primary amine having one more carbon atom than the original halide. Amide reduction provides a method for converting a carboxylic acid into an amine having the same number of carbon atoms.

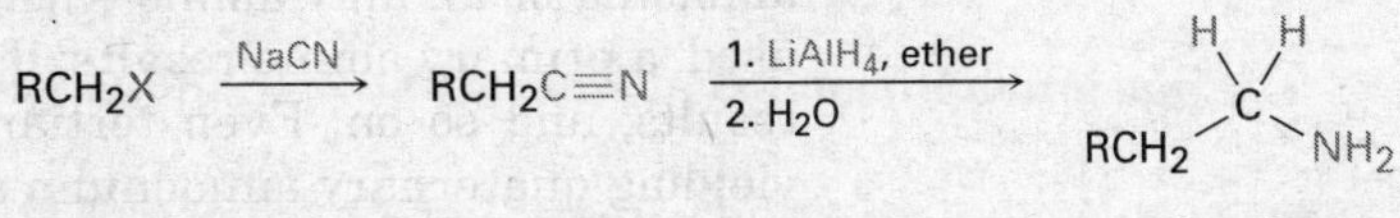

Alkyl halide **1° amine**

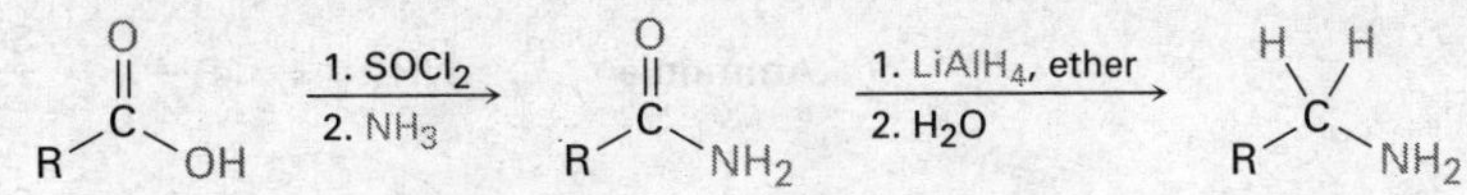

Carboxylic acid **1° amine**

Worked Example 12.1 Synthesizing an Amine from an Amide

What amide would you use to prepare *N*-ethylcyclohexylamine?

Strategy Reduction of an amide with $LiAlH_4$ yields an amine in which the amide carbonyl group has been replaced by a methylene (CH_2), $RCONR_2 \rightarrow RCH_2NR_2$. Since *N*-ethylcyclohexylamine has only one CH_2 attached to nitrogen (in the ethyl group), the product must come from reduction of *N*-cyclohexylacetamide.

Solution

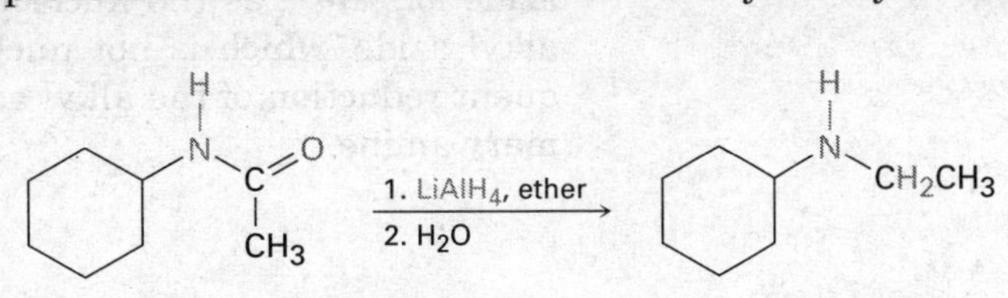

***N*-Cyclohexylacetamide** ***N*-Ethylcyclohexylamine**

Problem 12.7 Propose structures for amides that might be precursors of the following amines:

(a) $CH_3CH_2CH_2NH_2$ **(b)** $NH(CH_2CH_2CH_3)_2$ **(c)**

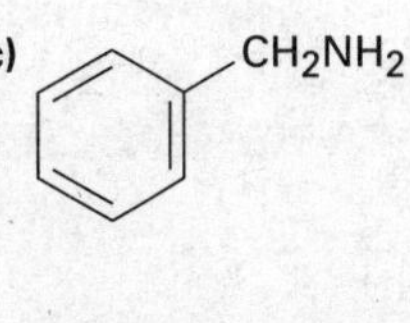

Problem 12.8 Propose structures for nitriles that might be precursors of the following amines:

(a) CH_3 (attached to C-2)
$CH_3CHCH_2CH_2NH_2$

(b) Benzylamine, $C_6H_5CH_2NH_2$

S_N2 Alkylation Reactions of Alkyl Halides

Ammonia and other amines are good nucleophiles in S_N2 reactions. As a result, the simplest method of alkylamine synthesis is by S_N2 alkylation of ammonia or an alkylamine with an alkyl halide (Section 7.5). If ammonia is used, a primary amine results; if a primary amine is used, a secondary amine results; and so on. Even tertiary amines react rapidly with alkyl halides, yielding quaternary ammonium salts, $R_4N^+ X^-$.

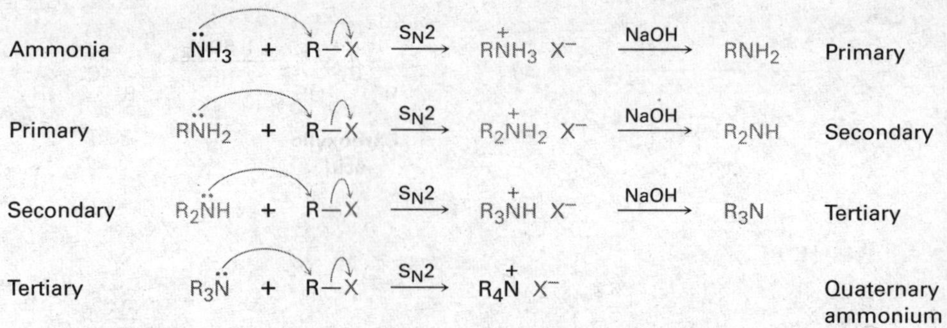

Unfortunately, none of these reactions stops cleanly after a single alkylation has occurred. Because ammonia and primary amines have similar reactivity, the initially formed monoalkylated amine often undergoes further reaction to yield a mixture of mono-, di-, and trialkylated products.

A better method for preparing primary amines from alkyl halides is to use azide ion, N_3^-, as the nucleophile rather than ammonia. The product is an alkyl azide, which is not nucleophilic, so overalkylation can't occur. Subsequent reduction of the alkyl azide with $LiAlH_4$ then leads to the desired primary amine.

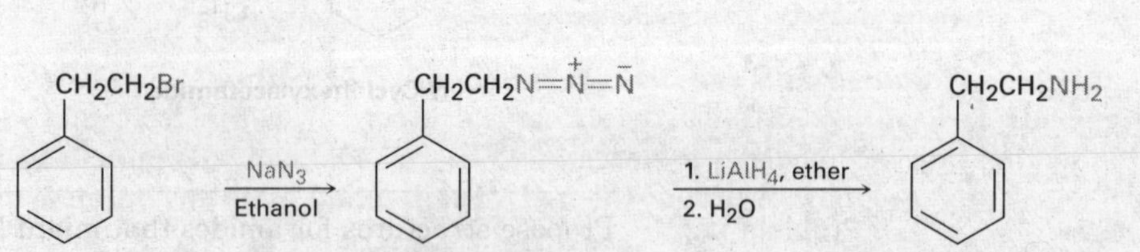

1-Bromo-2-phenylethane **2-Phenylethyl azide** **2-Phenylethylamine (89%)**

Worked Example 12.2 Synthesizing an Amine from an Alkyl Halide

How could you prepare diethylamine from ammonia and an alkyl halide?

Strategy Look at the starting material (NH_3) and the product $(CH_3CH_2)_2NH$, and note the difference. Since two ethyl groups have become bonded to the nitrogen atom, the reaction must involve ammonia and 2 equivalents of an ethyl halide.

Solution

$$2\ CH_3CH_2Br + NH_3 \longrightarrow (CH_3CH_2)_2NH$$

Problem 12.9 How could you prepare the following amines from ammonia and appropriate alkyl halides?

(a) Triethylamine **(b)** Tetramethylammonium bromide

Problem 12.10 What alkyl halide would you use to synthesize dopamine, a neurotransmitter involved in regulation of the central nervous system?

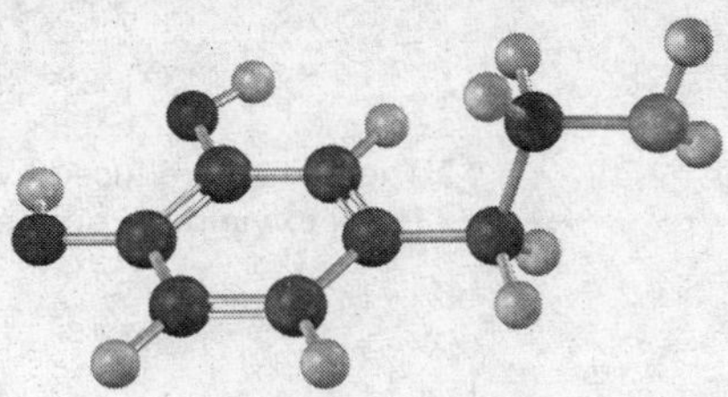

Dopamine

Reductive Amination of Aldehydes and Ketones

Amines can be synthesized from an aldehyde or ketone in a single step by reaction with ammonia or an amine in the presence of a reducing agent, a process called **reductive amination**. For example, amphetamine, a central nervous system stimulant, is prepared commercially by reductive amination of phenylpropan-2-one with ammonia, using hydrogen gas over a nickel catalyst as the reducing agent. In the laboratory, $NaBH_4$ is often used as the reducing agent rather than H_2 and nickel.

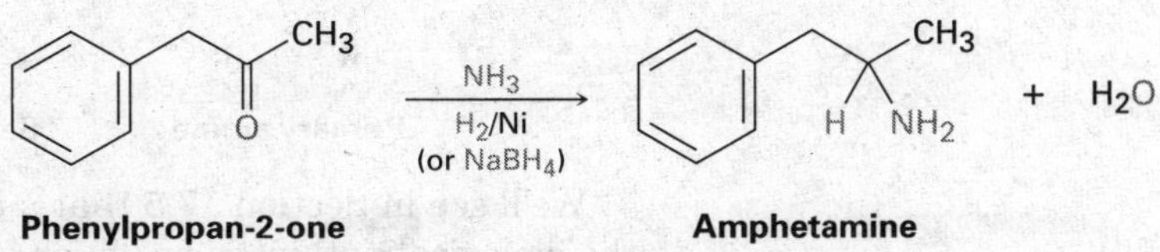

Reductive amination takes place by the pathway shown in Figure 12.2. An imine intermediate is first formed by a nucleophilic addition reaction (Section 9.9), and the C=N bond of the imine is then reduced to the amine, much as the C=O bond of a ketone can be reduced to an alcohol.

MECHANISM

Figure 12.2 Mechanism of reductive amination of a ketone to yield an amine. The imine-forming step was discussed in Section 9.9.

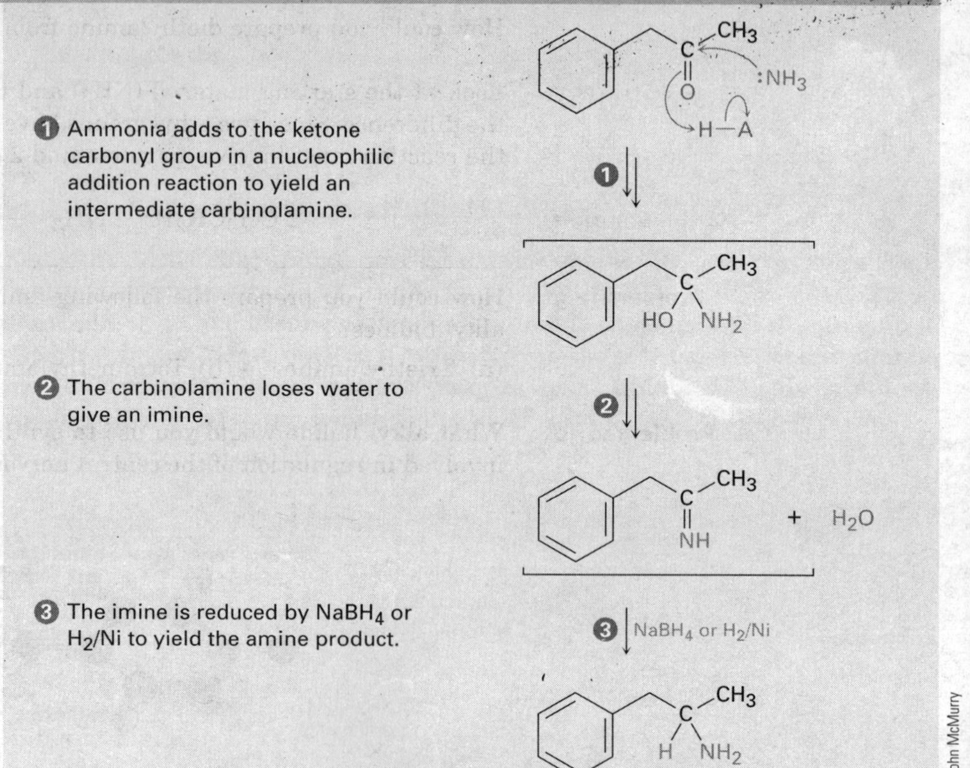

Ammonia, primary amines, and secondary amines can all be used in the reductive amination reaction, yielding primary, secondary, and tertiary amines, respectively.

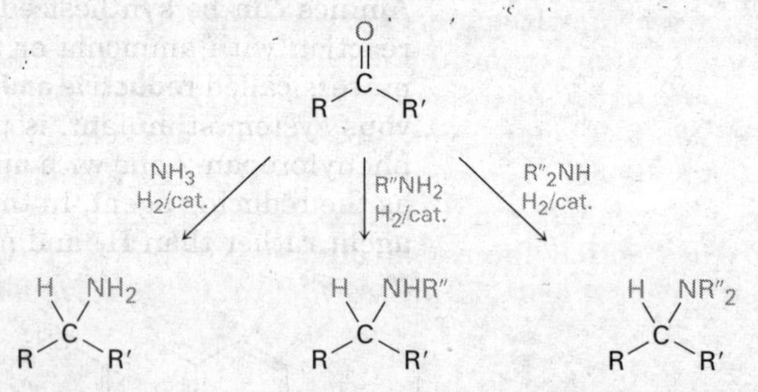

We'll see in Section 17.5 that a closely related reductive amination occurs in the biological pathways by which some amino acids are synthesized. The amino acid alanine, for instance, arises by reaction of pyruvic acid and ammonia to give an intermediate imine that is then enzymatically reduced.

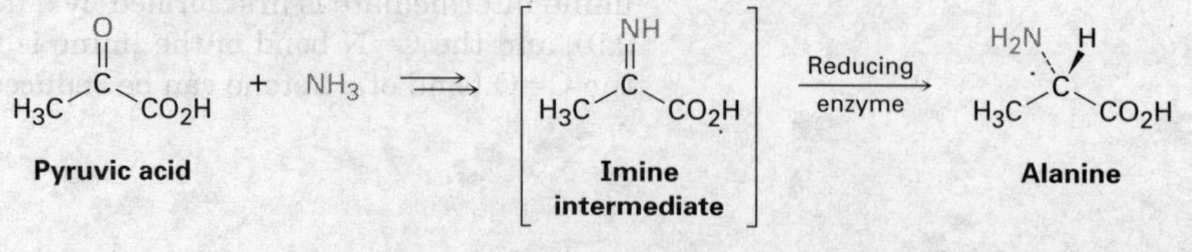

Worked Example 12.3 Synthesizing an Amine Using a Reductive Amination

How might you prepare *N*-methyl-2-phenylethylamine using a reductive amination reaction?

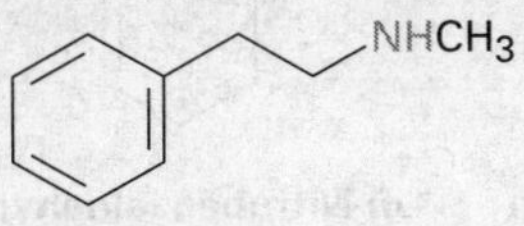

N-Methyl-2-phenylethylamine

Strategy Look at the target molecule, and identify the groups attached to nitrogen. One of the groups must be derived from the aldehyde or ketone component, and the other must be derived from the amine component. In the case of *N*-methyl-2-phenylethylamine, there are two combinations that can lead to the product: phenylacetaldehyde plus methylamine or formaldehyde plus 2-phenylethylamine. In general, it's better to choose the combination with the simpler amine component—methylamine in this case—and to use an excess of that amine as reactant.

Solution

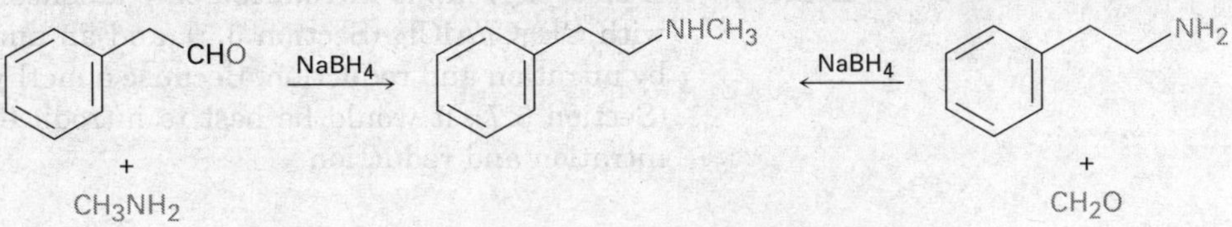

Problem 12.11 How could you prepare the following amines using reductive amination reactions? Show all precursors if more than one is possible.

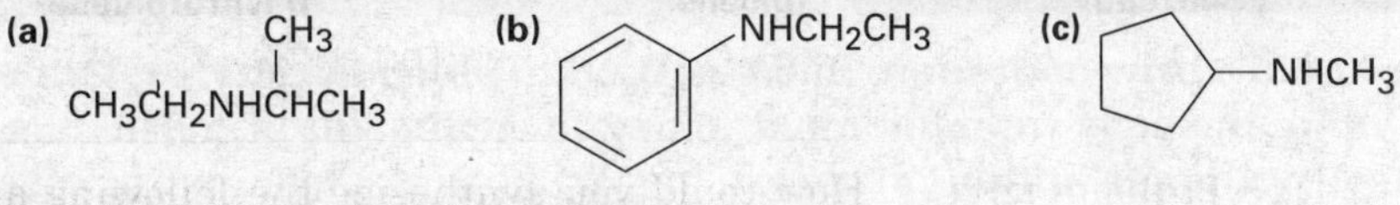

Problem 12.12 How could you prepare the following amine using a reductive amination reaction?

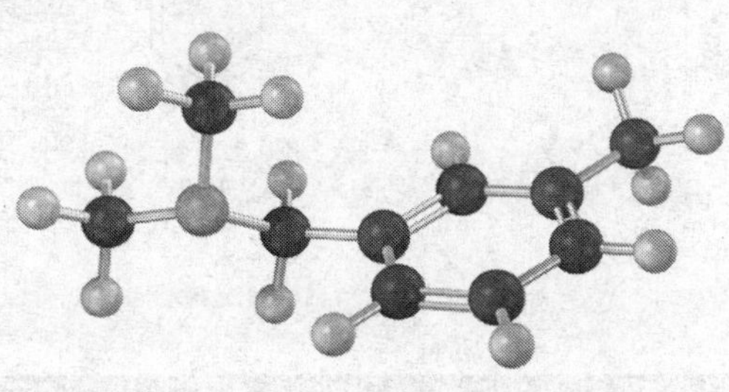

Reduction of Nitrobenzenes

Arylamines are prepared by nitration of an aromatic starting material, followed by reduction of the nitro group (Section 5.4). The reduction step can be carried out in different ways, depending on the circumstances. Catalytic hydrogenation over platinum works well but is sometimes incompatible with the presence elsewhere in the molecule of other reducible groups, such as C=C bonds. Iron, tin, and stannous chloride ($SnCl_2$) in aqueous acid are also effective.

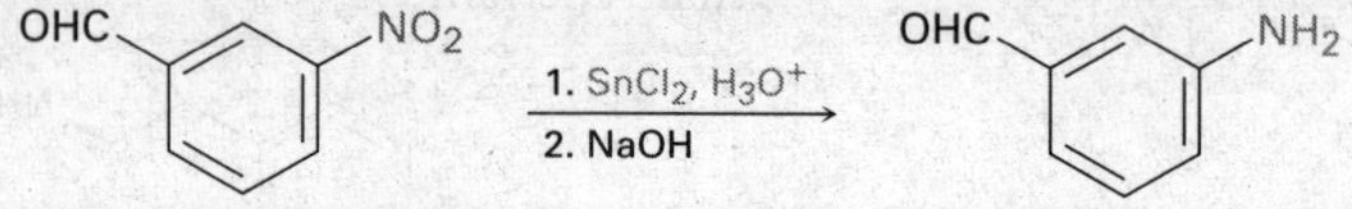

***m*-Nitrobenzaldehyde** **m-Aminobenzaldehyde (90%)**

Worked Example 12.4 Synthesizing an Aromatic Amine

How could you synthesize *p*-methylaniline from benzene? More than one step is needed.

Strategy A methyl group is introduced onto a benzene ring by a Friedel–Crafts reaction with $CH_3Cl/AlCl_3$ (Section 5.5), and an amino group is introduced onto a ring by nitration and reduction. Because a methyl group is ortho- and para-directing (Section 5.7), it would be best to introduce the methyl group first followed by nitration and reduction.

Solution

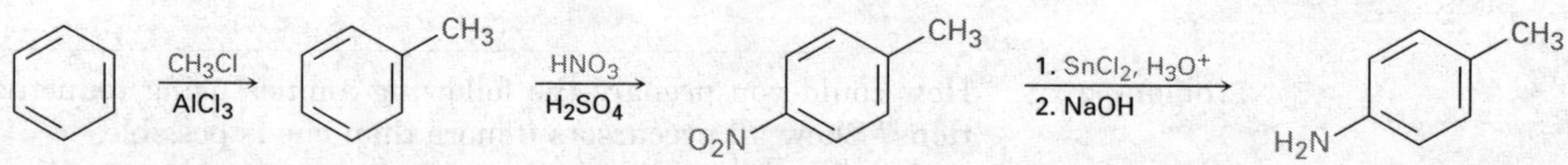

Benzene **Toluene** ***p*-Nitrotoluene** ***p*-Methylaniline**

Problem 12.13 How could you synthesize the following amines from benzene? More than one step is required in each case.

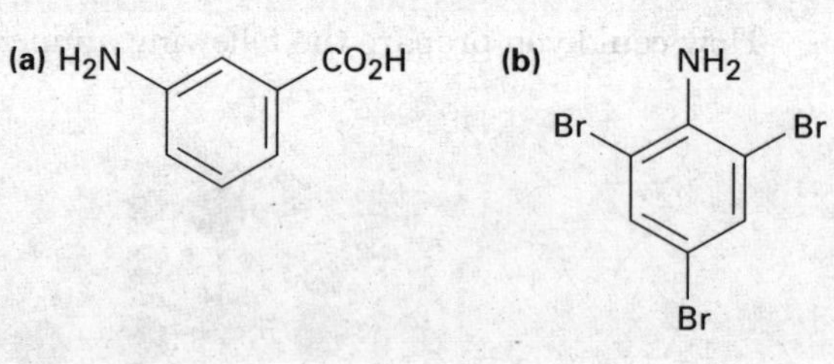

12.5 Reactions of Amines

We've already discussed the two most general reactions of alkylamines—alkylation and acylation. As we saw in the previous section, primary, secondary, and tertiary amines can be alkylated by reaction with alkyl halides. Primary and secondary (but not tertiary) amines can also be acylated by nucleophilic acyl substitution reactions with acid chlorides (Section 10.7) or acid anhydrides (Section 10.8) to give amides.

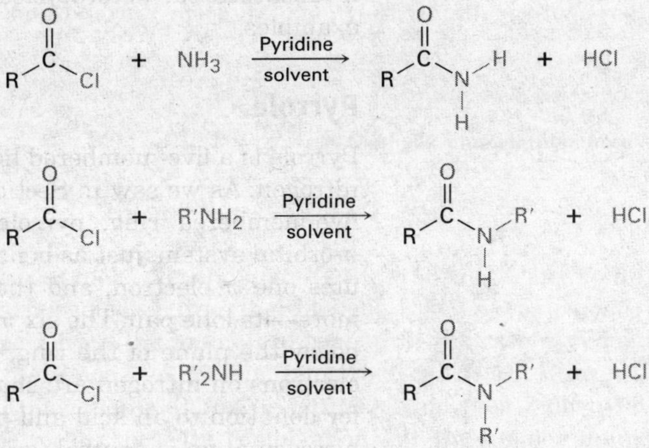

In addition, we've seen that primary amines react with aldehydes and ketones to give imines (Section 9.9).

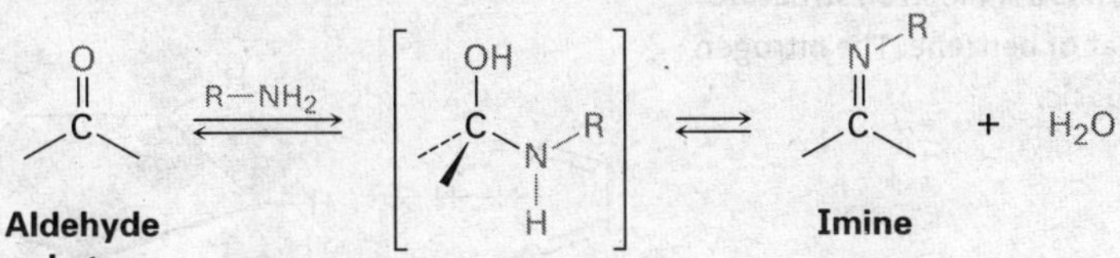

Aldehyde or ketone

Imine

12.6 Heterocyclic Amines

A **heterocycle**, as noted in Section 5.9 in connection with a discussion of aromaticity, contains atoms of two or more different elements in its ring, usually carbon along with nitrogen, oxygen, or sulfur. Heterocyclic amines are particularly common in organic chemistry, and many have important biological properties. Pyridoxal phosphate, a coenzyme; sildenafil (Viagra), a well-known pharmaceutical; and heme, the oxygen carrier in blood, are examples.

Pyridoxal phosphate (a coenzyme)

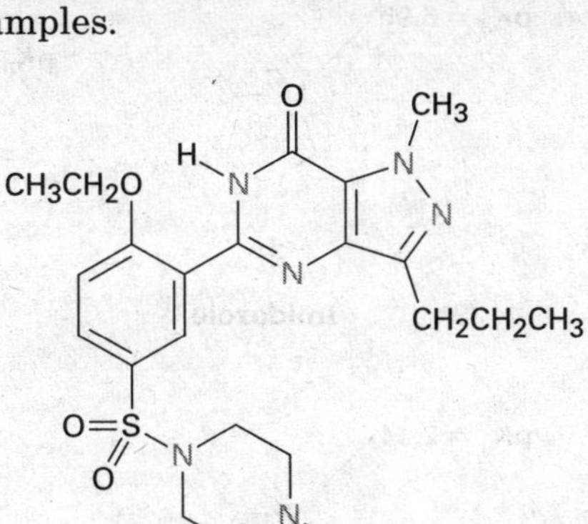

Sildenafil (Viagra)

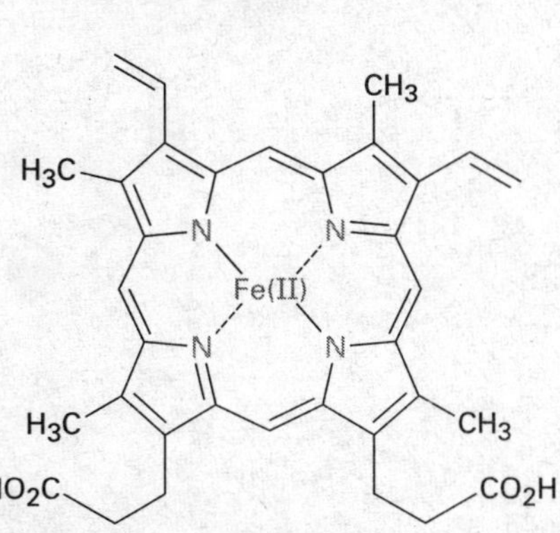

Heme

For the most part, heterocyclic amines have the same chemistry as their open-chain counterparts. In certain cases, though, particularly when the ring is unsaturated, heterocycles have unique properties. Let's look at several examples.

Pyrrole

Pyrrole is a five-membered heterocyclic amine with two double bonds and one nitrogen. As we saw in Section 5.9, pyrrole is aromatic. Even though it has a five-membered ring, pyrrole has six π electrons in a cyclic, conjugated π-orbital system, just as benzene does. Each of the four carbon atoms contributes one π electron, and the sp^2-hybridized nitrogen atom contributes two more—its lone pair. The six π electrons occupy p orbitals with lobes above and below the plane of the ring, as shown in Figure 12.3. Because the lone-pair electrons on nitrogen are shared in the aromatic ring, they are not available for donation to an acid and pyrrole is nonbasic. Note in Figure 12.3 how the nitrogen atom is neutral (green) rather than electron-rich (red).

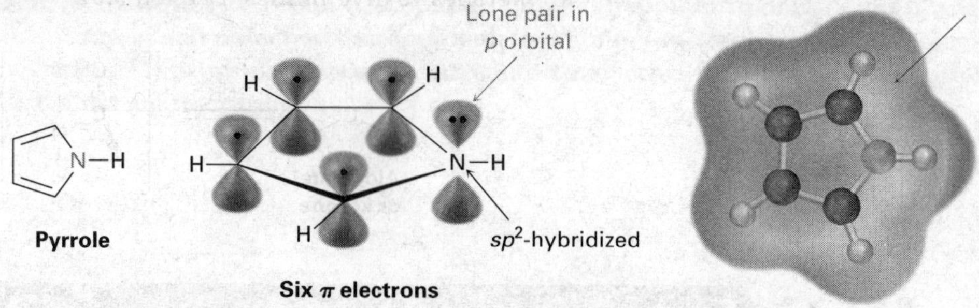

Figure 12.3 Pyrrole, an aromatic heterocycle, has a π electron structure similar to that of benzene. The nitrogen atom is nonbasic.

Other common five-membered aromatic heterocycles include imidazole and thiazole. Imidazole, a constituent of the amino acid histidine, has two nitrogens, only one of which is basic. Thiazole, the five-membered ring system on which the structure of thiamin (vitamin B_1) is based, also contains a basic nitrogen that is alkylated in thiamin to form a quaternary ammonium ion.

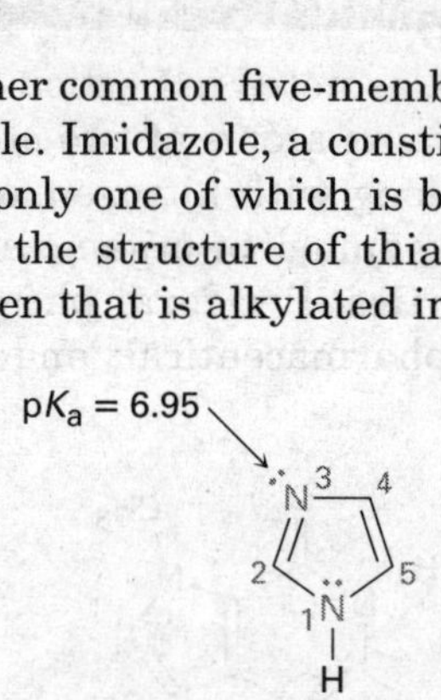

Imidazole

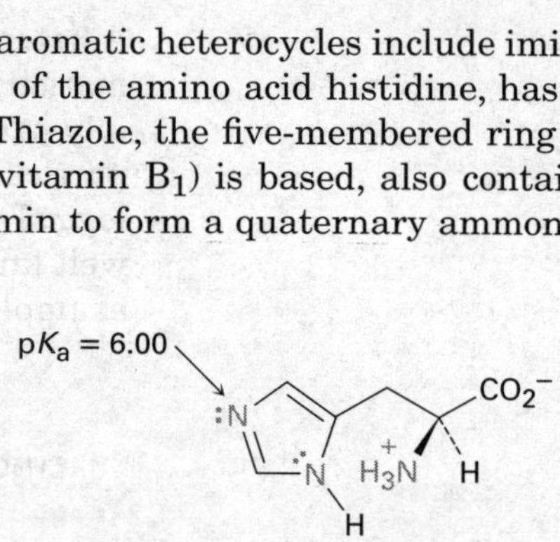

Histidine

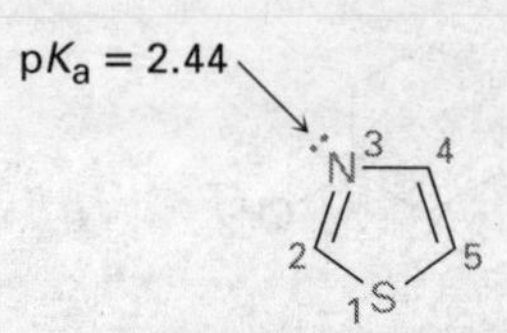

Thiazole

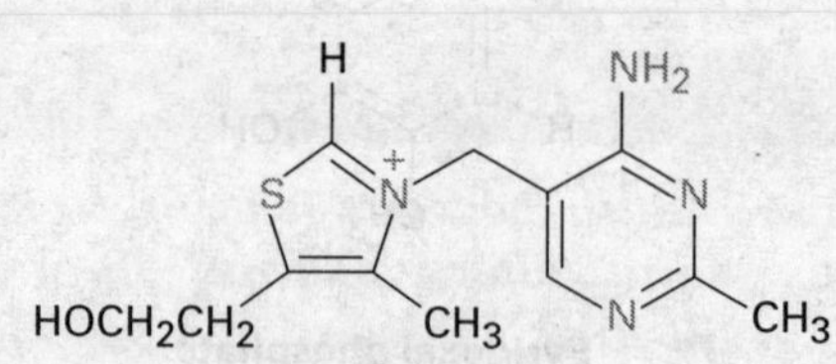

Thiamin (vitamin B_1)

Problem 12.14 Pyrrole undergoes typical electrophilic substitution reactions on the carbon next to nitrogen in the ring. What products would you expect to obtain from reaction of *N*-methylpyrrole with the following reagents?

(a) Br_2 **(b)** CH_3Cl, $AlCl_3$ **(c)** CH_3COCl, $AlCl_3$

Problem 12.15 Review the mechanism of the nitration of benzene (Section 5.4), and propose a mechanism for the nitration of pyrrole.

Pyridine

Pyridine is the nitrogen-containing heterocyclic analog of benzene. Like benzene, pyridine is a flat, aromatic molecule with bond angles of approximately 120°. Also like benzene, pyridine is aromatic with six π electrons in a cyclic, conjugated π-orbital system. The sp^2-hybridized nitrogen atom and the five carbon atoms each contribute one π electron to the cyclic, conjugated *p* orbitals of the ring.

Unlike the situation in pyrrole, the lone-pair electrons on the pyridine nitrogen atom are not part of the π orbital system but instead occupy an sp^2 orbital in the plane of the ring (Figure 12.4). As a result, the pyridine lone-pair electrons are available for donation to an acid and pyridine is therefore basic. Compare the electrostatic potential maps of pyrrole (Figure 12.3) and pyridine (Figure 12.4) to see this difference in basicity.

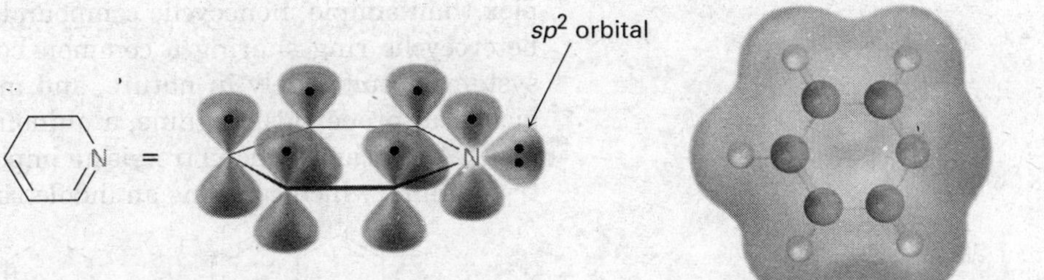

Pyridine

Figure 12.4 Electronic structure of pyridine, a nitrogen-containing analog of benzene.

Although less basic than typical alkylamines, pyridine (pK_b = 8.75) is nevertheless used in a variety of organic reactions when a base catalyst is required. You might recall, for instance, that the reaction of an acid chloride with an alcohol to yield an ester is commonly done in the presence of pyridine (Section 10.7).

Substituted pyridines, such as the B_6 complex vitamins pyridoxal and pyridoxine, are important biologically. Present in yeast, cereal, and other foodstuffs, the B_6 vitamins are necessary for the synthesis of some amino acids.

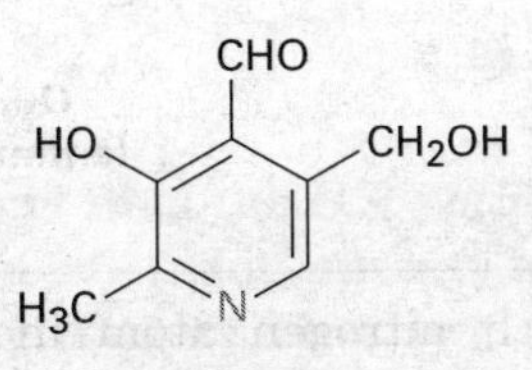

Pyridoxal

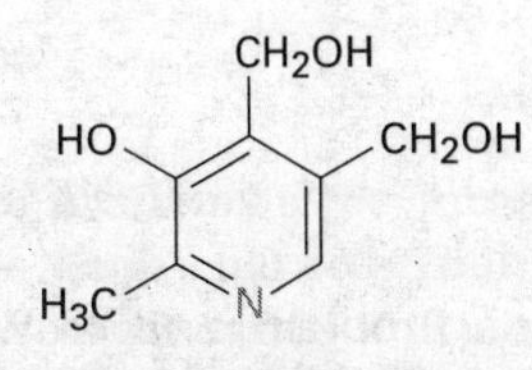

Pyridoxine

Problem 12.16 The five-membered heterocycle imidazole contains two nitrogen atoms, one "pyrrole-like" and one "pyridine-like." Draw an orbital picture of imidazole, and indicate the orbital in which each nitrogen has its electron lone pair.

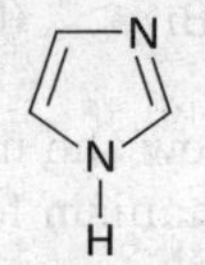

Imidazole

Problem 12.17 Which nitrogen atom in imidazole (Problem 12.16) is more basic according to the following electrostatic potential map? Why?

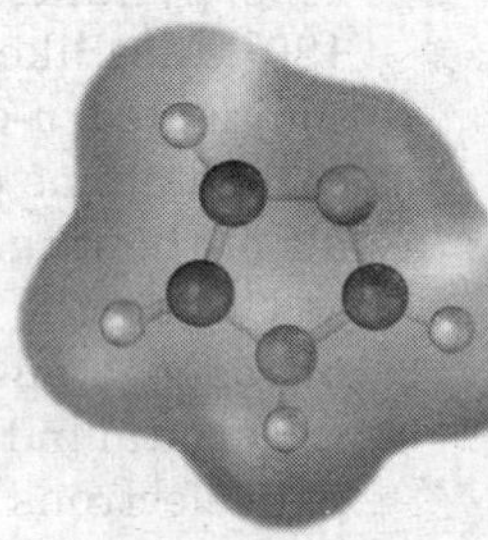

Fused-Ring Aromatic Heterocycles

Fused-ring heterocycles like quinoline, isoquinoline, and indole are more complex than simple monocyclic compounds. All three have a benzene ring and a heterocyclic ring sharing a common bond. These and many other fused-ring systems occur widely in nature, and many members of the class have useful biological properties. Quinine, a quinoline derivative found in the bark of the South American cinchona tree, is an important antimalarial drug. *N,N*-Dimethyltryptamine, which contains an indole ring, is a powerful hallucinogen.

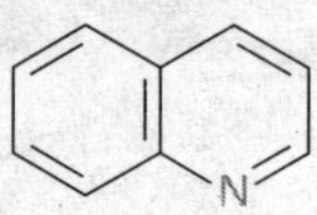

Quinoline

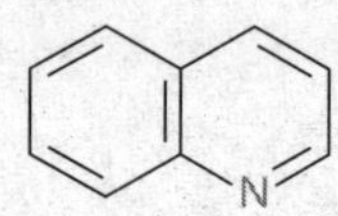

Isoquinoline

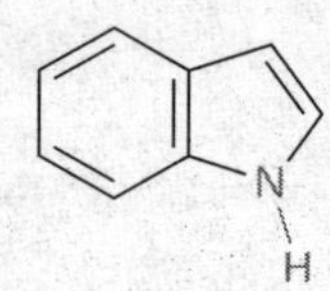

Indole

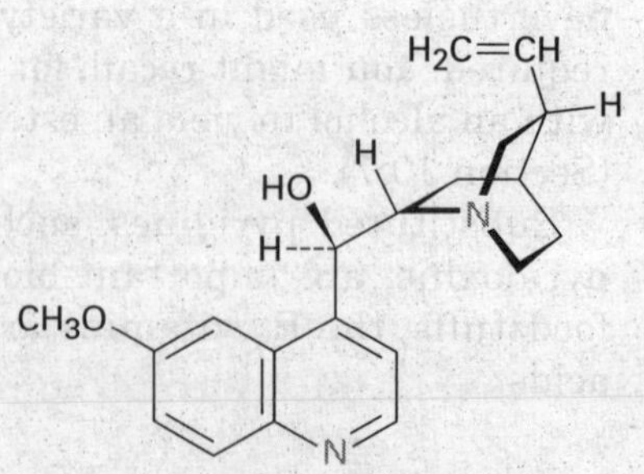

Quinine (antimalarial)

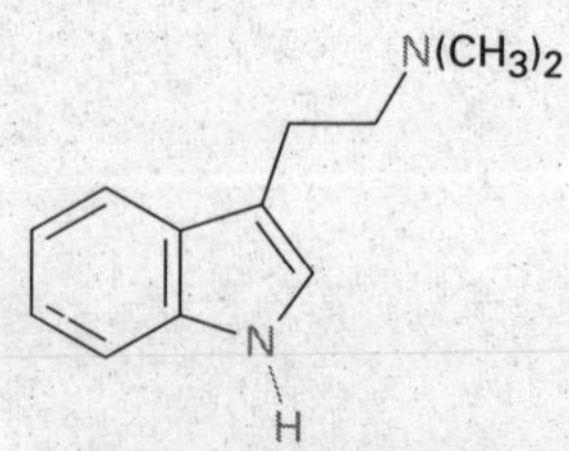

***N,N*-Dimethyltryptamine (hallucinogenic)**

Problem 12.18 Which nitrogen atom in the hallucinogenic indole alkaloid *N,N*-dimethyltryptamine do you think is more basic? Explain.

12.7 Alkaloids: Naturally Occurring Amines

Naturally occurring amines derived from plant sources were once known as "vegetable alkali" because their aqueous solutions are slightly basic, but they are now referred to as **alkaloids**. The study of alkaloids provided much of the impetus for the growth of organic chemistry in the 19th century and remains today a fascinating area of research.

Approximately 30,000 different alkaloids are known, varying widely in structure from the simple to the enormously complex. The odor of rotting fish, for example, is largely caused by the simplest alkaloid, methylamine. In fact, the use of acidic lemon juice to mask fish odors is just an acid–base reaction of the citric acid in lemons with methylamine to form the nonvolatile methylammonium salt.

Many alkaloids have pronounced biological properties, and many of the pharmaceutical agents used today are alkaloids from natural sources. As only a few examples, atropine, an antispasmodic agent used for the treatment of colitis, is obtained from the flowering plant *Atropa belladonna,* commonly called the deadly nightshade. Ephedrine, a bronchodilator and decongestant, is obtained from the Chinese plant *Ephedra sinica.* Reserpine, a tranquilizer and antihypertensive, comes from powdered roots of the semitropical plant *Rauwolfia serpentina.*

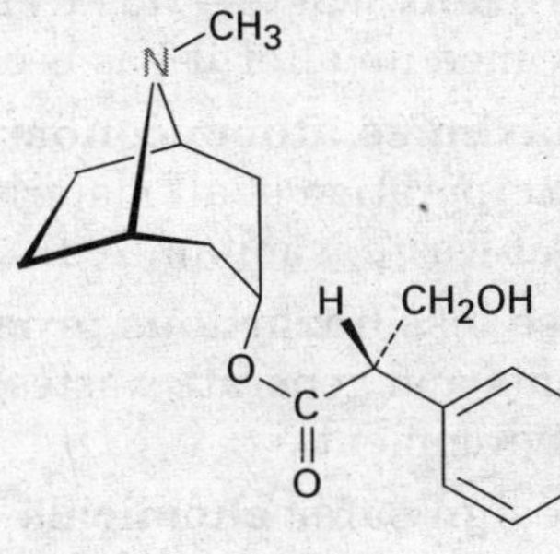

Atropine

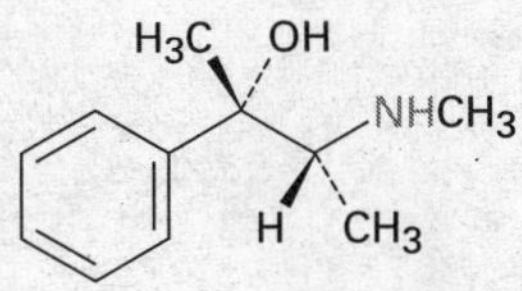

Ephedrine

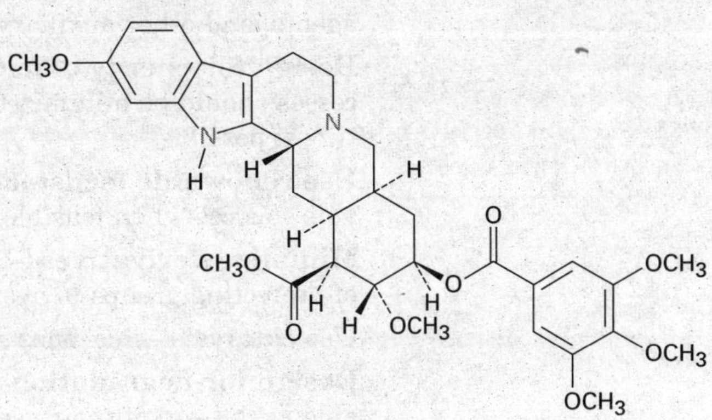

Reserpine

A recent report from the U.S. National Academy of Sciences estimates that less than 1% of all living species have been characterized. Thus, alkaloid chemistry today remains an active area of research, and innumerable substances with potentially useful properties remain to be discovered.

INTERLUDE

Green Chemistry

Let's hope disasters like this are never repeated.

Organic chemistry in the 20th century changed the world, giving us new medicines, insecticides, adhesives, textiles, dyes, building materials, composites, and all manner of polymers. But these advances did not come without a cost: every chemical process produces wastes that must be dealt with, including reaction solvents and toxic by-products that might evaporate into the air or be leached into groundwater if not disposed of properly. Even apparently harmless by-products must be safely buried or otherwise sequestered. As always, there's no such thing as a free lunch; with the good also comes the bad.

It may never be possible to make organic chemistry completely benign, but awareness of the environmental problems caused by many chemical processes has grown dramatically in recent years, giving rise to a movement called *green chemistry.* Green chemistry is the design and implementation of chemical products and processes that reduce waste and attempt to eliminate the generation of hazardous substances. There are 12 principles of green chemistry.

Prevent waste—Waste should be prevented rather than treated or cleaned up after it has been created.

Maximize atom economy—Synthetic methods should maximize the incorporation of all materials used in a process into the final product so that waste is minimized.

Use less hazardous processes—Synthetic methods should use reactants and generate wastes with minimal toxicity to health and the environment.

Design safer chemicals—Chemical products should be designed to have minimal toxicity.

Use safer solvents—Minimal use should be made of solvents, separation agents, and other auxiliary substances in a reaction.

Design for energy efficiency—Energy requirements for chemical processes should be minimized, with reactions carried out at room temperature if possible.

Use renewable feedstocks—Raw materials should come from renewable sources when feasible.

Minimize derivatives—Syntheses should be designed with minimal use of protecting groups to avoid extra steps and reduce waste.

Use catalysis—Reactions should be catalytic rather than stoichiometric.

Design for degradation—Products should be designed to be biodegradable at the end of their useful lifetimes.

Monitor pollution in real time—Processes should be monitored in real time for the formation of hazardous substances.

Prevent accidents—Chemical substances and processes should minimize the potential for fires, explosions, or other accidents.

continued

INTERLUDE

The foregoing 12 principles won't all be met in most real-world applications, but they provide a worthy goal to aim for and they can make chemists think more carefully about the environmental implications of their work. Real success stories are already occurring, and more are in progress. Approximately 7 million pounds per year of ibuprofen (6 billion tablets!) is now made by a "green" process that produces approximately 99% less waste than the process it replaces. Only three steps are needed, the anhydrous HF solvent used in the first step is recovered and reused, and the second and third steps are catalytic.

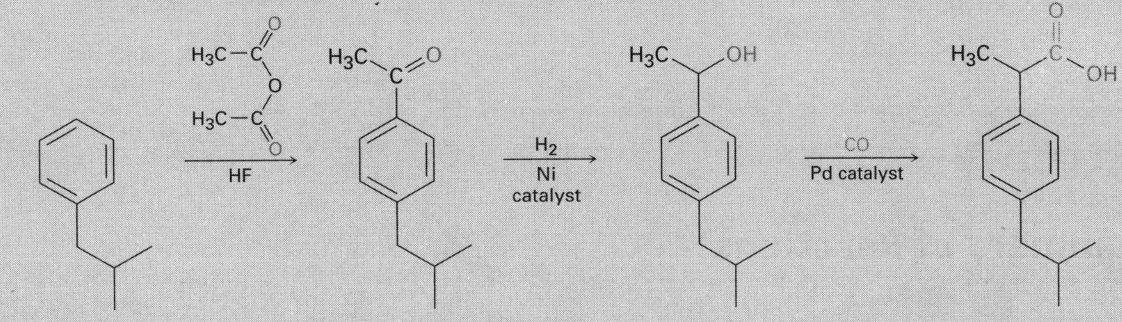

Isobutylbenzene → Ibuprofen

Summary and Key Words

alkaloid 421
alkylamine 405
amine 404
arylamine 405
basicity constant, K_b 408
heterocycle 417
quaternary ammonium salt 405
reductive amination 413

We've now seen all the common functional groups that occur throughout organic and biological chemistry. Of those groups, amines are among the most abundant. In addition to proteins and nucleic acids, the majority of pharmaceutical agents contain amine functional groups and many of the common coenzymes necessary for biological catalysis are amines.

Amines are organic derivatives of ammonia. They are named in the IUPAC system either by adding the suffix *-amine* to the name of the alkyl substituent or by considering the amino group as a substituent on a more complex parent molecule. Bonding in amines is similar to that in ammonia: the nitrogen atom is sp^3-hybridized, the three substituents are directed to three corners of a regular tetrahedron, and the lone pair of electrons occupies the fourth corner of the tetrahedron.

The chemistry of amines is dominated by the presence of the lone-pair electrons on nitrogen, which make amines both basic and nucleophilic. **Arylamines** are generally weaker bases than **alkylamines** because their lone-pair electrons are shared by orbital overlap with the aromatic π electron system.

The simplest method of amine synthesis involves S_N2 reaction of ammonia or an amine with an alkyl halide. Alkylation of ammonia yields a primary amine; alkylation of a primary amine yields a secondary amine; and so on. Amines can also be prepared from amides and nitriles by reduction with $LiAlH_4$ and from aldehydes and ketones by **reductive amination** with ammonia and a reducing agent. Arylamines are prepared by nitration of an aromatic ring followed by reduction of the nitro group. Many of the reactions that amines undergo are familiar from previous chapters. Thus, amines react with alkyl halides in S_N2 reactions and with acid chlorides in nucleophilic acyl substitution reactions.

Heterocyclic amines, compounds in which the nitrogen atom is in a ring, have a great diversity in their structures and properties. Pyrrole, pyridine, indole, and quinoline all show aromatic properties.

Summary of Reactions

1. Synthesis of amines (Section 12.4)

(a) S_N2 reaction of alkyl halides

Ammonia	$\ddot{N}H_3 + R{-}X \longrightarrow$	$R\overset{+}{N}H_3\ X^-$	$\xrightarrow{NaOH}$	RNH_2	Primary
Primary	$R\ddot{N}H_2 + R{-}X \longrightarrow$	$R_2\overset{+}{N}H_2\ X^-$	$\xrightarrow{NaOH}$	R_2NH	Secondary
Secondary	$R_2\ddot{N}H + R{-}X \longrightarrow$	$R_3\overset{+}{N}H\ X^-$	$\xrightarrow{NaOH}$	R_3N	Tertiary
Tertiary	$R_3\ddot{N} + R{-}X \longrightarrow$	$R_4\overset{+}{N}\ X^-$			Quaternary ammonium

(b) Reduction of azides

$$RCH_2{-}X \xrightarrow[\text{ethanol}]{Na^+\ ^-N_3} RCH_2{-}N{=}\overset{+}{N}{=}\overset{-}{N} \xrightarrow[2.\ H_2O]{1.\ LiAlH_4,\ \text{ether}} R{-}NH_2$$

(c) Reductive amination of aldehydes and ketones

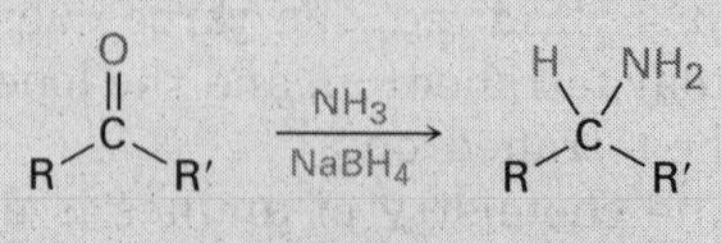

(d) Reduction of nitrobenzenes

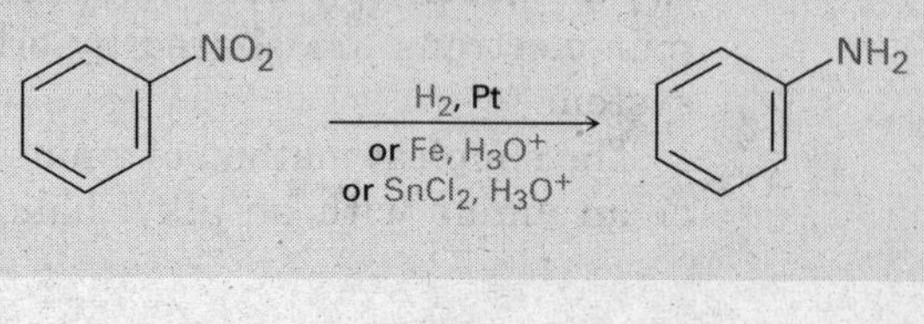

Exercises

Visualizing Chemistry

(Problems 12.1–12.18 appear within the chapter.)

OWL

Interactive versions of these problems are assignable in OWL.

12.19 Which nitrogen atom in the alkaloid tryptamine is more basic? Explain.

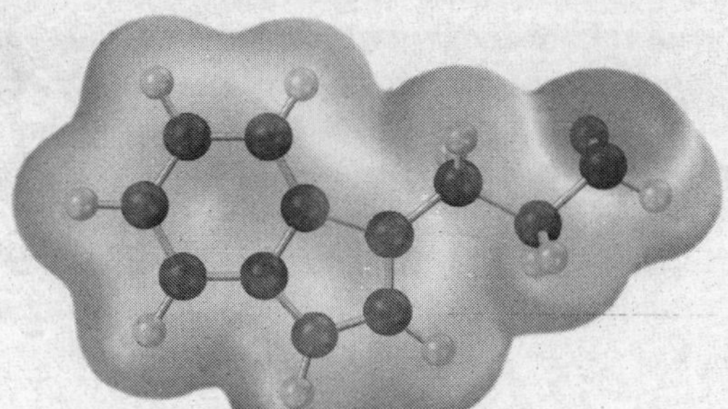

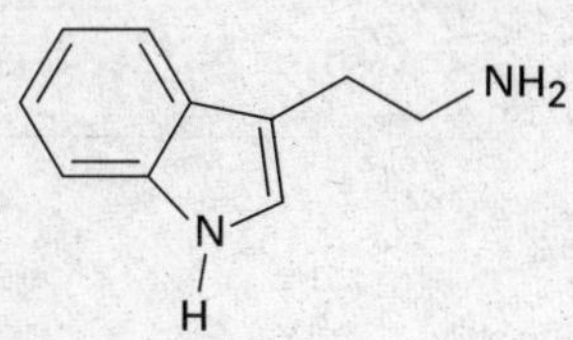

Tryptamine

12.20 Name the following amines, and identify each as primary, secondary, or tertiary:

(a)

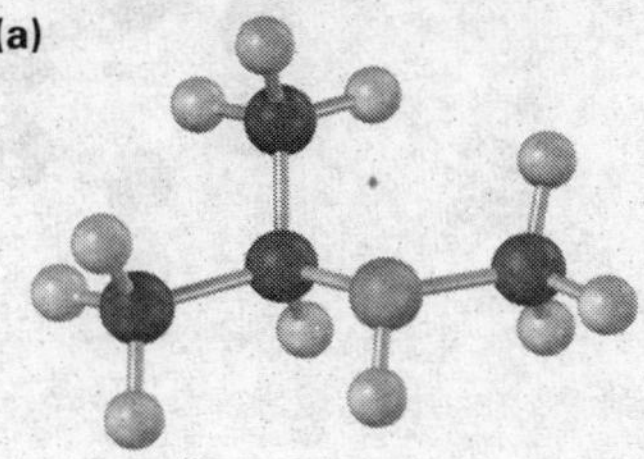

(b)

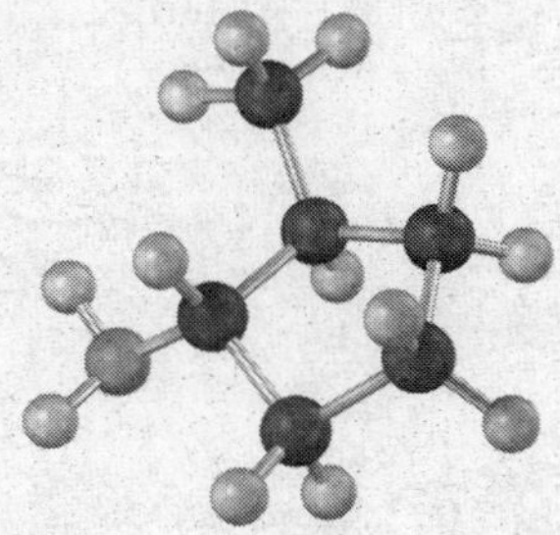

(c)

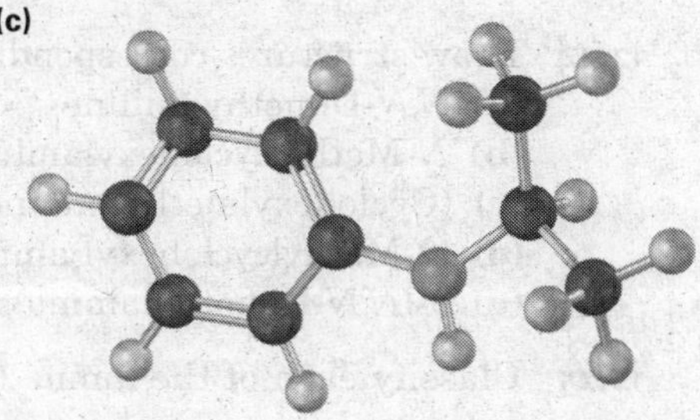

12.21 The following compound contains three nitrogen atoms. Rank them in order of increasing basicity.

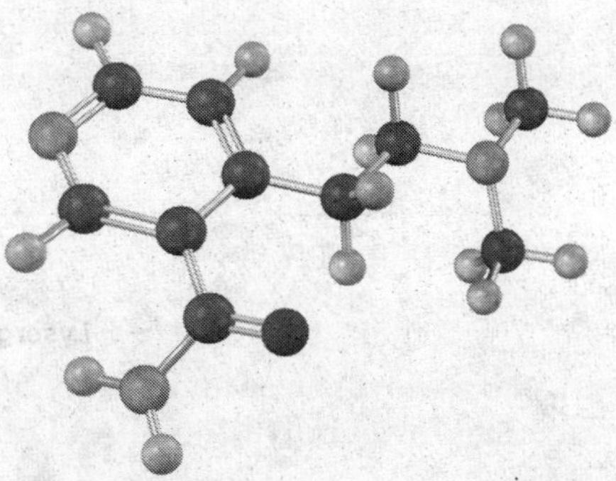

12.22 The following molecule can be prepared by reaction between a primary amine and a *dihalide.* Identify the two reactants, and write the reaction.

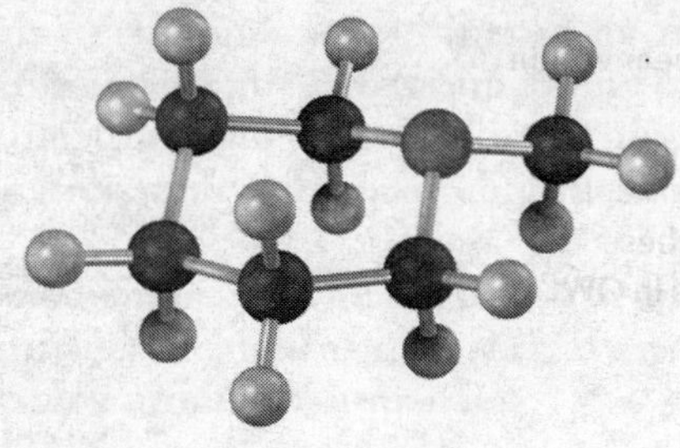

12.23 Name the following amine, including *R,S* stereochemistry, and draw the product of its reaction with (i) CH_3CH_2Br and (ii) CH_3COCl.

Additional Problems

Naming Amines

12.24 Draw structures corresponding to the following IUPAC names:
(a) *N,N*-Dimethylaniline
(b) *N*-Methylcyclohexylamine
(c) (Cyclohexylmethyl)amine
(d) (2-Methylcyclohexyl)amine
(e) 3-(*N,N*-Dimethylamino)propanoic acid

12.25 Classify each of the amine (not amide) nitrogen atoms in the following substances as primary, secondary, or tertiary:

(a)

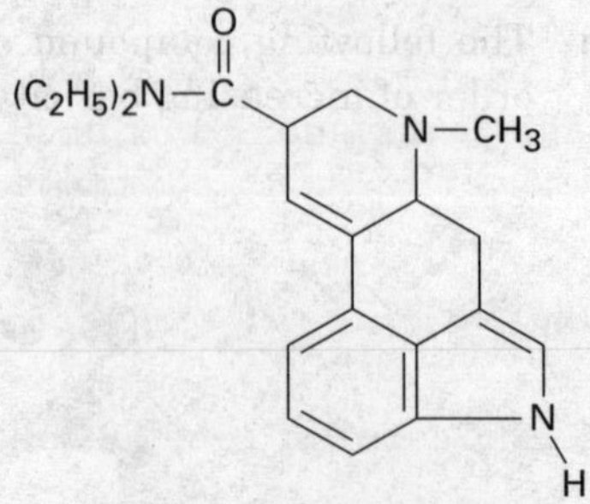

Lysergic acid diethylamide

(b)

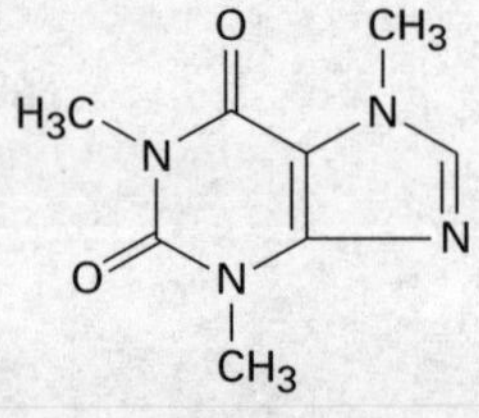

Caffeine

12.26 Mescaline, a powerful hallucinogen derived from the peyote cactus, has the systematic name 2-(3,4,5-trimethoxyphenyl)ethylamine. Draw its structure.

12.27 Propose structures for amines that fit the following descriptions:
(a) A secondary arylamine
(b) A 1,3,5-trisubstituted arylamine
(c) An achiral quaternary ammonium salt
(d) A five-membered heterocyclic amine

12.28 There are eight isomeric amines with the formula $C_4H_{11}N$. Draw them, name them, and classify each as primary, secondary, or tertiary.

12.29 Name the following compounds:

(a) 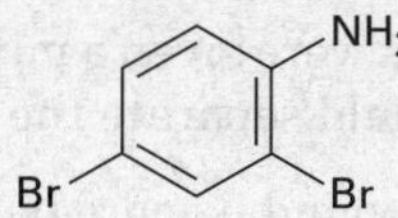

(b) 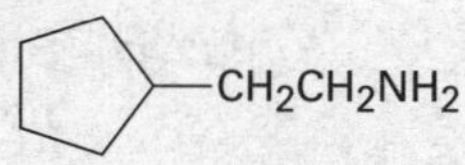

(c) 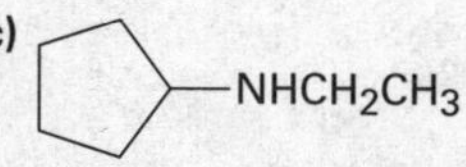

(d) 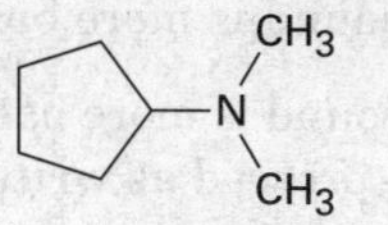

(e) 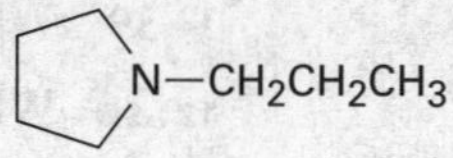

(f) $H_2NCH_2CH_2CH_2CN$

REACTIONS AND SYNTHESIS

12.30 How might you prepare the following amines from 1-bromobutane?
(a) Butylamine **(b)** Dibutylamine **(c)** Pentylamine

12.31 How might you prepare each of the amines in Problem 12.30 from butan-1-ol?

12.32 How might you prepare the following amines from ammonia and any alkyl halides needed?

(a) $CH_3CH_2CH_2CH_2CH_2CH_2NH_2$ (b) $(CH_3)_4N^+ \ I^-$

(c) 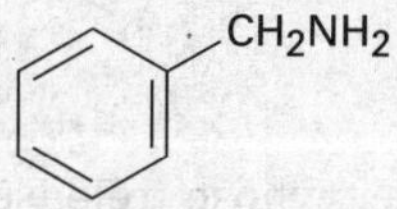

(d) 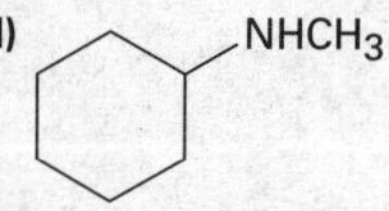

12.33 Show the products of the following reactions:

(a) $CH_3CH_2CH_2NH_2 \xrightarrow{CH_3Br} ?$

(b) 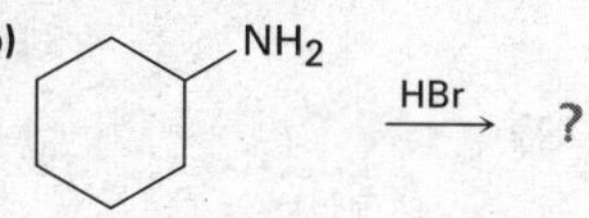

(c) 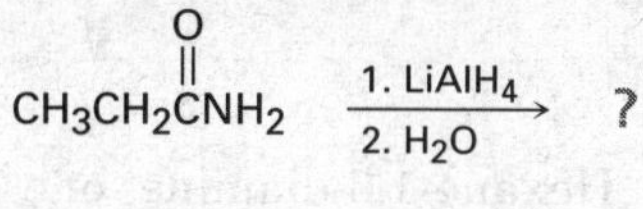

(d) 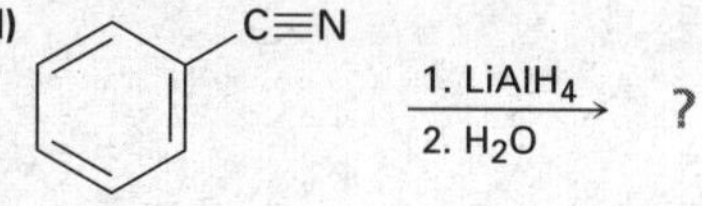

12.34 How might you prepare pentylamine from the following starting materials?
(a) Pentanamide (b) Pentanenitrile (c) Pentanoic acid

12.35 How would you prepare benzylamine, $C_6H_5CH_2NH_2$, from each of the following starting materials?

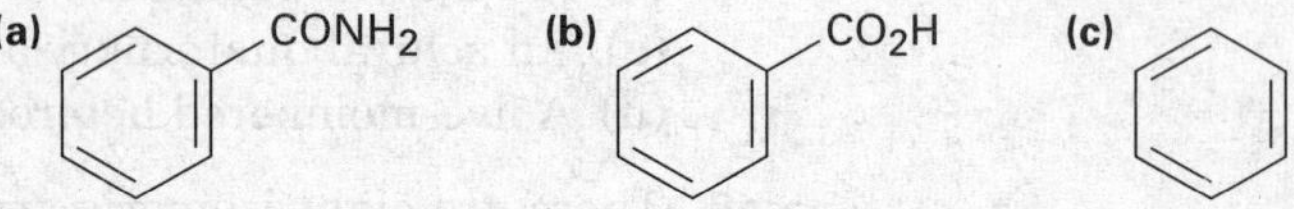

BASICITY

12.36 Would you expect diphenylamine to be more basic or less basic than aniline? Explain.

12.37 Suppose you were given a mixture of toluene, aniline, and phenol. Describe how you would separate the mixture into its three pure components.

12.38 Which compound is more basic, $CH_3CH_2NH_2$ or $CF_3CH_2NH_2$? Explain.

12.39 Which compound is more basic, *p*-aminobenzaldehyde or aniline?

12.40 Which compound is more basic, triethylamine or aniline? Does the following reaction proceed as written?

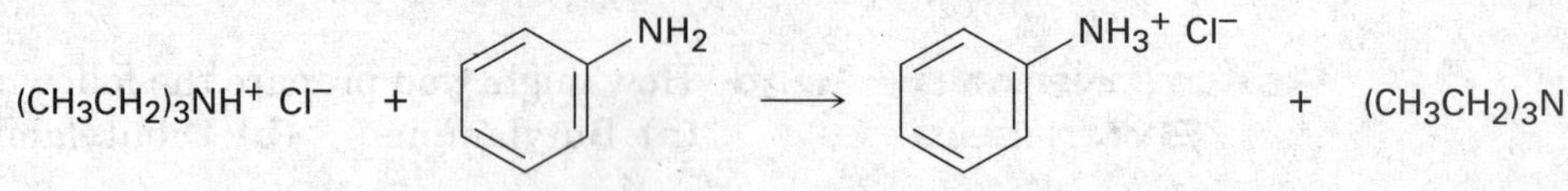

GENERAL PROBLEMS

12.41 Protonation of an amide using strong acid occurs on oxygen rather than on nitrogen. Explain, using resonance structures of *O*-protonated and *N*-protonated products.

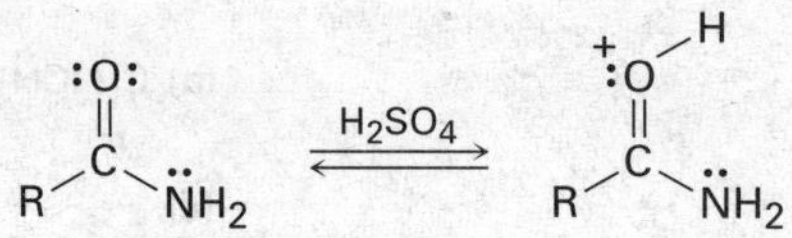

12.42 Histamine, whose release in the body triggers nasal secretions and constricted airways, has three nitrogen atoms—one "pyrrole-like," one "pyridine-like," and one alkylamine. List them in order of increasing basicity, and explain your ordering.

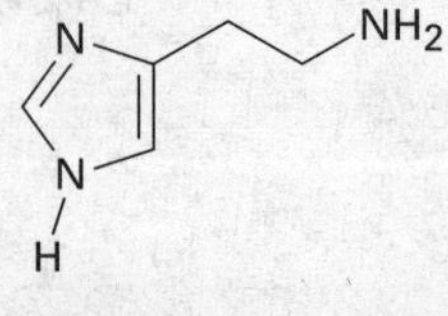

Histamine

12.43 Hexane-1,6-diamine, one of the starting materials used for the manufacture of nylon 66, can be synthesized by a route that begins with the addition of Cl_2 to buta-1,3-diene (Section 4.8). How would you carry out the complete synthesis?

12.44 Another method for making hexane-1,6-diamine (see Problem 12.43) starts from adipic acid (hexanedioic acid). How would you carry out the synthesis?

12.45 How can you explain the observation that *p*-nitroaniline is less basic than aniline by a factor of 40,000? (See Section 5.7.)

12.46 In light of your answer to Problem 12.45, which would you expect to be more basic, aniline or *p*-methoxyaniline? Explain.

12.47 Give the structures of the major organic products you would obtain from the reaction of *m*-methylaniline with the following reagents:
(a) Br_2 (1 mol) **(b)** CH_3I (excess) **(c)** CH_3COCl, pyridine

12.48 Fill in the missing reagents **a** through **d** in the following synthesis of racemic methamphetamine from benzene.

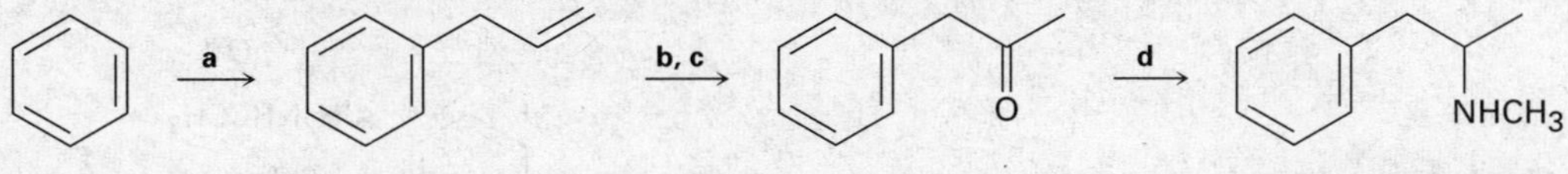

(R,S)-Methamphetamine

12.49 Atropine, $C_{17}H_{23}NO_3$, is a poisonous alkaloid isolated from the leaves and roots of the deadly nightshade, *Atropa belladonna*. In small doses, atropine acts as a muscle relaxant: 0.5 ng (1 nanogram = 10^{-9} g) is sufficient to cause pupil dilation. On reaction with aqueous NaOH, atropine yields tropic acid, $C_6H_5CH(CH_2OH)CO_2H$, and tropine, $C_8H_{15}NO$. Tropine, an optically inactive alcohol, yields tropidene on dehydration. Propose a structure for atropine.

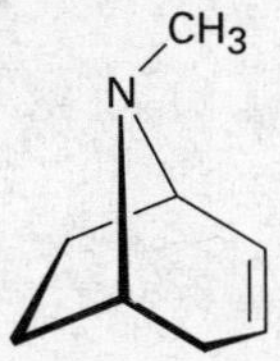

Tropidene

12.50 We've seen that amines are basic and amides are neutral. *Imides*, compounds with two carbonyl groups flanking an N–H, are actually acidic. Show by drawing resonance structures of the anion resulting from deprotonation why imides are acidic.

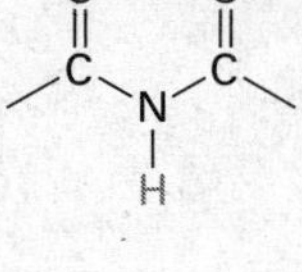

An imide

12.51 Furan, the oxygen-containing analog of pyrrole, is aromatic in the same way that pyrrole is. Draw an orbital picture of furan, and show how it has six electrons in its cyclic conjugated π orbitals.

Furan

12.52 Fill in the missing reagents **a** through **f** in the following scheme:

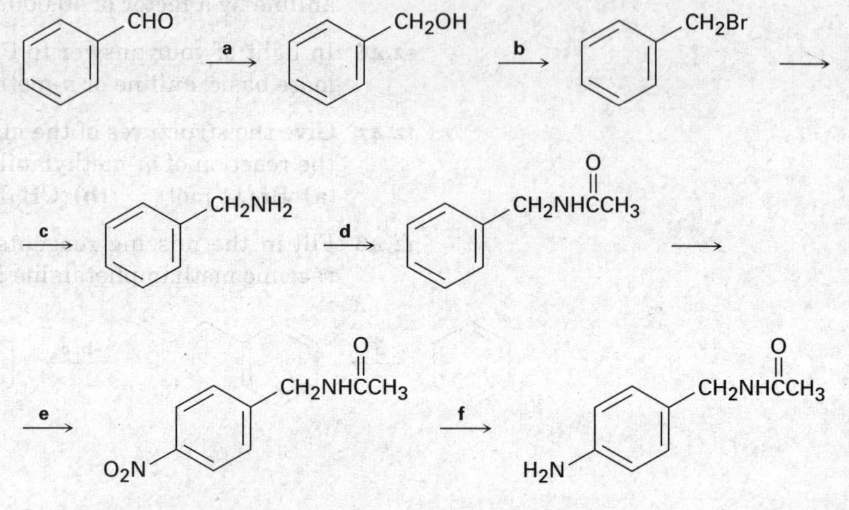

12.53 The amino acid proline is biosynthesized from glutamate semialdehyde by the following transformation, where NADH is the biological reducing agent nicotinamide adenine dinucleotide. What is the likely structure of intermediate **A**, and how is it formed?

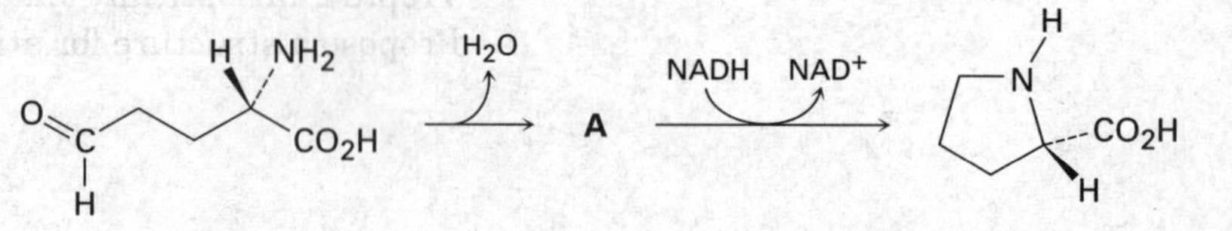

Glutamate 5-semialdehyde

Proline

12.54 The following transformation involves a conjugate nucleophilic addition reaction (Section 9.10) followed by an intramolecular nucleophilic acyl substitution reaction (Section 10.5). Show the mechanism.

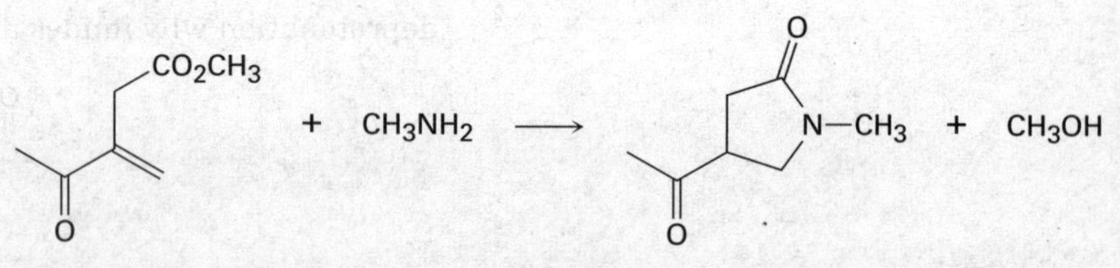

12.55 Choline, a component of the phospholipids in cell membranes, can be prepared by S_N2 reaction of trimethylamine with ethylene oxide. Show the structure of choline, and propose a mechanism for the reaction.

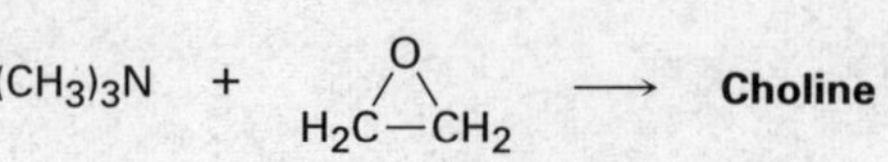

IN THE MEDICINE CABINET

12.56 The anti-inflammatory drug celecoxib, marketed as Celebrex, is widely used for treatment of rheumatoid arthritis. Draw a mechanism for the following step used in celecoxib synthesis.

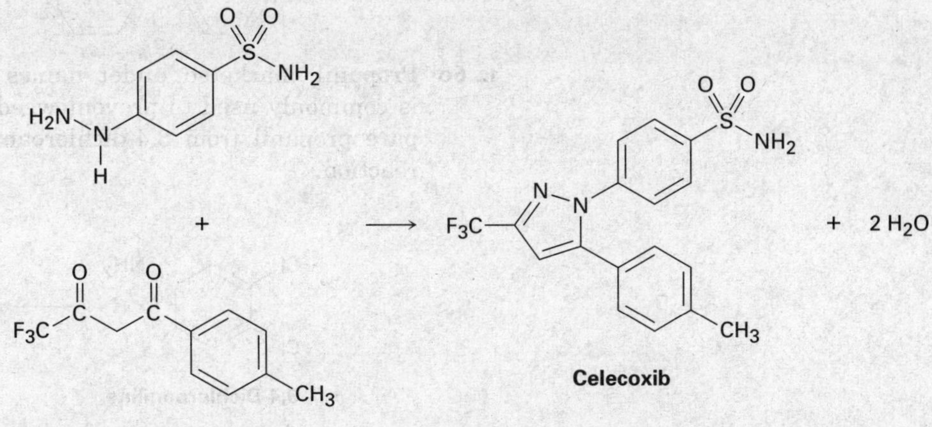

Celecoxib

12.57 How might you use a reductive amination to synthesize ephedrine, an amino alcohol that is widely used for the treatment of bronchial asthma?

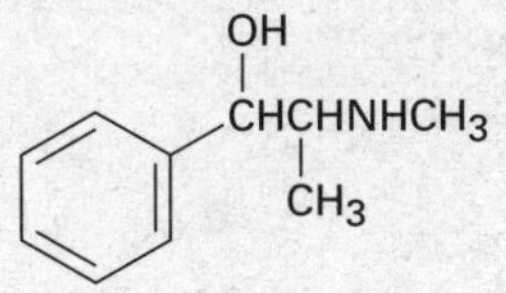

Ephedrine

12.58 Tetracaine, a substance used as a spinal anesthetic, can be prepared from benzene by the following route. Show how you could accomplish each of the transformations **(a)** through **(d)**.

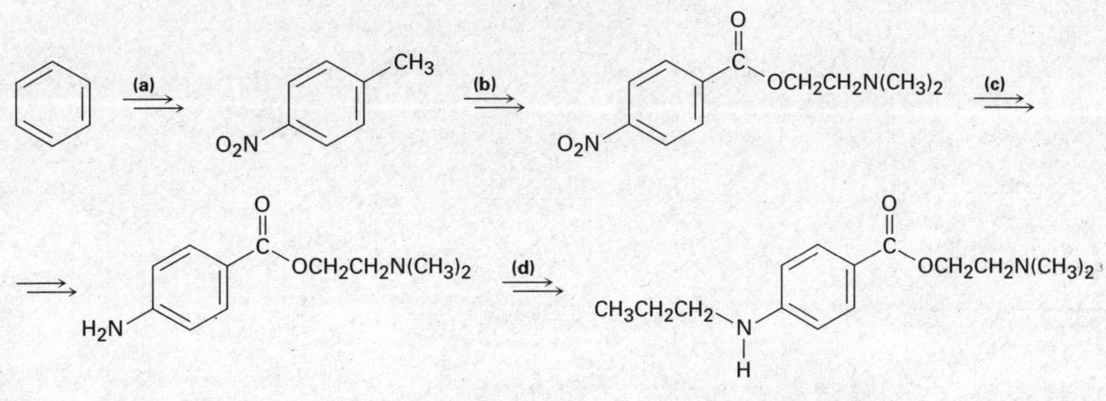

Tetracaine

IN THE FIELD

12.59 Paraquat, a broad-spectrum weed killer, is prepared using the following step. What is the structure of paraquat?

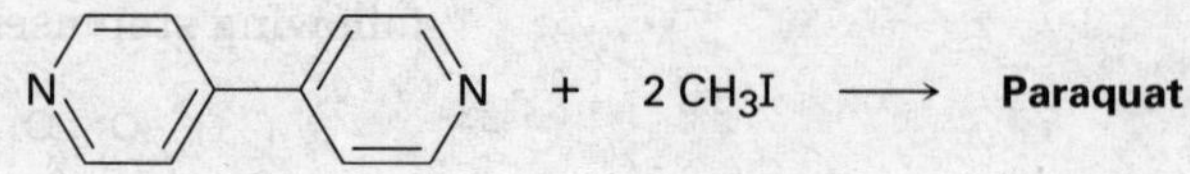

12.60 Propanil, marketed under names such as Stampede and Chem-Rice, is commonly used to prevent weeds in rice fields. How would you prepare propanil from 3,4-dichloroaniline? Show the mechanism of your reaction.

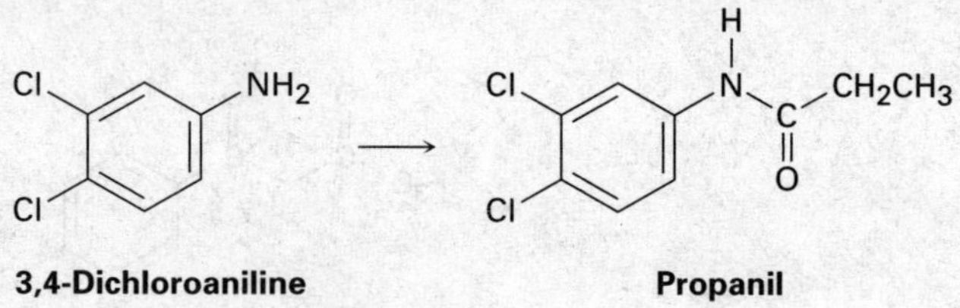

CHAPTER

13

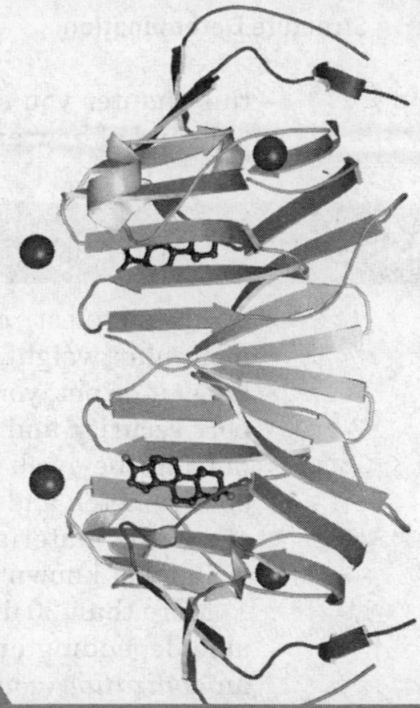

The anabolic steroids sometimes taken by athletes are transported to their target tissues in the body by binding to a protein called human sex hormone–binding globulin. The chemical structures of this and other complex organic molecules are generally determined by x-ray crystallography.

Structure Determination

Every time a reaction is run, the products must be identified, and every time a new compound is found in nature, its structure must be determined. Determining the structure of an organic molecule was a difficult and time-consuming process in the early days of chemistry, but powerful techniques and specialized instruments developed in the mid-1900s greatly simplified the task. We'll look at four of the most useful such techniques—mass spectrometry (MS), infrared spectroscopy (IR), ultraviolet spectroscopy (UV), and nuclear magnetic resonance spectroscopy (NMR)—each of which yields a different kind of structural information.

Mass Spectrometry	What is the molecular formula?
Infrared (IR) spectroscopy	What functional groups are present?
Ultraviolet (UV) spectroscopy	Is a conjugated π electron system present?
Nuclear magnetic resonance (NMR) spectroscopy	What is the carbon–hydrogen framework?

WHY THIS CHAPTER

Finding the structures of new molecules, whether small ones synthesized in the laboratory or large proteins and nucleic acids found in living organisms, is central to progress in chemistry and biochemistry. We'll only scratch the surface of structure determination in this book, but after reading

OWL

Online homework for this chapter can be assigned in OWL, an online homework assessment tool.

this chapter you should have a good idea of some techniques that are available and of how and when each is used.

13.1 Mass Spectrometry

At its simplest, **mass spectrometry (MS)** is a technique for measuring the molecular weight, and therefore the formula, of a molecule. Whether you know it or not, you see a mass spectrometer every time you pass through airport security and watch a security person swabbing a piece of luggage and putting the swab into an instrument. The instrument is a specialized mass spectrometer, and it can detect explosives by measuring the molecular weight of volatile materials on the swab and comparing the result to the molecular weights of known explosives.

More than 20 different kinds of commercial mass spectrometers are available depending on the intended application, but all have three basic parts: an *ionization source* in which sample molecules are given an electrical charge; a *mass analyzer* in which ions are separated by their mass-to-charge ratio, *m/z;* and a *detector* in which the separated ions are observed and counted. Since the number of charges *z* on each ion is usually 1, the value of *m/z* for an ion is simply its mass, *m*. Masses up to approximately 2500 atomic mass units (amu) can be analyzed with an accuracy up to four decimal places.

A typical mass spectrum, like that of hexane in Figure 13.1, is normally presented as a bar graph with *m/z* on the horizontal axis and relative abundance of ions on the vertical axis. The tallest peak, assigned an intensity of 100%, is called the *base peak,* and the peak that corresponds to the unfragmented ion is called the *parent peak,* or the *molecular ion (M^+)*. Hexane, for instance, shows $M^+ = 86$, corresponding to a formula of C_6H_{14}.

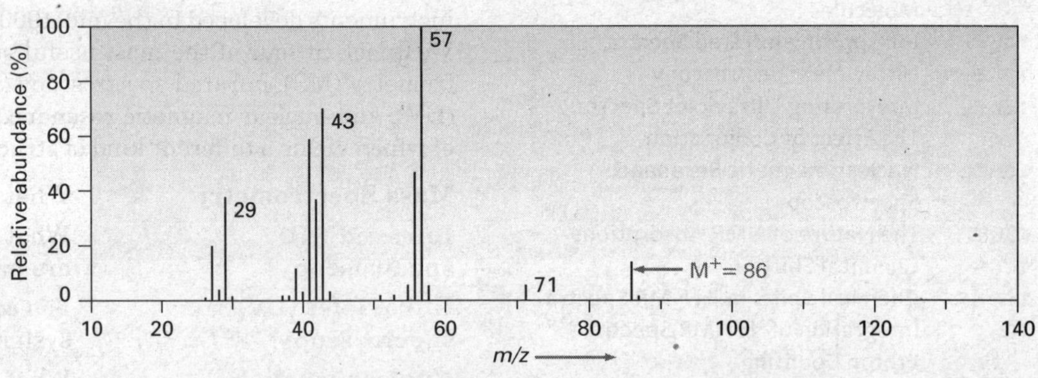

Figure 13.1 Mass spectrum of hexane (C_6H_{14}; MW = 86). The molecular ion is at $m/z = 86$, the base peak is at $m/z = 57$, and numerous other ions are present.

In addition to giving a molecular ion, most molecules fragment in the mass spectrometer, giving rise to numerous other ions that can provide structural information when interpreted. Hexane, for instance, shows peaks at $m/z = 71$ corresponding to loss of a CH_3- group, $m/z = 57$ corresponding to loss of an CH_3CH_2- group, $m/z = 43$ corresponding to loss of a $CH_3CH_2CH_2-$ group, and so forth.

You might also note in Figure 13.1 that in addition to the peak for the molecular ion at $m/z = 86$, there is also a small peak at $m/z = 87$ because of the presence of different isotopes in molecules. Although ^{12}C is the most abundant carbon isotope, a small amount (1.10% natural abundance) of ^{13}C is also present. Thus, a certain percentage of the molecules analyzed in the mass spectrometer contain a ^{13}C atom, giving rise to the observed M+1 peak.

Problem 13.1 The mass spectrum of the product formed in the addition of a halogen X_2 to but-2-ene showed a peak at m/z = 310. What halogen was used?

Problem 13.2 Oxidation of butan-1-ol to give butanal gives a by-product whose mass spectrum shows peaks at m/z = 88 and m/z = 72. What do you think the impurity might be?

13.2 Spectroscopy and the Electromagnetic Spectrum

Visible light, X rays, microwaves, radio waves, and so forth, are all different kinds of *electromagnetic radiation*. Collectively, they make up the **electromagnetic spectrum**, shown in Figure 13.2. The electromagnetic spectrum is arbitrarily divided into regions, with the familiar visible region accounting for only a small portion of the overall spectrum, from 3.8×10^{-7} to 7.8×10^{-7} m in wavelength. The visible region is flanked by the infrared and ultraviolet regions.

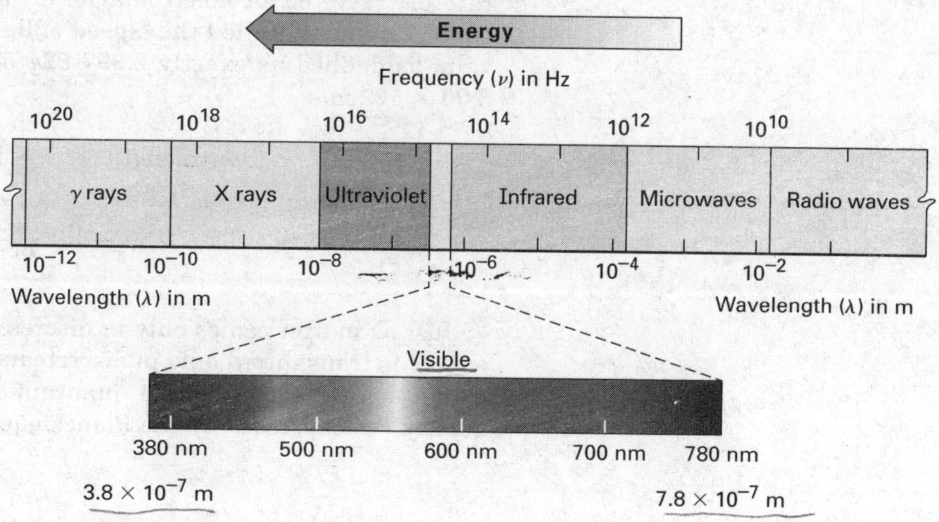

Figure 13.2 The electromagnetic spectrum covers a continuous range of wavelengths and frequencies, from radio waves at the low-frequency end to gamma (γ) rays at the high-frequency end. The familiar visible region accounts for only a small portion near the middle of the spectrum.

Electromagnetic radiation is often said to have dual behavior. In some respects it has the properties of a particle, called a *photon*, yet in other respects it behaves as an energy wave. Like all waves, electromagnetic radiation is characterized by a frequency, a wavelength, and an amplitude (Figure 13.3). The **wavelength, λ** (Greek lambda), is the distance from one wave maximum to the next. The **frequency, ν** (Greek nu), is the number of waves that pass by a fixed point per unit time, usually given in reciprocal seconds (s^{-1}), or **hertz, Hz** (1 Hz = 1 s^{-1}). The **amplitude** is the height of a wave, measured from midpoint to peak. The intensity of radiant energy, whether a feeble glow or a blinding glare, is proportional to the square of the wave's amplitude.

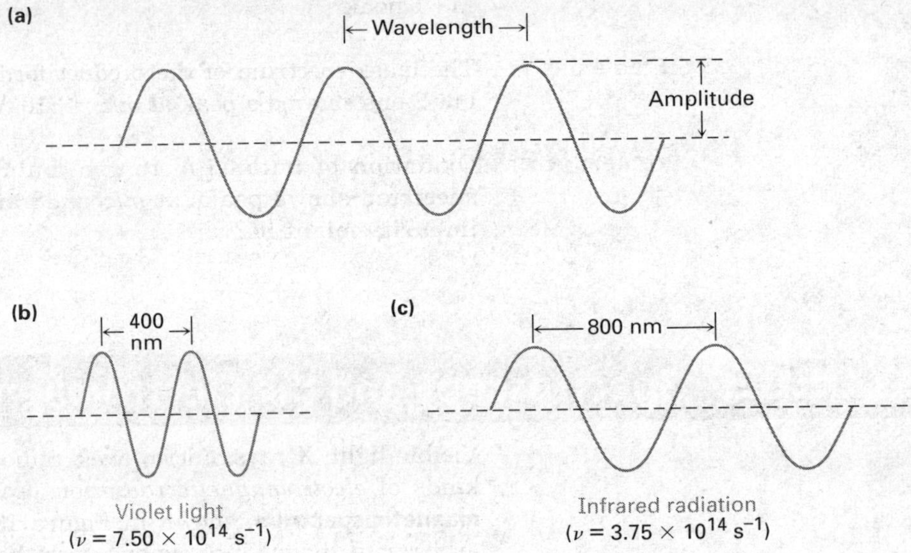

Figure 13.3 Electromagnetic waves are characterized by a wavelength, a frequency, and an amplitude. **(a)** Wavelength (λ) is the distance between two successive wave maxima. Amplitude is the height of the wave measured from the center. **(b), (c)** What we perceive as different kinds of electromagnetic radiation are simply waves with different wavelengths and frequencies.

Multiplying the length of a wave in meters (m) by its frequency in reciprocal seconds (s^{-1}) gives the speed of the wave in meters per second (m/s). The rate of travel of all electromagnetic radiation in a vacuum is a constant value, commonly called the "speed of light" and abbreviated *c*. Its numerical value is defined as exactly 2.997 924 58 × 10^8 m/s, usually rounded off to 3.00 × 10^8 m/s.

$$\text{Wavelength} \times \text{Frequency} = \text{Speed}$$

$$\lambda\,(\text{m}) \times \nu\,(\text{s}^{-1}) = c\,(\text{m/s})$$

$$\lambda = \frac{c}{\nu} \quad \text{or} \quad \nu = \frac{c}{\lambda}$$

Just as matter comes only in discrete units, called atoms, electromagnetic energy is transmitted only in discrete amounts, called *quanta*. The amount of energy, ϵ, corresponding to 1 quantum of energy, or 1 *photon*, of a given frequency ν is expressed by the Planck equation

$$\varepsilon = h\nu = \frac{hc}{\lambda}$$

where h = Planck's constant [6.62×10^{-34} (J · s) = 1.58×10^{-34} (cal · s)]
c = speed of light (3.00×10^8 m/s)
λ = wavelength in meters

The Planck equation says that the energy of a given photon varies directly with its frequency ν but inversely with its wavelength λ. High frequencies and short wavelengths correspond to high-energy radiation such as gamma rays; low frequencies and long wavelengths correspond to low-energy radiation such as radio waves. Multiplying ϵ by Avogadro's number, N_A, gives the same equation in more familiar units, where E represents the energy of Avogadro's number (one "mole") of photons of wavelength λ:

$$E = \frac{N_A hc}{\lambda} = \frac{1.20 \times 10^{-4}\ \text{kJ/mol}}{\lambda\ \text{(in meters)}} \quad \text{or} \quad \frac{2.86 \times 10^{-5}\ \text{kcal/mol}}{\lambda\ \text{(in meters)}}$$

Infrared radiation with a wavelength 1.0×10^{-5} m, for instance, has an energy of 12 kJ/mol (2.9 kcal/mol).

When an organic compound is exposed to electromagnetic radiation, it absorbs energy of certain wavelengths but passes, or transmits, energy of other wavelengths. If we irradiate an organic compound with energy of many wavelengths and determine which are absorbed and which are transmitted, we can determine the **absorption spectrum** of the compound. The results are displayed on a plot of wavelength versus the amount of radiation transmitted.

An example of an absorption spectrum—that of ethanol exposed to infrared radiation—is shown in Figure 13.4. The horizontal axis records the wavelength, and the vertical axis records the intensity of the various energy absorptions in percent transmittance. The baseline corresponding to 0% absorption (or 100% transmittance) runs along the top of the chart, so a downward spike means that energy absorption has occurred at that wavelength.

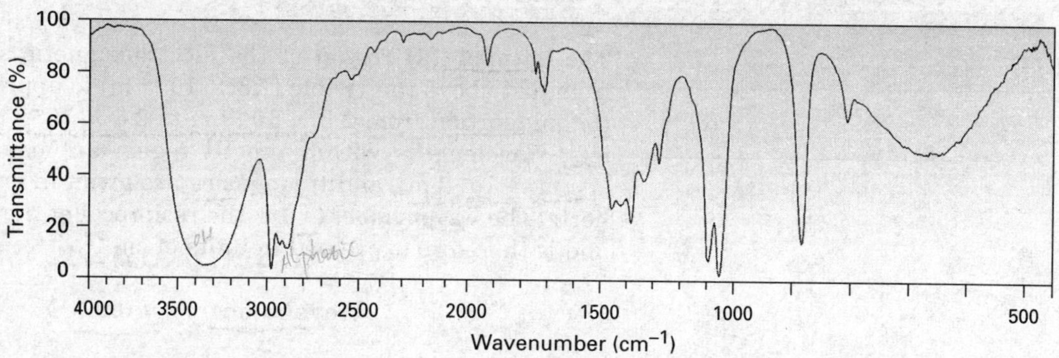

Figure 13.4 The infrared absorption spectrum of ethanol, CH_3CH_2OH. A transmittance of 100% means that all the energy is passing through the sample. A lower transmittance means that some energy is being absorbed. Thus, each downward spike corresponds to an energy absorption.

Worked Example 13.1

Converting from Frequency to Wavelength

What is the wavelength in meters of visible light with a frequency of 4.5×10^{14} Hz?

Strategy Frequency and wavelength are related by the equation $\lambda = c/\nu$, where c is the speed of light (3.0×10^8 m/s).

Solution

$$\lambda = \frac{3.0 \times 10^8\ \text{m/s}}{4.5 \times 10^{14}\ \text{s}^{-1}} = 6.7 \times 10^{-7}\ \text{m}$$

Worked Example 13.2

Calculating the Energy of Electromagnetic Waves

Which is higher in energy, FM radio waves with a frequency of 1.015×10^8 Hz (101.5 MHz) or visible light with a frequency of 5×10^{14} Hz?

Strategy Remember the equations $\epsilon = h\nu$ and $\epsilon = hc/\lambda$, which say that energy increases as frequency increases and as wavelength decreases.

Solution Because visible light has a higher frequency than radio waves, it is higher in energy.

Problem 13.3 How does the energy of infrared radiation with $\lambda = 1.0 \times 10^{-6}$ m compare with that of an X ray having $\lambda = 3.0 \times 10^{-9}$ m?

Problem 13.4 Which is higher in energy, radiation with $\nu = 4.0 \times 10^9$ Hz or radiation with $\lambda = 9.0 \times 10^{-6}$ m?

Problem 13.5 Cellphones use microwave radiation of around 1.9×10^9 Hz.

(a) What is the wavelength of cellphone radiation?

(b) Use the equation $E = 1.20 \times 10^{-4}$ (kJ/mol)/λ to calculate the amount of energy in cellphone radiation in kJ/mol. How does the amount of energy you calculated for a cellphone compare with the energy of visible red light (200 kJ/mol).

13.3 Infrared Spectroscopy of Organic Molecules

The **infrared (IR)** region of the electromagnetic spectrum covers the range from just above the visible (7.8×10^{-7} m) to approximately 10^{-4} m, but only the midportion from 2.5×10^{-6} m to 2.5×10^{-5} m is used by organic chemists. Wavelengths within the IR region are usually given in micrometers (1 μm = 10^{-6} m), and frequencies are given in *wavenumbers* rather than in hertz. The **wavenumber** ($\tilde{\nu}$) is the reciprocal of the wavelength in centimeters and is therefore expressed in units of cm^{-1}.

$$\text{Wavenumber:} \quad \tilde{\nu}\,(\text{cm}^{-1}) = \frac{1}{\lambda\ (\text{cm})}$$

Thus, the useful IR region is from 4000 to 400 cm^{-1}, corresponding to energies of 48.0 kJ/mol to 4.80 kJ/mol (11.5–1.15 kcal/mol).

Why does an organic molecule absorb some wavelengths of IR radiation but not others? All molecules have a certain amount of energy and are in constant motion. Their bonds stretch and contract, atoms wag back and forth, and other molecular vibrations occur. Some of the kinds of allowed vibrations are shown.

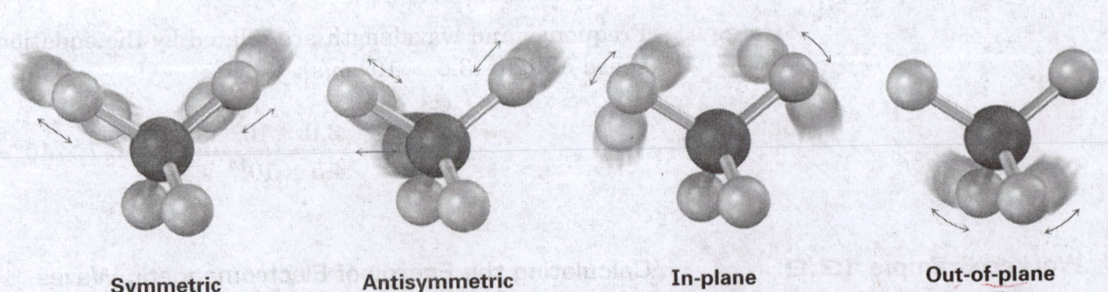

The amount of energy a molecule contains is not continuously variable but is *quantized.* That is, a molecule can stretch or bend only at specific frequencies corresponding to specific energy levels. Take bond stretching, for instance. Although we usually speak of bond lengths as if they were fixed, the numbers given are really averages. In fact, a typical C–H bond with an average bond length of 110 pm is actually vibrating at a specific frequency, alternately stretching and contracting as if there were a spring connecting the two atoms.

When the molecule is irradiated with electromagnetic radiation, *energy is absorbed if the frequency of the radiation matches the frequency of the vibration.* The result of this energy absorption is an increased amplitude for the vibration; in other words, the "spring" connecting the two atoms stretches and compresses a bit further. Since each frequency absorbed by a molecule corresponds to a specific molecular motion, we can find what kinds of motions a molecule has by measuring its IR spectrum. By then interpreting those motions, we can find out what kinds of bonds (functional groups) are present in the molecule.

IR spectrum → What molecular motions? → What functional groups?

13.4 Interpreting Infrared Spectra

The full interpretation of an IR spectrum is difficult because most organic molecules are so large that they have dozens of different molecular motions and thus have dozens of absorptions. Fortunately, we don't need to interpret an IR spectrum fully to get useful information because most functional groups have characteristic IR absorptions that don't change from one compound to another. The C=O absorption of a ketone is almost always in the range 1670 to 1750 cm^{-1}, the O–H absorption of an alcohol is almost always in the range 3400 to 3650 cm^{-1}, the C=C absorption of an alkene is almost always in the range 1640 to 1680 cm^{-1}, and so forth. By learning where characteristic functional-group absorptions occur, it's possible to get structural information from IR spectra. Table 13.1 lists the characteristic IR bands of some common functional groups.

Table 13.1 Characteristic Infrared Absorptions of Some Functional Groups

Functional Group	Absorption (cm^{-1})	Intensity	Functional Group	Absorption (cm^{-1})	Intensity
Alkane			Amine		
C–H	2850–2960	Medium	N–H	3300–3500	Medium
Alkene			C–N	1030–1230	Medium
=C–H	3020–3100	Medium	Carbonyl compound		
C=C	1640–1680	Medium	C=O	1670–1780	Strong
Alkyne			Aldehyde	1725	Strong
≡C–H	3300	Strong	Ketone	1715	Strong
C≡C	2100–2260	Medium	Ester	1735	Strong
Alkyl halide			Amide	1690	Strong
C–Cl	600–800	Strong	Carboxylic acid	1710	Strong, broad
C–Br	500–600	Strong	Carboxylic acid		
Alcohol			O–H	2500–3100	Strong, broad
O–H	3400–3650	Strong, broad	Nitrile		
C–O	1050–1150	Strong	C≡N	2210–2260	Medium
Arene			Nitro		
C–H	3030	Weak	NO_2	1540	Strong
Aromatic ring	1660–2000	Weak			
	1450–1600	Medium			

Look at the spectra of cyclohexanol and cyclohexanone in Figure 13.5 to see how IR spectroscopy can be used. Although both spectra contain many peaks, the characteristic absorptions of the different functional groups allow the compounds to be distinguished. Cyclohexanol shows a characteristic alcohol O–H absorption at 3300 cm^{-1} and a C–O absorption at 1060 cm^{-1}; cyclohexanone shows a characteristic ketone C=O peak at 1715 cm^{-1}.

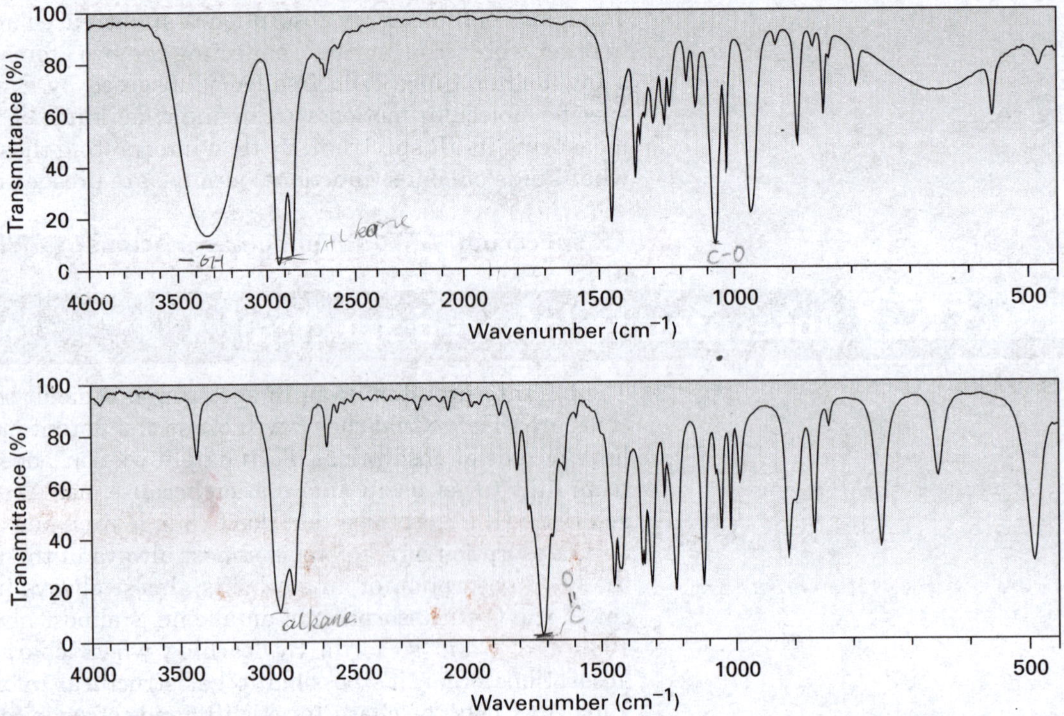

Figure 13.5 Infrared spectra of **(a)** cyclohexanol and **(b)** cyclohexanone. Spectra like these are easily obtained in minutes with milligram amounts of material.

One further point about infrared spectroscopy: it's also possible to obtain structural information from an IR spectrum by noticing which absorptions are *not* present. If the spectrum of an unknown does *not* have an absorption near 3400 cm^{-1}, the unknown is not an alcohol; if the spectrum does not have an absorption near 1715 cm^{-1}, the unknown is not a ketone; and so on.

It helps in remembering the positions of various IR absorptions to divide the infrared range from 4000 to 200 cm^{-1} into four parts, as shown in Figure 13.6.

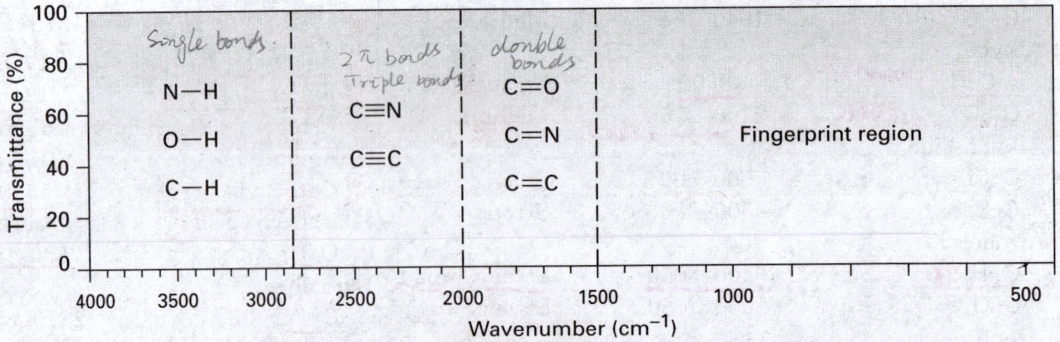

Figure 13.6 The four regions of the infrared spectrum—single bonds to hydrogen, triple bonds, double bonds, and fingerprint.

- The region from 4000 to 2500 cm^{-1} corresponds to absorptions caused by N–H, C–H, and O–H single-bond stretching motions. N–H and O–H bonds absorb in the 3300 to 3600 cm^{-1} range; C–H bond stretching occurs near 3000 cm^{-1}.

- The region from 2500 to 2000 cm^{-1} is where triple-bond stretching occurs. Both C≡N and C≡C bonds absorb here.
- The region from 2000 to 1500 cm^{-1} is where double bonds (C=O, C=N, and C=C) absorb. Carbonyl groups generally absorb in the range 1670 to 1780 cm^{-1}, and alkene stretching normally occurs in the narrow range 1640 to 1680 cm^{-1}.
- The region below 1500 cm^{-1} is the fingerprint portion of the IR spectrum. A large number of absorptions due to a variety of C–C, C–O, C–N, and C–X single-bond vibrations occur here, forming a unique pattern that acts as an identifying fingerprint of each organic compound.

Worked Example 13.3 Predicting Infrared Absorptions

Refer to Table 13.1 and make educated guesses about the functional groups that cause the following IR absorptions:

(a) 1735 cm^{-1} **(b)** 3500 cm^{-1}

Solution **(a)** An absorption at 1735 cm^{-1} is in the carbonyl-group region of the IR spectrum, probably an ester.

(b) An absorption at 3500 cm^{-1} is in the –OH (alcohol) region.

Worked Example 13.4 Using Infrared Spectroscopy

Acetone and prop-2-en-1-ol ($H_2C{=}CHCH_2OH$) are isomers. How could you distinguish them by IR spectroscopy?

Strategy Identify the functional groups in each molecule, and refer to Table 13.1.

Solution Table 13.1 shows that acetone has a strong C=O absorption at 1715 cm^{-1}, while prop-2-en-1-ol has an –OH absorption at 3500 cm^{-1} and a C=C absorption at 1660 cm^{-1}.

Problem 13.6 What functional groups might molecules contain if they show IR absorptions at the following frequencies?

(a) 1715 cm^{-1}
(b) 1540 cm^{-1}
(c) 2210 cm^{-1}
(d) 1720 and 2500–3100 cm^{-1}
(e) 3500 and 1735 cm^{-1}

Problem 13.7 How might you use IR spectroscopy to help distinguish between the following pairs of isomers?

(a) CH_3CH_2OH and CH_3OCH_3

(b) 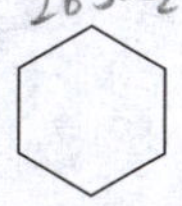 and $CH_3CH_2CH_2CH_2CH{=}CH_2$

(c) $CH_3CH_2\overset{O}{\overset{\|}{C}}OH$ and $HOCH_2CH_2\overset{O}{\overset{\|}{C}}H$

Problem 13.8 Where might the following compound have IR absorptions?

(a)
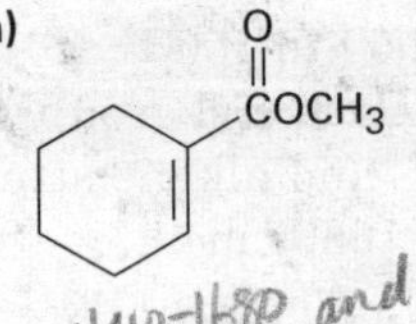

(b)
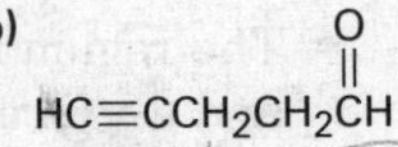

(c)
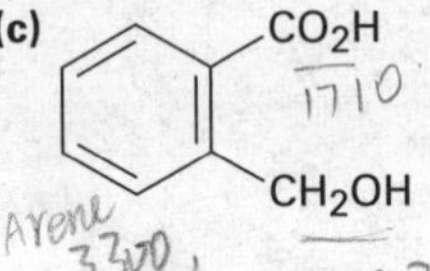

13.5 Ultraviolet Spectroscopy

The **ultraviolet (UV)** region of the electromagnetic spectrum extends from the low-wavelength end of the visible region (4×10^{-7} m) to the long-wavelength end of the X-ray region (10^{-8} m). The part of greatest interest to organic chemists, though, is the narrow range from 2×10^{-7} m to 4×10^{-7} m. Absorptions in this region are measured in *nanometers* (nm), where 1 nm = 10^{-9} m = 10^{-7} cm. Thus, the ultraviolet range of interest is from 200 to 400 nm.

We saw in the previous section that when a molecule is subjected to IR irradiation, the energy absorbed corresponds to the amount necessary to increase molecular vibrations. When UV radiation is used, the energy absorbed corresponds to the amount necessary to raise the energy level of a π electron in a conjugated molecule.

A typical UV spectrum—that of buta-1,3-diene—is shown in Figure 13.7. Unlike IR spectra, which generally have many peaks, UV spectra are usually quite simple. Often, there is only a single broad peak, which is identified by noting the wavelength at the very top, indicated as λ_{max}. For buta-1,3-diene, λ_{max} = 217 nm. Note that UV spectra differ from IR spectra in the way they are presented. For historical reasons, IR spectra are usually displayed so that the baseline corresponding to zero absorption runs across the top of the chart and a valley indicates an absorption, whereas UV spectra are displayed with the baseline at the bottom of the chart so that a peak indicates an absorption.

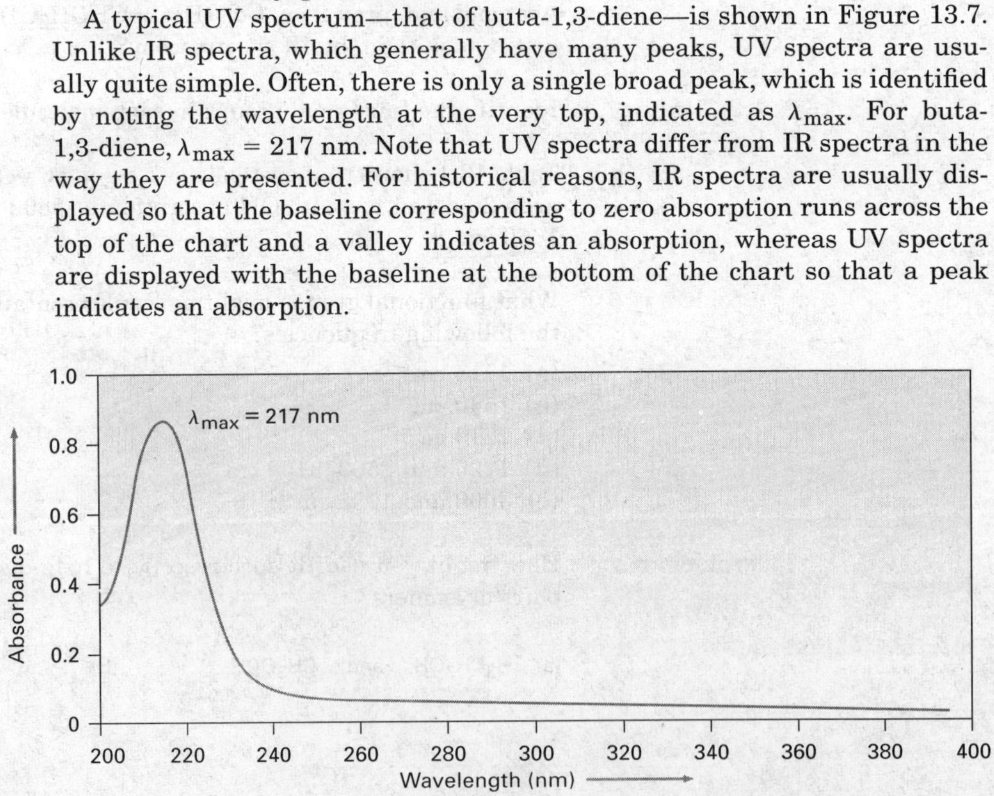

Figure 13.7 Ultraviolet spectrum of buta-1,3-diene.

The amount of UV light absorbed by a sample is expressed as the sample's absorbance (A), defined by the equation

$$A = \epsilon \cdot c \cdot l$$

where ϵ = the *molar absorptivity* of the molecule in units of L/(mol · cm) The higher the molar absorptivity, the greater the amount of radiation absorbed by the compound.

c = the concentration of a solution of sample in mol/L

l = the sample pathlength in cm

A particularly important use of this equation comes from rearranging it to the form $c = A/(\epsilon \cdot l)$, which lets us measure the concentration of a sample in solution when A, ϵ, and l are known. As an example, β-carotene, the pigment responsible for the orange color of carrots, has ϵ = 138,000 L/(mol · cm). If a sample of β-carotene is placed in a cell with a pathlength of 1.0 cm and the UV absorbance reads 0.37, then the concentration of β-carotene in the sample is

$$c = \frac{A}{\varepsilon l} = \frac{0.37}{\left(1.38 \times 10^5 \dfrac{\text{L}}{\text{mol} \cdot \text{cm}}\right)(1.00 \text{ cm})}$$
$$= 2.7 \times 10^{-6} \text{ M}$$

Problem 13.9 What is the concentration of cytosine, a constituent of nucleic acids, if its molar absorptivity is 6.1×10^3 L/(mol · cm) and its absorbance is 0.20 in a cell with a 1.0 cm pathlength?

13.6 Interpreting Ultraviolet Spectra: The Effect of Conjugation

The wavelength of radiation necessary to raise the energy of a π electron in a conjugated molecule depends on the nature of the molecule's π electron system. One of the most important factors is the extent of conjugation (Section 4.8). Thus, by measuring the UV spectrum of an unknown, we can derive structural information about the nature of any conjugated π electron system present in a molecule.

It turns out that the energy required for an electronic transition decreases as the extent of conjugation increases. Thus, buta-1,3-*di*ene absorbs at λ_{max} = 217 nm, hexa-1,3,5-*tri*ene absorbs at λ_{max} = 258 nm, and octa-1,3,5,7-*tetra*ene absorbs at λ_{max} = 290 nm. (Remember: longer wavelength means lower energy.)

Other kinds of conjugated π electron systems besides dienes and polyenes also show ultraviolet absorptions. Conjugated enones, such as but-3-en-2-one, and aromatic molecules, such as benzene, also have characteristic UV absorptions that aid in structure determination. The UV absorption maxima of some representative conjugated molecules are given in Table 13.2.

Table 13.2 Ultraviolet Absorption Maxima of Some Conjugated Molecules

Name	Structure	λ_{max} (nm)
2-Methybuta-1,3-diene	$H_2C{=}C(CH_3){-}CH{=}CH_2$	220
Cyclohexa-1,3-diene	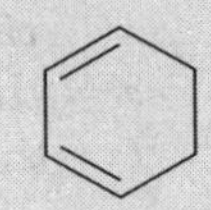	256
Hexa-1,3,5-triene	$H_2C{=}CH{-}CH{=}CH{-}CH{=}CH_2$	258
Octa-1,3,5,7-tetraene	$H_2C{=}CH{-}CH{=}CH{-}CH{=}CH{-}CH{=}CH_2$	290
But-3-en-2-one	$H_2C{=}CH{-}C({=}O){-}CH_3$	219
Benzene	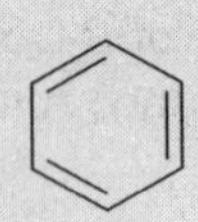	203

Worked Example 13.5 Using Ultraviolet Spectroscopy

Hexa-1,5-diene and hexa-1,3-diene are isomers. How can you distinguish them by UV spectroscopy?

$H_2C{=}CHCH_2CH_2CH{=}CH_2$ **Hexa-1,5-diene**

$CH_3CH_2CH{=}CHCH{=}CH_2$ **Hexa-1,3-diene**

Strategy Remember that only conjugated molecules show UV absorptions.

Solution Hexa-1,3-diene is a conjugated diene, but hexa-1,5-diene is nonconjugated. Only the conjugated isomer shows a UV absorption above 200 nm.

Problem 13.10 Which of the following compounds show UV absorptions in the range 200 to 400 nm?

(a) 

(b) 

(c) $H_2C{=}CHCOCH_3$ (with C=O)

(d) 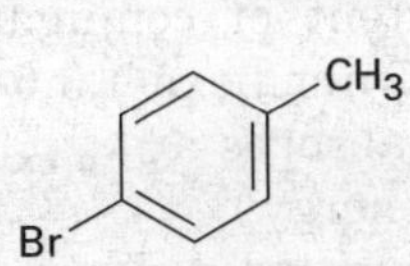

(e) 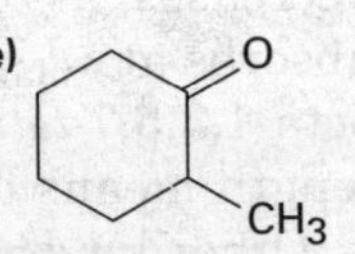

(f) 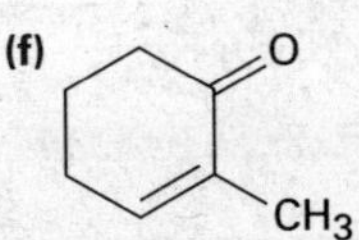

Problem 13.11 How can you distinguish between hexa-1,3-diene and hexa-1,3,5-triene by UV spectroscopy?

13.7 Nuclear Magnetic Resonance Spectroscopy

We've seen up to this point that IR spectroscopy provides information about a molecule's functional groups and that UV spectroscopy provides information about a molecule's conjugated π electron system. **Nuclear magnetic resonance (NMR) spectroscopy** complements these techniques by providing a "map" of the carbon–hydrogen framework in an organic molecule. Taken together, IR, UV, and NMR spectroscopies often make it possible to find the structures of even very complex molecules.

How does NMR spectroscopy work? Many kinds of nuclei, including ^1H and ^{13}C, behave as if they were spinning about an axis. Because they're positively charged, these spinning nuclei act like tiny magnets and interact with an external magnetic field (denoted B_0). In the absence of an external magnetic field, the nuclear spins of magnetic nuclei are oriented randomly. When a sample containing these nuclei is placed between the poles of a strong magnet, however, the nuclei adopt specific orientations, much as a compass needle orients in the earth's magnetic field.

A spinning ^1H or ^{13}C nucleus can orient so that its own tiny magnetic field is aligned either with (parallel to) or against (antiparallel to) the external field. The two orientations don't have the same energy and therefore aren't equally likely. The parallel orientation is slightly lower in energy, making this spin state slightly favored over the antiparallel orientation (Figure 13.8).

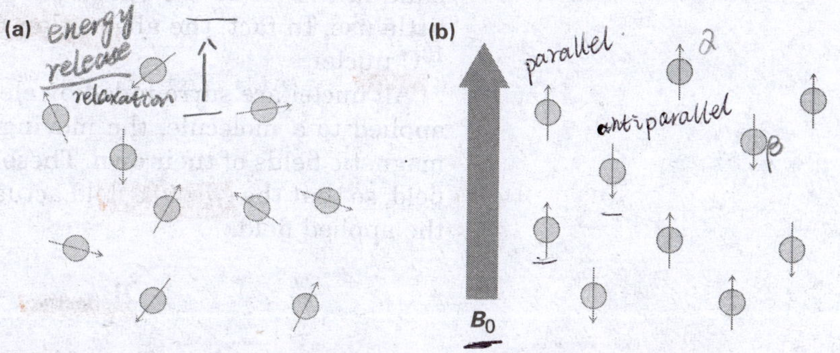

Figure 13.8 **(a)** Nuclear spins are oriented randomly in the absence of an external magnetic field but **(b)** have a specific orientation in the presence of an external field B_0. Some of the spins (red) are aligned parallel to the external field and others (blue) are antiparallel. The parallel spin state is lower in energy and therefore favored.

If the oriented nuclei are now irradiated with electromagnetic radiation of the right frequency, energy absorption occurs and the lower-energy state "spin-flips" to the higher-energy state. When this spin-flip occurs, the nuclei are said to be in resonance with the applied radiation—hence the name *nuclear magnetic resonance.*

The exact frequency necessary for resonance depends both on the strength of the external magnetic field and on the identity of the nuclei. If a very strong field is applied, the energy difference between the two spin states is larger so that higher-energy (higher-frequency) radiation is required. If a weaker magnetic field is applied, less energy is needed to effect the transition between nuclear spin states.

In practice, superconducting magnets that produce enormously powerful fields up to 21.2 tesla (T) are sometimes used, but field strengths in the range of 4.7 to 7.0 T are more common. At a magnetic field strength of 4.7 T, so-called radiofrequency (rf) energy in the 200 MHz range (1 MHz = 10^6 Hz) brings a ^{1}H nucleus into resonance, and rf energy of 50 MHz brings a ^{13}C nucleus into resonance. At the highest field strength currently available in commercial instruments (21.2 T), 900 MHz energy is required for ^{1}H spectroscopy.

Problem 13.12 NMR spectroscopy uses electromagnetic radiation with a frequency of 1×10^8 Hz. Is this a greater or lesser amount of energy than that used by IR spectroscopy?

13.8 The Nature of NMR Absorptions

From the description thus far, you might expect all ^{1}H nuclei in a molecule to absorb energy at the same frequency and all ^{13}C nuclei to absorb at the same frequency. If so, we would observe only a single NMR absorption band in the ^{1}H or ^{13}C spectrum of a molecule, a situation that would be of little use. In fact, the absorption frequency is not the same for all ^{1}H or all ^{13}C nuclei.

All nuclei are surrounded by electrons. When an external magnetic field is applied to a molecule, the moving electrons around nuclei set up tiny local magnetic fields of their own. These local fields act in opposition to the applied field, so that the *effective* field actually felt by the nucleus is a bit weaker than the applied field.

$$\boldsymbol{B}_{\text{effective}} = \boldsymbol{B}_{\text{applied}} - \boldsymbol{B}_{\text{local}}$$

In describing this effect of local fields, we say that the nuclei are **shielded** from the full effect of the applied field by their surrounding electrons. Because each specific nucleus in a molecule is in a slightly different electronic environment, each nucleus is shielded to a slightly different extent and the effective magnetic field felt by each is slightly different. These slight differences can be detected, and we therefore see a different NMR signal for each chemically distinct ^{1}H or ^{13}C nucleus in a molecule.

Figure 13.9 shows both the ^{1}H and the ^{13}C NMR spectra of methyl acetate, $CH_3CO_2CH_3$. The horizontal axis shows the effective field strength felt by the nuclei, and the vertical axis indicates the intensity of absorption of rf energy. Each peak in the NMR spectrum corresponds to a chemically distinct ^{1}H or ^{13}C nucleus in the molecule. Note, though, that ^{1}H and ^{13}C spectra can't be observed at the same time on the same spectrometer because different amounts of energy are required to spin-flip the different kinds of nuclei. The two spectra must be recorded separately.

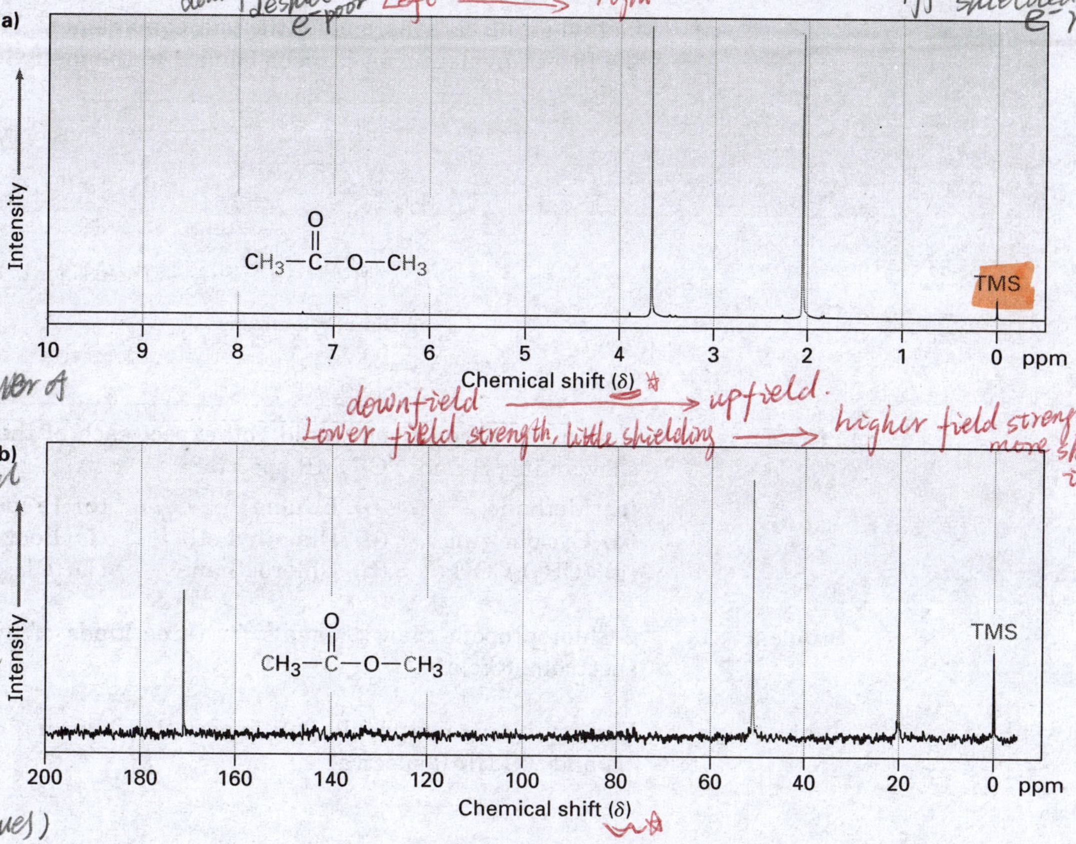

Figure 13.9 **(a)** The ^1H NMR spectrum and **(b)** the ^{13}C NMR spectrum of methyl acetate, $CH_3CO_2CH_3$. The small peak marked "TMS" at the far right of each spectrum is a calibration peak, as explained shortly.

The ^{13}C spectrum of methyl acetate in Figure 13.9b has three peaks, one for each of the three chemically distinct carbons in the molecule. The ^1H spectrum shows only *two* peaks, however, even though methyl acetate has *six* hydrogens. One peak is due to the $CH_3C{=}O$ hydrogens and the other to the $-OCH_3$ hydrogens. Because the three hydrogens in each methyl group have the same chemical (and magnetic) environment, they are shielded to the same extent and are said to be *equivalent.* Chemically equivalent nuclei always show a single absorption. The two methyl groups themselves, however, are nonequivalent, so the two sets of hydrogens absorb at different positions.

Worked Example 13.6 **Predicting the Number of NMR Absorptions**

How many signals would you expect *p*-dimethylbenzene to show in its ^1H and ^{13}C NMR spectra?

Strategy Look at the structure of the molecule, and count the number of kinds of chemically distinct ^1H and ^{13}C nuclei.

Solution Because of the molecule's symmetry, the two $-CH_3$ groups in *p*-dimethylbenzene are equivalent and all four ring hydrogens are equivalent. Thus, there are only two absorptions in the ^1H NMR spectrum. Also because of symmetry, there are only three absorptions in the ^{13}C NMR spectrum: one for the two equivalent methyl-group carbons, one for the four equivalent C–H ring carbons, and one for the two equivalent ring carbons bonded to the methyl groups.

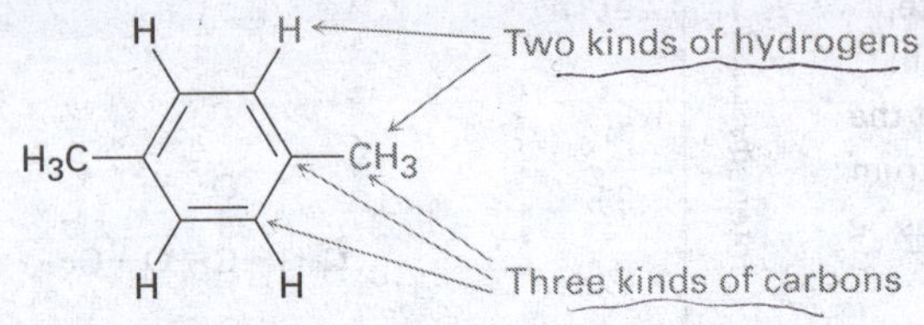

***p*-Dimethylbenzene**

Problem 13.13 How many absorptions would you expect each of the following compounds to show in its ^{1}H and ^{13}C NMR spectra?

(a) Methane **(b)** Ethane **(c)** Propane
(d) Cyclohexane **(e)** Dimethyl ether **(f)** Benzene
(g) $(CH_3)_3COH$ **(h)** Chloroethane **(i)** $(CH_3)_2C{=}C(CH_3)_2$

Problem 13.14 2-Chloropropene shows signals for three kinds of hydrogens in its ^{1}H NMR spectrum. Explain.

Problem 13.15 How many signals would you expect the following compound to show in its ^{1}H and ^{13}C NMR spectra?

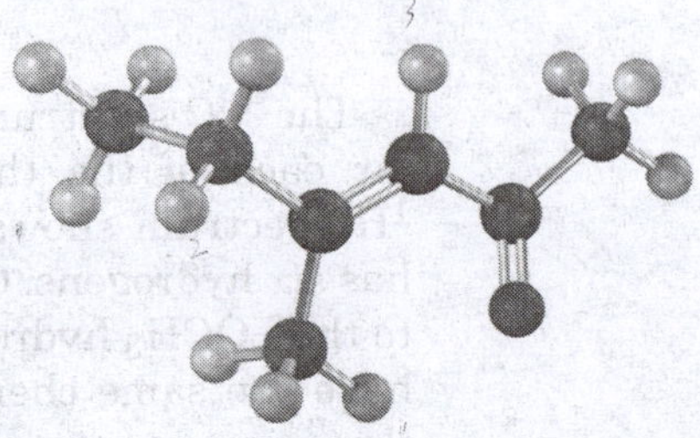

13.9 Chemical Shifts

NMR spectra are displayed on charts that show the applied field strength increasing from left to right (Figure 13.9). Thus, the left side of the chart is the low-field (or **downfield**) side, and the right side is the high-field (or **upfield**) side. Nuclei that absorb on the downfield side of the chart require a lower field strength for resonance, implying that they have relatively little shielding. Nuclei that absorb on the upfield side require a higher field strength for resonance, implying that they have more shielding.

To define the position of an absorption, the NMR chart is calibrated and a reference point is used. In practice, a small amount of tetramethylsilane [TMS; $(CH_3)_4Si$] is added to the sample so that a reference absorption peak is produced when the spectrum is run. TMS is used as reference for both ^{1}H and ^{13}C spectra because it produces in both a single peak that occurs upfield of most other absorptions in organic compounds. The ^{1}H and ^{13}C spectra of methyl acetate shown previously in Figure 13.9 have the TMS reference peak indicated.

The exact position on the chart at which a nucleus absorbs is called its **chemical shift**. The chemical shift of TMS is set as the zero point, and other peaks normally occur downfield, to the left on the chart. NMR charts are calibrated using an arbitrary scale called the **delta (δ) scale**, where 1 δ equals 1 part per million (ppm) of the spectrometer operating frequency. For example, if we were measuring the ^1H NMR spectrum of a sample using an instrument operating at 200 MHz, 1 δ would be 1 millionth of 200 million Hz, or 200 Hz. If we were measuring the spectrum using a 500 MHz instrument, 1 δ would be 500 Hz.

Although this method of calibrating NMR charts may seem complex, there's a good reason for it. As we saw earlier, the rf frequency required to bring a given nucleus into resonance depends on the spectrometer's magnetic field strength. But because there are many different kinds of spectrometers with many different magnetic field strengths, chemical shifts given in frequency units (Hz) vary greatly from one instrument to another. Thus, a resonance that occurs at 120 Hz downfield from TMS on one spectrometer might occur at 600 Hz downfield from TMS on another spectrometer with a more powerful magnet.

By using a system of measurement in which NMR absorptions are expressed in relative terms (parts per million relative to spectrometer frequency) rather than absolute terms (Hz), it's possible to compare spectra obtained on different instruments. *The chemical shift of an NMR absorption in δ units is constant, regardless of the operating frequency of the spectrometer.* A ^1H nucleus that absorbs at 2.0 δ on a 200 MHz instrument also absorbs at 2.0 δ on a 500 MHz instrument.

Worked Example 13.7 Converting between Hertz and δ Units

Cyclohexane shows an absorption at 1.43 δ in its ^1H NMR spectrum. How many hertz away from TMS is this on a spectrometer operating at 200 MHz? On a spectrometer operating at 300 MHz?

Strategy Remember that 1 δ = 1 ppm of spectrometer operating frequency.

Solution On a 200 MHz spectrometer, 1 δ = 200 Hz. Thus, 1.43 δ = 286 Hz away from the TMS reference peak. On a 300 MHz spectrometer, 1 δ = 300 Hz and 1.43 δ = 429 Hz.

Problem 13.16 When the ^1H NMR spectrum of acetone is recorded on a 100 MHz instrument, a single sharp resonance line at 2.1 δ is observed.

(a) How far away from TMS (in hertz) does the acetone absorption occur? 210 HZ

(b) What is the position of the acetone absorption in δ units on a 220 MHz instrument? 2.1 δ

(c) How many hertz away from TMS does the absorption in the 220 MHz spectrum correspond to? 460 HZ

Problem 13.17 The following ^1H NMR resonances were recorded on a spectrometer operating at 300 MHz. Convert each into δ units.

(a) $CHCl_3$, 2180 Hz 7.27 δ **(b)** CH_3Cl, 915 Hz 3.05 δ

(c) CH_3OH, 1040 Hz 3.47 δ **(d)** CH_2Cl_2, 1590 Hz 5.3 δ

13.10 Chemical Shifts in ^{1}H NMR Spectra

Everything we've said thus far about NMR spectroscopy applies to both ^{1}H and ^{13}C spectra. Now, let's focus only on ^{1}H NMR spectroscopy to see how it can be used in organic structure determination.

Most ^{1}H NMR absorptions occur in the range 0 to 10 δ, which can be divided into the five regions shown in Table 13.3. By remembering the positions of these regions, it's possible to tell at a glance what kinds of protons a molecule contains. (In speaking about NMR, the ^{1}H nucleus is often referred to as a *proton*.) In general, protons bonded to saturated, *sp*^{3}-hybridized carbons absorb at higher fields, whereas protons bonded to *sp*^{2}-hybridized carbons absorb at lower fields. Protons on carbons that are bonded to electronegative atoms, such as N, O, or halogen, also absorb at lower fields.

Table 13.3 Correlation of ^{1}H Chemical Shift with Environment

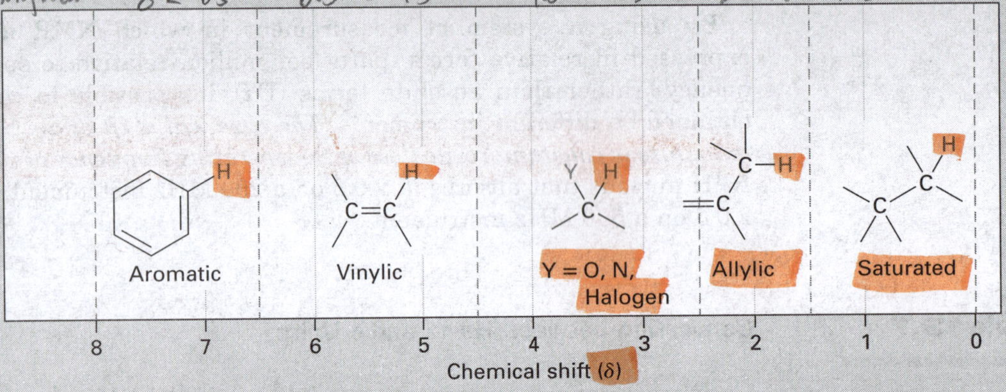

Type of hydrogen		Chemical shift (δ)
Reference	$Si(CH_3)_4$	0
Alkyl (primary)	$—CH_3$	0.7–1.3
Alkyl (secondary)	$—CH_2—$	1.2–1.6
Alkyl (tertiary)	—CH—	1.4–1.8
Allylic	C=C—C(H)—	1.6–2.2
Methyl ketone	$—C(=O)—CH_3$	2.0–2.4
Aromatic methyl	$Ar—CH_3$	2.4–2.7
Alkynyl	—C≡C—H	2.5–3.0
Alkyl halide	H—C—Hal	2.5–4.0

Type of hydrogen		Chemical shift (δ)
Alcohol	—C—O—H	2.5–5.0
Alcohol, ether	H—C—O—	3.3–4.5
Vinylic	C=C—H	4.5–6.5
Aryl	Ar—H	6.5–8.0
Aldehyde	—C(=O)—H	9.7–10.0
Carboxylic acid	—C(=O)—O—H	11.0–12.0

Worked Example 13.8 Predicting Chemical Shifts in ¹H NMR Spectra

Methyl 2,2-dimethylpropanoate $(CH_3)_3CCO_2CH_3$ has two peaks in its ¹H NMR spectrum. At what approximate chemical shifts do they come?

Strategy Identify the types of hydrogens in the molecule, and note whether each is alkyl, vinylic, or next to an electronegative atom. Then predict where each absorbs, using Table 13.3 as necessary.

Solution The $-OCH_3$ protons absorb around 3.5 to 4.0 δ because they are on carbon bonded to oxygen. The $(CH_3)_3C-$ protons absorb near 1.0 δ because they are typical alkane-like protons.

Problem 13.18 Each of the following compounds exhibits a single ¹H NMR peak. Approximately where would you expect each to absorb?

(a) Ethane **(b)** Acetone **(c)** Benzene **(d)** Trimethylamine

13.11 Integration of ¹H NMR Spectra: Proton Counting

Look at the ¹H NMR spectrum of methyl 2,2-dimethylpropanoate in Figure 13.10. There are two peaks, corresponding to the two kinds of protons, but the peaks aren't the same size. The peak at 1.20 δ, due to the $(CH_3)_3C-$ protons, is larger than the peak at 3.65 δ, due to the $-OCH_3$ protons.

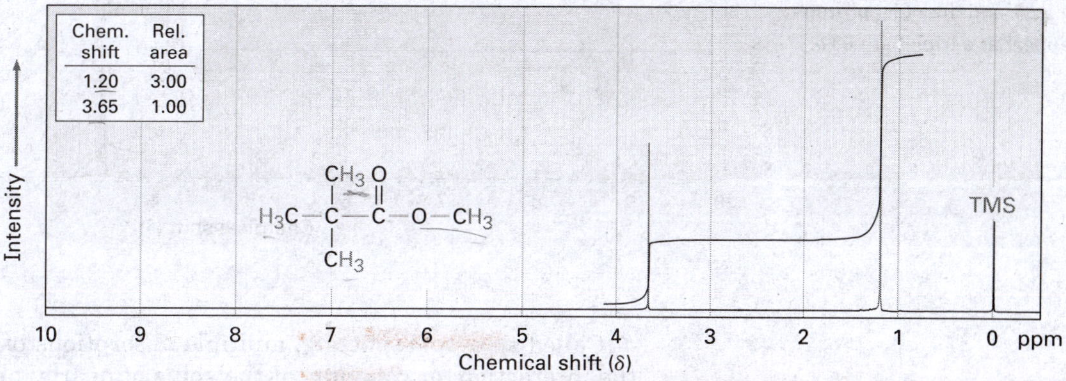

Figure 13.10 The ¹H NMR spectrum of methyl 2,2-dimethylpropanoate. Integrating the two peaks in a stair-step manner shows that they have a 1:3 ratio, corresponding to the 3:9 ratio of protons responsible. Modern instruments give a direct digital readout of relative peak areas.

The area under each peak is proportional to the number of protons causing that peak. By electronically measuring, or **integrating**, the area under each peak, it's possible to measure the relative numbers of the different kinds of protons in a molecule.

Modern NMR instruments provide a digital readout of relative peak areas, but an older, more visual method displays the integrated peak areas as a stair-step line, with the height of each step proportional to the area under the peak, and therefore proportional to the relative number of protons causing the peak. To compare the size of one peak against another, simply take a ruler and measure the heights of the various steps. For example, the two steps for the peaks in methyl 2,2-dimethylpropanoate are found to have a 1:3 (or 3:9) area ratio when integrated—exactly what we expect since the three $-OCH_3$ protons are equivalent and the nine $(CH_3)_3C-$ protons are equivalent.

Problem 13.19 How many peaks would you expect in the ^1H NMR spectrum of *p*-dimethylbenzene (*p*-xylene)? What ratio of peak areas would you expect to find on integration of the spectrum? Refer to Table 13.3 for approximate chemical shift values, and sketch what the spectrum might look like.

13.12 Spin–Spin Splitting in ^1H NMR Spectra

In the ^1H NMR spectra we've seen thus far, each chemically different proton in a molecule has given rise to a single peak. It often happens, though, that the absorption of a proton splits into *multiple* peaks, called a **multiplet**. For example, the ^1H NMR spectrum of bromoethane in Figure 13.11 indicates that the $-CH_2Br$ protons appear as four peaks (a *quartet*) centered at 3.42 δ and the $-CH_3$ protons appear as three peaks (a *triplet*) centered at 1.68 δ.

Figure 13.11 The ^1H NMR spectrum of bromoethane, CH_3CH_2Br. The $-CH_2Br$ protons appear as a quartet at 3.42 δ, and the $-CH_3$ protons appear as a triplet at 1.68 δ.

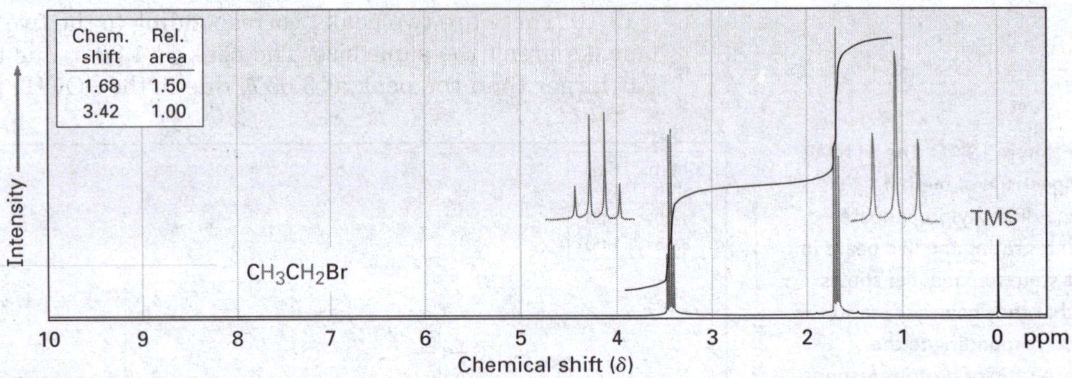

Called **spin–spin splitting**, multiple absorptions of a nucleus are caused by the interaction, or **coupling**, of the spins of nearby nuclei. In other words, the tiny magnetic field produced by one nucleus affects the magnetic field felt by neighboring nuclei. Look at the $-CH_3$ protons in bromoethane, for example. The three equivalent $-CH_3$ protons are neighbored by two other magnetic nuclei—the two protons on the adjacent $-CH_2Br$ group. Each of the neighboring $-CH_2Br$ protons has its own nuclear spin, which can align either with or against the applied field, producing a tiny effect that is felt by the $-CH_3$ protons.

There are three ways in which the spins of the two $-CH_2Br$ protons can align, as shown in Figure 13.12. If both proton spins align with the applied field, the total effective field felt by the neighboring $-CH_3$ protons is slightly larger than it would otherwise be. Consequently, the applied field necessary to cause resonance is slightly reduced. Alternatively, if one of the $-CH_2Br$ proton spins aligns with the field and one aligns against the field, there is no effect on the neighboring $-CH_3$ protons. (There are two ways this arrangement can occur, depending on which of the two proton spins aligns which way.) Finally, if both $-CH_2Br$ proton spins align against the applied field, the effective field felt by the $-CH_3$ protons is slightly smaller than it would otherwise be and the applied field needed for resonance is slightly increased.

Figure 13.12 The origin of spin–spin splitting in bromoethane. The nuclear spins of neighboring protons, indicated by horizontal arrows, align either with or against the applied field, causing the splitting of absorptions into multiplets.

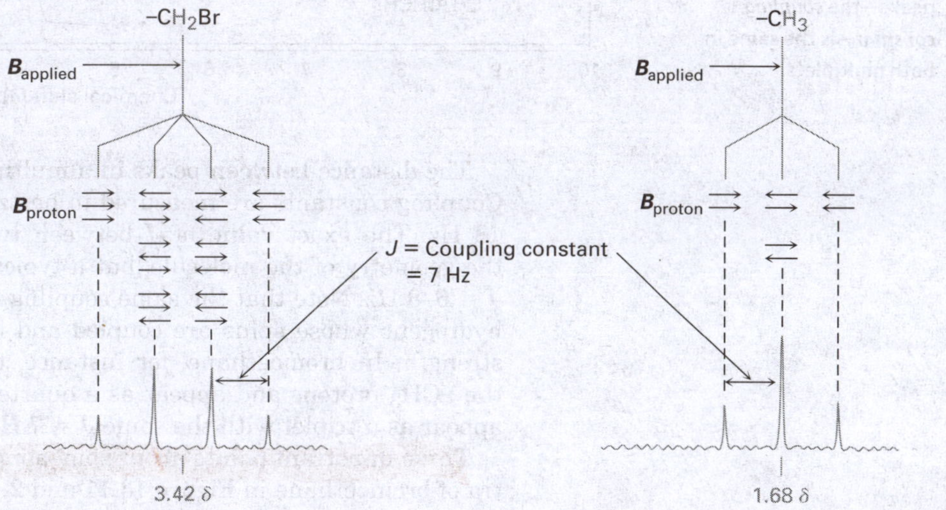

Any given molecule has only one of the three possible alignments of $-CH_2Br$ spins, but in a large collection of molecules, all three spin states are represented in a 1:2:1 statistical ratio. We therefore find that the neighboring $-CH_3$ protons come into resonance at three slightly different values of the applied field, and we see a 1:2:1 triplet in the NMR spectrum. One resonance is a little above where it would be without coupling, one is at the same place it would be without coupling, and the third resonance is a little below where it would be without coupling.

In the same way that the $-CH_3$ absorption of bromoethane is split into a triplet, the $-CH_2Br$ absorption is split into a quartet. The three spins of the neighboring $-CH_3$ protons can align in four possible combinations: all three with the applied field, two with and one against (three ways), one with and two against (three ways), or all three against. Thus, four peaks are produced for the $-CH_2Br$ protons in a 1:3:3:1 ratio.

As a general rule, called the **n + 1 rule**, protons that have n equivalent neighboring protons show $n + 1$ peaks in their NMR spectrum. For example, the spectrum of 2-bromopropane in Figure 13.13 shows a doublet at 1.71 δ and a seven-line multiplet, or *septet,* at 4.28 δ. The septet is caused by splitting of the –CHBr– proton signal by six equivalent neighboring protons on the two methyl groups ($n = 6$ leads to $6 + 1 = 7$ peaks). The doublet is due to signal splitting of the six equivalent methyl protons by the single –CHBr– proton ($n = 1$ leads to 2 peaks).

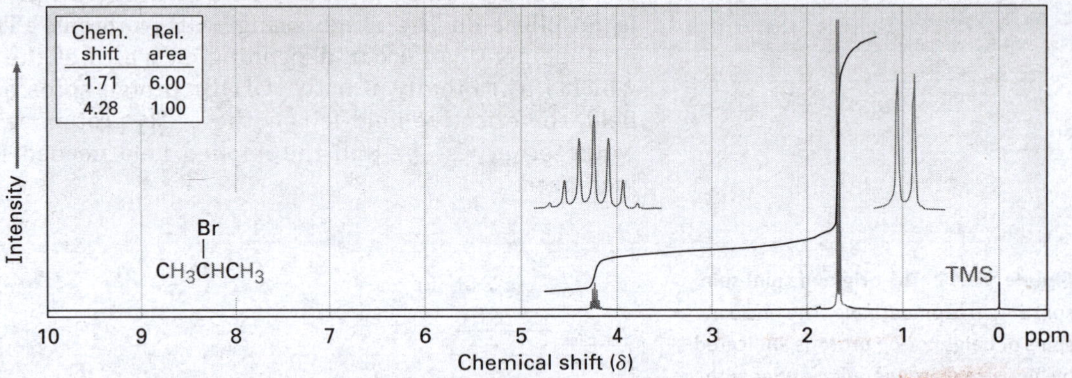

Figure 13.13 The ^{1}H NMR spectrum of 2-bromopropane. The $-CH_3$ proton signal at 1.71 δ is split into a doublet, and the –CHBr– proton signal at 4.28 δ is split into a septet. Note that the distance between peaks—the *coupling constant*—is the same in both multiplets.

The distance between peaks in a multiplet is called the **coupling constant, *J***. Coupling constants are measured in hertz and generally fall in the range 0 to 18 Hz. The exact value of J between two neighboring protons depends on the geometry of the molecule, but a typical value for an open-chain alkane is J = 6–8 Hz. Note that the same coupling constant is shared by both groups of hydrogens whose spins are coupled and is independent of spectrometer field strength. In bromoethane, for instance, the $-CH_2Br$ protons are coupled to the $-CH_3$ protons and appear as a quartet with J = 7 Hz. The $-CH_3$ protons appear as a triplet with the same J = 7 Hz coupling constant.

Three important points about spin–spin splitting are illustrated by the spectra of bromoethane in Figure 13.11 and 2-bromopropane in Figure 13.13.

- **Chemically equivalent protons don't show spin–spin splitting.** The equivalent protons may be on the same carbon or on different carbons, but their signals don't split.

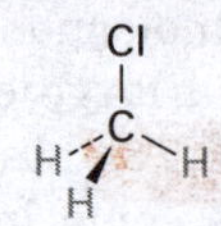

Three C–H protons are chemically equivalent; no splitting occurs.

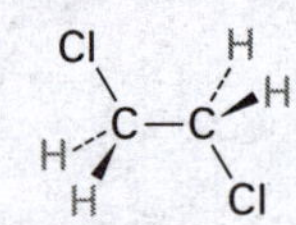

Four C–H protons are chemically equivalent; no splitting occurs.

- **The signal of a proton with *n* equivalent neighboring protons is split into a multiplet of *n* + 1 peaks.** Protons that are more than two carbon atoms apart usually don't split each other's signals.

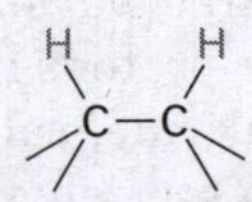

Splitting observed

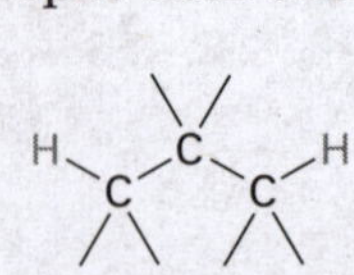

Splitting not usually observed

- **Two groups of protons coupled to each other have the same coupling constant *J*.**

Worked Example 13.9 Predicting Spin–Spin Splitting Patterns

Predict the splitting pattern for each kind of hydrogen in isopropyl propanoate, $CH_3CH_2CO_2CH(CH_3)_2$.

Strategy First, find how many different kinds of protons are present (there are four). Then, determine how many neighboring protons each kind has and apply the $n + 1$ rule.

Solution

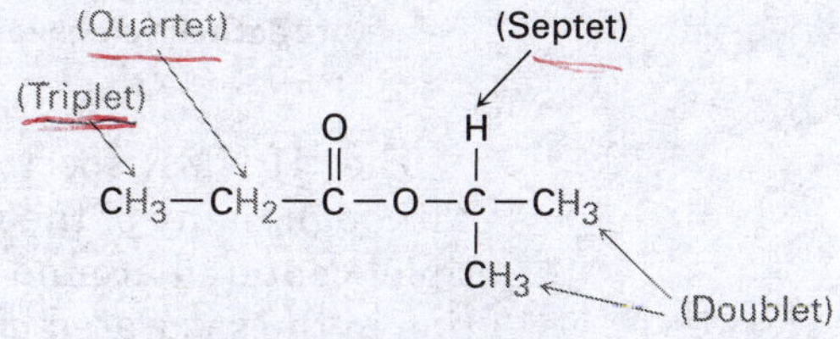

Isopropyl propanoate

Problem 13.20 Predict the splitting patterns for each proton in the following molecules:

(a) $CHBr_2CH_3$ (b) $CH_3OCH_2CH_2Br$ (c) $ClCH_2CH_2CH_2Cl$

(d)
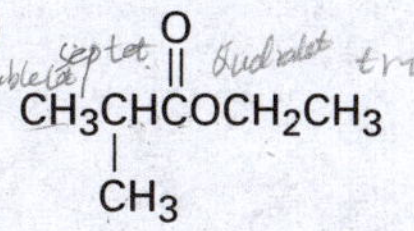

(e)
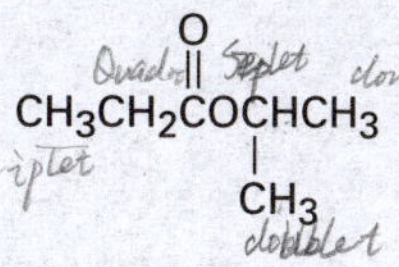

(f)
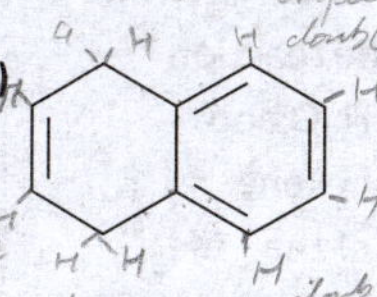

Problem 13.21 Propose structures for compounds that show the following ¹H NMR spectra:

(a) C_2H_6O; one singlet
(b) $C_3H_6O_2$; two singlets
(c) C_3H_7Cl; one doublet and one septet

Problem 13.22 Predict the splitting patterns for each kind of chemically distinct proton in the following molecule:

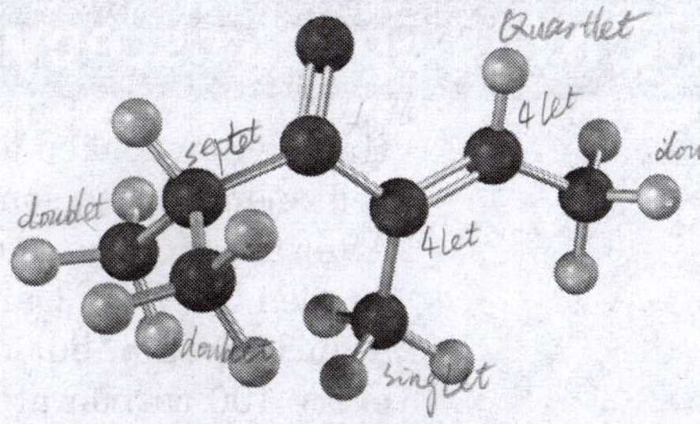

13.13 Uses of ¹H NMR Spectra

¹H NMR spectroscopy is used to help identify the product of nearly every reaction run in the laboratory. For example, we said in Section 4.3 that acid-catalyzed addition of H_2O to an alkene occurs with Markovnikov orientation

to give the more highly substituted alcohol. With the help of ^1H NMR, we can now prove this statement.

Does addition of H_2O to 1-methylcyclohexene yield 1-methylcyclohexanol or 2-methylcyclohexanol?

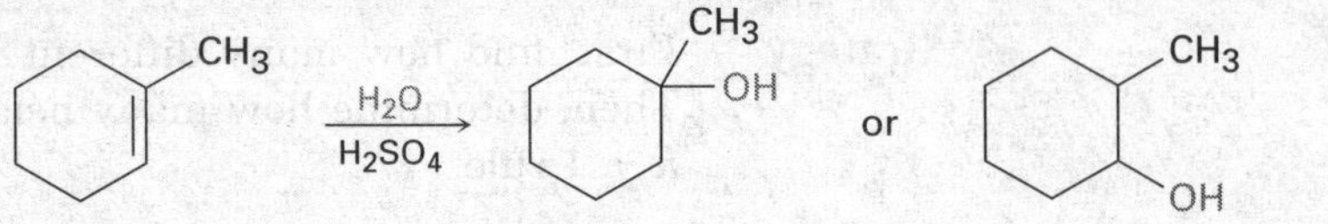

The ^1H NMR spectrum of the reaction product is shown in Figure 13.14. Although many of the ring protons overlap into a broad, poorly defined multiplet centered around 1.5 δ, the spectrum also shows a large singlet absorption in the saturated methyl region at 1.19 δ, indicating that the product has a methyl group with no neighboring hydrogens ($R_3C{-}CH_3$). Furthermore, the spectrum shows no absorptions around 4 δ, where we would expect the signal of an $R_2C\mathbf{H}OH$ proton to occur. Thus, the reaction product is 1-methylcyclohexanol.

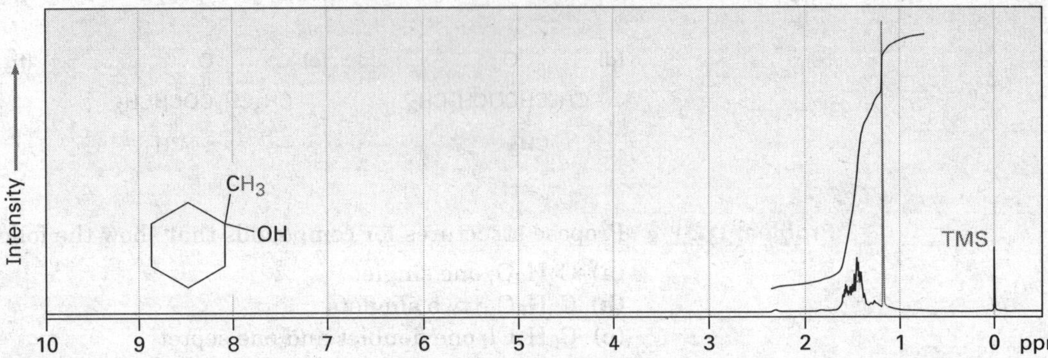

Figure 13.14 The ^1H NMR spectrum of the reaction product from H_2O and 1-methylcyclohexene. The presence of the $-CH_3$ absorption at 1.19 δ and the absence of any absorptions near 4 δ identify the product as 1-methylcyclohexanol.

13.14 ^{13}C NMR Spectroscopy

Having now looked at ^1H NMR spectroscopy, let's take a brief look at ^{13}C NMR before ending. In some ways, it's surprising that carbon NMR is even possible. After all, ^{12}C, the most abundant carbon isotope, has no nuclear spin and can't be seen by NMR. Carbon-13 is the only naturally occurring carbon isotope with a nuclear spin, but its natural abundance is only 1.1%. Thus, only about 1 of every 100 carbon atoms in an organic molecule is observable by NMR. Fortunately, the technical problems caused by this low abundance have been overcome by computer techniques, and ^{13}C NMR is a routine structural tool.

At its simplest, ^{13}C NMR makes it possible to count the number of carbon atoms in a molecule. Look at the ^{13}C NMR spectrum of methyl acetate shown previously in Figure 13.9b, for instance. Methyl acetate has three nonequivalent carbon atoms and thus has three peaks in its ^{13}C NMR spectrum.

(Coupling between adjacent carbon atoms isn't seen, because the low natural abundance of ^{13}C makes it unlikely that two such nuclei will be adjacent in a molecule.)

Most ^{13}C resonances are between 0 and 220 δ downfield from the TMS reference line, with the exact chemical shift of each ^{13}C resonance dependent on that carbon's electronic environment within the molecule. Figure 13.15 shows the correlation of chemical shift with environment.

Figure 13.15 Chemical shift correlations for ¹³C NMR.

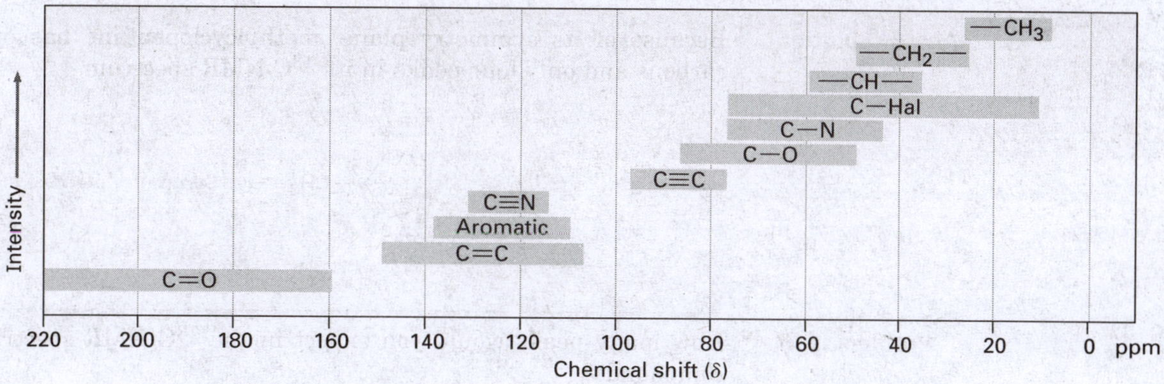

The factors that determine chemical shifts are complex, but it's possible to make some generalizations. One trend is that a carbon's chemical shift is affected by the electronegativity of nearby atoms. Carbons bonded to oxygen, nitrogen, or halogen absorb downfield (to the left) of typical alkane carbons. Another trend is that sp^3-hybridized carbons absorb in the range 0 to 90 δ, while sp^2 carbons absorb in the range 110 to 220 δ. Carbonyl carbons (C=O) are particularly distinct in the ^{13}C NMR spectrum and are always found at the extreme low-field end of the chart, in the range 160 to 220 δ. For example, the ^{13}C NMR spectrum of *p*-bromoacetophenone in Figure 13.16 shows an absorption for the carbonyl carbon at 197 δ.

The ^{13}C NMR spectrum of *p*-bromoacetophenone is interesting for another reason as well. Note that only six absorptions are observed, even though the molecule has eight carbons. *p*-Bromoacetophenone has a symmetry plane that makes carbons 4 and 4′, and carbons 5 and 5′, equivalent. Thus, the six ring carbons show only four absorptions in the range 128 to 137 δ. In addition, the $-CH_3$ carbon absorbs at 26 δ.

Figure 13.16 The ¹³C NMR spectrum of *p*-bromoacetophenone, $BrC_6H_4COCH_3$.

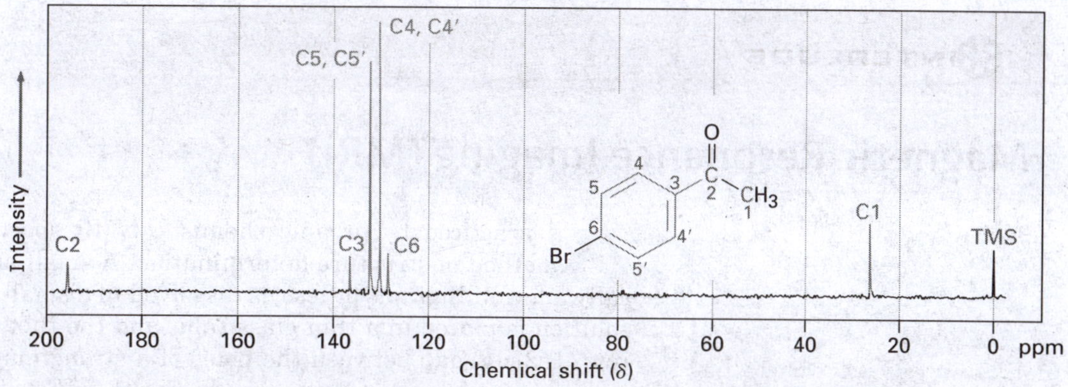

Worked Example 13.10 **Predicting the Number of ^{13}C Absorptions**

How many peaks would you expect in the ^{13}C NMR spectrum of methylcyclopentane?

Strategy Find the number of distinct types of carbons in the molecule, taking symmetry into account.

Solution Because of its symmetry plane, methylcyclopentane has only four kinds of carbons and only four peaks in its ^{13}C NMR spectrum.

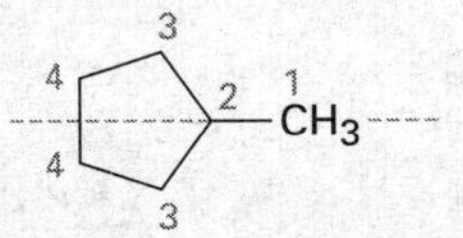

Symmetry plane

Problem 13.23 How many peaks would you expect in the ^{13}C NMR spectra of the following compounds?

(a) 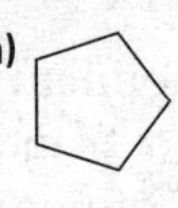

(b) 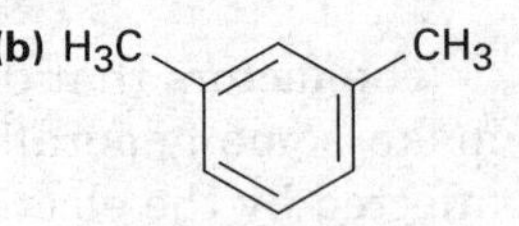

(c) 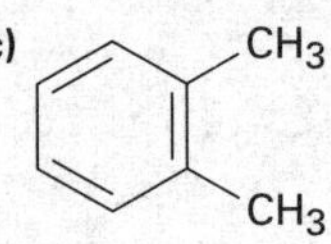

(d) 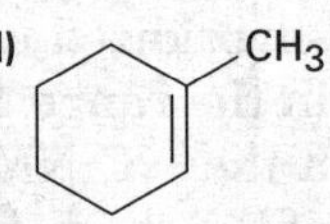

(e) 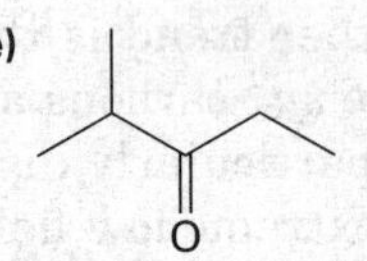

(f) 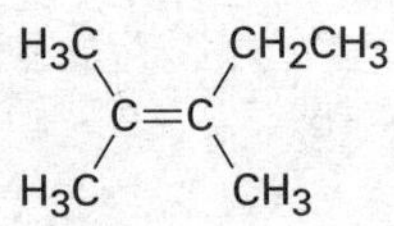

Problem 13.24 Propose structures for compounds whose ^{13}C NMR spectra fit the following descriptions:

(a) A hydrocarbon with seven peaks in its spectrum
(b) A six-carbon compound with only five peaks in its spectrum
(c) A four-carbon compound with three peaks in its spectrum

INTERLUDE

Magnetic Resonance Imaging (MRI)

As practiced by organic chemists, NMR spectroscopy is a powerful method of structure determination. A small amount of sample, typically a few milligrams or less, is dissolved in a small amount of solvent, the solution is placed in a thin glass tube, and the tube is placed into the narrow (1–2 cm) gap between the poles of a strong magnet. Imagine, though,

continued

INTERLUDE

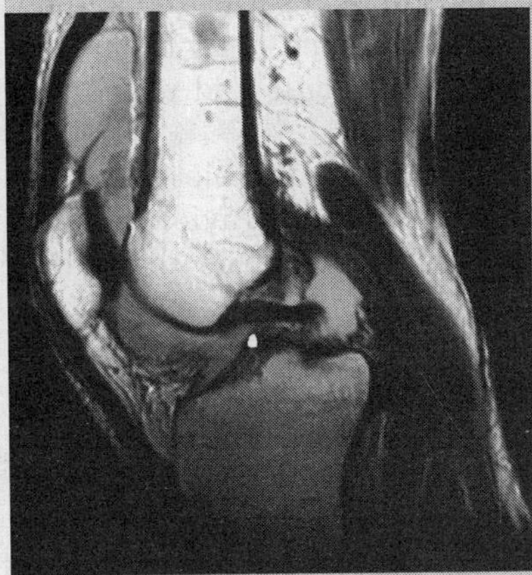

Living Art Enterprises/Photo Researchers, Inc.

This person won't be running again soon. The MRI shows a torn posterior cruciate ligament in the knee.

that a much larger NMR instrument were available. Instead of a few milligrams, the sample size could be tens of kilograms; instead of a narrow gap between magnet poles, the gap could be large enough for a whole person to climb into so that an NMR spectrum of body parts could be obtained. That large instrument is exactly what's used for *magnetic resonance imaging (MRI),* a diagnostic technique of enormous value to the medical community.

Like NMR spectroscopy, MRI takes advantage of the magnetic properties of certain nuclei, typically hydrogen, and of the signals emitted when those nuclei are stimulated by radiofrequency energy. Unlike what happens in NMR spectroscopy, though, MRI instruments use data manipulation techniques to look at the three-dimensional *location* of magnetic nuclei in the body rather than at the chemical nature of the nuclei. As noted, most MRI instruments currently look at hydrogen, present in abundance wherever there is water or fat in the body.

The signals detected by MRI vary with the density of hydrogen atoms and with the nature of their surroundings, allowing identification of different types of tissue and even allowing the visualization of motion. For example, the volume of blood leaving the heart in a single stroke can be measured, and heart motion can be observed. Soft tissues that don't show up well on X-ray films can be seen clearly, allowing diagnosis of brain tumors, strokes, and other conditions. The technique is also valuable in diagnosing damage to knees or other joints and is a noninvasive alternative to surgical explorations.

Several types of atoms in addition to hydrogen can be detected by MRI, and the applications of images based on ^{31}P atoms are being explored. The technique holds great promise for studies of metabolism.

Summary and Key Words

absorption spectrum 437
amplitude 435
chemical shift 449
coupling 452
coupling constant (J) 454
delta (δ) scale 449
downfield 448
electromagnetic spectrum 435
frequency (ν) 435
hertz (Hz) 435
infrared (IR) spectroscopy 438
integration 451
mass spectrometry 434
multiplet 452
$n + 1$ rule 453
nuclear magnetic resonance (NMR) spectroscopy 445
shielding 446
spin–spin splitting 452

Finding the structure of a new molecule, whether a small one synthesized in the laboratory or a large one found in living organisms, is central to progress in chemistry and biochemistry. Four main spectroscopic methods are used to determine the structures of organic molecules. Each gives a different kind of information.

Mass Spectrometry	What is the molecular formula?
Infrared spectroscopy	What functional groups are present?
Ultraviolet spectroscopy	Is a conjugated π electron system present?
Nuclear magnetic resonance spectroscopy	What carbon–hydrogen framework is present?

In **mass spectrometry**, molecules are first ionized and then sorted according to their mass-to-charge ratio (m/z). The ionized sample molecule is called the *molecular ion,* M^+, and measurement of its mass gives its molecular weight. Clues about molecular structure can be obtained by interpreting the fragmentation pattern of the molecular ion.

When an organic molecule is irradiated with **infrared (IR)** energy, frequencies of light corresponding to the energies of various molecular stretching and bending motions are absorbed. Each functional group has a characteristic set of IR absorptions that allows the group to be identified. For example, an alkene C=C bond absorbs in the range 1640 to 1680 cm^{-1}, a saturated ketone absorbs near 1715 cm^{-1}, and a nitrile absorbs near 2230 cm^{-1}. By observing which frequencies of IR radiation are absorbed by a molecule and which are not, the functional groups in a molecule can be identified.

Ultraviolet (UV) spectroscopy is applicable to conjugated π electron systems. When a conjugated molecule is irradiated with ultraviolet light, energy absorption occurs, leading to excitation of π electrons to higher energy levels. The greater the extent of conjugation, the longer the wavelength needed for excitation.

Nuclear magnetic resonance (NMR) spectroscopy is the most valuable of the common spectroscopic techniques. When ^{1}H and ^{13}C nuclei are placed in a magnetic field, their spins orient either with or against the field. On irradiation with radiofrequency (rf) waves, energy is absorbed and the nuclear spins flip from the lower-energy state to the higher-energy state. This absorption of energy is detected, amplified, and displayed as an NMR spectrum. NMR spectra display four general features.

- **Number of peaks.** Each nonequivalent kind of ^{1}H or ^{13}C nucleus in a molecule gives rise to a different peak.
- **Chemical shift.** The exact position of each peak is correlated to the chemical environment of each ^{1}H or ^{13}C nucleus. Most ^{1}H absorptions are in the range 0 to 10 δ downfield from the TMS reference signal, and most ^{13}C absorptions are in the range 0 to 220 δ.
- **Integration.** The area under each peak can be electronically integrated to determine the relative number of protons responsible for each peak.
- **Spin–spin splitting.** Neighboring nuclear spins can **couple**, splitting an NMR absorption into a **multiplet**. The NMR signal of a ^{1}H nucleus neighbored by n adjacent protons splits into $n + 1$ peaks with coupling constant J.

Exercises

Visualizing Chemistry

(Problems 13.1–13.24 appear within the chapter.)

OWL

Interactive versions of these problems are assignable in OWL.

13.25 Into how many peaks would you expect the ^{1}H NMR signal of the indicated protons to be split? (Yellow-green = Cl.)

(a)

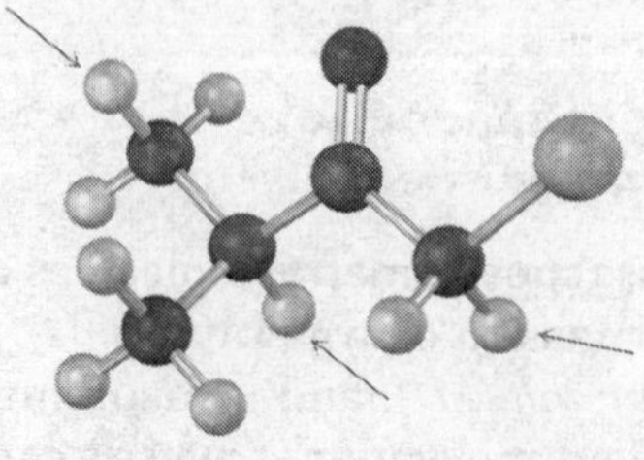

(b)

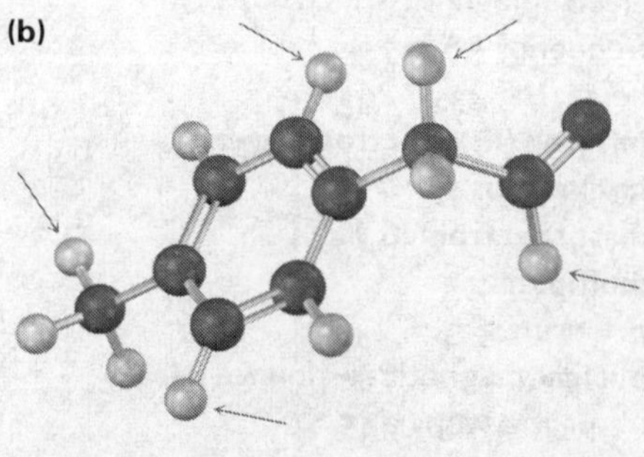

13.26 How many absorptions would you expect the following compound to have in its ^{1}H and ^{13}C NMR spectrum?

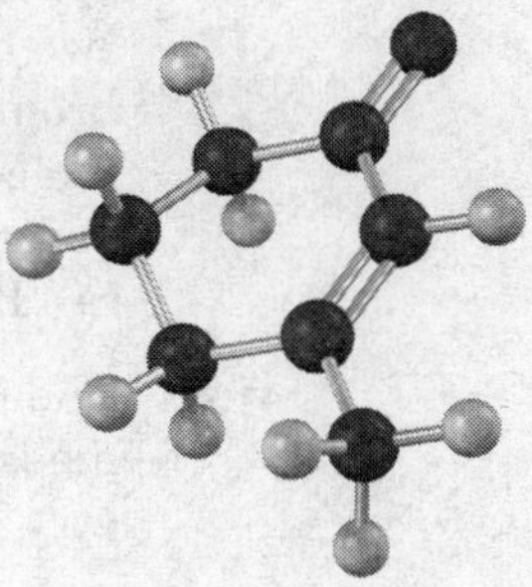

13.27 Where in the infrared spectrum would you expect each of the following compounds to absorb?

(a)

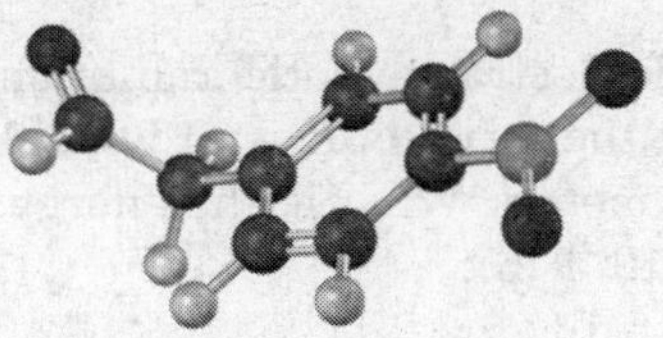

(b)

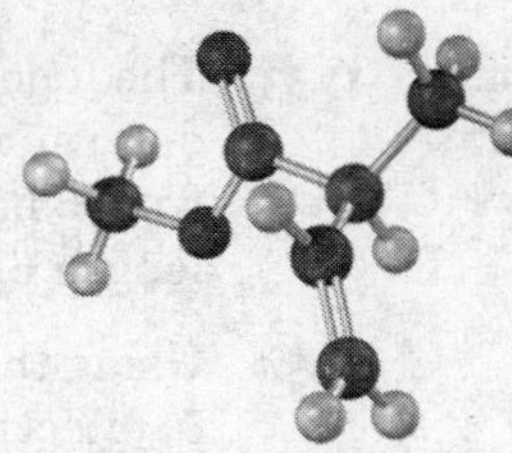

(c)

13.28 Sketch what you might expect the ^{1}H and ^{13}C NMR spectra of the following compound to look like (yellow-green = Cl):

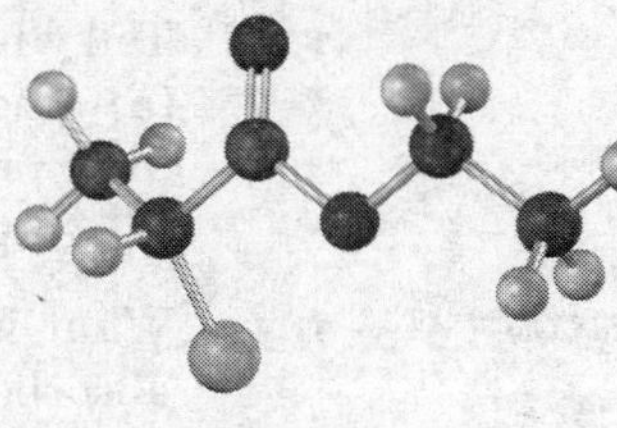

13.29 How many absorptions would you expect the following compound to have in its ^{13}C NMR spectrum?

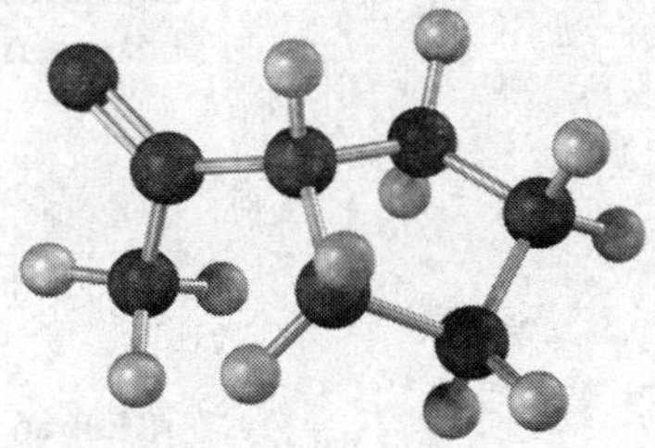

Additional Problems

Mass Spectrometry

13.30 Halogenated compounds are particularly easy to identify by their mass spectra because both chlorine and bromine occur naturally as mixtures of two abundant isotopes. Chlorine occurs as ^{35}Cl (75.8%) and ^{37}Cl (24.2%); bromine occurs as ^{79}Br (50.7%) and ^{81}Br (49.3%). At what masses do the molecular ions occur for the following formulas? What are the relative percentages of each molecular ion?
(a) Bromomethane, CH_3Br **(b)** 1-Chlorohexane, $C_6H_{13}Cl$

13.31 The male sex hormone testosterone contains only C, H, and O and has a mass of 288.2089 amu, as determined by high-resolution mass spectrometry. What is the likely molecular formula of testosterone?

13.32 Propose structures for compounds that fit the following mass-spectral data:
(a) A hydrocarbon with $M^+ = 132$
(b) A hydrocarbon with $M^+ = 166$
(c) A hydrocarbon with $M^+ = 84$

Electromagnetic Radiation and Spectroscopy

13.33 The energy of electromagnetic radiation in units of kJ/mol, can be determined by the formula $E = (1.20 \times 10^{-4}\ \text{kJ/mol})/\lambda$, where λ is the wavelength in meters. What is the energy of infrared radiation of wavelength 6.55×10^{-6} m?

13.34 Using the equation given in Problem 13.33, calculate the energy required to effect the electronic excitation of buta-1,3-diene ($\lambda_{max} = 217$ nm).

13.35 Using the equation given in Problem 13.33, calculate the amount of energy required to spin-flip a proton in a spectrometer operating at 100 MHz. Does increasing the spectrometer frequency from 100 MHz to 220 MHz increase or decrease the amount of energy necessary for resonance?

13.36 Tell what is meant by each of the following terms:
(a) Chemical shift **(b)** Coupling constant **(c)** λ_{max}
(d) Spin–spin splitting **(e)** Wavenumber
(f) Applied magnetic field

UV Spectroscopy

13.37 What is the concentration of a sample of abacavir in mg/mL if the UV absorbance measured in a cell with 1.00 cm pathlength is $A = 0.93$ and the molar absorptivity is 13,260 L/(mol · cm).

13.38 The pathway for metabolism of a drug depends on the species doing the metabolizing. Coumarin, an anticoagulant, metabolizes to a toxic compound in rats but not in humans. How would the UV spectra differ between the different metabolites?

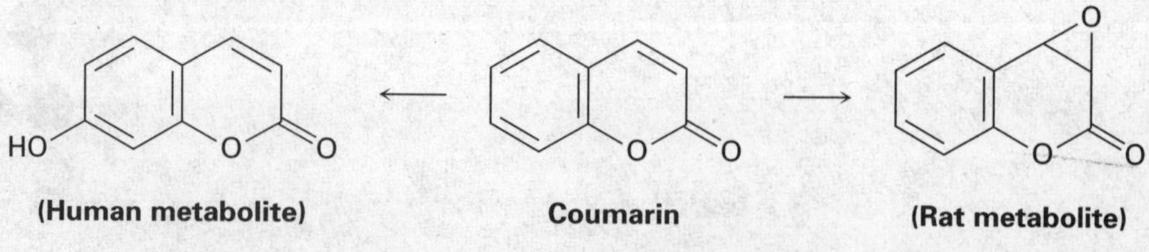

(Human metabolite) **Coumarin** **(Rat metabolite)**

13.39 The active metabolite of the antiviral drug abacavir has a UV molar absorptivity of 13,260 L/(mol · cm). What absorbance would you expect for a sample in a cell with a pathlength of 1.00 cm at a concentration of 42 μM? (1 μM = 10^{-6} M)

13.40 The mosquito attractant oct-1-en-3-ol can be prepared by reduction of oct-1-en-3-one. How can UV spectroscopy be used to show when the reaction is complete?

IR Spectroscopy

13.41 If C–O single-bond stretching occurs at 1000 cm^{-1} and C=O double-bond stretching occurs at 1700 cm^{-1}, which of the two requires more energy? How does your answer correlate with the relative strengths of single and double bonds?

13.42 What kinds of functional groups might compounds contain if they show the following IR absorptions?
(a) 1670 cm^{-1}
(b) 1735 cm^{-1}
(c) 1540 cm^{-1}
(d) 1715 cm^{-1} and 2500–3100 cm^{-1} (broad)

13.43 Propose structures for compounds that meet the following descriptions:
(a) C_5H_8, with IR absorptions at 3300 and 2150 cm^{-1}
(b) C_4H_8O, with a strong IR absorption at 3400 cm^{-1}
(c) C_4H_8O, with a strong IR absorption at 1715 cm^{-1}
(d) C_8H_{10}, with IR absorptions at 1600 and 1500 cm^{-1}

13.44 At what approximate positions might the following compounds show IR absorptions?

(a) 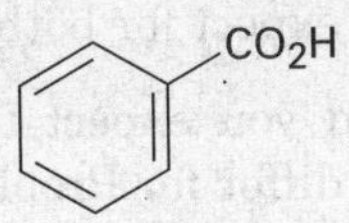

(b) 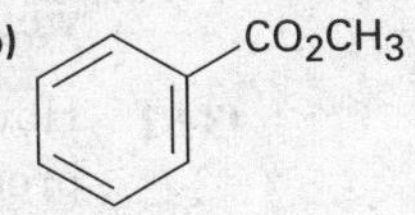

(c) 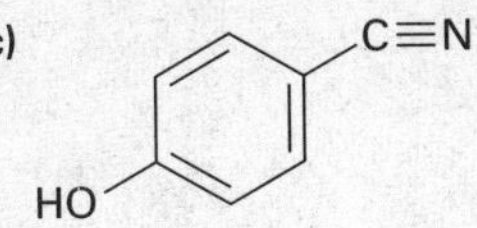

(d) 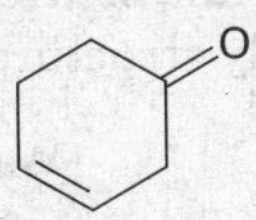

(e) $CH_3\overset{O}{\overset{\|}{C}}CH_2CH_2\overset{O}{\overset{\|}{C}}OCH_3$

NMR Spectroscopy

13.45 The following NMR absorptions, given in δ units, were obtained on a spectrometer operating at 300 MHz. Convert the chemical shifts from δ units into hertz downfield from TMS.
(a) 2.1 δ (b) 3.45 δ (c) 6.30 δ

13.46 When measured on a spectrometer operating at 100 MHz, chloroform ($CHCl_3$) shows a single sharp absorption at 7.3 δ.
(a) How many parts per million downfield from TMS does chloroform absorb?
(b) How many hertz downfield from TMS does chloroform absorb if the measurement is carried out on a spectrometer operating at 360 MHz?
(c) What is the position of the chloroform absorption in δ units measured on a 360 MHz spectrometer?

13.47 The following ^1H NMR absorptions, determined on a spectrometer operating at 100 MHz, are given in hertz downfield from the TMS standard. Convert the absorptions to δ units.
(a) 218 Hz (b) 478 Hz (c) 751 Hz

13.48 At what positions, in hertz downfield from the TMS standard, would the NMR absorptions in Problem 13.47 appear on a spectrometer operating at 220 MHz?

13.49 Describe the ^{1}H NMR spectra you would expect for the following compounds:
(a) CH_3CHCl_2 (b) $CH_3CO_2CH_2CH_3$ (c) $(CH_3)_3CCH_2CH_3$

13.50 The following compounds all show a single peak in their ^{1}H NMR spectra. List them in order of expected increasing chemical shift: CH_4, CH_2Cl_2, cyclohexane, CH_3COCH_3, $H_2C{=}CH_2$, benzene.

13.51 How many absorptions would you expect each of the following compounds to have in its ^{1}H and ^{13}C NMR spectra?

(a)
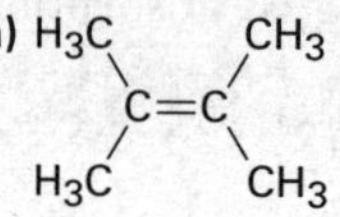

(b)
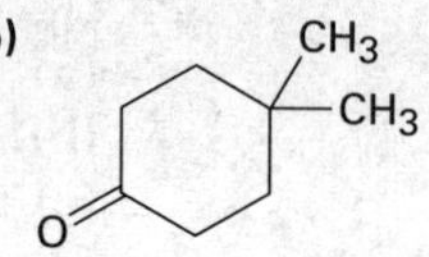

(c)
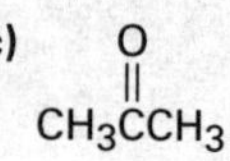

(d)
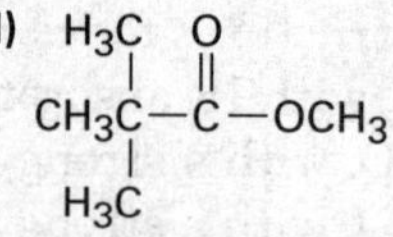

(e)
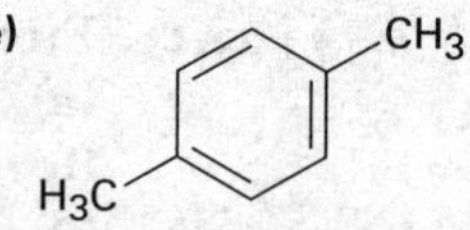

(f)
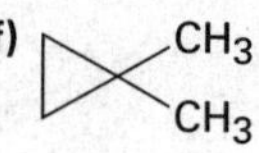

GENERAL PROBLEMS

13.52 Assume you are carrying out the dehydration of 1-methylcyclohexanol to yield 1-methylcyclohexene. How could you use IR spectroscopy to determine when the reaction is complete? What characteristic absorptions would you expect for both starting material and product?

13.53 How would you expect the mass spectra of the starting material and product to differ for Problem 13.52?

13.54 The IR spectrum of a compound with the formula C_7H_6O is shown. Propose a likely structure.

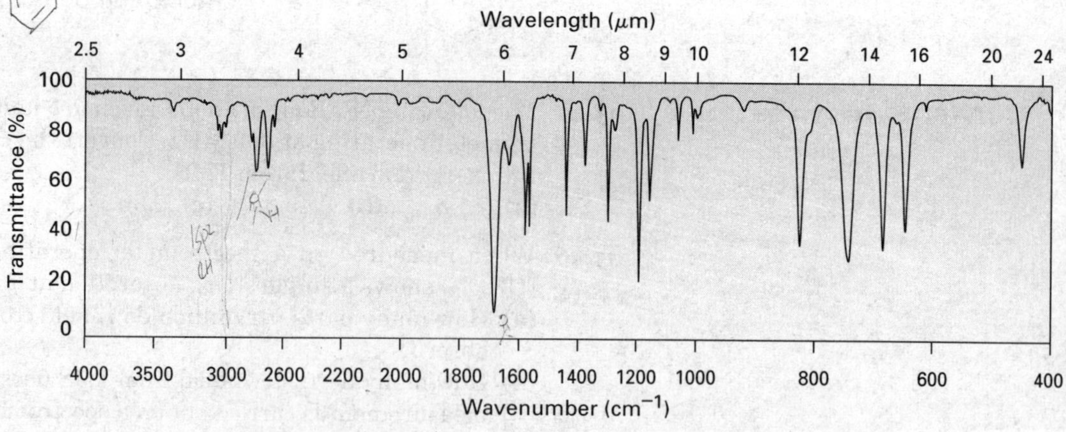

13.55 How would you use IR spectroscopy to distinguish between the following pairs of isomers?
(a) $(CH_3)_3N$ and $CH_3CH_2NHCH_3$
(b) CH_3COCH_3 and $CH_2{=}CHCH_2OH$
(c) CH_3COCH_3 and CH_3CH_2CHO

13.56 How would you use ^{1}H NMR spectroscopy to distinguish between the isomer pairs shown in Problem 13.55?

13.57 How could you use ^{13}C NMR spectroscopy to distinguish between the isomer pairs shown in Problem 13.55?

13.58 3,4-Dibromohexane can undergo base-induced loss of 2 HBr to yield either hex-3-yne or hexa-2,4-diene. How could you use UV spectroscopy to help identify the product? How could you use ^1H NMR spectroscopy?

13.59 Propose structures for compounds with the following formulas that show only one peak in their ^1H NMR spectra:
(a) C_5H_{12} **(b)** C_5H_{10} **(c)** $C_4H_8O_2$

13.60 At what approximate positions might the following compounds show IR absorptions?

(a) 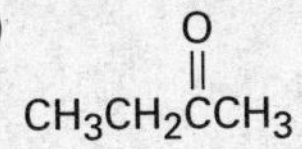

(b) 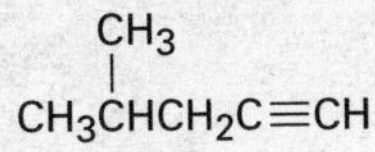

(c) 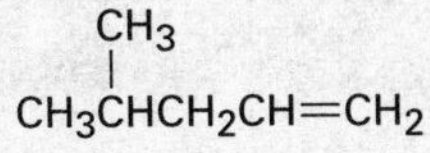

(d) 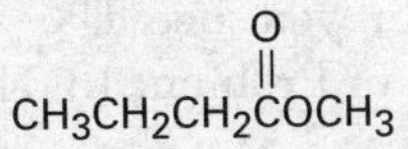

(e) 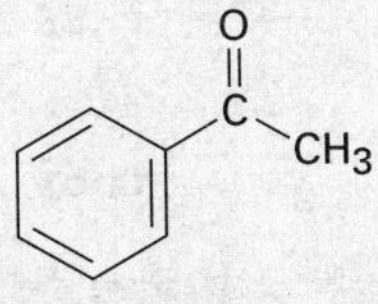

(f) 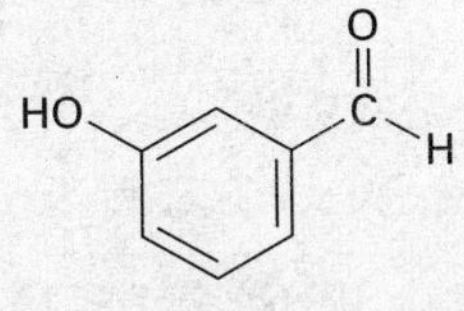

13.61 How many types of nonequivalent protons are present in each of the following compounds?

(a) 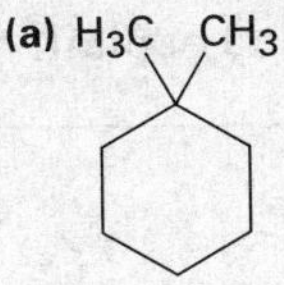

(b) $CH_3CH_2CH_2OCH_3$

(c) 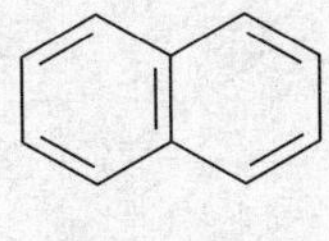

Naphthalene

(d) 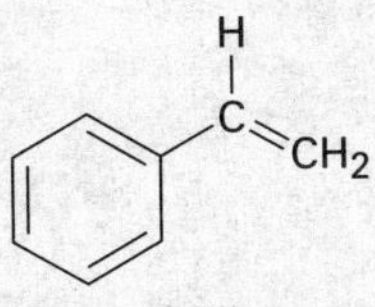

Styrene

(e) 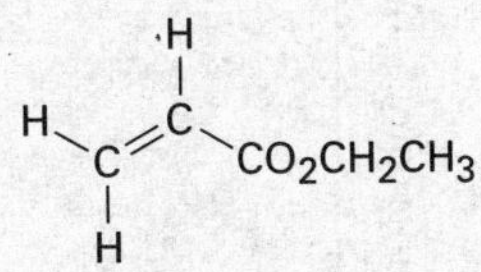

Ethyl acrylate

13.62 Dehydration of 1-methylcyclohexanol might lead to either of two isomeric alkenes, 1-methylcyclohexene or methylenecyclohexane. How could you use NMR spectroscopy (both ^1H and ^{13}C) to determine the structure of the product?

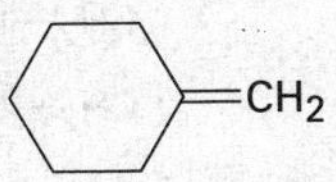

Methylenecyclohexane

13.63 Describe the ^1H and ^{13}C NMR spectra you expect for the following compounds:
(a) $ClCH_2CH_2CH_2Cl$ **(b)** $CH_3COCH_2CH_2Cl$

13.64 How can you use ^{1}H NMR to help distinguish between the following isomers?

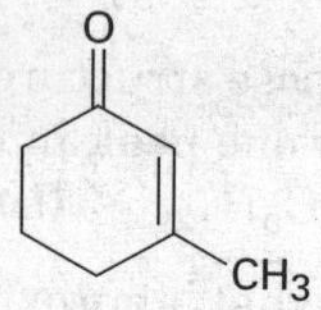

3-Methylcyclohex-2-enone

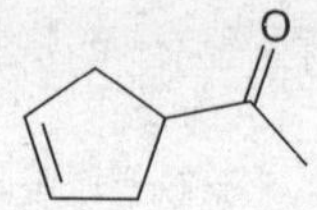

Cyclopent-3-enyl methyl ketone

13.65 How can you use ^{13}C NMR to help distinguish between the isomers in Problem 13.64?

13.66 How can you use UV spectroscopy to help distinguish between the isomers in Problem 13.64?

13.67 Propose structures for compounds that fit the following ^{1}H NMR data:

(a) $C_5H_{10}O$
6 H doublet at 0.95 δ, $J = 7$ Hz
3 H singlet at 2.10 δ
1 H multiplet at 2.43 δ

(b) C_3H_5Br
3 H singlet at 2.32 δ
1 H singlet at 5.25 δ
1 H singlet at 5.54 δ

13.68 The ^{1}H NMR spectrum of compound **A**, $C_3H_6Br_2$, is shown. Propose a structure for **A**, and explain how the spectrum fits your structure.

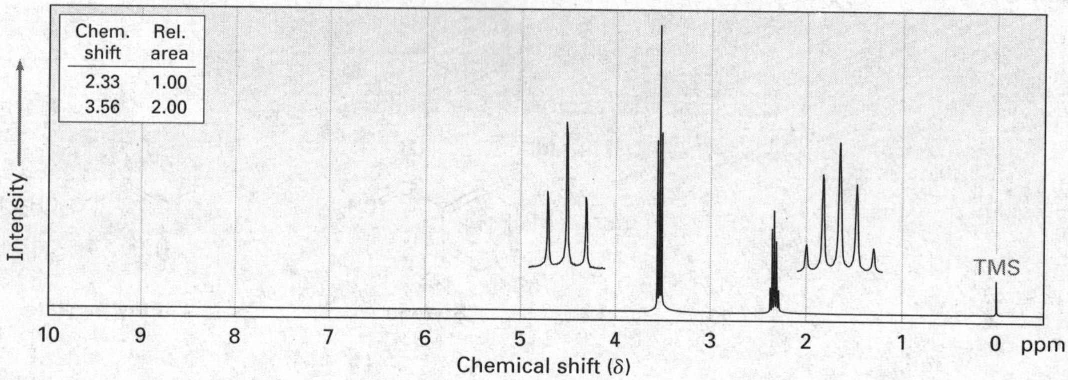

13.69 How can you use ^{1}H and ^{13}C NMR to help distinguish among the following four isomers?

$$\begin{array}{c}CH_2-CH_2\\ |\quad\quad |\\ CH_2-CH_2\end{array}\qquad H_2C{=}CHCH_2CH_3\qquad CH_3CH{=}CHCH_3\qquad \begin{array}{c}CH_3\\ |\\ CH_3C{=}CH_2\end{array}$$

13.70 Assume you have a compound with formula C_3H_6O.

(a) Propose as many structures as you can that fit the molecular formula (there are seven).

(b) If your compound has an IR absorption at 1715 cm^{-1}, what can you conclude?

(c) If your compound has a single ^{1}H NMR absorption at 2.1 δ, what is its structure?

13.71 Nitriles ($RC \equiv N$) react with Grignard reagents (RMgBr). The reaction product from 2-methylpropanenitrile with methylmagnesium bromide has the following spectroscopic properties. Propose a structure.

Molecular weight = 86

IR: 1715 cm^{-1}

^1H NMR: 1.05 δ (6 H doublet); 2.12 δ (3 H singlet); 2.67 δ (1 H septet)

^{13}C NMR: 18.2, 27.2, 41.6, 211.2 δ

13.72 Propose structures for compounds that fit the following ^1H NMR data:

(a) $C_4H_6Cl_2$
3 H singlet at 2.18 δ
2 H doublet at 4.16 δ, $J = 7$ Hz
1 H triplet at 5.71 δ, $J = 7$ Hz

(b) $C_{10}H_{14}$
9 H singlet at 1.30 δ
5 H singlet at 7.30 δ

13.73 The compound whose ^1H NMR spectrum is shown has the formula $C_4H_7O_2Cl$ and has an IR absorption peak at 1740 cm^{-1}. Propose a structure.

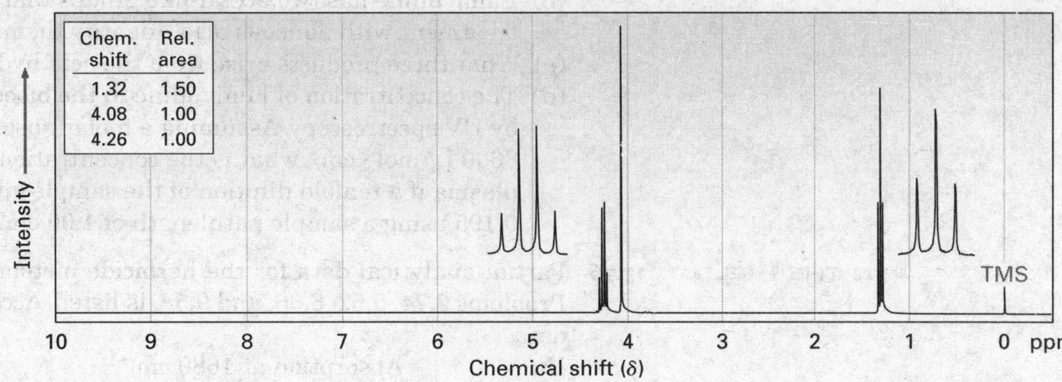

13.74 Propose a structure for a compound with formula C_4H_9Br that has the following ^1H NMR spectrum:

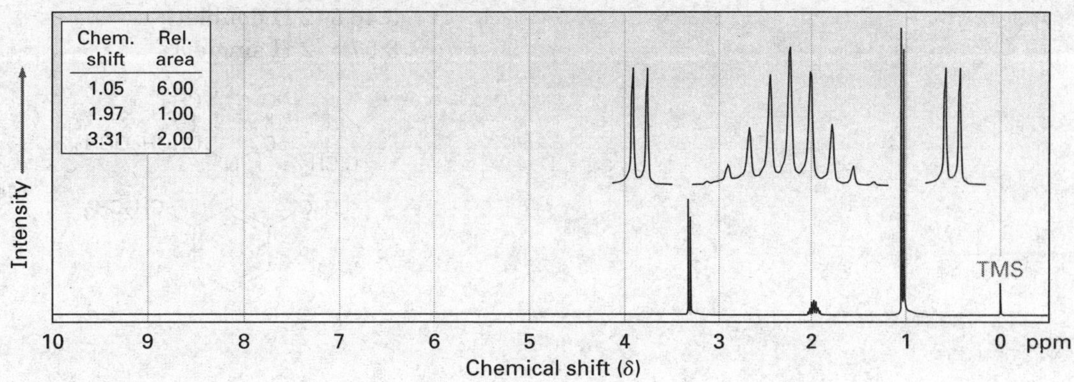

IN THE MEDICINE CABINET

13.75 Lamivudine, a drug used in the management of acquired immunodeficiency syndrome (AIDS), structurally resembles the nucleoside cytidine found in nucleic acids. Lamivudine is also referred to as 3TC, an acronym for 3-thiocytosine.

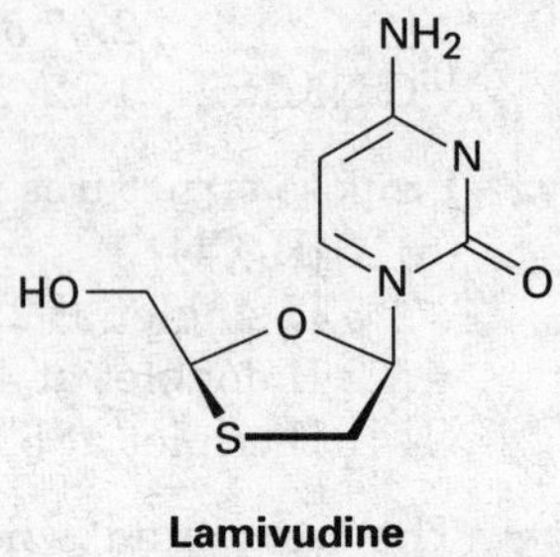

Lamivudine

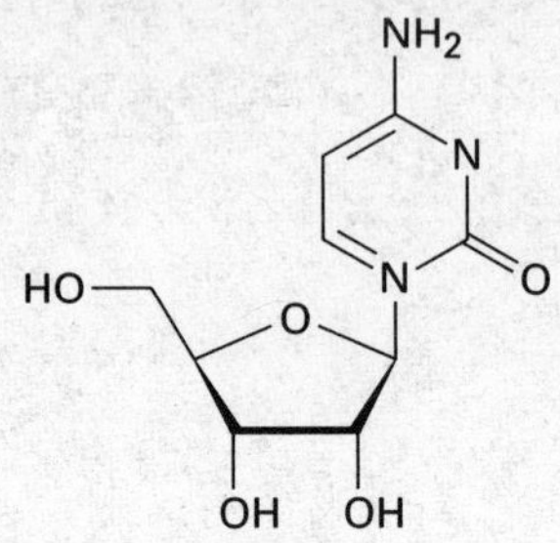

Cytidine

(a) How many chirality centers does lamivudine have?
(b) Lamivudine has two acetal-like groups that will hydrolyze on treatment with aqueous acid. Identify them.
(c) What three products arise from the acid hydrolysis of lamivudine?
(d) The concentration of lamivudine in the blood can be measured by UV spectroscopy. Assuming a molar absorptivity of 8600 L/(mol · cm), what is the concentration of drug in blood plasma if a tenfold dilution of the sample gives an absorbance of 0.195 using a sample pathlength of 1.00 cm?

IN THE FIELD

13.76 Partial analytical data for the herbicide metolachlor, seen previously in Problems 2.74, 5.62, 6.66, and 9.58, is listed. Account for each piece of the data.

IR: Absorption at 1680 cm^{-1}
UV: Broad absorption at 310 nm
^{1}H NMR: 1.11 δ (3 H doublet); 1.23 δ (3 H triplet); 2.21 δ (3 H singlet); 2.57 δ (2 H quartet); 3.25 δ (3 H singlet); 3.46 δ (2 H doublet); 3.57 δ (2 H singlet)

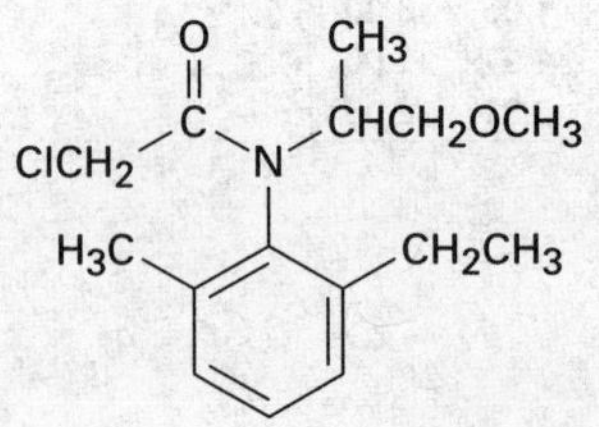

Metolachlor

CHAPTER

14

Produced by honeybees from the nectar of flowers, honey is primarily a mixture of the two simple sugars fructose and glucose.

Biomolecules: Carbohydrates

Carbohydrates occur in every living organism. The sugar and starch in food, and the cellulose in wood, paper, and cotton are nearly pure carbohydrate. Modified carbohydrates form part of the coating around living cells, other carbohydrates are part of the nucleic acids that carry our genetic information, and still others are used as medicines.

The word **carbohydrate** derives historically from the fact that glucose, the first carbohydrate to be obtained pure, has the molecular formula $C_6H_{12}O_6$ and was originally thought to be a "hydrate of carbon," $C_6(H_2O)_6$. This view was soon abandoned, but the name persisted. Today, the term *carbohydrate* is used to refer loosely to the broad class of polyhydroxylated aldehydes and ketones commonly called *sugars.* Glucose, also known as *dextrose* in medical work, is the most familiar example.

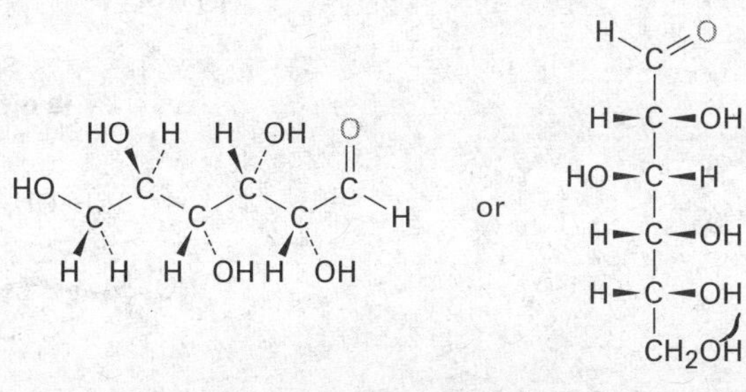

Glucose (dextrose), a pentahydroxyhexanal

OWL

Online homework for this chapter can be assigned in OWL, an online homework assessment tool.

Carbohydrates are synthesized by green plants during photosynthesis, a complex process in which sunlight provides the energy to convert CO_2 and H_2O into glucose plus oxygen. Many molecules of glucose are then chemically linked for storage by the plant in the form of either cellulose or starch. It has been estimated that more than 50% of the dry weight of the earth's biomass—all plants and animals—consists of glucose polymers. When eaten and metabolized, carbohydrates then provide animals with a source of readily available energy. Thus, carbohydrates act as the chemical intermediaries by which solar energy is stored and used to support life.

$$6\,CO_2 + 6\,H_2O \xrightarrow{\text{Sunlight}} 6\,O_2 + \underset{\textbf{Glucose}}{C_6H_{12}O_6} \longrightarrow \text{Cellulose, starch}$$

WHY THIS CHAPTER?

We've now seen all the common functional groups and reaction types that occur in organic and biological chemistry. In the few remaining chapters, we'll focus on the major classes of biological molecules, beginning in this chapter with a look at the structures and primary biological functions of carbohydrates. In Chapter 17, we'll return to the subject to see how carbohydrates are both synthesized and degraded by organisms.

14.1 Classification of Carbohydrates

Carbohydrates are generally classed as either *simple* or *complex*. **Simple sugars**, or **monosaccharides**, are carbohydrates like glucose and fructose that can't be converted into smaller sugars by hydrolysis. **Complex carbohydrates** are made of two or more simple sugars linked together. Sucrose (table sugar), for example, is a *disaccharide* made up of one glucose linked to one fructose. Similarly, cellulose is a *polysaccharide* made up of several thousand glucose molecules linked together. Enzyme-catalyzed hydrolysis of polysaccharides breaks them down into their constituent monosaccharides.

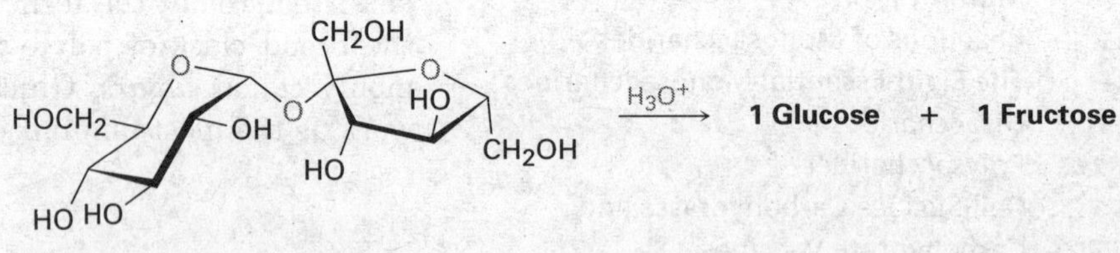

Sucrose (a disaccharide)

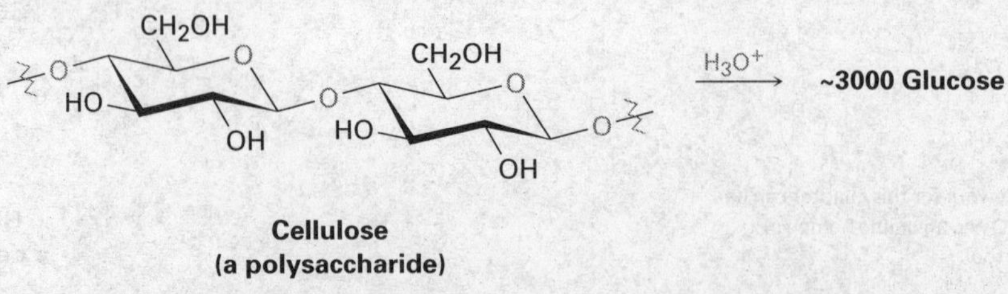

Cellulose (a polysaccharide)

Monosaccharides are further classified as either **aldoses** or **ketoses**. The *-ose* suffix designates a carbohydrate, and the *aldo-* and *keto-* prefixes identify the kind of carbonyl group in the molecule, whether aldehyde or ketone. The number of carbon atoms in the monosaccharide is indicated by the appropriate numerical prefix *tri-*, *tetr-*, *pent-*, *hex-*, and so forth. Thus, glucose is an *aldohexose,* a six-carbon aldehydo sugar; fructose is a *ketohexose,* a six-carbon keto sugar; and ribose is an *aldopentose,* a five-carbon aldehydo sugar. Most of the common simple sugars are either aldopentoses or aldohexoses.

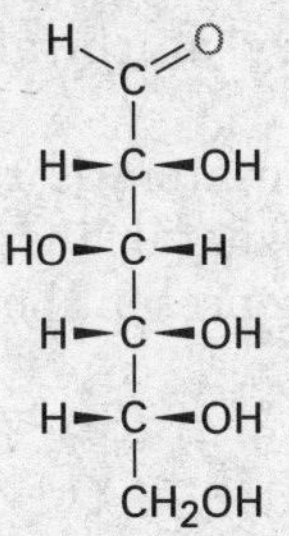

Glucose
(an aldohexose)

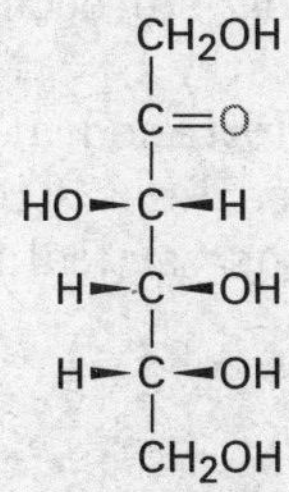

Fructose
(a ketohexose)

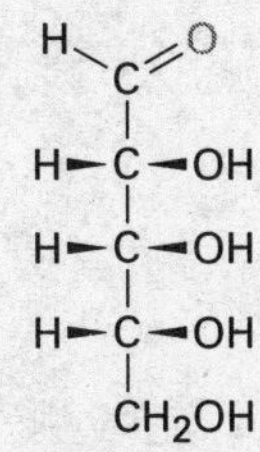

Ribose
(an aldopentose)

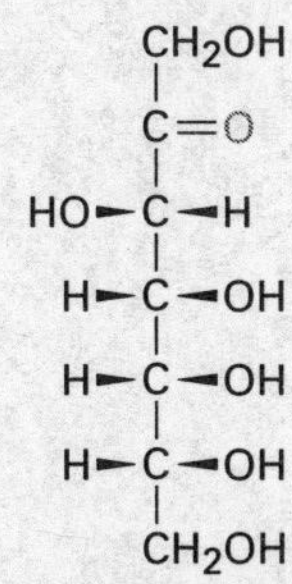

Sedoheptulose
(a ketoheptose)

Worked Example 14.1

Classifying Monosaccharides

Classify the following monosaccharide:

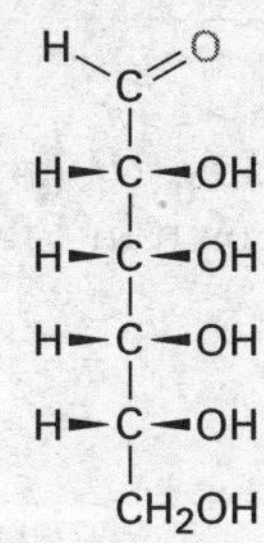

Allose

Solution Since allose has six carbons and an aldehyde carbonyl group, it is an aldohexose.

Problem 14.1 Classify each of the following monosaccharides:

(a)

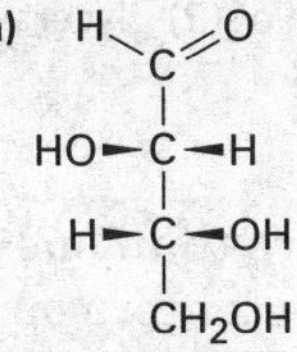

Threose

(b)

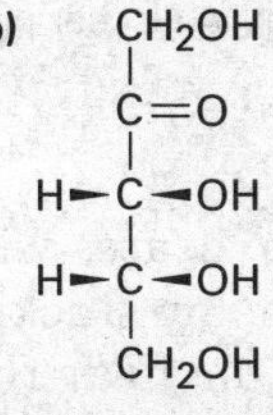

Ribulose

(c)

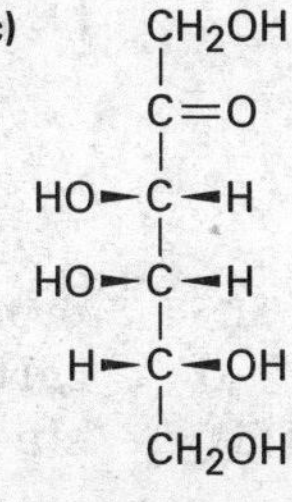

Tagatose

(d)

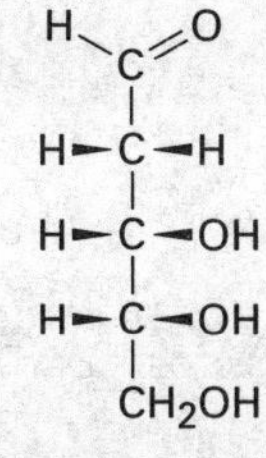

2-Deoxyribose

14.2 Depicting Carbohydrate Stereochemistry: Fischer Projections

Because carbohydrates usually have numerous chirality centers, it was recognized long ago that a quick method for representing stereochemistry is needed. In 1891, the German chemist Emil Fischer suggested a method based on the projection of a tetrahedral carbon atom onto a flat surface. These **Fischer projections** were soon adopted and are now a common means of representing stereochemistry at chirality centers, particularly in carbohydrate chemistry.

A tetrahedral carbon atom is represented in a Fischer projection by two crossed lines. The horizontal lines represent bonds coming out of the page, and the vertical lines represent bonds going into the page.

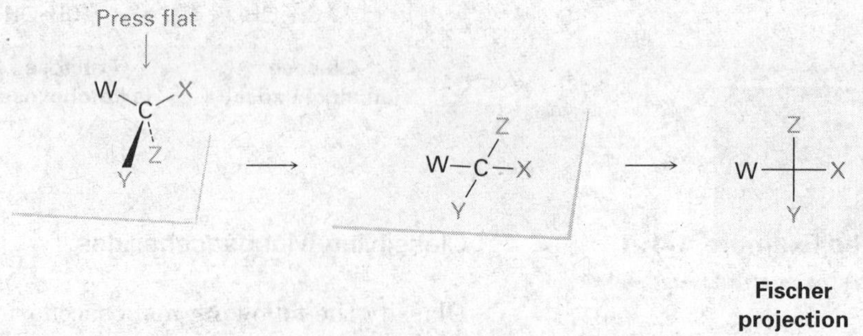

Thus, (*R*)-glyceraldehyde, the simplest monosaccharide, is represented as shown in Figure 14.1.

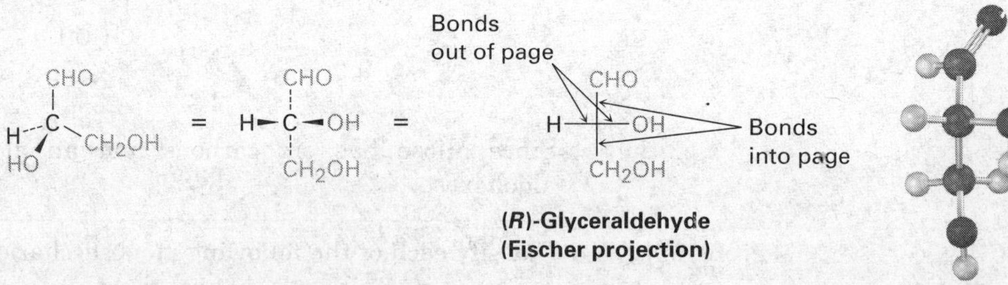

Figure 14.1 A Fischer projection of (*R*)-glyceraldehyde.

Carbohydrates with more than one chirality center are shown in Fischer projections by stacking the centers on top of one another, with the carbonyl carbon at or near the top. Glucose, for example, has four chirality centers stacked on top of one another in a Fischer projection. Such representations don't, however, give an accurate picture of a molecule's true three-dimensional conformation, which is curled around on itself like a bracelet.

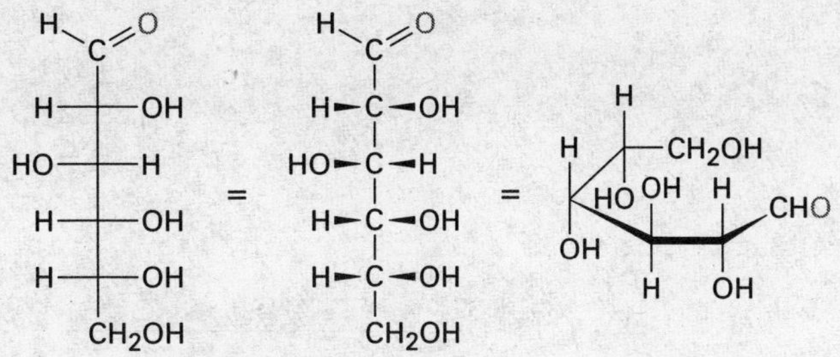

Glucose
(carbonyl group at top)

Worked Example 14.2

Drawing a Fischer Projection

Convert the following tetrahedral representation of (*R*)-butan-2-ol into a Fischer projection:

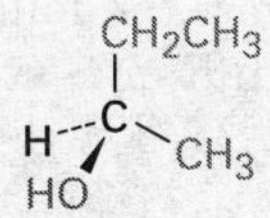

(*R*)-Butan-2-ol

Strategy Orient the molecule so that two horizontal bonds are facing you and two vertical bonds are receding away from you. Then press the molecule flat into the paper, indicating the chirality center as the intersection of two crossed lines.

Solution

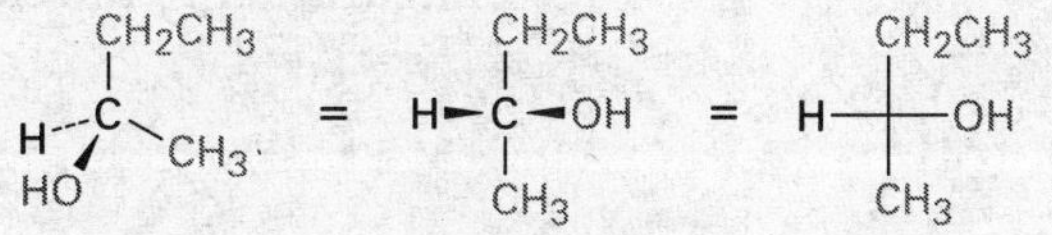

(*R*)-Butan-2-ol

Worked Example 14.3

Interpreting a Fischer Projection

Convert the following Fischer projection of lactic acid into a tetrahedral representation, and indicate whether the molecule is (*R*) or (*S*):

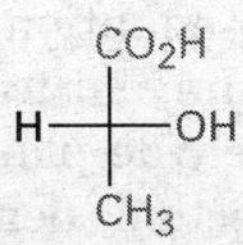

Lactic acid

Strategy Place a carbon atom at the intersection of the two crossed lines, and imagine that the two horizontal bonds are coming toward you and the two vertical bonds are receding away from you. The projection represents (R)-lactic acid.

Solution

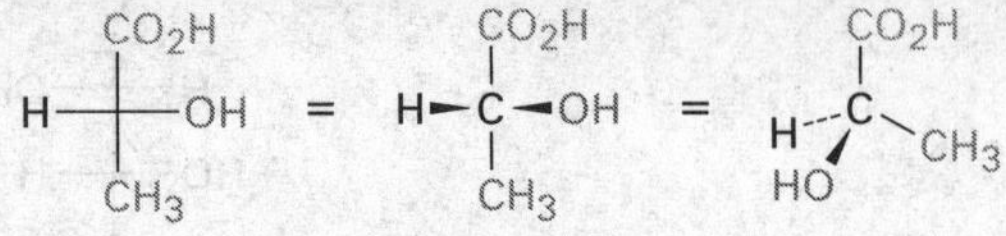

(R)-Lactic acid

Problem 14.2 Convert the following tetrahedral representation of (S)-glyceraldehyde into a Fischer projection:

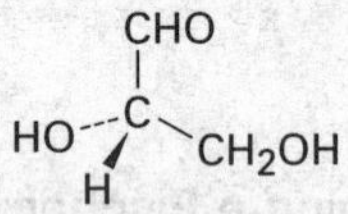

(S)-Glyceraldehyde

Problem 14.3 Draw Fischer projections of both (R)-2-chlorobutane and (S)-2-chlorobutane.

Problem 14.4 Convert the following Fischer projections into tetrahedral representations, and assign R or S stereochemistry to each:

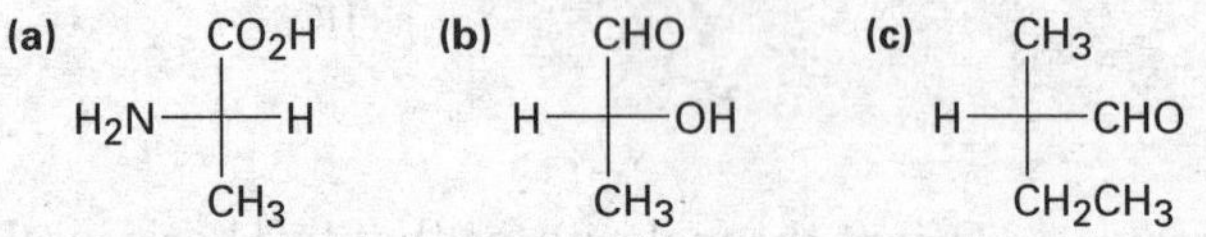

Problem 14.5 Redraw the following molecule as a Fischer projection, and assign R or S configuration to the chirality center (yellow-green = Cl):

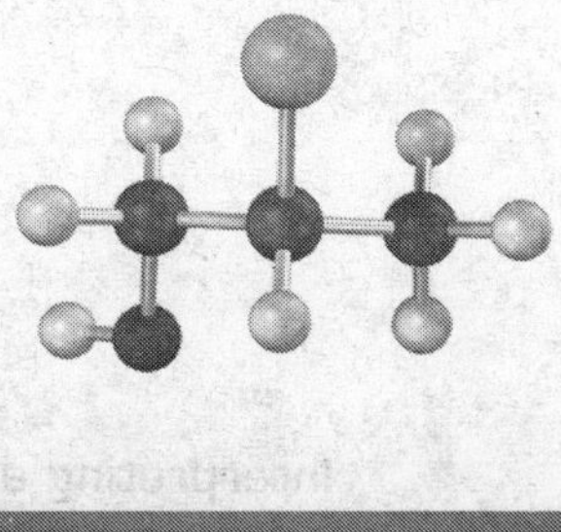

14.3 D,L Sugars

Glyceraldehyde, the simplest aldose, has only one chirality center and thus has two enantiomeric (mirror-image) forms. Only the dextrorotatory enantiomer occurs naturally, however. That is, a sample of naturally occurring glyceraldehyde placed in a polarimeter rotates plane-polarized light in a clockwise direction, denoted (+). Since (+)-glyceraldehyde has been found to have an R configuration at C2, it can be represented as in Figure 14.2. For historical reasons dating from long before the adoption of the R,S system, (R)-(+)-glyceraldehyde is also referred to as *D-glyceraldehyde* (D for dextrorotatory). The other enantiomer, (*S*)-(−)-glyceraldehyde, is known as L-*glyceraldehyde* (L for levorotatory).

Because of the way that monosaccharides are synthesized in nature, glucose, fructose, ribose, and most other naturally occurring monosaccharides have the same *R* stereochemical configuration as D-glyceraldehyde at the chirality center farthest from the carbonyl group. In Fischer projections, therefore, most naturally occurring sugars have the −OH group at the bottom chirality center pointing to the right (Figure 14.2). Such compounds are referred to as **D sugars.**

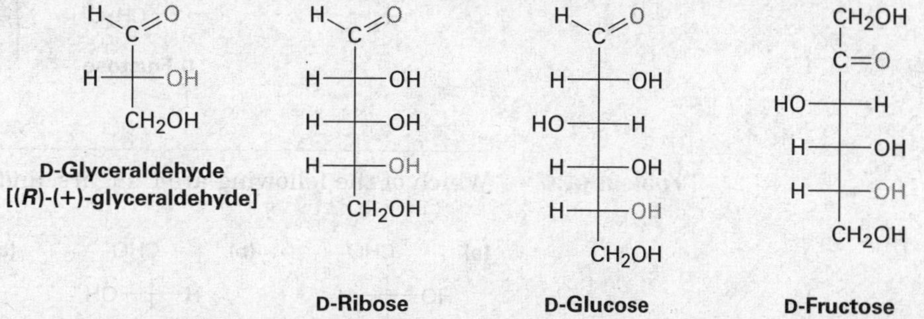

Figure 14.2 Some naturally occurring D sugars. The −OH group at the chirality center farthest from the carbonyl group has the same configuration as (*R*)-(+)-glyceraldehyde and points toward the right in Fischer projections.

In contrast to D sugars, all **L sugars** have an *S* configuration at the lowest chirality center, with the bottom −OH group pointing to the *left* in Fischer projections. Thus, an L sugar is the mirror image (enantiomer) of the corresponding D sugar and has the opposite configuration at all chirality centers.

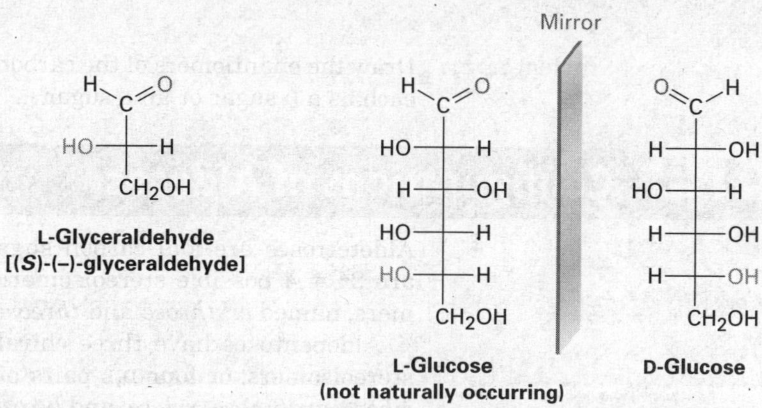

Note that the D and L notations have no relation to the direction in which a given sugar rotates plane-polarized light. A D sugar may be either dextrorotatory or levorotatory. The prefix D indicates only that the stereochemistry of the lowest chirality center is the same as that of D-glyceraldehyde and that the −OH group points to the right when the molecule is drawn in the standard way in a Fischer projection. Note also that the D,L system of carbohydrate nomenclature describes the configuration at only one chirality center and says nothing about the configuration of other chirality centers that may be present.

Worked Example 14.4 Drawing the Fischer Projection of an Enantiomer

Look at the Fischer projection of D-fructose in Figure 14.2, and draw a Fischer projection of L-fructose.

Strategy Since L-fructose is the enantiomer of D-fructose, simply take the structure of D-fructose and reverse the configuration at every chirality center.

Solution

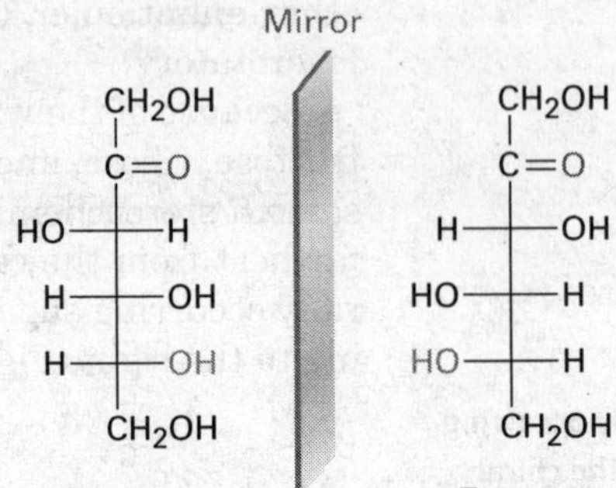

D-Fructose L-Fructose

Problem 14.6 Which of the following are L sugars, and which are D sugars?

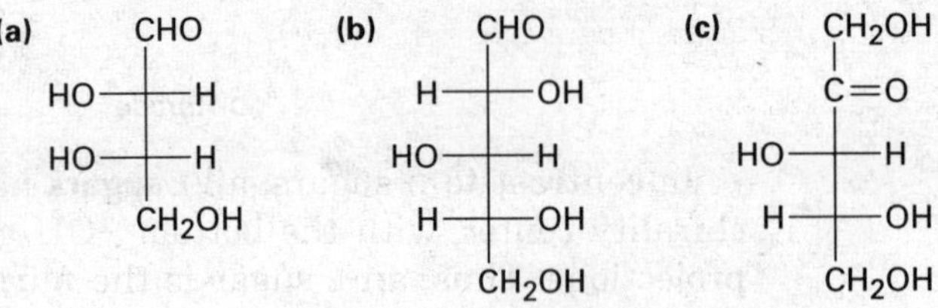

Problem 14.7 Draw the enantiomers of the carbohydrates shown in Problem 14.6, and identify each as a D sugar or an L sugar.

14.4 Configurations of Aldoses

Aldotetroses are four-carbon sugars with two chirality centers. Thus, there are $2^2 = 4$ possible stereoisomeric aldotetroses, or two D,L pairs of enantiomers, named *erythrose* and *threose.*

Aldopentoses have three chirality centers and a total of $2^3 = 8$ possible stereoisomers, or four D,L pairs of enantiomers. These four pairs are named *ribose, arabinose, xylose,* and *lyxose.* All except lyxose occur widely. D-Ribose is an important part of RNA (ribonucleic acid), L-arabinose is found in many plants, and D-xylose is found in both plants and animals.

Aldohexoses have four chirality centers and a total of $2^4 = 16$ possible stereoisomers, or eight D,L pairs of enantiomers. The names of the eight are *allose, altrose, glucose, mannose, gulose, idose, galactose,* and *talose.* Only D-glucose, from starch and cellulose, and D-galactose, from gums and fruit pectins, are widely distributed in nature. D-Mannose and D-talose also occur naturally, but in lesser abundance.

Fischer projections of the four-, five-, and six-carbon D aldoses are shown in Figure 14.3. Starting from D-glyceraldehyde, we can imagine constructing the two D aldotetroses by inserting a new chirality center just below the aldehyde carbon. Each of the two D aldotetroses then leads to two D aldopentoses (four total), and each of the four D aldopentoses leads to two D aldohexoses (eight total). In addition, each of the D aldoses in Figure 14.3 has a mirror-image L enantiomer, which is not shown.

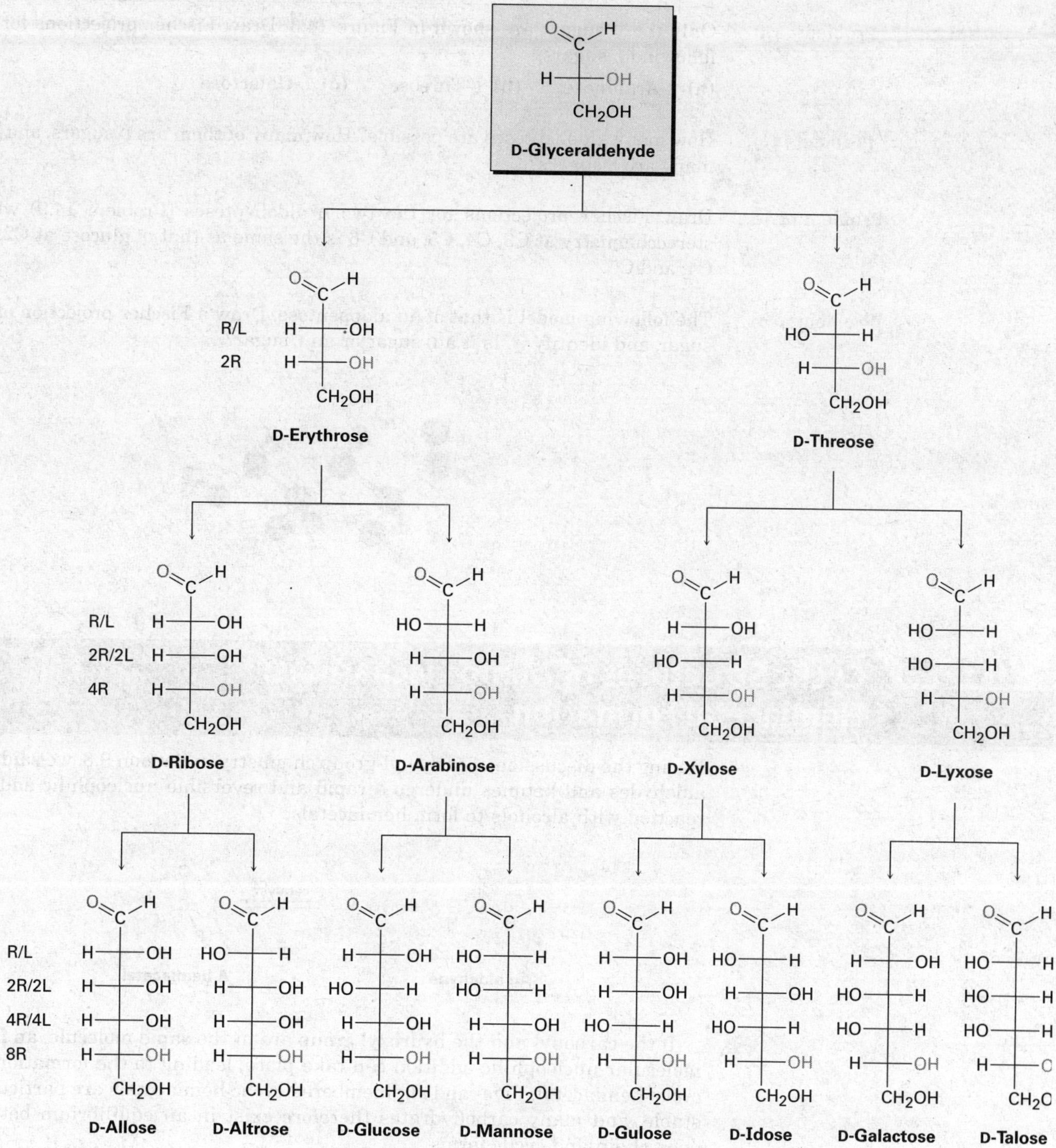

Figure 14.3 Configurations of D aldoses. The structures are arranged from left to right so that the –OH groups on C2 alternate right/left (R/L) in going across a series. Similarly, the –OH groups at C3 alternate two right/two left (2R/2L), the –OH groups at C4 alternate 4R/4L, and the –OH groups at C5 are to the right in all eight (8R). Each D aldose has a mirror-image L enantiomer that is not shown.

Problem 14.8 Only the D sugars are shown in Figure 14.3. Draw Fischer projections for the following L sugars:

(a) L-Arabinose **(b)** L-Threose **(c)** L-Galactose

Problem 14.9 How many aldoheptoses are possible? How many of them are D sugars, and how many are L sugars?

Problem 14.10 Draw Fischer projections for the two D aldoheptoses (Problem 14.9) whose stereochemistry at C3, C4, C5, and C6 is the same as that of glucose at C2, C3, C4, and C5.

Problem 14.11 The following model is that of an aldopentose. Draw a Fischer projection of the sugar, and identify it. Is it a D sugar or an L sugar?

14.5 Cyclic Structures of Monosaccharides: Hemiacetal Formation

During the discussion of carbonyl-group chemistry in Section 9.8, we said that aldehydes and ketones undergo a rapid and reversible nucleophilic addition reaction with alcohols to form hemiacetals.

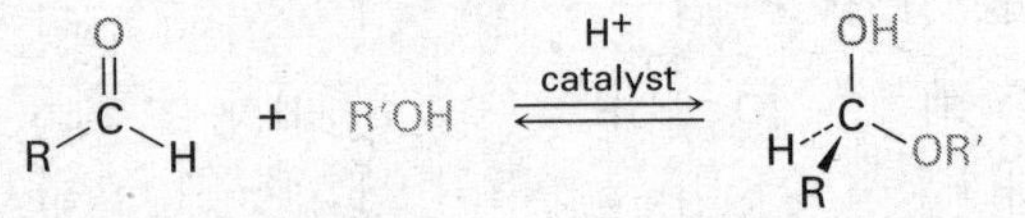

If the carbonyl and the hydroxyl group are in the same molecule, an intramolecular nucleophilic addition can take place, leading to the formation of a *cyclic* hemiacetal. Five- and six-membered cyclic hemiacetals are particularly stable, and many carbohydrates therefore exist in an equilibrium between open-chain and cyclic forms.

Glucose exists in aqueous solution primarily in the six-membered, **pyranose** form resulting from intramolecular nucleophilic addition of the –OH group at C5 to the C1 carbonyl group. Fructose, on the other hand, exists to the extent of about 72% in the pyranose form and about 28% in the five-membered **furanose** form resulting from addition of the –OH group at C5 to the C2 ketone. (The names *pyranose* for a six-membered ring and *furanose* for a five-membered ring are derived from the names of the simple cyclic ethers pyran and furan.) The cyclic forms of glucose and fructose are shown in Figure 14.4.

Figure 14.4 Glucose and fructose in their cyclic pyranose and furanose forms.

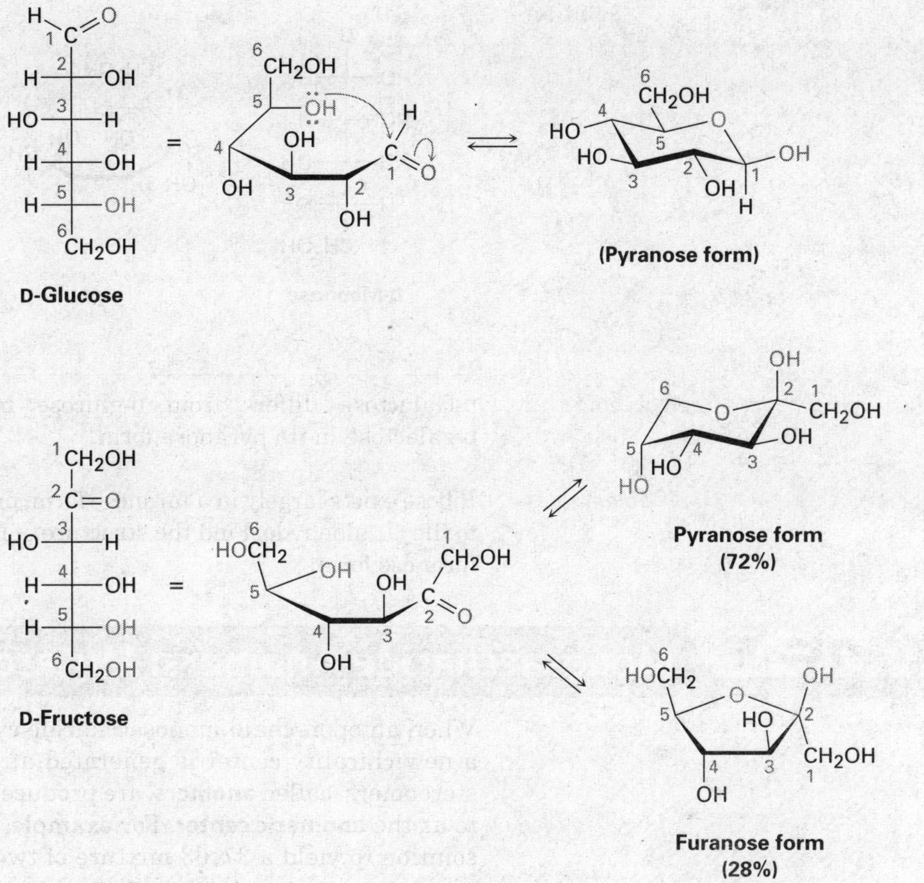

Like cyclohexane rings (Section 2.10), pyranose rings have a chair-like geometry with axial and equatorial substituents. By convention, the rings are usually drawn by placing the hemiacetal oxygen atom at the right rear, as shown in Figure 14.4. Note that an –OH group on the *right* in a Fischer projection is on the *bottom* face of the pyranose ring, and an –OH group on the *left* in a Fischer projection is on the *top* face of the ring. For D sugars, the terminal $-CH_2OH$ group is on the top of the ring, whereas for L sugars, the $-CH_2OH$ group is on the bottom.

Worked Example 14.5 — Drawing the Pyranose Form of a Monosaccharide

D-Mannose differs from D-glucose in its stereochemistry at C2. Draw D-mannose in its pyranose form.

Strategy First draw a Fischer projection of D-mannose. Then lay it on its side, and curl it around so that the –CHO group (C1) is on the right front and the $-CH_2OH$ group (C6) is toward the left rear. Now, connect the –OH at C5 to the

C1 carbonyl group to form the pyranose ring. In drawing the chair form, raise the leftmost carbon (C4) up and drop the rightmost carbon (C1) down.

Solution

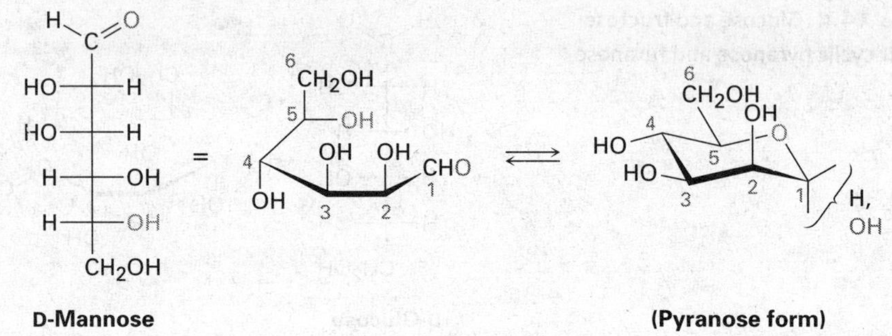

Problem 14.12 D-Galactose differs from D-glucose in its stereochemistry at C4. Draw D-galactose in its pyranose form.

Problem 14.13 Ribose exists largely in a furanose form, produced by addition of the C4 –OH group to the C1 aldehyde. Find the structure of D-ribose in Figure 14.3, and draw it in its furanose form.

14.6 Monosaccharide Anomers: Mutarotation

When an open-chain monosaccharide cyclizes to a pyranose or furanose form, a new chirality center is generated at the former carbonyl carbon. Two diastereomers, called **anomers**, are produced, with the hemiacetal carbon referred to as the **anomeric center**. For example, glucose cyclizes reversibly in aqueous solution to yield a 37:63 mixture of two anomers (Figure 14.5).

Figure 14.5 Structures of the alpha and beta anomers of glucose.

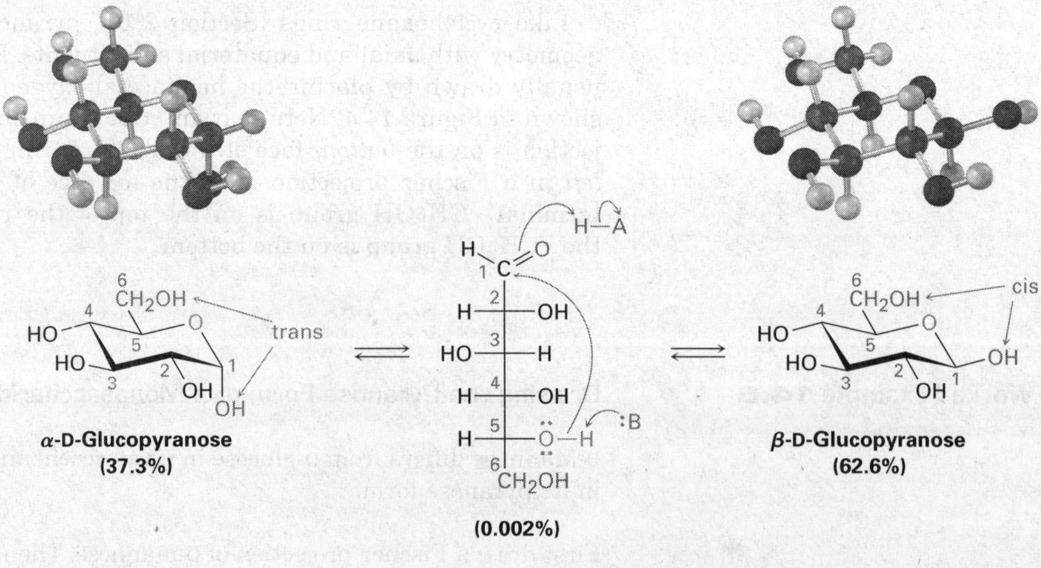

The minor anomer, which has the C1 –OH group trans to the $-CH_2OH$ substituent at C5, is called the **alpha (α) anomer**; its full name is α-D-glucopyranose.

The major anomer, which has the C1 –OH group cis to the $-CH_2OH$ substituent at C5, is called the **beta (β) anomer**; its full name is β-D-glucopyranose. Note that β-D-glucopyranose has all the substituents on the ring equatorial. Thus, β-D-glucopyranose is the least sterically crowded and most stable of the eight D aldohexoses.

Both anomers of D-glucopyranose can be crystallized and purified. Pure α-D-glucopyranose has a melting point of 146 °C and a specific rotation $[\alpha]_D = +112.2$; pure β-D-glucopyranose has a melting point of 148–155 °C and a specific rotation $[\alpha]_D = +18.7$. When a sample of either pure anomer is dissolved in water, however, the optical rotation slowly changes and ultimately reaches a constant value of +52.6. That is, the specific rotation of the α anomer solution decreases from +112.2 to +52.6.°, and the specific rotation of the β anomer solution increases from +18.7 to +52.6.°. Called **mutarotation**, this spontaneous change in optical rotation is caused by the slow interconversion of the pure α and β enantiomers to give the 37:63 equilibrium mixture.

Mutarotation occurs by a reversible ring opening of each anomer to the open-chain aldehyde form, followed by reclosure. Although equilibration is slow at neutral pH, it is catalyzed by both acid and base.

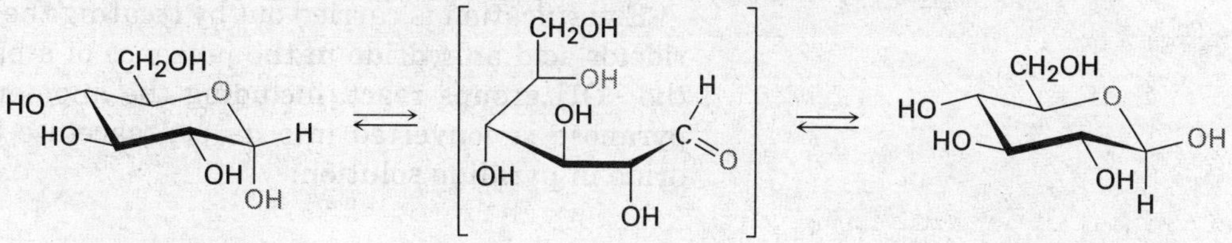

α-D-Glucopyranose $[\alpha]_D = +112.2$

β-D-Glucopyranose $[\alpha]_D = +18.7$

Worked Example 14.6

Drawing Pyranose Anomers

Draw the two pyranose anomers of D-galactose, and identify each as α or β.

Solution The α anomer has the –OH group at C1 pointing down, trans to the CH_2OH, and the β anomer has the –OH group at C1 pointing up, cis to the CH_2OH.

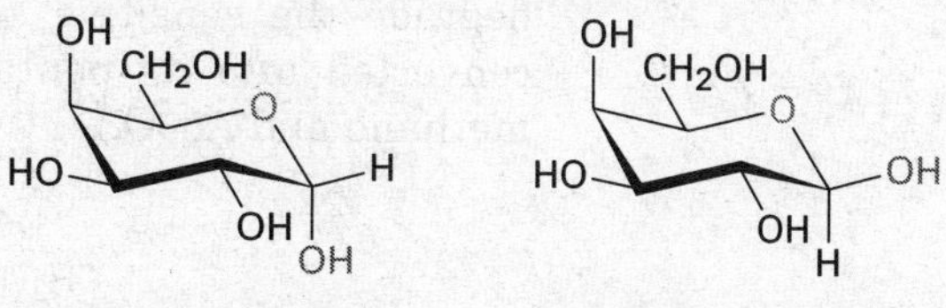

α-D-Galactopyranose β-D-Galactopyranose

Problem 14.14 At equilibrium in aqueous solution, D-fructose consists of 70% β-pyranose, 2% α-pyranose, 23% β-furanose, and 5% α-furanose forms. Draw all four.

Problem 14.15 Draw β-D-mannopyranose in its chair conformation, and label all substituents as axial or equatorial. Which would you expect to be more stable, mannose or galactose (Worked Example 14.6)?

14.7 Reactions of Monosaccharides

Because monosaccharides contain only two kinds of functional groups, hydroxyls and carbonyls, most of the chemistry of monosaccharides is the now-familiar chemistry of these two groups. As we've seen, alcohols can be converted to esters and ethers and can be oxidized; carbonyl compounds can react with nucleophiles and can be reduced.

Ester and Ether Formation

Monosaccharides behave as simple alcohols in much of their chemistry. For example, carbohydrate –OH groups can be converted into esters and ethers, which are often easier to work with than the free sugars. Because of their many hydroxyl groups, monosaccharides are usually soluble in water but insoluble in organic solvents such as ether. They are also difficult to purify and have a tendency to form syrups rather than crystals when water is removed. Ester and ether derivatives, however, are soluble in organic solvents and are easily purified and crystallized.

Esterification is carried out by treating the carbohydrate with an acid chloride or acid anhydride in the presence of a base (Sections 10.7 and 10.8). All the –OH groups react, including the anomeric one. For example, β-D-glucopyranose is converted into its pentaacetate by treatment with acetic anhydride in pyridine solution.

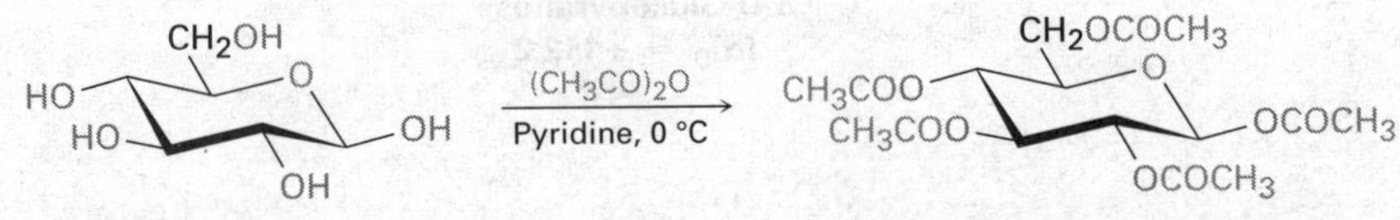

β-D-Glucopyranose → **Penta-O-acetyl-β-D-glucopyranose (91%)**

Carbohydrates are converted into ethers by treatment with an alkyl halide in the presence of base (Section 8.4). Silver oxide is a particularly mild and useful base for this reaction because hydroxide and alkoxide bases tend to degrade the sensitive sugar molecules. For example, α-D-glucopyranose is converted into its pentamethyl ether in 85% yield on reaction with iodomethane and Ag_2O.

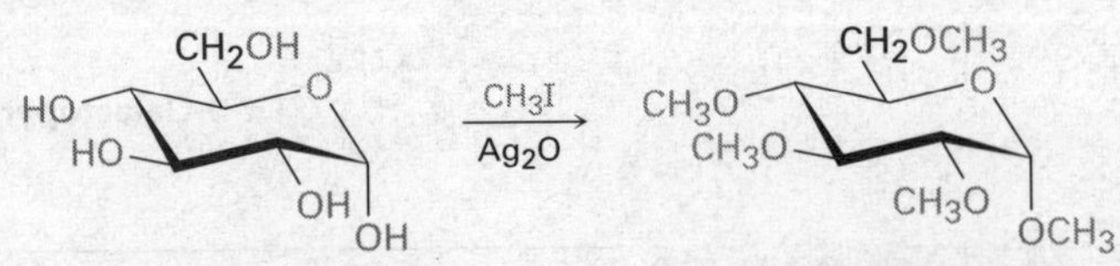

α-D-Glucopyranose → **α-D-Glucopyranose pentamethyl ether (85%)**

Problem 14.16 Draw the products you would obtain by reaction of β-D-ribofuranose with the following:
(a) CH_3I, Ag_2O **(b)** $(CH_3CO)_2O$, pyridine

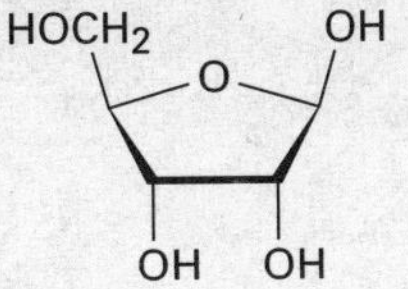

β-D-Ribofuranose

Glycoside Formation

We saw in Section 9.8 that treatment of a hemiacetal with an alcohol and an acid catalyst yields an acetal.

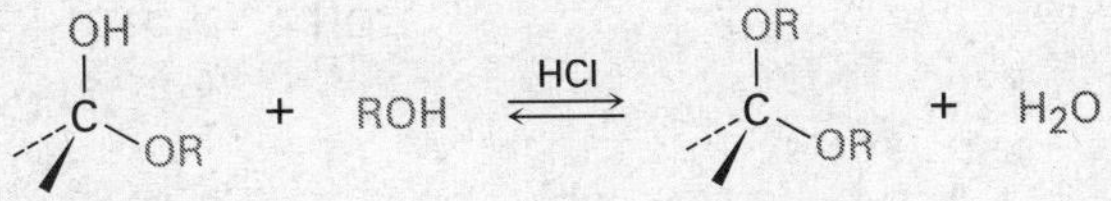

A hemiacetal An acetal

In the same way, treatment of a monosaccharide hemiacetal with an alcohol and an acid catalyst yields an acetal in which the anomeric –OH group has been replaced by an –OR group. For example, reaction of glucose with methanol gives a 2:1 mixture of α and β methyl D-glucopyranosides.

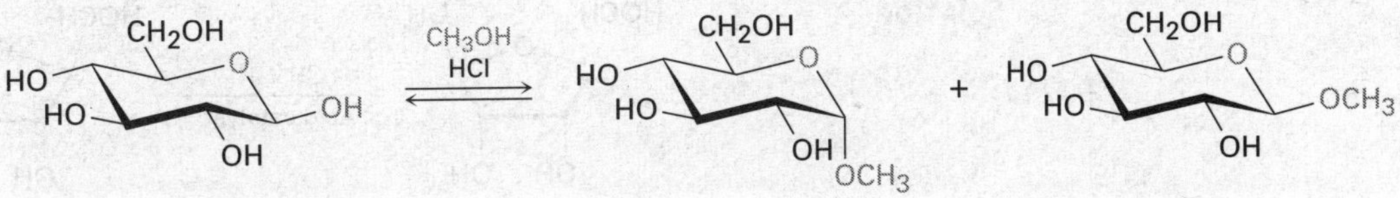

β-D-Glucopyranose (a cyclic hemiacetal) Methyl α-D-glucopyranoside (66%) Methyl β-D-glucopyranoside (33%)

Called **glycosides**, carbohydrate acetals are named by first citing the alkyl group and then replacing the *-ose* ending of the sugar with *-oside.* Like all acetals, glycosides are stable to water. They aren't in equilibrium with an open-chain form, and they don't show mutarotation. They can, however, be hydrolyzed to give back the free monosaccharide plus alcohol on treatment with aqueous acid.

Glycosides are abundant in nature, and many biologically important molecules contain glycosidic linkages. For example, digitoxin, the active component of the digitalis preparations used for treatment of heart disease, is a glycoside consisting of a steroid alcohol linked to a trisaccharide. Note also that the three sugars are linked to one another by glycoside bonds.

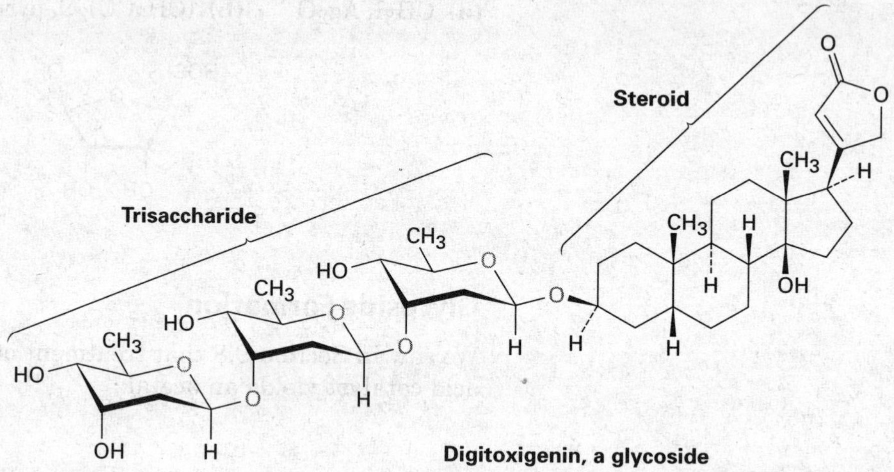

Digitoxigenin, a glycoside

Worked Example 14.7 **Predicting the Structure of a Glycoside**

What product would you expect from the acid-catalyzed reaction of β-D-ribofuranose with methanol?

Strategy The acid-catalyzed reaction of a monosaccharide with an alcohol yields a glycoside in which the anomeric –OH group is replaced by the –OR group of the alcohol.

Solution

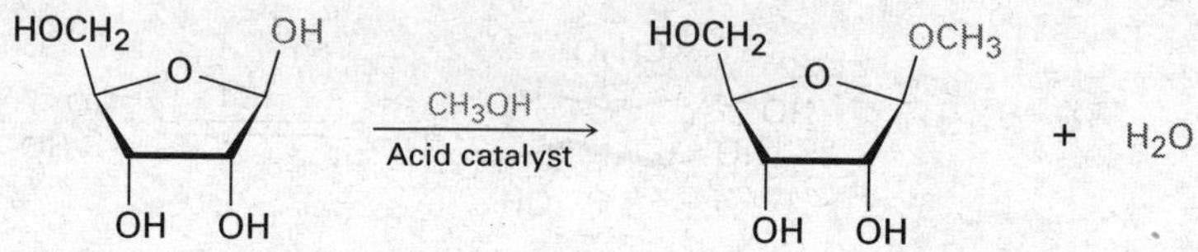

β-D-Ribofuranose **Methyl β-D-ribofuranoside**

Problem 14.17 Draw the product you would obtain from the acid-catalyzed reaction of β-D-galactopyranose with ethanol.

Biological Ester Formation: Phosphorylation

In living organisms, carbohydrates occur not only in their free form but also linked through their anomeric center to other biological molecules such as lipids (*glycolipids*) or proteins (*glycoproteins*). Collectively called *glycoconjugates*, these sugar-linked molecules are components of cell walls and are crucial to the mechanism by which different cell types recognize one another.

Glycoconjugate formation occurs by reaction of the lipid or protein with a glycosyl nucleoside diphosphate. This diphosphate is itself formed by initial reaction of a monosaccharide with adenosine triphosphate (ATP) to give a glycosyl monophosphate, followed by reaction with uridine triphosphate (UTP). (We'll see the structures of nucleoside phosphates in Section 16.5.) The purpose of the phosphorylation is to activate the anomeric –OH group of the sugar and make it a better leaving group in a nucleophilic substitution reaction with a protein or lipid (Figure 14.6).

Figure 14.6 Glycoprotein formation occurs by initial reaction of the starting carbohydrate with ATP to give a glycosyl monophosphate, followed by reaction with UTP to form a glycosyl uridine 5′-diphosphate. Nucleophilic substitution by an –OH (or $-NH_2$) group on a protein then gives the glycoprotein.

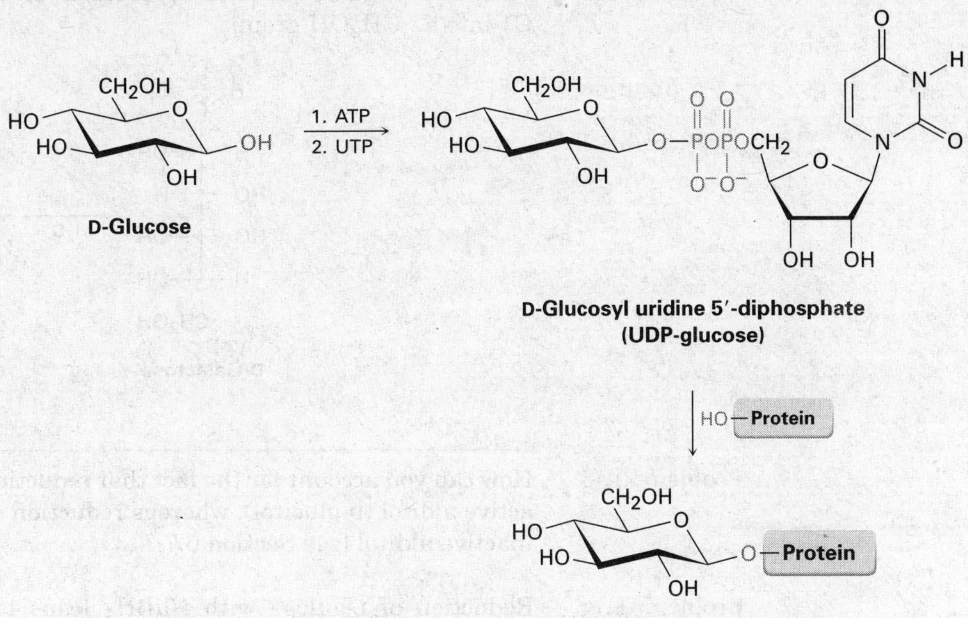

Reduction of Monosaccharides

Treatment of an aldose or a ketose with $NaBH_4$ reduces it to a polyalcohol called an **alditol**. The reaction occurs by reaction of the open-chain form present in the aldehyde ⇄ hemiacetal equilibrium.

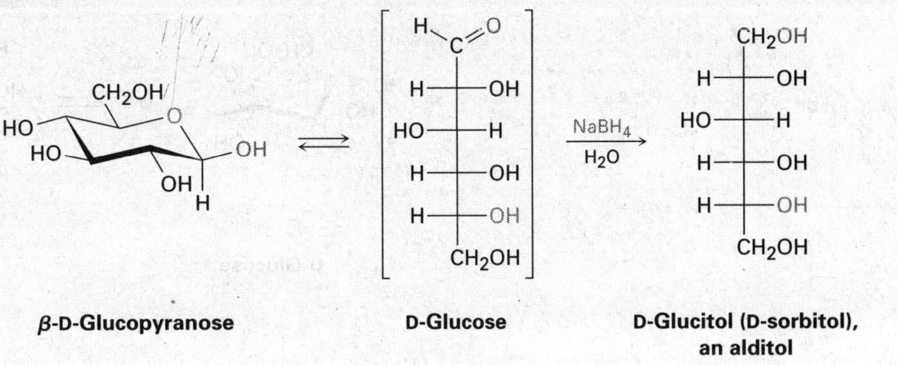

D-Glucitol, the alditol produced on reduction of D-glucose, is itself a naturally occurring substance that has been isolated from many fruits and berries. It is used under the name D-sorbitol as a sweetener and sugar substitute in many foods.

Worked Example 14.8 **Drawing the Structure of an Alditol**

Show the structure of the alditol you would obtain from reduction of D-galactose.

Strategy First draw D-galactose in its open-chain form. Then convert the –CHO group at C1 into a $-CH_2OH$ group.

Solution

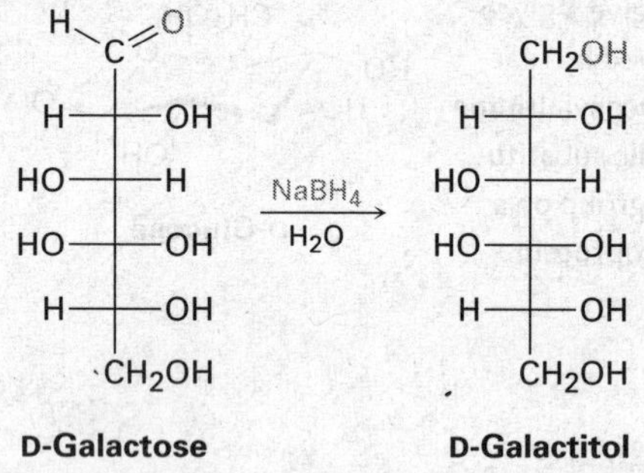

Problem 14.18 How can you account for the fact that reduction of D-glucose leads to an optically active alditol (D-glucitol), whereas reduction of D-galactose leads to an optically inactive alditol (see Section 6.7)?

Problem 14.19 Reduction of L-gulose with $NaBH_4$ leads to the same alditol (D-glucitol) as reduction of D-glucose. Explain.

Oxidation of Monosaccharides

Like other aldehydes, aldoses can be oxidized to yield the corresponding carboxylic acids, called **aldonic acids**. For laboratory purposes, a buffered solution of aqueous Br_2 is often used.

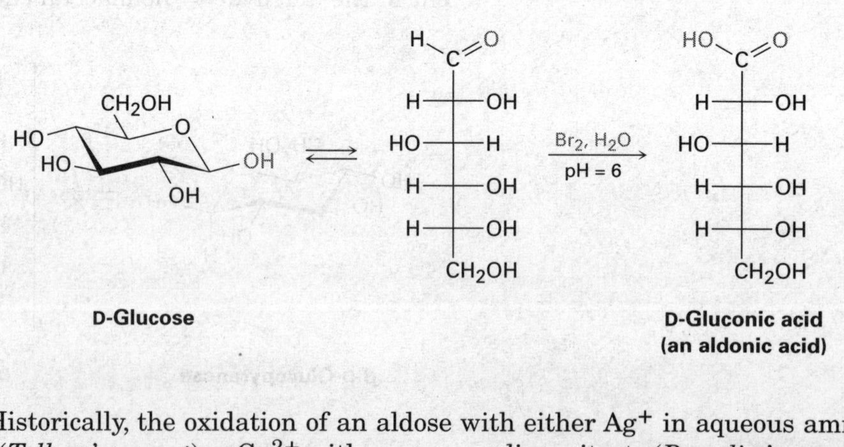

Historically, the oxidation of an aldose with either Ag^+ in aqueous ammonia (*Tollens' reagent*) or Cu^{2+} with aqueous sodium citrate (*Benedict's reagent*) formed the basis of simple tests for what are called **reducing sugars** (*reducing* because the aldose reduces the metal oxidizing agent). Some simple diabetes self-test kits sold in drugstores still use Benedict's reagent to detect glucose in urine, but more modern methods have largely replaced it.

All aldoses are reducing sugars because they contain aldehyde carbonyl groups, but glycosides are nonreducing because the acetal group can't open to an aldehyde under basic conditions.

If warm dilute HNO_3 (nitric acid) is used as the oxidizing agent, an aldose is oxidized to a dicarboxylic acid called an **aldaric acid**. Both the aldehyde carbonyl and the terminal $-CH_2OH$ group are oxidized in this reaction.

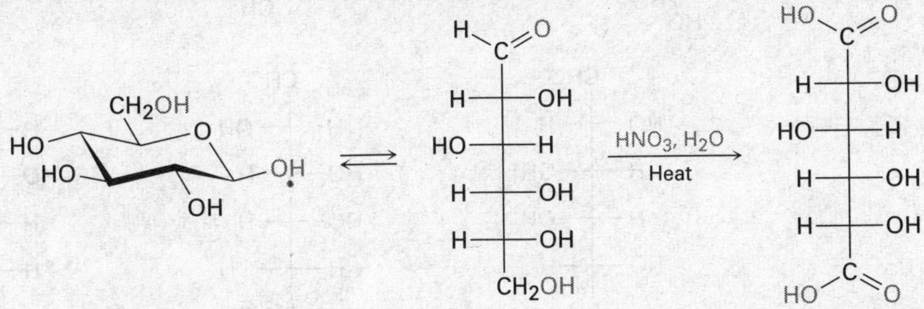

D-Glucose

D-Glucaric acid (an aldaric acid)

Finally, if only the $-CH_2OH$ end of the aldose is oxidized without affecting the $-CHO$ group, the product is a monocarboxylic acid called a **uronic acid**. The reaction must be done enzymatically; no chemical reagent is known that can accomplish this selective oxidation in the laboratory.

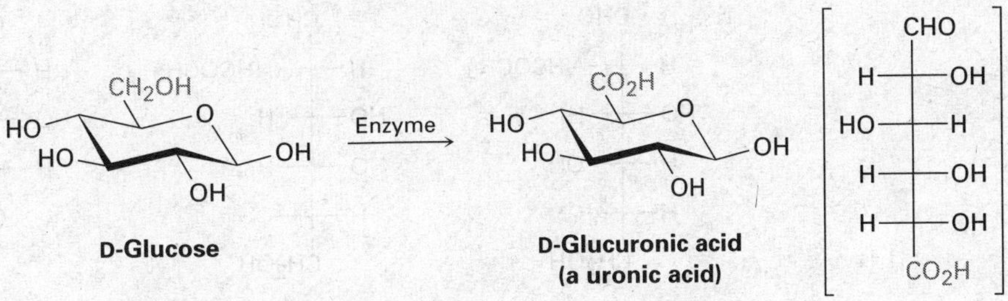

D-Glucose

D-Glucuronic acid (a uronic acid)

Problem 14.20 D-Glucose yields an optically active aldaric acid on treatment with nitric acid, but D-allose yields an optically inactive aldaric acid. Explain.

Problem 14.21 Which of the other six D aldohexoses yield optically active aldaric acids, and which yield optically inactive aldaric acids? (See Problem 14.20.)

14.8 The Eight Essential Monosaccharides

Humans need to obtain eight monosaccharides for proper functioning. Although all can be biosynthesized in the body from simpler precursors if necessary, it's more energetically efficient to obtain them from the diet. The eight are L-fucose (6-deoxy-L-galactose), D-galactose, D-glucose, D-mannose, *N*-acetyl-D-glucosamine, *N*-acetyl-D-galactosamine, D-xylose, and *N*-acetyl-D-neuraminic acid

(Figure 14.7). All are used for the synthesis of the glycoconjugate components of cell walls, and glucose is also the body's primary source of energy.

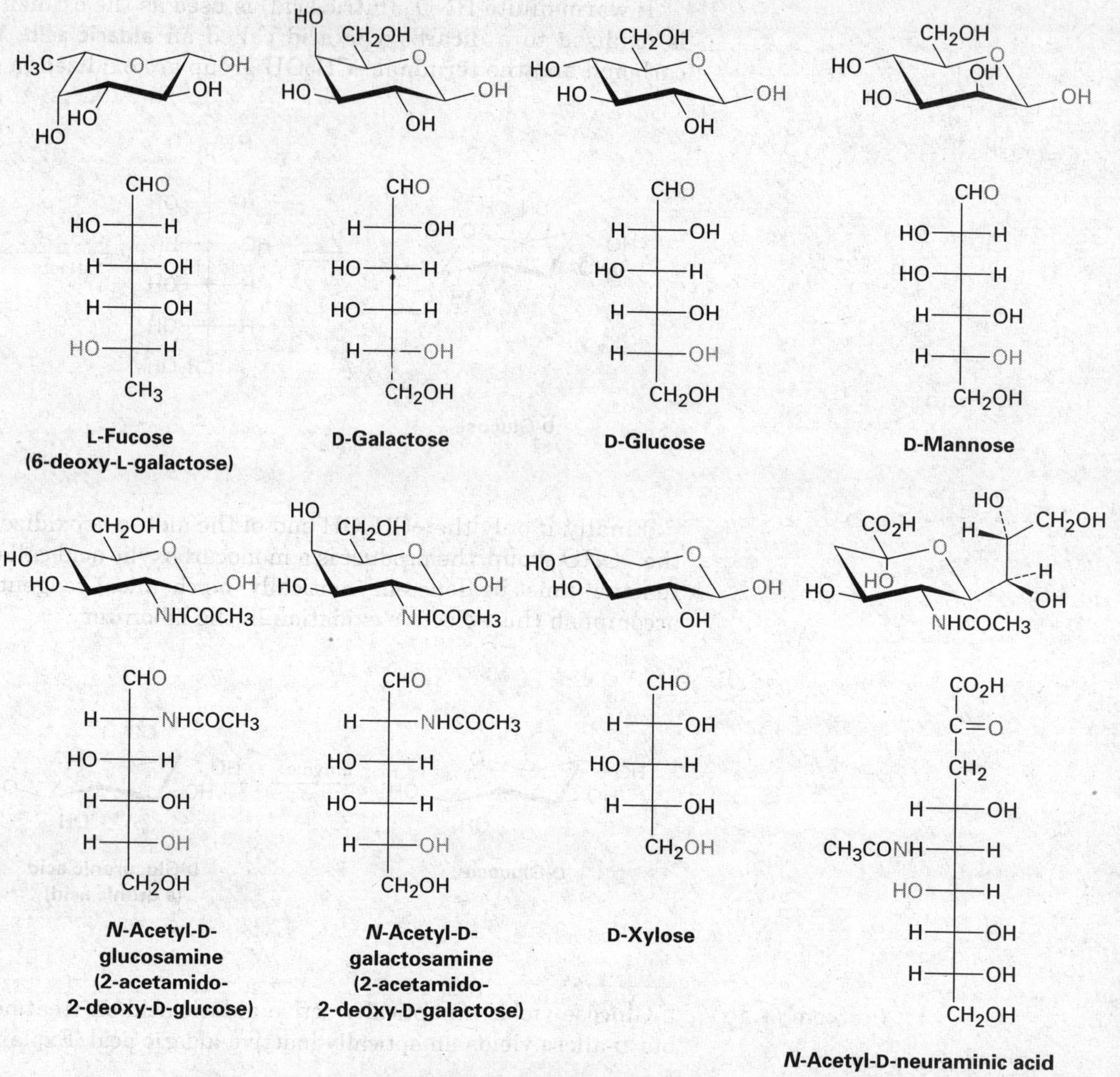

Figure 14.7 Structures of the eight monosaccharides essential to humans.

Of the eight essential monosaccharides, galactose, glucose, and mannose are simple aldohexoses, while xylose is an aldopentose. Fucose is a **deoxy sugar**, meaning that it has an oxygen atom "missing." That is, an –OH group (the one at C6) is replaced by an –H. *N*-Acetylglucosamine and *N*-acetylgalactosamine are amide derivatives of **amino sugars** in which an –OH (the one at C2) is replaced by an $-NH_2$ group. *N*-Acetylneuraminic acid is the parent compound of the *sialic acids,* a group of more than 30 substances with different modifications, including various oxidations, acetylations, sulfations, and methylations. All the essential monosaccharides are biosynthesized from glucose.

14.9 Disaccharides

We saw in Section 14.7 that reaction of a monosaccharide hemiacetal with an alcohol yields a glycoside in which the anomeric –OH group is replaced by an –OR. If the alcohol is itself a sugar, the glycoside product is a *disaccharide.*

Maltose and Cellobiose

Disaccharides contain a glycoside acetal bond between the anomeric carbon of one sugar and an –OH group at any position on the other sugar. A glycoside link between C1 of the first sugar and C4 of the second sugar is particularly common. Such a bond is called a **1→4 link**.

A glycoside bond to the anomeric carbon can be either α or β. Maltose, the disaccharide obtained by enzyme-catalyzed hydrolysis of starch, consists of two α-D-glucopyranose units joined by a 1→4-α-glycoside bond. Cellobiose, the disaccharide obtained by partial hydrolysis of cellulose, consists of two β-D-glucopyranose units joined by a 1→4-β-glycoside bond.

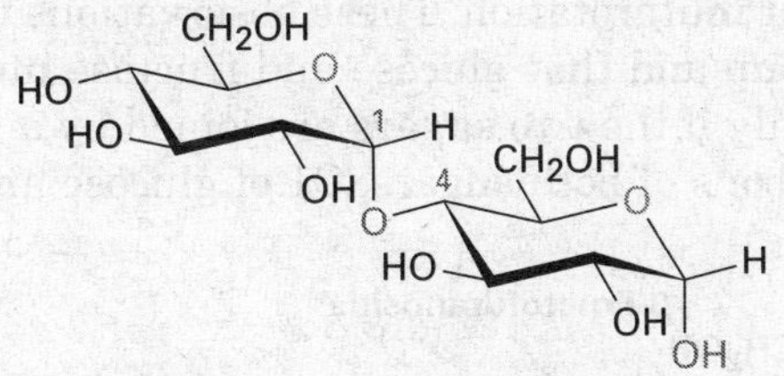

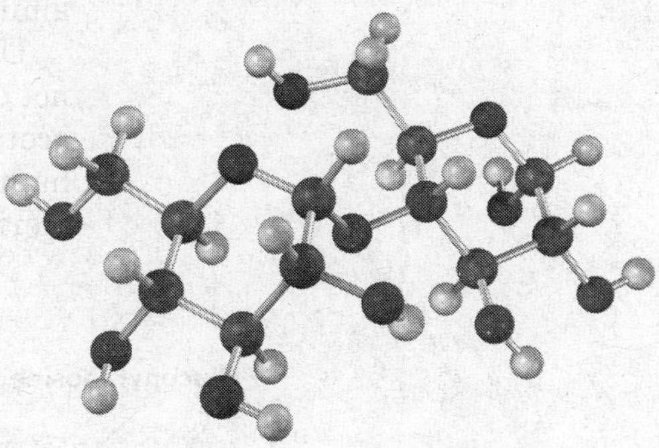

Maltose, a 1→ 4-α-glycoside
[4-*O*-(α-D-glucopyranosyl)-α-D-glucopyranose]

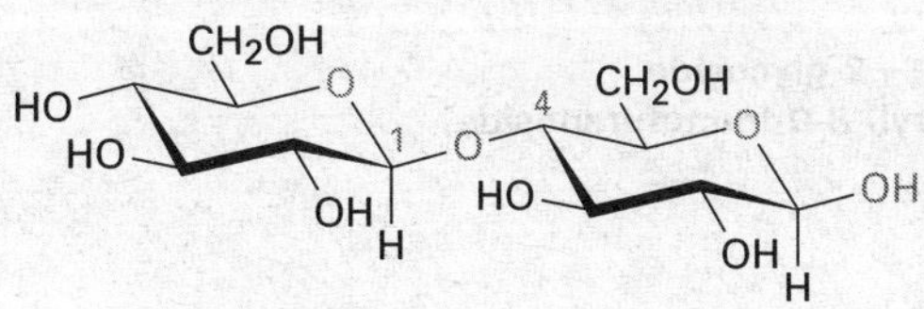

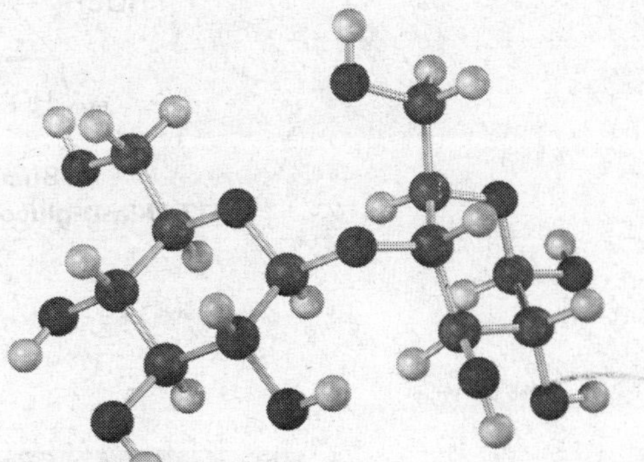

Cellobiose, a 1→ 4-β-glycoside
[4-*O*-(β-D-glucopyranosyl)-β-D-glucopyranose]

Maltose and cellobiose are both reducing sugars because the anomeric carbons on the right-hand glucopyranose units have hemiacetal groups and are in equilibrium with aldehyde forms. For a similar reason, both maltose and cellobiose show mutarotation of α and β anomers of the glucopyranose unit on the right.

Despite the similarities of their structures, cellobiose and maltose have dramatically different biological properties. Cellobiose can't be digested by humans and can't be fermented by yeast. Maltose, however, is digested without difficulty and is fermented readily.

Problem 14.22 Draw the structures of the products obtained from reaction of cellobiose with the following:
(a) $NaBH_4$ **(b)** Br_2, H_2O

Sucrose

Sucrose, or ordinary table sugar, is probably the most abundant pure organic chemical in the world. Whether from sugar cane (20% by weight) or from sugar beets (15% by weight), and whether raw or refined, all table sugar is sucrose.

Sucrose is a disaccharide that yields 1 equivalent of glucose and 1 equivalent of fructose on hydrolysis. This 1:1 mixture of glucose and fructose is often referred to as *invert sugar* because the sign of optical rotation changes, or inverts, during the hydrolysis of sucrose ($[\alpha]_D = +66.5$) to a glucose/fructose mixture ($[\alpha]_D = -22.0$). Some insects, such as honeybees, have enzymes called *invertases* that catalyze the hydrolysis of sucrose. Honey, in fact, is primarily a mixture of glucose, fructose, and sucrose.

Unlike most other disaccharides, sucrose is not a reducing sugar and does not exhibit mutarotation. These observations imply that sucrose has no hemiacetal group and that glucose and fructose must *both* be glycosides. This can happen only if the two sugars are joined by a glycoside link between the anomeric carbons of both sugars, C1 of glucose and C2 of fructose.

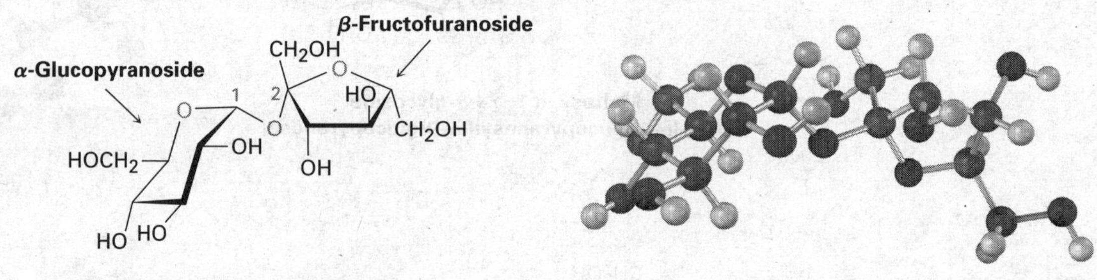

Sucrose, a 1→2-glycoside
[2-*O*-(α-D-glucopyranosyl)-β-D-fructofuranoside]

14.10 Polysaccharides

Polysaccharides are complex carbohydrates in which tens, hundreds, or even thousands of simple sugars are linked together through glycoside bonds. Because they have only the one free anomeric –OH group at the end of a very long chain, polysaccharides are not reducing sugars and don't show noticeable mutarotation. Cellulose and starch are the two most widely occurring polysaccharides.

Cellulose

Cellulose consists of several thousand D-glucose units linked by 1→4-β-glycoside bonds like those in cellobiose. Different cellulose molecules then interact to form a large aggregate structure held together by hydrogen bonds.

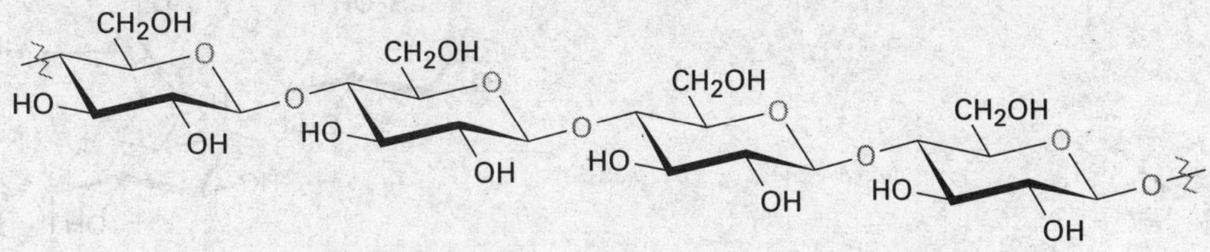

Cellulose, a 1→ 4-O-(β-D-glucopyranoside) polymer

Nature uses cellulose primarily as a structural material to impart strength and rigidity to plants. Leaves, grasses, and cotton are primarily cellulose. Cellulose also serves as a raw material for the manufacture of cellulose acetate, known commercially as *acetate rayon,* and cellulose nitrate, known as *guncotton.* Guncotton is the major ingredient in smokeless powder, the explosive propellant used in artillery shells and ammunition for firearms.

Starch and Glycogen

Potatoes, corn, and cereal grains contain large amounts of *starch,* a polymer of glucose in which the monosaccharide units are linked by 1→4-α-glycoside bonds like those in maltose. Starch can be separated into two fractions: *amylose* and *amylopectin*. Amylose accounts for about 20% by weight of starch and consists of several hundred glucose molecules linked together by 1→4-α-glycoside bonds.

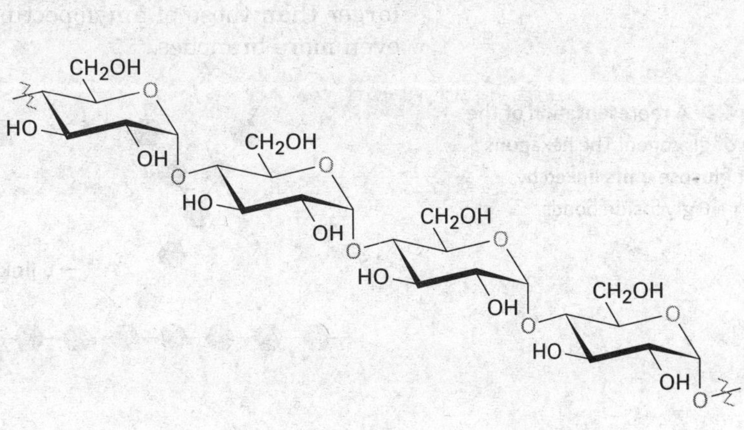

Amylose, a 1→ 4-O-(α-D-glucopyranoside) polymer

Amylopectin accounts for the remaining 80% of starch and is more complex in structure than amylose. Unlike cellulose and amylose, which are linear polymers, amylopectin contains 1→6-α-glycoside branches approximately every 25 glucose units.

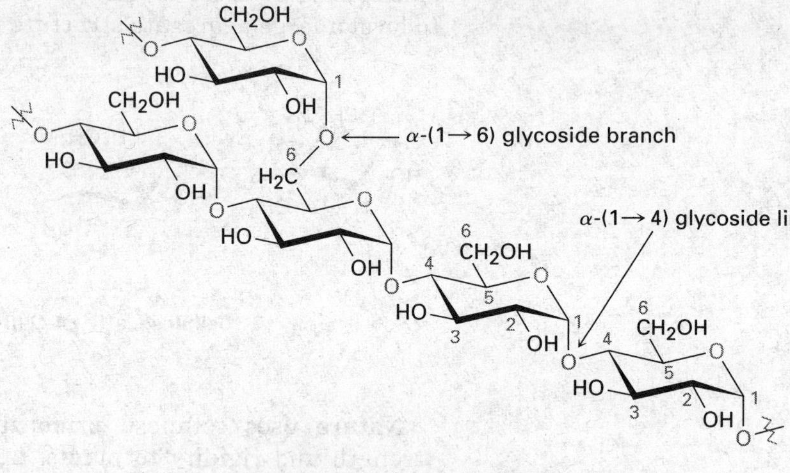

Amylopectin: α-(1→4) links with α-(1→6) branches

Starch is digested in the mouth and stomach by α-glycosidases, which catalyze the hydrolysis of glycoside bonds and release individual molecules of glucose. Like most enzymes, α-glycosidases are highly selective in their action. They hydrolyze only the α-glycoside links in starch and leave the β-glycoside links in cellulose untouched. Thus, humans can digest potatoes and grains but not grass and leaves.

Glycogen is a polysaccharide that serves the same energy storage function in animals that starch serves in plants. Dietary carbohydrates not needed for immediate energy are converted by the body to glycogen for long-term storage. Like the amylopectin found in starch, glycogen contains a complex branching structure with both 1→4 and 1→6 links (Figure 14.8). Glycogen molecules are larger than those of amylopectin—up to 100,000 glucose units—and contain even more branches.

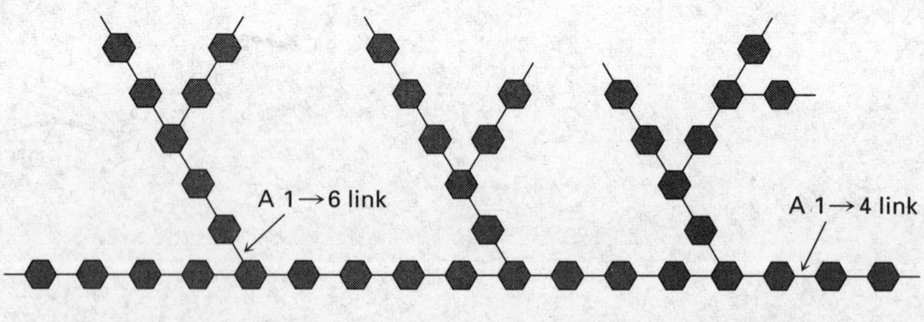

Figure 14.8 A representation of the structure of glycogen. The hexagons represent glucose units linked by 1→4 and 1→6 glycoside bonds.

14.11 Cell-Surface Carbohydrates and Carbohydrate Vaccines

It was once thought that carbohydrates were useful in nature only as structural materials and energy sources. Although carbohydrates do indeed serve these purposes, they have many other important biochemical functions as well. As noted previously, for instance, glycoconjugates are centrally involved in cell–cell recognition, the critical process by which one type of cell distinguishes another. Small polysaccharide chains, covalently bound by glycosidic links to –OH or $-NH_2$ groups on proteins, act as biochemical markers on cell surfaces, as illustrated by the human blood group antigens.

It has been known for more than a century that human blood can be classified into four blood group types (A, B, AB, and O) and that blood from a donor of one type can't be transfused into a recipient with another type unless the two types are compatible (Table 14.1). If an incompatible mix is made, the red blood cells clump together, or *agglutinate.*

Table 14.1 Human Blood Group Compatibilities

Donor blood type	A	B	AB	O
A	o	x	o	x
B	x	o	o	x
AB	x	x	o	x
O	o	o	o	o

The agglutination of incompatible red blood cells, which indicates that the body's immune system has recognized the presence of foreign cells in the body and has formed antibodies against them, results from the presence of polysaccharide markers on the surface of the cells. Types A, B, and O red blood cells each have their own unique markers, called *antigenic determinants;* type AB cells have both type A and type B markers. The structures of all three blood group determinants are shown in Figure 14.9. Note that the monosaccharide constituents of each marker are among the eight essential sugars shown previously in Figure 14.7.

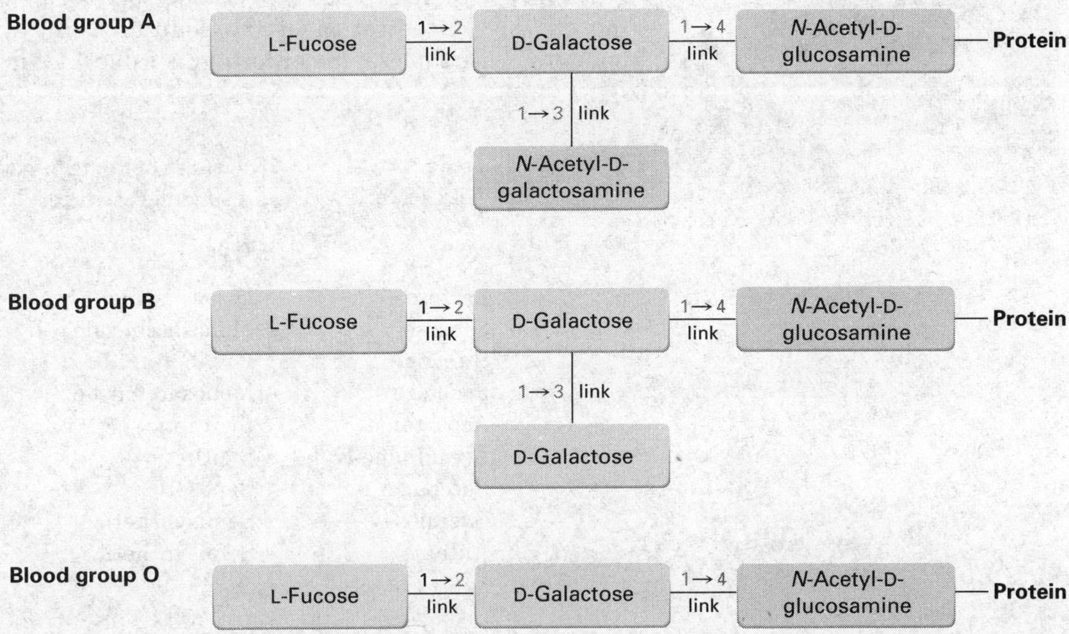

Figure 14.9 Structures of the A, B, and O blood group antigenic determinants.

All three blood group antigenic determinants contain *N*-acetylamino sugars as well as the unusual monosaccharide L-fucose.

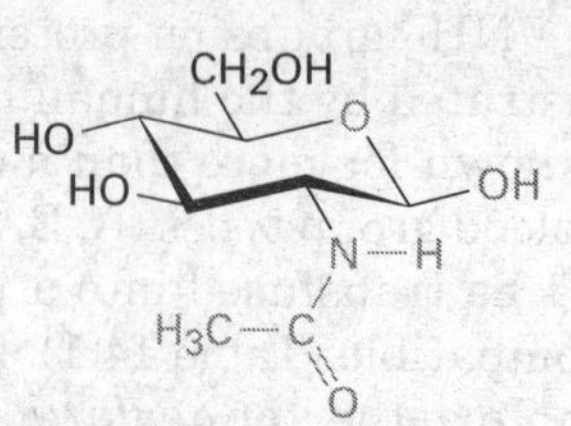

β-D-*N*-Acetylglucosamine

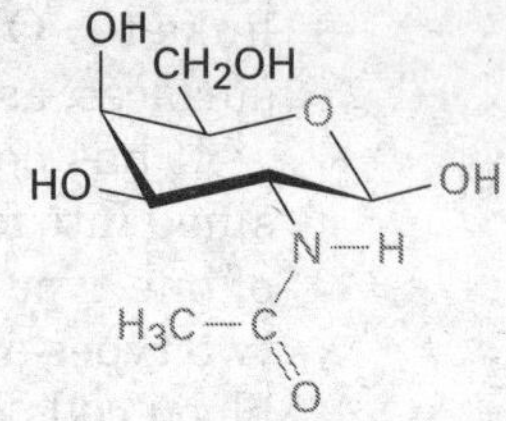

β-D-*N*-Acetylgalactosamine

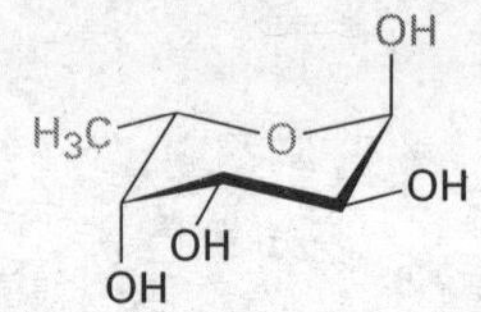

α-L-Fucose

Elucidation of the role of carbohydrates in cell recognition is a vigorous area of current research that offers hope of breakthroughs in treating a wide range of diseases from bacterial infections to cancer. Particularly exciting is the possibility of developing carbohydrate-based vaccines to help mobilize the body's immune system. Diseases currently being studied for vaccine development include pneumonia, malaria, several cancers, and AIDS.

INTERLUDE

Sweetness

"Dietary disaster!"

Say the word *sugar* and most people immediately think of sweet-tasting candies, desserts, and such. In fact, most simple carbohydrates *do* taste sweet, but the degree of sweetness varies greatly from one sugar to another. With sucrose (table sugar) as a reference point, fructose is nearly twice as sweet, but lactose is only about one-sixth as sweet. Comparisons are difficult, though, because perceived sweetness varies depending on the concentration of the solution being tasted and on personal opinion. Nevertheless, the ordering in Table 14.2 is generally accepted.

Table 14.2 Sweetness of Some Sugars and Sugar Substitutes

Name	Type	Sweetness
Lactose	Disaccharide	0.16
Glucose	Monosaccharide	0.75
Sucrose	**Disaccharide**	**1.00**
Fructose	Monosaccharide	1.75
Aspartame	Synthetic	180
Acesulfame-K	Synthetic	200
Saccharin	Synthetic	350
Sucralose	Semisynthetic	600
Alitame	Semisynthetic	2000

continued

INTERLUDE

The desire of many people to cut their caloric intake has led to the development of synthetic sweeteners such as saccharin, aspartame, acesulfame, and sucralose. All are far sweeter than natural sugars, so the choice of one or another depends on personal taste, government regulations, and (for baked goods) heat stability. Saccharin, the oldest synthetic sweetener, has been used for more than a century, although it has a somewhat metallic aftertaste. Doubts about its safety and potential carcinogenicity were raised in the early 1970s, but it has now been cleared of suspicion.

Acesulfame potassium, one of the most recently approved sweeteners, is proving to be extremely popular in soft drinks because it has little aftertaste. Sucralose, another recently approved sweetener, is particularly useful in baked goods because of its stability at high temperatures. Alitame, marketed in some countries under the name Aclame, is not approved for sale in the United States. It is some 2000 times as sweet as sucrose and, like acesufame-K, has no aftertaste. Of the five synthetic sweeteners listed in Table 14.2, only sucralose has clear structural resemblance to a carbohydrate, although it differs dramatically in containing three chlorine atoms. Aspartame and alitame are both dipeptides.

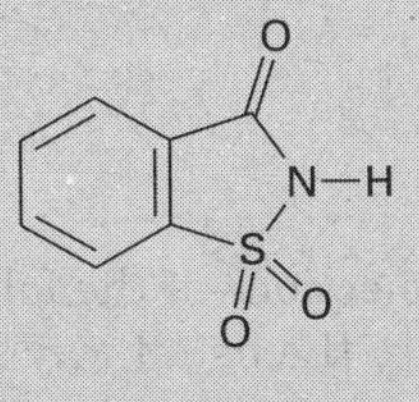

Saccharin

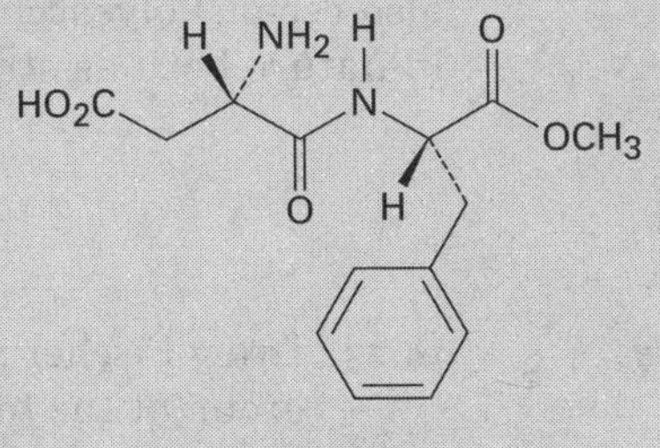

Aspartame

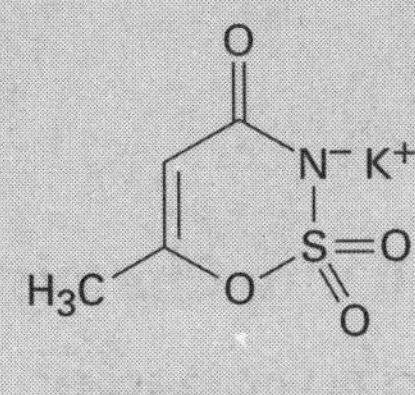

Acesulfame potassium

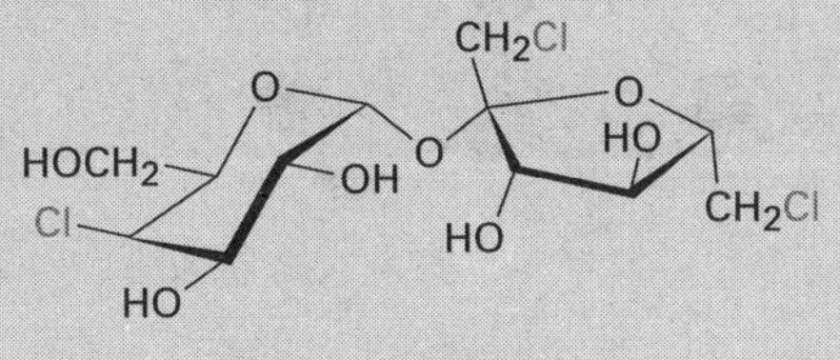

Sucralose

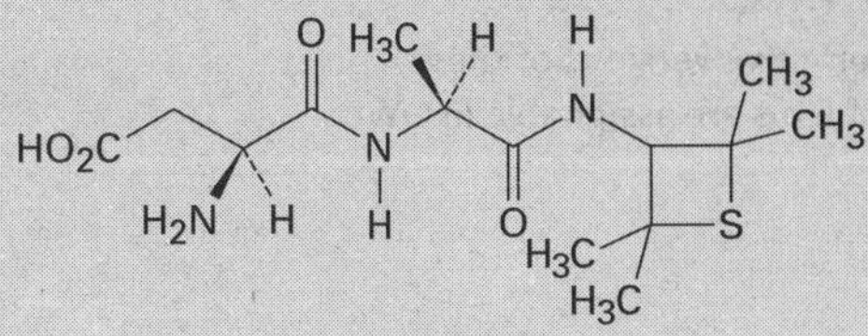

Alitame

Summary and Key Words

aldaric acid 487
alditol 485
aldonic acid 486
aldose 471
alpha (α) anomer 480
amino sugar 488
anomeric center 480
beta (β) anomer 481

Now that we've now seen all the common functional groups and reaction types, our focus has changed to looking at the major classes of biological molecules. **Carbohydrates** are polyhydroxy aldehydes and ketones. They are classified according to the number of carbon atoms and the kind of carbonyl group they contain. Glucose, for example, is an aldohexose, a six-carbon aldehydo sugar. **Monosaccharides** are further classified as either **D** or **L sugars**, depending on the stereochemistry of the chirality center farthest from the carbonyl group. Carbohydrate stereochemistry is frequently shown

carbohydrate 469
complex carbohydrate 470
D sugar 475
deoxy sugar 488
Fischer projection 472
furanose 478
glycoside 483
ketose 471
L sugar 475
1→4 link 489
monosaccharide 470
mutarotation 481
pyranose 478
reducing sugar 486
simple sugar 470
uronic acid 487

using **Fischer projections**, which represent a chirality center as the intersection of two crossed lines.

Monosaccharides normally exist as cyclic hemiacetals rather than as open-chain aldehydes or ketones. The hemiacetal linkage results from reaction of the carbonyl group with an –OH group three or four carbon atoms away. A five-membered cyclic hemiacetal is a **furanose**, and a six-membered cyclic hemiacetal is a **pyranose**. Cyclization leads to the formation of a new chirality center called the **anomeric center** and the production of two diastereomeric hemiacetals called **alpha (α)** and **beta (β) anomers**.

Much of the chemistry of monosaccharides is the familiar chemistry of alcohol and carbonyl functional groups. Thus, the –OH groups of carbohydrates form esters and ethers. The carbonyl group of a monosaccharide can be reduced with $NaBH_4$ to yield an **alditol**, oxidized with aqueous Br_2 to yield an **aldonic acid**, oxidized with warm HNO_3 to yield an **aldaric acid**, oxidized enzymatically to form a **uronic acid**, or treated with an alcohol in the presence of acid to form a **glycoside**.

Disaccharides are complex carbohydrates in which two simple sugars are linked by a glycoside bond between the anomeric carbon of one unit and an –OH of the second unit. The two sugars can be the same, as in maltose and cellobiose, or different, as in sucrose. The glycoside bond can be either α (maltose) or β (cellobiose) and can involve any –OH of the second sugar. A 1→4 link is most common (cellobiose, maltose), but other links, such as 1→2 (sucrose), also occur. Polysaccharides, such as cellulose, starch, and glycogen, are used in nature both as structural materials and for long-term energy storage.

Exercises

Visualizing Chemistry

(Problems 14.1–14.22 appear within the chapter.)

OWL
Interactive versions of these problems are assignable in OWL.

14.23 Draw Fischer projections of the following molecules, placing the carbonyl group at the top in the usual way. Identify each as a D or L sugar.

(a)

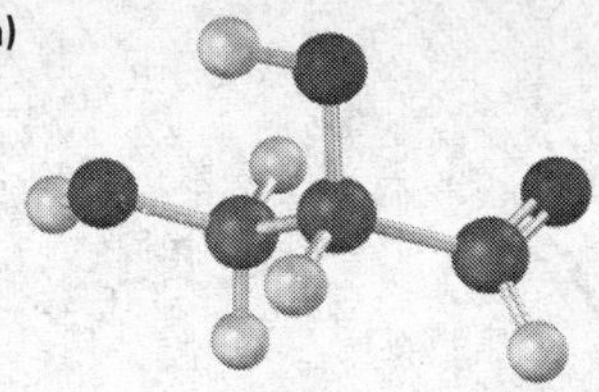

(b)

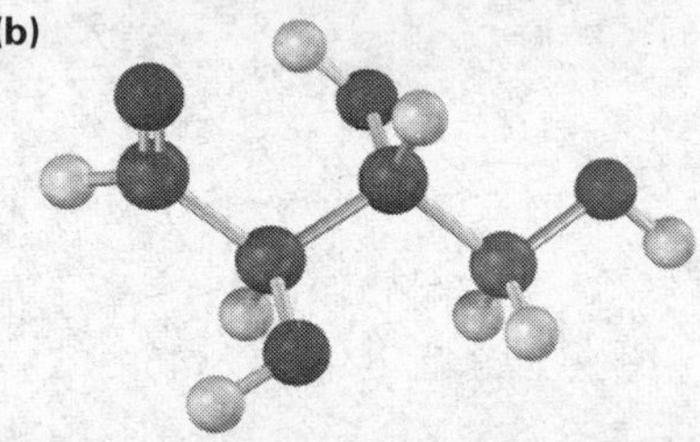

14.24 Identify the following aldoses (see Figure 14.3), and indicate whether each is a D or L sugar.

(a)

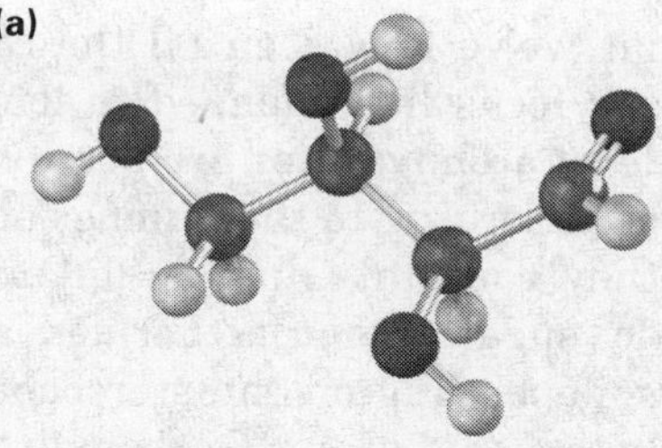

(b)

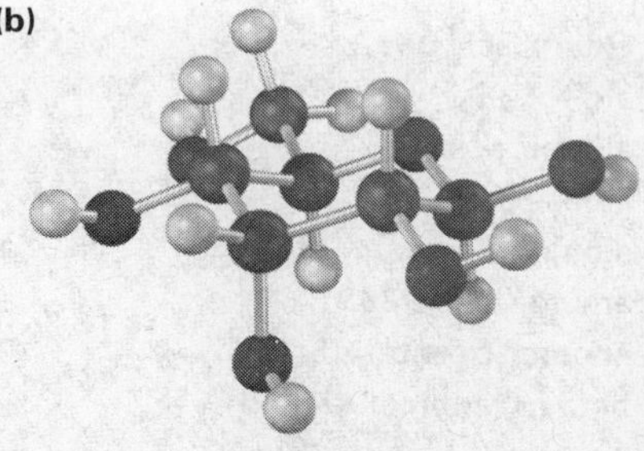

14.25 The following model is that of an aldohexose:

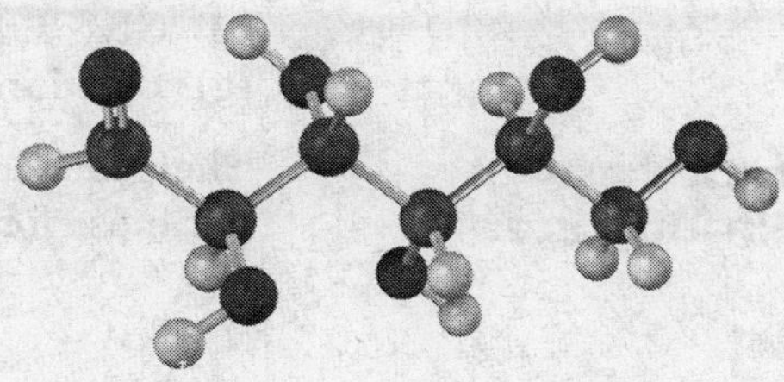

(a) Draw Fischer projections of the sugar, its enantiomer, and a diastereomer.
(b) Is this a D sugar or an L sugar? Explain.
(c) Draw the β anomer of the sugar in its furanose form.

14.26 The following structure is that of an L aldohexose in its pyranose form. Identify it (see Figure 14.3).

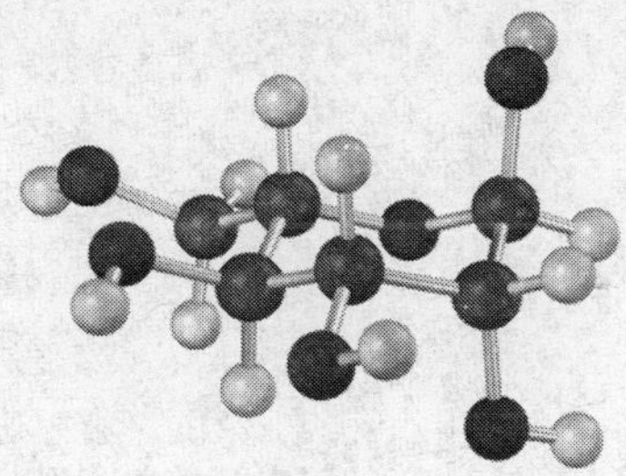

Additional Problems

CLASSIFYING MONOSACCHARIDES

14.27 Write an open-chain structure for a deoxyaldohexose.

14.28 Write an open-chain structure for a five-carbon amino sugar.

14.29 Classify the following sugars by type (for example, glucose is an aldohexose):

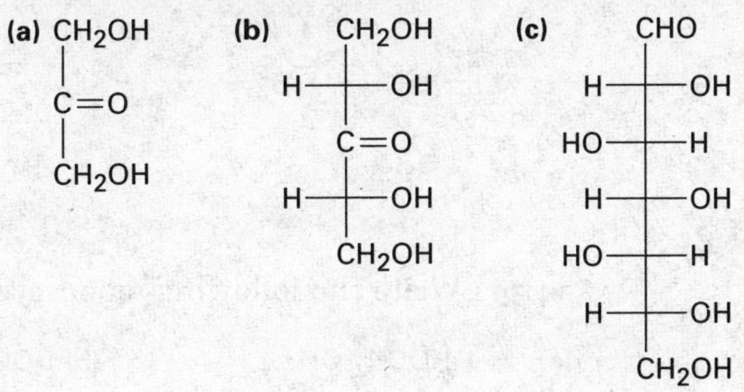

14.30 Write open-chain structures for a ketotetrose and a ketopentose.

14.31 Define the following terms, and give an example of each:
(a) Monosaccharide **(b)** Anomeric center **(c)** Fischer projection
(d) Glycoside **(e)** Reducing sugar **(f)** Pyranose form
(g) 1→4 Link **(h)** D-Sugar

Carbohydrate Structures

14.32 The following cyclic structure is that of allose. Is this a furanose or pyranose form? Is it an α or β anomer? Is it a D sugar or L sugar?

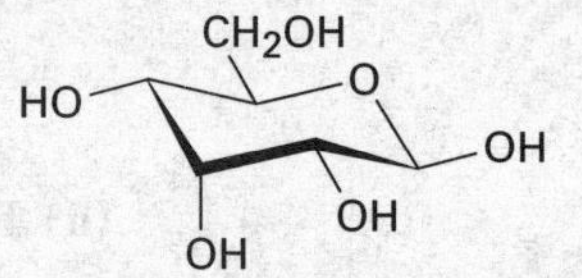

14.33 Uncoil allose (see Problem 14.32), and write it in its open-chain form.

14.34 The structure of ascorbic acid (vitamin C) is shown. Does ascorbic acid have a D or L configuration?

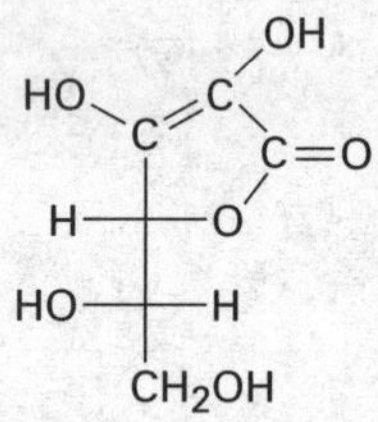
Ascorbic acid

14.35 Assign R or S stereochemistry to each chirality center in ascorbic acid (Problem 14.34).

14.36 Look up the structures of maltose and sucrose in Section 14.9, and explain why maltose is reduced by $NaBH_4$ but sucrose is not.

14.37 Look up the structure of D-talose in Figure 14.3, and draw the β anomer in its pyranose form. Identify the ring substituents as axial or equatorial.

14.38 Draw D-ribulose in its five-membered cyclic β hemiacetal form.

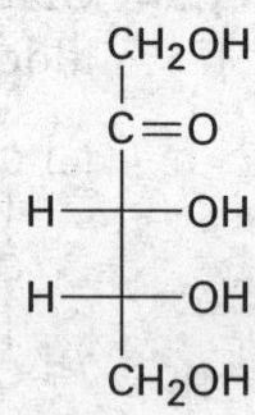
Ribulose

14.39 Write the following monosaccharides in their open-chain forms:

(a)
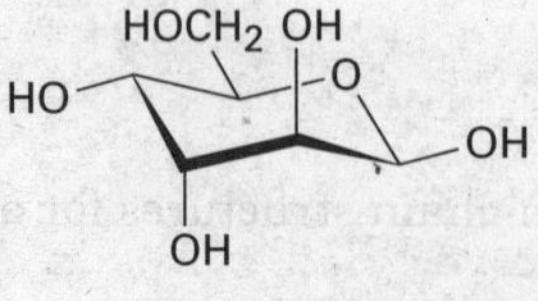

(b)
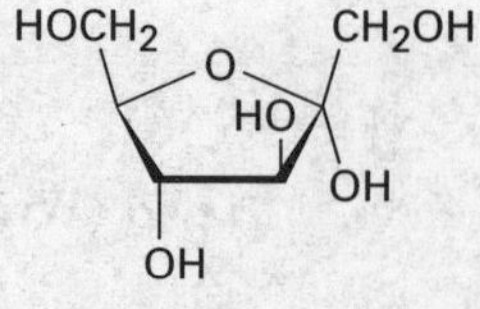

(c)
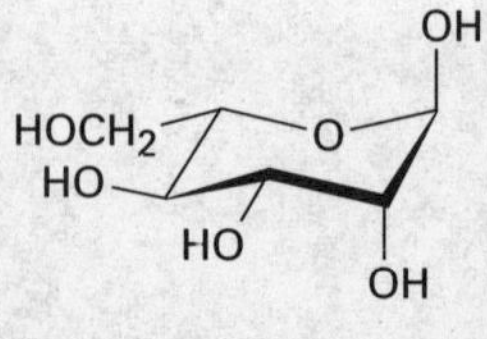

14.40 How many D-2-ketohexoses are there? Draw them.

14.41 What is the stereochemical relationship of D-allose to L-allose? What generalizations can you make about the following properties of the two sugars?
(a) Melting point **(b)** Solubility in water
(c) Specific rotation **(d)** Density

14.42 What is the stereochemical relationship of D-ribose to L-xylose? What generalizations can you make about the following properties of the two sugars?
(a) Melting point **(b)** Solubility in water
(c) Specific rotation **(d)** Density

Carbohydrate Reactions

14.43 One of the D-2-ketohexoses (see Problem 14.40) is called *sorbose.* On treatment with $NaBH_4$, sorbose yields a mixture of gulitol and iditol. What is the structure of D-sorbose? (Gulitol and iditol are the alditols obtained by reduction of gulose and idose.)

14.44 Another D-2-ketohexose, *psicose,* yields a mixture of allitol and altritol when reduced with $NaBH_4$ (see Problem 14.43). What is the structure of psicose?

14.45 Draw structures for the products you would expect to obtain from the reaction of β-D-talopyranose (see Problem 14.37) with each of the following reagents:
(a) $NaBH_4$ **(b)** Warm dilute HNO_3
(c) aqueous Br_2 **(d)** CH_3CH_2OH, H^+
(e) CH_3I, Ag_2O **(f)** $(CH_3CO)_2O$, pyridine

General Problems

14.46 Which of the eight D aldohexoses yield optically inactive (meso) alditols on reduction with $NaBH_4$?

14.47 What other D aldohexose gives the same alditol as D-talose? (See Problem 14.46.)

14.48 Which of the other three D aldopentoses gives the same aldaric acid as D-lyxose?

14.49 Which of the eight D aldohexoses give the same aldaric acids as their L enantiomers?

14.50 Convert the following Fischer projections into tetrahedral representations:

(a)

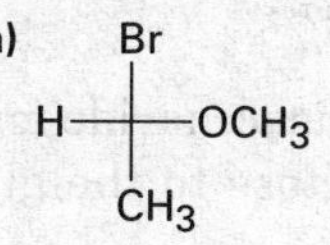

(b)

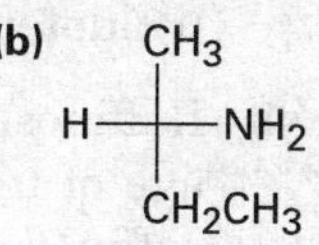

14.51 Draw Fischer projections of the following substances:
(a) (*R*)-2-Methylbutanoic acid **(b)** (*S*)-3-Methylpentan-2-one

14.52 The *Ruff degradation* is a method used to shorten an aldose chain by one carbon atom. The original C1 carbon atom is cut off, and the original C2 carbon atom becomes the aldehyde of the chain-shortened aldose. For example, D-glucose, an aldohexose, is converted by Ruff degradation into D-arabinose, an aldopentose. What other D aldohexose would also yield D-arabinose on Ruff degradation?

14.53 D-Galactose and D-talose yield the same aldopentose on Ruff degradation (Problem 14.52). What does this tell you about the stereochemistry of galactose and talose? Which D aldopentose is obtained?

14.54 D-Fructose and D-glucose are interconverted on treatment with dilute acid or base. Propose a mechanism for this interconversion. (See Section 11.1.)

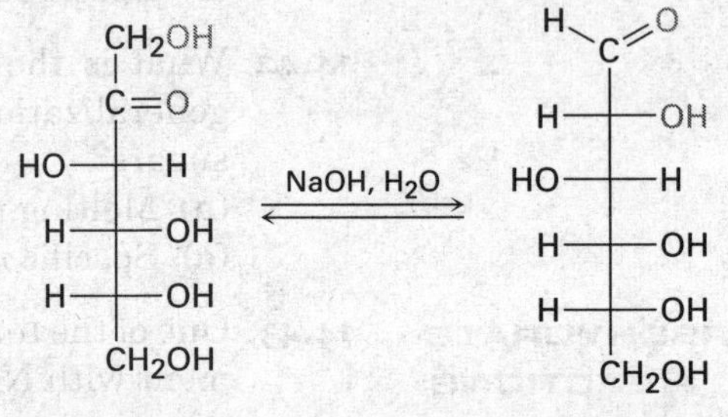

D-Fructose D-Glucose

14.55 What other aldohexose besides glucose is likely to be interconvertible with fructose when treated with dilute aqueous base? (See Problem 14.54.)

14.56 Mannose, one of the eight essential monosaccharides (Section 14.8), is biosynthesized as its 6-phosphate derivative from fructose 6-phosphate in three steps. The initial step is hemiacetal ring-opening, the second step is keto–enol equilibration, and the third step is hemiacetal formation. Propose a mechanism. (See Problem 14.54.)

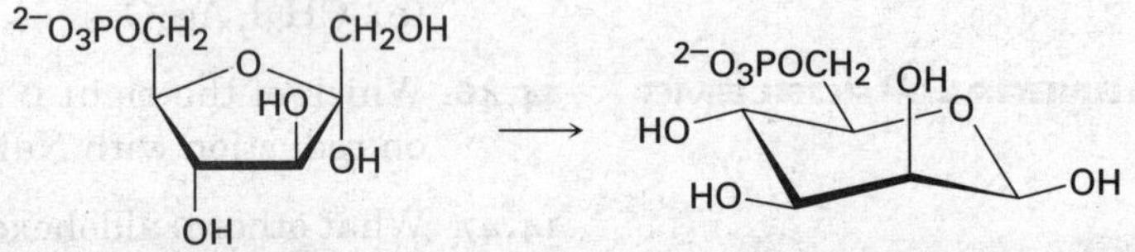

Fructose 6-phosphate Mannose 6-phosphate

14.57 The aldaric acid obtained by nitric acid oxidation of D-erythrose, one of the D aldotetroses, is optically inactive. The aldaric acid obtained from oxidation of the other D aldotetrose, D-threose, however, is optically active. How does this information allow you to assign structures to the two D aldotetroses?

14.58 Raffinose, a trisaccharide found in sugar beets, is formed by a 1→6 α linkage of D-galactose to the glucose unit of sucrose. Draw the structure of raffinose.

14.59 Is raffinose (see Problem 14.58) a reducing sugar? Explain.

14.60 Gentiobiose is a rare disaccharide found in saffron and gentian. It is a reducing sugar and forms only glucose on hydrolysis with aqueous acid. If gentiobiose contains a 1→6-β-glycoside link, what is its structure?

IN THE MEDICINE CABINET

14.61 Zanamivir, sold under the trade name Relenza, is one of the very few compounds active against the H5N1 avian flu virus. The virus attacks cells by recognizing a terminal sialic acid on a cell-surface glycoprotein and hydrolyzing it from the cell surface. Zanamivir is able to prevent the hydrolysis by mimicking the oxonium ion intermediate in the hydrolysis and blocking the enzyme necessary for removal. Why is Relenza a good mimic of the intermediate oxonium ion?

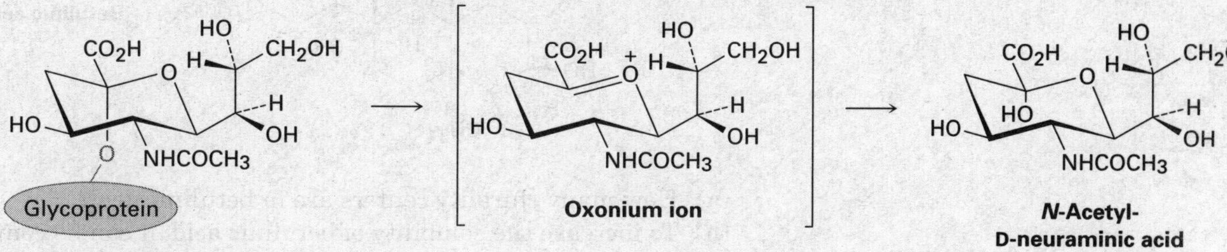

N-Acetyl-D-neuraminic acid glycoconjugate → **Oxonium ion** → **N-Acetyl-D-neuraminic acid**

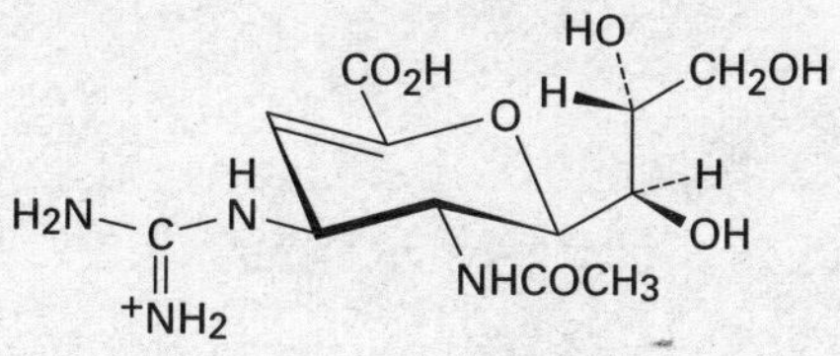

Zanamivir (Relenza)

14.62 *N*-Acetylneuraminic acid, one of the eight essential monosaccharides and the simplest sialic acid (Problem 14.61), is biosynthesized by reaction of *N*-acetylmannosamine with pyruvate, $CH_3COCO_2^-$. What kind of reaction takes place? Show the mechanism.

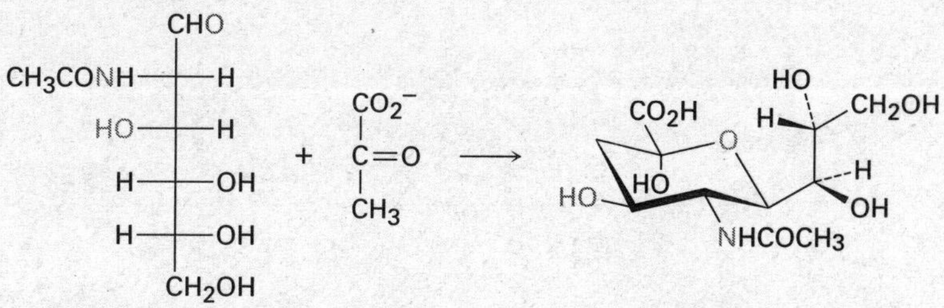

N-Acetylmannosamine + **Pyruvate** → **N-Acetyl-D-neuraminic acid**

IN THE FIELD

14.63 Betulinic acid is found in the bark of the white birch tree and is useful against melanomas, dangerous forms of skin cancer.

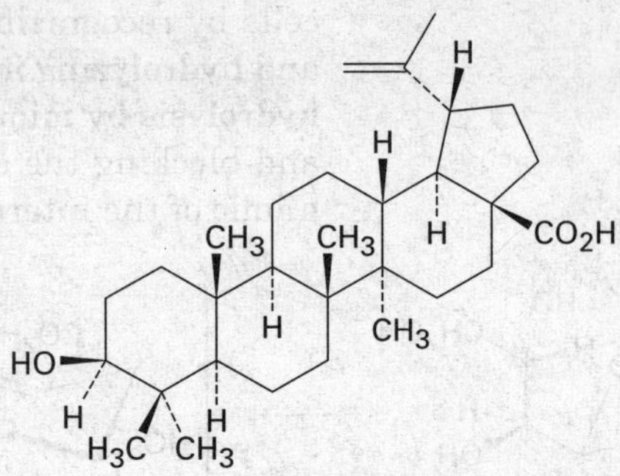

Betulinic acid

(a) How many chirality centers are in betulinic acid?

(b) To increase the solubility of betulinic acid, it can be converted into a glycoside by letting it react with a carbohydrate. Draw a mechanism for the formation of a glycoside bond between betulinic acid and glucose.

CHAPTER

15

The building blocks of life that we call proteins are aptly named after Proteus, the early Greek sea-god whose name means "first" or "primordial."

Proteus, Old Man of the Sea, from Thessalonika (mosaic), Greek, (1st century)/National Archaeological Museum, Athens, Greece/Ancient Art and Architecture Collection Ltd./The Bridgeman Art Library

Biomolecules: Amino Acids, Peptides, and Proteins

Proteins occur in every living organism, are of many different types, and have many different biological functions. The keratin of skin and fingernails, the fibroin of silk and spider webs, and the estimated 50,000 or so enzymes that catalyze the biological reactions in our bodies are all proteins. Regardless of their function, all proteins have a fundamentally similar structure and are made up of many *amino acids* linked together in a long chain.

Amino acids, as their name implies, are difunctional. They contain both a basic amino group and an acidic carboxyl group.

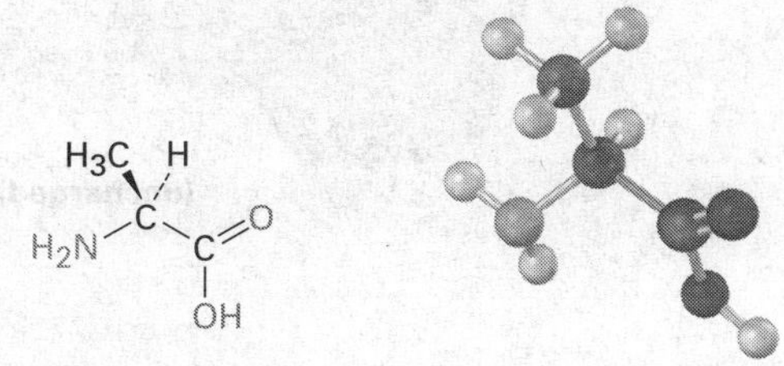

Alanine, an amino acid

OWL

Online homework for this chapter can be assigned in OWL, an online homework assessment tool.

Their value as building blocks to make proteins stems from the fact that amino acids can join together into long chains by forming amide bonds between the $-NH_2$ of one amino acid and the $-CO_2H$ of another. For classification purposes, chains with fewer than 50 amino acids are often called **peptides**, while the term **protein** is generally used for larger chains.

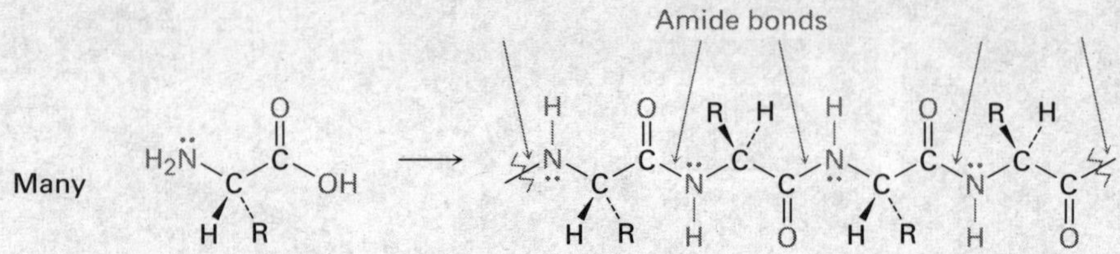

WHY THIS CHAPTER?

Continuing our look at the main classes of biomolecules, we'll focus in this chapter on amino acids, the fundamental building blocks from which the 150,000 or so proteins in our bodies are made. We'll then see how amino acids are incorporated into proteins, a topic crucial to any understanding of biological chemistry.

15.1 Structures of Amino Acids

Because amino acids contain both basic amino and acidic carboxyl groups, they undergo an intramolecular acid–base reaction and exist in aqueous solution primarily in the form of dipolar ions, called **zwitterions** (from the German *zwitter,* meaning "hybrid").

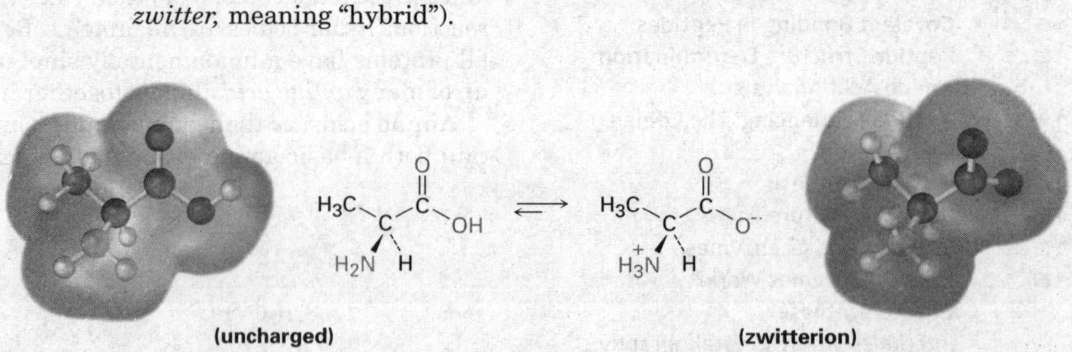

Alanine

Amino acid zwitterions are internal salts and therefore have many of the physical properties associated with salts. They are relatively soluble in water but insoluble in hydrocarbons and are crystalline substances with relatively high melting points. In addition, amino acids are *amphiprotic,* meaning that they can react either as acids or as bases, depending on the circumstances. In aqueous acid solution, an amino acid zwitterion is a base that accepts a proton onto its $-CO_2^-$ group to yield a cation; in aqueous base solution, the zwitterion is an acid that loses a proton from its $-NH_3^+$ group to form an anion.

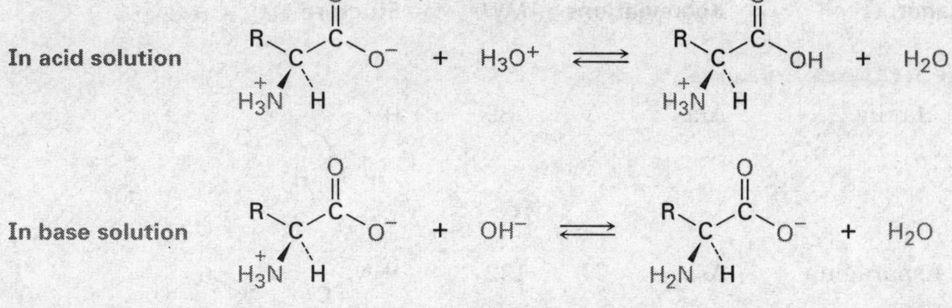

The structures, abbreviations (both one-letter and three-letter), and pK_a values of the 20 amino acids commonly found in proteins are shown in Table 15.1. All 20 are **α-amino acids**, meaning that the amino group in each is a substituent on the α carbon—the one next to the carbonyl group. Nineteen of the twenty are primary amines, RNH_2, and differ only in the identity of the **side chain**—the substituent attached to the α carbon. Proline is a secondary amine whose nitrogen and α carbon atoms are part of a five-membered pyrrolidine ring.

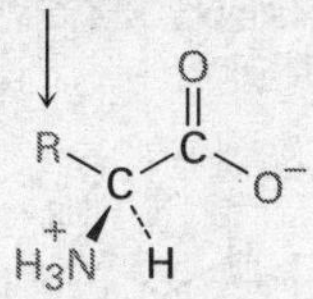

A primary α-amino acid

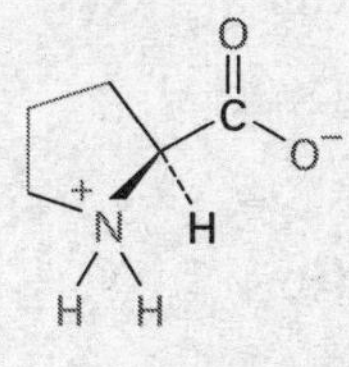

Proline, a secondary α-amino acid

In addition to the 20 amino acids commonly found in proteins, 2 others—selenocysteine and pyrrolysine—are found in some organisms, and more than 700 nonprotein amino acids are also found in nature. γ-Aminobutyric acid (GABA), for instance, is found in the brain and acts as a neurotransmitter; homocysteine is found in blood and is linked to coronary heart disease; and thyroxine is found in the thyroid gland, where it acts as a hormone.

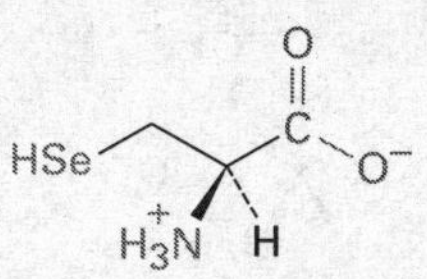

Selenocysteine

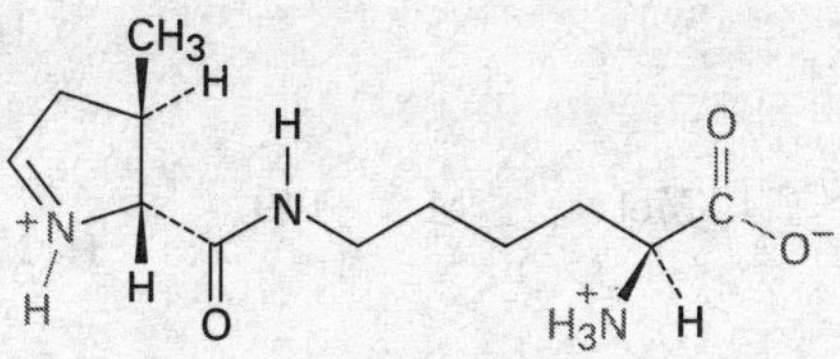

Pyrrolysine

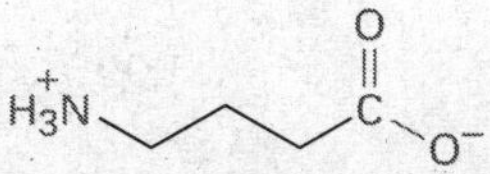

γ-Aminobutyric acid

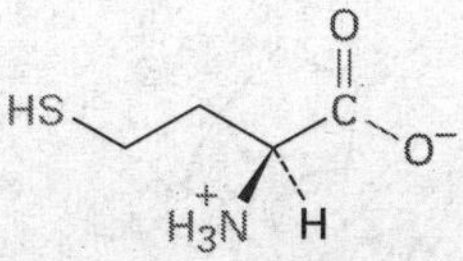

Homocysteine

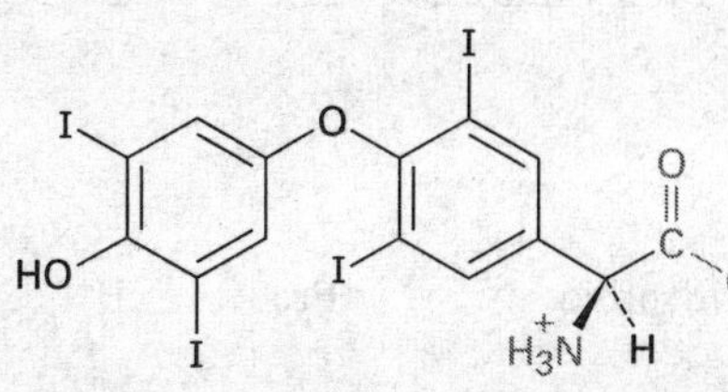

Thyroxine

Table 15.1 **Structures of the 20 Common Amino Acids Found in Proteins**

Name	Abbreviations		MW	Stucture	pK_a α-CO_2H	pK_a α-NH_3^+	pK_a side chain	pI
Neutral Amino Acids								
Alanine	Ala	A	89	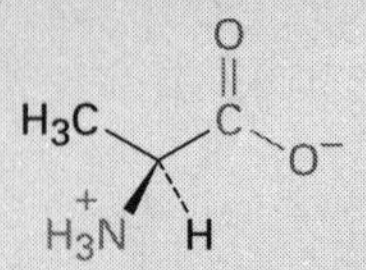	2.34	9.69	—	6.01
Asparagine	Asn	N	132	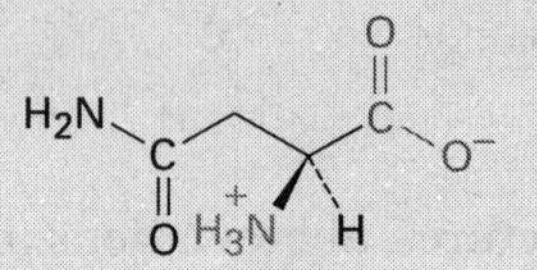	2.02	8.80	—	5.41
Cysteine	Cys	C	121	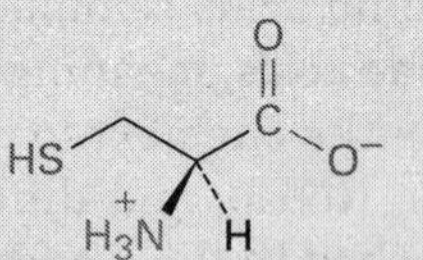	1.96	10.28	8.18	5.07
Glutamine	Gln	Q	146	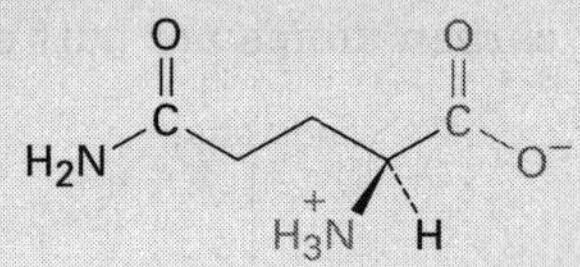	2.17	9.13	—	5.65
Glycine	Gly	G	75	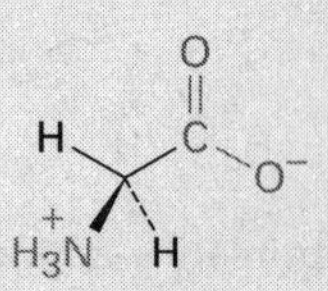	2.34	9.60	—	5.97
Isoleucine	Ile	I	131	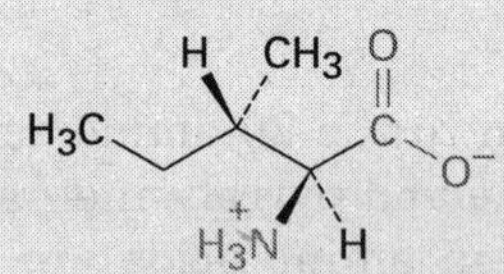	2.36	9.60	—	6.02
Leucine	Leu	L	131	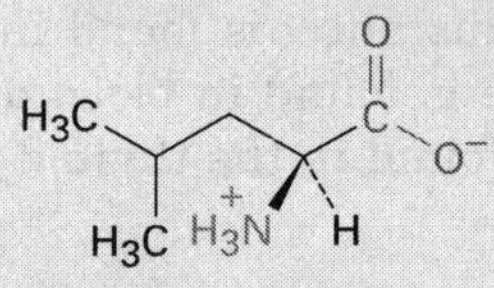	2.36	9.60	—	5.98
Methionine	Met	M	149	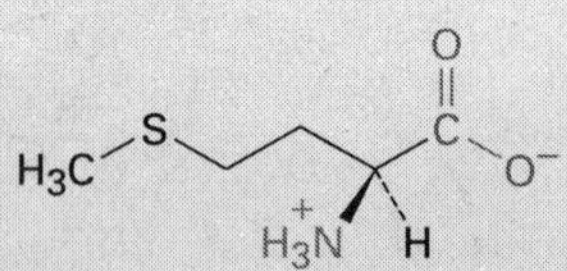	2.28	9.21	—	5.74
Phenylalanine	Phe	F	165	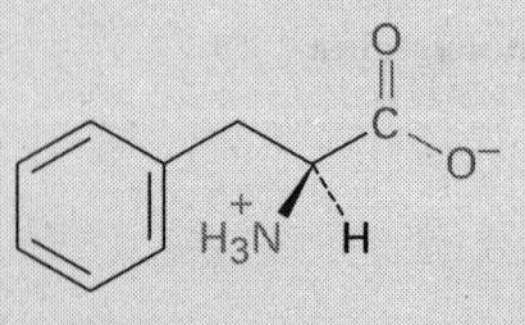	1.83	9.13	—	5.48
Proline	Pro	P	115	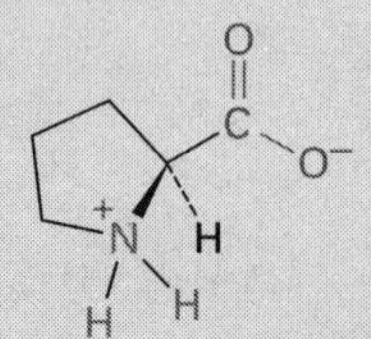	1.99	10.60	—	6.30

continued

Table 15.1 **Structures of the 20 Common Amino Acids Found in Proteins *(continued)***

Name	Abbreviations		MW	Stucture	pK_a α-CO_2H	pK_a α-NH_3^+	pK_a side chain	pI
Neutral Amino Acids *continued*								
Serine	Ser	S	105	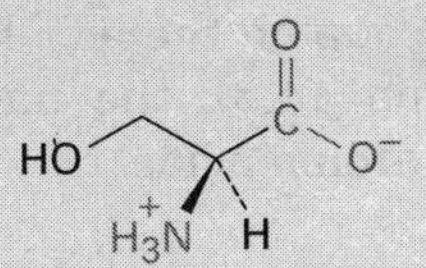	2.21	9.15	—	5.68
Threonine	Thr	T	119	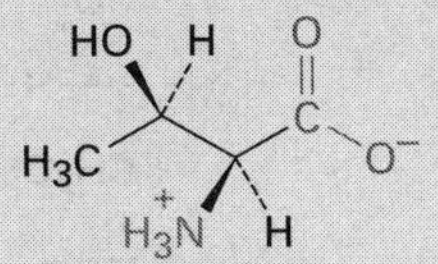	2.09	9.10	—	5.60
Tryptophan	Trp	W	204	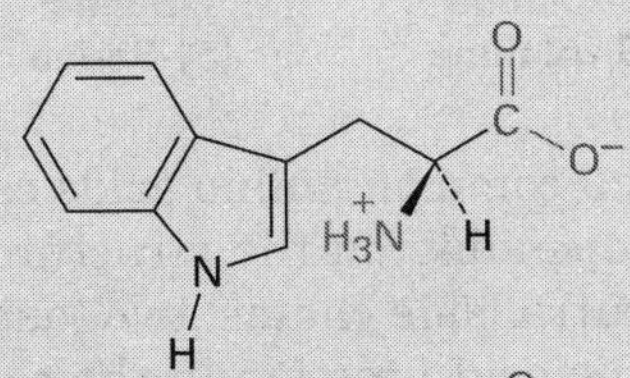	2.83	9.39	—	5.89
Tyrosine	Tyr	Y	181	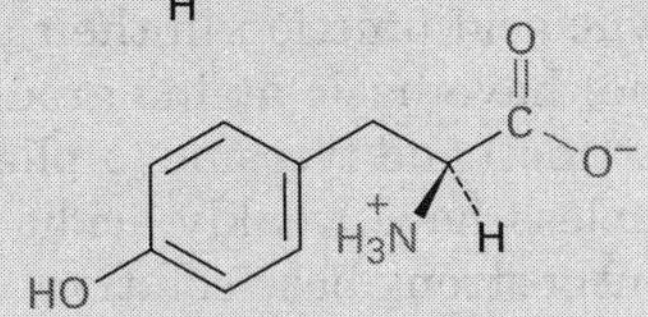	2.20	9.11	10.07	5.66
Valine	Val	V	117	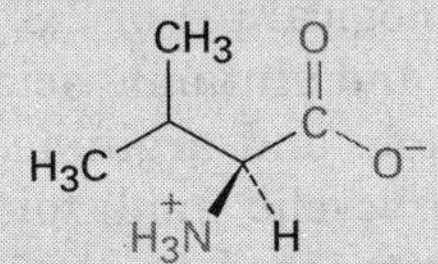	2.32	9.62	—	5.96
Acidic Amino Acids								
Aspartic acid	Asp	D	133	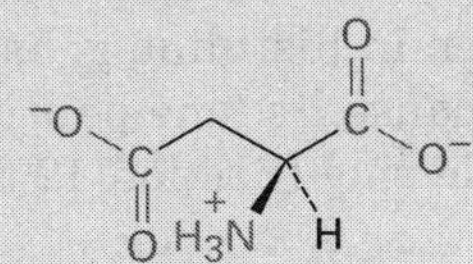	1.88	9.60	3.65	2.77
Glutamic acid	Glu	E	147	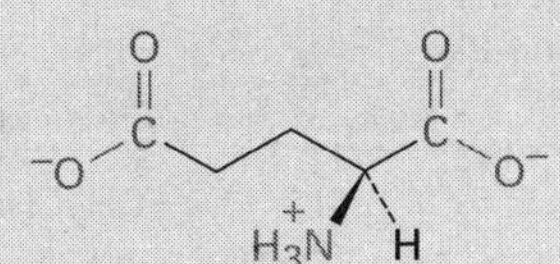	2.19	9.67	4.25	3.22
Basic Amino Acids								
Arginine	Arg	R	174	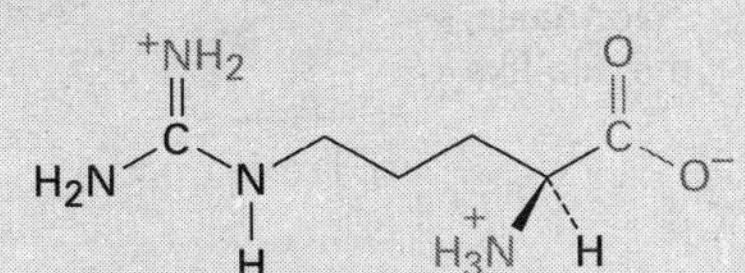	2.17	9.04	12.48	10.76
Histidine	His	H	155	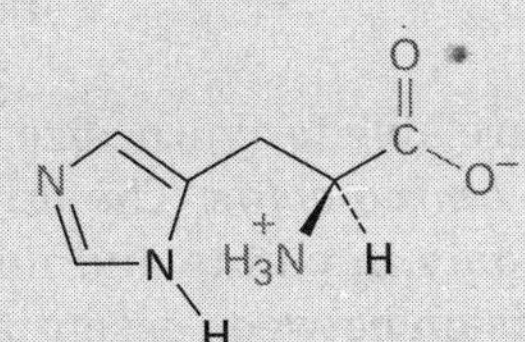	1.82	9.17	6.00	7.59
Lysine	Lys	K	146	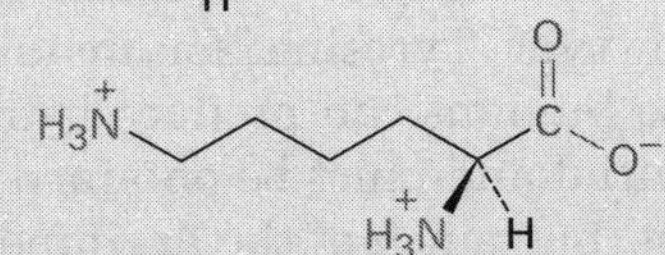	2.18	8.95	10.53	9.74

Except for glycine, $^{+}H_3NCH_2CO_2^{-}$, the α carbons of amino acids are chirality centers. Two enantiomers of each are therefore possible, but nature uses only one to build proteins. In Fischer projections, naturally occurring amino acids are represented by placing the carboxyl group at the top and the side chain at the bottom as if drawing a carbohydrate (Section 14.2) and then placing the amino group on the left. Because of their stereochemical similarity to L sugars (Section 14.3), the naturally occurring α-amino acids are often referred to as L amino acids.

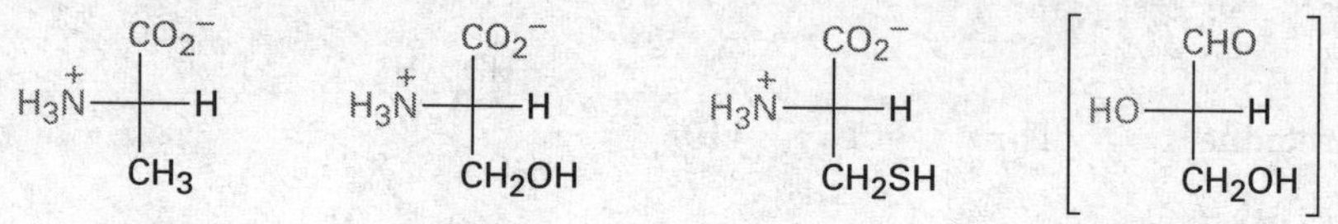

L-Alanine (*S*)-Alanine — L-Serine (*S*)-Serine — L-Cysteine (*R*)-Cysteine — L-Glyceraldehyde

The 20 common amino acids can be further classified as neutral, acidic, or basic, depending on the structure of their side chains. Fifteen of the twenty have neutral side chains, two (aspartic acid and glutamic acid) have an extra carboxylic acid function in their side chains, and three (lysine, arginine, and histidine) have basic amino groups in their side chains. Note that both cysteine (a thiol) and tyrosine (a phenol), although usually classified as neutral, nevertheless have weakly acidic side chains that can be deprotonated in a sufficiently strong base solution.

At the physiological pH of 7.3 within cells, the side-chain carboxyl groups of aspartic acid and glutamic acid are deprotonated and the basic side-chain nitrogens of lysine and arginine are protonated. Histidine, however, which contains a heterocyclic imidazole ring in its side chain, is not quite basic enough to be protonated at pH 7.3. Note that only the pyridine-like, doubly bonded nitrogen in histidine is basic. The pyrrole-like singly bonded nitrogen is nonbasic because its lone pair of electrons is part of the six-π-electron aromatic imidazole ring (Section 12.6).

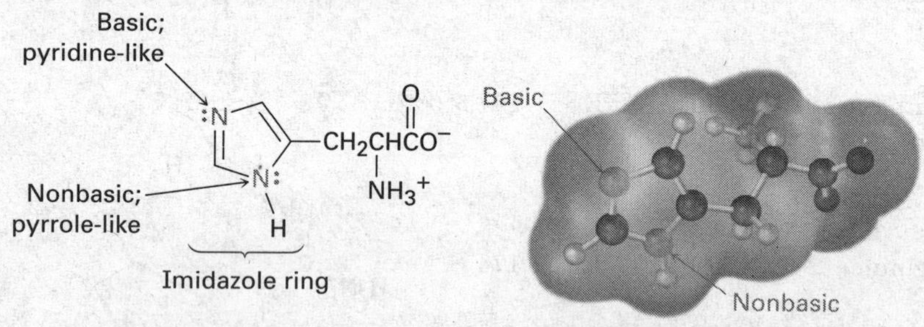

Histidine

Humans are able to synthesize only 11 of the 20 protein amino acids, called *nonessential amino acids.* The other 9, called *essential amino acids,* are biosynthesized only in plants and microorganisms and must be obtained in our diet. The division between essential and nonessential amino acids is not clear-cut, however. Tyrosine, for instance, is sometimes considered nonessential because humans can produce it from phenylalanine, yet phenylalanine itself is essential and must be obtained in the diet. Arginine can be synthesized by humans, but much of the arginine we need also comes from our diet.

Worked Example 15.1 **Acid–Base Reactions of Amino Acids**

Write an equation for the reaction of glycine hydrochloride, $Cl^{-}\,{}^{+}H_3NCH_2CO_2H$, with **(a)** 1 equivalent of NaOH and **(b)** 2 equivalents of NaOH.

Solution **(a)** Reaction with the first equivalent of NaOH removes the acidic $-CO_2H$ proton to give the neutral zwitterion.

$$Cl^{-}\ H_3\overset{+}{N}CH_2\overset{O}{\overset{\|}{C}}OH \ + \ NaOH \longrightarrow H_3\overset{+}{N}CH_2\overset{O}{\overset{\|}{C}}O^{-} \ + \ H_2O \ + \ NaCl$$

(b) Once the zwitterion has formed, reaction with a second equivalent of NaOH removes the remaining acidic proton from the $-NH_3^{+}$ group to give an amino carboxylate anion.

$$H_3\overset{+}{N}CH_2\overset{O}{\overset{\|}{C}}O^{-} \ + \ NaOH \longrightarrow H_2NCH_2\overset{O}{\overset{\|}{C}}O^{-}\ Na^{+} \ + \ H_2O$$

Problem 15.1 How many of the α-amino acids in Table 15.1 contain aromatic rings? How many contain sulfur? How many are alcohols? How many have hydrocarbon side chains?

Problem 15.2 Of the 19 L amino acids, 18 have the *S* configuration at the α carbon. Cysteine is the only L amino acid that has an *R* configuration. Explain.

Problem 15.3 Draw L-alanine in the standard three-dimensional format using solid, wedged, and dashed lines.

Problem 15.4 Write the products of the following reactions:

(a) Phenylalanine + 1 equiv NaOH → ?
(b) Product of **(a)** + 1 equiv HCl → ?
(c) Product of **(a)** + 2 equiv HCl → ?

15.2 Isoelectric Points

In acid solution, an amino acid is protonated and exists primarily as a cation. In base solution, an amino acid is deprotonated and exists primarily as an anion. Thus, at some intermediate pH, the amino acid must be exactly balanced between anionic and cationic forms and exist primarily as the neutral zwitterion. This pH is called the amino acid's **isoelectric point, p*I***.

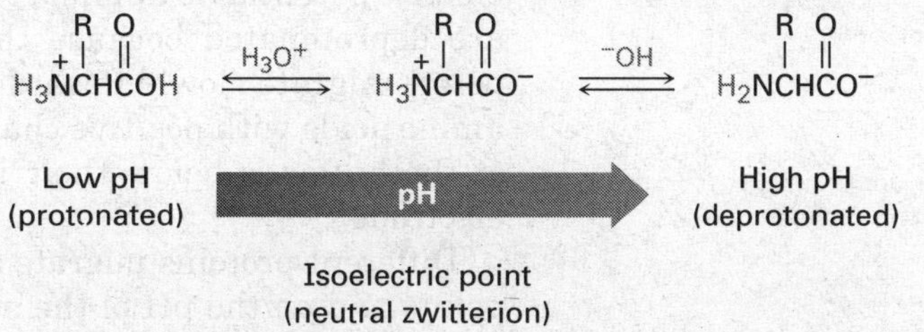

The isoelectric point of an amino acid depends on its structure, with values for the 20 common amino acids given in Table 15.1. The 15 neutral amino acids have isoelectric points near neutrality, in the pH range 5.0 to 6.5. The two acidic amino acids have isoelectric points at lower pH so that deprotonation of the side-chain $-CO_2H$ is suppressed, and the three basic amino acids have isoelectric points at higher pH so that protonation of the side-chain amino group is suppressed.

More specifically, the p*I* of any amino acid is the average of the two acid-dissociation constants that involve the neutral zwitterion. For the 13 amino acids with a neutral side chain, p*I* is the average of pK_{a1} and pK_{a2}. For the four amino acids with either a strongly or weakly acidic side chain, p*I* is the average of the two *lowest* pK_a values. For the three amino acids with a basic side chain, p*I* is the average of the two *highest* pK_a values.

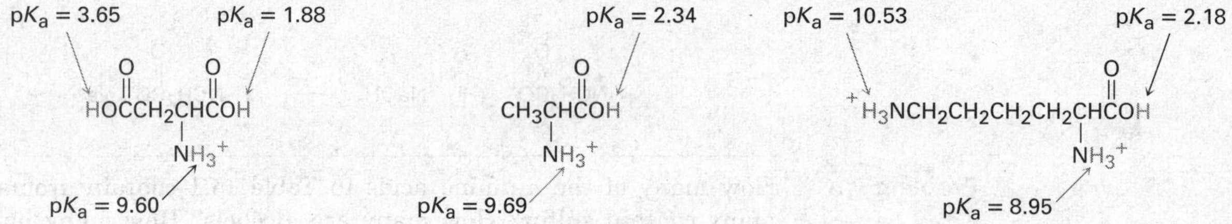

$$pI = \frac{1.88 + 3.65}{2} = 2.77 \qquad pI = \frac{2.34 + 9.69}{2} = 6.01 \qquad pI = \frac{8.95 + 10.53}{2} = 9.74$$

Acidic amino acid / **Aspartic acid** — **Neutral amino acid** / **Alanine** — **Basic amino acid** / **Lysine**

Just as individual amino acids have isoelectric points, entire proteins have an overall p*I* because of the cumulative effect of all the acidic or basic amino acids they may contain. The enzyme lysozyme, for instance, has a preponderance of basic amino acids and thus has a high isoelectric point ($pI = 11.0$). Pepsin, however, has a preponderance of acidic amino acids and a low isoelectric point ($pI \sim 1.0$). Not surprisingly, the solubilities and properties of proteins with different p*I*'s are strongly affected by the pH of the medium. Solubility in water is usually lowest at the isoelectric point, where the protein has no net charge, and is higher both above and below the p*I*, where the protein is charged.

We can take advantage of the differences in isoelectric points to separate a mixture of proteins into its pure constituents. Using a technique known as *electrophoresis,* a mixture of proteins is placed near the center of a strip of paper or gel. The paper or gel is then moistened with an aqueous buffer of a given pH, and electrodes are connected to the ends of the strip. When an electric potential is applied, those proteins with negative charges (those that are deprotonated because the pH of the buffer is above their isoelectric point) migrate slowly toward the positive electrode. At the same time, those amino acids with positive charges (those that are protonated because the pH of the buffer is below their isoelectric point) migrate toward the negative electrode.

Different proteins migrate at different rates, depending on their isoelectric points and on the pH of the aqueous buffer, thereby effecting a separation of the mixture into its components. Figure 15.1 illustrates the separation for a mixture containing basic, neutral, and acidic components.

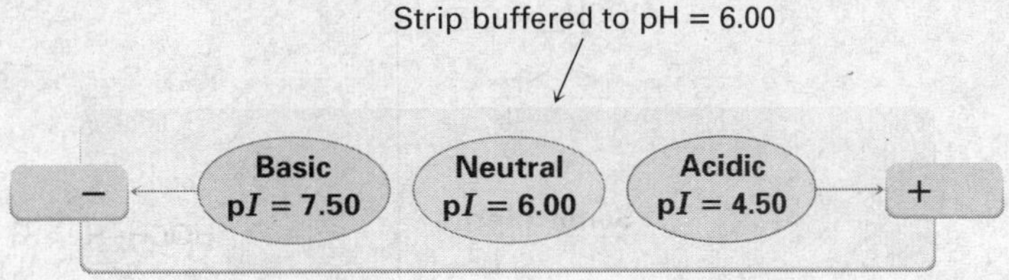

Figure 15.1 Separation of a protein mixture by electrophoresis. At pH = 6.00, a neutral protein does not migrate, a basic protein is protonated and migrates toward the negative electrode, and an acidic protein is deprotonated and migrates toward the positive electrode.

Problem 15.5 For the mixtures of amino acids indicated, predict the direction of migration of each component (toward the positive or negative electrode) and the relative rate of migration during electrophoresis.

(a) Valine, glutamic acid, and histidine at pH 7.6
(b) Glycine, phenylalanine, and serine at pH 5.7
(c) Glycine, phenylalanine, and serine at pH 6.0

Problem 15.6 Hemoglobin has pI = 6.8. Does hemoglobin have a net negative charge or net positive charge at pH = 5.3? At pH = 7.3?

15.3 Peptides and Proteins

Proteins and peptides are amino acid polymers in which the individual amino acids, called **residues**, are joined together by amide bonds, or *peptide bonds.* An amino group from one residue forms an amide bond with the carboxyl of a second residue, the amino group of the second forms an amide bond with the carboxyl of a third, and so on. For example, alanylserine is the dipeptide that results when an amide bond forms between the alanine carboxyl and the serine amino group.

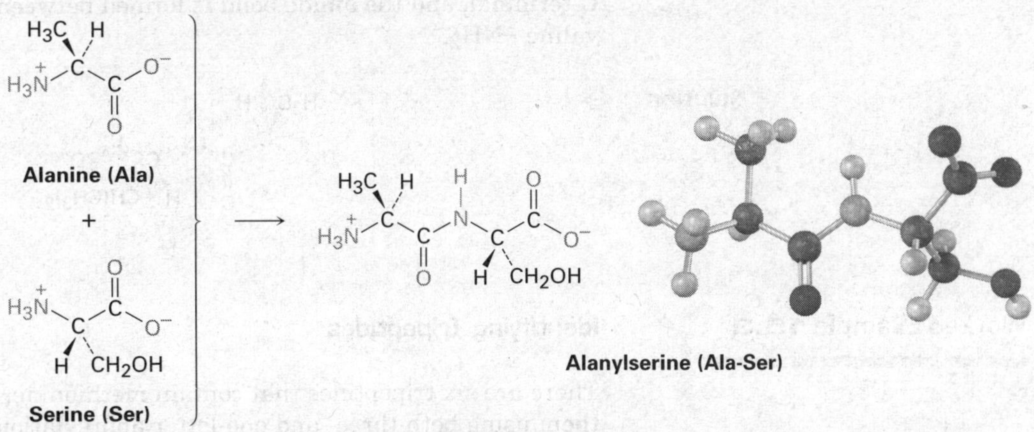

Note that two dipeptides can result from reaction between alanine and serine, depending on which carboxyl group reacts with which amino group.

If the alanine amino group reacts with the serine carboxyl, serylalanine results.

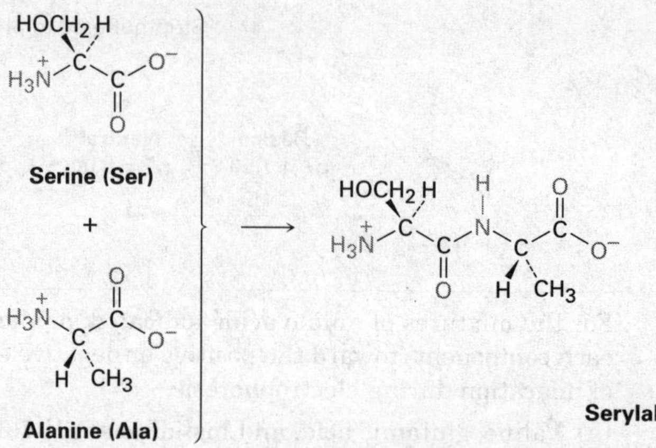

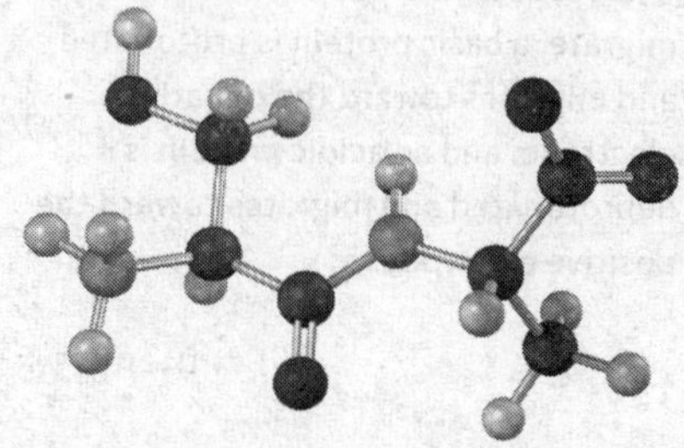

Serylalanine (Ser-Ala)

The long, repetitive sequence of –N–CH–CO– atoms that makes up a continuous chain is called the protein's **backbone**. By convention, peptides are written with the **N-terminal amino acid** (the one with the free $-NH_3^+$ group) on the left and the **C-terminal amino acid** (the one with the free $-CO_2^-$ group) on the right. The name of the peptide is indicated by using the abbreviations listed in Table 15.1 for each amino acid. Thus, alanylserine is abbreviated Ala-Ser or A-S, and serylalanine is abbreviated Ser-Ala or S-A. The one-letter abbreviations are more convenient, though less immediately recognizable, than the three-letter abbreviations.

Worked Example 15.2 Drawing the Structure of a Dipeptide

Draw the structure of Ala-Val.

Strategy By convention, the N-terminal amino acid is written on the left and the C-terminal amino acid on the right. Thus, alanine is N-terminal, valine is C-terminal, and the amide bond is formed between the alanine $-CO_2H$ and the valine $-NH_2$.

Solution

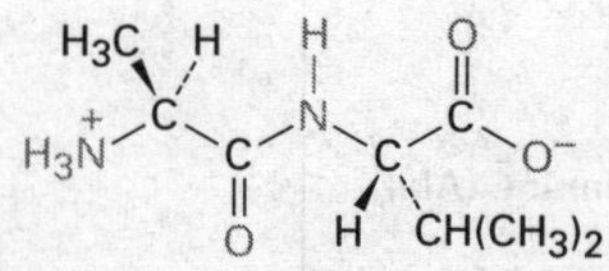

Ala-Val

Worked Example 15.3 Identifying Tripeptides

There are six tripeptides that contain methionine, lysine, and isoleucine. Name them using both three- and one-letter abbreviations.

Solution

Met-Lys-Ile (M-K-I)	Lys-Met-Ile (K-M-I)	Ile-Met-Lys (I-M-K)
Met-Ile-Lys (M-I-K)	Lys-Ile-Met (K-I-M)	Ile-Lys-Met (I-K-M)

Problem 15.7 Draw structures of the two dipeptides made from leucine and cysteine.

Problem 15.8 Using both three- and one-letter notations for each amino acid, name the six possible isomeric tripeptides that contain valine, tyrosine, and glycine.

Problem 15.9 Draw the structure of Met-Pro-Val-Gly, and indicate the positions of the amide bonds.

15.4 Covalent Bonding in Peptides

The amide bond that links different amino acids together in peptides is no different from any other amide bond (Section 10.10). An amide nitrogen is nonbasic because its unshared electron pair is delocalized by interaction with the carbonyl group. This overlap of the nitrogen *p* orbital with the *p* orbitals of the carbonyl group imparts a certain amount of double-bond character to the C–N bond and restricts rotation around it. The amide bond is therefore planar, and the N–H is oriented 180° to the C=O.

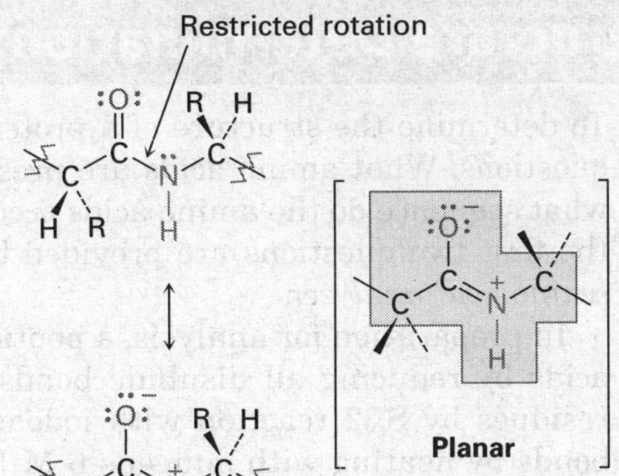

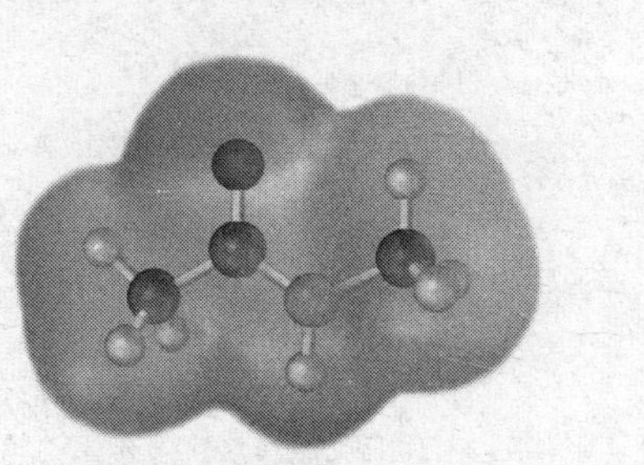

A second kind of covalent bonding in peptides occurs when a disulfide linkage, RS—SR, is formed between two cysteine residues. As we saw in Section 8.8, a disulfide is formed by mild oxidation of a thiol, RSH, and is cleaved by mild reduction.

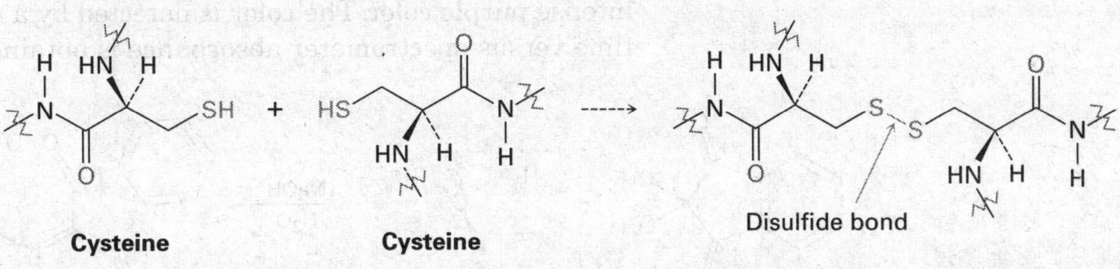

A disulfide bond between cysteine residues in different peptide chains links the otherwise separate chains together, while a disulfide bond between cysteine residues in the same chain forms a loop. Insulin, for instance, is composed of two chains that total 51 amino acids and are linked by two cysteine disulfide bridges.

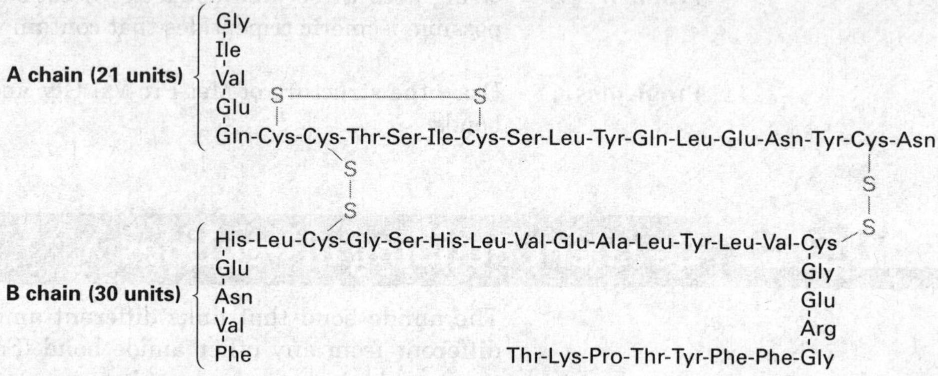

Insulin

15.5 Peptide Structure Determination: Amino Acid Analysis

To determine the structure of a protein or peptide, we need to answer three questions: What amino acids are present? How much of each is present? In what sequence do the amino acids occur in the peptide chain? The answers to the first two questions are provided by an automated instrument called an *amino acid analyzer.*

In preparation for analysis, a peptide is broken into its constituent amino acids by reducing all disulfide bonds, capping the –SH groups of cysteine residues by S_N2 reaction with iodoacetic acid, and hydrolyzing the amide bonds by heating with aqueous 6 M HCl at 110 °C for 24 hours. The resultant amino acid mixture is then separated into its constituents by a technique called ion-exchange chromatography. The mixture is placed at the top of a glass column filled with a special absorbent material, and a series of aqueous buffers is pumped through the column. The various amino acids migrate down the column at different rates depending on their structures and are thus separated.

As each amino acid exits the column, it is mixed with a solution of a substance called *ninhydrin* and undergoes a rapid reaction that produces an intense purple color. The color is detected by a spectrometer, and a plot of exit time versus spectrometer absorbance is obtained.

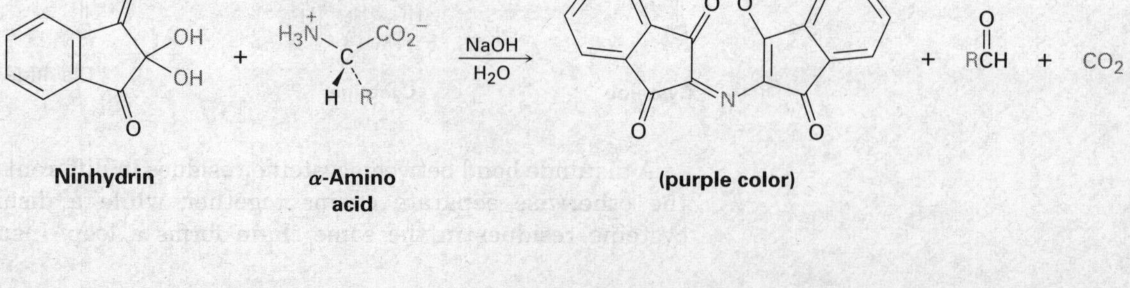

Because the time required for a given amino acid to pass down a standard column is reproducible, the identities of the amino acids in a peptide can be determined. The amount of each amino acid in the sample is determined by measuring the intensity of the purple color resulting from its reaction with ninhydrin. Figure 15.2 shows the results of amino acid analysis of a standard equimolar mixture of 17 α-amino acids. Typically, amino acid analysis requires about 100 picomoles (2–3 μg) of sample for a protein containing about 200 residues.

Figure 15.2 Amino acid analysis of an equimolar mixture of 17 amino acids.

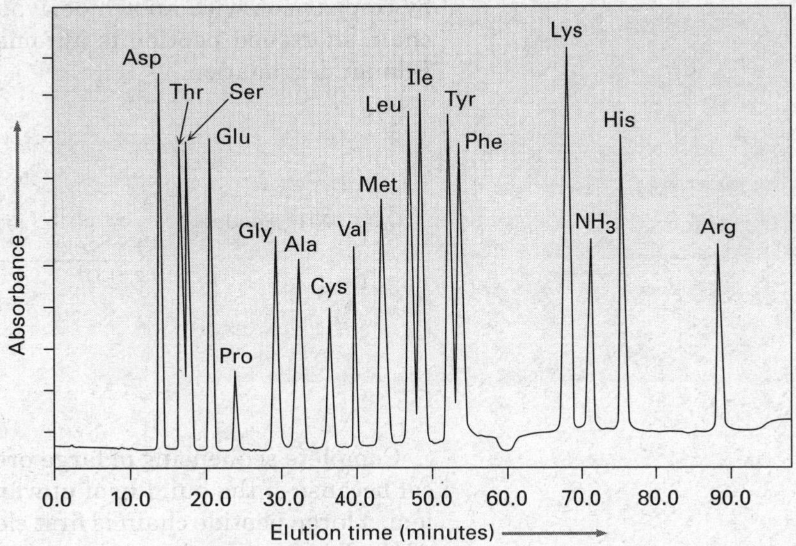

Problem 15.10 Show the structure of the product you would expect to obtain by S_N2 reaction of a cysteine residue with iodoacetic acid.

Problem 15.11 Show the structures of the products obtained on reaction of valine with ninhydrin.

15.6 Peptide Sequencing: The Edman Degradation

With the identities and relative amounts of amino acids known, the peptide is then *sequenced* to find out in what order the amino acids are linked together. Much peptide sequencing is now done by mass spectrometry (Section 13.1), but a chemical method of peptide sequencing called the *Edman degradation* is also used.

The general idea of peptide sequencing by Edman degradation is to cleave one amino acid at a time from the N terminus of the peptide chain. That terminal amino acid is then separated and identified, and the cleavage reactions are repeated on the chain-shortened peptide until the entire peptide sequence is known. Automated protein sequencers are available that allow as many as 50 repetitive sequencing cycles, with an efficiency that allows sample sizes of 1 to 5 picomoles—less than 0.1 μg of peptide.

Edman degradation involves treatment of a peptide with phenyl isothiocyanate (PITC), $C_6H_5—N{=}C{=}S$, followed by mild acid hydrolysis. PITC first attaches to the $-NH_2$ group of the N-terminal amino acid, and the N-terminal residue then splits from the chain giving a *phenylthiohydantoin* derivative (PTH) along with chain-shortened peptide. The PTH is identified by comparison with known derivatives of the common amino acids, and the chain-shortened peptide is automatically resubmitted to another round of Edman degradation.

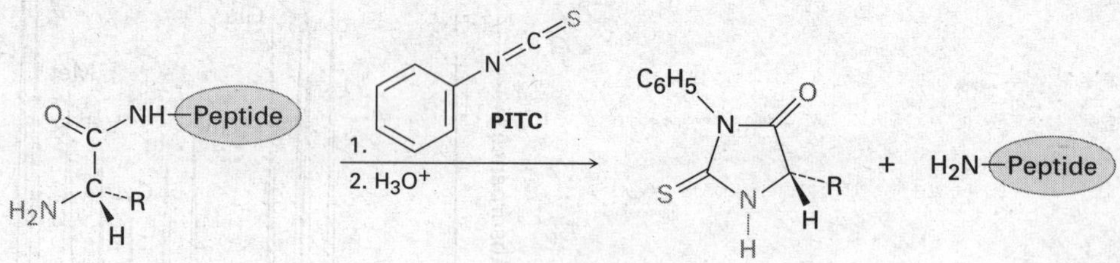

A phenylthiohydantoin (PTH)

Complete sequencing of large proteins by Edman degradation is impractical because of the buildup of unwanted by-products. To get around the problem, a large peptide chain is first cleaved by partial hydrolysis into a number of smaller fragments, the sequence of each fragment is determined, and the individual fragments are fitted together by matching the overlapping ends. In this way, protein chains with more than 400 amino acids have been sequenced.

Partial hydrolysis of a protein can be carried out either chemically with aqueous acid or enzymatically. Acid hydrolysis is unselective and gives a more-or-less random mixture of small fragments, but enzymatic hydrolysis is quite specific. The enzyme trypsin, for instance, catalyzes hydrolysis of peptides only at the carboxyl side of the basic amino acids arginine and lysine; chymotrypsin cleaves only at the carboxyl side of the aryl-substituted amino acids phenylalanine, tyrosine, and tryptophan.

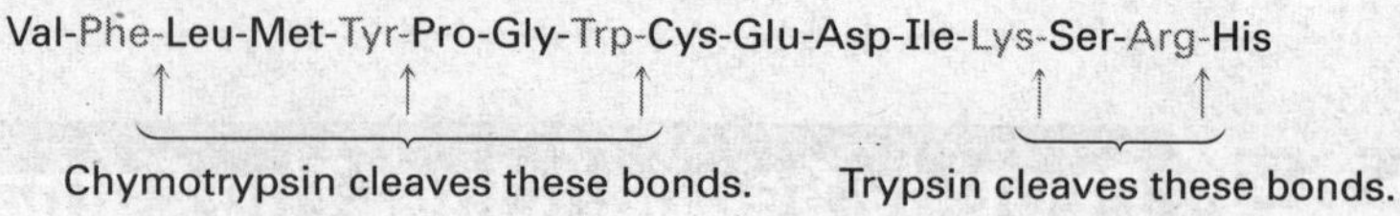

Worked Example 15.4

Sequencing a Simple Peptide

Angiotensin II is a hormonal octapeptide involved in controlling hypertension by regulating the sodium–potassium salt balance in the body. Amino acid analysis shows the presence of eight different amino acids in equimolar amounts: Arg,

Asp, His, Ile, Phe, Pro, Tyr, and Val. Partial hydrolysis of angiotensin II with dilute hydrochloric acid yields the following fragments:

Asp-Arg-Val-Tyr, Ile-His-Pro, Pro-Phe, Val-Tyr-Ile-His

What is the sequence of angiotensin II?

Strategy Line up the fragments to identify the overlapping regions, and then write the sequence.

```
Asp-Arg-Val-Tyr
        :   :
        Val-Tyr-Ile-His
                :   :
                Ile-His-Pro
                        :
                        Pro-Phe
```

Solution The sequence is Asp-Arg-Val-Tyr-Ile-His-Pro-Phe.

Problem 15.12 What fragments would result if angiotensin II (Worked Example 15.4) were cleaved with trypsin? With chymotrypsin?

Problem 15.13 Give the amino acid sequence of a hexapeptide containing Arg, Gly, Ile, Leu, Pro, and Val that produces the following fragments on partial acid hydrolysis: Pro-Leu-Gly, Arg-Pro, Gly-Ile-Val.

Problem 15.14 What is the N-terminal residue on a peptide that gives the following PTH derivative on Edman degradation?

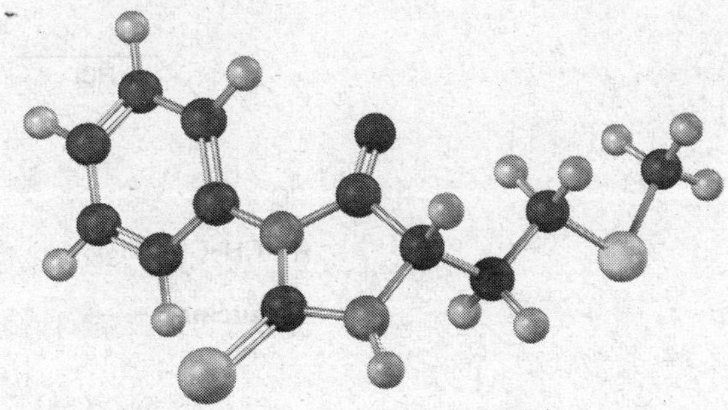

15.7 Peptide Synthesis

Once the structure of a peptide is known, synthesis is often the next goal—perhaps to obtain a larger amount for biological evaluation. A simple amide can be formed by treating an amine and a carboxylic acid with dicyclohexylcarbodiimide (DCC; Section 10.6), but peptide synthesis is more difficult because of the need for specificity. Many different amide links must be formed in a precise order rather than at random.

The solution to the specificity problem is *protection* (Section 9.8). If we wanted to couple alanine with leucine to synthesize Ala-Leu, for instance, we could protect the $-NH_2$ group of alanine and the $-CO_2H$ group of leucine to shield them from reacting, then form the desired Ala-Leu amide bond by reaction with DCC, and then remove the protecting groups.

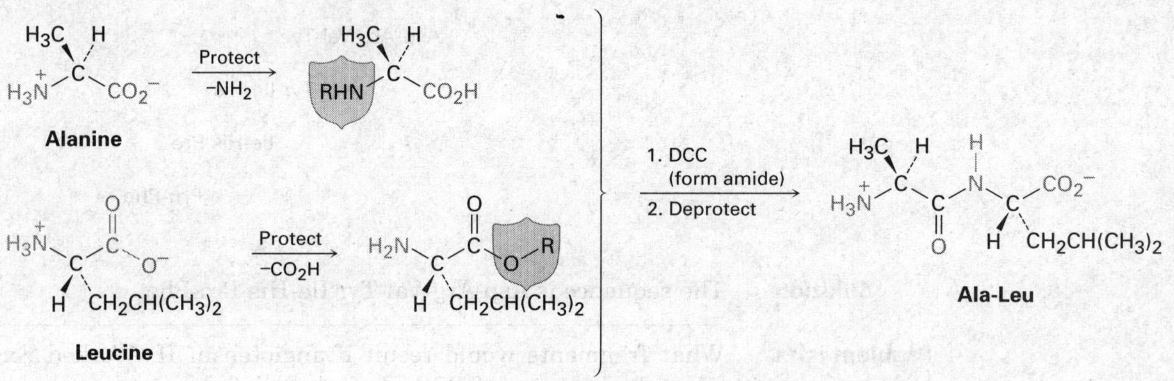

Carboxyl groups are often protected simply by converting them into methyl or benzyl esters. Esters are easily made from carboxylic acids (Section 10.6) and are easily hydrolyzed by mild treatment with aqueous NaOH.

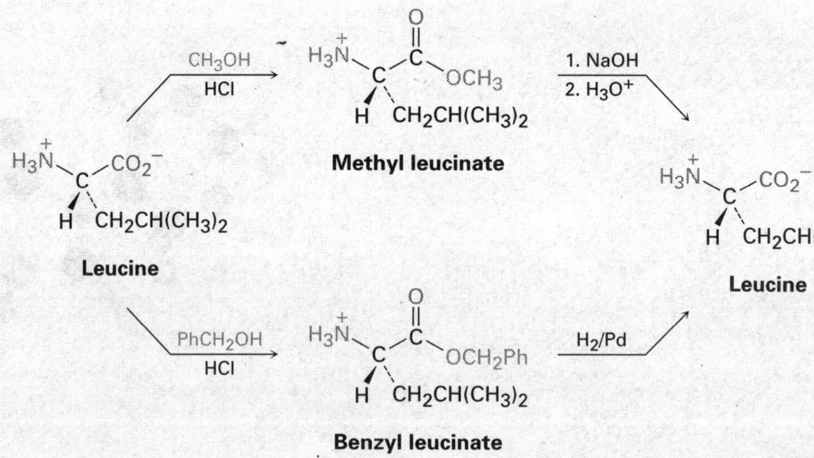

Amino groups are often prctected as their *tert*-butyloxycarbonyl amide (Boc) or fluorenylmethyloxycarbonyl amide (Fmoc) derivatives. The Boc protecting group is introduced by reaction of the amino acid with di-*tert*-butyl dicarbonate in a nucleophilic acyl substitution reaction and is removed by brief treatment with a strong acid such as trifluoroacetic acid, CF_3CO_2H. The

Fmoc protecting group is introduced by reaction with an acid chloride and is removed by treatment with base.

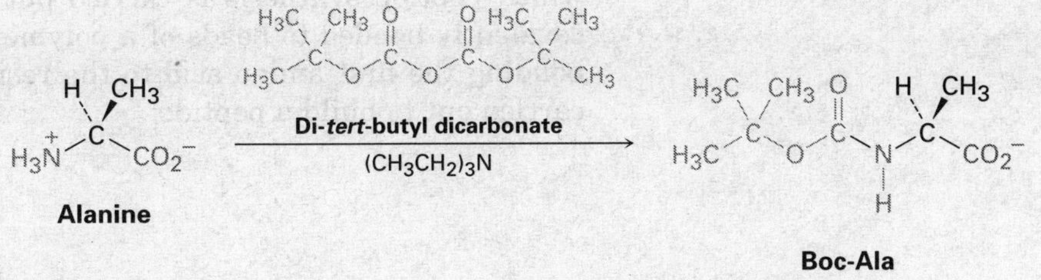

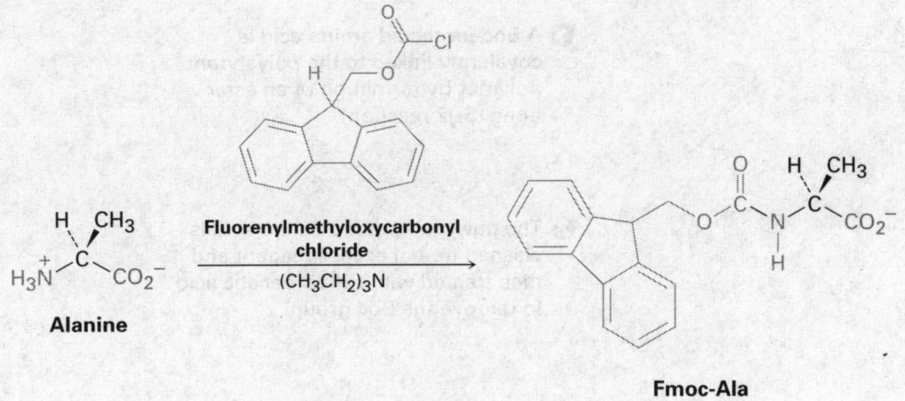

In summary, five steps are needed to synthesize a dipeptide such as Ala-Leu (Figure 15.3).

Figure 15.3 The five steps needed for peptide synthesis.

1. The amino group of alanine is protected as the Boc derivative, and
2. the carboxyl group of leucine is protected as the methyl ester.
3. The two protected amino acids are coupled using DCC.
4. The Boc protecting group is removed by acid treatment.
5. The methyl ester is removed by basic hydrolysis.

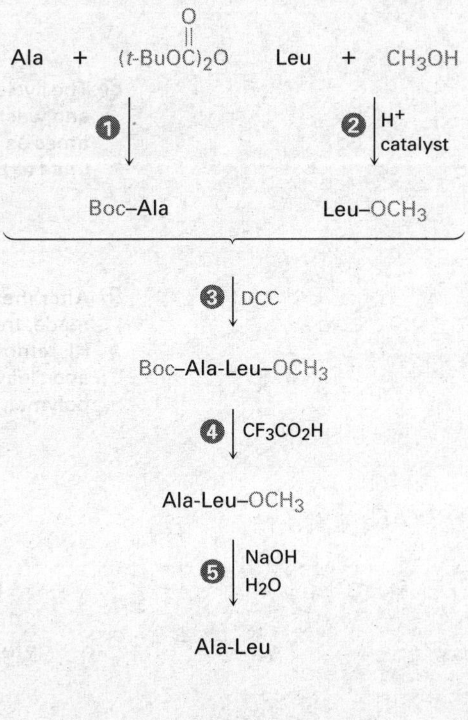

Although the steps just shown can be repeated to add one amino acid at a time to a growing chain, the synthesis of a large peptide by this sequential addition is arduous. Much simpler is the *Merrifield solid-phase* method, in which peptide synthesis is carried out with the growing amino acid chain covalently bonded to beads of a polymer resin rather than in solution. After bonding the first amino acid to the resin, a repeating series of four steps is carried out to build a peptide:

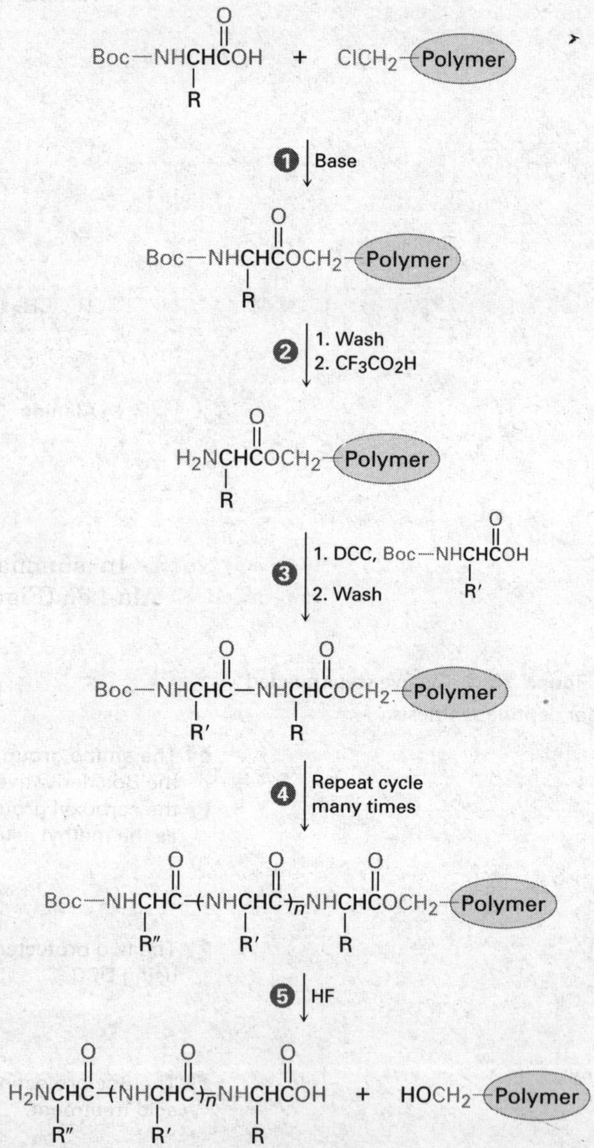

1. A Boc-protected amino acid is covalently linked to the polystyrene polymer by formation of an ester bond (S_N2 reaction).
2. The polymer-bonded amino acid is washed free of excess reagent and then treated with trifluoroacetic acid to remove the Boc group.
3. A second Boc-protected amino acid is coupled to the first by reaction with DCC. Excess reagents are removed by washing them from the insoluble polymer.
4. The cycle of deprotection, coupling, and washing is repeated as many times as desired to add amino acid units to the growing chain.
5. After the desired peptide has been made, treatment with anhydrous HF removes the final Boc group and cleaves the ester bond to the polymer, yielding the free peptide.

Robotic peptide synthesizers automatically repeat the coupling, washing, and deprotection steps with different amino acids. Each step occurs in high yield, and mechanical losses are minimized because the peptide intermediates are never removed from the insoluble polymer until the final step. Using this procedure, up to 30 mg of a peptide with 20 amino acids can be routinely prepared in a few hours.

Worked Example 15.5

Reactions of Amino Acids

Write equations for the reaction of methionine with:
(a) CH_3OH, HCl **(b)** Di-*tert*-butyl dicarbonate

Solution **(a)**

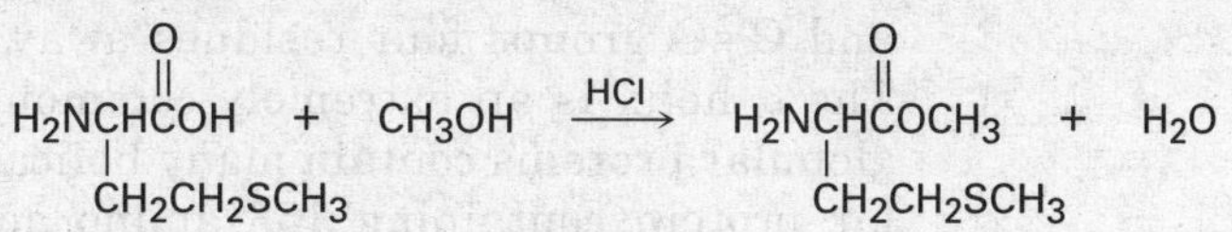

(b)

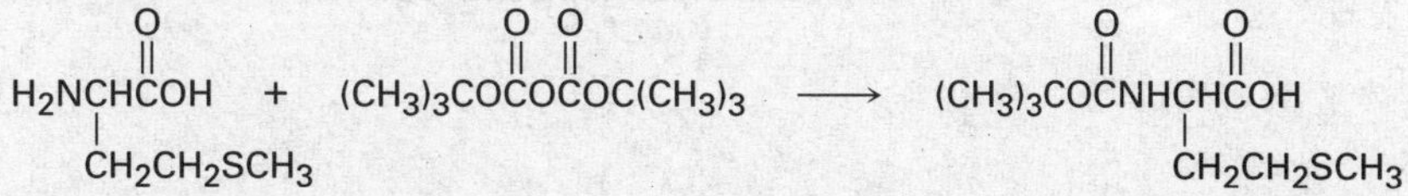

Problem 15.15 Write the structures of the intermediates in the five-step synthesis of Leu-Ala from alanine and leucine.

Problem 15.16 Show the mechanism for formation of a Boc derivative by reaction of an amino acid with di-*tert*-butyl dicarbonate.

15.8 Protein Structure

Proteins are usually classified as either *fibrous* or *globular*, according to their three-dimensional shapes. **Fibrous proteins**, such as the collagen in tendons and connective tissue and the myosin in muscle tissue, consist of polypeptide chains arranged side by side in long filaments. Because these proteins are tough and insoluble in water, they are used in nature for structural materials. **Globular proteins**, by contrast, are usually coiled into compact, roughly spherical shapes. These proteins are generally soluble in water and are mobile within cells. Most of the 3000 or so enzymes that have been characterized to date are globular proteins.

Proteins are so large that the word *structure* takes on a broader meaning than it does with simpler organic compounds. In fact, chemists speak of four different levels of structure when describing proteins:

- The **primary structure** of a protein is simply the amino acid sequence.
- The **secondary structure** of a protein describes how *segments* of the peptide backbone orient into a regular pattern.
- The **tertiary structure** describes how the *entire* protein molecule coils into an overall three-dimensional shape.
- The **quaternary structure** describes how different protein molecules come together to yield large aggregate structures.

Primary structure is determined, as we've seen, by sequencing the protein. Secondary, tertiary, and quaternary structures are determined either by NMR (Sections 13.7 through 13.13) or by a technique called X-ray crystallography, which is described in the *Interlude* at the end of this chapter.

The most common secondary structures are the α helix and the β-pleated sheet. An **α helix** is a right-handed coil of the protein backbone, much like the coil of a spiral staircase (Figure 15.4a). Each turn of the helix contains 3.6 amino acid residues, with a distance between coils of 540 pm, or 5.4 Å. The structure is stabilized by hydrogen bonds between amide N–H groups and C=O groups four residues away, with an N–H····O distance of 2.8 Å. The α helix is an extremely common secondary structure, and almost all globular proteins contain many helical segments. Myoglobin, a small globular protein containing 153 amino acid residues in a single chain, is an example (Figure 15.4b).

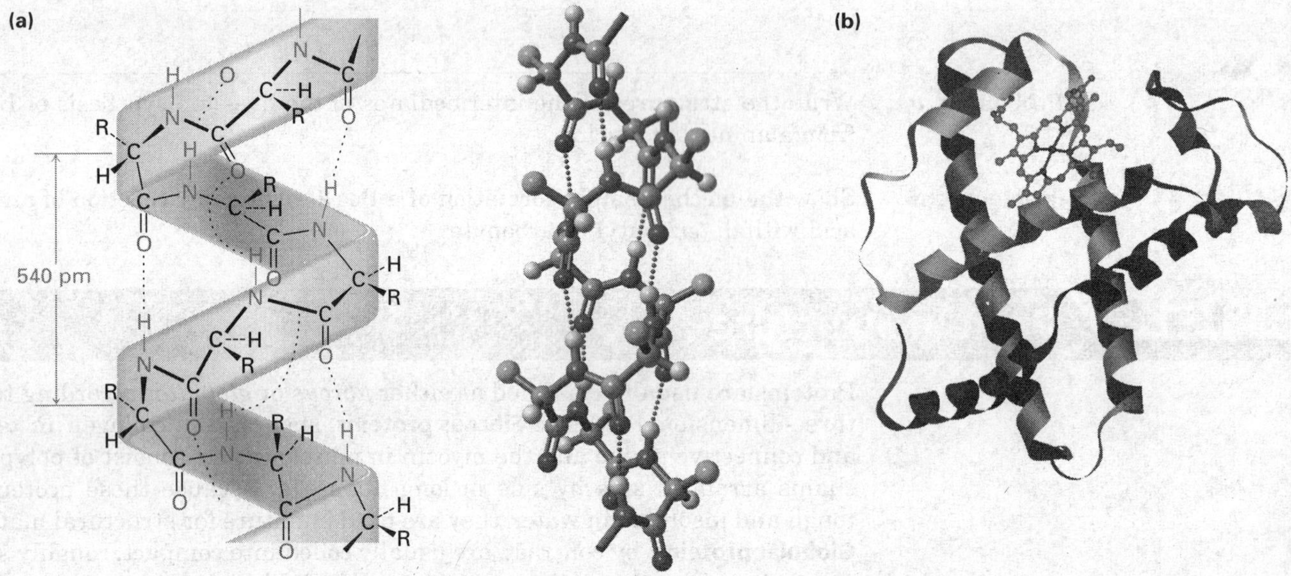

Figure 15.4 **(a) The α-helical secondary structure of proteins is stabilized by hydrogen bonds between the N–H group of one residue and the C=O group four residues away. (b) The structure of myoglobin, a globular protein with extensive helical regions that are shown as ribbons in this representation.**

A **β-pleated sheet** differs from an α helix in that the peptide chain is fully extended rather than coiled and the hydrogen bonds occur between residues in adjacent chains (Figure 15.5a). The neighboring chains can run either in the same direction (parallel) or in opposite directions (antiparallel), although the antiparallel arrangement is more common and energetically somewhat more favorable. Concanavalin A, for instance, consists of two identical chains of 237 residues with extensive regions of antiparallel β sheets (Figure 15.5b).

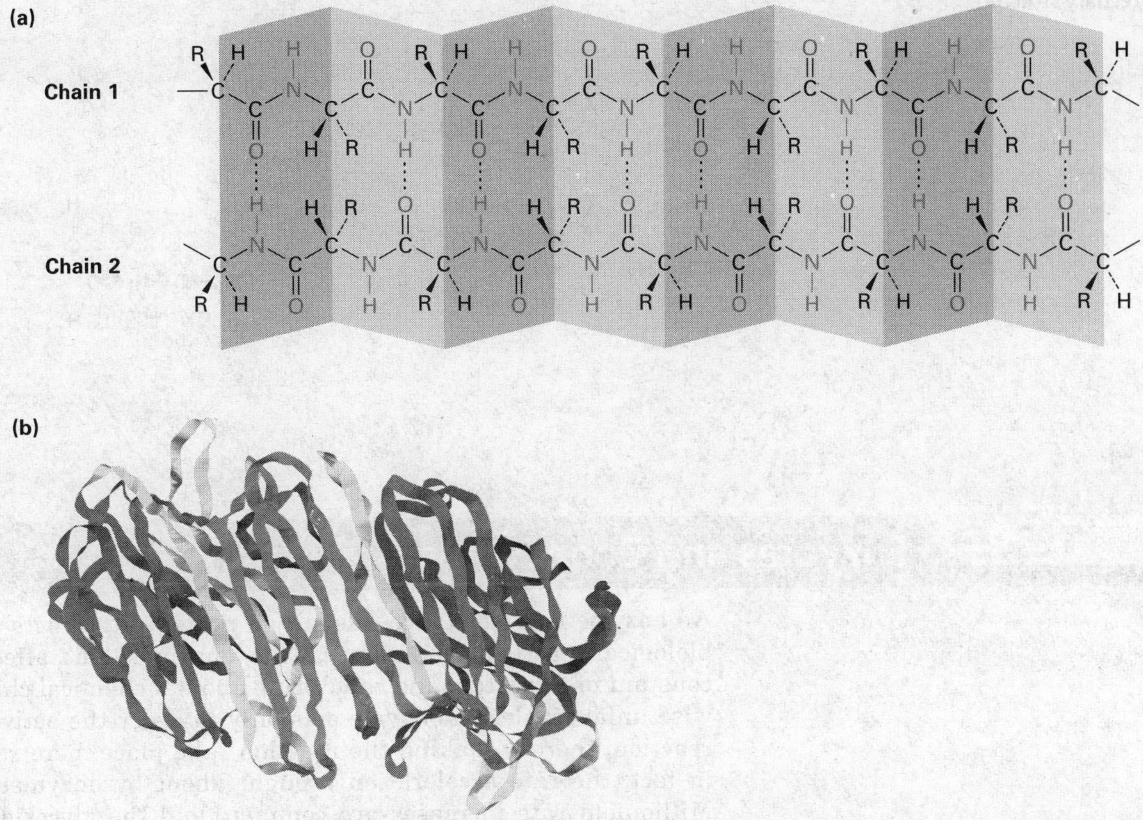

Figure 15.5 **(a)** The β-pleated sheet secondary structure of proteins is stabilized by hydrogen bonds between parallel or antiparallel chains. **(b)** The structure of concanavalin A, a protein with extensive regions of antiparallel β sheets, shown as ribbons.

What about tertiary structure? Why does a protein adopt the shape it does? The forces that determine the tertiary structure of a protein are the same forces that act on all molecules, regardless of size, to provide maximum stability. Particularly important are the hydrophilic (water-loving) interactions of the polar side chains on acidic or basic amino acids and the hydrophobic (water-fearing) interactions of nonpolar side chains. Those acidic or basic amino acids with charged side chains tend to congregate on the exterior of the protein, where they can be solvated by water. Those amino acids with neutral, nonpolar side chains tend to congregate on the hydrocarbon-like interior of a protein molecule, away from the aqueous medium.

Also important for stabilizing a protein's tertiary structure are the formation of disulfide bridges between cysteine residues, the formation of hydrogen bonds between nearby amino acid residues, and the presence of ionic attractions, called *salt bridges,* between positively and negatively charged sites on various amino acid side chains within the protein. The various kinds of stabilizing forces are summarized in Figure 15.6.

Figure 15.6 Kinds of interactions among amino acid side chains that stabilize a protein's tertiary structure.

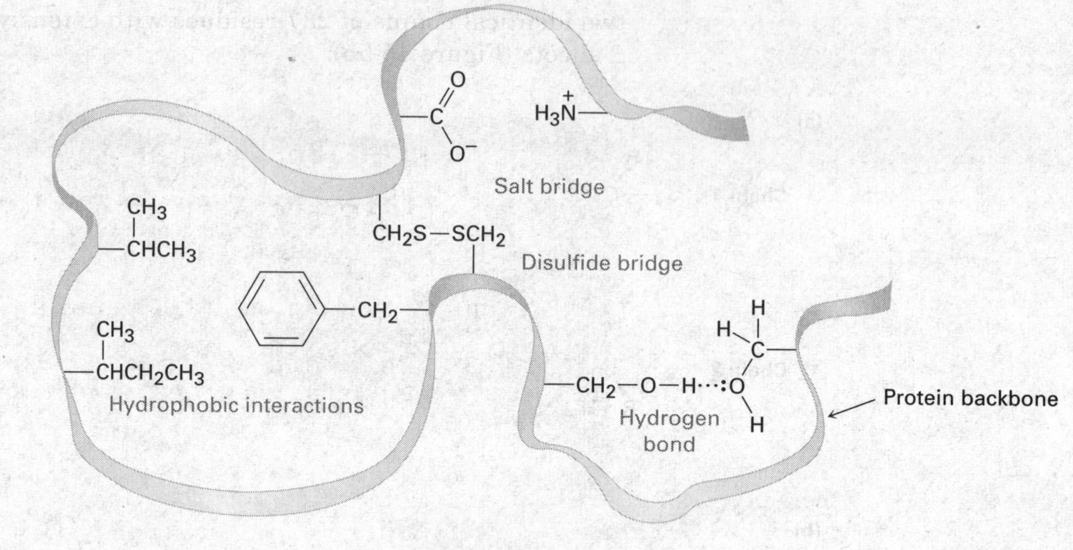

15.9 Enzymes and Coenzymes

An **enzyme** is a substance—usually a protein—that acts as a catalyst for a biological reaction. Like all catalysts, enzymes don't affect the equilibrium constant of a reaction and can't bring about a chemical change that is otherwise unfavorable. An enzyme acts only to lower the activation energy for a reaction, thereby making the reaction take place more rapidly. Sometimes, in fact, the rate acceleration brought about by enzymes is extraordinary. Millionfold rate increases are common, and the glycosidase enzymes that hydrolyze polysaccharides increase the reaction rate by a factor of more than 10^{17}, changing the time required for the reaction from millions of years to milliseconds.

Unlike many of the catalysts that chemists use in the laboratory, enzymes are usually specific in their action. Often, in fact, an enzyme will catalyze only a single reaction of a single compound, called the enzyme's *substrate.* For example, the enzyme amylase found in the human digestive tract catalyzes only the hydrolysis of starch to yield glucose; cellulose and other polysaccharides are untouched by amylase.

Different enzymes have different specificities. Some, such as amylase, are specific for a single substrate, but others operate on a range of substrates. Papain, for instance, a globular protein of 212 amino acids isolated from papaya fruit, catalyzes the hydrolysis of many kinds of peptide bonds. In fact, it's this ability to hydrolyze peptide bonds that makes papain useful as a meat tenderizer and a cleaner for contact lenses.

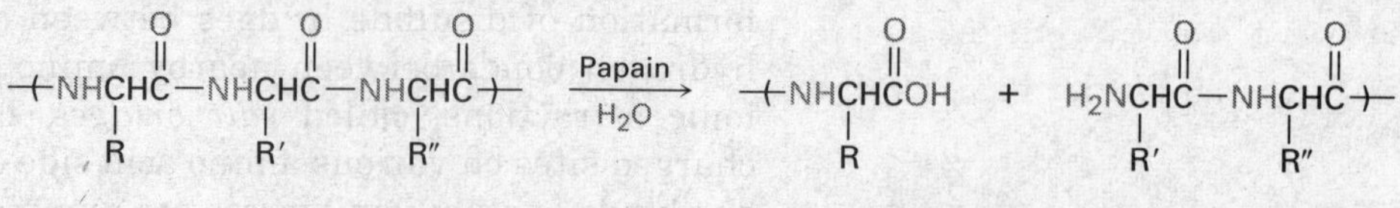

Enzymes are classified into six categories depending on the kind of reaction they catalyze, as shown in Table 15.2. *Oxidoreductases* catalyze oxidations and reductions; *transferases* catalyze the transfer of a group from one substrate to another; *hydrolases* catalyze hydrolysis reactions of esters, amides, and related substrates; *lyases* catalyze the elimination or addition of a small molecule such as H_2O from or to a substrate; *isomerases* catalyze isomerizations; and *ligases* catalyze the bonding together of two molecules, often coupled with the hydrolysis of ATP. The systematic name of an enzyme has two parts, ending with *-ase.* The first part identifies the enzyme's substrate, and the second part identifies its class. For example, hexose kinase is a transferase that catalyzes the transfer of a phosphate group from ATP to a hexose sugar.

Table 15.2 Classification of Enzymes

Class	Some subclasses	Function
Oxidoreductases	Dehydrogenases	Introduction of double bond
	Oxidases	Oxidation
	Reductases	Reduction
Transferases	Kinases	Transfer of phosphate group
	Transaminases	Transfer of amino group
Hydrolases	Lipases	Hydrolysis of ester
	Nucleases	Hydrolysis of phosphate
	Proteases	Hydrolysis of amide
Lyases	Decarboxylases	Loss of CO_2
	Dehydrases	Loss of H_2O
Isomerases	Epimerases	Isomerization of chirality center
Ligases	Carboxylases	Addition of CO_2
	Synthetases	Formation of new bond

In addition to their protein part, most enzymes also contain a small nonprotein part called a *cofactor.* A **cofactor** can be either an inorganic ion, such as Zn^{2+}, or a small organic molecule, called a **coenzyme**. A coenzyme is not a catalyst but is a reactant that undergoes chemical change during the reaction and requires an additional step or series of steps to return it to its initial state.

Many, although not all, coenzymes are derived from *vitamins*—substances that an organism requires for growth but is unable to synthesize and must receive in its diet. Coenzyme A from pantothenate (vitamin B_3), NAD^+ from niacin, FAD from riboflavin (vitamin B_2), tetrahydrofolate from folic acid, pyridoxal phosphate from pyridoxine (vitamin B_6), and thiamin diphosphate from thiamin (vitamin B_1) are examples. Table 15.3 shows the structures of some common coenzymes.

Table 15.3 Structures of Some Common Coenzymes

Adenosine triphosphate—ATP (phosphorylation)

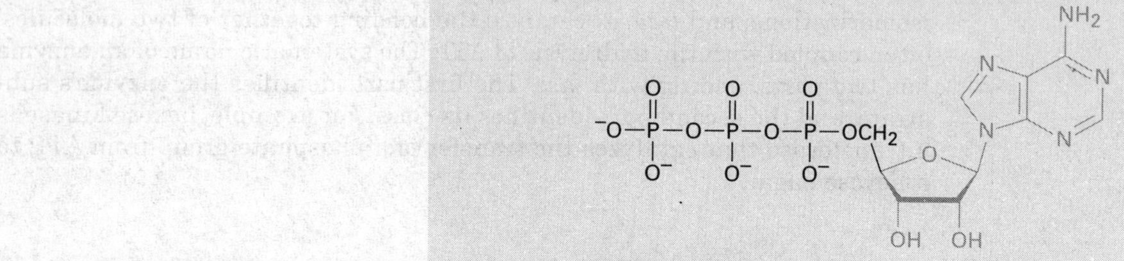

Coenzyme A (acyl transfer)

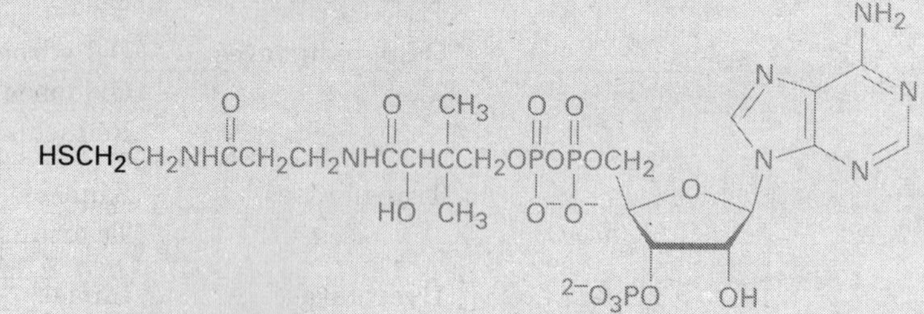

Nicotinamide adenine dinucleotide—NAD$^+$ (oxidation/reduction)
(NADP$^+$)

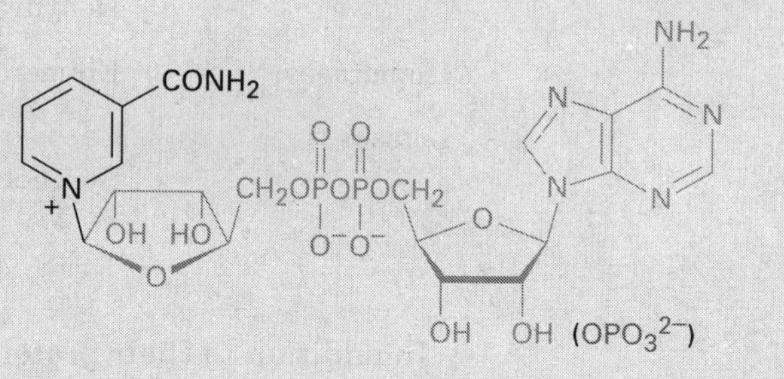

Flavin adenine dinucleotide—FAD (oxidation/reduction)

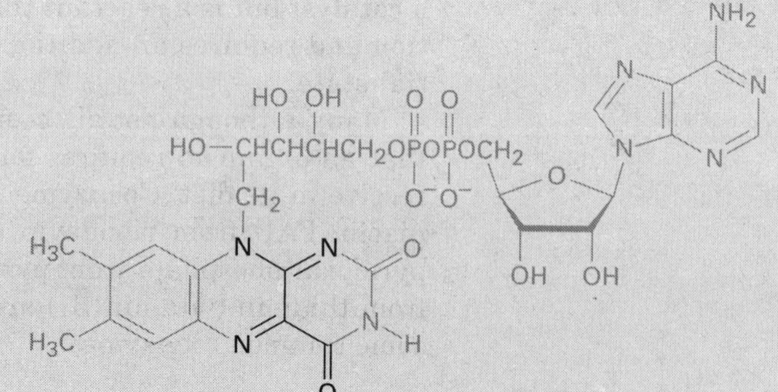

continued

Table 15.3 Structures of Some Common Coenzymes *(continued)*

Tetrahydrofolate (transfer of C_1 units)

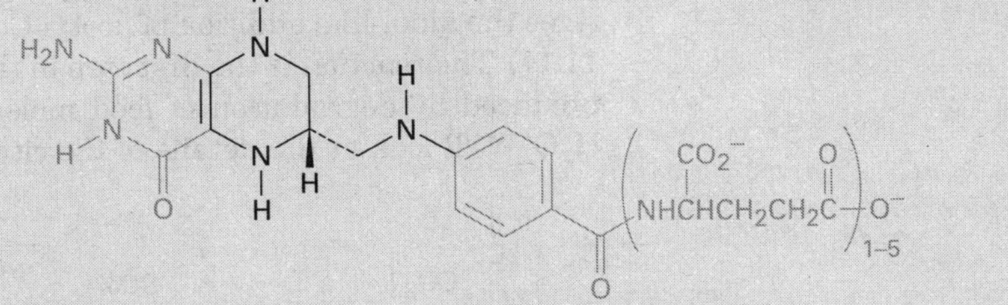

***S*-Adenosylmethionine (methyl transfer)**

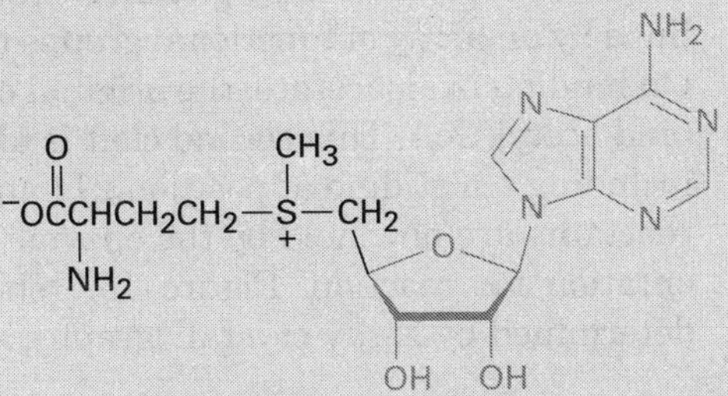

Lipoic acid (acyl transfer)

Pyridoxal phosphate (amino acid metabolism)

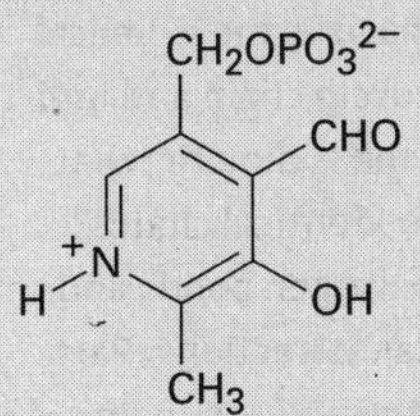

Thiamin diphosphate (decarboxylation)

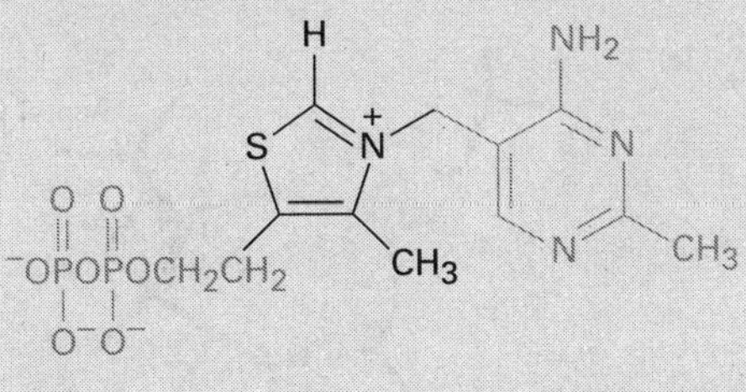

Biotin (carboxylation)

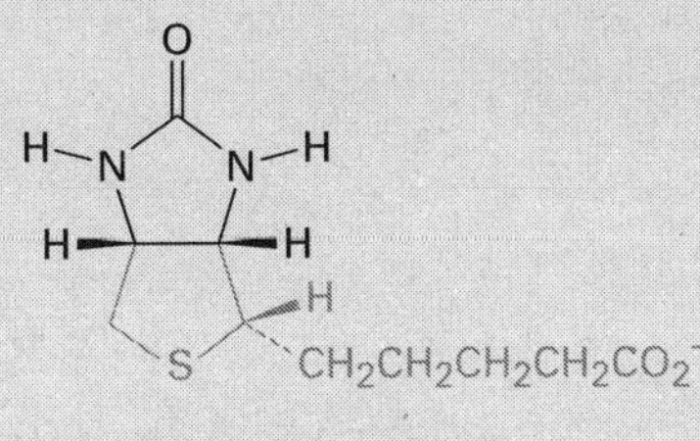

Problem 15.17 To what classes do the following enzymes belong?

(a) Pyruvate decarboxylase **(b)** Chymotrypsin

(c) Alcohol dehydrogenase

15.10 How Do Enzymes Work? Citrate Synthase

Enzymes work by bringing reactant molecules together, holding them in the orientation necessary for reaction, and providing any acidic or basic sites needed for catalysis. Let's look, for example, at citrate synthase, an enzyme that catalyzes the aldol-like addition of acetyl CoA to oxaloacetate to give citrate (Section 11.11). The reaction is the first step in the citric acid cycle, in which acetyl groups produced by degradation of food molecules are metabolized to yield CO_2 and H_2O. We'll look at the details of the citric acid cycle in Section 17.4.

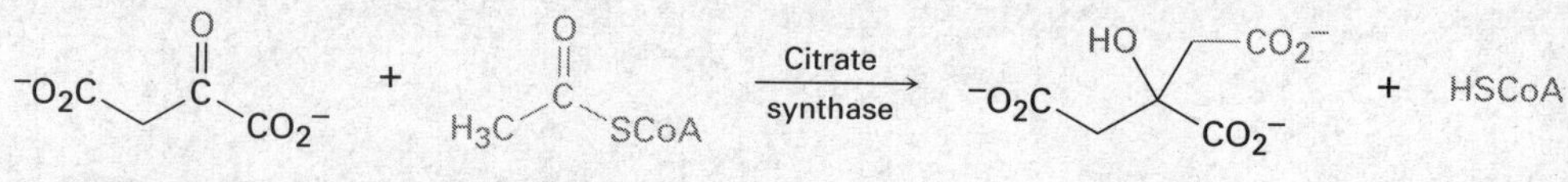

Oxaloacetate **Acetyl CoA** **Citrate**

Citrate synthase is a globular protein of 433 amino acids with a deep cleft lined by an array of functional groups that can bind to the substrate oxaloacetate. On binding oxaloacetate, the original cleft closes and another opens up nearby to bind acetyl CoA. This second cleft is also lined by appropriate functional groups, including a histidine at position 274 and an aspartic acid at position 375. The two reactants are now held by the enzyme in close proximity and with a suitable orientation for reaction. Figure 15.7 shows the structure of citrate synthase as determined by X-ray crystallography, along with a close-up of the active site.

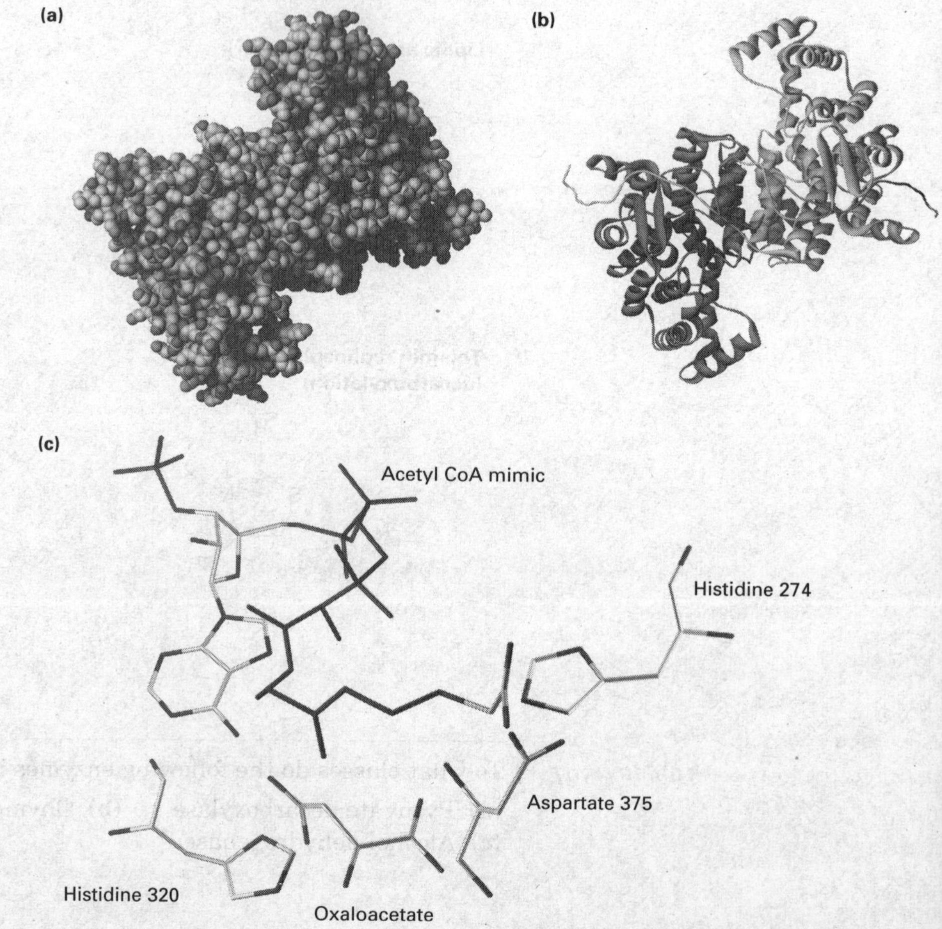

Figure 15.7 X-ray crystal structure of citrate synthase. Part **(a)** is a space-filling model and part **(b)** is a ribbon model, which emphasizes the α-helical segments of the protein chain and indicates that the enzyme is dimeric; that is, it consists of two identical chains held together by hydrogen bonds and other intermolecular attractions. Part **(c)** is a close-up of the active site, in which oxaloacetate and an unreactive acetyl CoA mimic are bound.

As shown in Figure 15.8, the first step in the aldol reaction is generation of the enol of acetyl CoA. The side-chain carboxyl of an aspartate residue acts as base to abstract an acidic α proton, while at the same time the side-chain imidazole ring of a histidine donates H^+ to the carbonyl oxygen. The enol thus produced then does a nucleophilic addition to the ketone carbonyl group of oxaloacetate. The first histidine acts as a base to remove the –OH hydrogen from the enol, while a second histidine residue simultaneously donates a proton to the oxaloacetate carbonyl group, giving citryl CoA. Water then hydrolyzes the thiol ester group in citryl CoA in a nucleophilic acyl substitution reaction, releasing citrate and coenzyme A as the final products.

MECHANISM

Figure 15.8 Mechanism of action of the enzyme citrate synthase.

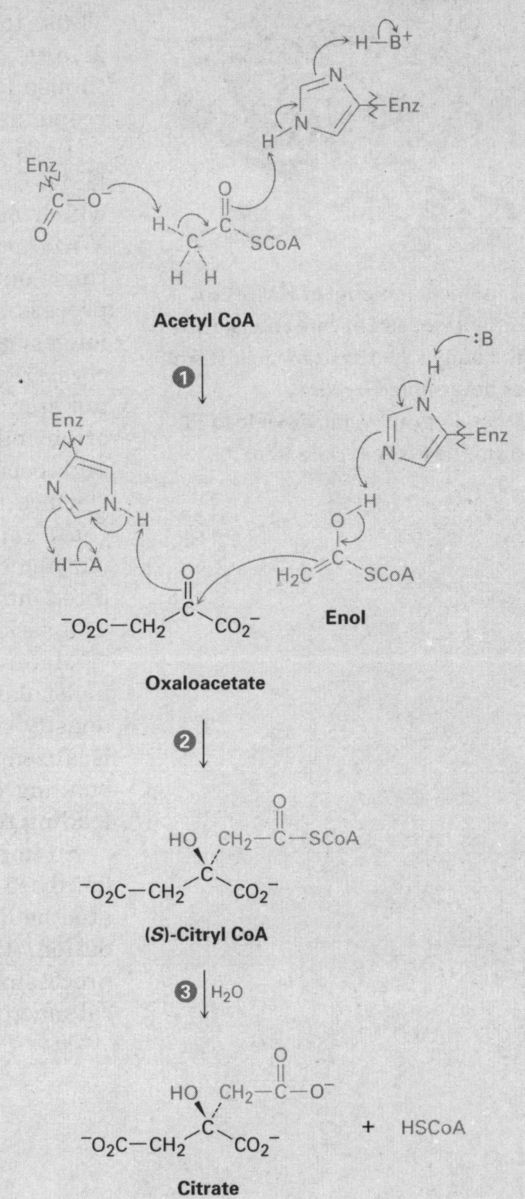

1 The side-chain carboxylate group of an aspartic acid acts as a base and removes an acidic α proton from acetyl CoA, while the N–H group on the side chain of a histidine acts as an acid and donates a proton to the carbonyl oxygen, giving an enol.

2 A histidine deprotonates the acetyl-CoA enol, which adds to the ketone carbonyl group of oxaloacetate in an aldol-like reaction. Simultaneously, an acid N–H proton of another histidine protonates the carbonyl oxygen, producing (*S*)-citryl CoA.

3 The thioester group of citryl CoA is hydrolyzed by a typical nucleophilic acyl substitution reaction to produce citrate plus coenzyme A.

INTERLUDE

X-Ray Crystallography

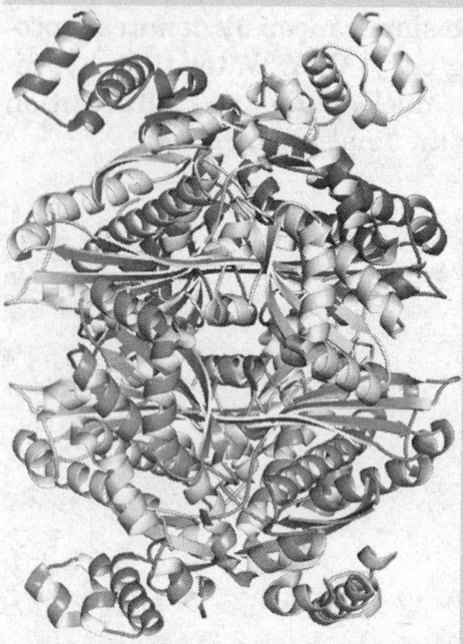

A molecular model of HMG-CoA reductase, an enzyme crucial to the body's synthesis of cholesterol, as determined by X-ray crystallography and downloaded from the Protein Data Bank.

Determining the three-dimensional shape of an object around you is easy—you look at it, your eyes focus the light rays reflected from the object, and your brain assembles the data into a recognizable image. If the object is small, you use a microscope and let the microscope lens focus the visible light. Unfortunately, there is a limit to what you can see, even with the best optical microscope. Called the *diffraction limit,* you can't see anything smaller than the wavelength of light you are using for the observation. Visible light has wavelengths of several hundred nanometers, but atoms in molecules have dimensions on the order of 0.1 nm. Thus, to "see" a molecule—whether a small one in the laboratory or a large, complex enzyme with a molecular weight in the hundreds of thousands—you need wavelengths in the 0.1 nm range, which corresponds to X rays.

Let's say that we want to determine the structure of the enzyme HMG-CoA reductase, which catalyzes a crucial step in the process by which our bodies synthesize cholesterol. The technique used is called *X-ray crystallography.* First, the molecule is crystallized (which often turns out to be the most difficult and time-consuming part of the entire process), and a small crystal with a dimension of 0.4 to 0.5 mm on its longest axis is glued to the end of a glass fiber. The fiber and attached crystal are then mounted in an instrument called an X-ray diffractometer, which consists of a radiation source, a sample positioning and orienting device that can rotate the crystal in any direction, a detector, and a controlling computer.

Once mounted, the crystal is irradiated with X rays, usually so-called Cu*Kα* radiation with a wavelength of 0.154 nm. When the X rays strike the enzyme crystal, they interact with electrons in the molecule and are scattered into a diffraction pattern that, when detected and visualized, appears as a series of bright spots against a null background.

Manipulation of the diffraction pattern to extract three-dimensional molecular data is a complex process, but the final result is that an electron-density map of the molecule is produced. Because electrons are largely localized around atoms, any two centers of electron density located within bonding distance of each other are assumed to represent bonded atoms, leading to a recognizable chemical structure.

So important is structural information for biochemistry that an online database of more than 60,000 biological structures has been created. Operated by Rutgers University and funded by the U.S. National Science Foundation, the Protein Data Bank (PDB) is a worldwide repository for processing and distributing three-dimensional structural data for biological macromolecules.

Summary and Key Words

α-amino acid 505
α helix 522
backbone 512
β-pleated sheet 522
C-terminal amino acid 512
coenzyme 525
cofactor 525
enzyme 524
fibrous protein 521
globular protein 521
isoelectric point, pI 509
N-terminal amino acid 512
peptide 504
primary structure 521
protein 504
quaternary structure 521
residue 511
secondary structure 521
side chain 505
tertiary structure 521
zwitterion 504

Proteins and **peptides** are large biomolecules made of **α-amino acid residues** linked together by amide bonds. Twenty α-amino acids are commonly found in proteins, and all except glycine have stereochemistry similar to that of L sugars.

Determining the structure of a peptide or protein begins with amino acid analysis. The peptide is first hydrolyzed to its constituent α-amino acids, which are then separated and identified. Next, the peptide is sequenced. Edman degradation by treatment with phenyl isothiocyanate (PITC) cleaves one residue from the **N terminus** of the peptide and forms an easily identifiable phenylthiohydantoin (PTH) derivative of that residue. An automated series of Edman degradations can sequence peptide chains up to 50 residues in length.

Peptide synthesis involves the use of protecting groups. An N-protected amino acid with a free $-CO_2H$ group is coupled using DCC to an O-protected amino acid with a free $-NH_2$ group. Amide formation occurs, the protecting groups are removed, and the sequence is repeated. Amines are usually protected as their *tert*-butyloxycarbonyl (Boc) or fluorenylmethyloxycarbonyl (Fmoc) derivatives; acids are usually protected as esters. The synthesis is often carried out by the Merrifield solid-phase method, in which the peptide is bonded to insoluble polymer beads.

Proteins have four levels of structure. **Primary structure** describes a protein's amino acid sequence; **secondary structure** describes how segments of the protein chain orient into regular patterns—either **α helix** or **β-pleated sheet**; **tertiary structure** describes how the entire protein molecule coils into an overall three-dimensional shape; and **quaternary structure** describes how individual protein molecules aggregate into larger structures.

Proteins are classified as either **globular** or **fibrous**. Fibrous proteins such as α-keratin are tough and water-insoluble; globular proteins such as myoglobin are water-soluble and mobile within cells. Most of the 3000 or so known enzymes are globular proteins.

Enzymes are biological catalysts that act by bringing reactant molecules together, holding them in the orientation necessary for reaction and providing any acidic or basic sites needed for catalysis. They are classified into six groups according to the kind of reaction they catalyze: *oxidoreductases* catalyze oxidations and reductions; *transferases* catalyze transfers of groups; *hydrolases* catalyze hydrolysis; *isomerases* catalyze isomerizations; *lyases* catalyze bond breakages; and *ligases* catalyze bond formations.

In addition to their protein part, many enzymes contain **cofactors**, which can be either metal ions or small organic molecules called **coenzymes**. Often, the coenzyme is a vitamin, a small molecule that must be obtained in the diet and is required in trace amounts for proper growth and functioning.

Exercises

Visualizing Chemistry

(Problems 15.1–15.17 appear within the chapter.)

OWL

Interactive versions of these problems are assignable in OWL.

15.18 Give the sequence of the following tetrapeptide (yellow = S):

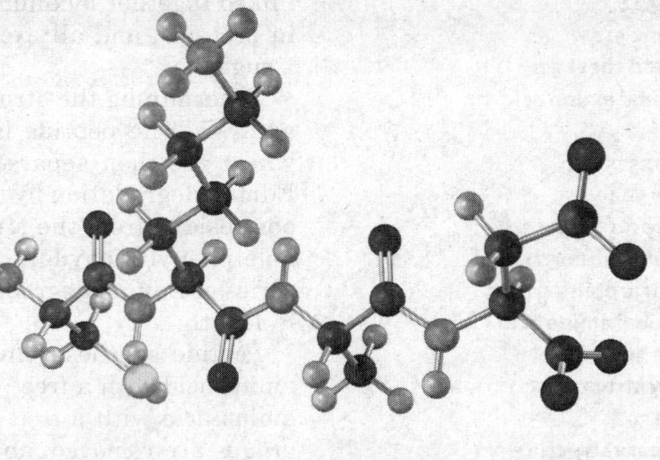

15.19 Isoleucine and threonine are the only two amino acids with two chirality centers. Assign *R* or *S* configuration to the methyl-bearing carbon atom of isoleucine.

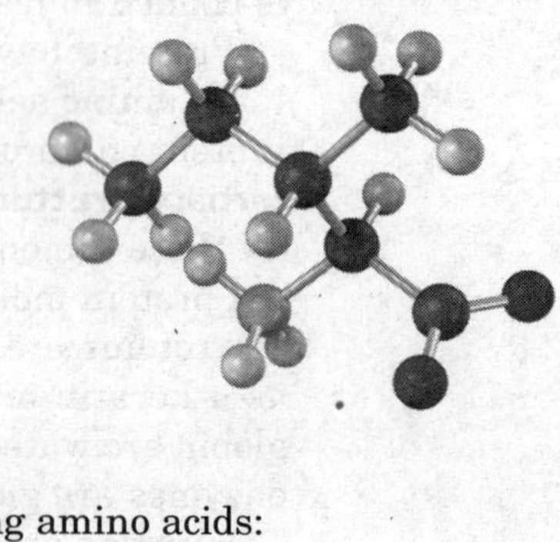

15.20 Identify the following amino acids:

(a)

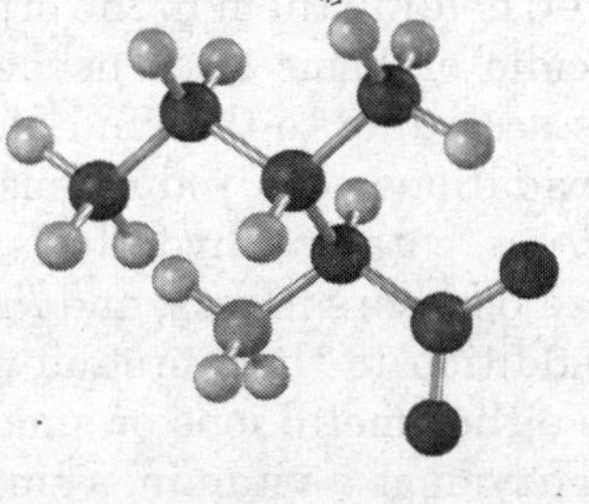

(b)

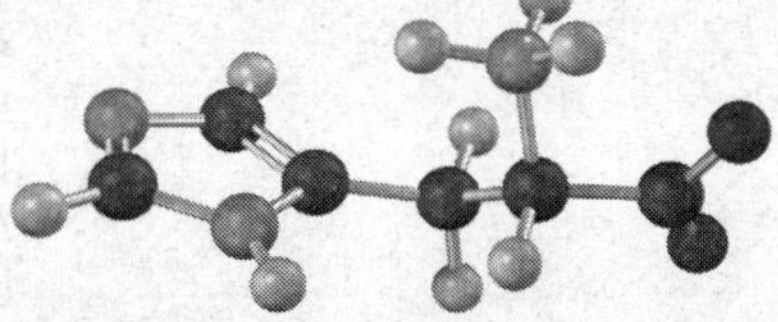

(c)

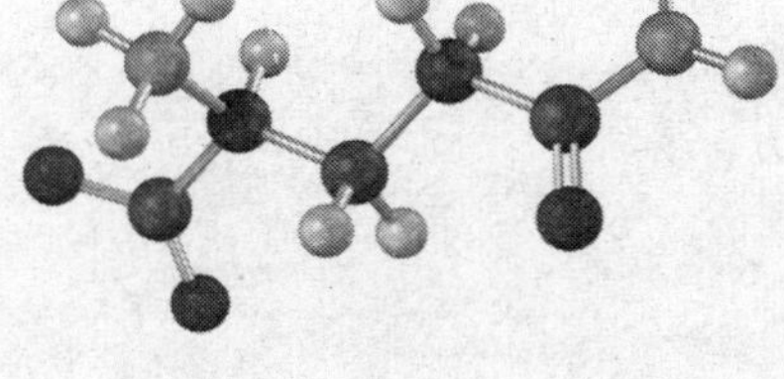

15.21 Is the following molecule a D amino acid or an L amino acid? Identify it.

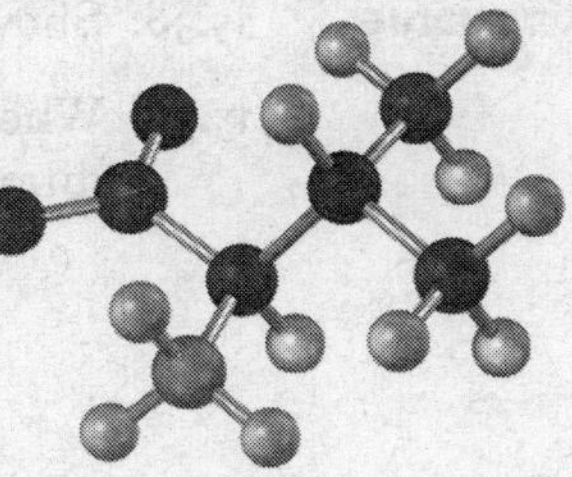

Additional Problems

STRUCTURE AND CHIRALITY

15.22 Why is cysteine such an important amino acid for determining the tertiary structure of a protein?

15.23 Draw a Fischer projection of (*S*)-proline, the only secondary amino acid.

15.24 The amino acid threonine, (2*S*,3*R*)-2-amino-3-hydroxybutanoic acid, has two chirality centers and a stereochemistry similar to that of the four-carbon sugar D-threose. Draw a Fischer projection of threonine.

15.25 Draw the Fischer projection of a diastereomer of threonine (see Problem 15.24).

15.26 What does the prefix "α" mean when referring to an α-amino acid?

15.27 Although only *S* amino acids occur in proteins, several *R* amino acids are found elsewhere in nature. For example, (*R*)-serine is found in earthworms and (*R*)-alanine is found in insect larvae. Draw Fischer projections of (*R*)-serine and (*R*)-alanine.

15.28 What amino acids do the following abbreviations stand for?
(a) Ser **(b)** Thr **(c)** Pro **(d)** F **(e)** Q **(f)** D

15.29 At what pH would you carry out an electrophoresis experiment if you wanted to separate a mixture of histidine, serine, and glutamic acid? Explain.

15.30 Draw the following amino acids in their zwitterionic forms:
(a) Serine **(b)** Tyrosine **(c)** Threonine

15.31 Write full structures for the following peptides, and indicate the positions of the amide bonds:
(a) Val-Phe-Cys **(b)** Glu-Pro-Ile-Leu

15.32 Draw structures of the predominant forms of lysine and aspartic acid at pH 3.0 and pH 9.7.

15.33 Using both one- and three-letter code names for each amino acid, write the structures of all the peptides containing the following amino acids:
(a) Val, Leu, Ser **(b)** Ser, Leu_2, Pro

15.34 The *endorphins* are a group of naturally occurring compounds in the brain that act to control pain. The active part of an endorphin is a pentapeptide called an *enkephalin,* which has the structure Tyr-Gly-Gly-Phe-Met. Draw the structure.

15.35 The amino acid analysis data in Figure 15.2 indicate that proline is not easily detected by reaction with ninhydrin. Suggest a reason.

REACTIONS

15.36 Show the steps involved in a Merrifield synthesis of Phe-Ala-Val.

15.37 When an unprotected α-amino acid is treated with dicyclohexylcarbodiimide (DCC), a 2,5-diketopiperazine results. Explain.

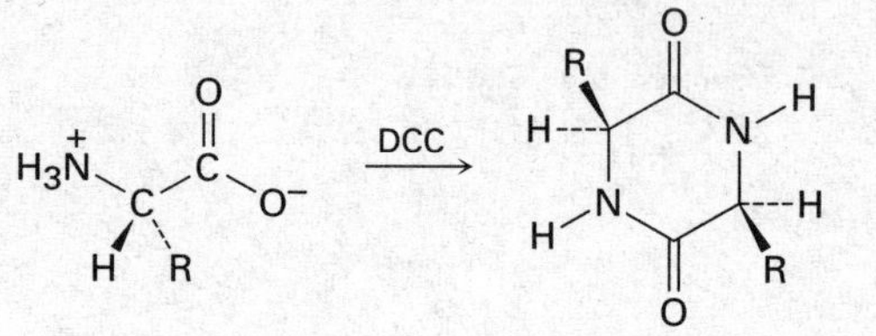

15.38 Predict the product of the reaction of valine with the following reagents:
(a) CH_3CH_2OH, H^+
(b) NaOH, H_2O
(c) Di-*tert*-butyl dicarbonate

15.39 Draw the structure of the phenylthiohydantoin product you would expect to obtain from Edman degradation of the following peptides:
(a) Val-Leu-Gly **(b)** Ala-Pro-Phe

PEPTIDE SEQUENCING

15.40 What is the structure of a nonapeptide that gives the following fragments when cleaved by chymotrypsin and by trypsin?

Trypsin cleavage: Val-Val-Pro-Tyr-Leu-Arg, Ser-Ile-Arg

Chymotrypsin cleavage: Leu-Arg, Ser-Ile-Arg-Val-Val-Pro-Tyr

15.41 Which amide bonds in the following polypeptide are cleaved by trypsin? By chymotrypsin?

Phe-Leu-Met-Lys-Tyr-Asp-Gly-Gly-Arg-Val-Ile-Pro-Tyr

15.42 Propose a structure for an octapeptide that shows the composition Asp, Gly_2, Leu, Phe, Pro_2, Val on amino acid analysis. Edman analysis shows a glycine N-terminal group, and leucine is the C-terminal group. Acidic hydrolysis gives the following fragments:

Val-Pro-Leu, Gly, Gly-Asp-Phe-Pro, Phe-Pro-Val

15.43 Give the amino acid sequence of hexapeptides that produce the following fragments on partial acid hydrolysis:
(a) Arg, Gly, Ile, Leu, Pro, Val gives Pro-Leu-Gly, Arg-Pro, Gly-Ile-Val
(b) Asp, Leu, Met, Trp, Val_2 gives Val-Leu, Val-Met-Trp, Trp-Asp-Val

GENERAL PROBLEMS

15.44 What kind of reaction does each of the following enzymes catalyze?
(a) A protease **(b)** A kinase **(c)** A carboxylase

15.45 How can you account for the fact that proline is never encountered in a protein α helix? The α-helical segments of myoglobin and other proteins stop when a proline residue is encountered in the chain.

15.46 What kinds of reactions do the following classes of enzymes catalyze?
(a) Hydrolases **(b)** Lyases **(c)** Transferases

15.47 The reaction of ninhydrin with an α-amino acid occurs in several steps.

(a) The first step is loss of water to give a triketone. Show the mechanism of the reaction and the structure of the triketone.

(b) The second step is formation of an imine by reaction of the amino acid with the triketone. Show its structure.

(c) The third step is a decarboxylation. Show the structure of the product and the mechanism of the decarboxylation reaction.

(d) The fourth step is hydrolysis of an imine to yield an amine and an aldehyde. Show the structures of both products.

(e) The final step is formation of the purple anion. Show the mechanism of the reaction.

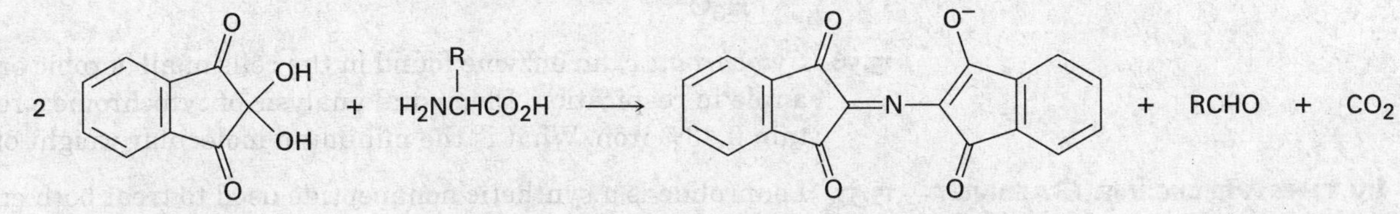

Ninhydrin

15.48 Draw as many resonance forms as you can for the purple anion obtained by reaction of ninhydrin with an amino acid (Problem 15.47).

15.49 Look up the structure of human insulin in Section 15.4, and indicate where in each chain the molecule is cleaved by trypsin and by chymotrypsin.

15.50 Arginine, which contains a *guanidino* group in its side chain, is the most basic of the 20 common amino acids. How can you account for this basicity, using resonance structures to see how the protonated guanidino group is stabilized?

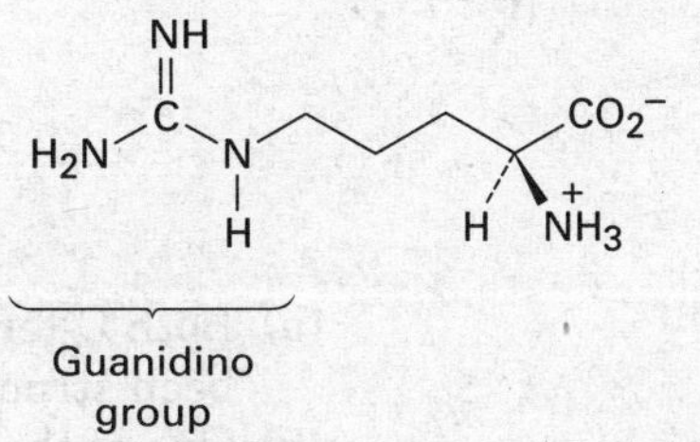

Arginine

15.51 Cysteine is the only amino acid that has L stereochemistry but an R configuration. Design another L amino acid of your own making that also has an R configuration.

15.52 Propose two structures for a tripeptide that gives Leu, Ala, and Phe on hydrolysis but does not react with phenyl isothiocyanate.

15.53 Which of the following amino acids are more likely to be found on the outside of a globular protein, and which on the inside? Explain.

(a) Valine **(b)** Aspartic acid **(c)** Isoleucine **(d)** Lysine

15.54 A hexapeptide with the composition Arg, Gly, Leu, Pro_3 has proline at both C-terminal and N-terminal positions. What is the structure of the hexapeptide if partial hydrolysis gives Gly-Pro-Arg, Arg-Pro, and Pro-Leu-Gly?

15.55 *Aspartame,* a nonnutritive sweetener marketed under the trade name NutraSweet, is the methyl ester of a simple dipeptide, Asp-Phe-OCH_3.

(a) Draw the full structure of aspartame.

(b) The isoelectric point of aspartame is 5.9. Draw the principal structure present in aqueous solution at this pH.

(c) Draw the principal form of aspartame present at physiological pH 7.6.

(d) Show the products of hydrolysis on treatment of aspartame with H_3O^+.

15.56 *Cytochrome c,* an enzyme found in the cells of all aerobic organisms, plays a role in respiration. Elemental analysis of cytochrome *c* reveals it to contain 0.43% iron. What is the minimum molecular weight of this enzyme?

IN THE MEDICINE CABINET

15.57 Leuprolide is a synthetic nonapeptide used to treat both endometriosis in women and prostate cancer in men.

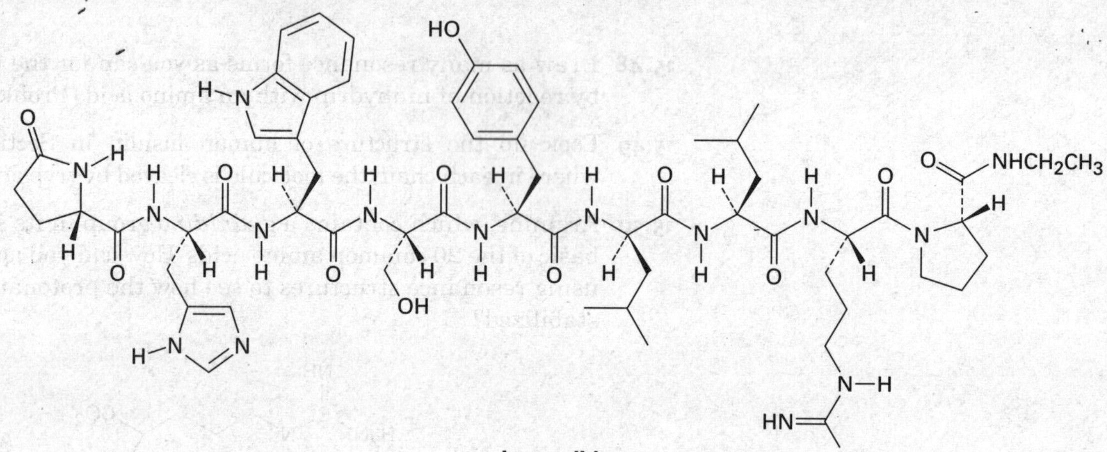

Leuprolide

(a) Both C-terminal and N-terminal amino acids in leuprolide have been structurally modified. Identify the modifications.

(b) One of the nine amino acids in leuprolide has D stereochemistry rather than the usual L. Which one?

(c) Write the structure of leuprolide using both one- and three-letter abbreviations.

(d) What charge would you expect leuprolide to have at neutral pH?

15.58 *Oxytocin,* a nonapeptide hormone secreted by the pituitary gland, stimulates uterine contraction and lactation during childbirth. Its sequence was determined from the following evidence:

1. Oxytocin is a cyclic peptide containing a disulfide bridge between two cysteine residues.
2. When the disulfide bridge is reduced, oxytocin has the constitution Asn, Cys_2, Gln, Gly, Ile, Leu, Pro, Tyr.
3. Partial hydrolysis of reduced oxytocin yields seven fragments:

   Asp-Cys, Ile-Glu, Cys-Tyr, Leu-Gly, Tyr-Ile-Glu, Glu-Asp-Cys, Cys-Pro-Leu

4. Gly is the C-terminal group.
5. Both Glu and Asp are present as their side-chain amides (Gln and Asn) rather than as free side-chain acids.

What is the amino acid sequence of reduced oxytocin? What is the structure of oxytocin?

IN THE FIELD **15.59** The herbicide atrazine binds to an enzyme in weeds by anchoring to two critical residues, a phenylalanine and a serine. Describe three interactions that occur between these residues and the herbicide (see Figure 15.6).

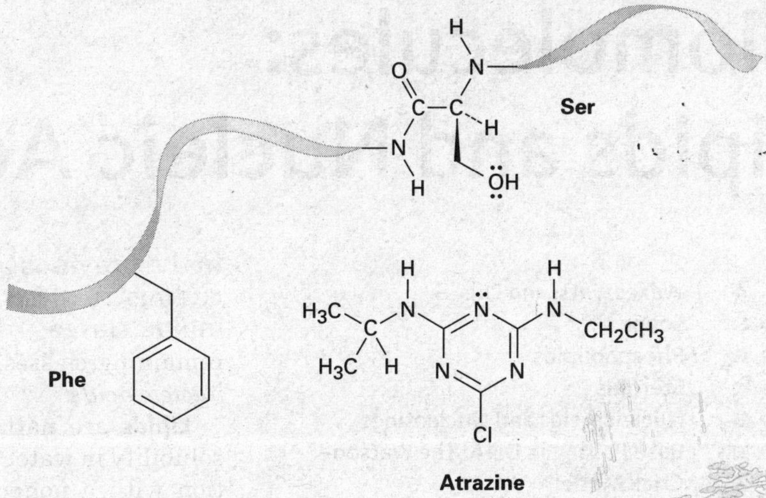

CHAPTER

16

Soap bubbles, so common yet so beautiful, are made from animal fat, a lipid.

Biomolecules: Lipids and Nucleic Acids

In the previous two chapters, we've discussed the organic chemistry of carbohydrates and proteins, two of the four major classes of biomolecules. Let's now look at the two remaining classes, beginning with *lipids* and continuing on to *nucleic acids.*

Lipids are naturally occurring molecules that have limited solubility in water and can be isolated from organisms by extraction with a nonpolar organic solvent. Fats, oils, waxes, many vitamins and hormones, and most nonprotein cell-membrane components are examples. Note that, unlike carbohydrates and proteins, lipids are defined by a physical property (solubility) rather than by structure.

Lipids are classified into two general types: those like fats and waxes, which contain ester linkages and can be hydrolyzed, and those like cholesterol and other steroids, which don't have ester linkages and can't be hydrolyzed.

OWL

Online homework for this chapter can be assigned in OWL, an online homework assessment tool.

Animal fat—a triester
(R, R′, R″ = C_{11}–C_{19} chains)

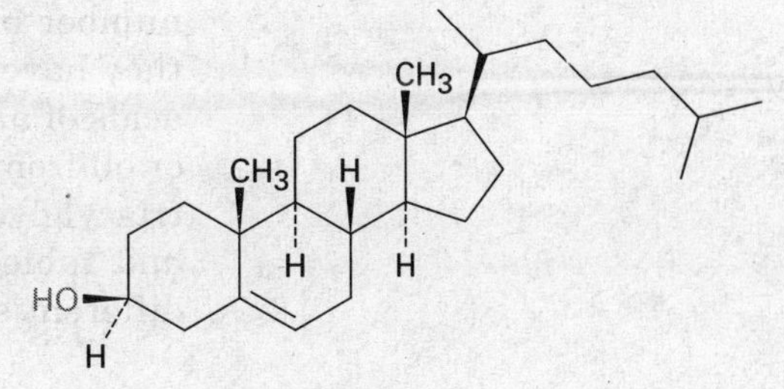

Cholesterol

16.1 Waxes, Fats, and Oils

Waxes are mixtures of esters of long-chain carboxylic acids with long-chain alcohols. The carboxylic acid usually has an even number of carbons from 16 through 36, while the alcohol has an even number of carbons from 24 through 36. One of the major components of beeswax, for instance, is triacontyl hexadecanoate, the ester of the C_{30} alcohol triacontan-1-ol and the C_{16} acid hexadecanoic acid. The waxy protective coatings on most fruits, berries, leaves, and animal furs have similar structures.

$$CH_3(CH_2)_{14}\overset{O}{\overset{\|}{C}}O(CH_2)_{29}CH_3$$

Triacontyl hexadecanoate (from beeswax)

Animal fats and vegetable oils are the most widely occurring lipids. Although they appear different—animal fats like butter and lard are solids, whereas vegetable oils like corn oil and peanut oil are liquids—their structures are closely related. Fats and oils are *triglycerides,* or **triacylglycerols**—triesters of glycerol with three long-chain carboxylic acids called **fatty acids**. Animals use fats for long-term energy storage because they are much less highly oxidized than carbohydrates and provide about six times as much energy as an equal weight of stored, hydrated glycogen.

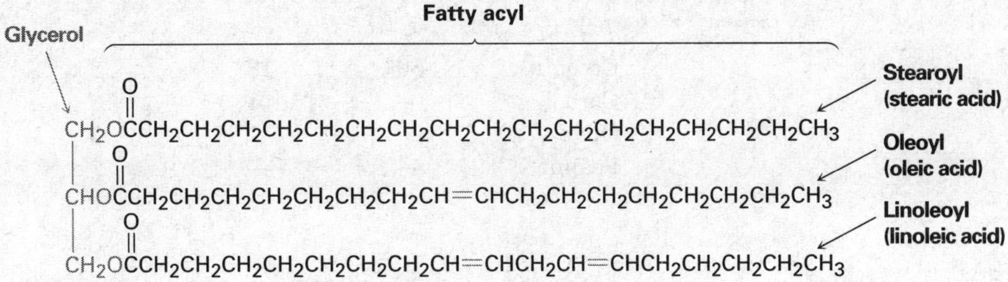

A triacylglycerol

Hydrolysis of a fat or oil with aqueous NaOH yields glycerol and three fatty acids. The fatty acids are generally unbranched and contain an even number of carbon atoms between 12 and 20. If double bonds are present, they have largely, although not entirely, *Z*, or cis, geometry. The three fatty acids of a specific triacylglycerol molecule need not be the same, and the fat or oil from a given source is likely to be a complex mixture of many different triacylglycerols. Table 16.1 lists some of the commonly occurring fatty acids, and Table 16.2 lists the approximate composition of some fats and oils from different sources.

Table 16.1 Structures of Some Common Fatty Acids

Name	No. of carbons	Melting point (°C)	Structure
Saturated			
Lauric	12	43.2	$CH_3(CH_2)_{10}CO_2H$
Myristic	14	53.9	$CH_3(CH_2)_{12}CO_2H$
Palmitic	16	63.1	$CH_3(CH_2)_{14}CO_2H$
Stearic	18	68.8	$CH_3(CH_2)_{16}CO_2H$
Arachidic	20	76.5	$CH_3(CH_2)_{18}CO_2H$
Unsaturated			
Palmitoleic	16	−0.1	(Z)-$CH_3(CH_2)_5CH{=}CH(CH_2)_7CO_2H$
Oleic	18	13.4	(Z)-$CH_3(CH_2)_7CH{=}CH(CH_2)_7CO_2H$
Linoleic	18	−12	(Z,Z)-$CH_3(CH_2)_4(CH{=}CHCH_2)_2(CH_2)_6CO_2H$
Linolenic	18	−11	(all Z)-$CH_3CH_2(CH{=}CHCH_2)_3(CH_2)_6CO_2H$
Arachidonic	20	−49.5	(all Z)-$CH_3(CH_2)_4(CH{=}CHCH_2)_4CH_2CH_2CO_2H$

Table 16.2 Composition of Some Fats and Oils

	Saturated fatty acids (%)				Unsaturated fatty acids (%)	
Source	C_{12} lauric	C_{14} myristic	C_{16} palmitic	C_{18} stearic	C_{18} oleic	C_{18} linoleic
Animal fat						
Lard	—	1	25	15	50	6
Butter	2	10	25	10	25	5
Human fat	1	3	25	8	46	10
Whale blubber	—	8	12	3	35	10
Vegetable oil						
Coconut	50	18	8	2	6	1
Corn	—	1	10	4	35	45
Olive	—	1	5	5	80	7
Peanut	—	—	7	5	60	20

More than 100 different fatty acids are known, and about 40 occur widely. Palmitic acid (C_{16}) and stearic acid (C_{18}) are the most abundant saturated fatty acids; oleic and linoleic acids (both C_{18}) are the most abundant unsaturated ones. Oleic acid is *monounsaturated* because it has only one double bond, whereas linoleic, linolenic, and arachidonic acids are **polyunsaturated fatty acids** because they have more than one double bond. Linoleic and linolenic acids occur in cream and are essential in the human diet; infants grow poorly and develop skin lesions if fed a diet of nonfat milk for prolonged periods.

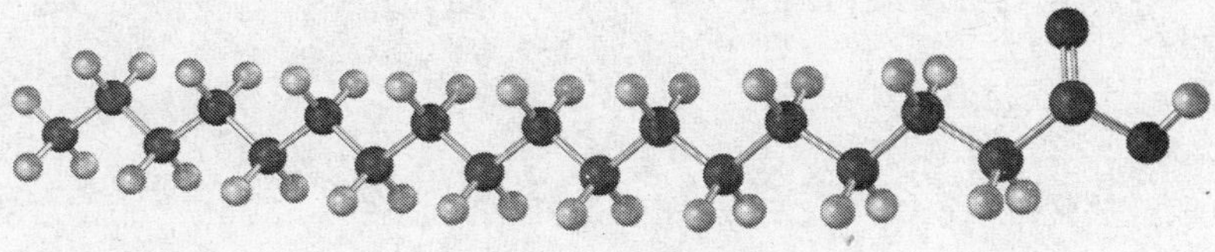

$$CH_3CH_2CH_2CH_2CH_2CH_2CH_2CH_2CH_2CH_2CH_2CH_2CH_2CH_2CH_2CH_2\overset{\overset{\displaystyle O}{\|}}{C}OH$$

Stearic acid

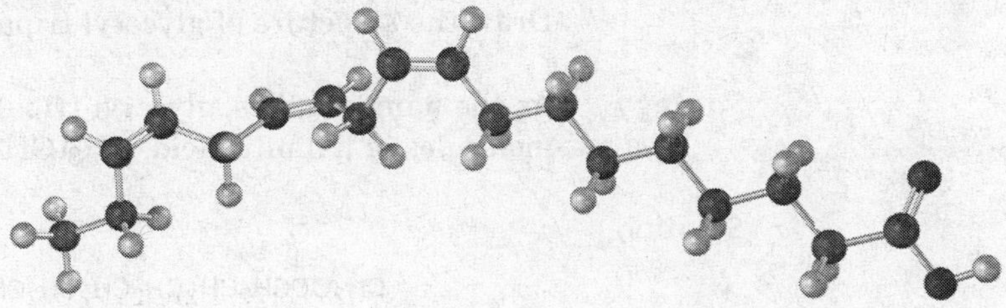

$$CH_3CH_2CH{=}CHCH_2CH{=}CHCH_2CH{=}CHCH_2CH_2CH_2CH_2CH_2CH_2CH_2\overset{\overset{\displaystyle O}{\|}}{C}OH$$

Linolenic acid, a polyunsaturated fatty acid

The data in Table 16.1 show that unsaturated fatty acids generally have lower melting points than their saturated counterparts, a trend that is also true for triacylglycerols. Since vegetable oils generally have a higher proportion of unsaturated to saturated fatty acids than animal fats (Table 16.2), they have lower melting points. The difference is a consequence of structure. Saturated fats have a uniform shape that allows them to pack together efficiently in a crystal lattice. In unsaturated vegetable oils, however, the C=C bonds introduce bends and kinks into the hydrocarbon chains, making crystal formation less favorable and lowering the melting point.

The C=C bonds in vegetable oils can be reduced by catalytic hydrogenation, typically carried out at high temperature using a nickel catalyst, to produce saturated solid or semisolid fats. Margarine and shortening are produced by hydrogenating soybean, peanut, or cottonseed oil until the proper consistency is obtained. Unfortunately, the hydrogenation reaction is accompanied by some cis–trans isomerization of the double bonds that remain, producing fats with about 10% to 15% trans unsaturated fatty acids. Dietary intake of trans fatty acids increases cholesterol levels in the blood, thereby increasing the risk for heart problems. The conversion of linoleic acid into elaidic acid is an example.

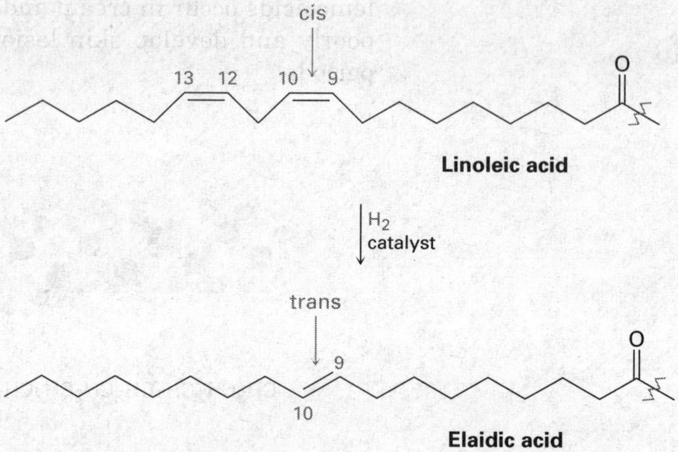

Worked Example 16.1 Drawing the Structure of a Fat

Draw the structure of glyceryl tripalmitate, a typical fat molecule.

Strategy As the name implies, glyceryl tripalmitate is the triester of glycerol with three molecules of palmitic acid, $CH_3(CH_2)_{14}CO_2H$.

Solution

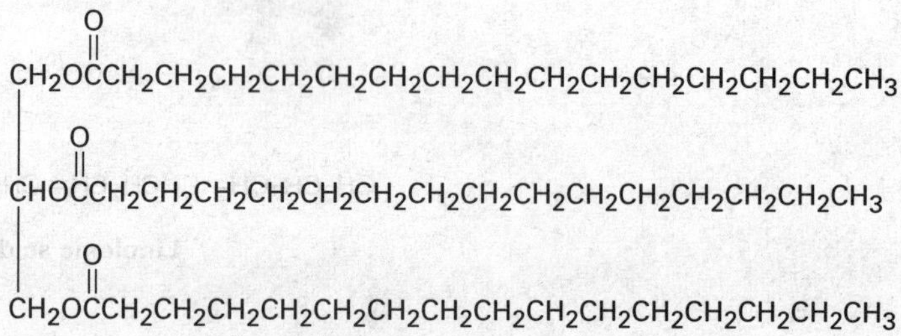

Glyceryl tripalmitate

Problem 16.1 Carnauba wax, used in floor and furniture polishes, contains an ester of a C_{32} straight-chain alcohol with a C_{20} straight-chain carboxylic acid. Draw its structure.

Problem 16.2 Draw structures of the following compounds. Which would you expect to have a higher melting point?

(a) Glyceryl trioleate **(b)** Glyceryl monooleate distearate

Problem 16.3 Fats and oils can be either optically active or optically inactive, depending on their structures. Draw the structure of an optically active fat that gives 2 equivalents of palmitic acid and 1 equivalent of stearic acid on hydrolysis. Draw the structure of an optically inactive fat that gives the same products on hydrolysis.

16.2 Soaps

Soap has been known since at least 600 BC, when the Phoenicians prepared a curdy material by boiling goat fat with extracts of wood ash. The cleansing properties of soap weren't generally recognized, however, and the use of soap didn't become widespread until the 18th century. Chemically, soap is a mixture of the sodium or potassium salts of long-chain fatty acids produced by hydrolysis (*saponification*) of animal fat with alkali.

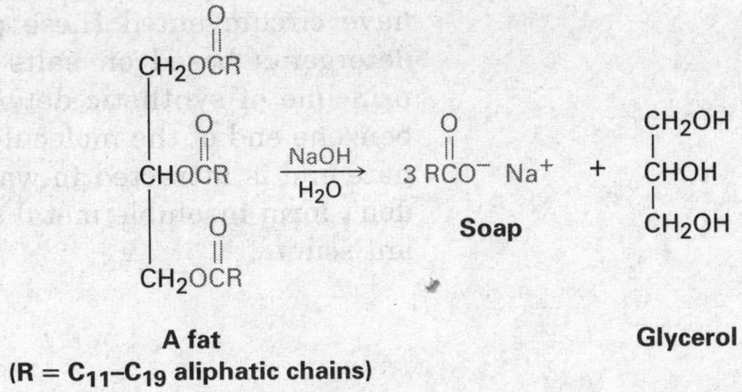

Crude soap curds contain glycerol and excess alkali as well as soap but can be purified by boiling with water and adding NaCl or KCl to precipitate the pure carboxylate salts. The smooth soap that precipitates is dried, perfumed, and pressed into bars for household use. Dyes are added to make colored soaps, antiseptics are added for medicated soaps, pumice is added for scouring soaps, and air is blown in for soaps that float.

Soaps act as cleansers because the two ends of a soap molecule are so different. The carboxylate end of the long-chain molecule is ionic and therefore *hydrophilic,* or attracted to water. The long hydrocarbon portion of the molecule, however, is nonpolar and *hydrophobic,* or water avoiding, and therefore more soluble in oils. The net effect of these two opposing tendencies is that soaps are attracted to both oils and water.

When soaps are dispersed in water, the long hydrocarbon tails cluster together on the inside of tangled, hydrophobic balls, while the ionic heads on the surface of the clusters stick out into the water layer. These spherical clusters, called **micelles**, are shown schematically in Figure 16.1. Grease and oil droplets are solubilized in water when they are coated by the nonpolar, hydrophobic tails of soap molecules in the center of micelles. Once solubilized, the grease and dirt can be rinsed away.

Figure 16.1 A soap micelle solubilizing a grease particle in water. An electrostatic potential map of a fatty-acid carboxylate shows how the negative charge is located in the head group.

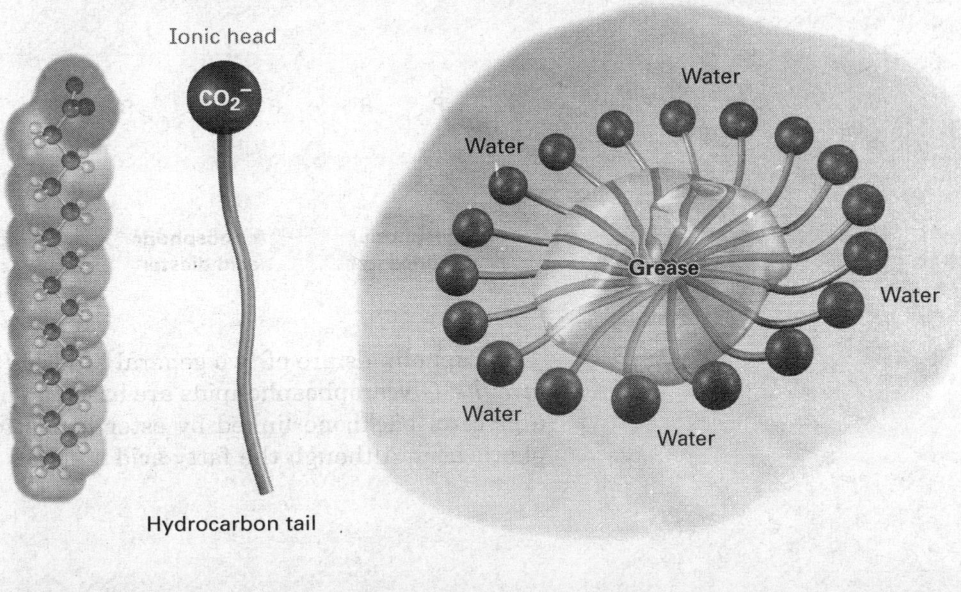

As useful as they are, soaps also have some drawbacks. In hard water, which contains metal ions such as Mg^{2+}, Ca^{2+}, and Fe^{3+}, soluble sodium carboxylates are converted into insoluble metal salts, leaving the familiar ring of scum around bathtubs and the gray tinge on white clothes. Chemists have circumvented these problems by synthesizing a class of synthetic detergents based on salts of long-chain alkylbenzenesulfonic acids. The principle of synthetic detergents is the same as that of soaps: the alkylbenzene end of the molecule is attracted to grease, while the anionic sulfonate end is attracted to water. Unlike soaps, though, sulfonate detergents don't form insoluble metal salts in hard water and don't leave an unpleasant scum.

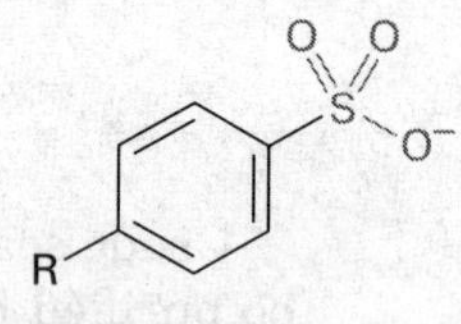

A synthetic detergent
(R = a mixture of C_{12} chains)

Problem 16.4 Draw the structure of magnesium oleate, one of the components of bathtub scum.

Problem 16.5 Write the saponification reaction of glyceryl dioleate monopalmitate with aqueous NaOH.

16.3 Phospholipids

Just as waxes, fats, and oils are esters of carboxylic acids, **phospholipids** are esters of phosphoric acid, H_3PO_4.

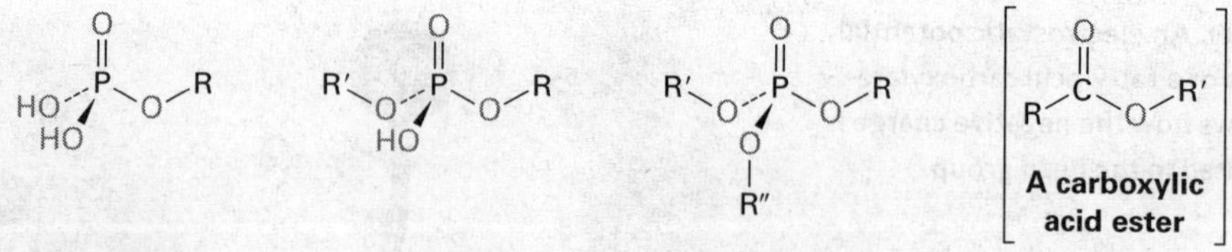

Phospholipids are of two general kinds: *glycerophospholipids* and *sphingomyelins.* Glycerophospholipids are based on phosphatidic acid, which contains a glycerol backbone linked by ester bonds to two fatty acids and one phosphoric acid. Although the fatty-acid residues can be any of the C_{12}–C_{20} units typically present in fats, the acyl group at C1 is usually saturated and the one at C2 is usually unsaturated. The phosphate group at C3 is also bonded to an amino alcohol such as choline $[HOCH_2CH_2N(CH_3)_3]^+$, ethanolamine $(HOCH_2CH_2NH_2)$, or serine $[HOCH_2CH(NH_2)CO_2H]$. The compounds are chiral and have an L, or *R*, configuration at C2.

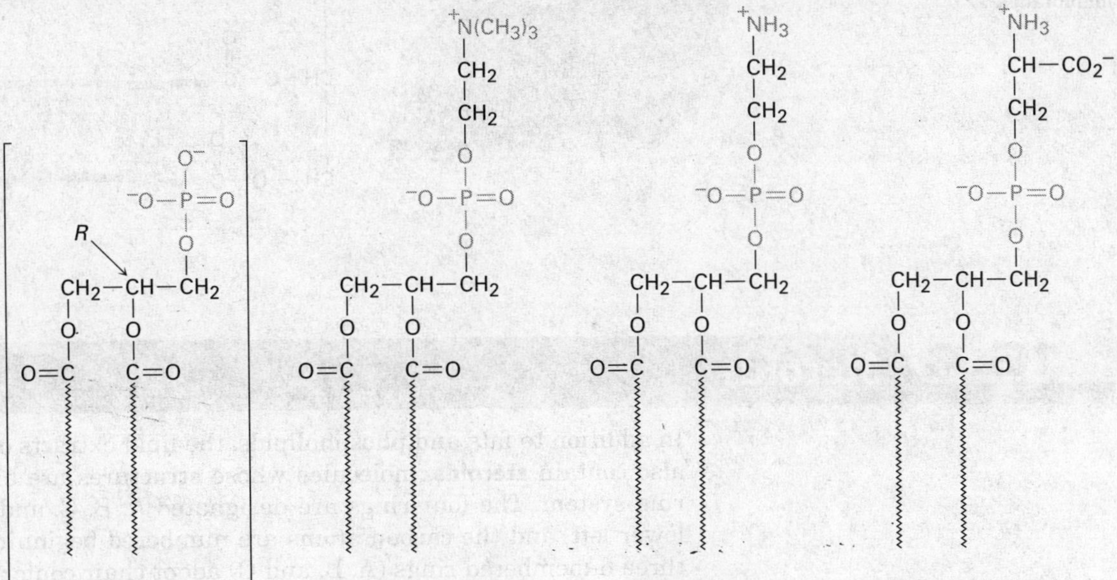

Phosphatidic acid **Phosphatidylcholine** **Phosphatidylethanolamine** **Phosphatidylserine**

Sphingomyelins are the second major group of phospholipids. These compounds have sphingosine or a related dihydroxyamine as their backbone and are particularly abundant in brain and nerve tissue, where they are a major constituent of the coating around nerve fibers.

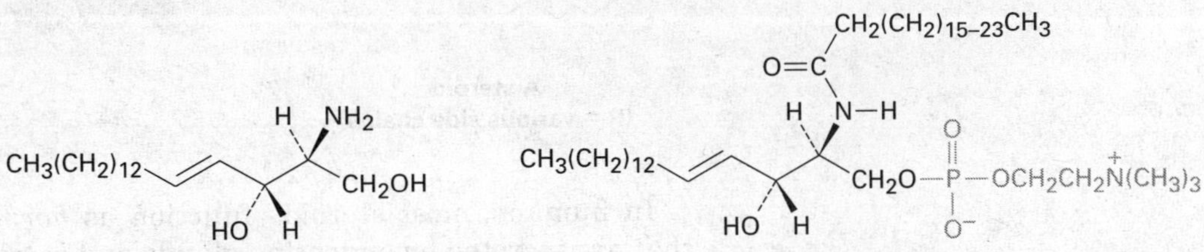

Sphingosine **A sphingomyelin**

Phospholipids are found widely in both plant and animal tissues and make up approximately 50% to 60% of cell membranes. Because they are like soaps in having a long, nonpolar hydrocarbon tail bound to a polar ionic head, phospholipids in the cell membrane organize into a **lipid bilayer** about 5.0 nm (50 Å) thick. As shown in Figure 16.2, the nonpolar tails aggregate in the center of the bilayer in much the same way that soap tails aggregate in the center of a micelle. This bilayer serves as an effective barrier to the passage of water, ions, and other components into and out of cells.

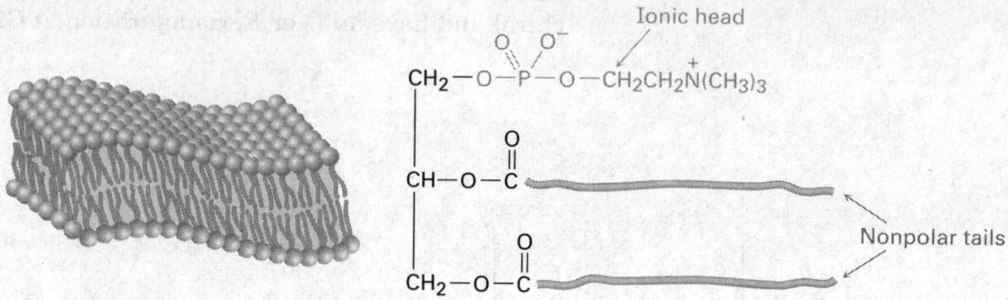

Figure 16.2 Aggregation of glycerophospholipids into the lipid bilayer that composes cell membranes.

16.4 Steroids

In addition to fats and phospholipids, the lipid extracts of plants and animals also contain **steroids**, molecules whose structures are based on a tetracyclic ring system. The four rings are designated A, B, C, and D, beginning at the lower left, and the carbon atoms are numbered beginning in the A ring. The three 6-membered rings (A, B, and C) adopt chair conformations but are constrained by their rigid geometry from undergoing the usual cyclohexane ring-flips (Section 2.11).

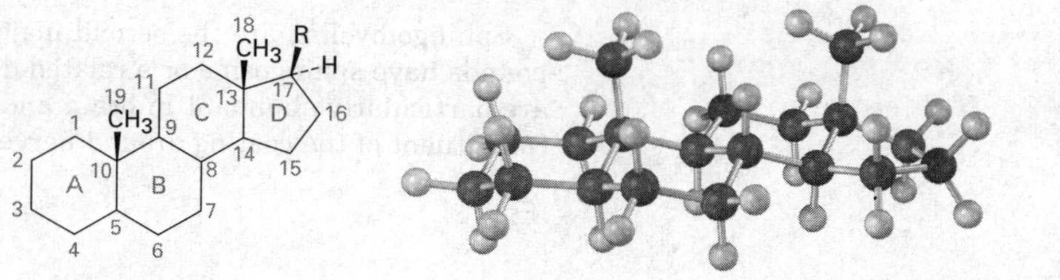

A steroid
(R = various side chains)

In humans, most steroids function as *hormones,* chemical messengers that are secreted by endocrine glands and carried through the bloodstream to target tissues. There are two main classes of steroid hormones: the *sex hormones,* which control maturation, tissue growth, and reproduction, and the *adrenocortical hormones,* which regulate a variety of metabolic processes.

Sex Hormones

Testosterone and androsterone are the two most important male sex hormones, or *androgens.* Androgens are responsible for the development of male secondary sex characteristics during puberty and for promoting tissue and muscle growth. Both are synthesized in the testes from cholesterol. Androstenedione is another minor hormone that has received particular attention because of its use by prominent athletes.

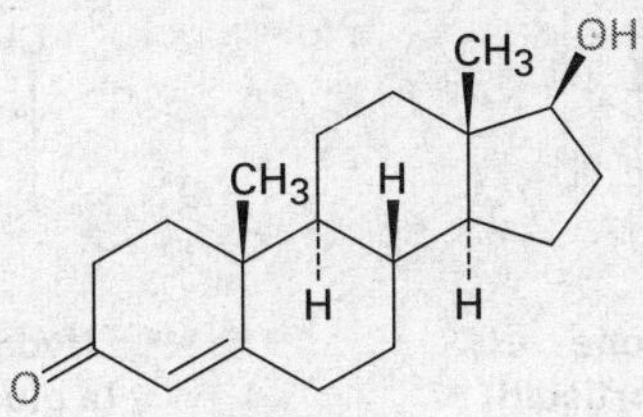

Testosterone

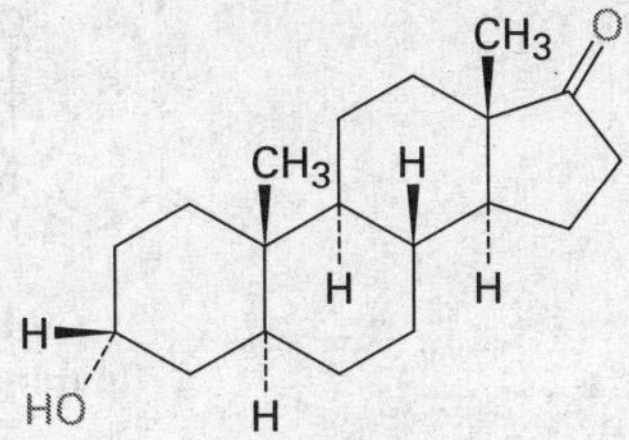

Androsterone

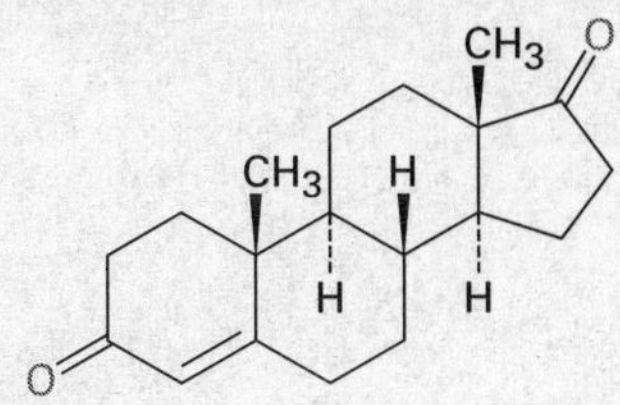

Androstenedione

(Androgens)

Estrone and estradiol are the two most important female sex hormones, or *estrogens.* Synthesized in the ovaries from testosterone, estrogenic hormones are responsible for the development of female secondary sex characteristics and for regulation of the menstrual cycle. Note that both have a benzene-like aromatic A ring. In addition, another kind of sex hormone, called a *progestin,* is essential for preparing the uterus for implantation of a fertilized ovum during pregnancy. *Progesterone* is the most important progestin.

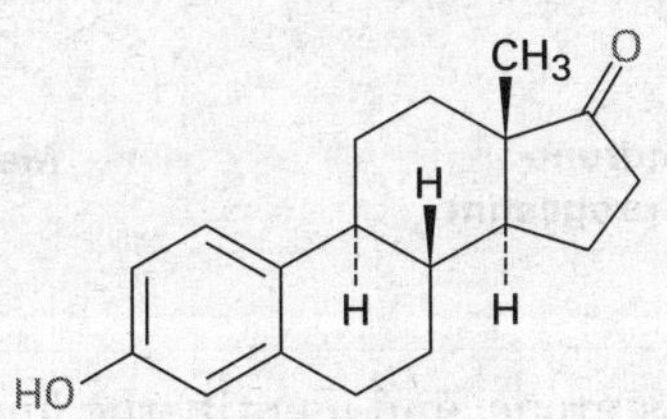

Estrone

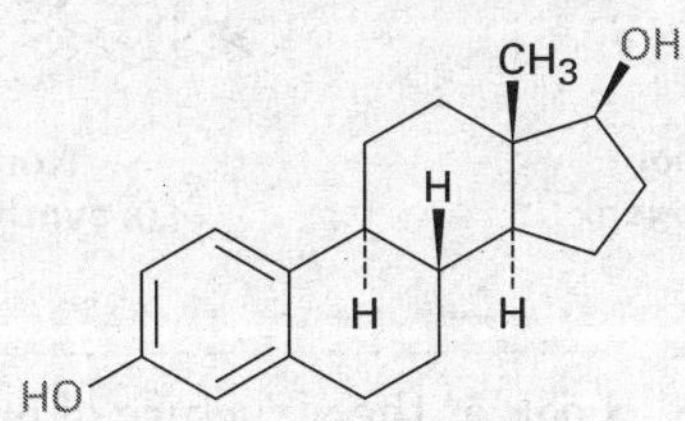

Estradiol

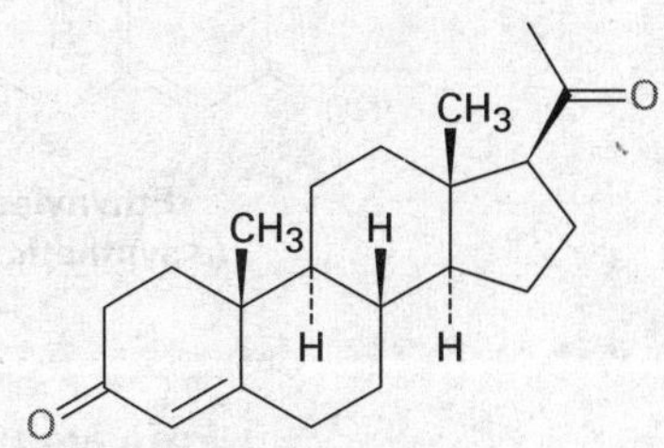

Progesterone
(a progestin)

(Estrogens)

Adrenocortical Hormones

Adrenocortical steroids are secreted by the adrenal glands, small organs located near the upper end of each kidney. There are two types of adrenocortical steroids, called *mineralocorticoids* and *glucocorticoids.* Mineralocorticoids, such as aldosterone, control tissue swelling by regulating cellular salt balance between Na^+ and K^+. Glucocorticoids, such as hydrocortisone, are involved in the regulation of glucose metabolism and in the control of inflammation. Glucocorticoid ointments are widely used to bring down the swelling from exposure to poison oak or poison ivy.

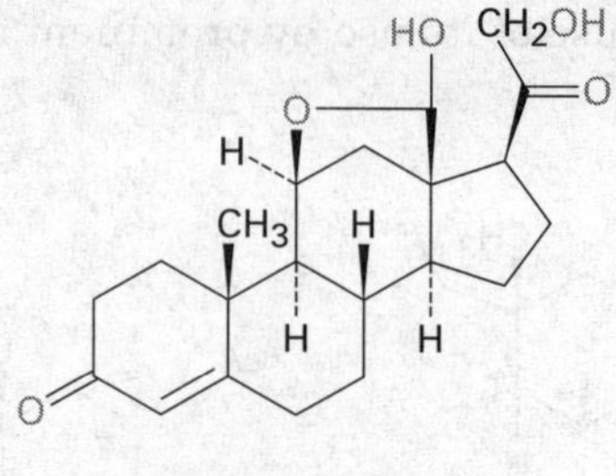

Aldosterone (a mineralocorticoid)

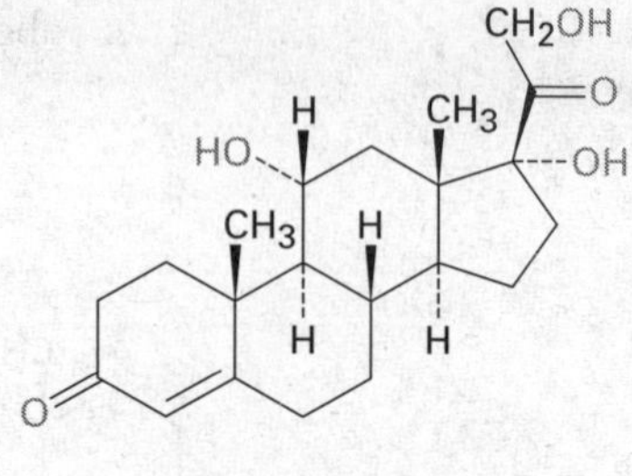

Hydrocortisone (a glucocorticoid)

Synthetic Steroids

In addition to the many hundreds of steroids isolated from plants and animals, thousands more have been synthesized in pharmaceutical laboratories in a search for new drugs. Among the best-known synthetic steroids are oral contraceptives and anabolic agents. Most birth-control pills are a mixture of two compounds, a synthetic estrogen, such as ethynylestradiol, and a synthetic progestin, such as norethindrone. Anabolic steroids, such as methandrostenolone (Dianabol), are synthetic androgens that mimic the tissue-building effects of natural testosterone.

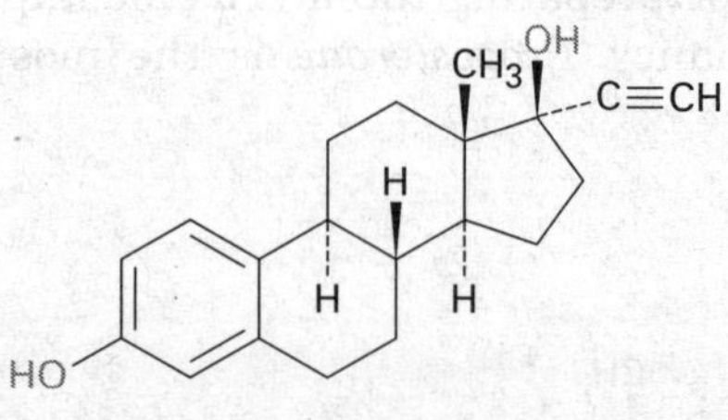

Ethynylestradiol (a synthetic estrogen)

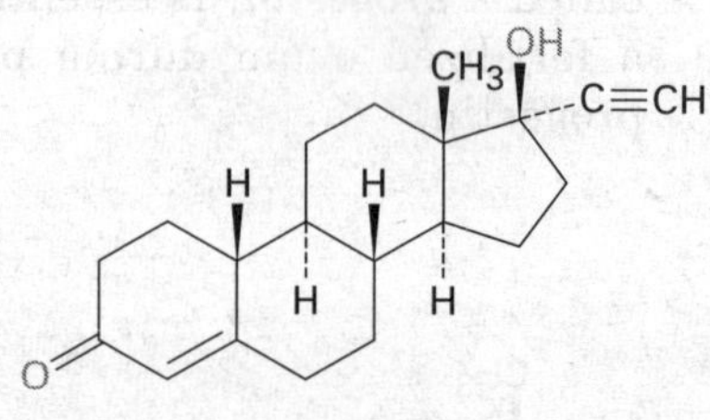

Norethindrone (a synthetic progestin)

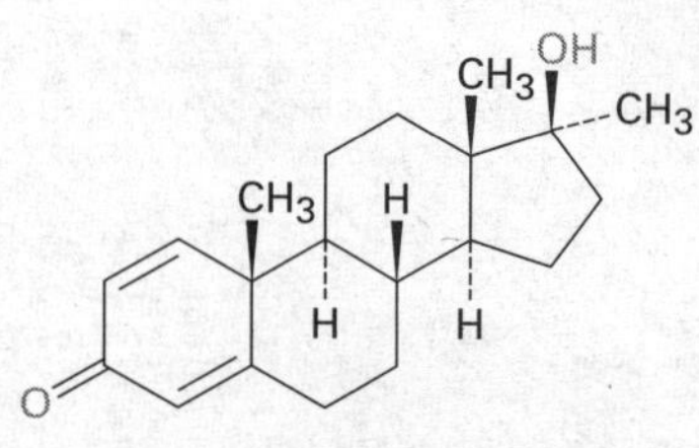

Methandrostenolone (Dianabol)

Problem 16.6 Look at the structure of progesterone, and identify the functional groups in the molecule.

Problem 16.7 Look at the structures of estradiol and ethynylestradiol, and point out the differences. What common structural feature do they share that makes both estrogens?

16.5 Nucleic Acids and Nucleotides

Nucleic acids are the last of the four major classes of biomolecules we'll consider. So much has been written and spoken about DNA in the media that you probably know the basics of DNA replication and transcription. The field is moving very rapidly, however, so there's probably a lot you may not be familiar with.

The **nucleic acids**, **deoxyribonucleic acid (DNA)** and **ribonucleic acid (RNA)**, are the carriers and processors of a cell's genetic information. Coded in a cell's DNA is all the information that determines the nature of the cell, controls the cell's growth and division, and directs biosynthesis of the enzymes and other proteins required for cellular functions. In addition to nucleic acids themselves, nucleic acid derivatives such as ATP are involved in many biochemical pathways, and several important coenzymes, including NAD^+, FAD, and coenzyme A, have nucleic acid components. See Table 15.3 on pages 526 and 527 for the structures.

Just as proteins are biopolymers made of amino acids, nucleic acids are biopolymers made of **nucleotides** joined together to form a long chain. Each nucleotide is composed of a **nucleoside** bonded to a phosphate group, and each nucleoside is composed of an aldopentose sugar linked through its anomeric carbon to the nitrogen atom of a heterocyclic amine base (Section 12.6).

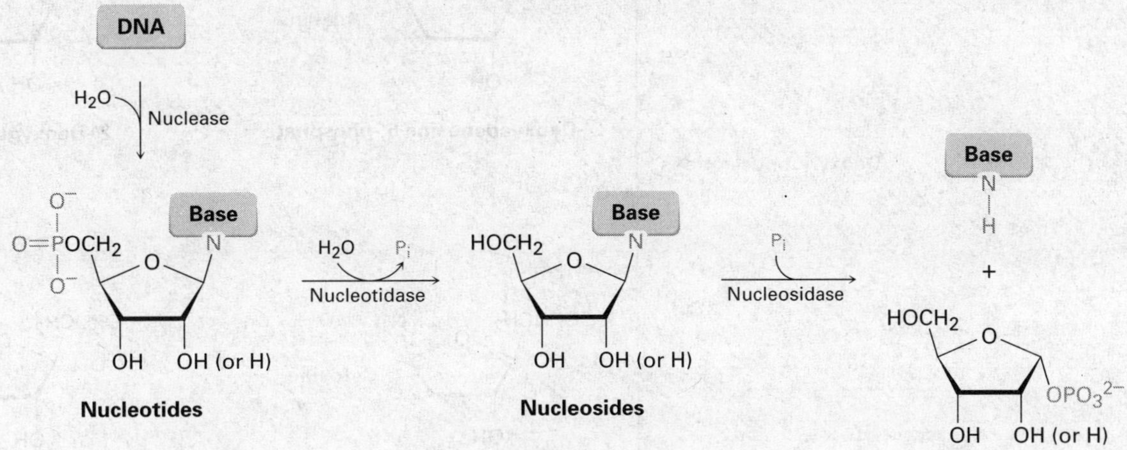

The sugar component in RNA is ribose, and the sugar in DNA is 2′-deoxyribose. (In naming and numbering nucleotides, numbers with a prime superscript refer to positions on the sugar, and numbers without a prime superscript refer to positions on the heterocyclic base. Thus, the prefix 2′-deoxy indicates that oxygen is missing from C2′ of ribose.) DNA contains four different amine bases, two substituted purines (adenine and guanine) and two substituted pyrimidines (cytosine and thymine). Adenine, guanine, and cytosine also occur in RNA, but thymine is replaced in RNA by a closely related pyrimidine base called uracil.

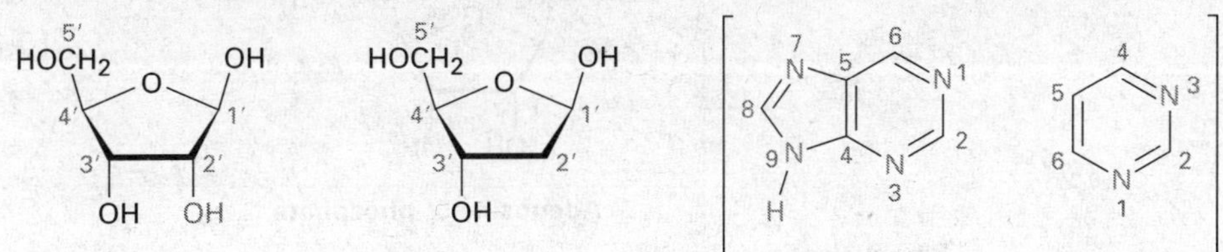

Ribose **2-Deoxyribose** **Purine** **Pyrimidine**

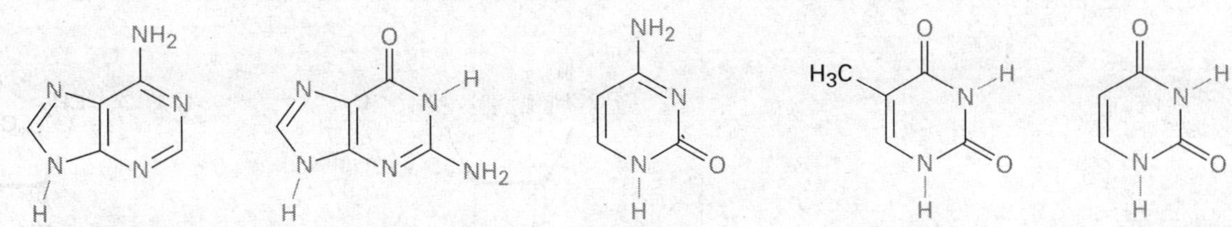

Adenine (A) DNA, RNA | **Guanine (G)** DNA, RNA | **Cytosine (C)** DNA, RNA | **Thymine (T)** DNA | **Uracil (U)** RNA

The structures of the four deoxyribonucleotides and the four ribonucleotides are shown in Figure 16.3. Although similar chemically, DNA and RNA differ dramatically in size. Molecules of DNA are enormous, containing as many as 245 million nucleotides and having molecular weights as high as 75 billion. Molecules of RNA, by contrast, are much smaller, containing as few as 21 nucleotides and having molecular weights as low as 7000.

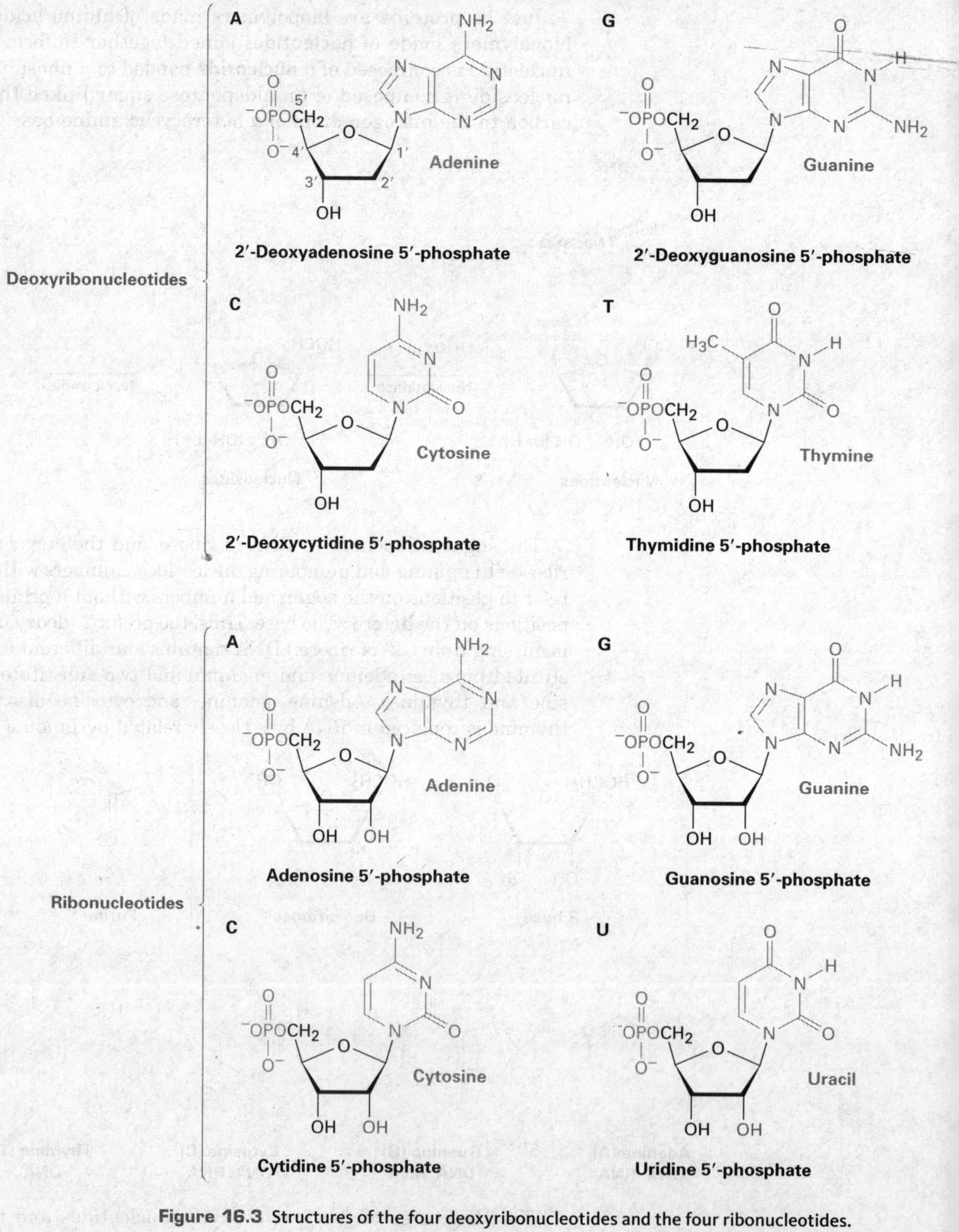

Figure 16.3 Structures of the four deoxyribonucleotides and the four ribonucleotides.

Nucleotides are linked together in DNA and RNA by *phosphodiester* bonds [RO—(PO_2^-)—OR′] between phosphate, the 5′ hydroxyl group on one nucleoside, and the 3′-hydroxyl group on another nucleoside. One end of the nucleic acid polymer has a free hydroxyl at C3′ (the *3′ end*), and the other end has a phosphate at C5′ (the *5′ end*). The sequence of nucleotides in a chain is described by starting at the 5′ end and identifying the bases in order of occurrence, using the abbreviations G, C, A, T (or U for RNA). Thus, a typical DNA sequence might be written as TAGGCT.

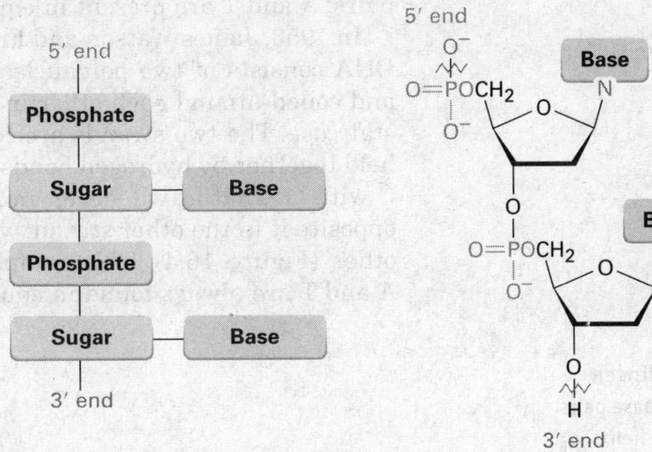

Worked Example 16.2 — Drawing the Structure of a Dinucleotide

Draw the full structure of the DNA dinucleotide CT.

Solution

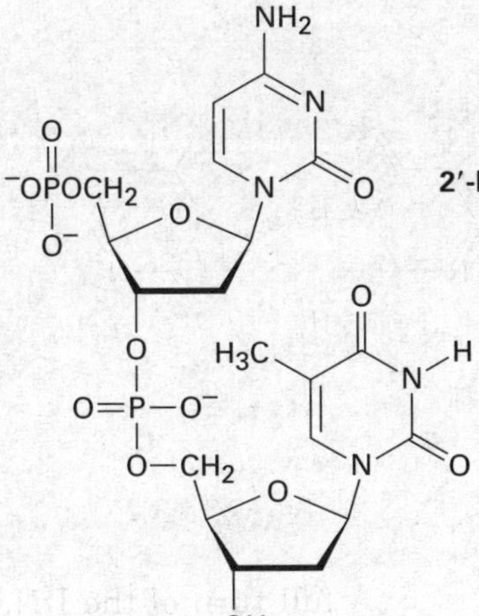

2′-Deoxycytidine (C)

Thymidine (T)

Problem 16.8 Draw the full structure of the DNA dinucleotide AG.

Problem 16.9 Draw the full structure of the RNA dinucleotide UA.

16.6 Base Pairing in DNA: The Watson–Crick Model

Samples of DNA isolated from different tissues of the same species have the same proportions of heterocyclic bases, but samples from different species often have greatly different proportions of bases. Human DNA, for example, contains about 30% each of A and T and about 20% each of G and C. The bacterium *Clostridium perfringens,* however, contains about 37% each of A and T and only 13% each of G and C. Note that in both examples, the bases occur in pairs: A and T are present in equal amounts, as are G and C. Why?

In 1953, James Watson and Francis Crick made their historic proposal that DNA consists of two polynucleotide strands, running in opposite directions and coiled around each other in a **double helix** like the handrails on a spiral staircase. The two strands are complementary rather than identical and are held together by hydrogen bonds between specific pairs of bases, A with T and C with G. That is, whenever an A base occurs in one strand, a T base occurs opposite it in the other strand; when a C base occurs in one, a G occurs in the other (Figure 16.4). This complementary base pairing thus explains why A and T are always found in equal amounts, as are G and C.

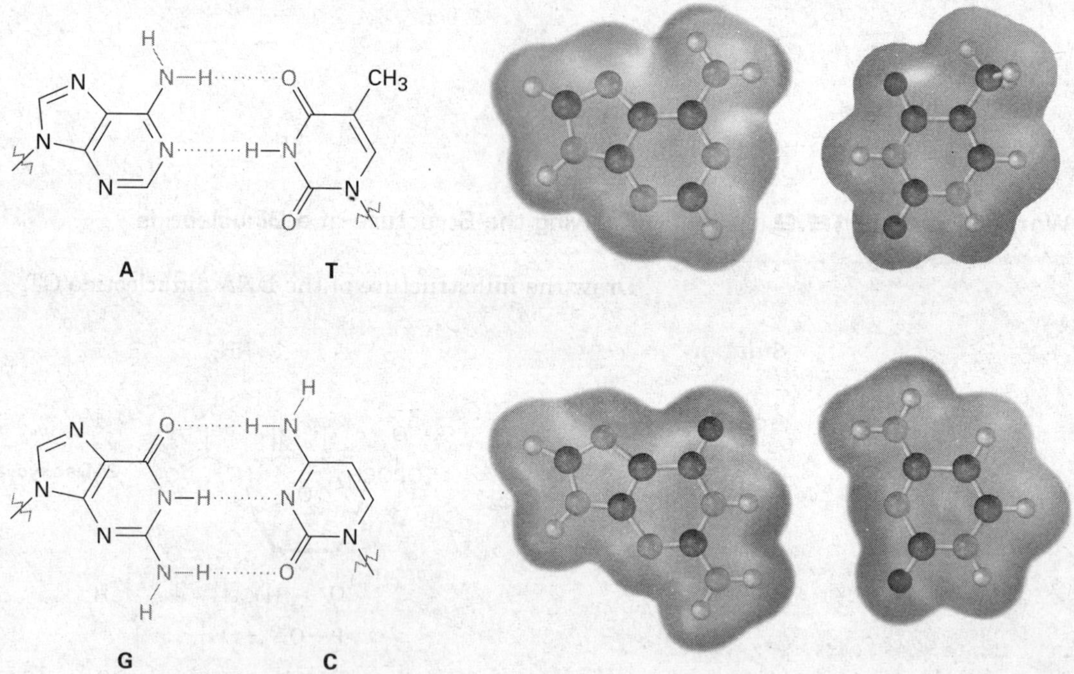

Figure 16.4 Hydrogen-bonding between base pairs in the DNA double helix. Electrostatic potential maps show that the faces of the bases are relatively neutral (green), while the edges have electron-poor (blue) and electron-rich (red) regions. Pairing G with C and A with T brings together the oppositely charged regions.

A full turn of the DNA double helix is shown in Figure 16.5. The helix is 20 Å wide, there are 10 base pairs per turn, and each turn is 34 Å in length. Notice that the two strands of the double helix coil in such a way that two kinds of "grooves" result, a *major groove* 12 Å wide and a *minor groove* 6 Å wide. The major groove is slightly deeper than the minor groove, and both are lined by hydrogen bond donors and acceptors. As a result, a variety of flat, polycyclic aromatic molecules are able to slip sideways, or *intercalate,* between the stacked bases. Many cancer-causing and cancer-preventing agents function by interacting with DNA in this way.

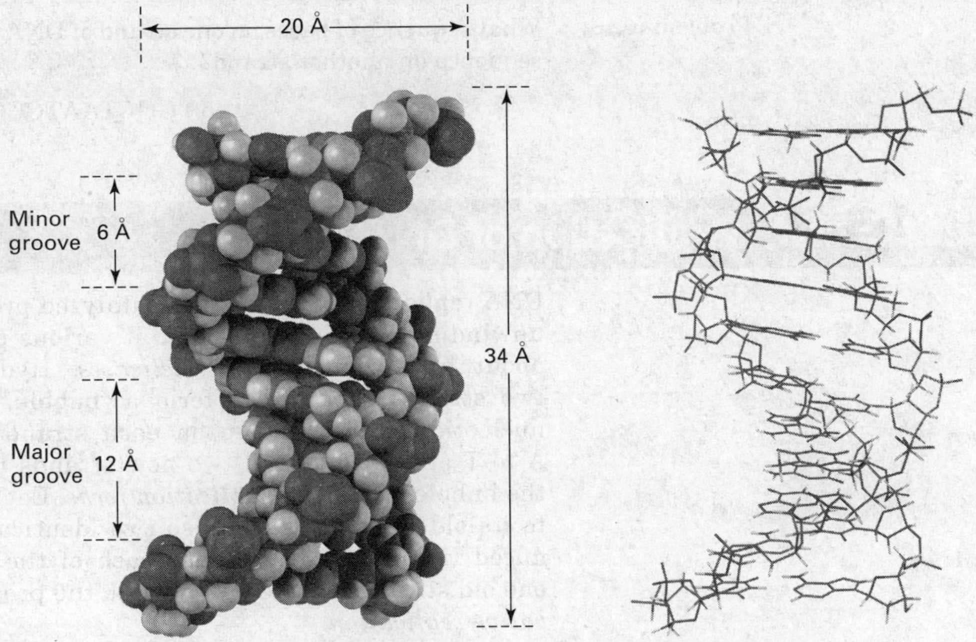

Figure 16.5 A turn of the DNA double helix in both space-filling and wire-frame formats. The sugar–phosphate backbone runs along the outside of the helix, and the amine bases hydrogen bond to one another on the inside. Both major and minor grooves are visible.

An organism's genetic information is stored as a sequence of deoxyribonucleotides strung together in the DNA chain. For the information to be preserved and passed on to future generations, a mechanism must exist for copying DNA. For the information to be used, a mechanism must exist for decoding the DNA message and implementing the instructions it contains. Three fundamental processes take place:

- **Replication**—the process by which identical copies of DNA are made so that information can be preserved and handed down to offspring.
- **Transcription**—the process by which the genetic messages are read and carried out of the cell nucleus to ribosomes, where protein synthesis occurs.
- **Translation**—the process by which the genetic messages are decoded and used to synthesize proteins.

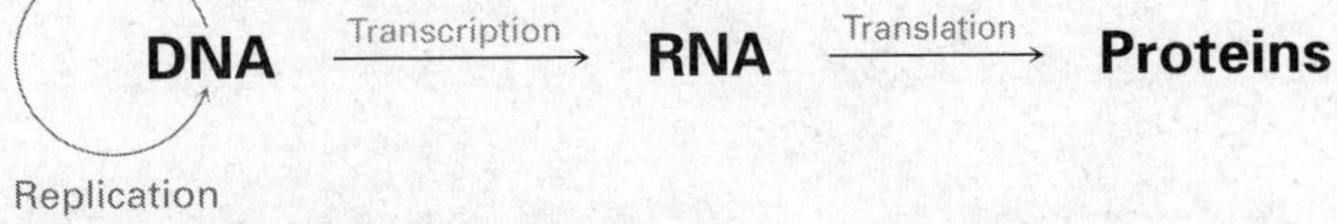

Worked Example 16.3 **Complementary DNA Sequences**

What sequence of bases on one strand of DNA is complementary to the sequence TATGCAT on another strand?

Strategy Remember that A and G form complementary pairs with T and C, respectively, and then go through the sequence replacing A by T, G by C, T by A, and C by G.

Solution Original: (5′) TATGCAT (3′)
Complement: (3′) ATACGTA (5′) or (5′) ATGCATA (3′)

Problem 16.10 What sequence of bases on one strand of DNA is complementary to the following sequence on another strand?

(5′) GGCTAATCCGT (3′)

16.7 Replication of DNA

DNA **replication** is an enzyme-catalyzed process that begins with a partial unwinding of the double helix at various points along the chain, brought about by enzymes called *helicases.* Hydrogen bonds are broken, the two strands separate to form a "bubble," and bases are exposed. New nucleotides then line up on each strand in a complementary manner, A to T and G to C, and two new strands begin to grow from the ends of the bubble, called the *replication forks.* Each new strand is complementary to its old template strand, so two identical DNA double helices are produced (Figure 16.6). Because each of the new DNA molecules contains one old strand and one new strand, the process is described as *semiconservative replication.*

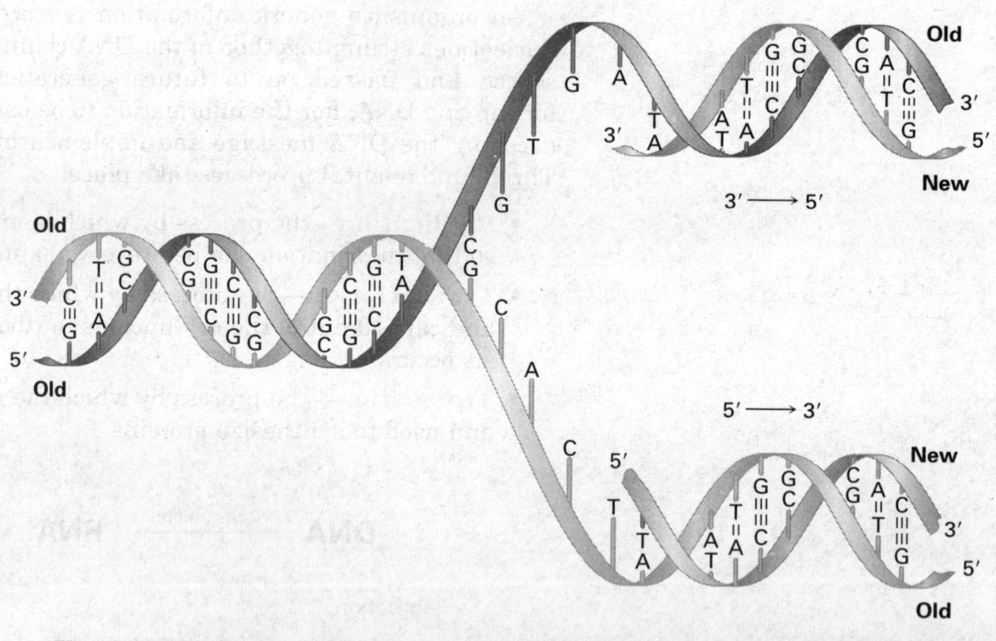

Figure 16.6 A representation of semiconservative DNA replication. The original double-stranded DNA partially unwinds, bases are exposed, nucleotides line up on each strand in a complementary manner, and two new strands begin to grow. Both strands are synthesized in the same 5′ → 3′ direction, one continuously and one in fragments.

Addition of nucleotides to a growing chain takes place in the 5′ → 3′ direction and is catalyzed by DNA polymerase. The key step is the addition of a nucleoside 5′-triphosphate to the free 3′-hydroxyl group of the growing chain, with loss of a diphosphate leaving group.

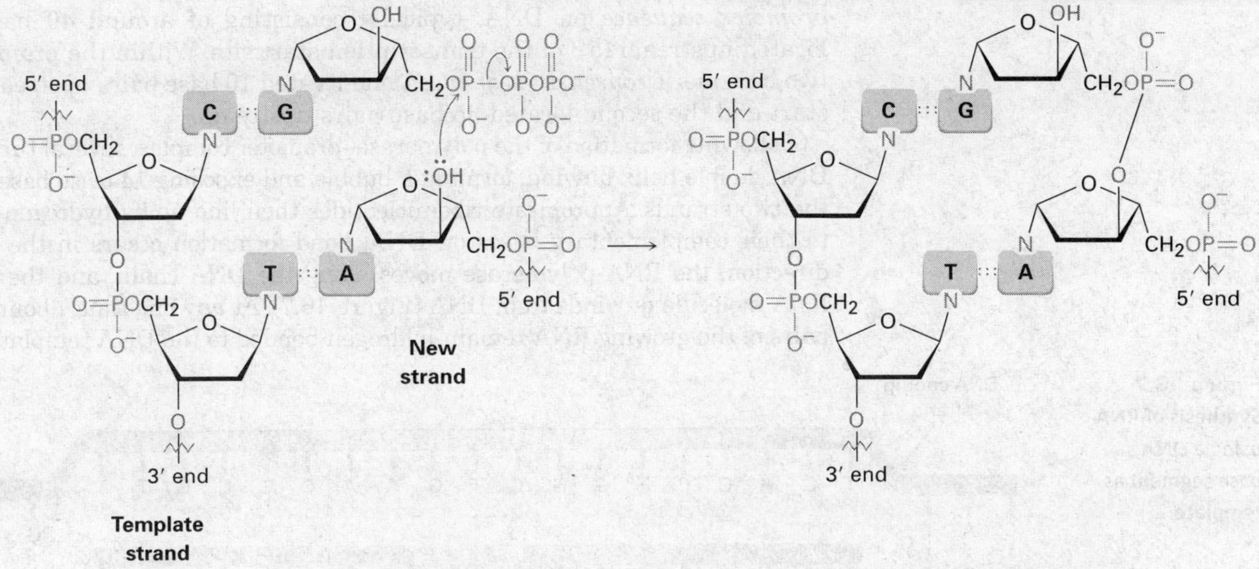

The magnitude of the replication process is staggering. The nucleus of every human cell contains two copies of 22 chromosomes plus an additional 2 sex chromosomes, for a total of 46. Each chromosome consists of one very large DNA molecule, and the sum of the DNA in each of the two sets of chromosomes is estimated to be 3.0 billion base pairs, or 6.0 billion nucleotides. Despite the size of these enormous molecules, their base sequence is faithfully copied during replication. The entire copying process takes only a few hours and, after proofreading and repair, an error gets through only about once each 10 to 100 billion bases.

16.8 Transcription of DNA

As noted previously, RNA is structurally similar to DNA but contains ribose rather than deoxyribose and uracil rather than thymine. There are three major kinds of RNA, each of which serves a specific purpose. In addition, there are a number of small RNAs that appear to control a wide variety of important cellular functions. All RNA molecules are much smaller than DNA, and all remain single-stranded rather than double-stranded.

- **Messenger RNA (mRNA)** carries genetic messages from DNA to ribosomes, small particles in the cytoplasm of a cell where protein synthesis occurs.
- **Ribosomal RNA (rRNA)** complexed with protein provides the physical makeup of the ribosomes.
- **Transfer RNA (tRNA)** transports amino acids to the ribosomes, where they are joined together to make proteins.
- **Small RNAs**, also called *functional RNAs,* have a variety of functions within the cell, including silencing transcription and catalyzing chemical modifications of other RNA molecules.

The genetic information in DNA is contained in segments called *genes,* each of which consists of a specific nucleotide sequence that encodes a specific protein. The conversion of that information from DNA into proteins begins in the nucleus of cells with the synthesis of mRNA by **transcription** of DNA. In bacteria, the process begins when RNA polymerase recognizes and binds to a *promoter sequence* on DNA, typically consisting of around 40 base pairs located upstream (5′) of the transcription start site. Within the promoter are two hexameric *consensus sequences,* one located 10 base pairs upstream of the start and the second located 35 base pairs upstream.

Following formation of the polymerase–promoter complex, several turns of the DNA double helix unwind, forming a bubble and exposing 14 or so base pairs of the two strands. Appropriate ribonucleotides then line up by hydrogen-bonding to their complementary bases on DNA, bond formation occurs in the 5′ → 3′ direction, the RNA polymerase moves along the DNA chain, and the growing RNA molecule unwinds from DNA (Figure 16.7). At any one time, about 12 base pairs of the growing RNA remain hydrogen-bonded to the DNA template.

Figure 16.7
Synthesis of RNA using a DNA base segment as template.

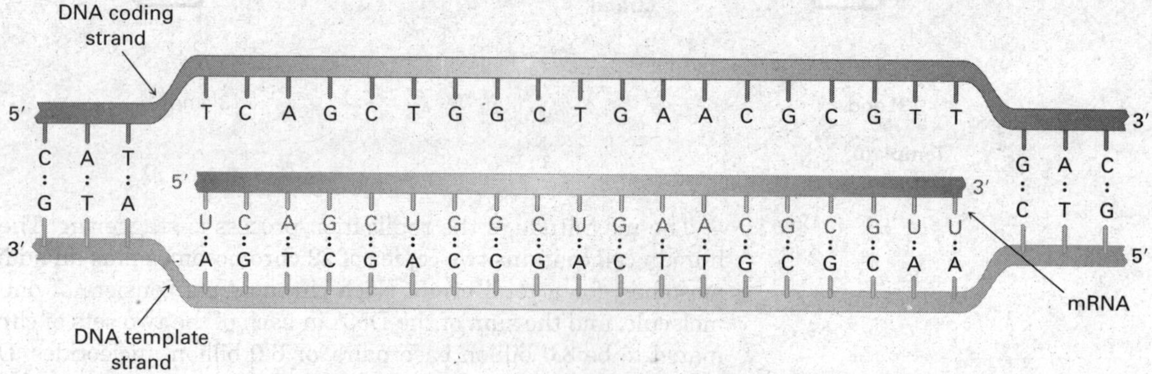

Unlike what happens in DNA replication, where both strands are copied, only one of the two DNA strands is transcribed into mRNA. The DNA strand that contains the gene is often called the **sense strand**, or *coding strand,* and the DNA strand that gets transcribed to give RNA is called the **antisense strand**, or *noncoding strand.* Because the sense strand and the antisense strand in DNA are complementary, and because the DNA antisense strand and the newly formed RNA strand are also complementary, *the RNA molecule produced during transcription is a copy of the DNA sense strand.* That is, the complement of the complement is the same as the original. The only difference is that the RNA molecule has a U everywhere the DNA sense strand has a T.

Another part of the picture in vertebrates and flowering plants is that genes are often not continuous segments of the DNA chain. Instead, a gene will begin in one small section of DNA called an *exon,* then be interrupted by a noncoding section called an *intron,* and then take up again farther down the chain in another exon. The final mRNA molecule results only after the noncoded sections are cut out of the transcribed mRNA and the remaining pieces are joined together by spliceosomes. The gene for triose phosphate isomerase in maize, for instance, contains eight noncoding introns accounting for approximately 70% of the DNA base pairs and nine coding exons accounting for only 30% of the base pairs.

Worked Example 16.4 RNA/DNA Complementary Sequences

What RNA base sequence is complementary to the following DNA base sequence?

(5′) TAAGCCGTG (3′)

Strategy Go through the sequence replacing A by U, G by C, T by A, and C by G.

Solution Original DNA: (5′) TAAGCCGTG (3′)
Complementary RNA: (3′) AUUCGGCAC (5′)

Problem 16.11 Show how uracil can form strong hydrogen bonds to adenine, just as thymine can.

Problem 16.12 What RNA base sequence is complementary to the following DNA base sequence?

(5′) GATTACCGTA (3′)

Problem 16.13 From what DNA base sequence was the following RNA sequence transcribed?

(5′) UUCGCAGAGU (3′)

16.9 Translation of RNA: Protein Biosynthesis

The primary cellular function of mRNA is to direct biosynthesis of the thousands of diverse peptides and proteins required by an organism—as many as 500,000 in a human. The mechanics of protein biosynthesis take place on ribosomes, small granular particles in the cytoplasm of a cell that consist of about 60% ribosomal RNA and 40% protein.

The specific ribonucleotide sequence in mRNA forms a message that determines the order in which amino acid residues are to be joined. Each "word," or **codon**, along the mRNA chain consists of a sequence of three ribonucleotides that is specific for a given amino acid. For example, the series UUC on mRNA is a codon directing incorporation of the amino acid phenylalanine into the growing protein. Of the $4^3 = 64$ possible triplets of the four bases in RNA, 61 code for specific amino acids and 3 code for chain termination. Table 16.3 shows the meaning of each codon.

Table 16.3 Codon Assignments of Base Triplets

		Third base (3′ end)			
First base (5′ end)	**Second base**	**U**	**C**	**A**	**G**
U	U	Phe	Phe	Leu	Leu
	C	Ser	Ser	Ser	Ser
	A	Tyr	Tyr	Stop	Stop
	G	Cys	Cys	Stop	Trp
C	U	Leu	Leu	Leu	Leu
	C	Pro	Pro	Pro	Pro
	A	His	His	Gln	Gln
	G	Arg	Arg	Arg	Arg
A	U	Ile	Ile	Ile	Met
	C	Thr	Thr	Thr	Thr
	A	Asn	Asn	Lys	Lys
	G	Ser	Ser	Arg	Arg
G	U	Val	Val	Val	Val
	C	Ala	Ala	Ala	Ala
	A	Asp	Asp	Glu	Glu
	G	Gly	Gly	Gly	Gly

The message embedded in mRNA is read by transfer RNA (tRNA) in a process called **translation**. There are 61 different tRNAs, one for each of the 61 codons that specifies an amino acid. A typical tRNA is single-stranded and has roughly the shape of a cloverleaf, as shown in Figure 16.8. It consists of about 70 to 100 ribonucleotides and is bonded to a specific amino acid by an ester linkage through the 3′ hydroxyl on ribose at the 3′ end of the tRNA. Each tRNA also contains on its middle leaf a segment called an **anticodon**, a sequence of three ribonucleotides complementary to the codon sequence. For example, the codon sequence UUC present on mRNA is read by a phenylalanine-bearing tRNA having the complementary anticodon base sequence GAA. [Remember that nucleotide sequences are written in the 5′ → 3′ direction, so the sequence in an anticodon must be reversed. That is, the complement to (5′) UUC (3′) is (3′) AAG (5′), which is written as (5′) GAA (3′).]

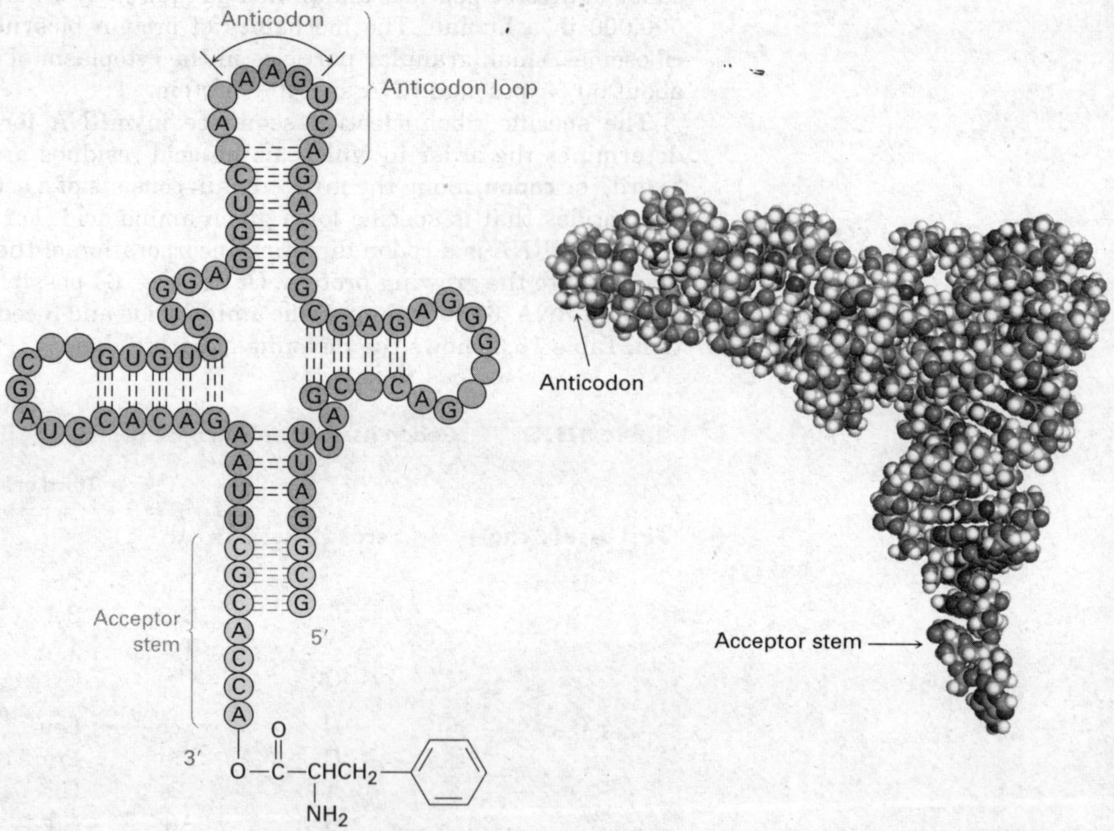

Figure 16.8 Structure of a tRNA molecule. The tRNA molecule is roughly cloverleaf-shaped and contains an anticodon triplet on one "leaf" and an amino acid unit attached covalently at its 3′ end. The example shown is a yeast tRNA that codes for phenylalanine. The nucleotides not specifically identified are chemically modified analogs of the four common ribonucleotides.

As each successive codon on mRNA is read, different tRNAs bring the correct amino acids into position for enzyme-mediated transfer to the growing peptide. When synthesis of the proper protein is completed, a "stop" codon signals the end and the protein is released from the ribosome. The process is illustrated in Figure 16.9.

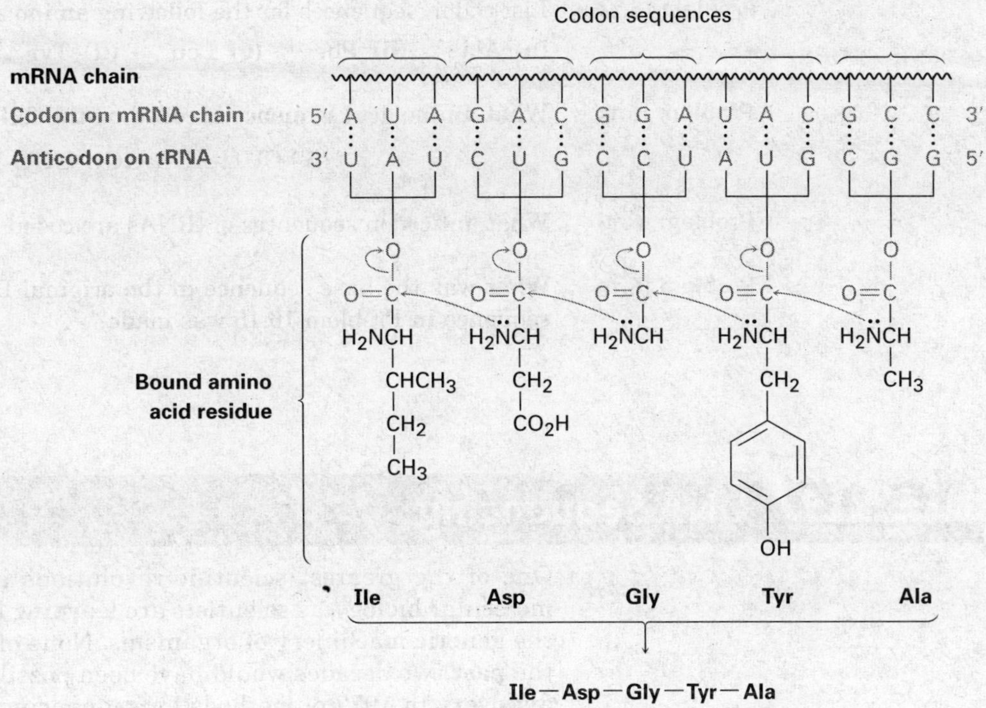

Figure 16.9 A representation of protein biosynthesis. The codon base sequences on mRNA are read by tRNAs containing complementary anticodon base sequences. Transfer RNAs assemble the proper amino acids into position for incorporation into the growing peptide.

Worked Example 16.5 Codon Sequences for Amino Acids

Give a codon sequence for valine.

Solution According to Table 16.3, there are four codons for valine: GUU, GUC, GUA, and GUG.

Worked Example 16.6 Finding the Amino Acid Sequence Transcribed from DNA

What amino acid sequence is coded by the following segment of a DNA sense strand?

(5′) CTA-ACT-AGC-GGG-TCG-CCG (3′)

Strategy The mRNA produced during translation is a copy of the DNA sense strand, with each T replaced by U. Thus, the mRNA has the sequence

(5′) CUA-ACU-AGC-GGG-UCG-CCG (3′)

Each set of three bases forms a codon, whose meaning can be found in Table 16.3.

Solution Leu-Thr-Ser-Gly-Ser-Pro

Problem 16.14 List codon sequences for the following amino acids:

(a) Ala (b) Phe (c) Leu (d) Tyr

Problem 16.15 What amino acid sequence is coded by the following mRNA base sequence?

(5′) CUU-AUG-GCU-UGG-CCC-UAA (3′)

Problem 16.16 What anticodon sequences of tRNAs are coded by the mRNA in Problem 16.15?

Problem 16.17 What was the base sequence in the original DNA strand on which the mRNA sequence in Problem 16.15 was made?

16.10 DNA Sequencing

One of the greatest scientific revolutions in history is now underway in molecular biology, as scientists are learning how to manipulate and harness the genetic machinery of organisms. None of the extraordinary advances of the past two decades would have been possible, however, were it not for the discovery in 1977 of methods for sequencing immense DNA chains.

The first step in DNA sequencing is to cleave the enormous chain at known points to produce smaller, more manageable pieces, a task accomplished by the use of *restriction endonucleases.* Each different restriction enzyme, of which more than 3500 are known and approximately 200 are commercially available, cleaves a DNA molecule at a point in the chain where a specific base sequence occurs. For example, the restriction enzyme *Alu*I cleaves between G and C in the four-base sequence AG-CT. Note that the sequence is a *palindrome,* meaning that the *sequence* (5′)-AGCT-(3′) is the same as its *complement* (3′)-TCGA-(5′) when both are read in the same 5′ → 3′ direction. The same is true for other restriction endonucleases.

If the original DNA molecule is cut with another restriction enzyme that has a different specificity for cleavage, still other segments are produced whose sequences partially overlap those produced by the first enzyme. Sequencing of all the segments, followed by identification of the overlapping regions, allows complete DNA sequencing.

Two methods of DNA sequencing are available. The *Maxam–Gilbert method* uses chemical techniques, while the **Sanger dideoxy method** uses enzymatic reactions. The Sanger method is the more commonly used of the two and is the method responsible for sequencing the entire human genome of 3.0 billion base pairs. In commercial sequencing instruments, the dideoxy method begins with a mixture of the following:

- The restriction fragment to be sequenced
- A small piece of DNA called a *primer,* whose sequence is complementary to that on the 3′ end of the restriction fragment
- The four 2′-deoxyribonucleoside triphosphates (dNTPs)
- Very small amounts of the four 2′,3′-*dideoxy*ribonucleoside triphosphates (ddNTPs), each of which is labeled with a fluorescent dye of a

different color. (A 2′,3′-dideoxyribonucleoside triphosphate is one in which both 2′ and 3′ –OH groups are missing from ribose.)

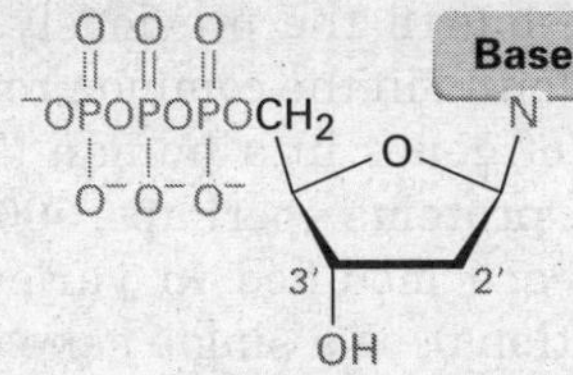

A 2′-deoxyribonucleoside triphosphate (dNTP)

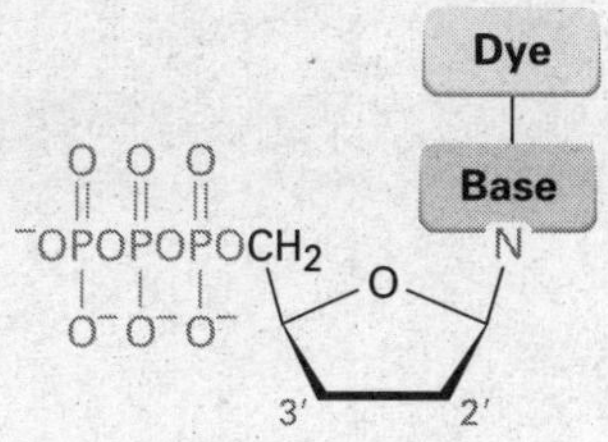

A 2′,3′-dideoxyribonucleoside triphosphate (ddNTP)

DNA polymerase is added to the mixture, and a strand of DNA complementary to the restriction fragment begins to grow from the end of the primer. Most of the time, only normal deoxyribonucleotides are incorporated into the growing chain because of their much higher concentration in the mixture, but every so often, a dideoxyribonucleotide is incorporated. When that happens, DNA synthesis stops because the chain end no longer has a 3′-hydroxyl group for adding further nucleotides.

When reaction is complete, the product consists of a mixture of DNA fragments of all possible lengths, each terminated by one of the four dye-labeled dideoxyribonucleotides. This product mixture is then separated according to the size of the pieces by gel electrophoresis (Section 15.2), and the identity of the terminal dideoxyribonucleotide in each piece—and thus the sequence of the restriction fragment—is determined by noting the color with which it fluoresces. Figure 16.10 shows a typical result.

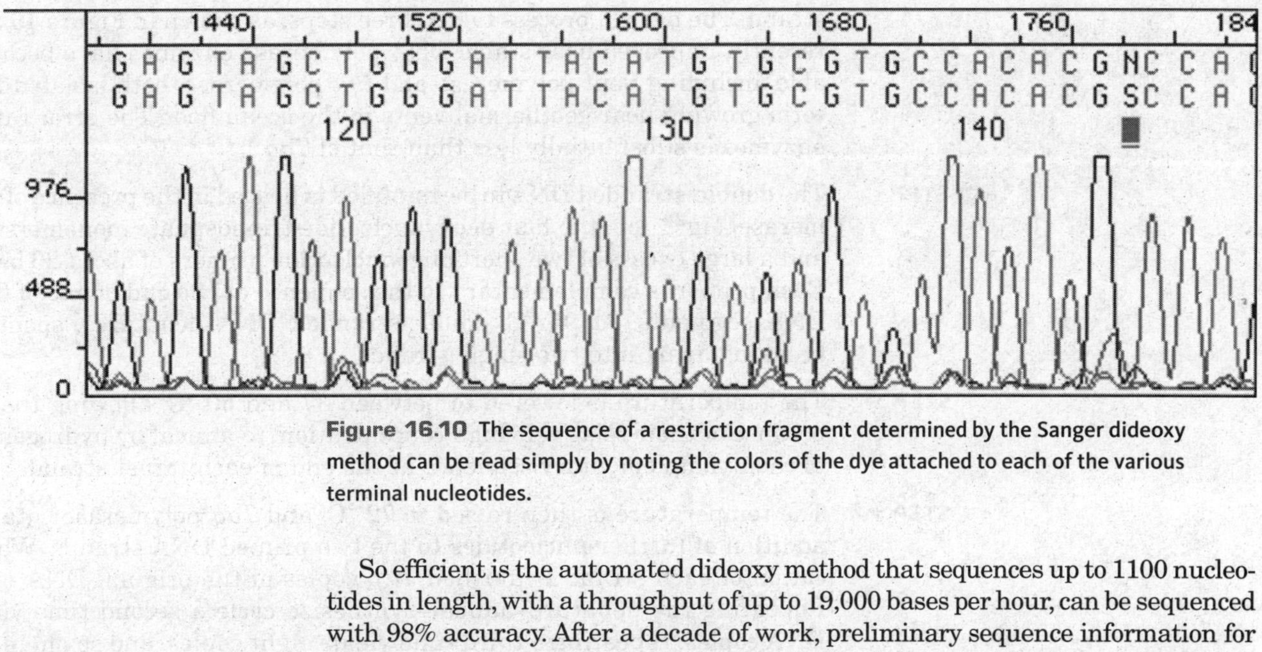

Figure 16.10 **The sequence of a restriction fragment determined by the Sanger dideoxy method can be read simply by noting the colors of the dye attached to each of the various terminal nucleotides.**

So efficient is the automated dideoxy method that sequences up to 1100 nucleotides in length, with a throughput of up to 19,000 bases per hour, can be sequenced with 98% accuracy. After a decade of work, preliminary sequence information for the entire human genome of 3.0 billion base pairs was announced early in 2001 and complete information was released in 2003. More recently, the genome sequencing of specific individuals, including that of James Watson, discoverer of the double helix, has been accomplished.

Remarkably, our genome appears to contain only about 21,000 genes, less than one-fourth the previously predicted number and only about twice the number found in the common roundworm. It's also interesting to note that the number of genes in a human (21,000) is much smaller than the number of kinds of proteins (perhaps 500,000). The discrepancy arises because most proteins are modified in various ways after translation (posttranslational modifications), so a single gene can ultimately give many different proteins.

16.11 The Polymerase Chain Reaction

It often happens that only a tiny amount of DNA can be obtained directly, as might occur at a crime scene, so methods for obtaining larger amounts are sometimes needed to carry out the sequencing and characterization. The invention of the **polymerase chain reaction (PCR)** in 1986 has been described as being to genes what Gutenberg's invention of the printing press was to the written word. Just as the printing press produces multiple copies of a book, PCR produces multiple copies of a given DNA sequence. Starting from less than 1 *picogram* of DNA with a chain length of 10,000 nucleotides (1 pg = 10^{-12} g; about 10^5 molecules), PCR makes it possible to obtain several micrograms (1 μg = 10^{-6} g; about 10^{11} molecules) in just a few hours.

The key to the polymerase chain reaction is *Taq* DNA polymerase, a heat-stable enzyme isolated from the thermophilic bacterium *Thermus aquaticus* found in a hot spring in Yellowstone National Park. *Taq* polymerase is able to take a single strand of DNA that has a short, primer segment of complementary chain at one end and then finish constructing the entire complementary strand. The overall process takes three steps, as shown in Figure 16.11. (More recently, improved heat-stable DNA polymerase enzymes have become available, including Vent polymerase and *Pfu* polymerase, both isolated from bacteria growing near geothermal vents in the ocean floor. The error rate of both enzymes is substantially less than that of *Taq*.)

STEP 1 The double-stranded DNA to be amplified is heated in the presence of *Taq* polymerase, Mg^{2+} ion, the four deoxynucleotide triphosphate monomers (dNTPs), and a large excess of two short oligonucleotide primers of about 20 bases each. Each primer is complementary to the sequence at the end of one of the target DNA segments. At 95 °C, double-stranded DNA denatures, spontaneously breaking apart into two single strands.

STEP 2 The temperature is lowered to between 37 and 50 °C, allowing the primers, because of their relatively high concentration, to anneal by hydrogen-bonding to their complementary sequence at the end of each target strand.

STEP 3 The temperature is then raised to 72 °C, and *Taq* polymerase catalyzes the addition of further nucleotides to the two primed DNA strands. When replication of each strand is finished, *two* copies of the original DNA now exist. Repeating the denature–anneal–synthesize cycle a second time yields four DNA copies, repeating a third time yields eight copies, and so on, in an exponential series.

PCR has been automated, and 30 or so cycles can be carried out in an hour, resulting in a theoretical amplification factor of 2^{30} ($\sim 10^9$). In practice, however, the efficiency of each cycle is less than 100%, and an experimental amplification of about 10^6 to 10^8 is routinely achieved for 30 cycles.

Figure 16.11 The polymerase chain reaction. Details are explained in the text.

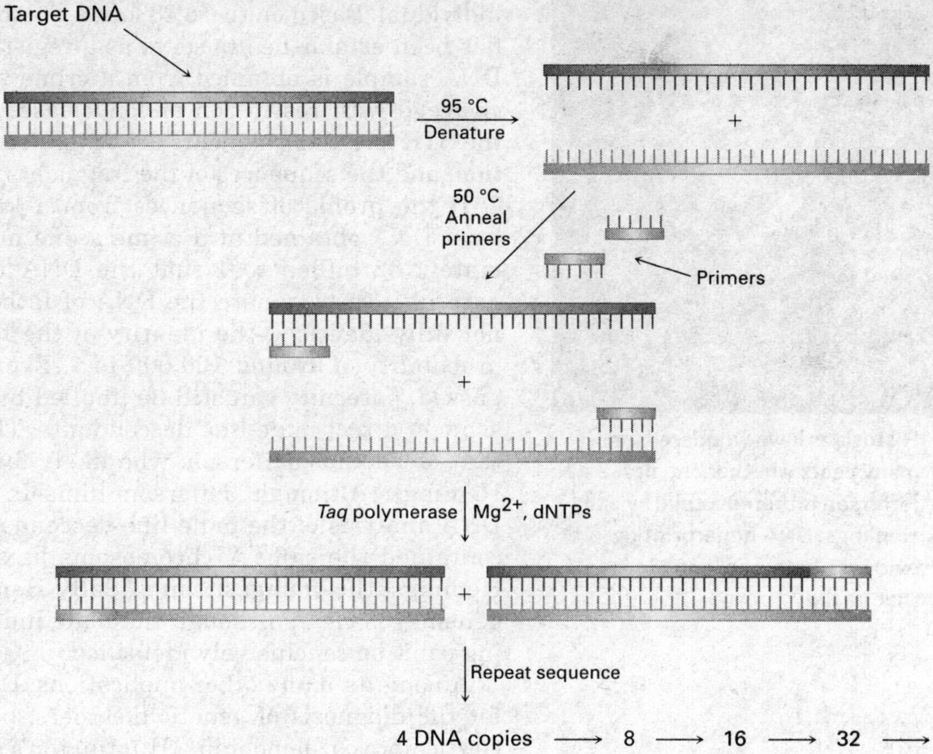

INTERLUDE

DNA Fingerprinting

The invention of DNA sequencing has affected society in many ways, few more dramatic than those stemming from the development of *DNA fingerprinting.* DNA fingerprinting arose from the discovery in 1984 that human genes contain short, repeating sequences of noncoding DNA, called *short tandem repeat* (STR) loci. Furthermore, the STR loci are slightly different for every individual, except identical twins. By sequencing these loci, a pattern unique to each person can be obtained.

Perhaps the most common and well-publicized use of DNA fingerprinting is that carried out by crime laboratories to link suspects to biological evidence—blood, hair follicles, skin, or semen—found at a crime scene. Thousands of court cases have now been decided based on DNA evidence.

continued

INTERLUDE

Rembrandt Peale/Getty Images

Historians have wondered for many years whether Thomas Jefferson fathered a child by Sally Hemings. DNA fingerprinting evidence obtained in 1998 suggests that he did.

For use in criminal cases, forensic laboratories in the United States have agreed on 13 core STR loci that are most accurate for identification of an individual. Based on these 13 loci, a Combined DNA Index System (CODIS) has been established to serve as a registry of convicted offenders. When a DNA sample is obtained from a crime scene, the sample is subjected to cleavage with restriction endonucleases to cut out fragments containing the STR loci, the fragments are amplified using the polymerase chain reaction, and the sequences of the fragments are determined.

If the profile of sequences from a known individual and the profile from DNA obtained at a crime scene match, the probability is approximately 82 billion to 1 that the DNA is from the same individual. In paternity cases, where the DNA of father and offspring are related but not fully identical, the identity of the father can be established with a probability of around 100,000 to 1. Even after several generations have passed, paternity can still be implied by DNA analysis of the Y chromosome of direct male-line descendants. The most well-known such case is that of Thomas Jefferson, who likely fathered a child by his slave Sally Hemings. Although Jefferson himself has no male-line descendants, DNA analysis of the male-line descendants of Jefferson's paternal uncle contained the same Y chromosome as a male-line descendant of Eston Hemings, the youngest son of Sally Hemings. Thus, a mixing of the two genomes is clear, although the male individual responsible for that mixing can't be conclusively identified.

Among its many other applications, DNA fingerprinting is widely used for the diagnosis of genetic disorders, both prenatally and in newborns. Cystic fibrosis, hemophilia, Huntington's disease, Tay–Sachs disease, sickle cell anemia, and thalassemia are among the many diseases that can be detected, enabling early treatment of an affected child. Furthermore, by studying the DNA fingerprints of relatives with a history of a particular disorder, it's possible to identify DNA patterns associated with the disease and perhaps obtain clues for eventual cure. In addition, the U.S. Department of Defense now requires blood and saliva samples from all military personnel. The samples are stored, and DNA is extracted if the need for identification of a casualty arises.

Summary and Key Words

anticodon 558
antisense strand 556
codon 557
deoxyribonucleic acid (DNA) 548
double helix 552
fatty acid 539
lipid 538
lipid bilayer 545
messenger RNA (mRNA) 555
micelle 543
continued

In this chapter, we've looked at lipids and nucleic acids, completing our coverage of the four main classes of biological molecules. **Lipids** are the naturally occurring substances isolated from plants and animals by extraction with organic solvents. Animal fats and vegetable oils are the most widely occurring lipids. Both are **triacylglycerols**—triesters of glycerol with long-chain **fatty acids**. **Phospholipids** are esters of phosphoric acid and have either glycerol or an amino alcohol for their backbone.

Steroids are plant and animal lipids with a characteristic tetracyclic skeleton. Steroids occur widely in body tissue and have many different kinds of physiological properties. Among the more important kinds of

nucleic acid 548
nucleoside 549
nucleotide 549
phospholipid 544
polymerase chain reaction (PCR) 562
polyunsaturated fatty acid 541
replication 554
ribonucleic acid (RNA) 548
ribosomal RNA (rRNA) 555
Sanger dideoxy method 560
sense strand 556
small RNAs 555
steroid 546
transcription 556
transfer RNA (tRNA) 555
translation 558
triacylglycerol 539
wax 539

steroids are the sex hormones (*androgens* and *estrogens*) and the adrenocortical hormones.

The **nucleic acids**, **DNA** (**deoxyribonucleic acid**) and **RNA** (**ribonucleic acid**), are biological polymers that act as chemical carriers of an organism's genetic information. Enzyme-catalyzed hydrolysis of nucleic acids yields **nucleotides**, the monomer units from which RNA and DNA are constructed. Further enzyme-catalyzed hydrolysis of the nucleotides yields **nucleosides** plus phosphate. Nucleosides, in turn, consist of a purine or pyrimidine base linked to C1 of an aldopentose sugar—ribose in RNA and 2-deoxyribose in DNA. The nucleotides are joined by phosphodiester bonds between the 5′ phosphate of one nucleotide and the 3′ hydroxyl on the sugar of another nucleotide.

Molecules of DNA consist of two complementary strands held together by hydrogen bonds between heterocyclic bases on the different strands and coiled into a **double helix**. Adenine (A) and thymine (T) form hydrogen bonds to each other, as do cytosine (C) and guanine (G).

Three processes take place in deciphering the genetic information of DNA:

- **Replication** of DNA is the process by which identical DNA copies are made. The DNA double helix unwinds, complementary deoxyribonucleotides line up in order, and two new DNA molecules are produced.
- **Transcription** is the process by which RNA is produced to carry genetic information from the nucleus to the ribosomes. A short segment of the DNA double helix unwinds, and complementary ribonucleotides line up to produce **messenger RNA (mRNA)**.
- **Translation** is the process by which mRNA directs protein synthesis. Each mRNA is divided into **codons**, ribonucleotide triplets that are recognized by small amino acid carrying molecules of **transfer RNA (tRNA)**, which deliver the appropriate amino acids needed for protein synthesis.

Sequencing of DNA is carried out by the **Sanger dideoxy method**. Small amounts of DNA can be amplified by a factor of 10^6 using the **polymerase chain reaction (PCR)**.

Exercises

Visualizing Chemistry

(Problems 16.1–16.17 appear within the chapter.)

OWL

Interactive versions of these problems are assignable in OWL.

16.18 Identify the following bases, and tell whether each is found in DNA, RNA, or both:

(a)

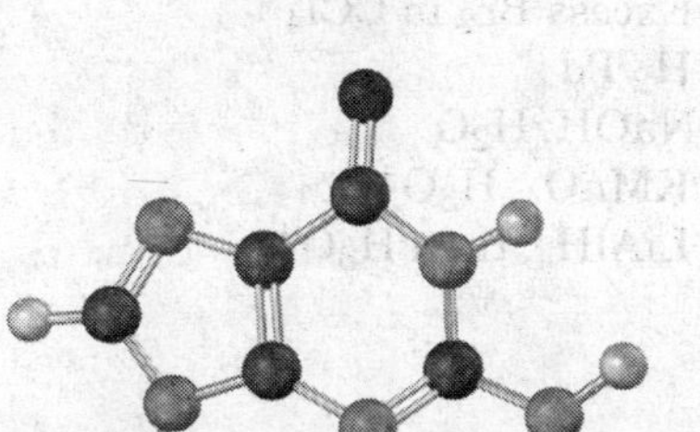

(b)

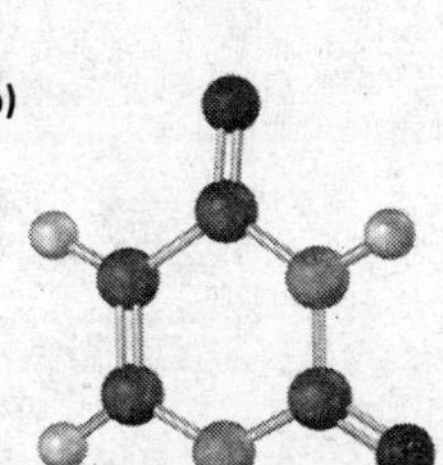

(c)

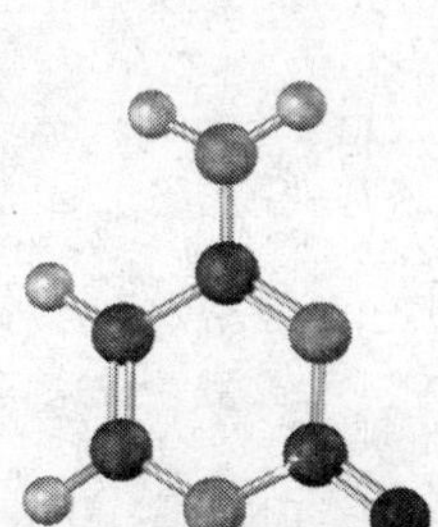

16.19 Cholesterol has the following structure. Tell whether the –OH group is axial or equatorial.

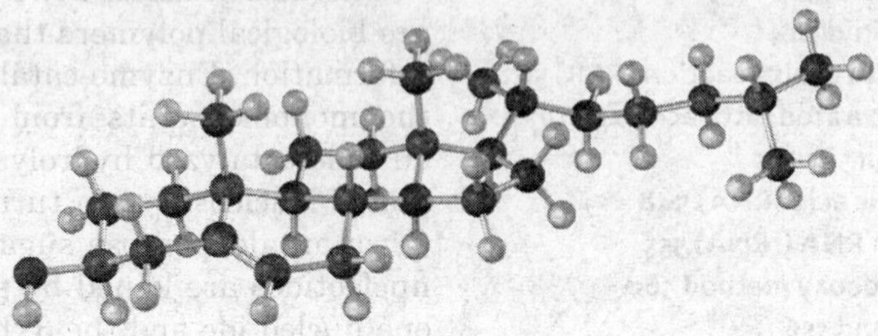

16.20 Identify the following nucleotide, and tell how it is used:

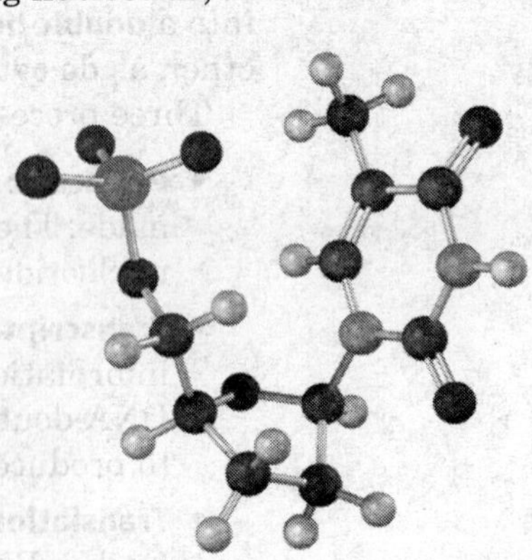

Additional Problems

LIPIDS

16.21 How would you convert oleic acid into the following substances?
(a) Methyl oleate **(b)** Methyl stearate **(c)** Nonanedioic acid

16.22 Write representative structures for the following:
(a) A fat **(b)** A vegetable oil **(c)** A steroid

16.23 Write the structures of the following molecules:
(a) Sodium stearate
(b) Ethyl linoleate
(c) Glyceryl dioleopalmitate

16.24 Show the products you would expect to obtain from the reaction of glyceryl trioleate with the following:
(a) Excess Br_2 in CCl_4
(b) H_2/Pd
(c) NaOH, H_2O
(d) $KMnO_4$, H_3O^+
(e) $LiAlH_4$, then H_3O^+

16.25 Spermaceti, a fragrant substance from sperm whales, was much used in cosmetics until it was banned in 1976 to protect whales from extinction. Chemically, spermaceti is cetyl palmitate, the ester of cetyl alcohol ($C_{16}H_{33}OH$) with palmitic acid. Draw its structure.

16.26 Draw the products you would obtain from treatment of cholesterol with the following reagents:
(a) Br_2 **(b)** H_2, Pd catalyst **(c)** CH_3COCl, pyridine

16.27 The plasmalogens are a group of lipids found in nerve and muscle cells. How do plasmalogens differ from fats?

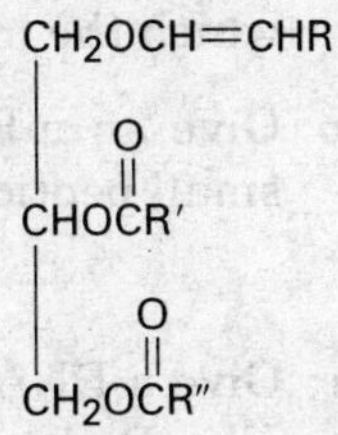

A plasmalogen

16.28 Eleostearic acid, $C_{18}H_{30}O_2$, is a rare fatty acid found in tung oil. On oxidation with $KMnO_4$, eleostearic acid yields 1 part pentanoic acid, 2 parts oxalic acid ($HO_2C—CO_2H$), and 1 part nonanedioic acid. Propose a structure for eleostearic acid.

16.29 Cardiolipins are a group of lipids found in heart muscles. What products would be formed if all ester bonds, including phosphates, were saponified by treatment with aqueous NaOH?

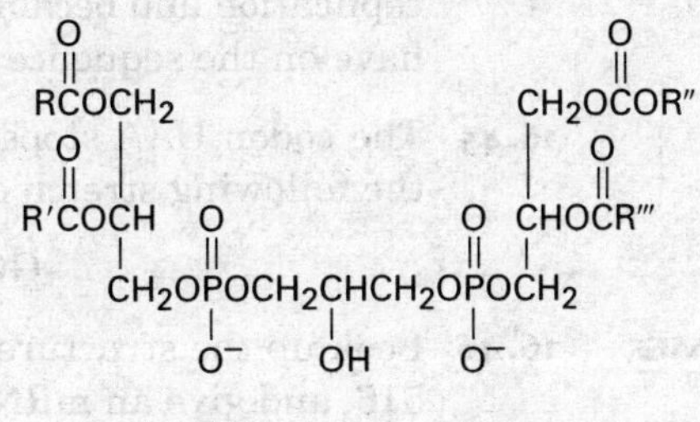

A cardiolipin

16.30 Stearolic acid, $C_{18}H_{32}O_2$, yields oleic acid on catalytic hydrogenation over the Lindlar catalyst. Propose a structure for stearolic acid.

16.31 If the average molecular weight of soybean oil is 1500, how many grams of NaOH are needed to saponify 5.00 g of the oil?

Nucleic Acids

16.32 What DNA sequence is complementary to the following sequence?

(5′) GAAGTTCATGC (3′)

16.33 Give codons for the following amino acids:
(a) Ile **(b)** Asp **(c)** Thr

16.34 The DNA from sea urchins contains about 32% A and about 18% G. What percentages of T and C would you expect in sea urchin DNA? Explain.

16.35 What amino acids do the following ribonucleotide codons code for?
(a) AAU **(b)** GAG **(c)** UCC **(d)** CAU **(e)** ACC

16.36 From what DNA sequences were each of the mRNA codons in Problem 16.35 transcribed?

16.37 What anticodon sequences of tRNAs are coded by each of the codons in Problem 16.35?

16.38 Draw the complete structure of the ribonucleotide codon UAC. For what amino acid does this sequence code?

16.39 Draw the complete structure of the deoxyribonucleotide sequence from which the mRNA codon in Problem 16.38 was transcribed.

16.40 Give an mRNA sequence that codes for synthesis of metenkephalin, a small peptide with morphine-like properties:

Tyr-Gly-Gly-Phe-Met

16.41 Give a DNA gene sequence (sense strand) that will code for metenkephalin (Problem 16.40).

16.42 Human and horse insulin both have two polypeptide chains with one chain containing 21 amino acids and the other containing 30 amino acids. How many nitrogen bases are present in the DNA to code for each chain?

16.43 Human and horse insulin (see Problem 16.42) differ in primary structure at two amino acids: at the 9th position in one chain (human has Ser and horse has Gly) and at the 30th position in the other chain (human has Thr and horse has Ala). How must the DNA differ?

16.44 If the gene sequence -TAA-CCG-GAT- on DNA were miscopied during replication and became -TGA-CCG-GAT-, what effect would the mutation have on the sequence of the protein produced?

16.45 The codon UAA stops protein synthesis. Why does the sequence UAA in the following stretch of mRNA not cause any problems?

-GCA-UUC-GAG-GUA-ACG-CCC-

General Problems

16.46 Look up the structure of angiotensin II in Worked Example 15.4 on page 516, and give an mRNA sequence that codes for its synthesis.

16.47 What amino acid sequence is coded by the following mRNA sequence?

CUA-GAC-CGU-UCC-AAG-UGA

16.48 What anticodon sequences of tRNAs are coded by the mRNA in Problem 16.47? What was the base sequence in the original DNA strand on which this mRNA was made? What was the base sequence in the DNA strand complementary to that from which this mRNA was made?

16.49 One of the steps in the metabolic degradation of guanine is hydrolysis to give xanthine. Propose a mechanism.

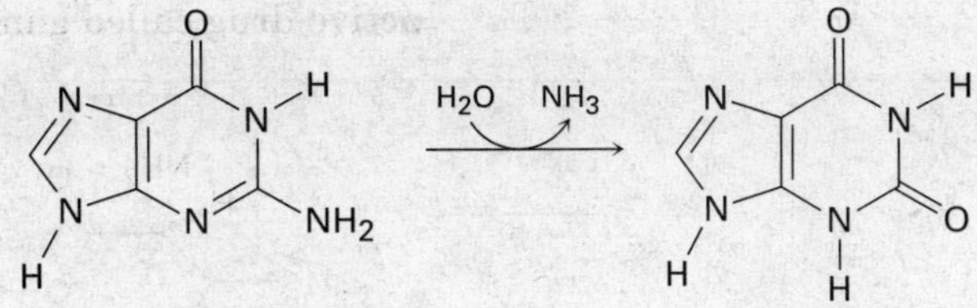

Guanine **Xanthine**

16.50 Nandrolone is an anabolic steroid sometimes taken by athletes to build muscle mass. Compare the structures of nandrolone and testosterone (page 547), and point out their structural similarities.

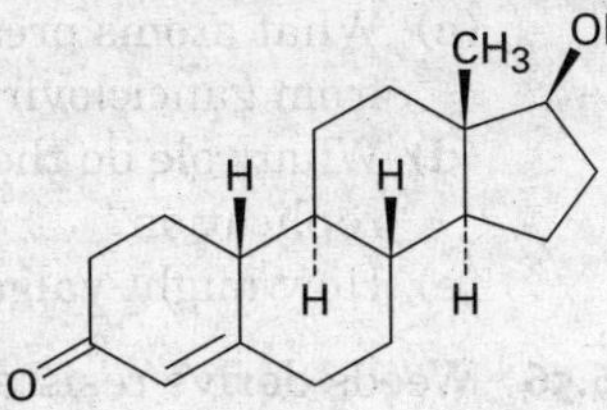

Nandrolone (an anabolic steroid)

16.51 Draw the structure of cyclic adenosine monophosphate (cAMP), a messenger involved in the regulation of glucose production in the body. Cyclic AMP has a phosphate ring connecting the 3′ and 5′ hydroxyl groups on adenosine.

16.52 Diethylstilbestrol (DES) exhibits estradiol-like activity even though it is structurally unrelated to steroids. Once used widely as an additive in animal feed, DES has been implicated as a causative agent in several types of cancers. Show how DES can be drawn so that it is sterically similar to estradiol.

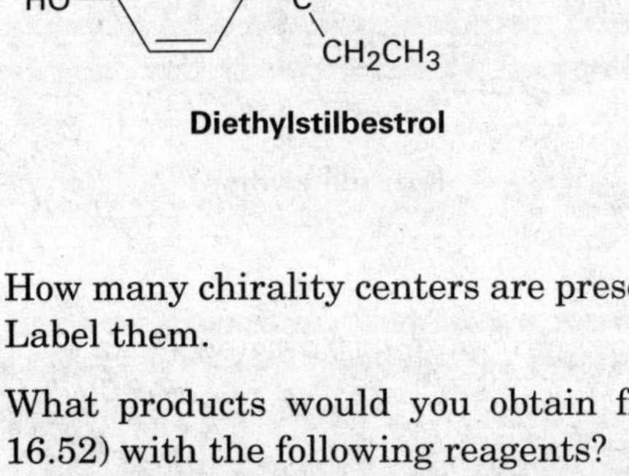

Diethylstilbestrol

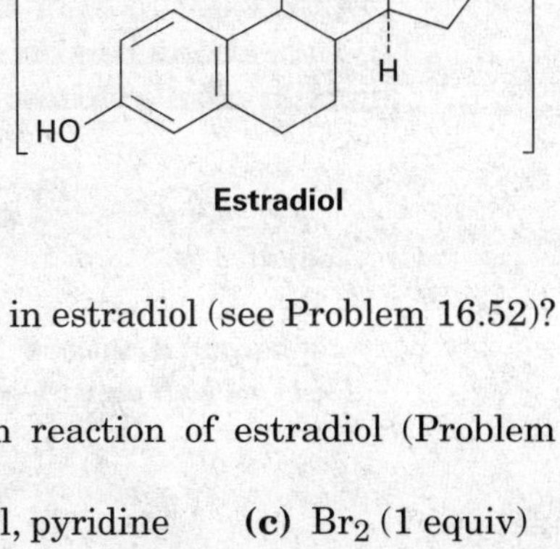

Estradiol

16.53 How many chirality centers are present in estradiol (see Problem 16.52)? Label them.

16.54 What products would you obtain from reaction of estradiol (Problem 16.52) with the following reagents?

(a) NaOH, then CH_3I **(b)** CH_3COCl, pyridine **(c)** Br_2 (1 equiv)

IN THE MEDICINE CABINET

16.55 Valganciclovir is an antiviral agent used for the treatment of cytomegalovirus. Called a *prodrug*, valganciclovir is inactive by itself but is rapidly converted in the intestine by hydrolysis of its ester bond to produce an active drug called ganciclovir, along with an amino acid.

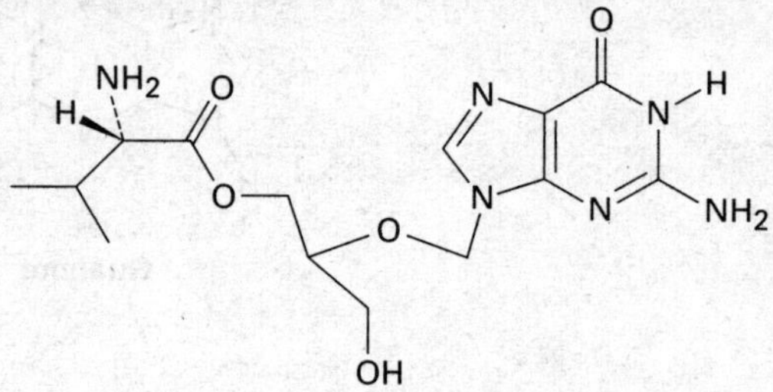

Valganciclovir

(a) What amino acid is produced by hydrolysis of the ester bond in valganciclovir?
(b) What is the structure of ganciclovir?
(c) What atoms present in the nucleotide deoxyguanine are missing from ganciclovir?
(d) What role do the atoms missing from deoxyguanine play in DNA replication?
(e) How might valganciclovir interfere with DNA synthesis?

IN THE FIELD

16.56 Weeds derive resistance to the herbicide atrazine by a mutation in which amino acid substitution of a serine residue by a glycine occurs.
(a) Identify the six codons that correspond to serine and the four codons that correspond to glycine.
(b) What is the minimum number of base mutations of DNA required for the evolution of resistance?

CHAPTER

17

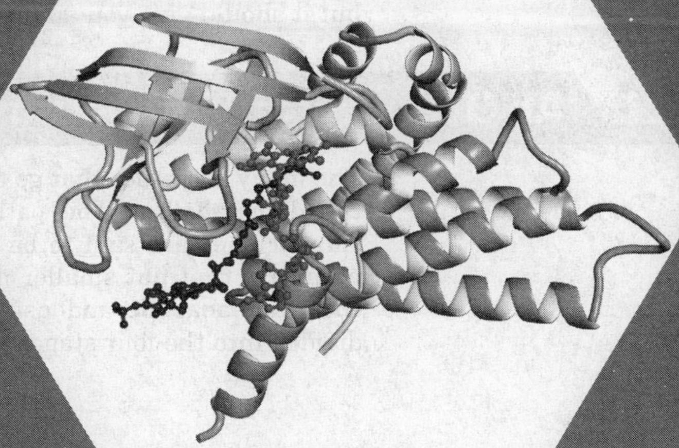

Acyl CoA dehydrogenase is an enzyme that catalyzes the introduction of a C=C double bond into fatty acids during their metabolism.

The Organic Chemistry of Metabolic Pathways

The organic chemical reactions that take place in even the smallest and simplest living organism are more complex than those carried out in any laboratory. Yet all those reactions in living organisms, regardless of their complexity, follow the same rules of reactivity and proceed by the same mechanisms we've seen in the preceding chapters.

A word of caution, though: biological molecules are often much larger and more complicated than the substances we've been dealing with thus far. But don't be intimidated. Keep your focus on the parts of the molecules where changes occur and ignore the parts where nothing changes. The reactions themselves are exactly the same additions, eliminations, substitutions, carbonyl condensations, and so forth, that we've been dealing with all along. By the end of this chapter, it should be clear that the chemistry of living organisms *is* organic chemistry.

WHY THIS CHAPTER?

OWL

Online homework for this chapter can be assigned in OWL, an online homework assessment tool.

To understand the chemistry of living organisms, knowing *what* occurs isn't enough. It's also necessary to understand *how* and *why* organisms use the chemistry they do. In this chapter, we'll look at some of the pathways by which organisms carry out their chemistry, focusing primarily on how they metabolize fats and carbohydrates. The treatment will be far from complete, but it should give you an idea of how biological processes occur.

17.1 An Overview of Metabolism and Biochemical Energy

The many processes that go on in the cells of living organisms are collectively called **metabolism**. The pathways that break down larger molecules into smaller ones are said to be **catabolic** and usually release energy, while the pathways that put smaller molecules together to synthesize larger biomolecules are **anabolic** and usually absorb energy. Catabolic processes can be divided into the four stages shown in Figure 17.1.

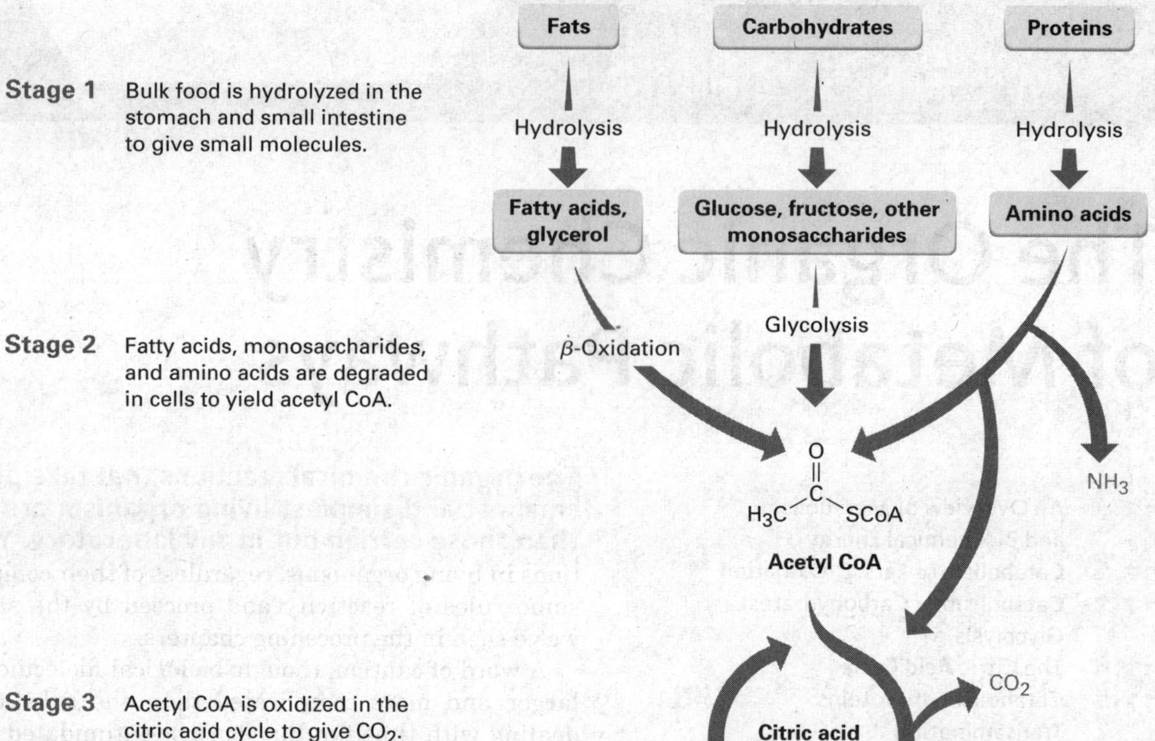

Figure 17.1 An overview of catabolic pathways for the degradation of food and the production of biochemical energy. The ultimate products of food catabolism are CO_2 and H_2O, with the energy released in the citric acid cycle used to drive the synthesis of adenosine triphosphate (ATP) from adenosine diphosphate (ADP) plus phosphate ion, $HOPO_3^{2-}$.

In the first catabolic stage, commonly called digestion, food is broken down in the mouth, stomach, and small intestine by hydrolysis of ester, acetal (glycoside), and amide (peptide) bonds to yield fatty acids, simple sugars, and amino acids. These small molecules are then absorbed by cells and further degraded in the second stage of catabolism to acetyl groups attached by a thioester bond to the large carrier molecule coenzyme A. The resultant compound, acetyl coenzyme A (acetyl CoA), is a key substance in the metabolism of food molecules and in many other biological pathways. As noted in Section 10.12, the acetyl group in acetyl CoA is linked to the sulfur atom of phosphopantetheine, which is itself linked to adenosine 3′,5′-bisphosphate.

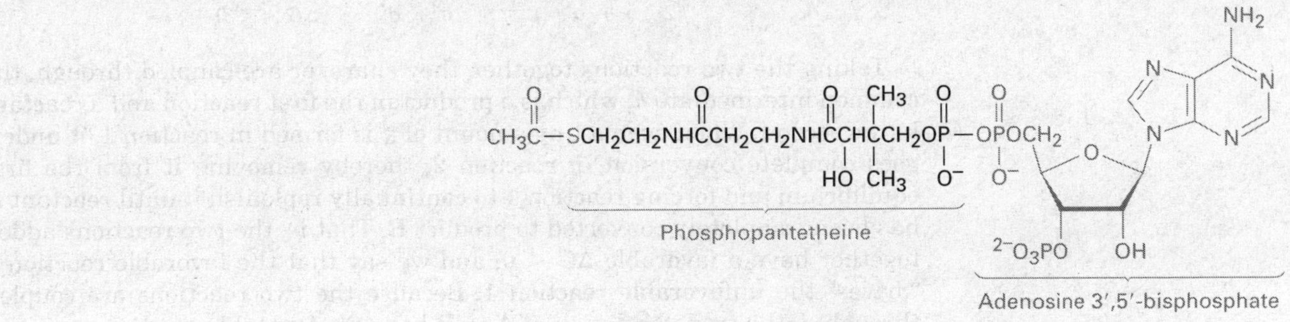

Acetyl CoA—a thioester

Acetyl CoA next enters the third stage of catabolism, the *citric acid cycle,* where it is oxidized to yield CO_2. Like most oxidations, this stage releases a large amount of energy, which is used in the fourth stage, the *electron-transport chain,* to make adenosine triphosphate, ATP.

ATP has been called the "energy currency" of the cell. Catabolic reactions buy ATP by using the energy they release to synthesize it from adenosine diphosphate (ADP) plus hydrogen phosphate ion, HPO_4^{2-}. Anabolic reactions then spend ATP by transferring a phosphate group to another molecule, thereby releasing energy and regenerating ADP. Energy production and use thus revolves around the ATP ⇄ ADP interconversion.

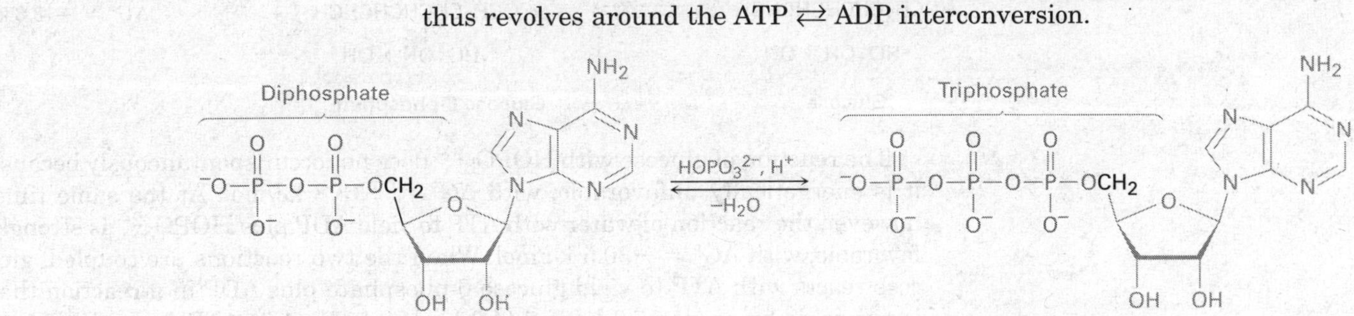

Adenosine diphosphate (ADP) **Adenosine triphosphate (ATP)**

How does the body use the ATP it spends? You may recall from your general chemistry course that a chemical reaction must have a favorable equilibrium constant and release energy (actually, *free energy, G*) to occur spontaneously. This means that the free-energy change for the reaction, ΔG, must be negative. If ΔG is positive, the reaction is unfavorable and can't occur spontaneously.

What typically happens for an energetically unfavorable reaction to occur is that it is "coupled" to an energetically favorable reaction so that the overall free-energy change for the two reactions together is favorable. Imagine, for instance, that reaction 1 does not occur to any reasonable extent because it has a small equilibrium constant and is energetically unfavorable; that is, the reaction has $\Delta G > 0$.

$$(1)\quad \mathbf{A} + m \rightleftarrows \mathbf{B} + n \qquad \Delta G > 0$$

where **A** and **B** are the biochemically "important" substances, while m and n are enzyme cofactors, H_2O, or other small molecules

Imagine also that product n can react with substance o to yield p and q in a second, energetically favorable reaction that has a large equilibrium constant and $\Delta G << 0$.

$$(2)\quad n + o \rightleftarrows p + q \qquad \Delta G << 0$$

Taking the two reactions together, they share, or are coupled through, the common intermediate n, which is a product in the first reaction and a reactant in the second. When even a tiny amount of n is formed in reaction 1, it undergoes complete conversion in reaction 2, thereby removing it from the first equilibrium and forcing reaction 1 to continually replenish n until reactant **A** has been completely converted to product **B**. That is, the two reactions added together have a favorable $\Delta G < 0$, and we say that the favorable reaction 2 "drives" the unfavorable reaction 1. Because the two reactions are coupled through n, the transformation of **A** to **B** becomes favorable.

$$(1)\quad \mathbf{A} + m \rightleftarrows \mathbf{B} + \cancel{n} \qquad \Delta G > 0$$

$$(2)\quad \cancel{n} + o \rightleftarrows p + q \qquad \Delta G << 0$$

$$\text{Net: } \mathbf{A} + m + o \rightleftarrows \mathbf{B} + p + q \qquad \Delta G < 0$$

As an example of two reactions that are coupled, look at the phosphorylation reaction of glucose to yield glucose 6-phosphate plus water, an important step in the breakdown of dietary carbohydrates.

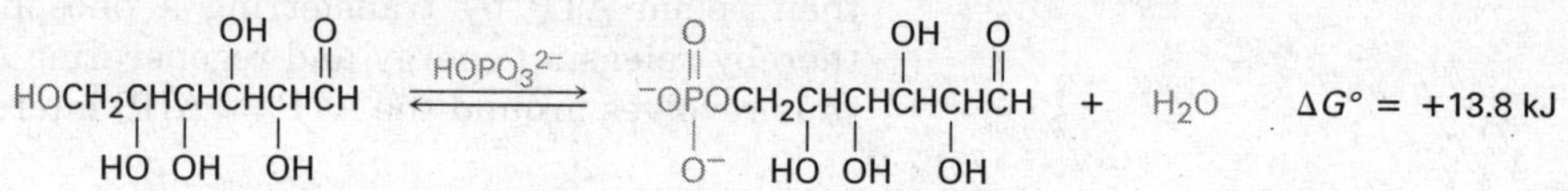

Glucose **Glucose 6-phosphate**

The reaction of glucose with $HOPO_3^{2-}$ does not occur spontaneously because it is energetically unfavorable, with $\Delta G° = +13.8$ kJ/mol. At the same time, however, the reaction of water with ATP to yield ADP plus $HOPO_3^{2-}$ is strongly favorable, with $\Delta G° = -30.5$ kJ/mol. When the two reactions are coupled, glucose reacts with ATP to yield glucose 6-phosphate plus ADP in a reaction that is favorable by about 16.7 kJ/mol (4.0 kcal/mol). That is, ATP drives the phosphorylation reaction of glucose.

$$\text{Glucose} + \cancel{HOPO_3^{2-}} \longrightarrow \text{Glucose 6-phosphate} + \cancel{H_2O} \qquad \Delta G° = +13.8 \text{ kJ/mol}$$

$$\text{ATP} + \cancel{H_2O} \longrightarrow \text{ADP} + \cancel{HOPO_3^{2-}} + H^+ \qquad \Delta G° = -30.5 \text{ kJ/mol}$$

$$\text{Net: Glucose} + \text{ATP} \longrightarrow \text{Glucose 6-phosphate} + \text{ADP} + H^+ \qquad \Delta G° = -16.7 \text{ kJ/mol}$$

It's this ability to drive otherwise unfavorable phosphorylation reactions that makes ATP so useful. The resultant phosphates are much more reactive as leaving groups in nucleophilic substitutions and eliminations than the alcohols they're derived from and are therefore much more chemically useful.

Problem 17.1 One of the steps in fat metabolism is the reaction of glycerol (propane-1,2,3-triol) with ATP to yield glycerol 1-phosphate. Write the reaction, and draw the structure of glycerol 1-phosphate.

17.2 Catabolism of Fats: β-Oxidation

Let's begin a study of some common metabolic pathways by looking at fat catabolism. The metabolic breakdown of fats and oils (triacylglycerols) begins with their hydrolysis to yield glycerol plus fatty acids. Glycerol is then phosphorylated by reaction with ATP and oxidized to yield dihydroxyacetone phosphate, which enters the carbohydrate catabolic pathway to be discussed in Section 17.3.

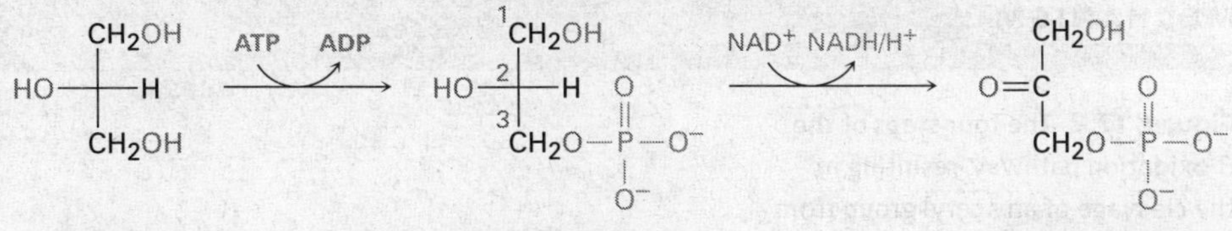

Glycerol   ***sn*-Glycerol 3-phosphate**   **Dihydroxyacetone phosphate (DHAP)**

Note how the preceding reactions are written. It's common practice when writing biochemical transformations to show only the structures of the reactant and product, while abbreviating the structures of coenzymes (Section 15.9) and other substances. Thus, the curved arrow intersecting the usual straight reaction arrow in the first step shows that ATP is also a reactant and that ADP is also a product. The coenzyme nicotinamide adenine dinucleotide (NAD^+) is required in the second step, and reduced nicotinamide adenine dinucleotide (NADH) plus a proton are products. We'll see shortly that NAD^+ is often involved as a biochemical oxidizing agent for converting alcohols to aldehydes or ketones.

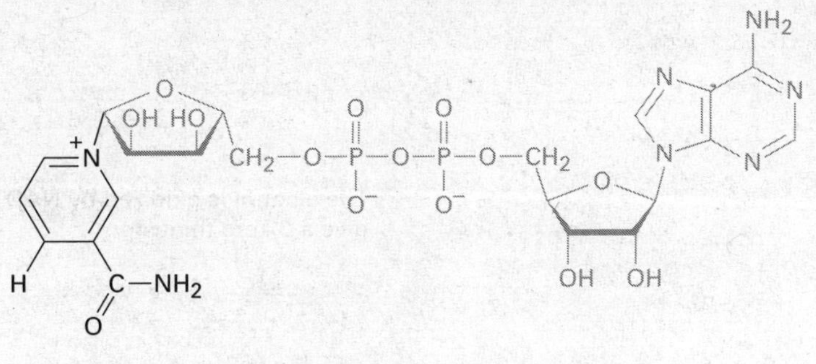

Nicotinamide adenine dinucleotide (NAD^+)

Reduce ⇅ Oxidize

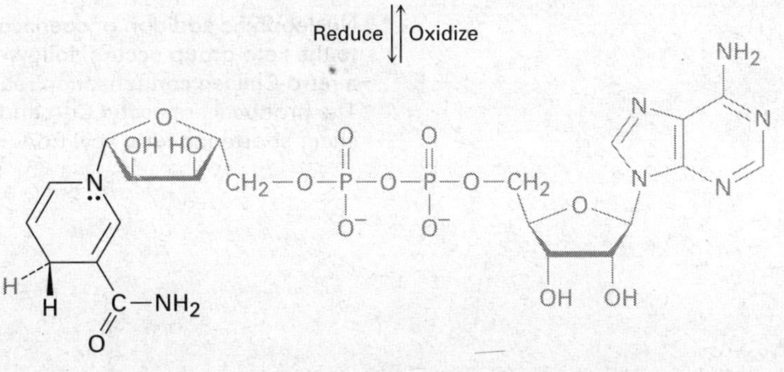

Reduced nicotinamide adenine dinucleotide (NADH)

Note also that glyceraldehyde 3-phosphate is written with its phosphate group dissociated, that is, $-OPO_3^{2-}$ rather than $-OPO_3H_2$. As remarked in Section 15.1, it's standard practice in writing biochemical structures to show carboxylic acids and phosphoric acids as their anions because they exist in this form at the physiological pH of 7.3 found in cells.

The fatty acids produced on triacylglycerol hydrolysis are catabolized by a repetitive four-step sequence of enzyme-catalyzed reactions called the **β-oxidation pathway**, shown in Figure 17.2. Each passage along the pathway results in the cleavage of a two-carbon acetyl group from the end of the fatty-acid chain, until the entire molecule is ultimately degraded. As each acetyl group is produced, it enters the citric acid cycle, which we'll see in Section 17.4.

MECHANISM

Figure 17.2 The four steps of the β-oxidation pathway, resulting in the cleavage of an acetyl group from the end of the fatty-acid chain. The chain-shortening step is a retro-Claisen reaction of a β-keto thioester. Individual steps are explained in the text.

1. A conjugated double bond is introduced by removal of hydrogens from C2 and C3 by the coenzyme flavin adenine dinucleotide (FAD).
2. Conjugate nucleophilic addition of water to the double bond gives a β-hydroxyacyl CoA.
3. The alcohol is oxidized by NAD^+ to give a β-keto thioester.
4. Nucleophilic addition of coenzyme A to the keto group occurs, followed by a retro-Claisen condensation reaction. The products are acetyl CoA and a chain-shortened fatty acyl CoA.

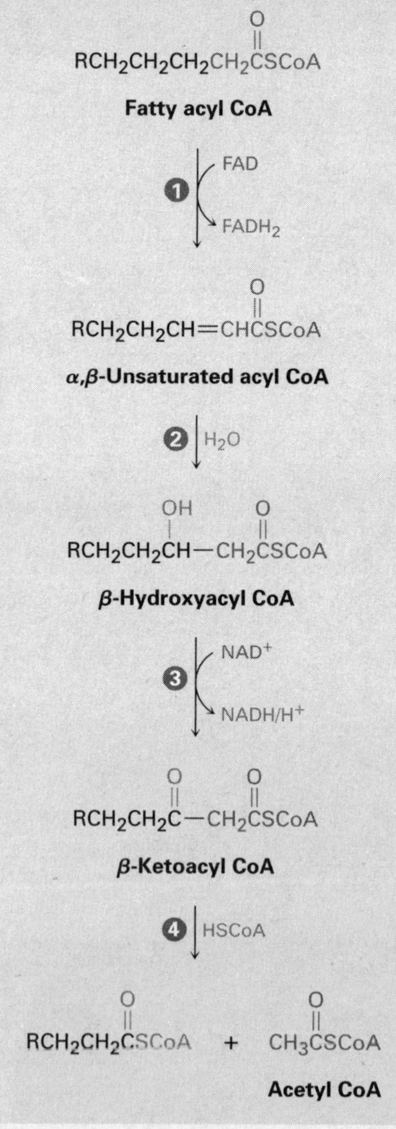

STEP 1 OF FIGURE 17.2: INTRODUCTION OF A DOUBLE BOND

The β-oxidation pathway begins when a fatty acid reacts with coenzyme A to give a fatty acyl CoA. Two hydrogen atoms are then removed from C2 and C3 by an acyl CoA dehydrogenase to yield an α, β-unsaturated acyl CoA. This kind of oxidation—the introduction of a conjugated double bond into a carbonyl compound—occurs frequently in biochemical pathways and is usually carried out by the coenzyme flavin adenine dinucleotide (FAD). Reduced $FADH_2$ is the by-product.

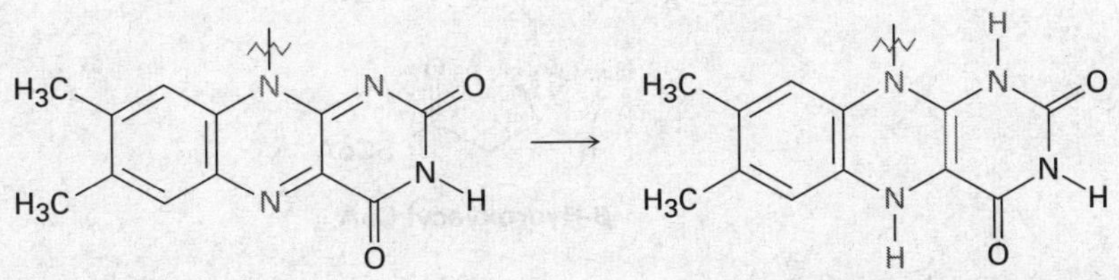

Flavin adenine dinucleotide (FAD) **Reduced flavin adenine dinucleotide ($FADH_2$)**

STEP 2 OF FIGURE 17.2: CONJUGATE ADDITION OF WATER

The α, β-unsaturated acyl CoA produced in step 1 reacts with water by a conjugate nucleophilic addition pathway (Section 9.10) to yield a β-hydroxyacyl CoA in a process catalyzed by enoyl-CoA hydratase. Water adds to the β carbon of the double bond, yielding an enolate ion intermediate, which is then protonated.

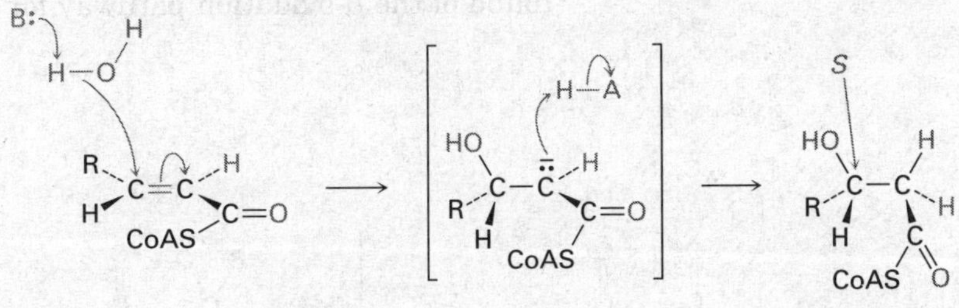

(3*S*)-Hydroxyacyl CoA

STEP 3 OF FIGURE 17.2: ALCOHOL OXIDATION

The β-hydroxyacyl CoA from step 2 is oxidized to a β-ketoacyl CoA in a reaction catalyzed by L-3-hydroxyacyl-CoA dehydrogenase. As in the oxidation of glycerol 1-phosphate to dihydroxyacetone phosphate mentioned at the beginning of this section, the alcohol oxidation requires NAD^+ as a coenzyme and yields reduced NADH as by-product.

The mechanism of the alcohol oxidation is similar in some respects to that of the conjugate nucleophilic addition reaction in step 2. Thus, a hydride ion expelled from the alcohol acts as a nucleophile and adds to the $C{=}C{-}C{=}N^+$

part of NAD^+ in much the same way that water acts as a nucleophile and adds to the C=C—C=O part of the unsaturated acyl CoA.

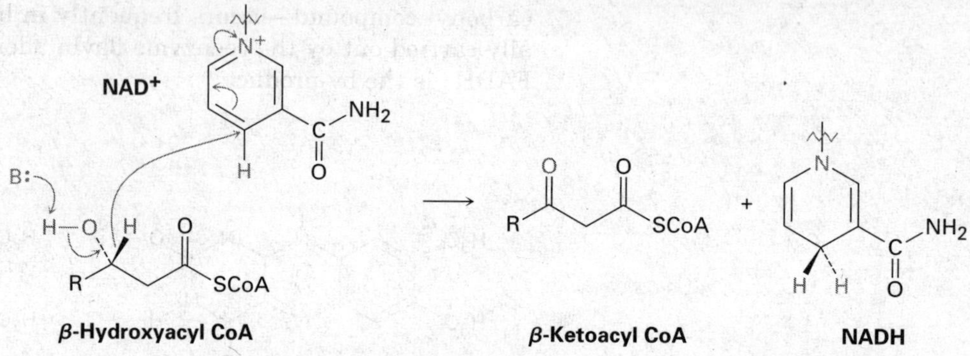

STEP 4 OF FIGURE 17.2: CHAIN CLEAVAGE Acetyl CoA is split off from the acyl chain in the final step of β-oxidation, leaving behind an acyl CoA that is two carbon atoms shorter than the original. The reaction is catalyzed by β-ketoacyl-CoA thiolase and is the exact reverse of a Claisen condensation reaction (Section 11.10). In the *forward* direction, a Claisen condensation joins two esters together to form a β-keto ester product. In the *reverse* direction, a retro-Claisen reaction splits a β-keto ester (or β-keto thioester in this case) to form two esters (or two thioesters).

The reaction occurs by nucleophilic addition of coenzyme A to the keto group of the β-keto acyl CoA to yield an alkoxide ion intermediate. Cleavage of the C2–C3 bond follows, with expulsion of an acetyl CoA enolate ion that is immediately protonated. The chain-shortened acyl CoA then enters another round of the β-oxidation pathway for further degradation.

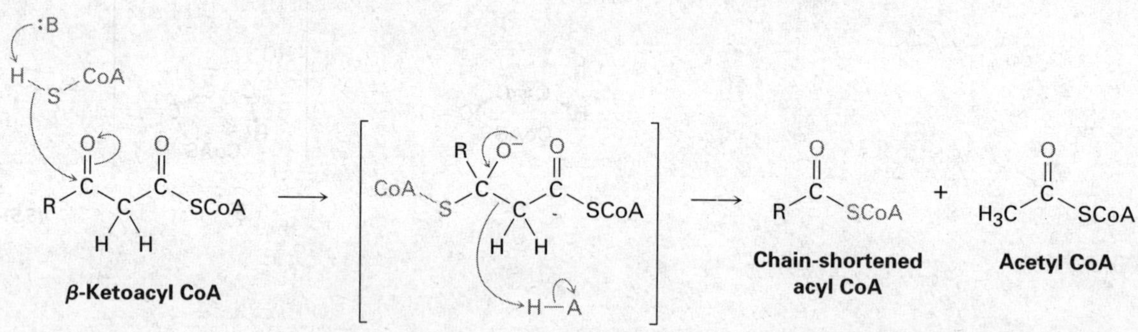

Look at the catabolism of myristic acid shown in Figure 17.3 to see the overall result of the β-oxidation pathway. The first passage converts the C_{14} myristoyl CoA into the C_{12} lauroyl CoA plus acetyl CoA; the second passage converts lauroyl CoA into the C_{10} caproyl CoA plus acetyl CoA; the third passage converts caproyl CoA into the C_8 capryloyl CoA; and so on. Note that the last passage produces two molecules of acetyl CoA because the precursor has four carbons.

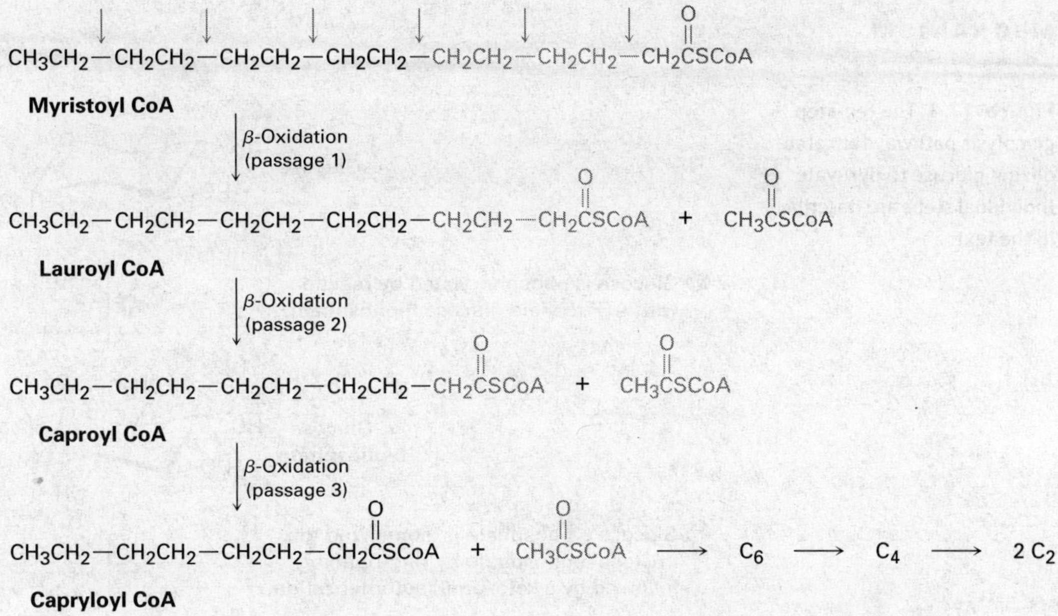

Figure 17.3 Catabolism of the C_{14} myristic acid in the β-oxidation pathway yields seven molecules of acetyl CoA after six passages.

Most fatty acids have an even number of carbon atoms, so none are left over after β-oxidation. Those fatty acids with an odd number of carbon atoms or with double bonds require additional steps for degradation, but all carbon atoms are ultimately released for further oxidation in the citric acid cycle.

Problem 17.2 Write the equations for the remaining passages of the β-oxidation pathway following those shown in Figure 17.3.

Problem 17.3 How many molecules of acetyl CoA are produced by catabolism of the following fatty acids, and how many passages of the β-oxidation pathway are needed?

(a) Palmitic acid, $CH_3(CH_2)_{14}CO_2H$

(b) Arachidic acid, $CH_3(CH_2)_{18}CO_2H$

17.3 Catabolism of Carbohydrates: Glycolysis

Glucose is the body's primary short-term energy source. Its catabolism begins with **glycolysis**, a series of ten enzyme-catalyzed reactions that break down glucose into 2 equivalents of pyruvate, $CH_3COCO_2^-$. The steps of glycolysis, also called the *Embden–Meyerhoff pathway* after its discoverers, are summarized in Figure 17.4.

MECHANISM

Figure 17.4 The ten-step glycolysis pathway for catabolizing glucose to pyruvate. Individual steps are described in the text.

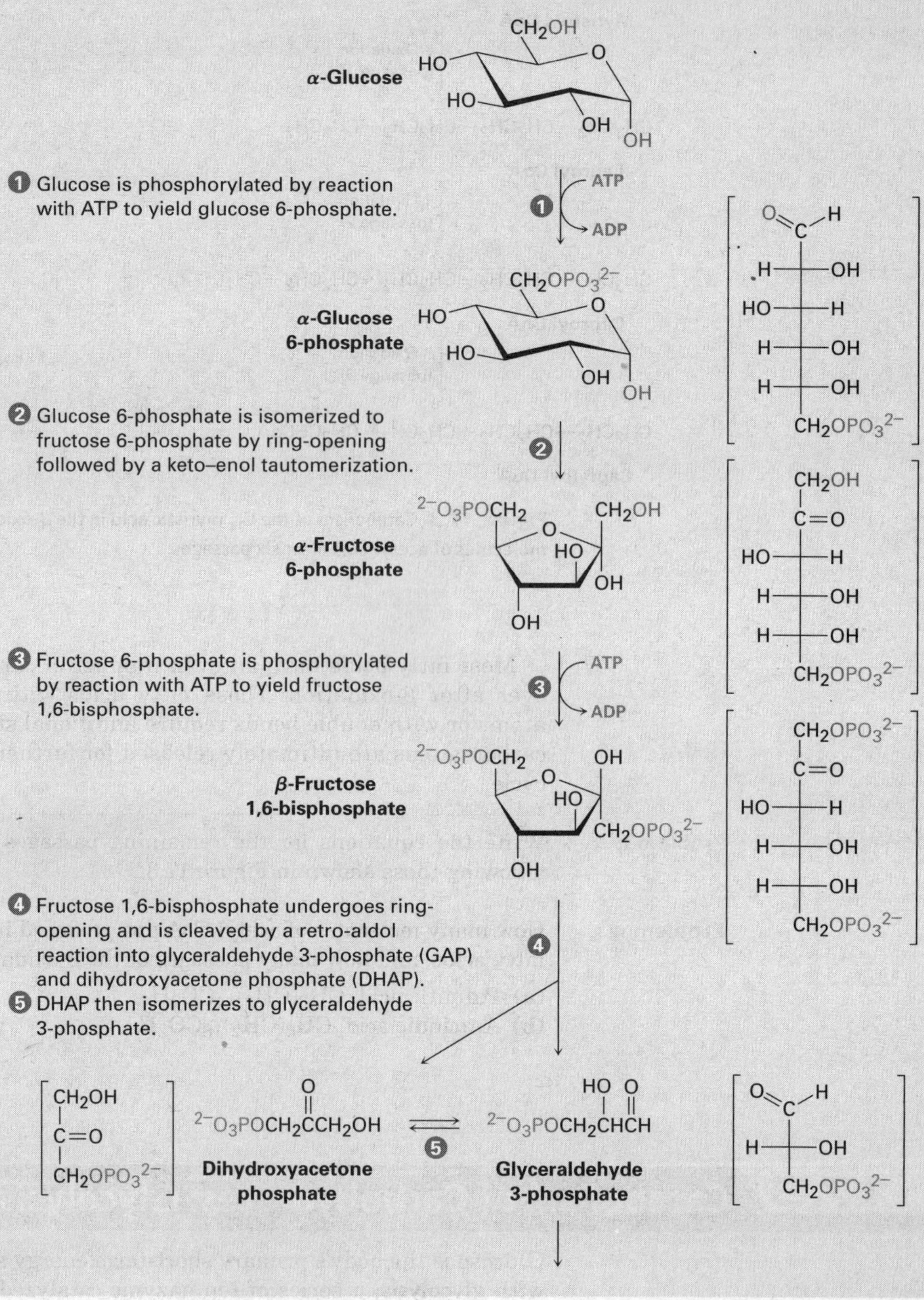

MECHANISM

Figure 17.4 Continued.

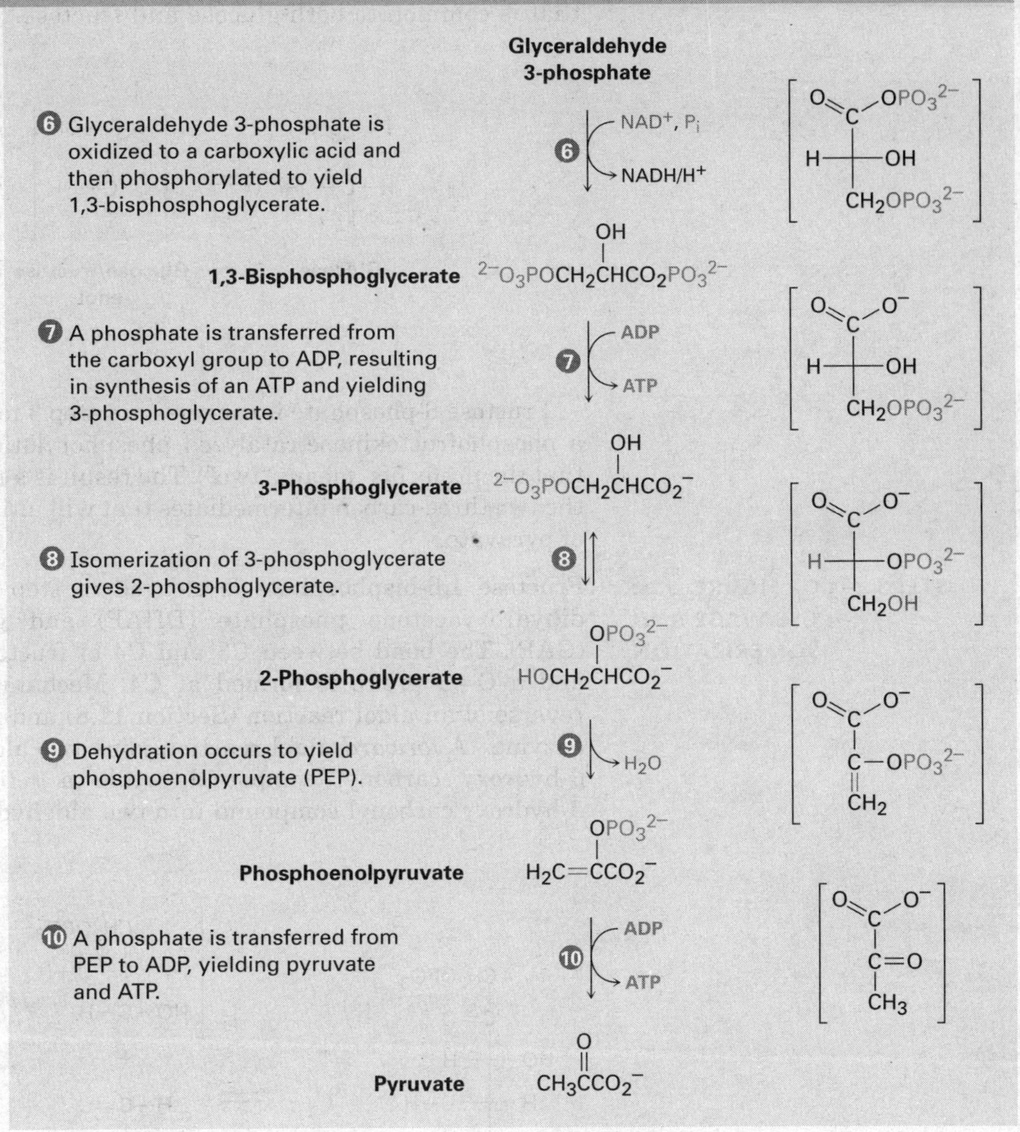

STEPS 1–3 OF FIGURE 17.4: PHOSPHORYLATION AND ISOMERIZATION Glucose, produced by digestion of dietary carbohydrates, is phosphorylated in step 1 by reaction with ATP in a reaction catalyzed by hexokinase. The glucose 6-phosphate that results is then isomerized in step 2 by glucose-6-phosphate isomerase to give fructose 6-phosphate. The isomerization

takes place by initial opening of the glucose hemiacetal ring to the open-chain form, followed by keto–enol tautomerization (Section 11.1) to an enol that is common to both glucose and fructose.

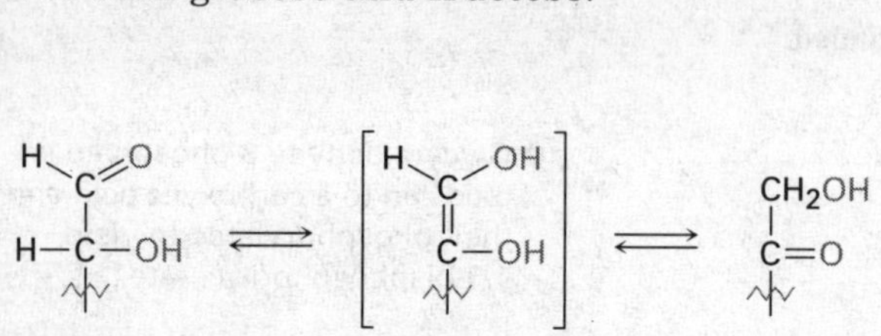

Glucose   Glucose/fructose enol   Fructose

Fructose 6-phosphate is converted in step 3 to fructose 1,6-bisphosphate by a phosphofructokinase-catalyzed phosphorylation reaction with ATP (recall that the prefix *bis-* means "two"). The result is a molecule ready to be split into the two three-carbon intermediates that will ultimately become two molecules of pyruvate.

STEPS 4–5 OF FIGURE 17.4: CLEAVAGE AND ISOMERIZATION Fructose 1,6-bisphosphate is cleaved in step 4 into two 3-carbon pieces, dihydroxyacetone phosphate (DHAP) and glyceraldehyde 3-phosphate (GAP). The bond between C3 and C4 of fructose 1,6-bisphosphate breaks, and a C=O group is formed at C4. Mechanistically, the cleavage is the reverse of an aldol reaction (Section 11.8) and is carried out by an aldolase enzyme. (A *forward* aldol reaction joins two aldehydes or ketones to give a β-hydroxy carbonyl compound, while a *retro*-aldol reaction cleaves a β-hydroxy carbonyl compound into two aldehydes or ketones.)

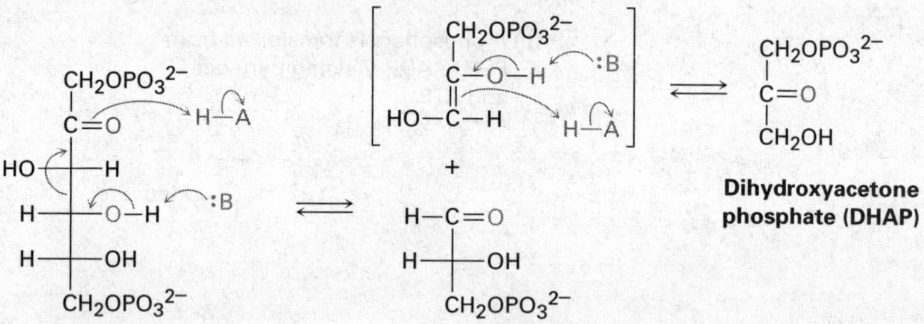

Fructose 1,6-bisphosphate   Glyceraldehyde 3-phosphate (GAP)   Dihydroxyacetone phosphate (DHAP)

Glyceraldehyde 3-phosphate continues on in the glycolysis pathway, but dihydroxyacetone phosphate is first isomerized in step 5 by triose phosphate isomerase. As in the glucose-to-fructose conversion of step 2, the isomerization of dihydroxyacetone phosphate to glyceraldehyde 3-phosphate takes place by keto–enol tautomerization through a common enol. The net result of steps 4 and 5 is the production of two glyceraldehyde 3-phosphate molecules, both of which pass down the rest of the pathway. Thus, each of the remaining five steps of glycolysis takes place twice for every glucose molecule that enters at step 1.

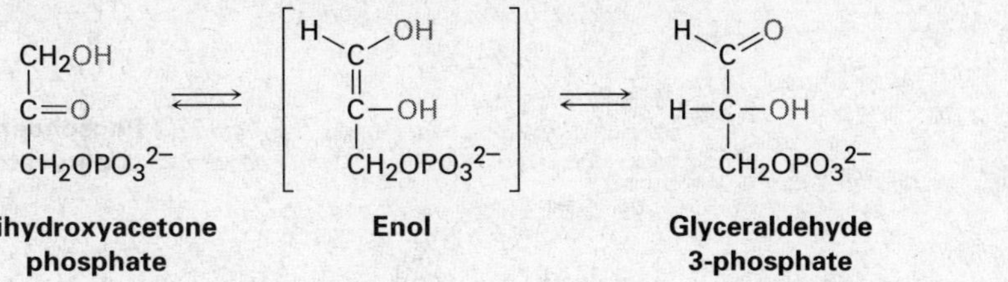

STEPS 6–8 OF FIGURE 17.4: OXIDATION AND PHOSPHORYLATION

Glyceraldehyde 3-phosphate is oxidized and phosphorylated in step 6 to give 1,3-bisphosphoglycerate. The reaction requires the coenzyme NAD^+ in the presence of phosphate ion (HPO_4^{2-}, abbreviated P_i) and is catalyzed by glyceraldehyde-3-phosphate dehydrogenase. Transfer of a phosphate group from the carboxyl of 1,3-bisphosphoglycerate to ADP in step 7 then yields 3-phosphoglycerate, which is isomerized to 2-phosphoglycerate in step 8. The phosphorylation is catalyzed by phosphoglycerate kinase, and the isomerization is catalyzed by phosphoglycerate mutase.

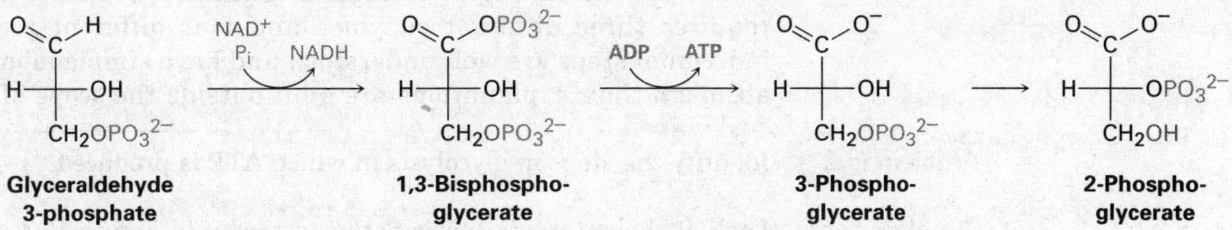

STEPS 9–10 OF FIGURE 17.4: DEHYDRATION AND DEPHOSPHORYLATION

Like most β-hydroxy carbonyl compounds produced in aldol reactions (Section 11.9), 2-phosphoglycerate undergoes a ready dehydration in step 9 by an E1cB mechanism (Section 7.8). The process is catalyzed by enolase, and the product is phosphoenolpyruvate, abbreviated PEP.

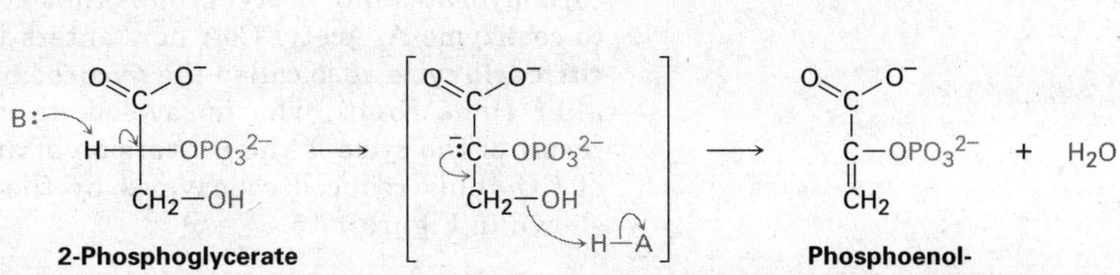

Transfer of the phosphate group to ADP in step 10 then generates ATP and gives pyruvate, a reaction catalyzed by pyruvate kinase.

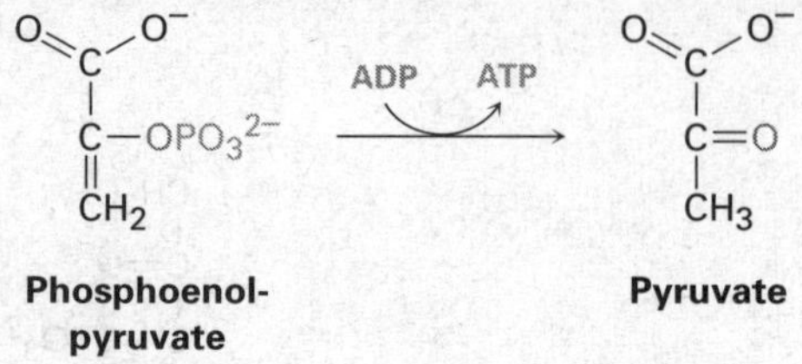

The overall result of glycolysis is

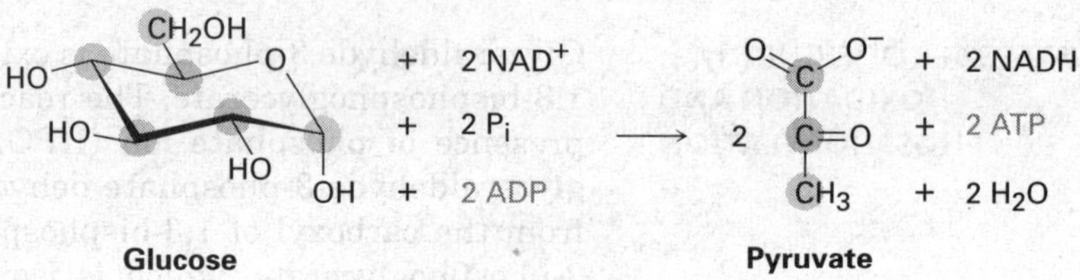

Pyruvate can undergo several further transformations, depending on the conditions and on the organism. Most commonly, pyruvate is converted to acetyl CoA through a complex, multistep sequence of reactions that requires three different enzymes and four different coenzymes. All the individual steps are well understood and have simple laboratory analogies, although their explanations are a bit outside the scope of this book.

Problem 17.4 Identify the steps in glycolysis in which ATP is produced.

Problem 17.5 Look at the entire glycolysis pathway shown in Figure 17.4 and make a list of the kinds of organic reactions that take place—nucleophilic acyl substitutions, aldol reactions, E1cB reactions, and so forth.

17.4 The Citric Acid Cycle

The first two stages of catabolism result in the conversion of fats and carbohydrates into acetyl groups that are bonded through a thioester link to coenzyme A. Acetyl CoA now enters the third stage of catabolism, the **citric acid cycle**, also called the *tricarboxylic acid (TCA) cycle* or *Krebs cycle* after Hans Krebs, who unraveled its complexities in 1937. The overall result of the cycle is the conversion of an acetyl group into two molecules of CO_2 plus reduced coenzymes by the eight-step sequence of reactions shown in Figure 17.5.

MECHANISM

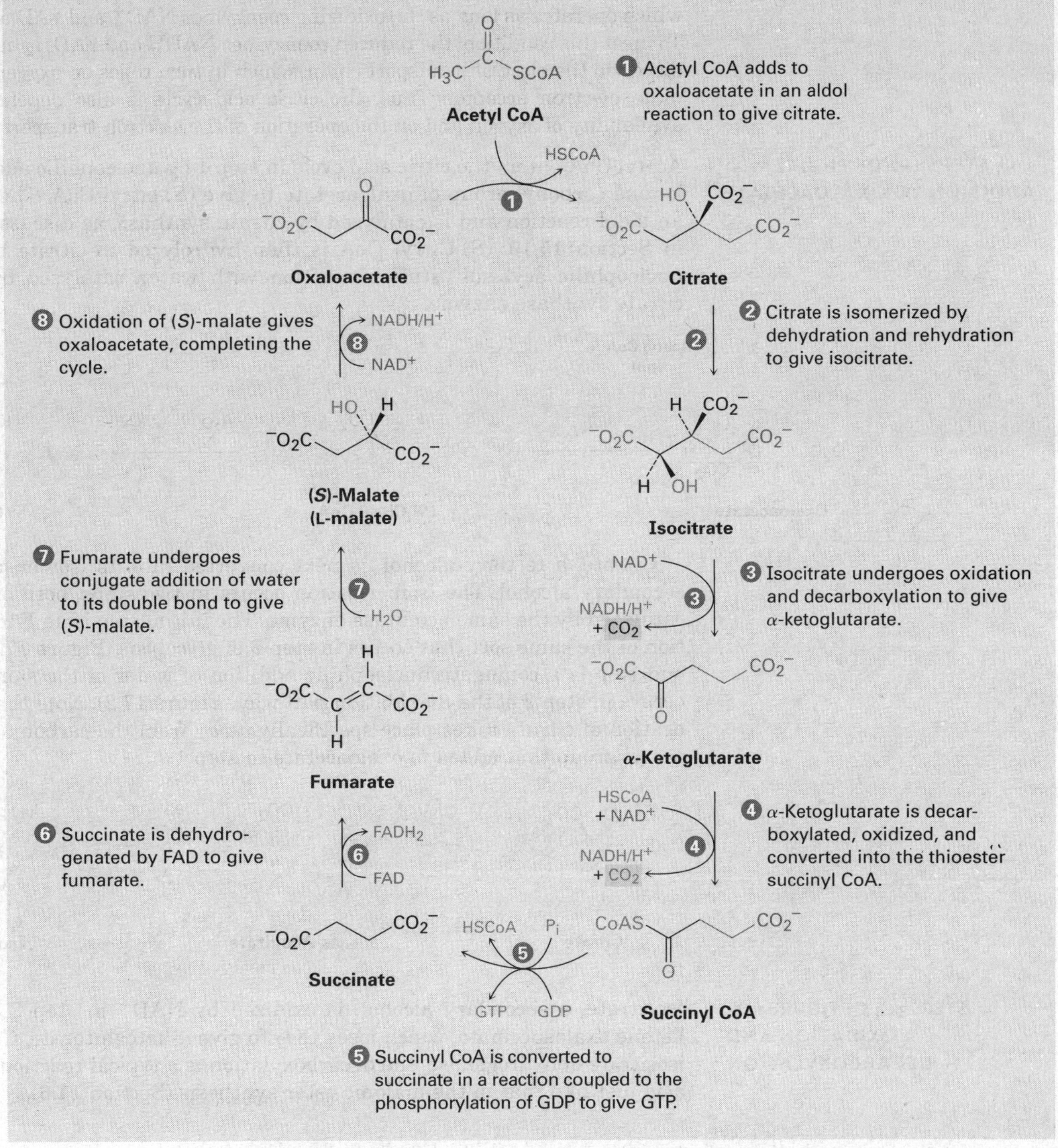

Figure 17.5 The citric acid cycle is an eight-step series of reactions that results in the conversion of an acetyl group into two molecules of CO_2 plus reduced coenzymes. Individual steps are explained in the text.

As its name implies, the citric acid *cycle* is a closed loop of reactions in which the product of the last step (oxaloacetate) is a reactant in the first step. The intermediates are constantly regenerated and flow continuously through the cycle, which operates as long as the oxidizing coenzymes NAD^+ and FAD are available. To meet this condition, the reduced coenzymes NADH and $FADH_2$ must be reoxidized via the electron-transport chain, which in turn relies on oxygen as the ultimate electron acceptor. Thus, the citric acid cycle is also dependent on the availability of oxygen and on the operation of the electron-transport chain.

STEPS 1–2 OF FIGURE 17.5: ADDITION TO OXALOACETATE

Acetyl CoA enters the citric acid cycle in step 1 by nucleophilic addition to the ketone carbonyl group of oxaloacetate to give (*S*)-citryl CoA. The addition is an aldol reaction and is catalyzed by citrate synthase, as discussed in detail in Section 15.10. (*S*)-Citryl CoA is then hydrolyzed to citrate by a typical nucleophilic acyl substitution reaction with water, catalyzed by the same citrate synthase enzyme.

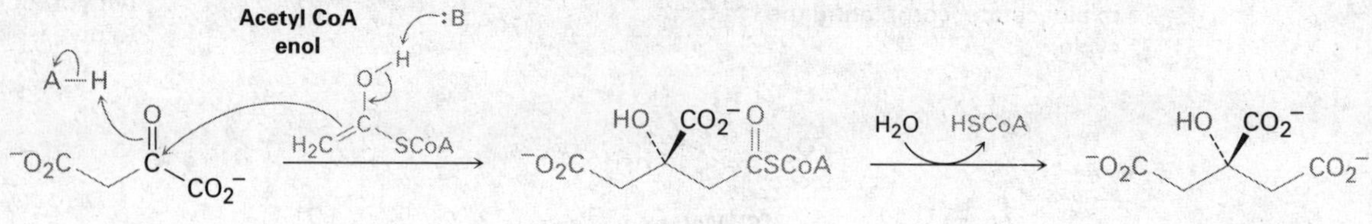

Oxaloacetate **(*S*)-Citryl CoA** **Citrate**

Citrate, a tertiary alcohol, is next converted into its isomer, isocitrate, a secondary alcohol. The isomerization occurs in two steps, both of which are catalyzed by the same aconitase enzyme. The initial step is an E1cB dehydration of the same sort that occurs in step 9 of glycolysis (Figure 17.4). The second step is a conjugate nucleophilic addition of water of the same sort that occurs in step 2 of the β-oxidation pathway (Figure 17.2). Note that the dehydration of citrate takes place specifically *away* from the carbon atoms of the acetyl group that added to oxaloacetate in step 1.

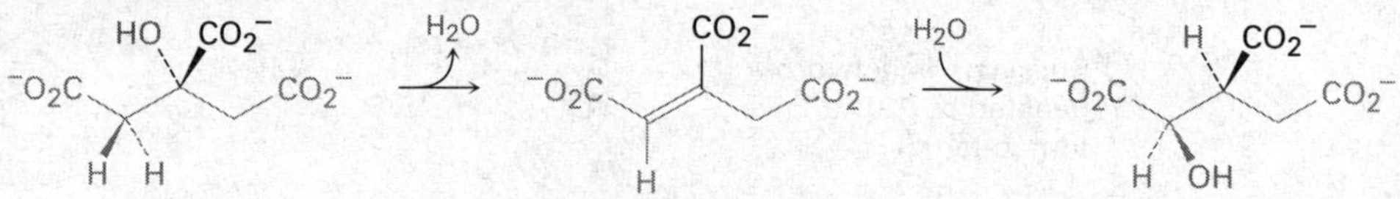

Citrate ***cis*-Aconitate** **Isocitrate**

STEPS 3–4 OF FIGURE 17.5: OXIDATION AND DECARBOXYLATION

Isocitrate, a secondary alcohol, is oxidized by NAD^+ in step 3 to give the ketone oxalosuccinate, which loses CO_2 to give α-ketoglutarate. Catalyzed by isocitrate dehydrogenase, the decarboxylation is a typical reaction of a β-keto acid just like that in the malonic ester synthesis (Section 11.6).

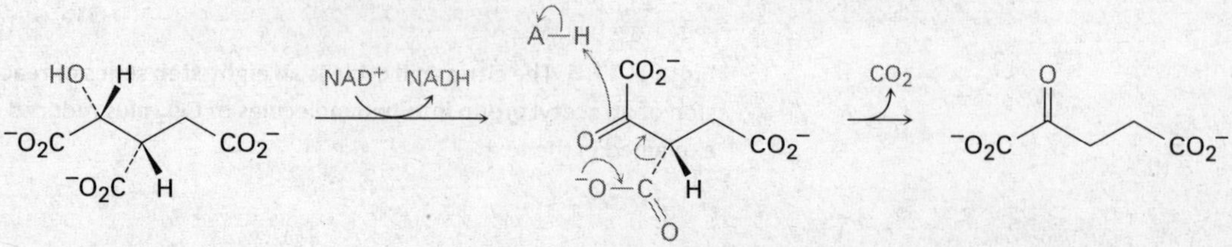

Isocitrate **Oxalosuccinate** **α-Ketoglutarate**

The transformation of α-ketoglutarate to succinyl CoA in step 4 is a multistep process analogous to the transformation of pyruvate to acetyl CoA that we saw in the previous section. Like the pyruvate conversion, the α-ketoglutarate conversion requires a number of different enzymes and coenzymes.

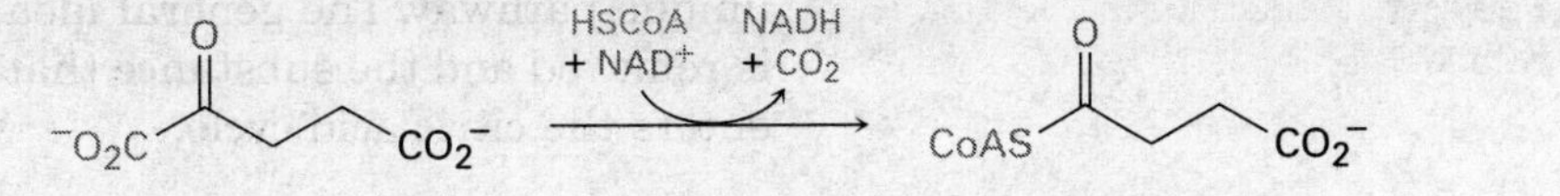

α-Ketoglutarate **Succinyl CoA**

STEPS 5–6 OF FIGURE 17.5: HYDROLYSIS AND DEHYDROGENATION

Succinyl CoA is hydrolyzed to succinate in step 5. The reaction is catalyzed by succinyl CoA synthetase and is coupled with phosphorylation of guanosine diphosphate (GDP) to give guanosine triphosphate (GTP). Succinate is then dehydrogenated by FAD and succinate dehydrogenase to give fumarate—a process analogous to that of step 1 in the β-oxidation pathway.

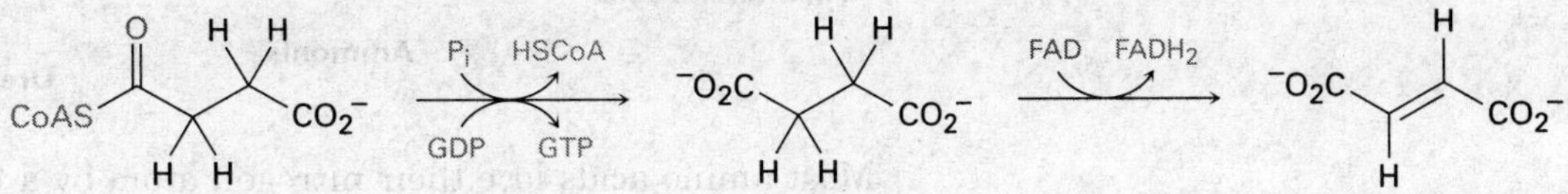

Succinyl CoA **Succinate** **Fumarate**

STEPS 7–8 OF FIGURE 17.5: REGENERATION OF OXALOACETATE

The final two steps in the citric acid cycle are the conjugate nucleophilic addition of water to fumarate to yield (*S*)-malate (L-malate) and the oxidation of malate by NAD^+ to give oxaloacetate. The addition is catalyzed by fumarase and is mechanistically similar to the addition of water to *cis*-aconitate in step 2. The oxidation is catalyzed by malate dehydrogenase. At this point, the citric acid cycle has returned to the beginning, ready to revolve again.

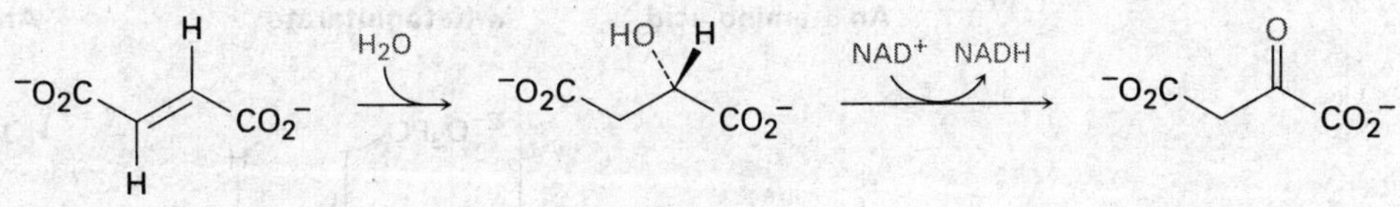

Fumarate **(*S*)-Malate** **Oxaloacetate**

The overall result of the citric acid cycle is

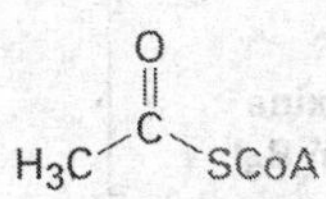

$$+\ 3\ NAD^+ + FAD + GDP + P_i + 2\ H_2O$$

Acetyl CoA

$$\longrightarrow\ 2\ CO_2 + HSCoA + 3\ NADH + 2\ H^+ + FADH_2 + GTP$$

Problem 17.6 Which of the substances in the citric acid cycle are tricarboxylic acids, thus giving the cycle its alternate name?

Problem 17.7 Write mechanisms for the reactions in step 2 of the citric acid cycle, the dehydration of citrate and the addition of water to aconitate.

17.5 Catabolism of Proteins: Transamination

The catabolism of proteins is more complex than that of fats and carbohydrates because each of the 20 amino acids is degraded through its own unique pathway. The general idea, however, is that the amino nitrogen atom is removed and the substance that remains is converted into a compound that enters the citric acid cycle.

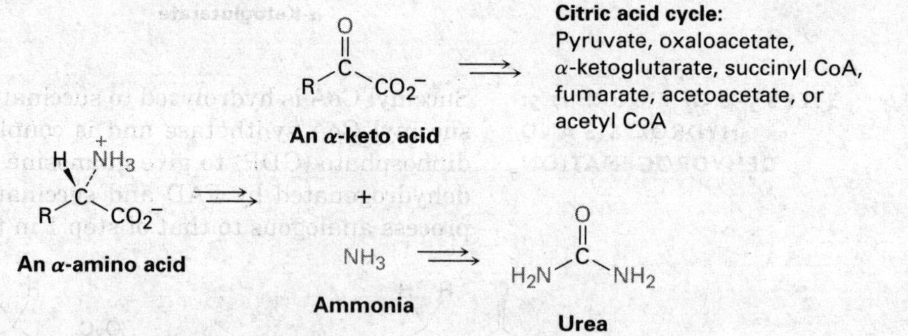

Most amino acids lose their nitrogen atom by a **transamination** reaction in which the $-NH_2$ group of the amino acid changes places with the keto group of α-ketoglutarate. The products are a new α-keto acid and glutamate. The overall process occurs in two parts, is catalyzed by various aminotransferases, and involves participation of the coenzyme pyridoxal phosphate (PLP), a derivative of pyridoxine (vitamin B_6).

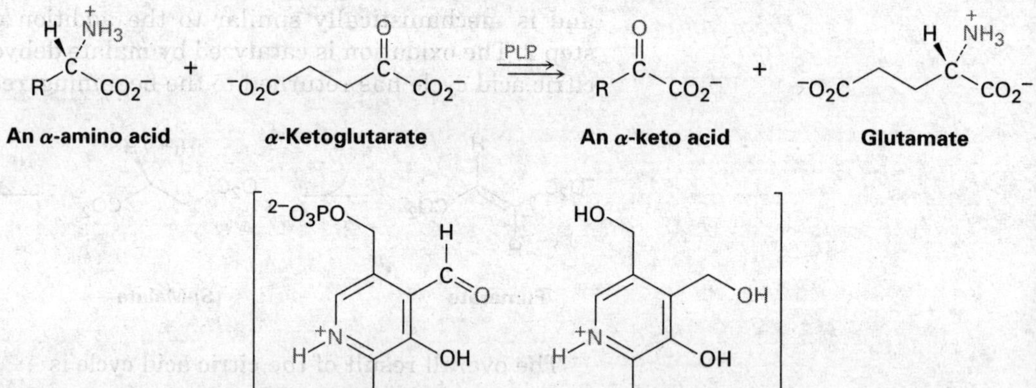

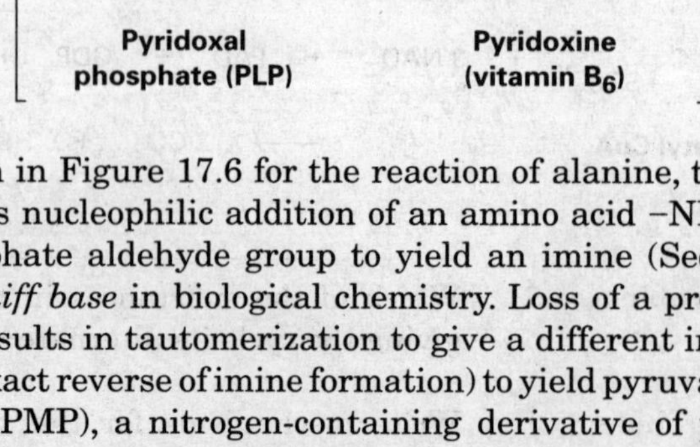

As shown in Figure 17.6 for the reaction of alanine, the key step in transamination is nucleophilic addition of an amino acid $-NH_2$ group to the pyridoxal phosphate aldehyde group to yield an imine (Section 9.9), frequently called a *Schiff base* in biological chemistry. Loss of a proton from the α position then results in tautomerization to give a different imine, which is hydrolyzed (the exact reverse of imine formation) to yield pyruvate and pyridoxamine phosphate (PMP), a nitrogen-containing derivative of pyridoxal phosphate. Pyruvate is then converted into acetyl CoA (Section 17.3), which enters the citric acid cycle for further catabolism.

MECHANISM

Figure 17.6 Mechanism of the transamination of alanine by pyridoxal phosphate (PLP) to give pyruvate.

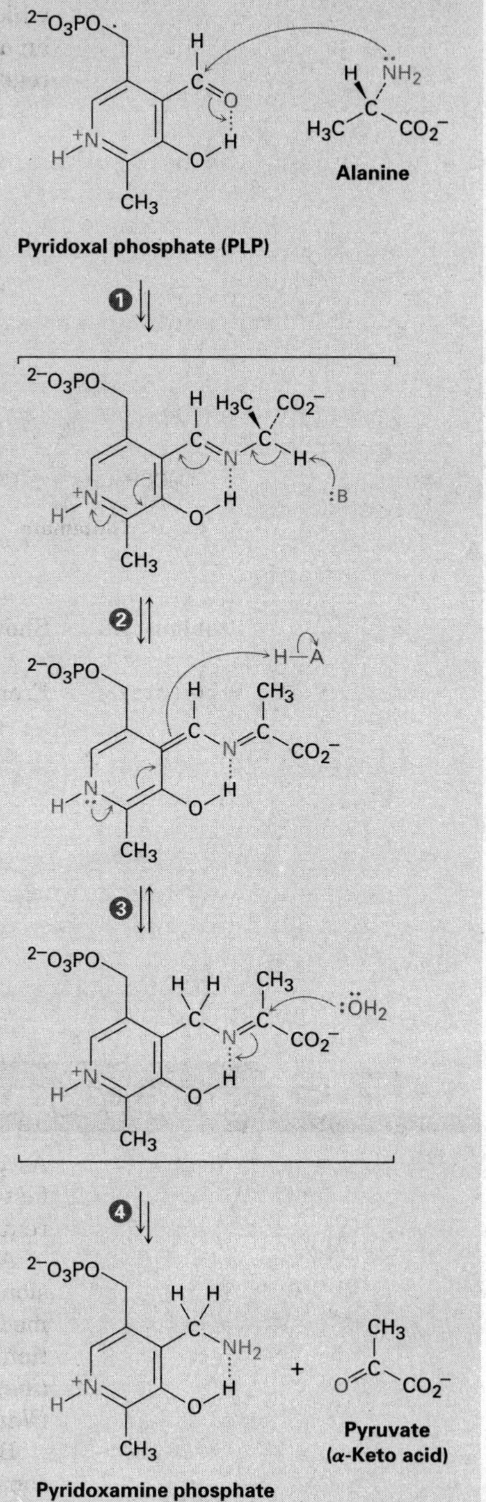

PLP–alanine imine

α-Keto acid imine

α-Keto acid imine tautomer

❶ Alanine reacts with pyridoxal phosphate by nucleophilic addition of its $-NH_2$ to the carbonyl group, giving an imine.

❷ Deprotonation of the acidic α carbon of alanine gives an α-keto acid imine . . .

❸ . . . that is reprotonated on carbon. The result of this deprotonation/reprotonation sequence is tautomerization of the imine C=N bond.

❹ Hydrolysis of the imine gives the transamination products pyridoxamine phosphate (PMP) and pyruvate, an α-keto acid.

The pyridoxamine phosphate formed by transamination transfers its nitrogen atom to α-ketoglutarate by the reverse of the steps in Figure 17.6, thereby forming glutamate and regenerating pyridoxal phosphate for further use. Glutamate, which now contains the nitrogen atom of the former amino acid, next undergoes an oxidation to an imine and subsequent hydrolysis to yield ammonium ion plus regenerated α-ketoglutarate. The oxidation of the amine to an imine is similar to the oxidation of a secondary alcohol to a ketone, and is carried out by NAD^+.

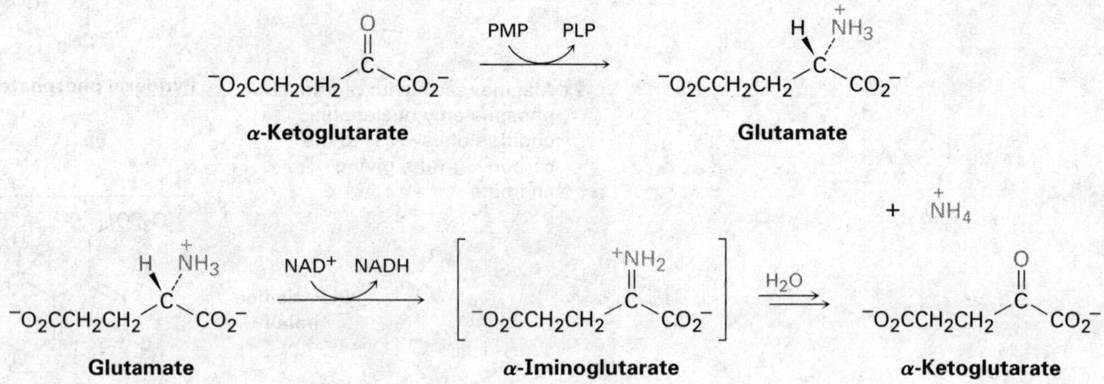

Problem 17.8 Show the structure of the α-keto acid formed by transamination of leucine.

Problem 17.9 From what amino acid was the following α-keto acid derived?

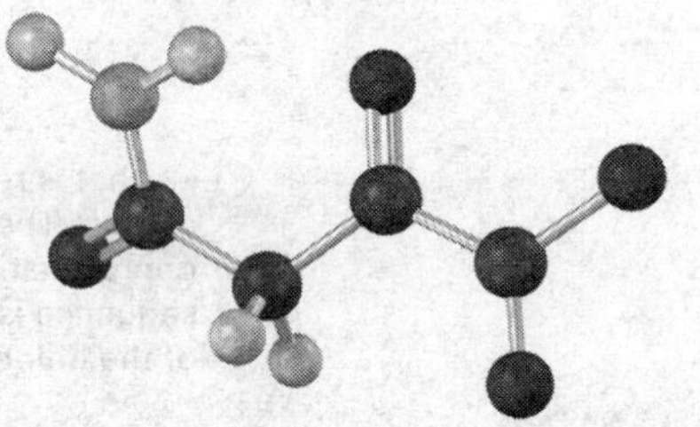

17.6 Some Conclusions about Biological Chemistry

As promised in the chapter introduction, the past few sections have been a fast-paced tour of a large number of reactions. Following it all undoubtedly required a lot of work and some page-turning to look at earlier sections.

After examining the various metabolic pathways, perhaps the main conclusion about biological chemistry is the remarkable similarity between the mechanisms of biological reactions and the mechanisms of laboratory reactions. In all the pathways described in this chapter, terms like *imine formation, aldol reaction, nucleophilic acyl substitution reaction, E1cB reaction,* and *Claisen reaction* appear constantly.

Biological reactions aren't mysterious; there are clear, understandable reasons for the reactions carried out within living organisms. Biological chemistry *is* organic chemistry. Understanding how nature works and how living organisms function is a fascinating field of study.

INTERLUDE

Statin Drugs

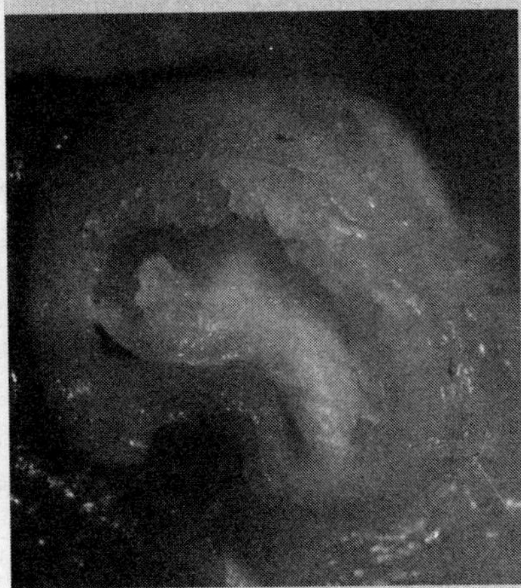

The buildup of cholesterol deposits inside arteries can cause coronary heart disease, a leading cause of death for both men and women.

Let's end this text by looking at a remarkable example of how organic and biological chemistry are changing modern medicine. Coronary heart disease—the buildup of cholesterol-containing plaques on the walls of heart arteries—is the leading cause of death for both men and women above age 20. It's estimated that up to one-third of women and one-half of men will develop the disease at some point in their lives.

The onset of coronary heart disease is directly correlated with blood cholesterol levels, and the first step in disease prevention is to lower those levels. It turns out that only about 25% of your blood cholesterol comes from what you eat; the remaining 75% (about 1000 mg each day) is biosynthesized in your liver from dietary fats and carbohydrates. Thus, any effective plan for lowering your cholesterol level means limiting the amount that your body makes, which in turn means understanding and controlling the metabolic pathway for cholesterol biosynthesis.

As you might imagine, that pathway is somewhat complex, but the chemical details of the more than 20 steps have all been worked out. Cholesterol arises from the simple precursor acetyl CoA, and the rate-limiting step in the pathway is the reduction of 3-hydroxy-3-methylglutaryl CoA (abbreviated HMG-CoA) to mevalonate, brought about by the enzyme HMG-CoA reductase. If that enzyme could be stopped from functioning, cholesterol biosynthesis would also be stopped.

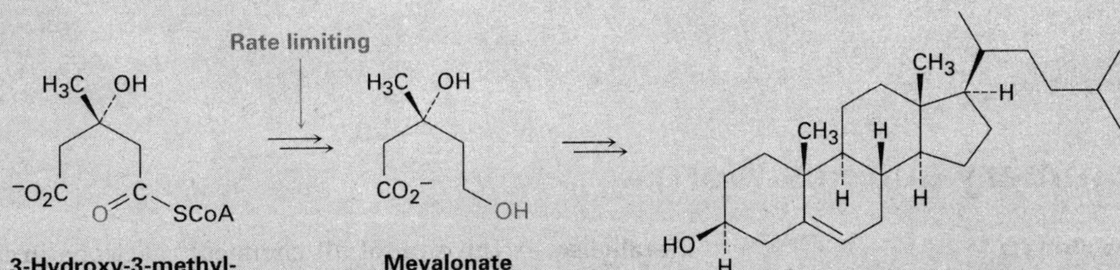

3-Hydroxy-3-methyl-glutaryl coenzyme A (HMG-CoA) **Mevalonate** **Cholesterol**

To find a drug that blocks HMG-CoA reductase, chemists did two experiments on a large number of potential drug candidates isolated from soil microbes. In one experiment, the drug candidate and mevalonate were added to liver extract; in the second experiment, only the drug candidate was added without mevalonate. If cholesterol was produced in the presence of mevalonate but was not produced in the absence of mevalonate, it meant that the drug candidate blocked mevalonate synthesis.

The drugs that block HMG-CoA reductase, and thus control cholesterol synthesis in the body, are called *statins*. They are the most widely prescribed drugs in the world, with an estimated $14.6 billion in annual sales. So effective are they that in the 10-year period following their introduction

continued

INTERLUDE

in 1994, the death rate from coronary heat disease decreased by 33% in the United States. Atorvastatin (Lipitor), simvastatin (Zocor), rosuvastatin (Crestor), pravastatin (Pravachol), and lovastatin (Mevacor) are examples. An X-ray crystal structure of the active site in the HMG-CoA reductase enzyme is shown in the accompanying graphic, along with a molecule of atorvastatin (blue) that is tightly bound in the active site and stops the enzyme from functioning. A good understanding of organic chemistry certainly paid off in this instance.

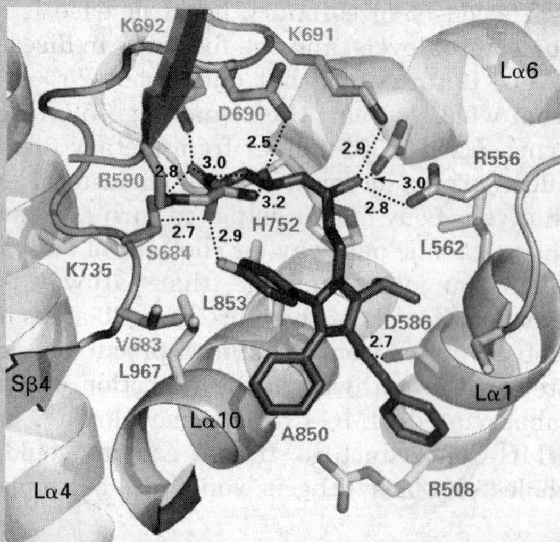

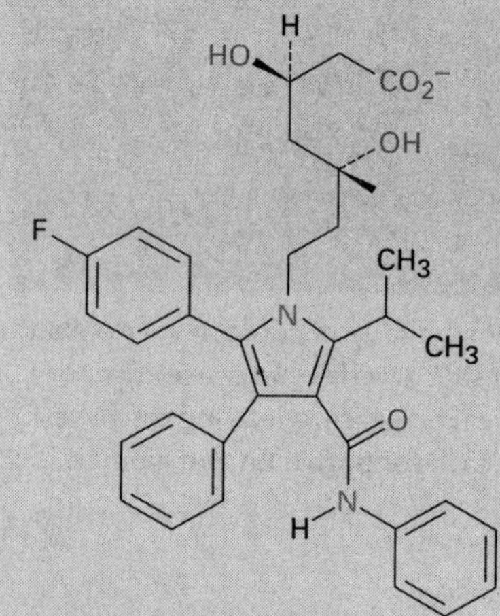

Atorvastatin (Lipitor)

Summary and Key Words

- **anabolism** 572
- **β-oxidation pathway** 576
- **catabolism** 572
- **citric acid cycle** 584
- **glycolysis** 579
- **metabolism** 572
- **transamination** 588

Metabolism is the sum of all chemical reactions in the body. Reactions that break down larger molecules into smaller ones are **catabolic** and reactions that build up larger molecules from smaller ones are called **anabolic**. Although the details of specific biochemical pathways are sometimes complex, all the reactions that occur follow the normal rules of organic chemical reactivity.

The catabolism of fats begins with hydrolysis to give glycerol and fatty acids. The fatty acids are degraded in the four-step **β-oxidation pathway** by removal of two carbons at a time, yielding acetyl CoA. Catabolism of carbohydrates begins with the hydrolysis of glycoside bonds to give glucose, which is degraded in the ten-step **glycolysis** pathway. Pyruvate, the initial product of glycolysis, is then converted into acetyl CoA. The acetyl groups produced by degradation of fats and carbohydrates next enter the eight-step **citric acid cycle**, where they are further degraded into CO_2. The cycle is a closed loop of reactions in which the product of the final step (oxaloacetate) is a reactant in the first step. The intermediates are constantly regenerated and flow continuously through the cycle, which operates as long as the oxidizing coenzymes NAD^+ and FAD are available.

Catabolism of proteins is more complex than that of fats or carbohydrates because each of the 20 different amino acids is degraded by its own unique pathway. In general, though, the amino nitrogen atoms are removed and the substances that remain are converted into compounds that enter the citric acid cycle. Most amino acids lose their nitrogen atom by **transamination**, a reaction in which the $-NH_2$ group of the amino acid changes places with the keto group of an α-keto acid such as α-ketoglutarate. The products are a new α-keto acid and glutamate.

The energy released in all catabolic pathways is used in the electron-transport chain to make molecules of adenosine triphosphate (ATP), the final result of food catabolism. ATP couples with and drives many otherwise unfavorable reactions.

Exercises

Visualizing Chemistry

(Problems 17.1–17.9 appear within the chapter.)

OWL

Interactive versions of these problems are assignable in OWL.

17.10 Identify the following intermediate in the citric acid cycle, and tell whether it has R or S stereochemistry:

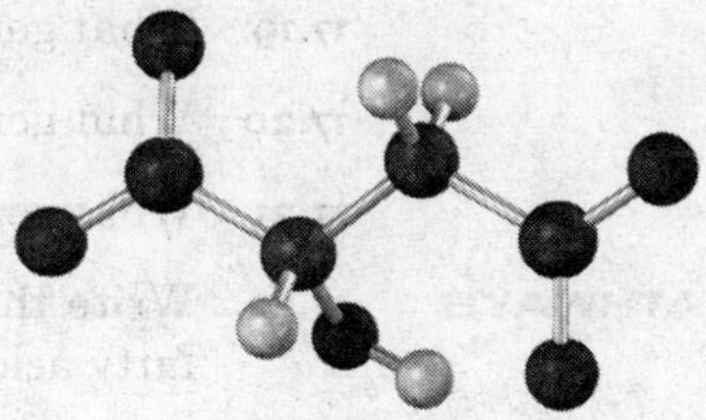

17.11 Identify the amino acid that is a catabolic precursor of each of the following α-keto acids:

(a)

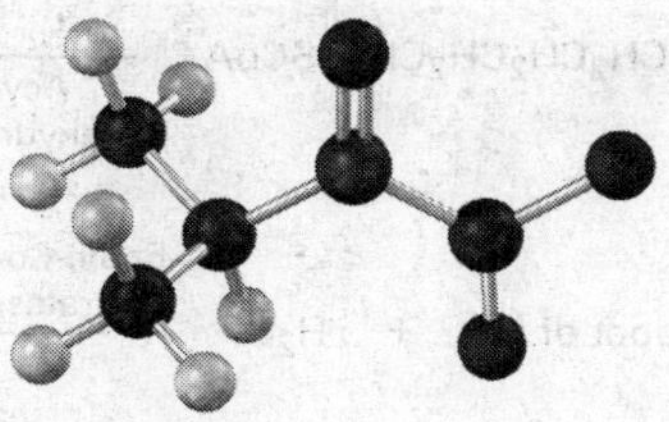

(b)

Additional Problems

General Metabolism

17.12 What is the difference between digestion and metabolism? Between anabolism and catabolism?

17.13 What chemical events occur during the digestion of the following kinds of food molecules?

(a) Fats **(b)** Complex carbohydrates **(c)** Proteins

17.14 How many grams of acetyl CoA (mol wt = 809.6 amu) are produced by catabolism of the following substances?
(a) 100.0 g glucose **(b)** 100.0 g palmitic acid **(c)** 100.0 g maltose

17.15 Which of the substances listed in Problem 17.14 is the most efficient precursor of acetyl CoA on a weight basis?

17.16 Lactate, a product of glucose catabolism in oxygen-starved muscles, can be converted into pyruvate by oxidation. What coenzyme do you think is needed? Write the equation in the normal biochemical format using a curved arrow.

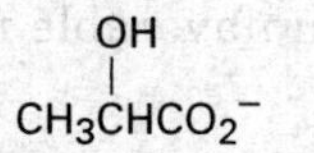
Lactate

17.17 Draw the structure of adenosine monophosphate (AMP), an intermediate in some biochemical pathways.

17.18 How many moles of acetyl CoA are produced by catabolism of the following substances?
(a) 1.0 mol glucose **(b)** 1.0 mol palmitic acid ($C_{15}H_{31}CO_2H$)
(c) 1.0 mol maltose

17.19 What general kind of reaction does ATP carry out?

17.20 What general kind of reaction does FAD carry out?

17.21 What general kind of reaction does NAD^+ carry out?

METABOLIC PATHWAYS

17.22 Write the equation for the final step in the β-oxidation pathway of any fatty acid with an even number of carbon atoms.

17.23 What substance is the starting point of the citric acid cycle, reacting with acetyl CoA in the first step and being regenerated in the last step?

17.24 Show the products of each of the following reactions:

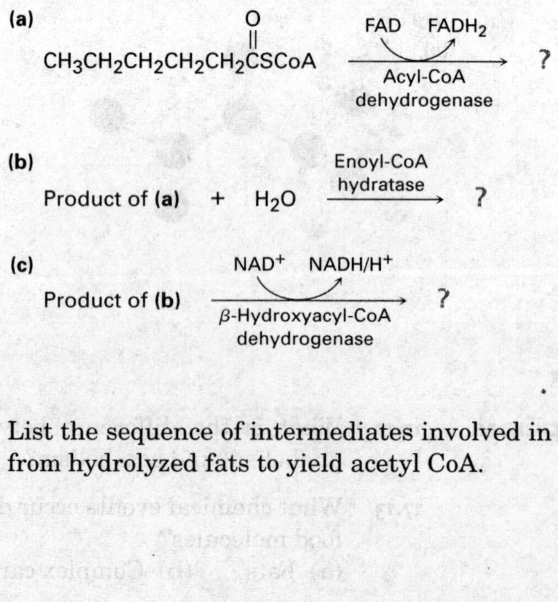

17.25 List the sequence of intermediates involved in the catabolism of glycerol from hydrolyzed fats to yield acetyl CoA.

17.26 What enzyme cofactor is associated with transamination?

17.27 What is the structure of the α-keto acid formed by transamination of each of the following amino acids?
(a) Valine **(b)** Phenylalanine **(c)** Methionine

General Problems

17.28 In the *pentose phosphate* pathway for degrading sugars, ribulose 5-phosphate is converted to ribose 5-phosphate. Propose a mechanism for the isomerization.

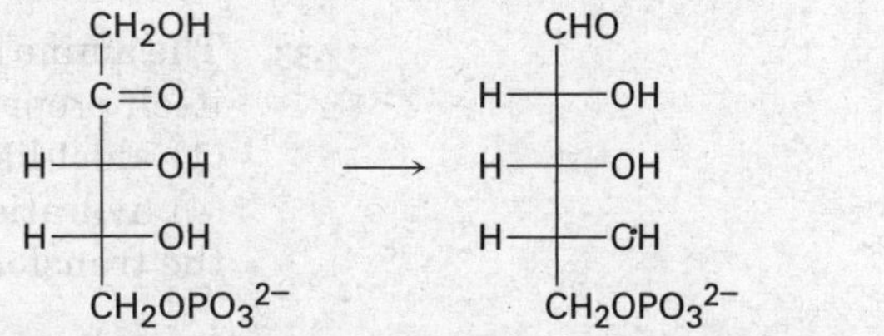

Ribulose 5-phosphate **Ribose 5-phosphate**

17.29 Another step in the pentose phosphate pathway for degrading sugars (see Problem 17.28) is the conversion of ribose 5-phosphate to glyceraldehyde 3-phosphate. What kind of organic process is occurring? Propose a mechanism for the conversion.

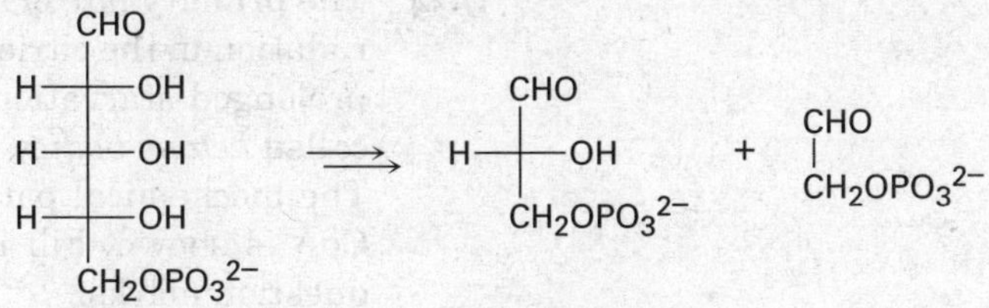

Ribose 5-phosphate **Glyceraldehyde 3-phosphate**

17.30 Fatty acids are synthesized in the body by a sequence that begins with acetyl CoA. The first step is the condensation of two acetyl CoA molecules to yield acetoacetyl CoA, which undergoes three further enzyme-catalyzed steps, yielding butyryl CoA. Based on the kinds of reactions that occur in the β-oxidation pathway, what do you think are the three further steps of fatty-acid biosynthesis?

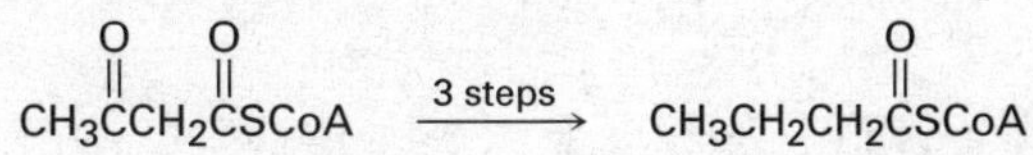

Acetoacetyl CoA **Butyryl CoA**

17.31 One of the steps in the *gluconeogenesis* pathway for synthesizing glucose in the body is the reaction of pyruvate with CO_2 to yield oxaloacetate. Tell what kind of reaction is occurring, and suggest a mechanism.

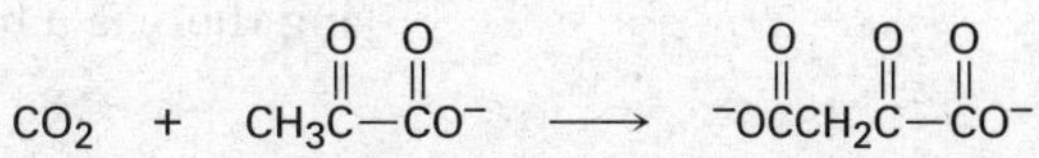

Pyruvate **Oxaloacetate**

17.32 Another step in gluconeogenesis (see Problem 17.31) is the conversion of oxaloacetate to phosphoenolpyruvate by decarboxylation and phosphorylation. Tell what kind of reaction is occurring, and suggest a mechanism.

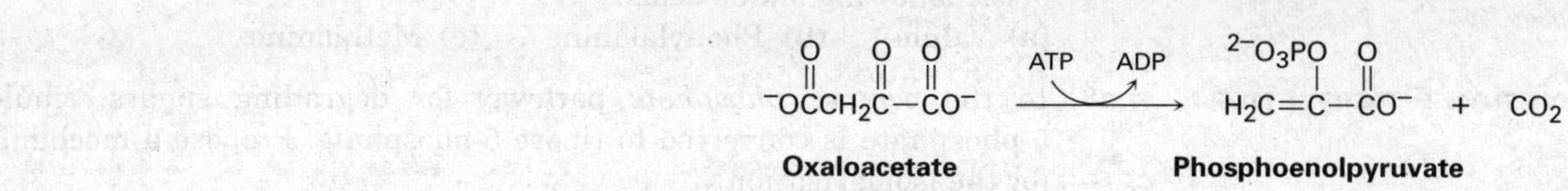

17.33 The amino acid leucine is biosynthesized from α-ketoisocaproate, which is itself prepared from α-ketoisovalerate by a multistep route that involves (1) aldol-like reaction with acetyl CoA, (2) hydrolysis, (3) dehydration, (4) hydration, (5) oxidation, and (6) decarboxylation. Show the steps in the transformation.

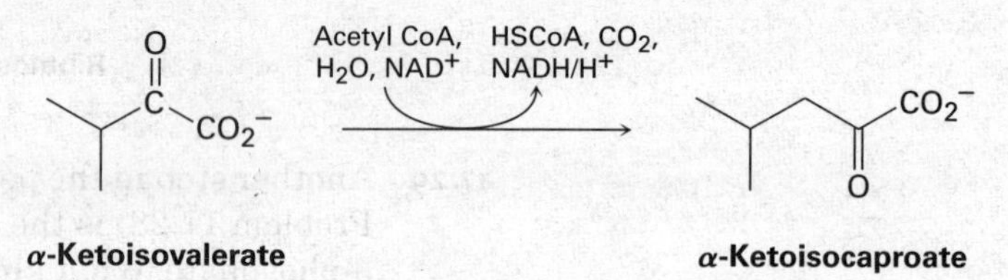

17.34 The primary fate of acetyl CoA under normal metabolic conditions is degradation in the citric acid cycle to yield CO_2. When the body is stressed by prolonged starvation, however, acetyl CoA is converted into compounds called *ketone bodies,* which can be used by the brain as a temporary fuel. The biochemical pathway for the synthesis of ketone bodies from acetyl CoA is shown. Fill in the missing information represented by the four question marks.

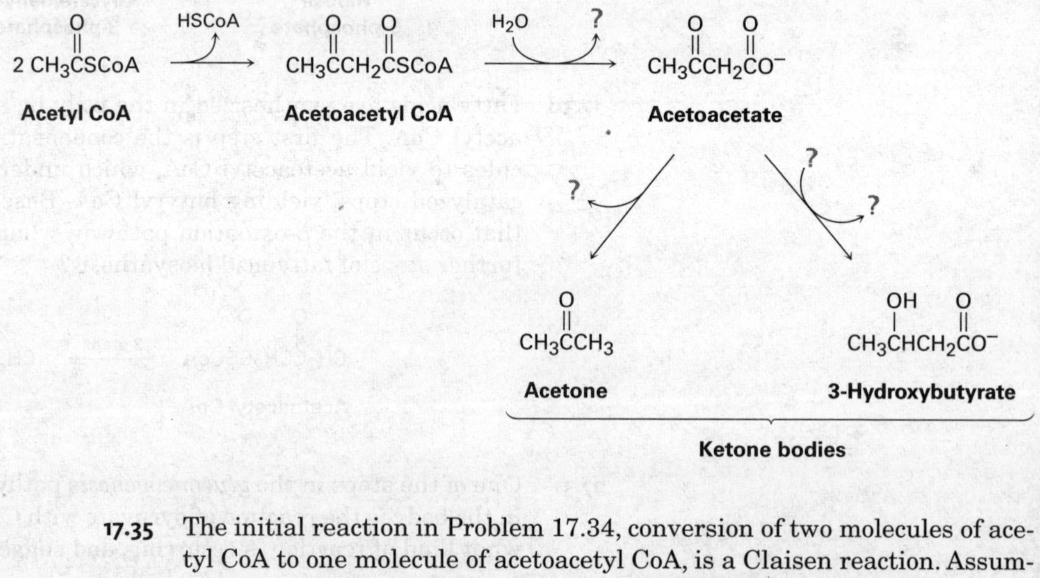

17.35 The initial reaction in Problem 17.34, conversion of two molecules of acetyl CoA to one molecule of acetoacetyl CoA, is a Claisen reaction. Assuming there is a base present, show the mechanism of the reaction.

17.36 The amino acid cysteine, $C_3H_7NO_2S$, is biosynthesized from a substance called cystathionine by a multistep pathway.

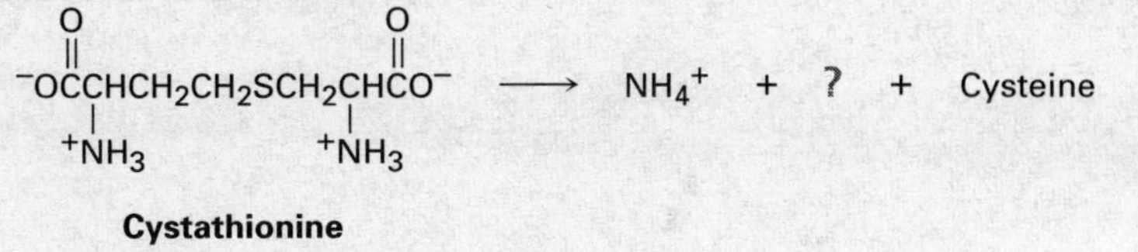

(a) The first step is a transamination. What is the product?

(b) The second step is an E1cB reaction. Show the products and the mechanism of the reaction.

(c) The final step is a double-bond reduction. What is the product represented by the question mark in the equation?

17.37 The amino acid tyrosine is metabolized by a series of steps that include the following transformations. Propose a mechanism for the conversion of fumaroylacetoacetate into fumarate plus acetoacetate.

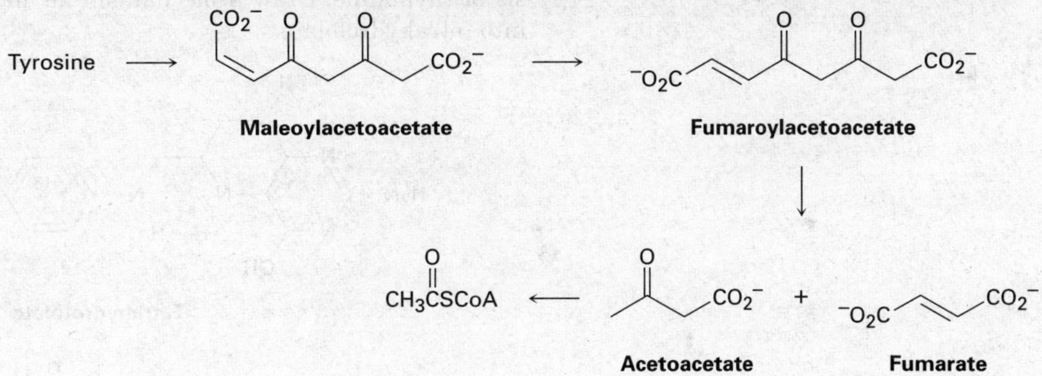

17.38 Propose a mechanism for the conversion of acetoacetate into acetyl CoA (Problem 17.37).

IN THE MEDICINE CABINET **17.39** Many drugs function by interfering with metabolic pathways. The anticancer drug methotrexate, for instance, inhibits the dihydrofolate reductase enzyme that catalyzes the following reactions:

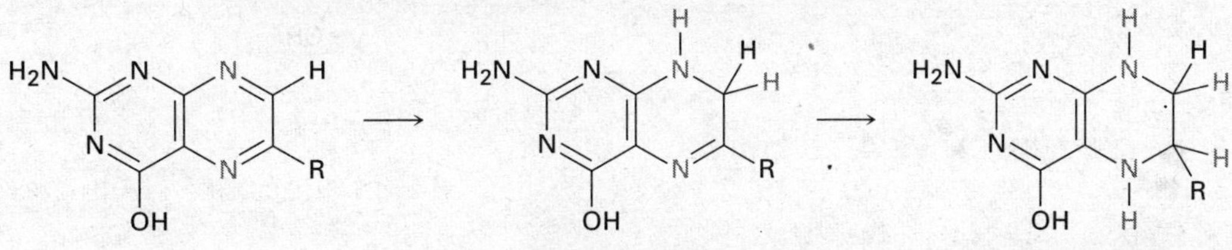

Folate Dihydrofolate Tetrahydrofolate

What cofactor is likely to be involved in these reductions? Propose a mechanism.

17.40 Methylation of tetrahydrofolate produces a cofactor called 5-methyltetrahydrofolate that is used in the conversion of homocysteine to methionine. Propose a mechanism for this reaction.

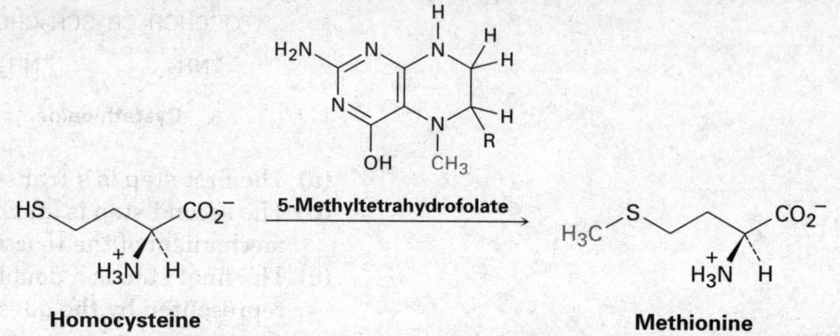

17.41 In addition to transferring a methyl group, tetrahydrofolate can also transfer a formaldehyde group, a process that is critical for the biosynthesis of thymidine. Draw a mechanism for incorporation of formaldehyde into tetrahydrofolate.

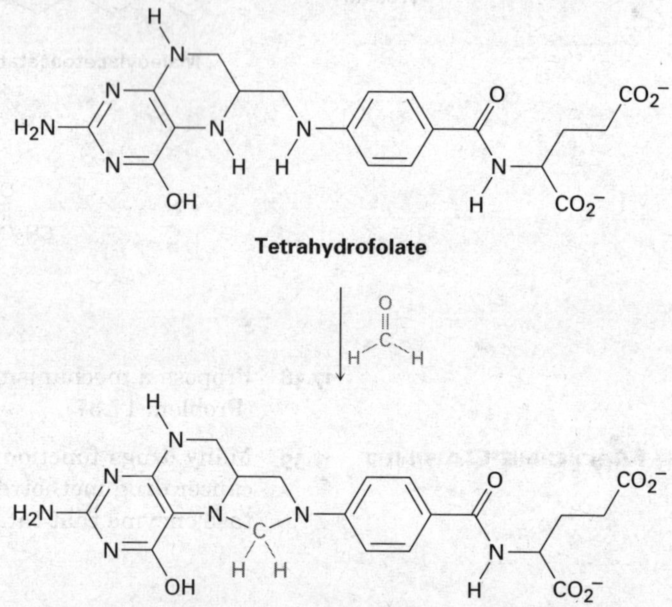

APPENDIX A

Nomenclature of Polyfunctional Organic Compounds

With more than 37 million organic compounds known and several thousand more being created daily, naming them all is a real problem. Part of the problem is due to the sheer complexity of organic structures, but part is also due to the fact that chemical names have more than one purpose. For Chemical Abstracts Service (CAS), which catalogs and indexes the worldwide chemical literature, each compound must have only one correct name. It would be chaos if half the entries for CH_3Br were indexed under "M" for methyl bromide and half under "B" for bromomethane. Furthermore, a CAS name must be strictly systematic so that it can be assigned and interpreted by computers; common names are not allowed.

People, however, have different requirements than computers. For people—which is to say chemists in their spoken and written communications—it's best that a chemical name be pronounceable and that it be as easy as possible to assign and interpret. Furthermore, it's convenient if names follow historical precedents, even if that means a particularly well-known compound might have more than one name. People can readily understand that bromomethane and methyl bromide both refer to CH_3Br.

As noted in the text, chemists overwhelmingly use the nomenclature system devised and maintained by the International Union of Pure and Applied Chemistry, or IUPAC. Rules for naming monofunctional compounds were given throughout the text as each new functional group was introduced, and a list of where these rules can be found is given in Table A.1.

Table A.1 Nomenclature Rules for Functional Groups

Functional group	Text section	Functional group	Text section
Acid anhydrides	10.1	Aromatic compounds	5.2
Acid halides	10.1	Carboxylic acids	10.1
Alcohols	8.1	Cycloalkanes	2.7
Aldehydes	9.2	Esters	10.1
Alkanes	2.3	Ethers	8.1
Alkenes	3.1	Ketones	9.2
Alkyl halides	7.1	Nitriles	10.1
Alkynes	3.1	Phenols	8.1
Amides	10.1	Sulfides	8.8
Amines	12.1	Thiols	8.8

Naming a monofunctional compound is reasonably straightforward, but even experienced chemists often encounter problems when faced with naming a complex polyfunctional compound. Take the following compound, for instance. It has three functional groups, ester, ketone, and C=C, but how should it be named? As an ester with an *-oate* ending, a ketone with an *-one* ending, or an alkene with an *-ene* ending? It's actually named methyl 3-(2-oxocyclohex-6-enyl)propanoate.

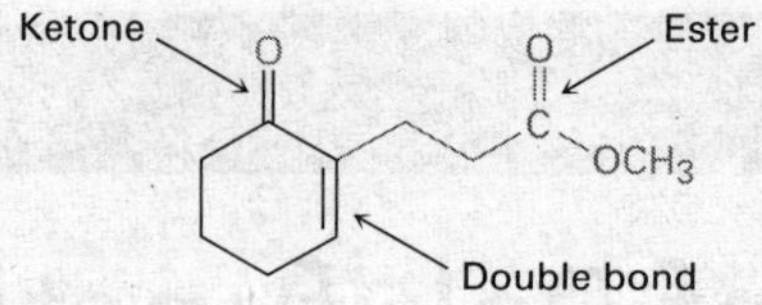

Methyl 3-(2-oxocyclohex-6-enyl)propanoate

The name of a polyfunctional organic molecule has four parts—suffix, parent, prefixes, and locants—which must be identified and expressed in the proper order and format. Let's look at each of the four.

Name Part 1. The Suffix: Functional-Group Precedence

Although a polyfunctional organic molecule might contain several different functional groups, we must choose just one suffix for nomenclature purposes. It's not correct to use two suffixes. Thus, keto ester **1** must be named either as a ketone with an *-one* suffix or as an ester with an *-oate* suffix, but it can't be named as an *-onoate.* Similarly, amino alcohol **2** must be named either as an alcohol (*-ol*) or as an amine (*-amine*), but it can't be named as an *-olamine* or *-aminol.*

1. 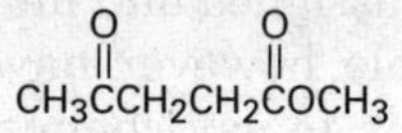

2. 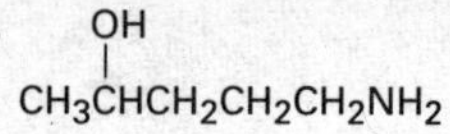

The only exception to the rule requiring a single suffix is when naming compounds that have double or triple bonds. Thus, the unsaturated acid $H_2C{=}CHCH_2CO_2H$ is but-3-enoic acid, and the acetylenic alcohol $HC{\equiv}CCH_2CH_2CH_2OH$ is pent-5-yn-1-ol.

How do we choose which suffix to use? Functional groups are divided into two classes, **principal groups** and **subordinate groups**, as shown in Table A.2. Principal groups can be cited either as prefixes or as suffixes, while subordinate groups are cited only as prefixes. Within the principal groups, an order of priority has been established, with the proper suffix for a given compound determined by choosing the principal group of highest priority. For example, Table A.2 indicates that keto ester **1** should be named as an ester rather than as a ketone because an ester functional group is higher in priority than a ketone. Similarly, amino alcohol **2** should be named as an alcohol rather than as an amine. Thus, the name of **1** is methyl 4-oxopentanoate, and the name of **2** is 5-aminopentan-2-ol. Further examples are shown:

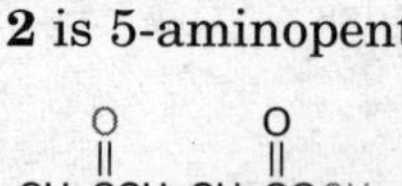

1. Methyl 4-oxopentanoate
(an ester with a ketone group)

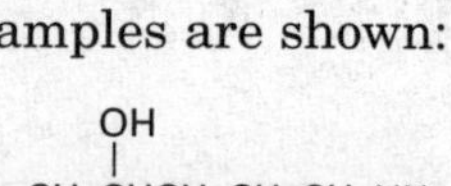

2. 5-Aminopentan-2-ol
(an alcohol with an amine group)

3. Methyl 5-methyl-6-oxohexanoate
(an ester with an aldehyde group)

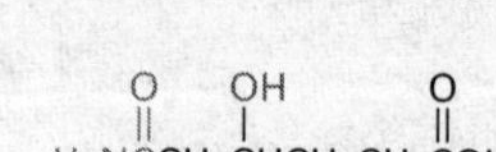

4. 5-Carbamoyl-4-hydroxypentanoic acid
(a carboxylic acid with amide and alcohol groups)

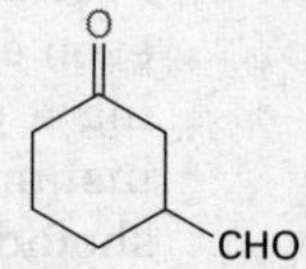

5. 3-Oxocyclohexanecarbaldehyde
(an aldehyde with a ketone group)

Table **A.2** Classification of Functional Groups[a]

Functional group	Name as suffix	Name as prefix
Principal groups		
Carboxylic acids	-oic acid -carboxylic acid	carboxy
Acid anhydrides	-oic anhydride -carboxylic anhydride	—
Esters	-oate -carboxylate	alkoxycarbonyl
Thioesters	-thioate -carbothioate	alkylthiocarbonyl
Acid halides	-oyl halide -carbonyl halide	halocarbonyl
Amides	-amide -carboxamide	carbamoyl
Nitriles	-nitrile -carbonitrile	cyano
Aldehydes	-al -carbaldehyde	oxo
Ketones	-one	oxo
Alcohols	-ol	hydroxy
Phenols	-ol	hydroxy
Thiols	-thiol	mercapto
Amines	-amine	amino
Imines	-imine	imino
Ethers	ether	alkoxy
Sulfides	sulfide	alkylthio
Disulfides	disulfide	—
Alkenes	-ene	—
Alkynes	-yne	—
Alkanes	-ane	—
Subordinate groups		
Azides	—	azido
Halides	—	halo
Nitro compounds	—	nitro

[a]Principal groups are listed in order of decreasing priority; subordinate groups have no priority order.

Name Part 2. The Parent: Selecting the Main Chain or Ring

The parent, or base, name of a polyfunctional organic compound is usually easy to identify. If the principal group of highest priority is part of an open chain, the parent name is that of the longest chain containing the largest number of principal groups. For example, compounds **6** and **7** are isomeric aldehydo amides, which must be named as amides rather than as aldehydes according to Table A.2. The longest chain in compound **6** has six carbons, and the substance is therefore named 5-methyl-6-oxohexanamide. Compound **7** also has a chain of six carbons, but the longest chain that contains both principal functional groups has only four carbons. The correct name of **7** is 4-oxo-3-propylbutanamide.

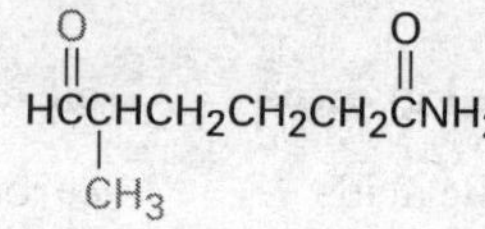

6. 5-Methyl-6-oxo**hexanamide**

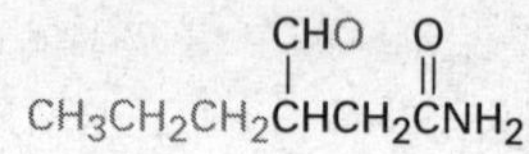

7. 4-Oxo-3-propyl**butanamide**

If the highest-priority principal group is attached to a ring, the parent name is that of the ring system. Compounds **8** and **9**, for instance, are isomeric keto nitriles and must both be named as nitriles according to Table A.2. Substance **8** is named as a benzonitrile because the –CN functional group is a substituent on the aromatic ring, but substance **9** is named as an acetonitrile because the –CN functional group is on an open chain. The correct names are 2-acetyl-(4-bromomethyl)benzonitrile (**8**) and (2-acetyl-4-bromophenyl)acetonitrile (**9**). As further examples, compounds **10** and **11** are both keto acids and must be named as acids, but the parent name in (**10**) is that of a ring system (cyclohexanecarboxylic acid) and the parent name in (**11**) is that of an open chain (propanoic acid). The full names are *trans*-2-(3-oxopropyl)cyclohexanecarboxylic acid (**10**) and 3-(2-oxocyclohexyl)propanoic acid (**11**).

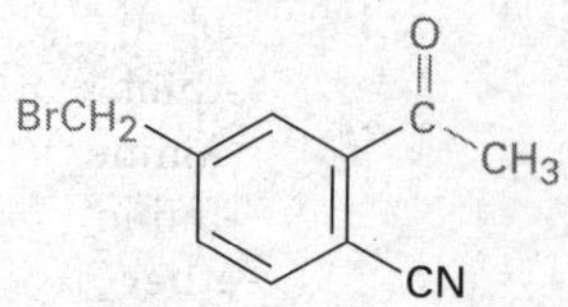

8. 2-Acetyl-(4-bromomethyl)**benzonitrile**

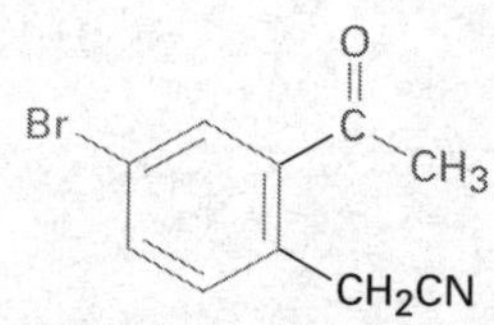

9. (2-Acetyl-4-bromophenyl)**acetonitrile**

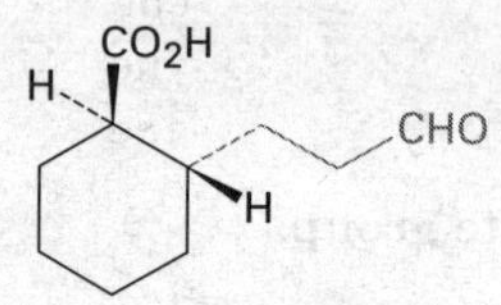

10. ***trans*-2-(3-oxopropyl)cyclohexanecarboxylic acid**

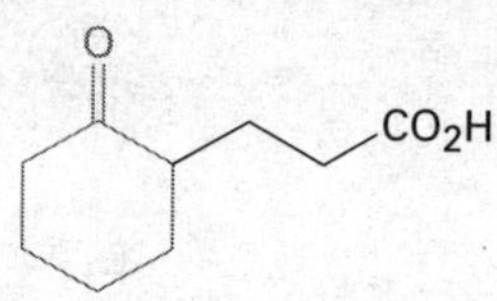

11. 3-(2-Oxocyclohexyl)**propanoic acid**

Name Parts 3 and 4. The Prefixes and Locants

With the parent name and the suffix established, the next step is to identify and give numbers, or *locants,* to all substituents on the parent chain or ring. These substituents include all alkyl groups and all functional groups other than the one cited in the suffix. For example, compound **12** contains three different functional groups (carboxyl, keto, and double bond). Because the carboxyl group is highest in priority and because the longest chain containing the functional groups has seven carbons, **12** is a heptenoic acid. In addition, the main chain has a keto (oxo) substituent and three methyl groups. Numbering from the end nearer the highest-priority functional group, **12** is named (*E*)-2,5,5-trimethyl-4-oxohept-2-enoic acid. Look back at some of the other compounds we've named to see other examples of how prefixes and locants are assigned.

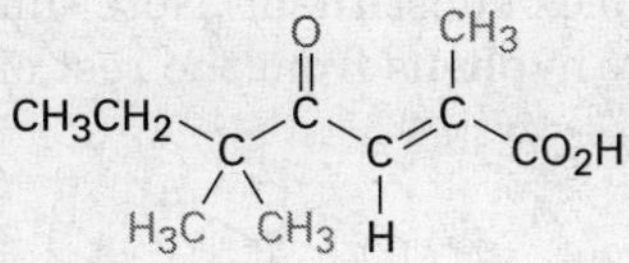

12. (*E*)-2,5,5-Trimethyl-4-oxo**hept-2-enoic acid**

Writing the Name

Once the name parts have been established, the entire name is written out. Several additional rules apply:

1. **Order of prefixes.** When the substituents have been identified, the main chain has been numbered, and the proper multipliers such as *di-* and *tri-* have been assigned, the name is written with the substituents listed in alphabetical, rather than numerical, order. Multipliers such as *di-* and *tri-* are not used for alphabetization purposes, but the prefix *iso-* is used.

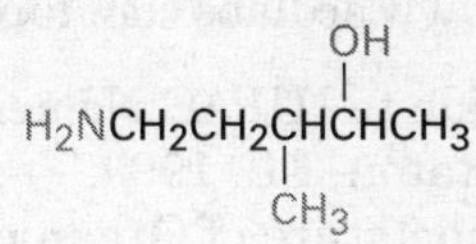

13. 5-Amino-3-methyl**pentan-2-ol**

2. **Use of hyphens; single- and multiple-word names.** The general rule is to determine whether the parent is itself an element or compound. If it is, then the name is written as a single word; if it isn't, then the name is written as multiple words. Methylbenzene is written as one word, for instance, because the parent—benzene—is itself a compound. Diethyl ether, however, is written as two words because the parent—ether—is a class name rather than a compound name. Some further examples follow:

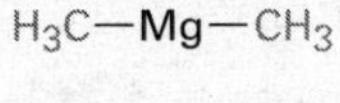

14. Dimethyl**magnesium** (one word, because magnesium is an element)

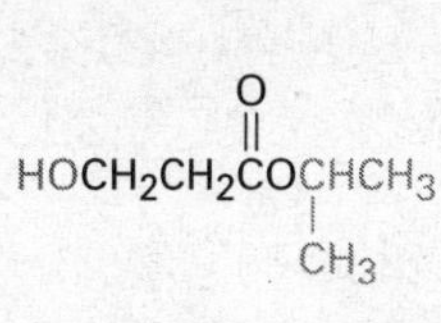

15. Isopropyl 3-hydroxy**propanoate** (two words, because "propanoate" is not a compound)

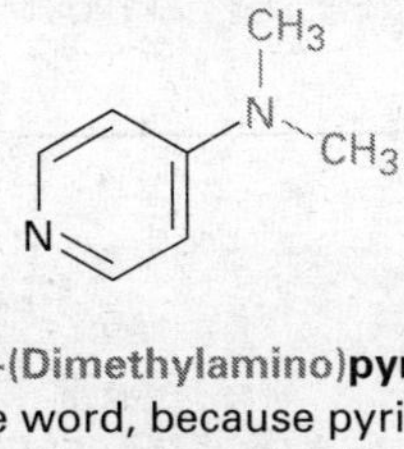

16. 4-(Dimethylamino)**pyridine** (one word, because pyridine is a compound)

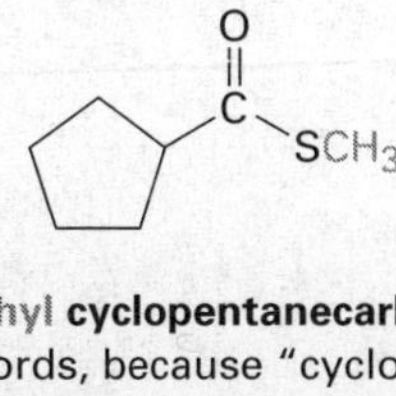

17. Methyl **cyclopentanecarbothioate** (two words, because "cyclopentanecarbothioate" is not a compound)

3. **Parentheses.** Parentheses are used to denote complex substituents when ambiguity would otherwise arise. For example, chloromethylbenzene has two substituents on a benzene ring, but (chloromethyl)benzene has only one complex substituent. Note that the expression in parentheses is not set off by hyphens from the rest of the name.

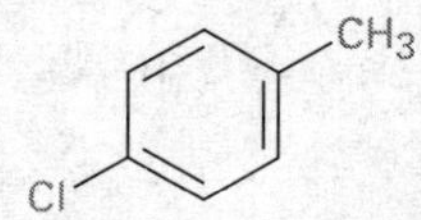

18. *p*-**Chloromethylbenzene**

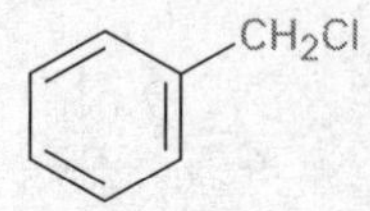

19. (Chloromethyl)**benzene**

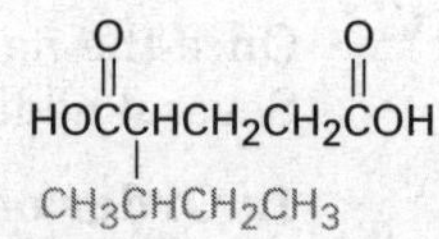

20. 2-(1-Methylpropyl)**pentanedioic acid**

Additional Reading

Further explanations of the rules of organic nomenclature can be found online at http://www.acdlabs.com/iupac/nomenclature/ and in the following references:

1. "A Guide to IUPAC Nomenclature of Organic Compounds," CRC Press, Boca Raton, FL, 1993.
2. "Nomenclature of Organic Chemistry, Sections A, B, C, D, E, F, and H," International Union of Pure and Applied Chemistry, Pergamon Press, Oxford, 1979.

APPENDIX B

Glossary

Absorbance (Section 13.5): In optical spectroscopy, the logarithm of the intensity of the incident light divided by the intensity of the light transmitted through a sample; $A = \log I_0/I$.

Absorption spectrum (Section 13.2): A plot of wavelength of incident light versus amount of light absorbed. Organic molecules show absorption spectra in both the infrared and ultraviolet regions of the electromagnetic spectrum.

Acetal (Section 9.8): A functional group consisting of two –OR groups bonded to the same carbon, $R_2C(OR')_2$. Acetals are often used as protecting groups for ketones and aldehydes.

Acetyl group (Section 9.2): The CH_3CO– group.

Acetylide anion (Section 4.11): The anion formed by removal of a proton from a terminal alkyne, $R{-}C{\equiv}C{:}^-$.

Achiral (Section 6.2): Lacking handedness. A molecule is achiral if it has a plane of symmetry and is thus superimposable on its mirror image.

Acid anhydride (Section 10.7): A functional group with two acyl groups bonded to a common oxygen atom, RCO_2COR'.

Acid halide (Section 10.7): A functional group with the general formula RCOX, where X is a halogen.

Acidity constant, K_a (Section 1.10): A measure of acid strength in water. For any acid HA, the acidity constant is given by the expression

$$K_a = \frac{[H_3O^+][A^-]}{[HA]}$$

Activating group (Section 5.7): An electron-donating group such as hydroxyl (–OH) or amino (–NH_2) that increases the reactivity of an aromatic ring toward electrophilic aromatic substitution.

Activation energy, E_{act} (Section 3.8): The difference in energy between ground state and transition state. The amount of activation energy required by a reaction determines the rate at which the reaction proceeds.

Active site (Section 15.10): The pocket in an enzyme where a substrate is bound and undergoes reaction.

Acyl group (Sections 5.5 and 9.2): A name for the –COR group.

Acyl phosphate (Section 10.12): A functional group with an acyl group bonded to a phosphate, $RCO_2PO_3^{2-}$.

Acylation (Section 5.5): The introduction of an acyl group, –COR, onto a molecule. For example, acylation of an aromatic ring yields a ketone, acylation of an alcohol yields an ester, and acylation of an amine yields an amide.

Acylium ion (Section 5.5): A resonance-stabilized carbocation in which the positive charge is located at a carbonyl-group carbon, $R{-}C^+{=}O \leftrightarrow R{-}C{\equiv}O^+$. Acylium ions are intermediates in Friedel–Crafts acylation reactions.

1,2-Addition (Sections 4.8 and 9.10): The addition of a reactant to the two ends of a double bond.

1,4-Addition (Sections 4.8 and 9.10): The addition of a reactant to atoms 1 and 4 of a conjugated diene or conjugated enone.

Addition reaction (Section 3.5): The reaction that occurs when two reactants combine to form a single new product with no atoms left over.

Adrenocortical hormone (Section 16.4): A steroid hormone secreted by the adrenal glands. There are two types of adrenocortical hormones: mineralocorticoids and glucocorticoids.

Alcohol (Chapter 8): A compound with an –OH group bonded to a saturated, sp^3-hybridized carbon atom.

Aldaric acid (Section 14.7): The dicarboxylic acid that results from oxidation of an aldose.

Aldehyde (Section 9.1): A compound containing the –CHO functional group.

Alditol (Section 14.7): The polyalcohol that results from reduction of the carbonyl group of a monosaccharide.

Aldol reaction (Section 11.8): A carbonyl condensation reaction between two ketones or aldehydes leading to a β-hydroxy carbonyl product.

Aldonic acid (Section 14.7): The monocarboxylic acid that results from mild oxidation of an aldose.

Aldose (Section 14.1): A simple sugar with an aldehyde carbonyl group.

Alicyclic (Section 2.7): An aliphatic cyclic hydrocarbon, or cycloalkane.

Aliphatic (Section 2.2): A nonaromatic hydrocarbon such as a simple alkane, alkene, or alkyne.

Alkaloid (Section 12.7): A naturally occurring compound that contains a basic amine functional group.

Alkane (Section 2.2): A compound that contains only carbon and hydrogen and has only single bonds.

Alkene (Chapter 3 Introduction): A hydrocarbon that contains a carbon–carbon double bond, $R_2C{=}CR_2$.

Alkoxide ion (Section 8.2): The anion RO^- formed by deprotonation of an alcohol.

Alkyl group (Section 2.2): The partial structure that remains when a hydrogen atom is removed from an alkane.

Alkyl halide (Chapter 7 Introduction): A compound with a halogen atom bonded to a saturated, sp^3-hybridized carbon atom.

Alkylamine (Section 12.1): An amino-substituted alkane, RNH_2, R_2NH, or R_3N.

Alkylation (Sections 5.5 and 11.6): The introduction of an alkyl group onto a molecule. For example, aromatic rings can be alkylated to yield arenes (ArH → ArR), and enolate anions can be alkylated to yield α-substituted carbonyl compounds.

Alkyne (Chapter 3 Introduction): A hydrocarbon that has a carbon–carbon triple bond.

Allylic (Section 4.8): The position next to a double bond.

α-Amino acid (Section 15.1): A compound with an amino group attached to the carbon atom next to the carboxyl group, $RCH(NH_2)CO_2H$.

α Anomer (Section 14.6): The cyclic hemiacetal form of a sugar that has the hemiacetal –OH group on the side of the ring opposite the terminal $-CH_2OH$.

α Helix (Section 15.8): A common secondary structure of a protein in which the chain coils into a spiral.

α Position (Chapter 11 Introduction): The position next to a carbonyl group.

α-Substitution reaction (Section 11.2): A reaction that results in substitution of a hydrogen on the α carbon of a carbonyl compound.

α, β-Unsaturated carbonyl compound (Section 9.10): A compound containing the C=C–C=O functional group.

Amide (Section 10.10): A compound containing the $-CONR_2$ functional group.

Amine (Section 12.1): An organic derivative of ammonia, RNH_2, R_2NH, or R_3N.

Amino acid (Section 15.1): *See* α-Amino acid.

Amino sugar (Section 14.8): A sugar with one of its –OH groups replaced by $-NH_2$.

Amphiprotic (Section 15.1): Capable of acting as either an acid or a base.

Amplitude (Section 13.2): The height of a wave from midpoint to peak.

Anabolism (Section 17.1): Metabolic reactions that synthesize larger molecules from smaller precursors.

Androgen (Section 16.4): A steroid male sex hormone such as testosterone.

Angle strain (Section 2.9): The strain introduced into a molecule when a bond angle is deformed from its ideal value.

Anomeric center (Section 14.6): The hemiacetal carbon atom in the cyclic pyranose or furanose form of a sugar.

Anomers (Section 14.6): Cyclic stereoisomers of sugars that differ only in their configurations at the hemiacetal (anomeric) carbon.

Anti stereochemistry (Section 4.4): The opposite of syn. An anti addition reaction is one in which the two ends of the double bond are attacked from different sides.

Anticodon (Section 16.9): A sequence of three bases on tRNA that read the codons on mRNA and bring the correct amino acids into position for protein synthesis.

Antisense strand (Section 16.8): The noncoding strand of double-helical DNA that does not contain the gene.

Aromatic (Section 5.1): The class of compounds that contain a benzene-like six-membered ring with three double bonds.

Aryl group (Section 5.2): An aromatic substituent group, Ar–.

Arylamine (Section 12.1): An amino-substituted aromatic compound, $ArNH_2$.

Axial position (Section 2.10): A bond to chair cyclohexane that lies along the ring axis perpendicular to the rough plane of the ring.

Backbone (Section 15.3): The repeating series of –N–CH–CO– atoms that make up a protein chain.

Base peak (Section 13.1): The most intense peak in a mass spectrum.

Basicity constant, K_b (Section 12.3): A value that expresses the strength of a base in water solution. The larger the K_b, the stronger the base.

Benzoyl group (Section 9.2): The C_6H_5CO– group.

Benzyl group (Section 5.2): The $C_6H_5CH_2$– group.

Benzylic position (Section 5.8): The position next to an aromatic ring.

***β* Anomer** (Section 14.6): The cyclic hemiacetal form of a sugar that has the hemiacetal –OH group on the same side of the ring as the terminal $-CH_2OH$.

***β*-Oxidation pathway** (Section 17.2): A series of four enzyme-catalyzed reactions that cleave two carbon atoms at a time from the end of a fatty-acid chain.

***β*-Pleated sheet** (Section 15.8): A protein secondary structure in which the chain folds back on itself so that two sections of the chain run parallel.

Bimolecular reaction (Section 7.5): A reaction whose rate-limiting step occurs between two reactants.

Boc derivative (Section 15.7): A butyloxycarbonyl N-protected amino acid.

Bond angle (Section 1.6): The angle formed between two adjacent bonds.

Bond length (Section 1.5): The equilibrium distance between the nuclei of two atoms that are bonded to each other.

Bond strength (Section 1.5): The amount of energy needed to break a bond to produce two radical fragments.

Branched-chain alkane (Section 2.2): An alkane that contains a branching connection of carbons as opposed to a straight-chain alkane.

Bromonium ion (Section 4.4): A species with a divalent, positively charged bromine, R_2Br^+.

Brønsted–Lowry acid (Section 1.10): A substance that donates a hydrogen ion (proton, H^+) to a base.

Brønsted–Lowry base (Section 1.10): A substance that accepts a hydrogen ion, H^+, from an acid.

C-Terminal amino acid (Section 15.3): The amino acid with a free $-CO_2H$ group at one end of a protein chain.

Cahn–Ingold–Prelog sequence rules (Sections 3.4 and 6.5): A series of rules for assigning relative rankings to substituent groups on a double-bond carbon atom or on a chirality center.

Carbanion (Section 7.3): A carbon-anion, or substance that contains a trivalent, negatively charged carbon atom ($R_3C:^-$).

Carbocation (Section 3.7): A carbon-cation, or substance that contains a trivalent, positively charged carbon atom having six electrons in its outer shell (R_3C^+).

Carbohydrate (Section 14.1): A polyhydroxy aldehyde or polyhydroxy ketone. Carbohydrates can be either simple sugars such as glucose or complex sugars such as cellulose.

Carbonyl condensation reaction (Section 11.7): A reaction between two carbonyl compounds in which the *α* carbon of one partner bonds to the carbonyl carbon of the other.

Carbonyl group (Section 9.1): The C=O functional group.

Carboxyl group (Section 10.1): The $-CO_2H$ group.

Carboxylate ion (Section 10.3): The anion of a carboxylic acid, RCO_2^-.

Carboxylic acid (Section 10.1): A compound containing the $-CO_2H$ functional group.

Carboxylic acid derivative (Chapter 10 Introduction): A compound in which an acyl group is bonded to an electronegative atom or substituent that can act as a leaving group in a substitution reaction. Esters, amides, and acid halides are examples.

Catabolism (Section 17.1): Metabolic reactions that break down large molecules.

Catalyst (Section 3.9): A substance that increases the rate of a chemical transformation by providing an alternative mechanism but is not itself changed in the reaction.

Chain-growth polymer (Section 10.13): A polymer produced by chain reaction of a monofunctional monomer.

Chair conformation (Section 2.9): A three-dimensional conformation of cyclohexane that resembles the rough shape of a chair. The chair form of cyclohexane has neither angle strain nor eclipsing strain.

Chemical shift (Section 13.9): The position on the NMR chart where a nucleus absorbs. By convention, the chemical shift of tetramethylsilane is set at zero and all other absorptions usually occur downfield (to the left on the chart).

Chiral (Section 6.2): Having handedness. A chiral molecule does not have a plane of symmetry, is not superimposable on its mirror image, and thus exists in right- and left-handed forms.

Chiral environment (Section 6.10): Chiral surroundings or conditions in which a molecule resides.

Chirality center (Section 6.2): An atom (usually carbon) that is bonded to four different groups. Also called a stereocenter.

Cis–trans isomers (Sections 2.8 and 3.3): Stereoisomers that differ in their stereochemistry about a double bond or a ring.

Citric acid cycle (Section 17.4): The metabolic pathway by which acetyl CoA is degraded to CO_2.

Claisen condensation reaction (Section 11.10): A carbonyl condensation reaction between two esters leading to formation of a β-keto ester product.

Coding strand (Section 16.8): The sense strand of double-helical DNA that contains the gene.

Codon (Section 16.9): A three-base sequence on the mRNA chain that encodes the genetic information necessary to cause specific amino acids to be incorporated into proteins.

Coenzyme (Section 15.9): A small organic molecule that acts as an enzyme cofactor.

Cofactor (Section 15.9): A small, nonprotein part of an enzyme necessary for biological activity.

Complex carbohydrate (Section 14.1): A carbohydrate composed of two or more simple sugars linked together by acetal bonds.

Condensed structure (Section 2.2): A shorthand way of drawing structures in which C–H and C–C bonds are understood rather than shown explicitly.

Configuration (Section 6.5): The three-dimensional arrangement of atoms bonded to a chirality center.

Conformation (Section 2.5): The exact three-dimensional shape of a molecule at any given instant, assuming that rotation around single bonds is frozen.

Conformers (Section 2.5): Conformational isomers that interconvert by bond rotation.

Conjugate acid (Section 1.10): The product that results when a base accepts H^+.

Conjugate (1,4) addition reaction (Section 9.10): The addition of a nucleophile to the β carbon atom of an α,β-unsaturated carbonyl compound.

Conjugate base (Section 1.10): The anion that results from dissociation of an acid.

Conjugation (Section 4.8): A series of alternating single and multiple bonds with overlapping *p* orbitals.

Constitutional isomers (Section 2.2): Isomers such as butane and 2-methylpropane, which have their atoms connected in a different order.

Coupled reactions (Section 17.1): Two reactions that share a common intermediate so that the energy released in the favorable step allows the unfavorable step to occur.

Coupling constant, *J* (Section 13.12): The magnitude of the spin–spin splitting interaction between nuclei whose spins are coupled.

Covalent bond (Section 1.4): A bond formed by sharing electrons between two nuclei.

Cycloalkane (Section 2.7): An alkane with a ring of carbon atoms.

D Sugar (Section 14.3): A sugar whose hydroxyl group at the chirality center farthest from the carbonyl group points to the right when the molecule is drawn in Fischer projection.

Deactivating group (Section 5.7): An electron-withdrawing substituent that decreases the reactivity of an aromatic ring toward electrophilic aromatic substitution.

Decarboxylation (Section 11.6): The loss of CO_2. β-Keto acids decarboxylate readily on heating.

Dehydration (Section 8.4): Elimination of water from an alcohol to yield an alkene.

Dehydrohalogenation (Section 7.7): Elimination of HX from an alkyl halide to yield an alkene on treatment with a strong base.

Delta (δ) scale (Section 13.9): The arbitrary scale used for defining the position of NMR absorptions; $1\ \delta = 1$ ppm of spectrometer frequency.

Deoxy sugar (Section 14.8): A sugar with an –OH group missing from one carbon.

Deoxyribonucleic acid (DNA) (Section 16.5): The biopolymer consisting of deoxyribonucleotide units linked together through phosphate–sugar bonds. Found in the nucleus of cells, DNA contains an organism's genetic information.

Deshielding (Section 13.9): An effect observed in NMR that causes a nucleus to absorb downfield because of a withdrawal of electron density from the nucleus.

Dextrorotatory (Section 6.3): An optically active substance that rotates the plane of polarization of plane-polarized light in a right-handed (clockwise) direction.

Diastereomers (Section 6.6): Non–mirror-image stereoisomers; diastereomers have the same configuration at one or more chirality centers but differ at other chirality centers.

1,3-Diaxial interaction (Section 2.11): The strain energy caused by a steric interaction between axial groups three carbon atoms apart in chair cyclohexane.

Digestion (Section 17.1): The first stage of catabolism, in which food molecules are hydrolyzed to yield fatty acids, amino acids, and monosaccharides.

Disaccharide (Section 14.9): A complex carbohydrate formed by linking two simple sugars through an acetal bond.

Disulfide (Section 8.8): A compound of the general structure RSSR′.

DNA (Section 16.5): *See* Deoxyribonucleic acid.

Double bond (Section 1.8): A covalent bond formed by sharing two pairs of electrons between atoms.

Double helix (Section 16.6): The structure of DNA in which two polynucleotide strands coil around each other.

Doublet (Section 13.12): A two-line NMR absorption caused by spin–spin splitting when the spin of the nucleus under observation couples with the spin of a neighboring magnetic nucleus.

Downfield (Section 13.9): The left-hand portion of the NMR chart.

***E* geometry** (Section 3.4): A term used to describe the stereochemistry of a carbon–carbon double bond in which higher-ranked groups on each carbon are on opposite sides of the double bond.

E1 reaction (Section 7.8): A unimolecular elimination reaction in which the substrate spontaneously dissociates to give a carbocation intermediate, which loses a proton in a separate step.

E1cB reaction (Section 7.8): A unimolecular elimination reaction in which a proton is first removed to give a carbanion intermediate, which then expels the leaving group in a separate step.

E2 reaction (Section 7.7): A bimolecular elimination reaction in which C–H and C–X bond cleavages are simultaneous.

Eclipsed conformation (Section 2.5): The geometric arrangement around a carbon–carbon single bond in which the bonds on one carbon are parallel to the bonds on the neighboring carbon as viewed in a Newman projection.

Edman degradation (Section 15.6): A method for selectively cleaving the N-terminal amino acid from a peptide.

Electromagnetic spectrum (Section 13.2): The range of electromagnetic energy, including infrared, ultraviolet, and visible radiation.

Electron-dot structure (Section 1.4): A representation of a molecule showing valence electrons as dots.

Electron-transport chain (Section 17.1): The final stage of catabolism, in which ATP is produced.

Electronegativity (Section 1.9): The ability of an atom to attract electrons in a covalent bond. Electronegativity generally increases from right to left and from bottom to top of the periodic table.

Electrophile (Section 3.6): An "electron-lover," or substance that accepts an electron pair from a nucleophile in a polar bond-forming reaction.

Electrophilic addition reaction (Section 3.7): The addition of an electrophile to an alkene to yield a saturated product.

Electrophilic aromatic substitution reaction (Section 5.3): The substitution of an electrophile for a hydrogen atom on an aromatic ring.

Electrophoresis (Sections 15.2 and 16.10): A technique for separating charged organic molecules, particularly proteins and amino acids, by placing them in an electric field.

Electrostatic potential map (Section 1.9): A molecular representation that uses color to indicate the calculated charge distribution in the molecule.

Elimination reaction (Section 3.5): The reaction that occurs when a single reactant splits apart into two products.

Embden–Meyerhof pathway (Section 17.3): An alternative name for glycolysis.

Enantiomers (Section 6.1): Stereoisomers that have a mirror-image relationship, with opposite configurations at all chirality centers.

Enantioselective synthesis (Chapter 6 *Interlude*): A method of synthesis from an achiral precursor that yields only a single enantiomer of a chiral product.

3′-End (Section 16.5): The end of a nucleic acid chain that has a free sugar hydroxyl group.

5′-End (Section 16.5): The end of a nucleic acid chain that has a phosphoric acid unit.

Energy diagram (Section 3.8): A graph depicting the energy changes that occur during a reaction.

Enol (Section 11.1): A vinylic alcohol, C=C—OH.

Enolate ion (Sections 9.10 and 11.1): The resonance-stabilized anion of an enol, $C{=}C{-}O^-$.

Enone (Section 11.9): An unsaturated ketone.

Entgegen (*E*) (Section 3.4): A term used to describe the stereochemistry of a carbon–carbon double bond in which higher-ranked groups on each carbon are on opposite sides of the double bond.

Enzyme (Section 15.9): A biological catalyst. Enzymes are large proteins that catalyze specific biochemical reactions.

Epoxide (Section 4.6): A three-membered ring ether functional group.

Epoxy resin (Chapter 8 *Interlude*): A polymer prepared by reaction of a bisphenol with epichlorohydrin.

Equatorial position (Section 2.10): A bond to cyclohexane that lies along the rough equator of the ring. (*See* Axial position.)

Essential amino acid (Section 15.1): An amino acid that must be obtained in the diet.

Essential monosaccharide (Section 14.8): One of eight monosaccharides essential for life and obtained in the diet.

Essential oil (Chapter 3 *Interlude*): The fragrant mixture of liquids extracted from many plants.

Ester (Section 10.9): A compound containing the $-CO_2R$ functional group.

Estrogen (Section 16.4): A female steroid sex hormone.

Ether (Section 8.1): A compound with two organic groups bonded to the same oxygen atom, R—O—R′.

Exon (Section 16.8): A section of DNA that contains genetic information.

Fat (Section 16.1): A solid triacylglycerol derived from an animal source.

Fatty acid (Section 16.1): A long straight-chain carboxylic acid found in fats and oils.

Fibrous protein (Section 15.8): A protein that consists of polypeptide chains arranged side by side in long threads.

Fingerprint region (Section 13.3): The complex region of the infrared spectrum from 1500 cm^{-1} to 400 cm^{-1}.

Fischer esterification reaction (Section 10.6): The acid-catalyzed reaction of an alcohol with a carboxylic acid to yield an ester.

Fischer projection (Section 14.2): A method for depicting the configuration of a chirality center using crossed lines. Horizontal bonds come out of the plane of the page, and vertical bonds go back into the plane of the page.

Fishhook arrow (Section 3.6): A half-headed curved arrow used to show the movement of a single electron in a radical reaction.

Fmoc derivative (Section 15.7): A fluorenylmethyloxycarbonyl N-protected amino acid.

Formyl group (Section 9.2): A –CHO group.

Frequency, ν (Section 13.2): The number of electromagnetic wave cycles that travel past a fixed point in a given unit of time, usually expressed in reciprocal seconds, s^{-1}, or hertz.

Friedel–Crafts reaction (Section 5.5): The introduction of an alkyl or acyl group onto an aromatic ring by an electrophilic substitution reaction.

Functional group (Section 2.1): An atom or group of atoms that is part of a larger molecule and has a characteristic chemical reactivity.

Furanose (Section 14.5): The five-membered ring structure of a simple sugar.

Geminal (Section 9.7): Referring to two groups attached to the same carbon atom.

Globular protein (Section 15.8): A protein that is coiled into a compact, nearly spherical shape.

Glycerophospholipid (Section 16.3): A lipid that contains a glycerol backbone linked to two fatty acids and a phosphoric acid.

Glycoconjugate (Section 14.7): A molecule in which a carbohydrate is linked through its anomeric center to another biological molecule such as a lipid or protein.

Glycol (Section 4.6): A diol, such as ethylene glycol, $HOCH_2CH_2OH$.

Glycolipid (Section 14.7): A biological molecule in which a carbohydrate is linked through its anomeric center to a lipid.

Glycolysis (Section 17.3): A series of ten enzyme-catalyzed reactions that break down a glucose molecule into two pyruvate molecules.

Glycoprotein (Section 14.7): A biological molecule in which a carbohydrate is linked through its anomeric center to a protein.

Glycoside (Section 14.7): A cyclic acetal formed by reaction of a sugar with another alcohol.

Green chemistry (Chapter 12 *Interlude*): The design and implementation of chemical products and processes that reduce waste and attempt to eliminate the generation of hazardous substances.

Grignard reagent (Section 7.3): An organomagnesium halide, RMgX.

Ground-state electron configuration (Section 1.2): The lowest-energy electron configuration of a molecule or atom.

Halogenation (Sections 4.4 and 5.3): The reaction of halogen with an alkene to yield a 1,2-dihalide addition product or with an aromatic compound to yield a substitution product.

Hemiacetal (Section 9.8): A functional group consisting of one –OR and one –OH group bonded to the same carbon.

Hertz (Hz) (Section 13.2): The standard unit for frequency; the number of waves that pass by a fixed point per second.

Heterocycle (Sections 5.9 and 12.6): A cyclic molecule whose ring contains more than one kind of atom.

Hormone (Section 16.4): A chemical messenger secreted by a specific gland and carried through the bloodstream to affect a target tissue.

Hybrid orbital (Section 1.6): An orbital derived from a combination of atomic orbitals. Hybrid orbitals, such as the sp^3, sp^2, and sp hybrids of carbon, are strongly directed and form stronger bonds than atomic orbitals do.

Hydration (Section 4.3): Addition of water to a molecule, such as occurs when alkenes are treated with strong aqueous acid.

Hydrocarbon (Section 2.2): A compound that has only carbon and hydrogen.

Hydrogen bond (Section 8.2): An attraction between a hydrogen atom bonded to an electronegative element and an electron lone pair on another atom.

Hydrogenation (Section 4.5): Addition of hydrogen to a double or triple bond to yield a saturated product.

Hydrophilic (Section 16.2): Water-loving; attracted to water.

Hydrophobic (Section 16.2): Water-fearing; not attracted to water.

Hydroquinone (Section 8.5): 1,4-dihydroxybenzene.

Hydroxylation (Section 4.6): The addition of one or more –OH groups to a molecule.

Imine (Section 9.9): A compound with a $R_2C{=}NR$ functional group; also called a Schiff base in biochemistry.

Inductive effect (Section 1.9): The electron-attracting or electron-withdrawing effect that is transmitted through σ bonds.

Infrared (IR) spectroscopy (Section 13.3): A kind of optical spectroscopy that uses infrared energy. IR spectroscopy is particularly useful in organic chemistry for determining the kinds of functional groups in molecules.

Integration (Section 13.11): A means of electronically measuring the ratios of the number of nuclei responsible for each peak in an NMR spectrum.

Intermediate (Section 3.8): A species that is formed during the course of a multistep reaction but is not the final product.

Intron (Section 16.8): A section of DNA that does not contain genetic information.

Ionic bond (Section 1.4): A bond between two ions due to the electrical attraction of unlike charges.

Isoelectric point, p*I* (Section 15.2): The pH at which the number of positive charges and the number of negative charges on a protein or amino acid are exactly balanced.

Isomers (Section 2.2): Compounds with the same molecular formula but different structures.

Isotopes (Section 1.1): Atoms of the same element that have different mass numbers.

IUPAC system of nomenclature (Section 2.3): Rules for naming compounds, devised by the International Union of Pure and Applied Chemistry.

Kekulé structure (Section 1.4): A representation of a molecule in which a line between atoms represents a covalent bond.

Keto–enol tautomerism (Section 11.1): The equilibration between a carbonyl form and vinylic alcohol form of a molecule.

Ketone (Section 9.1): A compound with two organic substituents bonded to a carbonyl group, $R_2C{=}O$.

Ketose (Section 14.1): A simple sugar with a ketone functional group.

Krebs cycle (Section 17.4): An alternative name for the citric acid cycle, by which acetyl CoA is degraded to CO_2.

L Sugar (Section 14.3): A sugar whose hydroxyl group at the chirality center farthest from the carbonyl group points to the left when the molecule is drawn in Fischer projection.

Lactam (Chapter 10 *Interlude*): A cyclic amide.

LD_{50} (Chapter 1 *Interlude*): The amount of a substance per kilogram body weight that is lethal to 50% of test animals.

Leaving group (Section 7.4): The group that is replaced in a substitution reaction.

Levorotatory (Section 6.3): An optically active substance that rotates the plane of polarization of plane-polarized light in a left-handed (counterclockwise) direction.

Lewis acid (Section 1.12): A substance with a vacant low-energy orbital that can accept an electron pair from a base.

Lewis base (Section 1.12): A substance that donates an electron lone pair to an acid.

Lewis structure (Section 1.4): A representation of a molecule showing covalent bonds as a pair of electron dots between atoms.

Lindlar catalyst (Section 4.11): A hydrogenation catalyst used to convert an alkyne to a cis alkene.

Line-bond structure (Section 1.4): A representation of a molecule showing covalent bonds as lines between atoms.

1→4 Link (Section 14.9): An acetal link between the C1 carbonyl group of one sugar and the C4 hydroxyl group of another sugar.

Lipid (Chapter 16 Introduction): A naturally occurring substance isolated from plants or animals by extraction with a nonpolar organic solvent.

Lipid bilayer (Section 16.3): The double layer of phospholipids that forms a cell membrane.

Locant (Sections 2.3 and 3.1): A number in the IUPAC name of a compound that specifies the point of attachment of a substituent to the parent chain or the position of a functional group in the chain.

Lone-pair electrons (Section 1.4): A nonbonding electron pair that occupies a valence orbital.

Magnetic resonance imaging, MRI (Chapter 13 *Interlude*): A medical diagnostic technique based on nuclear magnetic resonance.

Major groove (Section 16.6): The larger of two grooves in double-helical DNA.

Malonic ester synthesis (Section 11.6): The synthesis of a carboxylic acid by alkylation of an alkyl halide, followed by hydrolysis and decarboxylation.

Markovnikov's rule (Section 4.1): A guide for determining the regiochemistry (orientation) of electrophilic addition reactions. In the addition of HX to an alkene, the hydrogen atom bonds to the alkene carbon that has fewer alkyl substituents.

Mass spectrometry (Section 13.1): A technique for measuring the mass, and therefore the molecular weight (MW), of ions.

Mechanism (Section 3.6): A complete description of how a reaction occurs. A mechanism accounts for all reactants and all products and describes the details of each individual step in the overall reaction process.

Mercapto group (Section 8.8): An alternative name for the thiol group, –SH.

Meso compound (Section 6.7): A compound that contains one or more chirality centers but is nevertheless achiral because it has a symmetry plane.

Messenger RNA (mRNA) (Section 16.8): The kind of RNA transcribed from DNA and used to carry genetic messages from DNA to ribosomes.

Meta, *m*- (Section 5.2): A naming prefix used for 1,3-disubstituted benzenes.

Metabolism (Section 17.1): A collective name for the many reactions that go on in the cells of living organisms.

Micelle (Section 16.2): A spherical cluster of soap-like molecules that aggregate in aqueous solution. The ionic heads of the molecules lie on the outside where they are solvated by water, and the organic tails bunch together on the inside of the micelle.

Minor groove (Section 16.6): The smaller groove in double-helical DNA.

Molar absorptivity (Section 13.5): A quantitative measure of the amount of UV light absorbed by a sample.

Molecular ion (Section 13.1): The cation produced in the mass spectrometer by loss of an electron from the parent molecule. The mass of the molecular ion corresponds to the molecular weight of the sample.

Molecule (Section 1.4): A neutral collection of atoms held together by covalent bonds.

Monomer (Sections 4.7 and 10.13): The starting unit from which a polymer is made.

Monosaccharide (Section 14.1): A simple sugar.

Monoterpene (Chapter 3 *Interlude*): A ten-carbon lipid.

Multiplet (Section 13.12): A pattern of peaks in an NMR spectrum that arises by spin–spin splitting of a single absorption because of coupling between neighboring magnetic nuclei.

Mutarotation (Section 14.6): The change in optical rotation observed when a pure anomer of a sugar is dissolved in water and equilibrates to an equilibrium mixture of anomers.

***n* + 1 rule** (Section 13.12): The signal of a proton with n neighboring protons splits into $n + 1$ peaks in the NMR spectrum.

N-Terminal amino acid (Section 15.3): The amino acid with a free $-NH_2$ group at one end of a protein chain.

Natural gas (Section 2.4): A naturally occurring hydrocarbon mixture consisting chiefly of methane, along with smaller amounts of ethane, propane, and butane.

Natural product (Chapter 2 *Interlude*): A catchall term generally taken to mean a small molecule found in bacteria, plants, and other living organisms.

New molecular entity, NME (Chapter 2 *Interlude*): A new biologically active chemical substance approved for sale as a drug by the U.S. Food and Drug Administration.

Newman projection (Section 2.5): A means of indicating stereochemical relationships between substituent groups on neighboring carbons by looking end-on at a carbon–carbon bond.

Nitration (Section 5.4): The substitution of a nitro group onto an aromatic ring.

Nitrile (Section 10.11): A compound with a $-C{\equiv}N$ functional group.

Node (Section 1.1): A surface of zero electron density within an orbital. For example, a p orbital has a nodal plane passing through the center of the nucleus, perpendicular to the axis of the orbital.

Nonbonding electron (Section 1.4): A valence electron not used for bonding.

Nonessential amino acid (Section 15.1): One of the eleven amino acids that are biosynthesized by humans.

Normal (*n*) alkane (Section 2.2): A straight-chain alkane, as opposed to a branched alkane.

NSAID (Chapter 5 *Interlude*): A nonsteroidal anti-inflammatory drug, such as aspirin or ibuprofen.

Nuclear magnetic resonance (NMR) spectroscopy (Section 13.7): A spectroscopic technique that provides information about the carbon–hydrogen framework of a molecule.

Nucleic acid (Section 16.5): A biopolymer, either DNA or RNA, made of nucleotides joined together.

Nucleophile (Section 3.6): An electron-rich species that donates an electron pair to an electrophile in a polar bond-forming reaction. Nucleophiles are also Lewis bases.

Nucleophilic acyl substitution reaction (Section 10.5): A reaction in which a nucleophile attacks a carbonyl compound and substitutes for a leaving group bonded to the carbonyl carbon.

Nucleophilic addition reaction (Section 9.5): A reaction in which a nucleophile adds to the electrophilic carbonyl group of a ketone or aldehyde to give an alcohol.

Nucleophilic substitution reaction (Section 7.4): A reaction in which one nucleophile replaces another attached to a saturated carbon atom.

Nucleoside (Section 16.5): A nucleic acid constituent, consisting of a sugar residue bonded to a heterocyclic purine or pyrimidine base.

Nucleotide (Section 16.5): A nucleic acid constituent, consisting of a sugar residue bonded both to a heterocyclic purine or pyrimidine base and to phosphoric acid.

Nylon (Section 10.13): A polyamide step-growth polymer, usually prepared by reaction between a diacid and a diamine.

Olefin (Chapter 3 Introduction): An alternative name for an alkene.

Optical activity (Section 6.3): The ability of a chiral molecule in solution to rotate plane-polarized light.

Optical isomers (Section 6.4): An older, alternative name for enantiomers. Optical isomers are isomers that have a mirror-image relationship.

Orbital (Section 1.1): A region of space occupied by a given electron or pair of electrons.

Organic chemistry (Chapter 1 Introduction): The chemistry of carbon compounds.

Organohalide (Chapter 7 Introduction): A compound that contains one or more halogen atoms bonded to carbon.

Organometallic compound (Section 7.3): A compound that contains a carbon–metal bond. Grignard reagents, RMgX, are examples.

Ortho, *o*- (Section 5.2): A naming prefix used for 1,2-disubstituted benzenes.

Oxidation (Section 4.6): The addition of oxygen to a molecule or removal of hydrogen from it.

Oxirane (Section 4.6): An alternative name for an epoxide.

Para, *p*- (Section 5.2): A naming prefix used for 1,4-disubstituted benzenes.

Paraffin (Section 2.4): A common name for an alkane.

Parent peak (Section 13.1): The peak in a mass spectrum corresponding to the molecular ion and thus representing the molecular weight of the compound.

Peptide (Chapter 15 Introduction): A short amino acid polymer in which the individual amino acid residues are linked by amide bonds. (*See* Protein.)

Peptide bond (Section 15.4): An amide bond in a peptide chain.

Peroxyacid (Section 4.6): A compound with the $-CO_3H$ functional group.

Petroleum (Section 2.4): A complex mixture of naturally occurring hydrocarbons derived from the decomposition of plant and animal matter.

Phenol (Section 8.1): A compound with an –OH group bonded to an aromatic ring, ArOH.

Phenoxide ion (Section 8.2): The anion of a phenol, ArO^-.

Phenyl group (Section 5.2): The $-C_6H_5$ group, often abbreviated as –Ph.

Phospholipid (Section 16.3): A lipid that contains a phosphate residue.

Phosphoric acid anhydride (Section 17.1): A substance that contains a PO_2PO link, analogous to the CO_2CO link in carboxylic acid anhydrides.

Phosphorylation (Sections 14.7 and 17.1): A reaction that transfers a phosphate group from a phosphoric anhydride to an alcohol.

Pi (π) bond (Section 1.8): A covalent bond formed by sideways overlap of two *p* orbitals.

p*K*$_a$ (Section 1.10): The negative common logarithm of the K_a; used to express acid strength.

Plane of symmetry (Section 6.2): A plane that bisects a molecule such that one half of the molecule is the mirror image of the other half. Molecules that contain a plane of symmetry are achiral.

Plane-polarized light (Section 6.3): Light that has its electric waves oscillating in a single plane rather than in random planes.

Plasticizer (Section 10.2): A small organic molecule added to polymers to act as a lubricant between polymer chains.

Polar covalent bond (Section 1.9): A covalent bond in which the electrons are shared unequally between the atoms.

Polar reaction (Section 3.6): A reaction in which bonds are made when a nucleophile donates two electrons to an electrophile, and in which bonds are broken when one fragment leaves with both electrons from the bond.

Polarity (Section 1.9): The unsymmetrical distribution of electrons in a molecule that results when one atom attracts electrons more strongly than another.

Polycyclic aromatic compound (Section 5.9): A molecule that has two or more benzene rings fused together.

Polyester (Section 10.13): A polymer prepared by reaction between a diacid and a dialcohol.

Polymer (Sections 4.7 and 10.13): A large molecule made up of repeating smaller units.

Polymerase chain reaction (PCR) (Section 16.11): A method for amplifying small amounts of DNA to prepare larger amounts.

Polysaccharide (Section 14.10): A complex carbohydrate that has many simple sugars bonded together by acetal links.

Polyunsaturated fatty acid (Section 16.1): A fatty acid with more than one double bond in its chain.

Primary, secondary, tertiary, quaternary (Sections 2.2, 8.1, and 12.1): Terms used to describe the substitution pattern at a specific site. A primary site has one organic substituent attached to it, a secondary site has two organic substituents, a tertiary site has three, and a quaternary site has four.

	Carbon	Carbocation	Hydrogen	Alcohol	Amine
Primary	RCH_3	RCH_2^+	$RC\mathit{H}_3$	RCH_2OH	RNH_2
Secondary	R_2CH_2	R_2CH^+	$R_2C\mathit{H}_2$	R_2CHOH	R_2NH
Tertiary	R_3CH	R_3C^+	$R_3C\mathit{H}$	R_3COH	R_3N
Quaternary	$R_4\mathit{C}$				

Primary structure (Section 15.8): The amino acid sequence of a protein.

Protecting group (Section 9.8): A group that is temporarily introduced into a molecule to protect a functional group from reaction elsewhere in the molecule.

Protein (Chapter 15 Introduction): A large biological polymer containing 50 or more amino acid residues.

Protein Data Bank (Chapter 15 *Interlude*): A worldwide online repository of X-ray and NMR structural data for biological macromolecules. To access the Protein Data Bank, go to http://www.rcsb.org/pdb/.

PTH (Section 15.6): A phenylthiohydantoin derived from a terminal amino acid during Edman degradation.

Pyranose (Section 14.5): The six-membered ring structure of a simple sugar.

Quartet (Section 13.12): A set of four peaks in an NMR spectrum, caused by spin–spin splitting of a signal by three adjacent nuclear spins.

Quaternary: *See* Primary.

Quaternary ammonium salt (Section 12.1): A compound with four organic substituents attached to a positively charged nitrogen, $R_4N^+ X^-$.

Quaternary structure (Section 15.8): The highest level of protein structure, involving a specific aggregation of individual proteins into a larger cluster.

Quinone (Section 8.5): A cyclohexa-2,5-diene-1,4-dione.

***R* configuration** (Section 6.5): The configuration at a chirality center as specified using the Cahn–Ingold–Prelog sequence rules.

R group (Section 2.2): A generalized abbreviation for an organic partial structure.

Racemic mixture (Section 6.8): A 50:50 mixture of the two enantiomers of a chiral substance.

Radical (Section 3.6): A species that has an odd number of electrons, such as the chlorine radical, Cl·.

Radical reaction (Section 3.6): A reaction in which bonds are made by donation of one electron from each of two reagents, and in which bonds are broken when each fragment leaves with one electron.

Reaction intermediate (Section 3.8): A substance formed transiently during the course of a multistep reaction.

Reaction mechanism (Section 3.6): A complete description of how a reaction occurs.

Rearrangement reaction (Section 3.5): The reaction that occurs when a single reactant undergoes a reorganization of bonds and atoms to give an isomeric product.

Reducing sugar (Section 14.7): A sugar that reduces Ag^+ in the Tollens test or Cu^{2+} in the Fehling or Benedict tests.

Reduction (Section 4.5): The addition of hydrogen to a molecule or the removal of oxygen from it.

Reductive amination (Section 12.4): A method for synthesizing amines by treatment of an aldehyde or ketone with ammonia or an amine and a reducing agent.

Refining (Section 2.4): The process by which petroleum is converted into gasoline and other useful products.

Regiospecific (Section 4.1): A term describing a reaction that occurs with a specific orientation to give a single product rather than a mixture of products.

Replication (Section 16.7): The process by which double-stranded DNA uncoils and is replicated to produce two new copies.

Replication fork (Section 16.7): The point of unraveling in a DNA chain where replication occurs.

Residue (Section 15.3): An amino acid in a protein chain.

Resolution (Section 6.8): Separation of a racemic mixture into its pure component enantiomers.

Resonance forms (Section 4.9): Structural representations of a molecule that differ only in where the bonding electrons are placed.

Resonance hybrid (Section 4.9): The composite structure of a molecule described by different resonance forms.

Restriction endonuclease (Section 16.10): An enzyme that is able to cut a DNA strand at a specific base sequence in the chain.

Ribonucleic acid (RNA) (Sections 16.5 and 16.8): The biopolymer found in cells that serves to transcribe the genetic information found in DNA and uses that information to direct the synthesis of proteins.

Ribosomal RNA (rRNA) (Section 16.8): A kind of RNA that makes up ribosomes.

Ring-flip (Section 2.11): The molecular motion that converts one chair conformation of cyclohexane into another chair conformation, thereby interconverting axial and equatorial bonds.

RNA (Sections 16.5 and 16.8): *See* Ribonucleic acid.

***S* configuration** (Section 6.5): The configuration at a chirality center as specified using the Cahn–Ingold–Prelog sequence rules.

Saccharide (Section 14.1): A sugar.

Salt bridge (Section 15.8): The ionic attraction between charged amino acid side chains that helps stabilize a protein's tertiary structure.

Sanger dideoxy method (Section 16.10): A method for sequencing DNA strands.

Saponification (Section 10.9): An old term for the base-induced hydrolysis of an ester to yield a carboxylic acid salt.

Saturated (Section 2.2): A compound that has only single bonds.

Sawhorse representation (Section 2.5): A manner of representing stereochemistry that uses a stick drawing and gives an oblique view of the conformation around a single bond.

Schiff base (Section 17.5): An alternative name for an imine, $R_2C{=}NR'$, used primarily in biochemistry.

Secondary: *See* Primary.

Secondary structure (Section 15.8): The level of protein substructure that involves organization of chain sections into ordered arrangements such as β-pleated sheets or α helices.

Semiconservative replication (Section 16.7): A description of DNA replication in which each new DNA molecules contains one old strand and one new strand.

Sense strand (Section 16.8): The coding strand of double-helical DNA that contains the gene.

Sequence rules (Sections 3.4 and 6.5): A series of rules for assigning relative rankings to substituent groups on a double-bond carbon atom or on a chirality center.

Sesquiterpene (Chapter 3 *Interlude*): A 15-carbon lipid.

Shielding (Section 13.8): An effect observed in NMR that causes a nucleus to absorb toward the right (upfield) side of the chart. Shielding is caused by donation of electron density to the nucleus.

Side chain (Section 15.1): The substituent bonded to the α carbon of an α-amino acid.

Sigma (σ) bond (Section 1.8): A covalent bond formed by head-on overlap of atomic orbitals.

Simple sugar (Section 14.1): A carbohydrate like glucose that can't be hydrolyzed to smaller sugars.

Skeletal structure (Section 2.6): A shorthand way of writing structures in which carbon atoms are assumed to be at each intersection of two lines (bonds) and at the end of each line.

S_N1 reaction (Section 7.6): A nucleophilic substitution reaction that takes place in two steps through a carbocation intermediate.

S_N2 reaction (Section 7.5): A nucleophilic substitution reaction that takes place in a single step by backside displacement of the leaving group.

Solid-phase synthesis (Section 15.7): A technique of synthesis whereby the starting material is covalently bound to a solid polymer bead and reactions are carried out on the bound substrate. After the desired transformations have been effected, the product is cleaved from the polymer.

***sp* Hybrid orbital** (Section 1.8): A hybrid orbital derived from the combination of an *s* and a *p* atomic orbital. The two *sp* orbitals that result from hybridization are oriented at an angle of 180° to each other.

sp^2 Hybrid orbital (Section 1.8): A hybrid orbital derived by combination of an *s* atomic orbital with two *p* atomic orbitals. The three sp^2 hybrid orbitals that result lie in a plane at angles of 120° to each other.

sp^3 Hybrid orbital (Section 1.6): A hybrid orbital derived by combination of an *s* atomic orbital with three *p* atomic orbitals. The four sp^3 hybrid orbitals that result are directed toward the corners of a regular tetrahedron at angles of 109° to each other.

Specific rotation, $[\alpha]_D$ (Section 6.3): The amount by which an optically active compound rotates plane-polarized light under standard conditions.

Sphingomyelin (Section 16.3): A phospholipid that has sphingosine as its backbone rather than glycerol.

Spin–spin splitting (Section 13.12): The splitting of an NMR signal into a multiplet because of an interaction between nearby magnetic nuclei whose spins are coupled. The magnitude of spin–spin splitting is given by the coupling constant, *J*.

Staggered conformation (Section 2.5): The three-dimensional arrangement of atoms around a carbon–carbon single bond in which the bonds on one carbon bisect the bond angles on the second carbon as viewed end-on.

Statins (Chapter 17 *Interlude*): A drug that blocks the ability of the body to synthesize cholesterol.

Step-growth polymer (Section 10.13): A polymer in which each bond is formed independently of the others. Polyesters and polyamides (nylons) are examples.

Stereocenter (Section 6.2): An atom in a molecule that is a cause of chirality. Also called a chirality center.

Stereochemistry (Sections 2.8 and 6.1): The branch of chemistry concerned with the three-dimensional arrangement of atoms in molecules.

Stereoisomers (Section 2.8): Isomers that have their atoms connected in the same order but have different three-dimensional arrangements. The term includes both enantiomers and diastereomers.

Steric strain (Section 2.11): The strain imposed on a molecule when two groups are too close together and try to occupy the same space.

Steroid (Section 16.4): A lipid whose structure is based on a characteristic tetracyclic carbon skeleton with three 6-membered and one 5-membered ring.

STR loci (Chapter 16 *Interlude*): Short tandem repeat sequences of noncoding DNA that are unique to every individual and allow DNA fingerprinting.

Straight-chain alkane (Section 2.2): An alkane whose carbon atoms are connected without branching.

Substitution reaction (Section 3.5): The reaction that occurs when two reactants exchange parts to give two products.

Sulfide (Section 8.8): A compound that has two organic groups bonded to the same sulfur atom, R—S—R′.

Symmetry plane (Section 6.2): A plane that bisects a molecule such that one half of the molecule is the mirror image of the other half. Molecules containing a plane of symmetry are achiral.

Syn stereochemistry (Section 4.5): The opposite of anti. A syn addition reaction is one in which the two ends of the double bond react from the same side.

Tautomers (Section 11.1): Isomers that interconvert spontaneously, usually with the change in position of a hydrogen.

Terpenoid (Chapter 3 *Interlude*): A lipid that is formally derived by head-to-tail polymerization of isoprene units.

Tertiary: *See* Primary.

Tertiary structure (Section 15.8): The level of protein structure that involves the manner in which the entire protein chain is folded into a specific three-dimensional arrangement.

Thioester (Section 10.12): The sulfur analog of an ester, RCOSR′.

Thiol (Section 8.8): A compound with the –SH functional group.

Thiolate ion (Section 8.8): The sulfur analog of an alkoxide ion, RS^-.

Thiophenol (Chapter 8 Introduction): The sulfur analog of a phenol, Ar—SH.

TMS (Section 13.9): Tetramethylsilane, used as an NMR calibration standard.

Transamination (Section 17.5): A reaction in which the $-NH_2$ group of an amine changes places with the keto group of an α-keto acid.

Transcription (Section 16.8): The process by which the genetic information encoded in DNA is read and used to synthesize RNA in the nucleus of the cell.

Transfer RNA (tRNA) (Section 16.8): A kind of RNA that transports amino acids to the ribosomes, where they are joined together to make proteins.

Transition state (Section 3.8): An activated complex between reactants, representing the highest energy point on a reaction curve.

Translation (Section 16.9): The process by which the genetic information transcribed from DNA onto mRNA is read by tRNA and used to direct protein synthesis.

Triacylglycerol (Section 16.1): A lipid, such as that found in animal fat and vegetable oil, that is a triester of glycerol with long-chain fatty acids.

Tricarboxylic acid cycle (Section 17.4): An alternative name for the citric acid cycle by which acetyl CoA is degraded to CO_2.

Triple bond (Section 1.8): A covalent bond formed by sharing three pairs of electrons between atoms.

Triplet (Section 13.12): A symmetrical three-line splitting pattern observed in the ^1H NMR spectrum when a proton has two equivalent neighbor protons.

Ultraviolet (UV) spectroscopy (Section 13.5): An optical spectroscopy employing ultraviolet irradiation. UV spectroscopy provides structural information about the extent of electron conjugation in organic molecules.

Unimolecular reaction (Section 7.6): A reaction step that involves only one molecule.

Unsaturated (Section 3.1): A molecule that has one or more double or triple bonds and thus has fewer hydrogens than the corresponding alkane.

Upfield (Section 13.9): The right-hand portion of the NMR chart.

Uronic acid (Section 14.7): The monocarboxylic acid formed by oxidizing the $-CH_2OH$ end of a sugar without affecting the $-CHO$ end.

Valence bond theory (Section 1.5): A theory of chemical bonding that describes bonds as resulting from overlap of atomic orbitals.

Valence shell (Section 1.4): The outermost electron shell of an atom.

Vegetable oil (Section 16.1): A liquid triacylglycerol derived from a plant source.

Vinyl monomer (Section 4.7): A substituted alkene monomer used to make a chain-growth polymer.

Vinylic (Section 4.11): Referring to a substituent directly attached to a double-bond carbon atom.

Vitamin (Section 15.9): A small organic molecule that must be obtained in the diet and that is required for proper growth.

Vulcanization (Chapter 4 *Interlude*): A technique for cross-linking and hardening a diene polymer by heating with a few percent by weight of sulfur.

Wave equation (Section 1.1): A mathematical expression that defines the behavior of an electron in an atom.

Wave function (Section 1.1): A solution to the wave equation for defining the behavior of an electron in an atom. The square of the wave function defines the shape of an orbital.

Wavelength, λ (Section 13.2): The length of a wave from peak to peak.

Wavenumber, $\tilde{\nu}$ (Section 13.2): A unit of frequency measurement equal to the reciprocal of the wavelength in centimeters, cm^{-1}.

Wax (Section 16.1): A mixture of esters of long-chain carboxylic acids with long-chain alcohols.

Williamson ether synthesis (Section 8.4): The reaction of an alkoxide ion with an alkyl halide to yield an ether.

X-ray crystallography (Chapter 15 *Interlude*): A technique using X rays to determine the structure of molecules.

Z geometry (Section 3.4): A term used to describe the stereochemistry of a carbon–carbon double bond in which the two higher-ranked groups on each carbon are on the same side of the double bond.

Zaitsev's rule (Section 7.7): A rule stating that E2 elimination reactions normally yield the more highly substituted alkene as major product.

Zusammen (Z) (Section 3.4): A term used to describe the stereochemistry of a carbon–carbon double bond in which the two higher-ranked groups on each carbon are on the same side of the double bond.

Zwitterion (Sections 1.11 and 15.1): A neutral dipolar molecule whose positive and negative charges are not adjacent. For example, amino acids exist as zwitterions, $H_3N^+—CHR—CO_2^-$.

APPENDIX C

Answers to Selected In-Chapter Problems

The following answers to in-chapter problems are meant only as a quick check. Full answers and explanations for all problems, both in-chapter and end-of-chapter, are provided in the accompanying *Study Guide and Solutions Manual*.

Chapter 1

1.1 (a) 1 (b) 2 (c) 3

1.2 (a) B: $1s^2\ 2s^2\ 2p$ (b) P: $1s^2\ 2s^2\ 2p^6\ 3s^2\ 3p^3$
(c) O: $1s^2\ 2s^2\ 2p^4$ (d) Ar: $1s^2\ 2s^2\ 2p^6\ 3s^2\ 3p^6$

1.3

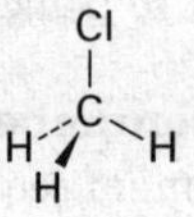

1.4

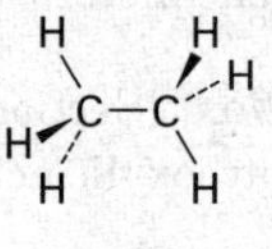

1.5 (a) CCl_4 (b) AlH_3 (c) CH_2Cl_2 (d) SiF_4

1.6

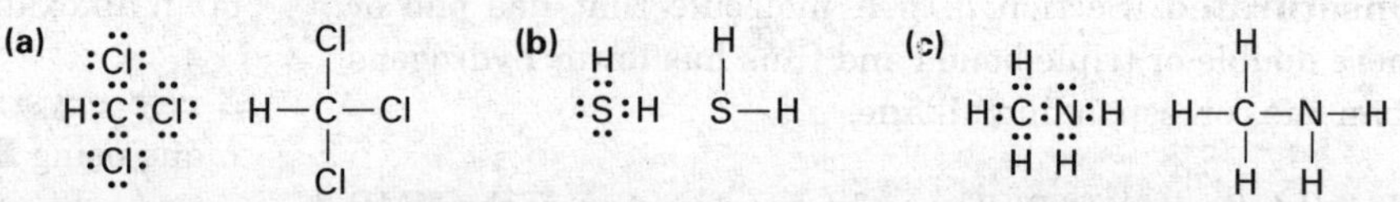

1.7 C_2H_7 has too many hydrogens for a compound with two carbons.

1.8

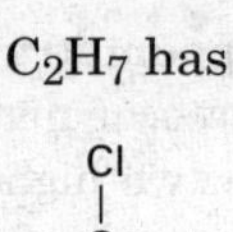

1.9 A carbon atom is larger than a hydrogen atom.

1.10 All bond angles are approximately 109°.

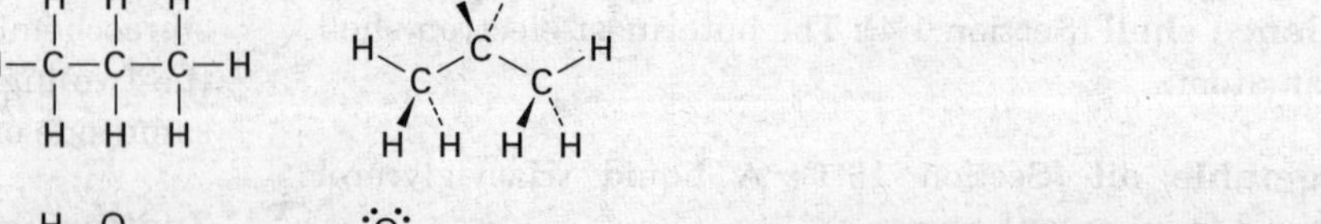

1.11

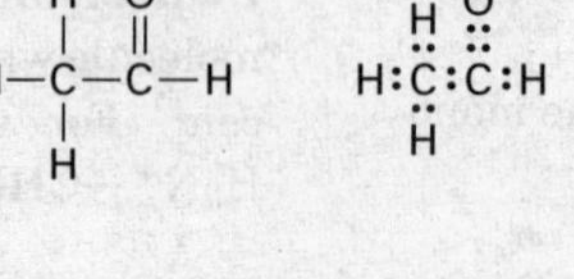

1.12 The CH_3 carbon is sp^3, the double-bond carbons are sp^2, and the C=C–C bond angle is approximately 120°.

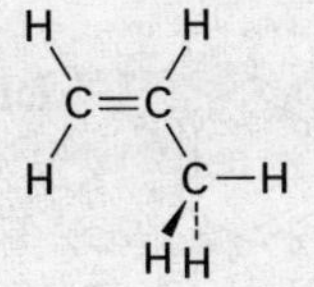

1.13 The CH_3 carbon is sp^3, the triple-bond carbons are sp, and the C≡C–C bond angle is approximately 180°.

1.14 All carbons are sp^2, and all bond angles are approximately 120°.

1.15

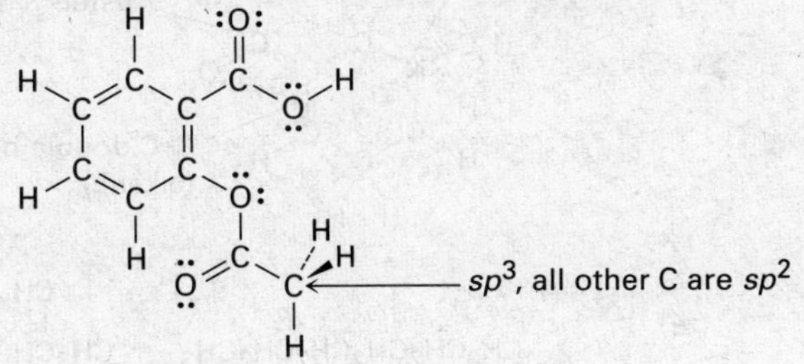

1.16 **(a)** H **(b)** Br **(c)** Cl

1.17 **(a)** C is $\delta+$, Br is $\delta-$ **(b)** C is $\delta+$, N is $\delta-$
(c) H is $\delta+$, N is $\delta-$ **(d)** C is $\delta+$, S is $\delta-$
(e) Mg is $\delta+$, C is $\delta-$ **(f)** C is $\delta+$, F is $\delta-$

1.18 CCl_4 and $Cl_2O < TiCl_3 < MgCl_2$

1.19

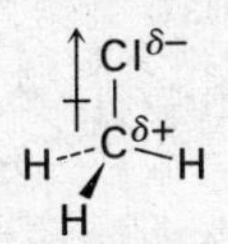

1.20 **(a)** Formic acid: $K_a = 1.8 \times 10^{-4}$; picric acid: $K_a = 0.42$
(b) Picric acid is stronger.

1.21 Water is the stronger acid.

1.22 **(a)** No **(b)** No

1.23 Lewis acids: (c), (d), (e); Lewis bases: (b), (f); both: (a)

1.24 **(a)** $CH_3CH_2OH + HCl \longrightarrow CH_3CH_2OH_2^+ \; Cl^-$; $(CH_3)_2NH + HCl \longrightarrow (CH_3)_2NH_2^+ \; Cl^-$; $(CH_3)_3P + HCl \longrightarrow (CH_3)_3PH^+ \; Cl^-$

(b) $HO^- + CH_3^+ \longrightarrow HO{-}CH_3$; $HO^- + B(CH_3)_3 \longrightarrow HO{-}B(CH_3)_3^-$;
$HO^- + MgBr_2 \longrightarrow HO{-}MgBr_2^-$

1.25

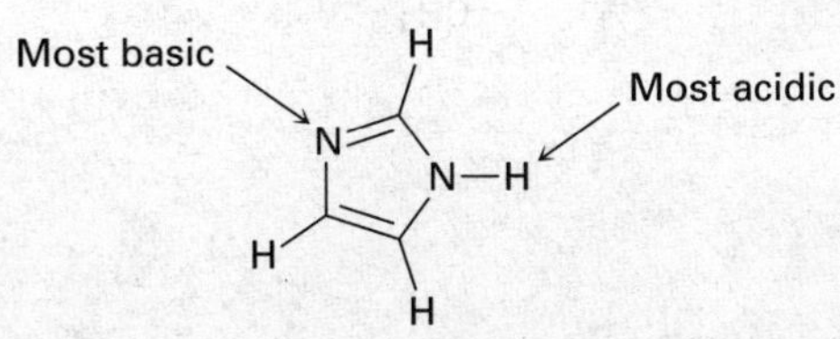

Chapter 2

2.1 **(a)** Carboxylic acid, double bond
(b) Carboxylic acid, aromatic ring, ester
(c) Aldehyde, alcohol

2.2 **(a)** CH_3OH **(b)** 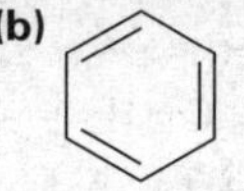 **(c)** 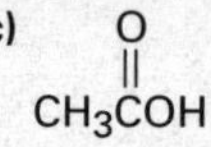 **(d)** CH_3NH_2

(e) 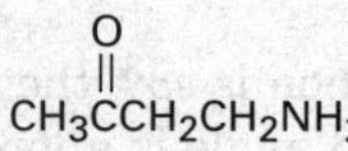 **(f)** $H_2C{=}CHCH{=}CH_2$

2.3

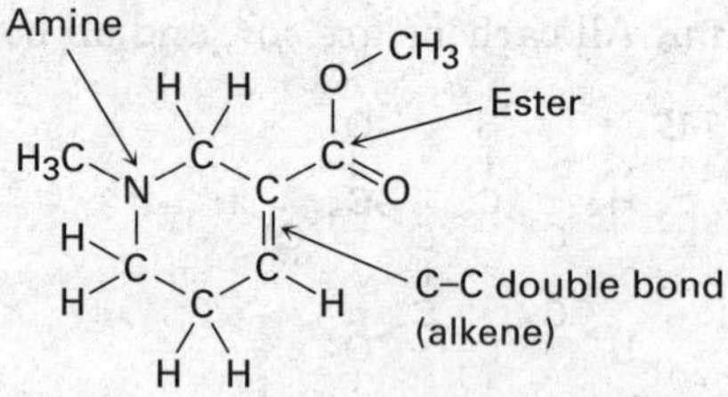

2.4

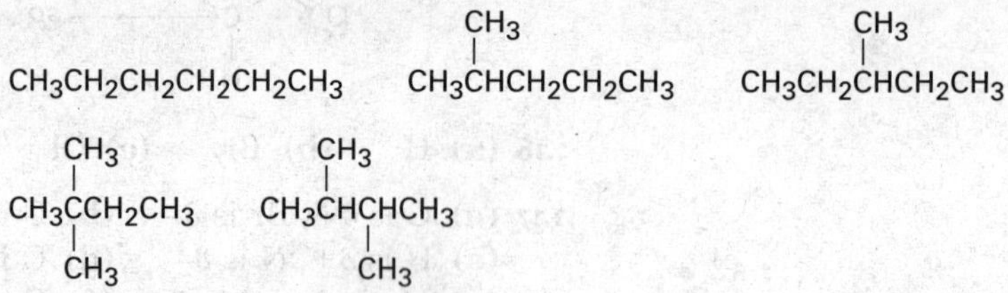

2.5 **(a)** $CH_3CH_2CH_2CH_2CH_2CH_2CH_2CH_3$
$(CH_3)_2CHCH_2CH_2CH_2CH_2CH_3$
$(CH_3)_3CCH_2CH_2CH_2CH_3$

(b)

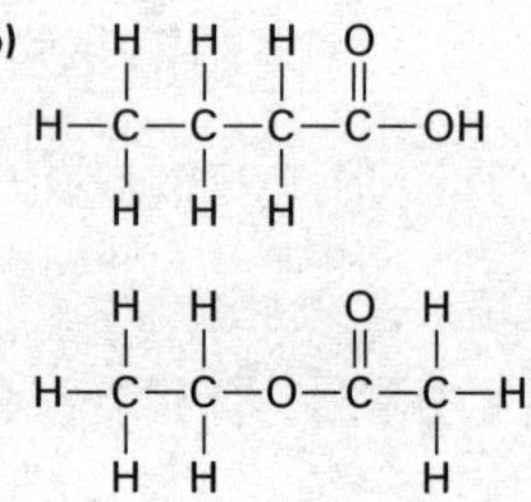

2.6

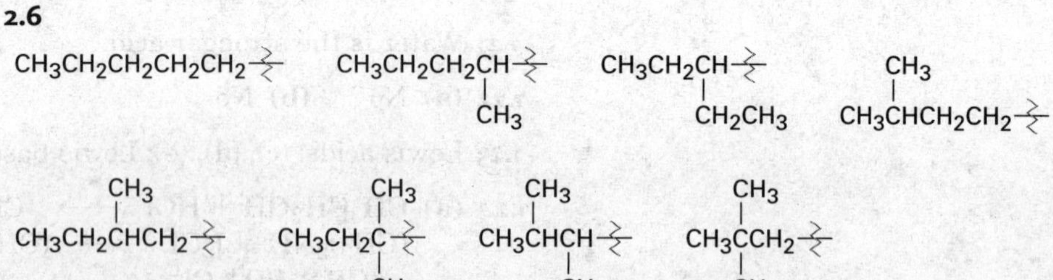

2.7 **(a)** 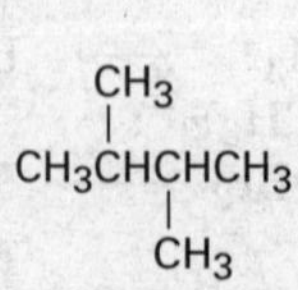 **(b)** 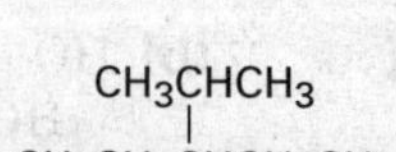 **(c)** 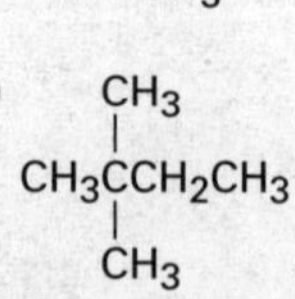

2.8 (a) 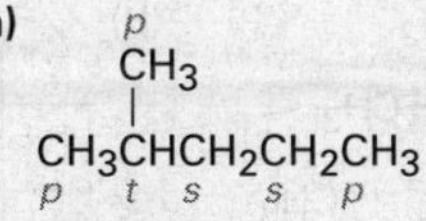 (b) 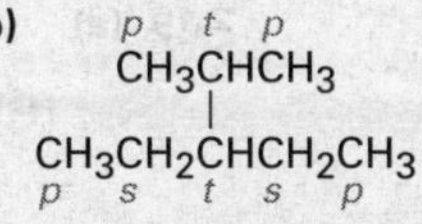 (c) 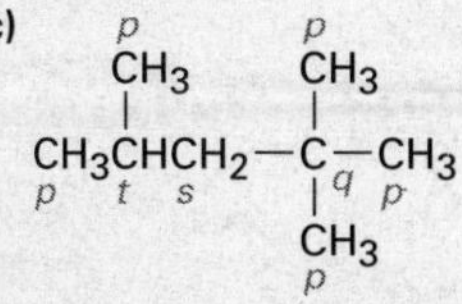

2.9 **(a)** Pentane, 2-methylbutane, 2,2-dimethylpropane
(b) 3,4-Dimethylhexane
(c) 2,4-Dimethylpentane
(d) 2,2,5-Trimethylheptane

2.10 (a) 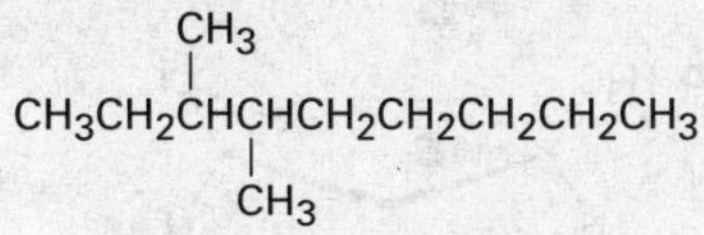 (b) 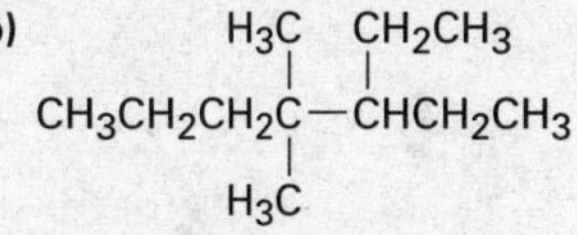

(c) 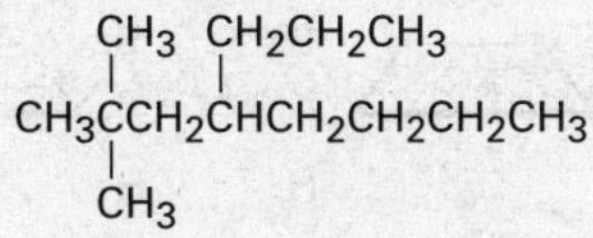 (d) 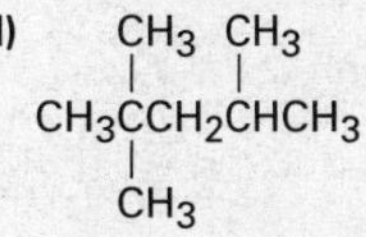

2.11 3,3,4,5-Tetramethylheptane

2.12

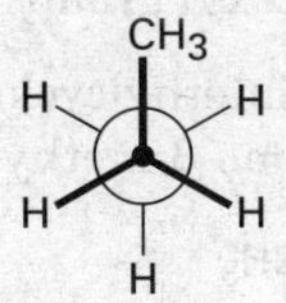

Most stable conformation (staggered)

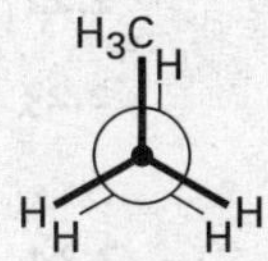

Least stable conformation (eclipsed)

2.13

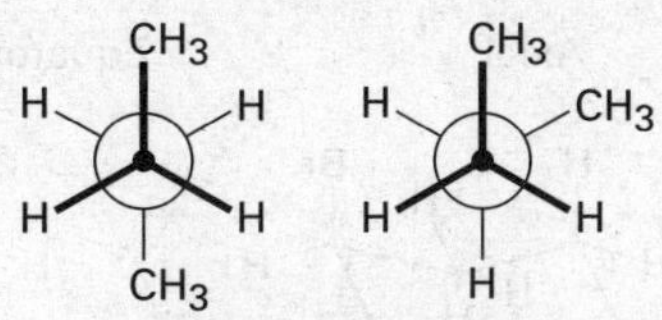

Staggered butane

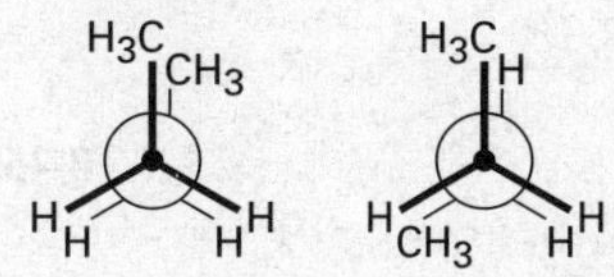

Eclipsed butane

2.14 The first staggered conformation of butane is the most stable.

2.15 **(a)** C_5H_5N **(b)** $C_6H_{10}O$ **(c)** C_8H_7N

2.16 (a) $CH_3CH_2CH{=}CH_2$ (b) 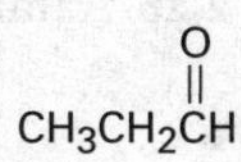 (c) 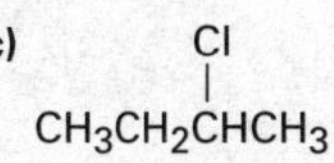

2.17

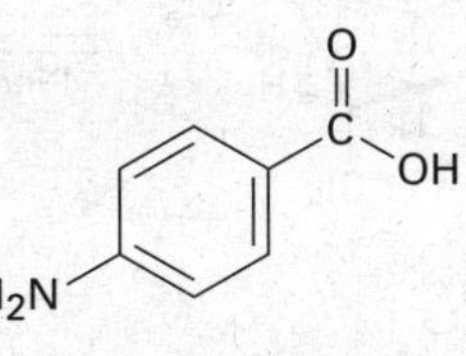

2.18 **(a)** 1,4-Dimethylcyclohexane **(b)** 1-Ethyl-3-methylcyclopentane
(c) Isopropylcyclobutane

2.19 **(a)**

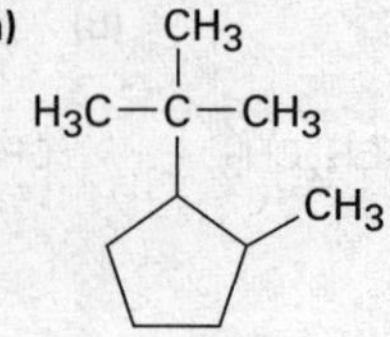

(b)

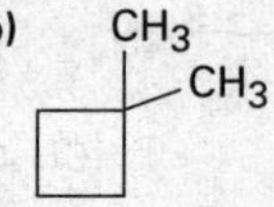

(c)

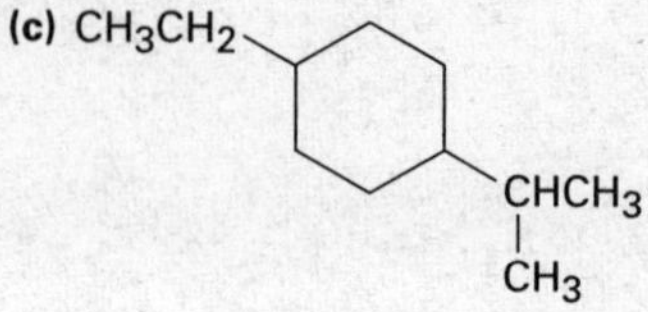

(d)

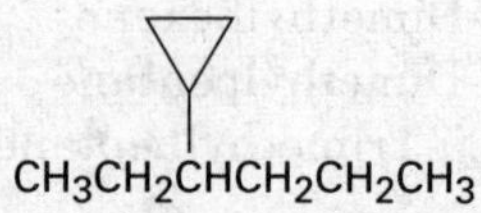

2.20

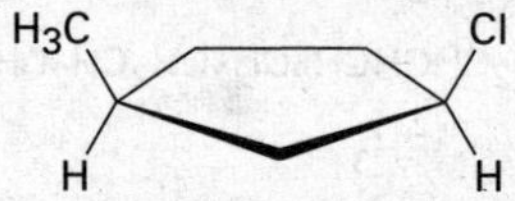

2.21

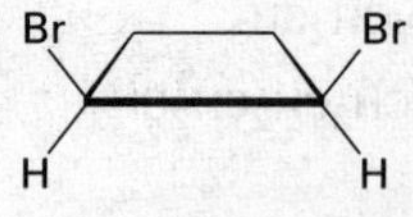

Cis

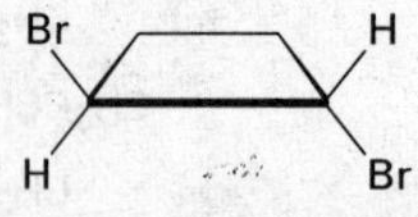

Trans

2.22 The two hydroxyl groups are cis; the two carbon chains are trans.

2.23 **(a)** *cis*-1,2-Dimethylcyclopentane
(b) *cis*-1-Bromo-3-methylcyclobutane

2.24

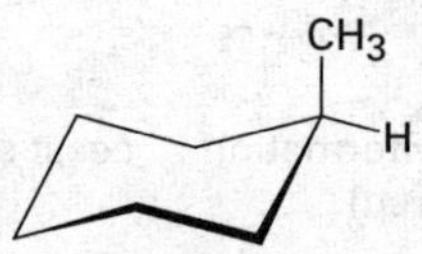

Axial

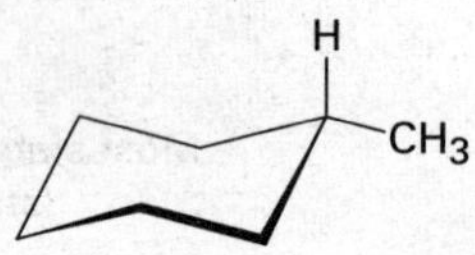

Equatorial

2.25

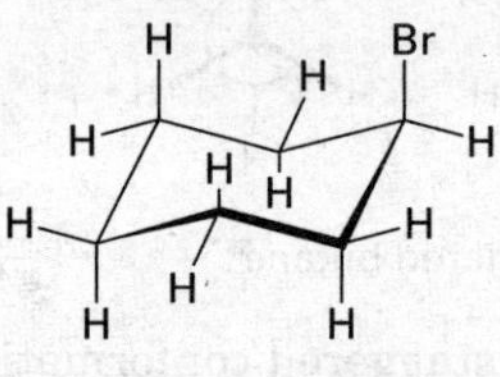

Axial (less stable)

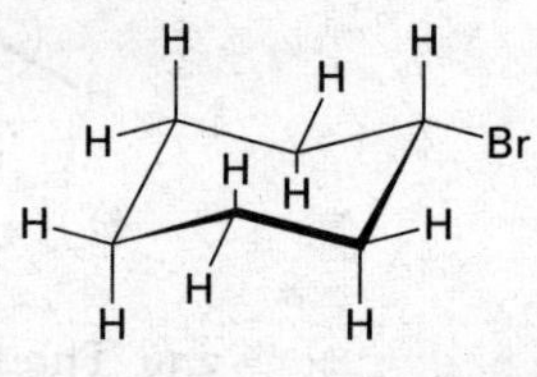

Equatorial (more stable)

2.26 Axial and equatorial positions alternate on each side of a ring.

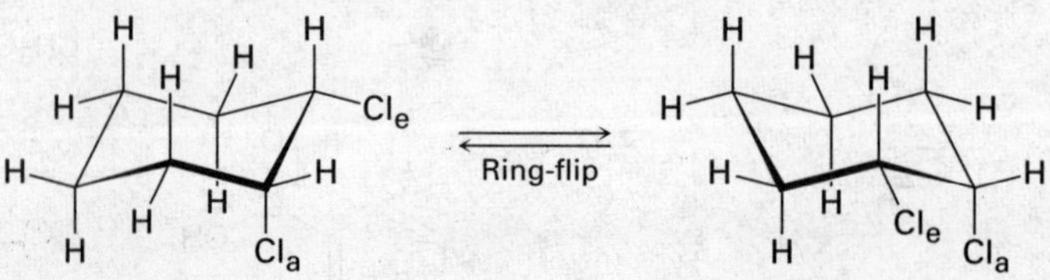

2.27 Axial and equatorial positions alternate on each side of a ring.

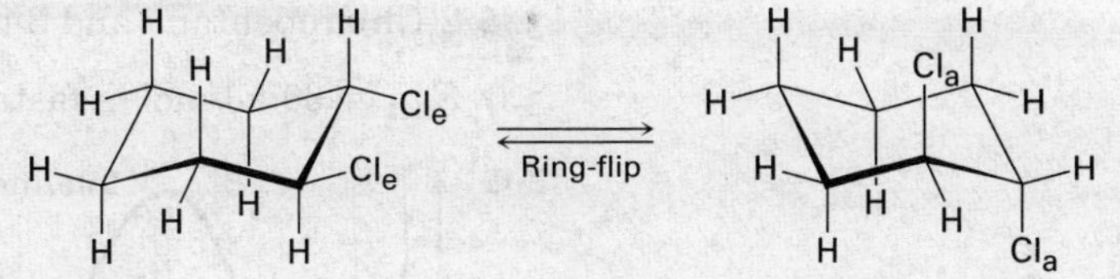

2.28 Less stable

Chapter 3

3.1 **(a)** 3,4,4-Trimethylpent-1-ene **(b)** 3-Methylhex-3-yne
(c) 4,7-Dimethylocta-2,5-diene **(d)** 6-Ethyl-7-methylnon-4-ene

3.2 **(a)** 1,2-Dimethylcyclohexene **(b)** 4,4-Dimethylcycloheptene
(c) 3-Isopropylcyclopentene

3.3 (a)

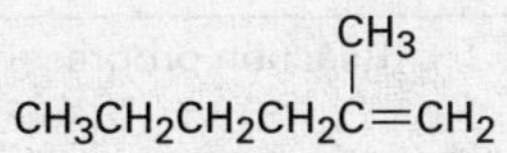

(b)

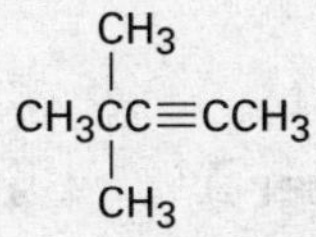

(c)

(d)

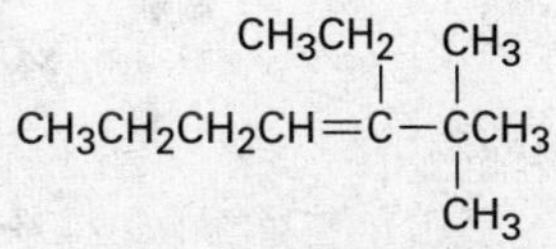

3.4 **(a)** 2,5-Dimethylhex-3-yne **(b)** 3,3-Dimethylbut-1-yne
(c) 3,3-Dimethyloct-4-yne **(d)** 2,5,5-Trimethylhept-3-yne

3.5 **(a)** 2,5,5-Trimethylhex-2-ene **(b)** 2,2-Dimethylhex-3-yne
(c) 2-Methylhepta-2,5-diene **(d)** 1-Methylcyclopenta-1,3-diene

3.6 Compounds (c), (d), (e), and (f) can exist as pairs of isomers.

3.7 **(a)** *cis*-3,4-Dimethylhex-2-ene **(b)** *trans*-6-Methylhept-3-ene

3.8 **(a)** –Br **(b)** –Br **(c)** $-CH_2CH_3$
(d) –OH **(e)** $-CH_2OH$ **(f)** $-CH{=}O$

3.9 **(a)** *Z* **(b)** *E* **(c)** *E*

3.10 *Z*

3.11 **(a)** Substitution **(b)** Elimination **(c)** Addition

3.12 (a)

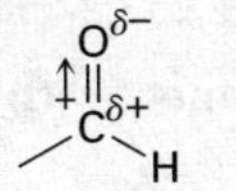

(b)

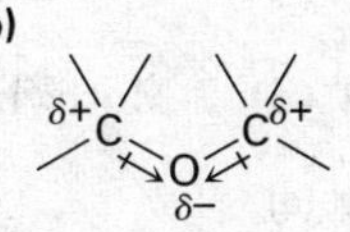

(c)

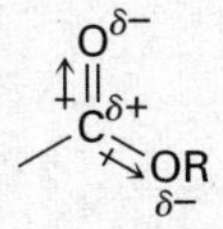

(d)

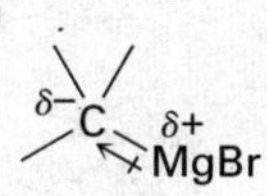

3.13 Electrophile: (a), (c); nucleophile: (b), (d), (e)

3.14 Boron is a Lewis acid/electrophile because it has only six outer-shell electrons.

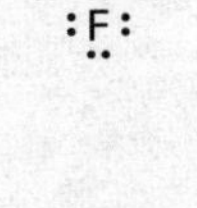

3.15 $(CH_3)_3C^+$ is the intermediate.

3.16 2-Chloropentane and 3-chloropentane

3.17 E_{act} = 60 kJ/mol is faster.

3.19

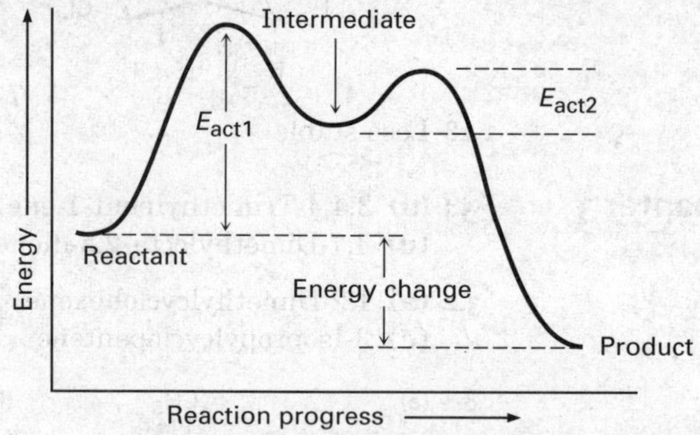

Chapter 4

4.1 **(a)** Chlorocyclohexane **(b)** 2-Bromo-2-methylpentane **(c)** 4-Methylpentan-2-ol **(d)** 1-Bromo-1-methylcyclohexane

4.2 **(a)** Cyclopentene **(b)** 1-Ethylcyclohexene or ethylidenecyclohexane **(c)** Hex-3-ene **(d)** Vinylcyclohexane (cyclohexylethylene)

4.3 (a) (b)

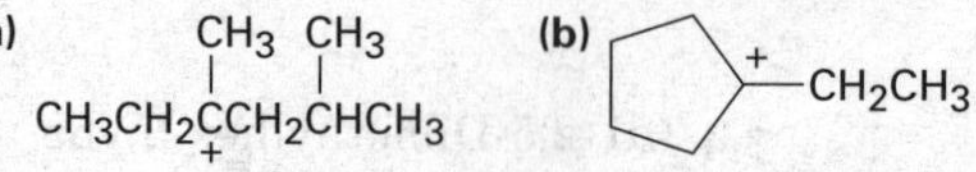

4.4 (a) (b) (c)

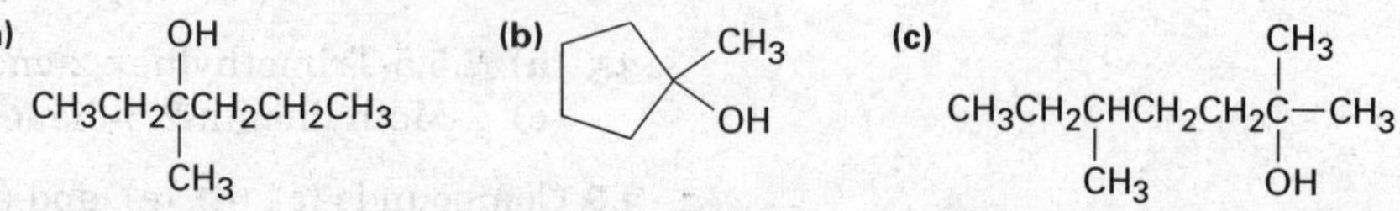

4.5 **(a)** But-1-ene or but-2-ene
(b) 3-Methylpent-2-ene or 2-ethylbut-1-ene
(c) 1,2-Dimethylcyclohexene or 2,3-dimethylcyclohexene

4.6 *trans*-1,2-Dibromo-1,2-dimethylcyclohexane

4.7

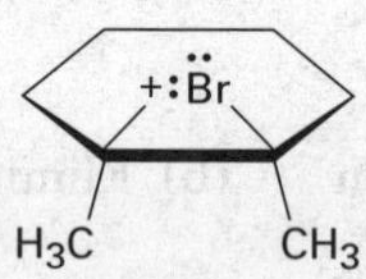

4.8 **(a)** 2-Methylpentane **(b)** 1,1-Dimethylcyclopentane

4.9 (a) (b)

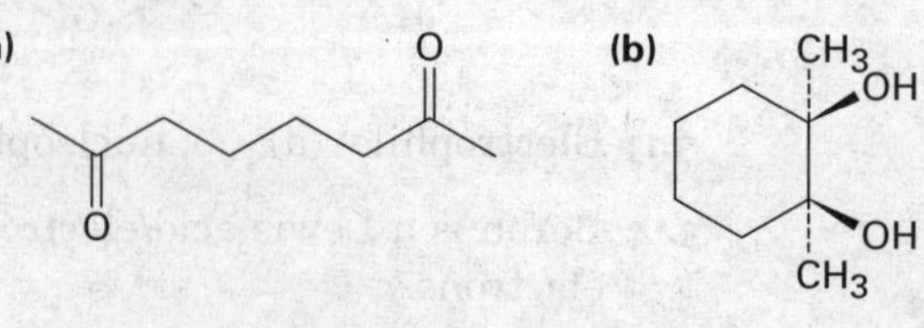

4.10 **(a)** 2-Methylpropene **(b)** Hex-3-ene

4.11 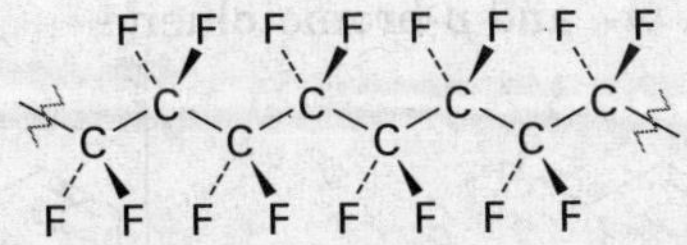

4.12 1,4-Dibromobut-2-ene and 3,4-dibromobut-1-ene

4.13 4-Chloropent-2-ene, 3-chloropent-1-ene, 1-chloropent-2-ene

4.14 $CH_3CH_2\overset{\delta+}{C}H{=}CH{=}\overset{\delta+}{C}H_2$ and $CH_3\overset{\delta+}{C}H{=}CH{=}\overset{\delta+}{C}HCH_3$

More stable

4.15

(a) 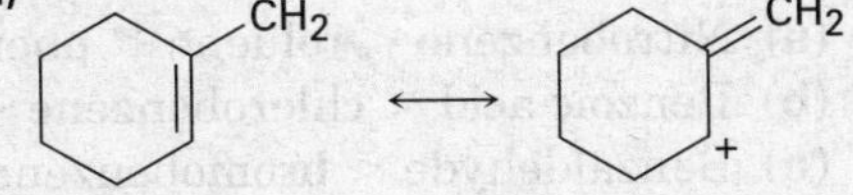

(b) 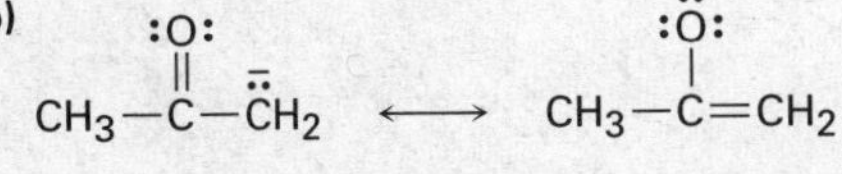

(c) 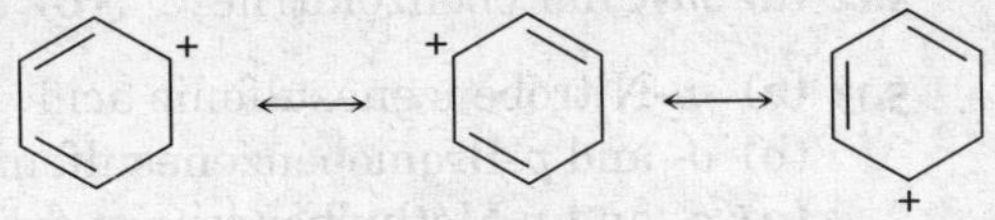

4.16 **(a)** 6-Methylhept-3-yne **(b)** 3,3-Dimethylbut-1-yne
(c) 5-Methylhex-2-yne **(d)** Hept-2-en-5-yne

4.17 **(a)** 1,2-Dichloropent-1-ene
(b) 4-Bromohept-3-ene and 3-bromohept-3-ene
(c) *cis*-6-Methylhept-3-ene

4.18 Octan-4-one

4.19 **(a)** Pent-1-yne **(b)** Hex-3-yne

4.20 **(a)** 1-Bromo-3-methylbutane + acetylene
(b) 1-Bromopropane + prop-1-yne, or bromomethane + pent-1-yne
(c) Bromomethane + 3-methylbut-1-yne

Chapter 5

5.1 The two structures are resonance forms, not isomers.

5.2 **(a)** meta **(b)** para **(c)** ortho

5.3 **(a)** *m*-Bromochlorobenzene **(b)** (3-Methylbutyl)benzene
(c) *p*-Bromoaniline **(d)** 2,5-Dichlorotoluene
(e) 1-Ethyl-2,4-dinitrobenzene **(f)** 1,2,3,5-Tetramethylbenzene

5.4 (a) 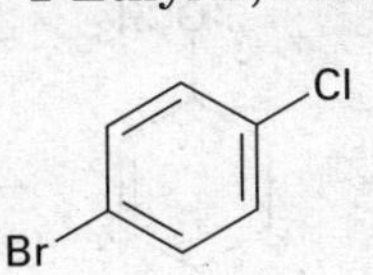

(b) 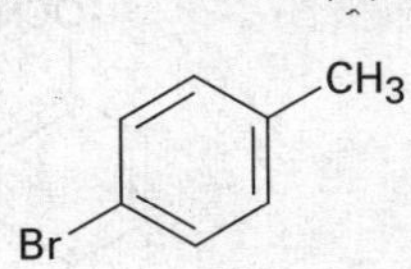

(c) 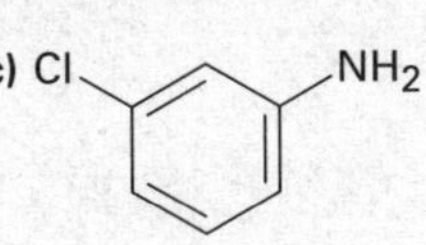

(d) 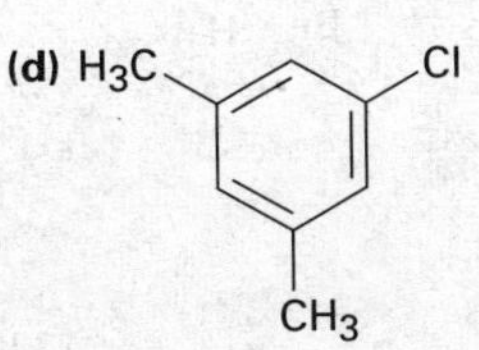

5.5 *o*-, *m*-, and *p*-bromotoluene

5.6

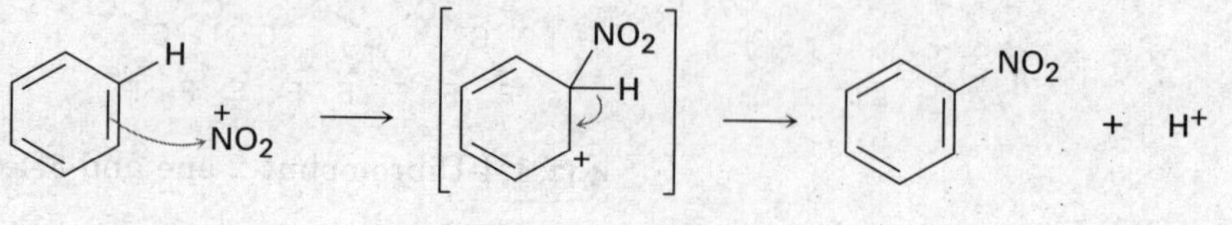

Carbocation intermediate

5.7 *p*-Xylene has one kind of ring position; *o*-xylene has two.

5.8 Three

5.9 **(a)** Ethylbenzene **(b)** 2-Ethyl-1,4-dimethylbenzene

5.10 **(a)** *tert*-Butylbenzene **(b)** Propanoylbenzene, $C_6H_5COCH_2CH_3$

5.11 **(a)** Nitrobenzene < toluene < phenol
(b) Benzoic acid < chlorobenzene < benzene < phenol
(c) Benzaldehyde < bromobenzene < benzene < aniline

5.12 **(a)** *m*-Chlorobenzonitrile **(b)** *o*- and *p*-Bromochlorobenzene

5.13 **(a)** *m*-Nitrobenzenesulfonic acid
(b) *o*- and *p*-Bromobenzenesulfonic acid
(c) *o*- and *p*-Methylbenzenesulfonic acid
(d) *m*-Carboxybenzenesulfonic acid
(e) *m*-Cyanobenzenesulfonic acid

5.14

Ortho

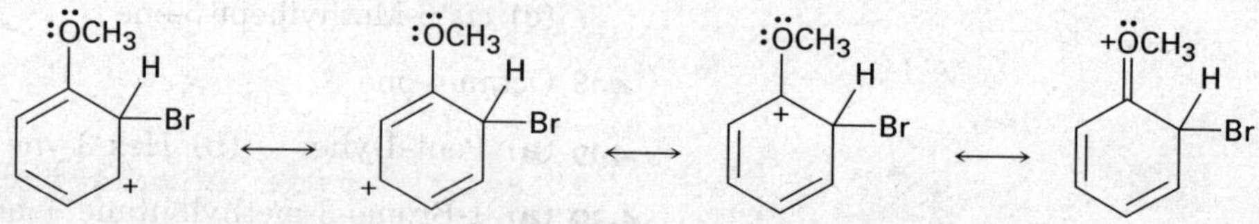

Meta

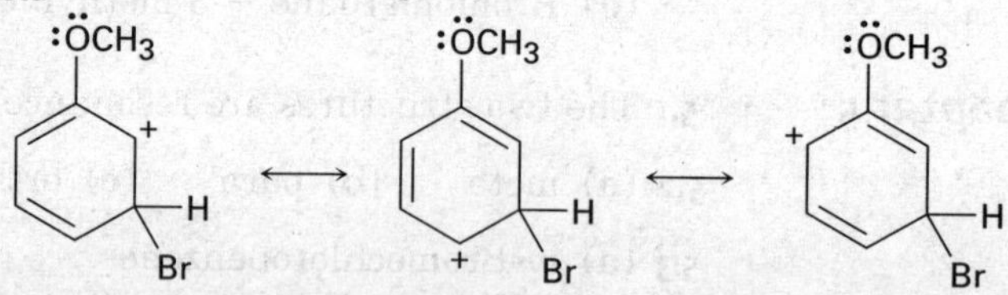

Para

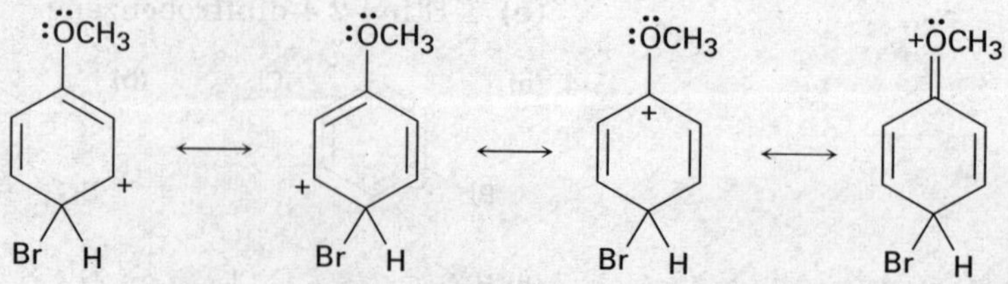

5.15 Ortho

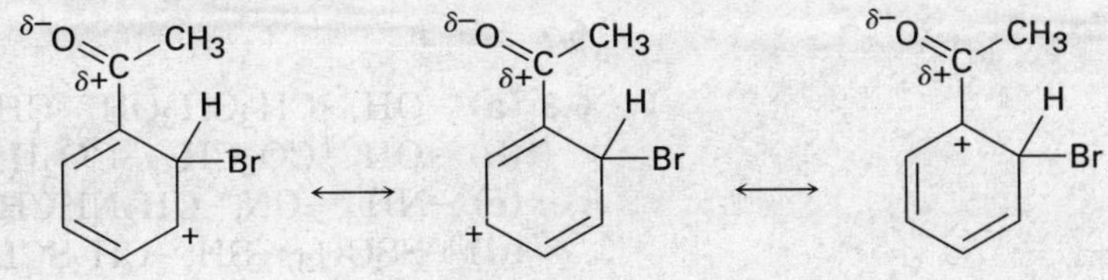

Meta

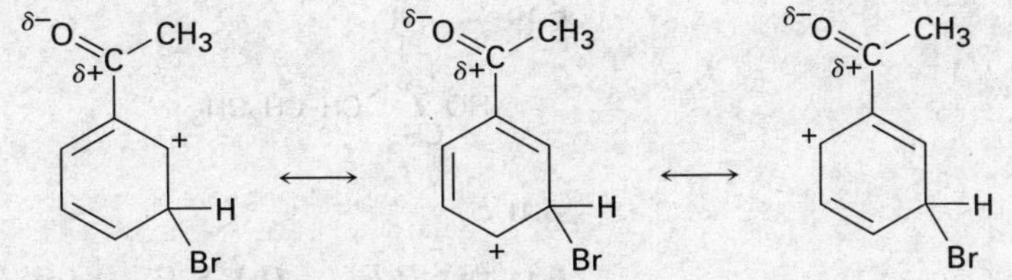

Para

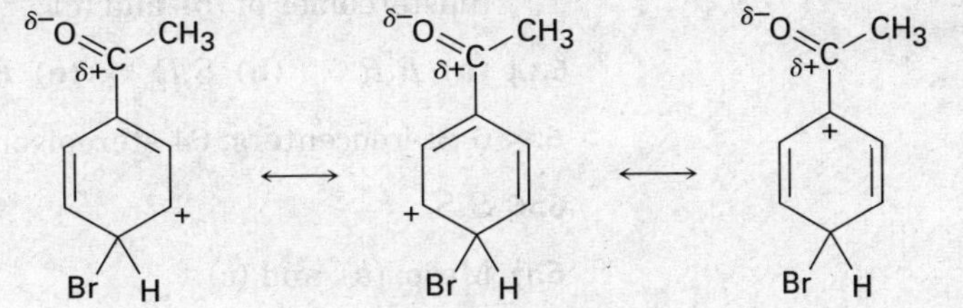

5.16 **(a)** *m*-Chlorobenzoic acid **(b)** *o*-Benzenedicarboxylic acid

5.17

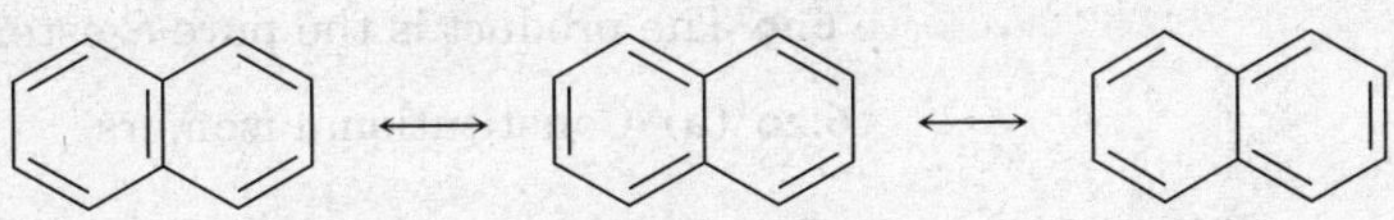

5.18 **(a)** 1. CH_3Cl, $AlCl_3$; 2. CH_3COCl, $AlCl_3$
(b) 1. Cl_2, $FeCl_3$; 2. HNO_3, H_2SO_4

5.19 **(a)** 1. Br_2, $FeBr_3$; 2. CH_3Cl, $AlCl_3$ **(b)** 1. 2 CH_3Cl, $AlCl_3$; 2. Br_2, $FeBr_3$

5.20 1. CH_3Cl, $AlCl_3$; 2. $KMnO_4$, H_2O; 3. Cl_2, $FeCl_3$

Chapter 6

6.1 Chiral: screw, shoe

6.2 Chiral: (b), (c)

6.3 Chiral: (b)

6.4

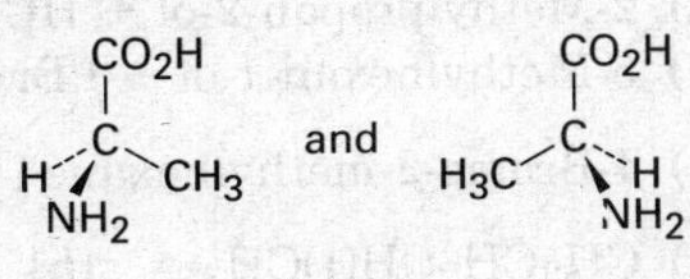

6.5 (a)

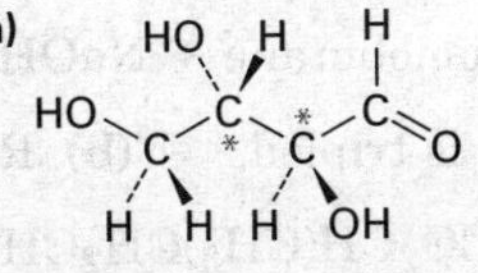

(b)

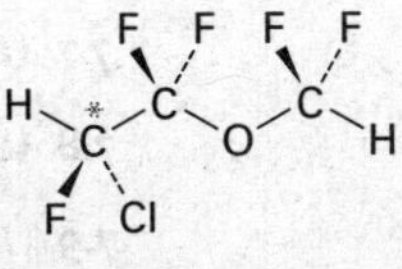

6.6 Levorotatory

6.7 +16.1

6.8 **(a)** $-OH, -CH_2CH_2OH, -CH_2CH_3, -H$
(b) $-OH, -CO_2CH_3, -CO_2H, -CH_2OH$
(c) $-NH_2, -CN, -CH_2NHCH_3, -CH_2NH_2$
(d) $-SSCH_3, -SH, -CH_2SCH_3, -CH_3$

6.9 **(a)** *S* **(b)** *S* **(c)** *R*

6.10

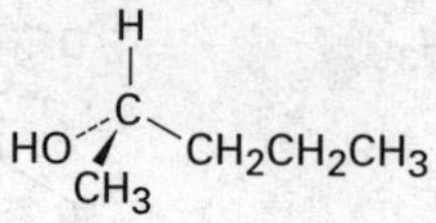

6.11 *S*

6.12 **(a)** *R,R* **(b)** *S,R* **(c)** *R,S*

6.13 Molecules (b) and (c) are enantiomers (mirror images). Molecule (a) is the diastereomer of (b) and (c).

6.14 **(a)** *R,R* **(b)** *S,R* **(c)** *R,S* **(d)** *S,S*

6.15 6 Stereocenters; 64 stereoisomers

6.16 *S,S*

6.17 Meso: (a) and (c)

6.18 Meso: (a) and (c)

6.19 The product is the pure *S* ester.

6.20 **(a)** Constitutional isomers **(b)** Diastereomers

Chapter 7

7.1 **(a)** 2-Bromobutane **(b)** 3-Chloro-2-methylpentane
(c) 1-Chloro-3-methylbutane **(d)** 1,3-Dichloro-3-methylbutane
(e) 1-Bromo-4-chlorobutane **(f)** 4-Bromo-1-chloropentane

7.2 **(a)** $CH_3CH_2CH_2C(CH_3)_2CH(Cl)CH_3$
(b) $CH_3CH_2CH_2C(Cl)_2CH(CH_3)_2$
(c) $CH_3CH_2C(Br)(CH_2CH_3)_2$
(d) $CH_3CH(Cl)CH_2CH(CH_3)CH(Br)CH_3$

7.3 1-Chloro-3-methylpentane, 2-chloro-3-methylpentane, 3-chloro-3-methylpentane, 3-(chloromethyl)pentane. The first two are chiral.

7.4 **(a)** 2-Methylpropan-2-ol + HCl **(b)** 4-Methylpentan-2-ol + PBr_3
(c) 5-Methylhexan-1-ol + PBr_3 **(d)** 2,4-Dimethylhexan-2-ol + HCl

7.5 **(a)** 4-Bromo-2-methylhexane **(b)** 1-Chloro-3,3-dimethylcyclopentane

7.6 **(a)** $CH_3CH_2CH(I)CH_3$ **(b)** $(CH_3)_2CHCH_2SH$ **(c)** $C_6H_5CH_2CN$

7.7 **(a)** 1-Bromobutane + NaOH **(b)** 1-Bromo-3-methylbutane + NaN_3

7.8 **(a)** Rate is tripled. **(b)** Rate is quadrupled.

7.9 (*R*) $CH_3CO_2CH(CH_3)CH_2CH_2CH_2CH_3$

7.10 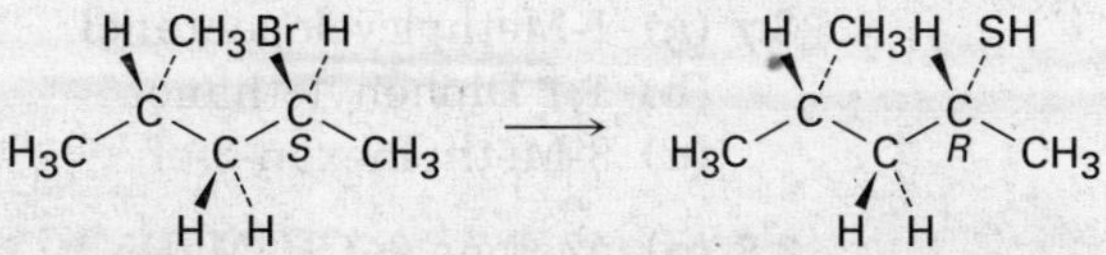

7.11 **(a)** Reaction with $CH_3CH_2CH_2Br$ is faster.
(b) Reaction with $(CH_3)_2CHCH_2Cl$ is faster.

7.12 $CH_3I > CH_3Br > CH_3F$

7.13 **(a)** Rate is unchanged. **(b)** Rate is doubled.

7.14 Racemic 3-bromo-3-methyloctane

7.15 The *S* substrate gives a racemic mixture of alcohols.

7.16 **(a)** 2-Methylpent-2-ene **(b)** 2,3,5-Trimethylhex-2-ene
(c) 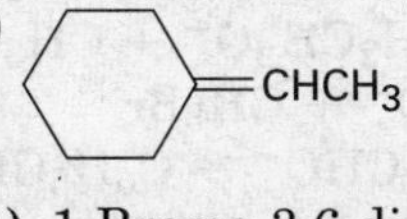

7.17 **(a)** 1-Bromo-3,6-dimethylheptane
(b) 1,2-Dimethyl-4-bromocyclopentane

7.18 The rate is tripled.

7.19 **(a)** S_N2 **(b)** E2 **(c)** S_N1 **(d)** E1cB

Chapter 8

8.1 **(a)** 5-Methylhexane-2,4-diol **(b)** 2-Methyl-4-phenylbutan-2-ol
(c) 4,4-Dimethylcyclohexanol **(d)** *trans*-2-Bromocyclopentanol
(e) 4-Bromo-3-methylphenol **(f)** 3-Methoxycyclopentene

8.2 Secondary: (a), (c), (d); tertiary: (b)

8.3

(a) 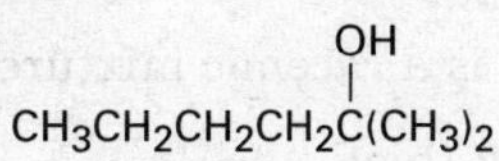 **(b)** 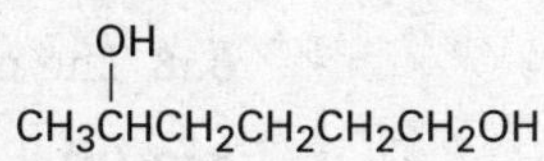 **(c)** 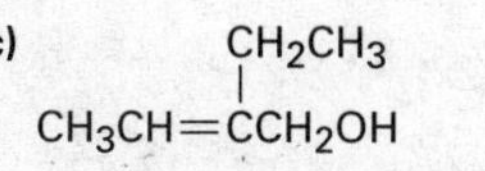

(d) 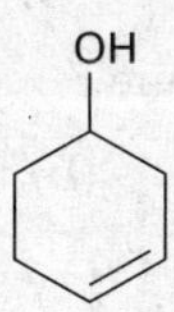 **(e)** 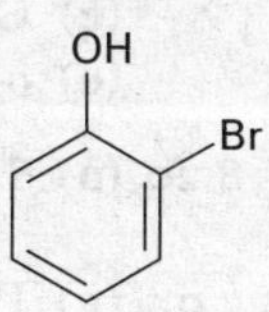 **(f)** 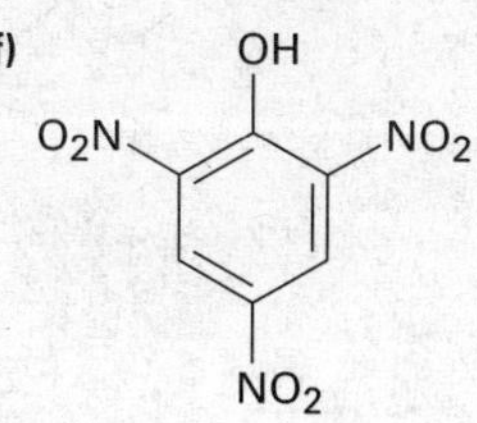

8.4 **(a)** Diisopropyl ether
(b) Cyclopentyl propyl ether
(c) *p*-Bromoanisole or 4-bromo-1-methoxybenzene
(d) Ethyl isobutyl ether

8.5 **(a)** $NaBH_4$ **(b)** $LiAlH_4$

8.6 **(a)** C_6H_5CHO, $C_6H_5CO_2H$, $C_6H_5CO_2R$
(b) $C_6H_5COCH_3$
(c) Cyclohexanone

8.7 **(a)** 1-Methylcyclopentanol
(b) 1,1-Diphenylethanol
(c) 3-Methylhexan-3-ol

8.8 **(a)** Acetone + CH_3MgBr
(b) Cyclohexanone + CH_3MgBr
(c) Pentan-3-one + CH_3MgBr, or butan-2-one + CH_3CH_2MgBr

8.9 **(a)** 2,3-Dimethylpent-2-ene **(b)** 2-Methylpent-2-ene

8.10 **(a)** 2,3-Dimethylcyclohexanol **(b)** Heptan-4-ol

8.11 **(a)** 1-Phenylethanol **(b)** 2-Methylpropan-1-ol **(c)** Cyclopentanol

8.12 **(a)** Cyclohexanone **(b)** Hexanoic acid **(c)** Hexan-2-one

8.13 **(a)** Cyclohexanone **(b)** Hexanal **(c)** Hexan-2-one

8.14 **(a)** $CH_3CH_2CH_2O^- + CH_3Br$
(b) $C_6H_5O^- + CH_3Br$
(c) $(CH_3)_2CHO^- + C_6H_5CH_2Br$

8.15 **(a)** Bromoethane > chloroethane > 2-bromopropane > 2-chloro-2-methylpropane

8.16 $CH_3CH_2COCH_2CH(CH_3)_2$;
(i) $CH_3CH_2CH(OCH_3)CH_2CH(CH_3)_2$
(ii) $CH_3CH_2CH(Cl)CH_2CH(CH_3)_2$
(iii) $CH_3CH_2COCH_2CH(CH_3)_2$

8.17 **(a)** 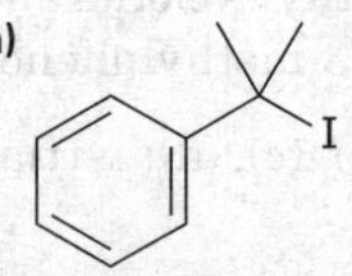 + CH_3OH **(b)** $CH_3CH_2CH(CH_3)OH$ + $CH_3CH_2CH_2I$

8.18 The product is a racemic mixture of *R*,*R* and *S*,*S* butane-1,2-diols.

8.19 **(a)** Butane-2-thiol **(b)** 2,2,6-Trimethylheptane-4-thiol
(c) Cyclopent-2-ene-1-thiol **(d)** Ethyl isopropyl sulfide
(e) *o*-Di(methylthio)benzene **(f)** 3-(Ethylthio)cyclohexanone

8.20 **(a)** 1. PBr_3; 2. $Na^+ \ ^-SH$ **(b)** 1. $LiAlH_4$; 2. PBr_3; 3. $Na^+ \ ^-SH$

Chapter 9

9.1 **(a)** Pentan-2-one **(b)** $CH_3CH_2CH_2CH{=}CHCHO$
(c) $CH_3CH_2COCH_2CH_2CHO$ **(d)** Cyclopentanone

9.2 **(a)** 2-Methylpentan-3-one
(b) 3-Phenylpropanal
(c) Octane-2,6-dione
(d) *trans*-2-Methylcyclohexanecarbaldehyde
(e) Pentanedial
(f) *cis*-2,5-Dimethylcyclohexanone

9.3 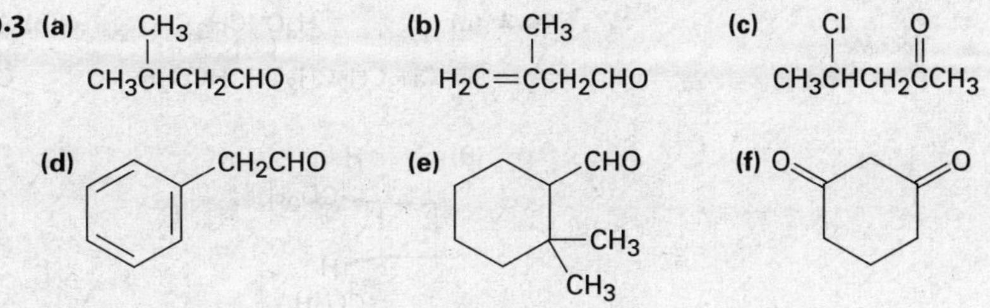

9.4 **(a)** Periodinane
(b) 1. $LiAlH_4$; 2. periodinane
(c) 1. $KMnO_4$; 2. $LiAlH_4$; 3. periodinane

9.5 **(a)** Periodinane **(b)** H_3O^+, $HgSO_4$ **(c)** $KMnO_4$, H_3O^+

9.6 **(a)** 1. H_3O^+; 2. periodinane **(b)** 1. CH_3COCl, $AlCl_3$; 2. $NaBH_4$

9.7 **(a)** Pentanoic acid
(b) 2,2-Dimethylhexanoic acid
(c) No reaction

9.8 $(CH_3)_2C(OH)CN$

9.9 $(CH_3)_2C(OH)OCH_3$

9.10 **(a)** $C_5H_9MgBr + (CH_3)_2CHCHO$ or $(CH_3)_2CHMgBr + C_5H_9CHO$ or $C_5H_9COCH(CH_3)_2 + NaBH_4$
(b) $PhCH_2CHO + NaBH_4$ or $PhCH_2CO_2R + LiAlH_4$ or $PhCH_2MgBr + CH_2O$
(c) $C_6H_{11}MgBr + (CH_3)_2C{=}O$ or $CH_3MgBr + C_6H_{11}CO_2R$ or $CH_3MgBr + C_6H_{11}COCH_3$

9.11 $C_5H_9COCH_3 + CH_3MgBr$ or $C_5H_9MgBr + (CH_3)_2C{=}O$

9.12 Labeled water adds reversibly to the carbonyl group.

9.13 The mechanism of acetal formation is shown in Figure 9.3.

9.14 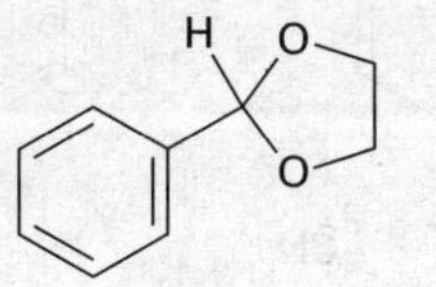

9.15 1. CH_3OH, acid catalyst; 2. CH_3MgBr; 3. H_3O^+

9.16 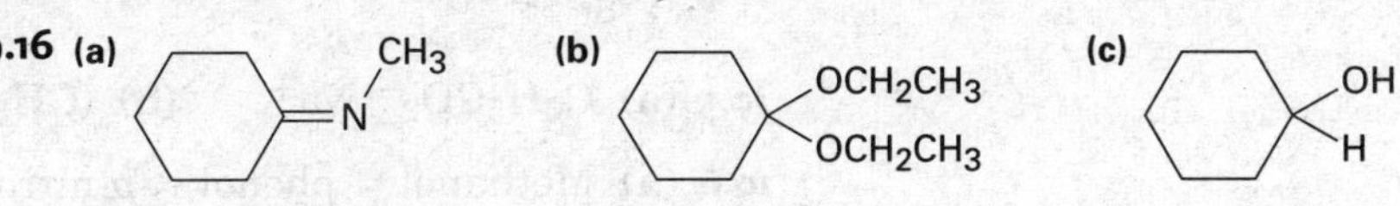

9.17 $(CH_3)_2CHCOCH_2CH_3 + CH_3NH_2$

9.18 6-Methylcyclohex-2-enone + $(CH_3)_2CHOH$

Chapter 10

10.1 **(a)** 3-Methylbutanoic acid
(b) 4-Bromopentanoic acid
(c) 2-Ethylpentanoic acid
(d) *cis*-Hex-4-enoic acid
(e) *cis*-Cyclopentane-1,3-dicarboxylic acid

10.2 (a) 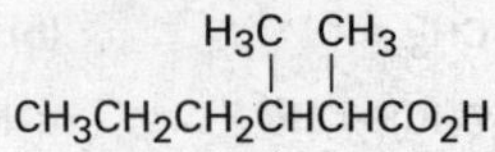 (b) 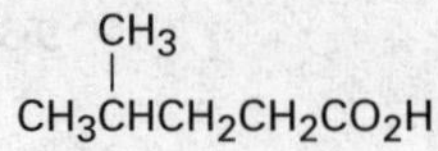 (c) 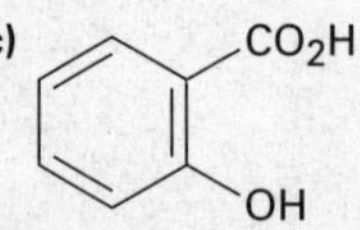

(d) 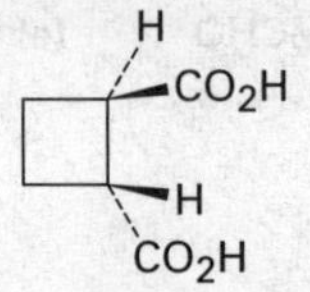

10.3 (a) 4-Methylpentanoyl chloride
(b) Cyclohexylacetamide
(c) Isopropyl 2-methylpropanoate
(d) Benzoic anhydride
(e) Isopropyl cyclopentanecarboxylate
(f) Cyclopentyl 2-methylpropanoate
(g) *N*-Methylpent-4-enamide
(h) 2-Methylbutanenitrile

10.4 (a)  (b) 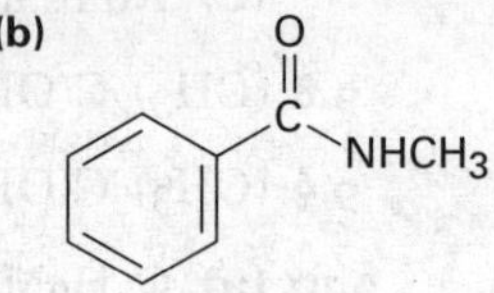 (c) 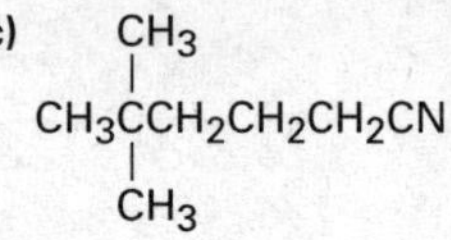

(d) 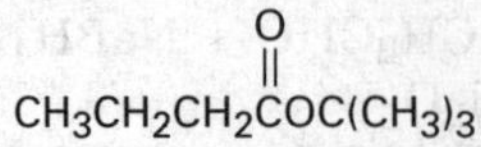 (e) 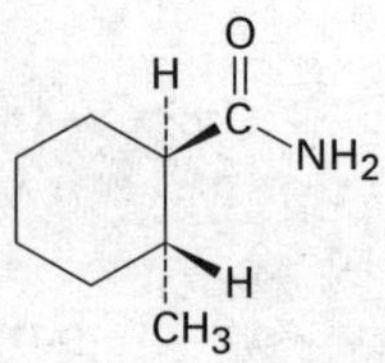

(f) 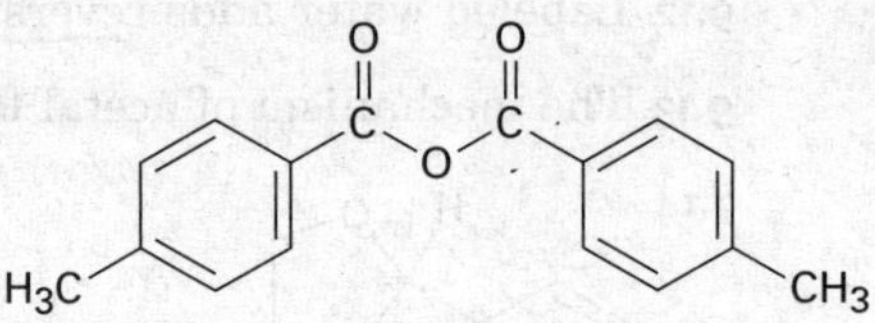 (g) 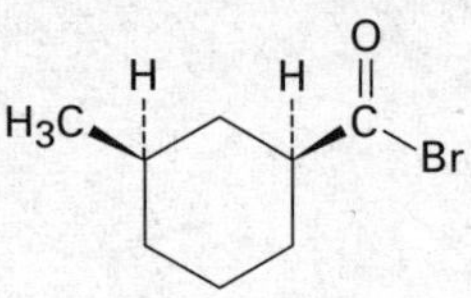

(h) 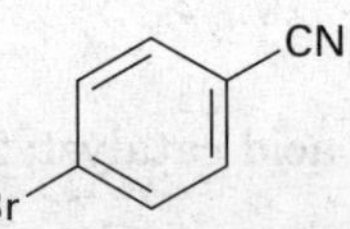

10.5 (a) $C_6H_5CO_2^-\ Na^+$ (b) $(CH_3)_3CCO_2^-\ K^+$

10.6 (a) Methanol < phenol < *p*-nitrophenol < acetic acid < sulfuric acid
(b) Ethanol < benzoic acid < *p*-cyanobenzoic acid

10.7 Lactic acid is stronger because of the electron-withdrawing effect of the –OH group.

10.8 1. NaCN; 2. NaOH, H_2O. Iodobenzene cannot be converted to benzoic acid by this method.

10.9 (a) CH_3COCl (b) $CH_3CH_2CO_2CH_3$
(c) $CH_3CO_2COCH_3$ (d) $CH_3CO_2CH_3$

10.10 **(a)** $CH_3CO_2^- Na^+$ **(b)** CH_3CONH_2
(c) $CH_3CO_2CH_3 + CH_3CO_2^- Na^+$ **(d)** $CH_3CONHCH_3$

10.11 **(a)** C_6H_5COCl **(b)** $C_6H_5CO_2CH_3$
(c) $C_6H_5CH_2OH$ **(d)** $C_6H_5CO_2^- Na^+$

10.12 **(a)** $CH_3CO_2H + CH_3CH_2CH_2CH_2OH$
(b) $CH_3CH_2CH_2CO_2H + CH_3OH$
(c) $PhCO_2H + (CH_3)_2CHOH$

10.13 **(a)** $CH_3CH_2COCl + CH_3OH$ **(b)** $CH_3COCl + CH_3CH_2OH$
(c) $CH_3COCl + C_6H_{11}OH$

10.14

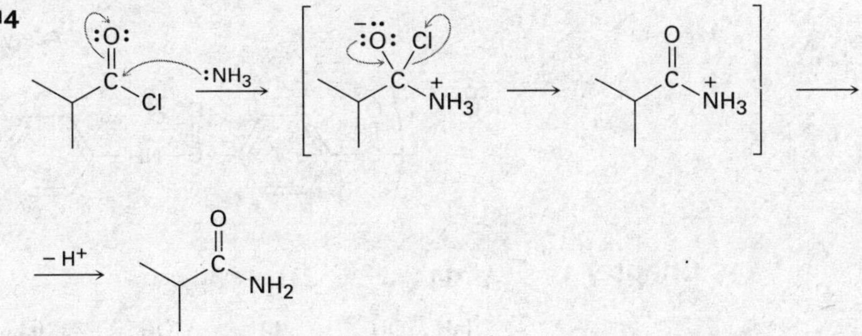

10.15 **(a)** $CH_3CH_2COCl + NH_3$ **(b)** $(CH_3)_2CHCH_2COCl + CH_3NH_2$
(c) $CH_3CH_2COCl + (CH_3)_2NH$ **(d)** $PhCOCl + (CH_3CH_2)_2NH$

10.17

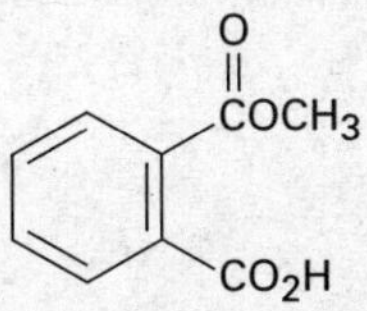

10.18 **(a)** $(CH_3)_2CHOH + CH_3CO_2H$
(b) $CH_3OH + C_6H_{11}CO_2H$

10.19 Reaction of an acid with an alkoxide ion gives the unreactive carboxylate ion.

10.20 **(a)** $CH_3CH_2CH_2CH(CH_3)CH_2OH + CH_3OH$
(b) $C_6H_5OH + C_6H_5CH_2OH$

10.21 **(a)** $C_5H_9CH_2CO_2R + CH_3CH_2MgBr$
(b) $CH_3CO_2R + H_2C{=}CHMgBr$

10.22 **(a)** H_2O, NaOH **(b)** 1. H_2O, NaOH; 2. $LiAlH_4$ **(c)** $LiAlH_4$

10.23

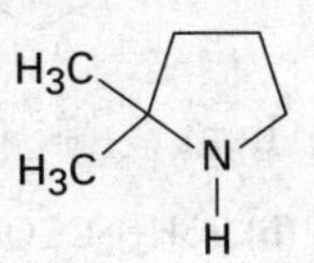

10.24 **(a)** $CH_3CH_2CN + CH_3CH_2MgBr$, then H_3O^+
(b) *p*-Nitrobenzonitrile + CH_3MgBr, then H_3O^+

10.25 1. NaCN; 2. CH_3CH_2MgBr, then H_3O^+

10.26

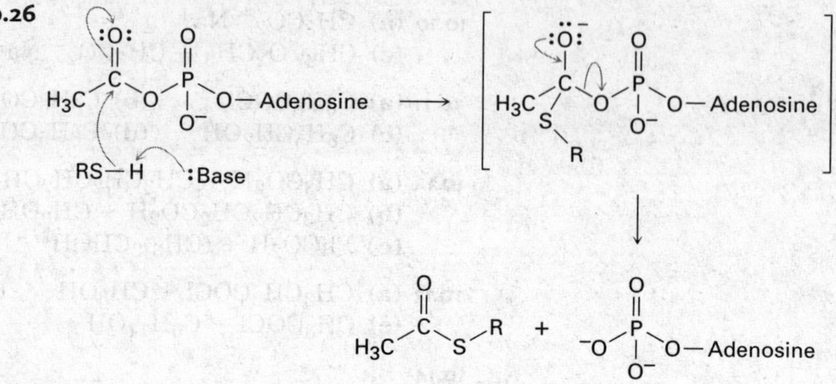

Acetyl CoA

10.27

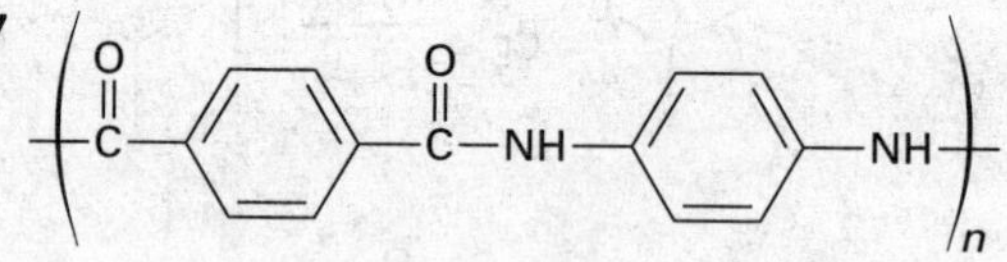

Chapter 11

11.1

(a)

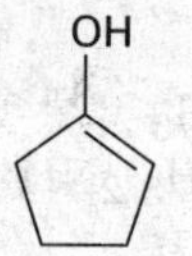

(b)

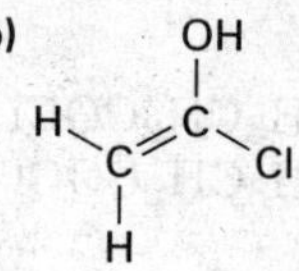

(c)

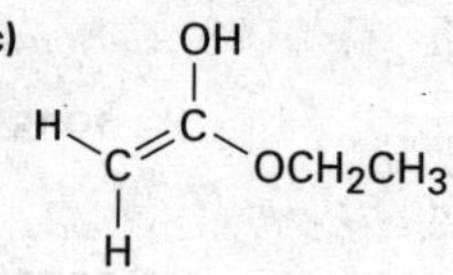

(d)

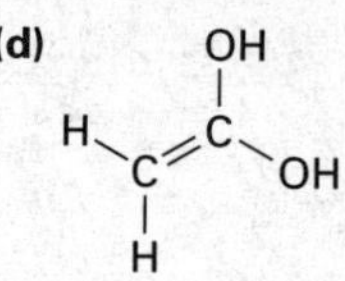

(e)

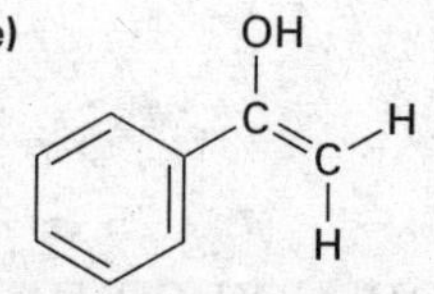

11.2 (a) 4 (b) 3 (c) 3 (d) 4 (e) 3

11.3

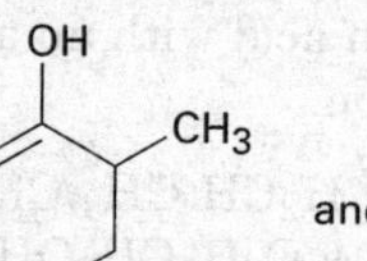

and

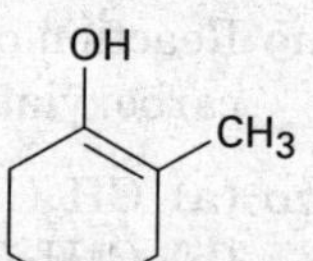

11.4 **(a)**

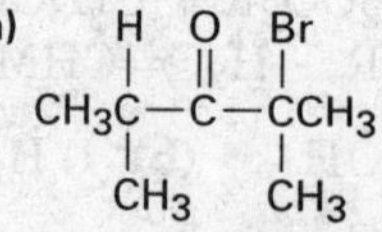

(b)

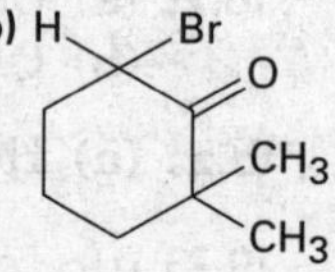

11.5 1. Br_2; 2. Pyridine, heat

11.6 **(a)** CH_3CH_2CHO **(b)** $(CH_3)_3CCOCH_3$

(c) CH_3CO_2H **(d)** $CH_3CH_2CH_2C{\equiv}N$

(e)

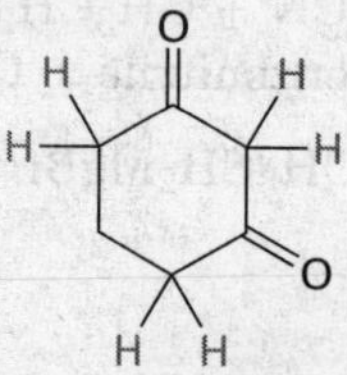

11.7 (a)

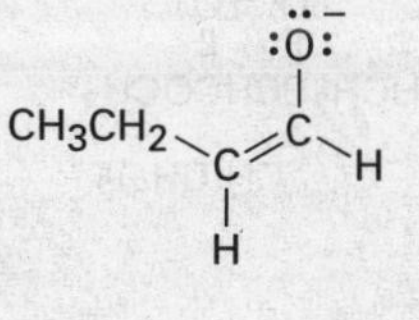

(b)

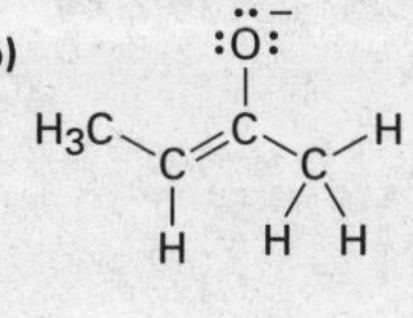

and

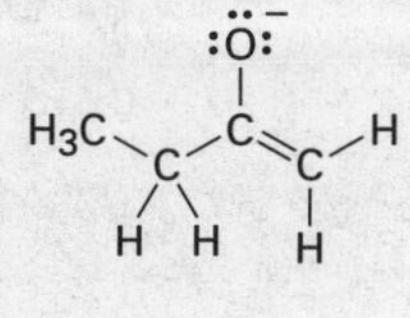

(c)

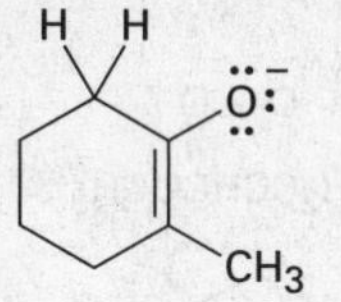

and

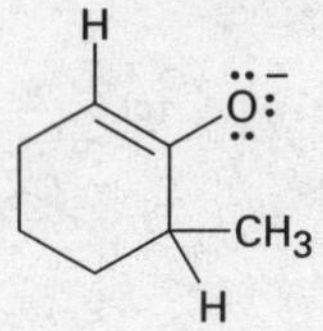

11.8

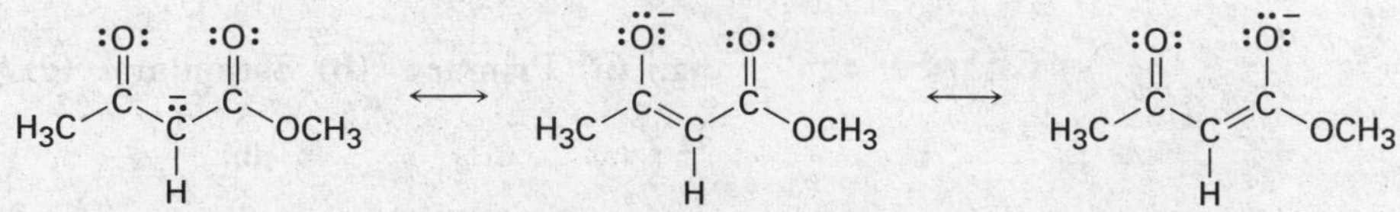

11.9 (a) CH_3CH_2Br (b) $C_6H_5CH_2Br$ (c) $(CH_3)_2CHCH_2CH_2Br$

11.10 (a) 1. Na^+ ^-OEt; 2. $(CH_3)_2CHCH_2Br$; 3. H_3O^+
(b) 1. Na^+ ^-OEt; 2. $CH_3CH_2CH_2Br$; 3. Na^+ ^-OEt; 4. CH_3Br; 5. H_3O^+

11.11 1. Na^+ ^-OEt; 2. $(CH_3)_2CHCH_2Br$; 3. Na^+ ^-OEt; 4. CH_3Br; 5. H_3O^+

11.12 Only (a) can undergo an aldol reaction.

11.13 (a)

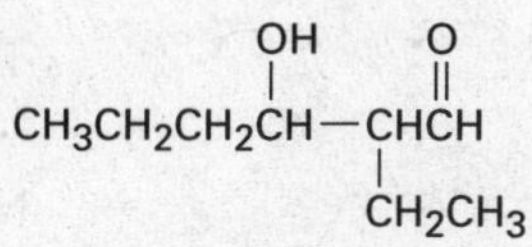

(b)

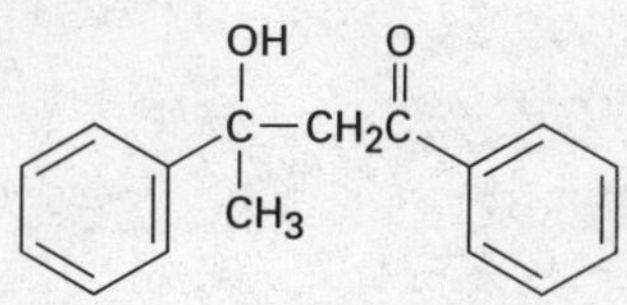

(c)

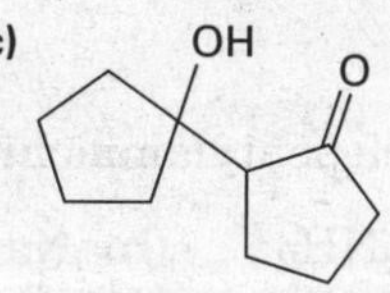

11.14 (a)

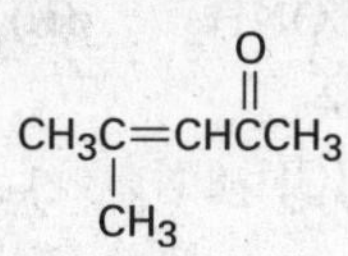

(b)

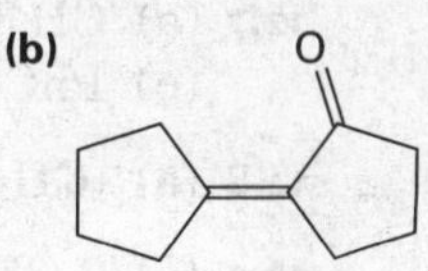

(c)

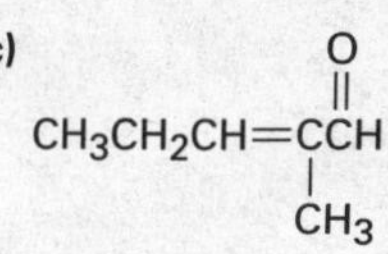

11.15

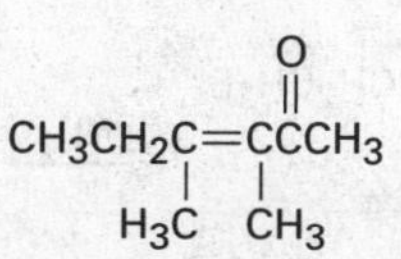

and

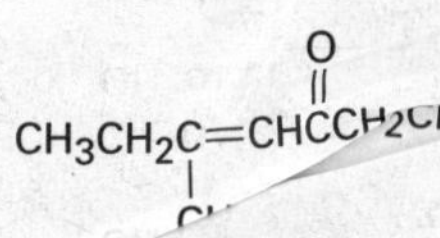

en reaction.

11.16 Only (c) undergoes

11.17 (a)
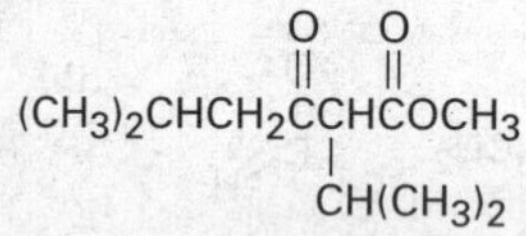
(b)
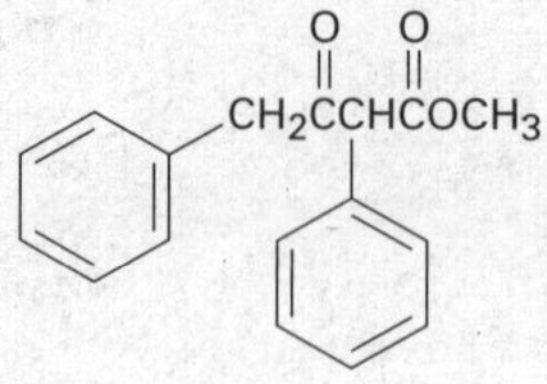

(c)
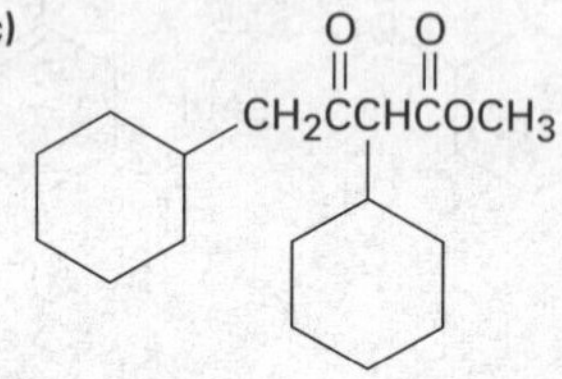

Chapter 12

12.1 **(a)** Primary **(b)** Secondary **(c)** Tertiary

12.2 (a)
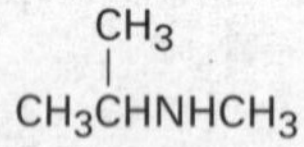
(b)
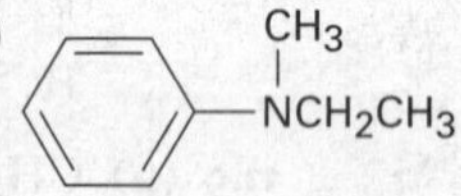
(c)
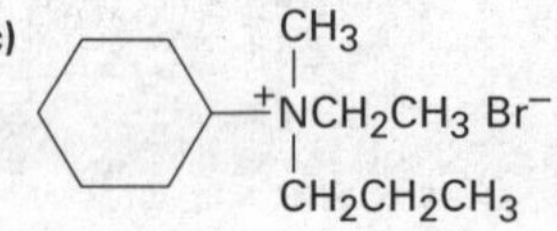

12.3 **(a)** Isopropylamine **(b)** Diethylamine
(c) *N*-Methylpyrrole **(d)** *N*-Methyl-*N*-ethylcyclohexylamine
(e) Diisopropylamine **(f)** Butane-1,3-diamine

12.4 (a) $(CH_3CH_2)_3N$ (b)
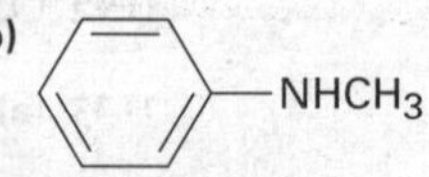
(c) $(CH_3CH_2)_4N^+\ Br^-$

(d)

(e)
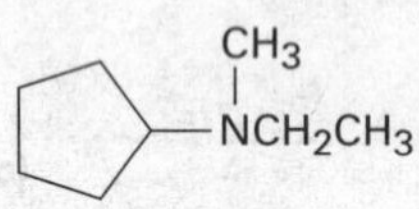

12.5 *N*-Methylcyclopentylammonium bromide

12.6 **(a)** $CH_3CH_2NH_2$ **(b)** NaOH **(c)** CH_3NHCH_3 **(d)** $(CH_3)_3N$

12.7 **(a)** $CH_3CH_2CONH_2$ **(b)** $CH_3CH_2CONHCH_2CH_2CH_3$
(c) $PhCONH_2$

12.8 **(a)** $(CH_3)_2CHCH_2CN$ **(b)** PhCN

12.9 **(a)** $3\ CH_3CH_2Br + NH_3$ **(b)** $4\ CH_3Br + NH_3$

12.10
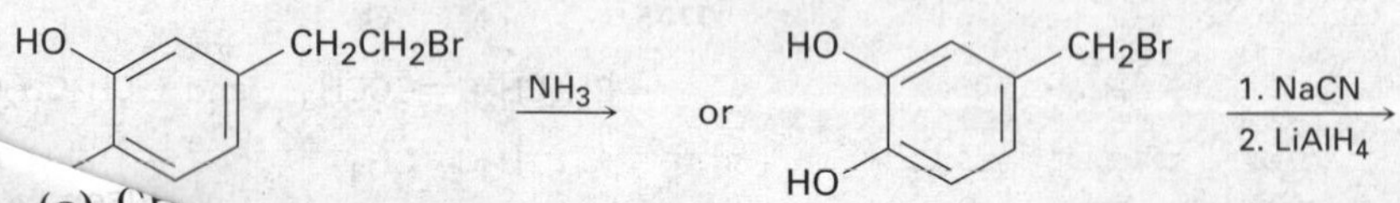

12.11 **(a)** CH...
(b) $C_6H_5NH_2$ + CH_3COCH_3 or $(CH_3)_2CHNH_2 + CH_3CHO$
(c) $C_5H_{11}NH_2$ + ...CHO

12.12 $(CH_3)_2NH$ + *o*-methylbenz...NH_2 + cyclopentanone

12.13 **(a)** 1. CH_3Cl, $AlCl_3$; 2. $KMnO_4$, H_2O, ...
(b) 1. HNO_3, H_2SO_4; 2. H_2/Pt catalyst; 3. ... SO_4; 4. H_2, Pt catalyst

12.14 **(a)** *N*-Methyl-2-bromopyrrole **(b)** *N*-Methyl-2-methylpyrrole
(c) *N*-Methyl-2-acetylpyrrole

12.15

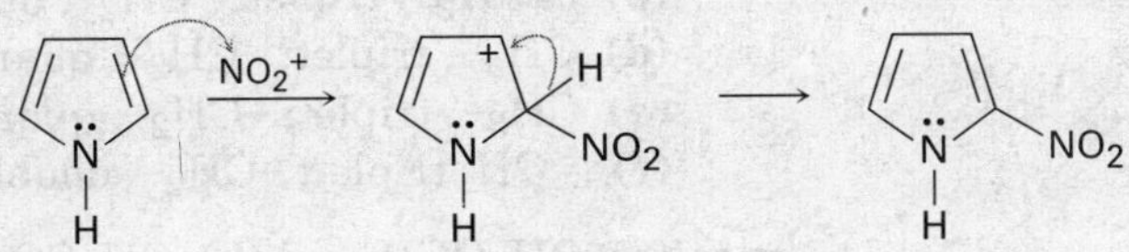

12.16

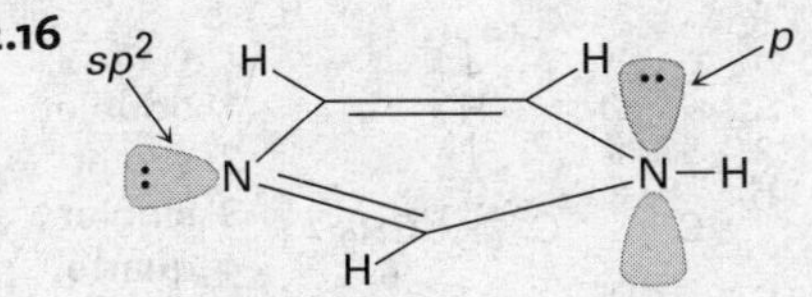

12.17 The pyridine-like doubly bonded nitrogen is more basic.

12.18 The side-chain nitrogen is more basic than the ring nitrogen.

Chapter 13

13.1 I_2

13.2 Butanoic acid

13.3 IR: $\epsilon = 2.0 \times 10^{-19}$ J; X ray: $\epsilon = 6.6 \times 10^{-17}$ J

13.4 $\lambda = 9.0 \times 10^{-6}$ m is higher in energy.

13.5 **(a)** 0.16 m **(b)** 7.5×10^{-4} kJ/mol; much less energy than light

13.6 **(a)** Ketone or aldehyde **(b)** Nitro
(c) Nitrile or alkyne **(d)** Carboxylic acid
(e) Alcohol and ester

13.7 **(a)** CH_3CH_2OH has an –OH absorption.
(b) Hex-1-ene has a double-bond absorption.
(c) Propanoic acid has a very broad –OH absorption.

13.8 Nitrile: 2210–2260 cm^{-1}; ketone: 1690 cm^{-1}; double bond: 1640 cm^{-1}

13.9 3×10^{-5} M

13.10 (a), (c), (d), and (f) have UV absorptions.

13.11 Hexa-1,3,5-triene absorbs at a longer wavelength.

13.12 The energy used by NMR spectroscopy is less than that used by IR spectroscopy.

13.13 **(a)** ^{1}H, 1; ^{13}C, 1 **(b)** ^{1}H, 1; ^{13}C, 1 **(c)** ^{1}H, 2; ^{13}C, 2
(d) ^{1}H, 1; ^{13}C, 1 **(e)** ^{1}H, 1; ^{13}C, 1 **(f)** ^{1}H, 1; ^{13}C, 1
(g) ^{1}H, 2; ^{13}C, 2 **(h)** ^{1}H, 2; ^{13}C, 2 **(i)** ^{1}H, 1; ^{13}C, 2

13.14 The vinylic C–H protons are nonequivalent.

13.15 ^{1}H, 5; ^{13}C, 7

13.16 **(a)** 210 Hz **(b)** 2.1 δ **(c)** 460 Hz

13.17 **(a)** 7.27 δ **(b)** 3.05 δ **(c)** 3.47 δ **(d)** 5.30 δ

13.18 **(a)** 0.88 δ **(b)** 2.17 δ **(c)** 7.17 δ **(d)** 2.22 δ

13.19 Two peaks; 3:2 ratio

13.20 **(a)** $-CHBr_2$, quartet; $-CH_3$, doublet
(b) CH_3O-, singlet; $-OCH_2-$, triplet; $-CH_2Br$, triplet
(c) $ClCH_2-$, triplet; $-CH_2-$, quintet
(d) CH_3-, triplet; $-CH_2-$, quartet; $-CH-$, septet; $(CH_3)_2$, doublet
(e) CH_3-, triplet; $-CH_2-$, quartet; $-CH-$, septet; $(CH_3)_2$, doublet
(f) $=CH$, triplet; $-CH_2-$, doublet; aromatic C–H, two multiplets

13.21 **(a)** CH_3OCH_3 **(b)** $CH_3CO_2CH_3$ **(c)** $(CH_3)_2CHCl$

13.22

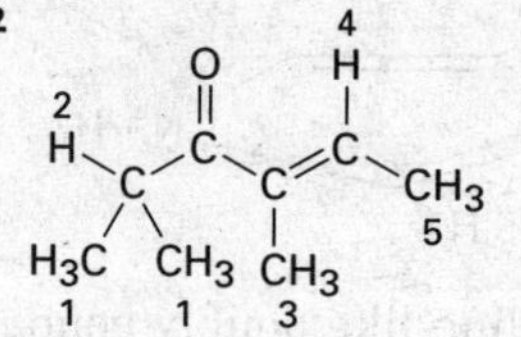

1 doublet
2 septet
3 singlet
4 quartet
5 doublet

13.23 **(a)** 1 **(b)** 5 **(c)** 4 **(d)** 7 **(e)** 5 **(f)** 7

13.24 **(a)** Hept-1-ene
(b) 2-Methylpentane
(c) 1-Chloro-2-methylpropane

Chapter 14

14.1 **(a)** Aldotetrose **(b)** Ketopentose
(c) Ketohexose **(d)** Aldopentose

14.2

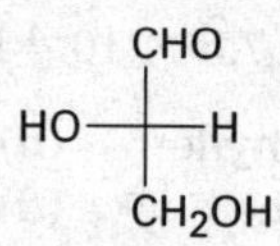

14.3

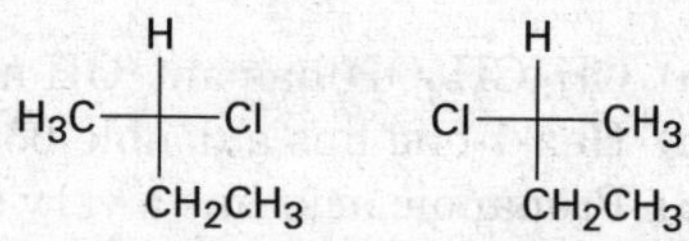

R *S*

14.4 **(a)** *S* **(b)** *R* **(c)** *S*

14.5

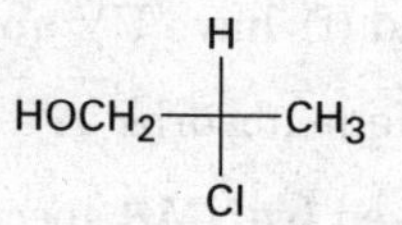

R

14.6 **(a)** L **(b)** D **(c)** D

14.7 (a)

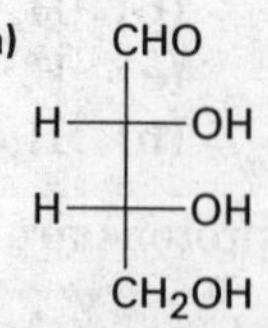

D

(b)

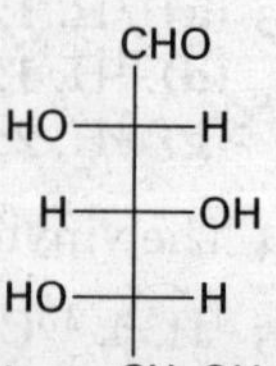

L

(c)

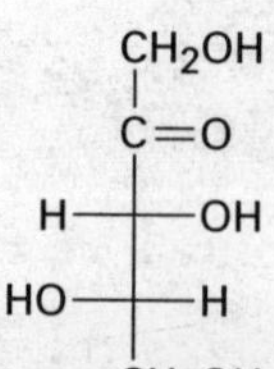

L

14.8 (a)

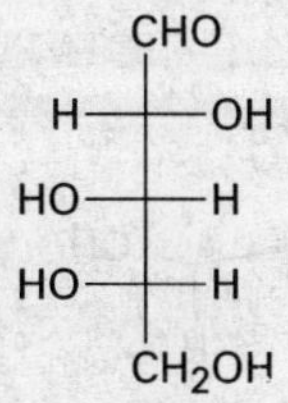

(b)

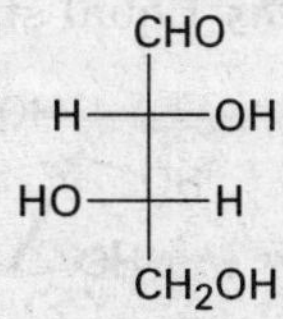

(c)

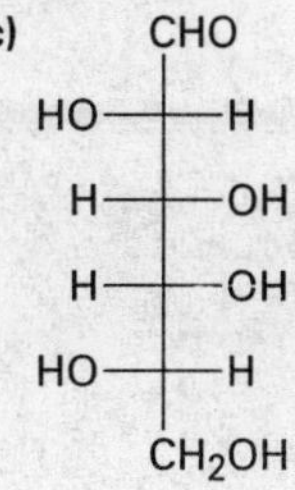

14.9 There are 16 D and 16 L aldoheptoses.

14.10

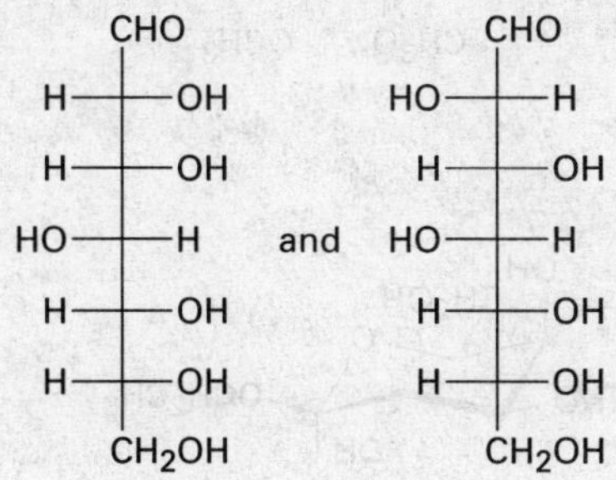

14.11

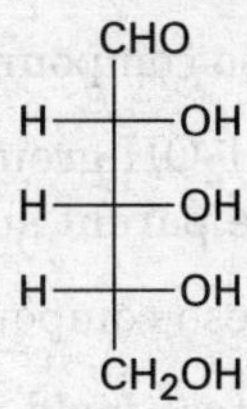

D-Ribose

14.12

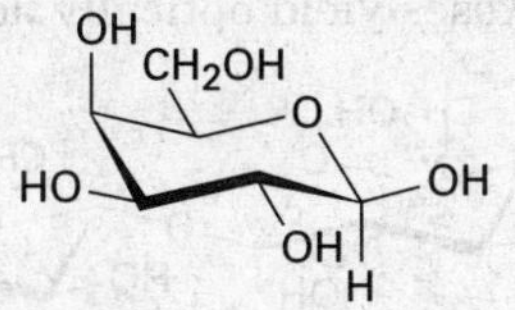

14.13

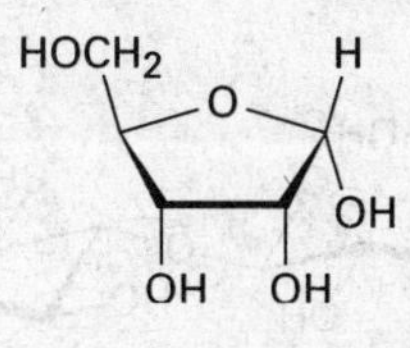

14.14

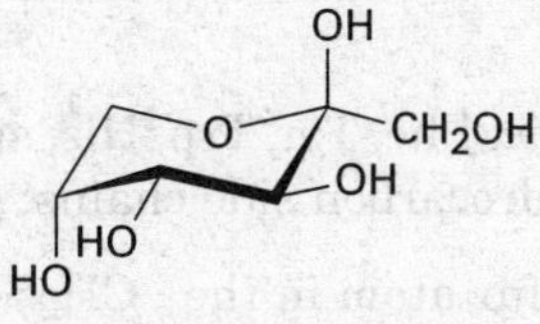

β-D-Fructopyranose

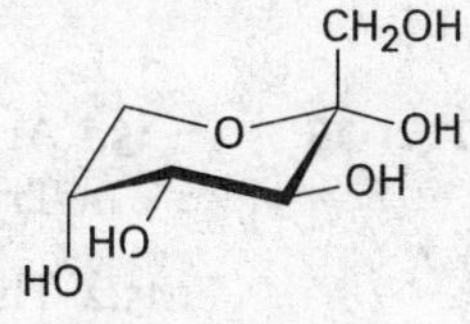

α-D-Fructopyranose

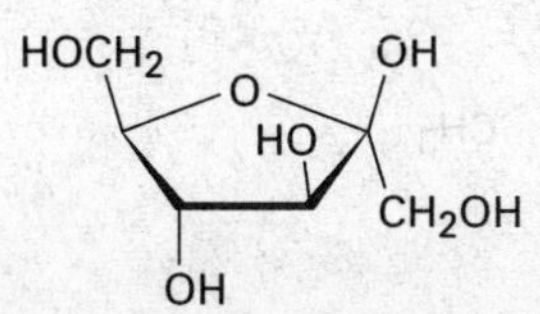

β-D-Fructofuranose

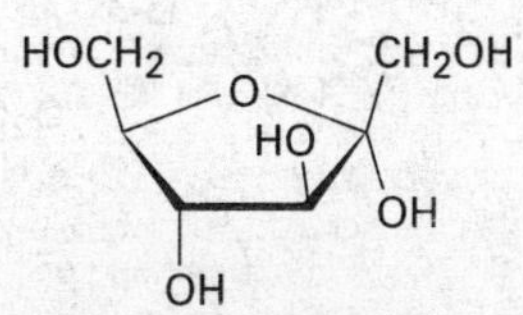

α-D-Fructofuranose

14.15 Equal stability

14.16 (a)

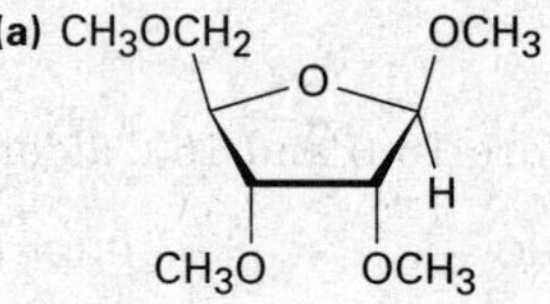

(b)

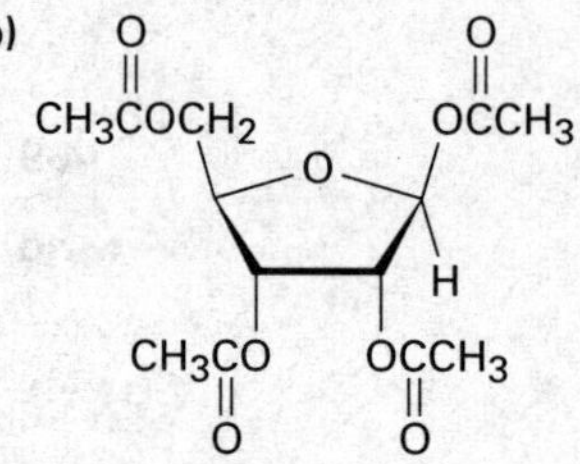

14.17

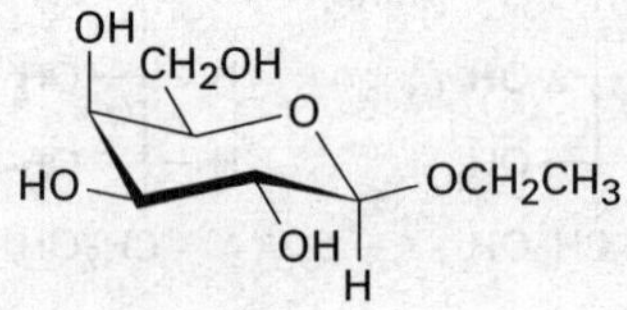

14.18 D-Galactitol is a meso compound.

14.19 An alditol has a $-CH_2OH$ group at both ends; either could have been a $-CHO$ group in the parent sugar.

14.20 D-Allaric acid is a meso compound; D-glucaric acid is not.

14.21 D-Allose and D-galactose yield meso aldaric acids; the other six D-aldohexoses yield optically active aldaric acids.

14.22 (a)

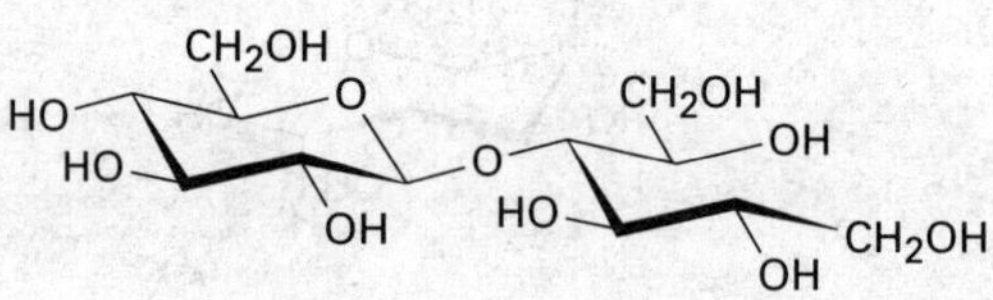

(b)

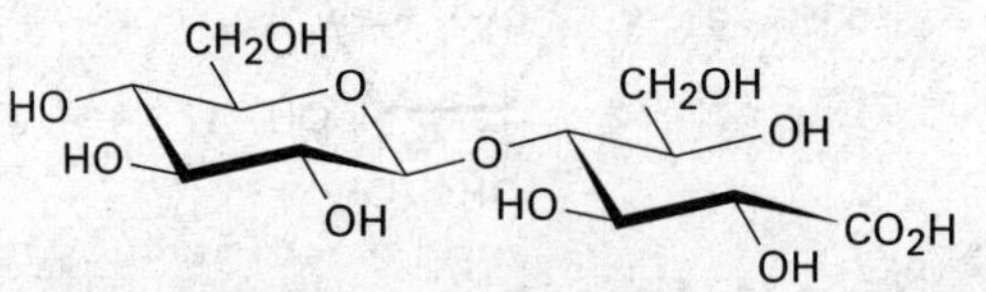

Chapter 15

15.1 Aromatic: Phe, Tyr, Trp, His; sulfur-containing: Cys, Met; alcohols: Ser, Trp; hydrocarbon side chains: Ala, Ile, Leu, Val.

15.2 The sulfur atom in the $-CH_2SH$ group of cysteine makes the side chain higher ranked than the $-CO_2H$ group.

15.3

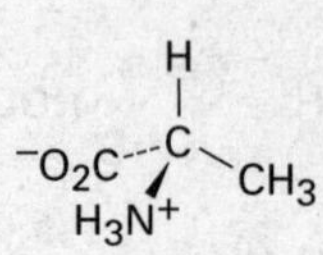

15.4

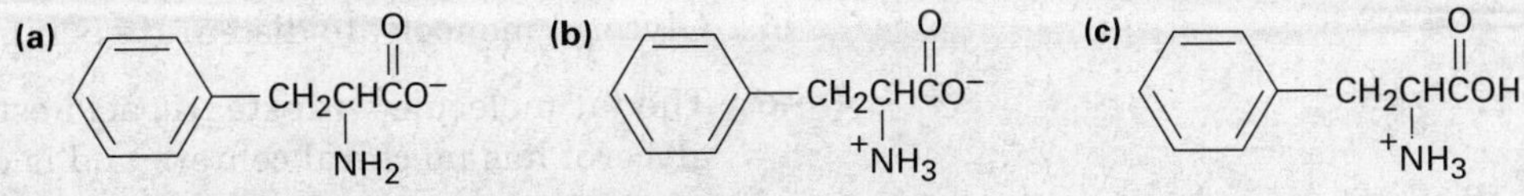

15.5 **(a)** Toward (+): Glu > Val; toward (–): none
(b) Toward (+): Phe; toward (–): Gly
(c) Toward (+): Phe > Ser; toward (–): none

15.6 Net positive at pH = 5.3; net negative at pH = 7.3

15.7

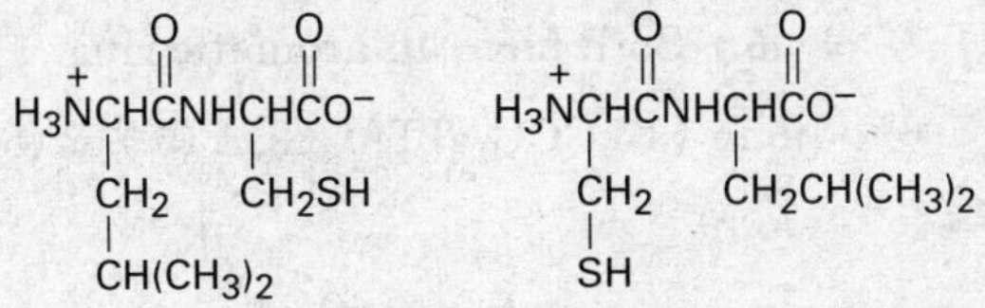

15.8 Val-Tyr-Gly (VYG), Tyr-Gly-Val (YGV), Gly-Val-Tyr (GVY), Val-Gly-Tyr (VGY), Tyr-Val-Gly (YVG), Gly-Tyr-Val (GYV)

15.9

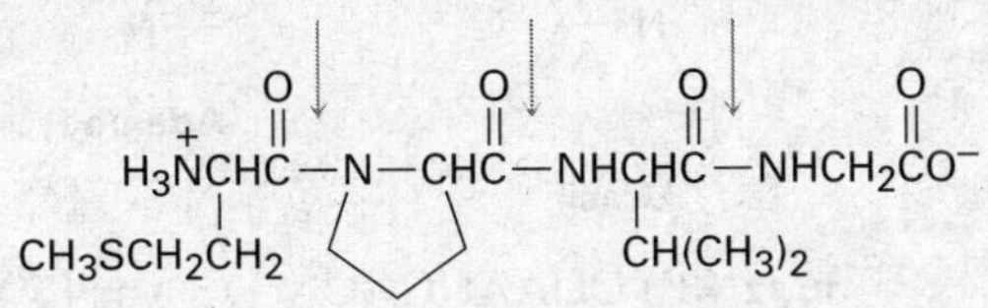

15.10

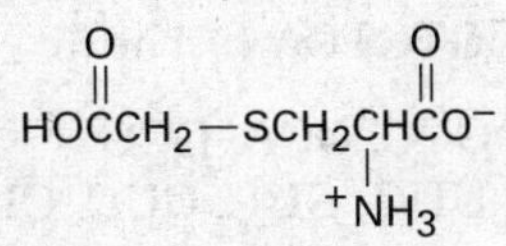

15.11

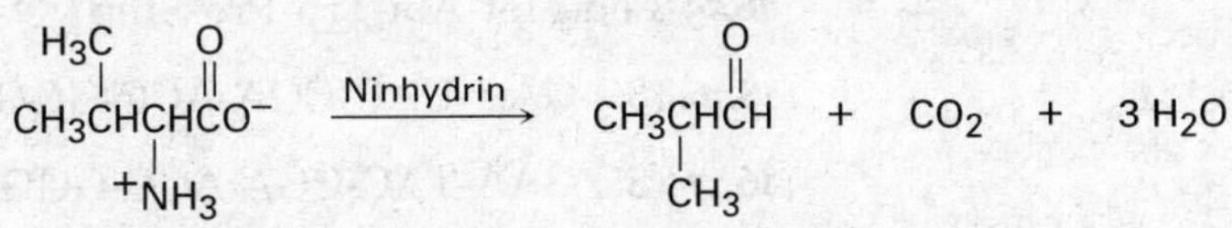

15.12 Trypsin: Asp-Arg + Val-Tyr-Ile-His-Pro-Phe
Chymotrypsin: Asp-Arg-Val-Tyr + Ile-His-Pro-Phe

15.13 Arg-Pro-Leu-Gly-Ile-Val

15.14 Methionine

15.15 (1) Protect the amino group of leucine.
(2) Protect the carboxylic acid group of alanine.
(3) Couple the protected amino acids with DCC.
(4) Remove the leucine protecting group.
(5) Remove the alanine protecting group.

15.16 This is a typical nucleophilic acyl substitution reaction, with the amine of the amino acid as the nucleophile and *tert*-butyl carbonate as the leaving group. The *tert*-butyl carbonate then loses CO_2 and gives *tert*-butoxide, which is protonated.

15.17 **(a)** Lyase **(b)** Hydrolase **(c)** Oxidoreductase

Chapter 16

16.1 $CH_3(CH_2)_{18}CO_2CH_2(CH_2)_{30}CH_3$

16.2 Glyceryl monooleate distearate is higher melting.

16.3 The fat molecule with stearic acid esterified to the central –OH group of glycerol has no chiral centers and is optically inactive.

16.4 $[CH_3(CH_2)_7CH{=}CH(CH_2)_7CO_2^-]_2\ Mg^{2+}$

16.5 Glyceryl dioleate monopalmitate → glycerol + 2 sodium oleate + sodium palmitate

16.6 Two ketones, double bond

16.7 Both have an aromatic ring.

16.10 (3′) CCGATTAGGCA (5′) or (5′) ACGGATTAGCC (3′)

16.11

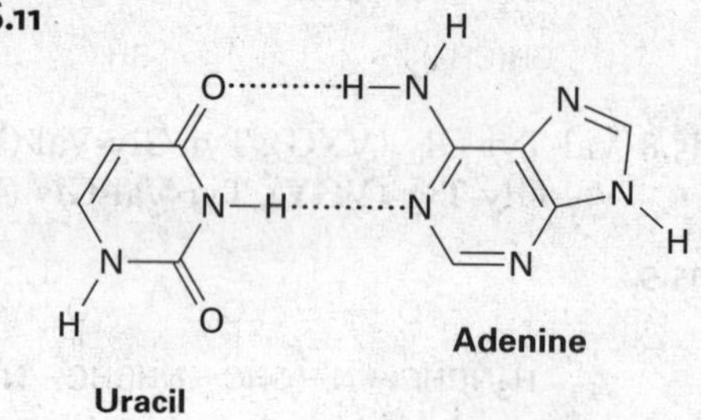

16.12 (3′) CUAAUGGCAU (5′) or (5′) UACGGUAAUC (3′)

16.13 (3′) AAGCGTCTCA (5′) or (5′) ACTCTGCGAA (3′)

16.14 **(a)** GCU, GCC, GCA, GCG **(b)** UUU, UUC
(c) UUA, UUG, CUU, CUC, CUA, CUG **(d)** UAU, UAC

16.15 Leu-Met-Ala-Trp-Pro-Stop

16.16 (3′) GAA-UAC-CGA-ACC-GGG-AUU (5′)

16.17 (3′) GAA-TAC-CGA-ACC-GGG-ATT (5′)

Chapter 17

17.1 $HOCH_2CH(OH)CH_2OH + ATP \rightarrow HOCH_2CH(OH)CH_2OPO_3^{2-} + ADP$

17.3 **(a)** 8 acetyl CoA; 7 passages **(b)** 10 acetyl CoA; 9 passages

17.4 Steps 7 and 10

17.5 Step 1: nucleophilic acyl substitution at phosphorus;
step 2: isomerization by keto–enol tautomerization;
step 3: like step 1;
step 4: retro aldol condensation;
step 5: like step 2;
step 6: oxidation;
step 7: like step 2;
step 8: isomerization;
step 9: E1cB reaction;
step 10: substitution at phosphorus, followed by tautomerization

17.6 Citrate and isocitrate

17.7

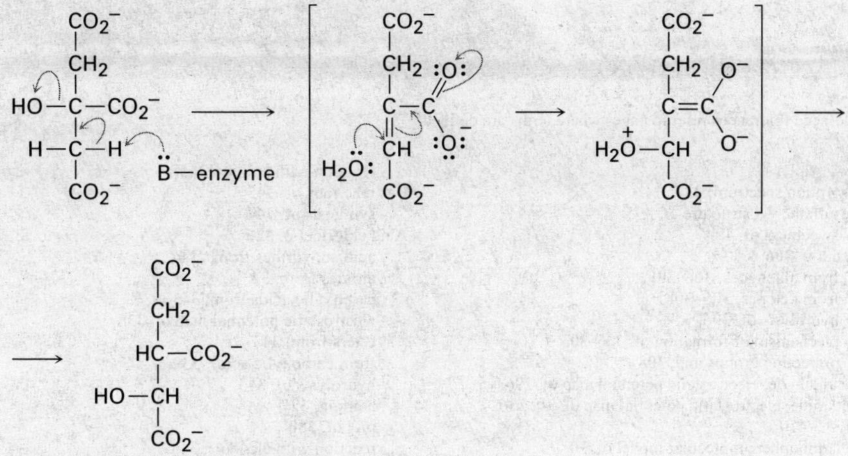

17.8 $(CH_3)_2CHCH_2COCO_2^-$

17.9 Asparagine

INDEX

Boldfaced references refer to pages where terms are defined.